The Etiology of Breast Cancer

ENDOGENOUS AND EXOGENOUS CAUSES

Fritz F. Parl

ISBN-13: 978-0615993737
ISBN-10: 0615993737

TABLE OF CONTENTS

When a German philosopher published a 500-page treatise 'About the Truth', the writer and physician Gottfried Benn commented: How can the truth be 500 pages long? I was reminded of this comment during the writing of this book on 'The Etiology of Breast Cancer'. How can the etiology be 500 pages long? Of course, the answer lies in the complexity of breast cancer, which appears to have multiple causes. In this book, which contains over 5000 references, I have made an effort to present the causes of breast cancer by reviewing the literature fairly and without bias. I have tried to express my opinions clearly without appearing dogmatic. Obviously, critics will find fault with mistakes, oversights, and opinions. To make it easier for the reader, I have highlighted sections in blue that are directly relevant to breast cancer.

I make this book available to all students anywhere in the world and dedicate it to all women with breast cancer, especially the "lay advocates". This remarkable group of women not only lives with the diagnosis of the disease but also makes an unselfish effort to understand and fight breast cancer for the benefit of their sisters, daughters, and future generations of women.

AhR	aryl hydrocarbon receptor
BER	base excision repair
CEE	conjugated equine estrogen
CI	confidence interval
COMT	catechol-O-methyltransferase
CYP	cytochrome P450
DBD	DNA binding domain
E_1	estrone
E_2	17β-estradiol
E_3	estriol
EGFR	epidermal growth factor receptor
ER	estrogen receptor
ERE	estrogen response element
GSH	glutathione
LBD	ligand binding domain
NER	nucleotide excision repair
PAH	polycyclic aromatic hydrocarbon
PR	progesterone receptor
ROS	reactive oxygen species
SNP	single-nucleotide polymorphism
SOD	superoxide dismutase
SULT	sulfotransferase

CHAPTER 1

Introduction

Undoubtedly, breast cancer is a complex disease. Traditional histopathological analysis distinguishes ductal and lobular phenotypes (1). Gene expression profiling led to the classification of at least four distinct subtypes (2-4). Recent whole-genome sequencing of breast cancers with integrative cluster analysis of gene copy number variation and gene expression data revealed 10 subgroups (5). Moreover, epidemiological studies indicate that about 10% of breast cancers are inherited or "familial", which leaves 90% unexplained or "sporadic" cancers. The apparent complexity makes it likely that breast cancer has multiple causes.

Breast cancers are histologically complex tissues composed of malignant epithelial cells and several benign cell types, i.e., endothelial, stromal, and adipose cells, as well as B and T lymphocytes and macrophages, each represented by distinct gene expression patterns (2, 6). The epithelial cell component itself is complex, derived from mammary stem cells that can both self-renew and propagate the full spectrum of cell types to generate functional lobulo-alveolar units during pregnancy (7, 8). Mammary epithelial cell development is thought to progress in a hierarchical process from undifferentiated stem cells into at least two differentiated cell types, basal/myoepithelial and luminal cells, which can be distinguished by cytokeratin markers. In turn, the basal/myoepithelial and luminal cells are progenitor cells, which give rise to at least four distinct tumor subtypes: basal-like cells expressing human epidermal growth factor receptor 1 (EGFR; HER1), human epidermal growth factor receptor 2 (HER2)-enriched cells, and luminal A and luminal B cells typically expressing estrogen receptor alpha (ERα) and progesterone receptor (PR) (3, 9). Interestingly, both mouse and human mammary stem cells exhibit a triple-negative phenotype for ERα, PR, and HER2 (10, 11).

According to the cancer stem cell hypothesis, a breast cancer can arise from transformation of a stem or progenitor cell resulting in a heterogeneous mixture of differentiated and stem/progenitor cells. Whatever the cell of tumor origin, the evolution of a normal cell toward a cancerous one is a complex process, accompanied by multiple steps of somatic mutations and epigenetic alterations that confer selective advantages upon the altered cells. Conceptually, the somatic mutations can be classified as 'driver' and 'passenger' mutations, the former conferring growth advantages and the latter arising in an ancestor of the cancer cell when it acquired one of its drivers (12). The alterations are thought to endow emerging tumor cells with evasion from programmed cell death, insensitivity to antigrowth signals, and unlimited replicative and proliferative potential (13, 14). A detailed genome analysis of 100 breast cancers for mutations in the coding exons of 21,416 genes revealed 6,964 single-base substitutions (68.0% missense, 6.1% nonsense, 2.3% splice site, <1.0% stop codon, 23.5% silent mutations) (15). The number of somatic mutations varied markedly between individual tumors but there was a strong correlation between the number of mutations, the age at which the cancer was diagnosed, and the histological grade of the tumor. Driver mutations were identified in at least 40 cancer genes with 73 different combinations among the 100 tumors highlighting the substantial genetic diversity of breast cancer.

The breast is a unique target organ for carcinogenesis because of the proximity of fat tissue and embedded epithelial cells (16). The fat tissue stores lipophilic procarcinogens and carcinogens whereas the epithelial cells are endowed with highly active enzymes capable of converting procarcinogens to carcinogens that can cause DNA damage (17). The lipophilic chemicals are endogenous (e.g., estrogens) or exogenous chemicals (e.g., heterocyclic amines, polycyclic aromatic hydrocarbons). Most of the potential carcinogens undergo several enzyme-catalyzed reactions to produce DNA-reactive metabolites. The first reaction is typically mediated by an oxidizing or "activating" phase I enzyme of the cytochrome P450 (CYP) family. Subsequent reactions are commonly catalyzed by conjugating phase II enzymes, which include several families such as glutathione S-transferases (GSTs), N-acetyltransferases (NATs), sulfotransferases (SULTs), UDP-glucuronosyltransferases (UGTs), and others. In addition to lipophilic carcinogens, there are reactive oxygen species (ROS), which are formed during normal metabolism of the cell and pathological stress conditions. ROS can cause oxidative DNA damage unless they are removed by non-enzymatic and enzymatic scavenger mechanisms. The anti-oxidant enzymes include superoxide dismutases (SODs), catalase, glutathione peroxidases (GPXs), and others. Finally, lipid peroxidation results in the formation of mutagenic and carcinogenic products such as malondialdehyde.

Each of the enzymes involved in carcinogen metabolism carries single nucleotide polymorphisms (SNPs), the most frequent type of genetic variation. Considering that there are approximately 10 million SNPs in the human genome they are likely to account for most of the genetic differences between individuals. Especially non-synonymous SNPs are of interest for disease susceptibility because the associated amino acid substitutions may result in altered protein function. Combining neighboring polymorphism into haplotypes may increase the association with disease. Like most common diseases, breast cancer arises from the interaction between multiple genetic variations, endogenous factors such as estrogens, and exogenous factors such as hormone replacement therapy or diet.

Abundant experimental and epidemiological evidence has implicated estrogens as prime risk factor for the development of breast cancer (18-22). Experiments on estrogen metabolism (23-25), formation of DNA adducts (26, 27), mutagenicity (28, 29), cell transformation (30, 31) and carcinogenicity (32, 33) have implicated certain estrogen metabolites, especially the catechol estrogen 4-hydroxyestradiol (4-OHE$_2$) to react with DNA via its quinone, causing mutations and initiating cancer. Estrogen-DNA adducts have been detected in normal and malignant human breast tissues (34-36) and we have provided direct experimental evidence that oxidative metabolism of 17β-estradiol (E$_2$) leads to the formation of 4-OHE$_2$ and deoxyribonucleoside adducts (37). Studies have also indicated that breast cancer risk is higher in women with early menarche and late menopause, who have longer exposure to estrogens. For this reason, current models of breast cancer risk prediction are mainly based on surrogates of cumulative estrogen exposure such as age, age at menarche, and age at first live birth (38-42). While traditional exposure data are valuable in risk calculation, they do not directly reflect mammary estrogen metabolism. Furthermore, current models do not address genetic variability between women in exposure to carcinogenic estrogen metabolites, including catechols and quinones.

The susceptibility of the mammary gland to carcinogens is markedly influenced by its physiological state at the time of exposure (43, 44). Around menarche, the human mammary gland begins to develop and differentiate into defined ducts and lobules. The primary lobules formed at this time are type 1 lobules, which differentiate further into type 2 and type 3 lobules during pregnancy (45). Normal human mammary epithelial cells (nHMEC) from type 1 lobules were more sensitive to the transforming effect of the carcinogen 7,12-dimethylbenz[α]anthracene (DMBA) than type 3 lobule cells (46). Virgin rats treated with DMBA developed significantly more mammary tumors than multiparous rats of the same strain. The difference in tumor incidence was attributed to the difference in the extent of mammary tissue differentiation, i.e., mammary glands of virgin rats contain more undifferentiated end buds with a larger number of proliferating epithelial cells than those of parous animals.

In the same manner as parity protects rodents from mammary tumorigenesis, full-term pregnancy at an early age decreases breast cancer risk in humans, most likely due to structural changes occurring during pregnancy and lactation (18, 47). An important requirement for the transformation of nHMEC is cell proliferation in the presence of carcinogen. The proliferating mammary epithelial cell needs to be able to metabolize the carcinogen and to replicate the damaged DNA, which carries the carcinogen-induced mutation to the subsequent cell population (48).

Tumor development or carcinogenesis is usually viewed as a stepwise process beginning with genotoxic effects (initiation) followed by enhanced cell proliferation (promotion) (49). The two major estrogens, 17ß-estradiol (E$_2$) and estrone (E$_1$), are substrates for CYP1A1 and CYP1B1 and ligands for the estrogen receptor. In their dual role of substrate and ligand, estrogens have been implicated in the development of breast cancer by simultaneously causing DNA damage via their oxidation products, the 2-OH and 4-OH catechol estrogens, and by stimulating cell proliferation and gene expression via the estrogen receptor (21, 22, 50). Thus, estrogens and their oxidative metabolites are unique carcinogens that affect both tumor initiation and promotion. However, the development of breast cancer is more complex than the two-stage process portrays. Initiation depends not only on DNA damage but also on DNA repair while promotion is a composite of cell proliferation and cell death or apoptosis. Moreover, the two stages are linked rather than separate. The maintenance of genomic integrity following DNA damage depends on the coordination of DNA repair with cell cycle checkpoint controls (51, 52). The integrity of the DNA damage response pathway is crucial for the prevention of neoplastic proliferation, as suggested by the fact that many proteins involved in these pathways are tumor suppressors such as BRCA1, BRCA2, ATM, and p53.

It is obvious that breast cancer is not caused by a single agent, protein or pathway but rather results from the interaction of multiple factors. In the broadest terms, breast cancer is caused by an imbalance of DNA damage, DNA repair, cell proliferation, and apoptosis allowing the following somewhat simplistic summary of its etiology:

$$(\text{DNA damage} - \text{DNA repair}) + (\text{cell proliferation} - \text{apoptosis}) = \text{breast cancer}$$

DNA DAMAGE

Testosterone

CYP19A1

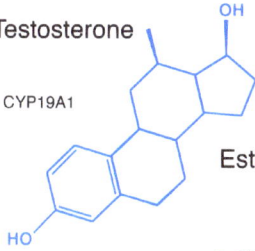

Estradiol

CYP1B1

4-OHE$_2$

E$_2$-3,4-Q

CAT	rs511895
COMT	rs4680
CYP17A1	rs743572
CYP19A1	rs700519, (TTTA)10
CYP1A1	rs1048943
CYP1B1	rs1056836, 1800440
GSTP1	rs1695
GSTM1	deletion
GSTT1	deletion
HSD17B1	rs605059, 676387
TXN1	rs4135179
TXNRD2	rs5748469, 756661

DNA REPAIR

4-OHE$_2$-N7-guanine

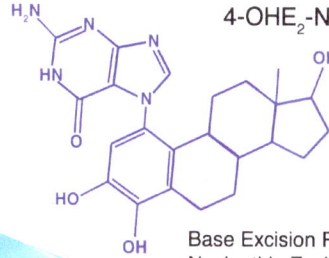

Base Excision Repair
Nucleotide Excision Repair
Homologous Recombination Repair

ATM	rs1800057
BABAM1	rs8170
ERCC2	rs13181
NBN	657del5
P53	rs12947788, 12951053, 17878362
RAD51L1	rs999737
WRN	rs1346044
XRCC3	rs861539

DNA
Adduct

G2

M

CELL PROLIFERATION

ER-Mediated Transcription

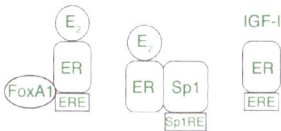

AURKA	rs1047972
CHEK2	rs17879961, 1100delC
ESR1	rs3020314, 1801132, 2234693
FGFR2	rs1219648, 2981582
IGF1	rs6220
MAP3K1	rs889312
NUMA1	rs3018301
TERT	rs10069690
TGFB1	rs1982073
TNF	rs1800629
TNRC9	rs3803662
TNP1	rs13387042

APOPTOSIS

Extrinsic Pathway

Caspase-8

CASP8	rs1045485, 1053485, 6723097
NQO1	rs1800566

Figure 1.1. Overview of mammary carcinogenesis as a combination of DNA damage, DNA repair, cell proliferation and apoptosis.

The overview emphasizes the predominant role of estrogens although other endogenous and exogenous genotoxic factors are likely to contribute to carcinogenesis as discussed in individual chapters. In the breast, the ovaries and adipose tissue, androgens (e.g., testosterone) are converted by aromatase (CYP19A1) to estrogens, in particular 17β-estradiol (E_2). In turn, E_2 is a substrate for several enzymes and a ligand for the estrogen receptor (ER). The main enzyme, CYP1B1, oxidizes E_2 to catechol estrogens and further to quinones, such as 4-OHE$_2$ and E_2-3,4-Q. The highly reactive E_2-3,4-Q forms addition products with deoxyribonucleosides. The resulting depurinating adducts, such as 4-OHE$_2$-N7-guanine, leave apurinic sites in the DNA. Unless repaired during G1 of the cell cycle, DNA replication during the S phase may lead to mutations that can be propagated into daughter cells during the M phase. The cell proliferation is driven by E_2-ER-mediated transcription of multiple genes while growth factors stimulate cell proliferation through binding to their membrane-bound tyrosine kinase receptors, which communicate the growth factor signals to pathways inside the cell, turning genes on and off in the process. Finally, cell proliferation is balanced by apoptosis, which can be initiated by a wide variety of intra- and extracellular factors. The suppression of apoptosis can lead to deregulated cell proliferation while deficient repair of damaged DNA permits the formation of mutations resulting in tumor formation. Thus, cancer is caused by a sustained imbalance of DNA damage, DNA repair, cell proliferation, and apoptosis. Nonsynonymous single nucleotide polymorphisms (SNPs) and other genetic variants have been identified in genes encoding key proteins in each of these four processes. Genome-wide association studies have linked specific SNPs and other genetic variants to breast cancer risk (see inserts with reference SNP ID numbers, rs#). Although each of these polymorphisms has a small effect individually it is likely that the large number of cumulative weak effects contributed by several variants in these linked pathways plays a major role in mammary carcinogenesis.

Figure 1.1 presents a conceptual rendition of how these four factors interact in mammary carcinogenesis. Each of these processes results from the interaction of multiple proteins, which are encoded by genes containing single nucleotide polymorphisms (SNPs) and other genetic variants. Genome-wide association studies have identified nonsynonymous SNPs and other genetic variants in many of these proteins that are linked to breast cancer risk (see inserts with reference SNP ID numbers, rs#). Although each of these polymorphisms has a small effect individually, the cumulative sustained effects contributed by several variants of proteins in these linked pathways plays a major role in mammary carcinogenesis. In the book, I attempt to address DNA damage, DNA repair, cell proliferation, and apoptosis and assess their contribution to the complex etiology of breast cancer.

References

1. Page DL, Anderson TJ. *Diagnostic Histopathology of the Breast.* London: Churchill Livingstone; 1987.

2. Perou CM, Sorlie T, Eisen MB, van de Rijn M, Jeffrey SS, Rees CA, et al. *Molecular portraits of human breast tumours.* Nature. 2000;406:747-52.

3. Prat A, Perou CM. *Mammary development meets cancer genomics.* Nat Med. 2009;15:842-4.

4. Sorlie T, Perou CM, Tibshirani R, Aas T, Geisler S, Johnsen H, et al. *Gene expression patterns of breast carcinomas distinguish tumor subclasses with clinical implications.* Proc Natl Acad Sci U S A. 2001;98:10869-74.

5. Curtis C, Shah SP, Chin SF, Turashvili G, Rueda OM, Dunning MJ, et al. *The genomic and transcriptomic architecture of 2,000 breast tumours reveals novel subgroups.* Nature. 2012;486:346-52.

6. Ronnov-Jessen L, Petersen OW, Bissell MJ. *Cellular changes involved in conversion of normal to malignant breast: importance of the stromal reaction.* Physiol Rev. 1996;76:69-125.

7. Shackleton M, Vaillant F, Simpson KJ, Stingl J, Smyth GK, Asselin-Labat ML, et al. *Generation of a functional mammary gland from a single stem cell.* Nature. 2006;439:84-8.

8. Stingl J, Eirew P, Ricketson I, Shackleton M, Vaillant F, Choi D, et al. *Purification and unique properties of mammary epithelial stem cells.* Nature. 2006;439:993-7.

9. Nielsen TO, Hsu FD, Jensen K, Cheang M, Karaca G, Hu Z, et al. *Immunohistochemical and clinical characterization of the basal-like subtype of invasive breast carcinoma.* Clin Cancer Res. 2004;10:5367-74.

10. Asselin-Labat ML, Shackleton M, Stingl J, Vaillant F, Forrest NC, Eaves CJ, et al. *Steroid hormone receptor status of mouse mammary stem cells.* J Natl Cancer Inst. 2006;98:1011-4.

11. Lim E, Vaillant F, Wu D, Forrest NC, Pal B, Hart AH, et al. *Aberrant luminal progenitors as the candidate target population for basal tumor development in BRCA1 mutation carriers.* Nat Med. 2009;15:907-13.

12. Stratton MR, Campbell PJ, Futreal PA. *The cancer genome.* Nature. 2009;458:719-24.

13. Hanahan D, Weinberg RA. *The hallmarks of cancer.* Cell. 2000;100:57-70.

14. Rivlin N, Brosh R, Oren M, Rotter V. *Mutations in the p53 Tumor Suppressor Gene: Important Milestones at the Various Steps of Tumorigenesis.* Genes Cancer. 2011;2:466-74.

15. Stephens PJ, Tarpey PS, Davies H, Van Loo P, Greenman C, Wedge DC, et al. *The landscape of cancer genes and mutational processes in breast cancer.* Nature. 2012;486:400-4.

16. Beer AE, Billingham RE. *Adipose tissue, a neglected factor in aetiology of breast cancer?* Lancet. 1978;2:296.

17. Phillips DH, Martin FL, Williams JA, Wheat LM, Nolan L, Cole KJ, et al. *Mutagens in human breast lipid and milk: the search for environmental agents that initiate breast cancer.* Environ Mol Mutagen. 2002;39:143-9.

18. MacMahon B, Cole P, Brown J. *Etiology of human breast cancer: a review.* J Natl Cancer Inst. 1973;50:21-42.

19. Parl FF. *Estrogens, Estrogen Receptor and Breast Cancer.* Amsterdam: IOS Press; 2000.

20. Pike MC, Krailo MD, Henderson BE, Casagrande JT, Hoel DG. *'Hormonal' risk factors, 'breast tissue age' and the age-incidence of breast cancer.* Nature. 1983;303:767-70.

21. Yager JD, Davidson NE. *Estrogen carcinogenesis in breast cancer.* New Engl J Med. 2006;354:270-82.

22. Yue W, Yager JD, Wang JP, Jupe ER, Santen RJ. *Estrogen receptor-dependent and independent mechanisms of breast cancer carcinogenesis.* Steroids. 2013;78:161-70.

23. Devanesan P, Santen RJ, Bocchinfuso WP, Korach KS, Rogan EG, Cavalieri E. *Catechol estrogen metabolites and conjugates in mammary tumors and hyperplastic tissue from estrogen receptor-α knock-out (ERKO)/Wnt-1 mice: implications for initiation of mammary tumors.* Carcinogenesis. 2001;22:1573-6.

24. Rogan EG, Badawi AF, Devanesan PD, Meza J, Edney JA, West WW, et al. *Relative imbalances in estrogen metabolism and conjugation in breast tissue of women with carcinoma: potential biomarkers of susceptibility to cancer.* Carcinogenesis. 2003;24:697-702.

25. Zhu BT, Conney AH. *Functional role of estrogen metabolism in target cells: review and perspectives.* Carcinogenesis. 1998;19:1-27.

26. Cavalieri EL, Stack DE, Devanesan PD, Todorvic R, Dwivedy I, Higginbotham S, et al. *Molecular origin of cancer: catechol estrogen-3,4-quinones as endogenous tumor initiators.* Proc Natl Acad Sci USA. 1997;94:10937-42.

27. Li KM, Todorovic R, Devanesan P, Higginbotham S, Kofeler H, Ramanathan R, et al. *Metabolism and DNA binding studies of 4-hydroxyestradiol and estradiol-3,4-quinone in vitro and in female ACI rat mammary gland in vivo.* Carcinogenesis. 2004;25:289-97.

28. Fernandez SV, Russo IH, Russo J. *Estradiol and its metabolites 4-hydroxyestradiol and 2-hydroxyestradiol induce mutations in human breast epithelial cells.* Int J Cancer. 2006;118:1862-8.

29. Zhao Z, Kosinska W, Khmelnitsky M, Cavalieri EL, Rogan EG, Chakravarti D, et al. *Mutagenic activity of 4-hydroxyestradiol, but not 2-hydroxyestradiol, in BB Rat2 embryonic cells, and the mutational spectrum of 4-hydroxyestradiol.* Chem Res Toxicol. 2006;19:475-9.

30. Russo J, Fernandez SV, Russo PA, Fernbaugh R, Sheriff FS, Lareef HM, et al. *17-Beta-estradiol induces transformation and tumorigenesis in human breast epithelial cells.* FASEB J. 2006;20:1622-34.

31. Shekhar MP, Nangia-Makker P, Wolman SR, Tait L, Heppner GH, Visscher DW. *Direct action of estrogen on sequence of progression of human preneoplastic breast disease.* Am J Pathol. 1998;152:1129-32.

32. Newbold RR, Liehr JG. *Induction of uterine adenocarcinoma in CD-1 mice by catechol estrogens.* Cancer Res. 2000;60:235-7.

33. Yue W, Santen RJ, Wang JP, Li Y, Verderame MF, Bocchinfuso WP, et al. *Genotoxic metabolites of estradiol in breast: potential mechanism of estradiol induced carcinogenesis.* J Steroid Biochem Mol Biol. 2003;86:477-86.

34. Embrechts J, Lemiere F, Van Dongen W, Esmans EL, Buytaert P, Van Marck E, et al. *Detection of estrogen DNA-adducts in human breast tumor tissue and healthy tissue by combined nano LC-nano ES tandem mass spectrometry.* J Am Soc Mass Spectrom. 2003;14:482-91.

35. Markushin Y, Zhong W, Cavalieri EL, Rogan EG, Small GJ, Yeung ES, et al. *Spectral characterization of catechol estrogen quinone (CEQ)-derived DNA adducts and their identification in human breast tissue extract.* Chem Res Toxicol. 2003;16:1107-17.

36. Zhang QA, R.L., Gross ML. *Estrogen carcinogenesis: specific identification of estrorgen-modified nucleobase in breast tissue from women.* Chem Res Toxicol. 2008;21:1509-13.

37. Belous AR, Hachey DL, Dawling S, Roodi N, Parl FF. *Cytochrome P450 1B1-mediated estrogen metabolism results in estrogen-deoxyribonucleoside adduct formation.* Cancer Res. 2007;67:812-7.

38. http://www.cancer.gov/bcrisktool.

39. Armstrong K, Eisen A, Weber B. *Assessing the risk of breast cancer.* New Engl J Med. 2000;342:564-71.

40. Claus EB, Risch N, Thompson WD. *Autosomal dominant inheritance of early-onset breast cancer. Implications for risk prediction.* Cancer. 1994;73:643-51.

41. Gail MH, Brinton LA, Byar DP, Corle DK, Green SB, Schairer C, et al. *Projecting individualized probabilities of developing breast cancer for white females who are being examined annually.* J Natl Cancer Inst. 1989;81:1879-86.

42. Tyrer J, Duffy SW, Cuzick J. *A breast cancer prediction model incorporating familial and personal risk factors.* Statist Med. 2004;23:1111-30.

43. Huggins C, Grand LC, Brillantes FP. *Critical significance of breast structure in the induction of mammary cancer in the rat.* Proc Natl Acad Sci U S A. 1959;45:1294-300.

44. Russo J. *Basis of cellular autonomy in the subsceptibility to carcinogenesis.* Toxicol Pathol. 1983;11:149-66.

45. Russo J, Hu YF, Yang X, Russo IH. *Developmental, cellular, and molecular basis of human breast cancer.* J Natl Cancer Inst Monogr. 2000:17-37.

46. Russo J, Reina D, Frederick J, Russo IH. *Expression of phenotypical changes by human breast epithelial cells treated with carcinogens in vitro.* Cancer Res. 1988;48:2837-57.

47. Colditz GA. *Epidemiology and prevention of breast cancer.* Cancer Epidemiol Biomarkers Prev. 2005;14:768-72.

48. Russo J, Russo IH. *Influence of differentiation and cell kinetics on the susceptibility of the rat mammary gland to carcinogenesis.* Cancer Res. 1980;40:2677-87.

49. Moolgavkar SH, Day NE, Stevens RG. *Two-stage model for carcinogenesis: Epidemiology of breast cancer in females.* J Natl Cancer Inst. 1980;65:559-69.

50. Parl FF, Dawling S, Roodi N, Crooke PS. *Estrogen metabolism and breast cancer: a risk model.* Ann NY Acad Sci. 2009;1155:68-75.

51. Kastan MB, Bartek J. *Cell-cycle checkpoints and cancer.* Nature. 2004;432:316-23.

52. Sherr CJ. *Cancer cell cycles.* Science. 1996;274:1672-7.

CHAPTER 2

Endogenous Causes

2.1. ESTROGENS
2.1.1. SYNTHESIS
2.1.1.1. INTRODUCTION

Estrogens have long been recognized as the prime risk factor for the development of breast cancer (1, 2).

Current models of breast cancer risk prediction are mainly based on cumulative estrogen exposure and incorporate such factors as age, age at menarche, and age at first live birth (3, 4, www.cancer.gov/bcrisktool). Defining the role of estrogens in the development of breast cancer requires an understanding of their biosynthesis, chemistry, and metabolism. All steroid hormones are derived from the parent compound, cholesterol, by sequential modifications of the carbon skeleton (5-7). The steroidogenic pathways in the adrenal cortex and gonads (testis and ovary) are shown in Figure 2.1.1.

References

1. MacMahon B, Feinleib M. *Breast cancer in relation to nursing and menopausal history.* J Natl Cancer Inst. 1960;24:733-53.

2. Pike MC, Krailo MD, Henderson BE, Casagrande JT, Hoel DG. *'Hormonal' risk factors, 'breast tissue age' and the age-incidence of breast cancer.* Nature. 1983;303:767-70.

3. Gail MH, Brinton LA, Byar DP, Corle DK, Green SB, Schairer C, et al. *Projecting individualized probabilities of developing breast cancer for white females who are being examined annually.* J Natl Cancer Inst. 1989;81:1879-86.

4. Rockhill B, Spiegelman D, Byrne C, Hunter DJ, Colditz GA. *Validation of the Gail et al. model of breast cancer risk prediction and implications for chemoprevention.* J Natl Cancer Inst. 2001;93:358-66.

5. Miller WL. *Minireview: regulation of steroidogenesis by electron transfer.* Endocrinology. 2005;146:2544-50.

6. Nebert DW, Dalton TP. *The role of cytochrome P450 enzymes in endogenous signalling pathways and environmental carcinogenesis.* Nature Rev. 2006;6:947-60.

7. Nebert DW, Russell DW. *Clinical importance of the cytochromes P450.* Lancet. 2002;360:1155-62.

2.1.1.2. CYTOCHROME P450 11A (CYP11A)

The conversion of cholesterol to pregnenolone is the first and rate-limiting step in the synthesis of all steroid hormones. This conversion involves three steps: 20-hydroxylation, 22-hydroxylation, and cleavage of the C_{20}-C_{22} bond to produce pregnenolone and isocaproic acid. These three steps are mediated by a single mitochondrial enzyme, cholesterol side-chain cleavage cytochrome P450 (P450scc), the product of the CYP11A gene (1). The CYP11A gene is comprised of nine exons spanning 30 kb at 15q24.1, encoding a 60 kDa protein composed of 521 amino acids. Deficiency of the CYP11A gene or mutations in the gene encoding the steroidogenic acute regulatory protein (StAR) cause congenital lipoid adrenal hyperplasia, in which the adrenal cortex becomes engorged with cholesterol (2, 3). The defect in the conversion of cholesterol to pregnenolone results in severe impairment of steroid biosynthesis in the adrenal gland and gonads. The deficient adrenal steroido-genesis leads to salt wasting, hyponatremia, hypovolemia, and death in infancy, although patients can survive to adulthood with appropriate mineralocorticoid- and gluco-corticoid replacement therapy. In genetic males (46XY), the deficient steroidogenesis in the fetal testis results in external female genitalia. Thorough examination of the CYP11A gene has revealed rare miscoding SNPs (4). The most common polymorphism involves a pentanucleotide repeat [(TAAAA)n] in the promoter region, located 529 bp upstream from the translation start site (487 bp upstream from exon 1) (5). The number of repeats ranges from 4 to 10 with varying frequencies across ethnic groups. The common alleles contain 4, 6, or 8 repeats, accounting for >90% of total alleles (4, 6-8). There was no association between the [(TAAAA)n] polymorphism and levels of circulating sex hormones in women (6-8). Several studies observed an association with the polycystic ovary syndrome (PCOS), which is characterized by hyperan-drogenemia and represents the most common cause of anovulatory infertility and hirsutism (5, 9-12). However, a large-scale analysis of over 2000 women did not detect any association of the polymorphism with either PCOS or serum testosterone concentrations (13).

Two studies have examined the association of the penta-nucleotide polymorphism with breast cancer risk. In the Shanghai Breast Cancer Study (1015 cases, 1082 controls) the frequency of the 8 repeat allele was more common in cases (12.6%) than controls (8.5%) (OR 1.6; 95% CI 1.3 - 1.9) (8). No association with the repeat polymorphism was found in the Multiethnic Cohort Study (1615 cases, 1962 controls) although a comprehensive haplotype analysis (36 SNPs) across the entire CYP11A gene showed a significant difference between cases and controls (p = 0.006) (4).

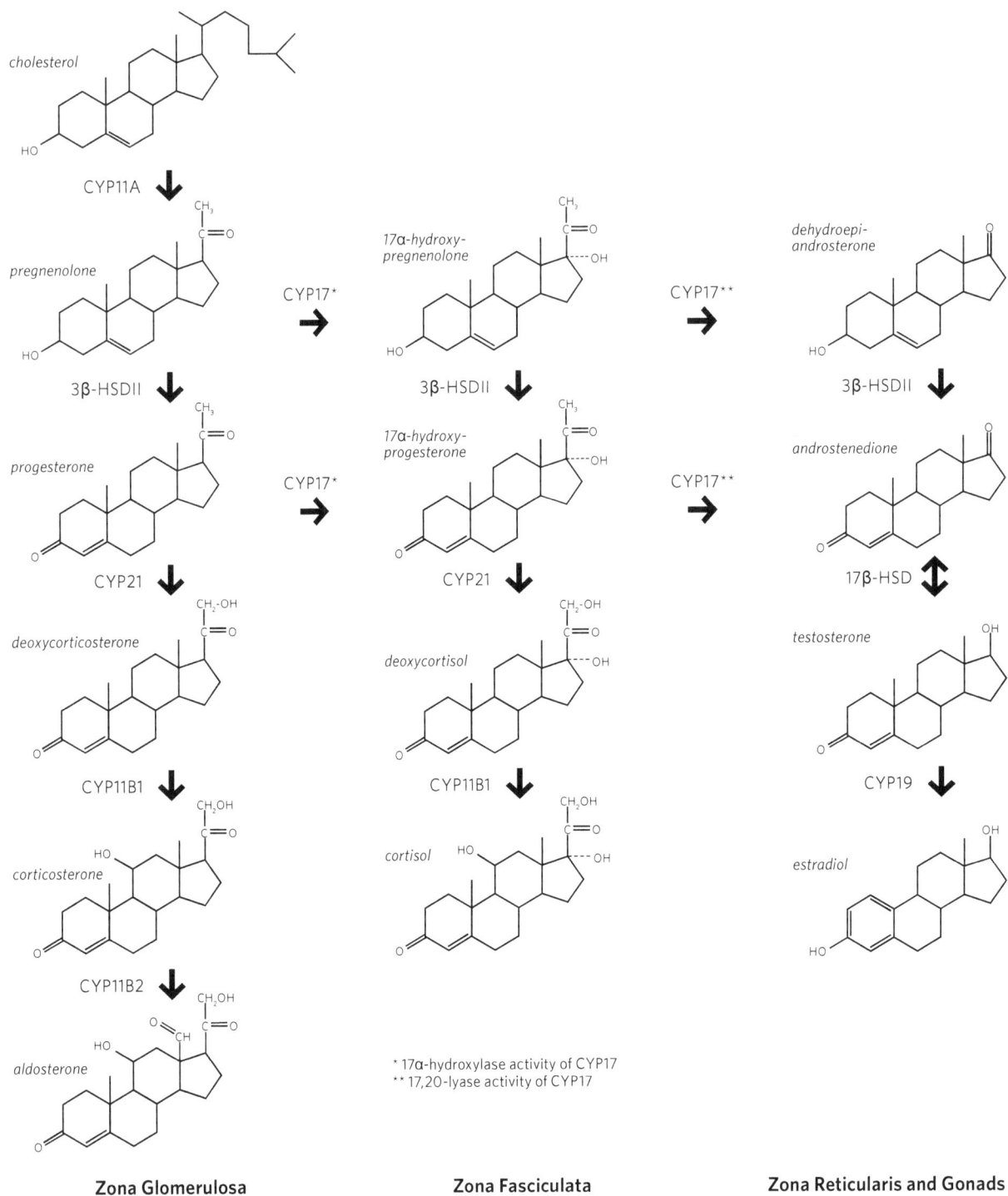

Figure 2.1.1. Overview of steroid biosynthesis in adrenal cortex and gonads.

References

1. Chung B, Matteson KJ, Voutilainen R, Mohandas TK, Miller WL. *Human cholesterol side-chain cleavage enzyme, P450scc: cDNA cloning, assignment of the gene to chromosome 15, and expression in the placenta.* Proc Natl Acad Sci USA. 1986;83:8962-6.

2. Bose HS, Sugawara T, Strauss JF, Miller WL. *The pathophysiology and genetics of congenital lipoid adrenal hyperplasia.* N Engl J Med. 1996;335:1870-8.

3. Hauffa BP, Miller WL, Grumbach MM, Conte FA, Kaplan SL. *Congenital adrenal hyperplasia due to deficient cholesterol side-chain cleavage activity (20, 22-desmolase) in a patient treated for 18 years.* Clin Endocrinol. 1985;23:481-93.

4. Setiawan VW, Cheng I, Stram DO, Giorgi E, Pike MC, Van Den Berg D, et al. *A systematic assessment of common genetic variation in CYP11A and risk of breast cancer.* Cancer Res. 2006;66:12019-25.

5. Gharani N, Waterworth DM, Batty S, White D, Gilling-Smith C, Conway GS, et al. *Association of the steroid synthesis gene CYP11a with polycystic ovary syndrome and hyperandrogenism.* Hum Mol Genet. 1997;6:397-402.

6. Garcia-Closas M, Herbstman J, Schiffman M, Glass A, Dorgan JF. *Relationship between serum hormone concentrations, reproductive history, alcohol consumption and genetic polymorphisms in pre-menopausal women.* Int J Cancer 2002;102:172-8.

7. San Millan JL, Sancho J, Calvo RM, Escobar-Morreale HF. *Role of the pentanucleotide (tttta)ₙ polymorphism in the promoter of the CYP11a gene in the pathogenesis of hirsutism.* Fertil Steril. 2001;75:797-802.

8. Zheng W, Gao YT, Shu XO, Wen W, Cai Q, Dai Q, et al. *Population-based case-control sudy of CYP11A gene polymorphism and breast cancer risk.* Cancer Epidemiol Biomarkers Prev. 2004;13:709-14.

9. Daneshmand S, Weitsman SR, Navab A, Jakimiuk AJ, Magoffin DA. *Overexpression of theca-cell messenger RNA in polycystic ovary syndrome does not correlate with polymorphisms in the cholesterol side-chain cleavage and 17α-hydroxylase/C₁₇₋₂₀ lyase promoters.* Fertil Steril. 2002;77:274-80.

10. Diamanti-Kandarakis E, Bartzis MI, Bergiele AT, Tsianateli TC, Kouli C. *Microsatellite polymorphism (tttta)ₙ at -528 base pairs of gene CYP11α influences hyperandrogenemia in patients with polycystic ovary syndrome.* Fertil Steril. 2000;73:735-41.

11. Franks S. *Polycystic ovary syndrome.* N Engl J Med. 1995;333:853-61.

12. Franks S, Gharani N, Waterworth D, Batty S, White D, Williamson R, et al. *The genetic basis of polycystic ovary syndrome.* Hum Reprod. 1997;12:2641-8.

13. Gaasenbeek M, Powell BL, Sovio U, Haddad L, Gharani N, Bennett A, et al. *Large-scale analysis of the relationship between CYP11A promoter variation, polycystic ovarian syndrome, and serum testosterone.* J Clin Endocrinol Metab. 2004;89:2408-13.

2.1.1.3. CYTOCHROME P450 17 (CYP17)

Pregnenolone can be dehydrogenated at the 3β-position by 3β-hydroxysteroid dehydrogenase (3β-HSDII) to form progesterone. The 21-carbon pregnenolone and progesterone then undergo two reactions in sequence (Figure 2.1.1.). First, they undergo 17α-hydroxylation, in which an oxygen is inserted into a C-H bond, yielding 17α-hydroxypregnenolone and 17α-hydroxyprogesterone, respectively. Second, 17α-OH-pregnenolone and 17α-OH-progesterone undergo cleavage of the C-17,20 carbon bond via the 17,20-lyase reaction to yield acetic acid and the 19-carbon, 17-ketosteroids dehydroepiandrosterone (DHEA) and androstenedione, respectively. The sequential 17α-hydroxylase and 17,20-lyase reactions are catalyzed by a single enzyme, cytochrome P450 17α (CYP17) (1, 2). CYP17 catalyzes both steroidal transformations on a single active site, but the 17,20-lyase activity is facilitated by cytochrome b5, which promotes the interaction with P450 oxidoreductase for electron transfer (3). Thus, the 17α-hydroxylase and 17,20-lyase activities of CYP17 can be regulated independently, which places the enzyme in a key branching point of human steroidogenesis (Figure 2.1.1.). In the gonads and the adrenal zona reticularis, the presence of both 17α-hydroxylase and 17,20-lyase activities results in the synthesis of DHEA and androstenedione, which serve as precursors for all androgens and estrogens. In the adrenal zona fasciculata, where the 17α-hydroxylase activity of CYP17 predominates, the glucocorticoid cortisol is produced. Finally, in the adrenal zona glomerulosa, where CYP17 is absent, steroidogenesis is directed toward the mineralocorticoid aldosterone (Figure 2.1.1.). The tissue-specific expression of CYP17 is unique among steroidogenic enzymes and obviously constitutes a major factor in determining the class of steroids produced.

The CYP17 gene at 10q24.3 consists of 8 exons encoding a 57 kDa protein containing 508 amino acids (4). The CYP17 gene is expressed in adrenal, testis, ovary, and adipose tissue (5, 6). Fat tissue also contains an isoform with a 156-bp in-frame deletion in the first exon (5). The expression of the CYP17 gene is stimulated by cAMP, which binds to the CYP17 promoter. Androgens are produced by CYP17 and repress its expression via a negative feedback mechanism. The latter involves binding of the androgen receptor to the cAMP-responsive region of the CYP17 promoter, which interferes with the binding of transcription factors (7). Sex steroid production requires both 17α-hydroxylase and 17,20-lyase activities and deficiencies in one or both actions due to inherited CYP17 mutations lead to ambiguity in male genital formation and lack of progression into puberty in females (2). Genetic females (46XX) have no pubertal development with primary amenorrhea, lack of axillary and pubic hair, and hypoplastic breasts. In genetic males (46XY), the fetal testis does not produce testosterone, resulting in absence of masculinization and female external genitalia. The clinical manifestations of CYP17 deficiency vary from the classical type to less symptomatic forms depending on the degree of the CYP17 gene alteration. CYP17 deficiency also causes an increase in ACTH levels, which provoke overproduction of mineralocorticoids, leading to low-renin hypertension, hypokalemia, and metabolic alkalosis.

The CYP17 gene contains three known polymorphisms (8-10): a G/A transition at nucleotide 47 in the 5'-untranslated region and a C/A transition at nucleotide 5471 in intron 6. The third polymorphism, a T/C substitution at nucleotide 27 in the 5'-untranslated region, is quite common. The CC genotype is most common in East Asian (32%) and Japanese (22%) populations and present in Caucasians (14%) and African Americans (13%) (11). Examination of the T and C alleles at +27 bp, designated A1 and A2, respectively, showed a significant association of the A2 allele with familial PCOS (8). The T/C substitution was hypothesized to create an additional Sp-1 binding site (CCACC box) with enhanced promoter activity and an increased rate of transcription (12). However, experiments with recombinant Sp-1 protein showed no evidence for binding to the +27 bp promoter region or for increased transcriptional activity of the C allele (10, 13). Since CYP17 is the rate-limiting enzyme in the biosynthesis of androgens and the latter are precursors of estrogens, the A2 allele has been studied extensively for possible associations with increased estrogen levels, earlier onset of puberty, or increased susceptibility to breast cancer.

Studies investigating the relationship between the CYP17 A2 allele and circulating E_2 levels in premenopausal women have yielded inconsistent results (14-19). In one study, E_2 levels measured around day 11 of the menstrual women with genotypes A1/A2 and A2/A2, respectively, compared with A1/A1 women (14). Another study found 19% and 42% higher E_2 levels for A1/A2 and A2/A2, respectively, but only among women with BMI values ≤25 kg/m^2 (18). A third study observed elevated levels of E_1 and DHEA (20). However, two other studies did not find statistically significant differences in E_2 levels among CYP17 genotypes (15, 19). One reason for the discrepant results may be that these studies relied on hormone measurements in a single blood sample from each woman. A study based on daily collected saliva samples for one entire menstrual cycle of 60 women found 54% higher E_2 levels in women with A2/A2 genotype compared to women with A1/A1 genotype (p = 0.0001) and 37% higher than women with A1/A2 genotype (p = 0.0008) (21). Studies in postmenopausal women of blood and urine estrogen levels also yielded inconsistent results (16, 19, 20, 22-25). Analysis of over 5000 postmenopausal women from the National Cancer Institute Breast and Prostate Cancer Cohort Consortium (BPC3) found no evidence for an association of the A2 allele with circulating hormones (26). The CYP17 C/T polymorphism did not show any association with the onset of breast development, one of the first signs of puberty (27). A review of 11 retrospective studies showed a modest association between the A2/A2 genotype and earlier menarche in seven studies (11). However, analysis of over 10,000 women in the BPC3 database found no evidence of interaction (26).

Multiple studies have examined the association of the A2 allele with breast cancer risk and obtained inconsistent results (10, 12, 20, 28-38). A meta-analysis of 15 case-control studies including 4,227 breast cancers and 4,730 controls showed no evidence of an association with increased risk (39). A systematic analysis of the CYP17 gene by dense genotyping across the entire 58 kb locus revealed several haplotype-tagging polymorphisms besides the +27 bp T/C substitution. Analysis of the latter in 5333 breast cancer cases and 7069 controls in the BPC3 data base found no evidence for an association of the A2 allele with increased breast cancer risk (26). Therefore, this comprehensive investigation does not support the hypothesis that common CYP17 germline variants contribute substantially to breast cancer risk.

References

1. Auchus RJ, Miller WL. *Molecular modeling of human P450c17 (17α-Hydroxylase/17,20-Lyase): Insights into reaction mechanisms and effects of mutations.* Mol Endocrinol. 1999;13:1169-82.

2. Yanase T, Simpson ER, Waterman MR. *17α-hydroxylase/17,20-lyase deficiency: From clinical investigation to molecular definition.* Endocr Rev. 1991;12:91-108.

3. Geller DH, Auchus RJ, Miller WL. *P450c17 mutations R347H and R358Q selectively disrupt 17,20-lyase activity by disrupting interactions with P450 oxidoreductase and cytochrome b5.* Mol Endocrinol. 1999;13:167-75.

4. Chung BC, Picado-Leonard J, Haniu M, Bienkowski M, Hall PF, Shively JE, et al. *Cytochrome P450c17 (steroid 17α-hydroxylase/17,20 lyase): Cloning of human adrenal and testis cDNAs indicates the same gene is expressed in both tissues.* Proc Natl Acad Sci USA. 1987;84:407-11.

5. Puche C, Jose M, Cabero A, Meseguer A. *Expression and enzymatic activity of the P450c17 gene in human adipose tissue.* Eur J Endocrinol. 2002;146:223-9.

6. Sasano H, Okamoto M, Mason JI, Simpson ER, Mendelson CR, Sasano N, et al. *Immunolocalization of aromatase, 17α-hydroxylase and side-chain-cleavage cytochromes P-450 in the human ovary.* J Reprod Fert. 1989;85:163-9.

7. Burgos-Trinidad M, Youngblood GL, Maroto MR, Scheller A, Robins DM, Payne AH. *Repression of cAMP-induced expression of the mouse P450 17α-hydroxylase/C17-20 lyase gene (Cyp17) by androgens.* Mol Endocrinol. 1997;11:87-96.

8. Carey AH, Waterworth D, Patel K, White D, Little J, Novelli P, et al. *Polycystic ovaries and premature male pattern baldness are associated with one allele of the steroid metabolism gene CYP17.* Hum Mol Genet. 1994;3:1873-6.

9. Crocitto LE, Feigelson HS, Yu MC, Kolonel LN, Henderson BE, Coetzee GA. *Short report on DNA marker at candidate locus.* Clin Genet. 1997;52:68-9.

10. Kristensen VN, Haraldsen EK, Anderson KB, Lonning PE, Erikstein B, Karesen R, et al. *CYP17 and breast cancer risk: The polymorphism in the 5' flanking area of the gene does not influence binding to Sp-1.* Cancer Res. 1999;59:2825-8.

11. Sharp L, Cardy AH, Cotton SC, Little J. *CYP17 gene polymorphisms: Prevalence and associations with hormone levels and related factors. A HuGE review.* Am J Epidemiol. 2004;160:729-40.

12. Feigelson HS, Coetzee GA, Kolonel LN, Ross RK, Henderson BE. *A polymorphism in the CYP17 gene increases the risk of breast cancer.* Cancer Res. 1997;57:1063-5.

13. Lin CJ, Martens JW, Miller WL. *NF-1C, Sp1, and Sp3 are essential for transcription of the human gene for P450c17 (steroid 17α-hydroxylase/17,20 lyase) in human adrenal NCI-H295A cells.* Mol Endocrinol. 2001;15:1277-93.

14. Feigelson HS, Shames LS, Pike MC, Coetzee GA, Stanczyk FZ, Henderson BE. *Cytochrome P450c17α gene (CYP17) polymorphism is associated with serum estrogen and progesterone concentrations.* Cancer Res. 1998;58:585-7.

15. Garcia-Closas M, Herbstman J, Schiffman M, Glass A, Dorgan JF. *Relationship between serum hormone concentrations, reproductive history, alcohol consumption and genetic polymorphisms in pre-menopausal women.* Int J Cancer 2002;102:172-8.

16. Hong CC, Thompson HJ, Jiang C, Hammond GL, Tritchler D, Yaffe M, et al. *Association between the T27C polymorphism in the cytochrome P450 c17α (CYP17) gene and risk factors for breast cancer.* Breast Cancer Res Treat. 2004;88:217-30.

17. Lurie G, Maskarinec G, Kaaks R, Stanczyk FZ, Marchand LL. *Association of genetic polymorphisms with serum estrogens measured multiple times during a 2-year period in premenopausal women.* Cancer Epidemiol Biomarkers Prev. 2005;14:1521-7.

18. Small CM, Marcus M, Sherman SL, Sullivan AK, Manatunga AK, Feigelson HS. *CYP17 genotype predicts serum hormone levels among pre-menopausal women.* Hum Reprod. 2005;20:2162-7.

19. Travis RC, Churchman M, Edwards SA, Smith G, Verkasalo PK, Wolf CR, et al. *No association of polymorphisms in CYP17, CYP19, and HSD17-B1 with plasma estradiol concentrations in 1,090 British women.* Cancer Epidemiol Biomarkers Prev. 2004;13:2282-4.

20. Haiman CA, Hankinson SE, Spiegelman D, Colditz GA, Willet WC, Speizer FE, et al. *The relationship between a polymorphism in CYP17 with plasma hormone levels and breast cancer.* Cancer Res. 1999;59:1015-20.

21. Jasienska G, Kapiszewska M, Ellison PT, Kalemba-Drozdz M, Nenko I, Thune I, et al. *CYP17 genotypes differ in salivary 17-α estradiol levels: A study based on hormonal profiles from entire menstrual cycles.* Cancer Epidemiol Biomarkers Prev. 2006;15:2132-5.

22. Haiman CA, Hankinson SE, Colditz GA, Hunter DJ, De Vivo I. *A polymorphism in CYP17 and endometrial cancer risk.* Cancer Res. 2001;61:3955-60.

23. Onland-Moret NC, Van Gils CH, Roest M, Grobbee DE, Peeters PH. *Cyp17, urinary sex steroid levels and breast cancer risk in postmenopausal women.* Cancer Epidemiol Biomarkers Prev. 2005;14:815-20.

24. Somner J, McLellan S, Cheung J, Mak YT, Frost ML, Knapp KM, et al. *Polymorphisms in the P450 c17 (17-hydroxylase/17,20-lyase) and P450 c19 (aromatase) genes: Association with serum sex steroid concentrations and bone mineral density in postmenopausal women.* J Clin Endocrinol Metab. 2004;89:344-51.

25. Tworoger SS, Chubak J, Aiello EJ, Ulrich CM, Atkinson C, Potter JD, et al. *Association of CYP17, CYP19, CYP1B1, and COMT polymorphisms with serum and urinary sex hormone concentrations in postmenopausal women.* Cancer Epidemiol Biomarkers Prev. 2004;13:94-101.

26. Setiawan VW, Schumacher FR, Haiman CA, Stram DO, Albanes D, Altshuler D, et al. *CYP17 genetic variation and risk of breast and prostate cancer from the National Cancer Institute Breast and Prostate Cancer Cohort Consortium (BPC3).* Cancer Epidemiol Biomarkers Prev. 2007;16:2237-46.

27. Kadlubar FF, Berkowitz GS, Delongchamp RR, Wang C, Green BL, Tang G, et al. *The CYP3A4*1B variant is related to the onset of puberty, a known risk factor for the development of breast cancer.* Cancer Epidemiol Biomarkers Prev. 2003;12:327-31.

28. Ambrosone CB, Moysich KB, Furberg H, Freudenheim JL, Bowman ED, Ahmed S, et al. *CYP17 genetic polymorphism, breast cancer, and breast cancer risk factors.* Breast Cancer Res. 2003;5:R45-R51.

29. Bergman-Jungestrom M, Gentile M, Lundin AC, Group SEBC, Wingren S. *Association between CYP17 gene polymorphism and risk of breast cancer in young women.* Int J Cancer. 1999;84:350-3.

30. Dunning AM, Healey CS, Pharoah PD, Foster NA, Lipscombe JM, Redman KL, et al. *No association between a polymorphism in the steroid metabolism gene CYP17 and risk of breast cancer.* Br J Cancer. 1998;77:2045-7.

31. Feigelson HS, McKean-Cowdin r, Coetzee GA, Stram DO, Kolonel LN, Henderson BE. *Building a multigenic model of breast cancer susceptibility: CYP17 and HSD17B1 are two important candidates.* Cancer Res. 2001;61:785-9.

32. Hamajima N, Iwatta H, Obata Y, Matsuo K, Mizutani M, Iwase T, et al. *No association of the 5' promoter region polymorphism of CYP17 with breast cancer risk in Japan.* Jpn J Cancer Res. 2000;91:880-5.

33. Helzlsouer KJ, Huang HY, Strickland PT, Hoffman S, Alberg AJ, Comstock GW, et al. *Association between CYP17 polymorphisms and the development of breast cancer.* Cancer Epidemiol Biomarkers Prev. 1998;7:945-9.

34. Huang CS, Chern HD, Chang KJ, Cheng CW, Hsu SM, Shen CY. *Breast cancer risk associated with genotype polymorphism of the estrogen-metabolizing genes CYP17, CYP1A1, and COMT: A multigenic study on cancer susceptibility.* Cancer Res. 1999;59:4870-5.

35. Mitrunen K, Jourenkova N, Kataja V, Eskelinen M, Kosma VM, Benhamou S, et al. *Steroid metabolism gene CYP17 polymorphism and the development of breast cancer.* Cancer Epidemiol Biomarkers Prev. 2000;9:1343-8.

36. Miyoshi Y, Iwao K, Ikeda N, Egawa C, Noguchi S. *Genetic polymorphism in CYP17 and breast cancer risk in Japanese women.* Eur J Cancer. 2000;36:2375-9.

37. Spurdle AB, Hopper JL, Dite GS, Chen X, Cui J, McCredie MR, et al. *CYP17 promoter polymorphism and breast cancer in australian women under age forty years.* J Natl Cancer Inst. 2000;92:1674-81.

38. Weston A, Pan CF, Bleiweiss IJ, Ksieski HB, Roy N, Maloney N, et al. *CYP17 genotype and breast cancer risk.* Cancer Epidemiol Biomarkers Prev. 1998;7:941-4.

39. Ye Z, Parry JM. *The CYP17 MspA1 polymorphism and breast cancer risk: a meta-analysis.* tagenesis. 2002;17:119-26.

2.1.1.4. ESTROGENS IN PLASMA AND BREAST TISSUE

In the circulation, estrogens are bound to albumin and sex hormone binding globulin (SHBG), with only a small fraction circulating unbound or free (between 1 and 5%). Since only the free hormone can enter cells, the unbound estrogen is considered biologically active. The binding affinity of SHBG for E_2 is orders of magnitude higher than the affinity of albumin for E_2 (1, 2). Thus, E_2 readily dissociates from albumin but remains tightly bound to SHBG. Consequently, the sum of free and albumin-bound fractions can be considered bioavailable E_2. Mathematical methods based on the law of mass action have been developed to calculate the various E_2 fractions using total hormone, SHBG, and albumin concentrations (3). Comparisons between calculated and measured values of both free and non-SHBG-bound E_2 have shown a high correlation (r > 0.91) (4). Thus, it is only necessary to measure total E_2 and SHBG concentrations to reliably calculate free E_2.

Only one enzyme, P450 aromatase, is capable of synthesizing estrogens. Since mammary epithelial cells do not express aromatase and therefore do not synthesize estrogens, an important question is how do estrogens gain access to mammary epithelium? The mode of supplying estrogens is different for pre- and postmenopausal women. The source of circulating estrogens in premenopausal women is predominantly the ovary, the main estrogen synthesized is E_2 and uptake of E_2 from the circulation is the primary mechanism for maintenance of E_2 concentrations

Table 2.1.1. Plasma estrogen concentrations in pre- and postmenopausal women

	E_2 (pg/ml)*	E_1 (pg/ml)	E_3 (ng/ml)
Pre-menopause	20 – 350	15 – 150	<2
Follicular phase	150 – 750	100 – 250	<2
Midcycle	30 – 450	15 – 200	<2
Luteal phase	0 – 30	15 – 55	<2
Post-menopause			

* Conversion to SI units of pmol/L may be made by using the equation pmol/L ≈ pg/ml x 3.7. E_1, estrone; E_2, 17β-estradiol; E_3, estriol.

in breast epithelium. This situation changes drastically in postmenopausal women in whom most of the estrogens are synthesized in peripheral tissues, especially in adipose tissue. Consequently, circulating estrogen levels are at least ten-fold lower after menopause than before (Table 2.1.1.). The reliable measurement of postmenopausal estrogen levels is below the sensitivity of current serum immunoassays, which have a lower level of quantitation of at best 10 pg/mL (5). Liquid chromatography-tandem mass spectrometry offers reliable quantitation in the 1 pg/mL range (6). After menopause, E_1 becomes the main estrogen resulting from the aromatization of the adrenal androgen androstenedione in adipose tissue (7) (Figure 2.1.2.). E_2 is produced through the aromatization of adrenal testosterone and through the reduction of E_1. Testosterone is produced in postmenopausal women in the ovaries and adrenals. Women who have had bilateral ovariectomies have lower concentrations of testosterone compared with women with intact ovaries (8).

Despite the decline in plasma estrogen levels after menopause, tumor E_2 levels in pre- and postmeno-pausal breast cancers do not differ significantly (9, 10). Similarly, the analysis of E_2 concentrations in nipple aspirate fluids from pre- and postmenopausal women showed no difference (11). The breast tissue/plasma E_2 ratio is 1:1 in premenopausal and 10 to 50:1 in post-menopausal women. Thus, E_2 levels in both normal and malignant breast tissues of postmenopausal women are one order of magnitude higher than in plasma from the same individual (12-15). How does the breast tissue of postmenopausal women maintain premenopausal levels in the presence of dramatically lower concentrations of estrogens in the plasma?

The main circulating estrogen in postmenopausal women is E_1 rather than E_2, and ~90% of E_1 is conjugated estrone sulfate, E_1S. By contrast, in breast tumors of postmeno-pausal women the reverse situation exists and E_2 is present in higher concentrations than E_1. The question arises whether plasma E_1S can serve as source for E_2 in

mammary epithelium of postmenopausal women. In order to examine the ability of breast tissue to take up E_1S, ^3H-E_1S or $E_1^{35}S$ were infused into postmenopausal women with advanced breast cancer (16). After infusion of ^3H-E_1S, significant levels of ^3H-E_1S were detected in normal and malignant breast tissues. The tissue:plasma ratios in a study group of five patients were 0.14 4 0.13 and 0.24 4 0.12, respectively. Similar ^3H-E_1S tissue:plasma ratios were detected after infusion of ^3H-E_1 indicating that the ^3H-E_1S detected in breast tissues after infusion of ^3H-E_1S may have originated from hydrolysis of ^3H-E_1S in tissues other than the breast, with subsequent uptake and sulfation in breast tissues. The sulfation is carried out by sulfotransferases (SULTs), whereas the hydrolysis is catalyzed by steroid sulfatase (STS). After infusion of $E_1^{35}S$, no significant levels of radioactivity were detectable in normal or malignant breast tissues demonstrating the high catalytic efficiency of STS and suggesting that E_1S is unable to cross cell membranes due to its polar nature (16). However, a superfamily of membrane transporter proteins has been identified that can mediate the specific uptake of organic anions, such as steroid sulfates (17). Immunohistochemical studies showed that one of these transporters, organic anion transporter polypeptide B (OATP-B), was strongly expressed in the majority of epithelial cells in invasive ductal carcinomas (18). Thus, the combined action of STS, SULTs, and membrane transporters appears necessary to explain intravascular E_1S levels and the high intracellular concentration of estrogens in postmenopausal women.

Can plasma E_1S serve as a source for E_2 in mammary epithelium? A third enzyme, namely 17β-hydroxysteroid dehydrogenase type 1 (17β-HSD1), is required for conversion of E_1 to the biologically more active E_2. MCF-7 breast cancer cells have been shown to convert physiological concentrations of E_1S to quantities of E_2 capable of stimulating cell growth (19). Metabolic studies with ^3H-E_1S showed the ability of normal breast tissue in pre- and postmenopausal women to convert E_1S into the active E_2 (20). The concentrations of E_1S, E_2S, and E_2 were higher in tumor tissue than in adjacent fat tissue (21). These findings suggest functional integration of STS, SULTs, and 17ß-HSD1 in mammary epithelial cells permitting intracellular production of E_2 from extracellular E_1S. In summary, a complex pathway is responsible for the presence of E_2 in mammary epithelial cells that involves the interaction of four different groups of enzymes, P450 aromatase, sulfotransferases, steroid sulfatase, and 17β-hydroxysteroid dehydrogenases (Figure 2.1.2.). Aromatase, STS, and 17β-HSD1 share an association with the membrane structure of the endoplasmic reticulum. The lipid bilayer may act as a solvent for the hydrophobic steroid and facilitate passage through the membrane to enter and leave the active enzyme site. The enzymes are described in the following chapters.

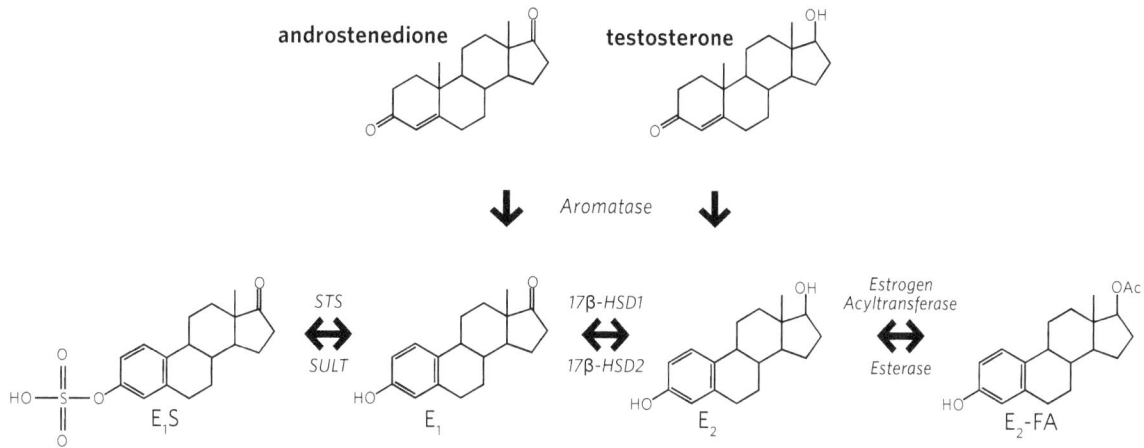

Figure 2.1.2. Overview of enzymes contributing to the presence of the main estrogens, 17β-estradiol (E_2) and estrone (E_1), in mammary epithelial cells.

Estrogens are synthesized by aromatase, which transforms the androgens androstenedione and testosterone into E_1 and E_2, respectively. E_1 and E_2 can be converted to sulfates and E_2 to 17β-fatty acid esters by sulfotransferase (SULT) and estrogen acyltransferase, respectively. The sulfate conjugates and fatty acid esters, in turn, can serve as a source for the parent estrogens, which are liberated by steroid sulfatase (STS) and esterase. 17β-hydroxysteroid dehydrogenases (17β-HSD) types 1 and 2 are responsible for the interconversion of E_1 and E_2.

References

1. Dunn JF, Nisula BC, Rodbard D. *Transport of steroid hormones: binding of 21 endogenous steroids to both testosterone-binding globulin and corticosteroid-binding globulin in human plasma.* J Clin Endocrinol Metab. 1981;53:58-68.

2. Pardridge WM. *Transport of protein-bound hormones into tissues in vivo.* Endocr Rev. 1981;2:103-23.

3. Sodergard R, Backstrom T, Shanbhag V, Carstensen H. *Calculation of free and bound fractions of testosterone and estradiol-17 beta to human plasma proteins at body temperature.* J Steroid Biochem. 1982;16:801-10.

4. Endogenous Hormones and Breast Cancer *Collaborative Group: Body Mass index, serum sex hormones, and breast cancer risk in postmenopausal women.* J Natl Cancer Inst. 2003;95:1218-26.

5. Santen RJ, Lee JS, Wang S, Demers LM, Mauras N, Wang H, et al. *Potential role of ultra-sensitive estradiol assays in estimating the risk of breast cancer and fractures.* Steroids. 2008;73:1318-21.

6. Blair IA. *Analysis of estrogens in serum and plasma from postmenopausal women: past present, and future.* Steroids. 2010;75:297-306.

7. Siiteri PK. *Adipose tissue as a source of hormones.* Am J Clin Nutr. 1987;45:277-82.

8. McTiernan A, Wu L, Barnabei VM, Chen C, Hendrix S, Modugno F, et al. *Relation of demographic factors, menstrual history, reproduction and medication use to sex hormone levels in postmenopausal women.* Breast Cancer Res Treat. 2008;108:217-31.

9. van Landeghem AA, Poortman J, Nabuurs M, Thijssen JH. *Endogenous concentration and subcellular distribution of estrogens in normal and malignant human breast tissue.* Cancer Res. 1985;45:2900-6.

10. Vermeulen A, Deslypere JP, Paridaens R, Leclercq G, Roy F, Heuson JC. *Aromatase, 17 beta-hydroxysteroid dehydrogenase and intratissular sex hormone concentrations in cancerous and normal glandular breast tissue in postmenopausal women.* Eur J Cancer Clin Oncol. 1986;22:515-25.

11. Chatterton RT, Jr., Khan SA, Heinz R, Ivancic D, Lee O. *Patterns of sex steroid hormones in nipple aspirate fluid during the menstrual cycle and after menopause in relation to serum concentrations.* Cancer Epidemiol Biomarkers Prev. 2010;19:275-9.

12. Edery M, Goussard J, Dehennin L, Scholler R, Reiffsteck J, Drosdowsky MA. *Endogenous oestradiol-17beta concentration in breast tumours determined by mass fragmentography and by radioimmunoassay: relationship to receptor content.* Eur J Cancer. 1981;17:115-20.

13. Geisler J, Lonning PE. *Aromatase inhibition: translation into a successful therapeutic approach.* Clin Cancer Res. 2005;11:2809-21.

14. Pasqualini JR, Chetrite G, Blacker C, Feinstein M-C, Delalonde L, Talbi M, et al. *Concentrations of estrone, estradiol, and estrone sulfate and evaluation of sulfatase and aromatase activities in pre- and postmenopausal breast cancer patients.* J Clin Endocrinol Metabol. 1996;81:1460-4.

15. Pasqualini JR, Chetrite G, Nguyen B-L. *Estrone sulfate-sulfatase and 17 beta-hydroxysteroid dehydrogenase activities: a hypothesis for their role in the evolution of human breast cancer from hormone-dependence to hormone-independence.* J Steroid Biochem Mol Biol. 1995;53:407-12.

16. Purohit A, Riaz AA, Ghilchik MW, Reed MJ. *The origin of oestrone sulphate in normal and malignant breast tissues in postmenopausal women.* Horm Metab Res. 1992;24:532-6.

17. Hagenbuch B, Meier PJ. *The superfamily of organic anion transporting polypeptides.* Biochim Biophys Acta. 2003;1609:1-18.

18. Pizzagalli F, Varga Z, Huber RD, Folkers G, Meier PJ, St-Pierre MV. *Identification of steroid sulfate transport processes in the human mammary gland.* J Clin Endocrinol Metab. 2003;88:3902-12.

19. Santner SJ, Ohlsson-Wilhelm B, Santen RJ. *Estrone sulfate promotes human breast cancer cell replication and nuclear uptake of estradiol in MCF-7 cell cultures.* Int J Cancer. 1993;54:119-24.

20. Soderqvist G, Olsson H, Wilking N, von Schoultz B, Carlstrom K. *Metabolism of estrone sulfate by normal breast tissue: influence of menopausal status and oral contraceptives.* J Steroid Biochem Mol Biol. 1994;48:221-4.

21. Blankenstein M, J. S, Daroszewski J, Milewicz A, Thijssen JH. *Estrogens in plasma and fatty tissue from breast cancer patients and women undergoing surgery for non-oncological reasons.* Gynecol Endocrinol. 1992;6:13-7.

2.1.1.5. AROMATASE

CYP 19A1 Gene Structure and Aromatase Enzyme Function. C19 steroids, i.e., androgens, are the biosynthetic precursors of estrogens. The biosynthesis of estrogens from androgens such as androstenedione, dehydroepian-drosterone and testosterone is mediated by the action of a specific cytochrome P450 enzyme called aromatase (Figure 2.1.3.) (1). Aromatase is encoded by the CYP19A1 gene which spans ~120 kb at 15q21.1 (2-5). The gene comprises a ~90 kb region that contains multiple untranslated first exons and a ~30 kb region, which contains nine coding exons, II – X, with the translation start site in the middle of exon II. Structure-function studies identified the catalytic site containing amino acids 298 to 313, which are encoded by exon VIII (6-9). Other amino acids participate in steroid binding, e.g., 133Ile in exon IV and 365Arg in exon IX (8, 9). Exon X encodes the heme-binding region with the conserved 437Cys serving as fifth coordinating ligand for the heme iron (10). Two alternative polyadenylation sites downstream from the stop codon give rise to two mRNA species of 3.4 and 2.9 kb (Figure 2.1.4.).

Aromatase catalyzes the series of reactions that leads to formation of the phenolic A ring characteristic of estrogens, with concomitant loss of the C-19 angular methyl group (Figure 2.1.3.). Associated with the aromatase is a flavoprotein, NADPH-cytochrome P450 reductase, a ubiquitous protein of the endoplasmic reticulum that is responsible for transferring reducing equivalents from NADPH to any microsomal form of cytochrome P450 with which it comes into contact (11). In a three-step process, each step requiring 1 mol of O_2, 1 mol of NADPH, and coupling with its redox partner NADPH-cytochrome P450 reductase, aromatase converts androstenedione and testosterone to estrone and 17β-estradiol, respectively (12). The C-19 angular methyl group is the site of attack by oxygen (13, 14). The steroid substrate presents its β-side to the active site of the aromatase and one of the three hydrogen atoms at C-19 becomes hydroxylated. This is followed by a second hydroxylation yielding the 19, 19-dihydroxysteroid, which may reversibly dehydrate to the 19-oxosteroid. The third oxidative reaction involves a peroxidative attack at C-19 that leads to a concerted conformational change with 1β-hydrogen elimination to give

a 1,10-double bond, followed by protonation of the 3-oxo group, and finally abstraction of the 2β-hydrogen (7). The newly formed estrogen is released from the active site of the enzyme. In summary, the first two steps are C19-methyl hydroxylation steps and the third involves the aromatization of the steroid A-ring, unique to aromatase. Since this is the only reaction in vertebrates capable of introducing an aromatic ring into a molecule, aromatase is the rate-limiting enzyme in estrogen biosynthesis.

Aromatase is highly selective for androgens, in contrast to many microsomal cytochrome P450 enzymes, which are not highly substrate specific. The structural basis for the androgen specificity and the unique catalytic mechanism is provided by an androgen-specific cleft that snugly binds androstenedione (12). The high-resolution x-ray structure of native aromatase purified from human placenta revealed that the catalytic cleft is comprised of 14 hydrophobic and hydrophilic amino acid residues. The hydrophobic residues and porphyrin rings of heme pack tightly against the steroid backbone, forming a cavity of no more than 400 $Å^3$ complementary in shape to the bound steroid. The C-19 methyl group, which undergoes oxidation during the hydroxylation reaction, is centered over the heme-bound iron atom, placing the C-2 carbon adjacent to 309Asp, which facilitates the proton extraction during the subsequent aromatization reaction (12, 15).

The CYP19A1 gene is unique among cytochrome P450 genes by having multiple promoters that drive the expression of different untranslated exons I in a tissue-specific manner (16). Tissue-specific promoter usage in syncytiotrophoblasts of the placenta, ovarian granulosa cells and corpus luteum, testicular Leydig cells, chondrocytes and osteoblasts of bone, various sites in the brain, as well as in adipose tissue gives rise to transcripts with unique 5'-noncoding termini (Figure 2.1.4.). For example, placental transcripts contain at their 5'-end the untranslated exon I.1 which is located 89 kb upstream from the translation start site in exon II (17). By contrast, transcripts in adipose tissue contain another distal untranslated exon, I.4, which is located 20 kb downstream from exon I.1 (18, 19). Transcripts in brain and bone contain the untranslated exons If and I.6, respectively. Finally, transcripts in the ovary contain sequence at their 5'-end that is immediately upstream of the translation start site in exon II (42). Several other untranslated exons have been characterized, e.g., I.2, I.5, 1.7, and 2a (16, 20). The corresponding transcripts are present at only trace levels in tissues and their significance is uncertain. Regardless of the distance, splicing of every 5'-untranslated exon to form the mature transcript occurs at a common, highly promiscuous 3'-acceptor site (AG/A\ GACT) that is upstream of the translation start site in exon II. This means that although transcripts in different

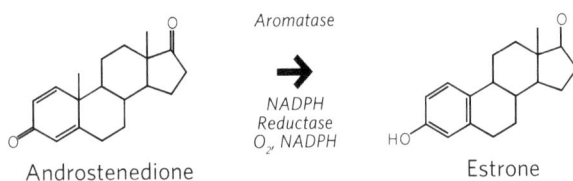

Figure 2.1.3. Aromatase reaction converts androstenedione to E_1 and, in a similar fashion, testosterone to E_2.

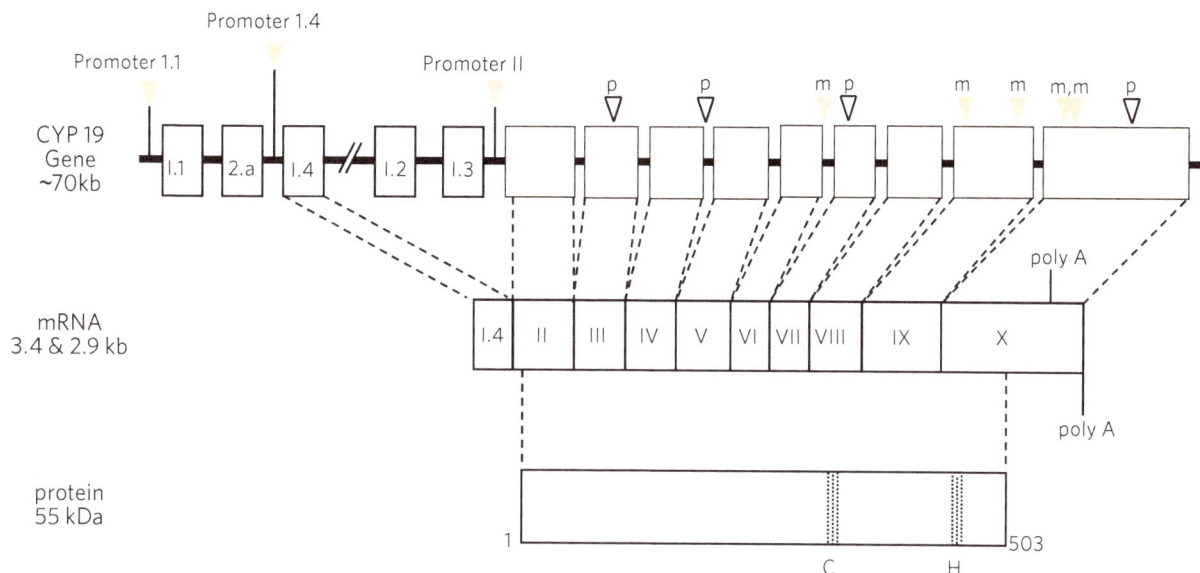

Figure 2.1.4. Overview of CYP19A1 gene which encodes aromatase mRNA and protein.
The CYP19A1 gene at 15q21.1 spans ~120 kb. The gene comprises nine coding exons, II – X (solid boxes) and at least ten untranslated exons I (open boxes), which are expressed from different promoters (solid arrows) in a tissue-specific manner. All exons I are spliced into the common 3' acceptor splice junction upstream from the translation start site in exon II. The exon I.4-specific transcript shown here is present in adipose tissue of the normal human breast. So-called promoter switching leads to PII-specific transcripts in adipose tissue adjacent to breast cancer. Two alternative polyadenylation sites (polyA) downstream from the stop codon in exon X give rise to two aromatase mRNA species of 3.4 and 2.9 kb. Regardless of the 5'-splicing and 3'-polyadenylation events, all transcripts contain the same open reading frame of 1509 nucleotides that are translated into a protein of 503 amino acids. The shaded areas indicate the catalytic domain (C) and the heme-binding region (H). The lengths of the coding exons II – X are drawn to scale, while those of the introns and non-coding exons I are not.

tissues have different 5'-termini, the nucleotide encoding the open reading frame of 503 amino acids and therefore the protein expressed in all tissues is always the same. In summary, tissue-specific expression of the CYP19A1 gene is determined by the use of tissue-specific promoters, which give rise to transcripts with unique 5'-noncoding termini all of which are translated into a single 55 kDa aromatase protein that is identical in all tissues (21). The enzyme is located primarily in the microsomal compartment of the cell but may also be expressed at the cell surface (22).

The tissue-specific regulation of aromatase expression is ultimately controlled by different response elements in the promoter regions of each of the untranslated first exons (Table 2.1.2.). For example, the ovarian promoter PII contains a cAMP-response element, CRE (TGCACGTCA), and a binding site for steroidogenic factor-1 (SF-1) (CAAGGTCA), an orphan nuclear receptor that was originally identified as a transcription factor coordinating expression of steroid hydroxylases in adrenocortical cells (23). Ovarian aromatase expression is primarily regulated by the gonadotropin FSH, whose activity is mediated by cAMP. Specifically, FSH binds to its receptor and activates adenylcyclase, which results in formation of cAMP and activation of protein kinase A (PKA). PKA then phosphory-

lates CREB, which binds to CRE to activate PII. The adipose promoter I.4 contains a glucocorticoid response element (GRE), a SP1 binding site, an interferon-gamma-activating sequence (GAS), and an activating protein-1 (AP-1) binding site (18, 19). Thus, the I.4 promoter induces expression of exon I.4-specific aromatase transcripts in fat tissue in the presence of glucocorticoids and members of the class I cytokine family, such as interleukin-6 and interleukin-11 (IL-6, IL-11) as well as tumor necrosis factor α (TNFα). The class I cytokines use the Jak/STAT signaling pathway to activate the GAS element, while TNFα probably acts through ceramide formation to stimulate binding of both c-fos and c-jun to the AP-1 element (18, 19). In this fashion, each tissue displays a unique way of regulating estrogen biosynthesis (Table 2.1.2.).

One mechanism by which obesity is linked to the increased incidence of estrogen-dependent breast cancer in post-menopausal women involves the adipokines leptin and adiponectin (24). The altered adipokine milieu associated with obesity leads to down-regulation of AMP-activated protein kinase (AMPK). One of the actions of AMPK is the phosphorylation and inhibition of cAMP-responsive element binding protein (CREB)-regulated transcription coactivator 2 (CRTC2), which regulates aromatase

Table 2.1.2. CYP19A1 expression is up-regulated via tissue-specific promoters and modulated by transcription factors

Tissue	Promoter	Regulatory Element	Transcription Factor	Regulator	References
Placenta	I.1	RXRE	Retinoid X receptor Mash-2	Retinoids hypoxia	(27, 28)
Adipose	I.4	GRE Sp1	GR	Glucocorticoids, IL-6, IL-11, TNFα	(18)
Ovary	II	CRE	CREB1	cAMP, FSH, IGF-1 PKA, PKC	(29-33)
Breast Cancer	II and I.3	S1, AP-1 AGGTCA	SF-1, ERRα-1 Leptin, LRH-1	PGE2	(34-37)

AP-1, activator protein 1; cAMP, cyclic adenosine monophosphate; CRE, cAMP responsive element; CREB1, cAMP responsive element binding protein 1; ERRα-1, estrogen-related receptor α; FSH, follicle-stimulating hormone; GR, glucocorticoid receptor; GRE, glucocorticoid response element; IGF-1, insulin-like growth factor-1; IL, interleukin; LRH-1, liver receptor homologue-1; PGE2, prostaglandin E2; PKA, protein kinase A; PKC, protein kinase C; RXRE, retinoid X DNA response element; SF-1, steroidogenic factor 1; Sp1, specificity protein 1; TNFα, tumor necrosis factor α.

promoter II activity (25). Since AMPK also inhibits the proliferation of both malignant and nonmalignant cells (26), the obesity-induced decrease of AMPK may lead to increased expression of aromatase and E2 production as well as increased cell proliferation.

Aromatase Expression in Adipose Tissue. The significance of adipose tissue as a source of estrogens was recognized in experiments comparing the fractional conversion of circulating androstenedione to E_1 in pre- and postmenopausal women (38, 39). There was a striking increase with obesity, suggesting that most of the extragonadal conversion occurred in adipose tissue, but also that there was an equally striking increase with age for any given body weight. Several studies have examined the aromatase expression in adipose tissue of women. In normal adipose tissue, aromatase expression was highest in buttocks, followed by thighs, and lowest in the abdomen (40). Expression increased in direct proportion to advancing age of the women. Regardless of site or age, the aromatase transcripts were predominantly exon I.4-specific with lower copy numbers of promoter II- and exon I.3-specific transcripts. Aromatase expression appears to occur primarily, if not exclusively, in adipose fibroblasts rather than in mature adipocytes (41). These fibroblasts, which surround the adipocytes, are mesenchymal elements (preadipocytes) believed to have the capacity to develop into adipocytes.

A morphometric analysis of adipose tissue in normal breast of premenopausal women examined the fibroblast to adipocyte ratio in relation to aromatase expression (42). Samples from three regions (outer, upper, and inner) of both breasts of 13 women undergoing reduction mammoplasty showed the highest fibroblast to adipocyte ratio and the highest aromatase transcript levels in the outer breast region. This distribution pattern corresponds to the most common site of breast cancer. Aromatase transcript levels were 3- to 4-fold higher in breast adipose

tissue of cancer patients than those of cancer-free individuals, even when the adipose tissue from the cancer patient was taken from a quadrant with no detectable tumor (43-45). The immunohistochemical analysis of breast cancers showed aromatase staining predominantly in stromal cells with negligible staining in malignant and benign epithelial cells, with intensities corresponding to the biochemically measured levels of basal and glucocorticoid- and cAMP-induced enzyme activity (46, 47). However, studies with different antibodies also visualized aromatase expression in malignant breast epithelial cells (48).

Aromatase Expression in Breast Cancer. A number of studies have attempted to relate the aromatase activity or expression in adipose tissue to the tumor site in breast cancer patients. Only one group did not find a higher activity in the affected than the non-affected quadrants (49). All other groups observed a direct relationship between tumor proximity and aromatase expression in the tumor-bearing quadrant (44, 50-52). These results suggest that tumors may secrete substances, which influence aromatase expression in adjacent stroma to establish a gradient of aromatase expression. IL-6 and IL-11 have been shown to be elaborated either by cells derived from adipose tissue, tumor-derived fibroblasts, or tumor tissue itself to stimulate aromatase expression in adipose fibroblasts and tumor-derived fibroblasts in culture (53, 54). Since aromatase expression is confined to the stromal component, one can envision the establishment of a local positive feedback loop whereby fibroblasts surrounding a developing tumor produce estrogens, which stimulate the tumor to produce a variety of cytokines and other factors. The second arm of the feedback loop was demonstrated in T47D breast cancer cells, in which expression of IL-11 transcripts was induced threefold by the addition of E_2 to the culture medium (53). Addition of E_2 to MCF-7 cells markedly increased the release of soluble IL-6 receptor, which is necessary for IL-6 responsiveness (55). Thus, a positive feedback loop is established via mesenchymal-epithelial interactions. Once the initiating events leading to tumor development have occurred, locally produced estrogens and tumor-derived growth factors may act by paracrine and autocrine mechanisms to sustain tumor growth (56).

Interestingly, the increased expression of aromatase in adipose tissue adjacent to breast cancer is associated with so-called promoter switching (43, 57, 58). Instead of exon I.4-specific transcripts found in normal adipose tissue, aromatase transcripts are primarily expressed from promoters I.3 and II. The molecular mechanism underlying the promoter switching involves two regulatory elements, S1 and CRE, located between promoters I.3 and II (59). S1 can function as an enhancer or a repressor element depending on the trans-factors that interact with the element. S1 behaves as an enhancer when ERRα-1 binds and as a repressor when EAR-2, COUP-TF, and RARγ bind (37, 60). The function of S1 depends on the expression levels of these nuclear receptors in cells, which differ between normal and malignant cells. In normal breast tissue, the three factors EAR-2, COUP-TF, or RARγ are expressed at high levels, directing S1 to function as a negative regulatory element that suppresses promoters I.3 and II. The converse is true in breast cancers, which express high levels of ERRα-1. Like S1, the CRE sequence can also function as enhancer or suppressor depending on its respective trans-factors, i.e., CREB1 and the zinc-finger protein Snail (59, 61). Levels of Snail are high in normal breast and low in breast cancer, while the converse applies to CREB1. The latter is stimulated via cAMP by prostaglandin E2 (PGE2), which is secreted by tumor epithelial cells, fibroblasts, and macrophages surrounding the tumor (19). PGE2 is a major driver of aromatase expression in breast cancer. PGE2 stimulates aromatase activity and expression in adipose stromal cells by activating prostaglandin receptor subtypes EP_1 and EP_2, linked to protein kinase C (PKC) and PKA, respectively (19, 62). Interestingly, treatment of adipose stromal cells with PGE2 induced expression of both aromatase and the orphan nuclear receptor, liver receptor homologue-1 (LRH-1) (63). LRH-1 was originally identified as a liver-specific transcription factor regulating bile acid synthesis and cholesterol metabolism (64). In breast tissue, LRH-1 was shown to stimulate aromatase expression by increasing promoter II activity (65). Moreover, LRH-1 expression paralleled aromatase expression, i.e., both exhibit low levels in normal breast tissue and markedly elevated levels in malignant breast and surrounding adipose tissue. Yet another hepatic protein, the bile acid receptor FXR (farnesoid X receptor), is also expressed in normal and malignant breast epithelial cells (67). FXR can downregulate LRH-1 by inducing the short heterodimer partner (SHP), an atypical nuclear receptor that lacks its own DNA binding

domain. The FXR-SHP-LHR-1 cascade, in turn, can be manipulated to inhibit aromatase expression (66, 67). A third regulatory element upstream of promoter I.3 was shown to bind the CCAAT/enhancer binding protein β (C/EBP β) in preadipocytes and C/EBPδ in breast cancer epithelial cells, resulting in upregulated aromatase expression (68, 69). p65, a subunit of nuclear factor-κB (NF-κB), prevented the upregulation by inhibiting the binding of C/EBPδ.

In normal and malignant breast tissue, aromatase expression appears to occur primarily in fibroblasts (preadipocytes) rather than in adipocytes or malignant epithelium. Aromatase expression in breast cancer cell lines such as MCF-7, T47D, and MDA-MB-231 is detectable at very low levels by mRNA analysis, protein immunoprecipitation, and activity measurement (70-72). The reduced expression may be due to different regulatory mechanisms of promoter activity in breast cancer cells compared to fibroblasts (73). Aromatase activity in MCF-7 and T47D cells can be induced by dibutyryl cAMP, which stimulates cAMP-dependent protein kinases and potentiated by phorbol ester, which stimulates PKC (74). The aromatase activity in MCF-7 cells is not high enough to produce sufficient estrogen for stimulating cell growth (74, 75). However, coculture of MCF-7 cells with stromal cells resulted in up-regulation of aromatase expression in the epithelial cell line (48). Aromatase-transfected MCF-7 cells inoculated into ovariectomized nude mice were able to aromatize androgens to increase local E_2 levels and stimulate cell proliferation (76). A combined xenograft of estrogen-dependent human breast cancer cells and fibroblasts obtained from a patient with benign breast disease also showed enhanced tumor growth following local testosterone administration, demonstrating *in situ* synthesis of estrogen by aromatization of androgen precursors in fibroblasts (77). Overexpression of aromatase in the mammary glands of transgenic mice resulted in morphological abnormalities including ductal hyperplasia, glandular dysplasia, and multinucleated epithelial cells, which are all indicative of preneoplastic changes (78). The morphological changes were accompanied by increased expression of ERα, PR, cyclin D1, and cyclin E (79, 80).

CYP19A1 Mutations. Several mutations of the CYP19A1 gene have been identified that give rise to either aromatase deficiency or overexpression. The resulting lack of estrogen leads to female pseudohermaphroditism and progressive virilization at puberty, primary amenorrhea, and polycystic ovaries. In one patient a GT/GC mutation in the consensus 5' splice acceptor sequence of intron 6 resulted in the use of a cryptic splice acceptor site and incorporation of 87 bases from intron 6, which were translated into an additional 29 amino acids inserted into the middle of the mature polypeptide (81). The patient was homozygous and the parents were obligate heterozygotes for the mutation. In two other patients homozygous non-synonymous mutations were identified in exon 9, one at bp 1094 (G/A) resulting in Arg365Gln (82) and the other at bp 1123 (C/T) resulting in Arg375Cys (83). A fourth patient was a compound heterozygote with a different single-base change on each allele, bp 1303 (C/T) and bp 1310 (G/A), both in exon 10, which resulted in substitutions Arg435Cys and Cys437Tyr, respectively (84). Both residues are in the heme-binding region, 435Arg being highly conserved and 437Cys comprising the fifth coordinating ligand of the heme iron. Studies of the corresponding recombinant mutated proteins confirmed that the latter mutations were associated with functional loss of enzyme activity. A rearrangement in chromosome15q21.2-3, which caused the coding region of the CYP19A1 gene to lie adjacent to constitutively active cryptic promoters that normally transcribe other genes, resulted in overexpression of aromatase, elevated plasma estrogen levels and severe gynecomastia of prepubertal onset (85).

CYP19A1 Polymorphisms, Aromatase Activity, and Plasma Estrogen Levels. Since aromatase is a critical enzyme in estrogen biosynthesis, the CYP19A1 gene has been scrutinized for functionally important polymorphisms that affect estrogen levels and influence breast cancer risk. A definitive investigation sequenced all coding exons, all upstream untranslated exons plus their presumed core promoter regions, all exon-intron splice junctions, and a portion of the 3'-untranslated region of CYP19A1 using 240 DNA samples from four ethnic groups, including 60 African-American, 60 Caucasian-American, 60 Han Chinese-American, and 60 Mexican-American subjects (86). The systematic study identified 88 polymorphisms resulting in 44 haplotypes including four non-synonymous polymorphisms: Trp39Arg in exon 2, Thr201Met in exon 5, Arg264Cys in exon 7 (rs700519), and Met364Thr in exon 9. The frequency of the variant alleles of the non-synonymous polymorphisms varied between ethnic groups. For example, only two of the four polymorphisms (Thr201Met and Arg264Cys) occur in Caucasian- and African-Americans (87). The frequency of 201Met was 0.05 in both groups and 264Cys occurred in 0.025 of Caucasian- and 0.225 of African-Americans. Three of the four polymorphisms were found in Han Chinese-Americans (Trp39Arg, Arg264Cys, and Met364Thr) with variant allele frequencies of 0.067, 0.117, and 0.008, respectively. The variants were expressed in COS-1 cells and the catalytic activity and protein level of the 201Met allozyme was similar to the wild-type enzyme (87). The 39Arg and 201Met allozymes were similar to the wild-type enzyme except that the immunoreactive protein level of 39Arg was decreased. The 264Cys variant showed significant decreases in levels of activity and immunoreactive protein, although another study obtained the same catalytic activity for the 264Arg

and 264Cys isoforms (88). Finally, 364Thr exhibited drastically reduced levels of immunoreactive protein and activity with a 4-fold increase in K_m value with androstenedione as substrate (86). Thus, the CYP19A1 coding region does not contain any non-synonymous SNPs with clear-cut functional differences, except Met364Thr, which was observed only in Han Chinese-Americans.

There are other silent polymorphisms, e.g., in exon 3 (nucleotide 240G/A) and exon 10 (nucleotide 1529C/T, 19 nucleotides downstream from the stop codon) in the 3' untranslated region of the gene. A tetranucleotide $(681TTTA)_n$ in intron 4 can be repeated 7 to 13 times, with 7 and 11 repeats occurring most commonly. Also in intron 4, 50 bp upstream from the tetranucleotide repeat is a 3-bp deletion, 630 delCTT, which is found in some individuals with the $(681TTTA)_7$ repeat. This generates two alleles, namely 7 repeats with the 3-bp deletion [7r(-3)] and 7 repeats without the deletion [7r]. One study observed that postmenopausal women carrying two [7r(-3)] alleles had about 20% lower plasma E_2 and E_1 concentrations than women with [7r] wild-type (89). However, another study found no association between the tetranucleotide repeat in intron 4 or the polymorphism in exon 10 and plasma E_2 concentration in postmenopausal women (90).

Large-scale haplotype-tagging SNPs and common haplotypes spanning the coding and 5'/3'-untranslated regions of CYP19A1 were shown to be significantly associated with a 10 to 20% increase in endogenous estrogen levels in postmenopausal women (91). The two most significantly associated SNPs, rs749292 (A allele) and rs727479 (A allele) were located in haplotype blocks spanning the 5'-untranslated region, presumably affecting promoter activity and thereby aromatase expression levels. Additional SNPs in either the 5'- or 3'-untranslated regions (rs6493497, rs7176005, rs10046) have also been associated with elevated plasma E_2 levels (92, 93).

CYP19A1 Polymorphisms and Breast Cancer Risk. Numerous association studies of SNPs or haplotypes have been published to elucidate the etiology of breast cancer (94, 95). Studies of individual nonsynonymous SNPs, such as Trp39Arg or Arg264Cys have yielded inconsistent results (88, 96-100). Similarly, the silent SNP in exon 3 (nucleotide 240G/A), the tetranucleotide repeat $(TTTA)_n$ in intron 4, and the polymorphism in exon 10 (nucleotide 1529C/T) showed no consistent association with breast cancer risk (96, 101-108).

A comprehensive haplotype analysis of the aromatase gene in relation to breast cancer risk was performed in the Multiethnic Cohort, a population-based study of women (African-American, Hawaiian, Japanese, Latina, Caucasian) living in Los Angeles or Hawaii (96).

The authors initially used 74 densely spaced SNPs (one every ~2.6 kb) spanning 190 kb of the CYP19A1 locus to characterize linkage disequilibrium and haplotype patterns among 69 or 70 healthy individuals from each ethnic population. They identified four genomic segments (blocks 1 – 4, SNPs 4 – 66) with a high degree of linkage disequilibrium, i.e., little evidence of historical recombination between SNPs. Within each block there was limited diversity of haplotypes, most of which were shared across populations. For example, the 50-kb block 4 (SNPs 44 – 66), which covers the entire coding region of the aromatase gene, contained 10 common haplotypes. Based on this database, the authors selected a reduced set of 25 haplotype-tagging SNPs (htSNPs) that provided high predictability of the common haplotypes within each block. Finally, they applied the htSNP set to a case-control study of the multiethnic cohort (1,355 cases, 2,580 controls) to search for differences in breast cancer risk according to the common haplotypes.

Modest associations were observed between haplotypes in block 2 and increased breast cancer risk, but the associations were not consistent across populations. Two specific haplotypes in blocks 2 and 3 were associated with increased breast cancer risk suggesting that women with this long-range haplotype 2b-3c may be carriers of a predisposing breast cancer susceptibility allele. In contrast, the large-scale haplotype-tagging SNP analysis (5,356 cases, 7,129 controls) that showed an association of rs749292 (A allele) and rs727479 (A allele) in the 5′-untranslated region with a 10 to 20% increase in endogenous estrogen levels in postmenopausal women found no significant association with breast cancer risk (91).

Clinical Correlation Studies. Between 40 and 79% of breast cancers are aromatase-positive as determined by enzymatic activity or immunohistochemical staining (46, 109-113). Aromatase expression was unrelated to menopausal status, tumor size, lymph node involvement, histological type and grade, DNA index, and S-phase fraction (44, 110, 111, 114). Aromatase mRNA levels were higher in cancers of elderly patients (80–99 years of age) than in normal breast tissue of the elderly or breast cancers of younger patients (37–70 years of age) (115). Several studies have examined the relationship between intratumoral aromatase activity and ER status but failed to find evidence for a consistent correlation. Aromatase positivity has been reported to be unrelated to ER expression (44, 52, 113), positively correlated to ER expression (109, 111), or inversely correlated to ER positivity (110). Similarly, studies of the prognostic value of aromatase expression have yielded inconsistent results. One study found an inverse relation between aromatase activity and time to tumor relapse while another observed the opposite (113, 116). Two additional studies found no correlation between aromatase

expression and breast cancer prognosis (117, 118).

Steroidal and Non-Steroidal Aromatase Inhibitors.
Aromatase is the main target for inhibition of estrogen production in endocrine treatment and prevention protocols of postmenopausal breast cancer. Various aromatase inhibitors have been developed that can be categorized into two types: steroidal and non-steroidal inhibitors (Figure 2.1.5.) (119). In general, steroidal inhibitors such as 4-hydroxyandrostenedione and exemestane are substrate analogues that compete with the normal androgen substrate for the enzyme binding-site (12, 120, 121). Non-steroidal inhibitors with imidazole or triazole structures interact directly with the prosthetic heme group of the enzyme; examples include aminoglutethimide, letrozole, and vorozole (122, 123). Treatment of postmenopausal breast cancer patients with aromatase inhibitors effectively reduces breast tissue exposure to estrogen. For example, treatment of ten postmenopausal women with breast cancer with the third generation steroidal inhibitor exemestane suppressed plasma levels of E_2, E_1, and E_1 sulfate by 92.2, 94.5, and 93.2%, respectively (121).

Treatment with the third generation non-steroidal inhibitor vorozole reduced the intratumoral aromatase activity 89% and the tumor levels of E_2 and E_1 80 and 64%, respectively (124). Two tightly linked SNPs (rs6493497, rs7176005) in the 5′-untranslated region of the CYP19A1 gene were associated with higher plasma E_2 levels and influenced the response of breast cancer patients to aromatase inhibitors (93). Although similar in their ability to inhibit estrogen synthesis, steroidal and non-steroidal inhibitors differ in other respects. For example, exemestane and its principal metabolite, 17-hydroexemestane, prevent bone loss and lower serum cholesterol and low-density lipoprotein levels in ovariectomized rats, which was not the case with the non-steroidal inhibitor letrozole (125).

Exemestane is structurally related to androstenedione and has some affinity for the androgen receptor. It appears that the protective effect of exemestane and 17-hydroexemestane on bone and lipid metabolism is mediated through their androgenicity rather than through conversion to estrogens. Aromatase inhibitors were more effective than tamoxifen in preventing contralateral cancers in women with early-stage breast cancer (126). In a randomized, placebo-controlled, double-blind study of 4,560 healthy postmenopausal women at high risk of breast cancer exemestane led to a 65% reduction in the diagnosis of invasive breast cancer at a median follow-up of three years (127). These results provide definitive evidence concerning the importance of estrogens in the pathogenesis of breast cancer.

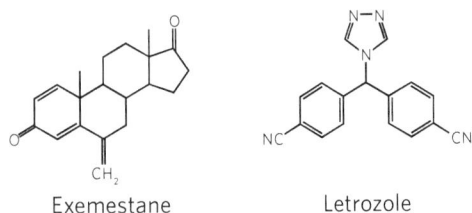

Exemestane   Letrozole

Figure 2.1.5. Structures of aromatase inhibitors. *Exemestane and letrozole are representative of steroidal and non-steroidal inhibitors, respectively.*

Phytoestrogens as Aromatase Inhibitors. Phytoestrogens, in particular flavones, have been shown to be competitive inhibitors of aromatase with respect to the androgen substrate, with K_i values at low micromolar concentrations (128, 129). Computer modeling and mutagenesis studies revealed that the flavones bind to the active site of aromatase in an orientation in which their A- and C-rings mimic rings D and C of the androgen substrate, respectively (130). *In vitro* and *in vivo* experiments indicate that grape seed extracts contain chemicals with a dual action of suppressing aromatase expression and inhibiting aromatase activity (131, 132). Oral ingestion of grape juice by nude mice suppressed the formation of tumors derived from MCF-7 cells that overexpressed aromatase.

References

1. Simpson ER, Mahendroo MS, Means GD, Kilgore MW, Hinshelwood MM, Graham-Lorence S, et al. *Aromatase cytochrome P450, the enzyme responsible for estrogen biosynthesis.* Endocr Rev. 1994;15:342-55.

2. Chen SA, Besman MJ, Sparkes RS, Zollman S, Klisak I, Mohandas T, et al. *Human aromatase: cDNA cloning, Southern blot analysis, and assignment of the gene to chromosome 15.* DNA. 1988;7:27-38.

3. Harada N, Yamada K, Saito K, Kibe N, Dohmae S, Takagi Y. *Structural characterization of the human estrogen synthetase (aromatase) gene.* Biochem Biophys Res Commun. 1990;166:365-72.

4. Means GD, Mahendroo MS, Corbin CJ, Mathis JM, Powell FE, Mendelson CR, et al. *Structural analysis of the gene encoding human aromatase cytochrome P-450, the enzyme responsible for estrogen biosynthesis.* J Biol Chem. 1989;264:19385-91.

5. Sebastian S, Bulun SE. *A highly complex organization of the regulatory region of the human CYP19 (aromatase) gene revealed by the human genome project.* J Clin Endocrinol Metabol. 2001;86:4600-2.

6. Chen S, Zhou D, Swiderek KM, Kadohama N, Osawa Y, Hall PF. *Structure-function studies of human aromatase.* J Steroid Biochem Mol Biol. 1993;44:347-56.

7. Graham-Lorence S, Khalil MW, Lorence MC, Mendelson CR, Simpson ER. *Structure-function relationships of human aromatase cytochrome P-450 using molecular modeling and site-directed mutagenesis.* J Biol Chem. 1991;266:11939-46.

8. Chen S, Zhou D. *Functional domains of aromatase cytochrome P450 inferred from comparative analyses of amino acid sequences and substantiated by site-directed mutagenesis experiments.* J Biol Chem. 1992;267:22587-94.

9. Conley A, Mapes S, Corbin CJ, Greger D, Graham S. *Structural determinants of aromatase cytochrome p450 inhibition in substrate recognition site-1.* Mol Endocrinol. 2002;16:1456-68.

10. Harada N. *Cloning of a complete cDNA encoding human aromatase: Immunochemical identification and sequence analysis.* Biochem Biophys Res Com. 1988;156:725-32.

11. Haniu M, McManus ME, Birkett DJ, Lee TD, Shively JE. *Structural and functional analysis of NADPH-cytochrome P-450 reductase from human liver: complete sequence of human enzyme and NADPH-binding sites.* Biochemistry. 1989; 28:8639-45.

12. Ghosh D, Griswold J, Erman M, Pangborn W. *Structural basis for androgen specificity and oestrogen synthesis in human aromatase.* Nature. 2009;457:219-23.

13. Amarneh B, Corbin CJ, Peterson JA, Simpson ER, Graham-Lorence S. *Functional domains of human aromatase cytochrome P450 characterized by linear alignment and site-directed mutagenesis.* Mol Endocrinol. 1993;7:1617-24.

14. Graham-Lorence S, Amarneh B, White RE, Peterson JA, Simpson ER. *A three-dimensional model of aromatase cytochrome P450.* Protein Sci. 1995;4:1065-80.

15. Chen S, Zhang F, Sherman MA, Kijima I, Cho M, Yuan Y-C, et al. *Structure-function studies of aromatase and its inhibitors: a progress report.* J Steroid Biochem Mol Biol. 2003;86:231-7.

16. Simpson ER, Clyne C, Rubin G, Boon WC, Robertson K, Britt K, et al. *Aromatase-a brief overview.* Ann Rev Physiol. 2002;64:93-127.

17. Kamat A, Alcorn JL, Kunczt C, Mendelson CR. *Characterization of the regulatory regions of the human aromatase (P450 arom) gene involved in placenta-specific expression.* Mol Endocrinol. 1998;12:1764-77.

18. Zhao Y, Mendelson CR, Simpson ER. *Characterization of the sequences of the human CYP19 (aromatase) gene that mediate regulation by glucocorticoids in adipose stromal cells and fetal hepatocytes.* Mol Endocrinol. 1995;9:340-9.

19. Zhao Y, Nichols JE, Valdez R, Mendelson CR, Simpson ER. *Tumor necrosis factor-A stimulates aromatase gene expression in human adipose stromal cells through use of an activating protein-1 binding site upstream of promoter 1.4.* Mol Endocrinol. 1996;10:1350-7.

20. Sebastian S, Takayama K, Shozu M, Bulun SE. *Cloning and characterization of a novel endothelial promoter of the human CYP19 (aromatase P450) gene that is up-regulated in breast cancer tissue.* Mol Endocrinol. 2002;16:2243-54.

21. Simpson ER, Michael MD, Agarwal VR, Hinshelwood MM, Bulun SW, Zhao Y. *Expression of the CYP19 (aromatase) gene: an unusual case of alternative promoter usage.* FASEB J. 1997;11:29-36.

22. Amarneh BA, Simpson ER. *Detection of aromatase cytochrome P450, 17 alpha-hydroxylase cytochrome P450 and NADPH:P450 reductase on the surface of cells in which they are expressed.* Mol Cell Endocrinol. 1996;119:69-74.

23. Parker KL, Rice DA, Lala DS, Ikeda Y, Luo X, Wong M, et al. *Steroidogenic factor 1: an essential mediator of endocrine development.* Rec Prog Horm Res. 2002;57:19-36.

24. Brown KA, Simpson ER. *Obesity and breast cancer: progress to understanding the relationship.* Cancer Res. 2010;70:4-7.

25. Brown KA, McInnes KJ, Hunger NI, Oakhill JS, Steinberg GR, Simpson ER. *Subcellular localization of cyclic AMP-responsive element binding protein-regulated transcription coactivator 2 provides a link between obesity and breast cancer in postmenopausal women.* Cancer Res. 2009;69:5392-9.

26. Motoshima H, Goldstein BJ, Igata M, Araki E. *AMPK and cell proliferation--AMPK as a therapeutic target for atherosclerosis and cancer.* J Physiol. 2006;574:63-71.

27. Jiang B, Kamat A, Mendelson CR. *Hypoxia prevents induction of aromatase expression in human trophoblast cells in culture: Potential inhibitory role of the hypoxia-inducible transcription factor Mash-2 (mammalian achaete-scute* homologous protein-2). Mol Endocrinol. 2000;14:1661-73.

28. Sun T, Zhao Y, Mangelsdorf DJ, Simpson ER. *Characterization of a region upstream of exon I.1 of the human CYP19 (aromatase) gene that mediates regulation by retinoids in human choriocarcinoma cells.* Endocrinology. 1998;139:1684-91.

29. Carlone DL, Richards JS. *Evidence that functional interactions of CREB and SF-1 mediate hormone regulated expression of the aromatase gene in granulosa cells and constitutive expression in R2C cells.* J Steroid Biochem Mol Biol. 1997;61:223-31.

30. Michael MD, Kilgore MW, Morohashi K, Simpson ER. *Ad4BP/SF-1 regulates cyclic AMP-induced transcription from the proximal promoter (PII) of the human aromatase P450 (CYP19) gene in the ovary.* J Biol Chem. 1995;270:13561-6.

31. Steinkampf MP, Mendelson CR, Simpson ER. *Regulation by follicle-stimulating hormone of the synthesis of aromatase cytochrome P-450 in human granulosa cells.* Mol Endocrinol. 1987;1:465-71.

32. Steinkampf MP, Mendelson CR, Simpson ER. *Effects of epidermal growth factor and insulin-like growth factor I on the levels of mRNA encoding aromatase cytochome P-450 of human ovarian granulosa cells.* Mol Cell Endocrinol. 1988;59:93-9.

33. Zeitoun k, Takayama K, Michael MD, Bulun SE. *Stimulation of aromatase P450 promoter (II activity in endometriosis and its inhibition in endometrium are regulated by competitive binding of steroidogenic factor-1 and chicken ovalbumin upstream promoter transcription factor to the same cis-acting element.* Mol Endocrinol. 1999;13:239-53.

34. Catalano S, Marsico S, Giordano C, Mauro L, Rizza P, Panno ML, et al. *Leptin enhances, via AP-1, expression of aromatase in the MCF-7-7 cell line.* J Biol Chem. 2003;278:28668-76.

35. Clyne CD, Speed CJ, Zhou J, Simpson ER. *Liver receptor homologue-1 (LRH-1) regulates expression of aromatase in preadipocytes.* J Biol Chem. 2002;277:20591-7.

36. Kinoshita Y, Chen S. *Induction of aromatase (CYP19) expression in breast cancer cells through a nongenomic action of estrogen receptor a.* Cancer Res. 2003;63:3546-55.

37. Yang C, Zhou D, Chen S. *Modulation of aromatase expression in the breas tissue by erra-1 orphan receptor.* Cancer Res. 1998;58:5695-700.

38. Grodin JM, Siiteri PK, MacDonald PC. *Source of estrogen production in postmenopausal women.* J Clin Endocrinol Metabol. 1973;36:207-14.

39. MacDonald PC, Edman CD, Hemsell DL, Porter JC, Siiteri PK. *Effect of obesity on conversion of plasma androstenedione to estrone in postmenopausal women with and without endometrial cancer.* Am J Obstet Gynecol. 1978;130:448-55.

40. Agarwal VR, Ashanullah CI, Simpson ER, Bulun SE. *Alternatively spliced transcripts of the aromatase cytochrome P450 (CYP19) gene in adipose tissue of women.* J Clin Endocrinol Metab. 1997;82:70-4.

41. Simpson ER, Zhao Y. *Estrogen biosynthesis in adipose.* Ann N Y Acad Sci. 1996;784:18-26.

42. Bulun SE, Sharda G, Rink J, Sharma S, Simpson ER. *Distribution of aromatase P450 transcripts and adipose fibroblasts in the human breast.* J Clin Endocrinol Metab. 1996;81:1273-7.

43. Agarwal VR, Bulun SE, Leitch M, Rohrich R, Simpson ER. *Use of alternative promoters to express the aromatase cytochrome P450 (CYP19) gene in breast adipose tissues of cancer-free and breast cancer patients.* J Clin Endocrinol Metab. 1996;81:3843-9.

44. Bulun SE, Price TM, Aitken J, Mahendroo MS, Simpson ER. *A link between breast cancer and local estrogen biosynthesis suggested by quantification of breast adipose tissue aromatase cytochrome P450 transcripts using competitive polymerase chain reaction after reverse transcription.* J Clin Endocrinol Metab. 1993;77:1622-8.

45. Miller WR. *Aromatase activity in breast tissue.* J Steroid Biochem Mol Biol. 1991;39:783-90.

46. Santen RJ, Martel J, Hoagland M, Naftolin F, Roa L, Harada N, et al. *Stromal spindle cells contain aromatase in human breast tumors.* J Clin Endocrinol Metab. 1994;79:627-32.

47. Santner SJ, Pauley RJ, Tait L, Kaseta J, Santen RJ. *Aromatase activity and expression in breast cancer and benign breast tissue stromal cells.* J Clin Endocrinol Metab. 1997;82:200-8.

48. Miki Y, Suzuki T, Tazawa C, Yamaguchi Y, Kitada K, Honma S, et al. *Aromatase localization in human breast cancer tissues: possible interactions between intratumoral stromal and parenchymal cells.* Cancer Res. 2007;67:3945-54.

49. Thijssen JH, Daroszewski J, Milewicz A, Blankenstein MA. *Local aromatase activity in human breast tissues.* J Steroid Biochem. 1993;44:577-82.

50. O'Neill JS, Elton RA, Miller WR. *Aromatase activity in adipose tissue from breast quadrants: a link with tumour site.* Br Med J Clin Res Ed. 1988;296:741-3.

51. Utsumi T, Harada N, Maruta M, Takagi Y. *Presence of alternatively spliced transcripts of aromatase gene in human breast cancer.* J Clin Endocrinol Metab. 1996;81:2344-9.

52. Vermeulen A, Deslypere JP, Paridaens R, Leclercq G, Roy F, Heuson JC. *Aromatase, 17 beta-hydroxysteroid dehydrogenase and intratissular sex hormone concentrations in cancerous and normal glandular breast tissue in postmenopausal women.* Eur J Cancer Clin Oncol. 1986;22:515-25.

53. Crichton MB, Nichols JE, Zaho Y, Bulun SE, Simpson ER. *Expression of transcripts of interleukin-6 and related cytokines by human breast tumors, breast cancer cells, and adipose stromal cells.* Mol Cell Endocrinol. 1996;118:215-20.

54. Purohit A, Ghilchik MW, Duncan L, Wang DY, Singh A, Walker MM, et al. *Aromatase activity and interleukin-6 production by normal and malignant breast tissues.* J Clin Endocrinol Metab. 1995;80:3052-8.

55. Singh A, Purohit A, Wang DY, Duncan LJ, Ghilchik MW, Reed MJ. *IL-6sR: release from MCF-7 breast cancer cells and role in regulating peripheral oestrogen synthesis.* J Endocrinol. 1995;147:R9-12.

56. Bulun SE, Simpson ER. *Regulation of aromatase expression in human tissues.* Breast Cancer Res Treat. 1994;30:19-29.

57. Harada N. *Aberrant expression of aromatase in breast cancer tissues.* J Steroid Biochem Mol Biol. 1997;61:175-84.

58. Harada N, Utsumi T, Takagi Y. *Molecular and epidemiological analyses of abnormal expression of aromatase in breast cancer.* Pharmacogenetics. 1995;5:S59-64.

59. Chen S, Itoh T, Wu K, Zhou D, Yang C. *Transcriptional regulation of aromatase expression in human breast tissue.* J Steroid Biochem Mol Biol. 2003;83:93-9.

60. Yang C, Yu B, Zhou D, Chen S. *Regulation of aromatase promoter activity in human breast tissue by nuclear receptors.* Oncogene. 2002;21:2854-63.

61. Okubo T, Truong TK, Yu B, Itoh T, Zhao J, Grube B, et al. *Down-regulation of promoter 1.3 activity of the human aromatase gene in breast tissue by zinc-finger protein, snail (SnaH).* Cancer Res. 2001;61:1338-46.

62. Richards JA, Bruggemeier RW. *Prostaglandin E2 regulates aromatase activity and expression in human adipose stromal cells via two disntinct receptor subtypes.* J Clin Endocrinol Metab. 2003;88:2810-6.

63. Zhou J, Suzuki T, Kovacic A, Saito R, Miki Y, Ishida T, et al. *Interactions between prostaglandin E2, liver receptor homologue-1, and aromatase in breast cancer.* Cancer Res. 2005;65:657-63.

64. Francis GA, Fayard E, Picard F, Auwerx J. *Nuclear receptors and the control of metabolism.* Ann Rev Physiol. 2003;65:261-311.

65. Clyne CD, Kovacic A, Speed CJ, Zhou J, Pezzi V, Simpson ER. *Regulation of aromatase expression by the nuclear receptor LRH-1 in adipose tissue.* Mol Cell Endocrinol. 2004;215:39-44.

66. Kovacic A, Speed CJ, Simpson ER, Clyne CD. *Inhibition of aromatase transcription via promoter II short heterodimer partner in human preadipocytes.* Mol Endocrinol. 2004;18:252-9.

67. Swales KE, Korbonits M, Carpenter R, Walsh DT, Warner TD, Bishop-Bailey D. *The farnesoid X receptor is expressed in breast cancer and regulates apoptosis and aromatase expression.* Cancer Res. 2006;66:10120-6.

68. Kijima I, Ye J, Glackin C, Chen S. *CCAAT/enhancer binding protein delta up-regulates aromatase promoters I.3/II in breast cancer epithelial cells.* Cancer Res. 2008;68:4455-64.

69. Zhou J, Gurates B, Yang S, Sebastian S, Bulun SE. *Malignant breast epithelial cells stimulate aromatase expression via promoter II in human adipose fibroblasts: an epithelial-stromal interaction in breast tumors mediated by CCAAT/enhancer binding protein beta.* Cancer Res. 2001;61:2328-34.

70. Hevir N, Trost N, Debeljak N, Rizner TL. *Expression of estrogen and progesterone receptors and estrogen metabolizing enzymes in different breast cancer cell lines.* Chem Biol Interact. 2011;191:206-16.

71. Zhou C, Zhou D, Esteban J, Murai J, Siiteri PK, Wilczynski S, et al. *Aromatase gene expression and its exon I usage in human breast tumors. Detection of aromatase messenger RNA by reverse transcription-polymerase chain reaction.* J Steroid Biochem Mol Biol. 1996;59:163-71.

72. Zhou D, Wang J, Chen E, Murai J, Siiteri PK, Chen S. *Aromatase gene is amplified in MCF-7 human breast cancer cells.* J Steroid Biochem Mol Biol. 1993;46:147-53.

73. Zhou D, Clarke P, Wang J, Chen S. *Identification of a promoter that controls aromatase expression in human breast cancer and adipose stromal cells.* J Biol Chem. 1996;271:15194-202.

74. Ryde CM, Nicholls JE, Dowsett M. *Steroid and growth factor modulation of aromatase activity in MCF7 and T47D breast carcinoma cell lines.* Cancer Res. 1992;52:1411-5.

75. Santner SJ, Chen S, Zhou D, Korsunsky Z, Martel J, Santen RJ. *Effect of androstenedione on growth of untransfected and aromatase-transfected MCF-7 cells in culture.* J Steroid Biochem Mol Biol. 1993;44:611-6.

76. Yue W, Wang J, Hamilton CJ, Demers LM, Santen RJ. *In situ aromatization enhances breast tumor estradiol levels and cellular proliferation.* Cancer Res. 1998;58:927-32.

77. Koh J, Kubota T, Sasano H, Hashimoto M, Hosoda Y, Kitajima M. *Stimulation of human tumor xenograft growth by local estrogen biosynthesis in stromal cells.* Anticancer Res. 1998;18:2375-80.

78. Tekmal RR, Ramachandra N, Gubba S, Durgam VR, Mantione J, Toda K, et al. *Overexpression of int-5/aromatase in mammary glands of transgenic mice results in the induction of hyperplasia and nuclear abnormalities.* Cancer Res. 1996;56:3180-5.

79. Diaz-Cruz ES, Sugimoto Y, Gallicano GI, Brueggemeier RW, Furth PA. *Comparison of increased aromatase versus ERalpha in the generation of mammary hyperplasia and cancer.* Cancer Res. 2011;71:5477-87.

80. Kirma N, Gill K, Mandava U, Tekmal RR. *Overexpression of aromatase leads to hyperplasia and changes in the expression of genes involved in apoptosis, cell cycle, growth, and tumor suppressor functions in the mammary glands of transgenic mice.* Cancer Res. 2001;61:1910-8.

81. Harada N, Ogawa H, Shozu M, Yamada K, Suhara K, Nishida E, et al. *Biochemical and molecular genetic analyses on placental aromatase (P450AROM) deficiency.* J Biol Chem. 1992;267:4781-5.

82. Carani C, Qin K, Simoni M, Faustini-Fustini M, Serpente S, Boyd J, et al. *Effect of testosterone and estradiol in a man with aromatase deficiency.* New Engl J Med. 1997;337:91-5.

83. Morishima A, Grumbach MM, Simpson ER, Fisher C, Qin K. *Aromatase deficiency in male and female siblings caused by a novel mutation and the physiological role of estrogens.* J Clin Endocrinol Metab. 1995;80:3689-98.

84. Ito Y, Fisher CR, Conte FA, Grumbach MM, Simpson ER. *Molecular basis of aromatase deficiency in an adult female with sexual infantilism and polycystic ovaries.* Proc Natl Acad Sci USA. 1993;90:11673-7.

85. Shozu M, Sebastian S, Takayama K, Hsu WT, Schultz RA, Neely K, et al. *Estrogen excess associated with novel gain-of-function mutations affecting the aromatase gene.* N Engl J Med. 2003;348:1855-65.

86. Ma CX, Adjei AA, Salavaggione OE, Coronel J, Pelleymounter L, Wang L, et al. *Human aromatase: gene resequencing and functional genomics.* Cancer Res. 2005;65:11071-82.

87. Geisler J, Lonning PE. *Aromatase inhibition: translation into a successful therapeutic approach.* Clin Cancer Res. 2005;11:2809-21.

88. Watanabe J, Harada N, Suemasu K, Higashi Y, Gotoh O, Kawajiri K. *Arginine-cysteine polymorphism at codon 264 of the human CYP19 gene does not affect aromatase activity.* Pharmacogenetics. 1997;7:419-24.

89. Tworoger SS, Chubak J, Aiello EJ, Ulrich CM, Atkinson C, Potter JD, et al. *Association of CYP17, CYP19, CYP1B1, and COMT polymorphisms with serum and urinary sex hormone concentrations in postmenopausal women.* Cancer Epidemiol Biomarkers Prev. 2004;13:94-101.

90. Travis RC, Churchman M, Edwards SA, Smith G, Verkasalo PK, Wolf CR, et al. *No association of polymorphisms in CYP17, CYP19, and HSD17-B1 with plasma estradiol concentrations in 1,090 British women.* Cancer Epidemiol Biomark Prev. 2004;13:2282-4.

91. Haiman CA, Dossus L, Setiawan VW, Stram DO, Dunning AM, Thomas G, et al. *Genetic variation at the CYP19A1 locus predicts circulating estrogen levels but not breast cancer risk in postmenopausal women.* Cancer Res. 2007;67:1893-7.

92. Dunning AM, Dowsett M, Healey CS, Tee L, Luben RN, Folkerd E, et al. *Polymorphisms associated with circulating sex hormone levels in postmenopausal women.* J Natl Cancer Inst. 2004;96:936-45.

93. Wang L, Ellsworth KA, Moon I, Pelleymounter LL, Eckloff BW, Martin YN, et al. *Functional genetic polymorphisms in the aromatase gene CYP19 vary the response of breast cancer patients to neoadjuvant therapy with aromatase inhibitors.* Cancer Res. 2010;70:319-28.

94. Pharoah PDP, Dunning AM, Ponder BAJ, Easton DF. *Association studies for finding cancer-susceptibility genetic variants.* Nature Rev Cancer. 2004;4:850-60.

95. Rebbeck TR, Ambrosone CB, Bell DA, Chanock SJ, Hayes RB, Kadlubar FF, et al. *SNPs, haplotypes, and cancer: Applications in molecular epidemiology.* Cancer Epidemiol Biomarkers Prev. 2004;13:681-7.

96. Haiman CA, Stram DO, Pike MC, Kolonel LN, Burtt NP, Altshuler D, et al. *A comprehensive haplotype analysis of CYP19 and breast cancer risk: the Multiethnic Cohort.* Human Mol Genet. 2003;12:2679-92.

97. Hirose K, Matsuo K, Toyama T, Iwata H, Hamajima N, Tajima K. *The CYP19 gene codon 39 Trp/Arg polymorphism increases breast cancer risk in subsets of premenopausal Japanese.* Cancer Epidemiol Biomarkers Prev. 2004;13:1407-11.

98. Lee K-M, Abel J, Ko Y, Harth V, Park W-Y, Seo J-S, et al. *Genetic polymorphisms of cytochrome P450 19 and 1B, alcohol use, and breast cancer risk in Korean women.* Br J Cancer. 2003;88:675-8.

99. Miyoshi Y, Iwao K, Ikeda N, Egawa C, Noguchi S. *Breast cancer risk associated with polymorphism in CYP19 in Japanese women.* Int J Cancer. 2000;89:325-8.

100. Nativelle-Serpentini C, Lambard S, Seralini GE, Sourdaine P. *Aromatase and breast cancer: W39R, an inactive protein.* Eur J Endocrinol. 2002;146:583-9.

101. Haiman CA, Hankinson SE, Spiegelman D, Brown M, Hunter DJ. *No association between a single nucleotide polymorphism in CYP19 and breast cancer risk.* Cancer Epidemiol Biomark Prev. 2002;11:215-6.

102. Haiman CA, Hankinson SE, Spiegelman D, De Vivo I, Colditz GA, Willett WC, et al. *A tetranucleotide repeat polymorphism in CYP19 and breast cancer risk.* Int J Cancer. 2000;87:204-10.

103. Healey CS, Dunning AM, Durocher F, Teare D, Pharoah PDP, Luben RN, et al. *Polymorphisms in the human aromatase cytochrome p450 gene (CYP19) and breast cancer risk.* Carcinogenesis. 2000;21:189-93.

104. Kristensen VN, Andersen TI, Lindblom A, Erikstein B, Magnus P, Borresen-Dale A-L. *A rare CYP19 (aromatase) variant may increase the risk of breast cancer.* Pharmacogenetics. 1998;8:43-8.

105. Polymeropoulos MH, Xiao H, Rath DS, Merril CR. *Tetranucleotide repeat polymorphism at the human aromatase cytochrome P-450 gene (CYP19).* Nucleic Acids Res. 1991;19:195.

106. Probst-Hensch NM, Ingles SA, Diep AT, Haile RW, Stanczyk FZ, Kolonel LN, et al. *Aromatase and breast cancer susceptibility.* Endocrine-Related Cancer. 1999;6:165-73.

107. Siegelmann-Danieli N, Buetow KH. *Constitutional genetic variation at the human aromatase gene (CYP19) and breast cancer risk.* Br J Cancer. 1999;79:456-63.

108. Sourdaine P, Parker MG, Telford J, Miller WR. *Analysis of the aromatase cytochrome P450 gene in human breast cancers.* J Mol Endocrinol. 1994;13:331-7.

109. Bolufer P, Ricart E, Lluch A, Vazquez C, Rodriguez A, Ruiz A, et al. *Aromatase activity and estradiol in human breast cancer: its relationship to estradiol and epidermal growth factor receptors and to tumor-node metastasis staging.* J Clin Oncol. 1992;10:438-46.

110. Esteban JM, Warsi Z, Haniu M, Hall P, Shively JE, Chen S. *Detection of intratumoral aromatase in breast carcinomas. An immunohistochemical study with clinico-pathologic correlation.* Am J Pathol. 1992;140(2):337-43.

111. Miller WR. *Oestrogens and breast cancer.* Br Med Bull. 1990;47:470-83.

112. Sasano H, Frost AR, Saitoh R, Harada N, Poutanen M, Vihko R, et al. *Aromatase and 17 beta-hydroxysteroid dehydrogenase type 1 in human breast carcinoma.* J Clin Endocrinol Metab. 1996;81:4042-6.

113. Silva MC, Rowlands MG, Dowsett M, Gusterson B, McKinna JA, Fryatt I, et al. *Intratumoral aromatase is a prognostic factor in human breast carcinoma.* Cancer Res. 1989;49:2588-91.

114. Lipton A, Santen RJ, Santner SJ, Harvey HA, White-Hershey D, Bartholomew MJ, et al. *Correlation of aromatase activity with histological differentiation of breast cancer - a morphometric analysis.* Breast Cancer Res Treat. 1988;12:31-5.

115. Honma N, Takubo K, Sawabe M, Arai T, Akiyama F, Sakamoto G, et al. *Estrogen-metabolizing enzymes in breast cancers from women over the age of 80 years.* J Clin Endocrinol Metab. 2006;91:607-13.

116. Yoshimura N, Harada N, Bukholm I, Karesen R, Borresen-Dale A-L, Kristensen VN. *Intratumoral mRNA expression of genes from the oestradiol metabolic pathway and clinical and histopathological parameters of breast cancer.* Breast Cancer Res. 2004;6:R46-R55.

117. Lipton A, Santen RJ, Santner SJ, Harvey HA, Sanders SI, Matthews YL. *Prognostic value of breast cancer aromatase.* Cancer. 1992;70:1951-5.

118. Miyoshi Y, Ando A, Hasegawa S, Ishitobi M, Taguchi T, Tamaki Y, et al. *High expression of steroid sulfatase mrRNA predicts poor prognosis in patients with estrogen receptor-positive breast cancer.* Clin Cancer Res. 2003;9:2288-93.

119. Smith IE, Dowsett M. *Aromatase inhibitors in breast cancer.* N Engl J Med. 2003;348:2431-42.

120. Dowsett M. *Aromatase inhibition: basic concepts, and the pharmacodynamics of formestane.* Ann Oncol. 1994;5 Suppl 7:S3-5.

121. Geisler J, King N, Anker G, Ornati G, Di Salle E, Lonning E, et al. *In vivo inhibition of aromatization by exemestane, a novel irreversible aromatase inhibitor, in postmenopausal breast cancer patients.* Clin Cancer Res. 1998;4:2089-93.

122. Kao Y, Cam LL, Laughton CA, Zhou D, Chen S. *Binding characteristics of seven inhibitors of human aromatase: A site-directed mutagenesis study.* Cancer Res. 1996;56:3451-60.

123. Miller WR. *Aromatase inhibitors-where are we now?* Br J Cancer. 1996;73:415-7.

124. de Jong PC, Ven JV, Nortier WR, Maitimu-Smeele I, Donker TH, Thijssen JH, et al. *Inhibition of breast cancer tissue aromatase activity and estrogen concentrations by the third-generation aromatase inhibitor vorozole.* Cancer Res. 1997;57:2109-11.

125. Goss PE, Qi S, Cheung AM, Hu H, Mendes M, Pritzker KPH. *Effects of the steroidal aromatase inhibitor exemestane and the nonsteroidal aromatase inhibitor letrozole on bone and lipid metabolism in ovariectomized rats.* Clin Cancer Res. 2004;10:5717-23.

126. Coombes RC, Hall E, Gibson LJ, Paridaens R, Jassem J, Delozier T, et al. *A randomized trial of exemestane after two to three years of tamoxifen therapy in postmenopausal women with primary breast cancer.* N Engl J Med. 2004;350:1081-92.

127. Goss PE, Ingle JN, Ales-Martinez JE, Cheung AM, Chlebowski RT, Wactawski-Wende J, et al. *Exemestane for breast-cancer prevention in postmenopausal women.* N Engl J Med. 2011;364:2381-91.

128. Ibrahim AR, Abul-Hajj YJ. *Aromatase inhibition by flavonoids.* J Steroid Biochem Mol Biol. 1990;37:257-60.

129. Kellis JT, Jr., Vickery LE. *Inhibition of human estrogen synthetase (aromatase) by flavones.* Science. 1984;225:1032-4.

130. Kao YC, Zhou C, Sherman M, Laughton CA, Chen S. *Molecular basis of the inhibition of human aromatase (estrogen synthetase) by flavone and isoflavone phytoestrogens: A site-directed mutagenesis study.* Environ Health Perspect. 1998;106:85-92.

131. Eng ET, Williams D, Mandava U, Kirma N, Tekmal RR, Chen S. *Suppression of aromatase (estrogen synthetase) by red wine phytochemicals.* Breast Cancer Res Treat. 2001;67:133-46.

132. Kijima I, Phung S, Hur G, Kwok SL, Chen S. *Grape seed extract is an aromatase inhibitor and a suppressor of aromatase expression.* Cancer Res. 2006;66:5960-7.

2.1.1.6. 17β-HYDROXYSTEROID DEHYDROGENASE

17β-Hydroxysteroid dehydrogenase (17β-HSD) enzymes are responsible for the interconversion of 17-ketosteroids and their 17β-hydroxysteroid counterparts, such as E1 and E2, androstenedione and testosterone, and androstanedione and dihydrotestosterone (Figure 2.1.6.). Thus, 17β-HSDs catalyze both oxidation and reduction reactions. They also participate in other biochemical pathways, e.g., the elongation of long chain fatty acids.

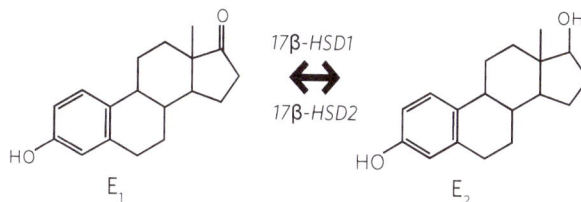

Figure 2.1.6. 17β-HSD1 and 17β-HSD2 catalyze the interconversion of E_1 and E_2.

Fourteen human genes encoding isozymes of 17β-HSD have been identified (1-3) (Table 2.1.3.). These isozymes are designated types 1, 2, 3, etc. or 17β-HSD1, HSD2, HSD3, etc. (4). They share the use of nicotinamide adenine dinucleotide [NAD(H)] or its phosphate [NADP(H)] as cofactors to catalyze the hydride transfer between 17-keto and 17β-hydroxysteroid pairs in a positional and stereospecific manner. Some of them also have secondary 20α-hydroxysteroid dehydrogenase activity, as demonstrated by the ability to convert 20α-dihydroprogesterone to progesterone (5). Since both estrogens and androgens possess the highest activity in the 17β-hydroxy form, i.e., E_2 and testosterone, the 17β–HSD enzymes regulate the biological activity of sex hormones and thereby contribute to the control of steroid hormone action. 17β-HSD1 to 4 and HSD6 belong to the short-chain dehydrogenase/reductase (SDR) protein family, whereas 17β-HSD5 is a member of the aldoketoreductase (AKR) family (6, 7). As can be expected on the basis of their dissimilar primary sequence, the 17β-HSD isozymes also differ in their substrate and cofactor specificity. In particular, 17β-HSD1 and 3 catalyze reductive reactions of estrogens and androgens, whereas 17β-HSD2, 4, and 5 are oxidative (4, 8). Certain 17β-HSD isozymes convert primarily estrogens, while others accept both phenolic and neutral steroids as substrates. Furthermore, the 17β-HSD isozymes display different patterns of expression in both classical steroidogenic and peripheral tissues. The enzyme activities are distributed in microsomal and cytosolic cell fractions.

17β-HSD1: The 17β-HSD1 gene at 17q11 spans 3.2 kb and contains 6 exons (Figure 2.1.7.). It is located within 13 kb from a 17β-HSD1 pseudogene that has a premature, in-frame stop codon (9, 10). Two transcription start points in the 17β-HSD1 gene result in two mRNAs, 1.3 and 2.3 kb in

Table 2.1.3. Human 17β-Hydroxysteroid Dehydrogenases: Catalytic Roles in Regulating Estrone – 17β-Estradiol Balance and Expression in Breast Epithelium

Gene	Locus	Protein (kDa)	Principal Reaction	Protein Expression in Breast Epithelium
HSD1	17q11	35	$E_1 \rightarrow E_2$	normal cells: yes malignant cells: variable
HSD2	16q24	43	$E_2 \rightarrow E_1$	normal cells: yes malignant cells: no (except ZR-75)
HSD3	9q22	35	$E_1 \rightarrow E_2$	no
HSD4	5q2	80	$E_2 \rightarrow E_1$	no
HSD5	10p15	34	$E_2 \rightarrow E_1$	normal and malignant cells
HSD7*	10p11.2	37	$E_1 \rightarrow E_2$	normal and malignant cells
HSD8	6p21.3		$E_2 \rightarrow E_1$	
HSD10	Xp11.2		$E_2 \rightarrow E_1$	normal and malignant cells
HSD11	1q32-41	44	$E_2 \rightarrow E_1$	
HSD12	11q11		$E_1 \rightarrow E_2$	normal and malignant cells
HSD14	19q13.33		$E_2 \rightarrow E_1$	normal and malignant cells

*17β-HSD6 and 17β-HSD9 have been identified in rodents but not in man (1). HSD, 17β-hydroxysteroid dehydrogenase; E1, estrone; E2, 17β-estradiol.

size, that differ in their 5' noncoding regions (11). However, only the 1.3 kb transcript correlates with 17β-HSD1 protein concentration and responds to progestin induction of the 17β-HSD1 gene (12). The 35 kDa 17β-HSD1 protein is composed of 327 amino acids. Determination of the three-dimensional structure revealed a NADPH cofactor-binding site near the N-terminal and a conserved catalytic domain with 142Ser, 155Tyr, and 159Lys as critically important residues for hydride transfer (13). Other residues, e.g., 152Asn, 153Asp, and 187Pro, determine substrate specificity (14-16). The active enzyme exists as homodimer consisting of noncovalently bound subunits (17). The equine estrogen equilin, which is present in Premarin®, competes with E_1 for binding to the active site of 17β-HSD1 and thereby inhibits reduction of E_1 to E_2 (18).

17β-HSD1 is abundantly expressed in ovarian granulosa cells and placental syncytiotrophoblasts (3). In the syncytiotrophoblasts, 17β-HSD1 not only converts E_1 to E_2 but also 16α-OHE$_1$ to 16α-OHE$_2$ (= estriol, E_3), accounting for the high levels of E_2 (10 – 15 mg/24 h) and E_3 (60 – 150 mg/24 h) secretion during pregnancy. 17β-HSD1 is also expressed in epithelial cells of the breast and endometrium. The 17β-HSD1 concentration in mammary epithelial cells is regulated by basic fibroblast growth factor, tumor necrosis factor α, progestin, and retinoic acid, the latter acting through a retinoic acid response element in the 17β-HSD1 promoter (12, 19-21). In a series of eight breast cancer cell lines only three (BT-20, MDA-MB-361, T47D) expressed 17β-HSD1, while five (MCF-7, ZR-75, HBL 100, MDA-MB-231, MDA-MB-468) failed to express the enzyme by Western blot analysis. Of the three cell lines expressing 17β-HSD1, only T47D responded to progestin treatment with an increase

in 17β-HSD1 mRNA and protein. Thus, expression and functional response of the 17β-HSD1 gene in breast cancer cells are highly variable. Progestins have a complex effect on 17β-HSD activity and can direct the $E_1 \leftrightarrow E_2$ interconversion in both directions (22). Studies with hormone-dependent breast cancer cells MCF-7 and T47D have shown that some progestins stimulate the reductive activity, E_1 to E_2, and thereby enhance cell proliferation (12). Other progestins favor the oxidative direction, E_2 to E_1, and thereby inhibit cell growth (23, 24).

Kinetic studies of 17β-HSD1 in breast cancer cells showed that the catalytic efficiency for the reductive activity from E_1 to E_2 was about threefold higher than that for the oxidative activity converting E_2 to E_1. Thus, 17β-HSD1 regulates the intracellular ligand supply for ER by favoring the predominance of E_2 over E_1 in mammary epithelial cells (25). Infusion of ^{3}H-E_2 or ^{3}H-E_1 into postmenopausal women with breast cancer showed preferential conversion of E_1 to E_2 in tumor tissue (26).

The degree of 17β-HSD1 expression reported in breast cancer tissue is quite variable. For example, the detection of 17β-HSD1 mRNA varies from 16 to 100% (27, 28) and the immunohistochemical staining of 17β-HSD1 ranges from 20 to 61% of cases (12, 28-30). Investigators observed a positive, inverse, or no correlation between 17β-HSD1 expression and ER-positive status in breast cancers (28-30). One study observed significantly higher expression of 17β-HSD1 in postmenopausal cancers while another study found no correlation with menopausal status (30, 31). Not surprisingly, correlation between 17β-HSD1 expression and prognosis has

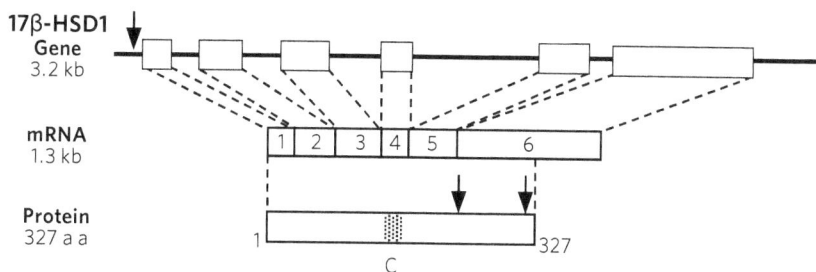

Figure 2.1.7. Overview of 17β-HSD1 gene, mRNA, and protein.
The 17β-HSD1 gene at 17q11 spans 3.2 kb and contains 6 exons. The 1.3 kb mRNA contains a short 5' untranslated segment (the first 10 nucleotides in exon 1) and a longer 3' untranslated segment (the last 329 nucleotides of exon 6). The open reading frame of 981 nucleotides encodes a protein of 327 amino acids with a molecular weight of 35 kDa. The shaded area indicates the catalytic domain (C). The lengths of the exons are drawn in scale, while those of the introns are not. The arrows indicate polymorphic sites in the TATA box and in codons Ala237Val and Ser312Gly.

been inconsistent, with investigators finding either no association or a shorter overall and disease-free survival in breast cancer patients (28, 30, 31). One reason for the discrepant 17β-HSD1 expression data could be the amplification of the 17β-HSD1 gene, which was observed in 14.5% of postmenopausal breast cancers (32).

The 17β-HSD1 gene contains several polymorphisms, including the common nonsynonymous Ser312Gly (rs605059) in exon 6 (37). Two meta-analyses including 10 studies (>11,000 cases, >14,000 controls in total) found no evidence of an association of this or other polymorphisms in the 17β-HSD1 gene with breast cancer risk (33, 34). However, several of the individual studies showed a reduced risk for the variant allele. For example, the GENICA study observed an OR of 0.83 (95% CI 0.64 – 1.09) and 0.73 (95% CI 0.52 – 1.01) for heterozygotes and minor allele homozygotes compared with major allele homozygotes (35). The conclusion of the meta-analysis was that the nonsynonymous Ser312Gly (rs605059) by itself does not have a major effect on breast cancer risk but that the possibility of a weak effect cannot be excluded (34). This is consistent with experimental data showing no difference in the catalytic activity of recombinant wild-type and variant 312 alleles (16). Nevertheless, one study observed higher plasma E_2 levels in lean women with the homozygous 312Gly/ Gly genotype (p = 0.01) (36). A rare polymorphism, Ala237Val also does not affect the catalytic activity of 17β-HSD1 (16, 37). A point mutation (nucleotide -27 A/C) was identified in the TATA box of the 17β-HSD1 gene that decreased the promoter activity approximately 45% as determined by reporter gene analysis. Although this mutation may affect transcriptional efficiency, it was observed with equal frequency (4%) in breast cancer patients and healthy individuals, making a role in tumorigenesis unlikely (38). Additional nucleotide variations of no apparent functional importance in the 5' upstream region were also identified with equal frequency in breast cancer patients and controls.

17β-HSD2: The 17β-HSD2 gene at 16q24 spans >40 kb and contains seven exons (5, 39) (Figure 2.1.8.). The gene encodes two alternatively spliced transcripts (40). The predominant 1.5-kb transcript comprises exons 1 – 3 and 6 – 7. Exon 1 contains a 167-nucleotide 5' and exon 7 a 96-nucleotide 3' untranslated region. The open reading frame of 1,161 nucleotides encodes a protein of 387 amino acids with a molecular weight of 43 kDa. The less abundant transcript comprises all exons 1 – 7. The inclusion of exons 4 and 5 introduces a change in the translational reading frame that results in a different carboxy-terminal segment and a shorter protein (241 amino acids) devoid of enzymatic activity. Although 17β-HSD types 1 and 2 have similar hydropathy patterns and share sequence similarity in the catalytic domain (amino acids 232 – 242 in 17β-HSD2), the overall amino acid identity is only ~20%. Localized in the microsomal fraction, 17β-HSD2 uses NAD^+ as cofactor to catalyze the conversion of E_2 to E_1 with 20-fold greater efficiency than the reverse E_1 to E_2 reaction (39). Interestingly, 17β-HSD2 also converts 2-MeOE$_2$ into MeOE$_1$ with similar efficiency (41, 42). This catalytic function has been shown to be important for the responsiveness of breast cancer cell lines to the antiproliferative action of 2-MeOE$_2$. Analysis of 17β-HSD2 expression in one ER-negative (MDA-MB435) and three ER-positive (MCF-7, T47D, ZR-75) breast cancer cell lines revealed expression only in ZR-75 cells. While MCF-7, T47D, and MDA-MB435 were highly sensitive to the antiproliferative effect of 2-MeOE$_2$, ZR-75 cells were resistant. The high level of 17β-HSD2 in ZR-75 cells was responsible for the rapid conversion of 2-MeOE$_2$ to inactive 2-MeOE1 and account for the selective insensitivity of ZR-75 cells to the anti-proliferative action of 2-MeOE$_2$ (41). 17β-HSD2 is regularly expressed in normal mammary epithelium but not always present in breast cancer cells (27, 28, 30, 43-45). 17β-HSD2 mRNA was detected in 10 to 69% of tumors but the protein was absent in all breast cancers.

The 17β-HSD3 gene at 9q22 encodes the 17β-HSD3 isozyme, which is predominantly expressed in the testis

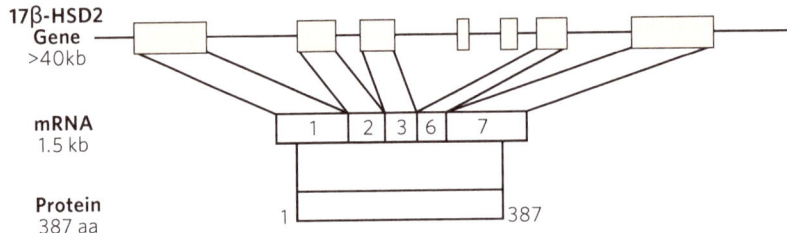

Figure 2.1.8. Overview of 17β-HSD2 gene, mRNA, and protein.
The 17β-HSD2 gene at 16q24 spans >40 kb and contains seven exons. The gene encodes two alternatively spliced mRNAs. The major 1.5-kb transcript contains exons 1 – 3 and 6 – 7. Exon 1 contains a 5' untranslated segment and exon 7 a 3' untranslated segment. The open reading frame of 1,161 nucleotides encodes a protein of 387 amino acids with a molecular weight of 43 kDa. The length of the exons is drawn to scale, while those of the introns are not.

(46). Although 17β-HSD3 is capable of converting E_1 to E_2, it is primarily responsible for catalyzing androstenedione to testosterone. Inherited deficiency of 17β-HSD3 causes male pseudohermaphroditism (47, 48). The 17β-HSD4 gene at 5q22 encodes the 17β-HSD4 isozyme, which is expressed in liver, kidney, ovary, and testis (49, 50). 17β-HSD4 is multifunctional including the oxidation of E_2 to E_1 (Table 2.1.3.). 17β-HSD4 expression was also observed in normal mammary epithelial cells (45). However, there was no correlation between the presence of 17β-HSD4 and oxidative activity. Instead, oxidative 17β-HSD activity appeared in cell lines where 17β-HSD2 was expressed, whereas reductive 17β-HSD activity was present in cells expressing 17β-HSD1. 17β-HSD5 (also known as aldo-keto reductase AKR1C3) is expressed in the prostate where it converts androstenedione to testosterone (51, 52). 17β-HSD5 is also expressed in breast with higher levels detected in breast cancer specimens than in normal breast tissue (28). Since 17β-HSD5 recognizes a wide range of substrates, including androgens, estrogens, progestins, and prostaglandins, its role in breast tissue is uncertain. Two polymorphisms, Glu77Gly and Lys183Arg, have been associated with decreased activity (53). The 17β-HSD12 gene at 11q11 resembles the 17β-HSD3 gene in its genomic structure, number of exons, and conserved active sites (54). The 17β-HSD12 isozyme converts E_1 into E_2 in the ovary and breast and regulates fatty acid synthesis, specifically the process of very long chain fatty acid elongation, such as arachidonic acid, which may subsequently be metabolized to prostaglandins (55). Tissue E_2 concentrations measured in 16 postmenopausal breast cancers showed no correlation with 17β-HSD12 expression. The 17β-HSD14 gene at 19q13.33 encodes an isozyme that decreased the level of E_2 in MCF-7 and SKBR3 cells by catalyzing the conversion of E_2 to E_1 (56). In summary, 17β-HSD1 and 17β-HSD2 are the main 17β-HSD isozymes in normal breast. They are evenly expressed in normal mammary epithelial cells thereby balancing oxidative and reductive activities. Their expression in malignant cells is variable.

References

1. Adamski J, Jakob FJ. *A guide to 17β-hydroxysteroid dehydrogenases.* Mol Cell Endocrinol. 2001;171:1-4.

2. Marchais-Oberwinkler S, Henn C, Moller G, Klein T, Negri M, Oster A, et al. *17beta-Hydroxysteroid dehydrogenases (17beta-HSDs) as therapeutic targets: protein structures, functions, and recent progress in inhibitor development.* J Steroid Biochem Mol Biol. 2011;125:66-82.

3. Peltoketo H, V. L-T, Simard J, Adamski J. *17β-hydroxysteroid dehydrogenase (HSD)/17-keto-steroid reductase (KSR) family; nomenclature and main characteristics of the 17HSD/KSR enzymes.* J Mol Endocrinology. 1999;23:1-11.

4. Peltoketo H, Vihko P, Vihko R. *Regulation of estrogen action: role of 17β-hydroxysteroid dehydrogenases.* Vitam Horm. 1999;55:35-98.

5. Casey ML, MacDonald PC, Andersson S. *17β-hydroxysteroid dehydrogenase type 2: chromosomal assignment and progestin regulation of gene expression in human endometrium.* J Clin Invest. 1994;94:2135-41.

6. Jornvall H, Persson B, Krook M, Atrian S, Gonzalez-Duarte R, Jeffery J, et al. *Short-chain dehydrogenases/reductases (SDR).* Biochemistry. 1995;34:6003-13.

7. Krozowski Z. *The short-chain alcohol dehydrogenase superfamily: variations on a common theme.* J Steroid Biochem Mol Biol. 1994;51:125-30.

8. Vihko P, Isomaa V, Ghosh D. *Structure and function of 17β-hydroxysteroid dehydrogenase type 1 and type 2.* Mol Cell Endocrinol. 2001;171:71-6.

9. Luu-The V, Labrie C, Simard J, Lachance Y, Zhao HF, Couet J, et al. *Structure of two in tandem human 17 beta-hydroxysteroid dehydrogenase genes.* Mol Endocrinol. 1990;4:268-75.

10. Peltoketo H, Isomaa V, Vihko R. *Genomic organization and DNA sequences of human 17 beta-hydroxysteroid dehydrogenase genes and flanking regions.* Eur J Biochem. 1992;209:459-66.

11. Labrie F, Luu-The V, Labrie C, Berube D, Couet J, Zhao HF, et al. *Characterization of two mRNA species encoding human estradiol 17 beta-dehydrogenase and assignment of the gene to chromosome 17.* J Steroid Biochem. 1989;34:189-97.

12. Poutanen M, Moncharmont B, Vihko R. *17 beta-hydroxysteroid dehydrogenase gene expression in human breast cancer cells: regulation of expression by a progestin.* Cancer Res. 1992;52:290-4.

13. Ghosh D, Pletnev VZ, Zhu DW, Wawrzak Z, Duax WL, Pangborn W, et al. *Structure of human estrogenic 17 beta-hydroxysteroid dehydrogenase at 2.20 A resolution.* Structure. 1995;3:503-13.

14. Huang Y-W, Pineau I, Chang H-J, Azzi A, Bellmare v, Laberge S, et al. *Critical residues for the specificity of cofactors and substrates in human estrogenic 17β-hydroxysteroid dehydrogenase 1: variants designed from the three-dimensional structure of the enzyme.* Mol Endocrinol. 2001;15:2010-20.

15. Puranen T, Poutanen M, Ghosh D, Vihko R, Vihko P. *Origin of substrate specificity of human and rat 17 β-hydroxysteroid dehydrogenase type 1, using chimeric enzymes and site-directed substitutions.* Endocrinology. 1997;138:3532-9.

16. Puranen TJ, Poutanen MH, Peltoketo HE, Vihko PT, Vihko RK. *Site-directed mutagenesis of the putative active site of human 17 beta-hydroxysteroid dehydrogenase type 1.* Biochem J. 1994;304:289-93.

17. Puranen T, Poutanen M, Ghosh D, Vihko P, Vihko R. *Characterization of structural and functional properties of human 17 beta-hydroxysteroid dehydrogenase type 1 using recombinant enzymes and site-directed mutagenesis.* Mol Endocrinol. 1997;11:77-86.

18. Sawicki MW, Erman M, Puranen T, Vihko P, Ghosh D. *Structure of the ternary complex of human 17β-hydroxysteroid dehydrogenase type 1 with 3-hydroxyestra-1,3,5,7-tetraen-17-one (equilin) and NADP+.* Proc Natl Acad Sci USA. 1999;96.

19. Duncan LJ, Reed MJ. *The role and proposed mechanism by which oestradiol 17 beta-hydroxysteroid dehydrogenase regulates breast tumour oestrogen concentrations.* J Steroid Biochem Mol Biol. 1995;55:565-72.

20. Peltoketo H, Isomaa V, Poutanen M, Vihko R. *Expression and regulation of 17 beta-hydroxysteroid dehydrogenase type 1.* J Endocrinol. 1996;150:S21-S30.

21. Piao YS, Peltoketo H, Oikarinen J, Vihko R. *Coordination of transcription of the human 17 beta-hydroxysteroid dehydrogenase type 1 gene (EDH17B2) by a cell-specific enhancer and a silencer: identification of a retinoic acid response element.* Mol Endocrinol. 1995;9:1633-44.

22. Pasqualini JR. *The selective estrogen enzyme modulators in breast cancer: a review.* Biochim Biophys Acta. 2004;1654:123-43.

23. Chetrite GS, Ebert C, Wright F, Philippe JC, Pasqualini JR. *Effect of medrogestone on 17 beta-hydroxysteroid dehydrogenase activity in the hormone-dependent MCF-7 and T-47D human breast cancer cell lines.* J Steroid Biochem Mol Biol. 1999;68:51-6.

24. Chetrite GS, Kloosterboer H, Philippe JC, Pasqualini JR. *Effects of ORG OD14 (Livial) and its metabolites on 17 beta-hydroxysteroid dehydrogenase activity in hormone-dependent MCF-7 and T-47D breast cancer cells.* Anticancer Res. 1999;19:261-7.

25. Miettinen MM, Mustonen MV, Poutanen MH, Isomaa VV, Vihko RK. *Human 17 beta-hydroxysteroid dehydrogenase type 1 and 2 isoenzymes have opposite activities in cultured cells and characteristic cell-and tissue-specific expression.* Biochem J. 1996;314:839-45.

26. McNeill JM, Reed MJ, Beranek PA, Bonney RC, Ghilchik MW, Robinson DJ, et al. *A comparison of the in vivo uptake and metabolism of 3H-oestrone and 3H-oestradiol by normal breast and breast tumour tissues in post-menopausal women.* Int J Cancer. 1986;38:193-6.

27. Gunnarsson C, Olsson BM, Stal O, Group MotSBC. *Abnormal expression of 17β-hydroxysteroid dehydrogenases in breast cancer predicts late recurrence.* Cancer Res. 2001;61:8448-51.

28. Oduwole OO, Li Y, Isomaa VV, Mantyniemi A, Pulkka AE, Soini Y, et al. *17β-hydroxysteroid dehydrogenase type 1 is an independent prognostic marker in breast cancer.* Cancer Res. 2004;64:7604-9.

29. Sasano H, Frost AR, Saitoh R, Harada N, Poutanen M, Vihko R, et al. *Aromatase and 17 beta-hydroxysteroid dehydrogenase type 1 in human breast carcinoma.* J Clin Endocrinol Metab. 1996;81:4042-6.

30. Suzuki T, Moriya T, Ariga N, Kaneko C, Kanazawa M, Sasano H. *17β-hydroxysteroid dehydrogenase type 1 and type 2 in human breast carcinoma: A correlation to clinicopathological parameters.* Br J Cancer. 2000;82:518-23.

31. Miyoshi Y, Ando A, Shiba E, Taguchi T, Tamaki Y, Noguchi S. *Involvement of up-regulation of 17β-hydroxysteroid dehydrogenase type 1 in maintenance of intratumoral high estradiol levels in postmenopausal breast cancers.* Int J Cancer. 2001;94:685-9.

32. Gunnarsson C, Ahnstrom M, Kirschner K, Olsson B, Nordenskjold B, Rutqvist LE, et al. *Amplification of hsd17b1 and erbb2 in primary breast cancer.* Oncogene. 2003;22:34-40.

33. Feigelson HS, Cox DG, Cann HM, Wacholder S, Kaaks R, Henderson BE, et al. *Haplotype analysis of the HSD17B1 gene and risk of breast cancer: a comprehensive approach to multicenter analyses of prospective cohort studies.* Cancer Res. 2006;66:2468-75.

34. Gaudet MM, Chanock S, Dunning A, Driver K, Brinton LA, Lissowska J, et al. *HSD17B1 genetic variants and hormone receptor-defined breast cancer.* Cancer Epidemiol Biomarkers Prev. 2008;17:2766-72.

35. Justenhoven C, Hamann U, Schubert F, Zapatka M, Pierl CB, Rabstein S, et al. *Breast cancer: a candidate gene approach across the estrogen metabolic pathway.* Breast Cancer Res Treat. 2008;108:137-49.

36. Setiawan VW, Hankinson SE, Colditz GA, Hunter DJ, De Vivo I. *HSD17B1 gene polymorphisms and risk of endometrial and breast cancer.* Cancer Epidemiol Biomarkers Prev. 2004;13:213-9.

37. Normand T, Narod S, Labrie F, Simard J. *Detection of polymorphisms in the estradiol 17 beta-hydroxysteroid dehydrogenase II gene at the EDH17B2 locus on 17q11-q21.* Hum Mol Genet. 1993;2:479-83.

38. Peltoketo H, Piao Y, Mannermaa A, Ponder BAJ, Isomaa V, Poutanen M, et al. *A point mutation in the putative TATA box, detected in nondiseased individuals and patients with hereditary breast cancer, decreases promoter activity of the 17 beta-hydroxysteroid dehydrogenase type 1 gene 2 (EDH17B2) in vitro.* Genomics. 1994;23:250-2.

39. Wu L, Einstein M, Geissler WM, Chan HK, Elliston KO, Andersson s. *Expression cloning and characterization of human 17β-hydroxysteroid dehydrogenase type 2, a microsomal enzyme possessing 20α-hydroxysteroid dehydrogenase activity.* J Biol Chem. 1993;268:12964-9.

40. Labrie Y, Durocher F, Lachance Y, Turgeon C, Simard J, Labrie C, et al. *The human type II 17β-hydroxysteroid dehydrogenase gene encodes two alternatively spliced mRNA species.* DNA Cell Biol. 1995;14:849-61.

41. Liu Z-J, Lee WJ, Zhu BT. *Selective insensitivity of ZR-75-1 human breast cancer cells to 2-methoxyestradiol: evidence for type II 17β-hydroxysteroid dehydrogenase as the underlying cause.* Cancer Res. 2005;65:5802-11.

42. Newman SP, Ireson CR, Tutill HJ, Day JM, Parsons MF, Leese MP, et al. *The role of 17beta-hydroxysteroid dehydrogenases in modulating the activity of 2-methoxyestradiol in breast cancer cells.* Cancer Res. 2006;66:324-30.

43. Ariga N, Moriya T, Suzuki T, Kimura M, Ohuchi N, Satomi S, et al. *17β-hydroxysteroid dehydrogenase type 1 and type 2 in ductal carcinoma in situ and intraductal proliferative lesions of the human breast.* Anticancer Res. 2000;20:1101-8.

44. Gunnarsson C, Hellqvist E, Stal O, Group SSBC. *17β-hydroxysteroid dehydrogenases involved in local oestrogen synthesis have prognostic significance in breast cancer.* Br J Cancer. 2005;92:547-52.

45. Miettinen M, Mustonen M, Poutanen M, Isomaa V, Wickman M, Soderqvist G, et al. *17β-hydroxysteroid dehydrogenases in normal human mammary epithelial cells and breast tissue.* Breast Cancer Res Treat. 1999;57:175-82.

46. Andersson S, Geissler WM, Patel S, Wu L. *The molecular biology of androgenic 17 beta-hydroxysteroid dehydrogenases.* J Steroid Biochem Mol Biol. 1995;53:37-9.

47. Andersson S, Geissler WM, Wu L, Davis DL, Grumbach MM, New MI, et al. *Molecular genetics and pathophysiology of 17 beta-hydroxysteroid dehydrogenase 3 deficiency.* J Clin Endocrinol Metab. 1996;81:130-6.

48. Geissler WM, Davis DL, Wu L, Bradshaw KD, Patel S, Mendonca BB, et al. *Male pseudohermaphroditism caused by mutations of testicular 17 beta-hydroxysteroid dehydrogenase 3.* Nature Genetics. 1994;7:34-9.

49. Adamski J, Normand T, Leenders F, Monte D, Begue A, Stehelin D, et al. *Molecular cloning of a novel widely expressed human 80 kDa 17 beta-hydroxysteroid dehydrogenase IV.* J Biochem. 1995;311:437-43.

50. de Launoit Y, Adamski J. *Unique multifunctional hsd17b4 gene product: 17β-hydroxysteroid dehydrogenase 4 and D-3-hydoxyacyl-coenzyme A dehydrogenase/hydratase involved in Zellweger syndrome.* J Mol Endocrinol. 1999;22:227-40.

51. Penning TM, Burczynski ME, Jez JM, Hung C-F, Lin H-K, Ma H, et al. *Human 3a-hydroxysteroid dehydrogenase isoforms (AKR1C1-AKR1C4) of the aldo-keto reductase superfamily: functional plasticity and tissue distribution reveals roles in the inactivation and formation of male and female sex hormones.* Biochem J. 2000;351:67-77.

52. Qiu W, Zhou M, Labrie F, Lin S-X. *Crystal structures of the multispecific 17β-hydroxysteroid dehydrogenase type 5: critical androgen regulation in human peripheral tissues.* Mol Endocrinol. 2004;18:1798-807.

53. Plourde M, Ferland A, Soucy P, Hamdi Y, Tranchant M, Durocher F, et al. *Analysis of 17beta-hydroxysteroid dehydrogenase types 5, 7, and 12 genetic sequence variants in breast cancer cases from French Canadian Families with high risk of breast and ovarian cancer.* J Steroid Biochem Mol Biol. 2009;116:134-53.

54. Luu-The V, Tremblay P, Labrie F. *Characterization of type 12 17beta-hydroxysteroid dehydrogenase, an isoform of type 3 17beta-hydroxysteroid dehydrogenase responsible for estradiol formation in women.* Mol Endocrinol. 2006;20:437-43.

55. Nagasaki S, Suzuki T, Miki Y, Akahira J, Kitada K, Ishida T, et al. *17Beta-hydroxysteroid dehydrogenase type 12 in human breast carcinoma: a prognostic factor via potential regulation of fatty acid synthesis.* Cancer Res. 2009;69:1392-9.

56. Jansson AK, Gunnarsson C, Cohen M, Sivik T, Stal O. *17beta-hydroxysteroid dehydrogenase 14 affects estradiol levels in breast cancer cells and is a prognostic marker in estrogen receptor-positive breast cancer.* Cancer Res. 2006;66:11471-7.

2.1.1.7. STEROID SULFATASE

A dozen mammalian sulfatase enzymes and their genes have been identified (1, 2). All members of the sulfatase family catalyze the hydrolysis of sulfate ester bonds (O-sulfatase activity) in a variety of substrates, ranging from complex molecules, such as glycosaminoglycans, to simple ones, such as 3-hydroxysteroid sulfates. Only a single steroid sulfatase (STS; steryl-sulfatase, EC 3.1.6.2) exists, in contrast to the many sulfotransferases (SULTs). STS and SULTs act in opposite directions. As described in Chapter 4.3.4, the various SULTs catalyze the formation of steroid sulfates, whereas STS catalyzes the hydrolysis of sulfated steroids, such as cholesterol sulfate, pregnenolone sulfate, deoxycorticosterone sulfate, dehydroepiandrosterone sulfate (DHEAS), and estrone sulfate (E_1S) (Figure 2.1.9.).

The formation and hydrolysis of steroid sulfates by SULTs and STS represents an important mechanism for regulating the biological activity of steroid hormones. Steroid sulfates cannot bind to steroid hormone receptors (e.g., E_1S to estrogen receptor) because of the sulfate group. Removal of the sulfate moiety by STS is required before the steroids can exert biological effects. Plasma concentrations of DHEAS and E_1S are several-fold higher than their unconjugated counterparts and the half-lives (~10 h) are considerably longer than those of the unconjugated steroids (~30 min). Due to the combination of high concentration and prolonged half-life, plasma E_1S represents an important reservoir for the formation of biologically active estrogens via the action of STS.

The STS gene at Xp22.3 – Xpter spans 146 kb and contains 10 exons (Figure 2.1.10.). The gene is located in the pseudo-autosomal region of the X-chromosome that escapes X-inactivation (3-5). It is part of a cluster of four sulfatase genes that have near identical genomic structure suggesting that they arose by duplication of a common ancestral gene. The duplication appears to have occurred before the divergence of X and Y because the Y-chromosome contains a STS pseudogene. The extent of sequence similarity between the X- and Y-genes suggests that they have been diverging for approximately 40 million years (6). The STS protein has been purified and crystallized (7, 8). Unlike the lysosomal sulfatases, which are active in an acidic environment, STS is localized in microsomes and active at neutral pH. The overall shape of the STS protein is mushroom-like with the crown protruding toward the

luminal side of the endoplasmic reticulum and the stalk traversing the lipid bilayer. The catalytic site of all sulfatases contains a conserved cysteine residue that undergoes posttranslational modification to a formylglycine, fGly, to produce the active enzyme (9, 10). In STS, the 75fGly residue forms a geminal diol, which is crucial to the catalytic reaction (Figure 2.1.11.). In the proposed reaction scheme, one of the hydroxyl groups attacks the sulfur atom of E1S leading to the release of E_1 by desulfonation and the formation of fGlyS. The attack of a water molecule then regenerates fGly from fGlyS (11). Other conserved residues lining the catalytic site include 35Asp, 36Asp, 342Asp, 126His, 290His, 134Lys, 368Lys, and 79Arg. The substrate recognition site contains hydrophobic residues 74Leu, 98Arg, 99Thr, 101Val, 103Leu, 167Leu, 177Val, 178Phe, 180Thr, 181Gly, 484Thr, 485His, 486Val, and 488Phe (8).

Deficiency of the STS enzyme due to complete deletion of the STS gene results in X-linked ichthyosis, a condition that is characterized by scaly skin caused by increased cholesterol sulfate levels in the stratum corneum (12). Catalytically inactive STS may also result from point mutations in the carboxyl region of the enzyme, which is important for substrate binding (13). Immunohisto-chemical studies have shown that STS is present in the cytoplasm of epithelial but not stromal cells in breast tissue (14). The level of STS mRNA determined by real time RT-PCR was significantly higher in malignant (1458 amol/mg RNA) than normal breast tissue (536 amol/mg) (15). This result is consistent with enzymatic studies, which also showed higher STS activity in malignant breast tissue (16). STS is present in both ER-positive (MCF-7, T47D) and ER-negative (MDA-MB-231) breast cancer cell lines as well as in tumors of pre- and postmeno-pausal women regardless of ER status (17-19). Several studies have shown that STS mRNA expression was an independent prognostic indicator with high levels associated with a poor prognosis (14, 20-22). One can speculate that the up-regulation of STS leads to high intratumoral estrogen concentrations and enhanced stimulation of tumor growth. Thus, STS-expressing tumor cells escaping surgical removal may proliferate rapidly

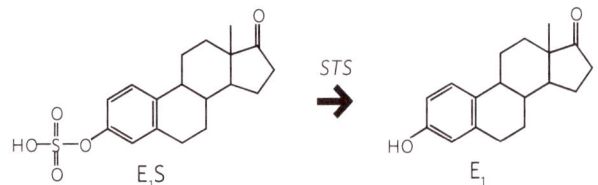

Figure 2.1.9. STS hydrolyzes the sulfate ester bond in E_1S to liberate E_1.
The enzyme can similarly convert E_2S to E_2.

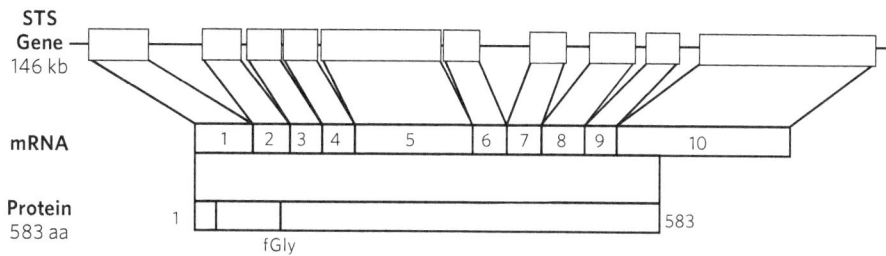

Figure 2.1.10. Overview of STS gene, mRNA, and protein.
The STS gene at Xp22.3 – Xpter spans 146 kb and contains 10 exons. Variable mRNA transcripts are due to the use of alternative polyadenylation sites within exon 10. The open reading frame of 1749 nucleotides encodes a protein of 583 amino acids with a N-terminal signal peptide of 21 amino acids (shaded zone) leaving the mature microsomal protein of 562 amino acids. The conserved residue 75 cysteine is posttranslationally modified to the catalytically active formylglycine, fGly. Variable glycosylation at asparagine residues 47 and 259 results in a protein of ~65 kDa. Complete deletion and point mutations of the STS gene cause X-linked ichthyosis.

and cause early relapse. Estrogens may be produced *in situ* in breast tissue either by aromatase or STS. The latter is quantitatively more important for estrogen production in breast tumors (23). Using the appropriate substrate concentrations, as much as 10-fold more E_1 could originate from E_1S via STS catalysis than from androstenedione by aromatase action (24). Thus, the action of STS makes a major contribution to the in situ production of estrogen in breast tissue.

Several studies have shown that progestins act as "selective estrogen enzyme modulators (SEEMs)" in hormone-responsive breast cancer cells (25). Specifically, several progestins exert an inhibitory effect on STS, which produces estrogen, in conjunction with a stimulatory effect on sulfotransferase, which forms the inactive estrogen sulfate. Collectively, these findings help to explain the anti-proliferative effect of progestins in breast tissue. Interestingly, the same investigators also showed in MCF-7 and T47D cells that E_2

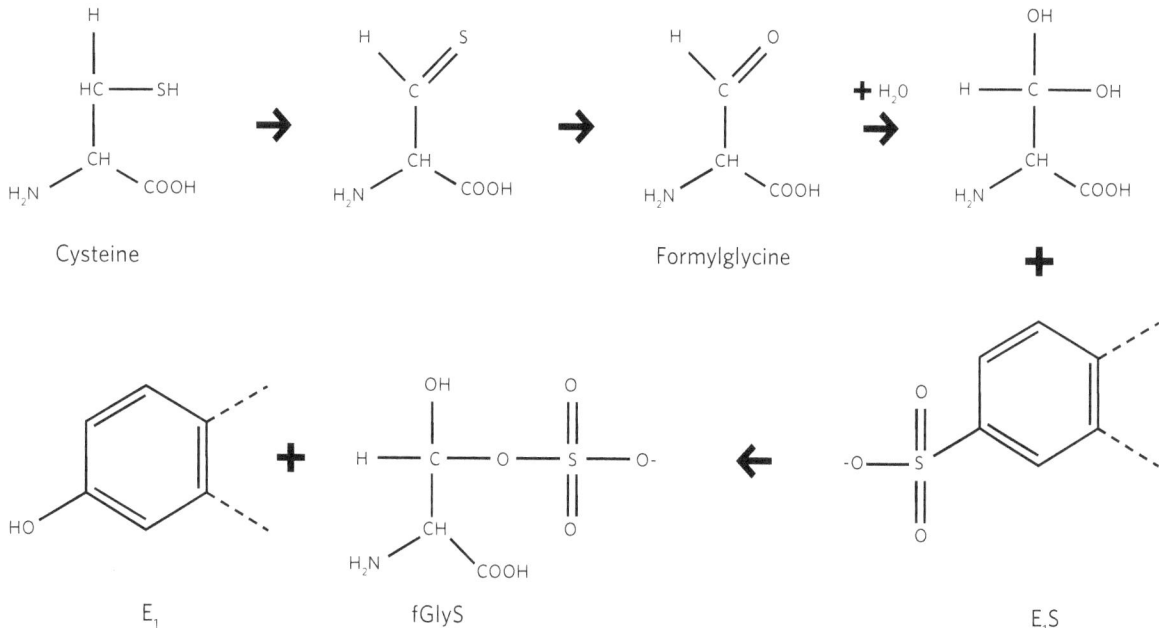

Figure 2.1.11. Reaction scheme for cleavage of E_1S by STS.
The catalytic site of STS contains a conserved cysteine, which undergoes posttranslational modification to formylglycine, fGly. The formyl group is converted to an aldehyde hydrate with two geminal hydroxyl groups. One of the hydroxyls attacks the sulfur atom of E_1S. The nucleophilic substitution of E_1S leads to the release of E_1 by desulfonation and the formation of fGlyS by transesterification. The second, non-esterified hydroxyl group of the covalent fGlyS intermediate induces the elimination of sulfate resulting in regeneration of the aldehyde for another catalytic cycle.

inhibited STS in a dose-dependent manner (IC50 1.8×10^{-9} and 8.8×10^{-10} M, respectively) and thereby decreased its own formation by blocking the conversion of E_1S to E_1, which is necessary for the 17β-HSD1-mediated conversion of E_1 to E_2 (26).

Since the conversion of E_1S to E_1 by STS represents a major source of estrogen in breast cancer cells, several groups of compounds have been designed to inhibit STS. These compounds are either synthetic steroid analogs that mimic the substrate E_1S or nonsteroidal agents such as p-O-sulfamoyl-N-alkanoyl tyramines (27-29). The most potent of the steroid analogs is estrone-3-O-sulfamate (EMATE), an active site-directed, irreversible inhibitor of STS that is highly estrogenic, presumably due to the release of the estrogenic steroid nucleus (30-32). Several analogs of EMATE have been developed that are less estrogenic than the parent while possessing similar or superior inhibitory activity (2). These analogs were modified in the A-ring, D-ring, or both, e.g., 2-methoxyestrone-3-O-sulfamate, and 2-methoxy-17α-p-tert-benzylestradiol-3-O-sulfamate (33) (Figure 2.1.12.).

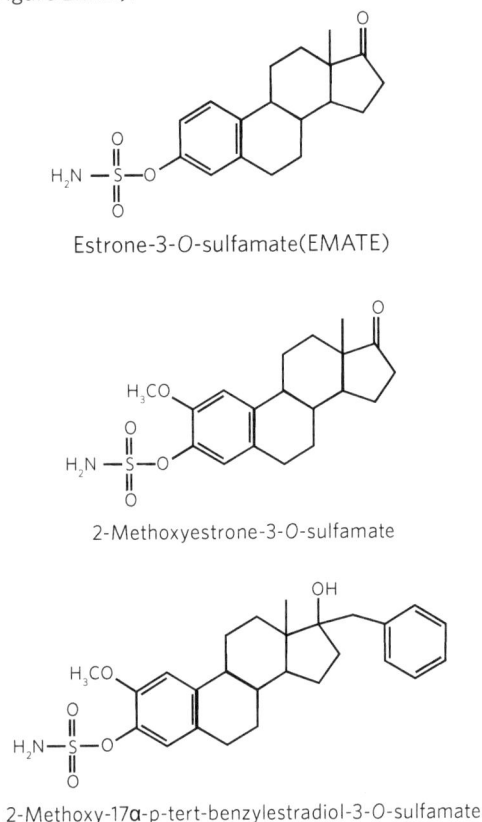

Estrone-3-O-sulfamate(EMATE)

2-Methoxyestrone-3-O-sulfamate

2-Methoxy-17α-p-tert-benzylestradiol-3-O-sulfamate

Figure 2.1.12. Estrone-3-O-sulfamate (EMATE) and other examples of irreversible STS inhibitors, 2-methoxyestrone-3-O-sulfamate, and 2-methoxy-17α-p-tert-benzylestradiol-3-O-sulfamate.

References

1. Parenti G, Meroni G, Ballabio A. *The sulfatase gene family. Curr Opin Genet Devel.* 1997;7:386-91.

2. Reed MJ, Purohit A, Woo WL, Newman SP, Potter BVL. *Steroid sulfatase: molecular biology, regulation, and inhibition.* Endocrine Rev. 2005;26:171-202.

3. Franco B, Meroni G, Parenti G, Levilliers J, Bernard L, Gebbia M, et al. *A cluster of sulfatase genes on Xp22.3: mutations in chrondrodysplasia punctata (CDPX) and implications for warfarin embryopathy.* Cell. 1995;81:15-25.

4. Meroni G, Franco B, Archidiacono N, Messali S, Andolfi G, Rocchi M, et al. *Characterization of a cluster of sulfatse genes on Xp22.3 suggests gene duplications in an ancestral pseudoautosomal region.* Hum Mol Genet. 1996;5:423-31.

5. Yen PH, Allen E, Marsh B, Mohandas T, Wang N, Taggart RT, et al. *Cloning and expression of steroid sulfatase cDNA and the frequent occurrence of deletions in STS deficiency: implications for X-Y interchange.* Cell. 1987;49:443-54.

6. Yen PH, Marsh B, Allen E, Tsai SP, Ellison J, Connolly L, et al. *The human X-linked steroid sulfatase gene and a Y-encoded pseudogene: evidence for an inversion of the Y chromosome during primate evolution.* Cell. 1988;55:1123-35.

7. Hernandez-Guzman FG, Higashiyama T, Osawa Y, Ghosh D. *Purification, characterization and crystallization of human placental estrone/dehydroepiandrosterone sulfatase, a membrane-bound enzyme of the endoplasmic reticulum.* J Steroid Biochem Mol Biol. 2001;78:441-50.

8. Hernandez-Guzman FG, Higashiyama T, Pangborn W, Osawa Y, Ghosh D. *Structure of human estrone sulfatase suggests functional roles of membrane association.* J Biol Chem. 2003;278:22989-97.

9. Dierks T, Schmidt B, von Figura K. *Conversion of cysteine to formylglycine: a protein modification in the endoplasmic reticulum.* Proc Natl Acad Sci USA. 1997;94:11963-8.

10. Schmidt B, Selmer T, Ingendoh A, von Figura K. *A novel amino acid modification in sulfatases that is defective in multiple sulfatase deficiency.* Cell. 1995;82:271-8.

11. Recksiek M, Selmer T, Dierks T, Schmidt B, von Figura K. *Sulfatases, trapping of the sulfated enzyme intermediate by substituting the active site formylglycine.* J Biol Chem. 1998;273:6096-103.

12. Shapiro LJ, Yen P, Pomerantz D, Martin E, Rolewic L, Mohandas T. *Molecular studies of deletions at the human steroid sulfatase locus.* Proc Natl Acad Sci USA. 1989;86:8477-81.

13. Gonzalez-Huerta LM, Riviera-Vega MR, Kofman-Alfero SH, Cuevas-Sovarrubias SA. *Novel missense mutation (Arg432Cys) in a patient with steroid sulphatase-deficiency.* Clin Endocrinol. 2003;59:263-5.

14. Suzuki T, Nakata T, Miki Y, Kaneko C, Moriya T, Ishida T, et al. *Estrogen sulfotransferase and steroid sulfatase in human breast carcinoma.* Cancer Res. 2003;63:2762-70.

15. Utsumi T, Yoshimura N, Takeuchi S, Maruta M, Maeda K, Harada N. *Elevated steroid sulfatase expression in breast cancers.* J Steroid Biochem Mol Biol. 2000;73:141-5.

16. Pasqualini JR, Chetrite G, Blacker C, Feinstein M-C, Delalonde L, Talbi M, et al. *Concentrations of estrone, estradiol, and estrone sulfate and evaluation of sulfatase and aromatase activities in pre- and postmenopausal breast cancer patients.* J Clin Endocrinol Metabol. 1996;81:1460-4.

17. Chetrite G, Paris J, Botella J, Pasqualini JR. *Effect of nomegestrol acetate on estrone-sulfatase and 17beta-hydroxysteroid dehydrogenase activities in human breast cancer.* J Steroid Biochem Mol Biol. 1996;58:525-31.

18. Prost O, Turrel MO, Dahan N, Craveur C, Adessi GL. *Estrone and dehydroepiandrosterone sulfatase activities and plasma estrone sulfate levels in human breast carcinoma.* Cancer Res. 1984;44:661-4.

19. Purohit A, Reed MJ. *Oestrogen sulphatase activity in hormone-dependent and hormone-independent breast-cancer cells: modulation by steroidal and non-steroidal therapeutic agents.* Int J Cancer. 1992;50:901-5.

20. Miyoshi Y, Ando A, Hasegawa S, Ishitobi M, Taguchi T, Tamaki Y, et al. *High expression of steroid sulfatase mrRNA predicts poor prognosis in patients with estrogen receptor-positive breast cancer.* Clin Cancer Res. 2003;9:2288-93.

21. Utsumi T, Yoshimura N, Takeuchi S, Ando J, Maruta M, Maeda K, et al. *Steroid sulfatase expression is an independent predictor* of recurrence in human breast cancer. Cancer Res. 1999;59:377-81.

22. Yoshimura N, Harada N, Bukholm I, Karesen R, Borresen-Dale A-L, Kristensen VN. *Intratumoral mrRNA expression of genes from the oestradiol metabolic pathway and clinical and histopathological parameters of breast cancer.* Breast Cancer Res. 2004;6:R46-R55.

23. Santner SJ, Feil PD, Santen RJ. *In situ estrogen production via the estrone sulfatase pathway in breast tumors: relative importance versus the aromatase pathway.* J Clin Endocrinol Metab. 1984;59:29-33.

24. Pasqualini JR, Chetrite G, Nestour EL. *Control and expression of oestrone sulphatase activities in human breast cancer.* J Endocrinol. 1996;150:S99-S105.

25. Pasqualini JR. *The selective estrogen enzyme modulators in breast cancer: a review.* Biochim Biophys Acta. 2004;1654:123-43.

26. Pasqualini JR, Chetrite GS. *Paradoxical effect of estradiol: it can block its own bioformation in human breast cancer cells.* J Steroid Biochem Mol Biol. 2001;78:21-4.

27. Anderson C, Freeman J, Lucas LH, Farley M, Dalhoumi H, Widlanski TS. *Estrone sulfatase: probing structural requirements for substrate and inhibitor recognition.* Biochemistry. 1997;36:2586-94.

28. Purohit A, Reed MJ, Morris NC, Williams GJ, Potter BVL. *Regulation and inhibition of steroid sulfatase activity in breast cancer.* Ann N Y Acad Sci. 1996;784:40-9.

29. Selcer KW, Hegde PV, Li P. *Inhibition of estrone sulfatase and proliferation of human breast cancer cells by nonsteroidal (p-O-Sulfamoyl)-N-alkanoyl tyramines.* Cancer Res. 1997;57:702-7.

30. Elger W, Schwarz S, Hedden A, Reddersen G, B. S. *Sulfamates of various estrogens are prodrugs with increased systemic and reduced hepatic estrogenicity at oral application.* J Steroid Biochem Mol Biol. 1995;55:395-403.

31. Hidalgo Aragones MI, Purohit A, Parish D, Sahm UG, Pouton CW, Potter BV, et al. *Pharmacokinetics of oestrone-3-0-sulphamate.* J Steroid Biochem Mol Biol. 1996;58:611-7.

32. James VH, Reed MJ, Purohit A. *Inhibition of oestrogen synthesis in postmenopausal women with breast cancer.* J Steroid Biochem Mol Biol. 1992;43:149-53.

33. Ciobanu LC, Luu-The V, Martel C, Labrie F, Poirier D. *Inhibition of estrogen sulfate-induced uterine growth by potent nonestrogenic steroidal inhibitors of steroid sulfatase.* Cancer Res. 2003;63:6442-6.

2.1.1.8. ESTROGEN FATTY ACID ESTERS

Several steroids, including E_2, have been shown to undergo esterification to long-chain fatty acids in a number of mammalian tissues (1-4). For example, incubation of microsomes from human breast cancer tissue with E_2 and fatty acyl-coenzyme A (CoA) resulted in fatty acid esterification of E_2 at C-17β but not at the phenolic C-3 (5, 6). The resulting E_2-17β-fatty acid esters (E_2-FA) are bulkier and even more lipophilic than the parent hormone (Figure 2.1.13.). The responsible enzyme, fatty acyl-CoA: estradiol-17β-acyltransferase, has a pH optimum of 5.5, which distinguishes it from the related enzyme, acyl-CoA: cholesterol acyltransferase (optimal pH ~7.0) (7). The fatty acyl-CoA: estradiol-17β-acyltransferase shows specificity for the D-ring, especially the C-17β group of the estrogen molecule. The vicinity of a bulky 16α-hydroxy group appears to hamper the accessibility to the C-17β hydroxyl, resulting in a reduced rate (28%) of esterification for E_3 compared to E_2 (8, 9). At the same time, the D-ring esterification

results in sterically hindered secondary C-17β and C-16α esters (E_2-17β-FA, E_3-17β-FA, E_3-16α-FA) that protect the estrogens from metabolic degradation and thereby prolongs their biological action. The synthesis and sequestration of these hydrophobic estrogen esters forms a readily available, preformed hormone pool that does not require de novo synthesis. Rather, only esterase activity is needed to liberate the hormone. Two types of esterases have been identified in MCF-7 breast cancer cells, a nonspecific esterase that acts upon short-chain esters of E_2 (e.g., E_2-17β-acetate) and a more specific esterase capable of hydrolyzing E_2-17β-stearate (Figure 2.1.13.) (10). The latter type of esterase has been isolated from human breast cyst fluid and bovine placenta with molecular weights of 90 and 84 kDa, respectively (11, 12). The placental esterase appeared to be identical to hormone-sensitive lipase.

The D-ring esterification of E_2 has two effects. (i) The fatty acid moiety shields the D-ring from oxidative metabolism to E_1 (13). (ii) The bulky fatty acid moiety prevents the binding of E_2-FA to the ER (14). Thus, E_2-FA may play a role in estrogen action by affecting the intracellular equilibrium between E_1 and E_2. The estrogenic potency of E_1 appears to be limited by its rapid clearance from the cell (15).

The formation of E_2-FA may slow the intracellular loss of estrogen by protecting the C-17 hydroxyl group of E_2 from 17β-HSD, the enzyme that converts E_2 to E_1. In this manner, E_2-FA could serve as a long-lived storage form that provides available E_2 through an enzymatic hydrolytic process. The E_2-FA are known to be potent, long-acting estrogens (4, 16). Intravenous administration of E_2-stearate to female rats produced a sustained and greater uterotropic response than a similar dose of E_2. The associated induction of uterine ER was also markedly sustained compared with induction by E_2, although the response was delayed and only about one third of the E_2 response (17). Since E_2-stearate does not bind to the ER, this delay is consistent with the requirement for hydrolysis of the ester before interaction with the ER (14). Neither of the ionic conjugates of E_2 (sulfate and glucuronate) produced an increase in ER concentration or uterotropic response. These findings suggest that E_2-FA represent a metabolically protected, slowly releasable pool of preformed estrogen. Since these fatty acid esters are highly lipophilic, they would be expected to accumulate in fatty tissue such as the mammary gland and serve as reservoir for slow esterase-mediated release of E_2. The availability of an intracellular pool of E_2 that does not require de novo synthesis to produce its stimulatory effect raises the possibility of local target tissue control of the extent of hormonal stimulation through conversion of the inactive esters to the active E_2.

The absolute extent of conversion of E_2 to E_2-FA is much smaller than the rate of other metabolic reactions such as sulfoconjugation. For instance, E_2 was converted to

E_2-sulfate at a 18-fold higher overall rate than to the E_2-FA over a 19-hr period in ZR-75 breast cancer cells (9). Thus, fatty acid esterification of E_2 is likely to represent only a minor fraction of the metabolic estrogen efflux observed at any time. Nevertheless, by virtue of the highly lipophilic character of E_2-FA, accumulation of fatty acid esters may represent a major fraction of estrogen derivatives present in the intracellular compartment.

Incubation of human breast cancer cells with [^{3}H]E_2 resulted in the synthesis of E_2-FA. E2-17β-oleate was the major component (~30%) among the fatty acid esters, while E_2-17β-arachidonate was present in much lower concentration, possibly due to more rapid hydrolysis of the latter (5, 13). E_2-FA is synthesized in breast cancer cells regardless of their hormonal sensitivity, i.e., the ER content. Addition of 10 nM [^{3}H]E_2 to the culture medium resulted in 2-hr E_2-FA levels of 270 fmol/mg DNA in ER-positive MCF-7 cells and ~900 fmol/mg DNA in ER-negative MDA-MB-231 and MDA-MB-330 cells (18). Upon subsequent withdrawal of E_2 from the medium, intracellular concentration of E_2-FA decreased rapidly in the first 5-hr period, then declined more slowly to ~50 fmol/mg DNA at 24 hours. Intracellular concentrations of E_2 were maintained over this time period. E_2-FA was not detectable in the medium. Thus, accumulation of E_2-FA in cells upon continuous exposure to E_2 represents the net result of esterification (acylation) and deesterification (hydrolysis) reactions (18).

In the circulation, the E_2-FAs are mostly bound by plasma lipoproteins with the majority (54%) recovered in the high-density lipoprotein and 28% in the low-density lipoprotein fractions (19). E_2-FAs are present in very small amounts in blood of premenopausal women (20, 21). The E_2-FA concentration increased 10-fold during pregnancy, from 40 pmol/L in early pregnancy to 400 pmol/L in late pregnancy (22). Higher concentrations of E_2-FA were measured in the fluid of ovarian follicles and breast cysts (21, 23). The highest levels were observed in subcutaneous fat and omentum, ranging from 957 ± 283 fmol/g fat in premenopausal women to 669 ± 158 fmol/g in women less than 12 years postmenopausal and 399 ± 146 fmol/g in women postmenopausal for over 15 years (21). Since postmenopausal women have very low blood levels of E_2

and undetectable circulating E_2-FA, it appears likely that E_2-FA is synthesized in adipose tissue from E_2 formed in adipocytes by aromatization.

Treatment of postmenopausal women with either oral or transdermal E_2 replacement for 12 weeks resulted in a differential effect on serum E_2-FA and nonesterified E_2. Both types of application led to an increase in nonesterified E_2 but only the oral therapy caused an increase (27%) in median serum E_2-FA (24). The change in serum E_2-FA concentrations during treatment, but not that of nonesterified E_2 correlated positively with enhanced forearm blood flow responses *in vivo*. These data suggest that an increase in serum E_2-FA may contribute to the beneficial effects of oral E_2 treatment, compared with an equipotent dose of transdermal E_2. Chronic treatment of ovariectomized rats with 0.5 or 5 nmol/day E_2-stearate for 10 or 23 days had a stronger stimulatory effect on mammary gland cell proliferation than treatment with equimolar doses of E_2 (25). E_2 monoesters at C-17β showed longer estrogenic effects than monoesters at C-3, while the E_2-3,17β-diesters exhibited the shortest effect (26).

Two commonly prescribed hypolipidemic drugs, clofibrate and gemfibrozil, increase the size and number of hepatic peroxisomes upon administration to rodents. Treatment of rats with clofibrate caused a multifold increase in the hepatic microsomal formation of E_2-FA (7). The stimulatory effect of clofibrate on hepatic fatty acid esterification of E_2 was paralleled by enhanced E_2-induced increases in the formation of lobules in the mammary gland and by increased incorporation of bromodeoxy-uridine, a marker of cell proliferation, into these lobules (27). These experimental findings may be relevant to explain clinical observations that men receiving clofibrate medication manifested side effects related to disturbed sex hormone function such as decreased libido and breast tenderness or enlargement.

References

1. Abul-Hajj YJ. *Formation of estradiol-17b fatty acyl 17-esters in mammary tumors.* Steroids. 1982;40:149-56.

2. Hochberg RB. *Biological esterification of steroids.* Endocrine Rev. 1998;19:331-48.

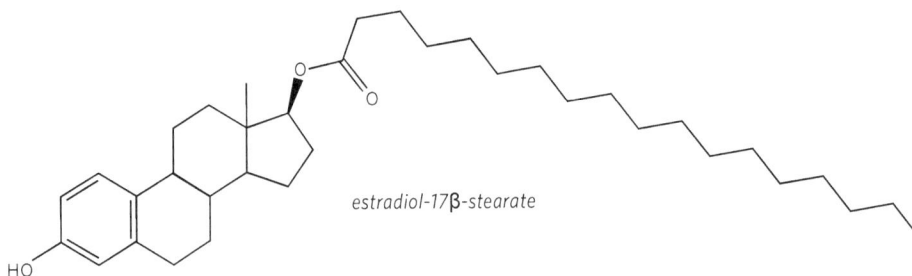

estradiol-17β-stearate

Figure 2.1.13. Estradiol-17β-stearate, a naturally occurring fatty acid ester of E_2.

3. Hochberg RB, Pajuja SL, Zielinski JE, Larner JM. *Steroidal fatty acid esters. J Steroid Biochem Mol Biol.* 1991;40:577-85.

4. Schatz F, Hochberg RB. *Lipoidal derivative of estradiol: the biosynthesis of a nonpolar estrogen metabolite.* Endocrinology. 1981;109:697-703.

5. Martyn P, Smith DL, Adams JB. *Selective turnover of the essential fatty acid ester components of estradiol-17b lipoidal derivatives formed by human mammary cancer cells in culture.* J Steroid Biochem. 1987;28:393-8.

6. Mellon-Nussbaum SH, Ponticorvo L, Schatz F, Hochberg RB. *Estradiol fatty acid esters. The isolation and identification of the lipoidal derivative of estradiol synthesized in the bovine uterus.* J Biol Chem. 1982;257:5678-84.

7. Xu S, Zhu BT, Conney AH. *Stimulatory effect of clofibrate and gemfibrozil administration on the formation of fatty acid esters of estradiol by rat liver microsomes.* J Pharmacol Exp Ther. 2001;296:188-97.

8. Pahuja SL, Zielinski JE, Giordano G, McMurray WJ, Hochberg RB. *The biosynthesis of D-ring fatty acid esters of estriol.* J Biol Chem. 1991;266:7410-6.

9. Poulin R, Poirier D, Theriault C, Couture J, Belanger A, Labrie F. *Wide spectrum of steroids serving as substrates for the formation of lipoidal derivatives in ZR-75-1 human breast cancer cells.* J Steroid Biochem. 1990;35:237-47.

10. Katz J, Finlay TH, Banerjee S, Levitz M. *An estrogen-dependent esterase activity in MCF-7 cells.* J Steroid Biochem. 1987;26:687-92.

11. Banerjee S, Katz J, Levitz M, Finlay TH. *Purification and properties of an esterase from human breast cyst fluid.* Cancer Res. 1991;51:1092-8.

12. Lee F-T, Adams JB, Garton AJ, Yeaman SJ. *Hormone-sensitive lipase is involved in the hydrolysis of lipoidal derivatives of estrogens and other steroid hormones.* Biochim Biophys Acta. 1988;963:258-64.

13. Hershcopf RJ, Bradlow HL, Fishman J, Swaneck GE, Larner JM, Hochberg RB. *Metabolism of estradiol fatty acid esters in man.* J Clin Endocrinol Metab. 1985;61:1071-5.

14. Janocko L, Larner JM, Hochberg RB. *The interaction of C-17 esters of of estradiol with the estrogen receptor.* Endocrinology. 1984;114:1180-6.

15. Tseng L, Stolee A, Gurpide E. *Quantitative studies on the uptake and metabolism of estrogens and progesterone by human endometrium.* Endocrinology. 1972;90:390-404.

16. Larner JM, MacLusky NJ, Hochberg RB. *The naturally occurring C-17 fatty acid esters of estradiol are long-acting estrogens.* J Steroid Biochem. 1985;22:407-13.

17. MacLusky NJ, Larner JM, Hochberg RB. *Actions of an estradiol-17-fatty acid ester in estrogen target tissues of the rat: comparison with other C-17 metabolites and a pharma-cological C-17 ester.* Endocrinology. 1989;124:318-24.

18. Adams JB, Hall RT, Nott S. *Esterification-deesterification of estradiol by human mammary cancer cells in culture.* J Steroid Biochem. 1986;24:1159-62.

19. Vihma V, Tiitinen A, Ylikorkala O, Tikkanen MJ. *Quantitative determination of estradiol fatty acid esters in lipoprotein fractions in human blood.* J Clin Endocrin Metab. 2003;88:2552-5.

20. Janocko L, Hochberg RB. *Estradiol fatty acid esters occur naturally in human blood.* Science. 1983;222:1334-6.

21. Larner JM, Shackleton CHL, Roitman E, Schwartz PE, Hochberg RB. *Measurement of estradiol -17-fatty acid esters in human tissues.* J Clin Endocrinol Metabolism. 1992;75:195-200.

22. Vihma V, Adlercreutz H, Tiitinen A, Kiuru P, Wahala K, Tikkanen MJ. *Quantitative determination of estradiol fatty acid esters in human pregnancy serum and ovarian follicular fluid.* Clin Chem. 2001;47:1256-62.

23. Larner JM, Pajuja SL, Shackleton CHL, McMurray WJ, Giordano G, Hochberg RB. *The isolation and characterization of estradiol-fatty acid esters in human ovarian follicular fluid.* J Biol Chem. 1993;268:13893-9.

24. Vihma V, Vehkavaara S, Yki-Jarvinen H, Hohtari H, Tikkanen MJ. *Differential effects of oral and transdermal estradiol treatment on circulating estradiol fatty acid ester concentrations in postmenopausal women.* J Clin Endocrinol Metab. 2003;88:588-93.

25. Mills LH, Lee AJ, Parlow AF, Zhu BT. *Preferential growth stimulation of mammary glands over uterine endometrium in female rats by a naturally occurring estradiol-17b-fatty acid ester.* Cancer Res. 2001;61:5764-70.

26. Vazquez-Alcantara MA, Menjivar M, Garcia GA, Diaz-Zagoya JC, Garza-Flores J. *Long-acting estrogenic responses of estradiol fatty acid esters.* J Steroid Biochem. 1989;33:1111-8.

27. Xu S, Zhu BT, Cai MX, Conney AH. *Stimulatory effect of clofibrate on the action of estradiol in the mammary gland but not in the uterus of rats.* J Pharmacol Exp Ther. 2001;297:50-6.

2.1.2. ESTROGEN METABOLISM
2.1.2.1. OXIDATION PRODUCTS
2.1.2.1.1. INTRODUCTION

All naturally occurring estrogens have an unsaturated (aromatic) A ring, a phenolic group at C-3, and a methyl group at C-13 (Figure 2.1.2.1.). The main estrogens, 17β-estradiol (E_2) and estrone (E_1) undergo extensive oxidative metabolism via the action of several cytochrome P450 (CYP) monooxygenases. Each CYP favors the hydroxylation of specific carbons and altogether the P450 enzymes can hydroxylate virtually all carbons in the steroid molecule with the exception of the inaccessible angular carbons 5, 8, 9, 10, and 13 (1-7). The remarkable hydroxylation activity of CYPs may be due to two binding orientations of the estrogens at the active site of each enzyme. It appears that steroid substrates can enter P450 enzymes with antiparallel orientation of the steroid long axis, positioning either ring A or D proximal to the heme group of the active site. In the case of E_2, either the 3-OH or 17-OH group may support docking into the active site. Evidence for multiple substrate binding modes is provided by x-ray analysis of CYP2C5-substrate complexes (8, 9). The generation of hydroxyl and keto functions at specific sites of the steroid nucleus markedly affects the biologic property of the respective estrogen metabolites, i.e., different hydroxylation reactions yield estrogenic, nonestrogenic or carcinogenic metabolites. Quantitatively and functionally, the most important reactions occur at C-2, 4, and 16. Both parent hormones and oxidative metabolites can be further modified by conjugation reactions resulting in over 30 metabolites identified in breast cancer tissue (10, 11).

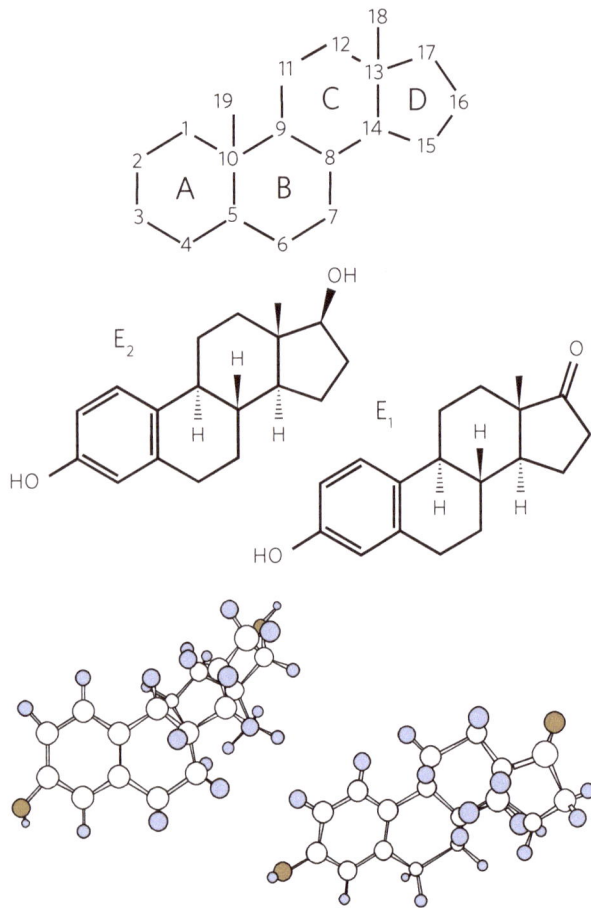

Figure 2.1.2.1. Steroid molecule with ring lettering and carbon numbering (top). The principal estrogens, 17β-estradiol (E_2) and estrone (E_1), differ only by the C-17 substituent as shown in the steric models (bottom). Both E_2 and E_1 undergo extensive oxidative metabolism by cytochrome P450 monooxygenases with the main reactions occurring at C-2, C-4, and C-16.

References

1. Badawi AF, Cavalieri EL, Rogan EG. *Role of human cytochrome P450 1A1, 1A2, 1B1, and 3A4 in the 2-, 4-, and 16a-hydroxylation of 17 b-estradiol.* Metabolism. 2001;50:1001-3.

2. Kisselev P, Schunck W-H, Roots I, Schwarz D. *Association of CYP1A1 polymorphisms with differential metabolic activation of 17b-estradiol and estrone.* Cancer Res. 2005;65:2972-8.

3. Lee AJ, Cai MX, Thomas PE, Conney AH, Zhu BT. *Characterization of the oxidative metabolites of 17β-estradiol and estrone formed by 15 selectively expressed*

human cytochrome p450 isoforms. Endocrinology. 2003;144:3382-98.

4. Lee AJ, Kosh JW, Conney AH, Zhu BT. *Characterization of the NADPH-dependent metabolism of 17β-estradiol to multiple metabolites by human liver microsomes and selectively expressed human cytochrome p450 3A4 and 3A5.* J Pharmacol Exp Ther. 2001;298:420-32.

5. Lee AJ, Mills LH, Kosh JW, Conney AH, Zhu BT. *NADPH-dependent metabolism of estrone by human liver microsomes.* J Pharmacol Exp Ther. 2002;300:838-49.

6. Martucci CP, Fishman J. *P450 enzymes of estrogen metabolism.* Pharmac Ther. 1993;57:237-57.

7. Zhu BT, Conney AH. *Functional role of estrogen metabolism in target cells: review and perspectives.* Carcinogenesis. 1998;19:1-27.

8. Wester MR, Johnson EF, Marques-Soarres C, Dansette PM, Mansuy D, Stout CD. *Structure of a substrate complex of mammalian cytochrome P450 2C5 at 2.3 Å resolution: Evidence for multiple substrate binding modes.* Biochemistry. 2003;42:6370-9.

9. Williams PA, Cosme J, Sridhar V, Johnson EF, McRee DE. *Mammalian microsomal cytochrome*

P450 monooxygenase: structural adaptations for membrane binding and functional diversity. Mol Cell. 2000;5:121-31.

10. Castagnetta LAM, Granata OM, Traina A, Ravazzolo B, Amoroso M, Miele M, et al. *Tissue content of hydroxyestrogens in relation to survival of breast cancer patients.* Clin Cancer Res. 2002;8:3146-55.

11. Rogan EG, Badawi AF, Devanesan PD, Meza J, Edney JA, West WW, et al. *Relative imbalances in estrogen metabolism and conjugation in breast tissue of women with carcinoma: potential biomarkers of susceptibility to cancer.* Carcinogenesis. 2003;24:697-702.

2.1.2.1.2. CATECHOL ESTROGENS

Overview. Hydroxylation at C-2 and C-4 produces catechol estrogens, which are characterized by hydroxyl groups at vicinal carbons (i.e., C-2 and C-3 or C-3 and C-4) of the aromatic A-ring (Figure 2.1.2.2.). Specifically, hydroxylation at the C-2 position of E_2 and E_1 results in formation of the 2-catechol estrogens, 2-hydroxyestradiol (2-OHE$_2$) and 2-hydroxyestrone (2-OHE$_1$), respectively. Similarly, hydroxylation at the C-4 position produces the corresponding 4-catechol estrogens, 4-hydroxyestradiol (4-OHE$_2$) and 4-hydroxyestrone (4-OHE$_1$). The mechanism of aromatic hydroxylation of estrogens involves hydrogen abstraction from the phenolic hydroxy group, electron delocalization of the phenoxy radical to a carbon-centered radical, and subsequent formation of catechol metabolites by hydroxy radical addition at C-2 or C-4 depending on steric or electronic constraints (1). Catechol estrogens are produced in various tissues (liver, breast, uterus, pituitary, hypothalamus, cerebral cortex and others) and released into the circulation where they may bind with low affinity to sex hormone-binding globulin (2). The concentration of catechol estrogens in the systemic circulation is very low, which is probably due to rapid conjugative metabolism (O-methylation, glucuronidation, sulfonation) followed by urinary excretion. The highest levels are produced during pregnancy, when urinary excretion steadily increases until reaching a plateau in the last trimester (3).

Physiological Function. Catechol estrogens have been shown to mediate the activation of dormant blastocysts for implantation into the receptive uterus (4). Specifically, 4-OHE$_2$ produced in the uterus from E_2 mediates blastocyst activation for implantation in a paracrine manner. The effect of 4-OHE$_2$ on blastocyst activation is not mediated by the ER but via prostaglandin synthesis (5).

Hormonal Activity. The oxidative metabolism of estrogens to catechol estrogens is generally thought to terminate the estrogenic signal although catechol estrogens retain some binding affinity to the ER (6-9). Treatment of MCF-7 cells with 2-OHE$_2$ and 4-OHE$_2$ increased the rate of cell proliferation and the expression of estrogen-inducible genes such as PR and pS2. Relative to E$_2$, the effects of 2-OHE$_2$ and 4-OHE$_2$ on proliferation rate, PR, and pS2 expression were 36 and 76%, 10 and 28%, 48 and 79%, respectively (8, 10). Thus, the ER is transcriptionally activated in breast cancer cells upon binding of catechol estrogens. The lower estrogenic potency of 2-OHE$_2$ compared to 4-OHE$_2$ may be due to more rapid dissociation of the former from ER (11).

Catechol-O-methyl transferase (COMT) Substrates.
COMT catalyzes the O-methylation of catechol estrogens forming monomethyl ethers at the 2-OH, 3-OH, and 4-OH groups. However, the enzyme generated two products from 2-OHE, but only one product from 4-OHE (12-14). In the case of 2-OHE$_2$ and 2-OHE$_1$, COMT catalyzed the methylation of the 2-OH and 3-OH groups, resulting in the formation of 2-MeOE$_2$ and 2-OH-3-MeOE$_2$, and 2-MeOE$_1$ and 2-OH-3-MeOE$_1$, respectively. In contrast, in the case of 4-OHE$_2$ and 4-OHE$_1$, methylation occurred only at the 4-OH group, resulting in the formation of 4-MeOE$_2$ and 4-MeOE$_1$, respectively. 3-MeO-4-OHE$_2$ and 3-MeO-4-OHE$_1$ were not produced by COMT.

CYP1A1 and CYP1B1 Substrates. The catechol estrogens are products as well as substrates of CYP1A1 and CYP1B1 (15, 16). Specifically, CYP1A1 converts E$_2$ to 2-OHE$_2$ and

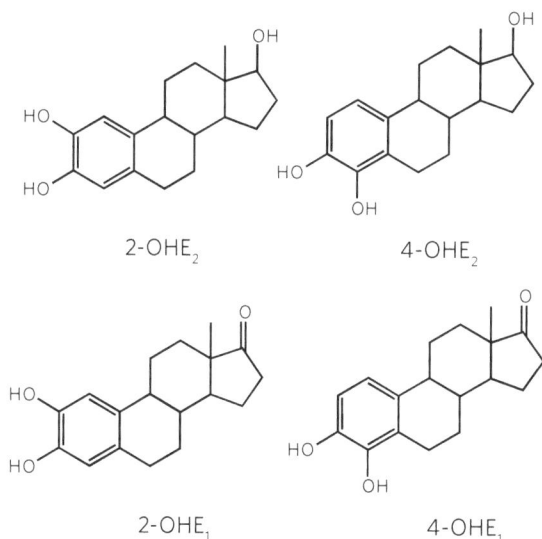

Figure 2.1.2.2. The parent estrogens, E$_2$ and E$_1$, are hydroxylated to the catechol estrogens 2-hydroxyestradiol (2-OHE$_2$) and 4-hydroxyestradiol (4-OHE$_2$) and 2-hydroxyestrone (2-OHE$_1$) and 4-hydroxyestrone (4-OHE$_1$), respectively.

further to the semiquinone E$_2$-2,3-SQ and quinone E$_2$-2,3-Q. CYP1B1 converts E$_2$ to 2-OHE2 as well as 4-OHE$_2$ and further to the corresponding semiquinones and quinones. E$_1$ is metabolized in a similar manner by CYP1A1 and CYP1B1. It appears that the catalytic process of CYP-mediated estrogen oxidation comprises several equilibria as shown for the C-2 oxidation in the following diagram (Figure 2.1.2.3.), which omits the semiquinone for the sake of clarity.

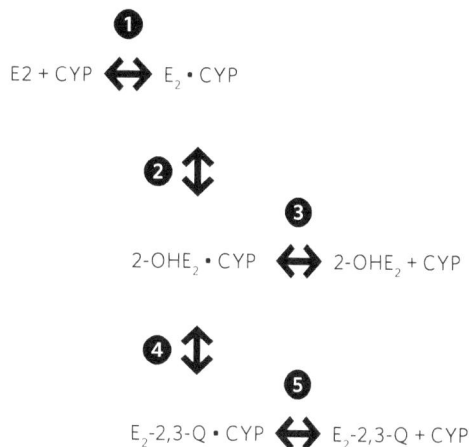

Figure 2.1.2.3. CYP-mediated oxidative metabolism of E$_2$ to 2-OHE$_2$ and E$_2$-2,3-Q.

Carcinogenic Activity. Experimental evidence indicates that 4-catechol estrogens are more carcinogenic than the 2-OH isomers. Treatment with 4-OHE$_2$, but not 2-OHE$_2$, induced renal cancer in Syrian hamster (17, 18). Analysis of renal DNA demonstrated that 4-OHE$_2$ significantly increased 8-hydroxyguanosine levels, whereas 2-OHE$_2$ did not cause oxidative DNA damage (19). In addition to the induction of renal cancer in the hamster model, 4-OHE$_2$ is capable of inducing uterine adenocarcinoma, a hormonally related cancer, in mice. Administration of E$_2$, 2-OHE$_2$, and 4-OHE$_2$ induced endometrial carcinomas in 7, 12, and 66%, respectively, of treated CD-1 mice (20). Examination of microsomal E$_2$ hydroxylation activity in human breast cancer showed significantly higher 4-OHE$_2$/2-OHE$_2$ ratios in tumor tissue than in adjacent normal breast tissue (6), while the latter tissue samples contained four-fold higher levels of 4-OHE$_2$ than normal tissue from benign breast biopsies (21). Comparison of intra-tissue concentrations of estrogens (E$_1$, E$_2$, E$_3$), hydroxyestrogens (16α-OHE$_1$, 2-OHE$_1$, 2-OHE$_2$, 4-OHE$_1$, 4-OHE$_2$) and methoxyestrogens (2-MeOE$_1$, 2-MeOE$_2$, 4-MeOE$_1$, 4-MeOE$_2$) in normal and malignant breast revealed the highest concentration of 4-OHE$_2$ in malignant tissue (22). The concentration (1.6 nmol/g tissue) determined by combined HPLC and GC/MS was more than twice as high as that of any other compound. Such high levels in neoplastic mammary tissue suggest a mechanistic role of 4-OHE$_2$ in tumor development.

References

1. Sarabia SF, Zhu BT, Kurosawa T, Tohma M, Liehr JG. *Mechanism of cytochrome P450-catalyzed aromatic hydroxylation of estrogens.* Chem Res Toxicol. 1997;10:767-71.

2. Ball P, Knuppen R. *Formation, metabolism, and physiologic importance of catecholestrogens.* Am J Obstet Gynecol. 1990;163:2163-70.

3. Berg FD, Kuss E. *Serum concentration and urinary excretion of "classical" estrogens, catecholestrogens and 2-methoxyestrogens in normal pregnancy.* Arch Gynecol Obstetrics. 1992;251:17-27.

4. Paria BC, Das SK, Dey SK. *Embryo implantation requires estrogen-directed uterine preparation and catecholestrogen-mediated embryonic activation.* Adv Pharmacol. 1998;42:840-3.

5. Paria BC, Lim H, Das SK, Reese J, Dey SK. *Molecular signaling in uterine receptivity for implantation.* Semin Cell Develop Biol. 2000;11:67-76.

6. Liehr JG, Ricci MJ. *4-Hydroxylation of estrogens as marker of human mammary tumors.* Proc Natl Acad Sci USA. 1996;93:3294-6.

7. Liehr JG, Ricci MJ, Jefcoate CR, Hannigan EV, Hokanson JA, Zhu BT. *4-Hydroxylation of estradiol by human uterine myometrium and myoma microsomes: implications for the mechanism of uterine tumorigenesis.* Proc Natl Acad Sci USA. 1995;92:9220-4.

8. Schutze N, Vollmer G, Tiemann I, Geiger M, Knuppen R. *Catecholestrogens are MCF-7 cell estrogen receptor agonists.* J Steroid Biochem Mol Biol. 1993;46:781-9.

9. Van Aswegen CH, Purdy RH, Wittliff JL. *Binding of 2-hydroxyestradiol and 4-hydroxyestradiol to estrogen receptors from human breast cancers.* J Steroid Biochem. 1989;32:485-92.

10. Schutze N, Vollmer G, Knuppen R. *Catecholestrogens are agonists of estrogen receptor dependent gene expression in MCF-7 cells.* J Steroid Biochem Mol Biol. 1994;48:453-61.

11. Barnea ER, MacLusky NJ, Naftolin F. *Kinetics of catechol estrogen-estrogen receptor dissociation: a possible factor underlying differences in catechol estrogen biological activity.* Steroids. 1983;41:643-56.

12. Dawling S, Roodi N, Mernaugh RL, Wang XY, Parl FF. *Catechol-O-methyltransferase (COMT)-mediated metabolism of catechol estrogens: comparison of wild-type and variant COMT isoforms.* Cancer Res. 2001;61:6716-22.

13. Goodman JE, Jensen LT, He P, Yager JD. *Characterization of human soluble high and low activity catechol-O-methyltransferase catalyzed catechol estrogen methylation.* Pharmacogenetics. 2002;12:517-28.

14. Lautala P, Ulmanen I, Taskinen J. *Molecular mechanisms controlling the rate and specificity of catechol O-methylation by human soluble catechol O-methyltransferase.* Mol Pharmacol. 2001;59:393-402.

15. Dawling S, Hachey DL, Roodi N, Parl FF. *In vitro model of mammary estrogen metabolism: Structural and kinetic differences in mammary metabolism of catechol estrogens 2- and 4-hydroxyestradiol.* Chem Res Toxicol. 2004;17:1258-64.

16. Hachey DL, Dawling S, Roodi N, Parl FF. *Sequential action of phase I and II enzymes cytochrome P450 1B1 and glutathione S-transferase P1 in mammary estrogen metabolism.* Cancer Res. 2003;63:8492 - 9.

17. Li JJ, Li SA. *Estrogen carcinogenesis in Syrian hamster tissues: role of metabolism.* Fed Proc. 1987;46:1858-63.

18. Liehr JG, Fang WF, Sirbasku DA, Ari-Ulubelen A. *Carcinogenicity of catechol estrogens in Syrian hamsters.* J Steroid Biochem. 1986;24:353-6.

19. Han X, Liehr JG. *Microsome-mediated 8-hydroxylation of guanine bases of DNA by steroid estrogens: correlation of DNA damage by free radicals with metabolic activation to quinones.* Carcinogenesis. 1995;16:2571-4.

20. Newbold RR, Liehr JG. *Induction of uterine adenocarcinoma in CD-1 mice by catechol estrogens.* Cancer Res. 2000;60:235-7.

21. Rogan EG, Badawi AF, Devanesan PD, Meza J, Edney JA, West WW, et al. *Relative imbalances in estrogen metabolism and conjugation in breast tissue of women with carcinoma: potential biomarkers of susceptibility to cancer.* Carcinogenesis. 2003;24:697-702.

22. Castagnetta LAM, Granata OM, Traina A, Ravazzolo B, Amoroso M, Miele M, et al. *Tissue content of hydroxyestrogens in relation to survival of breast cancer patients.* Clin Cancer Res. 2002;8:3146-55.

2.1.2.1.3. 16α-HYDROXYESTROGENS

Overview. 16α-hydroxylation of E_1 and E_2 leads to 16α-hydroxyestrone (16α-OHE$_1$) and 16α-hydroxyestradiol, i.e., estriol (16α-OHE$_2$ = E_3), respectively (Figure 2.1.2.3).

It has been widely held for years that 16α-OHE$_2$ = E_3 is not the product of direct 16α-hydroxylation of E_2, but instead is formed via 16α-hydroxylation of E_1 followed by enzymatic or non-enzymatic 17β-reduction (1, 2). However, experiments with liver microsomes of male and female subjects showed that 16α-hydroxylation was not a major pathway for E_1 hydroxylation (3). An analysis of 15 CYP isoenzymes showed that CYP1A1, 3A4, 3A5, and 2C8 catalyzed the 16α-hydroxylation of both E_1 and E_2, further disproving the view that 16α-hydroxylation only occurs with E_1 as substrate (4, 5). Another isoenzyme, CYP3A7, distinguished the two estrogen substrates with >100 times higher V_{max}:K_m ratio for the 16α-hydroxylation of E_1 than E_2. The difference in reaction rates is most likely due to the difference in structure at the C-17 position of E_1 and E_2. The presence of the 17-ketogroup in E_1 appears to be essential for substrate recognition and 16α-hydroxylation by CYP3A7 (6). 16α-hydroxylation is induced by E_2 in ER-positive MCF-7 cells but not in ER-negative MDA-MB-231 breast cancer cells, suggesting that the 16α-hydroxylation pathway of E_2 metabolism is regulated via ER (7). However, other factors must be involved since E_2 caused a decrease in 16α-hydroxylation in ER-positive T47D cells (8).

Hormonal Activity. Like the catechol estrogens, the 16α-hydroxylated estrogens are hormonally active, chemically reactive, and potentially mutagenic. Both 16α-OHE$_1$ and E_3 are hormonally active by binding to ER. Moreover, 16α-OHE$_1$ possesses the unique property of

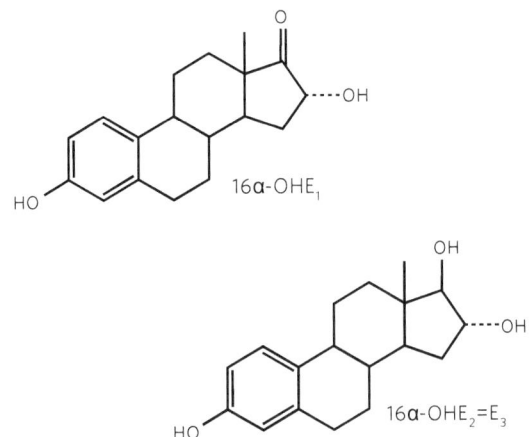

Figure 2.1.2.3. 16α-hydroxylation of E_1 and E_2 leads to 16α-hydroxyestrone (16α-OHE$_1$) and 16α-hydroxyestradiol, i.e., estriol (16α-OHE$_2$ = E_3), respectively.

covalently binding to ER and other nuclear proteins, such as histones (9, 10). Mechanistically, a Schiff base is formed from 16α-OHE_1 by reacting with amino groups in proteins. The Schiff base, in turn, undergoes Heyns rearrangement, resulting in formation of a stable 16-keto-17β-amino estrogen adduct (11, 12). The ability of 16α-OHE_1 to covalently bind ER and thereby cause a persistent activation of the ER has been proposed as a mechanism for prolonged estrogenic effects of this metabolite (9).

Carcinogenic Activity. Bradlow and Fishman have proposed that increased formation of 16α-OHE_1 and E_3 may be associated with increased risk of developing breast cancer (13). They cite several observations in support of linking 16α-hydroxyestrogen metabolites to increased breast cancer risk. (i) 16α-OHE_1 can bind covalently to ER, allowing a prolonged stimulatory effect on cell proliferation (9). (ii) 16α-OHE_1 induced genotoxic damage and aberrant cell proliferation in mouse mammary epithelial cells (14, 15). (iii) Explants from human breast cancer tissues showed four- to fivefold higher levels of 16α-hydroxylation than explants from reduction mammoplasties (16). Based on these observations and clinical studies (17), Bradlow and Fishman presented the hypothesis that the ratio of the two urinary metabolites 2-OHE_1/16α-OHE_1 is inversely correlated with the risk for breast cancer. They chose the numerator 2-OHE_1 to reflect the "good" C-2 and the denominator 16α-OHE_1 to reflect the "bad" C-16α hydroxylation pathways of estrogen metabolism (13, 14, 18). Enzyme immunoassays for simultaneous quantitation of 2-OHE_1 and 16α-OHE_1 levels in urine have been developed and improved to correlate with results obtained by gas chromatography-mass spectrometry (19-21).

Results from epidemiological studies on the association between 2- and 16α-hydroxylation and breast cancer are mixed. Several case-control studies found an increased risk of breast cancer associated with a lower 2-OHE_1/16α-OHE_1 ratio (17, 22-25). Other groups did not observe a difference in the 2-OHE_1/16α-OHE_1 ratios between controls and patients (26, 27). All of these studies measured metabolites after breast cancer diagnosis, leaving open the possibility that the results may have been affected by the tumor. Two prospective studies addressed this issue but yielded inconsistent results. The first prospective study was carried out in women on the island of Guernsey, United Kingdom. Urine samples were collected and stored in the 1970s when all women were healthy. Almost 20 years later, reanalysis of the samples yielded a median of 1.6 for the 2-OHE_1/16α-OHE_1 ratio in 42 postmenopausal women who had developed breast cancer and 1.7 in 139 matched control subjects from the Guernsey III cohort follow-up (28). Compared with women in the lowest tertile category of 2-OHE_1/16α-OHE_1, women in the highest tertile had an odds ratio for breast cancer of 0.71, but the 95% confidence

interval was wide and not statistically significant (95% CI, 0.29 – 1.75). Analysis of premenopausal women in the Guernsey cohort showed no difference between cases and controls. The second prospective study of Italian women had a shorter average follow-up of 5.5 years (29). The odds ratio in postmenopausal women was 1.29 (95% CI = 0.53 – 3.10). In the premenopausal group, women in the highest quintile of the 2/16α-OHE_1 ratio had an odds ratio of 0.58 (95% CI 0.25 – 1.34). A third type of epidemiological study examined urinary metabolites in women of different ethnic groups known to have different rates of breast cancer. Two studies compared 2- and 16α-hydroxylated urinary metabolites in healthy Asian and non-Asian women to determine whether a different 2-OHE_1/16α-OHE_1 ratio paralleled the difference in breast cancer risk in these populations. The first study compared 13 Asian premenopausal women (mostly Vietnamese) living in Hawaii with 12 omnivorous Finnish premenopausal women (30). Compared with the Asian women, the Finnish women had higher 2-OHE_1, but similar 16α-OHE_1 levels, resulting in four- to fivefold higher 2-OHE_1/16α-OHE_1 ratios (p < 0.001). The second study examined healthy postmenopausal women randomly selected from the Singapore Chinese Health Study (n = 67) and the Los Angeles Multiethnic Cohort Study (n = 58) (31). Although breast cancer is substantially lower in Singapore than American women, there were no significant differences between the groups in urinary 16α-OHE_1 levels or 2-OHE_1/16α-OHE_1 ratios. Finally, no differences were found in premenopausal women with or without family history of breast cancer (32). Several other observations further weaken the argument for the etiological role of 16α-hydroxylated estrogens in breast cancer development. Two studies searched for 16α-hydroxylation activity in breast cancers and detected such activity in less than one third of tumors (33, 34). In contrast, 2- and 4-hydroxylase activity has been identified in most breast cancers (33-35). 16α-OHE_1 and E_3 were only weak carcinogens in the estrogen-induced hamster kidney tumor model under experimental conditions that produced a 100% tumor incidence in animals treated with 4-OHE_2 (36-38). In the discussion of 2- and 16α-hydroxylations, the C-4 hydroxylation pathway has often been overlooked for technical reasons associated with the difficulty in accurately quantitating individual catechol estrogens by HPLC. Since 4-OHE_2 and 4-OHE_1 have been shown to be carcinogenic, determination of the 2-OHE_1/16α-OHE_1 ratio or measurement of total catechol estrogens without distinguishing 2- and 4-hydroxylated metabolites may be misleading. Moreover, levels of estrogen metabolites in urine most likely reflect metabolic activity in the liver.

Associations between urinary metabolites and cancer risk in extrahepatic tissues, such as the breast where the expression of P450 isozymes may be quite different, must be interpreted with caution. Finally, pregnant women produce large amounts of 16α-OHE$_1$ and E$_3$, yet full-term pregnancy decreases rather than increases breast cancer risk (39-41). An estrogen that is unique to pregnancy is 15α-hydroxyestriol (15α,16α-dihydroxyestradiol), also called estetrol, which has low affinity for the ER and exhibits weak biological activity (42). In the last trimester, urinary estetrol is excreted in amounts that exceed those of all other estrogens except E$_3$ and 16α-OHE$_1$ (43).

In summary, the 2-OHE$_1$-to-16α-OHE$_1$ ratio goes back to the 1980s, when several studies examined the 2-OHE$_1$-to-16α-OHE$_1$ ratio in relation to breast cancer risk. Unfortunately, the HPLC techniques in the 1980s did not allow separation of the 2-OH from the 4-OH metabolites (they were lumped together as 2-OH and the 4-OH peak was hidden behind the 2-OH peak and thereby went unrecognized). The importance of 4-OH estrogens became only apparent after technical improvements allowed their distinct identification and biological characterization. The methodological limitations of the early HPLC studies were not improved by the Estramet® enzyme immunoassay (developed in the early 1990s), which was used for many subsequent studies. The assay was developed to measure 2-OHE$_1$ but actually does not distinguish between 2-OHE$_1$ and 2-OHE$_2$. Moreover, the assay measures only three estrogen metabolites (16α-OHE$_1$, 2-OHE$_1$ + 2-OHE$_2$) while serum, urine, and tissues contain a dozen or more metabolites, several of which are present in higher concentrations. These metabolites can be measured by HPLC, GC/MS, and LC/MS assays, all of which are technically superior to the immunoassay. Importantly, 4-OHE$_1$ and especially 4-OHE$_2$ are more carcinogenic than 2-OHE$_1$/2 based on their propensity to form DNA adducts that result in mutations.

References

1. Fishman J, Bradlow HL, Fukushima DK, O'Connor J, Rosenfeld RS, Graepel GJ, et al. *Abnormal estrogen conjugation in women at risk for familial breast cancer at the periovulatory stage of the menstrual cycle.* Cancer Res. 1983;43:1884-90.

2. Nebert DW. *Elevated Estrogen 16alpha- hydroxylase activity: is this a genotoxic or nongenotoxic biomarker in human breast cancer risk?* J Natl Cancer Inst. 1993;85:1888-91.

3. Lee AJ, Mills LH, Kosh JW, Conney AH, Zhu BT. *NADPH-dependent metabolism of estrone by human liver microsomes.* J Pharmacol Exp Ther. 2002;300:838-49.

4. Huang Z, Guengerich FP, Kaminsky LS. *16α-hydroxylation of estrone by human cytochrome p4503A4/5.* Carcinogenesis. 1998;19:867-72.

5. Lee AJ, Cai MX, Thomas PE, Conney AH, Zhu BT. *Characterization of the oxidative metabolites of 17β-estradiol and estrone formed by 15 selectively expressed human cytochrome p450 isoforms.* Endocrinology. 2003;144:3382-98.

6. Lee AJ, Conney AH, Zhu BT. *Human cytochrome p450 3A7 has a distinct high catalytic activity for the 16α-hydroxylation of estrone but not 17b-estradiol.* Cancer Res. 2003;63:6532-6.

7. Niwa T, Bradlow HL, Fishman J, Swaneck GE. *Determination of estradiol 2-and 16 alpha-hydroxylase activities in MCF-7 human breast cancer cells in culture using radiometric analysis.* J Steroid Biochem. 1989;33:311-4.

8. Niwa T, Bradlow HL, Fishman J, Swaneck GE. *Induction and inhibition of estradiol hydroxylase activities in MCF-7 human breast cancer cells in culture.* Steroids. 1990;55:297-302.

9. Swaneck GE, Fishman J. *Covalent binding of the endogenous estrogen 16a-hydroxyestrone to estradiol receptor in human breast cancer cells: characterization and intranuclear localization.* Proc Natl Acad Sci USA. 1988;85:7831-5.

10. Yu SC, Fishman J. *Interaction of histones with estrogens. Covalent adduct formation with 16 alpha-hydroxyestrone.* Biochemistry. 1985;24:8017-21.

11. Miyairi S, Ichikawa T, Nambara T. *Structure of the adduct of 16 alpha-hydroxyestrone with a primary amine: evidence for the Heyns rearrangement of steroidal D-ring alpha-hydroxyimines.* Steroids. 1991;56:361-6.

12. Miyairi S, Maeda K, Oe T, Kato T, Naganuma A. *Effect of metal ions on the stable adduct formation of 16alpha-hydroxyestrone with a primary amine via the Heyns rearrangement.* Steroids. 1999;64:252-8.

13. Bradlow HL, Hershcopf R, Martucci C, Fishman J. *16 alpha-hydroxylation of estradiol: a possible risk marker for breast cancer.* Ann NewYork Acad Sci. 1986;464:138-51.

14. Fishman J, Osborne MP, Telang NT. *The role of estrogen in mammary carcinogenesis.* Ann N Y Acad Sci. 1995;768:91-100.

15. Telang NT, Suto A, Wong GY, Osborne MP, Bradlow HL. *Induction by estrogen metabolite 16 alpha-hydroxyestrone of genotoxic damage and aberrant proliferation in mouse mammary epithelial cells.* J Natl Cancer Inst. 1992;84:634-8.

16. Osborne MP, Bradlow HL, Wong GYC, Telang NT. *Upregulaton of estradiol C16alpha-hydroxylation in human breast tissue: a potential biomarker of breast cancer risk.* J Natl Cancer Inst. 1993;85:1917-20.

17. Schneider J, Kinne D, Fracchia A, Pierce V, Anderson KE, Bradlow HL, et al. *Abnormal oxidative metabolism of estradiol in women with breast cancer.* Proc Natl Acad Sci USA. 1982;79:3047-51.

18. Bradlow HL, Telang NT, Sepkovic DW, Osborne MP. *2-hydroxyestrone: the 'good' estrogen.* J Endocrinol. 1996;150 Suppl:S259-S65.

19. Bradlow HL, Sepkovic DW, Klug T, Osborne MP. *Application of an improved ELISA assay to the analysis of urinary estrogen metabolites.* Steroids. 1998;63:406-13.

20. Pasagian-Macaulay A, Meilahn EN, Bradlow HL, Sepkovic DW, Buhari AM, Simkin-Silverman L, et al. *Urinary markers of estrogen metabolism 2- and 16 alpha-hydroxylation in premenopausal women.* Steroids. 1996;61:461-7.

21. Ziegler RG, Rossi SC, Fears TR, Bradlow HL, Adlercreutz H, Sepkovic D, et al. *Quantifying estrogen metabolism: an evaluation of the reproducibility and validity of enzyme immunoassays for 2-hydroxyestrone and 16alpha-hydroxyestrone in urine.* Environ Health Perspect. 1997;105:607-14.

22. Coker AL, Crane MM, Sticca RP, Sepkovic DW. *Ethnic differences in estrogen metabolism in healthy women.* J Natl Cancer Inst. 1997;89:89-90.

23. Ho GH, Luo XW, Ji CY, Foo SC, Ng EH. *Urinary 2/16 alpha-hydroxyestrone ratio: correlation with serum insulin-like growth factor binding protein-3 and a potential biomarker of breast cancer risk.* Ann Acad Med Singapore. 1998;27:294-9.

24. Kabat GC, Chang CJ, Sparano JA, Sepkovic DW, Hu X, Khalil A, et al. *Urinary estrogen metabolites and breast cancer: A case-control study.* Cancer Epidemiol Biomarkers Prev. 1997;6:505-9.

25. Zheng W, Dunning L, Jin F, Holtzman J. *Urinary estrogen metabolites and breast cancer: A case-control study.* Cancer Epidemiol Biomarkers Prev. 1998;7:85-6.

26. Adlercreutz H, Fotsis T, Hockerstedt K, Hamalainen E, Bannwart C, Bloigu S, et al. *Diet and urinary estrogen profile in permenopausal omnivorous and vegetarian women and in premenopausal women with breast cancer.* J Steroid Biochem. 1989;34:527-30.

27. Ursin G, London S, Stanczyk FZ, Gentzchein E, Paganini-Hill A, Ross RK, et al. *Urinary 2-hydroxyestrone/16alpha-hydroxyestrone ratio and risk of breast cancer in postmenopausal women.* J Natl Cancer Inst. 1999;91:1067-72.

28. Meilahn EN, De Stavola B, Allen DS, Fentiman I, Bradlow HL, Sepkovic DW, et al. *Do urinary oestrogen metabolites predict breast cancer? Guernsey III cohort follow-up.* Br J Cancer. 1998;78:1250-5.

29. Muti P, Bradlow HL, Micheli A, Krogh V, Freudenheim JL,

Schunemann HJ, et al. *Estrogen metabolism and risk of breast cancer: a prospective study of the 2:16a-hydroxy-estrone ratio in premenopausal and postmenopausal women.* Epidemiology. 2000;11:635-40.

30. Adlercreutz H, Gorbach SL, Goldin BR, Woods MN, Dwyer JT, Hamalainen E. *Estrogen metabolism and excretion in Oriental and Caucasian women.* J Natl Cancer Inst. 1994;86:1076-82.

31. Ursin G, Wilson M, Henderson BE, Kolonel LN, Monroe K, Lee H-P, et al. *Do urinary estrogen metabolites reflect the differences in breast cancer risk between Singapore Chinese and United States African-American and white women?* Cancer Res. 2001;61:3326-9.

32. Ursin G, London S, Yang D, Tseng CC, Pike MC, Bernstein L, et

al. *Urinary 2-hydroxestrone/16alpha-hydroxyestrone ratio and family history of breast cancer in premenopausal women.* Breast Cancer Res Treat. 2002;72:139-43.

33. Abul-Hajj YJ, Thijssen JH, Blankenstein MA. *Metabolism of estradiol by human breast cancer.* Eur J Cancer Clin Oncol. 1988;24:1171-8.

34. Imoto S, Mitani F, Enomoto K, Fujiwara K, Ikeda T, Kitajima M, et al. *Influence of estrogen metabolism on proliferation of human breast cancer.* Breast Cancer Res Treat. 1997;42:57-64.

35. Liehr JG, Ricci MJ. *4-Hydroxylation of estrogens as marker of human mammary tumors.* Proc Natl Acad Sci USA. 1996;93:3294-6.

36. Li JJ, Li SA. *Estrogen carcino-*

genesis in Syrian hamster tissues: role of metabolism. Fed Proc. 1987;46:1858-63.

37. Li JJ, Li SA, Oberley TD, Parsons JA. *Carinogenic activities of various steroidal and nonsteroidal estrogens in the hamster kidney: Relation to hormonal activity and cell proliferation.* Cancer Res. 1995;55:4347-51.

38. Liehr JG, Fang WF, Sirbasku DA, Ari-Ulubelen A. *Carcinogenicity of catechol estrogens in Syrian hamsters.* J Steroid Biochem. 1986;24:353-6.

39. MacMahon B, Cole P, Lin TM, Lowe CR, Mirra AP, Ravnihar B, et al. *Age at first birth and breast cancer risk.* Bull World Health Org. 1970;43:209-21.

40. Rosner B, Colditz GA, Willett WC. *Reproductive risk factors in a*

prospective study of breast cancer: the Nurses' Health Study. Am J Epidemiol. 1994;139:819-35.

41. Trichopoulos D, Hsieh CC, MacMahon B, Lin TM, Lowe CR, Mirra AP, et al. *Age at any birth and breast cancer risk.* Int J Cancer. 1983;31:701-4.

42. Martucci CP, Fishman J. *P450 enzymes of estrogen metabolism.* Pharmac Ther. 1993;57:237-57.

43. Taylor NF, Shackleton CH. *15alpha-hydroxyoestriol and other polar oestrogens in pregnancy monitoring.* Ann Clin Biochem. 1978;15:1-11.

2.1.2.1.4. ESTROGEN QUINONES

Overview. The P450 enzymes oxidize estrogens to catechol estrogens and further to semiquinones (i.e., E_2-2,3-SQ, E_2-3,4-SQ, E_1-2,3-SQ, and E_1-3,4-SQ) and quinones (i.e., E_2-2,3-Q, E_2-3,4-Q, E_1-2,3-Q, and E_1-3,4-Q) (Figure 2.1.2.4.).

Other enzyme systems can also oxidize in vitro the catechol estrogens to quinones, e.g., horseradish peroxidase or lactoperoxidase/H_2O_2, catalase/H_2O_2, prostaglandin H-synthetase/arachidonic acid, and tyrosinase/O_2 (1-3). The oxidative pathway catechol → semiquinones → quinone is reversible in the presence of reducing enzymes, such as NADPH-dependent P450 reductase, NAD(P)H:quinone oxidoreductase (DT-diaphorase), cytochrome b_5 reductase, and xanthine oxidase. Thus, we have a complete redox cycle, in which the intermediary semiquinones can be formed by oxidation of the catechols or by reduction of the quinones. However, the semiquinone radicals are transient (msec half-life), decaying rapidly via disproportionation to yield the corresponding catechols and quinones (Figure 2.1.2.5.). Experimentally, the semiquinone radicals can be stabilized by chelation with diamagnetic divalent cations (e.g., Mg^{2+}, Zn^{2+}) and detected by electron spin resonance spectroscopy (2, 4-6). Depending on the cofactor used, microsomal preparations containing P450 enzymes catalyze either the oxidation or reduction of estrogen metabolites. For example, when 2-OHE_2 was incubated with hepatic microsomes and cumene hydroperoxide, the oxidation to E_2-2,3-Q was favored. In contrast, in the presence of NADPH as cofactor, the reduction of E_2-2,3-Q to 2-OHE_2 was preferred (7). Not surprisingly, the oxidation to quinones can also be promoted by virtually any chemical oxidant (e.g., molecular oxygen, metal ions, and metal

oxides) (8). On the other hand, in the presence of NADH or NADPH, the estrogen quinones are rapidly reduced to the catechol estrogens, which do not autoxidize, i.e., once the quinone was reduced, the reaction terminated (9). Thus, quantitative studies of catechol estrogens generally include reducing agents (e.g., ascorbate) to reduce the quinones to catechols (10, 11). Experimentally, the estrogen quinones can also be converted into radical species, e.g., E_2-3,4-Q radical anion and cation (12-14) (Figure 2.1.2.6.). The extent to which quinone radicals contribute to the formation of in vivo products is uncertain.

It has been shown *in vitro* that the *o*-quinones isomerize nonenzymatically to hydroxylated *p*-quinone methides (8, 10, 15). For example, E_2-2,3-Q isomerized to two quinone methides, one with only one alkyl substituent in the B ring (2-OHE_2-QM1) and the other stabilized by two alkyl substituents on the methylene group in the C ring (2-OHE_2-QM2) (Figure 2.1.2.7.). In contrast, E2-3,4-Q isomerized to the potentially more stable C-ring *p*-quinone methide (4-OHE_2-QM2), whereas the B-ring *o*-quinone methide (4-OHE_2-QM1) was not detected (15). Quinone methides differ structurally from quinones, as one of the carbonyl oxygens is replaced by a methylene or substituted methylene group. This substitution makes the methide an even more potent electrophile than the *o*-quinone, but results in a reduced capacity for redox chemistry (8). Consequently, reactions of quinone methides in biological systems are characterized by nonenzymatic Michael additions at the exocyclic methylene carbon, generating benzylic adducts of proteins and nucleic acids.

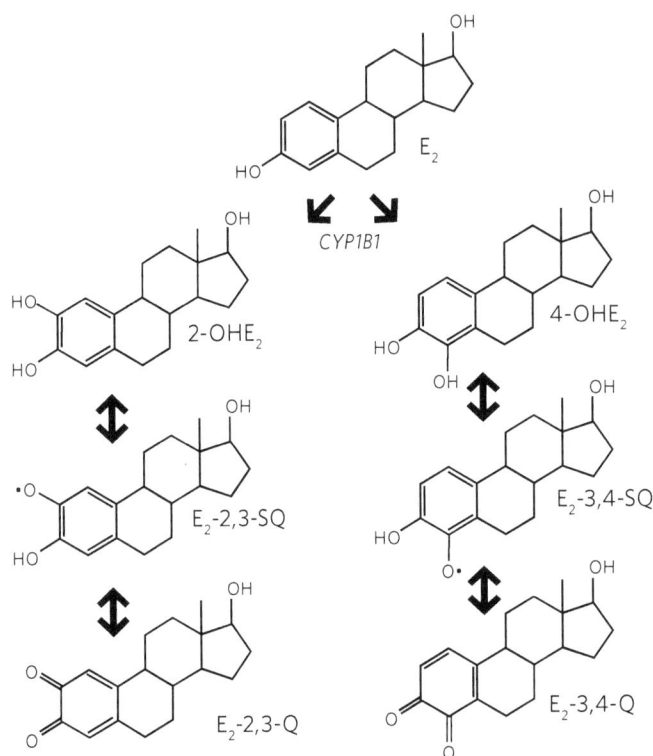

Figure 2.1.2.4. Cytochrome P450 1B1-mediated oxidation of E_2 to catechol estrogens (2-OHE$_2$, 4-OHE$_2$) and further to semiquinones (E$_2$-2,3-SQ, E$_2$-3,4-SQ) and quinones (E$_2$-2,3-Q, E$_2$-3,4-Q). The same oxidative pathway applies to E$_1$.

Chemical reactivity. Like other quinones the estrogen quinones are highly reactive compounds. The chemical reactivity of individual *o*-quinones is poorly understood although it is clear that, as a group, they are more electrophilic than their *p*-quinone structural isomers due to the strained 1,2-diketone functionality. Thus, the *o*-quinone formed from 2-OHE$_1$ has a half-life of 42 sec at pH 7.4 and 37°C and the o-quinone formed from 4-OHE$_1$ has a half-life of 12 min under identical conditions (11). Experimentally, the amount of *o*-quinone formation from catechol estrogens cannot be determined reliably because of the short half-life of the estrogen quinones and their indiscriminate reaction with nucleophilic residues *in vitro* and *in vivo*, including microsomal proteins such as P450 enzymes (8).

The estrogen replacement drug Premarin® contains the equine estrogens equilin and equilenin, which differ structurally from E$_1$ and E$_2$ by having an unsaturated B ring. The principal metabolite of equilin and equilenin is the catechol 4-hydroxyequilenin (4-OHEN), which contains aromatic A and B rings (8, 16) (Figure 2.1.2.8.). The aromatization of the B ring in 4-OHEN affects the formation of its o-quinone. In contrast to 4-OHE$_1$ and 4-OHE$_2$,

4-OHEN is readily oxidized to an *o*-quinone without the need for cytochrome P450 catalysis. The autoxidation rate constant of 4-OHEN was determined to be 1.9×10^{-2} sec^{-1} at 25°C and pH 7.4 (9). It appears that the adjacent aromatic ring stabilizes 4-OHEN-*o*-quinone through extended π conjugation, leading to reduction in ground state energy and a corresponding increased ease of formation from the catechol. At the same time, 4-OHEN-o-quinone is much more stable (half-life 2.3 h) compared to E$_1$-3,4-Q (12 min) (9).

Like other quinones, the estrogen quinones exert dual actions as oxidants and electrophiles (8, 17, 18). As electrophiles, the estrogen quinones are highly reactive, labile compounds, which readily undergo Michael addition to a variety of nucleophiles, ranging from amino acids such as lysine and cysteine to the tripeptide glutathione, as well as nucleic acids and DNA (12, 19-25). The Michael addition is characterized by the addition of a nucleophile to an α,β-unsaturated carbonyl compound, the electrophile acceptor. For example, the reaction of lysine with E$_2$-3,4-Q results in several products (24). One reaction pathway involves 1,2-addition of the α-amino group of lysine to the C-3 carbonyl group of E$_2$-3,4-Q (Figure 2.1.2.9.).

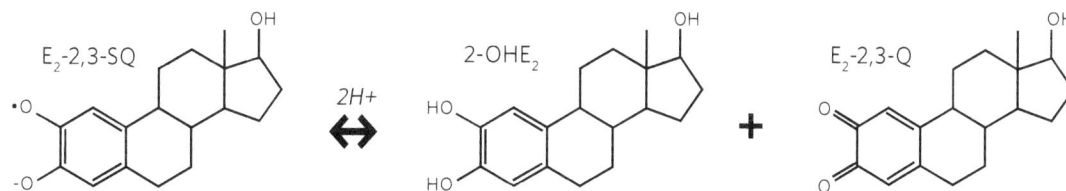

Figure 2.1.2.5. Semiquinone radicals, such as E_2-2,3-SQ, are transient and decay rapidly via disproportionation to yield the corresponding catechols and quinones, 2-OHE_2 and E_2-2,3-Q.

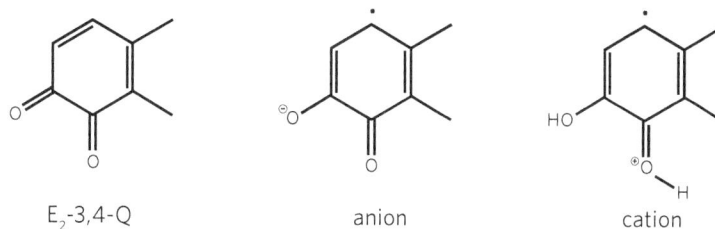

Figure 2.1.2.6. Ring A of the estrogen quinone E_2-3,4-Q and its radical anion and cation.

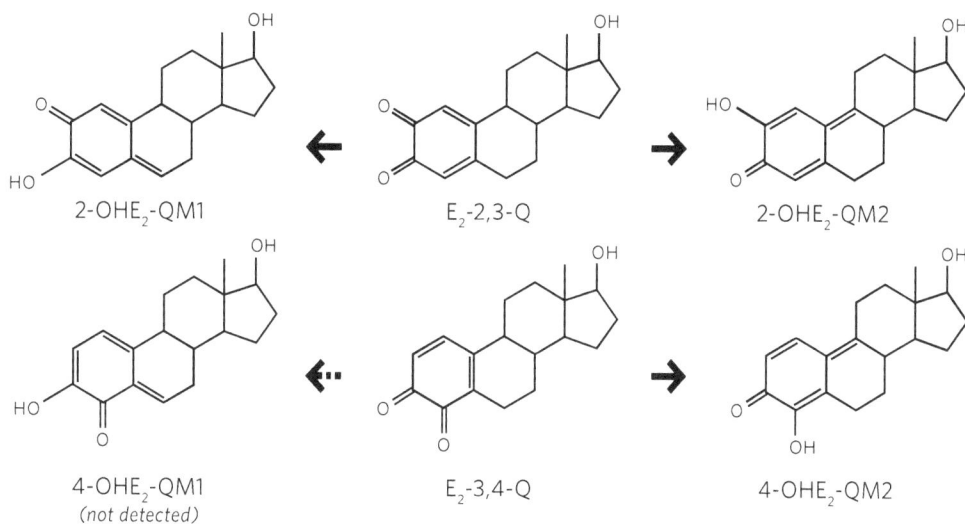

Figure 2.1.2.7. Isomerization of *o*-estrogen quinones to *p*-quinone methides.

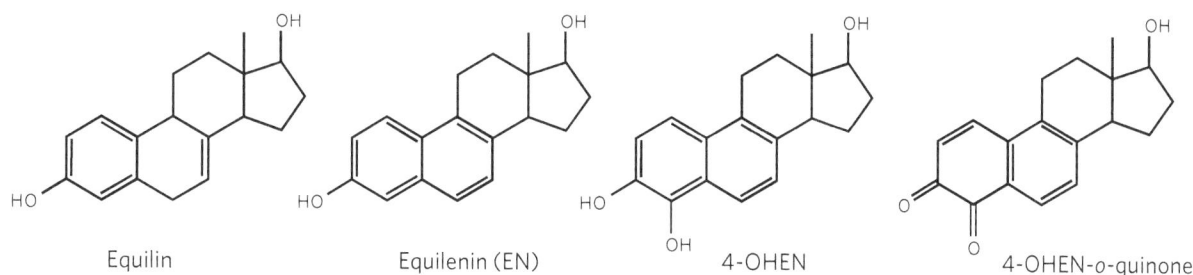

Figure 2.1.2.8. The equine estrogens equilin and equilenin (EN) are metabolized to 4-hydroxyequilenin (4-OHEN) and the 4-OHEN-*o*-quinone.

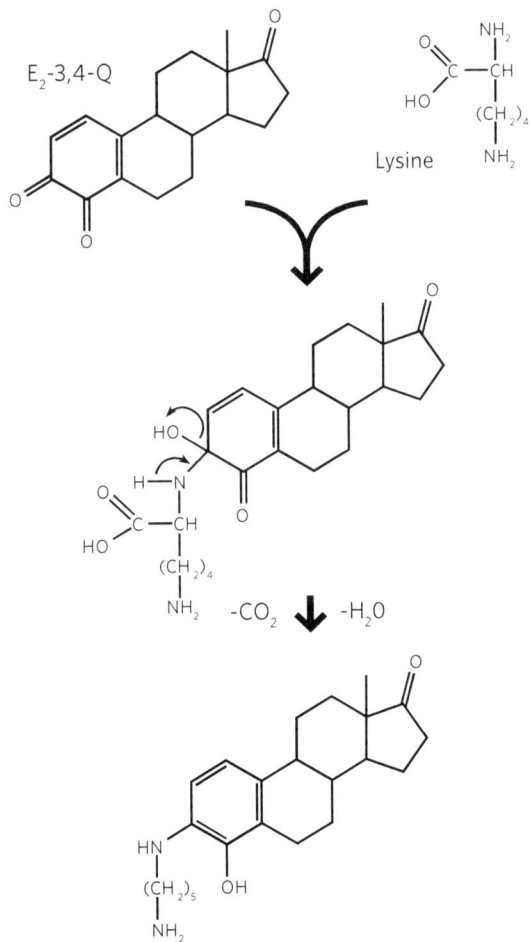

E$_2$-3,4-Q

Lysine

$-CO_2$ $-H_2O$

Figure 2.1.2.9. The reaction of E$_2$-3,4-Q with lysine proceeds via 1,2-Michael addition of the α-amino group to the C-3 carbonyl to yield several products, of which only one is shown.

Analysis of breast tissue from women with and without breast cancer revealed significantly higher concentrations of estrogen quinone conjugates with cysteine and N-acetylcysteine in tumor tissue than in normal breast (26). Altogether, the level of estrogen quinone conjugates in cases was three times that of controls. In the same study, the levels of 4-OHE$_2$ and 4-OHE$_1$ were nearly four times higher in cases. Thus, the levels of 3,4-catechol estrogens and quinone conjugates were highly significant predictors of breast cancer (26).

References

1. Cavalieri EL, Stack DE, Devanesan PD, Todorvic R, Dwivedy I, Higginbotham S, et al. *Molecular origin of cancer: catechol estrogen-3,4-quinones as endogenous tumor initiators.* Proc Natl Acad Sci USA. 1997;94:10937-42.

2. Kalyanaraman B, Felix CC, Sealy RC. *Semiquinone anion radicals of catechol(amine)s, catechol estrogens, and their metal ion complexes.* Environ Health Perspect. 1985;64:185-98.

3. Liehr JG, Roy D. *Free radical generation by redox cycling of estrogens.* Free Radical Biol Med. 1990;8:415-23.

4. Kalyanaraman B, Sealy RC, Liehr JG. *Characterization of semiquinone free radicals formed from stilbene catechol estrogens.* J Biol Chem. 1989;264:11014-9.

5. Kalyanaraman B, Sealy RC, Sivarajah K. *An electron spin resonance study of o-semiquinones formed during the enzymatic and autoxidation of catechol estrogens.* J Biol Chem. 1984;259:14018-22.

6. Seacat AM, Kuppusamy P, Zweier JL, Yager JD. *ESR identification of free radicals formed from the oxidation of catechol estrogens by Cu2+.* Arch Biochem Biophys. 1997;347:45-52.

7. Liehr JG, Ulubelen AA, Strobel HW. *Cytochrome P-450-mediated redox cycling of estrogens.* J Biol Chem. 1986;261:16865-70.

8. Bolton JL, Pisha E, Zhang F, Qiu S. *Role of quinoids in estrogen carcinogenesis.* Chem Res Toxicol. 1998;11:1113-27.

9. Shen L, Pisha E, Huang Z, Pezzuto JM, Krol E, Alam Z, et al. *Bioreductive activation of catechol estrogen-ortho-quinones: aromatization of the B ring in 4-hydroxyequilenin markedly alters quinoid formation and reactivity.* Carcinogenesis. 1997;18:1096-101.

10. Butterworth M, Lau SS, Monks TJ. *17b-estradiol metabolism by hamster hepatic microsomes: Comparison of catechol estrogen o-methylation with catechol estrogen oxidation and glutathione conjugation.* Chem Res Toxicol. 1996;9:793-9.

11. Iverson SL, Shen L, Anlar N, Bolton JL. *Bioactivation of estrone and its catechol metabolites to quinoid-glutathione conjugates in rat liver microsomes.* Chem Res Toxicol. 1996;9:492-9.

12. Akanni A, Abul-Hajj YJ. *Estrogen-nucleic acid adducts: reaction of 3,4-estrone o-quinone radical anion with deoxyribonucleosides.* Chem Res Toxicol. 1997;10:760-6.

13. Akanni A, Abul-Hajj YJ. *Estrogen-nucleic acid adducts: Dissection of the reaction of 3,4-estrone quinone and its radical anion and radical cation with deoxynucleosides and DNA.* Chem Res Toxicol. 1999;12:1247-53.

14. Akanni A, Tabakovic K, Abul-Hajj YJ. *Estrogen-nucleic acid adducts: reaction of 3,4-estrone o-quinone with nucleic acid bases.* Chem Res Toxicol. 1997;10:477-81.

15. Bolton JL, Shen L. *p-Quinone methides are the major decomposition products of catechol estrogen o-quinones.* Carcinogenesis. 1996;17:925-9.

16. Zhang F, Swanson SM, Van Breemen RB, Liu X, Yang Y, Gu C, et al. *Equine estrogen metabolite of 4-hydroxyequilenin induces DNA damage in the rat mammary tissues: Formation of single-strand breaks, apurinic sites, stable adducts, and oxidized bases.* Chem Res Toxicol. 2001;14:1654-9.

17. Bolton JL, Trush MA, Penning TM, Dryhurst G, Monks TJ. *Role of quinones in toxicology.* Chem Res Toxicol. 2000;13:135-60.

18. Cavalieri EL, Li KM, Balu N, Saeed M, Devanesan P, Higginbotham S, et al. *Catechol ortho-quinones: the electrophilic compounds that form depurinating DNA adducts and could initiate cancer and other diseases.* Carcinogenesis. 2002;23:1071-7.

19. Abul-Hajj YJ, Tabakovic K, Gleason WB, Ojala WH. *Reactions of 3,4-estrone quinone with mimics of amino acid side chains.* Chem Res Toxicol. 1996;9:434-8.

20. Bolton JL, Yu L, Thatcher RJ. *Quinoids formed from estrogens and antiestrogens.* Methods Enzymol. 2004;378:110-23.

21. Cao K, Devanesan PD, Ramanathan R, Gross ML, Rogan EG, Cavalieri EL. *Covalent binding of catechol estrogens to glutathione catalyzed by horseradish peroxidase, lactoperoxidase, or rat liver microsomes.* Chem Res Toxicol. 1998;11:917-24.

22. Khasnis D, Abul-Hajj YJ. *Estrogen quinones: Reaction with propylamine.* Chem Res Toxicol. 1994;7:68-72.

23. Stack DE, Byun J, Gross ML, Rogan EG, Cavalieri EL. *Molecular characteristics of catechol estrogen quinones in reactions with deoxyribonucleosides.* Chem Res Toxicol. 1996;9:851-9.

24. Tabakovic K, Abul-Hajj YJ. *Reaction of lysine with estrone 3,4-o-quinone.* Chem Res Toxicol. 1994;7:696-701.

25. Wang MY, Liehr JG. *Induction by estrogens of lipid peroxidation and lipid peroxide-derived malonaldehyde-DNA adducts in male Syrian hamsters: role of lipid peroxidation in estrogen-induced kidney carcinogenesis.* Carcinogenesis. 1995;16:1941-5.

26. Rogan EG, Badawi AF, Devanesan PD, Meza J, Edney JA,

West WW, et al. *Relative imbalances in estrogen metabolism and conjugation in breast tissue of women with carcinoma: potential biomarkers of susceptibility to cancer.* Carcinogenesis. 2003;24:697-702.

2.1.2.2. CONJUGATION PRODUCTS
2.1.2.2.1. METHOXYESTROGENS

Methoxyestrogens are methyl ether metabolites of catechol estrogens (Figure 2.1.2.10.). Methoxyestrogens are products of the COMT-mediated O-methylation of catechol estrogens. At the same time, methoxyestrogens are substrates for CYP1A1 and CYP1B1, which catalyze their O-demethylation to catechol estrogens, effectively reversing the COMT reaction (1). Specifically, both CYP1A1 and CYP1B1 demethylated 2-MeOE$_2$ and 2-OH-3-MeOE$_2$ to 2-OHE$_2$, and CYP1B1 additionally demethylated 4-MeOE$_2$ to 4-OHE$_2$. Thus, CYP1A1 and CYP1B1 recognize both the parent hormone E$_2$ and the methoxyestrogens 2-MeOE$_2$, 2-OH-3-MeOE$_2$, and 4-MeOE$_2$ as substrates.

Kinetic analysis showed that E$_2$ and the methoxyestrogens are alternate substrates, each catalyzed by the same enzyme, albeit by a different type of reaction (1). Because they are converted to identical catechol estrogen products, each inhibits formation of 2-OHE$_2$ and 4-OHE$_2$ from the other substrate in a noncompetitive manner. This means that methoxyestrogens exert feedback inhibition on CYP1A1 and CYP1B1, which affects the entire oxidative estrogen metabolism pathway in several ways (Figure 2.1.2.11.). First, CYP1A1 and CYP1B1 generate catechol estrogen substrates for COMT, and at the same time, they compete with COMT by converting the catechol estrogens to estrogen quinones. In turn, the methoxyestrogens generated by COMT are alternate substrates for CYP1A1 and CYP1B1 and inhibit oxidation of the parent hormone E$_2$ (and most likely the catechol estrogens as well) by CYP1A1 and CYP1B1. Second, the inhibition occurs at a strategic point in the pathway where it branches into 2-OH and 4-OH catechol estrogens. This may be important in view of the apparent difference in carcinogenicity of 2-OH and 4-OH catechol estrogens (2-5). Third, all three products of the COMT-mediated reaction (i.e., 2-MeOE$_2$, 2-OH-3-MeOE$_2$, and 4-MeOE$_2$) act as inhibitors, thereby maximizing the feedback regulation (1). Although not examined, it is likely that the methoxyestrogens produced by COMT from E$_1$ (i.e., 2-MeOE$_1$, 2-OH-3-MeOE$_1$, and 4-MeOE$_1$) similarly act as inhibitors, thereby doubling the number of physiologically inhibiting substrates. Fourth, the feedback regulation occurs at the step in the pathway preceding the conversion to estrogen semiquinones and quinones, thereby reducing the formation of reactive oxygen species during semiquinone – quinone

redox cycling and the potential for estrogen-induced DNA damage. Fifth, we see the joint involvement of phase I (CYP1A1 and CYP1B1) and phase II enzymes (COMT) in a biologically important feedback loop.

Since CYP1A1 and CYP1B1 play key roles as phase I enzymes in metabolizing various xenobiotic chemicals to procarcinogens in extrahepatic tissues (6), their inhibition by endogenous methoxyestrogens may represent a physiologic defense mechanism against carcinogenesis. CYP1A1 and CYP1B1 are highly inducible by numerous xenobiotics (7, 8) and the associated multifold increase in enzyme action may be attenuated by the methoxyestrogen feedback inhibition produced by the constitutively expressed COMT.

In mice and humans, 2-MeOE$_2$ is generated by COMT in the placenta and increases in the circulation during pregnancy. Treatment of pregnant mice with a selective COMT inhibitor (Ro 41-0960) reduced the number of preterm births (9). Women who developed severe eclampsia had significantly lower plasma levels of 2-MeOE$_2$ due to lower levels of COMT (10).

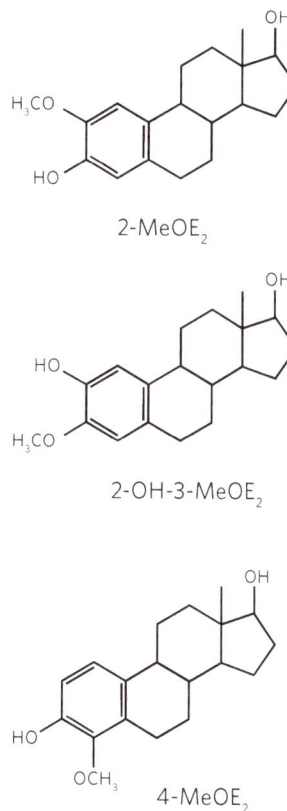

Figure 2.1.2.10. Methoxyestrogens 2-methoxyestradiol (2-MeOE$_2$), 2-hydroxy-3-methoxyestradiol (2-OH-3-MeOE$_2$) and 4-methoxyestradiol (4-MeOE$_2$).

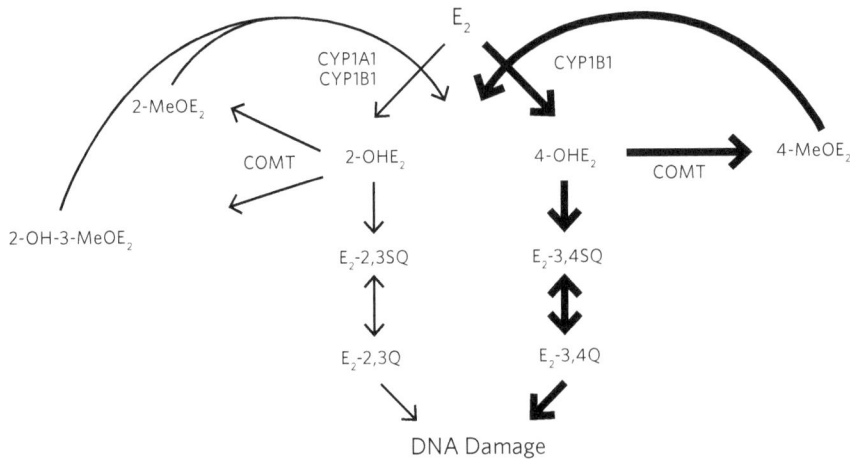

Figure 2.1.2.11. Oxidative metabolism of E_2 to 2-OH and 4-OH catechol estrogens (2-OHE$_2$ and 4-OHE$_2$) is catalyzed by CYP1A1 and CYP1B1 and inhibited by methoxyestrogens (2-MeOE$_2$, 2-OH-3MeOE$_2$, and 4-MeOE$_2$), which are produced by COMT. *This feedback inhibition reduces the formation of estradiol semiquinones (E_2-2,3SQ and E_2-3,4SQ) and quinones (E_2-2,3Q and E_2-3,4Q), which may form quinone-DNA adducts or cause oxidative DNA damage by reactive oxygen species arising during redox cycling. The same pathway applies to estrone. Straight arrows represent enzymes catalyzing the indicated conversions. Curved arrows represent feedback loops. The 4-OH catechol estrogens induce more DNA damage than the 2-OH catechol estrogens as indicated by the thicker arrow.*

Antiproliferative Action. Interestingly, 2-MeOE$_2$ is not just a byproduct of estrogen metabolism but is also endowed with antiproliferative activity (11). 2-MeOE$_2$ has been shown to inhibit the proliferation of both hormone-dependent and hormone-independent breast cancer cells (12, 13). Oral administration of 2-MeOE$_2$ (75 mg/kg) for one month suppressed the growth of a human breast cancer cell line (estrogen receptor negative) in immunodeficient mice by 60% without toxicity (14). The antiproliferative effect is not limited to breast cancer cells but extends to leukemia, pancreas, and lung cancer cells (15, 16). Human xenograft studies in animal models demonstrated oral bioavailability and a high therapeutic index of methoxyestrogens without signs of systemic toxicity. These features and their broad antitumor activity against a variety of tumor cells make methoxyestrogens attractive as potential therapeutic agents that are currently being tested in clinical trials (16, 17). Several synthetic analogs were equally effective as 2-MeOE$_2$ or even more potent than the endogenous compound (18-20) (Figure 2.1.2.12.).

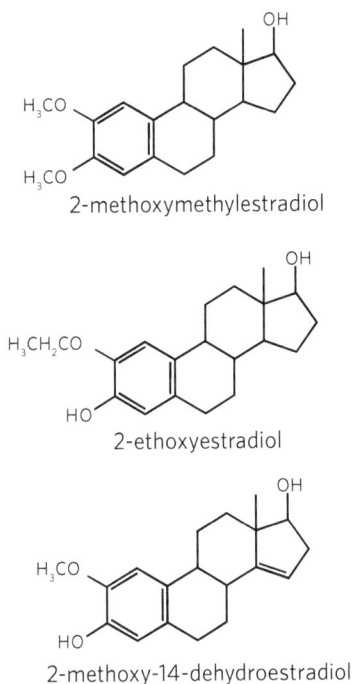

The antiproliferative effect of 2-MeOE$_2$ appears to be concentration-dependent and involve several mechanisms. At nano- and micromolar concentrations, 2-MeOE$_2$ caused disruption of microtubule function, induction of apoptosis, and inhibition of angiogenesis (12, 21, 22). At millimolar concentrations, 2-MeO$_2$ caused chromosome breaks and aneuploidy (23, 24). A comparison of 30 estrogens and estrogen derivatives to induce disruption of the cytoplasmic microtubule network in cultured Chinese Hamster V79 cells showed the strongest activity (EC$_{50}$ 2 µM) for 2-MeOE$_2$, followed by 2-OH-3-MeOE$_2$, 17β-estradiol, and 17β-estradiol. 4-MeOE$_2$, 2-MeOE$_1$, 2-OH-3-MeOE$_1$, and 4-MeOE$_1$ were less active than the 3-alkyl ethers 2-OH-3-ethylestradiol and 2-OH-3-propylestradiol (25). The EC$_{50}$ values of compounds correlated with their inhibitory activity on cell growth.

2-methoxymethylestradiol

2-ethoxyestradiol

2-methoxy-14-dehydroestradiol

Figure 2.1.2.12. Synthetic analogs of 2-methoxyestradiol

References

1. Dawling S, Roodi N, Parl FF. *Methoxyestrogens exert feedback inhibition on cytochrome P450 1A1 and 1B1.* Cancer Res. 2003;63:3127-32.

2. Cavalieri EL, Stack DE, Devanesan PD, Todorvic R, Dwivedy I, Higginbotham S, et al. *Molecular origin of cancer: catechol estrogen-3,4-quinones as endogenous tumor initiators.* Proc Natl Acad Sci USA. 1997;94:10937-42.

3. Han X, Liehr JG. *Microsome-mediated 8-hydroxylation of guanine bases of DNA by steroid estrogens: correlation of DNA damage by free radicals with metabolic activation to quinones.* Carcinogenesis. 1995;16:2571-4.

4. Li JJ, Li SA. *Estrogen carcinogenesis in Syrian hamster tissues: role of metabolism.* Fed Proc. 1987;46:1858-63.

5. Liehr JG, Fang WF, Sirbasku DA, Ari-Ulubelen A. *Carcinogenicity of catechol estrogens in Syrian hamsters.* J Steroid Biochem. 1986;24:353-6.

6. Guengerich FP. *Common and uncommon cytochrome P450 reactions related to metabolism and chemical toxicity.* Chem Res Toxicol. 2001;14:611-50.

7. Nebert DW, Nelson DR, Coon MJ, Estabrook RW, Feyereisen R, Fujii-Kuriyama Y, et al. *The P450 superfamily: update on new sequences, gene mapping, and recommended nomenclature.* DNA Cell Biol. 1991;10:1-14.

8. Whitlock JP. *Induction of cytochrome P4501A1.* Ann Rev Pharm Toxicol. 1999;39:103-25.

9. Wentz MJ, Shi SQ, Shi L, Salama SA, Harirah HM, Fouad H, et al. *Treatment with an inhibitor of catechol-O-methyltransferase activity reduces preterm birth and impedes cervical resistance to stretch in pregnant rats.* Reproduction. 2007;134:831-9.

10. Kanasaki K, Palmsten K, Sugimoto H, Ahmad S, Hamano Y, Xie L, et al. *Deficiency in catechol-o-methyltransferase and 2-methoxyoestradiol is associated with pre-eclampsia.* Nature. 2008;453:1117-21.

11. Zhu BT, Conney AH. *Is 2-methoxyestradiol an endogenous estrogen metabolite that inhibits mammary carcinogenesis?* Cancer Res. 1998;58:2269-77.

12. LaVallee TM, Zhan XH, Herbstritt CJ, Kough EC, Green SJ, Pribluda VS. *2-methoxyestradiol inhibits proliferation and induces apoptosis independently of estrogen receptors a and b.* Cancer Res. 2002;62:3691-7.

13. Lottering ML, Haag M, Seegers JC. *Effects of 17b-estradiol metabolites on cell cycle events in MCF-7 cells.* Cancer Res. 1992;52:5926-32.

14. Klauber N, Parangi S, Flynn E, Hamel E, D'Amato RJ. *Inhibition of angiogenesis and breast cancer in mice by the microtubule inhibitors 2-methoxyestradiol and taxol.* Cancer Res. 1997;57:81-6.

15. Schumacher G, Kataoka M, Roth JA, Mukhopadhyay T. *Potent antitumor activity of 2-methoxyestradiol in human pancreatic cancer cell lines.* Clin Cancer Res. 1999;5:493-9.

16. Schumacher G, Neuhaus P. *The physiologic estrogen metabolite 2-methoxyestradiol reduces tumor growth and induces apoptosis in human solid tumors.* J Cancer Res Clin Oncol. 2001;127:405 - 10.

17. Pribluda VS, Gubish J, E.R. , LaVallee TM, Treston A, Swartz GM, Green SJ. *2-methoxyestradiol: an endogenous antiangiogenic and antiproliferative drug candidate.* Cancer Metastasis Rev. 2000;19:173-9.

18. Brueggemeier RW, Bhat AS, Lovely CJ, Coughenour HD, Joomprabutra S, Weitzel DH, et al. *2-methoxymethylestradiol: a new 2-methoxy estrogen analog that exhibits antiproliferative activity and alters tubulin dynamics.* J Steroid Biochem Mol Biol. 2001;78:145-56.

19. Cushman M, He H-M, Katzenellenbogen JA, Varma RK, Hamel E, Lin CM, et al. *Synthesis of analogs of 2-methoxyestradiol with enhanced inhibitory effects of tubulin polymerization and cancer cell growth.* J Medicinal Chem. 1997;40:2323-34.

20. Tinley TL, Leal RM, Randall-Hlubek DA, Cessac JW, Wilkens LR, Rao PN, et al. *Novel 2-methoxystradiol analogues with antitumor activity.* Cancer Res. 2003;63:1538-49.

21. D'Amato RJ, Lin CM, Flynn E, Folkman J, Hamel E. *2-Methoxyestradiol, an endogenous mammalian metabolite, inhibits tubulin polymerization by interacting at the colchicine site.* Proc Natl Acad Sci USA. 1994;91:3964-8.

22. Fotsis T, Zhang Y, Pepper MS, Adlercreutz H, Montesano R, Nawroth PP, et al. *The endogenous oestrogen metabolite 2-methoxyoestradiol inhibits angiogenesis and suppresses tumour growth.* Nature. 1994;368:237-9.

23. Tsutsui T, Tamura Y, Hagiwara M, Miyachi T, Hikiba H, Kubo C, et al. *Induction of mammalian cell transformation and genotoxicity by 2-methoxyestradiol, an endogenous metabolite of estrogen.* Carcinogenesis. 2000;21:735-40.

24. Tsutsui T, Tamura Y, Yagi E, Barrett JC. *Involvement of genotoxic effects in the initiation of estrogen-induced cellular transformation: studies using Syrian hamster embryo cells treated with 17b-estradiol and eight of its metabolites.* Int J Cancer. 2000;86:8-14.

25. Aizu-Yokota E, Susaki A, Sato Y. *Natural estrogens induce modulation of microtubules in Chinese hamster V79 cells in culture.* Cancer Res. 1995;55:1863-8.

2.1.2.2.2. ESTROGEN GLUTATHIONE CONJUGATES

The labile estrogen quinones react with a variety of physiological compounds, including amino acids, such as lysine and cysteine, and the tripeptide glutathione (γ-glutamyl-cysteinyl-glycine, GSH) (1, 2). In these reactions, the quinones form Michael addition products with nucleophilic groups, typically yielding a mixture of several stable compounds (3, 4). Mass spectrometric analysis of the GSH-estrogen isomers revealed that the catechol estrogen attachment is at the cysteine moiety of GSH, with the cysteine sulfur binding to an A-ring carbon vicinal to the catechol carbons, i.e., C-1 or C-4 in 2-OHE$_2$ and C-2 in 4-OHE$_2$ (1, 5). Experiments using C-1 tritiated E$_1$-3,4-Q proved the regiospecific attachment of GSH at C-2 and demonstrated that the conjugation did not eliminate tritium (3, 6). Thus, the point of attachment of the –SG moiety is always directly adjacent to an oxygen bearing carbon, in line with all other known quinone-GSH conjugates (7) (Figure 2.1.2.13.).

An *in vitro* study with GSH observed a faster rate of estrogen quinone conjugation in the presence of GSTP1 than in the absence of the enzyme (8). Specifically, GSTP1 yielded two conjugates, 2-OHE2-1-SG and 2-OHE$_2$-4-SG, from the corresponding quinone E$_2$-2,3-Q, and one conjugate, 4-OHE$_2$-2-SG, from E$_2$-3,4-Q. The rate of conjugation was in the order 4-OHE$_2$-2-SG > 2-OHE$_2$-4-SG >> 2-OHE$_2$-1-SG, indicating a difference in regiospecific reactivity of the two quinones. The same study also showed that the enzymatic conversion of catechol estrogens to estrogen quinones by CYP1B1 is a necessary step for the subsequent GSH conjugation reaction, providing proof that the phase I oxidizing enzyme and the phase II conjugating enzyme act in sequence in metabolizing estrogens. The enzymatic reaction with GSTP1 yielded only mono-conjugates, i.e., 2-OHE$_2$-1-SG, 4-OHE$_2$-4-SG, and 4-OHE$_2$-2-SG. There was no evidence of bis-conjugates, such as 2-OHE$_2$-1,4-bisSG, which can be synthesized chemically by using much higher concentrations of GSH.

All GSH conjugates are catabolized via the mercapturic acid pathway (9-11) (Fig. 2.1.2.14.). First, the glutamyl moiety is removed from the GSH conjugate by transpeptidation, which is catalyzed by γ-glutamyl transpeptidase. Then, the cysteinylglycine is hydrolyzed by cysteinyl-dipeptidase to yield the cysteine conjugate. The final step consists of acetylation to the N-acetylcysteine conjugate, a mercapturic acid compound. Intramammillary injection of female ACI rats with 4-OHE$_2$ or E$_2$-3,4-Q resulted in the production of GSH-, cysteine-, and N-acetylcysteine-estrogen conjugates in the mammary gland (12). Thus, estrogen quinones are detoxified in tissues by GST-mediated GSH conjugation and the resultant GSH conjugates are catabolized to N-acetyl-cysteine conjugates. The GSH conjugates are excreted in bile and urine mostly as N-acetylcysteine conjugates but also as cysteine conjugates (1, 13-16).

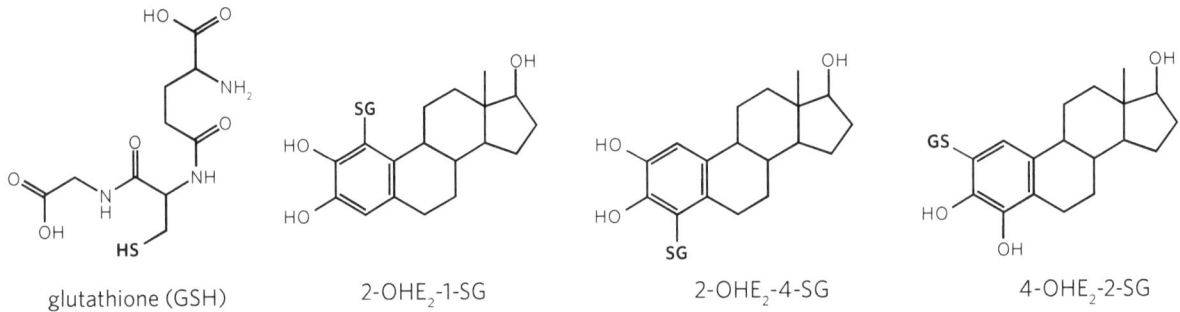

glutathione (GSH)   2-OHE$_2$-1-SG   2-OHE$_2$-4-SG   4-OHE$_2$-2-SG

Figure 2.1.2.13. The tripeptide glutathione (γ-glutamyl-cysteinyl-glycine, GSH) forms Michael addition products with estrogen quinones, resulting in the conjugates 2-OHE$_2$-1-SG, 2-OHE$_2$-4-SG, and 4-OHE$_2$-2-SG.

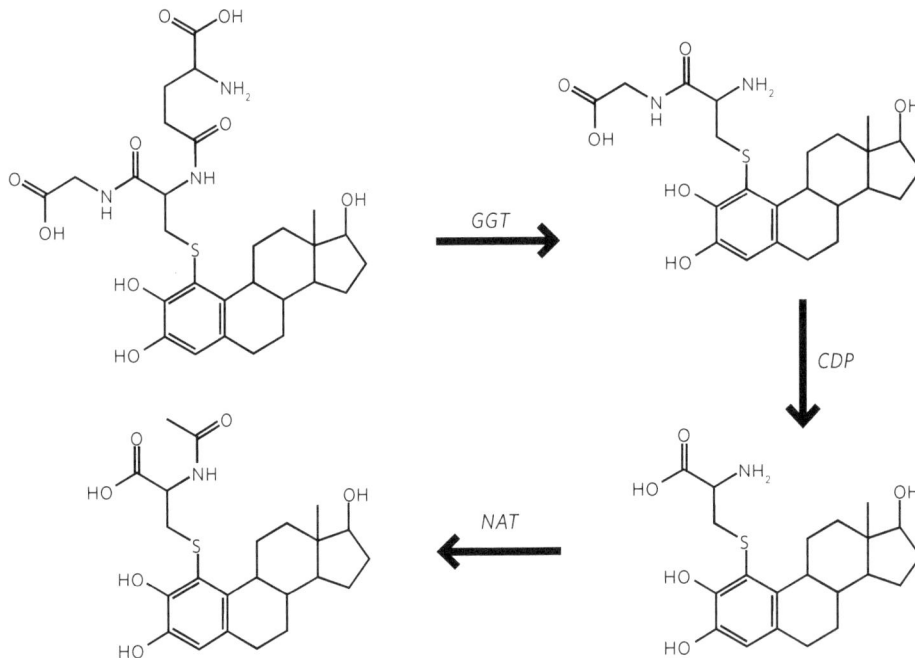

Figure 2.1.2.14. Estrogen-GSH conjugates, such as 2-OHE$_2$-1-SG, are catabolized via the mercapturic acid pathway. *The latter consists of sequential removal of glutamate and glycine by γ-glutamyl transpeptidase (GGT) and cysteinyl-dipeptidase (CDP), followed by acetylation of cysteine by N-acetyltransferase (NAT) to the N-acetylcysteine conjugate.*

References

1. Cao K, Devanesan PD, Ramanathan R, Gross ML, Rogan EG, Cavalieri EL. *Covalent binding of catechol estrogens to glutathione catalyzed by horseradish peroxidase, lactoperoxidase, or rat liver microsomes.* Chem Res Toxicol. 1998;11:917-24.

2. Tabakovic K, Abul-Hajj YJ. *Reaction of lysine with estrone 3,4-o-quinone.* Chem Res Toxicol. 1994;7:696-701.

3. Abul-Hajj YJ, Cisek PL. *Regioselective reaction of thiols with catechol estrogens and estrogen-O-quinones.* J Steroid Biochem. 1986;25:245-7.

4. Abul-Hajj YJ, Tabakovic K, Gleason WB, Ojala WH. *Reactions of 3,4-estrone quinone with mimics of amino acid side chains.* Chem Res Toxicol. 1996;9:434-8.

5. Ramanathan R, Cao K, Cavalieri E, Gross ML. *Mass spectrometric methods for distinguishing structural isomers of glutathione conjugates of estrone and estradiol.* J Am Soc Mass Spectrom. 1998;9:612-9.

6. Abul-Hajj YJ, Cisek PL. *Catechol estrogen adducts.* J Steroid Biochem. 1988;31:107-10.

7. Bolton JL, Trush MA, Penning TM, Dryhurst G, Monks TJ. *Role of quinones in toxicology.* Chem Res Toxicol. 2000;13:135-60.

8. Hachey DL, Dawling S, Roodi N, Parl FF. *Sequential action of phase I and II enzymes cytochrome P450 1B1 and glutathione S-transferase P1 in mammary estrogen metabolism.* Cancer Res. 2003;63:8492 - 9.

9. Boyland E, Chasseaud LF. *The role of glutathione and glutathione S-transferases in mercapturic acid biosynthesis.* Adv Enzymol Rel Areas Mol Biol. 1969;32:173-219.

10. Reed DJ. *Glutathione: toxicological implications.* Annu Rev Pharmacol Toxicol. 1990;30:603-31.

11. Taylor GW, Donnelly LE, Murray S, Rendell NB. *Excursions in biomedical mass spectrometry.* Br J Clin Pharmacol. 1996;42:119-26.

12. Li KM, Todorovic R, Devanesan P, Higginbotham S, Kofeler H, Ramanathan R, et al. *Metabolism and DNA binding studies of 4-hydroxyestradiol and estradiol-3,4-quinone in vitro and in female ACI rat mammary gland in vivo.* Carcinogenesis. 2004;25:289-97.

13. Cao K, Stack DE, Ramanathan R, Gross ML, Rogan EG, Cavalieri EL. *Synthesis and structure elucidation of estrogen quinones conjugated with cysteine, N-acetylcysteine, and glutathione.* Chem Res Toxicol. 1998;11:909-16.

14. Devanesan P, Santen RJ, Bocchinfuso WP, Korach KS, Rogan EG, Cavalieri E. *Catechol estrogen metabolites and conjugates in mammary tumors and hyperplastic tissue from estrogen receptor-α knock-out (ERKO)/Wnt-1 mice: implications for initiation of mammary tumors* Carcinogenesis. 2001;22:1573-6.

15. Devanesan P, Todorovic R, Zhao J, Gross ML, Rogan EG, Cavalieri EL. *Catechol estrogen conjugates and DNA adducts in the kidney of male syrian golden hamsters treated with 4-hydroxyestradiol: potential biomarkers for estrogen-initiated cancer.* Carcinogenesis. 2001;22:489-97.

16. Markushin Y, Zhong W, Cavalieri EL, Rogan EG, Small GJ, Yeung ES, et al. *Spectral characterization of catechol estrogen quinone (CEQ)-derived DNA adducts and their identification in human breast tissue extract.* Chem Res Toxicol. 2003;16:1107-17.

for the estrogen receptor due to steric hindrance (2). By increasing polarity, glucuronidation of estrogens facilitates partitioning of the lipophilic steroids into the aqueous compartment and their ultimate excretion in bile and urine.

The parent hormones, E_2 and E_1, and their respective catechols are recognized as substrates but individual UGT isoforms display distinct differences in substrate specificity and conjugation efficiency. Comparison of UGT1A1 and 2B7 showed regioselective conjugation of E_2, i.e., UGT1A1 only conjugated the C-3 hydroxyl group of the A-ring, whereas UGT2B7 conjugated the 17β-hydroxyl in the D-ring, yielding E_2-3Glu and E_2-17βGlu, respectively (3, 4). The catechol estrogens can also be substrates and, in fact, the UGT1A1, 1A9, and 2B7 isoforms were more active toward the catechol estrogens than the parent hormones. On the other hand, comparison of catechol estrogen substrates revealed that UGT1A1 and 1A3 were more active toward 2-OHE$_2$, while UGT1A9 and 2B7 conjugated 4-OHE$_2$ more efficiently (5, 6). Although the catechols derived from E_2 and E_1 are generally metabolized with similar efficiencies, UGT2B7 displayed 7- to 12-fold higher activity (1300 pmol/min/mg microsomal protein) toward 4-OHE$_1$ than 4-OHE$_2$, in spite of similar apparent Km values (6, 7). However, the highest activity was recorded for the UGT1A9-mediated conjugation of 4-OHE$_2$ (2500 pmol/min/mg) (5).

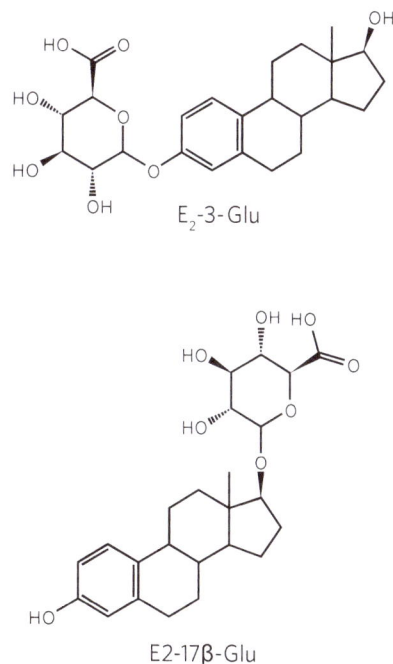

E$_2$-3- Glu

E2-17β-Glu

Figure 2.1.2.15. UDP-glucuronyltransferases catalyze the conjugation of glucuronic acid to either the C-3 or C-17β hydroxyl group of E$_2$, yielding E$_2$-3-glucuronide (E$_2$-3-Glu) and E$_2$-17β-glucuronide (E$_2$-17β-Glu).

2.1.2.2.3. ESTROGEN GLUCURONIDES

The uridine diphosphate (UDP)-glucuronyltransferase (UGT) gene family encodes microsomal enzymes that catalyze the transfer of the polar D-glucuronic acid moiety from UDP-glucuronic acid to a wide variety of endogenous lipophilic compounds, including steroid hormones (1). Glucuronidation of E$_2$ can occur at either the C-3 or C-17β hydroxyl group (Figure 2.1.2.15.), abolishing its affinity

References

1. Mackenzie PI, Mojarrabi B, Meech R, Hansen A. *Steroid UDP glucuronosyltransferases: characterization and regulation.* J Endocrinol. 1996;150:S79-S86.

2. Roy AK. *Regulation of steroid hormone action in target cells by specific hormone-inactivating enzymes.* Proc Soc Exp Biol Med. 1992;199:265-72.

3. Gall WE, Zawada G, Mojarrabi B, Tephly TR, Green MD, Coffman BL, et al. *Differential glucuronidation of bile acids, androgens and estrogens by human UGT1A3 and 2B7.* J Steroid Biochem Mol Biol. 1999;70:101-8.

4. Senafi SB, Clarke DJ, Burchell B. *Investigation of the substrate specificity of a cloned expressed human bilirubin UDP-glucuronosyltransferase: UDP-sugar specificity and involvement in steroid and xenobiotic glucuronidation.* Biochem J. 1994;303:233-40.

5. Albert C, Vallee M, Guillaume B, Belanger A, Hum DW. *The monkey and human uridine diphosphate-glucuronosyltransferase UGT1A9, expressed in steroid target tissues, are estrogen-conjugating enzymes.* Endocrinology. 1999;140:3292-302.

6. Cheng Z, Rios GR, King CD, Coffman BL, Green MD, Mojarrabi B, et al. *Glucuronidation of catechol estrogens by expressed human UDP-glucuronosyltransferases (UGTs) 1A1, 1A3, and 2B7.* Toxicol Sci. 1998;45:52-7.

7. Turgeon D, Carrier J-S, Levesque E, Hum DW, Belanger A. *Relative enzymatic activity, protein stability, and tissue distribution of human steroid-metabolizing UGT2B subfamily members.* Endocrinology. 2001;142:778-87.

2.1.2.2.4. ESTROGEN SULFATES

Several members of the sulfotransferase (SULT) gene family are capable of sulfating hydroxysteroids, including estrogens. The SULTs transfer the sulfonate group from the cofactor, 3'-phosphoadenosine-5'-phosphosulfate (PAPS), to the hydroxyl group of the steroid to produce the steroid sulfate. In the case of estrogens, several SULT isoforms were shown to be capable of conjugating E_1 and E_2, yielding E_1-3-sulfate (E_1-3-S) and E_2-3-sulfate (E_2-3-S), respectively (Figure 2.1.2.16.). The catechol estrogens, e.g., 2-OHE$_2$ and 4-OHE$_2$, can also be conjugated to 2-OHE$_2$-3-S and 4-OHE$_2$-3-S, respectively (Figure 2.1.2.16.). However, the isoforms differed significantly in their affinity, i.e., only SULT1E1 conjugated these substrates at nanomolar concentrations, in contrast to micromolar concentrations observed for SULT1A1, 1A2, 1A3, and 2A1 (1, 2). Methoxyestrogens may also be substrates.

An analysis of estrogen metabolites in MCF-7 culture media by mass spectrometry showed that 96% of 2-MeOE$_2$ was conjugated to 2-MeOE$_2$-3-S (Figure 2.1.2.16.), but only 27% of 4-MeOE$_2$, apparently by the combined action of COMT and SULT (3). SULT1E1 is the principal isoform for the sulfation of E_2, 2-OHE$_2$, and 4-OHE$_2$, but there is presently uncertainty whether SULT1A1 or 1E1 is responsible for the sulfation of methoxyestrogens (1, 3).

The importance of SULTs for estrogen conjugation is demonstrated by the observation that a major component of circulating estrogen in humans is sulfate conjugated,

i.e., E_1-3-S. The addition of the charged sulfate group to E_2 or E_2 prevents binding of the hormones to the estrogen receptor and thereby ameliorates their mitogenic action (4). Consequently, SULTs play a major role in the control of estrogen action in target tissues by catalyzing the conversion of E_1 and E_2 to their inactive sulfated forms. The importance of the sulfation of catechol estrogens and methoxyestrogens is uncertain. The resulting sulfated metabolites are more hydrophilic and may be excreted.

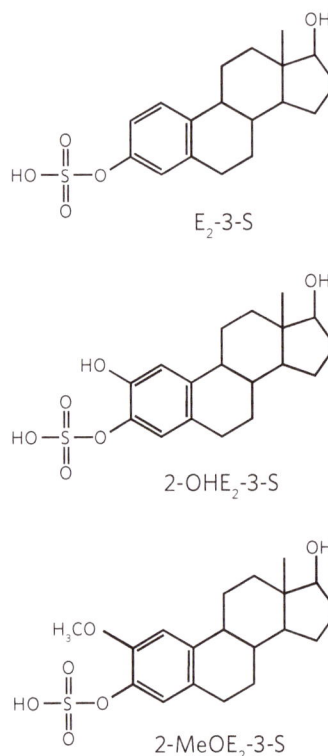

Figure 2.1.2.16. Sulfotransferases catalyze the conjugation of the sulfate group to hydroxysteroids, producing estrogen sulfate conjugates such as 17β-estradiol-3-sulfate (E_2-3-S), 2-hydroxyestradiol-3-sufate (2-OHE$_2$-3-S), and 2-methoxyestradiol-3-sulfate (2-MeOE$_2$-3-S).

References

1. Adjei AA, Weinshilboum RM. *Catecholestrogen sulfation: possible role in carcinogenesis.* Biochem Biophys Res Comm. 2002;292:402-8.

2. Falany JL, Falany CN. *Expression of cytosolic sulfotransferases in normal mammary epithelial cells and breast cancer cell lines.* Cancer Res. 1996;56:1551-5.

3. Spink BC, Katz BH, Hussain MM, Pang S, Connor SP, Aldous KM, et al. *SULT1A1 catalyzes 2-methoxyestradiol sulfonation in MCF-7 breast cancer cells.* Carcinogenesis. 2000;21:1947-57.

4. Santner SJ, Feil PD, Santen RJ. *In situ estrogen production via the estrone sulfatase pathway in breast tumors: relative importance versus the aromatase pathway.* J Clin Endocrinol Metab. 1984;59:29-33.

2.1.3. DNA ADDUCTS

2.1.3.1. ESTROGEN-DNA ADDUCTS

The estrogen quinones are the ultimate mutagenic metabolites of the catechol estrogen pathway because they can react directly with DNA bases to form covalent bonds. The nature of these estrogen-DNA adducts has been examined by reacting the quinones (E_1-2,3-Q, E_2-2,3-Q, E_1-3,4-Q, E_2-3,4-Q) with individual nucleic acid bases (adenine, guanine, cytosine, thymine) and deoxyribonucleosides (dA, dG, dC, dT) (1-8). Each base and nucleoside can form several adducts, depending on the reaction conditions, such as pH and ionic state of the quinone. In general, reactivity of the electrophilic quinone with potential nucleophilic sites of the nucleosides is greater with nitrogen than carbon sites (Table 2.1.3.1.). Purine bases are more reactive toward estrogen quinones than pyrimidine bases as demonstrated by the reaction of E_2-2,3-Q with T-rich model oligonucleotides containing one (TTTTTATTTTTT, TTTTTGTTTTTT) or two (TTTTTATTTATT, TTTTATTTGTTTT) reactive bases. Adducts were formed only with A and G but not T (9).

Table 2.1.3.1. Estrogen-DNA Adducts

Quinone	Nucleoside or Base	Adduct
E-2,3-Q	dA	6(α,β)*-N6-dA
	dG	6(α,β)-N2-dG
E-3,4-Q	dA	1(α,β)-C8-Ade
	Ade	1(α,β)-N3-Ade
	dG	1(α,β)-N7-Gua
		1(α,β)-C8-Gua
		2(α,β)-C8-Gua
	dC	1(α,β)-N4-dC
		2(α,β)-N4-dC
	dT	1(α,β)-N3-dT

The diversity of estrogen-DNA adducts is the result of several types of reactions. For example, incubation of E_1-3,4-Q with dG in an acidic environment results in Michael addition of N7 of guanine to C1 of the estrogen quinone to form 4-OHE$_1$-1(α,β)-N7-Gua (7) (Table 2.1.3.1., Figure 2.1.3.1.). The substitution of N7 destabilizes the glycosidic bond, which leads to loss of the deoxyribose moiety and release of the modified base from the DNA backbone, a process referred to as depurination. The generation of two conformational isomers, α and β, is the result of a rotational barrier about the C1-N7 bond. Thus, in one isomer, the guanine moiety is located primarily on the "α" side of the estrogen ring system, 4-OHE$_1$-1α-N7-Gua, while the other isomer has the guanine moiety located on the "β" side, 4-OHE$_1$-1β-N7-Gua. NMR analysis showed that the two isomers were formed in a ratio of approximately 60% β and

40% α (7). Since the rotation is not completely hindered, the two isomers cannot be separated analytically. Two minor, alternate reactions can occur between E_1-3,4-Q and dG, namely the attack of C8 of guanine at C1 or C2 of the estrogen quinone, resulting in 4-OHE$_1$-1(α,β)-C8-Gua and 4-OHE1-2(α,β)-C8-Gua, both of which also lose the deoxyribose moiety (Table 2.1.3.1., Figure 2.1.3.2.).

The reaction of 2'-deoxyribonucleosides with E_1-2,3-Q differs from E_1-3,4-Q in two aspects. First, the deoxyribose is not lost when E_1-2,3-Q is incubated with dG or dA. Second, the reaction does not occur with the quinone but its tautomer. The 2,3-*o*-quinone isomerizes rapidly into the more electrophilic *p*-quinone methide, which then undergoes the Michael addition reaction. The linkage occurs between C6 of the steroid B ring and the exocyclic amino group of either dG or dA (Table 2.1.3.1., Figure 2.1.3.3.). The resulting adducts, such as 2-OHE$_1$-6(α,β)-

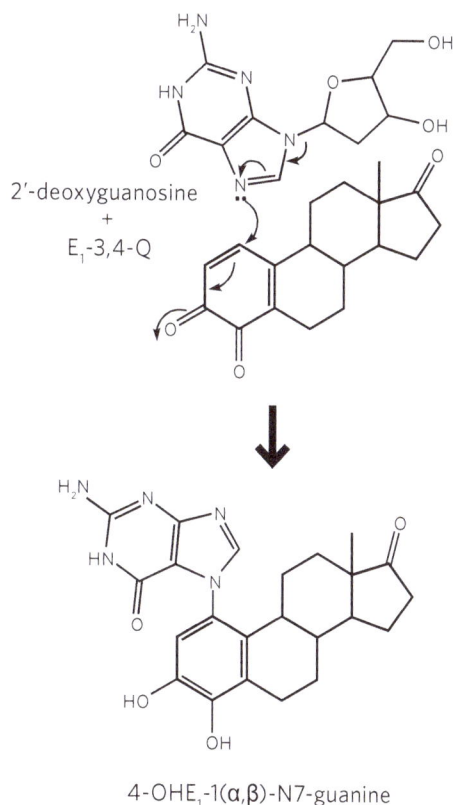

2'-deoxyguanosine
+
E_1-3,4-Q

4-OHE$_1$-1(α,β)-N7-guanine

Figure 2.1.3.1. Reaction of nucleophile 2'-deoxyguanosine with electrophile E_1-3,4-Q results in Michael addition of N7 of guanine to C1 of estrogen to form 4-OHE$_1$-1(α,β)-N7-guanine, a depurinating adduct that is released from DNA by destabilization of the glycosyl bond.

N2-dG and 2-OHE$_1$-6(α,β)-N6-dA, are stable and retain the deoxyribose (7). A minor reaction occurs at C9 of the B ring (9, 10). When E$_1$-3,4-Q was reacted with adenine in dimethylformamide under cathodic conditions, coupling between C8 of adenine and C1 of the estrogen quinone resulted in 4-OHE$_1$-1(α,β)-C8-Ade (Table 2.1.3.1., Figure 2.1.3.2.) (1, 2, 4). The mechanism of formation indicated that the semiquinone radical was the alkylating species and not the quinone or the quinone methide. Thus, estrogen-DNA adducts can be formed as Michael addition products by reacting with the quinone or quinone methide or via a radical coupling reaction with the semiquinone.

The 2,3-quinones preferentially form stable adducts whereas the 3,4-quinones are more likely to produce depurinating adducts. This was shown experimentally in reactions with calf thymus DNA. E$_1$-2,3-Q and E$_2$-2,3-Q produced 8.6 and 1.1 ×mol stable adduct/mol DNA-phosphate, respectively, whereas E$_1$-3,4-Q and E$_2$-3,4-Q generated 0.11 and 0.07 ×mol stable adduct/mol DNA-phosphate, respectively (11, 12). However, E$_1$-3,4-Q and E$_2$-3,4-Q produced two to three orders of magnitude higher levels of depurinating adducts than E$_2$-2,3-Q and E$_1$-2,3-Q (6, 13). Thus, when both stable and depurinating adducts were considered together, the 3,4-quinones produced much higher levels of DNA adducts (12).

When 4-OHE$_2$ was oxidized by horseradish peroxidase, lactoperoxidase, or tyrosinase in the presence of calf thymus DNA and H$_2$O$_2$, the depurinating adducts 4-OHE$_2$-1(α,β)-N7-guanine and 4-OHE$_2$-1(α,β)-N3-adenine were formed in concentrations ranging from 6.5 to 363 mmol/mol DNA-phosphate (14). The depurinating adducts constituted >99% of the total adduct concentration. Similar results were obtained upon incubation of DNA with phenobarbital-induced rat liver microsomes and cumene hydroperoxide, CuOOH (14). In incubations of E$_1$-3,4-Q with the mitochondrial cytochrome oxidase subunit III gene, the quinone was covalently linked mainly to guanine (15). The *in vitro* replication of the COIII template was obstructed indicating an arrest of DNA polymerase by the estrogen-guanine adducts. The formation of depurinating estrogen-DNA adducts could be inhibited by selected natural antioxidants, such as resveratrol, reduced lipoic acid, melatonin, and N-acetylcysteine (13).

Catechol estrogens and estrogen quinones have also been administered to cells and animals to elicit DNA damage. The immortalized breast epithelial cells MCF-10F were successively treated with 1 μM 4-OHE$_2$ for a 24-h period twice weekly for two weeks and the culture medium collected at 72, 120, 192 and 240 h post-plating (16). The culture medium was analyzed, rather than the cell pellet,

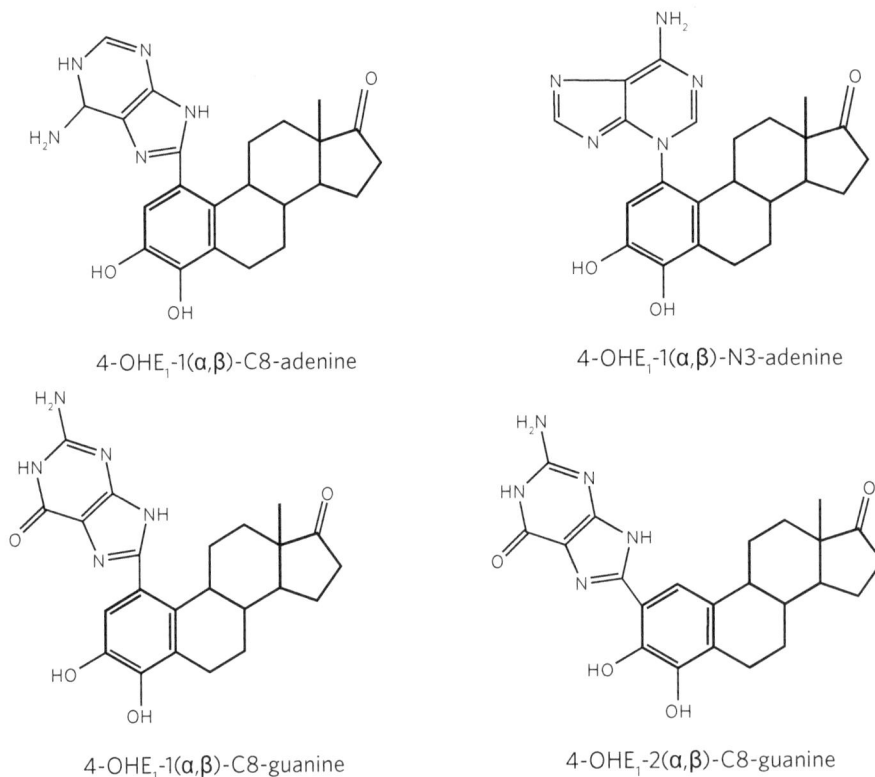

4-OHE$_1$-1(α,β)-C8-adenine

4-OHE$_1$-1(α,β)-N3-adenine

4-OHE$_1$-1(α,β)-C8-guanine

4-OHE$_1$-2(α,β)-C8-guanine

Figure 2.1.3.2. E$_1$-3,4-Q and E$_2$-3,4-Q react with deoxynucleosides to form apurinic estrogen-DNA adducts such as 4-OHE$_1$-1(α,β)-C8-adenine, 4-OHE$_1$-1(α,β)-N3-adenine, 4-OHE$_1$-1(α,β)-C8-guanine, and 4-OHE$_1$-2(α,β)-C8-guanine.

because it contains virtually all the estrogen metabolites, conjugates, and depurinating DNA adducts. The major adducts were 4-OHE$_2$-1-N3-Ade and 4-OHE$_1$-1-N3-Ade, which were highest after the third treatment. Smaller amounts of 4-OHE$_1$(E$_2$)-1-N7-Gua were also detected. Thus, MCF-10F cells oxidize 4-OHE$_2$ to E$_1$(E$_2$)-3,4-Q, which react with DNA to form the depurinating N3-Ade and N7-Gua adducts. Exposure of MCF-7 cells to 50 µM E1-3,4-Q for 2 h yielded detectable levels of 4-OHE$_1$-1-N7-Gua and 4-OHE$_1$-1-C8-Gua in the culture medium (3). This is a remarkable finding because it proves release of adducts from the DNA backbone and demonstrates their hydrophilicity, which permits exit from the cells within a short time frame. The result was confirmed in another study with 0.92 pg/ml 4-OHE1-1-N7-Gua and 4-OHE$_2$-1-N7-Gua in culture media of MCF-7 cells treated with 20 µM 4-OHE$_2$ for 24 h (17). Treatment of ERα-positive MCF-7 breast cancer cells with 10 µM E$_1$-3,4-Q for 1 h resulted in DNA single-strand breaks (18). A lower dose (100 µM) of E$_2$ (24 h) or 4-OHE$_2$ (2 h) was shown to elicit single-strand breaks in both MCF-7 and ERα-negative MDA-MB-231 cells, indicating that estrogen receptor status was not relevant to these genomic effects (19).

Oxidative DNA adducts are generated by reactive oxygen species resulting from redox cycling of catechol estrogens and estrogen quinones. In contrast to oxidative DNA adducts, apurinic estrogen-DNA adducts such as 4-OHE$_1$-1(α,β)-N7-Gua are unique among DNA adducts for three reasons (i) it contains the estrogen metabolite as a "fingerprint" of estrogen exposure, (ii) it is released from the DNA backbone by cleavage of the deoxyribose glycosyl bond, and (iii) by virtue of its hydrophilicity it is capable of exiting the cell as shown by its appearance in the culture medium of MCF-7 cells incubated with 4-OHE$_1$ or E$_1$-3,4-Q (3, 17).

Implantation of E$_1$ or E$_2$ in male Syrian golden hamsters induced renal carcinomas in 100% of the animals, but does not induce liver tumors (20). Examination of kidney and liver extracts of hamsters treated for 2 h with 8 µmol E$_2$ per 100 g body weight did not reveal any adduct formation. However, pretreatment with an inhibitor of GSH synthesis, L-buthionine (SR)-sulfoximine, to deplete GSH levels in the hamsters, resulted in the production of similar levels of 4-OHE$_1$-1(α,β)-N7-Gua and 4-OHE$_2$-1(α,β)-N7-Gua (0.27 nmol/g tissue) in renal but not hepatic tissues (21). Animal experiments have also been carried out with estrogen

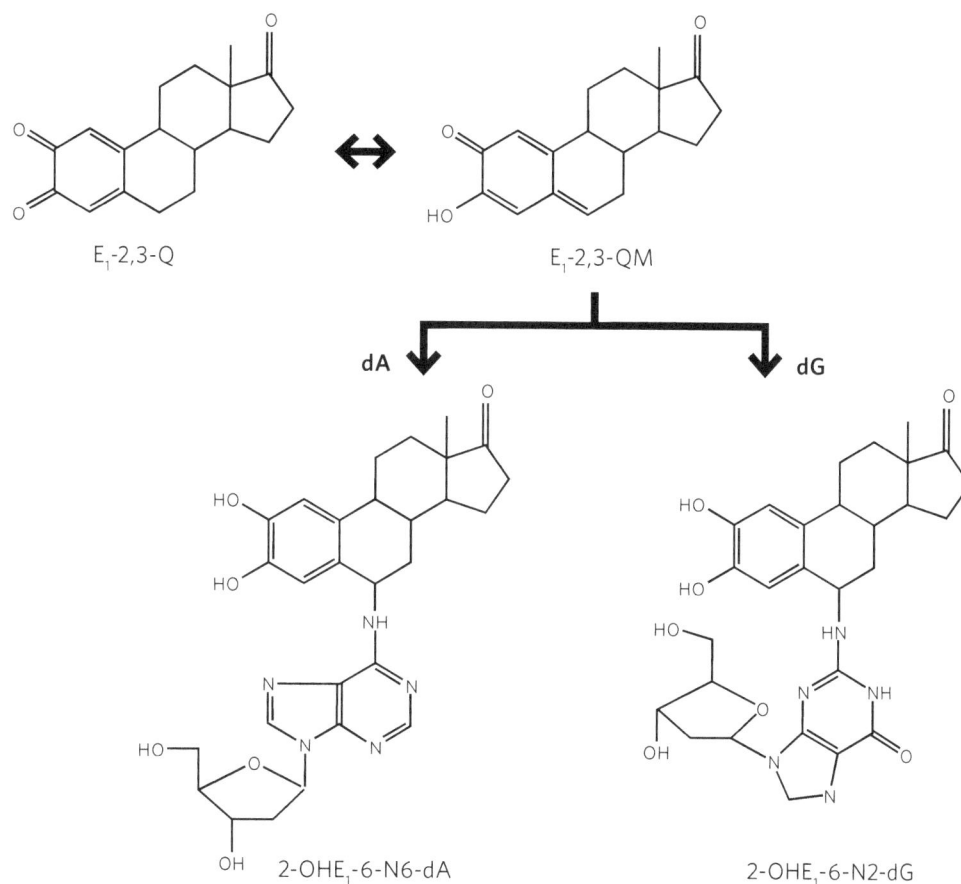

Figure 2.1.3.3. E$_1$-2,3-Q isomerizes to the more electrophilic E$_1$-2,3-QM, which reacts with deoxynucleosides to form stable estrogen-DNA adducts, such as 2-OHE$_1$-6-N6-dA and 2-OHE$_1$-6-N2-dG.

quinones, e.g., intramammillary injections of 200 nmol E_2-3,4-Q in 20 µl DMSO (= 10 mM E_2-3,4-Q) into female Sprague-Dawley rats (6). The breast tissue was collected 2 h post injection and 4-OHE$_2$-1-N7-Gua was detected at a concentration of 2.3 µmol adduct/mol DNA-phosphate. Identical treatment of female ACI rats followed by excision of mammary tissue 1 h later revealed the depurinating adducts 4-OHE$_2$-1(α,β)-N7-Gua and 4-OHE$_2$-1(α,β)-N3-Ade at 66 and 61 pmol/g tissue, respectively (14). The corresponding E_1 isomers, 4-OHE$_1$-1(α,β)-N7-Gua and 4-OHE$_1$-1(α,β)-N3-Ade, were also detected at 24 and 20 pmol/g tissue, respectively. The level of stable adducts was <0.1% of the total adduct concentration. Unexpectedly, injection of the ACI rats with 4-OHE$_2$ resulted in 40% higher adduct levels than in the E_2-3,4-Q-treated animals. This is presumably due to the high reactivity of the quinone, which can react indiscriminately with various cellular molecules. In contrast, 4-OHE$_2$ can be oxidatively metabolized to E_2-3,4-Q in regions close to DNA. In a third study, female SENCAR mice were treated topically with 200 nmol E_2-3,4-Q in 50 fl acetone/DMSO in a shaved area of dorsal skin (22). The skin tissue was excised 1 h after the application and approximately equal concentrations (12 µmol/mol DNA-phosphate) of 4-OHE$_2$-1(α,β)-N7-Gua and 4-OHE$_2$-1(α,β)-N3-Ade were detected. Again, the concentration of stable adducts was <0.1% of total adducts. Finally, in vitro experiments have demonstrated that CYP1B1-mediated oxidation of E_2 in the presence of deoxyguanosine caused the formation of the 4-OHE$_2$-N7-Gua adduct, providing direct evidence that metabolism of the parent hormone can initiate DNA damage (23).

Equine Estrogen Adducts. Equine 4-hydroxyequilenin (4-OHEN) also forms adducts with dA, dG, and dC, but not dT. These adducts are unusual by forming a ring structure between 4-OHEN and base, involving the exocyclic amino group as shown in Figure 2.1.3.4 (24). The major structural difference between thymidine and the other deoxynucleosides is the lack of an exocyclic amino group in the former, which confirms that coupling between 4-OHEN and dA, dG, and dC involves this group.

Mechanistic studies on the rate of formation of 4-OHEN-dG implicate the semiquinone radical as the alkylating species. Conditions that enhance the rate of autoxidation (i.e., base, reducing agents) lead to a corresponding increase in the rate of adduct formation. In contrast, inhibitors of autoxidation (acid, methanol, metal chelators) prevent formation of 4-OHEN-dG adducts. Finally, experiments with scavengers of reactive oxygen species suggest that hydrogen peroxide and not superoxide or free radical is required for adduct formation. Thus, it appears that autoxidation of 4-OHEN generates the semiquinone radical which abstracts a hydrogen atom from the exocyclic amino group of dNs generating dN-nH•. This radical combines with a 4-OHEN-semiquinone radical at the C1 position in the case of

adenine and the C3 position of guanine and cytosine. Cell culture experiments showed that 4-OHEN induced DNA damage in breast cancer cell lines and cellular transformation in vitro (25, 26). Intramammillary injection of 4-OHEN (200 or 2000 nmol in 20 µl DMSO) into rats resulted in single-strand breaks and oxidized bases (27). Only the higher dose (equivalent to 100 mM) caused the formation of depurinating (4-OHEN-Gua) and stable (4-OHEN-dA, 4-OHEN-dG) adducts.

Estrogen Quinone-Protein Adducts. As described in Section 2.1.2.1.4, the highly reactive estrogen quinones can form adducts with proteins via lysine or cysteine residues. Both E_2-3,4-Q and E_2-2,3-Q were shown to form adducts with serum albumin in a concentration-dependent manner (28). The reaction occurred with cysteine residues, reached maximum levels within 10 min and remained constant thereafter for up to 24 h. The analysis of serum albumin by GC/MS revealed the presence of cysteinyl adducts of E_2-2,3-Q-1-S-Alb, E_2-2,3-Q-4-S-Alb, and E_2-3,4-Q-2-S-Alb in blood samples from healthy women (n = 10) and breast cancer patients (n = 20). Levels of E_2-2,3-Q-derived adducts were 2-fold greater in controls than those of E_2-3,4-Q-2-S-Alb whereas the converse was true in patients.

Summary. The estrogen quinones can damage DNA directly via Michael addition, resulting in two classes of DNA adducts (i) stable adducts that remain in DNA unless removed by repair and (ii) depurinating adducts that are released from DNA by destabilization of the glycosyl bond (6). Regardless of mechanism of formation, all estrogen quinone-induced DNA lesions are potentially mutagenic (8, 22, 29-31). However, only the 3,4-quinones are carcinogenic. The difference in carcinogenicity appears to be due to the difference in reactivity and stability of the quinones. The 2,3-quinones undergo clearance more rapidly than the 3,4-quinones, which are more stable and therefore more likely to diffuse from the site of formation to the DNA target (32). In addition, apurinic sites are repaired by a different mechanism (base excision repair) than stable adducts (nucleotide excision repair). Table 2.1.3.2 summarizes the various factors that may contribute to the difference in carcinogenic potency of 2,3- and 3,4-quinones.

Table 2.1.3.2. Differences between 2,3- and 3,4-estrogen quinones and their respective DNA adducts

Property	E-2,3-Q	E-3,4-Q
Half-life	42 sec	12 min
Reactive form	p-quinone methide	o-quinone
Adduction site	B ring	A ring
Adduct stability	stable	unstable
Loss of 2'-deoxyribose	no	yes
Repair mechanism	nucleoside excision	base excision

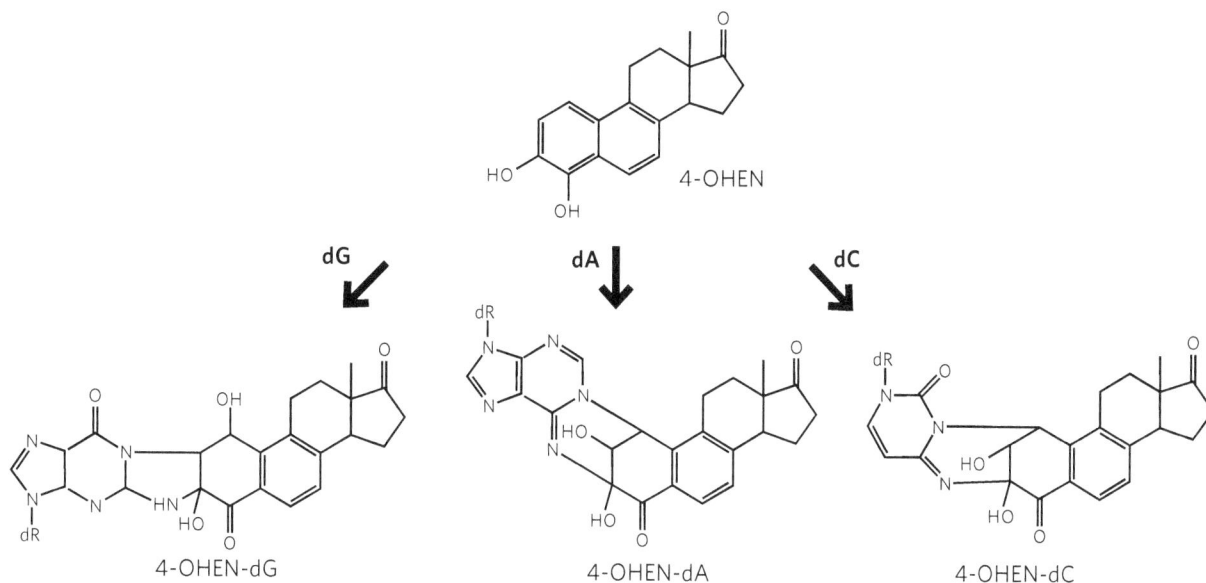

Figure 2.1.3.4. 4-OHEN reacts with dG, dA, and dC but not thymidine to form unusual ring-structure adducts (dR = deoxyribose).

References

1. Abul-Hajj YJ, Tabakovic K, Tabakovic I. *An estrogen-nucleic acid adduct. Electroreductive intermolecular coupling of 3,4-estrone-o-quinone and adenine.* J Am Chem Soc. 1995;117:6144-5.

2. Akanni A, Abul-Hajj YJ. *Estrogen-nucleic acid adducts: reaction of 3,4-estrone o-quinone radical anion with deoxyribonucleosides.* Chem Res Toxicol. 1997;10:760-6.

3. Akanni A, Abul-Hajj YJ. *Estrogen-nucleic acid adducts: Dissection of the reaction of 3,4-estrone and its radical anion and radical cation with deoxynucleosides and DNA.* Chem Res Toxicol. 1999;12:1247-53.

4. Akanni A, Tabakovic K, Abul-Hajj YJ. *Estrogen-nucleic acid adducts: reaction of 3,4-estrone o-quinone with nucleic acid bases.* Chem Res Toxicol. 1997;10:477-81.

5. Cavaleri E, Frenkel K, Liehr JG, Rogan E, Roy D. *Estrogens as endogenous genotoxic agents-DNA adducts and mutations.* J Natl Cancer Inst Monogr. 2000;27:75-93.

6. Cavalieri EL, Stack DE, Devanesan PD, Todorvic R, Dwivedy I, Higginbotham S, et al. *Molecular origin of cancer: catechol estrogen-3,4-quinones as endogenous tumor initiators.* Proc Natl Acad Sci USA. 1997;94:10937-42.

7. Stack DE, Byun J, Gross ML, Rogan EG, Cavalieri EL. *Molecular characteristics of catechol estrogen quinones in reactions with deoxyribonucleosides.* Chem Res Toxicol. 1996;9:851-9.

8. Terashima I, Suzuki N, Dasaradhi L, Tan CK, Downey KM, Shibutani S. *Translesional synthesis on DNA templates containing an estrogen quinone-derived adduct: N²-(2-hydroxyestron-6-yl)-2'-deoxyguanosine and N⁶-(2-hydroxyestron-6-yl)-2'-deoxyadenosine.* Biochemistry. 1998;37:13807-15.

9. Debrauwer L, Rathahao E, Couve C, Poulain S, Pouyet C, Jouanin I, et al. *Oligonucleotide covalent modifications by estrogen quinones evidenced by use of liquid chromatography coupled to negative electrospray ionization tandem mass spectrometry.* J Chromatogr A. 2002;976:123-34.

10. Convert O, Van Aerden C, Debrauwer L, Rathahao E, Molines H, Fournier F, et al. Reactions of estradiol-2,3-quinone with deoxyribonucleosides: possible insights in the reactivity of estrogen quinones with DNA. Chem Res Toxicol. 2002;15:754-64.

11. Dwivedy I, Devanesan P, Cremonesi P, Rogan E, Cavalieri E. *Synthesis and characterization of estrogen 2,3-and 3,4-quinones. Comparison of DNA adducts formed by the quinones versus horseradish peroxidase-activated catechol estrogens.* Chem Res Toxicol. 1992;5:828-33.

12. Stack DE, Cavalieri EL, Rogan EG. *Catecholestrogens procarcinogens: depurinating adducts and tumor initiation.* Adv Pharmacol. 1998;42:833-6.

13. Zahid M, Gaikwad NW, Rogan EG, Cavalieri EL. *Inhibition of depurinating estrogen-DNA adduct formation by natural compounds.* Chem Res Toxicol. 2007;20:1947-53.

14. Li KM, Todorovic R, Devanesan P, Higginbotham S, Kofeler H, Ramanathan R, et al. *Metabolism and DNA binding studies of 4-hydroxyestradiol and estradiol-3,4-quinone in vitro and in female ACI rat mammary gland in vivo.* Carcinogenesis. 2004;25:289-97.

15. Roy D, Abul-Hajj YJ. *Estrogen-nucleic acid adducts: guanine is major site for interaction between 3,4-estrone quinone and COIII gene.* Carcinogenesis. 1997;18:1247-9.

16. Saeed M, Rogan E, Fernandez SV, Sheriff F, Russo J, Cavalieri E. *Formation of depurinating N3Adenine and N7Guanine adducts by MCF-10F cells cultured in the presence of 4-hydroxyestradiol.* Int J Cancer. 2007;120:1821-4.

17. Yue W, Santen RJ, Wang JP, Li Y, Verderame MF, Bocchinfuso WP, et al. *Genotoxic metabolites of estradiol in breast: potential mechanism of estradiol induced carcinogenesis.* J Steroid Biochem Mol Biol. 2003;86:477-86.

18. Nutter LM, Ngo EO, Abul-Hajj YJ. *Characterization of DNA damage induced by 3,4-estrone-o-quinone in human cells.* J Biol Chem. 1991;25:16380-6.

19. Rajapakse N, Butterworth M, Kortenkamp A. *Detection of DNA strand breaks and oxidized DNA bases at the single-cell level resulting from exposure to estradiol and hydroxylated metabolites.* Environ Mol Mutagen. 2005;45:397-404.

20. Li JJ, Li SA. *Estrogen carcinogenesis in Syrian hamster tissues: role of metabolism.* Fed Proc. 1987;46:1858-63.

21. Cavalieri EL, Kumar S, Todorovic R, Higginbotham S, Badawi AF, Rogan EG. *Imbalance of estrogen homeostasis in kidney and liver of hamsters treated with estradiol: implications for estrogen-induced initiation of renal tumors.* Chem Res Toxicol. 2001;14:1041-50.

22. Chakravarti D, Mailander PC, Li KM, Higginbotham S, Zhang HL, Gross ML, et al. *Evidence that a burst of DNA depurination in SENCAR mouse skin induces error-prone repair and forms mutations in the H-ras gene.* Oncogene. 2001;20:7945-53.

23. Belous AR, Hachey DL, Dawling S, Roodi N, Parl FF. *Cytochrome P450 1B1-mediated estrogen metabolism results in estrogen-deoxyribonucleoside adduct formation.* Cancer Res. 2007;67:812 - 7.

24. Shen L, Pisha E, Huang Z, Pezzuto JM, Krol E, Alam Z, et al. *Bioreductive activation of catechol estrogen-ortho-quinones: aromatization of the B ring in 4-hydroxyequilenin markedly alters quinoid formation and reactivity.* Carcinogenesis. 1997;18:1096-101.

25. Chen Y, Lieu X, Pisha E, Constantinou AI, Hua Y, Shen L, et al. *A metabolite of equine estrogens, 4-hydroxyequilenin, induces DNA damage and apoptosis in breast cancer cell lines.* Chem Res Toxicol. 2000;13:342-50.

26. Pisha E, Lui X, Constantinou AI, Bolton JL. *Evidence that a metabolite of equine estrogens, 4-hydroxyequilenin, induces cellular transformation in vitro.* Chem Res Toxicol. 2001;14:82-90.

27. Zhang F, Swanson SM, Van Breemen RB, Liu X, Yang Y, Gu C, et al. *Equine estrogen metabolite of 4-hydroxyequilenin induces DNA damage in the rat mammary tissues: Formation of single-strand breaks, apurinic sites, stable adducts, and oxidized bases.* Chem Res Toxicol. 2001;14:1654-9.

28. Chen DR, Chen ST, Wang TW, Tsai CH, Wei HH, Chen GJ, et al. *Characterization of estrogen quinone-derived protein adducts and their identification in human serum albumin derived from breast cancer patients and healthy controls.* Toxicol Lett. 2011;202:244-52.

29. Kuchino Y, Mori F, Kasai H, Inoue H, Iwai S, Miura K, et al. *Misreading of DNA templates containing 8-hydroxydeoxyguanosine at the modified base and at adjacent residues.* Nature. 1987;327:77-9.

30. Shibutani S, Takeshita M, Grollman A. *Insertion of specific bases during DNA synthesis past the oxidation-damaged base 8-oxodG.* Nature. 1991;349:431-4.

31. Terashima I, Suzuki N, Shibutani S. *Mutagenic properties of estrogen-quinone derived DNA adducts in Simian kidney cells.* Biochemistry. 2001;40:166 - 72.

32. Tabakovic K, Gleason WB, Ojala WH, Abul-Hajj YJ. *Oxidative transformation of 2-hydroxyestrone. Stability and reactivity of 2,3-estrone quinone and its relationship to estrogen carcinogenicity.* Chem Res Toxicol. 1996;9:860-5.

2.1.3.2. MUTAGENIC AND CARCINOGENIC POTENTIAL

Mutagenic Potential. Generally, the quinones of E_1 and E_2 form the same adducts, e.g., 4-OHE$_1$-1-N7-Gua and 4-OHE$_2$-1-N7-Gua, consistent with similar free energies calculated for both reactions (1). However, there is a fundamental difference between 2,3-quinones (E_1-2,3-Q, E_2-2,3-Q) and 3,4-quinones (E_1-3,4-Q, E_2-3,4-Q). 3,4-quinone-derived adducts, such as 4-OHE$_1$-1-N7-Gua, 4-OHE$_1$-1-C8-Gua, 4-OHE$_2$-2-C8-Gua, 4-OHE$_1$-1-C8-Ade, and 4-OHE$_1$-1-N3-Ade, are lost from DNA by cleavage of the glycosidic bond, leaving apurinic sites in the DNA backbone. Unrepaired apurinic sites can lead to miscoding during DNA replication and result in mutations. For example, when an adenine adduct is lost by depurination, adenine is the most likely base to be inserted opposite the apurinic site in the next round of replication. When the coding strand of DNA is then replicated, a thymine is inserted opposite the new adenine, resulting in an A/T transversion. On the other hand, when a guanine adduct is lost by depurination, leaving an apurinic site in the DNA, the preferential insertion of adenine in the opposite DNA strand leads to a G/T transversion at the site of the adduct (2, 3).

The mutagenicity of estrogen quinones has been demonstrated in cell and animal experiments (4). Transfection of Simian kidney COS-7 cells with oligonucleotides containing 2-OHE$_1$-6-N2-dG resulted in G/T transversion with a mutation frequency of 18.2% opposite the adduct (5). 2-OHE$_1$-6-N6-dA caused A/T transversion and A/G transition with frequencies of 3.3% and 2.7%, respectively. Treatment of BB rat2 cells with multiple low-dose exposures of 4-OHE$_2$ resulted in mutations of the cII gene, especially at A:T base pairs, consistent with the known ability of E_2-3,4-Q to form a significant fraction of DNA adducts at adenines (6). Topical application of E_2-3,4-Q to the skin of female SENCAR mice resulted in the formation of 4-OHE$_2$-1-N7-Gua and 4-OHE$_2$-1-N3-Ade (7). Examination of specific genes, such as the H-*ras* oncogene,

revealed the occurrence of mutations in H-*ras* as early as 6 h after treatment. Although the level of depurinating adducts was similar, i.e., 12.5 μmol 4-OHE$_2$-1-N3-Ade and 12.1 μmol 4-OHE$_2$-1-N7-Gua per mol DNA-phosphate, the majority of mutations originated opposite the depurinated N3-Ade site in form of A/G transitions. The N3-Ade adduct is released from DNA instantaneously and generates A/G by misrepair and/or misreplication. In contrast, the N7-Gua adduct, which depurinates with a half-life of ~3 h, did not induce mutations, presumably because the slow rate of depurination allows the cell to correctly repair the DNA damage (8). In another study, female ACI rats, which are susceptible to estrogen-induced mammary tumors, were treated with E_2-3,4-Q by intramammillary injection. As in the mouse skin, depurinating N7-Gua and N3-Ade adducts as well as H-ras mutations were detected in the mammary tissue (8, 9). Again, the majority of mutations were A/G transitions.

Carcinogenic Potential. Treatment of ERα-negative MCF-10F breast epithelial cells twice a week for two weeks with E_2, 2-OHE$_2$, and 4-OHE$_2$ induced phenotypic changes indicative of neoplastic transformation, such as anchorage independent growth, colony formation in agar, inability to form ductules in collagen matrix and invasiveness in Matrigel invasion chambers (10-12). Doses of 70 and 0.007 nM E_2 and 4-OHE$_2$ also induced loss of heterozygosity at 13q12.3 and 17p13.1, respective sites of the BRCA1 and p53 genes (13). The genomic alterations were not abrogated by the anti-estrogen ICI182780, suggesting that ERα is not involved in the neoplastic transformation in this model. 2-OHE$_2$ induced genomic changes only at a higher dose, 3.6 μM. The most highly invasive cells in the Matrigel invasion chambers induced tumor formation in female SCID mice (10). The induced tumors were ERα–negative adenocarcinomas that exhibited genomic alterations similar to those reported in primary human breast cancer. Estrogens have also been shown to induce tumors in various organs of rats, mice, and hamsters (14). Implantation of E_1 and E_2 has been found to induce kidney carcinomas in male Syrian golden hamsters (15). In the same animal model, 4-OHE_1 and 4-OHE_2 were also carcinogenic, whereas 2-OHE_2 was not (15, 16). Administration of estrogen quinones showed that only 3,4-quinones were carcinogenic while no carcinogenic effects were detected with 2,3-quinone treatment. Intra-mammillary injection of 5 μmol E_1-3,4-Q/100 μl DMSO (= 50 mM E_1-3,4-Q) into 21-day old female CD-1 rats did not result in mammary tumor formation (17). However, the concentration of 50 mM E_1-3,4-Q is severely cytotoxic (18). In newborn CD-1 mice, repeated intraperitoneal injection of 4-OHE_2, 2-OHE_2, or E_2 produced uterine adenocar-cinomas (19). The strongest carcinogen was 4-OHE_2, which caused cancer formation in 66% of treated animals, followed by 2-OHE_2 (12%) and E_2 (7%). It is interesting that 2,3-quinones produced stable adducts but did not induce kidney tumors in hamsters (15, 16). In contrast, 3,4-quinones formed predominantly 4-OHE_1-1-N7-Gua and 4-OHE_2-1-N7-Gua, and induced kidney tumors in hamsters and liver cancer in mice (15, 16, 20). These data support the view that E_1-3,4-Q and E_2-3,4-Q are endogenous ultimate carcinogens.

Estrogen-DNA Adducts in Human Urine. Depurinating estrogen-DNA adducts are released from DNA and have been shown to exit cells and appear in culture media (21, 22). Thus, there is the potential that these adducts are present in the bloodstream and excreted in urine. One study confirmed the presence of estrogen-DNA adducts by HPLC/tandem mass spectrometry and detected 4-OHE_2-N7-Gua in all 10 female human urine samples analyzed (23). A quantitative study examined estrogen metabolites, conjugates and adducts in urine samples from 46 healthy women, 12 high-risk women and 17 women with breast cancer (24). The predominant adducts were 4-$OHE_{2/1}$-1-N3-Ade and 4-$OHE_{2/1}$-1-N7-Gua. The ratio of these adducts to their respective estrogen metabolites and conjugates were significantly higher in high-risk women and women with breast cancer than in control subjects (p < 0.001). There was no difference between the high-risk and cancer groups.

Estrogen-DNA Adducts in Human Breast Tissue. Few studies have measured estrogen-DNA adducts in human breast tissue. One study searched for estrogen-DNA adducts in normal and malignant human breast tissues by performing a qualitative liquid chromatography/mass spectrometry analysis (25). Comparison of 5 tumor and adjacent normal tissue samples revealed the presence of 4-OHE_2-dG in 4 tumor and 3 normal samples. 4-OHE_2-dG was also detected in 6 of 8 additional tumors. Other adducts identified in order of frequency were 4-OHE_1-dG, 4-OHE_2-dT, 4-OHE_1-dA, and 4-OHE_2-dC. Interestingly, other adducts contained 4-OH-equilenin and 4-OH-17α-ethinylestradiol, reflecting usage of Premarin and oral contraceptives. Another exogenously

derived adduct, benzo[a]pyrene-7,8-diol-9.10-epoxide-dG, was found in one normal and four tumor samples. Another study combined capillary electrophoresis with low temperature phosphorescence spectroscopy to quantify 4-OHE_1-1-N3-Ade, 4-OHE_1-1-N7-Gua, 4-OHE_2-1-N3-Ade, and 4-OHE_2-1-N7-Gua in one normal breast biopsy and one breast cancer (26). The normal sample contained 4-OHE_2-1-N3-Ade, 4-OHE_1-1-N3-Ade, and 4-OHE_2-1-N7-Gua at 1.18, 0.25, and 0.11 pmol/g tissue, respectively. The tumor contained 4-OHE_1-1-N3-Ade and 4-OHE_2-1-N3-Ade at 8.4 and 0.9 pmol/g tissue, respectively. A third study used liquid chromatography tandem mass spectrometry and detected 4-OHE_1-1-N3-Ade levels in the range of 20 – 70 fmol/g tissue in six breast tissue samples, two from control women and four normal samples from women who underwent mastectomy for breast cancer (27). The concentra-tions measured were close to the detection limit of the instrument and did not permit distinction between cancer and control tissues. The concentrations measured in the third study were significantly lower than those reported in the second study (20 – 70 versus 8,400 fmol/g tissue). Experimental data suggest that the conversion of catechol estrogens to estrogen-DNA adducts is on the order of 1% of the amount of estrogen present in tissue (28, 29). Other data indicate that the concentration of estrogens in breast tissue of pre- and postmenopausal women is on the order of 1 pmol/g tissue with similar proportions of parent hormones and catechol estrogens (30). The production of 20 – 70 fmol/g tissue would require a conversion rate of 2 – 7% of catechol estrogens to estrogen-DNA adducts whereas the production of 8,400 fmol/g would exceed the amount of estrogens in breast tissue.

Antiestrogens, such as tamoxifen, are also oxidized by cytochrome P450 enzymes to catechols and quinones that are capable of forming DNA adducts (31-34). These adducts possess mutagenic potential and have been identified in the endometrium treated with tamoxifen in the course of breast cancer therapy (35, 36). This genotoxic mechanism may explain the increased risk of endometrial cancer in women treated with tamoxifen for breast cancer (37).

References

1. Huetz P, Kamarulzaman EE, Wahab HA, Mavri J. *Chemical reactivity as a tool to study carcino-genicity: reaction between estradol and estrone 3,4 quinones ultimate carcinogens and guanine.* J Chem Inf Comput Sci. 2004;44:310-4.

2. Cavalieri EL, Rogan EG. *Role of aromatic hydrocarbons in disclosing how catecholestrogens initiate cancer.* Adv Pharmacol. 1998;42:837-40.

3. Loeb LA, Preston BD. *Mutagenesis by apurinic/apyrimidinic sites.* Ann Rev Genetics. 1986;20:201-30.

4. Cavaleri E, Frenkel K, Liehr JG, Rogan E, Roy D. *Estrogens as endogenous genotoxic agents-DNA adducts and mutations.* J Natl Cancer Inst Monogr. 2000;27:75-93.

5. Terashima I, Suzuki N, Shibutani S. *Mutagenic properties of estrogen-*

quinone derived DNA adducts in Simian kidney cells. Biochemistry. 2001;40:166-72.

6. Zhao Z, Kosinska W, Khmelnitsky M, Cavalieri EL, Rogan EG, Chakravarti D, et al. *Mutagenic activity of 4-hydroxyestradiol, but not 2-hydroxyestradiol, in BB rat2 embryonic cells, and the mutational spectrum of 4-hydroxyestradiol.* Chem Res Toxicol. 2006;19:475-9.

7. Chakravarti D, Mailander PC, Li KM, Higginbotham S, Zhang HL, Gross ML, et al. *Evidence that a burst of DNA depurination in SENCAR mouse skin induces error-prone repair and forms mutations in the H-ras gene.* Oncogene. 2001;20:7945-53.

8. Li KM, Todorovic R, Devanesan P, Higginbotham S, Kofeler H, Ramanathan R, et al. *Metabolism and DNA binding studies of 4-hydroxyestradiol and estradiol-3,4-quinone in vitro and in female ACI rat mammary gland in vivo.* Carcinogenesis. 2004;25:289-97.

9. Cavalieri EL, Rogan EG. *A unifying mechanism in the iniation of cancer and other diseases by catechol quinones.* Ann NY Acad Sci. 2004;1028:247-57.

10. Russo J, Fernandez SV, Russo PA, Fernbaugh R, Sheriff FS, Lareef HM, et al. *17-Beta-estradiol induces transformation and tumorigenesis in human breast epithelial cells.* FASEB J. 2006;20:1622-34.

11. Russo J, Lareef MH, Balogh G, Guo S, Russo IH. *Estrogen and its metabolites are carcinogenic agents in human breast epithelial cells.* J Steroid Biochem Mol Biol. 2003;87:1-25.

12. Russo J, Lareef MH, Tahin Q, Hu Y-F, Slater C, Ao X, et al. *17b-estradiol is carcinogenic in human breast epithelial cells.* J Steroid Biochem Mol Biol. 2002;80:149-62.

13. Fernandez SV, Russo IH, Russo J. *Estradiol and its metabolites 4-hydroxyestradiol and 2-hydroxyestradiol induce mutations*

in humman breast epithelial cells. International Journal of Cancer. 2006;118:1862-8.

14. Liehr JG. *Is estradiol a genotoxic mutagenic carcinogen?* Endocrine Rev. 2000;21:40-54.

15. Li JJ, Li SA. *Estrogen carcinogenesis in Syrian hamster tissues: role of metabolism.* Fed Proc. 1987;46:1858-63.

16. Liehr JG, Ulubelen AA, Strobel HW. *Cytochrome P-450-mediated redox cycling of estrogens.* J Biol Chem. 1986;261:16865-70.

17. El-Bayoumy K, Ji BY, Upadhyaya P, Chae YH, Kurtzke C, Rivenson A, et al. *Lack of tumorigenicity of cholesterol epoxides and estrone-3,4-quinone in the rat mammary gland.* Cancer Res. 1996;56:1970-3.

18. Nutter LM, Ngo EO, Abul-Hajj YJ. *Characterization of DNA damage induced by 3,4-estrone-o-quinone in human cells.* J Biol Chem. 1991;25:16380-6.

19. Newbold RR, Liehr JG. *Induction of uterine adenocarcinoma in CD-1 mice by catechol estrogens.* Cancer Res. 2000;60:235-7.

20. Cavalieri EL, Stack DE, Devanesan PD, Todorvic R, Dwivedy I, Higginbotham S, et al. *Molecular origin of cancer: catechol estrogen-3,4-quinones as endogenous tumor initiators.* Proc Natl Acad Sci USA. 1997;94:10937-42.

21. Akanni A, Abul-Hajj YJ. *Estrogen-nucleic acid adducts: Dissection of the reaction of 3,4-estrone quinone and its radical anion and radical cation with deoxynucleosides and DNA.* Chem Res Toxicol. 1999;12:1247-53.

22. Yue W, Santen RJ, Wang JP, Li Y, Verderame MF, Bocchinfuso WP, et al. *Genotoxic metabolites of estradiol in breast: potential mechanism of estradiol induced carcinogenesis.* J Steroid Biochem Mol Biol. 2003;86:477-86.

23. Bransfield LA, Rennie A, Visvanathan K, Odwin SA, Kensler TW, Yager JD, et al. *Formation of two novel estrogen guanine adducts and HPLC/MS detection of 4-hydroxyestradiol-N7-guanine in human urine.* Chem Res Toxicol. 2008;21:1622-30.

24. Gaikwad NW, Yang L, Muti P, Meza JL, Pruthi S, Ingle JN, et al. *The molecular etiology of breast cancer: evidence from biomarkers of risk.* International Journal of Cancer. 2008;122:1949-57.

25. Embrechts J, Lemiere F, Van Dongen W, Esmans EL, Buytaert P, Van Marck E, et al. *Detection of estrogen DNA-adducts in human breast tumor tissue and healthy tissue by combined nano LC-nano ES tandem mass spectrometry.* J Am Soc Mass Spectrom. 2003;14:482-91.

26. Markushin Y, Zhong W, Cavalieri EL, Rogan EG, Small GJ, Yeung ES, et al. *Spectral characterization of catechol estrogen quinone (CEQ)-derived DNA adducts and their identification in human breast tissue extract.* Chem Res Toxicol. 2003;16:1107-17.

27. Zhang QA, R.L., Gross ML. *Estrogen carcinogenesis: specific identification of estrorgen-modified nucleobase in breast tissue from women.* Chem Res Toxicol. 2008;21:1509-13.

28. Belous AR, Hachey DL, Dawling S, Roodi N, Parl FF. *Cytochrome P450 1B1-mediated estrogen metabolism results in estrogen-deoxyribonucleoside adduct formation.* Cancer Res. 2007;67:812 - 7.

29. Saeed M, Rogan E, Fernandez SV, Sheriff F, Russo J, Cavalieri E. *Formation of depurinating N3Adenine and N7Guanine adducts by MCF-10F cells cultured in the presence of 4-hydroxyestradiol.* International Journal of Cancer. 2007;120:1821-4.

30. Rogan EG, Badawi AF, Devanesan PD, Meza J, Edney JA, West WW, et al. *Relative imbalances*

in estrogen metabolism and conjugation in breast tissue of women with carcinoma: potential biomarkers of susceptibility to cancer.* Carcinogenesis. 2003;24:697-702.

31. Bolton JL. *Quinoids, quinoid radicals, and phenoxyl radicals formed from estrogens and antiestrogens.* Toxicology. 2002;177:55-65.

32. Bolton JL, Yu L, Thatcher RJ. *Quinoids formed from estrogens and antiestrogens.* Methods Enzymol. 2004;378:110-23.

33. Kim SY, Suzuki N, Laxmi YRS, Rieger R, Shibutani S. *a-Hydroxylation of tamoxifen and toremifene by human and rat cytochrome p450 subfamily enzymes.* Chem Res Toxicol. 2003;16:1138-44.

34. Moorthy B, Sriram P, Pathak DN, Bodell WJ, Randerath K. *Tamoxifen metabolic activation: comparison of DNA adducts formed by microsomal and chemical activation of tamoxifen and 4-hydroxytamoxifen with DNA adducts formed in vivo.* Cancer Res. 1996;56:53-7.

35. Shibutani S, Ravindernath A, Suzuki N, Terashima I, Sugarman SM, Grollman AP, et al. *Identification of tamoxifen-DNA adducts in the endometrium of women treated with tamoxifen.* Carcinogenesis. 2000;21:1461-7.

36. Terashima I, Suzuki N, Shibutani S. *Mutagenic potential of a-(N2-deoxyguanosinyl)tamoxifen lesions, the major DNA adducts detected in endometrial tissues of patients treated with tamoxifen.* Cancer Res. 1999;59:2091-5.

37. Bernstein L, Deapen D, Cerhan JR, Schwartz SM, Liff J, McGann-Maloney E, et al. *Tamoxifen therapy for breast cancer and endometrial cancer risk.* J Natl Cancer Inst. 1999;91:1654-62.

2.1.4. ENDOGENOUS ESTROGENS
2.1.4.1. INTRODUCTION

Numerous epidemiological and experimental studies have implicated estrogens in the development of breast cancer (1-8). For example, a pooled analysis of nine prospective studies of endogenous sex hormones in 2,428 postmenopausal women (663 cases, 1765 controls) revealed a strong association of serum total and free E_2 concentrations with breast cancer risk (9). The relative risk of breast cancer for women whose free E_2 levels were in the top quintile was 2.58 compared with 1.00 for those women whose levels were in the bottom quintile. Since the nine studies employed different methods to measure E_2, there were considerable differences in the median E_2 values reported. In spite of this variability, the median serum E_2 concentrations in seven of the nine studies were higher in the case patients than in the control subjects. In Table 2.1.4.1., we summarized the median E_2 values for the nine studies as well as the corresponding cases/controls E_2 ratios, which ranged from 0.91 to 1.34, with an average of 1.126. These ratios appear rather narrow and are of unknown biological significance but, at the least, suggest that small differences in E_2 concentration acting over years are important for mammary carcinogenesis. Subsequent studies observed strong associations of E_2 levels with breast hyperplasia and *in situ* carcinoma (10, 11).

Several studies of postmenopausal women have examined not only E_2 and its fractions [free, bioavailable; (12)], but also E_1 and SHBG as well as precursor androgens (e.g., DHEA, androstenedione, testosterone). Since estrogens can increase SHBG hepatic synthesis and blood levels while androgens have the opposite effect, the complex interaction of all these compounds may be difficult to analyze (13). Nevertheless, the majority of studies showed increased postmenopausal breast cancer risk for women in the upper compared to women in the lower quintiles of all sex steroids examined (DHEA, androstenedione, testosterone, E_1) (9). In contrast, high SHBG levels were associated with reduced risk, consistent with the role of the protein to bind E_2 in the circulation and restrict its biological activity. Any contribution of androgens to risk appears to be largely through their role as substrates for estrogen production (14, 15). There was no association between postmenopausal breast cancer risk and progesterone levels (10). Progesterone, androgens, and SHBG are discussed separately in Section 2.2. A prospective case-control study of postmenopausal women nested within the Prostate, Lung, Colorectal, and Ovarian Cancer Screening Trial (PLCO; 277 breast cancer cases, 423 controls) examined serum concentrations of 15 estrogens and estrogen metabolites (16). The parent estrogens, E_2 and E_1, and their hydroxylation and methylation products at the C-2, C-4, and C-16 positions were measured by liquid chromatography-tandem mass spectrometry. An increased ratio of the 4-hydroxylation pathway catechols to their inactive methylated derivatives was associated with an increased risk of breast cancer (1.34, 95% CI 1.04-1.72). On the other hand, the ratio of the 2-hydroxylation pathway to parent estrogens (E_1/E_2) was associated with a decreased risk of breast cancer (0.66, 95% CI 0.51-0.87). Estrogen and progesterone measurements in premenopausal women are difficult to interpret because serum levels change with the menstrual cycle and the cycle length varies between women. Studies in premenopausal women have yielded inconsistent associations of breast cancer risk with circulating steroid hormone levels (17-19).

Endogenous estrogen exposure is determined by several reproductive factors including age at menarche and menopause, parity and age at first birth, as well as obesity and genetic factors. Ovarian hormones are produced during both phases of the menstrual cycle. In the follicular phase, progesterone levels are low and estrogen levels increase toward ovulation. In the luteal phase both progesterone and estrogen are elevated. Breast epithelial cell proliferation varies with the menstrual cycle and DNA synthesis is significantly higher in the luteal than the follicular phase (20-22). Since mitotically active cells are susceptible to mutagenic events, one would expect a higher probability of neoplastic transformation in breast tissue with prolonged exposure to ovarian hormones (23). Indeed, numerous epidemiological studies have shown that the larger the number of ovulatory cycles and the associated cumulative estrogen exposure during the reproductive years, the higher the risk of breast cancer (2, 3). The reproduction-related exposure tended to be associated with increased risk of ER- and/or PR-positive but not receptor-negative tumors (24, 25).

Table 2.1.4.1. Correlation of postmenopausal serum E_2 concentration by study and case-control status with breast cancer risk (9).

Study; Country	Median Estradiol (pmol/L)		
	Cases	Controls	Ratio
Columbia, Missouri; United States	55.1	51.4	1.07
Guernsey; United Kingdom	45.5	35.0	1.30
Nurses' Health Study: United States	29.4	25.7	1.14
NYU WHS; United States	134	101	1.33
ORDET; Italy	21.9	21.7	1.01
Rancho Bernardo; United States	36.7	40.4	0.908
RERF; Japan	63.1	64.5	0.978
SOF; United States	29.4	22.0	1.34
Washington County; United States	62.4	58.7	1.06
Average of all centers			1.126

NYU WHS = New York University Women's Health Study
ORDET = Study of Hormones and Diet in the Etiology of Breast Tumors
RERF = Radiation Effects Research Foundation
SOF = Study of Osteoporotic Fractures

References

1. Armstrong K, Eisen A, Weber B. *Assessing the risk of breast cancer.* New Engl J Med. 2000;342:564-71.

2. Bernstein L, Ross RK. *Endogenous hormones and breast cancer risk.* Epidemiol Rev. 1993;15:48-65.

3. Colditz GA. *Epidemiology and prevention of breast cancer.* Cancer Epidemiol Biomarkers Prev. 2005;14:768-72.

4. Henderson BE, Feigelson HS. *Hormonal carcinogenesis.* Carcinogenesis. 2000;21:427-33.

5. Kelsey JL, Berkowitz GS. *Breast cancer epidemiology.* Cancer Res. 1988;48:5615-23.

6. Kelsey JL, Gammon MD, John EM. *Reproductive and hormonal risk factors.* Epidemiol Rev. 1993;15:36-47.

7. Key TJ, Pike MC. *The role of oestrogens and progestagens in the epidemiology and prevention of breast cancer.* Eur J Cancer Clin Oncol. 1988;24:29-43.

8. Russo J, Russo IH. *Toward a physiological approach to breast cancer prevention.* Cancer Epidemiol Biomarkers Prev. 1994;3:353-64.

9. The Endogenous Hormones and Breast Cancer Collaborative Group. *Endogenous sex hormones and breast cancer in postmenopausal women: Reanalysis of nine prospective studies.* J Natl Cancer Inst. 2002;94:606-16.

10. Missmer SA, Eliassen AH, Barbieri RL, Hankinson SE. *Endogenous estrogen, androgen, and progesterone concentrations and breast cancer risk among postmenopausal women.* J Natl Cancer Inst. 2004;96:1856-65.

11. Schairer C, Hill D, Sturgeon SR, Fears T, Mies C, Ziegler RG, et al. *Serum concentrations of estrogens, sex hormone binding globulin, and androgens and risk of breast hyperplasia in postmenopausal women.* Cancer Epidemiol Biomarkers Prev. 2005;14:1660-5.

12. Sodergard R, Backstrom T, Shanbhag V, Carstensen H. *Calculation of free and bound fractions of testosterone and estradiol-17 beta to human plasma proteins at body temperature.* J Steroid Biochem. 1982;16:801-10.

13. Pasquali R, Vicennati V, Bertazzo D, Casimirri F, Pascal G, Tortelli O, et al. *Determinants of sex hormone-binding globulin blood concentrations in premenopausal and postmenopausal women with different estrogen status.* Virgilio-Menopause-Health Group. Metabolism. 1997;46:5-9.

14. Kaaks R, Rinaldi S, Key TJ, Berrino F, Peeters PH, Biessy C, et al. *Postmenopausal serum androgens, oestrogens and breast cancer risk: the European prospective investigation into cancer and nutrition.* Endocr Relat Cancer. 2005;12:1071-82.

15. Zeleniuch-Jacquotte A, Shore RE, Koenig KL, Akhmedkhanov A, Afanasyeva Y, Kato I, et al. *Postmenopausal levels of oestrogen, androgen, and SHBG and breast cancer: long-term results of a prospective study.* Br J Cancer. 2004;90:153-9.

16. Fuhrman BJ, Schairer C, Gail MH, Boyd-Morin J, Xu X, Sue LY, et al. *Estrogen metabolism and risk of breast cancer in postmenopausal women.* Journal of the National Cancer Institute. 2012;104:326-39.

17. Eliassen AH, Missmer SA, Tworoger SS, Spiegelman D, Barbieri RL, Dowsett M, et al. *Endogenous steroid hormone concentrations and risk of breast cancer among premenopausal women.* J Natl Cancer Inst. 2006;98:1406-15.

18. Kaaks R, Berrino F, Key T, Rinaldi S, Dossus L, Biessy C, et al. *Serum sex steroids in premenopausal women and breast cancer risk within the European Prospective Investigation into Cancer and Nutrition (EPIC).* J Natl Cancer Inst. 2005;97:755-65.

19. Sturgeon SR, Potischman N, Malone KE, Dorgan JF, Daling J, Schairer C, et al. *Serum levels of sex hormones and breast cancer risk in premenopausal women: a case-control study (USA).* Cancer Causes Control. 2004;15:45-53.

20. Ferguson DJ, Anderson TJ. *Morphological evaluation of cell turnover in relation to the menstrual cycle in the "resting" human breast.* Br J Cancer. 1981;44:177-81.

21. Masters JR, Drife JO, Scarisbrick JJ. *Cyclic Variation of DNA synthesis in human breast epithelium.* J Natl Cancer Inst. 1977;58:1263-5.

22. Meyer JS. *Cell proliferation in normal human breast ducts, fibroadenomas and other ductal hyperplasias measured by nuclear laeling with tritiated thymidine. Effects of menstrual phase, age, and oral contraceptive hormones.* Hum Pathol. 1977;8:67-81.

23. Pike MC, Spicer DV, Dahmoush L, Press MF. *Estrogens, progestogens, normal breast cell proliferation and breast cancer risk.* Epidemiol Rev. 1993;15:17-35.

24. Althuis MD, Fergenbaum JH, Garcia-Closas M, Brinton LA, Madigan MP, Sherman ME. *Etiology of hormone receptor-defined breast cancer: a systematic review of the literature.* Cancer Epidemiol Biomarkers Prev. 2004;13:1558-68.

25. Colditz GA, Rosner BA, Chen WY, Holmes MD, Hankinson SE. *Risk factors for breast cancer according to estrogen and progesterone receptor status.* J Natl Cancer Inst. 2004;96:218-28.

2.1.4.2. REPRODUCTIVE FACTORS
2.1.4.2.1. MENARCHE

Early age at menarche is a well established positive risk factor of breast cancer, i.e., the younger a woman's age at menarche, the higher her risk (1-3). For each two-year delay in onset of menstrual activity, risk is reduced by 10 – 20%. Thus, the relative risk is approximately 1.2 for women in whom menarche occurred before the age of 12 compared with women in whom it occurred at the age of 14 (4).

Late age tended to be associated with a reduction in ER+/PR+ but not receptor-negative tumors as well as basal-like but not luminal A tumor subtypes (5-7). The age at menarche is influenced by nutrition and genetic factors implicated in energy homeostasis and hormonal regulation (8). Prenatal exposures may also play a role, e.g., maternal cigarette smoking during pregnancy was associated with an earlier age at menarche of the daughter while physical activity during pregnancy was related to a delay of menarche in the offspring (9, 10).

The effect of age at menarche on breast cancer risk may be mediated simply by the prolonged exposure of breast epithelium to estrogens produced by regular ovulatory cycles (11). In addition, some studies reported that women with early menarche have higher estrogen levels than women with later menarche for several years after menarche (12). A longitudinal study of 200 schoolgirls showed that girls with early menarche establish regular ovulatory cycles more quickly and have higher serum E_2 levels than girls with a later age at menarche (13). Follow-up of 44 participants during a 13-year span from menarche into the third decade of life showed that women who had their menarche before age 12 maintained significantly higher serum E_2 and lower sex hormone-binding globulin (SHBG) concentrations during the follicular phase than women who had their menarche at age 13 or more (14). These data suggest that women with early menarche are subject to a high degree of estrogen stimulation at least until 30 years of age. A more recent study also observed an inverse association between age at menarche and E_2 (15). However, several other studies of pre- and postmenopausal women found no correlation between age at menarche and serum E_2 or SHBG levels (16-20). Ovarian hormone concentrations are highest in the luteal phase of the menstrual cycle. The length of the luteal phase remains relatively constant, whereas thelength of the follicular phase can vary significantly. Since women with irregular or longer cycles spend relatively less of their reproductive years in luteal phase, they might be expected to have a lower risk of breast cancer. However, there was no clear relation between breast cancer risk and irregular menstrual cycles or menstrual cycle length (21, 22).

The age at which a woman reaches maximum height is linked to the onset of puberty, i.e., a later age at attained height is a marker of having a later pubertal growth spurt. One study showed that women who reached their maximum height at ≤12 years of age had a 1.4-fold (95% CI 1.0 – 1.8) increased risk of breast cancer compared to women who reached their maximum height at ≥17 years of age (p_{trend} = 0.04) (23). The age of onset of puberty frequently differs between dizygotic and even monozygotic twins. In monozygotic twins in which breast cancer developed in both members of the pair, the twin with earlier puberty was much more likely to develop the disease first (odds ratio 5.4; 95% CI 2.0 – 14.5) (24).

2.1.4.2.2. PARITY

Nulliparity and a late age at first birth both increase the lifetime incidence of breast cancer, especially the risk of ER-positive tumors (5, 25, 26).

The risk of breast cancer among women who have their first child before the age of 20 is about half as high as that of women who have their first child after the age of 30 years (4, 27). Subsequent births, especially at an early age, further reduce the risk of breast cancer (28-30). The spacing of births is also related to risk, i.e., the closer the births are the lower the risk (25). Combining the information on timing of menarche and parity has shown that the length of the interval between age at menarche and age at first birth is positively associated with breast cancer risk (25, 26). Women who had an interval of ≥16 years between the ages of menarche and first birth had a 1.5-fold higher risk of breast cancer compared to women with ≤5 years between these ages (31).

While pregnancies clearly convey protection, there is epidemiological evidence of a transient increase in breast cancer risk after giving birth (32). Comparing uniparous with nulliparous women, the transient increase peaked five years following delivery (odds ratio 1.49; 95% CI 1.01 – 2.20) (33). The excess risk was most pronounced among women who were older at the time of their first delivery. Among women who were 35 years old at first delivery, the odds ratio five years after delivery was 1.26 (95% CI 1.10 – 1.44) compared to nulliparous women (32). Thus, there is a crossover in the effect of a first pregnancy around the age of 35, beyond which the transient increase in breast cancer risk overshadows the protection conveyed by pregnancies in general (34).

How can we explain the dual effect of pregnancy on breast cancer risk, i.e., a short-term increase in risk followed by a long-term risk reduction? A plausible biologic interpretation is that pregnancy increases the short-term risk of breast cancer by stimulating the growth of cells that have undergone the early stages of malignant transformation but that it confers long-time protection by inducing the differentiation of normal mammary stem cells (32). Animal experiments have shown that pregnancy results in the differentiation of the terminal end buds of the breast into secretory units, which are characterized by lower proliferative activity (35, 36). The differentiation has been postulated to shift a subset of epithelial cells susceptible to carcinogenesis, called stem cells 1, to stem cells 2, which are refractory to carcinogenesis (37). Both types of stem cells have characteristic gene expression profiles. Since breast tissue in a nulliparous woman is relatively undifferentiated compared with a parous woman, it is susceptible to carcinogenic insults, particularly between the ages at menarche and first birth. Other mechanisms may contribute to the protective effect of pregnancy, in particular hormonal changes in estrogen and insulin-like growth factor (IGF) levels (38). A comparison of plasma and urinary estrogen levels of nuns with those of their parous sisters showed that the latter had approximately 20% lower total and free plasma E_2 concentrations which was reflected in lower urinary estrogen levels (39). SHBG concentrations were higher in parous than in nulliparous premenopausal women, leading to a relatively lower level of free hormone (39-41). These results suggest that parity may provide a net overall benefit with respect to the endogenous estrogen profile which permanently reduces risk. However, subsequent studies found no difference in E_2 or SHBG levels between parous and nulliparous pre- or postmenopausal women (17, 19, 42). Other pregnancy-related hormones including prolactin and human chorionic gonadotropin may also play a role, albeit less well defined (43, 44).

If a full-term pregnancy protects against breast cancer by allowing complete differentiation of breast cells, an interrupted pregnancy might increase a woman's risk of breast cancer because of incomplete differentiation. Epidemiologic investigations of the association between abortion and subsequent risk of breast cancer have yielded inconsistent results in case-control studies (45, 46). However, case-control interview studies have the weakness of differential reporting of abortions and founder if patients with breast cancer are either more or less likely than other women to recall or report their history accurately (47). A large population study avoided this weakness by using data from the Danish National Registry of Induced Abortions, which is based on mandatory reporting of such abortions in Denmark (48). In this cohort of 1.5 million women (28.5 million person-years) there were 10,246 women with breast cancer. In the subgroup of 280,965 women (2.7 million person-years) with 370,715 induced abortions 1338 cases of breast cancer were diagnosed. Overall, induced abortion was not associated with an increased risk of breast cancer (relative risk 1.00; 95% CI 0.94 – 1.06).

2.1.4.2.3. LACTATION

A meta-analysis of 47 epidemiological studies established that breastfeeding had a protective effect against breast cancer (49).

The reduction in breast cancer risk was independent of parity, age at first delivery, and menopausal status. A multi-center study involving 5,878 cases and 8,216 controls determined a relative risk of 0.78 (95% CI 0.66 - 0.91) compared with women who were parous but had never breast-fed (50). The risk of breast cancer decreased progressively with increasing cumulative duration of lactation, i.e., the cumulative number of months of breastfeeding from all pregnancies (49, 51). Breastfeeding protects against both ER+/PR+ and ER-/PR- tumors, whereas parity and age at first birth only protect against ER+/PR+ cancers, suggesting different underlying mechanisms (5, 52, 53). Several mechanisms have been implicated in the protective effect of breastfeeding, e.g., delaying the re-establishment of ovulatory cycles, increased prolactin and decreased estrogen production, terminal differentiation of mammary epithelium, and elimination of carcinogens during lactation (54).

Mean E_2 and E_1 levels in nipple aspirates of breast fluid were lower in premenopausal parous women than in nulliparous women (55).

2.1.4.2.4. PRE- AND PERINATAL FACTORS

Pre- and perinatal factors may also influence the daughter's risk of developing breast cancer. In particular, high concentrations of intrauterine estrogen or other pregnancy hormones may influence the largely undifferentiated mammary gland perinatally leading to increased breast cancer risk in adulthood (34, 56). Consistent with this hypothesis is the observation that breast cancer risk was reduced in women whose mothers experienced preeclampsia/eclampsia, a condition associated with lower circulating levels of estrogen compared to those in healthy pregnant women (57-60). Several studies have examined the relation between birthweight and breast cancer because high birthweight is associated with high intrauterine exposure to estrogens, IGFs, and leptin (61-64). A nested case-control study within the Nurses' Health cohort ascertained birthweight information from the mothers of 582 nurses with invasive breast cancer and the mothers of 1569 nurses who did not have breast cancer and served as controls (65). The risk of breast cancer among women who weighed less than 2,500 g at birth was about half that of women who weighed 4,000 g or more at birth. A Danish study that included 91,601 women and 2,074 cases of breast cancer showed a relative risk of 1.17 (95% CI 1.02 – 1.33) for the highest birthweight group (median 4,000 g) compared to the lowest group (median 2,500 g) (66). A meta-analysis of these and other studies determined a relative breast cancer risk of 1.25 (95% CI 1.14 – 1.38) among premenopausal women with high birthweight (67-69). The meta-analysis showed no association of birthweight with breast cancer in postmenopausal women (69). Adjustment for gestational age, which is strongly correlated with birthweight did not change the results for pre- or postmenopausal women.

Studies of breast cancer risk in twins offer the unique opportunity to control for the role of the intrauterine environment and maternal factors (69). Monozygotic twins are genetically identical and share a single placenta, whereas dizygotic twins are genetically dissimilar and develop in two placentas. Due to the presence of two placentas, estrogen levels are typically higher in dizygotic twins than in single-placenta (i.e., monozygotic twin and singleton) pregnancies (70). A complicating factor is the lower birth weight of twins compared to singletons.

Overall, it appears that monozygotic twins have the highest risk (standardized incidence ratio [SIR] 4.4, 95% CI 3.6 – 5.6) compared to dizygotic twins who have a higher risk (SIR 1.7; 95% 1.1 – 2.6) than singleton pregnancies (71, 72). One large study observed that twins were at an increased risk of developing breast cancer compared with singletons (relative risk 1.62; 95% CI 1.0 - 2.7), particularly women with a twin brother, which is a proxy for being dizygotic (relative risk 2.06) (73).

However, not all studies yielded consistent results and the underlying mechanisms may not only involve intrauterine factors but also hormonal exposure later in life, i.e., at puberty (24, 57, 59, 74-76).

Finally, breast cancer risk may be affected by hormonal treatment during pregnancy. Women who took the synthetic estrogen diethylstilbestrol (DES) during pregnancy to prevent miscarriage had a significantly higher breast cancer risk (relative risk 1.35; CI 1.05 - 1.74) than women who did not (77). Later studies provided definitive support that prenatal DES exposure was associated with an increased risk of breast cancer after age 40 years (78).

Animal experiments have shown that environmental hormone-mimicking or endocrine-disrupting compounds can influence rodent mammary gland development and lead to an increase in the incidence of mammary tumors (79).

2.1.4.2.5. MENOPAUSE

The menopausal status is defined by amenorrhea for at least 12 months with E_2 levels <40 pg/mL (150 pmol/L) and FSH levels >15 IU/L or a history of bilateral ovariectomy (80). Numerous epidemiological studies have established that late age at menopause is associated with an increased risk of breast cancer (81-83). For every one-year increase in age at menopause, the risk of breast cancer increases by approximately 3% (84-86). Bilateral oophorectomy before natural menopause brings about a reduction in breast cancer risk. Bilateral oophorectomy before age 40 conferred the strongest protective effect, decreasing the lifetime risk to 45% in comparison with having a natural menopause at age 50 (2, 22). These results are consistent with the hypothesis that reduced estrogen production and exposure would result in decreased breast cancer risk.

The increased risks associated with early age at menarche and late age at menopause indicate that the larger the number of ovulatory cycles and the associated cumulative estrogen exposure during the reproductive years, the higher the risk of breast cancer (87). A large European study examined the risk of breast cancer in postmenopausal women in relation to their lifetime cumulative number of menstrual cycles (88). Using ≤415 cycles as reference, the hazard ratio for 416 – 453, 454 – 490, and ≥491 cycles was 1.88 (95% CI 1.14 – 3.12), 1.74 (95% CI 1.05 – 2.87), and 1.80 (95% CI 1.09 – 2.96), respectively.

There are two questions concerning menopause and circulating estrogen levels that have received inconsistent answers. Does the age when a woman enters menopause influence her postmenopausal hormone levels? Most studies found no association between age at menopause and serum E_2 or SHBG concentrations in postmenopausal women (15, 17, 19, 40, 89). Does the time since menopause influence serum estrogen levels, i.e., do hormone levels decline with postmenopausal age? Some studies have shown a modest decline in serum E_2 concentrations with increasing time since menopause, whereas others did not find any significant change in E_2 levels with increasing post-menopausal age (15, 17, 40).

The end of the reproductive period is associated with a gradual cessation of ovarian function with fluctuating estrogen and progesterone levels. The fluctuation and five- to tenfold decrease in E_2 levels causes menopausal symptoms which vary considerably between women. The spectrum of manifestations includes vasomotor symptoms (hot flushes, night sweats, migraine), urogenital atrophy (vaginal dryness), and psychological changes (insomnia, mood changes). Two studies observed that women with hot flushes and migraine had a reduced risk of breast cancer (90, 91). Since both symptoms are associated with falling E_2 levels they may serve as surrogate markers for perimeno-pausal hormonal changes relevant for breast cancer risk.

References

1. Armstrong K, Eisen A, Weber B. *Assessing the risk of breast cancer.* New Engl J Med. 2000;342:564-71.

2. Brinton LA, Schairer C, Hoover RN, Fraumeni JF. *Menstrual factors and risk of breast cancer.* Cancer Invest. 1988;6:245-54.

3. Kampert JB, Whittemore AS, Paffenbarger RS. *Combined effect of childbearing, menstrual events, and body size on age-specific breast cancer risk.* Am J Epidemiol. 1988;128:962-79.

4. Gail MH, Brinton LA, Byar DP, Corle DK, Green SB, Schairer C, et al. *Projecting individualized probabilities of developing breast cancer for white females who are being examined annually.* J Natl Cancer Inst. 1989;81:1879-86.

5. Althuis MD, Fergenbaum JH, Garcia-Closas M, Brinton LA, Madigan MP, Sherman ME. *Etiology of hormone receptor-defined breast cancer: a systematic review of the literature.* Cancer Epidemiol Biomarkers Prev. 2004;13:1558-68.

6. Cotterchio M, Kreiger N, Theis B, Sloan M, Bahl S. *Hormonal factors and the risk of breast cancer according to estrogen- and progesterone-receptor subgroup.* Cancer Epidemiol Biomarkers Prev. 2003;12:1053-60.

7. Yang XR, Sherman ME, Rimm DL, Lissowska J, Brinton LA, Peplonska B, et al. *Differences in risk factors for breast cancer molecular subtypes in a population-based study.* Cancer Epidemiol Biomarkers Prev. 2007;16:439-43.

8. Elks CE, Perry JR, Sulem P, Chasman DI, Franceschini N, He C, et al. *Thirty new loci for age at menarche identified by a meta-analysis of genome-wide association studies.* Nat Genet. 2010;42:1077-85.

9. Colbert LH, Graubard BI, Michels KB, Willett WC, Forman MR. *Physical activity during pregnancy and age at menarche of the daughter.* Cancer Epidemiol Biomarkers Prev. 2008;17:2656-62.

10. Windham GC, Bottomley C, Birner C, Fenster L. *Age at menarche in relation to maternal use of tobacco, alcohol, coffee, and tea during pregnancy.* Am J Epidemiol. 2004;159:862-71.

11. MacMahon B, Trichopoulos D, Brown J, Andersen AP, Aoki K, Cole P, et al. *Age at menarche, probability of ovulation and breast cancer risk.* Int J Cancer. 1982;29:13-6.

12. MacMahon B, Trichopoulos D, Brown J, Andersen AP, Cole P, deWaard F, et al. *Age at menarche, urine estrogens and breast cancer risk.* Int J Cancer. 1982;30:427-31.

13. Apter D, Vihko R. *Early menarche, a risk factor for breast cancer, indicates early onset of ovulatory cycles.* J Clin Endocrinol Metab. 1983;57:82-6.

14. Apter D, Reinila M, Vihko R. *Some endocrine characteristics of early menarche, a risk factor for breast cancer, are preserved into adulthood.* Int J Cancer. 1989;44:783-7.

15. Madigan MP, Troisi R, Potischman N, Dorgan JF, Brinton LA, Hoover RN. *Serum hormone levels in relation to reproductive and lifestyle factors in postmenopausal women.* Cancer Causes Control. 1998;9:199-207.

16. Bernstein L, Pike MC, Ross RK, Henderson BE. *Age at menarche and estrogen concentrations of adult women.* Cancer Causes Control. 1991;2:221-5.

17. Lamar CA, Dorgan JF, Longcope C, Stanczyk FZ, Falk RT, Stephenson HE, Jr. *Serum sex hormones and breast cancer risk factors in postmenopausal women.* Cancer Epidemiol Biomarkers Prev. 2003;12:380-3.

18. Moore JW, Key TJ, Wang DY, Bulbrook RD, Hayward JL, Takatani O. *Blood concentrations of estradiol and sex hormone-binding globulin in relation to age at menarche in premenopausal British and Japanese women.* Breast Cancer Res Treat. 1991;18 Suppl:S47-S50.

19. Verkasalo PK, Thomas HV, Appleby PN, Davey GK, Key TJ. *Circulating levels of sex hormones and their relation to risk factors for breast cancer: a cross-sectional study in 1092 pre- and postmenopausal women (United Kingdom).* Cancer Causes Control. 2001;12:47-59.

20. Westhoff C, Gentile G, Lee J, Zacur H, Helbig D. *Predictors of ovarian steroid secretion in reproductive-age women.* Am J Epidemiol. 1996;144:381-8.

21. Terry KL, Willett WC, Rich-Edwards JW, Hunter DJ, Michels KB. *Menstrual cycle characteristics and incidence of premenopausal breast cancer.* Cancer Epidemiol Biomarkers Prev. 2005;14:1509-13.

22. Titus-Ernstoff L, Longnecker MP, Newcomb PA, Dain B, Greenberg ER, Mittendorf R, et al. *Menstrual factors in relation to breast cancer risk.* Cancer Epidemiol Biomarkers Prev. 1998;7:783-9.

23. Li CI, Littman AJ, White E. *Relationship between age maximum height is attained, age at menarche, and age at first full-term birth and breast cancer risk.* Cancer Epidemiol Biomarkers Prev. 2007;16:2144-9.

24. Hamilton AS, Mack TM. *Puberty and genetic susceptibility to breast cancer in a case-control study in twins.* N Engl J Med. 2003;348:2313-22.

25. Colditz GA. *Epidemiology and prevention of breast cancer.* Cancer Epidemiol Biomarkers Prev. 2005;14:768-72.

26. Kelsey JL, Gammon MD, John EM. *Reproductive and hormonal risk factors.* Epidemiol Rev. 1993;15:36-47.

27. MacMahon B, Cole P, Lin TM, Lowe CR, Mirra AP, Ravnihar B, et al. *Age at first birth and breast cancer risk.* Bull World Health Org. 1970;43:209-21.

28. Albrektsen G, Heuch I, Tretli S, Kvale G. *Breast cancer incidence before age 55 in relation to parity and age at first and last births: a prospective study of one million Norwegian women.* Epidemiology. 1994;5:604-11.

29. Rosner B, Colditz GA, Willett WC. *Reproductive risk factors in a prospective study of breast cancer: the Nurses' Health Study.* Am J Epidemiol. 1994;139:819-35.

30. Trichopoulos D, Hsieh CC, MacMahon B, Lin TM, Lowe CR, Mirra AP, et al. *Age at any birth and breast cancer risk.* Int J Cancer. 1983;31:701-4.

31. Li CI, Malone KE, Daling JR, Potter JD, Bernstein L, Marchbanks PA, et al. *Timing of menarche and first full-term birth in relation to breast cancer risk.* Am J Epidemiol. 2008;167:230-9.

32. Lambe M, Hsieh C, Trichopoulos D, Ekbom A, Pavia M, Adami HO. *Transient increase in the risk of breast cancer after giving birth.* N Engl J Med. 1994;331:5-9.

33. Liu Q, Wuu J, Lambe M, Hsieh SF, Ekbom A, Hsieh CC. *Transient increase in breast cancer risk after giving birth: postpartum period with the highest risk (Sweden).* Cancer Causes Control. 2002;13:299-305.

34. Trichopoulos D, Lagiou P, Adami HO. *Towards an integrated model for breast cancer etiology: the crucial role of the number of mammary tissue-specific stem cells.* Breast Cancer Res. 2005;7:13-7.

35. Russo J, Hu YF, Yang X, Russo IH. *Developmental, cellular, and molecular basis of human breast cancer.* J Natl Cancer Inst Monogr. 2000:17-37.

36. Russo J, Romero AL, Russo IH. *Architectural pattern of the normal cancerous breast under the influence of parity.* Cancer Epidemiol Biomarkers Prev. 1994;3:219-24.

37. Russo J, Moral R, Balogh GA, Mailo D, Russo IH. *The protective role of pregnancy in breast cancer.* Breast Cancer Res. 2005;7:131-42.

38. Russo J, Hu YF, Silva ID, Russo IH. *Cancer risk related to mammary gland structure and development.* Microsc Res Tech. 2001;52:204-23.

39. Bernstein L, Pike MC, Ross RK, Judd HL, Brown JB, Henderson BE. *Estrogen and sex hormone-binding globulin levels in nulliparous and parous women.* J Natl Cancer Inst. 1985;74:741-5.

40. Chubak J, Tworoger SS, Yasui Y, Ulrich CM, Stanczyk FZ, McTiernan A. *Associations between reproductive and menstrual factors and postmenopausal sex hormone concentrations.* Cancer Epidemiol Biomarkers Prev. 2004;13:1296-301.

41. Moore JW, Key TJ, Bulbrook RD, Clark GM, Allen DS, Wang DY, et al. *Sex hormone binding globulin and risk factors for breast cancer in a population of normal women who had never used exogenous sex hormones.* Br J Cancer. 1987;56:661-6.

42. Hankinson SE, Colditz GA, Hunter DJ, Manson JE, Willett WC, Stampfer MJ, et al. *Reproductive factors and family history of breast cancer in relation to plasma estrogen and prolactin levels in postmenopausal women in the Nurses' Health Study (United States).* Cancer Causes Control. 1995;6:217-24.

43. Musey VC, Collins DC, Musey PI, Martino-Saltzman D, Preedy JR. *Long-term effect of a first pregnancy on the secretion of prolactin.* New Engl J Med. 1987;316:229-34.

44. Toniolo P, Grankvist K, Wulff M, Chen T, Johansson R, Schock H, et al. *Human chorionic gonadotropin in pregnancy and maternal risk of breast cancer.* Cancer Res. 2010;70:6779-86.

45. Daling JR, Malone KE, Voigt LF, White E, Weiss NS. *Risk of breast cancer among young women: relationship to induced abortion.* J Natl Cancer Inst. 1994;86:1584-92.

46. Newcomb PA, Storer BE, Longnecker MP, Mittendorf R, Greenberg ER, Willett WC. *Pregnancy termination in relation to risk of breast cancer.* JAMA. 1996;275:283-7.

47. Hartge P. *Abortion, breast cancer, and epidemiology.* N Engl J Med. 1997;336:127-8.

48. Melbye M, Wohlfahrt J, Olsen JH, Frisch M, Westergaard T, Helweg-Larsen K, et al. *Induced abortion and the risk of breast cancer.* N Engl J Med. 1997;336:81-5.

49. Beral V. *Breast cancer and breastfeeding: collaborative reanalysis of individual data from 47 epidemiological studies in 30 countries, including 50302 women with breast cancer and 96973 women without the disease.* Lancet. 2002;360:187-95.

50. Newcomb PA, Storer BE, Longnecker MP, Mittendorf R, Greenberg ER, Clapp RW, et al. *Lactation and a reduced risk of premenopausal breast cancer.* New Engl J Med. 1994;330:81-7.

51. Layde PM, Webster LA, Baughman AL, Wingo PA, Rubin GL, Ory HW. *The independent associations of parity, age at first full term pregnancy, and duration of breastfeeding with the risk of breast cancer.* Cancer and Steroid Hormone Study Group. J Clin Epidemiol. 1989;42:963-73.

52. Lord SJ, Bernstein L, Johnson KA, Malone KE, McDonald JA, Marchbanks PA, et al. *Breast cancer risk and hormone receptor status in older women by parity, age of first birth, and breastfeeding: a case-control study.* Cancer Epidemiol Biomarkers Prev. 2008;17:1723-30.

53. Ma H, Wang Y, Sullivan-Halley J, Weiss L, Burkman RT, Simon MS, et al. *Breast cancer receptor status: do results from a centralized pathology laboratory agree with SEER registry reports?* Cancer Epidemiol Biomarkers Prev. 2009;18:2214-20.

54. Kelsey JL, John EM. *Lactation and the risk of breast cancer.* New Engl J Med. 1994;330:136-7.

55. Petrakis NL, Wrensch MR, Ernster VL, Miike R, Murai J, Simberg N, et al. *Influence of pregnancy and lactation on serum and breast fluid estrogen levels: implications for breast cancer risk.* Int J Cancer. 1987;40:587-91.

56. Trichopoulos D. *Hypothesis: does breast cancer originate in utero?* Lancet. 1990;335:939-40.

57. Ekbom A, Hsieh C, Lipworth L, Adami H, Trichopoulos D. *Intrauterine environment and breast cancer risk in women: a population-based study.* J Natl Cancer Inst. 1997;88:71-6.

58. Ekbom A, Trichopoulos D, Adami H, Hsieh C, Lan S. *Evidence of prenatal influences on breast cancer risk.* Lancet. 1992;340:1015-8.

59. Potischman N, Troisi R. *In-utero and early life exposures in relation to risk of breast cancer.* Cancer Causes Control. 1999;10:561-73.

60. Rosing U, Carlstrom K. *Serum levels of unconjugated and total oestrogens and dehydroepiandrosterone, progesterone and urinary oestriol excretion in pre-eclampsia.* Gynecol Obstet Invest. 1984;18:199-205.

61. Hills FA, English J, Chard T. *Circulating levels of IGF-I and IGF-binding protein-1 throughout pregnancy: relation to birthweight and maternal weight.* J Endocrinol. 1996;148:303-9.

62. Petridou E, Panagiotopoulou K, Katsouyanni K, Spanos E, Trichopoulos D. *Tobacco smoking, pregnancy estrogens, and birth weight.* Epidemiology. 1990;1:247-50.

63. Reece EA, Wiznitzer A, Le E, Homko CJ, Behrman H, Spencer EM. *The relation between human fetal growth and fetal blood levels of insulin-like growth factors I and II, their binding proteins, and receptors.* Obstet Gynecol. 1994;84:88-95.

64. Vatten LJ, Nilsen ST, Odegard RA, Romundstad PR, Austgulen R. *Insulin-like growth factor I and leptin in umbilical cord plasma and infant birth size at term.* Pediatrics. 2002;109:1131-5.

65. Michels KB, Trichopoulos D, Robins JM, Rosner BA, Manson JE, Hunter DJ, et al. *Birthweight as a risk factor for breast cancer.* Lancet. 1996;348:1542-6.

66. Ahlgren M, Melbye M, Wohlfahrt J, Sorensen TI. *Growth patterns and the risk of breast cancer in women.* N Engl J Med. 2004;351:1619-26.

67. dos Santos Silva I, De Stavola BL, Hardy RJ, Kuh DJ, McCormack VA, Wadsworth ME. *Is the association of birth weight with premenopausal breast cancer risk mediated through childhood growth?* Br J Cancer. 2004;91:519-24.

68. McCormack VA, dos Santos Silva I, Koupil I, Leon DA, Lithell HO. *Birth characteristics and adult cancer incidence: Swedish cohort of over 11,000 men and women.* Int J Cancer. 2005;115:611-7.

69. Michels KB, Xue F. *Role of birthweight in the etiology of breast cancer.* Int J Cancer. 2006;119:2007-25.

70. Thomas HV, Murphy MF, Key TJ, Fentiman IS, Allen DS, Kinlen LJ. *Pregnancy and menstrual hormone levels in mothers of twins compared to mothers of singletons.* Ann Hum Biol. 1998;25:69-75.

71. Mack TM, Hamilton AS, Press MF, Diep A, Rappaport EB. *Heritable breast cancer in twins.* Br J Cancer. 2002;87:294-300.

72. Peto J, Mack TM. *High constant incidence in twins and other relatives of women with breast cancer.* Nat Genet. 2000;26:411-4.

73. Weiss HA, Potischman NA, Brinton LA, D. B, Coates RJ, Gammon MD, et al. *Prenatal and perinatal risk factors for breast cancer in young women.* Epidemiology. 1997;8:181-7.

74. Braun MM, Ahlbom A, Floderus B, Brinton LA, Hoover RN. *Effect of twinship on incidence of cancer of the testis, breast, and other sites (Sweden).* Cancer Causes Control. 1995;6:519-24.

75. Hsieh CC, Lan SJ, Ekbom A, Petridou E, Adami HO, Trichopoulos D. *Twin membership and breast cancer.* Am J Epidemiol. 1992;136:1321-6.

76. Hubinette A, Lichtenstein P, Ekbom A, Cnattingius S. *Birth characteristics and breast cancer risk: a study among like-sexed twins.* Int J Cancer. 2001;91:248-51.

77. Colton T, Greenberg ER, Noller K, Resseguie L, Van Bennekom C, Heeren T, et al. *Breast cancer in mothers prescribed diethylstilbestrol in pregnancy.* J Am Med Assoc. 1993;269:2096-100.

78. Palmer JR, Wise LA, Hatch EE, Troisi R, Titus-Ernstoff L, Strohsnitter W, et al. *Prenatal diethylstilbestrol exposure and risk of breast cancer.* Cancer Epidemiol Biomarkers Prev. 2006;15:1509-14.

79. Fenton SE. *Endocrine-disrupting compounds and mammary gland development: early exposure and later life consequences.* Endocrinology. 2006;147:S18-24.

80. Pasquali R, Vicennati V, Bertazzo D, Casimirri F, Pascal G, Tortelli O, et al. *Determinants of sex hormone-binding globulin blood concentrations in premenopausal and postmenopausal women with different estrogen status.* Virgilio-Menopause-Health Group. Metabolism. 1997;46:5-9.

81. Hsieh CC, Trichopoulos D, Katsouyanni K, Yuasa S. *Age at menarche, age at menopause, height and obesity as risk factors for breast cancer: associations and interactions in an international case-control study.* Int J Cancer. 1990;46:796-800.

82. Lilienfeld AM. *The relationship of cancer of the female breast to artificial menopause and marital status.* Cancer. 1956;9:927-34.

83. Trichopoulos D, MacMahon B, Cole P. *Menopause and breast cancer risk.* J Natl Cancer Inst. 1972;48:605-13.

84. Collaborative Group on Hormonal Factors in Breast Cancer. *Breast cancer and hormone replacement therapy: collaborative reanalysis of data from 51 epidemiological studies of 52 705 women with breast cancer and 108 411 women without breast cancer.* Lancet. 1997;350:1047-59.

85. Pike MC, Krailo MD, Henderson BE, Casagrande JT, Hoel DG. *'Hormonal' risk factors, 'breast tissue age' and the age-incidence of breast cancer.* Nature. 1983;303:767-70.

86. Rosner B, Colditz GA. *Nurses' Health Study: log-incidence mathematical model of breast cancer incidence.* J Natl Cancer Inst. 1996;88:359-64.

87. Henderson BE, Ross RK, Judd HL, Drailo MD, Pike MC. *Do regular ovulatory cycles increase breast cancer risk?* Cancer. 1985;56:1206-8.

88. Chavez-MacGregor M, Elias SG, Onland-Moret NC, van der Schouw YT, Van Gils CH, Monninkhof E, et al. *Postmenopausal breast cancer risk and cumulative number of menstrual cycles.* Cancer Epidemiol Biomarkers Prev. 2005;14:799-804.

89. Meldrum DR, Davidson BJ, Tataryn IV, Judd HL. *Changes in circulating steroids with aging in postmenopausal women.* Obstet Gynecol. 1981;57:624-8.

90. Huang Y, Malone KE, Cushing-Haugen KL, Daling JR, Li CI. *Relationship between menopausal symptoms and risk of postmenopausal breast cancer.* Cancer Epidemiol Biomarkers Prev. 2011;20:379-88.

91. Mathes RW, Malone KE, Daling JR, Davis S, Lucas SM, Porter PL, et al. *Migraine in postmenopausal women and the risk of invasive breast cancer.* Cancer Epidemiol Biomarkers Prev. 2008;17:3116-22.

2.1.4.3. OBESITY AND BONE MASS
2.1.4.3.1. OBESITY AND BODY MASS INDEX (BMI)

Obesity is defined as an excess of body fat. The percentage of body fat can be estimated in several ways including the use of calipers to measure the thickness of subcutaneous fat in multiple places on the body and bioelectric impedance analysis, which uses the resistance of electrical flow through the body. A related measure of obesity is the body mass index (BMI), which is defined as weight/height^2 and measured in kg/m^2. In 1999 – 2004, 62% of women in the US were overweight or obese, with a BMI >25 kg/m^2 or >30 kg/m^2, respectively (1). BMI reflects both lean body mass and adipose tissue, whereas weight gain throughout adult life reflects primarily the accumulation of peripheral adipose tissue. Many studies have shown a J-shaped relationship between BMI and mortality from cardiovascular disease and certain cancers including breast cancer (2, 3). In white adults, overweight and obesity and possible underweight are associated with increased all-cause mortality, which is generally lowest with a BMI of 20.0 to 24.9.

Numerous studies have demonstrated that overweight and obese postmenopausal women have an increased risk of breast cancer compared to normal-weight, age-matched postmenopausal women (4-7). One study reported an odds ratio of 1.27 (CI 1.11 – 1.45) for the upper quintile group of height relative to the lowest and a corresponding odds ratio of 1.57 (CI 1.37 – 1.79) for weight (8). The relative risk in proportion to BMI increased by 3.1% per 1 kg/m^2 increase (9).

The elevated risk has been explained by higher levels of circulating estrogens secondary to increased conversion of androgen to estrogen in adipose tissue and a higher proportion of bioavailable estrogen due to lower levels of SHBG in obese women (10-14). A detailed analysis of serum estrogen fractions in relation to BMI showed a doubling of total and free E_2 across BMI tertiles from 5.1 to 10.5 pg/ml and 0.07 to 0.16 pg/ml, respectively, and a near doubling of E_1 and E_1 sulfate from 27.1 to 40.1 pg/mL and 271 to 458 pg/mL, respectively (15). The positive and negative correlations of serum estrogen and SHBG levels, respectively, with BMI support the hypothesis that the increased breast cancer risk of overweight postmenopausal women is mediated through elevated endogenous estrogens (16, 17).

Data collected by the Endogenous Hormones and Breast Cancer Collaborative Group were examined to determine whether estrogen levels could explain the association of BMI with breast cancer risk (18). Adjusting for free E_2 reduced the relative risk for breast cancer associated with a 5 kg/m^2 increase in BMI from 1.19 (95% CI 1.04 – 1.34) to 1.02 (95% CI 0.89 – 1.17). The increased risk was also substantially reduced after adjusting for total E_2, E_1, and E_1 sulfate whereas adjustment for androgens (testosterone,

androstenedione, DHEA, DHEA sulfate) had little effect on risk. In summary, the increase in breast cancer risk with increasing BMI in postmenopausal women is largely the result of the increase in estrogens, particularly free E_2.

An analysis of data from 155,723 postmenopausal women enrolled in the Women's Health initiative showed that women in the highest versus lowest BMI quartile had a 1.39-fold (95% CI 1.22 – 1.58) increased risk of ER-positive breast cancer (19).

These findings are consistent with other studies showing an association with ER-positive but not ER-negative tumors, which support the hypothesis of an estrogen-mediated effect of body fat on postmenopausal breast cancer risk (20, 21). Since obesity has been associated with increased risk of postmenopausal breast cancer due to increased production of endogenous estrogens in peripheral fat tissue, it is important to determine whether exogenous estrogen in form of menopausal estrogen therapy has a potentiating effect on breast cancer risk.

The majority of studies shows that among HRT users the estrogenic effect of obesity is imperceptible and does not further increase the risk of breast cancer already associated with HRT. In other words, both lean and heavy women have high levels of circulating estrogens by virtue of their HRT (4, 22-25). The increased breast cancer risk associated with adult weight gain and obesity was limited to postmenopausal women who never used HRT.

Since BMI does not differentiate between fat and lean body mass, dual-energy X-ray absorptiometry was used to precisely measure adiposity but did not provide better correlation with serum estrogen levels than BMI (26). The association between body mass and breast cancer risk is not as clear in premenopausal women.

A systematic review and meta-analysis of prospective observational studies from North America, Europe, Australia, and the Asia-Pacific region showed a positive association between increased BMI and premenopausal breast cancer in Asia-Pacific populations, but an inverse association in the other regions (27). Several studies reported a reversal in risk compared to postmeno-pausal women, i.e., obese premenopausal women have a significantly lower risk than their thin counterparts (28, 29). However, other studies failed to find such a correlation (7, 8). A meta-analysis of data from 23 studies examined the association of BMI and incidence of premenopausal breast cancer (30). The odds ratios from four cohort and 19 case-control studies were 0.70 (CI 0.54 – 0.91) and 0.88 (CI 0.76 – 1.02), respectively.

Overall, the data support a modest inverse association between obesity and premenopausal breast cancer risk which may be explained by two mechanisms. One possibility is that obese premenopausal women may exhibit a greater degree of anovulation resulting in lower levels of both estrogen and progesterone, lower breast cell division rates and, consequently, lower breast cancer risk (31, 32). Another possibility is that E_2 may be sequestered in adipose tissue or eliminated by the liver due to the larger albumin-bound pool of E_2 in obese premenopausal women (33).

An analysis of serum samples collected in the follicular phase of the menstrual cycles revealed a decrease in total and free E_2 across BMI tertiles from 137 to 76 pg/mL and 1.9 to 1.2 pg/mL, respectively (15). The SHBG concentration also decreased as BMI increased. Other studies also observed lower serum E_2 values in obese than non-obese premenopausal women (34, 35). Thus, differences in circulating estrogen concentrations may explain the opposite roles obesity plays as risk factor for breast cancer in pre- and postmenopausal women.

Another potential mechanism linking excess weight and breast cancer risk involves insulin and insulin-like growth factors (36, 37). Chronic hyperinsulinemia decreases concentrations of IGF binding proteins, which leads to an increase in free IGF-1 to stimulate tumor formation via enhanced mitogenesis and anti-apoptosis (38).

Levels of circulating total IGF-1 were associated with an increased risk of premenopausal but not postmenopausal breast cancer (38). Obesity is a risk factor not only for postmenopausal breast cancer but also for type 2 diabetes.

A double-transgenic mouse model of type 2 diabetes with a nonfunctional IGF-I receptor and a PyVmT oncogene displayed an accelerated development of ductal hyperplasia (39). However, an analysis of 31 common genetic markers for type 2 diabetes and obesity showed no association with breast cancer (40). The estrogen and IGF mechanisms may also interact since chronic hyperinsulinemia reduces serum SHBG and thereby increases bioavailable estrogen (41-43). A third mechanism by which obesity is linked to the increased incidence of estrogen-dependent breast cancer in postmenopausal women involves the adipokines leptin and adiponectin (44). The altered adipokine milieu associated with obesity leads to down-regulation of AMP-activated protein kinase (AMPK). One of the actions of AMPK is the phosphorylation and inhibition of cAMP-responsive element binding protein (CREB)-regulated transcription coactivator 2 (CRTC2), which regulates aromatase promoter II activity (45). Since AMPK also inhibits the proliferation of both malignant and nonmalignant cells (46), the obesity-induced decrease of AMPK may lead to increased expression of aromatase and E_2 production as well as increased cell proliferation.

2.1.4.3.2. MAMMOGRAPHIC DENSITY

The breast is composed of fat, connective tissue, and epithelial cells, which form ducts and glands. Breast density is determined as percentage of epithelial and connective tissues relative to fat (47). High percent density was strongly related to younger age, lower BMI, nulliparity, late age at first delivery, and pre/perimenopausal status (48).

A meta-analysis of 42 studies showed that mammographic density is one of the strongest risk factors for breast cancer (49). For percentage density measured among prediagnostic mammograms, combined relative risks of incident breast cancer in the general population were 1.79 (95% CI 1.48 - 2.16), 2.11 (95% CI 1.70 - 2.63), 2.92 (95% CI 2.49 - 3.42), and 4.64 (95% CI 3.64 - 5.91) for categories 5 - 24%, 25 - 49%, 50 - 74%, and ≥75% relative to <5%. Thus, compared with women with no visible breast density, women with a density of ≥75% had an almost fivefold increased risk of breast cancer. No differences were observed by age or menopausal status at mammography or by ethnicity. Paradoxically, breast density declines gradually with age, whereas breast cancer risk increases (47).

The lower epithelial proportion in postmenopausal women reflects the involution of lobules after menopause (50). It is unresolved whether technical advances, such as measuring the volume of mammographic density instead of two-dimensional estimation of percent dense area, will improve risk prediction (48, 51).

Breast cancer has a slight predilection for the left breast, which tends to be the larger breast in many women (52, 53). In general, breast size is not a risk factor for breast cancer because large breast size mostly reflects adipose tissue. Only in a subgroup of women who were especially lean before their first pregnancy has breast size been positively associated with postmenopausal breast cancer risk (29, 54). It is likely that breast size is a more reliable proxy of mammary gland mass in thin women. Women who had undergone surgical reduction of their breasts subsequently had reduced breast cancer risk (55, 56).

The mechanisms through which breast density affects breast cancer risk are not fully understood (49). Since breast density reflects the percentage of epithelial cells and breast cancer arises in epithelial cells, the mammary gland mass is a plausible risk factor for breast cancer. The larger the mammary gland mass the greater the number of epithelial cells, including stem cells that are at risk of carcinogenesis and/or an increased rate of proliferation (57, 58). One study showed a positive association of mammographic density with plasma IGF-I levels and an inverse association with IGFBP-3 levels in premenopausal women, suggesting an influence of growth factors on breast density (59). However, there was no association among postmenopausal women. Twin studies have demonstrated that mammographic density has a strong genetic component (60). Breast cancer risk is associated with mammographic density in BRCA1 and BRCA2 carriers (61). However, mammographic density is not higher in carriers, indicating that genes influencing density act independently of BRCA1/2 mutation status. It is uncertain whether breast cancer susceptibility loci, such as LSP1 (rs3817198) on 11p, are associated with breast density (62, 63). Physical inactivity and high mammographic breast density have both been associated with increased breast cancer risk raising the question whether these two risk factors are related. However, most studies observed no association between physical activity and breast density (64, 65).

2.1.4.3.3. BONE MASS

E_2 affects bone mass via the ERα in bone-forming osteoblasts and bone-resorbing osteoclasts (66-69). An analysis of serum E_2 concentrations and bone mineral density (BMD) assessed by dual x-ray absorptiometry showed that postmenopausal women with serum E_2 levels between 10 and 25 pg/ml have 5 to 10% greater BMD than women with E_2 levels below 5 pg/mL (70). A prospective study showed that the reduction in estrogen levels after menopause was correlated with loss of bone mineral density (71). Postmenopausal women with undetectable serum E_2 concentrations and high SHBG levels have an increased risk of hip and vertebral fractures (72). Direct and indirect evidence indicates that BMD reflects estrogen exposure in middle-aged and elderly women and serves as a strong predictor of the risk of postmenopausal breast cancer. Two studies reported that women with hip fractures had a 16% lower risk of developing breast cancer and women with forearm fractures an even greater reduction of 58% compared to women without fractures (73, 74). Direct measurement of BMD by absorptiometry revealed a positive association with breast cancer risk. Postmenopausal women enrolled in the Study of Osteoporotic Fractures who had a BMD above the 25th percentile were at 2.0 to 2.5 times increased risk of breast cancer compared to women below the 25th percentile (75). Women enrolled in the Framingham Study who were in the highest BMD quartile had a rate ratio of developing breast cancer of 3.5 (CI 1.8 - 6.8) compared to women in the lowest quartile (76). Postmenopausal women with high BMD who are receiving hormone replacement therapy could have an even higher risk of breast cancer (77).

2.1.4.3.4. PHYSICAL ACTIVITY

Higher levels of physical activity have been associated with a reduction in breast cancer risk (78, 79). This association is most consistent among postmenopausal women and each additional hour of recreational activity per week reported to reduce risk by 6% (80). A prospective study of Norwegian women showed a reduced risk in both pre- and postmenopausal women (81). Postmenopausal women who engaged in regular physical activity (≥3 h/week) had a reduced risk for breast cancer compared with inactive women (82, 83). Compared with women who reported no recreational activities, those with more than five weekly hours of vigorous activity had a relative risk of 0.62 (95% CI 0.49 - 0.78) (84). The decrease was still observed among women who were overweight, nulliparous, had a family history of breast cancer, or used hormone replacement therapy.

The question whether exercise early in life reduces the risk of postmenopausal breast cancer was addressed in a population-based case-control study of female residents of Wisconsin, Massachusetts, Maine, and New Hampshire (85). Frequency of participation in strenuous physical activity when 14 - 22 years of age, weight at age 18 and 5 years before interview where ascertained through telephone interviews.

Compared to women with no adolescent physical activity and little adult weight gain, those with frequent strenuous physical activity and little weight gain showed a reduction up to 45% in risk of postmenopausal breast cancer. The benefit of early-life physical activity was lost in postmenopausal women who gained substantive adult weight. A 2008 review of 34 case-control and 28 cohort studies found evidence for a risk reduction associated with increased physical activity in 47 studies (76%) with an average risk decrease of 25 - 30% (78). A dose-response effect existed in 28 of 33 studies, i.e., vigorous activity was associated with a greater risk reduction. Stronger decreases in risk were observed for recreational activity and lifetime or later life activity. Moreover, stronger decreases were found among postmenopausal women and women with normal BMI. There was a greater risk reduction for ER-negative tumors although other studies observed reduced risk for ER-positive breast cancer (19).

Several mechanisms have been suggested for the protective effect of exercise on breast cancer. In premenopausal women, physical activity is hypothesized to reduce risk by decreasing the cumulative exposure to estrogens by altering the normal cycle of ovulation and menses (79). In postmenopausal women, the protective effect of exercise may occur via the reduction of circulating estrogen levels mediated by its effect on overweight and obesity (86). Other mechanisms include a reduction in insulin resistance and chronic inflammation (86).

A 12-month moderate-intensity exercise intervention study of 173 sedentary, overweight postmenopausal women examined the effect on serum E_1, E_2, and free E_2 (87). After 3 months, exercisers experienced E_1, E_2, and free E_2 declines of 3.8, 7.7, and 8.2%, respectively, versus no change or increased concentrations in controls (p = 0.03, 0.07, and 0.02, respectively). At 12 months, the direction of effect remained the same, although the differences were no longer statistically significant, except in women whose percentage of body fat decreased by ≥2%. Their E_1, E_2, and free E_2 levels declined 11.9, 13.7, and 16.7%, respectively, at 12 months. Physical activity did not affect the 2-hydroxylation pathway of estrogen metabolism or the 2-OHE_1/16α-OHE_1 ratio (88).

Two interventional studies in premenopausal women yielded null results with regard to circulating hormone levels and body weight. In one study of 391 premenopausal sedentary women, 16 weeks of 150 minutes per week of moderate aerobic exercise did not significantly alter serum levels of E_2, E_1 sulfate, progesterone, testosterone, or SHBG (89). There were no significant changes in body weight, although lean body mass increased while percent body fat decreased. The second study with a similar exercise regimen also found no change in hormone levels or body weight (90). However, exercise combined with a caloric restrictive diet (20 – 35% of baseline energy requirement) resulted in significant reductions in body weight and serum E_2. Most cross-sectional studies are consistent with these null results although some have observed a reduction in total or free E_2 with increasing activity (14, 91, 92). One study analyzed free E_2 levels in saliva samples collected daily for one complete menstrual cycle in 139 regularly menstruating women 24 – 37 years of age (93). There was a negative relation between E_2 concentrations and habitual physical activity, i.e., typical daily activity in form of housework, walking, exercise, and occupational work with mean E_2 levels of 5.7 pg/mL in the low and 4.9 pg/mL in the high activity group (p < 0.009).

References

1. Ogden CL, Carroll MD, Curtin LR, McDowell MA, Tabak CJ, Flegal KM. *Prevalence of overweight and obesity in the United States, 1999-2004*. JAMA. 2006;295:1549-55.

2. Allison DB, Faith MS, Heo M, Kotler DP. *Hypothesis concerning the U-shaped relation between body mass index and mortality*. Am J Epidemiol. 1997;146:339-49.

3. Berrington de Gonzalez A, Hartge P, Cerhan JR, Flint AJ, Hannan L, MacInnis RJ, et al. *Body-mass index and mortality among 1.46 million white adults*. N Engl J Med. 2010;363:2211-9.

4. Feigelson HS, Jonas CR, Teras LR, Thun MJ, Calle EE. *Weight gain, body mass index, hormone replacement therapy, and postmenopausal breast cancer in a large prospective study*. Cancer Epidemiol Biomarkers Prev. 2004;13:220-4.

5. Lahmann PH, Lissner L, Gullberg B, Olsson H, Berglund G. *A prospective study of adiposity and postmenopausal breast cancer risk: the Malmo Diet and Cancer Study*. Int J Cancer. 2003;103:246-52.

6. Tretli S. *Height and weight in relation to breast cancer morbidity and mortality. A prospective study of 570,000 women in Norway*. Int J Cancer. 1989;44:23-30.

7. Yong LC, Brown CC, Schatzkin A, Schairer C. *Prospective study of relative weight and risk of breast cancer: the Breast Cancer Detection Demonstration Project follow-up study, 1979 to1987-1989*. Am J Epidemiol. 1996;143:985-95.

8. Trentham-Dietz A, Newcomb PA, Storer BE, Longnecker MP, Baron J, Greenberg ER, et al. *Body size and risk of breast cancer*. Am J Epidemiol. 1997;145:1011-9.

9. Collaborative Group on Hormonal Factors in Breast Cancer. *Breast cancer and hormone replacement therapy: collaborative reanalysis of data from 51 epidemiological studies of 52 705 women with breast cancer and 108 411 women without breast cancer*. Lancet. 1997;350:1047-59.

10. Hankinson SE, Manson JE, Spiegelman D, Willet WC, Longcope C, Speizer FE. *Reproducibility of plasma hormone levels in postmenopausal wonem over a 2-3- year period*. Cancer, Epidemiol Biomarkers Prev. 1995;4:649-54.

11. Key TJ, Pike MC. *The role of oestrogens and progestagens in the epidemiology and prevention of breast cancer*. Eur J Cancer Clin Oncol. 1988;24:29-43.

12. MacDonald PC, Edman CD, Hemsell DL, Porter JC, Siiteri PK. *Effect of obesity on conversion of plasma androstenedione to estrone in postmenopausal women with and without endometrial cancer*. Am J Obstet Gynecol. 1978;130:448-55.

13. Moore JW, Key TJ, Bulbrook RD, Clark GM, Allen DS, Wang DY, et al. *Sex hormone binding globulin and risk factors for breast cancer in a population of normal women who had never used exogenous sex hormones*. Br J Cancer. 1987;56:661-6.

14. Verkasalo PK, Thomas HV, Appleby PN, Davey GK, Key TJ. *Circulating levels of sex hormones and their relation to risk factors for breast cancer: a cross-sectional study in 1092 pre- and postmenopausal women (United Kingdom)*. Cancer Causes Control. 2001;12:47-59.

15. Potischman N, Swanson CA, Siiteri P, Hoover RN. *Reversal of relation between body mass and endogenous estrogen concentrations with menopausal status*. J Natl Cancer Inst. 1996;88:756-8.

16. Lipworth L, Adami HO, Trichopoulos D, Carlstrom K, Mantzoros C. *Serum steroid hormone levels, sex hormone-binding globulin, and body mass index in the etiology of postmenopausal breast cancer*. Epidemiology. 1996;7:96-100.

17. Madigan MP, Troisi R, Potischman N, Dorgan JF, Brinton LA, Hoover RN. *Serum hormone levels in relation to reproductive and lifestyle factors in postmenopausal women*. Cancer Causes Control. 1998;9:199-207.

18. Key T. *Free estradiol and breast cancer risk in postmenopausal women: comparison of measured and calculated values*. Cancer Epidemiol Biomarkers Prev. 2003;12:1457-61.

19. Phipps AI, Chlebowski RT, Prentice R, McTiernan A, Stefanick ML, Wactawski-Wende J, et al. *Body size, physical activity, and risk of triple-negative and estrogen receptor-positive breast cancer*. Cancer Epidemiol Biomarkers Prev. 2011;20:454-63.

20. Enger SM, Ross RK, Paganini-Hill A, Carpenter CL, Bernstein L. *Body size, physical activity, and breast cancer hormone receptor status: results from two case-control studies*. Cancer Epidemiol Biomarkers Prev. 2000;9:681-7.

21. Potter JD, Cerhan JR, Sellers TA, McGovern PG, Drinkard C, Kushi LR, et al. *Progesterone and estrogen receptors and mammary neoplasia in the Iowa Women's Health Study: How many kinds of breast cancer are there?* Cancer Epidemiol Biomarkers Prev. 1995;4:319-26.

22. Colditz GA, Stampfer MJ, Willett WC, Hunter DJ, Manson JE, Hennekens CH, et al. *Type of postmenopausal hormone use and risk of breast cancer: 12 year follow-up from the Nurses' Health Study*. Cancer Causes & Control. 1992;3:433-9.

23. Huang Z, Hankinson SE, Colditz GA, Stampfer MJ, Hunter DJ, Manson JE, et al. *Dual effects of weight and weight gain on breast cancer risk*. J Am Med Assoc. 1997;278:1407-11.

24. Newcomb PA, Longnecker MP, Storer BE, Mittendorf R, Baron J, Clapp RW, et al. *Long-term hormone replacement therapy and risk of breast cancer in postmenopausal women*. Am J Epidemiol. 1995;142:788-95.

25. Stanford JL, Weiss NS, Voigt LF, Daling JR, Habel LA, Rossing MA. *Combined estrogen and progestin hormone replacement therapy in relation to risk of breast cancer in middle-aged women.* JAMA. 1995;274:137-42.

26. Mahabir S, Baer DJ, Johnson LL, Hartman TJ, Dorgan JF, Campbell WS, et al. *Usefulness of body mass index as a sufficient adiposity measurement for sex hormone concentration associations in postmenopausal women.* Cancer Epidemiol Biomarkers Prev. 2006;15:2502-7.

27. Renehan AG, Tyson M, Egger M, Heller RF, Zwahlen M. *Body-mass index and incidence of cancer: a systematic review and meta-analysis of prospective observational studies.* Lancet. 2008;371:569-78.

28. London SJ, Colditz GA, Stampfer MJ, Willett WC, Rosner B, Speizer FE. *Prospective study of relative weight, height, and risk of breast cancer.* J Am Med Assoc. 1989;262:2853-8.

29. Swanson CA, Coates RJ, Schoenberg JB, Malone KE, Gammon MD, Stanford JL, et al. *Body size and breast cancer risk among women under age 45 years.* Am J Epidemiol. 1996;143:698-706.

30. Ursin G, Longnecker MP, Haile RW, Greenland S. *A meta-analysis of body mass index and risk of premenopausal breast cancer.* Epidemiology. 1995;9:137-41.

31. Henderson BE, Ross RK, Pike MC. *Hormonal chemoprevention of cancer in women.* Science. 1993;259:633-8.

32. Pike MC, Spicer DV, Dahmoush L, Press MF. *Estrogens, progestogens, normal breast cell proliferation and breast cancer risk.* Epidemiol Rev. 1993;15:17-35.

33. Siiteri PK, Murai JT, Hammond GL, Nisker JA, Raymoure WJ, Kuhn RW. *The serum transport of steroid hormones.* Rec Prog Horm Res. 1982;38:457-510.

34. Grenman S, Ronnemaa T, Irjala K, Gronroos M. *Sex steroid, gonadotropin, cortisol and prolactin levels in healthy, massively obese women: correlation with abdominal fat size and effect of weight reduction.* J Clin Endocrinol Metab. 1986;63:1257-61.

35. Kopelman PG, Pilkington TR, White N, Jeffcoate SL. *Abnormal sex steroid secretion and binding in massively obese women.* Clin Endocrinol. 1980;12:363-9.

36. Kaaks R. *Nutrition, insulin, IGF-1 metabolism and cancer risk: a summary of epidemiological evidence.* Novartis Found Symp. 2004;262:247-60; discussion 60-68.

37. Key TJ, Appleby PN, Reeves GK, Roddam AW. *Insulin-like growth factor 1 (IGF1), IGF binding protein 3 (IGFBP3), and breast cancer risk: pooled individual data analysis of 17 prospective studies.* Lancet Oncol. 2010;11:530-42.

38. Renehan AG, Frystyk J, Flyvbjerg A. *Obesity and cancer risk: the role of the insulin-IGF axis.* Trends Endocrinol Metab. 2006;17:328-36.

39. Novosyadlyy R, Lann DE, Vijayakumar A, Rowzee A, Lazzarino DA, Fierz Y, et al. *Insulin-mediated acceleration of breast cancer development and progression in a nonobese model of type 2 diabetes.* Cancer Res. 2010;70:741-51.

40. Chen F, Wilkens LR, Monroe KR, Stram DO, Kolonel LN, Henderson BE, et al. *No association of risk variants for diabetes and obesity with breast cancer: the Multiethnic Cohort and PAGE studies.* Cancer Epidemiol Biomarkers Prev. 2011;20:1039-42.

41. Calle EE, Kaaks R. *Overweight, obesity and cancer: epidemiological evidence and proposed mechanisms.* Nat Rev Cancer. 2004;4:579-91.

42. Kaaks R. *Nutrition, hormones, and breast cancer: is insulin the missing link?* Cancer Causes Control. 1996;1996:605-25.

43. Lorincz AM, Sukumar S. *Molecular links between obesity and breast cancer.* Endocr Relat Cancer. 2006;13:279-92.

44. Brown KA, Simpson ER. *Obesity and breast cancer: progress to understanding the relationship.* Cancer Res. 2010;70:4-7.

45. Brown KA, McInnes KJ, Hunger NI, Oakhill JS, Steinberg GR, Simpson ER. *Subcellular localization of cyclic AMP-responsive element binding protein-regulated transcription coactivator 2 provides a link between obesity and breast cancer in postmenopausal women.* Cancer Res. 2009;69:5392-9.

46. Motoshima H, Goldstein BJ, Igata M, Araki E. *AMPK and cell proliferation--AMPK as a therapeutic target for atherosclerosis and cancer.* J Physiol. 2006;574:63-71.

47. Boyd NF, Lockwood GA, Byng JW, Tritchler DL, Yaffe MJ. *Mammographic densities and breast cancer risk.* Cancer Epidemiol Biomarkers Prev. 1998;7:1133-44.

48. Lokate M, Kallenberg MG, Karssemeijer N, Van den Bosch MA, Peeters PH, Van Gils CH. *Volumetric breast density from full-field digital mammograms and its association with breast cancer risk factors: a comparison with a threshold method.* Cancer Epidemiol Biomarkers Prev. 2010;19:3096-105.

49. McCormack VA, dos Santos Silva I. *Breast density and parenchymal patterns as markers of breast cancer risk: a meta-analysis.* Cancer Epidemiol Biomarkers Prev. 2006;15:1159-69.

50. Gertig DM, Stillman IE, Byrne C, Spiegelman D, Schnitt SJ, Connolly JL, et al. *Association of age and reproductive factors with benign breast tissue composition.* Cancer Epidemiol Biomarkers Prev. 1999;8:873-9.

51. Shepherd JA, Kerlikowske K, Ma L, Duewer F, Fan B, Wang J, et al. *Volume of mammographic density and risk of breast cancer.* Cancer Epidemiol Biomarkers Prev. 2011;20:1473-82.

52. Senie RT, Saftlas AF, Brinton LA, Hoover RN. *Is breast size a predictor of breast cancer risk or the laterality of the tumor?* Cancer Causes Control. 1993;4:203-8.

53. Weiss HA, Devesa SS, Brinton LA. *Laterality of breast cancer in the United States.* Cancer Causes Control. 1996;7:539-43.

54. Egan KM, Newcomb PA, Titus-Ernstoff L, Trentham-Dietz A, Baron JA, Willett WC, et al. *The relation of breast size to breast cancer risk in postmenopausal women (United States).* Cancer Causes Control. 1999;10:115-8.

55. Brinton LA, Persson I, Boice JD, Jr., McLaughlin JK, Fraumeni JF, Jr. *Breast cancer risk in relation to amount of tissue removed during breast reduction operations in Sweden.* Cancer. 2001;91:478-83.

56. Brown MH, Weinberg M, Chong N, Levine R, Holowaty E. *A cohort study of breast cancer risk in breast reduction patients.* Plast Reconstr Surg. 1999;103:1674-81.

57. Adami HO, Signorello LB, Trichopoulos D. *Towards an understanding of breast cancer etiology.* Semin Cancer Biol. 1998;8:255-62.

58. Trichopoulos D, Lipman RD. *Mammary gland mass and breast cancer risk.* Epidemiology. 1992;3:523-6.

59. Byrne C, Colditz GA, Willett WC, Speizer FE, Pollak M, Hankinson SE. *Plasma insulin-like growth factor (IGF) I, IGF-binding protein 3, and mammographic density.* Cancer Res. 2000;60:3744-8.

60. Boyd NF, Dite GS, Stone J, Gunasekara A, English DR, McCredie MR, et al. *Heritability of mammographic density, a risk factor for breast cancer.* N Engl J Med. 2002;347:886-94.

61. Mitchell G, Antoniou AC, Warren R, Peock S, Brown J, Davies R, et al. *Mammographic density and breast cancer risk in BRCA1 and BRCA2 mutation carriers.* Cancer Res. 2006;66:1866-72.

62. Lee E, Haiman CA, Ma H, van den Berg D, Bernstein L, Ursin G. *The role of established breast cancer susceptibility loci in mammographic density in young women.* Cancer Epidemiol Biomark Prev. 2008;17:258-60.

63. Odefrey F, Stone J, Gurrin LC, Byrnes GB, Apicella C, Dite GS, et al. *Common genetic variants associated with breast cancer and mammographic density measures that predict disease.* Cancer Res. 2010;70:1449-58.

64. Peters TM, Ekelund U, Leitzmann M, Easton D, Warren R, Luben R, et al. *Physical activity and mammographic breast density in the EPIC-Norfolk cohort study.* Am J Epidemiol. 2008;167:579-85.

65. Reeves KW, Gierach GL, Modugno F. *Recreational physical activity and mammographic breast density characteristics.* Cancer Epidemiol Biomarkers Prev. 2007;16:934-42.

66. Denger S, Reid G, Kos M, Flouriot G, Parsch D, Brand H, et al. *ERalpha gene expression in human primary osteoblasts: evidence for the expression of two receptor proteins.* Mol Endocrinol. 2001;15:2064-77.

67. Eriksen EF, Colvard DS, Berg NJ, Graham ML, Mann KG, Spelsberg TC, et al. *Evidence of estrogen receptors in normal human osteoblast-like cells.* Science. 1988;241:84-6.

68. Korach KS. *Insights from the study of animals lacking functional estrogen receptor.* Science. 1994;266:1524-7.

69. Spelsberg TC, Subramaniam M, Riggs BL, Khosla S. *The actions and interactions of sex steroids and growth factors/cytokines on the skeleton.* Mol Endocrinol. 1999;13:819-28.

70. Ettinger B, Pressman A, Sklarin PB, D.C., Cauley JA, Cummings SR. *Associations between low levels* of serum estradiol, bone density, and fractures among elderly women: the Study of Osteoporotic Fractures. J Clin Endocrinol Metab. 1998;83:2239-43.

71. Ahlborg HG, Johnell O, Turner CH, Rannevik G, Karlsson MK. *Bone loss and bone size after menopause.* N Engl J Med. 2003;349:327-34.

72. Cummings SR, Browner WS, Bauer D, Stone K, Ensrud K, Jamal S, et al. *Endogenous hormones and the risk of hip and vertebral fractures among older women. Study of Osteoporotic Fractures Research Group.* N Engl J Med. 1998;339:733-8.

73. Olsson H, Hagglund G. *Reduced cancer morbidity and mortality in a prospective cohort of women with distal forearm fractures.* Am J Epidemiol. 1992;136:422-7.

74. Persson I, Adami HO, McLaughlin JK, Naessen T, Fraumeni JF. *Reduced risk of breast and endometrial cancer among women with hip fractures.* Cancer Causes Control. 1994;5:523-8.

75. Cauley JA, Lucas FL, Kuller LH, Vogt MT, Browner WS, Cummings SR. *Bone mineral density and risk of breast cancer in older women: the study of osteoporotic fractures.* Study of Osteoporotic Fractures Research Group. J Am Med Assoc. 1996;276:1404-8.

76. Zhang Y, Kiel DP, Kreger BE, Cupples LA, Ellison RC, Dorgan JF, et al. *Bone mass and the risk of breast cancer among postmeno-pausal women.* New Engl J Med. 1997;336:611-7.

77. Kuller LH, Cauley JA, Lucas L, Cummings S, Browner WS. *Sex steroid hormones, bone mineral density, and risk of breast cancer.* Environ Health Perspect. 1997;105 Suppl:593-9.

78. Friedenreich CM, Cust AE. *Physical activity and breast cancer risk: impact of timing, type and dose of activity and population subgroup effects.* Br J Sports Med. 2008;42:636-47.

79. Gammon MD, John EM, Britton JA. *Recreational and occupational physical activities and risk of breast cancer.* J Natl Cancer Inst. 1998;90:100-17.

80. Monninkhof EM, Elias SG, Vlems FA, van der Tweel I, Schuit AJ, Voskuil DW, et al. *Physical activity and breast cancer: a systematic review.* Epidemiology. 2007;18:137-57.

81. Thune I, Brenn T, Lund E, Gaard M. *Physical activity and the risk of breast cancer.* N Engl J Med. 1997;336:1269-75.

82. Friedenreich CM. *Physical activity and cancer prevention: from observational to intervention research.* Cancer Epidemiol Biomarkers Prev. 2001;10:287-301.

83. McTiernan A, Kooperberg C, White E, Wilcox S, Coates R, Adams-Campbell LL, et al.

Recreational physical activity and the risk of breast cancer in postmeno-pausal women: the Women's Health Initiative Cohort Study. JAMA. 2003;290:1331-6.

84. Tehard B, Friedenreich CM, Oppert JM, Clavel-Chapelon F. *Effect of physical activity on women at increased risk of breast cancer: results from the E3N cohort study.* Cancer Epidemiol Biomarkers Prev. 2006;15:57-64.

85. Shoff SM, Newcomb PA, Trentham-Dietz A, Remington PL, Mittendorf R, Greenberg ER, et al. *Early-life physical activity and postmenopausal breast cancer: effect of body size and weight change.* Cancer Epidemiol Biomarkers Prev. 2000;9:591-5.

86. Neilson HK, Friedenreich CM, Brockton NT, Millikan RC. *Physical activity and postmenopausal breast cancer: proposed biologic mechanisms and areas for future research.* Cancer Epidemiol Biomarkers Prev. 2009;18:11-27.

87. McTiernan A, Tworoger SS, Ulrich CM, Yasui Y, Irwin ML, Rajan KB, et al. *Effect of exercise on serum estrogens in postmenopausal women: A 12-month randomized clinical trial.* Cancer Res. 2004;64:2923-8.

88. Atkinson C, Lampe JW, Tworoger SS, Ulrich CM, Bowen D, Irwin ML, et al. *Effects of a moderate intensity exercise intervention on estrogen metabolism in postmeno-pausal women.* Cancer Epidemiol Biomarkers Prev. 2004;13:868-74.

89. Smith AJ, Phipps WR, Arikawa AY, O'Dougherty M, Kaufman B, Thomas W, et al. *Effects of aerobic exercise on premenopausal sex hormone levels: results of the WISER study, a randomized clinical trial in healthy, sedentary, eumenorrheic women.* Cancer Epidemiol Biomarkers Prev. 2011;20:1098-106.

90. Williams NI, Reed JL, Leidy HJ, Legro RS, De Souza MJ. *Estrogen and progesterone exposure is reduced in response to energy deficiency in women aged 25-40 years.* Hum Reprod. 2010;25:2328-39.

91. Schmitz KH, Lin H, Sammel MD, Gracia CR, Nelson DB, Kapoor S, et al. *Association of physical activity with reproductive hormones: the Penn Ovarian Aging Study.* Cancer Epidemiol Biomarkers Prev. 2007;16:2042-7.

92. Tworoger SS, Missmer SA, Eliassen AH, Barbieri RL, Dowsett M, Hankinson SE. *Physical activity and inactivity in relation to sex hormone, prolactin, and insulin-like growth factor concentrations in premenopausal women - exercise and premenopausal hormones.* Cancer Causes Control. 2007;18:743-52.

93. Jasienska G, Ziomkiewicz A, Thune I, Lipson SF, Ellison PT. *Habitual physical activity and estradiol levels in women of reproductive age.* Eur J Cancer Prev. 2006;15:439-45.

2.2. OTHER HORMONES

2.2.1. INTRODUCTION

The mammary gland is the only organ that undergoes most of its development postnatally. A rudimentary ductal system forms during embryogenesis but the bulk of the development occurs in two growth phases at the onset of puberty and pregnancy. At puberty, the combined action of estrogen and locally produced growth factors regulates proliferation of the terminal end buds to promote ductal elongation and dichotomous branching (1, 2). Until pregnancy, the mammary gland remains relatively quiescent although the ductal system gradually becomes more complex through the growth of side branches, which occurs during the luteal phase of repeated estrous cycles when ovarian steroids reach peak serum levels. During pregnancy, exposure to progesterone and prolactin results in extensive epithelial proliferation. The ductal side-branching becomes more extensive and alveoli differentiate to become the site of milk production. Since hormones other than estrogen are involved in mammary gland development, in particular progesterone and prolactin, the question has been raised whether these hormones play a role in breast cancer development (3, 4).

References

1. Conneely OM, Jericevic BM, Lydon JP. *Progesterone receptors in mammary gland development and tumorigenesis.* J Mammary Gland Biol Neoplasia. 2003;8:205-14.

2. Mallepell S, Krust A, Chambon P, Brisken C. *Paracrine signaling through the epithelial estrogen receptor alpha is required for proliferation and morphogenesis in the mammary gland.* Proc Natl Acad Sci U S A. 2006;103:2196-201.

3. Clevenger CV, Furth PA, Hankinson SE, Schuler LA. *The role of prolactin in mammary carcinoma.* Endocr Rev. 2003;24:1-27.

4. Ismail PM, Amato P, Soyal SM, DeMayo FJ, Conneely OM, O'Malley BW, et al. *Progesterone involvement in breast development and tumorigenesis--as revealed by progesterone receptor "knockout" and "knockin" mouse models.* Steroids. 2003;68:779-87.

2.2.2. PROGESTERONE

Progesterone plays a key role in breast development with mammary epithelial cells as the primary target. The ductal system in the mammary glands of young female mice gradually becomes more complex through the growth of side branches, which occurs during the luteal phase of repeated estrous cycles when serum progesterone reaches peak levels. The ductal side-branching becomes more extensive during early pregnancy when alveolar buds form and differentiate into lobular units with alveolar cells that become the site of milk production. The mitogenic action of progesterone on mammary epithelial cells is mediated via the progesterone receptor (PR) and paracrine signaling (1-5). The PR gene encodes two protein isoforms, PR-A and PR-B, by transcription at two distinct promoters and translation initiation at two alternative AUG signals (3, 6). The A and B isoforms are structurally identical with the exception of an N-terminal extension in the PR-B protein that allows differential co-regulator recruitment and cell-specific transactivation of PR-B (7). Knockout experiments have shown that PR-A-/- mice have severe abnormalities in ovarian and uterine development leading to infertility while mammary gland morphogenesis and response to progesterone are normal. In contrast, PR-B-/- mice show normal ovarian and uterine responses to progesterone but lack mammary ductal side-branching and alveologenesis during pregnancy. Thus, in the developing mouse mammary gland, ERα is required for ductal morphogenesis and PR-B for subsequent side-branching and lobulo-alveolar development (2, 8, 9).

A downstream molecular target that mediates the effect of progesterone is <u>r</u>eceptor <u>a</u>ctivator of <u>n</u>uclear factor <u>κ</u>B <u>l</u>igand (RANKL) that interacts with its mammary epithelial receptor RANK to induce transcriptional upregulation of cyclin D (10, 11). RANKL belongs to the tumor necrosis factor (TNF) family and is also known as osteoprotegerin-ligand (OPGL), a key factor in osteoclast differentiation that is essential for bone remodeling (11). The expression of RANKL/OPGL is regulated by multiple hormones involved in mammary gland and bone morphogenesis, e.g., progesterone (but not E_2), prolactin, vitamin D, calcitonin, and TNFα. Ablation of RANKL or components of this signaling pathway results in defective lobulo-alveolar development which is secondary to reduced proliferative expansion of alveolar buds and enhanced apoptosis of alveolar epithelial cells. At the same time, ectopic expression of RANKL in PR-/- mammary epithelial cells rescues the PR-/- phenotype and permits lobular-alveolar development (1).

The epithelial cell component of the mammary gland is complex with a heterogeneous population of cells that are derived from mammary stem cells, which can both self-renew and propagate the full spectrum of cell types to generate functional lobulo-alveolar units during pregnancy (12, 13). The epithelial cell development is thought to progress in a hierarchical process from undifferentiated stem cells into at least two differentiated cell types, basal/myoepithelial and luminal cells, which can be distinguished by cytokeratin markers. In turn, the basal/myoepithelial and luminal cells are progenitor cells, which give rise to at least four distinct tumor subtypes: basal-like cells expressing human epidermal growth factor receptor 1 (EGFR; HER1), human epidermal growth factor receptor 2 (HER2)-enriched cells, and luminal A and luminal B cells typically expressing ERα and PR (14, 15). Interestingly, both mouse and human mammary stem cells exhibit a triple-negative phenotype for ERα, PR, and HER2 (16, 17). Progesterone treatment results in an expansion of stem cells through paracrine signaling via RANKL (18-20). Transgenic mice overexpressing RANK develop mammary tumors after multiparity or combined progestin/carcinogen treatment, medroxyprogesterone acetate plus dimethylbenz[a]anthracene (MPA plus DMBA) (21, 22). Conversely, pharmacological inhibition of RANKL attenuated mammary tumor development. These results indicate that the RANKL/RANK system plays an important role in progestin-induced mammary tumorigenesis.

Experimental studies have provided conflicting evidence regarding the possible role of progesterone in breast cancer development. Depending on the experimental model system, the cell context, and the duration of treatment, progesterone can elicit either proliferative or anti-proliferative effects on breast epithelial cell growth (23). A mouse model showed that implantation of a slow-release progesterone pellet at lactation into a single mammary gland inhibited involution of the gland and apoptosis of epithelial cells (24). A murine model of DMBA-induced mammary tumorigenesis showed that the progesterone signal contributed to mammary tumor susceptibility (4).

A clinical study of 40 postmenopausal women demonstrated the anti-proliferative activity of progesterone on normal breast epithelial cells (25). The women received daily topical application of a gel containing a placebo, E_2, progesterone, or a combination of both during the 14 days preceding aesthetic breast surgery or excision of a benign lesion. Increasing the E_2 concentration enhanced the number of cycling epithelial cells, whereas increasing the progesterone concentration significantly limited the number of cycling cells. On the other hand, mitotic rates of breast epithelial cells are highest in the luteal phase of the menstrual cycle when progesterone levels peak (26).

Progesterone and synthetic progestins may induce different responses. For example, the 19-nor-testosterone-derived progestins that are present in second- and

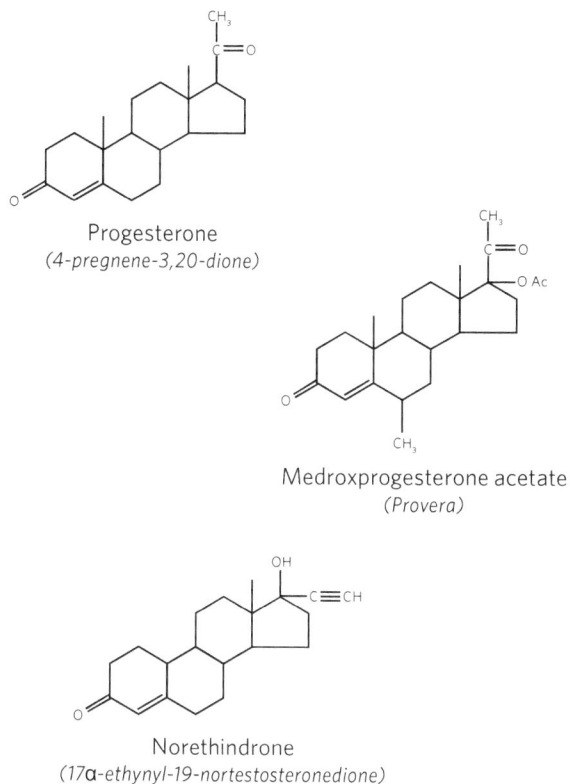

Progesterone
(4-pregnene-3,20-dione)

Medroxprogesterone acetate
(Provera)

Norethindrone
(17α-ethynyl-19-nortestosteronedione)

Figure 2.2.1. Chemical structure of progesterone and synthetic progestins.

third-generation oral contraceptives display estrogenic activity by binding to the ER (27). Progestins also exert an effect on several enzymes involved in estrogen metabolism and have been defined as "selective estrogen enzyme modulators (SEEMs)" in hormone-responsive breast cancer cells (28). Specifically, progestins have a complex effect on enzymes regulating the interconversion of the inactive E_1-S to E_1, i.e., steroid sulfatase and sulfotransferases (Figure 2.1.2.). Several progestins exert an inhibitory effect on estrone sulfatase, which produces E_1, in conjunction with a stimulatory effect on sulfotransferase, which forms E_1-S. Progestins also have a complex effect on 17β-hydroxysteroid dehydrogenase (17β-HSD) activity and can direct the $E_1 \leftrightarrow E_2$ interconversion in both directions (28). Studies with MCF-7 and T47D cells have shown that some progestins stimulate the reductive activity, E_1 to E_2, and thereby enhance cell proliferation (29). Other progestins favor the oxidative direction, E_2 to E_1, and thereby inhibit cell growth (30, 31). Thus, various progestins caused a significant decrease of E_2 formation (fmol E_2/mg DNA) when physiological concentrations of E_1-S, a source for E_2, were incubated with MCF-7 and T47D cells (32). Collectively, these data help to explain the anti-proliferative effect of progestins in breast tissue.

Epidemiological studies of the association between endogenous progesterone and breast cancer risk have yielded inconsistent results with positive, inverse, and no associations being reported (33-35). The Italian ORDET study observed an inverse association of serum progesterone with breast cancer risk with a relative risk of 0.40 for the highest versus lowest tertile (95% CI 0.15 – 1.08) (36). Similarly, the European Prospective Investigation into Cancer and Nutrition (EPIC) observed a reduced risk with elevated concentrations of progesterone (odds ratio for highest versus lowest quartile 0.61; 95% CI 0.38 – 0.98) (37). There was no association between breast cancer risk and progesterone levels in postmenopausal women (38).

In contrast, exogenous progestins in form of hormone replacement therapy (HRT) clearly increase breast cancer risk. The increase in risk is greater for estrogen-progestin combination therapy (e.g., conjugated equine estrogen plus MPA) than for estrogen replacement therapy alone (39-43).

One reason for the discrepant epidemiological results between endogenous and exogenous progestogens is the difference in chemical structure and biological activity. For example, MPA binds to the PR like progesterone but induces the expression of different proteins. Moreover, MPA has androgenic and glucocorticoid activity, whereas progesterone has minimal activity (44).

References

1. Beleut M, Rajaram RD, Caikovski M, Ayyanan A, Germano D, Choi Y, et al. *Two distinct mechanisms underlie progesterone-induced proliferation in the mammary gland.* Proc Natl Acad Sci U S A. 2010;107:2989-94.

2. Brisken C, Park S, Vass T, Lydon JP, O'Malley BW, Weinberg RA. *A paracrine role for the epithelial progesterone receptor in mammary gland development.* Proc Natl Acad Sci U S A. 1998;95:5076-81.

3. Conneely OM, Jericevic BM, Lydon JP. *Progesterone receptors in mammary gland development and tumorigenesis.* J Mammary Gland Biol Neoplasia. 2003;8:205-14.

4. Ismail PM, Amato P, Soyal SM, DeMayo FJ, Conneely OM, O'Malley BW, et al. *Progesterone involvement in breast development and tumorigenesis--as revealed by progesterone receptor "knockout" and "knockin" mouse models.* Steroids. 2003;68:779-87.

5. Williams SP, Sigler PB. *Atomic structure of progesterone complexed with its receptor.* Nature. 1998;393:392-6.

6. Kastner P, Krust A, Turcotte B, Stropp U, Tora L, Gronemeyer H, et al. *Two distinct estrogen-regulated promoters generate transcripts encoding the two functionally different human progesterone receptor forms A and B.* EMBO J. 1990;9:1603-14.

7. Giangrande PH, McDonnell DP. *The A and B isoforms of the human progesterone receptor: two functionally different transcription factors encoded by a single gene.* Recent Prog Horm Res. 1999;54:291-313.

8. Mallepell S, Krust A, Chambon P, Brisken C. *Paracrine signaling through the epithelial estrogen receptor alpha is required for proliferation and morphogenesis in the mammary gland.* Proc Natl Acad Sci U S A. 2006;103:2196-201.

9. Mulac-Jericevic B, Lydon JP, DeMayo FJ, Conneely OM. *Defective mammary gland morphogenesis in mice lacking the progesterone receptor B isoform.* Proc Natl Acad Sci U S A. 2003;100:9744-9.

10. Cao Y, Bonizzi G, Seagroves TN, Greten FR, Johnson R, Schmidt EV, et al. *IKKalpha provides an* essential link between RANK signaling and cyclin D1 expression during mammary gland development. Cell. 2001;107:763-75.

11. Fata JE, Kong YY, Li J, Sasaki T, Irie-Sasaki J, Moorehead RA, et al. *The osteoclast differentiation factor osteoprotegerin-ligand is essential for mammary gland development.* Cell. 2000;103:41-50.

12. Shackleton M, Vaillant F, Simpson KJ, Stingl J, Smyth GK, Asselin-Labat ML, et al. *Generation of a functional mammary gland from a single stem cell.* Nature. 2006;439:84-8.

13. Stingl J, Eirew P, Ricketson I, Shackleton M, Vaillant F, Choi D, et al. *Purification and unique properties of mammary epithelial stem cells.* Nature. 2006;439:993-7.

14. Nielsen TO, Hsu FD, Jensen K, Cheang M, Karaca G, Hu Z, et al. *Immunohistochemical and clinical characterization of the basal-like subtype of invasive breast carcinoma.* Clin Cancer Res. 2004;10:5367-74.

15. Prat A, Perou CM. *Mammary development meets cancer genomics.* Nat Med. 2009;15:842-4.

16. Asselin-Labat ML, Shackleton M, Stingl J, Vaillant F, Forrest NC, Eaves CJ, et al. *Steroid hormone receptor status of mouse mammary stem cells.* J Natl Cancer Inst. 2006;98:1011-4.

17. Lim E, Vaillant F, Wu D, Forrest NC, Pal B, Hart AH, et al. *Aberrant luminal progenitors as the candidate target population for basal tumor development in BRCA1 mutation carriers.* Nat Med. 2009;15:907-13.

18. Asselin-Labat ML, Vaillant F, Sheridan JM, Pal B, Wu D, Simpson ER, et al. *Control of mammary stem cell function by steroid hormone signalling.* Nature. 2010;465:798-802.

19. Joshi PA, Jackson HW, Beristain AG, Di Grappa MA, Mote PA, Clarke CL, et al. *Progesterone induces adult mammary stem cell expansion.*

Nature. 2010;465:803-7.

20. Lydon JP. *Stem cells: Cues from steroid hormones.* Nature. 2010;465:695-6.

21. Gonzalez-Suarez E, Jacob AP, Jones J, Miller R, Roudier-Meyer MP, Erwert R, et al. *RANK ligand mediates progestin-induced mammary epithelial proliferation and carcinogenesis.* Nature. 2010;468:103-7.

22. Schramek D, Leibbrandt A, Sigl V, Kenner L, Pospisilik JA, Lee HJ, et al. *Osteoclast differentiation factor RANKL controls development of progestin-driven mammary cancer.* Nature. 2010;468:98-102.

23. Lange CA, Richer JK, Horwitz KB. *Hypothesis: Progesterone primes breast cancer cells for cross-talk with proliferative or antiproliferative signals.* Mol Endocrinol. 1999;13:829-36.

24. Feng Z, Marti A, Jehn B, Altermatt HJ, Chicaiza G, Jaggi R. *Glucocorticoid and progesterone inhibit involution and programmed cell death in the mouse mammary gland.* J Cell Biol. 1995;131:1095-103.

25. Foidart JM, Colin C, Denoo X, Desreux J, Beliard A, Fournier S, et al. *Estradiol and progesterone regulate the proliferation of human breast epithelial cells.* Fertil Steril. 1998;69:963-9.

26. Pike MC, Spicer DV, Dahmoush L, Press MF. *Estrogens, progestogens, normal breast cell proliferation, and breast cancer risk.* Epidemiol Rev. 1993;15:17-35.

27. Jeng M, Parker CJ, Jordan VC. *Estrogenic potential of progestins in oral contraceptives to stimulate human breast cancer cell proliferation.* Cancer Res. 1992;52:6539-46.

28. Pasqualini JR. *The selective estrogen enzyme modulators in breast cancer: a review.* Biochim Biophys Acta. 2004;1654:123-43.

29. Poutanen M, Moncharmont B, Vihko R. *17 beta-hydroxysteroid*

dehydrogenase gene expression in human breast cancer cells: regulation of expression by a progestin. Cancer Res. 1992;52:290-4.

30. Chetrite GS, Ebert C, Wright F, Philippe JC, Pasqualini JR. *Effect of medrogestone on 17beta-hydroxysteroid dehydrogenase activity in the hormone-dependent MCF-7 and T-47D human breast cancer cell lines.* J Steroid Biochem Mol Biol. 1999;68:51-6.

31. Chetrite GS, Kloosterboer H, Philippe JC, Pasqualini JR. *Effects of ORG OD14 (Livial) and its metabolites on 17 beta-hydroxysteroid dehydrogenase activity in hormone-dependent MCF-7 and T-47D breast cancer cells.* Anticancer Res. 1999;19:261-7.

32. Pasqualini JR. *Differential effects of progestins on breast tissue enzymes.* Maturitas. 2003;46S1:S45-S54.

33. Eliassen AH, Missmer SA, Tworoger SS, Spiegelman D, Barbieri RL, Dowsett M, et al. *Endogenous steroid hormone concentrations and risk of breast cancer among premenopausal women.* J Natl Cancer Inst. 2006;98:1406-15.

34. Sturgeon SR, Potischman N, Malone KE, Dorgan JF, Daling J, Schairer C, et al. *Serum levels of sex hormones and breast cancer risk in premenopausal women: a case-control study (USA).* Cancer Causes Control. 2004;15:45-53.

35. Thomas HV, Key TJ, Allen DS, Moore JW, Dowsett M, Fentiman IS, et al. *A prospective study of endogenous serum hormone concentrations and breast cancer risk in premenopausal women on the island of Guernsey.* Br J Cancer. 1997;75:1075-9.

36. Micheli A, Muti P, Secreto G, Krogh V, Meneghini E, Venturelli E, et al. *Endogenous sex hormones and subsequent breast cancer in premenopausal women.* Int J Cancer. 2004;112:312-8.

37. Kaaks R, Berrino F, Key T, Rinaldi S, Dossus L, Biessy C, et al. *Serum sex steroids in premenopausal women and breast cancer risk within the European Prospective Investigation into Cancer and Nutrition (EPIC).* J Natl Cancer Inst. 2005;97:755-65.

38. Missmer SA, Eliassen AH, Barbieri RL, Hankinson SE. *Endogenous estrogen, androgen, and progesterone concentrations and breast cancer risk among postmenopausal women.* J Natl Cancer Inst. 2004;96:1856-65.

39. Anderson GL, Limacher M, Assaf AR, Bassford T, Beresford SA, Black H, et al. *Effects of conjugated equine estrogen in postmenopausal women with hysterectomy: the Women's Health Initiative randomized controlled trial.* JAMA. 2004;291:1701-12.

40. Beral V. *Breast cancer and hormone-replacement therapy in the Million Women Study.* Lancet. 2003;362:419-27.

41. Colditz GA, Hankinson SE, Hunter DJ, Willett WC, Manson JE, Stampfer MJ, et al. *The use of estrogens and progestins and the risk of breast cancer in postmenopausal women.* N Engl J Med. 1995;332:1589-93.

42. Rossouw J, GAnderson G, Prentice R, LaCroix A, Kooperberg C, Stefanick M, et al. *Risk and benefits of estrogen plus progestin in healthy postmenopausal women.* JAMA. 2002;288:321-33.

43. Schairer C, Lubin J, Troisi R, Sturgeon S, Brinton L, Hoover R. *Menopausal estrogen and estrogen-progestin replacement therapy and breast cancer risk.* JAMA. 2000;283:485-91.

44. Turgeon JL, McDonnell DP, Martin KA, Wise PM. *Hormone therapy: physiological complexity belies therapeutic simplicity.* Science. 2004;304:1269-73.

2.2.3. ANDROGENS

Women produce androgens in the adrenals and ovaries, including androstenedione, testosterone and the inactive precursors dehydroepiandrosterone (DHEA) and DHEA sulfate (Figure 2.1.1.). Circulating levels of testosterone are significantly lower in women than in men with normal reference ranges of 0.1 – 1.0 ng/mL and 2 – 10 ng/mL, respectively. Androgen concentrations are highest around 20 years of age, when concentration levels begin a steady

decline with a 50% parallel decrease in serum DHEA, DHEA sulfate, and testosterone by 40 years of age (1). Thus, the decline of serum androgens occurs well before menopause. In premenopausal women, theca-interstitial cells surrounding the ovarian follicles secrete about 50% of circulating androstenedione and 25% of testosterone and DHEA (2). The adrenals contribute about 50% of circulating androstenedione and DHEA, up to 25% of testosterone and virtually all of DHEA sulfate, which can be converted into DHEA, itself convertible into androstenedione. The remaining percentages of circulating testosterone and DHEA are derived from the peripheral conversion of androstenedione and DHEA sulfate, respectively, in adipose tissue, liver and kidneys. After menopause, most of the androgens are synthesized in peripheral tissues from DHEA and DHEA sulfate of adrenal origin (1). However, a fraction of androgens continues to be produced by the post-menopausal ovaries since women who have had bilateral ovariectomies have lower concentrations of testosterone compared with women with intact ovaries (3).

There is conflicting evidence regarding the possible role of androgens in breast cancer development. Androgens have been observed to have both stimulatory and inhibitory effects on the proliferation of normal and malignant mammary epithelial cells *in vitro* and on the growth of experimentally induced mammary tumors in animals (1, 4). The affinity of testosterone for ERα is several hundred-fold lower than that of E_2 and physiological concentrations of testosterone in the circulation are too low to exert any effect on ERα in either pre- or post-menopausal women. However, plasma concentrations of DHEA are 1 – 20 nM and levels of DHEA sulfate, the most abundant adrenal androgen, are in the micromolar range. At these physiologically relevant concentrations, DHEA has been shown to act as an estrogen agonist and stimulate ERα and cell growth (5). Thus, androgens may act directly, promoting growth via binding to the androgen receptor or the ER (4). Three studies involving a combined total of over 1500 primary breast cancers revealed a significant correlation between AR and ER expression (6-8). The concentration of AR in tumor cytosol was generally lower than that of ER. However, a subtype of breast cancer called apocrine carcinoma is characterized by lack of ER expression in the presence of high levels of AR expression (9, 10). More likely is an indirect mechanism via conversion to estrogens, either peripherally in adipose or in breast tissue (11). That is to say that any contribution of androgens to breast cancer risk appears to be largely through their role as substrates for estrogen production (12-14).

Several prospective epidemiological studies have examined androgens in pre- and postmenopausal women. These studies also measured E_2 and its fractions, E_1, and SHBG as well as the androgens DHEA, androstenedione, and testosterone. Such a complex set of data frequently leads to complicated interpretations, especially in studies with multiple additional variables that could potentially influence breast cancer risk.

Four prospective studies in premenopausal women found that increased serum levels of one or more of the androgens were associated with increased breast cancer risk. The European Prospective Investigation into Cancer and Nutrition (EPIC) examined serum sex steroids in premenopausal women and found no significant association of risk with serum estrogen levels but an increased risk with elevated concentrations of testosterone (odds ratio for highest versus lowest quartile 1.73; 95% CI 1.16 – 2.57), androstenedione (OR 1.56; 95% CI 1.05 – 2.32), and DHEA sulfate (OR 1.48; 95% CI 1.02 – 2.14) (15). A case-control study nested within the Nurses Health Study observed a significantly increased risk of breast cancer associated with higher follicular total and free E_2 levels but a modest, non-significant increase for total and free testosterone and androstenedione (12). However, when the analysis was restricted to invasive cancers, there was a significant association with high levels of luteal total testosterone levels (RR 2.0; 95% CI 1.1 – 3.6). A third study found no risk association of E_2, testosterone, or androstenedione but an elevated risk ratio for DHEA (RR 2.42; 95% CI 1.1 – 5.2) for DHEA (16). The Italian ORDET study observed an association of free testosterone with breast cancer risk with a relative risk of 2.85 for the highest versus lowest tertile (95% CI 1.11 – 7.33) (17). Most prospective studies of androgens in postmenopausal women have shown consistent associations between increased breast cancer risk and elevated androgen levels, similar to the observations made with estrogen levels (13, 14, 18).

In the pooled analysis of 9 prospective studies the relative risk of the highest versus lowest quintile was 2.22 for testosterone (95% CI 1.59 – 3.10), androstenedione 2.15 (95% CI 1.44 – 3.21), DHEA 2.04 (95% CI 1.21 – 3.45), and DHEA sulfate 1.75 (95% CI 1.26 – 2.43) (19). The increased serum concentrations of DHEA sulfate in women who develop breast cancer suggest increased adrenal androgen secretion since virtually all of DHEA sulfate is produced by the adrenals. Interestingly, serum concentrations of testosterone, androstenedione, DHEA, and androstenediol were not associated with risk of breast hyperplasia in postmenopausal women (20).

References

1. Labrie F, Luu-The V, Labrie C, Belanger A, Simard J, Lin SX, et al. *Endocrine and intracrine sources of androgens in women: inhibition of breast cancer and other roles of androgens and their precursor dehydroepiandrosterone.* Endocr Rev. 2003;24:152-82.

2. Lukanova A, Kaaks R. *Endogenous hormones and ovarian cancer: epidemiology and current hypotheses.* Cancer Epidemiol Biomarkers Prev. 2005;14:98-107.

3. McT.iernan A, Wu L, Barnabei VM, Chen C, Hendrix S, Modugno F, et al. *Relation of demographic factors,*

menstrual history, reproduction and medication use to sex hormone levels in postmenopausal women. Breast Cancer Res Treat. 2008;108:217-31.

4. Liao DJ, Dickson RB. *Roles of androgens in the development, growth, and carcinogenesis of the mammary gland.* J Steroid Biochem Mol Biol. 2002;80:175-89.

5. Maggiolini M, Donze O, Jeannin E, Ando S, Picard D. *Adrenal androgens stimulate the proliferation of breast cancer cells as direct activators of estrogen receptor alpha.* Cancer Res. 1999;59:4864-9.

6. Allegra JC, Lippman ME, Thompson EB, Simon R, Barlock A, Green L, et al. *Distribution, frequency, and quantitative analysis of estrogen, progesterone, androgen, and glucocorticoid receptors in human breast cancer.* Cancer Res. 1979;39:1447-454.

7. Brentani MM, Franco EL, Oshima CT, Pacheco MM. *Androgen, and progesterone receptor levels in malignant and benign breast tumors: a multivariate analysis approach.* Int J Cancer. 1986;38:637-42.

8. Lea OA, Kvinnsland S, Thorsen T. *Improved measurement of androgen receptors in human breast cancer.* Cancer Res. 1989;49:7162-7.

9. Doane AS, Danso M, Lal P, Donaton M, Zhang L, Hudis C, et al. *An estrogen receptor-negative breast cancer subset characterized by a hormonally regulated transcriptional program and response to androgen.* Oncogene. 2006;25:3994-4008.

10. Farmer P, Bonnefoi H, Becette V, Tubiana-Hulin M, Fumoleau P, Larsimont D, et al. *Identification of molecular apocrine breast tumours by microarray analysis.* Oncogene. 2005;24:4660-71.

11. Siiteri PK. *Adipose tissue as a source of hormones.* Am J Clin Nutr. 1987;45:277-82.

12. Eliassen AH, Missmer SA, Tworoger SS, Spiegelman D, Barbieri RL, Dowsett M, et al. *Endogenous steroid hormone concentrations and risk of breast cancer among premenopausal women.* J Natl Cancer Inst. 2006;98:1406-15.

13. Kaaks R, Rinaldi S, Key TJ, Berrino F, Peeters PH, Biessy C, et al. *Postmenopausal serum androgens, oestrogens and breast cancer risk: the European prospective investigation into cancer and nutrition.* Endocr Relat Cancer. 2005;12:1071-82.

14. Zeleniuch-Jacquotte A, Shore RE, Koenig KL, Akhmedkhanov A, Afanasyeva Y, Kato I, et al. *Postmenopausal levels of oestrogen, androgen, and SHBG and breast cancer: long-term results of a prospective study.* Br J Cancer. 2004;90:153-9.

15. Kaaks R, Berrino F, Key T, Rinaldi S, Dossus L, Biessy C, et al. *Serum sex steroids in premenopausal women and breast cancer risk within the European Prospective Investigation into Cancer and Nutrition (EPIC).* J Natl Cancer Inst. 2005;97:755-65.

16. Sturgeon SR, Potischman N, Malone KE, Dorgan JF, Daling J, Schairer C, et al. *Serum levels of sex hormones and breast cancer risk in premenopausal women: a case-control study (USA).* Cancer Causes Control. 2004;15:45-53.

17. Micheli A, Muti P, Secreto G, Krogh V, Meneghini E, Venturelli E, et al. *Endogenous sex hormones and subsequent breast cancer in premenopausal women.* Int J Cancer. 2004;112:312-8.

18. Missmer SA, Eliassen AH, Barbieri RL, Hankinson SE. *Endogenous estrogen, androgen, and progesterone concentrations and breast cancer risk among postmenopausal women.* J Natl Cancer Inst. 2004;96:1856-65.

19. The Endogenous Hormones and Breast Cancer Collaborative Group. *Endogenous sex hormones and breast cancer in postmenopausal women: Reanalysis of nine prospective studies.* J Natl Cancer Inst. 2002;94:606-16.

20. Schairer C, Hill D, Sturgeon SR, Fears T, Mies C, Ziegler RG, et al. *Serum concentrations of estrogens, sex hormone binding globulin, and androgens and risk of breast hyperplasia in postmenopausal women.* Cancer Epidemiol Biomarkers Prev. 2005;14:1660-5.

2.2.4. SEX HORMONE BINDING GLOBULIN

The Sex Hormone Binding Globulin (SHBG) gene at 17p12-p13 encodes a 373-amino acid protein, which is produced mainly by the liver as a homodimeric glycoprotein with O-linked oligosaccharides at 7Thr and N-linked oligosaccharides at 351Asn and 367Asn (1, 2). SHBG binds and transports estrogens and androgens in the blood with high affinity (3). Estrogens are also bound to albumin with only a small fraction circulating free or not bound to SHBG or albumin (between 1 and 5%). The binding affinity of SHBG for E_2 is orders of magnitude higher than the affinity of albumin for E_2 (4, 5). Thus, E_2 readily dissociates from albumin but remains tightly bound to SHBG. Since only the free hormone can enter cells, the unbound estrogen is considered biologically active and SHBG may play a role in mammary carcinogenesis by regulating the availability of free E_2 in the circulation.

In addition to its role as transport protein, SHBG can inhibit E_2-mediated induction of progesterone receptor expression, cell growth and anti-apoptosis in estrogen-dependent MCF-7 breast cancer cells but not in ER-negative MDA-MB-231 cells (6). This action is mediated by specific binding sites for SHBG, which have been identified on membranes of MCF-7 cells although no data are available on receptor structure and a gene encoding the putative receptor has not been identified (6). However, it has been shown that the N-terminal glycosylation at 7Thr is required for correct binding of SHBG to the cell membrane and its subsequent biological action (2). There is evidence that E_2 interacts with its membrane receptor, ERm, activating the MAP kinase ERK and inhibiting Jun kinase, which results in the induction of bcl-2 and inhibition of apoptosis in MCF-7 cells (7). There is evidence that E_2 also binds to membrane-bound SHBG, with the complex activating adenyl cyclase, which generates cAMP that, in turn, inhibits ERK and apoptosis (8). Thus, there are separate membrane-initiated pathways involving SHBG and E_2 that have opposite effects on proliferation and apoptosis in MCF-7 cells (6).

The concentration of SHBG in the circulation is influenced by hormonal, nutritional, and genetic factors. For example, estrogens increase SHBG hepatic synthesis and blood levels while androgens have the opposite effect (9). The SHBG gene contains a common nonsynonymous polymorphism in exon 8, Asp327Asn (rs6259), which introduces an extra site for N-glycosylation. The amino acid substitution does not alter the E_2 binding property of SHBG or its biologic activity in MCF-7 cells. However, the 327Asn allele was associated with an increased half-life of SHBG in an animal model as well as elevated serum SHBG levels and a reduced E_2:SHBG ratio in women (10-12). A G/T substitution (rs1799941) in the 5'-untranslated region was also associated with raised SHBG levels in postmenopausal women (12). A genome-wide association study reported an association of yet another SNP, rs6761, with SHBG levels (13). Finally, a pentanucleotide repeat polymorphism, $(TAAAA)_n$, at approximately -800 bp from the transcription start site also influenced SHBG levels; women homozygous for six repeats had significantly higher levels than nine-repeat homozygous individuals, possibly as a consequence of linkage disequilibrium with rs1799941 (10, 14).

Several prospective epidemiological studies have examined SHBG in pre- and postmenopausal women together with serum estrogens and androgens. The studies of premenopausal women have not shown any association of circulating SHBG levels and breast cancer risk (15-17). In contrast, the studies of postmenopausal women have shown that blood SHBG levels are inversely associated with breast cancer risk (18). Thus, high SHBG levels were associated with reduced risk, consistent with the role of the protein to bind E_2 in the circulation and restrict its biological activity. However, serum concentrations of SHBG were not associated with risk of breast hyperplasia in postmenopausal women (19). The Shanghai Breast Cancer Study also observed a protective association of the variant 327Asn allele (rs6259) with breast cancer risk in postmenopausal women (odds ratio 0.73; 95% CI 0.53 – 0.99) but not in premenopausal women (20). In contrast, a large European study of 11 tagging SNPs in and around the SHBG gene found no association of the 327Asn allele with breast cancer risk in postmenopausal women (14). Instead, a T/C substitution (rs6257) in intron 1 was significantly associated with breast cancer risk (0.88; 95% CI 0.82 – 0.95).

References

1. Berube D, Seralini GE, Gagne R, Hammond GL. *Localization of the human sex hormone-binding globulin gene (SHBG) to the short arm of chromosome 17 (17p12---p13).* Cytogenet Cell Genet. 1990;54:65-7.

2. Raineri M, Catalano MG, Hammond GL, Avvakumov GV, Frairia R, Fortunati N. *O-Glycosylation of human sex hormone-binding globulin is essential for inhibition of estradiol-induced MCF-7 breast cancer cell proliferation.* Mol Cell Endocrinol. 2002;189:135-43.

3. Hammond GL. *Potential functions of plasma steroid-binding proteins.* Trends Endocrinol Metab. 1995;6:298-304.

4. Dunn JF, Nisula BC, Rodbard D. *Transport of steroid hormones: binding of 21 endogenous steroids to both testosterone-binding globulin and corticosteroid-binding globulin in human plasma.* J Clin Endocrinol Metab. 1981;53:58-68.

5. Pardridge WM. *Transport of protein-bound hormones into tissues in vivo.* Endocr Rev. 1981;2:103-23.

6. Fortunati N, Catalano MG. *Sex hormone-binding globulin (SHBG) and estradiol cross-talk in breast cancer cells.* Horm Metab Res. 2006;38:236-40.

7. Razandi M, Pedram A, Greene GL, Levin ER. *Cell membrane and nuclear estrogen receptors (ERs) originate from a single transcript: studies of ERa and ERb expressed in chinese hamster ovary cells.* Mol Endocrinol. 1999;13:307-19.

8. Catalano MG, Frairia R, Boccuzzi G, Fortunati N. *Sex hormone-binding globulin antagonizes the anti-apoptotic effect of estradiol in breast cancer cells.* Mol Cell Endocrinol. 2005;230:31-7.

9. Pasquali R, Vicennati V, Bertazzo D, Casimirri F, Pascal G, Tortelli O, et al. *Determinants of sex hormone-binding globulin blood concentrations in premenopausal and postmenopausal women with different estrogen status.* Virgilio-Menopause-Health Group. Metabolism. 1997;46:5-9.

10. Cousin P, Calemard-Michel L, Lejeune H, Raverot G, Yessaad N, Emptoz-Bonneton A, et al. *Influence of SHBG gene pentanucleotide TAAAA repeat and D327N polymorphism on serum sex hormone-binding globulin concentration in hirsute women.* J Clin Endocrinol Metab. 2004;89:917-24.

11. Cousin P, Dechaud H, Grenot C, Lejeune H, Hammond GL, Pugeat M. *Influence of glycosylation on the clearance of recombinant human sex hormone-binding globulin from rabbit blood.* J Steroid Biochem Mol Biol. 1999;70:115-21.

12. Dunning AM, Dowsett M, Healey CS, Tee L, Luben RN, Folkerd E, et al. *Polymorphisms associated with circulating sex hormone levels in postmenopausal women.* J Natl Cancer Inst. 2004;96:936-45.

13. Melzer D, Perry JR, Hernandez D, Corsi AM, Stevens K, Rafferty I, et al. *A genome-wide association study identifies protein quantitative trait loci (pQTLs).* PLoS Genet. 2008;4:e1000072.

14. Thompson DJ, Healey CS, Baynes C, Kalmyrzaev B, Ahmed S, Dowsett M, et al. *Identification of common variants in the SHBG gene affecting sex hormone-binding globulin levels and breast cancer risk in postmenopausal women.* Cancer Epidemiol Biomarkers Prev. 2008;17:3490-8.

15. Eliassen AH, Missmer SA, Tworoger SS, Spiegelman D, Barbieri RL, Dowsett M, et al. *Endogenous steroid hormone concentrations and risk of breast cancer among premenopausal women.* J Natl Cancer Inst. 2006;98:1406-15.

16. Kaaks R, Rinaldi S, Key TJ, Berrino F, Peeters PH, Biessy C, et al. *Postmenopausal serum androgens, oestrogens and breast cancer risk: the European prospective investigation into cancer and nutrition.* Endocr Relat Cancer. 2005;12:1071-82.

17. Sturgeon SR, Potischman N, Malone KE, Dorgan JF, Daling J, Schairer C, et al. *Serum levels of sex hormones and breast cancer risk in premenopausal women: a case-control study (USA).* Cancer Causes Control. 2004;15:45-53.

18. The Endogenous Hormones and Breast Cancer Collaborative Group. *Endogenous sex hormones and breast cancer in postmenopausal women: Reanalysis of nine prospective studies.* J Natl Cancer Inst. 2002;94:606-16.

19. Schairer C, Hill D, Sturgeon SR, Fears T, Mies C, Ziegler RG, et al. *Serum concentrations of estrogens, sex hormone binding globulin, and androgens and risk of breast hyperplasia in postmenopausal women.* Cancer Epidemiol Biomarkers Prev. 2005;14:1660-5.

20. Cui Y, Shu XO, Cai Q, Jin F, Cheng JR, Cai H, et al. *Association of breast cancer risk with a common functional polymorphism (Asp327Asn) in the sex hormone-binding globulin gene.* Cancer Epidemiol Biomarkers Prev. 2005;14:1096-101.

2.2.5. PROLACTIN

Prolactin is a 23-kDa peptide composed of 199 amino acids, which is produced by the lactotropic cells of the pituitary as well as extrapituitary sites, including the breast, uterus, and prostate (1-3). Prolactin is essential for mammary gland development and lactation. The actions of prolactin are endocrine as well as autocrine/paracrine, mediated by six prolactin receptor isoforms (4). The classic type 1 transmembrane receptor stimulates Janus kinases, which, in turn, phosphorylate members of the STAT (signal transducers and activators of transcription) family to induce the expression of the milk protein β-casein (4, 5). The synthesis of prolactin is regulated by E_2 via the estrogen receptor and peak levels are produced in the third trimester of pregnancy (3, 6, 7). During lactation the mammary gland is nonresponsive to E_2 and in vitro E_2 has been shown to inhibit prolactin-induced milk protein production (8, 9). The inhibitory effect of E_2 during lactation may be due to negative cross-talk between ERα and STAT with ERα-repressive action on STAT5-regulated milk protein expression (5). Xenoestrogens, such as 2-amino-3-methylimidazo[4,5-b] pyridine are also capable of stimulating prolactin synthesis and secretion (10). Prolactin and prolactin receptor expression have been detected in normal and malignant breast tissues and in many breast cancer cell lines (11-13).

Prolactin acts as a cocarcinogen in rat mammary tumorigenesis (14). The growth of mammary cancers induced by dimethylbenz[a]anthracene (DMBA) was partly controlled by prolactin. Both hypophysectomy and prolactin-lowering ergot drugs were found to prevent or inhibit the growth of mammary tumors in this experimental model, whereas elevations in prolactin stimulated growth (15). Transgenic mice overexpressing the prolactin gene developed mammary tumors (16). Thus, there is clear evidence that prolactin plays a role in mammary carcinogenesis in animals.

Prolactin can act as an autocrine/paracrine growth factor in human breast cancer cells (1, 17-19). For example, prolactin and E_2 exhibited an additive stimulatory effect on T-47D cell proliferation (20). The mitogenic activity of locally produced prolactin was suppressed by prolactin antagonists, anti-prolactin antibodies, or prolactin antisense oligonucleotides (1, 20). Tumors derived from T47D breast cancer cells inoculated into nude mice grew larger upon treatment with prolactin, whereas treatment with a prolactin antagonist inhibited tumor growth (20).

Whether prolactin influences the risk of human breast cancer is uncertain (4). A comparison of age-matched parous and nulliparous women showed that pregnancy has a subsequent, long-term inhibitory effect on prolactin secretion (21). Several prospective studies have been performed in pre- and postmenopausal women. None of the premenopausal studies showed an association between circulating prolactin levels and breast cancer risk (22-24). Analysis of the Nurses' Health Cohort showed an association between high plasma prolactin concentrations and increased risk of postmenopausal breast cancer (25, 26). However, population-based prospective cohort studies in the UK, Sweden, and Japan found no association between prolactin levels and risk of post-menopausal breast cancer (23, 24, 27). An editorial in 1987 found little evidence that patients with pathological-ly or pharmacologically elevated serum prolactin levels are at a higher risk for developing mammary cancer (15). That statement appears correct today and whether high normal levels are associated with breast cancer is debatable.

References

1. Ben-Jonathan N, Liby K, McFarland M, Zinger M. *Prolactin as an autocrine/paracrine growth factor in human cancer.* Trends Endocrinol Metab. 2002;13:245-50.

2. Binart N, Ormandy CJ, Kelly PA. *Mammary gland development and the prolactin receptor.* Adv Exp Med Biol. 2000;480:85-92.

3. Freeman ME, Kanyicska B, Lerant A, Nagy G. *Prolactin: structure, function, and regulation of secretion.* Physiol Rev. 2000;80:1523-631.

4. Clevenger CV, Furth PA, Hankinson SE, Schuler LA. *The role of prolactin in mammary carcinoma.* Endocr Rev. 2003;24:1-27.

5. Faulds MH, Pettersson K, Gustafsson JA, Haldosen LA. *Cross-talk between ERs and signal transducer and activator of transcription 5 is E_2 dependent and involves two functionally separate mechanisms.* Mol Endocrinol. 2001;15:1929-40.

6. Maurer RA, Notides AC. *Iden-tification of an estrogen-responsive element from the 5'-flanking region of the rat prolactin gene.* Mol Cell Biol. 1987;7:4247-54.

7. Schaufele F. *Regulation of estrogen receptor activation of the prolactin enhancer/promoter by antagonistic activation function-2-interacting proteins.* Mol Endocrinol. 1999;13:935-45.

8. Kleinberg DL, Todd J, Babitsky G, Greising J. *Estradiol inhibits prolactin induced alpha-lactalbumin production in normal primate mammary tissue in vitro.* Endocrinology. 1982;110:279-81.

9. Shyamala G, Ferenczy A. *The nonresponsiveness of lactating mammary gland to estradiol.* Endocrinology. 1982;110:1249-56.

10. Lauber SN, Gooderham NJ. *The cooked meat derived genotoxic carcinogen 2-amino-3-methylimidazo[4,5-b]pyridine has potent hormone-like activity: mechanistic support for a role in breast cancer.* Cancer Res. 2007;67:9597-602.

11. Shaw-Bruha CM, Pirrucello SJ, Shull JD. *Expression of the prolactin gene in normal and neoplastic human breast tissues and human mammary cell lines: promoter usage and alternative mRNA splicing.* Breast Cancer Res Treat. 1997;44:243-53.

12. Touraine P, Martini JF, Zafrani B, Durand JC, Labaille F, Malet C, et al. *Increased expression of prolactin receptor gene assessed by quantitative polymerase chain reaction in human breast tumors versus normal breast tissues.* J Clin Endocrinol Metab. 1998;83:667-74.

13. Vonderhaar BK. *Prolactin: the forgotten hormone of human breast cancer.* Pharmacol Ther. 1998;79:169-78.

14. Welsch CW, Nagasawa H. *Prolactin and murine mammary tumorigenesis: a review.* Cancer Res. 1977;37:951-63.

15. Kleinberg DL. *Prolactin and breast cancer.* N Engl J Med. 1987;316:269-72.

16. Wennbo H, Gebre-Medhin M, Gritli-Linde A, Ohlsson C, Isaksson OG, Tornell J. *Activation of the prolactin receptor but not the growth hormone receptor is important for induction of mammary tumors in transgenic mice.* J Clin Invest. 1997;100:2744-51.

17. Biswas R, Vonderhaar BK. *Role of serum in the prolactin responsiveness of MCF-7 human breast cancer cells in

long-term tissue culture.* Cancer Res. 1987;47:3509-14.

18. Liby K, Neltner B, Mohamet L, Menchen L, Ben-Jonathan N. *Prolactin overexpression by MDA-MB-435 human breast cancer cells accelerates tumor growth.* Breast Cancer Res Treat. 2003;79:241-52.

19. Simon WE, Albrecht M, Trams G, Dietel M, Holzel F. *In vitro growth promotion of human mammary carcinoma cells by steroid hormones, tamoxifen, and prolactin.* J Natl Cancer Inst. 1984;73:313-21.

20. Chen WY, Ramamoorthy P, Chen N, Sticca R, Wagner TE. *A human prolactin antagonist, hPRL-G129R, inhibits breast cancer cell proliferation through induction of apoptosis.* Clin Cancer Res. 1999;5:3583-93.

21. Musey VC, Collins DC, Musey PI, Martino-Saltzman D, Preedy JR. *Long-term effect of a first pregnancy on the secretion of prolactin.* New Engl J Med. 1987;316:229-34.

22. Helzlsouer KJ, Alberg AJ, Bush TL, Longcope C, Gordon GB, Comstock GW. *A prospective study of endogenous hormones and breast cancer.* Cancer Detect Prev. 1994;18:79-85.

23. Kabuto M, Akiba S, Stevens RG, Neriishi K, Land CE. *A prospective study of estradiol and breast cancer in Japanese women.* Cancer Epidemiol Biomarkers Prev. 2000;9:575-9.

24. Wang DY, De Stavola BL, Bulbrook RD, Allen DS, Kwa HG, Fentiman IS, et al. *Relationship of blood prolactin levels and the risk of subsequent breast cancer.* Int J Epidemiol. 1992;21:214-21.

25. Hankinson SE, Willett WC, Michaud DS, Manson JE, Colditz GA, Longcope C, et al. *Plasma prolactin levels and subsequent risk of breast cancer in postmeno-pausal women.* J Natl Cancer Inst. 1999;91:629-34.

26. Tworoger SS, Eliassen AH, Rosner B, Sluss P, Hankinson SE. *Plasma prolactin concentrations and risk of postmenopausal breast cancer.* Cancer Res. 2004;64:6814-9.

27. Manjer J, Johansson R, Berglund G, Janzon L, Kaaks R, Agren A, et al. *Postmenopausal breast cancer risk in relation to sex steroid hormones, prolactin and SHBG (Sweden).* Cancer Causes Control. 2003;14:599-607.

2.3. REACTIVE OXYGEN SPECIES
2.3.1. INTRODUCTION

When we oxidize molecules with atmospheric oxygen, O_2, the oxygen molecule itself becomes reduced and is transformed into water. The reduction of oxygen occurs by the stepwise addition of electrons, resulting in the formation of intermediates collectively called "reactive oxygen species" (ROS) to convey their higher reactivities relative to molecular O_2 (Figure 2.3.1.) (1-3). One of the intermediates, the hydroperoxyl radical, rapidly dissociates at pH 7.4 to produce the superoxide anion radical: $HO_2\bullet \rightarrow H^+ + O_2\bullet-$. A **free radical** is defined as any chemical species capable of independent existence and containing one or more unpaired electrons, i.e., electrons that occupy atomic or molecular orbitals by themselves. Free radicals can be formed when a non-radical either gains or loses a single electron. Thus, the reduction of oxygen yields H_2O_2 and two free radicals, $O_2\bullet-$ and $HO\bullet$. Under physiological conditions, the steady-state levels of these ROS are typically kept low, i.e., 10^{-7} to 10^{-9} M H_2O_2, 10^{-10} to 10^{-11} M $O_2\bullet-$, and 10^{-15} to 10^{-20} M $HO\bullet$ (4). ROS are formed during normal metabolism of the cell, especially in mitochondria and peroxisomes, and by the activities of several enzyme systems, e.g., cytochromes P450, lipoxygenases, and plasma membrane associated oxidases (5). Mitochondria constitute the greatest source of ROS, as the mitochondrial electron-transport system consumes over 80% of the oxygen utilized by the cell (6). In normal cells, as much as 1% of the electrons flowing through this electron-transport chain are thought to undergo one-electron reductions of O_2 and form ROS. The production of ROS is increased in pathological stress conditions. For example, proliferating cells may be exposed to hypoxia and lack of glucose, both of which cause a progressive elevation in mitochondrial ROS production (7). A moderate increase in ROS can promote cell proliferation and differentiation (8). However, excessive amounts of ROS can cause oxidative damage to proteins, lipids, and DNA.

The steady-state levels of ROS ($O_2\bullet-$ and $HO\bullet$) were several-fold higher in breast cancer cells (HCT116, MB231) than in normal human mammary epithelial cells (9). In contrast, normal mammary epithelial stem cells contain lower concentrations of ROS than their more mature progeny cells and this difference may be critical for maintaining stem cell function (10).

The **superoxide anion radical ($O_2\bullet-$)** is formed when one electron enters one of the π^*2p orbitals of oxygen. The resulting molecule is both an anion and free radical, called superoxide, or more correctly, the superoxide anion radical. The addition of an extra electron to the oxygen molecule weakens the double bond, producing a more reactive molecule with only one and one-half bonds. It is relatively unstable and has a half-life of 10^{-5} s at 37°C. $O_2\bullet-$ is a weak oxidizing agent able to oxidize molecules such as ascorbic acid and glutathione. However, $O_2\bullet-$ is a much stronger reducing agent and can reduce iron complexes such as Fe^{3+}-EDTA. The enzyme superoxide dismutase (SOD) catalyzes the conversion of $O_2\bullet-$ to hydrogen peroxide and oxygen and keeps the cellular level of $O_2\bullet-$ to <10^{-10} M.

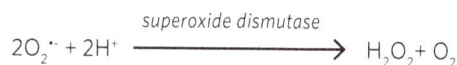

$$2O_2\bullet- + 2H^+ \xrightarrow{\text{superoxide dismutase}} H_2O_2 + O_2$$

Hydrogen peroxide (H_2O_2) has an uncharged covalent structure. It readily mixes with water and is treated as a water molecule by the body, rapidly diffusing across all membranes. It is a weak oxidant and a weak reducing agent that is relatively stable in the absence of transition metal ions. H_2O_2 is produced by SOD in any system that generates superoxide. Other enzymes, such as glucose oxidase and urate oxidase, produce H_2O_2 directly by the transfer of two electrons to oxygen. H_2O_2 is removed from cells by the action of catalase or glutathione peroxidase.

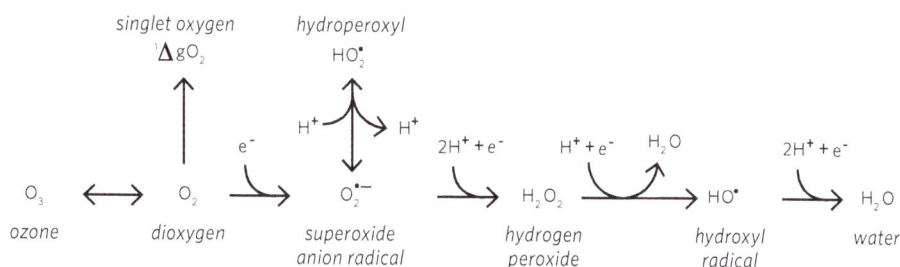

Figure 2.3.1. In the cell, oxygen can undergo several types of reaction. O_2 can be excited to form singlet oxygen or become partially reduced by transfer of one, two or three electrons to form, respectively, superoxide anion radical ($O_2\bullet-$), hydrogen peroxide (H_2O_2), or hydroxyl radical ($HO\bullet$). Collectively, these forms of oxygen are called reactive oxygen species (ROS).

$$2H_2O_2 \xrightarrow{\text{catalase}} 2H_2O + O_2$$

$$2H_2O_2 + 2GSH \xrightarrow{\text{glutathione peroxidase}} 2H_2O + GSSG$$

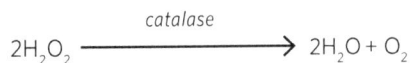

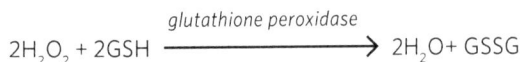

The **hydroxyl radical (HO•)** is the most aggressive ROS formed in vivo and can attack any biological molecule at an almost diffusion-controlled rate, causing damage within a few molecular diameters from its site of production. Unlike superoxide and hydrogen peroxide, which are mainly controlled enzymatically (by superoxide dismutase, catalase, and glutathione peroxidase, respectively), the hydroxyl radical is far too reactive to be restricted in such a way and will even attack antioxidant enzymes. It has a half-life of 10^{-9} to 10^{-10} s (1). HO• can be generated by several reactions, such as the high-energy ionization of water (radiolysis):

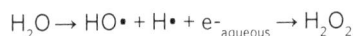

$$H_2O \rightarrow HO• + H• + e^-_{\text{aqueous}} \rightarrow H_2O_2$$

In the so-called Fenton reaction between iron salts and hydrogen peroxide, an electron transfer from Fe^{2+} to the weak O-O bond of H_2O_2 results in the formation of Fe^{3+} and the hydroxyl radical (11, 12):

$$Fe^{2+} + H_2O_2 \rightarrow Fe^{3+} + HO- + HO•$$

In the so-called Haber-Weiss reaction, the electron transfer occurs from the superoxide to H_2O_2, resulting in the formation of oxygen and hydroxyl radical:

$$O_2•- + H_2O_2 \rightarrow O_2 + HO- + HO•$$

Several metals besides iron are capable of undergoing changes in oxidation status. These metals, e.g., copper, chromium, can reduce hydrogen peroxide to hydroxyl radical in a Fenton-type reaction:

$$Cu^+ + H_2O_2 \rightarrow Cu^{2+} + HO- + HO•$$

Singlet oxygen. The input of appropriate energy can excite the unpaired electrons of the oxygen molecule, thereby forming singlet oxygen. Two forms of singlet oxygen can arise that differ in electron arrangement, energy, and half-life. $^1\Delta gO_2$ has two electrons with opposite spins in a common π^*2p orbital, an energy of 94 kJ mol^{-1}, and a half-life of 2×10^{-6} s at 37ºC. $^1\Sigma^+O_2$ has two electrons with opposite spins in different π^*2p orbitals, an energy of 157 kJ mol^{-1}, and a half-life of 10^{-11} s. Since $^1\Sigma^+O_2$ decays rapidly to the $^1\Delta gO_2$ state in biological systems, it is usually ignored. Thus, singlet oxygen is generally equated with $^1\Delta gO_2$, which is not a free radical because it does not contain an unpaired

electron. Nevertheless, it is a highly reactive form of oxygen in which the spin restriction (two unpaired electrons with parallel spins) is removed, thereby increasing its oxidizing ability. The formation of singlet oxygen is important in photochemical reactions. Absorption of light of the correct energy can excite certain chemical agents known as photosensitizers into a higher energy state. This energy can be transferred to an oxygen molecule in close proximity, exciting it to its singlet state. The photosensitizer simultaneously returns to its ground state. Many endogenous compounds can act as photosensitizing agents, e.g., heme, bilirubin, retinal, and flavins. Similarly, there are exogenous photosensitizers, such as tetracycline antibiotics and acridine orange.

Ozone (O_3) is a pale blue gas with a distinctive odor that is formed from its allotrope oxygen by silent electrical discharge: $3O_2 \rightarrow 2O_3$ (allotropy is the ability of a substance to exist in two or more physical forms). Unlike all the other ROS, ozone is not produced endogenously. Ozone is extremely toxic to living organisms because it is a much more powerful oxidant than oxygen, capable of oxidizing DNA, lipids, and proteins. However, ozone is thermodynamically unstable with respect to oxygen and explosive at high concentrations.

Solar ultraviolet light consists of UVC (<280 nm), UVB (280 - 320 nm), and UVA (>320 - 400 nm) radiations, all of which are mutagenic. Oxygen is generated during photosynthesis in plants and released into the atmosphere where it absorbs the entire solar UVC fraction in the stratosphere. In the process, oxygen decomposes and recombines to form ozone which, in turn, absorbs most of the solar UVB radiation. The remaining mutagenic UV light to which living organisms on earth are exposed is mainly (~95%) composed of UVA (13). Thus, ozone is produced in the stratosphere by the action of sunlight on atmospheric oxygen and the stratospheric ozone layer is beneficial by absorbing mutagenic solar UV radiation. In contrast, in the lower troposphere ozone is a major pollutant and principal component of photochemical smog. Ozone can be formed in the vicinity of electrical machinery and may cause health problems in poorly ventilated areas.

In summary, ROS are generated during normal metabolism and pathological stress conditions. A moderate increase in ROS can promote cell proliferation and differentiation (8). However, excessive amounts of ROS can cause oxidative damage to proteins, lipids, and DNA. In response to the destructive nature of ROS, mammalian cells have evolved both nonenzymatic and enzymatic mechanisms to scavenge ROS. The nonenzymatic mechanism mainly involved small molecule antioxidants, such as reduced glutathione (GSH) and ascorbate (vitamin C). However, the more efficient clearance of ROS required the coordinate actions of anti-oxidant enzymes such as superoxide dismutase and catalase (see Chapter 4.4. Anti-oxidant Enzymes).

Reactive Nitrogen Species (RNS). Nitric oxide (NO•) is a reactive radical produced from arginine by nitric oxide synthase. Nitric oxide has a very short half-life and can react with superoxide to form peroxynitrite (ONOO⁻) (14). Peroxynitrite is one of the most reactive RNS that is capable of affecting protein function by modifying essential reactive thiols or tyrosine residues. For example, peroxynitrite has been shown to irreversibly inactivate phase I and II enzymes, including cytochromes P450, glutathione S-transferase, and N-acetyltransferase (15). Just like ROS, RNS play a role in physiological and pathophysiological processes, such as vasorelaxation, apoptosis, inflammation, and cancer through the oxidative modification of proteins, lipids, and DNA.

References

1. Acworth IN. *The Handbook of Redox Biochemistry.* Chelmsford, MA: ESA, Inc.; 2003.

2. Gutteridge JMC. *Lipid peroxidation and antioxidants as biomarkers of tissue damage.* Clin Chem. 1995;41:1819-28.

3. Halliwell B, Whiteman M. *Measuring reactive species and oxidative damage in vivo and in cell culture: how should you do it and what do the results mean?* Br J Pharmacol. 2004;142:231-55.

4. Chance B, Sies H, Boveris A. *Hydroperoxide metabolism in mammalian organs.* Physiol Rev. 1979;59:527-605.

5. Thannickal VJ, Fanburg BL. *Reactive oxygen species in cell signaling.* Am J Physiol Lung Cell Mol Physiol. 2000;279:L1005-L28.

6. Shigenaga MK, Aboujaoude EN, Chen Q, Ames BN. *Assays of oxidative DNA damage biomarkers 8-oxo-2'-deoxyguanosine and 8-oxoguanine in nuclear DNA and biological fluids by high-performance liquid chromatography and electrochemical detection.* Methods Enzymol. 1994;234:16-33.

7. Ralph SJ, Rodriguez-Enriquez S, Neuzil J, Saavedra E, Moreno-Sanchez R. *The causes of cancer revisited: "mitochondrial malignancy" and ROS-induced oncogenic transformation - why mitochondria are targets for cancer therapy.* Mol Aspects Med. 2010;31:145-70.

8. Trachootham D, Alexandre J, Huang P. *Targeting cancer cells by ROS-mediated mechanisms: a radical therapeutic approach?* Nat Rev Drug Discov. 2009;8:579-91.

9. Aykin-Burns N, Ahmad IM, Zhu Y, Oberley LW, Spitz DR. *Increased levels of superoxide and H2O2 mediate the differential susceptibility of cancer cells versus normal cells to glucose deprivation.* Biochem J. 2009;418:29-37.

10. Diehn M, Cho RW, Lobo NA, Kalisky T, Dorie MJ, Kulp AN, et al. *Association of reactive oxygen species levels and radioresistance in cancer stem cells.* Nature. 2009;458:780-3.

11. Fenton HJH. *Oxidation of tartaric acid in presence of iron.* J Chem Soc Trans. 1894;65:899-905.

12. Haber F, Weiss JJ. *The catalytic decomposition of hydrogen peroxide by iron salts.* Proc Roy Soc London A. 1934;147:332.

13. de Gruijl FR. *Photocarcinogenesis: UVA vs UVB radiation.* Skin Pharmacol Appl Skin Physiol. 2002;15:316-20.

14. Nordberg J, Arner ES. *Reactive oxygen species, antioxidants, and the mammalian thioredoxin system.* Free Radic Biol Med. 2001;31:1287-312.

15. Dairou J, Atmane N, Rodrigues-Lima F, Dupret JM. *Peroxynitrite irreversibly inactivates the human xenobiotic-metabolizing enzyme arylamine N-acetyltransferase 1 (NAT1) in human breast cancer cells.* J Biol Chem. 2004;279:7708-14.

2.3.2. REDOX CYCLING

ROS can also be produced as a byproduct of certain biochemical reactions, such as β-oxidation of fatty acids in peroxisomes, prostaglandin synthesis, and detoxification reactions by cytochrome P450 enzymes in form of redox cycling (1). Aromatic compounds with oxygen-containing substituents, such as phenols, hydroquinones, and catechols, can be converted to quinones enzymatically, by metal ions, and in some cases molecular oxygen. Quinones are commonly named as derivatives of their aromatic parent system, e.g., estrogen quinones, benzo[a]pyrene quinones. Several quinones undergo redox cycling, i.e., they form oxidation-reduction couples with their hydroquinones and semiquinone radicals, which are all interconvertible by reversible one-electron redox steps (2). The one-electron step redox cycles are coupled with molecular oxygen to form ROS, resulting in superoxide, hydrogen peroxide, and hydroxyl radical. These cycles can operate under physiological conditions in the presence of enzyme systems, such as cytochrome P450 and NADPH-dependent cytochrome P450 reductase, as shown for the oxidative metabolism of estrogen (Figure 2.3.2.).

As oxidants, estrogen quinones redox cycle with their semiquinones, e.g., 4-OHE_2 is oxidized to its semiquinone, $E_2\text{-3,4-SQ}$, and quinone, $E_2\text{-3,4-Q}$, followed by reduction of the quinone back to the semiquinone or catechol. The one- and two-electron reductions of the quinone and semiquinone are catalyzed by NADPH cytochrome P450 reductase and other reducing enzymes. Likewise, the oxidation of the catechol to the semiquinone and quinone is catalyzed by cytochrome P450, e.g., CYP1B1, and other oxidative enzymes. The enzyme-mediated redox cycling produces NOS. Alternatively, non-enzymatic redox couples between copper ions and catechol estrogens can also generate ROS (3-5). Together, the metal ion-catalyzed and enzyme-mediated redox cycling can continually form ROS from catalytic amounts of catechol estrogens that are reused in the process. This cycling reaction may go on indefinitely, depending on the availability of catechol substrate, appropriate enzymes, and metal ions. The ROS can cause several types of DNA damage including oxidized bases and single- and double-strand breaks (6-10).

Quinone metabolites of exogenous compounds, such as the polycyclic aromatic hydrocarbons (PAHs, see Chapter 3.3), can also undergo redox cycling with their semiquinone radicals and hydroquinones to generate ROS (11). An example is benzo[a]pyrene (BP), which is converted into a radical cation by one-electron oxidation resulting from the removal of a π electron (12). This reaction, which is catalyzed by cytochrome P450 and peroxidase enzymes, leads to the formation of the phenol 6-hydroxy-benzo[a]pyrene (6-OH-BP) (13). 6-OH-BP is a major reactive metabolite of BP, which rapidly autoxidizes by a one-electron mechanism to the isomeric quinone metabolites BP-1,6-dione, BP-3,6-dione, and BP-6,12-dione (Figure 2.3.3.). All three BP-diones are easily reduced by NAD(P)H, glutathione, or copper to intermediate semiquinone radicals and further to the corresponding BP-diols (14). The BP-diols, in turn, are rapidly autoxidized to the BP-diones upon exposure to oxygen. Thus, catalytic amounts of these BP metabolites,

together with respiratory enzymes such as NADH dehydrogenase, function as cyclic oxidation-reduction couples that link NADH and oxygen in the continuous production of ROS (15). In summary, both endogenous and exogenous hydroquinone/catechol – quinone metabolites undergo reversible, univalent oxidation-reduction cycles, which produce ROS capable of causing oxidative DNA damage.

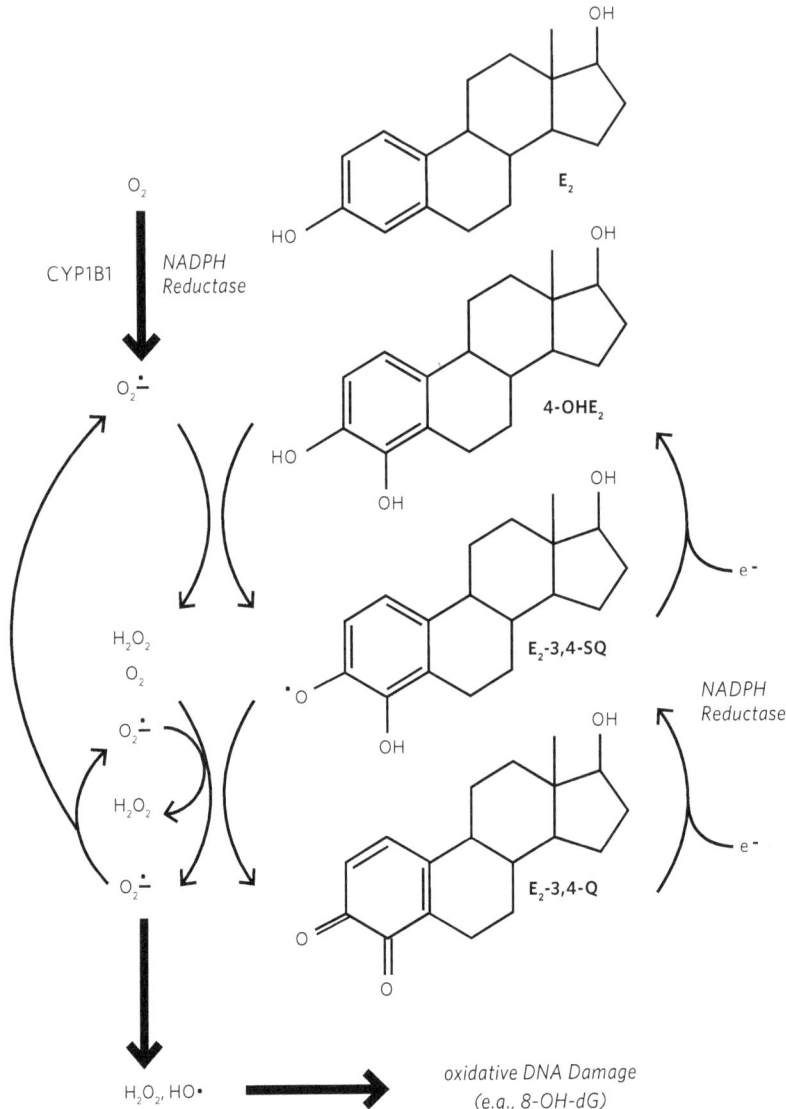

Figure 2.3.2. The oxidative metabolism of estrogens leads to the formation of catechol estrogens and estrogen quinones. *The oxidation enables redox cycling between catechol estrogens and estrogen quinones and results in the formation of ROS. Cytochrome P450 1B1 (CYP1B1) catalyzes the sequential oxidation of E_2 to the catechol estrogen, 4-OHE$_2$, its semiquinone, E_2-3,4-SQ, and quinone, E_2-3,4-Q. NADPH-dependent cytochrome P450 reductase catalyzes the one-electron reductions of the quinone and semiquinone thereby enabling the redox cycle. The resulting ROS cause damage to macromolecules including DNA, resulting in oxidative DNA adducts, e.g., 8-hydroxy-2'-deoxy-guanosine (8-OH-dG).*

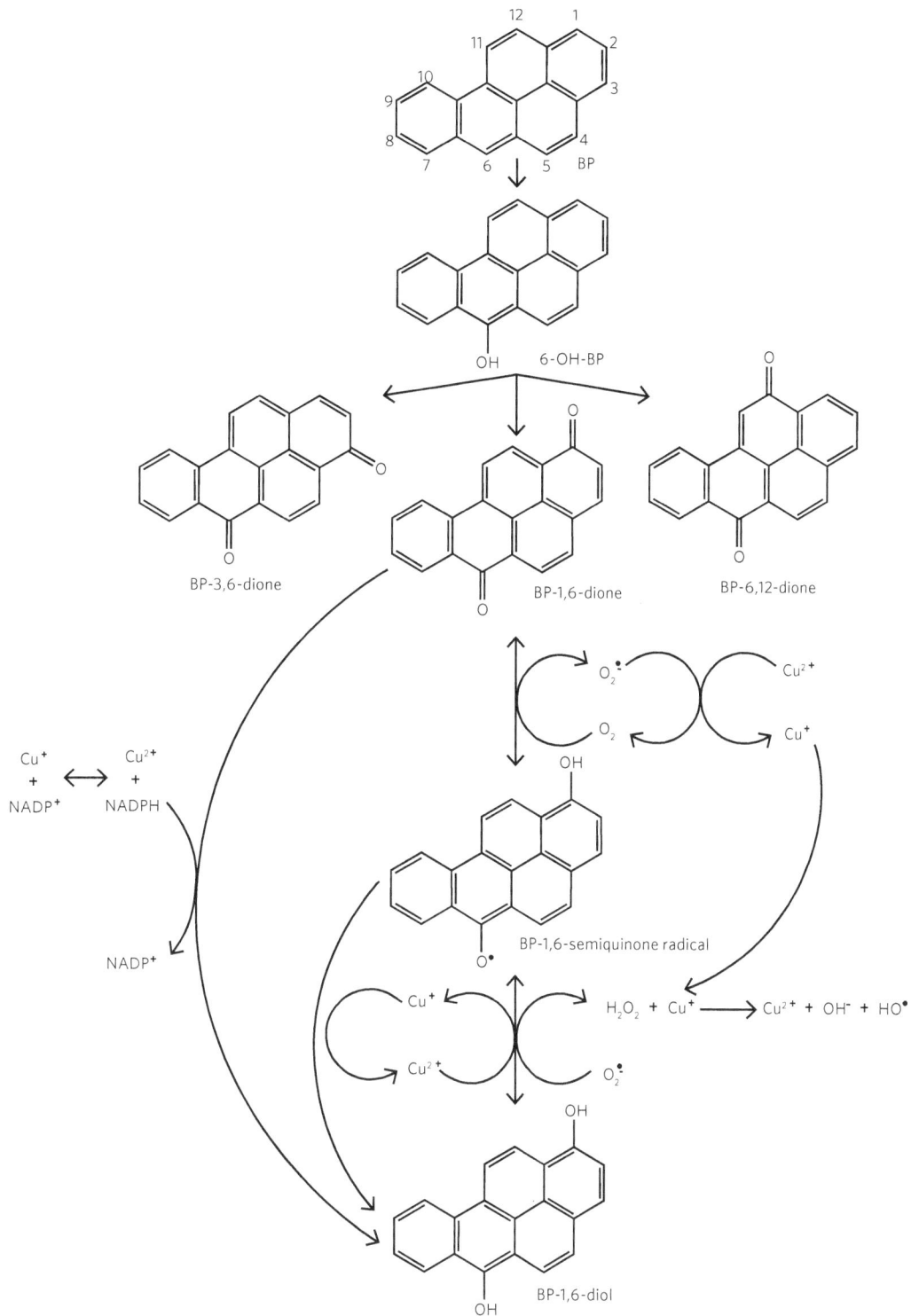

Figure 2.3.3. Oxidative metabolism of benzo[a]pyrene (BP) leads to 6-hydroxy-benzo[a]pyrene (6-OH-BP) and three isomeric quinones, BP-1,6-dione, BP-3,6-dione, and BP-6,12-dione.
All three quinones undergo redox cycling in the presence of NAHPH and cupric chloride resulting in the formation of ROS. Only the BP-1,6-dione/ diol cycle is shown.

References

1. Trachootham D, Alexandre J, Huang P. *Targeting cancer cells by ROS-mediated mechanisms: a radical therapeutic approach?* Nature Reviews Drug Discovery. 2009;8:579-91.

2. Bolton JL, Trush MA, Penning TM, Dryhurst G, Monks TJ. *Role of quinones in toxicology.* Chem Res Toxicol. 2000;13:135-60.

3. Hiraku Y, Yamashita N, Nishiguchi M, Kawanishi S. *Catechol estrogens induce oxidative DNA damage and estradiol enhances cell proliferation.* Int J Cancer. 2001;92:333-7.

4. Li Y, Trush MA, Yager JD. *DNA damage caused by reactive species originating from a copper-dependent oxidation of the 2-hydroxy catechol of estradiol.* Carcinogenesis. 1994;15:1421-7.

5. Mobley JA, Bhat AS, Brueggemeier RW. *Measurement of oxidative DNA damage by catechol*

estrogens and analogues in vitro. Chem Res Toxicol. 1999;12:270-7.

6. Han X, Liehr JG. *8-Hydroxylation of guanine bases in kidney and liver DNA of hamsters treated with estradiol: role of free radicals in estrogen-induced carcinogenesis.* Cancer Res. 1994;54:5515-7.

7. Han X, Liehr JG. *DNA single-strand breaks in kidneys of Syrian hamsters treated with steroidal estrogens: hormone-induced free radical damage preceding renal malignancy.* Carcinogenesis. 1994;15:997-1000.

8. Han X, Liehr JG. *Microsome-mediated 8-hydroxylation of guanine bases of DNA by steroid estrogens: correlation of DNA damage by free radicals with metabolic activation to quinones.* Carcinogenesis. 1995;16:2571-4.

9. Nutter LM, Ngo EO, Abul-Hajj YJ. *Characterization of DNA damage induced by 3,4-estrone-o-quinone*

in human cells. J Biol Chem. 1991;25:16380-6.

10. Roy D, Floyd RA, Liehr JG. *Elevated 8-Hydroxydeoxyguanosine levels in DNA of diethylstilbestrol-treated Syrian hamsters: covalent DNA damage by free radicals generated by redox cycling of diethylstilbestrol.* Cancer Res. 1991;51:3882-5.

11. Burdick AD, Davis JW, Liu KJ, Hudson LG, Shi H, Monske ML, et al. *Benzo(a)pyrene quinones increase cell proliferation, generate reactive oxygen species, and transactivate the epidermal growth factor receptor in breast epithelial cells.* Cancer Res. 2003;63:7825-33.

12. Cavalieri EL, Rogan EG. *Central role of radical cations in metabolic activation of polycyclic aromatic hydrocarbons.* Xenobiotica. 1995;25:677-88.

13. Cavalieri EL, Rogan EG, Cremonesi P, Devanesan PD. *Radical cations as precursors in the metabolic formation of quinones from benzo[a]pyrene and 6-fluorobenzo[a]pyrene. Fluoro substitution as a probe for one-electron oxidation in aromatic substrates.* Biochem Pharmacol. 1988;37:2173-82.

14. Lorentzen RJ, Ts'o POP. *Benzo[a]pyrenedione/benzo[a]pyrenediol oxidation-reduction couples and the generation of reactive reduced molecular oxygen.* Biochemistry. 1977;16:1467-73.

15. Flowers L, Ohnishi ST, Penning TM. *DNA strand scission by polycyclic aromatic hydrocarbon o-quinones: role of reactive oxygen species, Cu(ii)/Cu(i) redox cycling, and o-semiquinone anion radicals.* Biochemistry. 1997;36:8640-8.

2.3.3. OXIDATIVE DNA DAMAGE

The interaction of ROS with DNA results in oxidative modifications of base and sugar, DNA-protein cross-links, abasic sites, single-strand breaks, and double-strand breaks (1). There are over 30 different types of oxidized bases that can be formed in DNA by ROS. Figure 2.3.4. shows examples of oxidative DNA adducts, which are generated by ROS resulting from any redox-cycling process, including redox cycling of estrogen and BP metabolites (Figure 2.3.2. and 2.3.3.). Most adducts formed by ROS are produced by the hydroxyl free radical, HO•, which attacks DNA in two ways: (i) hydrogen abstraction from the deoxyribose sugar units can lead to cleavage of the sugar—phosphate backbone of the DNA and (ii) addition to the double bonds of heterocyclic DNA bases at rate constants of 3 - 10 x 10^9 $M^{-1}s^{-1}$ (2). The addition to the π bonds of purines forms C4-OH-, C5-OH-, and C8-OH-adduct radical intermediates, which undergo different reactions. The C4-OH- and C5-OH-adduct radicals undergo dehydration and yield oxidizing purine (-H)• radicals, which convert back to the starting purine upon reduction. In contrast, the C8-OH radical intermediate can give rise to two different nonradical adducts as shown in Figure 2.3.5. for 2'-deoxy-guanosine (dG) (2-4). The C8-OH N7-radical is redox ambivalent depending on the redox status of the DNA. A shift in the redox status favoring reduction leads to the ring-opened 2,6-diamino-4-hydroxy-5-formamidopyrim-idine 2'-deoxyguanosine (FAPy-dG), whereas the shift in favor of oxidation results in the formation of 8-hydroxy-2'-deoxyguanosine (8-OH-dG). The 8-OH-dG can exist

in four tautomeric forms and two alternative molecular conformations between the base and deoxyribose moieties, anti or syn N9-C1'-linkage (5). While guanine normally binds to cytosine, the 6,8-diketo tautomer of 8-OH-dG, 8-oxo-dG in the syn conformation can form hydrogen bonds with adenine. Specifically, 8-oxoG differs from G by the oxo group at C8 and an NH at N7, which allows 8-oxoG to base pair with either C or A. Thus, formation of 8-oxo-dG causes DNA misreplication, which can lead to mutation, particularly G/T transversion. In vitro replication of 8-oxo-dG-containing templates by mammalian DNA polymerases α and β caused mutations in nearly 100% (6). However, the mutation frequency observed in vivo in different cultured cells was only 17% because of the efficient removal of 8-oxo-dG by excision repair (7-11). The mutation frequency of 8-oxo-dA was at least four times lower than that of 8-oxo-dG (10).

8-oxo-dG is highly reactive toward further oxidation and can give rise to several other oxidative compounds including 5-hydroxy-8-oxo-dG, an intermediate that forms two different hydantoin products, 5-guanidinohydantoin (Gh) and spiroiminodihydantoin (Sp) (Figure 2.3.5.) (12-16). Hydration and decarboxylation of 5-hydroxy-8-oxo-dG leads to the open-ring product Gh, which predominates in double-stranded DNA. Alternatively, 5-hydroxy-8-oxo-dG isomerizes via acyl migration to the bulky heterocycle Sp, which predominates in single-stranded DNA and in the cellular nucleoside pool. Both Sp and Gh contain a

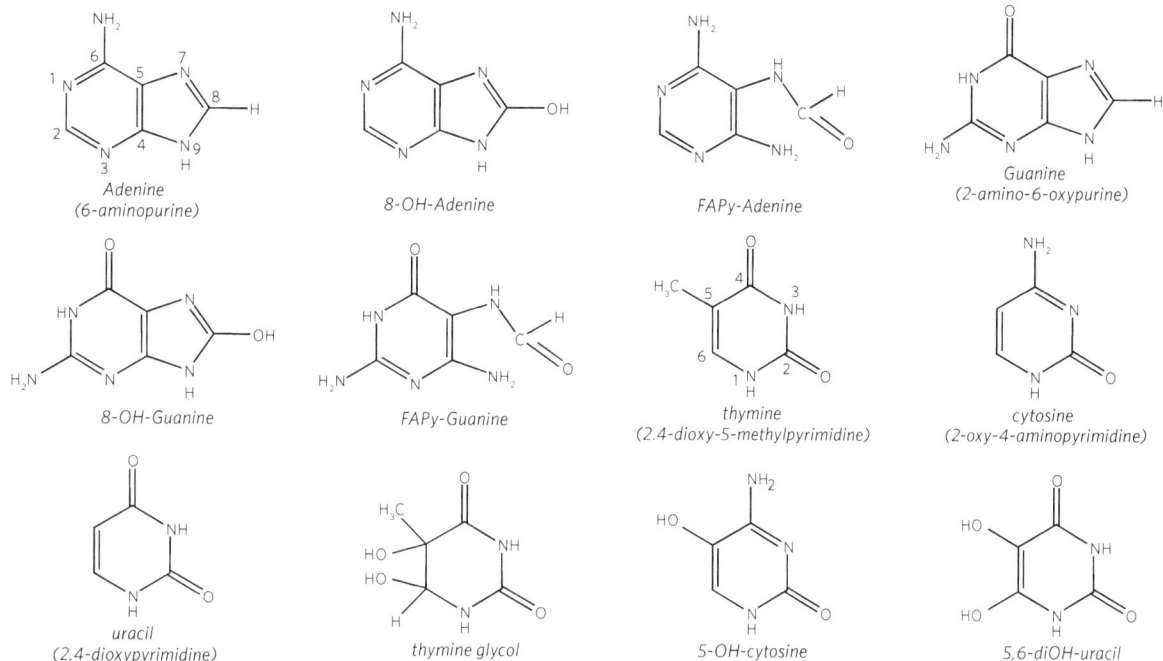

Figure 2.3.4. Nucleic acid bases and examples of oxidative adducts (4,6-diamino-5-formamidopyrimidine = FAPy-adenine; 2,6-diamino-4-hydroxy-5-formamidopyrimidine = FAPy-guanine).

tetrahedral, sp3, carbon within the normally planar nucleic acid base ring structure. This conformational change would be expected to disrupt normal base pair stacking in duplex DNA and thereby distort the DNA helix. Otherwise, the structures of the hydantoins resemble 8-oxoG and FapyG in retaining hydrogen-bonding functionality that can mimic T. Both Gh and Sp are an order of magnitude more mutagenic than 8-oxo-dG and cause G/T and/or G/C transversions (17). While most investigations have focused on oxidative base damage, the ROS can also modify the sugar moiety of DNA (18). For example, the hydroxyl radical, HO•, can abstract an H-atom from any carbon atom (except C2') of 2'-deoxyribose at rate constants of ~2 x 10^9 $M^{-1}s^{-1}$ (3, 4). The carbon-centered sugar radicals undergo further reactions that can lead to DNA strand breaks and base-free sites by a variety of mechanisms.

Oxidative DNA adducts can be measured by several techniques, such as HPLC with or without electrochemical detection, gas or liquid chromatography with mass spectrometry, capillary electrophoresis, ^{32}P postlabeling, and antibody-based methods (4, 19-21). Each technique offers specific analytical advantages but all face the risk of measuring artifacts when applied to body fluids or tissue samples (20, 22-25). The problem of accurately quantifying levels of oxidative DNA lesions in cells is compounded by the fact the process of isolation and characterization of oxidized DNA can lead not just to more oxidation during

manipulation but to oxidative destruction as well (12). Guanine is the most electron-rich of the four DNA bases and therefore highly susceptible to attack by electrophiles including oxidizing agents. Hence, guanine is the most easily oxidized of the nucleic acid bases and 8-oxo-dG the most commonly detected oxidative adduct found in normal human cells at a concentration of one adduct per 10^6 guanines and at higher levels in malignant cells (24). 8-oxo-dG is produced by many carcinogens including BP and oxidative estrogen metabolites (26).

The catechol estrogen pathway can induce DNA mutations via two mechanisms (i) direct estrogen quinone adduction (Chapter 2.1.3.) and (ii) ROS from redox cycling (27). However, there is disagreement about the relative importance of each process (28-30). In vitro experiments show that catechol estrogens induce oxidative adduct formation and strand breaks (31). Both types of DNA damage are dose-dependent, require the presence of copper, and are enhanced by NADH. Similar concentrations of 2-OHE$_2$ and 4-OHE$_2$ (10 to 100 nM) in the presence of physiological concentrations of copper (Cu II; 20 µM) and NADH (100 µM) resulted in the formation of 8-oxo-dG in calf thymus DNA (32). In contrast, 4-OHE$_2$ caused more extensive strand breaks than 2-OHE$_2$ (33, 34). 8-oxo-dG concentrations were also increased over control values in DNA incubated with 4-OHE$_2$ but not 2-OHE$_2$ and a microsomal activating system (35). Similarly,

DNA analysis of catechol estrogen-induced renal cancer in Syrian hamster showed that 4-OHE$_2$ significantly increased 8-oxo-dG levels, whereas 2-OHE$_2$ did not cause oxidative DNA adduction (36). 8-oxo-dG and 8-oxo-dA were also produced in breast tissue of rats following the intramammillary injection of 200 nmol equine 4-OHEN (37).

Oxidized bases are removed from the DNA by base excision repair (see Chapter 5.2.). The base excision repair is initiated by a lesion-specific DNA glycosylase, which recognizes and removes the damaged base. For example, 8-oxo-dG is recognized and removed by 8-oxoG-DNA glycosylase (OGG1), yielding 8-oxoG. Urine contains 8-oxoG but also 8-oxo-dG and the latter frequently measured as

a general biomarker of oxidative stress although the DNA repair source of the modified 2'-deoxyribonucleoside is uncertain (38). The examination of oxidatively modified DNA lesions in urine, such as 8-oxo-dG, allows noninvasive assessment of oxidative stress, which circumvents tissue extraction and the resultant artifacts (38). However, there remains a lack of consensus between methods and reference ranges for healthy and diseased individuals have not been established.

Urinary 8-oxo-dG levels have been measured in breast cancer patients. The Long Island Breast Cancer Study Project (400 cases, 401 controls) and the Shanghai Women's Health Study (327 cases, 654 controls) did

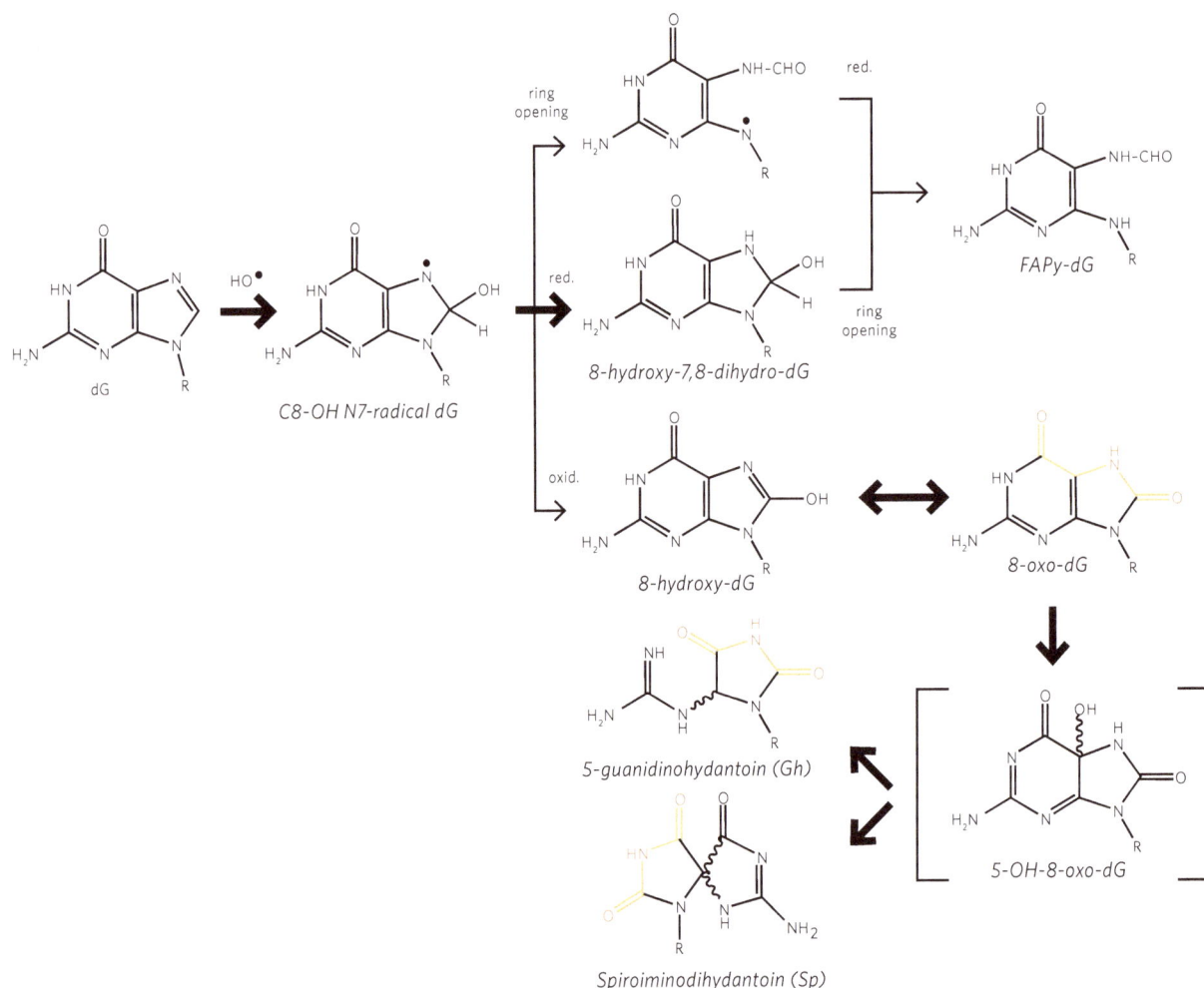

Figure 2.3.5. Oxidation products of guanine.
The attack of HO• on dG leads to the formation of the redox ambivalent C8-OH N7-radical. Reduction of this intermediate results in the ring-opened 2,6-diamino-4-hydroxy-5-formamidopyrimidine 2'-deoxyguanosine (FAPy-dG), whereas oxidation results in the formation of 8-OH-dG. There are at least four tautomeric forms of 8-OH-dG involving the N1, C6, N7, and C8 positions and two alternative molecular conformations between the base and deoxyribose moieties, anti or syn. The 6,8-diketo tautomer (8-oxo-dG) in the syn conformation is the predominant form. 8-oxo-dG can be further oxidized via the intermediate 5-OH-8-oxo-dG to two hydantoin compounds, 5-guanidinohydantoin (Gh) and spiroiminodihydantoin (Sp). Since they result from two sequential dG oxidation events, Gh and Sp are so-called hyperoxidized products.

not observe any association of breast cancer risk with 8-oxo-dG levels measured by ELISA (39, 40). However, a significant inverse trend (p = 0.04) was observed in the Long Island Breast Cancer Study Project when 77 cases were removed from the analysis that had received radiation treatment at some point before providing the urine sample. A smaller study used HPLC with electrochemical detection to compare 60 patients with cancer at stages I - III with 60 age-matched controls and observed significantly higher levels among the cancer subjects (41). Unexpectedly, the concentrations of 8-oxo-dG decreased with each stage of breast cancer, even after adjustment by creatinine and body weight. Concentrations of urinary 8-oxo-dG in stage I were 1.5 times higher than 8-oxo-dG levels in stage II, and 3.8 times higher than levels in stage III. However, in another study concentrations of several other urinary nucleosides were higher in patients with metastatic than primary breast cancers (42). Oxidative DNA adducts have also been identified in breast tissues. One study employed GC/MS to measure FAPy-A, FAPy-G, 8-oxo-A, and 8-oxo-G adducts in breast tissue samples from reduction mammoplasties, invasive ductal carcinomas, and microscopically normal breast adjacent to cancer (43). Seventy tissue samples obtained from 15 reduction mammoplasties contained a high ratio of FAPy-A + FAPy-G / 8-oxo-A + 8-oxo-G. In contrast, 22 malignant and normal appearing tissue samples from cancer mastectomies contained much lower levels of FAPy-A + FAPy-G and significantly higher concentrations of 8-oxo-dG. These findings agree with the notion that the open-ring FAPy adducts block DNA replication but do not cause mutations, whereas 8-oxo-A and 8-oxo-G lead to misincorporation of bases and mutations. The same group observed an age-related peak of 8-oxo-A and 8-oxo-G levels in breast stromal DNA of women aged 32 – 46 compared to either younger or older women (44). There was no difference in 8-oxo-A and 8-oxo-G levels of matched stromal, epithelial, and myoepithelial DNA from the same individual suggesting a random attack of ROS on DNA in all mammary cells. Additional studies reported a 2.3-fold increase in 8-oxo-dG concentration in breast cancer with lymph node metastases compared to non-metastatic cancer (45). Another group used HPLC with electrochemical detection and detected no difference in 8-oxo-dG levels between 22 breast cancers and adjacent non-cancerous breast tissues (46). The use of an immunoblot method showed a 9.8-fold higher 8-oxo-dG concentration in breast cancers than in normal mammary tissue and a 12.9-fold higher level in cultured breast cancer cells than in normal breast epithelial cells (47).

References

1. Nakamura J, Purvis ER, Swenberg JA. *Micromolar concentrations of hydrogen peroxide induce oxidative DNA lesions more efficiently than millimolar concentrations in mammalian cells.* Nucleic Acids Res. 2003;31:1790-5.

2. Steenken S. *Purine bases, nucleosides, and nucleotides: aqueous solution redox chemistry and transformation reactions of their radical cations and e- and OH adducts.* Chem Rev. 1989;89:503-20.

3. Breen AP, Murphy JA. *Reactions of oxyl radicals with DNA.* Free Rad Biol Med. 1995;18:1033-77.

4. Dizdaroglu M, Jaruga P, Birincioglu M, Rodriguez H. *Free radical-induced damage to DNA: mechanisms and measurement.* Free Rad Biol Med. 2002;32:1102-15.

5. Culp SJ, Cho BP, Kadlubar FF, Evans FE. *Structural and conformational analyses of 8-hydroxy-2'-deoxyguanosine.* Chem Res Toxicol. 1989;2:416-22.

6. Shibutani S, Takeshita M, Grollman A. *Insertion of specific bases during DNA synthesis past the oxidation-damaged base 8-oxodG.* Nature. 1991;349:431-4.

7. Bruner SD, Norman DPG, Verdine GL. *Structural basis for recognition and repair of the endogenous mutagen 8-oxoguanine in DNA.* Nature. 2000;403:859-66.

8. Cheng KC, Cahill DS, Kasai H, Nishimura S, Loeb LA. *8-hydroxyguanine, an abundant form of oxidative DNA damage, causes G → T and A → C substitutions.* J Biol Chem. 1992;267:166-72.

9. Klein JC, Bleeker MJ, Saris CP, Roelen HCPF, Brugghe HF, van den Elst H, et al. *Repair and replication of plasmids with site-specific 8-oxodG and 8-AAFdG residue in normal and repair-deficient human cells.* Nucleic Acids Res. 1992;20:4437-43.

10. Tan X, Grollman AP, Shibutani S. *Comparison of the mutagenic properties of 8-oxo-7,8-dihydro-2'-deoxyadenosine and 8-oxo-7,8-dihydro-2'-deoxyguanosine DNA lesions in mammalian cells.* Carcinogenesis. 1999;20:2287-92.

11. Tchou J, Kasai H, Shibutani S, Chung M-H, Laval J, Grollman AP, et al. *8-oxoguanine (8-hydroxyguanine) DNA glycosylase and its substrate specificity.* Proc Natl Acad Sci USA. 1991;88:4690-4.

12. Burrows CJ, Muller JG, Kornyushyna O, Luo W, Duarte V, Leipold MD, et al. *Structure and potential mutagenicity of new hydantoin products from guanosine and 8-Oxo-7,8-Dihydroguanine oxidation by transition metals.* Environ Health Perspect. 2002;110:713-7.

13. Luo W, Muller JG, Rachlin EM, Burrows CJ. *Characterization of hydantoin products from one-electron oxidation of 8-oxo-7,8-dihydroguanosine in a nucleoside model.* Chem Res Toxicol. 2001;14:927-38.

14. Neeley WL, Essigmann JM. *Mechanisms of formation, genotoxicity, and mutation of guanine oxidation products.* Chem Res Toxicol. 2006;19:491-505.

15. White B, Tarun MC, Gathergood N, Rusling JF, Smyth MR. *Oxidised guanidinohydantoin (Ghox) and spiroiminodihydantoin (Sp) are major products of iron- and copper-mediated 8-oxo-7,8-dihydroguanine and 8-oxo-7,8-dihydro-2'-deoxyguanosine oxidation.* Mol BioSyst. 2005;1:373-81.

16. Yu H, Venkatarangan L, Wishnok JS, Tannenbaum SR. *Quantitation of four guanine oxidation products from reaction of DNA with varying doses of peroxynitrite.* Chem Res Toxicol. 2005;18:1849-57.

17. Henderson PT, Delaney JC, Muller JG, Neeley WL, Tannenbaum SR, Burrows CJ, et al. *The hydantoin lesions formed from oxidation of 7,8-dihydro-8-oxoguanine are potent sources of replication errors in vivo.* Biochemistry. 2003;42:9257-62.

18. Demple B, DeMott MS. *Dynamics and diversions in base excision DNA repair of oxidized abasic lesions.* Oncogene. 2002;21:8926-34.

19. Dizdaroglu M. *Chemical determination of oxidative DNA damage by gas chromatography-mass spectrometry.* Methods Enzymol. 1994;234:3-16.

20. Halliwell B, Whiteman M. *Measuring reactive species and oxidative damage in vivo and in cell culture: how should you do it and what do the results mean?* Br J Pharmacol. 2004;142:231-55.

21. Shigenaga MK, Aboujaoude EN, Chen Q, Ames BN. *Assays of oxidative DNA damage biomarkers 8-oxo-2'-deoxyguanosine and 8-oxoguanine in nuclear DNA and biological fluids by high-performance liquid chromatography and electrochemical detection.* Methods Enzymol. 1994;234:16-33.

22. Cadet J, D'Ham C, Douki T, Pouget JP, Ravanat JL, Sauvaigo S. *Facts and artifacts in the* measurement of oxidative base damage to DNA. Free Rad Res. 1998;29:541-50.

23. Collins A, Cadet J, Epe B, Gedik C. *Problems in the measurement of 8-oxoguanine in human DNA.* Report of a workshop, DNA oxidation, held in Aberdeen, UK, 19-21 January, 1997. Carcinogenesis. 1997;18:1833-6.

24. Collins AR, Cadet J, Moller L, Poulsen HE, Vina J. *Are we sure we know how to measure 8-oxo-7,8-dihydroguanine in DNA from human cells.* Arch Biochem Biophys. 2004;423:57-65.

25. Weiss JM, Goode EL, Ladiges WC, Ulrich CM. *Polymorphic variation in hOGG1 and risk of cancer: A review of the functional and epidemiologic literature.* Mol Carcinog. 2005;42:127-41.

26. De Bont R, van Larebeke N. *Endogenous DNA damage in humans: a review of quantitative data.* Mutagenesis. 2004;19:169-85.

27. Chang M. *Dual roles of estrogen metabolism in mammary carcinogenesis.* BMB reports. 2011;44:423-34.

28. Cavalieri EL, Rogan EG. *A unified mechanism in the initiation of cancer.* Ann NY Acad Sci. 2002;959:341-54.

29. Cavalieri EL, Rogan EG. *A unifying mechanism in the iniation of cancer and other diseases by catechol quinones.* Ann NY Acad Sci. 2004;1028:247-57.

30. Lin PH, Nakamura J, Yamaguchi S, Asakura S, Swenberg JA. *Aldehydic DNA lesions induced by catechol estrogens in calf thymus DNA.* Carcinogenesis. 2003;24:1133-41.

31. Mobley JA, Bhat AS, Brueggemeier RW. *Measurement of oxidative DNA damage by catechol estrogens and analogues in vitro.* Chem Res Toxicol. 1999;12:270-7.

32. Hiraku Y, Yamashita N, Nishiguchi M, Kawanishi S. *Catechol estrogens induce oxidative DNA damage and estradiol enhances cell proliferation.* Int J Cancer. 2001;92:333-7.

33. Thibodeau PA, Kachadourian R, Lemay R, Bisson M, Day BJ, Paquette B. *In vitro pro- and antioxidant properties of estrogens.* J Steroid Biochem Mol Biol. 2002;81:227-36.

34. Thibodeau PA, Paquette B. *DNA damage induced by catecholestrogens in the presence of copper (ii): generation of reactive oxygen species and enhancement by NADH.* Free Radic Biol Med. 1999;27:1367-77.

35. Han X, Liehr JG. *Microsome-mediated 8-hydroxylation of guanine bases of DNA by steroid estrogens: correlation of DNA damage by free radicals with metabolic activation to quinones.* Carcinogenesis. 1995;16:2571-4.

36. Han X, Liehr JG. *8-Hydroxylation of guanine bases in kidney and liver DNA of hamsters treated with estradiol: role of free radicals in estrogen-induced carcinogenesis.* Cancer Res. 1994;54:5515-7.

37. Zhang F, Swanson SM, Van Breemen RB, Liu X, Yang Y, Gu C, et al. *Equine estrogen metabolite of 4-hydroxyequilenin induces DNA damage in the rat mammary tissues: Formation of single-strand breaks, apurinic sites, stable adducts, and oxidized bases.* Chem Res Toxicol. 2001;14:1654-9.

38. Cooke MS, Olinski R, Loft S. *Measurement and meaning of oxidatively modified DNA lesions in urine.* Cancer Epidemiol Biomarkers Prev. 2008;17:3-14.

39. Lee KH, Shu XO, Gao YT, Ji BT, Yang G, Blair A, et al. *Breast cancer and urinary biomarkers of polycyclic aromatic hydrocarbon and oxidative stress in the Shanghai Women's Health Study.* Cancer Epidemiol Biomarkers Prev. 2010;19:877-83.

40. Rossner P, Terry MB, Gammon MD, Zhang FF, Teitelbaum SL, Eng SM, et al. *OGG1 polymorphisms and breast cancer risk.* Cancer Epidemiol Biomarkers Prev. 2006;15:811-5.

41. Kuo HW, Chou SY, Hu TW, Wu FY, Chen DJ. *Urinary 8-hydroxy-2'-deoxyguanosine (8-OHdG) and genetic polymorphisms in breast cancer patients.* Mutation Research. 2007;631:62-8.

42. Zheng YF, Kong HW, Xiong JH, Lv S, Xu GW. *Clinical significance and prognostic value of urinary nucleosides in breast cancer patients.* Clin Biochem. 2005;38:24-30.

43. Malins DC, Holmes EH, Polissar NL, Gunselman SJ. *The etiology of breast cancer. Characteristic alteration in hydroxyl radical-induced DNA base lesions during oncogenesis with potential for evaluating incidence risk.* Cancer. 1993;71:3036-43.

44. Malins DC, Anderson KM, Jaruga P, Ramsey CR, Gilman NK, Green VM, et al. *Oxidative changes in the DNA of stroma and epithelium from the female breast: potential implications for breast cancer.* Cell Cycle. 2006;5:1629-32.

45. Malins DC, Polissar NL, Gunselman SJ. *Progression of human breast cancers to the meststatic state is linked to hydroxyl radical-induced DNA damage.* Proc Natl Acad Sci USA. 1996;93:2557-63.

46. Nagashima M, Tsuda H, Takenoshita S, Nagamachi Y, Hirohashi S, Yokota J, et al. *8-hydroxydeoxyguanosine levels in DNA of human breast cancer are not significantly different from those of non-cancerous breast tissues by the HPLC-ECD method.* Cancer Lett. 1995;90:157-62.

47. Musarrat J, Arezina-Wilson J, Wani AA. *Prognostic and aetiological relevance of 8-hydroxyguanosine in human breast carcinogenesis.* Eur J Cancer. 1996;32A:1209-14.

2.3.4. ROS MODULATION OF GENE EXPRESSION

In addition to oxidative damage to DNA, ROS can also affect protein functions through multiple mechanisms, including regulation of protein expression, post-translational modifications and alteration of protein stability (1). The most significant effects of ROS on signaling pathways have been observed in the nuclear factor erythroid 2-related factor (NF-E2/rf2 or Nrf2), mitogen-activated protein (MAP) kinase/AP-1, hypoxia-inducible transcription factor 1α (HIF-1α), and NF-κB (2-4). The resultant altered gene expression patterns evoked by ROS may contribute to carcinogenesis.

ARE-Nrf2 Pathway. The nuclear factor erythroid 2-related factor (NF-E2/rf2 or Nrf2) is a transcription factor, which heterodimerizes with members of the small Maf family of transcription factors and binds to anti-oxidant response elements (AREs with the consensus sequence TGACnnnGC) in promoters leading to the induction of ARE-regulated phase 2 genes (5, 6). Among these genes are several that encode antioxidant enzymes, such as superoxide dismutase, catalase, glutathione peroxidase, glutathione reductase, thioredoxin, and thioredoxin reductase (7, 8). Under basal unstressed conditions, the activity of Nrf2 is regulated by a cytosolic repressor protein, Keap1 (Kelch-like ECH-associated protein 1). Keap1 binds tightly to Nrf2 and targets it for ubiquitination and proteasomal degradation through Cullin 3-based E3 ligase, thereby repressing the ability of Nrf2 to induce phase 2 genes. Upon exposure to ROS, redox-sensitive cysteine residues in Keap1 are oxidized resulting in conformational change that leads to the disruption of the Nrf2-Keap1 interaction and translocation of the liberated Nrf2 to the nucleus (6, 9). Thus, pro-oxidants trigger the ARE-Nrf2 pathway by modifying Keap1, causing dissociation of Nrf2, which translocates into the nucleus where it heterodimerizes with small Maf proteins and activates ARE-driven transcription (6, 9, 10). The ROS-induced induction of Nrf2 expression promotes the production of anti-oxidant enzymes that control ROS levels as part of a physiological detoxification program in normal cells. The subcellular localization studies have been questioned because of nonspecific cross-reactivity of the anti-Nrf2 antibody and the observation that immunohistochemical studies with a more specific antibody detected Nrf2 in the nucleus in the absence of stress inducers (11).

Somatic mutations of both Nrf2 and Keap1 genes have been identified in lung, gall bladder, and head and neck cancers leading to constitutive activation of Nrf2 (5). This presents an apparent conundrum because Nrf2 protects normal cells against carcinogenesis. It appears that its constitutive activation in certain tumors contributes to the survival and growth of cancer cells. Interestingly, certain oncogenic alleles (e.g., K-Ras Gly12Asp) can also increase Nrf2 transcription to stably elevate the basal Nrf2 anti-oxidant program and thereby lower intracellular ROS (12). This mechanism confers a sustained reduced intracellular environment that enables tumor initiation.

One study reported a Keap1 mutation, Cys23Tyr, in the breast cancer cell line Q293 resulting in its inability to repress Nrf2 (13).

References

1. Trachootham D, Alexandre J, Huang P. *Targeting cancer cells by ROS-mediated mechanisms: a radical therapeutic approach?* Nature Rev Drug Discovery. 2009;8:579-91.

2. Finkel T, Holbrook NJ. *Oxidants, oxidative stress and the biology of ageing.* Nature. 2000;408:239-47.

3. Klaunig JE, Kamendulis LM, Hocevar BA. *Oxidative stress and oxidative damage in carcinogenesis.* Toxicol Pathol. 2010;38:96-109.

4. Verschoor ML, Wilson LA, Singh G. *Mechanisms associated with mito-chondrial-generated reactive oxygen species in cancer.* Canadian J Physiol Pharmacol. 2010;88:204-19.

5. Hayes JD, McMahon M. *NRF2 and KEAP1 mutations: permanent activation of an adaptive response in cancer.* Trends in Biochem Sciences. 2009;34:176-88.

6. Itoh K, Wakabayashi N, Katoh Y, Ishii T, Igarashi K, Engel JD, et al. *Keap1 represses nuclear activation of antioxidant responsive elements by Nrf2 through binding to the amino-terminal Neh2 domain.* Genes Development. 1999;13:76-86.

7. Osburn WO, Kensler TW. *Nrf2 singaling: an adaptive response pathway for protection against environmental toxic insults.* Mutation Res. 2008;659:31-9.

8. Talalay P, Dinkova-Kostova AT, Holtzclaw WD. *Importance of phase 2 gene regulation in protection against electrophile and reactive oxygen toxicity and carcinogenesis.* Advances in Enzyme Regulation. 2003;43:121-34.

9. Wakabayashi N, Dinkova-Kostova AT, Holtzclaw WD, Kang MI, Kobayashi A, Yamamoto M, et al. *Protection against electrophile and oxidant stress by induction of the phase 2 response: Fate of cysteines of the Keap1 sensor modified by inducers.* Proc Natl Acad Sci USA. 2004;101:2040-5.

10. Li W, Kong AN. *Molecular mechanisms of Nrf2-mediated antioxidant response.* Mol Carcino-genesis. 2009;48:91-104.

11. Nguyen T, Nioi P, Pickett CB. *The Nrf2-antioxidant response element signaling pathway and its activation by oxidative stress.* J Biol Chem. 2009 15;284:13291-5.

12. DeNicola GM, Karreth FA, Humpton TJ, Gopinathan A, Wei C, Frese K, et al. *Oncogene-induced Nrf2 transcription promotes ROS detoxification and tumorigenesis.* Nature. 2011;475:106-9.

13. Nioi P, Nguyen T. *A mutation of Keap1 found in breast cancer impairs its ability to repress Nrf2 activity.* Biochem Biophys Res Commun. 2007;362:816-21.

2.4. LIPID PEROXIDATION
2.4.1. INTRODUCTION

Nearly all fatty acids occurring in animal cells have chain lengths of 14 to 26 carbon atoms. By and large, they are even-numbered, unbranched, saturated or unsaturated. The latter are mono- or polyunsaturated fatty acids (PUFAs). All double bonds in naturally produced fatty acids are in the cis-configuration (Table 2.4.1.). Chemists count the chain from the carbonyl carbon, whereas physiologists count in the opposite direction starting from the n (ω) methyl carbon. Thus, the term n-3 (also called ω-3 or omega-3) fatty acid signifies that the first double bond exists as the third carbon-carbon bond from the terminal methyl end of the carbon chain. Similarly, omega-6 and omega-9 fatty acids have final carbon-carbon bonds in the n-6 and n-9 positions, respectively (Table 2.4.1.).

Oxygen radicals attack not only DNA but other cellular macromolecules as well, including lipid components, such as PUFAs (1). PUFAs of phospholipid membranes are the most abundant subcellular target for such attack in a complex process termed lipid peroxidation (2, 3). PUFAs contain one or more methylene groups positioned between cis double bonds. The methylene-interrupted double bonds are highly reactive to oxidizing agents and facilitate free radical oxidation by abstraction of their bisallylic hydrogens. Abstraction of a hydrogen leads to a resonance-stabilized free radical intermediate and subsequent addition of oxygen to form peroxyl radicals.

Regeneration of a free carbon-centered radical allows propagation of the radical chain reaction (Figure 2.4.1.). The oxidizability of PUFAs is linearly dependent on the number of bisallylic methylenes present in a fatty acid (4). Doco-sahexaenoic acid (DHA) is the most highly unsaturated fatty acid in biological systems. DHA is typically esterified to the sn-2 position of a phospholipid glycerol backbone. The autoxidative addition of one oxygen molecule to DHA can yield up to ten positionally isomeric hydroperoxides containing a conjugated diene.

The fate of the peroxyl radical depends on its position in the carbon chain of the fatty acid. If the peroxyl radical exists at one of the two ends of the double bond system, it is reduced to a hydroperoxide. If the peroxyl radical is at an internal position, cyclization to an adjacent double bond produces a cyclic peroxide adjacent to a carbon-centered radical, which can undergo a second cyclization to form a bicyclic peroxide. Thus, lipid peroxidation generates a mixture of peroxides, which decompose into a wide array of oxidative degradation products including 4-hydroxyalkenals and malondialdehyde (MDA) (5, 6). MDA can also arise from prostaglandin metabolism by various cytochrome P450 enzymes, which convert the prostaglandin endoperoxide intermediate PGH2 to hydroxyheptadecatrienoic acid and

Table 2.4.1. Common Saturated and Unsaturated Fatty Acids

Common Name	No. of carbon atoms	No. of double bonds	Position of final double bond (ω)	Chemical Name
Palmitic acid	16	0	0	hexadecanoic acid
Stearic acid	18	0	0	octadecanoic acid
Oleic acid	18	1	1	9-octadecanoic acid
Linoleic acid	18	2	2	9,12-octadecanoic acid
α-Linolenic acid (ALA)	18	3	3	all-cis-9,12,15-octadecatrienoic acid
Arachidonic acid	20	4	4	5,8,11,14-eicosatetraenoic acid
Eicosapentaenoic acid (EPA)	20	5	5	all-cis-5,8,11,14,17-eicosapentaenoic acid
Docosahexaenoic acid (DHA)	22	6	6	all-cis-4,7,10,13,16-docosahexaenoic acid

MDA (7). In turn, the 4-hydroxyalkenals and MDA can form a variety of mutagenic DNA adducts (8, 9).

4-Hydroxyalkenals. Lipid peroxides undergo a variety of secondary reactions including cleavage to form 4-hydroxy-alkenals, such as 4-hydroxyhexenal, 4-hydroxynonenal, and 4-hydroxydecenal (5, 10). 4-hydroxynonenal is the major aldehyde formed during peroxidation of the omega-6 fatty acids, linoleic acid and arachidonic acid. While 4-hydroxynonenal is derived from oxidized omega-6 fatty acids, 4-hydroxyhexenal is derived from oxidized omega-3 fatty acids. The 4-hydroxyalkenals elicit a variety of biological effects, such as inhibition of enzymes, inhibition of calcium sequestration by microsomes, exhibition of chemotactic activity toward neutrophils, and inhibition of protein synthesis (11). The 4-hydroxyalkenals are thought to exert these effects by binding to the sulfhydryl groups of proteins. These bifunctional electrophiles also react with nucleosides to form several etheno-adducts, such as 1,N6-etheno-dA and 1,N2-etheno-dG (9, 12). For example, 4-hydroxy-2-nonenal reacts with the N2 group of guanine followed by ring closure at N1, resulting in the cyclic adduct trans-4-hydroxy-2-nonenal-deoxyguanosine (Figure 2.4.2.). The cyclization allows the formation of several stereoisomers (13). The etheno-adducts have mutagenic potential by inducing both transitions and transversions (14-17). It appears that the mutagenic potential of etheno-adducts is greater than that resulting from oxidative DNA damage (18-20). Etheno-adducts are formed as a result of lipid peroxidation and 1,N6-etheno-dA and 1,N2-etheno-dG are identified in normal human tissue DNA (9, 21). Adduct levels are influenced by dietary fat intake, the ratio of omega-6 fatty acids to other fatty acids and antioxidants such as vitamin E (α-tocopherol) consumed in the diet (22).

Malondialdehyde (MDA) is a naturally occurring product of lipid peroxidation and prostaglandin biosynthesis that is mutagenic and carcinogenic (1, 6, 10). MDA reacts with nucleic acid bases to form multiple adducts. Both its carbonyl equivalents adduct to N2 and N1 of dG with loss of two water molecules to form pyrimido[1,2α]purin-10(3H)-one, a pyrimidopurinone, designated M1G (Figure

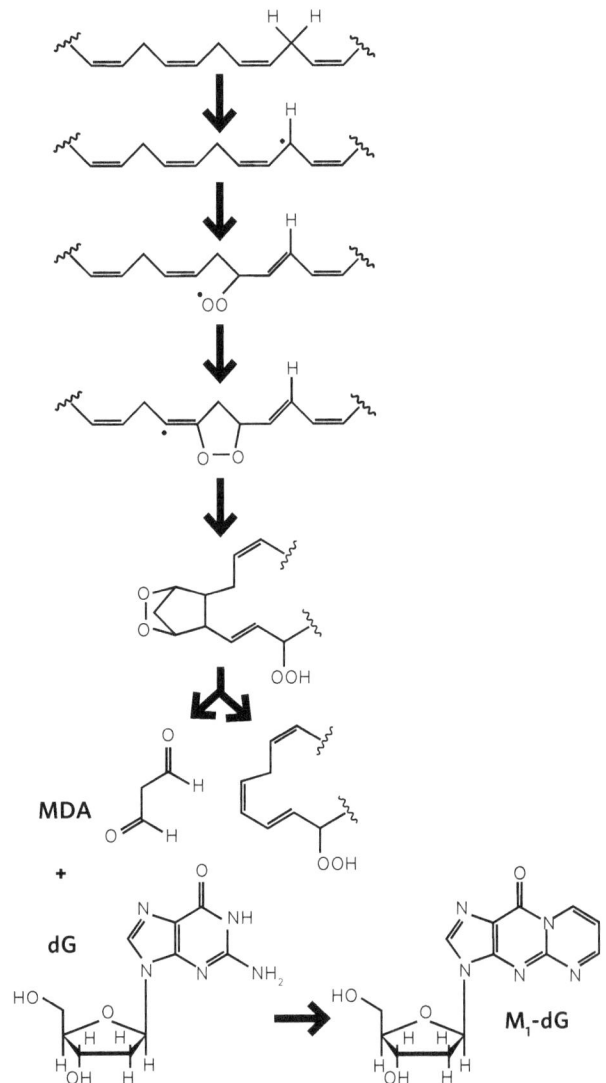

Figure 2.4.1. Pathway of MDA formation via lipid peroxidation of PUFA in phospholipid.
Radical-induced autoxidation of a tetraenoic PUFA results in bicyclic endoperoxide as MDA precursor. MDA forms adduct with dG resulting in pyrimido[1,2α]purin-10(3H)-one, designated M1G.

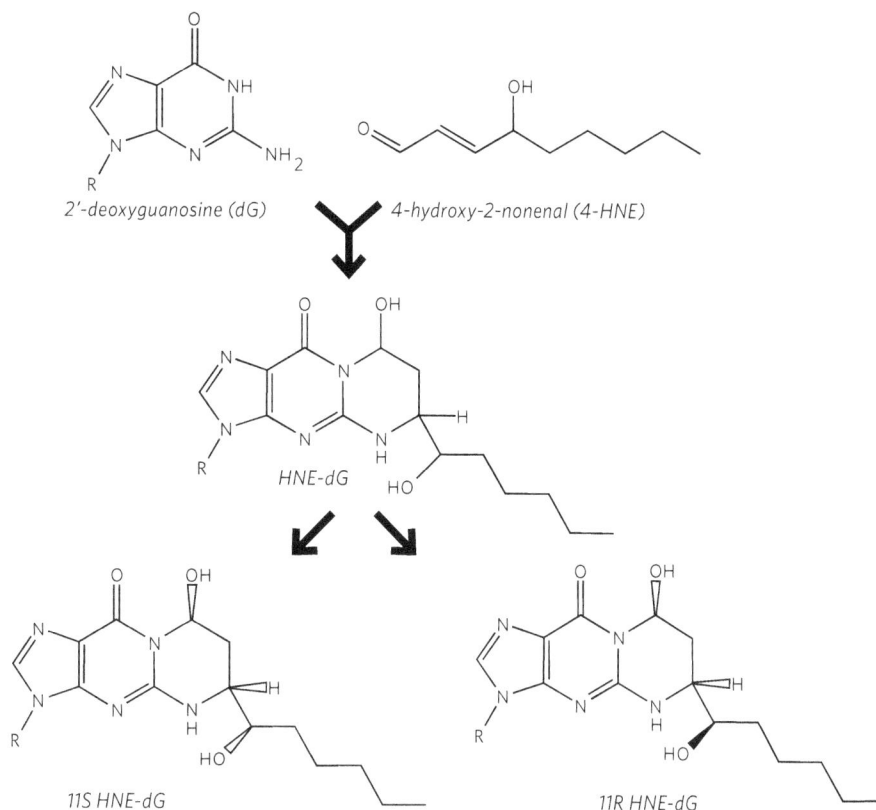

Figure 2.4.2. *4-hydroxynonenal-dG (HNE-dG) is formed from the reaction of 4-hydroxy-2-nonenal with 2'-deoxyguanosine. The exocyclic adduct consists of stereoisomers such as 11S and 11R HNE-dG.*

2.4.1.). The condensation products with dA and dC arise by addition of one of the carbonyl equivalents of MDA with the exocyclic amino groups to form oxopropenyl derivatives [N^6-(3-oxopropenyl)deoxyadenosine, M_1A] and [N^4-(3-oxo-propenyl)deoxycytidine, M_1C], respectively. M1G induced sequence-dependent frameshift mutations and base pair substitutions in bacteria and mammalian cells (23).

The extent of MDA formation from lipid peroxidation is determined by multiple factors including the composition of the parent lipid (e.g., the degree of PUFA unsaturation), the stimulus for oxygenation, and the reaction environment in which oxygenation takes place. Although MDA has been isolated as an end product of PUFA autoxidation in lipoproteins, membranes, and cells, constituents of these complex entities can modulate the extent of MDA formation. For example, iron, an abundant transition metal in living systems, promotes fatty hydroperoxide decomposition to MDA at physiological pH and temperature (3). Vitamin E, the principal peroxyl-radical-trapping antioxidant, influences the amount and types of decomposition products formed from fatty hydroperoxides (24).

F2-isoprostanes. Another group of lipid peroxidation markers are the F2-isoprostanes, which are generated by nonenzymatic free-radical-catalyzed peroxidation of arachidonic acid. Depending on which of the labile hydrogen atoms of arachidonic acid is first abstracted by free radicals, four different regioisomers are formed, which give rise to theoretically 64 diastereoisomers. Most studies have focused on 15-F_{2t}-isoprostane (10, 25). Isoprostanes are formed primarily *in situ* on phospholipids at sites of free radical generation. They are released from cell membranes into the circulation by phospholipases and excreted into the urine as free isoprostanes.

References

1. Marnett LJ. *Oxyradicals and DNA damage. Carcinogenesis.* 2000;21:361-70.

2. Esterbauer H, Schaur RJ, Zollner H. *Chemistry and biochemistry of 4-hydroxynonenal, malonaldehyde and related aldehydes.* Free Radical Biol Med. 1991;11:81-128.

3. Janero DR. *Malondialdehyde and thiobarbituric acid-reactivity as diagnostic indices of lipid peroxidation and peroxidative tissue injury.* Free Radical Biol Med. 1990;9:515-40.

4. Cosgrove JP, Church DF, Pryor WA. *The kinetics of the autoxidation of polyunsaturated fatty acids.* Lipids. 1987;22:299-304.

5. Lee SH, Oe T, Blair IA. *Vitamin C-induced decomposition of lipid hydroperoxides to endogenous genotoxins.* Science. 2001;292:2083-6.

6. Marnett LJ. *Lipid peroxidation-DNA damage by malondialdehyde.* Mutat Res. 1999;424:83-95.

7. Plastaras JP, Guengerich FP, Nebert DW, Marnett LJ. *Xenobiotic-metabolizing cytochromes P450* convert prostaglandin endoperoxide to hydroxyheptadecatrienoic acid and the mutagen, malondialdehyde. J Biol Chem. 2000;275:11784-90.

8. Chaudhary AK, Reddy GR, Blair IA, Marnett LJ. Characterization of an N6-oxopropenyl-2'-dexyadenosine adduct in malondialdehyde-modified DNA using liquid chromatography/ electrospray ionization tandem mass spectrometry. Carcinogenesis. 1996;17:1167-70.

9. Doerge DR, Churchwell MI, Fang JL, Beland FA. Quantification of etheno-DNA adducts using liquid chromatography, on-line sample processing, and electrospray tandem mass spectrometry. Chem Res Toxicol. 2000;13:1259-64.

10. Dalle-Donne I, Rossi R, Colombo R, Giustarini D, Milzani A. Biomarkers of oxidative damage in human disease. Clin Chem. 2006;52:601-23.

11. van Kuijk FJGM, Thomas DW, Stephens RJ, Dratz EA. Gas chromatography-mass spectometry of 4-hydroxynonenal in tissues. Methods Enzymol. 1990;186:399-406.

12. Sodum RS, Chung FL. 1,N2-ethenodeoxyguanosine as a potential marker for DNA adduct formation by trans-4-hydroxy-2-nonenal. Cancer Res. 1988;48:320-3.

13. Wolfle WT, Johnson RE, Minko IG, Lloyd RS, Prakash S, Prakash L. Replication past a trans-4-hydroxynonenal minor-groove adduct by the sequential action of human DNA polymerases iota and kappa. Mol Cell Biol. 2006;26:381-6.

14. Bartsch H. DNA adducts in human carcinogenesis: etiological relevance and structure-activity relationship. Mutat Res. 1996;340:67-79.

15. Langouet S, Mican AN, Muller M, Fink SP, Marnett LJ, Muhle SA, et al. Misincorporation of nucleotides opposite five-membered exocyclic ring guanine derivatives by Escherichia coli polymerases in vitro and in vivo: 1, N2-ethenoguanine, 5,6,7,9-tetrahydro-9-oxoimidazo [1,2-a] purine. Biochemistry. 1998;37:5184-93.

16. Pandya GA, Moriya M. 1,N6-ethenodeoxyadenosine, a DNA adduct highly mutagenic in mammalian cells. Biochemistry. 1996;35:11487-92.

17. Singer B, Kusmierek JT, Folkman W, Chavez F, Dosanjh MK. Evidence for the mutagenic potential of the vinyl chloride induced adduct, N2,3-etheno-deoxyguanosine, using a site-directed kinetic assay. Carcinogenesis. 1991;12:745-7.

18. Cadet J, D'Ham C, Douki T, Pouget JP, Ravanat JL, Sauvaigo S. Facts and artifacts in the measurement of oxidative base damage to DNA. Free Rad Res. 1998;29:541-50.

19. Levine RL, Yang IY, Hossain M, Pandya GA, Grollman AP, Moriya M. Mutagenesis induced by a single 1,N6-ethenodeoxyadenosine adduct in human cells. Cancer Res. 2000;60:4098-104.

20. Pang B, Zhou X, Yu H, Dong M, Taghizadeh K, Wishnok JS, et al. Lipid peroxidation dominates the chemistry of DNA adduct formation in a mouse model of inflammation. Carcinogenesis. 2007;28:1807-13.

21. Chung FL, Nath RG, Ocando J, Nishikawa A, Zhang L. Deoxyguanosine adducts of t-4-hydroxy-2-nonenal are endogenous DNA lesions in rodents and humans: detection and potential sources. Cancer Res. 2000;60:1507-11.

22. Hagenlocher T, Nair J, Becker N, Korfmann A, Bartsch H. Influence of dietary fatty acid, vegetable, and vitamin intake on etheno-DNA adducts in white blood cells of healthy female volunteers: a pilot study. Cancer Epidemiol Biomarkers Prev. 2001;10:1187-91.

23. VanderVeen LA, Hashim MF, Shyr Y, Marnett LJ. Induction of frameshift and base pair substitution mutations by the major DNA adduct of the endogenous carcinogen malondialdehyde. Proc Natl Acad Sci U S A. 2003;100:14247-52.

24. Gerber M, Richardson S, Crastes de Paulet P, Pujol H, Crastes de Paulet A. Relationship between vitamin E and polyunsaturated fatty acids in breast cancer. Nutritional and metabolic aspects. Cancer. 1989;64:2347-53.

25. Cracowski JL, Durand T, Bessard G. Isoprostanes as a biomarker of lipid peroxidation in humans: physiology, pharmacology and clinical implications. Trends Pharmacol Sci. 2002;23:360-6.

2.4.2. EXPERIMENTAL AND CLINICAL RESULTS

The mutagenicity of MDA was measured in the $lacZ\alpha$ forward mutation assay using a recombinant M13 phage, M13MB102 (1). Single-stranded M13MB102 DNA was reacted with MDA at neutral pH and the modified DNA transformed into strains of E. coli induced for the UV irradiation response. Increasing concentrations of MDA led to an increase in $lacZ\alpha$-mutations coincident with an increase in the level of the M1G adduct. Analysis of the $lacZ\alpha$- mutants by DNA sequencing revealed a range of mutations. The most common sequence changes induced by MDA were base-pair substitutions (76%). Of these, 43% (29/68) were transversions, most of which were G/T. Transitions (57%) were comprised exclusively of C/T (22/39) and A/G (17/39). Frameshift mutations were identified in 16% of the induced mutants and were comprised of mainly single base additions occurring in runs of reiterated bases (11/14). The ability of MDA to induce a variety of mutations at low levels of DNA modification suggests that it may also play a role in spontaneous mutagenesis. However, it is not possible to unequivocally establish the involvement of MDA in the etiology of a particular genetic disease because some of the MDA-induced mutations are also induced by other endogenous agents (e.g., G/T transversions by oxygen radicals and C/T transitions by nitric oxide and spontaneous deamination). Nevertheless, the major MDA-DNA adduct, M_1G, occurs endogenously at significant levels in rodent liver DNA and its formation is stimulated in animal models by lipid peroxidation (2).

M_1G has been detected in liver, white blood cells, and breast epithelium from healthy individuals at steady-state levels ranging from 1 - 120 per 10^8 nucleotides (3-5). The amount of M_1G is approximately five times that of M_1A, whereas M_1C is found only in trace amounts (6). Examination of MDA-DNA adducts in human breast showed significantly higher levels of MDA-dA and MDA-dG in non-cancerous tissue adjacent to breast cancer than in normal tissue obtained from reduction mammoplasty. However, these results are complicated by confounding variables such as patient age, body mass, smoking status, and other benzo[a]pyrene-like DNA adducts. In fact, in 11 patients with breast cancer, MDA-DNA adduct levels were actually lower in the tumor tissue than in the surrounding normal tissue (7). Fine-needle aspirate samples of 22 patients with breast cancer contained significantly higher MDA-dG levels than samples from 13 controls (8). Poorly-differentiated tumors had higher levels than well-differentiated tumors.

In several studies, serum/plasma levels of MDA were found to be higher among breast cancer patients than controls (9-14). Urinary MDA levels were measured in premenopausal women at different risks for breast cancer as determined by the appearance of the breast parenchyma on mammography (15). Levels in 24-h urine samples in 30 women with mammographic dysplasia were approximately double that of women without these radiologic changes. Urinary MDA and F2-isoprostances were also measured in breast cancer case-control studies. The Shanghai Women's Health Study (327 cases, 654 controls) did not observe any association of breast cancer risk with MDA levels (16). On the other hand, urinary F2-isoprostances were positively associated with breast cancer risk among overweight women (17). The Long Island Breast Cancer Study Project (400 cases, 401 controls) observed a positive association of increasing quartiles of $15\text{-}F_{2t}$-isoprostane levels with breast cancer risk ($p_{trend} = 0.002$) (18).

The prevailing view is that lipid peroxidation contributes to malignant transformation and carcinogenesis via MDA and isoprostane formation. However, the opposing view has been advanced that lipid peroxidation might protect against breast cancer by inducing apoptosis and growth arrest (19). Estrogens, by virtue of their oxidative metabolism, may play a role in stimulating lipid peroxidation and MDA formation. In an animal model of estrogen-induced carcinogenesis, E_2 treatment of male Syrian hamsters increased renal concentrations of both lipid peroxides and MDA-DNA adducts (5). If oxidative estrogen metabolism can induce lipid peroxidation, the question arises whether lipid peroxidation, in turn, can facilitate oxidative estrogen metabolism. As illustrated in the following Figure 2.4.3., one can envision interaction of the oxidative estrogen metabolism and lipid peroxidation pathways, which may increase the potential to cause DNA damage. Specifically, several types of DNA adducts are formed simultaneously by two classes of endogenous compounds, catechol estrogens, which are formed during the oxidative metabolism of estrogens, and MDA, which is formed enzymatically as a byproduct of the cyclooxygenase pathway of arachidonic acid metabolism as well as nonenzymatically during the peroxidation of PUFAs.

The subcellular distribution of CYP1B1 and its cohort enzyme, P450 NADPH reductase, may affect the disposition of E_2 and its metabolites. The hydrophobicity of E_2 permits its preferential partitioning into lipid membranes such as the endoplasmic reticulum, which facilitates access to CYP1B1 and metabolism to catechol estrogens. The resultant catechols are also lipophilic in nature and may be retained in the hydrophobic environment of the endoplasmic reticulum, where they may undergo oxidation to the corresponding estrogen quinones. Thus, oxidative estrogen metabolism takes place in the immediate vicinity of the phospholipid bilayer membranes of the endoplasmic reticulum. In fact, the phospholipid plays an essential role for CYP enzyme activity. *In vitro* experiments show that recombinant, purified CYP is inactive without the presence of both P450 NADPH reductase and phospholipid (20). As a matter of

experimental convenience, the phospholipid employed in these *in vitro* assays of CYP activity is typically composed of phosphatidyl choline esterified with two saturated fatty acids, e.g., L-α-dilauroyl-sn-glycero-3-phosphocholine. Yet, *in vivo*, phospholipids are generally esterified at the sn-2 position with PUFAs, such as arachidonic acid (20:4ω6) or docosahexaenoic acid (22:6ω3). Thus, PUFAs are part of the ternary system CYP1B1 – P450 NADPH reductase – polyunsaturated phospholipid that estrogens encounter physiologically in the cellular environment.

Thus, oxidative estrogen metabolism and lipid peroxidation may be linked increasing the potential to cause DNA damage. As described above, E_2 treatment of male Syrian hamsters increased renal concentrations of both lipid peroxides and MDA-DNA adducts (5). Although these findings suggest a connection between estrogen metabolism and lipid peroxidation, the molecular basis for such a relation cannot be ascertained in animal experiments. In the outlined ternary system, the hypothesis could be tested experimentally whether the CYP1B1-mediated metabolism of E_2 directly causes lipid peroxidation and whether lipid peroxides, in turn, can contribute to DNA damage in two ways, i.e., as cofactors in catechol estrogen quinone-semiquinone redox cycling and as precursors of MDA.

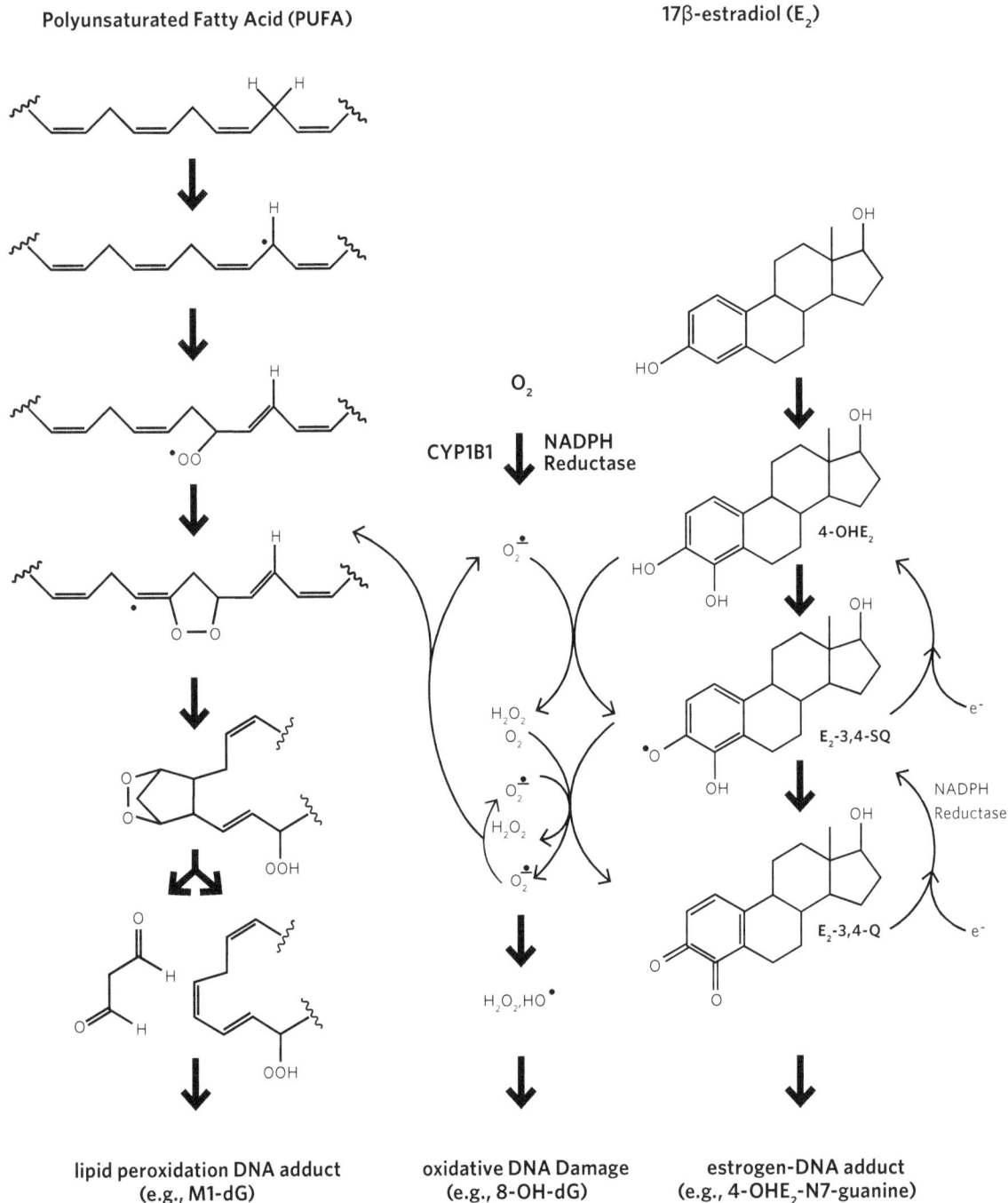

Figure 2.4.3. Interaction of oxidative estrogen metabolism and lipid peroxidation pathways.
Both pathways produce reactive peroxyl radicals during redox cycling or propagation of the radical chain reaction. Lipid peroxides may act as oxidants in the CYP1B1-mediated oxidation of catechol estrogens to semiquinones and quinones. The quinones can be reduced via semiquinone intermediates to catechol estrogens by NADPH-dependent P450 reductase. The semiquinones can also react with molecular oxygen to form superoxide radicals, which are converted to hydrogen peroxide by superoxide dismutase. In the presence of metal ions, hydrogen peroxide is reduced to hydroxyl radicals that may initiate lipid peroxidation. Overstoichiometric amounts of lipid peroxides may be formed by a continuous propagation reaction of lipid radicals with molecular oxygen. In turn, lipid peroxides can support and fuel the metabolic redox cycling of estrogen substrates.

References

1. Benamira M, Johnson K, Chaudhary A, Bruner K, Tibbetts C, Marnett LW. *Induction of mutations by replication of malondialdehyde-modified M13 DNA in Escherichia coli: determination of the extent of DNA modification, genetic requirements for mutagenesis, and types of mutations induced.* Carcinogenesis. 1995;16:93-9.

2. Chaudhary AK, Nokubo M, Marnett LJ, Blair IA. *Analysis of the malondialdehyde-2'-deoxyguanosine adduct in rat liver DNA by gas chromatography/electron capture negative chemical ionization mass spectrometry.* Biol Mass Spectrom. 1994;23:457-64.

3. Chaudhary AK, Nokubo M, Reddy GR, Yeola SN, Morrow JD, Blair IA, et al. *Detection of endogenous malondialdehyde-deoxyguanosine adducts in human liver.* Science. 1994;265:1580-2.

4. Vaca CE, Fang JL, Mutanen M, Valsta L. *32P-postlabelling determination of DNA adducts of malonaldehyde in humans: total white blood cells and breast tissue.* Carcinogenesis. 1995;16:1847-51.

5. Wang MY, Liehr JG. *Induction by estrogens of lipid peroxidation and lipid peroxide-derived malonaldehyde-DNA adducts in male Syrian hamsters: role of lipid peroxidation in estrogen-induced kidney carcinogenesis.* Carcinogenesis. 1995;16:1941-5.

6. Chaudhary AK, Reddy GR, Blair IA, Marnett LJ. *Characterization of an N6-oxopropenyl-2'-dexyadenosine adduct in malondialdehyde-modified DNA using liquid chromatography/electrospray ionization tandem mass spectrometry.* Carcinogenesis. 1996;17:1167-70.

7. Wang M, Dhingra K, Hittelman WN, Liehr JG, de Andrade M, Li D. *Lipid peroxidation-induced putative malondialdehyde-DNA adducts in human breast tissues.* Cancer Epidemiol Biomarkers Prev. 1996;5:705-10.

8. Peluso M, Munnia A, Risso GG, Catarzi S, Piro S, Ceppi M, et al. *Breast fine-needle aspiration malondialdehyde deoxyguanosine adduct in breast cancer.* Free Radic Res. 2011;45:477-82.

9. Akbulut H, Akbulut KG, Icli F, Buyukcelik A. *Daily variations of plasma malondialdehyde levels in patients with early breast cancer.* Cancer Detect Prev. 2003;27:122-6.

10. do Val Carneiro JL, Nixdorf SL, Mantovani MS, da Silva do Amaral Herrera AC, Aoki MN, Amarante MK, et al. *Plasma malondialdehyde levels and CXCR4 expression in peripheral blood cells of breast cancer patients.* J Cancer Res Clin Oncol. 2009;135:997-1004.

11. Gonenc A, Ozkan Y, Torun M, Simsek B. *Plasma malondialdehyde (MDA) levels in breast and lung cancer patients.* J Clin Pharm Ther. 2001;26:141-4.

12. Khanzode SS, Muddeshwar MG, Khanzode SD, Dakhale GN. *Antioxidant enzymes and lipid peroxidation in different stages of breast cancer.* Free Radic Res. 2004;38:81-5.

13. Ray G, Batra S, Shukla NK, Deo S, Raina V, Ashok S, et al. *Lipid peroxidation, free radical production and antioxidant status in breast cancer.* Breast Cancer Res Treat. 2000;59:163-70.

14. Sener DE, Gonenc A, Akinci M, Torun M. *Lipid peroxidation and total antioxidant status in patients with breast cancer.* Cell Biochem Funct. 2007;25:377-82.

15. Boyd NF, McGuire V. *The possible role of lipid peroxidation in breast cancer risk.* Free Radic Biol Med. 1991;10:185-90.

16. Lee KH, Shu XO, Gao YT, Ji BT, Yang G, Blair A, et al. *Breast cancer and urinary biomarkers of polycyclic aromatic hydrocarbon and oxidative stress in the Shanghai Women's Health Study.* Cancer Epidemiol Biomarkers Prev. 2010;19:877-83.

17. Dai Q, Gao YT, Shu XO, Yang G, Milne G, Cai Q, et al. *Oxidative stress, obesity, and breast cancer risk: results from the Shanghai Women's Health Study.* J Clin Oncol. 2009;27:2482-8.

18. Rossner P, Jr., Gammon MD, Terry MB, Agrawal M, Zhang FF, Teitelbaum SL, et al. *Relationship between urinary 15-F2t-isoprostane and 8-oxodeoxyguanosine levels and breast cancer risk.* Cancer Epidemiol Biomarkers Prev. 2006;15:639-44.

19. Gago-Dominguez M, Castelao JE, Pike MC, Sevanian A, Haile RW. *Role of lipid peroxidation in the epidemiology and prevention of breast cancer.* Cancer Epidemiol Biomarkers Prev. 2005;14:2829-39.

20. Shaw PM, Hosea NA, Thompson DV, Lenius JM, Guengerich FP. *Reconstitution premixes for assays using purified recombinant human cytochrome P450, NADPH-cytochrome P450 reductase, and cytochrome b5.* Arch Biochem Biophys. 1997;348:107-15.

CHAPTER 3

Exogenous Causes

3.1. INTRODUCTION

In 1990, worldwide incidence rates for breast cancer ranged from 11.8 per 100,000 women in Eastern China to 86.3 per 100,000 in North America (1, 2). The 8-fold difference in geographic incidence rates and the change over time and among migrants suggest that exogenous or environmental factors can influence breast cancer risk.

Geographic differences notwithstanding, there has been a worldwide increase in incidence rates since the 1960s in low- and high-incidence countries alike. For example, the incidence rate of breast cancer in Japan rose more than two-fold from 1959-60 to 1983-87 (3). In the United States, a steady rise in the annual age-adjusted incidence of breast cancer by an average of about 0.5% per year reached a historic high of over 300 breast cancers per 100,000 women in the 1990s (4-6). For the first time in decades, recent data showed a decline in incidence rates beginning in 2002 (5, 7, 8). The decrease was evident only in women who were 50 years of age or older and appears temporally related to the reduced use of hormone replacement therapy among postmenopausal women since 2002. There are many exogenous factors with potential relevance for breast cancer, ranging from ionizing radiation and electromagnetic fields to environmental pollutants (e.g., chemical carcinogens, xenoestrogens), dietary constituents (e.g., fat intake, coffee, tea), drugs (e.g., oral contraceptives, hormone replacement therapy), and lifestyle choices (e.g., smoking, alcohol). Ionizing radiation has been unequivocally associated with increased breast cancer risk (9, 10). The increased risk is a lifelong concern in women exposed in childhood to chest radiotherapy or radiation doses equivalent to multiple computed tomography (CT) procedures (11). In 2005, the International Agency for Research in Cancer (IARC) also classified oral contraceptives and hormone replacement therapy as carcinogenic to humans (group 1 carcinogen) with an increased breast cancer risk based on the review of epidemiological data (12, 13). The risk associated with exposure to other factors, such as environmental chemicals or xenobiotics, remains uncertain (14). Nevertheless, all these factors are of interest because of their potential role in the prevention of breast cancer. It is worth mentioning that some factors, which are known to influence cancer in other organs, such as occupation and lung cancer, do not play a role in breast cancer (15).

References

1. Muir C, Waterhouse J, Mack T, Powell J, Whelan S. *Cancer incidence in five continents.* International Agency for Research on Cancer. 1987;5; Lyon, France.

2. Parkin DM, Pisani P, Ferlay J. *Estimates of the worldwide incidence of 25 major cancers in 1990.* Int J Cancer. 1999;80:827-41.

3. Nagata C, Kawakami N, Shimizu H. *Trends in the incidence rate and risk factors for breast cancer in Japan.* Breast Cancer Res Treat. 1997;44:75-82.

4. Li C, Anderson B, Daling J, Moe R. *Trends in incidence rates of invasive lobular and ductal breast carcinoma.* JAMA. 2003;289:1421-4.

5. Ravdin P, Cronin K, Howlader N, Berg C, Chlebowski R, Feuer E, et al. *The decrease in breast-cancer incidence in 2003 in the United States.* N Engl J Med. 2007;356:1670-4.

6. Ries L, Harkins D, Krapcho M. *SEER Cancer Statistics Review, 1975-2003.* http://seercancergov/crs/1975_2003/results_merged/sect_01_overviewpdf. 2006.

7. Eheman C, Shaw K, Ryerson A, Miller J, Ajani U, White M. *The changing incidence of in situ and invasive ductal and lobular breast carcinomas: United States, 1999-2004.* Cancer Epidemiol Biomarkers Prev. 2009;18:1763-9.

8. Glass AG, Lacey JV, Jr., Carreon JD, Hoover RN. *Breast cancer incidence, 1980-2006: combined roles of menopausal hormone therapy, screening mammography, and estrogen receptor status.* J Natl Cancer Inst. 2007;99:1152-61.

9. Kelsey JL. *Breast cancer epidemiology: Summary and future directions.* Epidemiol Rev. 1993;15:256-63.

10. Preston D, Mattsson A, Holmberg E, Shore R, Hildreth N, Boice J. *Radiation effects on breast cancer risk: A pooled analysis of eight cohorts.* Radiat Res. 2002;158:220-35.

11. Adams M, Dozier A, Shore R, Lipshultz S, Schwartz R, Constine L, et al. *Breast cancer risk 55+ years after irradiation for an enlarged thymus and its implications for early childhood medical irradiation today.* Cancer Epidemiol Biomarkers Prev. 2010;19:48-58.

12. Anderson GL, Autier P, Beral V, Bosland MC, Fernandez E, Haslam SZ, et al. *Combined estrogen-progestogen contraceptives and combined estrogen-progestogen menopausal therapy.* IARC Monographs on the Evaluation of Carcinogenic Risks to Humans; 2007; Lyon, France: International Agency for Research on Cancer; 2007.

13. Cogliano V, Grosse Y, Baan R, Straif K, Secretan B, El Ghissassi F. *Carcinogenicity of combined oestrogen-progestagen contraceptives and menopausal treatment.* Lancet Oncol. 2005;6:552-3.

14. Brody JG, Rudel RA. *Environmental pollutants and breast cancer.* Environ Health Perspect. 2003;111:1007-19.

15. Ji BT, Blair A, Shu XO, Chow WH, Hauptmann M, Dosemeci M, et al. *Occupation and breast cancer risk among Shanghai women in a population-based cohort study.* Am J Ind Med. 2008;51:100-10.

3.2. EXOGENOUS ESTROGENS
3.2.1. INTRODUCTION

Breast cancer risk has been strongly linked to ages at menarche, first childbirth, and menopause, suggesting an important role for ovarian hormones in the development of the disease. Further support for the involvement of ovarian hormones derives from the protective effect of ovariectomy on the occurrence and prognosis of breast cancer, the presence of estrogen receptors in breast cancers, and the efficacy of antiestrogens in the treatment of breast cancer. Furthermore, both *in vitro* and *in vivo* studies have shown that estrogens have mitogenic effects on breast epithelial cells conceivably resulting in an increase in the number of normal cells, in the proliferation of initiated cells, or in altered growth control of malignant or premalignant cells (1-5).

All these considerations have raised concern regarding the effect of exogenous estrogens on breast cancer risk. This concern is magnified because of the widespread use of exogenous estrogens in form of oral contraceptives and menopausal replacement regimens. It is estimated that over 60 million women worldwide use oral contraceptives and that about 20% of postmenopausal American women receive hormone replacement therapy (6, 7). The potential impact of dietary and environmental estrogens on breast cancer risk has also generated considerable interest (www.cfe.cornell.edu/bcerf; www.tmc.tulane.edu/ecme). These agents include phytoestrogens as well as a variety of xenoestrogens. The following four sections address oral contraceptives, estrogen replacement therapy, phytoestrogens, and xenoestrogens in relation to breast cancer.

References

1. Anderson TJ, Battersby S, King RJ, McPherson K, Going JJ. *Oral contraceptive use influences resting breast proliferation.* Hum Pathol. 1989;20:1139-44.

2. Feigelson HS, Henderson BE. *Estrogens and breast cancer.* Carcinogenesis. 1996;17:2279-84.

3. King RJ. William L. McGuire Memorial Symposium. *Estrogen and progestin effects in human breast carcinogenesis.* Breast Cancer Res Treat. 1993;27:3-15.

4. Preston-Martin S, Pike MC, Ross RK, Jones PA, Henderson BE. *Increased cell division as a cause of human cancer.* Cancer Res. 1990;50:7415-21.

5. Spicer DV, Pike MC. *Sex steroids and breast cancer prevention.* J Natl Cancer Inst. 1994;16:139-47.

6. Diczfalusy E. *The worldwide use of steroidal contraception.* Int J Fertil. 1989;34 Suppl:56-63.

7. Jolleys JV, Olesen F. *A comparative study of prescribing of hormone replacement therapy in USA and Europe.* Maturitas. 1996;23:47-53.

3.2.2. ORAL CONTRACEPTIVES

Hormonal contraception has been used since the early 1960s by millions of women worldwide, providing them with a reliable method of regulating their fertility. Although progesterone is the active ingredient that blocks ovulation, contraceptives containing only progestogen are less effective than combined contraceptives (about 3.0 versus 0.1 to 1.0 pregnancies per 100 woman-years of use) (1). Combined oral contraceptives (OCs) are a mixture of a synthetic estrogen and one of several C-19 or C-21 steroids with progestational activity. Orally ingested E_2 is oxidized by 17ß-hydroxysteroid dehydrogenase to the less active E_1 within the intestinal mucosa and liver, increasing serum E_1 concentrations (2). Synthetic 17α-alkylated estrogens, such as ethinyl estradiol, are not oxidized by 17ß-estradiol dehydrogenase. For this reason, ethinyl estradiol has been the principal estrogen in almost all OCs (Figure 3.2.1.). The concentration of ethinyl estradiol in OC formulations has been decreased over the years from 50 to 100 ×g in the 1960s to 35 µg or less today (3). Synthetic progestogens are used because progesterone is very poorly absorbed. These progestogens have some androgenic activity, especially 19-nortestosterone derivatives such as norgestrel, norethynodrel, and norethindrone. Evidence has been presented that 19-nortestosterone derivatives also have estrogenic activity based on their ability to stimulate proliferation of ER-positive breast cancer cells and to induce ERE-CAT activity and PR synthesis (4). Newer progestogens, such as desogestrel, gestodene, and norgestimate, have less androgenic and estrogenic activity (5-8). Oral administration of estrogen leads to hormone concentrations in hepatic sinusoidal blood that are four to five times higher than those in peripheral blood. This so-called first-pass effect promotes the hepatic synthesis and secretion of several coagulation factors and renin substrates that may be detrimental and of lipid apolipoproteins that may be beneficial. Parenteral routes of estrogen administration (e.g., subcutaneous implant, vaginal ring) offer the advantage of avoiding the first pass through the liver, thereby allowing lower doses of hormone and reducing metabolic side effects. The effect of OCs on proliferative and secretory activity of normal breast epithelium has been examined. The immunohistochemical staining of secretory markers, such as immunoglobulin A, secretory component, and α-lactalbumin, was more frequently positive during artificial than natural menstrual cycles (9, 10) . A comparison of benign breast biopsies obtained from nulliparous OC users and control subjects revealed significantly higher ^{3}H-thymidine labeling of normal lobular units in the user group (11). The rise in labeling index was related to the estrogen rather than the progestogen concentration. A similar study (12) of fibroadenomas and reduction mammoplasties also showed a higher proliferative rate of normal epithelial cells in OC users than control women. In comparison to the natural cycle, OC use also reduced the percentage of ER-positive cells during the second week of the cycle. However, there was no OC effect on PR expression and ER expression was not different during the first, third, and fourth weeks of the cycle (12).

Many epidemiological studies have investigated whether OC use affects breast cancer risk (13). In 1996, the Collaborative Group on Hormonal Factors in Breast Cancer reanalyzed the worldwide epidemiological evidence based on individual data of 53,297 women with invasive breast cancer and 100,239 women without breast cancer from 54 studies conducted in 25 countries (14, 15). The pooled analysis included prospective as well as case-control studies with controls chosen from either the general or hospital populations. The breast cancer risk associated with OC use was determined relative to never-users, stratified by study, age at diagnosis, and parity. Overall, ever-users had a slightly, but significantly increased risk of 1.07 ($p = 0.00005$). Ever-use is only a crude measure of exposure that encompasses different patterns of OC use, such as current and past

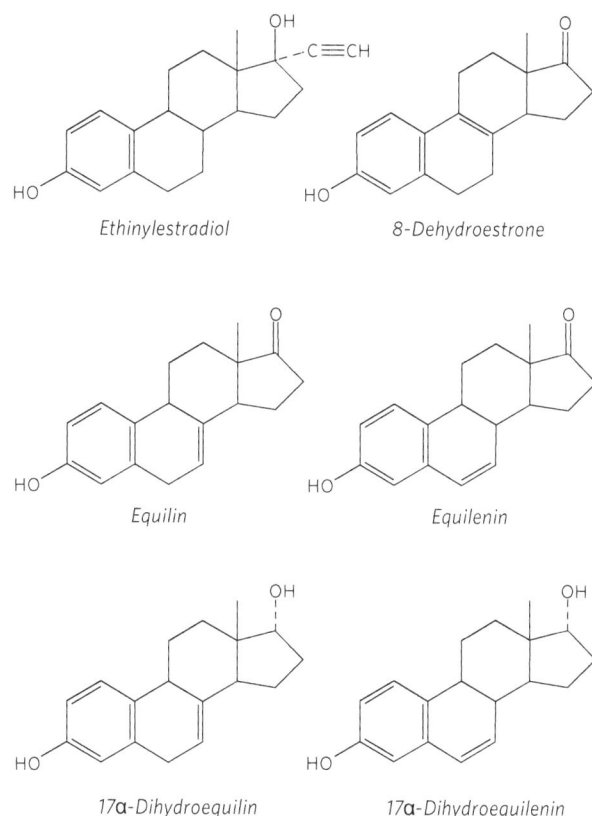

Ethinylestradiol

8-Dehydroestrone

Equilin

Equilenin

17α-Dihydroequilin

17α-Dihydroequilenin

Figure 3.2.1. Chemical structures of estrogens present in oral contraceptives (ethinyl estradiol) and Premarin (8-dehydroestrone, equilin, equilenin, 17α-dihydroequilin, and 17α-dihydroequilenin).

use. For this reason, breast cancer risk was considered in relation to timing of exposure, i.e., total duration of use, age at first use, time since first use, and time since last use. The analysis revealed that the time since last use was the most important of the four exposure indices. Current users had the highest risk (RR 1.24; 95% CI 1.15 - 1.33). The risk remained elevated for ten years after stopping OCs, but gradually declined during this period, i.e., one to four years after stopping the risk was 1.16 (95% CI 1.08 - 1.23), and five to nine years after stopping the risk was 1.07 (95% CI 1.02 - 1.13). For women who stopped OC use ten or more years ago, the relative risk did not differ significantly from that of never-users (RR 1.01; 95% CI 0.96 - 1.05). Although the 1996 Collaborative Group Study had the strength of large size (54 studies, 25 countries), the studies varied in the quality of their design, 50% of the cases were diagnosed before 1984 and only 40% of the women had ever used OCs (16). In the 2002 Women's Contraceptive and Reproductive Experiences (Women's CARE) population-based study of 4575 women with breast cancer and 4682 controls more than 75% of the women had used OCs (17). The risk of of breast cancer among women who had ever used any type of OC, as compared with those who had never used OCs, was 0.9 (95% CI, 0.8 – 1.0). Separate analysis of current and previous users revealed relative risks of 1.0 (95% CI 0.8 – 1.3) and 0.9 (95% CI 0.8 – 1.0), respectively. Thus, there was no association between past or present OC use and breast cancer in the Women's CARE study. In contrast, the 2002 Norwegian-Swedish Women's Lifestyle and Health Cohort Study observed an increased risk among women who were current or recent OC users (RR, 1.6; 95% CI, 1.2 – 2.1) (18). Long-term users of OCs were at higher risk of breast cancer than never users (test for trend, P = 0.005).

The 1996 Collaborative Group Study assessed women's use of OC between 1960 and 1990 and roughly half of the breast cancer cases were diagnosed before 1984. Since the 1960s, OC formulations have changed significantly including lower levels of estrogens and new progestogens. In terms of use, a greater number of younger women are using OCs, and they are using them with increasing duration (17). A 2006 meta-analysis evaluated 39 case-control studies that had most cases diagnosed after 1980 to better reflect more contemporary use patterns (19). The analysis revealed an increased risk of premenopausal breast cancer (OR, 1.19; 95% CI, 1.09 - 1.29). Among parous women, the assocaition was stonger when OCs were used before the first full-term pregnancy (OR, 1.44; 95% CI 1.28 - 1.62) than after the pregnancy (OR, 1.15; 95% CI, 1.06 - 1.26). Another study observed a small, but significant increase in risk of ductal carcinoma associated with OC use (OR, 1.15; 95% CI, 1.01 – 1.31) (20). Interestingly, OC use was associated with a 2.5-fold increased risk for ER-/PR-/HER2- (triple-negative) breast cancer and no significantly increased risk for non-triple negative breast cancer (21).

The 1996 Collaborative Group Study observed a slightly increased risk associated with current or recent OC use at all ages. However, the risk appeared highest for women who began use before age 20 and declined with increasing age at diagnosis (14). Several other studies found that long-term OC use in young women was associated with an excess of early-onset breast cancer (22-24). In contrast, the Women's CARE study observed no association of breast cancer risk with initiation of OC use at a young age (17). There is consensus that OC use among women 45 years or older was not associated with increased breast cancer risk (25, 26).

The reproductive history of a woman affects both her use of OCs and her risk of breast cancer. Nulliparous women are a special group in that there is no opportunity for the effects of OC use to be modified by childbearing. In the Collaborative Group Study, the pattern of risk with respect to time since last use was similar in nulliparous and parous women (14). Moreover, among parous women, the pattern of risk was similar irrespective of whether OC use began before or after birth of the first child. Similar patterns of risk with respect to time since last use were also observed for women of different parity and for women who had their first child at different ages (15).

The majority of studies have found little evidence of increased breast cancer risk related to prolonged duration of OC use (13, 27, 28). The Collaborative Group Study noted a weak trend of increasing risk with increasing duration ($p = 0.05$) while the Women's CARE study observed no increase in risk with longer periods of use or with higher doses of estrogen (14, 17). In most studies, there was also no relation between dose and type of hormone within the contraceptive, once recency of use had been taken into account. Multiple other factors, including family history of breast cancer, ethnic origin, height, and weight, also did not significantly modify the relative risks associated with current or recent OC use (14, 17, 29). Several studies have examined OC use in carriers of BRCA1 and BRCA2 mutations and found no evidence for increased breast cancer risk except possibly for BRCA2 carriers using OC for at least 5 years (30-32).

In summary, large scale studies such as the 1996 Collaborative Group analysis found a slightly increased breast cancer risk associated with OC use. The relation observed in the 1996 Collaborative Group Study is unusual, because the risk increases soon after first exposure, does not increase with duration of exposure, and returns to normal ten years after cessation. Such a pattern seems incompatible with a genotoxic effect of OCs. Instead, the pattern is consistent with the classic concept of the promotion of tumors that have already been initiated. Thus, OCs would be considered late-stage promotional agents of tumor growth. However, the 2006 meta-analysis revealed that OC use before first full-term pregnancy was more strongly associated with breast cacner risk than use after pregnancy suggesting that breast tissue is particularly susceptible to carcinogenic insult before differentiation during the first full-term pregnancy (33, 34). On the basis of the accumulated data, the International Agency for Research in Cancer (IARC) classified OCs as carcinogenic to humans (group 1 carcinogen) in 2005, which is a higher classification than the 1999 IARC evaluation (35, 36). While the risk of breast cancer appears to be increased, IARC concluded that OC use decreased the risk of ovarian and endometrial cancer. Additionally, OC use is associated with other adverse events, such as deep venous thrombosis, especially in women with factor V Leiden, and myocardial infarction among women over 35 years of age who smoke.

References

1. Baird DT, Glasier AF. *Hormonal contraception.* N Engl J Med. 1993;328:1543-9.

2. Kuhl H. *Pharmacokinetics of oestrogens and progestogens.* Maturitas. 1990;12:171-97.

3. Piper JM, Kennedy DL. *Oral contraceptives in the United States: trends in content and potency.* Int J Epidemiol. 1987;16:215-21.

4. Jeng M, Parker CJ, Jordan VC. *Estrogenic potential of progestins in oral contraceptives to stimulate human breast cancer cell proliferation.* Cancer Res. 1992;52:6539-46.

5. Burkman R. *Oral contraceptives: Current status Clin Obstet Gynecol.* 2001;44:62-72.

6. Kuhl H. *Comparative pharmacology of newer progestogens.* Drugs. 1996;51:188-215.

7. Phillips A, Hahn DW, McGuire JL. *Preclinical evaluation of norgestimate, a progestin with minimal androgenic activity.* Am J Obstet Gynecol. 1992;167:1191-6.

8. Rebar RW, Zeserson K. *Characteristics of the new progestogens in combination oral contraceptives.* Contraception. 1991;44:1-10.

9. Ciolino HP, Daschner PJ, Yeh GC. *Resveratrol inhibits transcription of CYP1A1 in vitro by preventing activation of the aryl hydrocarbon receptor.* Cancer Res. 1998;58:5707-12.

10. Going JJ, Anderson TJ, Battersby S, MacIntyre CC. *Proliferative and secretory activity in human breast during natural and artificial menstrual cycles.* Am J Pathol. 1988;130:193-204.

11. Anderson TJ, Battersby S, King RJ, McPherson K, Going JJ. *Oral contraceptive use influences resting breast proliferation.* Hum Pathol. 1989;20:1139-44.

12. Williams G, Anderson E, Howell A, Watson R, Coyne J, Roberts SA, et al. *Oral contraceptive (OCP) use increases proliferation and decreases oestrogen receptor content of epithelial cells in the normal human breast.* Int J Cancer. 1991;48:206-10.

13. Malone KE, Daling JR, Weiss NS. *Oral contraceptives in relation to breast cancer.* Epidemiol Rev. 1993;15:80-97.

14. Collaborative Group on Hormonal Factors in Breast Cancer. *Breast cancer and hormonal contraceptives: collaborative reanalysis of individual data on 53 297 women with breast cancer and 100 239 women without breast cancer from 54 epidemiological studies.* Lancet. 1996;347:1713-27.

15. Collaborative Group on Hormonal Factors in Breast Cancer. *Breast cancer and hormonal contraceptives: further results.* Contraception. 1996;54:1S-106S.

16. Davidson N, Helzlsouer K. *Good news about oral contraceptives.* N Engl J Med. 2002;346:2078-80.

17. Marchbanks P, McDonald J, Hoyt G, Folger S, Mandel M, Daling J, et al. *Oral contraceptives and the risk of breast cancer.* N Engl J Med. 2002;346:2025-32.

18. Kumle M, Weiderpass E, Braaten T, Persson I, Adami H, Lund E. *Use of oral contraceptives and breast cancer risk: The Norwegian-Swedish Women's Lifestyle and Health Cohort Study.* Cancer Epidemiol Biomarkers Prev. 2002;11:1375-81.

19. Kahlenborn C, Modugno F, Potter D, Severs W. *Oral contraceptive use as a risk factor for premenopausal breast cancer:*

A meta-analysis. Mayo Clin Proc. 2006;81:1290-302.

20. Nichols H, Trentham-Dietz A, Egan K, Titus-Ernstoff L, Hampton J, Newcomb P. *Oral contraceptive use and risk of breast carcinoma in situ.* Cancer Epidemiol Biomarkers Prev. 2007;16:2262-8.

21. Dolle J, Daling J, White E, Brinton L, Doody D, Porter P, et al. *Risk factors for triple-negative breast cancer in women under the age of 45 years.* Cancer Epidemiol Biomarkers Prev. 2009;18:1157-66.

22. Brinton LA, Gammon MD, Malone KE, Schoenberg JB, Daling JR, Coates RJ. *Modification of oral contraceptive relationships on breast cancer risk by selected factors among younger women.* Contraception. 1997;55:197-203.

23. Rosenberg L, Palmer JR, Rao S, Zauber AG, Strom BL, Warshauer ME, et al. *Case-control study of oral contraceptive use and risk of breast cancer.* Am J Epidemiol. 1996;143:25-37.

24. White E, Malone KE, Weiss NS, Daling JR. *Breast cancer among young U.S. women in relation to oral contracetpive use.* J Natl Cancer Inst. 1994;86:505-14.

25. Brinton LA, Daling JR, Liff JM, Schoenberg JB, Malone KE, Stanford JL, et al. *Oral contraceptives and breast cancer risk among younger women.* J Natl Cancer Inst. 1995;87:827-35.

26. Hankinson SE, Colditz GA, Manson JE, Willett WC, Hunter DJ, Stampfer MJ, et al. *A prospective study of oral contraceptive use and risk of breast cancer (Nurses' Health Study, United States).* Cancer Causes Control. 1997;8:65-72.

27. Newcomb PA, Longnecker MP, Storer BE, Mittendorf R, Baron J, Clapp RW, et al. *Recent oral contraceptive use and risk of breast cancer.* Cancer Causes Control. 1996;7:525-32.

28. Primic-Zakelj M, Evstifeeva T, Ravnihar B, Boyle P. *Breast-cancer risk and oral contraceptive use in slovenian women aged 25-54.* Int J Cancer. 1995;62:414-20.

29. Palmer JR, Rosenberg L, Rao RS, Strom BL, Warshauer ME, Harlap S, et al. *Oral contraceptive use and breast cancer risk among African-American women.* Cancer Causes Control. 1995;6:321-31.

30. Haile R, Thomas D, McGuire V, Felberg A, John E, Milne R, et al. *BRCA1 and BRCA2 mutation carriers, oral contraceptive use, and breast cancer before age 50.* Cancer Epidemiol Biomarkers Prev. 2006;15:1863-70.

31. Lee E, Ma H, McKean-Cowdin R, Van Den Berg D, Bernstein L, Henderson B, et al. *Effect of reproductive factors and oral contraceptives on breast cancer risk in BRCA1/2 mutation carriers and noncarriers: Results from a population based study.* Cancer Epidemiol Biomarkers Prev. 2008;17:3170-8.

32. Milne R, Knight J, John E, Dite G, Balbuena R, Ziogas A, et al. *Oral contraceptive use and risk of early-onset breast cancer in carriers and noncarriers of BRCA1 and BRCA 2 mutations.* Cancer Epidemiol Biomarkers Prev. 2005;14:350-6.

33. Cerhan J. *Oral contraceptive use and breast cancer risk: Current status.* Mayo Clin Proc. 2006;81:1287-9.

34. Colditz GA, Hankinson SE, Hunter DJ, Willett WC, Manson JE, Stampfer MJ, et al. *The use of estrogens and progestins and the risk of breast cancer in postmenopausal women.* N Engl J Med. 1995;332:1589-93.

35. Anderson GL, Autier P, Beral V, Bosland MC, Fernandez E, Haslam SZ, et al., editors. *Combined estrogen-progestogen contraceptives and combined estrogen-progestogen menopausal therapy.* IARC Monographs on the Evaluation of Carcinogenic Risks to Humans; 2007; Lyon, France: International Agency for Research on Cancer.

36. Cogliano V, Grosse Y, Baan R, Straif K, Secretan B, El Ghissassi F. *Carcinogenicity of combined oestrogen-progestagen contraceptives and menopausal treatment.* Lancet Oncol. 2005;6:552-3.

3.2.3. ESTROGEN REPLACEMENT THERAPY

Plasma estrogen levels in postmenopausal women are about 10-fold lower than those in premenopausal women. Thus, the postmenopausal period may simplistically be considered an endocrine-deficiency state, and hormone replacement therapy can provide effective relief from menopausal vasomotor symptoms and prevent osteoporosis and osteoarthritis (1, 2). On the other hand, unopposed estrogen therapy was shown to be associated with a three- to fourfold increased risk of endometrial cancer (3-5). The relation between estrogen replacement therapy and breast cancer risk was less striking. Nevertheless, any associated risk is a concern because of the widespread use of hormone replacement therapy affecting millions of women. Estimates indicated that over 20% of postmenopausal American women received hormone replacement therapy in the 1990s, reaching a peak of about 26 million visits with menopausal hormone therapy prescriptions in 2001 (6, 7). In 2002, the randomized trial of the Women's Health Initiative showed a significant increase in the risk of coronary heart disease associated with estrogen-progestin combination therapy (8). Since that time utilization has declined and analysis of two prescription-based databases indicated a drop to about 17 million visits by 2003 (6, 9). A comprehensive analysis of women enrolled in the large Kaiser Permanente Northwest health plan related breast cancer incidence between 1980 and 2006 to menopausal hormone therapy and screening mammography (10). While incidence rates for women under age 45 remained stable, the rates for women aged 45 – 59 years and women aged 60 years or older both rose about 50% from the early 1980s to 2001. This rise seemed to occur in two phases, the first during the 1980s, coinciding with the progressive adoption of screening mammography by over 75% of eligibile women in the health plan, and the second corresponding to increases in menopausal hormone therapy throughout the 1990s. The incidence rates for both older age groups dropped dramatically in 2003 – 2006 in parallel with a profound decline in menopausal hormone therapy prescriptions.

In the United States, the most commonly prescribed estrogen consists of conjugated equine estrogen (CEE; primarily Premarin®, Wyeth-Ayerst Laboratories, Philadelphia, Pennsylvania), a complex mixture containing E_1 (42%), 8-dehydroestrone (18%), equilin (17%), 17α-dihydroequilenin (10%), and the remainder consisting of small amounts (<5% each) of equilenin, 17α-estradiol, 17α-dihydroequilin, 17β-dihydroequilin, and 17β-dihydroequilenin, all in the form of sulfate esters (Figure 3.2.1.) (11, 12). Interestingly, E_2 accounts for only 1.5% of estrogens present in Premarin®. In Europe, the preferred estrogens are estradiol valerate and mixtures of E_2, E_1, and E_3 (13). Although plasma concentrations of E_2 and E_1 are similar after daily oral administration of estradiol valerate or CEE, physiological effects may differ depending on the type of estrogen used, most likely reflecting differences in pharmacokinetics (14, 15). "Equine" estrogens (equilin and equilenin) may differ from "classical" estrogens (E_2, E_1, and E_3) in their ability to serve as substrates for estrogen-metabolizing enzymes in humans (12). However, like E_1 and E_2, equilin and equilenin are carcinogenic in the hamster kidney model and the catechol metabolite of equilenin, 4-hydroxyequilenin, has been shown to produce DNA adducts (11, 16). Not surprisingly, estradiol-based replacement therapy resulted in 12- and 9-fold higher urinary levels of 2-OHE$_1$/$_2$ and 16α-OHE1, respectively, in users compared with nonusers (17). CEE in combination with medroxyprogesterone are prescribed in the US as Prempro®. When estrogen is the only component, the term estrogen replacement therapy (ERT) is used. When progestogens are added along with estrogen, the combination is referred to as hormone replacement therapy (HRT). Unfortunately, a clear distinction between ERT and HRT is not made in all epidemiological studies and frequently the type of estrogen is not specified.

Both dosage and hormone composition of replacement regimens have changed over the years. The currently preferred CEE dose of 0.625 mg/d for 25 days in a 28-day treatment cycle is lower than those commonly prescribed during the 1970s. Studies that examined the relation of dosage and breast cancer risk generally found no marked variation in risk according to dose (18-22). The hormonal regimen in studies before 1985 consisted exclusively of estrogens, but nowadays includes combinations of estrogens and progestogens. The addition of progestogens has unequivocally decreased the risk of endometrial cancer that was associated with unopposed estrogen therapy (4, 13). With regard to a beneficial effect of combination therapy on breast cancer risk, several studies did not find a difference in relative risk between women taking estrogen alone and women taking combination therapy (22-25). Hence, these data do not support the need for progestogen addition for women who have undergone hysterectomy. The lack of a protective effect of progestogens for breast cancer suggests that mammary epithelial cell respond differently than endometrial cells. Several studies have examined the relation between HRT and estrogen receptor status, reporting a statistically significant association with ER+ but not ER- breast cancers (10, 26, 27). The rising incidence of breast cancer between 1980 and 2000 involved all histologic subgroups of ER+ tumors, especially low-grade and mixed ductal-lobular cancers (28). Interestingly, the recent decline in breast cancer incidence observed after 2002 affected both ER+ and ER- tumors (10, 28).

Many epidemiological studies have investigated whether HRT affects breast cancer risk and several reviews and meta-analyses have summarized the literature (20, 21, 29-32). By far the most comprehensive evaluation of HRT use and breast cancer risk was carried out by the Collaborative Group on Hormonal Factors in Breast Cancer that also investigated OC use in relation to breast cancer. The Collaborative Group HRT study brought together and reanalyzed the worldwide epidemiological evidence based on individual data of 52,705 women with invasive breast cancer and 108,411 women without breast cancer from 51 studies conducted in 21 countries (18).

The pooled analysis included prospective studies such as the Nurses' Health Study as well as case-control studies with controls chosen from either the general or hospital populations. The breast cancer risk associated with HRT was determined relative to never-users, stratified by study, age at diagnosis, time since menopause, BMI, parity, and the age a woman was when her first child was born. The relative risk of breast cancer among current users of HRT increased with increasing duration of use. The risk increased by a factor of 1.023, i.e., by 2.3%, for each year of use (95% CI 1.011 - 1.036; 2p = 0.0002). This means that women who used HRT for five years or longer had a relative risk of 1.35 (95% CI 1.21 - 1.49; 2p = 0.00001).

The relation with increasing duration of use did not differ significantly between current and recent users, i.e., women whose use ceased one to four years previously. By contrast, past users who stopped HRT five or more years previously had no significant increase in breast cancer risk either overall or in relation to duration of use. After duration of use and time since last use had been taken into account, no residual effects remained for any other index of the timing of exposure to HRT, including age at first use and the related measure time between menopause and first use. Compared with breast cancers in never-users, those in ever-users were less likely to have spread to axillary lymph nodes (2p = 0.02) or to more distant sites (2p = 0.01) than to be localized to the breast (18). It is uncertain whether the less advanced clinical stage is due to the biological effects of HRT, the exclusion of women with previously undiagnosed breast cancer before they began HRT, the earlier diagnosis of breast cancer in current or recent users than in never-users, or a combination of factors.

Both cohort and case-control studies have attempted to identify whether certain subgroups of menopausal hormone users are at increased risk of breast cancer. Results have been difficult to interpret, primarily because numbers of users within subgroups have usually been limited. In addition, subgroups are even more likely to selection bias known to affect all studies of menopausal hormone use and breast cancer risk. Women who use hormone therapy differ from nonusers both prior to initiating therapy and during its use. Those electing hormone treatment tend to be at lower risk of breast cancer because of early menopause, to have no family history of breast cancer, and to have negative mammograms (33, 34). On the other hand, women using HRT have more physician contact, including physical examinations and mammography, and therefore are more likely to have breast cancer diagnosed (19, 26). Thus, it is not surprising that some results are contradictory and that consensus has not always been reached regarding potential high-risk subsets of users.

Between 1987 and 1999, incidence rates of ductal carcinoma have remained essentially constant whereas the rates of lobular cancer increased 1.5-fold (35). Thus, the proportion of breast cancers with a lobular component increased from 9.5% in 1987 to 15.6% in 1999. Several studies have shown that long-term HRT has been associated with a 2.0- to 5.6-fold increased risk of invasive lobular cancer but a much smaller or no increase of invasive ductal breast cancer (35-37). These data indicate that the increased utilization of HRT in the 1980s and 90s coincides with the increase in incidence rate of lobular cancer observed during the same period. Interestingly, a recent analysis of National Cancer Institute Surveillance, Epidemiology and End Results (SEER) data from 1999 through 2004 revealed a 20.5% decrease of invasive lobular cancer, which may reflect the decline in HRT during this period (38). If the decrease in lobular breast cancer incidence is indeed associated with discontinuation of HRT, the rapidity of change would suggest that clinically occult breast cancers stopped progressing or even regressed soon after discontinuation of the therapy (39).

Obesity has been associated with increased risk of postmenopausal breast cancer due to increased production of endogenous estrogens in peripheral fat tissue (40, 41). In view of the rising incidence of obesity in the American population, it is important to determine whether exogenous estrogen in form of menopausal estrogen therapy has a potentiating effect on breast cancer risk. Earlier studies found an elevated risk associated with estrogen use in obese women (42, 43), while several recent studies did not find any increase in relative risk (22, 24, 44-47). In fact, data from the Nurses' Health Study showed that the increased breast cancer risk associated with adult weight gain was limited to postmenopausal women who never used hormone replacement (48). Among these women, the relative risk was 1.99 (95% CI 1.43 - 2.76) for weight gain more than 20 kg versus unchanged weight since age 18. Other large studies support the observation that there is no statistically significant interaction between HRT use and BMI or weight gain (49). These findings indicate that among HRT users, both lean and heavy women have high levels of circulating estrogens by virtue of their HRT; against this background, the estrogenic effect of obesity is imperceptible and does not further increase the risk of breast cancer. Nevertheless, both total and free E_2 levels were positively associated with breast cancer risk among women older than 60 years and among women with a BMI <25 kg/m^2 (50).

Demographic/clinical predictors of breast cancer. The type and age at onset of menopause are strong predictors of both breast cancer and probability that estrogens will be prescribed. Women who have a surgical menopause resulting from ovarian ablation at a young age (<45 years) experience approximately a halving of their breast cancer risk compared with women who have a natural menopause

at ages 50 to 55 (51). At the same time, oophorectomized women are most likely to receive estrogen replacement therapy. Thus, the effect of hormone therapy might be quite different depending on the presence or absence of ovaries. However, the Collaborative Group HRT study found virtually the same increase in breast cancer risk associated with HRT use in women with natural menopause and with bilateral oophorectomy (18).

Family history of breast cancer. In evaluating studies of this subgroup, one should keep in mind that women with a family history of breast cancer are more likely to be never users, but are at increased risk of breast cancer, which may result in bias toward null. One meta-analysis found a significantly higher risk among women with family history of breast cancer, whereas another observed no difference in breast cancer risk for ever-use of hormone therapy between women with and without family history (19, 32). Two additional case-control studies also did not find any increased risk associated with family history of breast cancer (22, 24). The Collaborative Group HRT study also found no evidence for an increased risk (18). One study found that the catalase genotype (rs1001179, C/T polymorphism at -262 bp in the promoter region) modified the effect of HRT on breast cancer risk (52). The catalase genotype alone was not associated with breast cancer risk while ever use of HRT was associated with increased risk (OR, 1.39; 95% CI, 1.11 - 1.75). The increase with ever use was more pronounced among those women with variant CT or TT genotype (OR, 1.88; 95%, 1.29 - 2.75) than among those with CC (OR, 1.15; 95% CI, 0.86 - 1.54).

History of benign breast disease. Most studies agree that estrogen therapy does not significantly elevate breast cancer risk in women with a history of benign breast disease (19, 20, 22, 24, 46, 47). Of interest are subgroups of women with specific types of benign breast disease, i.e., atypical hyperplasia, proliferative disease without atypia, and complex fibroadenoma that are associated with up to four-fold elevations in breast cancer risk (53). A retrospective cohort study of these subgroups determined that ERT does not significantly elevate the risk of breast cancer in women with a history of any type of benign breast disease, including atypical hyperplasia (54). Women who have had carcinoma of the breast are usually disqualified from receiving estrogen therapy. Small scale studies suggest, however, that breast carcinoma may not be reactivated by estrogen therapy, and that a history of breast cancer may not be an absolute contraindication to estrogen therapy, especially for women with severe menopausal symptoms (55, 56). Since HRT is usually prescribed to "replace" the falling levels of circulating ovarian hormones at the menopause, one might expect that while women are using such therapy the effects of menopause on breast cancer risk will be delayed. Indeed, the findings of the Collaborative Group HRT study are in agreement with this expectation. Current or recent use of HRT was estimated to increase the relative risk of breast cancer by 2.3% for each year of use, which may be seen as comparable to the 2.8% increase in the relative risk of breast cancer that normally occurs for each year that menopause is delayed (18).

References

1. Cauley JA, Seeley DG, Ensrud K, Ettinger B, Black D, Cummings SR. *Estrogen replacement therapy and fractures in older women.* Ann Intern Med. 1995;122:9-16.

2. Nevitt MC, Cummings SR, Lane NE, Hochberg MC, Scott JC, Pressman AR, et al. *Association of estrogen replacement therapy with the risk of osteoarthritis of the hip in elderly white women. Study of osteoporotic fractures research group.* Arch Intern Med. 1996;156:2073-80.

3. Antunes CM, Strolley PD, Rosenshein NB, Davies JL, Tonascia JA, Brown C, et al. *Endometrial cancer and estrogen use. report of a large case-control study.* N Engl J Med. 1979;300:9-13.

4. Persson I, Yuen J, Bergkvist L, Schairer C. *Cancer incidence and mortality in women receiving estrogen and estrogen-progestin replacement therapy-long-term follow-up of a Swedish cohort.* Int J Cancer. 1996;67:327-32.

5. Shapiro S, Kelly JP, Rosenberg L, Kaufman DW, Helmrich SP, Rosenshein NB, et al. *Risk of localized and widespread endometrial cancer in relation to recent and discontinued use of conjugated estrogens.* N Engl J Med. 1985;313:969-72.

6. Hing E, Brett K. *Changes in U.S. prescribing patterns of menopausal hormone therapy, 2001-2003.* Obstet Gynecol. 2006;108:33-40.

7. Jolleys JV, Olesen F. *A comparative study of prescribing of hormone replacement therapy in USA and Europe.* Maturitas. 1996;23:47-53.

8. Rossouw J, GAnderson G, Prentice R, LaCroix A, Kooperberg C, Stefanick M, et al. *Risk and benefits of estrogen plus progestin in healthy postmenopausal women.* JAMA. 2002;288:321-33.

9. Solomon CG, Dluhy RG. *Rethinking postmenopausal hormone therapy.* N Engl J Med. 2003;348:579-80.

10. Glass AG, Lacey JV, Jr., Carreon JD, Hoover RN. *Breast cancer incidence, 1980-2006: combined roles of menopausal hormone therapy, screening mammography, and estrogen receptor status.* J Natl Cancer Inst. 2007;99:1152-61.

11. Li JJ, Li SA, Oberley TD, Parsons JA. *Carinogenic activities of various steroidal and nonsteroidal estrogens in the hamster kidney: Relation to hormonal activity and cell proliferation.* Cancer Res. 1995;55:4347-51.

12. Shen L, Pisha E, Huang Z, Pezzuto JM, Krol E, Alam Z, et al. *Bioreductive activation of catechol estrogen-ortho-quinones: aromatization of the B ring in 4-hydroxyequilenin markedly alters quinoid formation and reactivity.* Carcinogenesis. 1997;18:1096-101.

13. Belchetz PE. *Hormonal treatment of postmenopausal women.* New Engl J Med. 1994;330:1062-71.

14. Hammond CB, Maxson WS. *Estrogen replacement therapy.* Clin Obstet Gynecol. 1986;29:407-30.

15. Wren BG, Brown LB, Routledge DA. *Differential clinical response to oestrogens after menopause.* Med J Aust. 1982;2:329-32.

16. Shen L, Qiu S, Chen Y, Zhang F, van Breemen RG, Nikolic D, et al. *Alkylation of 2'-deoxynucleosides and DNA by the Premarin metabolite 4-hydroxyequilenin semiquinone radical.* Chem Res Toxicol. 1998;11:94-101.

17. Wellejus A, Olsen A, Tjonneland A, Thomsen B, Overvad K, Loft S. *Urinary hydroxyestrogens and breast cancer risk among postmenopausal women: A prospective study.* Cancer Epidemiol Biomarkers Prev. 2005;14:2137-42.

18. Collaborative Group on Hormonal Factors in Breast Cancer. *Breast cancer and hormone replacement therapy: collaborative reanalysis of data from 51 epidemiological studies of 52 705 women with breast cancer and 108 411 women without breast cancer.* Lancet. 1997;350:1047-59.

19. Colditz GA, Egan KM, Stampfer MJ. *Hormone replacement therapy and risk of breast cancer: Results from epidemiologic studies.* Am J Obstet Gynecol. 1993;168:1473-80.

20. Dupont WD, Page DL. *Menopausal estrogen replacement therapy and breast cancer.* Arch Intern Med. 1991;151:67-72.

21. Sillero-Arenas M, Delgado-Rodriguez M, Rodigues-Canteras R, Bueno-Cavanillas A, Galvez-Vargas R. *Menopausal hormone replacement therapy and breast cancer: A meta-analysis.* Obstet Gynecol. 1992;79:286-94.

22. Stanford JL, Weiss NS, Voigt LF, Daling JR, Habel LA, Rossing MA. *Combined estrogen and progestin hormone replacement therapy in relation to risk of breast cancer in middle-aged women.* JAMA. 1995;274:137-42.

23. Colditz GA, Hankinson SE, Hunter DJ, Willett WC, Manson JE, Stampfer MJ, et al. *The use of estrogens and progestins and the risk of breast cancer in postmenopausal women.* N Engl J Med. 1995;332:1589-93.

24. Newcomb PA, Longnecker MP, Storer BE, Mittendorf R, Baron J, Clapp RW, et al. *Long-term hormone replacement therapy and risk of breast cancer in postmenopausal women.* Am J Epidemiol 1995;142:788-95.

25. Yang CP, Daling JR, Band PR, Gallagher RP, White E, Weiss NS. *Noncontraceptive hormone use and risk of breast cancer.* Cancer Causes Control. 1992;3:475-9.

26. Daling J, Malone K, Doody D, Voigt L, Bernstein L, Marchbanks P, et al. *Association of regimens of hormone replacement therapy to prognostic factors among women diagnosed with breast cancer aged 50-64 years.* Cancer Epidemiol Biomarkers Prev. 2003;12:1175-81.

27. Li C, Malone K, Porter P, Weiss N, Tang M, Cushing-Haugen K, et al. *Relationship between long durations and different regimens of hormone therapy and risk of breast cancer.* JAMA. 2003;289:3254-63.

28. Brinton LA, Richesson D, Leitzmann MF, Gierach GL, Schatzkin A, Mouw T, et al. *Menopausal hormone therapy and breast cancer risk in the NIH-AARP Diet and Health Study Cohort.* Cancer Epidemiol Biomarkers Prev. 2008;17:3150-60.

29. Brinton LA, Schairer C. *Estrogen replacement therapy and breast cancer risk.* Epidemiology. 1993;15:66-79.

30. Colditz GA. *Relationship between estrogen levels, use of hormone replacement therapy, and breast cancer.* J Natl Cancer Inst. 1998;90:814-23.

31. Feigelson HS, Henderson BE. *Estrogens and breast cancer.* Carcinogenesis. 1996;17:2279-84.

32. Steinberg KK, Thacker SB, Smith SJ, Stroup DF, Zack MM, Flanders WD, et al. *A meta-analysis of the effect of estrogen replacement therapy on the risk of brest cancer.* JAMA. 1991;265:1985-90.

33. Barrett-Conner E. *Postmenopausal estrogen and prevention bias.* Ann Intern Med. 1991;115:455-6.

34. Bergkvist L, Adami HO, Persson I, Hoover R, Schairer C. *The risk of breast cancer after estrogen and estrogen-progestin replacement.* N Engl J Med. 1989;321:293-7.

35. Li C, Anderson B, Daling J, Moe R. *Trends in incidence rates of invasive lobular and ductal breast carcinoma.* JAMA. 2003;289:1421-4.

36. Chen C, Weiss N, Newcomb P, Barlow W, White E. *Hormone replacement therapy in relation to breast cancer.* JAMA. 2002;287:734-41.

37. Rosenberg L, Magnusson C, Lindstrom E, Wedren S, Hall P, Dickman P. *Menopausal hormone therapy and other breast cancer risk factors in relation to the risk of different histological subtypes of breast cancer: a case-control study.* Breast Cancer Res. 2006;8:R11.

38. Eheman C, Shaw K, Ryerson A, Miller J, Ajani U, White M. *The changing incidence of in situ and invasive ductal and lobular breast carcinomas: United States, 1999-2004.* Cancer Epidemiol Biomarkers Prev. 2009;18:1763-9.

39. Ravdin P, Cronin K, Howlader N, Berg C, Chlebowski R, Feuer E, et al. *The decrease in breast-cancer incidence in 2003 in the United States.* N Engl J Med. 2007;356:1670-4.

40. Key TJ, Pike MC. *The role of oestrogens and progestagens in the epidemiology and prevention of breast cancer.* Eur J Cancer Clin Oncol. 1988;24:29-43.

41. Trentham-Dietz A, Newcomb PA, Storer BE, Longnecker MP, Baron J, Greenberg ER, et al. *Body size and risk of breast cancer.* Am J Epidemiol. 1997;145:1011-9.

42. Mills PK, Beeson WL, Phillips RL, Fraser GE. *Prospective study of exogenous hormone use and breast cancer in seventh-day adventists.* Cancer. 1989;64:591-7.

43. Sherman B, Wallace R, Bean J. *Estrogen use and breast cancer. Interaction with body mass.* Cancer. 1983;51:1527-31.

44. Colditz GA, Stampfer MJ, Willett WC, Hunter DJ, Manson JE, Hennekens CH, et al. *Type of postmenopausal hormone use and* *risk of breast cancer: 12 year follow-up from the Nurses' Health Study.* Cancer Causes Control. 1992;3:433-9.

45. Harris JR, Lippman ME, Veronesi U, Willett W. *Breast cancer.* New Engl J Med. 1992;327:319-28.

46. Kaufman DW, Palmer JR, de Mouzon J, Rosenberg L, Stolley PD, Warshauer ME, et al. *Estrogen replacement therapy and the risk of breast cancer: results from the case-control surveillance study.* Am J Epidemiol. 1991;134:1375-85.

47. Palmer JR, Rosenberg L, Clarke EA, Miller DR, Shapiro S. *Breast cancer risk after estrogen replacement therapy: results from the Toronto Breast Cancer Study.* Am J Epidemiol. 1991;134:1386-95.

48. Huang Z, Hankinson SE, Colditz GA, Stampfer MJ, Hunter DJ, Manson JE, et al. *Dual effects of weight and weight gain on breast cancer risk.* JAMA. 1997;278:1407-11.

49. Feigelson HS, Jonas CR, Teras LR, Thun MJ, Calle EE. *Weight gain, body mass index, hormone replacement therapy, and post-menopausal breast cancer in a large prospective study.* Cancer Epidemiol Biomarkers Prev. 2004;13:220-4.

50. Tworoger SS, Missmer SA, Barbieri RL, Willett WC, Colditz GA, Hankinson SE. *Plasma sex hormone concentrations and subsequent risk of breast cancer among women using postmenopausal hormones.* J Natl Cancer Inst. 2005;97:595-602.

51. Trichopoulos D, MacMahon B, Cole P. *Menopause and breast cancer risk.* J Natl Cancer Inst. 1972;48:605-13.

52. Quick S, Shields P, Nie J, Platek M, McCann S, Hutson A, et al. *Effect modificaton by catalase genotype suggests a role for oxidative stress in the association of hormone replacement therapy with postmenopausal breast cancer risk.* Cancer Epidemiol Biomarkers Prev. 2008;17:1082-7.

53. Dupont WD, Page DL, Parl FF, Vnencak-Jones CL, Plummer WD, Rados MS, et al. *Long-term risk of breast cancer in women with fibroadenoma.* N Engl J Med. 1994;331:10-5.

54. Dupont WD, Page DL, Parl FF, Plummer WD, Schuyler PA, Kasami M, et al. *Estrogen replacement therapy in women with a history of proliferative breast disease.* Cancer. 1999;85: 1277- 1283.

55. Creasman WT. *Estrogen replacement therapy: is previously treated cancer a contraindication.* Obstet Gynecol. 1991;77:308-12.

56. Wren BG. *Hormone therapy following breast and uterine cancer.* Baillieres Clin Endocrinol Metab. 1993;7:225-42.

3.2.4. PHYTOESTROGENS

Classification and Chemical Structure. Phytoestrogens are dietary components of plant origin with weak estrogenic activity. They are not related to brassinosteroids, a group of steroid hormones that are widely distributed throughout the plant kingdom and cause biological effects on plant growth when applied exogenously (1, 2). Instead, they are non-steroidal compounds classified as isoflavones, lignans, coumestans, and hydroxystilbenes, all of which share a diphenolic structure that conveys exceptional stability (3) (Figure 3.2.2). Phytoestrogens occur in plants as glycoside conjugates and serve primarily as bactericidal and fungicidal agents for the plants. During digestion in the human intestinal tract, colonic bacteria hydrolyze and convert the phytoestrogens to biologically active mammalian forms that are absorbed into the systemic and enterohepatic circulation and excreted in urine and feces. For example, the main plant lignans, matairesinol and secoisolaricinol, are converted by bacteria in the proximal colon to enterolactone and enterodiol, respectively (Figure 3.2.2). Thus, the two

main mammalian lignans are bacterial metabolites of plant precursors. Similarly, the isoflavones biochanin A and formononetin are converted by colonic bacteria to the active compounds genistein and daidzein, respectively (Figure 3.2.2). Daidzein can be further metabolized to O-desmethylangolensin and equol. Related phytoestrogens are the flavones (e.g., flavone, luteolin) and flavonols (e.g., quercetin). Oral administration of antibiotics destroys the intestinal microflora and thereby prevents both the hydrolysis of glycosides and the conversion of plant precursors to active phytoestrogens.

Dietary Sources. Isoflavones occur in high concentrations in soybeans, soy flour, soy flakes, and nonfermented (tofu, soymilk) and fermented (miso) soy products (4). Lower

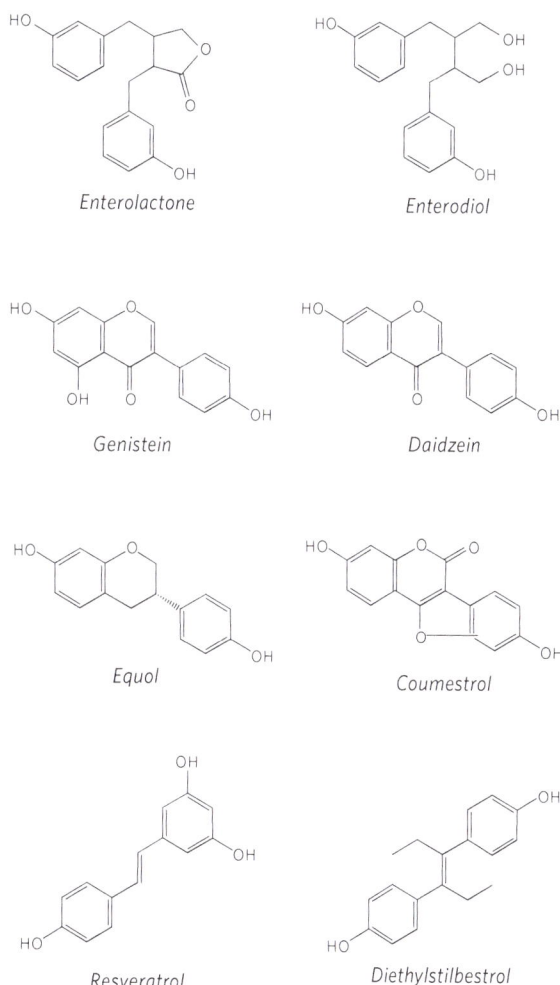

Enterolactone

Enterodiol

Genistein

Daidzein

Equol

Coumestrol

Resveratrol

Diethylstilbestrol

Figure 3.2.2. Chemical structures of selected phytoestrogens include lignans (enterolactone, enterodiol), isoflavonoids (genistein, daidzein, equol), and coumestans such as coumestrol. Resveratrol (3,5,4'-trihydroxy-trans-stilbene) is shown with the synthetic diethylstilbestrol (DES).

concentrations are present in legumes, such as peas (e.g., chickpeas, green split peas), beans (e.g., kidney, black-eyed, lima), and lentils. Beer also contains significant amounts of isoflavones, including formononetin and daidzein (5). Lignans are widely distributed in cereals, vegetables, legumes, and fruits. Relatively high levels of matairesinol and secoisolaricinol are also present in black and green teas (6). However, the highest concentration of lignans is found in oilseeds, such as flaxseed. Coumestrol is present in sprouts of alfalfa, clover, mungbeans, and soybeans. (3) Resveratrol occurs in grapes and a variety of medicinal plants (7). Lignans and isoflavones can be measured in plasma as well as urine and feces. Quantitation of individual phytoestrogens is carried out by gas chromatography-mass spectrometry, high performance liquid chromatography, and radioimmunoassays (8-11). Phytoestrogen excretion in the urine varies by type of diet (12, 13). Vegetarian diets are associated with higher levels of lignan excretion and the consumption of soy-based food influences the amount of isoflavone excretion (14-17).

Molecular Biology, Cell Culture, and Animal Studies.
It has long been recognized that phytoestrogens exert hormonal effects in cell culture systems and animals. What is the basis for the estrogenic activity of phytoestrogens? Experimental evidence indicates that the estrogenic activity of phytoestrogens is mediated by binding to intracellular estrogen receptors, including the estrogen receptors ERα and ERβ, the estrogen-related receptors ERRα, ERRβ, and ERRγ, as well as the so-called nuclear type II estrogen-binding sites, EBS-II (18, 19). For example, genistein and E_2 have approximately equal affinity of binding to ERβ (20, 21). Comparative structural studies have shown that genistein, like E_2, is completely buried within the hydrophobic core of the ligand-binding domain of ERβ (21). However, the alignment of genistein towards the transactivation AF-2 helix of the protein is sub-optimal compared to the orientation of E_2, resulting in only partial agonist activity of the phytoestrogen. Nevertheless, phytoestrogens compete with E_2 for binding to ER, activate transcription of estrogen-responsive reporter genes transfected into breast cancer cells, and stimulate expression of estrogen-regulated genes, such as pS2 (7, 22-24). Coumestrol, a fluorescent phytoestrogen, localized ER in situ in cultured cells (25). Estrogen-related receptors, ERRs, share a high amino acid homology with ERs, especially in the DNA-binding domain (26). Surprisingly, ERRs do not bind estrogens but are transcriptionally active in the absence of estrogen ligands, fulfilling the criteria of so-called orphan nuclear receptors (27). Structural analysis showed that the ligand-binding domain of ERRα is too small for E_2, but can accommodate isoflavone molecules, such as genistein, which are agonists of ERRs (19). Finally, phytoestrogens also bind to so-called nuclear type II estrogen-binding sites (EBS-II), which have greater capacity and lower affinity for E_2 than ER (28-30). Physiological concentrations of phytoestrogens stimulate the proliferation of breast cancer cells in vitro and in vivo in ovariectomized athymic mice (22, 31, 32). Administration of phytoestrogens to immature or ovariectomized rats significantly increased uterine weight and reduced bone loss (30, 33).

Phytoestrogens have other effects besides ER-mediated processes, some of which are antiestrogenic in the intact animal. For instance, phytoestrogens inhibit two key enzymes in estrogen biosynthesis, i.e., aromatase and 17β-hydroxysteroid dehydrogenase type 1 (34-36). By reducing the conversions of androstenedione to E_1 and E_1 to E_2, phytoestrogens effectively lower the concentration of endogenous estrogens, thereby reducing the risk of estrogen-dependent cancer. In addition, lignans and isoflavones stimulate SHBG synthesis in the liver. The increase in circulating SHBG reduces the relative amount of free E_2 and thereby decreases the bioavailability of E_2 for target cells (28).

Phytoestrogens, such as quercetin and genistein, inhibit cell cycle progression at the G1/S and G2/M boundaries, respectively (37, 38). The inhibitory effect of genistein appears to be mediated via inhibition of tyrosine phosphorylation by protein tyrosine kinase. (39) Phytoestrogens also inhibit angiogenesis and suppress the expression of stress response-related genes (40, 41). Together, these effects account for the anticancer action of phytoestrogens that leads to suppression of mammary cancer development in animal models (42). In vitro and in vivo experiments demonstrate both estrogenic and antiproliferative effects of phytoestrogens in ER-positive and -negative breast cancer cell lines (43-45).

An endogenous ligand for EBS-II, methyl p-hydroxyphenyl-lactate (MeHPLA), has been identified which is derived from flavonoid metabolism (46). Although the precise mechanism involved in MeHPLA regulation of cell growth remains to be determined, occupancy of EBS-II inhibits estrogen stimulation of rat uterine growth and breast cancer cell proliferation in vitro and in vivo (47, 48). Levels of MeHPLA are decreased in breast cancer due to degradation by MeHPLA esterase (49). An esterase-resistant analogue of MeHPLA, 2,6-bis[(3,4-dihydroxyphenyl)-methylene] cyclohexanone (BDHPC), bound irreversibly to EBS-II and inhibited growth of breast cancer cells. The antiproliferative effect of BDHPC was due to inhibition of the G1 phase of the cell cycle and shown to act synergistically with the inhibitory effect of genistein on the G2/M transition (50).

The red wine constituent, resveratrol (Figure 3.2.2), has attracted attention because of its ability to inhibit diverse cellular events associated with tumor initiation, promotion, and progression (51). For example, resveratrol prevented or delayed the development of carcinogen-induced (i.e., N-methyl-N-nitrosourea, 7,12-dimethylbenz[a]anthracene) mammary tumors in rats (52-54). Resveratrol was found to act as an antioxidant and inhibit cyclooxygenase andhy-droperoxidase functions (51). Although resveratrol binds to the estrogen receptor, it also exerted antiestrogenic acitivity in cells at submicromolar concentrations (53, 55, 56). Furthermore, resveratrol inhibits the expression of cytochrome P450 enzymes, such as CYP1A1, CYP1B1, and CYP3A4, which are involved in carcinogen activation (57, 58). Mechanistic studies revealed that resveratrol prevents the binding of the aryl hydrocarbon receptor to promoter sequences that regulate CYP1A1 expression (59). Interestingly, resveratrol is also metabolized by CYP1B1 to piceatannol (3,4,3',5'-tetrahydroxystilbene), which is more potent than resveratrol in inducing apoptosis (60). Other red wine flavonoids, the procyanidin B dimers, were shown to suppress estrogen biosynthesis by inhibiting aromatase (61).

Human Studies. Despite the growing body of data, effects of phytoestrogen consumption on endogenous plasma hormones have been inconsistent, probably as a result of methodological differences in subject characteristics, study design, and isoflavone form and dose (62). Plasma isoflavone (e.g., equol) levels may depend not only on the amount of ingested soy but also on the composition of the intestinal microflora (63, 64). Most reported studies have used randomized crossover or parallel-arm designs. Soy has been provided as isolated soy protein, soy milk or flour. Isoflavones have been consumed at levels of 7 – 200 mg/day, and the diet periods have ranged from 2 weeks to 6 months. Overall, these studies confirm that soy consumption alters the endocrine equilibrium of premenopausal women. For example, women with regular ovulatory cycles, who received daily supplementary soy protein for one month, experienced significantly increased follicular or reduced luteal phase lengths (65, 66). Circulating E_2 concentrations decreased after ingestion of a soy milk-supplemented diet (about 200 mg/day isoflavones) for one month (67). Ingestion of half the amount of soy milk for three consecutive menstrual cycles resulted in decreased serum levels of E_2 (27%) and E_1 (23%) (68). Two intervention studies using lower amounts of isoflavones reported either no change or an increase in circulating E_2 levels (69, 70). Thus, the effect of isoflavones on endogenous estrogen levels appears to be dose-dependent and requires a high intake of isoflavones to cause a decrease in circulating estrogen levels (68). The inverse effect is assumed to be mediated by a reduction of gonadotropin concentrations. A decrease in LH and FSH levels was observed in some studies but not others suggesting that additional factors mediate the soy-induced reduction of circulating estrogens in premenopausal women (66). The decrease in circulating estrogen levels was associated with a decrease in daily urinary excretion of parent hormones (E_1, E_2, E_3) and purported genotoxic metabolites (4-OHE$_1$, 4-OHE$_2$) (71). The 2-OH-E$_1$/16α-OHE$_1$ ratio was increased (72). Overall, these data indicate that

isoflavone consumption decreases estrogen synthesis and alters estrogen metabolism away from genotoxic toward inactive metabolites. Few hormonal effects have been observed in postmenopausal women, such as a decrease in the frequency and severity of hot flashes after receiving supplementary soy flour (73-75). Soy consumption also altered the estrogen metabolism in postmenopausal women with an increase in urinary 2-OHE$_1$/16α-OHE$_1$ ratio (71).

The detailed analysis of other chemopreventive phyto-chemicals, such as the polyphenol resveratrol, has also revealed complex relationships between mechanism, bioavailability, and preclinical efficacy (76). Resveratrol is present in red wine raising the question about the quantity of wine ingestion required to reach biologically effective doses. Resveratrol reaches peak plasma levels 1.5 h post-oral administration but has plasma elimination half-life of 3 – 9 h (77). However, the peak concentration of resveratrol may be insufficient to be chemopreventive. On the other hand, resveratrol conjugate metabolites (i.e., glucuronides, sulfates) reach high systemic levels commensurate with chemopreventive efficacy.

In summary, the dietary studies reported hormonal effects of phytoestrogens in both pre- and postmenopausal women, lowering endogenous estrogen levels in the high estrogen milieu of the former and acting as estrogen agonists in the low estrogen milieu of the latter. Thus, one would hypothesize that the antiestrogenic effect of soy isoflavone consumption may lower breast cancer risk in premenopausal women, whereas estrogenic effects may benefit the bone and vasomotor systems in postmenopausal women (62).

Epidemiological Studies of Phytoestrogen Consumption and Breast Cancer Risk. Epidemiological studies on the relationship between soy consumption and risk of breast cancer have yielded discordant results (78, 79). A meta-analysis of 18 studies published between 1978 and 2004 examined 12 case-control and six cohort or nested case-control studies (80). In a pooled analysis of all women, high soy intake was modestly associated with reduced breast cancer risk (OR 0.86; 95% CI 0.75 – 0.99). Among the 10 studies that stratified by menopausal status, the inverse association between soy exposure and breast cancer risk was somewhat stronger in premenopausal women (OR 0.70; 95% CI 0.58 – 0.85) than in postmenopausal women (OR 0.77; 95% CI 0.60 – 0.98). However, eight studies did not provide menopause-specific results, six of which did not support an association.

Studies on geographical differences in breast cancer incidence suggest a possible preventive role for soy products (28). Populations at low risk for breast cancer, e.g., women in Japan and other Asian countries, generally consume diets rich in soy products and legumes and have high plasma and urine levels of phytoestrogens (81). Conversely, Western populations at high risk for breast cancer, e.g., omnivorous women in Finland and Boston, Massachusetts, excrete low amounts of lignans and isoflavones (14). Throughout Asia, the consumption of legumes is estimated to supply 25 to 45 mg of total isoflavones in the diet each day compared to Western countries where less than 5 mg/day is consumed (82). However, a large prospective Japanese study of nearly 35,000 women revealed no significant association between soy intake during adulthood and breast cancer incidence (83).

Case-control studies have also used urinary excretion levels as surrogate for phytoestrogen intake but again observed inconsistent results. In a retrospective case-control study, urinary lignans levels were significantly lower in postmenopausal women with breast cancer than in age-matched, omnivorous or vegetarian controls (14). Conversely, high excretion levels of equol and enterolactone were associated with a significant reduction in breast cancer risk (84). A prospective nested case-control study of a cohort of postmenopausal Dutch women found no association between the urinary excretion of genistein and enterolactone and breast cancer risk (85). The European Prospective Investigation of Cancer and Nutrition-Norfolk study even observed the opposite effect, namely a significant association between high equol and daizein levels and increased breast cancer risk (86). Chinese women enrolled in the Shanghai Breast Cancer Study showed a reduced risk of breast cancer when the urinary excretion of isoflavonoids and lignans was in the upper 50% compared with those in the lower 50% (odds ratio 0.28; 95% CI 0.15 – 0.50) (87, 88). The inverse association was more pronounced among women with high plasma E_2 and low E_1 sulfate and SHBG levels (89). Women with benign breast conditions, including proliferative lesions, also had higher isoflavone excretion levels than age-matched controls (90).

A study in Asian Americans examined the intake of soy and green tea in relation to breast cancer risk (91). Each had a significant, independent protective effect on breast cancer risk. The benefit of soy was primarily observed among subjects who were nondrinkers of green tea. Similarly, the protective effect of green tea was primarily observed among subjects who were low soy consumers. On the other hand, a Chinese study found no association of breast cancer risk with soy consumption but a strong inverse relation with fruit and vegetable intake illustrating the difficulty in obtaining consistent results from complex dietary studies (92).

With the growing awareness that HRT is a risk factor for breast cancer, an increasing number of women in Western countries are turning to herbal preparations to manage their menopausal symptoms. A wide range of herbal preparations are offered for that purpose, including not only phyoestrogens but also mixed polyphenolic substances with less specified pharmaceutical agents from plant species, e.g. black cohosh (Actaea or Cimicifuga racemosa), St. John's Wort (Hypericum perforatum), chasteberry (Vitex agnus castus), Asian ginseng (Panax ginseng). Many of these bind competitively to ERα and ERβ and exert low estrogenic and/or antiestrogenic activity (93, 94). Some were shown to interfere with enzymes involved in estrogen synthesis and metabolism or promote hormonally independent antioxidative actions and apoptosis. The variability and complexity of herbal preparations makes it difficult to design epidemiological studies to assess their long-term effect on breast cancer risk. One study observed that ever-use of herbal preparations was inversely associated with invasive breast cancer risk in a dose-dependent manner (OR 0.74; 95% CI 0.63 – 0.87) (95).

References

1. Ecker JR. *BRI-ghtening the pathway to steroid hormone signaling events in plants.* Cell. 1997;90:825-7.

2. Mandava NB. *Plant growth-promoting brassinosteroids.* Ann Rev Plant Physiol Plant Mol Biol. 1988;39:23-52.

3. Adlercreutz H, Mazur W. *Phyto-oestrogens and Western diseases.* Ann Med. 1997;29:95-120.

4. Reinli K, Block G. *Phytoestrogen content of foods-a compendium of literature values.* Nutr Cancer. 1996;26:123-48.

5. Lapcik O, Hill M, Hampl R, Wahala K, Adlercreutz H. *Identification of isoflavonoids in beer.* Steroids. 1998;63:14-20.

6. Mazur WM, Wahala K, Rasku S, Salakka A, Hase T, Adlercreutz H. *Lignan and isoflavonoid concentrations in tea and coffee.* Br J Nutr. 1998;79:37-45.

7. Gehm BC, McAndrews JM, Chien P, Jameson JL. *Resveratrol, a polyphenolic compound found in grapes and wine, is an agonist for the estrogen receptor.* Proc Natl Acad Sci USA. 1997;94:14138-43.

8. Adlercreutz H, Fotsis T, Bannwart C, Wahala K, Brunow G, Hase T. *Isotope dilution gas chromatographic-mass spectrometric method for the determination of lignans and isoflavonoids in human urine, including identification of genistein.* Clin Chim Acta. 1991;3:263-78.

9. Adlercreutz H, Fotsis T, Watanabe S, Lampe J, Wahala K, Makela T, et al. *Determination of legnans and isoflavonoids in plasma by isotope dilution gas chromatography-mass spectrometry.* Cancer Detect Prev. 1994;18:259-71.

10. Lapcik O, Hampl R, al-Maharik N, Salakka A, Wahala K, Adlercreutz H. *A novel radioim-munoassay for daidzein.* Steroids. 1997;62:315-20.

11. Zeleniuch-Jacquotte A, Adlercreutz H, Akhmedkhanov A, Toniolo P. *Reliability of serum measurements of lignans and isoflavonoid phytoestrogens over a two-year period.* Cancer Epidemiol Biomarkers Prev. 1998;7:885-9.

12. Adlercreutz H, Fotsis T, Bannwart C, Wahala K, Makela T, Brunow G, et al. *Determination of urinary ligans and phytoestrogen metabolites, potential antiestrogens and anticarcinogens, in urine of women on various habitual diets.* J Steroid Biochem. 1986;25:791-7.

13. Horn-Ross PL, Barnes S, Kirk M, Coward L, Parsonnet J, Hiatt RA. *Urinary phytoestrogen levels in young women from a multiethnic population.* Cancer Epidemiol Biomarkers Prev. 1997;6:339-45.

14. Adlercreutz H, Fotsis T, Heikkinen R, Dwyer JT, Woods M, Goldin BR, et al. *Excretion of the lignans enterolactone and enterodiol and of equol in omnivorous and vegetarian postmenopausal women* and in women with breast cancer. Lancet. 1982;2:1295-9.

15. Hutchins AM, Lampe JW, Martini MC, Campbell DR, Slavin JL. *Vegetables, fruits and legumes: effect on urinary isoflavonoid phytoestrogen and lignan excretion.* J Am Diet Assoc. 1995;95:769-74.

16. Kirkman LM, Lampe JW, Campbell DR, Martini MC, Slavin JL. *Urinary lignan and isoflavonoid excretion in men and women consuming vegetable and soy diets.* Nutr Cancer. 1995;24:1-12.

17. Lampe JW, Gustafson DR, Hutchins AM, Martini MC, Li S, Wahala K, et al. *Urinary isoflavonoid and lignan excretion on a Western diet: relation to soy, vegetable, and fruit intake.* Cancer Epidemiol Biomarkers Prev. 1999;8:699-707.

18. Rollerova E, Urbancikova M. *Intracellular estrogen receptors, their characterization and function (Review).* Endocr Regul. 2000;34:203-18.

19. Suetsugi M, Su L, Karlsberg K, Yuan YC, Chen S. *Flavone and isoflavone phytoestrogens are agonists of estrogen-related receptors.* Mol Cancer Res. 2003;1:981-91.

20. Kuiper GJM, Carlsson B, Grandien K, Enmark E, Haggblad J, Nilsson S, et al. *Comparison of the ligand binding specificity and transcript tissue distribution of estrogen receptors alpha and beta.* Endocrinology. 1997;138:863-70.

21. Pike AC, Brzozowski AM, Hubbard RE, Bonn T, Thorsell AG, Engstrom O, et al. *Structure of the ligand-binding domain of oestrogen receptor beta in the presence of a partial agonist and a full antagonist.* EMBO J. 1999;18:4608-18.

22. Martin PM, Horwitz KB, Ryan DS, McGuire WL. *Phytoestrogen interaction with estrogen receptors in human breast cancer cells.* Endocrinology. 1978;103:1860-7.

23. Tamir S, Eizenberg M, Somjen D, Stern N, Shelach R, Kaye A, et al. *Estrogenic and antiproliferative properties of glabridin from licorice in human breast cancer cells.* Cancer Res. 2000;60:5704-9.

24. Wang TT, Sathyamoorthy N, Phang JM. *Molecular effects of genistein on estrogen receptor mediated pathways.* Carcinogenesis. 1996;17:271-5.

25. Miksicek RJ. *In situ localization of the estrogen receptor in living cells with the fluorescent phytoestrogen coumestrol.* J Histochem Cytochem. 1993;41:801-10.

26. Giguere V, Yang N, Segui P, Evans RM. *Identification of a new class of steroid hormone receptors.* Nature. 1988;331:91-4.

27. Hong H, Yang L, Stallcup MR. *Hormone-independent transcriptional activation and coactivator binding by novel orphan nuclear receptor ERR3.* J Biol Chem. 1999;274:22618-26.

28. Adlercreutz H, Mousavi Y, Clark J, Hockerstedt K, Hamalaninen E, Wahala K, et al. *Dietary phytoestrogens and cancer: in vitro and in vivo studies.* J Steroid Biochem Mol Biol. 1992;41:331-7.

29. Markaverich BM, Clark JH. *Two binding sites for estradiol in rat uterine nuclei: relationship to uterotropic response.* Endocrinology. 1979;105:1458-62.

30. Markaverich BM, Webb B, Densmore CL, Gregory RR. *Effects of coumestrol on estrogen receptor function and uterine growth in ovariectomized rats.* Environ Health Perspect. 1995;103:574-81.

31. Hsieh C, Santell RC, Haslam SZ, Helferich WG. *Estrogenic effects of genistein on the growth of estrogen receptor-positive human breast cancer (MCF-7) cells in vitro and in vivo.* Cancer Res. 1998;58:3833-8.

32. Welshons WV, Murphy CS, Koch R, Calaf G, Jordan VC. *Stimulation of breast cancer cells in vitro by the environmental estrogen enterolactone and the phytoestrogen equol.* Breast Cancer Res Treat. 1987;10:169-75.

33. Draper CR, Edel MJ, Dick IM, Randall AG, Martin GB, Prince RL. *Phytoestrogens reduce bone loss and bone resorption in oophorectomized rats.* J Nutr. 1997;127:1795-9.

34. Adlercreutz H, Bannwart C, Wahala K, Makela T, Brunow G, Hase T, et al. *Inhibition of human aromatase by mammalian lignans and isoflavonoid phytoestrogens.* J Steroid Biochem Mol Biol. 1993;44:147-53.

35. Makela S, Poutanen M, Kostian ML, Lehtimaki N, Strauss L, Santti R, et al. *Inhibition of 17beta-hydroxysteroid oxidoreductase by flavonoids in breast and prostate cancer cells.* Proc Soc Exp Biol Med. 1998;217:310-6.

36. Makela S, Poutanen M, Lehtimaki J, Kostian ML, Santti R, Vihko R. *Estrogen-specific 17 beta-hydroxysteroid oxidoreductase type1 (E.C.1.1.1.62) as a possible target for the action of phytoestrogens.* Proc Soc Exp Biol Med. 1995;208:51-9.

37. Matsukawa Y, Marui N, Sakai T, Satomi Y, Yoshida M, Matsumoto K, et al. *Genistein arrests cell cycle progression at G2-M.* Cancer Res. 1993;53:1328-31.

38. Yoshida M, Sakai T, Hosokawa N, Marui N, Matsumoto K, Fujioka A, et al. *The effect of quercetin on cell cycle progression and growth of human gastric cancer cells.* FEBS Lett. 1990;260:10-3.

39. Pagliacci MC, Smacchia M, Migliorati G, Grignani F, Riccardi C, Nicoletti I. *Growth-inhibitory effects of the natural phyto-oestrogen genistein in MCF-7 human breast cancer cells.* Eur J Cancer. 1994;1994:1675-82.

40. Fotsis T, Pepper MS, Aktas E, Breit S, Rasku S, Adlercreutz H, et al. *Flavonoids, dietary-derived inhibitors of cell proliferation and in vitro angiogenesis.* Cancer Res. 1997;57:2916-21.

41. Zhou Y, Lee AS. *Mechanism for the suppression of the mammalian stress response by genistein, an anticancer phytoestrogen from soy.* J Natl Cancer Inst. 1998;90:381-8

42. Lamartiniere CA, Moore JB, Brown NM, Thompson R, Hardin MJ, Barnes S. *Genistein suppresses mammary cancer in rats.* Carcinogenesis. 1995;16:2833-40.

43. Hirano T, Fukuoka K, Oka K, Naito T, Hosaka K, Mitsuhashi H, et al. *Antiproliferative activity of mammalian lignan derivatives against the human breast carcinoma cell line, ZR-75-1.* Cancer Invest. 1990;8:595-602.

44. Shao A, Wu J, Shen Z, Barsky SH. *Genistein exerts multiple suppressive effects on human breast carcinoma cells.* Cancer Res. 1998;58:4851-7.

45. Zava DT, Duwe G. *Estrogenic and antiproliferative properties of genistein and other flavonoids in human breast cancer cells in vitro.* Nutr Cancer. 1997;27:31-40.

46. Markaverich BM, Gregory RR, Alejandro MA, Clark JH, Johnson GA, Middleditch BS. *Methyl p-hydroxyphenyllactate. An inhibitor of cell growth and proliferation and an endogenous ligand for nuclear type II binding sites.* J Biol Chem. 1988;263:7203-10.

47. Markaverich BM, Gregory RR, Alejandro M, Kittrell FS, Medina D, Clark JH, et al. *Methyl p-hydroxyphenyllactate and nuclear type II binding sites in malignant cells: metabolic fate and mammary tumor growth.* Cancer Res. 1990;50:1470-8.

48. Scambia G, Ranelletti FO, Benedetti PP, Pianteli M, Bonanno G, De Vincenzo R, et al. *Quercetin inhibits the growth of a multidrug-resistant estrogen-receptor-negative MCF-7 human breast-cancer cell line expressing type II estrogen-binding sites.* Cancer Chemother Pharmacol. 1991;28:255-8.

49. Carbone A, Serra FG, Ferrandina G, Scambia G, Terribile D, Bellantone R, et al. *Methyl-p-hydroxyphenyllactate-esterase activity in breast cancer: a potentially new prognostic factor in short-term follow-up.* Cancer Res. 1997;57:5406-9.

50. Attalla H, Makela TP, Wahala K, Rasku S, Andersson LC, Adlercreutz H. *2,6-(Bis((3,4-dihydroxyphenyl)-methylene) cyclohexanone (BDHPC)-induced apoptosis and p53-independent growth inhibition: synergism with genistein.* Biochem Biophys Res Commun. 1997;239:467-72.

51. Jang M, Cai L, Udeani GO, Slowing KV, Thomas CF, Beecher CW, et al. *Cancer chemopreventive activity of resveratrol, a natural product derived from grapes.* Science. 1997;275:218-20.

52. Banerjee S, Bueso-Ramos C, Aggarwal BB. *Suppression of 7,12-dimethylbenz(a)anthracene-induced mammary carcinogenesis in rats by resveratrol: role of nuclear factor-kappaB, cyclooxygenase 2, and matrix metalloprotease 9.* Cancer Res. 2002;62:4945-54.

53. Bhat KP, Lantvit D, Christov K, Mehta RG, Moon RC, Pezzuto JM. *Estrogenic and antiestrogenic properties of resveratrol in mammary tumor models.* Cancer Res. 2001;61:7456-63.

54. Soleas GJ, Grass L, Josephy PD, Goldberg DM, Diamandis EP. *A comparison of the anticarcinogenic properties of four red wine polyphenols.* Clin Biochem. 2002;35:119-24.

55. Basly JP, Marre-Fournier F, Le Bail JC, Habrioux G, Chulia AJ. *Estrogenic/antiestrogenic and scavenging properties of (E)- and (Z)-resveratrol.* Life Sci. 2000;66:769-77.

56. Bowers JL, Tyulmenkov VV, Jernigan SC, Klinge CM. *Resveratrol acts as a mixed agonist/antagonist for estrogen receptors alpha and beta.* Endocrinology. 2000;141:3657-67.

57. Chan WK, Delucchi AB. *Resveratrol, a red wine constituent, is a mechanism-based inactivator of cytochrome P450 3A4.* Life Sci. 2000;67:3103-12.

58. Chang TK, Lee WB, Ko HH. *Trans-resveratrol modulates the catalytic activity and mRNA* expression of the procarcinogen-activating human cytochrome P450 1B1. Can J Physiol Pharmacol. 2000;78:874-81.

59. Ciolino HP, Daschner PJ, Yeh GC. *Resveratrol inhibits transcription of CYP1A1 in vitro by preventing activation of the aryl hydrocarbon receptor.* Cancer Res. 1998;58:5707-12.

60. Potter GA, Patterson LH, Wanogho E, Perry PJ, Butler PC, Ijaz T, et al. *The cancer preventative agent resveratrol is converted to the anticancer agent piceatannol by the cytochrome P450 enzyme CYP1B1.* Br J Cancer. 2002;86:774-8.

61. Eng ET, Ye J, Williams D, Phung S, Moore RE, Young MK, et al. *Suppression of estrogen biosynthesis by procyanidin dimers in red wine and grape seeds.* Cancer Res. 2003;63:8516-22.

62. Kurzer MS. *Hormonal effects of soy isoflavones: studies in premenopausal and postmenopausal women.* J Nutr. 2000;130:660S-1S.

63. Duncan AM, Merz-Demlow BE, Xu X, Phipps WR, Kurzer MS. *Premenopausal equol excretors show plasma hormone profiles associated with lowered risk of breast cancer.* Cancer Epidemiol Biomarkers Prev. 2000;9:581-6.

64. Xu X, Harris KS, Wang HJ, Murphy PA, Hendrich S. *Bioavailability of soybean isoflavones depends upon gut microflora in women.* J Nutr. 1995;125:2307-15.

65. Cassidy A, Bingham S, Setchell KD. *Biological effects of a diet of soy protein rich in isoflavones on the menstrual cycle of premenopausal women.* Am J Clin Nutr. 1994;60:333-40.

66. Lu LJ, Anderson KE, Grady JJ, Kohen F, Nagamani M. *Decreased ovarian hormones during a soya diet: implications for breast cancer prevention.* Cancer Res. 2000;60:4112-21.

67. Lu LJ, Anderson KE, Grady JJ, Nagamani M. *Effects of soya consumption for one month on steroid hormones in premenopausal women: implications for breast cancer risk reduction.* Cancer Epidemiol Biomarkers Prev. 1996;5:63-70.

68. Nagata C, Takatsuka N, Inaba S, Kawakami N, Shimizu H. *Effect of soymilk consumption on serum estrogen concentrations in premenopausal Japanese women.* J Natl Cancer Inst. 1998;90:1830-5.

69. Cassidy A, Bingham S, Setchell K. *Biological effects of isoflavones in young women: importance of the chemical composition of soyabean products.* Br J Nutr. 1995;74:587-601.

70. Petrakis NL, Barnes S, King EB, Lowenstein J, Wiencke J, Lee MM, et al. *Stimulatory influence of soy protein isolate on breast secretion in pre-and postmenopausal women.* Cancer Epidemiol Biomarkers Prev. 1996;5:785-94.

71. Xu X, Duncan AM, Wangen KE, Kurzer MS. *Soy consumption alters endogenous estrogen metabolism in postmenopausal women.* Cancer Epidemiol Biomarkers Prev. 2000;9:781-6.

72. Lu LJ, Cree M, Josyula S, Nagamani M, Grady JJ, Anderson KE. *Increased urinary excretion of 2-hydroxyestrone but not 16alpha-hydroxyestrone in premenopausal women during a soya diet containing isoflavones.* Cancer Res. 2000;60:1299-305.

73. Duncan AM, Underhill KE, Xu X, Lavalleur J, Phipps WR, Kurzer MS. *Modest hormonal effects of soy isoflavones in postmenopausal women.* J Clin Endocrinol Metab. 1999;84:3479-84.

74. Murkies AL, Lombard C, Strauss BJ, Wilcox G, Burger HG, Morton MS. *Dietary flour supplementation decreases post-menopausal hot flushes: effect of soy and wheat.* Maturitas. 1995;21:189-95.

75. Washburn S, Burke GL, Morgan T, Anthony M. *Effect of soy protein supplementation on serum lipoproteins, blood pressure, and menopausal symptoms in perimenopausal women.* Menopause. 1999;6:7-13.

76. Gescher AJ, Steward WP. *Relationship between mechanisms, bioavailibility, and preclinical chemopreventive efficacy of resveratrol: a conundrum.* Cancer Epidemiol Biomarkers Prev. 2003;12:953-7.

77. Boocock DJ, Faust GE, Patel KR, Schinas AM, Brown VA, Ducharme MP, et al. *Phase I dose escalation pharmacokinetic study in healthy volunteers of resveratrol, a potential cancer chemopreventive agent.* Cancer Epidemiol Biomarkers Prev. 2007;16:1246-52.

78. Caserta D, Maranghi L, Mantovani A, Marci R, Maranghi F, Moscarini M. *Impact of endocrine disruptor chemicals of gynaecology.* Hum Reproduction Update. 2008;14:59-72.

79. Messina M, McCaskill-Stevens W, Lampe JW. *Addressing the soy and breast cancer relationship: review, commentary, and workshop proceedings.* J Natl Cancer Inst. 2006;98:1275-84.

80. Trock BJ, Hilakivi-Clarke L, Clarke R. *Meta-analysis of soy intake and breast cancer risk.* J Natl Cancer Inst. 2006;98:459-71.

81. Adlercreutz H, Gorbach SL, Goldin BR, Woods MN, Dwyer JT, Hamalainen E. *Estrogen metabolism and excretion in Oriental and Caucasian women.* J Natl Cancer Inst. 1994;86:1076-82.

82. Coward L, Barnes NC, Setchell KDR, Barnes S. *The isoflavones genistein and daidzein in soybean foods from American and Asian diets.* J Agric Food Chem. 1993;41:1961-7.

83. Key TJ, Sharp GB, Appleby PN, Beral V, Goodman MT, Soda M, et al. *Soya foods and breast cancer risk: a prospective study in Hiroshima and Nagasaki, Japan.* Br J Cancer. 1999;81:1248-56.

84. Ingram D, Sanders K, Kolybaba M, Lopez D. *Case-control study of phyto-oestrogens and breast cancer.* Lancet. 1997;350:990-4.

85. den Tonkelaar I, Keinan-Boker L, Veer PV, Arts CJ, Adlercreutz H, Thijssen JH, et al. *Urinary phytoestrogens and postmenopausal breast cancer risk.* Cancer Epidemiol Biomarkers Prev. 2001;10:223-8.

86. Grace PB, Taylor JI, Low YL, Luben RN, Mulligan AA, Botting NP, et al. *Phytoestrogen concentrations in serum and spot urine as biomarkers for dietary phytoestrogen intake and their relation to breast cancer risk in European prospective investigation of cancer and nutrition-norfolk.* Cancer Epidemiol Biomarkers Prev. 2004;13:698-708.

87. Dai Q, Franke AA, Jin F, Shu XO, Hebert JR, Custer LJ, et al. *Urinary excretion of phytoestrogens and risk of breast cancer among Chinese women in Shanghai.* Cancer Epidemiol Biomarkers Prev. 2002;11:815-21.

88. Zheng W, Dai Q, Custer LJ, Shu XO, Wen WQ, Jin F, et al. *Urinary excretion of isoflavonoids and the risk of breast cancer.* Cancer Epidemiol Biomarkers Prev. 1999;8:35-40.

89. Dai Q, Franke AA, Yu H, Shu XO, Jin F, Hebert JR, et al. *Urinary phytoestrogen excretion and breast cancer risk: evaluating potential effect modifiers endogenous estrogens and anthropometrics.* Cancer Epidemiol Biomarkers Prev. 2003;12:497-502.

90. Lampe JW, Nishino Y, Ray RM, Wu C, Li W, Lin MG, et al. *Plasma isoflavones and fibrocystic breast conditions and breast cancer among women in Shanghai, China.* Cancer Epidemiol Biomarkers Prev. 2007;16:2579-86.

91. Wu AH, Yu MC, Tseng C, Hankin J, Pike MC. *Green tea and risk of breast cancer in Asian Americans.* Int J Cancer. 2003;106:574-9.

92. Shannon J, Ray R, Wu C, Nelson Z, Gao DL, Li W, et al. *Food and botanical groupings and risk of breast cancer: a case-control study in Shanghai, China.* Cancer Epidemiol Biomarkers Prev. 2005;14:81-90.

93. Liu J, Burdette JE, Xu H, Gu C, van Breemen RB, Bhat KP, et al. *Evaluation of estrogenic activity of plant extracts for the potential treatment of menopausal symptoms.* J Agric Food Chem. 2001;49:2472-9.

94. Rice S, Whitehead SA. *Phytoestrogens oestrogen synthesis and breast cancer.* J Steroid Biochem Mol Biol. 2008;108:186-95.

95. Obi N, Chang-Claude J, Berger J, Braendle W, Slanger T, Schmidt M, et al. *The use of herbal preparations to alleviate climacteric disorders and risk of postmenopausal breast cancer in a German case-control study.* Cancer Epidemiol Biomarkers Prev. 2009;18:2207-13.

3.2.5. XENOESTROGENS

Background. Large quantities of pesticides and industrial chemicals have been released into the environment since World War II. Although these chemicals are unrelated to steroid hormones, a surprisingly large number possesses weak estrogen-like effects and accordingly has been labeled as xenoestrogens or environmental estrogens. To date, more than 500 chemicals have been identified as xenoestrogens, including many chemicals in common use, such as constituents of detergents, pesticides, and plastics (1, 2). Xenoestrogens have also been called 'endocrine disruptors' and implicated in a variety of health effects in humans and wildlife by mimicking or interfering with the actions of endogenous estrogen (3, 4). For example, male reproductive abnormalities including declining semen quality and an increasing incidence of hypospadia, cryptorchidism, and testicular cancer have been related to xenoestrogens (5, 6). Female fertility problems and a worldwide decline in the duration of lactation were similarly related to xenoestrogens (7, 8).

Xenoestrogens have widespread consumer uses and are detectable in indoor air and dust. An analysis of residential environments in 120 homes detected up to 28 and 42 compounds per home in air and dust, respectively (9). The most abundant compounds were phthalates (plasticizers, emulsifiers), o-phenylphenol (disinfectant), 4-tert-butylphenol (adhesive) and polybrominated diphenyl ethers (flame retardants). The presence of xenoestrogens is often unexpected. For example, the plastics ingredient bisphenol-A is present in dental sealant and lacquer coating of food cans and has been detected in saliva and preserved food, respectively (10, 11). Ironically, the presence of xenoestrogens can give rise to spurious results in endocrine research laboratories. For example, commercial preparations of phenolsulfonphthalein (phenol red), a pH indicator dye commonly added to cell culture media was found to have estrogenic activity resulting from a minor lipophilic contaminant (about 0.002%) that was characterized as bis(4-hydroxyphenyl)[2-(phenoxysulfo-nyl)phenyl]methane (12). This compound was shown to bind to ER and to stimulate the proliferation of ER-positive MCF-7 breast cancer cells (13). MCF-7 cells grown in media prepared in water autoclaved in polycarbonate flasks exhibited higher PR levels than cells grown in media prepared with water autoclaved in glass. The estrogenic compound that leached out of the polycarbonate plastic during the autoclaving process was identified as bisphenol-A by nuclear magnetic resonance spectroscopy and mass spectrometry (14). Similarly, the xenoestrogen p-nonylphenol is released from polystyrene plastic tubes and can induce PR and stimulate MCF-7 cell proliferation (15).

The age-adjusted incidence rate of breast cancer in the United States has increased at an annual rate of about

one percent since formal tracking of cases through cancer registries began in the 1930s (16). The question has been raised whether the secular increase in breast cancer could be explained by increased exposure to xenoestrogens (17-22).

Chemical Structure. Xenoestrogens include a wide range of chemicals with diverse structures (Figure 3.2.3.). Organochlorines constitute the largest group with pesticides such as DDT (dichlorodiphenyltrichloroethane or 1,1,1-trichloro-2,2-bis[p-chlorophenyl] ethane) and its metabolite DDE (dichlorodiphenyldichloroethene or 1,1-dichloro-2,2-bis[p-chlorophenyl]ethylene), methoxychlor and its metabolite HPTE (2,2-bis[p-hydroxyphenyl]-1,1,1-trichlo roethane), and polychlorinated biphenyls (PCBs; e.g., dichlorobiphenyls [DCBs], tetrachlorobiphenyls [TCBs], and hexachlorobi-phenyls [HCBs]) and hydroxylated biphenyls, which are used in electrical insulators, lubricants, and plasticizers (23). Polybrominated diphenyl ethers (PBDEs) are used as flame retardants (9). Other xenoestrogens are carbamate

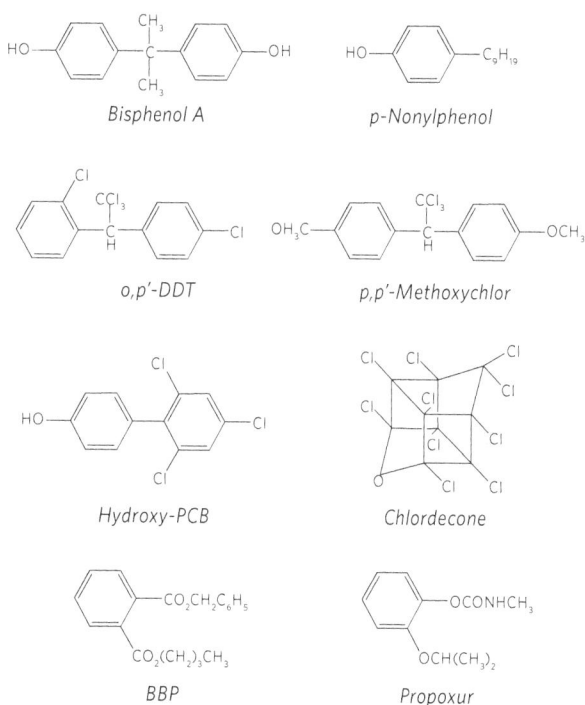

Bisphenol A   *p-Nonylphenol*

o,p'-DDT   *p,p'-Methoxychlor*

Hydroxy-PCB   *Chlordecone*

BBP   *Propoxur*

Figure 3.2.3. Chemical structures of selected xenoestrogens, many of which are phenols with a hydrophobic moiety at the para-position, such as bisphenol-A and p-nonylphenol. *Organochlorines include o,p'-DDT (dichlorodiphenyltrichloroethane), p,p'-methoxychlor, and polychlorinated biphenyls (PCBs), such as hydroxy-PCB and chlordecone (kepone). Additional examples are phthalates, such as BBP (benzylbutylphthalate), and carbamates, such as propoxur.*

insecticides (e.g., propoxur) and a variety of chemicals used in the manufacture of plastics including alkylphenols (e.g., nonylphenol, octylphenol), bisphenol-A, and phthalate esters (e.g., benzylbutylphthalate [BBP]) (24, 25).

Molecular Biology, Cell Culture, and Animal Studies. If one considers the range of chemical structures that bind to ER, it is virtually impossible to predict which structure will be the best ligand and which modification will lead to a complete loss in ligand-binding properties. For example, the insecticide chlordecone (kepone; Figure 3.2.3.), with its chlorinated, box-like structure, binds as an agonist to ER and in rats has been shown to exhibit estrogenic activity on uterine growth, embryo implantation, and the hypothalamo-pituitary axis (26-28). ER binding studies have to consider not only ERα but also ERβ, since both are capable of binding xenoestrogens, e.g., methoxychlor and bisphenol-A, albeit with very low affinity (29). At the same time, the parent xenoestrogen as well as its metabolite may bind to ER, as is true for neutral organochlorines and their hydroxylated metabolites (30). In addition to the classic nuclear ER pathway, xenoestrogens may act via the estrogen-related receptor α-1 (ERRα1), nonclassical cell membrane-linked signal transduction pathways, or even through epigenetic mechanisms (31-34). Therefore, the characterization of a chemical as xenoestrogen requires assessment of estrogenic activity by both *in vitro* and *in vivo* assays (35). Bioassays utilize activation of estrogen-regulated genes such as PR and pS2, stimulation of ER-positive breast cancer cell proliferation, vaginal cell cornification, and gain in uterine wet weight in the rodent uterotropic assay to assess the estrogenic activity of test chemicals (2, 36). The effect of SHBG on the access of xenoestrogens to ER can be determined by comparing the proliferative rate of MCF-7 cells grown in serum-free medium and in the presence of human serum (37).

Several groups developed yeast screening assays in which yeast cells were transformed with plasmids encoding the human ERα and an ERE linked to a reporter gene (38-41). The relative activity of xenoestrogens is determined by reference to E$_2$. In the yeast screen, p-nonylphenol, bisphenol-A, methoxychlor, and DDT were 5,000- to 1,000,000-fold less active than E$_2$. The yeast screen facilitates systematic structural analysis of classes of compounds such as alkylphenols, in which both the position (para > meta > ortho) and the branching (tertiary > secondary = normal) of the alkyl group affect estrogenicity (1, 42). Such studies allow the development of three-dimensional models that may aid in the experimental identification of xenoestrogens (43). Nevertheless, combination of *in vitro* and *in vivo* assays is recommended to determine the estrogenic activity of putative xenoestrogens since their potency often varies between different assays (44, 45). Xenoestrogens can act together to produce an effect even when each individual component of the mixture is

present below a threshold for effect (46). Thus, hazard assessments that ignore the possibility of synergistic action of xenoestrogens will lead to underestimations of risk.

DDT was shown to bind to ER, induce PR, and exert estrogenic activity in the mammalian uterus and avian oviduct (47-50). DDT also promotes maturation of the undifferentiated mammary gland and growth of mammary tumors in rats (51-53). In addition, DDT stimulates growth-arrested MCF-7 cells to enter the cell cycle by enhancing cyclin D1 synthesis and cdk2 activity. The relative potency of DDT in inducing cell cycle progression is 100 to 300 times less than that of E$_2$ when measured in the presence of insulin (54). Female Her2/neu transgenic mice were treated with the organochlorine dieldrin (0.45, 2.25, and 4.45 µg/g body weight) throughout gestation and lactation and the effect of the dieldrin exposure on mammary tumor formation in their offspring compared to corn oil treated controls (55). The three dieldrin concentrations were selected to produce serum levels representative of human background body burdens, occupational exposure, and overt toxicity. Compared to the control treatment only the highest dose caused increased tumor formation in the pups. It is noteworthy that the 4.45 µg/g dose also significantly reduced pregnancy rate in the mothers by >50% with no effect observed in the lower dose treatment groups.

Human Studies. Although DDT and PCBs have been banned in the United States and most Western nations since the 1970s, their residues persist and can be detected up to several decades later in body fluids and tissues of most residents of industrialized countries (56-58). Because organochlorines are highly lipophilic and metabolically resistant, they undergo lifelong sequestration in adipose tissue. Plasma DDE levels were shown to be positively correlated with body mass index and increase with age, reflecting exposure that occurred over a period of many years (59, 60). Nevertheless, interindividual differences in absorption, distribution, metabolism, and elimination result in wide ranges of adipose tissue/serum or plasma coefficients and make it difficult to determine lifetime intake for specific organo-chlorine compounds (60-62).

Because of its high fat content, breast milk secretion is the major route of excretion for organochlorines (63). Breast milk concentrations of organochlorines decrease with cumulative breast feeding duration (64). Hence, organo-chlorine body burden in young women differs markedly according to their lactation history. The apparent protective effect of breast feeding on breast cancer could be mediated by increased excretion and therefore decreased body burden of carcinogenic organochlorines (65).

Epidemiological Studies of Xenoestrogen Exposure and Breast Cancer Risk. The best studied xenoestrogens in relation to breast cancer risk are organochlorines such as DDT, its metabolite DDE, and PCBs, which have been measured in blood or adipose tissue. A few early studies reported higher levels of DDE among women with breast cancer than among controls, but a later review noted statistically significant associations in only 6 of 27 studies (66-68). Two large, prospective studies involving nearly 400 American women with breast cancer did not show a significant association of plasma DDE and PCB levels with breast cancer risk (59, 69). The combined analysis of five studies consisting of 1400 patients with breast cancer and 1642 control women in the northeastern United States found no association of breast cancer risk with plasma/serum concentrations of DDE or PCBs (70). The case-control study of the Long Island Breast Cancer Study Project assessed organochlorines in blood samples drawn near the time of diagnosis (cases) or interview (controls) (71). There were no significant differences in lipid-adjusted blood levels of DDE, the termiticide chlordane, the pesticide dieldrin, and four frequently occurring polychlorinated biphenyls. No dose-response relations were apparent. Women in North Carolina who lived or worked on farms with reported exposure to pesticides did not show an increase in breast cancer risk (72). European case-control studies of over 1000 women also found no difference in adipose DDE concentrations between case and control subjects (73-75). A Danish study of blood samples collected in the 1970s found an association of dieldrin with increased breast cancer risk (19). However, a follow-up study of Danish women observed no difference in dieldrin concentrations in adipose tissue samples obtained in the 1990s from 409 breast cancer patients and 409 control women (73). Neither of the Danish studies found an association with DDT, DDE, or other organochlorines. Studies in Mexico and Vietnam, countries which still utilize DDT to control malaria, also did not observe a difference in blood DDE levels between breast cancer patients and control subjects (76, 77). Two case-control studies have examined the interaction of enzyme polymorphisms and organochlorines in relation to breast cancer risk. PCB congeners were shown to induce cytochrome P450 enzymes, which, in turn, metabolize PCB leading to the formation of DNA adducts (78, 79). Although there was no independent association of either PCB exposure or a cytochrome P450 1A1 (CYP1A1) variant (codon Ile462Val) with breast cancer, the relative risk among postmenopausal women with plasma PCB levels in the highest third of the distribution in the control group and at least one 462Val allele was significantly increased compared with women who were homozygous 462Ile/Ile and had PCB levels in the lowest third (80, 81).

Much less is known about other xenoestrogens and their potential association with breast cancer. Occupational exposure to octylphenol, nonylphenol, bisphenol-A,

4-tert-butylphenol, and butylbenzylphthalate was not associated with a statistically significant increase in breast cancer (82). On the other hand, accidental exposure to 2,3,7,8-tetrachlorodibenzo-p-dioxin (TCDD or dioxin) was associated with increased breast cancer risk. A study of dioxin in women who were infants to 40 years of age in 1976 at the time of an industrial explosion in Seveso, Italy, revealed a 2.1-fold increase (95% CI 1.0 – 4.6) in breast cancer risk among women with a 10-fold increase in serum dioxin levels (83). The PCE (perchloroethylene) Study on Cape Cod, Massa-chussetts observed higher risk in women who were accidentally exposed to PCE leaching from the vinyl lining of improperly prepared drinking water distribution pipes from the late 1960s through the early 1980s (84, 85). The odds ratio for women with exposures >75 percentile compared to none was 1.6 (95% CI 1.1 – 2.4). Drinking water contaminated by wastewater is a potential source of exposure to mammary carcinogens and xenoestrogens from commercial products and excreted natural and pharmaceutical hormones. A meta-analysis of case-control studies of chlorinated drinking water reported an increased risk for bladder and rectal cancers but not for breast cancer (relative risk 1.18; 95% CI 0.90 – 1.54) (86). A study of Cape Cod women (824 cases, 745 controls) also found no evidence for an association between breast cancer and drinking water contaminated by wastewater (87).

Overall, these data do not support the hypothesis that exposure to DDT and PCBs increase the risk of breast cancer (88). Since organochlorine pesticides have been banned several decades ago, any potential threat of DDT or PCBs on breast cancer development is diminishing with every passing year. On balance, exposure to xenoestrogens is far outweighed by dietary intake of phytoestrogens, making it unlikely that xenoestrogens contribute significantly to a woman's overall lifetime exposure to estrogens (89, 90). Thus, whatever adverse effect xenoestrogens may have on human health, an increase in breast cancer risk at exposure levels of the general population is not supported by existing data.

References

1. Nishihara T, Nishikawa J, Kanayama T, Dakeyama F, Saito K, Imagawa M, et al. *Estrogenic Activities of 517 Chemicals by Yeast Two-Hybrid Assay.* J Health Sci 2000;46:282-98.

2. Soto AM, Sonnenschein C, Chung KL, Fernandez MF, Olea N, Serrano FO. *The E-SCREEN assay as a tool to identify estrogens: an update on estrogenic environmental pollutants.* Environ Health Perspect. 1995;103 Suppl 7:113-22.

3. Colborn T, vom Saal FS, Soto AM.

Developmental effects of endocrine-disrupting chemicals in wildlife and humans. Environ Health Perspect. 1993;101:378-84.

4. Maffini MV, Rubin BS, Sonnenschein C, Soto AM. *Endocrine disruptors and reproductive health: the case of bisphenol-A.* Mol Cell Endocrinol. 2006;254-255:179-86.

5. Sharpe RM, Skakkebaek NE. *Are oestrogens involved in falling sperm counts and disorders of the male reproductive tract?* Lancet. 1993;341:1392-5.

6. Toppari J, Larsen JC, Christiansen P, Giwercman A, Grandjean P, Guillette LJJ, et al. *Male reproductive health and environmental xenoestrogens.* Environ Health Perspect. 1996;104:741-803.

7. Gladen BC, Rogan WJ. *DDE and shortened duration of lactation in a northern Mexican town.* Am J Public Health. 1995;85:504-8.

8. Golden RJ, Noller KL, Titus-Ernstoff L, Kaufman RH, Mittendorf R, Stillman R, et al. *Environmental endocrine modulators and human bealth: an assessment of the biological evidence.* Crit Rev Toxicol. 1998;28:109-227.

9. Rudel RA, Camann DE, Spengler JD, Korn LR, Brody JG. *Phthalates, alkylphenols, pesticides, polybrominated diphenyl ethers, and other endocrine-disrupting compounds in indoor air and dust.* Environ Sci Technol. 2003;37:4543-53.

10. Brotons JA, Olea-Serrano MF, Villalobos M, Pedraza V, Olea N. *Xenoestrogens released from lacquer coatings in food cans.* Environ Health Perspect. 1995;103:608-12.

11. Olea N, Pulgar R, Perez P, Olea-Serrano F, Rivas A, Novillo-Fertrell A, et al. *Estrogenicity of resin-based composites and sealants used in dentistry.* Environ Health Perspect. 1996;104:298-305.

12. Bindal RD, Katzenellenbogen JA. *Bis (4-hydroxyphenyl) [2-phenoxysulfonyl) phenyl] methane: isolation and structure elucidation of a novel estrogen from commercial preparations of phenol red (phenosulfonphthalein).* J Med Chem. 1988;31:1978-83.

13. Berthois Y, Katzenellenbogen JA, Katzenellenbogen BS. *Phenol red tissue culture media is a weak estrogen: implications concerning the study of estrogen-responsive cells in culture.* Proc Natl Acad Sci USA. 1986;83:2496-500.

14. Krishnan AV, Stathis P, Permuth SF, Tokes L, Feldman D. *Bisphenol-A: an estrogenic substance is released from polycarbonate flasks during autoclaving.* Endocrinology. 1993;132:2279-86.

15. Soto AM, Justicia H, Wray JW, Sonnenschein C. *p-Nonyl-phenol: an estrogenic xenobiotic released from "modified" polystyrene.* Environ Health Perspect. 1991;92:167-73.

16. Harris JR, Lippman ME, Veronesi U, Willett W. *Breast cancer.* New Engl J Med. 1992;327:319-28.

17. Brody JG, Rudel RA. *Environmental pollutants and breast cancer.* Environ Health Perspect. 2003;111:1007-19.

18. Davis DL, Bradlow HL, Wolff M, Woodruff T, Hoel DG, Anton-Culver H. *Medical hypothesis: Xenoestrogens as preventable causes of breast cancer.* Environ Health Perspect. 1993;101:372-7.

19. Hoyer AP, Grandjean P, Jorgensen T, Brock JW, Hartvig HB. *Organochlorine exposure and risk of breast cancer.* Lancet. 1998;352:1816-20.

20. Hunter DJ, Kelsey KT. *Pesticide residues and breast cancer: The harvest of a silent spring.* J Natl Cancer Inst. 1993;85:598-9.

21. Key T, Reeves G. *Organochlorines in the environment and breast cancer.* Br Med J. 1994;308:1520-1.

22. Wolff MS, Collman GW, Barrett JC, Huff J. *Breast cancer and environmental risk factors: Epidemiological and experimental findings.* Annu Rev Pharmacol Toxicol. 1996;36:573-96.

23. Kutz FW, Wood PH, Bottimore DP. *Organochlorine pesticides and polychlorinated biphenyls in human adipose tissue.* Rev Environ Contam Toxicol. 1991;120:1-82.

24. Jobling S, Reynolds T, White R, Parker MG, Sumpter JP. *A variety of environmentally persistent chemicals, including some phthalate plasticizers, are weakly estrogenic.* Environ Health Perspect. 1995;103:582-7.

25. Sonnenschein C, Soto AM. *An updated review of environmental estrogen and androgen mimics and antagonists.* J Steroid Biochem Mol Biol. 1998;65:143-50.

26. Hudson PM, Yoshikawa K, Ali SF, Lamb JCI, Reel JR, Hong JS. *Estrogen-like activity of chlordecone (kepone) on the hypothalamo-pituitary axis: effects on the pituitary enkephalin system.* Toxicol Appl Pharmacol. 1984;74:383-9.

27. Johnson DC. *Estradiol-chlordecone (Kepone) interactions: additive effect of combinations for uterotropic and embryo implantation.* Toxicol Lett. 1996;89:57-64.

28. Palmiter RD, Mulvihill ER. *Estrogenic activity of the insecticide kepone on the chicken oviduct.* Science. 1978;201:356-8.

29. Kuiper GJM, Carlsson B, Grandien K, Enmark E, Haggblad J, Nilsson S, et al. *Comparison of the ligand binding specificity and transcript tissue distribution of estrogen receptors alpha and beta.* Endocrinology. 1997;138:863-70.

30. Gaido KW, Maness SC, McDonnell DP, Dehal SS, Kupfer D, Safe S. *Interaction of methoxychlor and related compounds with estrogen receptor alpha and beta, and androgen receptor: structure-activity studies.* Mol Pharmacol. 2000;58:852-8.

31. Ho SM, Tang WY, Belmonte de Frausto J, Prins GS. *Developmental exposure to estradiol and bisphenol A increases susceptibility to prostate carcinogenesis and epigenetically regulates phosphodiesterase type 4 variant 4.* Cancer Res. 2006;66:5624-32.

32. Hsu PY, Deatherage DE, Rodriguez BA, Liyanarachchi S, Weng YI, Zuo T, et al. *Xenoestrogen-induced epigenetic repression of microRNA-9-3 in breast epithelial cells.* Cancer Res. 2009;69:5936-45.

33. Steinmetz R, Young PC, Caperell-Grant A, Gize EA, Madhukar BV, Ben-Jonathan N, et al. *Novel estrogenic action of the pesticide residue beta-hexachlorocyclohexane in human breast cancer cells.* Cancer Res. 1996;56:5403-9.

34. Yang C, Chen S. *Two organochlorine pesticides, toxaphene and chlordane, are antagonists for estrogen-related receptor alpha-1 orphan receptor.* Cancer Res. 1999;59:4519-24.

35. Shelby MD, Newbold RR, Tully DB, Chae K, Davis VL. *Assessing environmental chemicals for estrogenicity using a combination of in vitro and in vivo assays.* Environ Health Perspect. 1996;104:1296-300.

36. Odum J, Lefevre PA, Tittensor S, Paton D, Routledge EJ, Beresford NA, et al. *The rodent uterotrophic asay: critical protocol features, studies with nonyl phenols, and comparison with a yeast estrogenicity assay.* Regulatory Toxicol Pharmacol. 1997;25:176-88.

37. Nagel SC, vom Saal FS, Thayer KA, Dhar MG, Boechler M, Welshons WV. *Relative binding affinity-serum modified access (RBA-SMA) assay predicts the relative in vivo bioactivity of the xenoestrogens bisphenol A and octylphenol.* Environ Health Perspect. 1997;105:70-6.

38. Arnold SF, Robinson MK, Notides AC, Guillette LJJ, McLachlan JA. *A yeast estrogen screen for examining the relative exposure of cells to natural and xenoestrogens.* Environ Health Perspect. 1996;104:544-8.

39. Coldham NG, Dave M, Sivapathasundaram S, McDonnell DP, Connor C, Sauer MJ. *Evaluation of a recombinant yeast cell estrogen screening assay.* Environ Health Perspect. 1997;105:734-42.

40. Gaido KW, Leonard LS, Lovell S, Gould JC, Babai D, Portier CJ, et al. *Evaluation of chemicals with endocrine modulating activity in a yeast-based steroid hormone receptor gene transcription assay.* Toxicol Applied Pharmacol. 1997;143:205-12.

41. Harris CA, Henttu P, Parker MG, Sumpter JP. *The estrogenic activity of phthalate esters in vitro.* Environ Health Perspect. 1997;105:802-11.

42. Routledge EJ, Sumpter JP. *Structural features of alkylphenolic chemicals associated with estrogenic activity.* J Biol Chem. 1997;272:3280-8.

43. Waller CL, Oprea TI, Chae K, Park HK, Korach KS, Laws SC, et al. *Ligand-based identification of environmental estrogens.* Chem Res Toxicol. 1996;9:1240-8.

44. Petit F, Le Goff P, Cravedi JP, Valotaire Y, Pakdel F. *Two complementary bioassays for screening the estrogenic potency of xenobiotics: recombinant yeast for trout estrogen receptor and trout hepatocyte cultures.* J Mol Endocrinol. 1997;19:321-5.

45. Zacharewski T. *Identification and assessment of endocrine disruptors: limitations of in vivo and in vitro assays.* Environ Health Perspect. 1998;106 (Suppl 2):577-82.

46. Silva E, Rajapakse N, Kortenkamp A. *Something from "nothing"--eight weak estrogenic chemicals combined at concentrations below NOECs produce significant mixture effects.* Environ Sci Technol. 2002;36:1751-6.

47. Bitman J, Cecil HC, Harris SJ, Fries GF. *Estrogenic activity of o,p'-DDT in the mammalian uterus and avian oviduct.* Science. 1968;162:371-2.

48. Das SK, Tan J, Johnson DC, Dey SK. *Differential spatiotemporal regulation of lactoferrin and progesterone receptor genes in the mouse uterus by primary estrogen, catechol estrogen, and xenoestrogen.* Endocrinology. 1998;139:2905-15.

49. Kupfer D, Bulger WH. *Interaction of o, p'DDT with the estrogen-binding protein (EBP) in human mammary and uterine tumors.* Res Commun Chem Pathol Pharmacol. 1977;16:451-62.

50. Nelson JA. *Effects of dichlorodiphenyltrichloroethane (DDT) analogs and polychlorinated biphenyl (PCB) mixtures on 17 beta-(3H) estradiol binding to rat uterine receptor.* Biochem Pharmacol. 1974;23:447-51.

51. Brown NM, Lamartiniere CA. *Xenoestrogens alter mammary gland differentiation and cell proliferation in the rat.* Environ Health Prospect. 1995;103:708-13.

52. Robison AK, Sirbasku DA, Stancel GM. *DDT supports the growth of an estrogen-responsive tumor.* Toxicol Lett. 1985;27:109-13.

53. Scribner JD, Mottet NK. *DDT acceleration of mmmary gland tumors induced in the male Sprague-Dawley rat by 2-acetamidophenanthrene.* Carcinogenesis. 1981;2:1235-9.

54. Dees C, Askari M, Foster JS, Ahamed S, Wimalasena J. *DDT mimicks estradiol stimulation of breast cancer cells to enter the cell cycle.* Mol Carcinogenesis. 1997;18:107-14.

55. Cameron HL, Foster WG. *Developmental and lactational exposure to dieldrin alters mammary tumorigenesis in Her2/neu transgenic mice.* PLoS One. 2009;4:e4303.

56. Hoyer AP, Jorgensen T, Grandjean P, Hartvig HB. *Repeated measurements of organochlorine exposure and breast cancer risk (Denmark).* Cancer Causes Control. 2000;11:177-84.

57. Jensen AA. *Polychlorobiphenyls (PCBs), polychlorodibenzo-p-dioxins (PCDDs) and polychlorodibenzofurans (PCDFs) in human milk, blood and adipose tissue.* Science Total Environ. 1987;64:259-93.

58. Longnecker MP, Rogan WJ, Lucier G. *The human health effects of DDT (dichlorodiphenyltrichloro-ethane) and PCBS (polychlorinated biphenyls) and an overview of organochlorines in public health.* Ann Rev Public Health. 1997;18:211-44.

59. Hunter DJ, Hankinson SE, Laden F, Colditz GA, Manson jE, Willett WC, et al. *Plasma organochlorine levels and the risk of breast cancer.* New Engl J Med. 1997;337:1253-8.

60. Wolff MS, Britton JA, Teitelbaum SL, Eng S, Deych E, Ireland K, et al. *Improving organochlorine biomarker models for cancer research.* Cancer Epidemiol Biomarkers Prev. 2005;14:2224-36.

61. Petreas M, Smith D, Hurley S, Jeffrey SS, Gilliss D, Reynolds P. *Distribution of persistent, lipid-soluble chemicals in breast and abdominal adipose tissues: lessons learned from a breast cancer study.* Cancer Epidemiol Biomarkers Prev. 2004;13:416-24.

62. Rusiecki JA, Matthews A, Sturgeon S, Sinha R, Pellizzari E, Zheng T, et al. *A correlation study of organochlorine levels in serum, breast adipose tissue, and gluteal adipose tissue among breast cancer cases in India.* Cancer Epidemiol Biomarkers Prev. 2005;14:1113-24.

63. Rogan WJ, Ragan NB. *Chemical contaminants, pharmacokinetics, and the lactating mother.* Environ Health Perspect. 1994;102 Suppl 11:89-95.

64. Dewailly E, Ayotte P, Laliberte C, Weber JP, Gingras S, Natel AJ. *Polychlorinated biphenyl (PCB) and dichlorodiphenyl dichloroethylene (DDE) concentrations in the breast milk of women in Quebec.* Am J Public Health. 1996;86:1241-6.

65. Dewailly E, Dodin S, Verreault R, Ayotte P, Sauve L, Morin J, et al. *High organochlorine body burden in women with estrogen receptor-positive breast cancer.* J Natl Cancer Inst. 1994;86:232-4.

66. Falck FJ, Ricci AJ, Wolff MS, Godbold J, Deckers P. *Pesticides and polychlorinated biphenyl residues in human breast lipids and their relation to breast cancer.* Arch Environ Health. 1992;47:143-6.

67. Snedeker SM. *Pesticides and breast cancer risk: a review of DDT, DDE, and dieldrin.* Environ Health Perspect. 2001;109 Suppl 1:35-47.

68. Wolff MS, Toniolo PG, Lee EW, Rivera M, Dubin N. *Blood levels of organochlorine residues and risk of breast cancer.* J Natl Cancer Inst. 1993;85:648-52.

69. Krieger N, Wolff MS, Hiatt RA, Rivera M, Vogelman J, Orentreich N. *Breast cancer and serum organochlorines: a prospective study among white, black and Asian women.* J Natl Cancer Inst. 1994;86:589-99.

70. Laden F, Collman G, Iwamoto K, Alberg AJ, Berkowitz GS, Freudenheim JL, et al. *1,1-Dichloro-2,2-bis(p-chlorophenyl)ethylene and polychlorinated biphenyls and breast cancer: combined analysis of five U.S. studies.* J Natl Cancer Inst. 2001;93:768-76.

71. Gammon MD, Wolff MS, Neugut AI, Eng SM, Teitelbaum SL, Britton JA, et al. *Environmental toxins and breast cancer on Long Island. II. Organochlorine compound levels in blood.* Cancer Epidemiol Biomarkers Prev. 2002;11:686-97.

72. Millikan R, DeVoto E, Duell EJ, Tse CK, Savitz DA, Beach J, et al. *Dichlorodiphenyldichloroethene, polychlorinated biphenyls, and breast cancer among African-American and white women in North Carolina.* Cancer Epidemiol Biomarkers Prev. 2000;9:1233-40.

73. Raaschou-Nielsen O, Pavuk M, Leblanc A, Dumas P, Philippe Weber J, Olsen A, et al. *Adipose organochlorine concentrations and risk of breast cancer among postmenopausal Danish women.* Cancer Epidemiol Biomarkers Prev. 2005;14:67-74.

74. van't Veer P, Lobbezoo IE, Martin-Moreno JM, Guallar E, Gomex-Aracena J, Kardinaal AF, et al. *DDT (dicophane) and postmenopausal breast cancer in Europe: case-control study.* Br Med J. 1997;315:81-5.

75. Ward EM, Schulte P, Grajewski B, Andersen A, Patterson DG, Jr., Turner W, et al. *Serum organochlorine levels and breast cancer: a nested case-control study of Norwegian women.* Cancer Epidemiol Biomarkers Prev. 2000;9:1357-67.

76. Lopez-Carrillo L, Blair A, Lopez-Cervantes M, Cebrian M, Rueda C, Reyes R, et al. *Dichlorodiphenyltrichloroethane serum levels and breast cancer risk: a case-control study from Mexico.* Cancer Res. 1997;57:3728-32.

77. Schecter A, Toniolo P, Dai LC, Thuy LT, Wolff MC. *Blood levels of DDT and breast cancer risk among women living in the north of Vietnam.* Arch Environ Contam Toxicol. 1997;33:453-6.

78. McLean MR, Robertson LW, Gupta RC. *Detection of PCB adducts by the 32P-postlabeling technique.* Chem Res Toxicol. 1996;9:165-71.

79. Oakley GG, Robertson LW, Gupta RC. *Analysis of polychlorinated biphenyl-DNA adducts by 32P-postlabeling.* Carcinogenesis. 1996;17:109-14.

80. Laden F, Ishibe N, Hankinson SE, Wolff MS, Gertig DM, Hunter DJ, et al. *Polychlorinated biphenyls, cytochrome P450 1A1, and breast cancer risk in the Nurses' Health Study.* Cancer Epidemiol Biomarkers Prev. 2002;11:1560-5.

81. Moysich KB, Shields PG, Freudenheim JL, Schisterman EF, Vena JE, Kostyniak P, et al. *Polychlorinated biphenyls, cytochrome P4501A1 polymorphism, and postmenopausal breast cancer risk.* Cancer Epidemiol Biomarkers Prev. 1999;8:41-4.

82. Aschengrau A, Coogan PF, Quinn MM, Cashins LJ. *Occupational exposure to estrogenic chemicals and the occurrence of breast cancer: an exploratory analysis.* Am J Ind Med. 1998;34:6-14.

83. Warner M, Eskenazi B, Mocarelli P, Gerthoux PM, Samuels S, Needham L, et al. *Serum dioxin concentrations and breast cancer risk in the Seveso Women's Health Study.* Environ Health Perspect. 2002;110:625-8.

84. Aschengrau A, Rogers S, Ozonoff D. *Perchloroethylene-contaminated drinking water and the risk of breast cancer: additional results from Cape Cod, Massachusetts, USA.* Environ Health Perspect. 2003;111:167-73.

85. Vieira V, Aschengrau A, Ozonoff D. *Impact of tetrachloroethylene-contaminated drinking water on the risk of breast cancer: using a dose model to assess exposure in a case-control study.* Environ Health. 2005;4:3.

86. Morris RD, Audet AM, Angelillo IF, Chalmers TC, Mosteller F. *Chlorination, chlorination by-products, and cancer: a meta-analysis.* Am J Public Health. 1992;82:955-63.

87. Brody JG, Aschengrau A, McKelvey W, Swartz CH, Kennedy T, Rudel RA. *Breast cancer risk and drinking water contaminated by wastewater: a case control study.* Environ Health. 2006;5:28.

88. Safe SH. *Xenoestrogens and breast cancer.* New Engl J Med. 1997;337:1303-4.

89. Safe SH. *Environmental and dietary estrogens and human health: Is there a problem?* Environ Health Perspect. 1995;103:346-51.

90. Safe SH. *Interactions between hormones and chemicals in breast cancer.* Annu Rev Pharmacol Toxicol. 1998;38:121-58.

3.3. CHEMICAL CARCINOGENS
3.3.1. INTRODUCTION

Results from animal studies show that many xenobiotics induce mammary carcinoma at high incidence (1-4). These chemicals include polycyclic aromatic hydrocarbons (PAHs) and heterocyclic and aromatic amines. Humans are exposed to several of these chemicals on a regular basis either in the diet or in the air in the form of tobacco smoke, auto exhaust, and industrial pollution. In fact, epidemiological and experimental evidence suggests that human breast cancer may be caused by exposure to each of these classes of carcinogens (5-13).

All these carcinogens are lipophilic compounds, which are expected to accumulate in mammary fat tissue. Therefore, it stands to reason that the adipose tissue surrounding mammary epithelial cells is the main source of mammary carcinogens. In 1978, Beer and Billingham proposed "the abundant adipose tissue that intimately invests both the alveolar and the ductal epithelium of the mammary gland may function as a slow-release depot for lipid-soluble carcinogenic agents of both endogenous and exogenous origin taken up from the bloodstream". (14) Indeed, extracts of mammary lipid from women undergoing reduction mammoplasty possess genotoxic activity in several in vitro assays (15-17). Similarly, mutagenic activity has been detected in nipple aspirates and breast cyst fluid (18-20). Furthermore, extracts of human breast milk have been shown to be genotoxic (21). The genotoxic activity was found to be less prevalent in milk samples from countries of lower breast cancer incidence (the Far East) compared with that in samples from the UK (22).

Experimental evidence implicates specific carcinogens in the induction of mammary DNA damage. One study quantified DNA adducts derived from one major member each of heterocyclic amines (2-amino-1-methyl-6-phenylimidazo[4,5-b]pyridine, PhIP), aromatic amines (4-aminobiphenyl, ABP), and polycyclic aromatic hydrocarbons (benzo[a]pyrene, B[a]P) (23). The DNA was isolated from exfoliated ductal epithelial cells in breast milk obtained from healthy, nonsmoking mothers. Of the 64 DNA samples analyzed, adducts were found in 31 samples. Thirty samples contained detectable levels of PhIP adducts, with a mean value of 4.7 adducts/10^7 nucleotides; 18 were positive for ABP adducts (mean 4.7 adducts/10^7 nucleotides); and 13 were found to contain B[a]P adducts (mean 1.9 adducts/10^7 nucleotides). Another study showed that PhIP and B[a]P induced DNA single-strand breaks in viable exfoliated cells isolated from breast milk (24). These data show that women are exposed to dietary and environmental carcinogens and that these carcinogens react with DNA in breast ductal epithelial cells, the cells from which breast cancers arise.

References

1. Bennett LM, Davis BJ. *Identification of mammary carcinogens in rodent bioassays.* Environ Mol Mutagen. 2002;39:150-7.

2. Dunnick JK, Elwell MR, Huff J, Barrett JC. *Chemically induced mammary gland cancer in the National toxicology Program's carcinogenesis bioassay.* Carcinogenesis. 1995;16:173-9.

3. El-Bayoumy K. *Environmental carcinogens that may be involved in human breast cancer etiology.* Chem Res Toxicol. 1992;5:585-90.

4. Josephy PD. *The role of peroxidase-catalyzed activation of aromatic amines in breast cancer.* Mutagenesis. 1996;11:3-7.

5. Dai Q, Franke AA, Jin F, Shu XO, Hebert JR, Custer LJ, et al. *Urinary excretion of phytoestrogens and risk of breast cancer among Chinese women in Shanghai.* Cancer Epidemiol Biomarkers Prev. 2002;11:815-21.

6. De Stefani E, Ronco A, Mendilaharsu M, Guidobono M, H. D-P. *Meat intake, heterocyclic amines, and risk of breast cancer: a case-control study in Uruguay.* Cancer Epidemiol Biomarkers Prev. 1997;6:573-81.

7. Eldridge SR, Gould MN, Butterworth BE. *Genotoxicity of environmental agents in human mammary epithelial cells.* Cancer Res. 1992;52:5617-21.

8. Gould MN, Cathers LE, Moore CJ. *Human breast cell-mediated mutagenesis of mammalian cells by polycyclic aromatic hydrocarbons.* Cancer Res. 1982;42:4619-24.

9. Luch A. *Nature and nurture - lessons from chemical carcinogenesis.* Nat Rev Cancer. 2005;5:113-25.

10. Sinha R, Gustafson DR, Kulldorff M, Wen W, Cerhan JR, Zheng W. *2-amino-1-methyl-6-phenylimidazo[4,5-b]pyridine, a carcinogen in high-temperature-cooked meat, and breast cancer risk.* J Natl Cancer Inst. 2000;92:1352-4.

11. Stampfer MR, Bartholomew JC, Smith HS, Bartley JC. *Metabolism of benzo[a]pyrene by human mammary epithelial cells: toxicity and DNA adduct formation.* Proc Natl Acad Sci USA. 1981;78:6251-5.

12. Sugimura T, Wakabayashi K, Nakagama H, Nagao M. *Heterocyclic amines: Mutagens/carcinogens produced during cooking of meat and fish.* Cancer Sci. 2004 ;95:290-9.

13. Zheng W, Gustafson DR, Sinha R, Cerhan JR, Moore D, Hong CP, et al. *Well-done meat intake and the risk of breast cancer.* J Natl Cancer Inst. 1998;90:1724-9.

14. Beer AE, Billingham RE. *Adipose tissue, a neglected factor in aetiology of breast cancer?* Lancet. 1978;2:296.

15. Martin FL, Carmichael PL, Crofton-Sleigh C, Venitt S, Phillips DH, Grover PL. *Genotoxicity of human mammary lipid.* Cancer Res. 1996;56:5342-6.

16. Martin FL, Venitt S, Carmichael PL, Crofton-Sleigh C, Stone EM, Cole KJ, et al. *DNA damage in breast epithelial cells: detection by the single-cell gel (comet) assay and induction by human mammary lipid extracts.* Carcinogenesis. 1997;18:2299-305.

17. Phillips DH, Martin FL, Williams JA, Wheat LM, Nolan L, Cole KJ, et al. *Mutagens in human breast lipid and milk: the search for environmental agents that initiate breast cancer.* Environ Mol Mutagen. 2002;39:143-9.

18. Petrakis NL, Maack CA, Lee RE, Lyon M. *Mutagenic activity in nipple aspirates of human breast fluid.* Cancer Res. 1980;40:188-9.

19. Scott WN, Miller WR. *The mutagenic activity of human breast secretions.* J Cancer Res Clin Oncol. 1990;116:499-502.

20. Scott WN, Miller WR. *Mutagens in human breast cyst fluid.* J Cancer Res Clin Oncol. 1991;117:254-8.

21. Martin FL, Cole KJ, Weaver G, Williams JA, Millar BC, Grover PL, et al. *Genotoxicity of human milk extracts and detection of DNA damage in exfoliated cells recovered from breast milk.* Biochem Biophys Res Commun. 1999;259:319-26.

22. Martin FL, Cole KJ, Weaver G, Hong GS, Lam BC, Balaram P, et al. *Genotoxicity of human breast milk from different countries.* Mutagenesis. 2001;16:401-6.

23. Gorlewska-Roberts K, Green B, Fares M, Ambrosone CB, Kadlubar FF. *Carcinogen-DNA adducts in human breast epithelial cells.* Environ Mol Mutagenesis. 2002;39:184-92.

24. Martin FL, Cole KJ, Williams JA, Millar BC, Harvey D, Weaver G, et al. *Activation of genotoxins to DNA-damaging species in exfoliated breast milk cells.* Mutation Res. 2000;470:115-24.

3.3.2. HETEROCYCLIC AND AROMATIC AMINES

INTRODUCTION

Heterocyclic Amines. All of us are constantly exposed to heterocyclic amines (HAs) and aromatic amines (also known as arylamines, AAs), dietary and industrial compounds with documented carcinogenic potential. The HAs 2-amino-3-methyl imidazo[4,5-f]quinoline (IQ), 2-amino-3,4-dimethylimidazo[4,5-f]quinoline (MeIQ), 2-amino-3,8-dimethylimidazo[4,5-f] quinoxaline (MeIQx), 2-amino-1,7-dimethylimidazo[4,5-g]quinoxaline (7-MeIgQx), and 2-amino-1-methyl-6-phenylimidazo[4,5-b]pyridine (PhIP) are the only mammary carcinogens known to be present in everyday foods (Figure 3.3.1.). HAs are food-derived chemicals that are formed under normal cooking conditions by pyrolysis of amino acids (1-6). PhIP is the most abundant HA detected in the browned surfaces of cooked meat, i.e., beef, pork, mutton, poultry, and fish. Cooking methods such as broiling, frying, and grilling yield concentrations of PhIP ranging from 0.5 – 70 ng/g of cooked meat (6-10). The PhIP production increases with rising cooking temperature.

Ingested HAs are metabolized by phase I and II enzymes and some of the resulting metabolites form DNA adducts in various tissues, including breast (11). The majority of metabolites are excreted in urine and feces. Quantitative analysis of urinary HA metabolites has been used to assess dietary exposure (12-15). HAs are mutagenic in bacterial and mammalian toxicity assays (1, 16). PhIP fed to rodents induces carcinomas of the colon, prostate, and mammary gland (17-20). Dietary exposure to HAs and PhIP, in particular, has been implicated in the etiology of several human malignancies, including colon, prostate, and breast cancer (5, 19, 21-23). The incremental cancer risk attributable to dietary intake of HAs in humans in the developed world has been estimated to be 1.1×10^{-4}, wherein PhIP contributes up to 46% of this incremental risk (24). Applied to the US population, this would mean that about 33,000 cancers (1.1×10^{-4} incremental cancer risk x 3×10^8 persons) are attributable to HA ingestion.

Aromatic amines, such as aniline and benzidine and their derivatives, are used as intermediates in the synthesis of numerous industrial compounds including azo dyes and antioxidants in consumer goods (25). They are also produced environmentally when plant and other organic materials are burned. Thus, AAs are present in dyes and

cigarette smoke but have also been detected in human milk (26). The prototype is 4-aminobiphenyl (ABP), a recognized human urinary bladder carcinogen (Figure 3.3.2.). Although no longer used in industry, ABP appears to be an environmental contaminant arising from combustion sources, such as cigarette smoke and synthetic fuels. ABP is found in significant quantities in cigarettes (1 – 100 ng/cigarette) and tobacco smoke and blood samples from cigarette smokers contained significantly higher hemoglobin adduct levels of ABP (mean value 154 pg/g Hb) compared to 28 pg/g Hb for nonsmokers (27). Studies on quitting smokers revealed that adduct levels declined over a period of 6 – 8 weeks to nonsmoker levels. The finding of ABP adducts in nonsmokers is consistent with low-level ubiquitous contamination of air, food, and water. ABP has been shown to be a potent urinary bladder carcinogen in mice, dogs, and humans (28). Metabolites of ABP, such as 4-acetylaminobiphenyl (AABP) and N-OH-AABP, cause the formation of mammary carcinomas in rats (29). These metabolites form adducts with proteins and DNA. For example, N-OH-ABP covalently binds as the sulfinic acid amide to amino acid 93 cysteine of human β-globin (30).

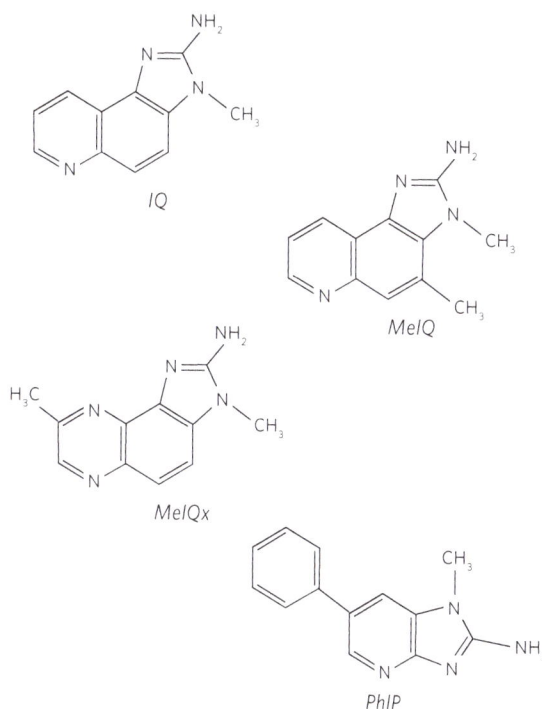

Figure 3.3.1. Structures of carcinogenic heterocyclic amines contain quinoline, quinoxaline, or pyridine moieties, such as 2-amino-3-methyl imidazo[4,5-f]quinoline (IQ), 2-amino-3,4-dimethylimidazo[4,5-f]quinoline (MeIQ), 2-amino-3,8-dimethylimidazo[4,5-f]quinoxaline (MeIQx), and 2-amino-1-methyl-6-phenylimidazo[4,5-b]pyridine (PhIP).

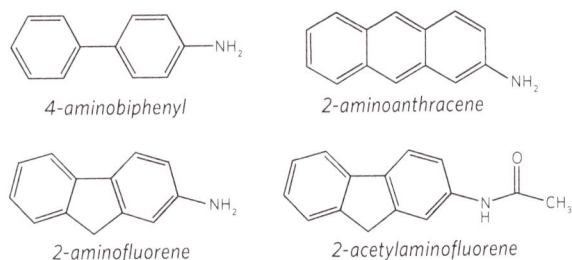

Figure 3.3.2. Structures of selected carcinogenic aromatic amines.

METABOLISM

High temperature pyrolysis of amino acids produces mutagenic HAs. For example, pyrolysis of phenylalanine produces the weak mutagen 2-amino-5-phenylpyridine (31). More potent HAs can be formed experimentally by heating a mixture of amino acids, sugar, and creatine (or creatinine), all of which are present in abundance in foods, such as meat and fish (32-35). Individual amino acids (e.g., glycine, alanine, glutamic acid, tryptophan, phenylalanine) react with the hexose (e.g., glucose, fructose) to produce heteroaromatic moieties and then a condensation reaction occurs with creatinine to produce the imidazole ring characteristic of HAs (Figure 3.3.3.).

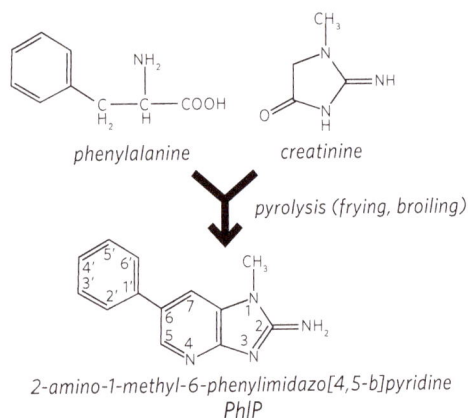

Figure 3.3.3. Formation of 2-amino-1-methyl-6-phenylimidazo[4,5-b]pyridine (PhIP) by pyrolysis of amino acids in the presence of creatinine and sugar, all of which are present in meat.

Heterocyclic Amines. HAs are metabolized by several cytochrome P450 enzymes, including CYP1A2 and 3A4 in liver and CYP1A1 and 1B1 in extrahepatic tissues (Figure

3.2.7.) (36-42). For example, PhIP is oxidized to two metabolites, the inactive 4'-hydroxy-2-amino-1-methyl-6-phenylimidazo[4,5-b]pyridine (4'-OH-PhIP) and the mutagenic 2-hydroxyamino-1-methyl-6-phenylimidazo[4,5-b]pyridine (N2-OH-PhIP) (16, 43). The latter is formed by oxidation at the exocyclic amino group of PhIP. The hydroxylamine N2-OH-PhIP can also be reverted to PhIP by microsomal NADH-dependent cytochrome b5 reductase, which is present in hepatocytes, e.g., HepG2 cells (44). The N-hydroxy products of some HAs can be further oxidized to nitroso derivatives, e.g., IQ-NHOH and IQ-N=O from IQ, which are capable of forming DNA adducts (45).

HAs and their N-hydroxy metabolites undergo conjugation by three classes of phase II enzymes, namely N-acetyltransferase (NAT, specifically NAT2), sulfotransferase (SULT), and UDP-glucuronosyltransferase (UGT). UGT forms glucuronides that are excreted through urine or bile, while NAT2 and SULT form highly electrophilic O-acetyl and O-sulfonyl esters, respectively, that are capable of binding to DNA and generate covalent DNA adducts (46-50).

Specifically, UGT esterifies N2-OH-PhIP and 4'-OH-PhIP in addition to forming stable glucuronide conjugates at the N2 and N3 positions of PhIP (Figure 3.3.4.). Thus, the list of glucuronides includes N2-OH-PhIP-N2-glucuronide, N2-OH-PhIP-N3-glucuronide, 4'-PhIP-glucuronide, PhIP-N2-glucuronide, and PhIP-N3-glucuronide, all of which are water-soluble and can be HAs and their N-hydroxy metabolites undergo conjugation by three classes of phase II enzymes, namely N-acetyltransferase (NAT, specifically NAT2), sulfotransferase (SULT), and UDP-glucuronosyltransferase (UGT). UGT forms glucuronides that are excreted through urine or bile, while NAT2 and SULT form highly electrophilic O-acetyl and O-sulfonyl esters, respectively, that are capable of binding to DNA and generate covalent DNA adducts (46-50).

Specifically, UGT esterifies N2-OH-PhIP and 4'-OH-PhIP in addition to forming stable glucuronide conjugates at the N2 and N3 positions of PhIP (Figure 3.3.4.). Thus, the list of glucuronides includes N2-OH-PhIP-N2-glucuronide, N2-OH-PhIP-N3-glucuronide, 4'-PhIP-glucuronide, PhIP-N2-glucuronide, and PhIP-N3-glucuronide, all of which are water-soluble and can be excreted through urine or bile. Comparison of UGT isoforms 1A1, 1A4, 1A6, and 1A9 expressed in microsomes of a human lymphoblastoid cell line revealed the following ranking of conjugation capacity for the substrate N2-OH-PhIP: UGT1A1 > UGT1A9 > UGT1A4 >> UGT1A6 (51). Over 10 metabolites of PhIP can be detected in human urine after administration of PhIP or after consuming a meal of well-cooked meat (21, 52, 53). N2-OH-PhIP-N2-glucuronide is the predominant metabolite accounting for 40 – 60% of all PhIP urinary metabolites, followed by N2-OH-PhIP-N3-glucuronide, PhIP-N2-glucuronide, and 4'-PhIP-sulfate (each <10%), whereas <1% was unmetabolized PhIP.

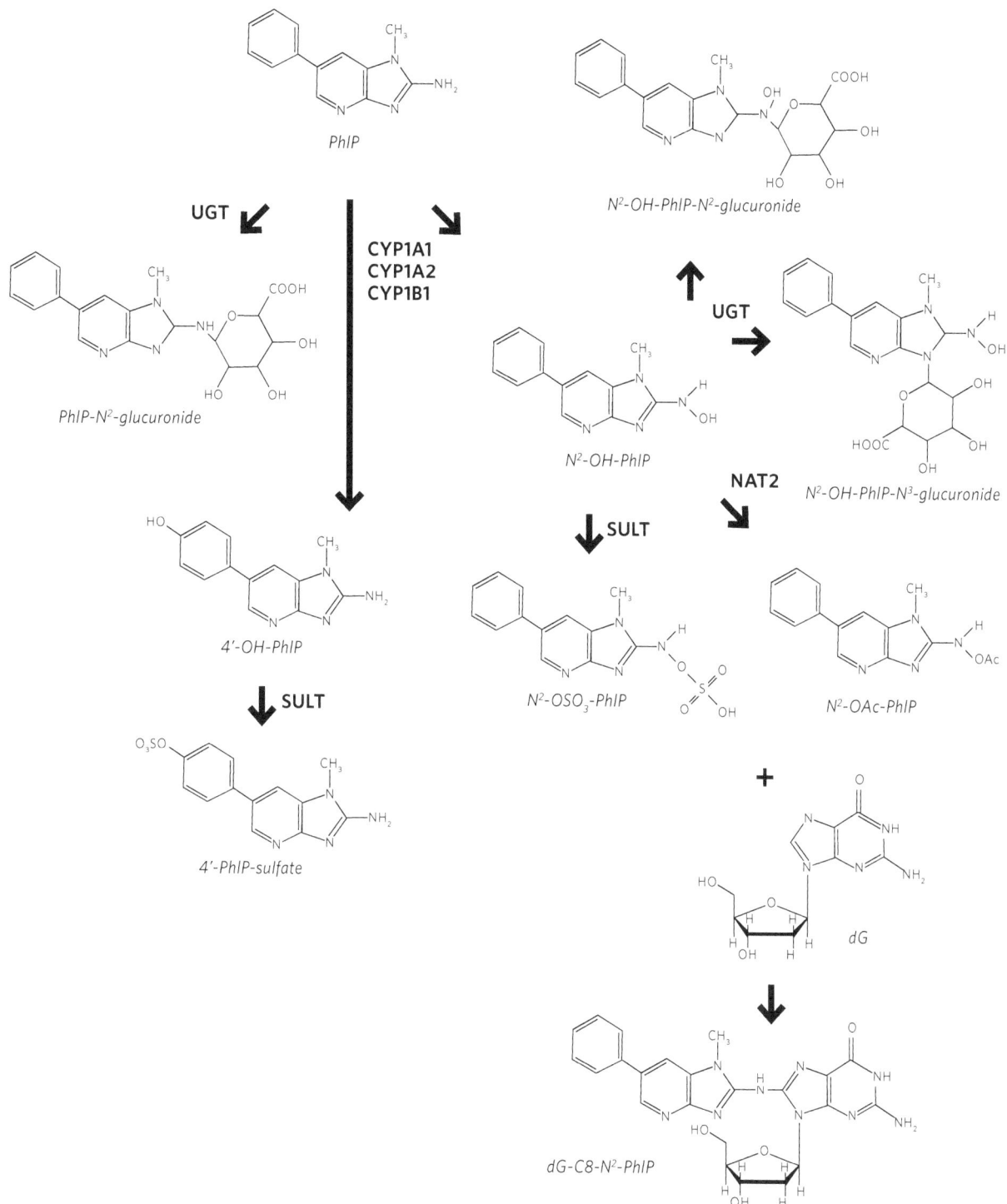

Figure 3.3.4. Metabolism of 2-amino-1-methyl-6-phenylimidazo[4,5-b]pyridine (PhIP).
The metabolism begins with the CYP-mediated oxidation to the N-hydroxy derivative, which is esterified by NAT2 or SULT, resulting in an unstable product that generates a nitrenium ion capable of attacking C8 of guanine bases to form a stable PhIP-DNA adduct. Alternate reactions lead to the formation of sulfate or glucuronide metabolites, which are water-soluble and can be excreted through urine or bile.

Aromatic Amines. The first step in the metabolic activation of AAs occurs by cytochrome P450-catalyzed N-hydroxylation (54). The resulting N-hydroxy derivatives readily react with proteins but are poorly reactive with DNA. However, they can be converted by N-acetyltransferase-mediated O-acetylation to esters that readily form DNA adducts (48). For example, following absorption of ABP, hepatic CYP1A2 catalyzes the N-hydroxylation of ABP to the N-hydroxylamine, N-OH-ABP (55). N-OH-ABP then enters the circulation where it either becomes bound to proteins, such as hemoglobin or undergoes renal filtration into the urinary bladder lumen (28). Upon reabsorption into the bladder epithelium, N-OH-ABP is converted by NAT1 to N-acetoxy-ABP ester, which is capable of reacting with DNA. Like heterocyclic amines, aromatic amines and their N-hydroxy metabolites are also good substrates for sulfotransferases and may therefore be activated in mammary tissues by similar pathways (56).

DNA Damage. Heterocyclic Amines. The acetylation or sulfation of N-hydroxy-HA species by NAT2 or SULT produces highly unstable esters, such as the N-acetoxy derivatives, which undergo heterolytic cleavage to produce the nitrenium ion-acetate ion pair. The reactive nitrenium ion-acetate ion pair. The reactive nitrenium ion intermediate then binds to DNA, forming adducts exclusively with 2'-deoxyguanosine (dG), mostly at the C8 position and fewer at the N2 position of guanine (57-59). For example, NAT2 generates the O-acetyl ester N2-OAc-PhIP, whereas SULT forms N2-SO$_4$-PhIP and 4'-SO$_4$-PhIP (Figure 3.4.4) (60, 61). Both N-acetoxy-PhIP and N-sulfonyloxy-PhIP reacted with dG, but not with the other deoxyribonucleosides, to form N2-(2'-deoxyguanosin-8-yl)-PhIP (dG-C8-PhIP; Figure 3.3.4.) (58, 62, 63). The reaction of N2-OH-PhIP with calf thymus DNA at pH 5.0 yielded only 0.313 ± 0.23 nmol of bound PhIP residues/mg DNA, whereas N2-acetoxy-PhIP yielded 5.38 ± 1.16 nmol under identical reaction conditions (64). NAT2-mediated conjugation of N2-OH-PhIP was tenfold more active in producing PhIP-DNA adducts than SULT-mediated esterification (60, 65). Structural analysis of dG-C8-PhIP has shown that the PhIP ligand intercalates into the DNA helix by denaturing and displacing the modified base pair (66). The location of the bulky PhIP-N-methyl and phenyl ring leads to widening of the minor groove and compression of the major groove at the lesion site with significant bending of the helix. The glycosidic torsion angle of dG-C8-PhIP exists predominantly in the abnormal syn conformation. Major adducts of IQ were identified as N-(deoxyguanosin-8-yl)-IQ and 5'-(deoxyguanosin-N2-yl)-IQ (59). The dG-C8 and dG-N2 isomeric adducts can be distinguished by yielding characteristic fragmentation patterns using mass spectrometry. The *syn* form is preferred for the dG-C8 compound while the *anti* form is preferred for the dG-N2 adduct of IQ.

It is generally accepted that the formation of PhIP-DNA adducts, which lead to mutations in critical genes, is central to the role of PhIP as a carcinogen (6, 57). PhIP-DNA adducts give rise to mutations either by misrepair of damaged DNA or as a consequence of errors during replication. Error-prone Y family polymerases (e.g., polymerases η, κ, and ι in humans) are capable of replicating past different types of DNA damage in a process termed translesion synthesis (67, 68). Y family polymerases have been shown structurally to accommodate various carcinogen adducts. The crystal structure of a ternary complex composed of Dpo4 (a bacterial homolog of human pol κ), a DNA substrate containing a dG-C8-PhIP adduct, and an incoming dNTP has been examined (69). The adduct was readily accommodated in the spacious major groove Dpo4 open pocket, with Dpo4 capable of incorporating dCTP, dTTP or dATP opposite the adduct reasonably well. Analysis of PhIP-induced rat mammary cancers revealed a characteristic mutational fingerprint, consisting of a guanine deletion from the 5'-GGGA-3' microsatellite sequence in four of five tumors and a predominance of G:C/T:A transversions, which are also the predominant type of mutation in rat prostate tumors (70-72).

In addition to its well-established genotoxic properties, PhIP has several other effects on the mammary gland. These include effects on mammary gland development and proliferation, alterations in levels of circulating hormones, and changes in cell signaling and gene expression (73). PhIP was also shown to have potent estrogenic activity mediated through direct binding to the ligand-binding domain of the estrogen receptor α, resulting in the transcription of ERα-regulated genes, stimulation of mitogen-activated protein kinase signaling, and proliferation of estrogen-dependent mammary epithelial cells, such as MCF-7 (74-77). The proliferative effect occurred at concentrations of PhIP (~10^{-9} to 10^{-11} M) that are likely to be equivalent to systemic human exposure via consumption of cooked meat (78). These data suggest that PhIP plays a role in cancer initiation and promotion.

Aromatic amines do not react directly with DNA but require CYP-mediated hydroxylation and then attachment of an acetyl- or sulfoxy-group to the oxygen of the hydroxylamine, which is catalyzed by NAT or SULT, respectively. Loss of the oxygen, acetyl- or sulfoxy-group leaves a nitrenium ion that then reacts with a nucleophile, such as DNA. For example, NAT1 and NAT2 play a major role in the bioactivation of ABP by converting N-OH-ABP to N-acetoxy-ABP ester, which is capable of reacting with DNA, forming adducts, mainly N-(deoxyguanosine-8-yl)-4-aminobiphenyl (dG-ABP) and dA-ABP (79). The adduct formation at the C8 position of guanine appears to involve a transient guanyl-N7 C8 position of guanine appears to involve a transient guanyl-N7 intermediate (80). Structural analysis of the C8-dG-ABP adduct revealed that the ABP ligand adopts primarily a major groove-binding structure with minimal perturbation of the DNA helix (81, 82).

MOLECULAR BIOLOGY, CELL CULTURE, AND ANIMAL STUDIES.

Heterocyclic Amines. Incubation of normal human mammary epithelial cells with IQ, MeIQ, MeIQx, or PhIP for 22 h resulted in the formation of carcinogen-DNA adducts demonstrating the ability of mammary epithelium to activate dietary HAs to DNA binding species (83). Normal human mammary epithelial cells express SULT1E1, which catalyzes the sulfation of the endogenous substrates E_2 and E_1 (84). Incubation of calf thymus DNA with human recombinant SULT1E1 and increasing levels of [^{3}H]N2-OH-PhIP resulted in a 3.5-fold increase in [^{3}H] N2-OH-PhIP covalently bound to DNA as quantified by liquid scintillation spectrometry (61). SULT1E1 did not catalyze the DNA binding of N2-OH-IQ and N2-OH-MeIQx, indicating that the enzyme only recognized N2-OH-PhIP as substrate. When normal mammary epithelial cells were cultured in the presence of 10 ×M [^{3}H]N2-OH-PhIP, binding to native DNA occurred at 60 – 240 pmol/mg DNA. Addition of the competing substrate E_1 reduced the binding to only half, suggesting that SULT1E1 contributes to bioactivation of PhIP in breast tissue (61). Treatment of human mammary epithelial cells with 20 µM N2-OH-PhIP with or without 50 µM resveratrol led to a decrease in PhIP-DNA adducts ranging from 31 to 69% in the presence of resveratrol (65). The decrease was due to inhibition of NAT2 and SULT by resveratrol. In vitro experiments showed that the DNA binding of N2-acetoxy-PhIP could be inhibited by glutathione S-transferase-mediated GSH conjugation (85). GSTA1 was more effective than GSTP1 (90% and 30% inhibition, respectively). These GSTs had no effect on the DNA binding of N2-acetoxy-IQ or –MeIQx. MeIQx treatment of DNA repair-deficient Chinese Hamster ovary cells transfected with human CYP1A1 and either NAT2*4 (rapid acetylator) or NAT2*5B (slow acetylator) resulted in significantly higher MeIQx-DNA adduct levels in CYP1A1/NAT2*4 than CYP1A1/NAT2*5B cells at all concentrations of MeIQx tested (86). These data and results obtained in primary cultures of human mammary epithelial cells (87) provide experimental support for epidemiologic studies reporting higher frequency of HA-related cancers in rapid NAT2 acetylators.

Chronic administration of PhIP in the diet has been shown to cause mammary gland cancer in rodents. A clear dose effect of PhIP was observed in dietary supplementation studies of female Fischer 344 (F344) and Sprague-Dawley (SD) rats. For example, supplementation of PhIP into the diet at a concentration of 400 parts per million (ppm) for 52 weeks induced mammary adenocarcinomas in 47% of F344 rats (Ito 1991). Lowering the dose but extending the exposure

time to 104 weeks yielded tumors in 47% (100 ppm) and 7% (25 ppm) animals (88). PhIP also induced mammary cancers in BALB/c mice and MeIQ and IQ induced tumors in F344 and SD rats (19, 23, 89). Additionally, trans-placental and trans-breast milk exposure to PhIP has also been shown to increase the risk of mammary tumors in female progeny of SD rats (90). The PhIP-induced mammary gland carcinogenesis in SD rats was shown to be associated with cell cycle deregulation (via overexpression of cyclin D1, Cdk4, phosphorylated retinoblastoma protein) and oxidative stress (suggested by overexpression of glutathione peroxidase 2, GPX2) (91, 92). The tumorigenic effect of HAs may be modified by dietary fat (17). Six-week-old female SD rats received ten doses of PhIP (75 mg/kg p.o., days 1 – 5 and 8 – 12). Two days after the last dose of PhIP, animals were placed on a high polyunsaturated-fat diet (23.5% corn oil) or a standard low-fat diet (5% corn oil). After 25 weeks on the defined diet, mammalian tumor incidence (average tumor mass ± SE) was 53% (5.7 ± 1.3 g) and 16% (2.4 ± 0.9 g) in rats on the high fat and standard low fat diet, respectively. In addition, there were striking differences in the histopathology of the induced tumors. Mammary gland tumors in PhIP-treated rats on the low fat diet were all histologically benign, containing hypertrophic and fibrocystic changes and sclerosing adenosis. In contrast, 80% of tumors in the high-fat group were histologically malignant including intraductal (papillary, cribriform, and comedo), tubular, and infiltrating ductal carcinomas. These data indicate that a high-fat diet in combination with a HA carcinogen derived from cooked meat may enhance the incidence and severity of mammary gland cancer.

Aromatic Amines. Many human cancers display genetic abnormalities in form of chromosomal or microsatellite instability. Exposure of genetically stable colorectal (HCT116) and bladder (RT112) cancer cells to sublethal doses of ABP generated resistant clones, which displayed chromosomal but not microsatellite instability (93). In contrast, treatment with another carcinogen, N-methyl-N'-nitro-N-nitrosoguanidine (MNNNG), induced microsatellite but not chromosomal instability suggesting that these two forms of genetic instability developed in a mutually exclusive fashion in these human cancer cells.

HUMAN STUDIES

Heterocyclic Amines. HAs are bioavailable in human populations as demonstrated by their detection in milk and urine from healthy women eating normal diets and charbroiled beef (15, 94). These results indicate that women are continually exposed to carcinogenic HAs in food. Dietary studies of volunteers reveal the time course of HA formation and excretion. In one study, PhIP metabolites were detectable in 35% (7 of 20) of urine samples collected from 10 healthy male volunteers who had refrained from eating smoked or broiled food for two weeks (95). The urinary concentration of PhIP metabolites increased 14- to 38-fold following consumption of an identical amount of broiled beef. The urinary metabolite concentration declined to pre-feed levels within 48 to 72 hours following cessation of broiled beef consumption. Dietary studies of volunteers consuming equal amounts of fried meat also revealed large inter-individual differences in urinary excretion of PhIP metabolites (96-98). The inter-individual variation results from differences in digestion or adsorption from the meat matrix, other foods in the diet, and genetic differences in PhIP metabolism (21, 99, 100). Members of the ABC family of drug transporters, such as BCRP and MRP2, may be involved in intestinal uptake of PhIP (101). The mouse homologue Bcrp1 was shown to restrict the exposure of mice to ingested PhIP by decreasing its uptake from the gut lumen and by mediating hepatobiliary and intestinal elimination (102).

PhIP-DNA adducts have been measured in normal and malignant tissues. Five human volunteers ingested a dietary-relevant dose of [^{14}C]PhIP (70 - 84 µg) 48 to 72 h before surgery for removal of colon tumors (103). Blood samples were collected at various time points revealing adduct formation with circulating proteins (albumin, hemoglobin) and DNA extracted from white blood cells. A comparison of DNA adducts in normal and malignant colon tissue revealed significantly higher levels in the latter. A study of ten patients with colon cancer observed an inverse relation of colon PhIP-DNA adducts with urinary levels of the N2-OH-PhIP-N2-glucuronide metabolite, suggesting that a urinary biomarker may predict the propensity of an individual to form colon DNA adducts (52). In another study, female patients undergoing breast surgery were administered [^{14}C]PhIP (20 µg per subject) and DNA adducts were identified in breast tissue, ranging from 26 to 480 adducts per 1012 DNA bases 24 h following treatment (11). Assuming that all of the radioactive material is bound covalently to DNA, the data indicate that PhIP is bioavailable to breast tissue to form DNA adducts after exposure to PhIP at dietary-relevant doses.

An immunohistochemical assay to detect the presence of PhIP-DNA adducts in breast tissue was validated against the ^{32}P-postlabeling method (mean spectrophotometric absorbance of 0.10 corresponded to 2.6 adducts/10^7 nucleotides) and then applied to normal breast tissues from 106 women with newly diagnosed breast cancer and 49 women undergoing reduction mammoplasty (104). PhIP-DNA adducts were detected in 82 and 71% of normal breast tissue sections from cancer and control patients, respectively. However, the median absorbance differed significantly between the cancer and control

populations, i.e., 0.18 (range 0 – 0.57) and 0.08 (0 – 0.38), respectively. Thus, normal mammary tissue in breast cancer patients displayed significantly higher levels of PhIP-DNA adducts than reduction mammoplasty tissue, indicating a positive association between PhIP-adduct formation and breast cancer risk.

Aromatic Amines. NAT1 and NAT2 convert N-OH-ABP to N-acetoxy-ABP ester, which is capable of reacting with DNA, forming dG-ABP and dA-ABP adducts in various tissues, e.g., urinary bladder and lung (28, 79, 105-107). In the human breast cell line MCF7, N-OH-ABP was converted by NAT1 to the N-acetoxy ester, which resulted in the formation of dG-ABP as the primary adduct and dA-ABP as minor adduct (108). A study of exfoliated ductal epithelial cells in human breast milk detected ABP-DNA adducts in 18 of 64 samples using ^{32}P post-labeling with levels ranging from 1 to 300/10^8 nucleotides (109). The investigation of environmental and inherited factors, such as smoking and NAT genotype on ABP-DNA adduct formation in breast, revealed higher ABP-DNA adduct levels in normal breast tissue adjacent to tumors than in the corresponding cancers (110). There was no correlation between the levels of ABP-DNA and NAT2 polymorphisms even when subjects were stratified by smoking status.

EPIDEMIOLOGICAL STUDIES OF HA AND AA EXPOSURE AND BREAST CANCER RISK

Several epidemiological studies have shown that intake of well done meat may be associated with an increased risk of breast cancer (111-113). However, such association was not observed in other studies (114-116). A 2009 meta-analysis of ten studies investigating the association between breast cancer risk and red meat consumption in premenopausal women observed an increased risk in seven case-control studies (summary relative risk 1.24; 95% CI 1.08 – 1.42) and none in three cohort studies (1.11; 95% CI 0.94 – 1.31) (117). Variation in cooking methods across study populations or discrepant dietary assessment may have contributed to the inconsistent findings. For example, the Shanghai Breast Cancer Study (1459 cases, 1556 controls) used a carefully designed food frequency questionnaire to assess meat consumption as well as cooking method (118). Each study participant was asked to report whether she usually deep-fried meat and fish to one of the following levels: level 1, the entire surface is brown with a slightly burnt flavor; level 2, the majority of the surface is brown; level 3, a small portion of surface is brown; level 4, virtually no surface is brown; or level 5, the meat still has bloody color. The data analysis showed a positive association of breast cancer risk with high intake of deep-fried, well-done red meat (OR 1.92; 95% CI 1.30 – 2.83 for highest versus lowest quintile; P_{trend} = 0.002). The positive association was more pronounced among women with a high body mass index. A "Charred Database" called Computerized Heterocyclic Amines Resource for Research in Epidemiology of Disease (CHARRED; http://www.charred.cancer.gov) was developed by Sinha and colleagues at the National Cancer Institute to estimate the daily intake of meat mutagens from a food frequency questionnaire validated against measured extracts of ~120 categories of meat samples prepared by different cooking methods with varying doneness levels (119). Application of this method to two large prospective studies, the NIH-AARP Diet and Health Study and the Nurses' Health Study, showed no association of meat intake and meat preparation with breast cancer risk (120, 121). Similarly, the prospective study of the Malmö Diet and Cancer Cohort (11,699 women) found no association between estimated intake of HAs and breast cancer risk (122). However, a significant positive association between higher estimated intake of MeIQx and risk of post-menopausal breast cancer was observed among 52,158 women in the prospective Prostate, Lung, Colorectal, and Ovarian (PLCO) Cancer Screening Trial (123). How can we explain these discrepant findings? The interaction with other dietary factors may have contributed to these inconsistent results. For example, the Long Island Breast Cancer Study Project (1508 cases, 1556 controls) observed no risk association with the food-frequency derived measure of HA intake but found an increased risk among postmenopausal women consuming the most grilled, barbecued, and smoked meats throughout life (OR 1.47; 95% CI 1.12 – 1.92 for highest versus lowest tertile of intake) (124). The risk was even higher for women in the top tertile with low fruit and vegetable intake (OR 1.74; 95%CI 1.20 – 2.50). In the Malmö Diet and Cancer Cohort, which did not find an overall risk association with HA intake, a subgroup of women in the bottom tertile who had the highest intake of omega-6 polyunsaturated fatty acids were at increased risk (OR 1.83; 95% CI 1.23 – 2.72 (122). Acrylamide is another potential carcinogen present in tobacco smoke, which can also be formed in foods during heating by reaction of amino acids with sugars (125, 126). The Maillard reaction products derived from glucose and asparagine, a major amino acid present in potatoes and cereals, are responsible for acrylamide levels up to 3.9 ppm in fried starchy foods, such as French fries, potato crisps, and cookies. Acrylamide is oxidized by CYP2E1 to its epoxide metabolite, glycidamide, which forms DNA adducts with guanine and adenine (127). A prospective study of dietary acrylamide intake observed increased risks of postmenopausal endometrial and ovarian cancer, particularly among never-smokers (128). However, there was no association of breast cancer risk with acrylamide intake.

NAT1 and NAT2 catalyze N- and O-acetylation of various arylamines, including the O-acetylation of N-hydroxy HAs to N-acetoxy HAs, the ultimate carcinogens of metabolic HA activation produced in well-done meat (14, 129). Epidemiologic studies of meat cooking in relation to NAT genotype have produced mixed results with respect to breast cancer risk. A NAT1 study reported a positive association of breast cancer risk with the NAT1*11 allele in individuals who consumed well-done meat (OR 5.6; 95% CI 0.5 - 62.7) (130). Another study observed a significant dose-response relation between breast cancer risk and consumption of well-done meat among women with the rapid/intermediate NAT2 genotype (trend test P = 0.003). The risk was nearly eightfold higher (OR 7.6; 95% confidence interval 1.1 - 50.4) compared to the consumption of rare or medium-done meats (131). However, the analysis of four other case-control and cohort populations, including the large Collaborative Breast CancerStudy (CBCS) and the Nurses' Health Study, found no statistically significant interaction of the NAT2 genotype with meat consumption and breast cancer risk (114-116, 132). Finally, a study of 106 women with breast cancer and 49 women undergoing reduction mammoplasty examined the association between level of PhIP-DNA adduct and various lifestyle and genetic factors (104). Stratified analysis did not reveal any association between PhIP-DNA adduct and age, ethnicity, smoking, well-done meat consumption, dietary intake of PhIP, or polymorphisms of CYP1A1, CYP1B1, NAT2, and GSTM1 genes. However, there was a borderline significant interaction (p = 0.047) between consumption of well-done meat, NAT2 genotype, and level of PhIP-DNA adducts.

References

1. Felton JS, Knize MG. *Occurence, identification, and bacterial mutagenicity of heterocyclic amines in cooked food.* Mutation Res. 1991;259:205-17.

2. Felton JS, Knize MG, Shen NH, Lewis PR, Andresen BD, Happe J, et al. *The isolation and identification of a new mutagen from fried ground beef: 2-amino-1-methyl-6-phenylimidazo[4,5-b] pyridine (PhIP).* Carcinogenesis. 1986;7:1081-6.

3. Skog K. *Cooking procedures and food mutagens: a literature review.* Food Chem Toxicol. 1993;31:655-75.

4. Skog K. *Problems associated with the determination of heterocyclic amines in cooked foods and human exposure.* Food Chem Toxicol. 2002;40:1197-203.

5. Sugimura T, Wakabayashi K, Nakagama H, Nagao M. *Heterocyclic amines: Mutagens/carcinogens produced during cooking of meat and fish.* Cancer Sci. 2004;95:290-9.

6. Turesky RJ. *Formation and biochemistry of carcinogenic heterocyclic aromatic amines in cooked meats.* Toxicol Lett. 2007;168:219-27.

7. Knize MG, Salmon CP, Mehta SS, Felton JS. *Analysis of cooked muscle meats for hetercyclic aromatic amine carcinogens.* Mutation Res. 1997;376:129-34.

8. Knize MG, Salmon CP, Pais P, Felton JS. *Food heating and the formation of heterocyclic aromatic amine and polycyclic aromatic hydrocarbon mutagens/carcinogens.* Adv Exp Med Biol. 1999;459:179-93.

9. Murray S, Lynch A, Knize MG, Gooderham NJ. *Quantifica-tion of the carcinogens 2-amino-3,8-dimethyl- and 2-amino-3,4,8-trimethylimidazo[4,5-f] quinoxaline and 2-amino-1-methyl-6-phenylimidazo[4,5-b]pyridine in food using a combined assay based on gas chromatography-negative ion mass spectrometry.* J Chromat. 1993;616:211-9.

10. Wakabayashi K, Nagao M, Esumi H, Sugimura T. *Food-derived mutagens and carcinogens.* Cancer Res. 1992;52:2092s-8s.

11. Lightfoot TJ, Coxhead JM, Cupid BC, Nicholson S, Garner RC. *Analysis of DNA adducts by accelerator mass spectrometry in human breast tissue after administration of 2-amino-1-methyl-6-phenylimidazo[4,5-b]pyridine and benzo[a]pyrene.* Mutation Res. 2000;472:119-27.

12. Davies DS, Gooderham NJ, Murray S, Lynch A, de la Torre R, Segura J, et al. *Chemical methods for assessing systemic exposure to dietary heterocyclic amines in man.* Arch Toxicol. 1996;18:251-8.

13. Friesen MD, Garren L, Bereziat J-C, Kadlubar F, Lin D. *Gas chromatography-mass spectrometry analysis of 2-amino-1-methyl-6-phenylimidazo[4,5-b]pyridine in urine and feces.* Environ Health Perspect. 1993;99:179-81.

14. Stillwell WG, Kidd LCR, Wishnok JS, Tannenbaum SR, Sinha R. *Urinary excretion of unmetabolized and phase II conjugates of 2-amino-1-methyl-6-phenylimidazo [4,5-b]pyridine and 2-amino-3,8-dimethylimidazo[4,5-f]quinoxaline in humans: relationship to cytochrome p4501A2 and N--acetyltransferase activity.* Cancer Res. 1997;57:3457-64.

15. Ushiyama H, Wakabayashi K, Hirose M, Itoh H, Sugimura T, Nagao M. *Presence of carcinogenic heterocyclic amines in urine of healthy volunteers eating normal diet, but not of inpatients receiving parenteral alimentation.* Carcinogenesis. 1991;12:1417-22.

16. Holme JA, Wallin H, Brunborg G, J.Søderlund E, K.Hongslo J, Alexander J. *Genotoxicity of the food mutagen 2-amino-1-methyl-6-phenylimidazo[4,5-b]pyridine (PhIP): formation of 2-hydroxamino-PhIP, a directly acting genotoxic metabolite.* Carcinogenesis. 1989;10:1389-96.

17. Ghoshal A, Preisegger KH, Takayama S, Thorgeirsson SS, Snyderwine EG. *Induction of mammary tumors in female Sprague-Dawley rats by food-derived carcinogen 2-amino-1-metyl-6-phenylimidazo[4,5-b]pyridine and effect of dietary fat.* Carcinogenesis. 1994;15:2429-33.

18. Ito N, Hasegawa R, Sano M, Tamano S, Esumi H, Takayama S, et al. *A new colon and mammary carcinogen in cooked food, 2-amino-1-methyl-6-phenylimidazo[4,5-b]pyridine (PhIP).* Carcinogenesis. 1991;12:1503-6.

19. Nagao M, Ushijima T, Wakabayashi K, Ochiai M, Kushida H, Sugimura T, et al. *Dietary carcinogens and mammary carcinogenesis; induction of rat mammary carcinomas by administration of heterocyclic amines in cooked foods.* Cancer. 1994;74:1063-9.

20. Shirai T, Sano M, Tamano S, Takahashi S, Hirose M, Futakuchi M, et al. *The prostate: a target for carcinogenicity of 2-amino-1-methyl-6-phenylimidazo[4,5-b]pyridine (PhIP) derived from cooked foods.* Cancer Res. 1997;57:195-8.

21. Felton JS, Knize MG, Salmon CP, Malfatti MA, Kulp KS. *Human exposure to heterocyclic amine food mutagens/carcinogens: relevance to breast cancer.* Environ Mol Mutagen. 2002;39:112-8.

22. Nelson CP, Kidd LC, Sauuageot J, Isaacs WB, De Marzo AM, Groopman JD, et al. *Protection against 2-amino-1-methyl-6-phenylimidazo[4,5-b]pyridine cytotoxicity and DNA adduct formation in human prostate by glutathione S-transferase P1.* Cancer Res. 2001;61:103-9.

23. Snyderwine EG. *Some perspectives on the nutritional aspects of breast cancer research. Food-derived heterocyclic amines as etiologic agents in human mammary cancer.* Cancer. 1994;74:1070-7.

24. Layton DW, Bogen KT, Knize MG, Hatch FT, Johnson VM, Felton JS. *Cancer risk of heterocyclic amines in cooked foods: an analysis and implications for research.* Carcinogenesis. 1995;16:39-52.

25. Radomski JL. *The primary aromatic amines: their biological properties and structure-activity relationships.* Annu Rev Pharmacol Toxicol. 1979;19:129-57.

26. DeBruin LS, Pawliszyn JB, Josephy PD. *Detection of monocyclic aromatic amines, possible mammary carcinogens, in human milk.* Chem Res Toxicology. 1999;12:78-82.

27. Bryant MS, Vineis P, Skipper PL, Tannenbaum SR. *Hemoglobin adducts of aromatic amines: associates with smoking status and type of tobacco.* Proc Natl Acad Sci USA. 1988;85:9788-91.

28. Kadlubar FF, Dooley KL, Teitel CH, Roberts DW, Benson RW, Butler MA, et al. *Frequency of urination and its effects on metabolism, pharmacokinetics, blood hemoglobin adduct formation, and liver and urinary bladder DNA adduct levels in beagle dogs given the carcinogen 4-aminobiphenyl.* Cancer Res. 1991;51:4371-7.

29. Miller JA, Wyatt CS, Miller EC, Hartmann HA. *The N-hydroxylation of 4-acetylaminobiphenyl by the rat and dog and the strong carcinogenicity of N-hydroxy-4-acetylaminobiphenyl in the rat.* Cancer Res. 1961;21:1465-73.

30. Bryant MS, Skipper PL, Tannenbaum SR, Maclure M. *Hemoglobin adducts of 4-aminobiphenyl in smokers and nonsmokers.* Cancer Res. 1987;47:602-8.

31. Kosuge T, Tsuji K, Wakabayashi K, Okamoto T, Shudo K, Iitaka Y, et al. *Isolation and structure studies of mutagenic principles in amino acid pyrolysates.* Chem Pharm Bull. 1978;26:611-9.

32. Grivas S, Nyhammar T, Olsson K, Jägerstad M. *Formation of a new mutagenic DiMeIQx compound in a model system by heating creatinine, alanine and fructose.* Mutation Res. 1985;151:177-83.

33. Jägerstad M, Olsson K, Grivas S, Negishi C, Wakabayashi K, Tsuda M, et al. *Formation of 2-amino-3,8-dimethylimidazo[4,5-f]quinoxaline in a model system by heating creatinine, glycine and glucose.* Mutation Res. 1984;126:239-44.

34. Skog KI, Johansson MA, Jagerstad MI. *Carcinogenic heterocyclic amines in model systems and cooked foods: a review on formation, occurrence and intake.* Food Chem Toxicol. 1998;36:879-96.

35. Taylor RT, Fultz E, Knize M. *Mutagen formation in a model beef supernatant fraction. IV. Properties of the system.* Environ Health Perspect. 1986;67:59-74.

36. Boobis AR, Gooderham NJ, Edwards RJ, Murray S, Lynch AM, Yadollahi-Farsani M, et al. *Enzymic and interindividual differences in the human metabolism of heterocyclic amines.* Arch Toxicol. 1996;S18:286-302.

37. Boobis AR, Lynch AM, Murray S, de la Torre R, Solans A, Farre M, et al. *CYP1A2-catalyzed conversion of dietary heterocyclic amines to their proximate carcinogens is their major route of metabolism in humans.* Cancer Res. 1994;54:89-94.

38. Buonarati MH, Roper M, Morris CJ, Happe JA, Knize MG, Felton JS. *Metabolism of 2-amino-1-methyl-6-phenylimidazo[4,5-b]pyridine (PhIP) in mice.* Carcinogenesis. 1992;13:621-7.

39. Crofts FG, Strickland PT, Hayes CL, Sutter TR. *Metabolism of 2-amino-1-methyl-6-phenylimidazo[4,5-b]pyridine (PhIP) by human cytochrome P4501B1.* Carcinogenesis. 1997;18:1793-8.

40. Crofts FG, Sutter TR, Strickland PT. *Metabolism of 2-amino-1-methyl-6-phenylimidazo[4,5-b]pyridine by human cytochrome p4501A1, p4501A2 and p4501B1.* Carcinogenesis. 1998;19:1969-73.

41. Snyderwine EG, Turesky RJ, Turteltaub KW, Davis CD, Sadrieh N, Schut N, et al. *Metabolism of food-derived heterocyclic amines in nonhuman primates.* Mutation Res. 1997;376:203-10.

42. Turesky RJ, Constable A, Richoz J, Varga N, Markovic J, Martin MV, et al. *Activation of heterocyclic aromatic amines by rat and human liver microsomes and by purified rat and human cytochrome P450 1A2.* Chem Res Toxicol. 1998;11:925-36.

43. Prabhu S, Lee M-J, Hu W-Y, Winnik B, Yang I, Buckley B, et al. *Determination of 2-amino-1-methyl-6-phenylimidazo[4,5-b]pyridine (PhIP) and its metabolite 2-hydroxy-amino-PhIP by liquid chromatography/electrospray ionization-ion trap mass spectrometry.* Anal Biochem. 2001;298:306-13.

44. King RS, Teitel CH, Shaddock JG, Casciano DA, Kadlubar FF. *Detoxification of carcinogenic aromatic and heterocyclic amines by enzymatic reduction of the N-hydroxy derivative.* Cancer Lett. 1999;143:167-71.

45. Kim D, Guengerich FP. *Cytochrome P450 activation of arylamines and heterocyclic amines.* Annu Rev Pharmacol Toxicol. 2005;45:27-49.

46. Buonarati MH, Felton JS. *Activation of 2-amino-1-methyl-6-phenylimidazo[4,5-b]pyridine (PhIP) to mutagenic metabolites.* Carcinogenesis. 1990;11:1133-8.

47. Buonarati MH, Turteltaub KW, Shen NH, Felton JS. *Role of sulfation and acetylation in the activation of 2-hydroxyamino-1-methyl-6-phenylimidazo[4,5-b]pyridine to intermediates which bind DNA.* Mutation Res. 1990;245:185-90.

48. Davis CD, Schut HA, Snyderwine EG. *Enzymatic phase II activation of the N-hydroxylamines of IQ, MeIQx and PhIP by various organs of monkeys and rats.* Carcinogenesis. 1993;14:2091-6.

49. Glatt H, Pabel U, Meinl W, Frederiksen H, Frandsen H, Muckel E. *Bioactivation of the heterocyclic aromatic amine 2-amino-3-methyl-9H-pyrido [2,3-b]indole (MeAalphaC) in recombinant test systems* expressing human xenobiotic-metabolizing enzymes. Carcinogenesis. 2004;25:801-7.

50. Wu RW, Panteleakos FN, Kadkhodayan S, Bolton-Grob R, McManus ME, Felton JS. *Genetically modified Chinese hamster ovary cells for investigating sulfotransferase-mediated cytotoxicity and mutation by 2-amino-1-methyl-6- phenylimidazo[4,5-b] pyridine.* Environ Mol Mutagen. 2000;35:57-65.

51. Malfatti MA, Felton JS. *N-glucuronidation of 2-amino-1-methyl-6-phenylimidazo [4,5-b]pyridine (PhIP) and N-hydroxy-PhIP by specific human UDP-glucuronosyltransferases.* Carcinogenesis. 2001;22:1087-93.

52. Malfatti MA, Dingley KH, Nowell-Kadlubar S, Ubick EA, Mulakken N, Nelson D, et al. *The urinary metabolite profile of the dietary carcinogen 2-amino-1-methyl-6-phenylimidazo[4,5-b]pyridine is predictive of colon DNA adducts after a low-dose exposure in humans.* Cancer Res. 2006;66:10541-7.

53. Malfatti MA, Kulp KS, Knize MG, Davis C, Massengill JP, Williams S, et al. *The identification of [2-14C]2-amino-1-methyl-6-phenylimidazo[4,5-b]pyridine metabolites in humans.* Carcinogenesis. 1999;20:705-13.

54. Kato R, Yamazoe Y. *Metabolic activation and covalent binding to nucleic acids of carcinogenic heterocyclic amines from cooked foods and amino acid pyrolysates.* Jpn J Cancer Res. 1987;78:297-311.

55. Butler MA, Iwasaki M, Guengerich FP, Kadlubar FF. *Human cytochrome P-450PA (P-450IA2), the phenacetin O-deethylase, is primarily responsible for the hepatic 3-demethylation of caffeine and N-oxidation of carcinogenic arylamines.* Proc Natl Acad Sci USA. 1989;86:7696-700.

56. Chou HC, Lang NP, Kadlubar FF. *Metabolic activation of the N-hydroxy derivative of the carcinogen 4-aminobiphenyl by human tissue sulfotransferases.* Carcinogenesis. 1995;16:413-7.

57. Schut HA, Snyderwine EG. *DNA adducts of heterocyclic amine food mutagens: implications for mutagenesis and carcinogenesis.* Carcinogenesis. 1999;20:353-68.

58. Goodenough AK, Schut HA, Turesky RJ. *Novel LC-ESI/MS/MS(n) method for the characterization and quantification of 2'-deoxyguanosine adducts of the dietary carcinogen 2-amino-1-methyl-6-phenylimidazo[4,5-b]pyridine by* 2-D linear quadrupole ion trap mass spectrometry. Chem Res Toxicol. 2007;20:263-76.

59. Turesky RJ, Vouros P. *Formation and analysis of heterocyclic aromatic amine-DNA adducts in vitro and in vivo.* J Chromatogr B Analyt Technol Biomed Life Sci. 2004;802:155-66.

60. Dubuisson JG, Gaubatz JW. *Bioactivation of the proximal food mutagen 2-hydroxyamino-1-methyl-6-phenylimidazo[4,5-b]pyridine (N-OH-PhIP) to DNA-binding species by human mammary gland enzymes.* Nutrition. 1998;14:683-6.

61. Lewis AJ, Walle UK, King RS, Kadlubar FF, Falany CN, Walle T. *Bioactivation of the cooked food mutagen N-hydroxy-2-amino-1-methyl-6-phenylimidazo[4,5-b]pyridine by estrogen sulfotransferase in cultured human mammary epithelial cells.* Carcinogenesis. 1998;19:2049-53.

62. Frandsen H, Grivas S, Andersson R, Dragsted L, Larsen JC. *Reaction of the N2-acetoxy derivative of 2-amino-1-methyl-6-phenylimidazo[4,5-b]pyridine (PhIP) with 2'-deoxyguanosine and DNA. Synthesis and identification of N2-(2'-deoxyguanosin-8-yl)-PhIP.* Carcinogenesis. 1992;13:629-35.

63. Lin D, Kaderlik KR, Turesky RJ, Miller DW, Lay JO, Kadlubar FF. *Identification of N-(Deoxyguanosin-8-yl)-2-amino-1-methyl-6-phenylimidazo[4,5-b]pyridine as the major adduct formed by the food-borne carcinogen, 2-amino-1-methyl-6-phenylimidazo[4,5-b]pyridine, with DNA.* Chem Res Toxicol. 1992;5:691-7.

64. Friesen MD, Kaderlik K, Lin D, Garren L, Bartsch H, Lang NP, et al. *Analysis of DNA adducts of 2-amino-1-methyl-6-phenylimidazo[4,5-b]pyridine in rat and human tissues by alkaline hydrolysis and gas chromatography/electron capture mass spectrometry: validation by comparison with 32P-postlabeling.* Chem Res Toxicol. 1994;7:733-9.

65. Dubuisson JG, Dyess DL, Gaubatz JW. *Resveratrol modulates human mammary epithelial cell O-acetyltransferase, sulfotransferase, and kinase activation of the heterocyclic amine carcinogen N-hydroxy-PhIP.* Cancer Lett. 2002;182:27-32.

66. Brown K, Hingerty BE, Guenther EA, Krishnan VV, Broyde S, Turteltaub KW, et al. *Solution structure of the 2-amino-1- methyl-6-phenylimidazo[4,5-b]pyridine C8-deoxyguanosine adduct in duplex DNA.* Proc Natl Acad Sci USA. 2001;98:8507-12.

67. Pages V, Fuchs RP. *How DNA lesions are turned into mutations within cells?* Oncogene. 2002;21:8957-66.

68. Prakash S, Johnson RE, Prakash L. *Eukaryotic translesion synthesis DNA polymerases: specificity of structure and function.* Annu Rev Biochem. 2005;74:317-53.

69. Zhang L, Rechkoblit O, Wang L, Patel DJ, Shapiro R, Broyde S. *Mutagenic nucleotide incorporation and hindered translocation by a food carcinogen C8-dG adduct in Sulfolobus solfataricus P2 DNA polymerase IV (Dpo4): modeling and dynamics studies.* Nucleic Acids Res. 2006;34:3326-37.

70. Nagao M, Ushijima T, Toyota M, Inoue R, Sugimura T. *Genetic changes induced by heterocyclic amines.* Mutation Res. 1997;376:161-7.

71. Nakai Y, Nelson WG, De Marzo AM. *The dietary charred meat carcinogen 2-amino-1-methyl-6-phenylimidazo[4,5-b]pyridine acts as both a tumor initiator and promoter in the rat ventral prostate.* Cancer Res. 2007;67:1378-84.

72. Yu M, Jones ML, Gong M, Sinha R, Schut HA, Snyderwine EG. *Mutagenicity of 2-amino-1-methyl-6-phenylimidazo[4,5-b]pyridine (PhIP) in the mammary gland of Big Blue rats on high- and low-fat diets.* Carcinogenesis. 2002;23:877-84.

73. Snyderwine EG, Venugopal M, Yu M. *Mammary gland carcinogenesis by food-derived heterocyclic amines and studies on the mechanisms of carcinogenesis of 2-amino-1-methyl-6-phenylimidazo[4,5-b]pyridine (PhIP).* Mutat Res. 2002;506-507:145-52.

74. Bennion BJ, Cosman M, Lightstone FC, Knize MG, Montgomery JL, Bennett LM, et al. *PhIP carcinogenicity in breast cancer: computational and experimental evidence for competitive interactions with human estrogen receptor.* Chem Res Toxicol. 2005;18:1528-36.

75. Gooderham NJ, Creton S, Lauber SN, Zhu H. *Mechanisms of action of the carcinogenic heterocyclic amine PhIP.* Toxicol Lett. 2007;168:269-77.

76. Lauber SN, Ali S, Gooderham NJ. *The cooked food derived carcinogen 2-amino-1-methyl-6-phenylimidazo[4,5-b] pyridine is a potent oestrogen: a mechanistic basis for its tissue-specific carcinogenicity.* Carcinogenesis. 2004;25:2509-17.

77. Lauber SN, Gooderham NJ. *The cooked meat derived genotoxic carcinogen 2-amino-3-methylimidazo[4,5-b]pyridine has potent hormone-like activity: mechanistic support for a role in breast cancer.* Cancer Res. 2007;67:9597-602.

78. Creton SK, Zhu H, Gooderham NJ. *The cooked meat carcinogen 2-amino-1-methyl-6-phenylimidazo[4,5-b]pyridine activates the extracellular signal regulated kinase mitogen-activated protein kinase pathway.* Cancer Res. 2007;67:11455-62.

79. Culp SJ, Roberts DW, Talaska G, Lang NP, Fu PP, Lay Jr JO, et al. *Immunochemical, 32P-postlabeling, and GC/MS detection of 4-amino-biphenyl-DNA adducts in human peripheral lung in relation to metabolic activation pathways involving pulmonary N-oxidation, conjugation, and peroxidation.* Mutation Res. 1997;378:97-112.

80. Guengerich FP, Mundkowski RG, Voehler M, Kadlubar FF. *Formation and reactions of N(7)-aminoguanosine and derivatives.* Chem Res Toxicol. 1999;12:906-16.

81. Cho BP, Beland FA, Marques MM. *NMR structural studies of a 15-mer DNA sequence from a ras protooncogene, modified at the first base of codon 61 with the carcinogen 4-aminobiphenyl.* Biochemistry. 1992;31:9587-602.

82. Patel DJ, Mao B, Gu Z, Hingerty BE, Gorin A, Basu AK, et al. *Nuclear magnetic resonance solution structures of covalent aromatic amine-DNA adducts and their mutagenic relevance.* Chem Res Toxicol. 1998;11:391-407.

83. Carmichael PL, Stone EM, Grover PL, Gusterson BA, Phillips DH. *Metabolic activation and DNA binding of food mutagens and other environmental carcinogens in human mammary epithelial cells.* Carcinogenesis. 1996;17:1769-72.

84. Falany JL, Falany CN. *Expression of cytosolic sulfotransferases in normal mammary epithelial cells and breast cancer cell lines.* Cancer Res. 1996;56:1551-5.

85. Lin D, Meyer DJ, Ketterer B, Lang NP, Kadlubar FF. *Effects of human and rat glutathione S-transferases on the covalent DNA binding of the N-acetoxy derivatives of heterocyclic amine carcinogens in vitro: a possible mechanism of organ specificity in their carcinogenesis.* Cancer Res. 1994;54:4920-6.

86. Bendaly J, Zhao S, Neale JR, Metry KJ, Doll MA, States JC, et al. *2-Amino-3,8-dimethylimidazo-[4,5-f]quinoxaline-induced DNA* adduct formation and mutagenesis in DNA repair-deficient Chinese hamster ovary cells expressing human cytochrome P4501A1 and rapid or slow acetylator N-acetyltransferase 2. Cancer Epidemiol Biomarkers Prev. 2007;16:1503-9.

87. Stone EM, Williams JA, Grover PL, Gusterson BA, Phillips DH. *Inter-individual variation in the metabolic activation of heterocyclic amines and their N-hydroxy derivatives in primary cultures of human mammary epithelial cells.* Carcinogenesis. 1998;19:873-9.

88. Hasegawa R, Sano M, Tamano S, Imaida K, Shirai T, Nagao M, et al. *Dose-dependence of 2-amino-1-methyl-6-phenylimidazo[4,5-b] pyridine (PhIP) carcinogenicity in rats.* Carcinogenesis. 1993;14:2553-7.

89. Nagao M, Ushijima T, Watanabe N, Okochi E, Ochiai M, Nakagama H, et al. *Studies on mammary carcinogenesis induced by a heterocyclic amine, 2-amino-1-methyl-6-phenylimidazo[4,5-b] pyridine, in mice and rats.* Environ Mol Mutagenesis. 2002;39:158-64.

90. Hasegawa R, Kimura J, Yaono M, Takahashi S, Kato T, Futakuchi M, et al. *Increased risk of mammary carcinoma development following transplacental and trans-breast milk exposure to a food-derived carcinogen, 2-amino-1-methyl-6-phenylimidazo[4,5-b]pyridine (PhIP), in Sprague-Dawley rats.* Cancer Res. 1995;55:4333-8.

91. Naiki-Ito A, Asamoto M, Hokaiwado N, Takahashi S, Yamashita H, Tsuda H, et al. *Gpx2 is an overexpressed gene in rat breast cancers induced by three different chemical carcinogens.* Cancer Res. 2007;67:11353-8.

92. Qiu C, Shan L, Yu M, Snyderwine EG. *Deregulation of the cyclin D1/Cdk4 retinoblastoma pathway in rat mammary gland carcinomas induced by the food-derived carcinogen 2-amino-1-methyl-6-phenylimidazo[4,5-b]pyridine.* Cancer Res. 2003;63:5674-8.

93. Saletta F, Matullo G, Manuguerra M, Arena S, Bardelli A, Vineis P. *Exposure to the tobacco smoke constituent 4-aminobiphenyl induces chromosomal instability in human cancer cells.* Cancer Res. 2007;67:7088-94.

94. DeBruin LS, Martos PA, Josephy PD. *Detection of PhIP (2-amino-1-methyl-6-phenylimidazo[4,5-b] pyridine) in the milk of healthy women.* Chem Res Toxicol. 2001;14:1523-8.

95. Friesen MD, Rothman N, Strickland PT. *Concentration of 2-amino-1-methyl-6-phenylimidazo(4,5-b)pyridine (PhIP) in urine and alkali-hydrolyzed urine after consumption of charbroiled beef.* Cancer Lett. 2001;173:43-51.

96. Knize MG, Kulp KS, Malfatti MA, Salmon CP, Felton JS. *Liquid chromatography-tandem mass spectrometry method of urine analysis for determining human variation in carcinogen metabolism.* J Chromat A. 2001;914:95-103.

97. Lynch AM, Knize MG, Boobis AR, Gooderham NJ, Davies DS, Murray S. *Intra- and inter-individual variability in systemic exposure in humans to 2-amino-3,8-dimethylimidazo[4,5-f] quinoxaline and 2-amino-1-methyl-6-phenylimidazo[4,5-b]pyridine, carcinogens present in cooked beef.* Cancer Res. 1993;52:6216-23.

98. Reistad R, Rossland DJ, Latub-Kala KJ, Rasmussen T, Vikse R, Becher G, et al. *Heterocyclic aromatic amines in human urine following a fried meat meal.* Food Chem Toxicol. 1997;35:945-55.

99. Baranczewski P, Moller L. *Relationship between content and activity of cytochrome P450 and induction of heterocyclic amine DNA adducts in human liver samples in vivo and in vitro.* Cancer Epidemiol Biomarkers Prev.2004;13:1071-8.

100. Kidd LCR, Stillwell WG, Yu MC, Wishnok JS, Skipper PL, Ross RK, et al. *Urinary excretion of 2-amino-1-methyl-6-phenylimidazo[4,5-b] pyridine (PhIP) in White, African-American, and Asian-American men in Los Angeles county.* Cancer Epidemiol Biomarkers Prev. 1999;8:439-45.

101. Dietrich CG, de Waart DR, Ottenhoff R, Schoots IG, Elferink RP. *Increased bioavailability of the food-derived carcinogen 2-amino-1-methyl-6-phenylimidazo [4,5-b] pyridine in MRP2-deficient rats.* Mol Pharmacol. 2001;59:974-80.

102. van Herwaarden AE, Jonker JW, Wagenaar E, Brinkhuis RF, Schellens JHM, Beijnen JH, et al. *The breast cancer resistance protein (Bcrp1/Abcg2) restricts exposure to the dietary carcinogen 2-amino-1-methyl-6-phenylimidazo[4,5-b]pyridine.* Cancer Res. 2003;63:6447-52.

103. Dingley KH, Curtis KD, Nowell S, Felton JS, Lang NP, Turteltaub KW. *DNA and protein adduct formation in the colon and blood of humans after exposure to a dietary-relevant dose of 2-amino-1-methyl-6-phenylimidazo[4,5-b]pyridine.* Cancer Epidemiol Biomarkers Prev. 1999;8:507-17.

104. Zhu J, Chang P, Bondy ML, Sahin AA, Singletary SE, Takahashi S, et al. *Detection of 2-amino-1-methyl-yl-6-phenylimidazo[4,5-b]-pyridine-DNA adducts in normal breast tissues and risk of breast cancer.* Cancer Epidemiol Biomarkers Prev. 2003;12:830-7.

105. Badawi AF, Hirvonen A, Bell DA, Lang NP, Kadlubar FF. *Role of aromatic amine acetyltransferases, NAT1 and NAT2, in carcinogen-DNA adduct formation in the human urinary bladder.* Cancer Res. 1995;55:5230-7.

106. Lin D, Lay Jr. JO, Bryant MS, Malaveille C, Friesen MD, Bartsch H, et al. *Analysis of 4-aminobiphenyl-DNA adducts in human urinary bladder and lung by alkaline hydrolysis and negative ion gas chromatography-mass spectrometry.* Environ Health Perspect. 1994;102:11-6.

107. Talaska G, al-Juburi AZ, Kadlubar FF. *Smoking related carcinogen-DNA adducts in biopsy samples of human urinary bladder: identification of N-(deoxyguanosin-8-yl)-4-aminobiphenyl as a major adduct.* Proc Natl Acad Sci USA. 1991;88:5350-4.

108. Swaminathan S, Frederickson SM, Hatcher JF. *Metabolic activation of N-hydroxy-4-acetylaminobiphenyl by cultured human breast epithelial cell line MCF 10A.* Carcinogenesis. 1994;15:611-7.

109. Gorlewska-Roberts K, Green B, Fares M, Ambrosone CB, Kadlubar FF. *Carcinogen-DNA adducts in human breast epithelial cells.* Environ Mol Mutagenesis. 2002;39:184-92.

110. Faraglia B, Chen SY, Gammon MD, Zhang Y, Teitelbaum SL, Neugut AI, et al. *Evaluation of 4-aminobiphenyl-DNA adducts in human breast cancer: the influence of tobacco smoke.* Carcinogenesis. 2003;24:719-25.

111. De Stefani E, Ronco A, Mendilaharsu M, Guidobono M, H. D-P. *Meat intake, heterocyclic amines, and risk of breast cancer: a case-control study in Uruguay.* Cancer Epidemiol Biomarkers Prev. 1997;6:573-81.

112. Sinha R, Gustafson DR, Kulldorff M, Wen W, Cerhan JR, Zheng W. *2-amino-1-methyl-6-phenylimidazo[4,5-b]pyridine, a carcinogen in high-temperature-cooked meat, and breast cancer risk.* J Natl Cancer Instit. 2000;92:1352-4.

113. Zheng W, Gustafson DR, Sinha R, Cerhan JR, Moore D, Hong CP, et al. *Well-done meat intake and the risk of breast cancer.* JJ Natl Cancer Instit. 1998;90:1724-9.

114. Ambrosone CB, Freudenheim JL, Sinha R, Graham S, Marshall JR, Vena JE, et al. *Breast cancer risk, meat consumption and N-acetyltransferase (NAT2) genetic polymorphisms.* Int J Cancer. 1998;75:825-30.

115. Delfino RJ, Sinha R, Smith C, West J, White E, Lin HJ, et al. *Breast cancer, heterocyclic aromatic amines from meat and N-acetyltransferase 2 genotype.* Carcinogenesis. 2000;21:607-15.

116. Gertig DM, Hankinson SE, Hough H, Spiegelman D, Colditz GA, Willett WC, et al. *N-acetyl transferase 2 genotypes, meat intake and breast cancer risk.* Int J Cancer. 1999;80:13-7.

117. Taylor VH, Misra M, Mukherjee SD. *Is red meat intake a risk factor for breast cancer among premenopausal women?* Breast Cancer Res Treat. 2009;117:1-8.

118. Dai Q, Shu X-o, Jin F, Gao Y-T, Ruan Z-X, Zheng W. *Consumption of animal foods, cooking methods, and risk of breast cancer.* Cancer Epidemiol Biomarkers Prev. 2002;11:801-8.

119. Sinha R, Cross A, Curtin J, Zimmerman T, McNutt S, Risch A, et al. *Development of a food frequency questionnaire module and databases for compounds in cooked and processed meats.* Mol Nutr Food Res. 2005;49:648-55.

120. Kabat GC, Cross AJ, Park Y, Schatzkin A, Hollenbeck AR, Rohan TE, et al. *Meat intake and meat preparation in relation to risk of postmenopausal breast cancer in the NIH-AARP diet and health study.* Int J Cancer. 2009;124:2430-5.

121. Wu K, Sinha R, Holmes MD, Giovannucci E, Willett W, Cho E. *Meat mutagens and breast cancer in postmenopausal women--a cohort analysis.* Cancer Epidemiol Biomarkers Prev.2010;19:1301-10.

122. Sonestedt E, Ericson U, Gullberg B, Skog K, Olsson H, Wirfalt E. *Do both heterocyclic amines and omega-6 polyunsaturated fatty acids contribute to the incidence of breast cancer in postmeno-pausal women of the Malmo diet and cancer cohort.* Int J Cancer. 2008;123:1637-43.

123. Ferrucci LM, Cross AJ, Graubard BI, Brinton LA, McCarty CA, Ziegler RG, et al. *Intake of meat, meat mutagens, and iron and the risk of breast cancer in the Prostate, Lung, Colorectal, and Ovarian Cancer Screening Trial.* Br J Cancer. 2009;101:178-84.

124. Steck SE, Gaudet MM, Eng SM, Britton JA, Teitelbaum SL, Neugut AI, et al. *Cooked meat and risk of breast cancer--lifetime versus recent dietary intake.* Epidemiology. 2007;18:373-82.

125. Dybing E, Farmer PB, Andersen M, Fennell TR, Lalljie SP, Muller DJ, et al. *Human exposure and internal dose assessments of acrylamide in food.* Food Chem Toxicol. 2005;43:365-410.

126. Vesper HW, Caudill SP, Osterloh JD, Meyers T, Scott D, Myers GL. *Exposure of the U.S. population to acrylamide in the National Health and Nutrition Examination Survey 2003-2004.* Environ Health Perspect. 2010;118:278-83.

127. Gamboa da Costa G, Churchwell MI, Hamilton LP, Von Tungeln LS, Beland FA, Marques MM, et al. *DNA adduct formation from acrylamide via conversion to glycidamide in adult and neonatal mice.* Chem Res Toxicol. 2003;16:1328-37.

128. Hogervorst JG, Schouten LJ, Konings EJ, Goldbohm RA, van den Brandt PA. *A prospective study of dietary acrylamide intake and the risk of endometrial, ovarian, and breast cancer.* Cancer Epidemiol Biomarkers Prev. 2007;16:2304-13.

129. Sadrieh N, Davis CD, Snyderwine EG. *N-acetyltransferase expression and metabolic activation of the food-derived heterocyclic amines in the human mammary gland.* Cancer Res. 1996;56:2683-7.

130. Zheng W, Deitz AC, Campbell DR, Wen W-Q, Cerhan JR, Sellers TA, et al. *N-acetyltransferase 1 genetic polymorphism, cigarette smoking, well-done meat intake, and breast cancer risk.* Cancer Epidemiol Biomarkers Prev. 1999;8:233-9.

131. Deitz AC, Zheng W, Leff MA, Gross M, Wen W-Q, Doll MA, et al. *N-acetyltransferase-2 genetic polymorphism, well-done meat intake, and breast cancer risk among postmenopausal women.* Cancer Epidemiol Biomarkers Prev.2000;9:905-10.

132. Mignone LI, Giovannucci E, Newcomb PA, Titus-Ernstoff L, Trentham-Dietz A, Hampton JM, et al. *Meat consumption, heterocyclic amines, NAT2, and the risk of breast cancer.* Nutr Cancer. 2009;61:36-46.

3.3.3. POLYCYCLIC AROMATIC HYDROCARBONS

INTRODUCTION

Polycyclic aromatic hydrocarbons (PAHs) are formed by the incomplete combustion of organic matter and produced whenever vegetation or fossil fuels are burned. Emitted from natural sources such as forest fires and industrial sources ranging from power plants to motor vehicle engines, PAHs find their way into ambient air as pollutants and into the food chain as contaminants (1, 2). Thus, human exposure is unavoidable. For example, the estimated intake of a single PAH, benzo[a]pyrene (B[a]P), for the US population ranges from 0.12 to 2.8 µg/day with 2.2 µg/day for the average non-smoking individual (3). Since B[A]P partitions mainly in the soil, the food chain is the dominant pathway of human exposure, accounting for 97% of the daily intake. B[a]P is present in virtually all food items, including produce, dairy products, and meat. Most of the ingested B[a]P is absorbed by the gastrointestinal tract due to its high lipophilicity. B[a]P intake from ambient air by inhalation and from water contributes about 2% and 1%, respectively, to the total daily intake in nonsmokers. Substantially higher quantities of PAH are present in tobacco smoke and in grilled meat. Consequently, ingestion of charbroiled meat and smoking are associated with higher levels of PAH exposure. For example, mainstream smoke of filter cigarettes yields about 10 ng B[a]P per cigarette, leading to a daily intake of about 200 ng B[a]P for a pack-a-day cigarette smoker (4).

Although PAHs are exogenous chemicals, they are recognized as ligands by the endogenous aryl hydrocarbon receptor (AhR) and as substrates by cytochrome P450 and other enzymes. The unoccupied AhR contains one or more molecules of the ligand-binding subunit and one or more copies of a 90-kDa heat shock protein (5-7). PAH binding leads to dissociation of the heat shock protein from the AhR and dimerization with the AhR nuclear translocator protein, ARNT (8). This heterodimeric complex stimulates the transcription of certain genes, including CYPs, by interacting with specific cis-acting nucleotide sequences located upstream of these genes, termed xenobiotic responsive elements (XREs) (9-11). With regard to carcinogenesis, it has been demonstrated that the hydrophobic, chemically inert PAHs are by themselves innocuous compounds. However, they become carcinogens by undergoing metabolic activation to electrophilic species, e.g., phenols, dihydrodiols, quinones, epoxides, and dihydrodiol-epoxides, which elicit DNA adduct formation and oxidative DNA damage. In turn, these genotoxic responses cause errors in DNA replication and mutations that initiate the carcinogenic process (12-15). The preferential formation of certain stereoisomers during metabolic activation can determine the level of DNA damage, the efficiency of DNA repair, and the carcinogenic potency of a PAH (16). Since PAHs induce CYPs via the AhR, they are capable of promoting their own metabolism (17). In fact, it appears that the carcinogenicity of PAHs is dependent on the presence of AhR as transcriptional regulator of CYP expression. Administration of B[a]P to AhR (+/+), (+/-), and (-/-) mice resulted in tumor formation in the first two groups but not the AhR-null animals (18).

Most carcinogenic PAHs possess a bay- or fjord-region (Fig. 3.3.5.). Bay-region PAHs are planar, rigid, and more extended than the sterically hindered fjord-region PAHs, which are curved and twisted. Despite structural differences, PAHs undergo similar types of metabolic reactions and the pentacyclic B[a]P can therefore be considered as a prototype PAH environmental carcinogen (19, 20). There are also nitrogenous analogues, polycyclic aza-aromatic hydrocarbons, such as 3-nitrobenzanthrone and 7-methylbenz[c]acridine (21-23).

METABOLISM

PAHs are metabolized by three oxidation pathways: (i) CYP-mediated monooxygenation with formation of diol-epoxides, (ii) one-electron oxidation to produce radical cations, and (iii) formation of o-quinones catalyzed by dihydrodiol dehydrogenases (Figures 3.3.6 and 3.3.7.)

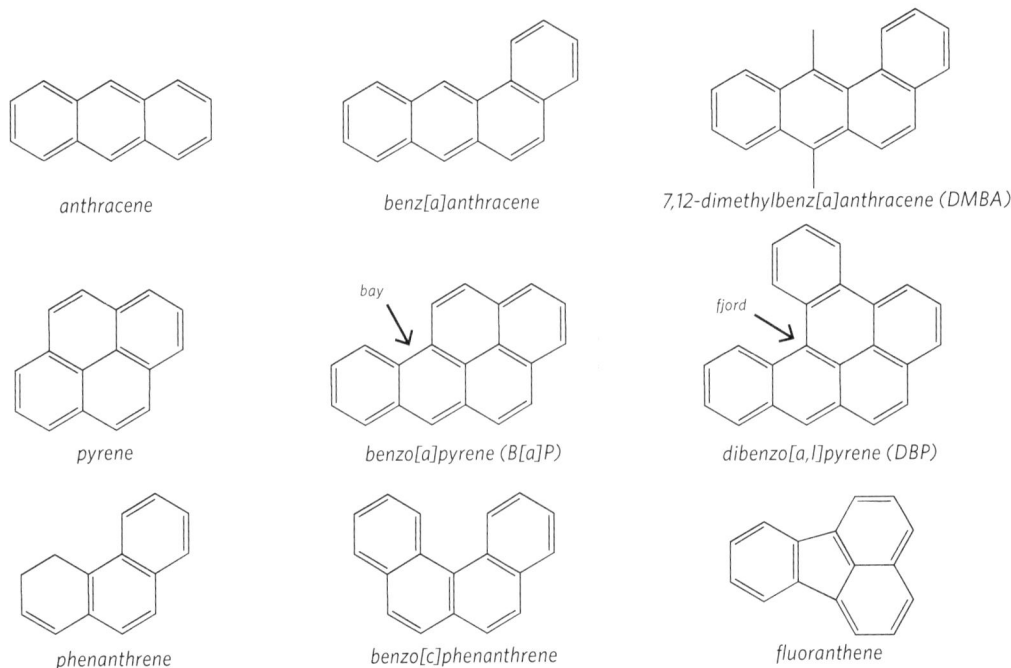

Figure 3.3.5. Structures of polycyclic aromatic hydrocarbons (PAHs).
Some of the PAHs have been shown to cause mammary cancers in animals, e.g., 7,12-dimethylbenz[a]anthracene (DMBA), benzo[a]pyrene (B[a]P), and dibenzo[a,l]pyrene (DBP).

(23). Some PAHs are activated exclusively by one of these mechanisms while others (e.g., B[a]P) are activated by a combination of reactions. Although PAHs have multiple carbon atoms available for reaction (e.g., the B[a]P molecule has 12), oxidations occur at select positions. For example, diol epoxides are preferentially formed in the bay-region of B[a]P and one-electron oxidation favors formation of the C6 radical cation due to the higher charge density at position 6. In addition to these principal activation pathways, a minor pathway involves sulfonation of methyl-substituted PAHs via electrophilic sulfuric acid ester formation catalyzed by sulfotransferases (24).

The first oxidative mechanism leads from B[a]P to vicinal dihydrodiol epoxides, which are sterically hindered but highly reactive metabolites capable of forming B[A]P-DNA adducts. As illustrated in Figure 3.3.6., the first reaction in this pathway is catalyzed by several cytochrome P450 enzymes, including CYP1A1 and CYP1B1 (19, 25, 26). B[a]P is oxygenated by the CYPs to three epoxides, namely 4,5-epoxide, 7,8-epoxide, and 9,10-epoxide. Another enzyme, the microsomal epoxide hydrolase, catalyzes the next reaction, the hydration of the epoxides to their corresponding dihydrodiols (27). Diols can exist as cis and trans configurational isomers. The epoxide hydrolase is both substrate- and product-stereoselective, i.e., the enzyme hydrates preferentially the 7,8-epoxide and produces only the *trans*-isomer, *trans*-7,8-diol. In the third reaction, the

CYPs oxygenate B[a]P *trans*-7,8-diol to 7,8-diol-9,10-epoxides (BPDEs). There are four possible stereoisomeric 9,10-epoxides that can be derived from the *trans*-7,8-diol. Different investigators have used different nomenclatures to designate each isomer. The benzylic carbon 7 of the diol function of B[a]P is commonly used as reference point. Where the benzylic hydroxyl and the epoxide function are *trans* or *cis*, they are referred to as *anti* or *syn* isomers, respectively. Both *anti*- and *syn*-BPDE isomers can exist in two optically active configurations, i.e., the (+)-*anti*-BPDE, (-)-*anti*-BPDE, (+)-*syn*-BPDE, and (-)-*syn*-BPDE enantiomers (Figure 3.3.6.). The oxidation of B[a]P-diols to BPDEs can also occur nonenzymatically by ROS (28). Several other enzymes, such as glutathione S-transferases, glucuronyl transferases, and sulfotransferases, may conjugate the reaction intermediates, converting them to water-soluble metabolites and thereby preventing the covalent binding of the BPDEs to DNA (29-33). In fact, cell culture and animal experiments have shown that more than half of the metabolites from culture media or body fluids could not be extracted with organic solvents. Therefore, the production of water-soluble metabolites plays an important role in the detoxification of B[a]P.

The second mechanism involves the removal of one electron from the π electron system of the PAH molecule through one-electron oxidation leading to the formation of a radical cation (Figure 3.3.7.) (23). The radical cation of PAH may

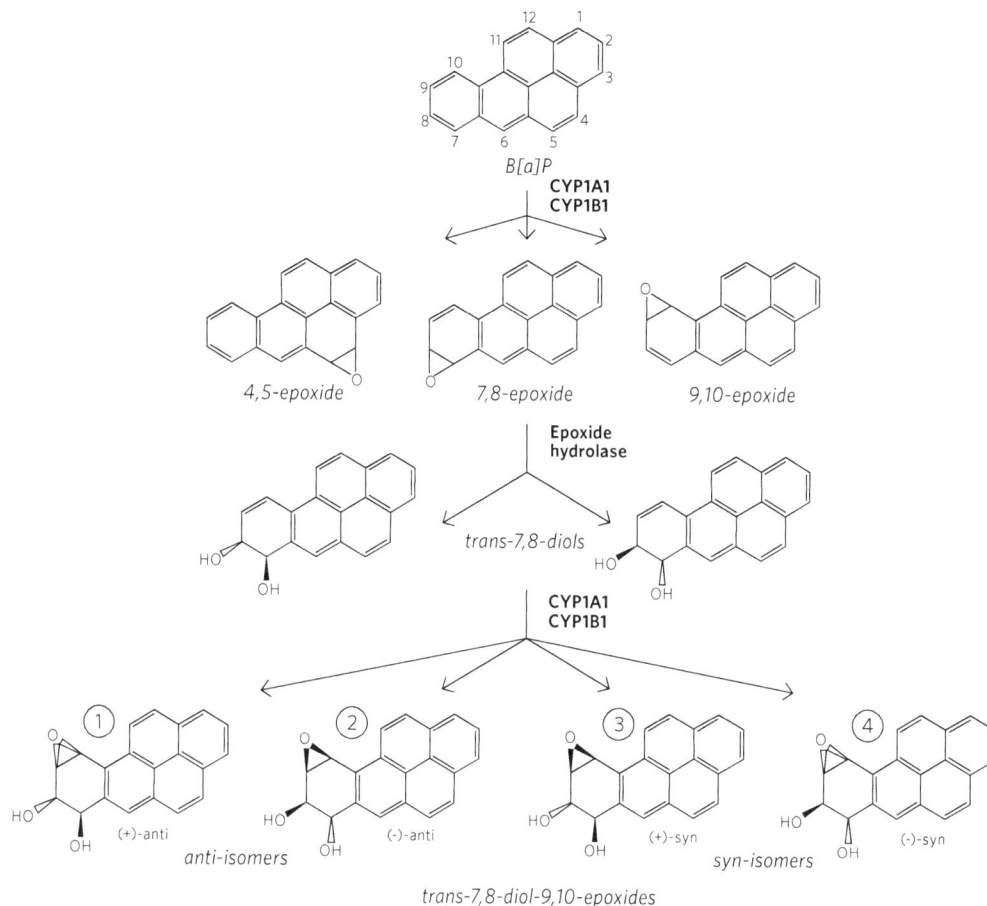

B[a]P

**CYP1A1
CYP1B1**

4,5-epoxide   *7,8-epoxide*   *9,10-epoxide*

**Epoxide
hydrolase**

trans-7,8-diols

**CYP1A1
CYP1B1**

(1) (+)-anti   (2) (-)-anti   (3) (+)-syn   (4) (-)-syn

anti-isomers   *syn-isomers*

trans-7,8-diol-9,10-epoxides

Figure 3.3.6. Oxidative B[a]P metabolism to diol epoxides (BPDEs)

be generated by chemical oxidants such as ferric ion III, enzymatically by peroxidases and CYPs, or photochemically by ultraviolet light (34-37). In the case of B[a]P, the C6 radical cation reacts with DNA to form depurinating adducts with guanine and adenine. The third mechanism involves the formation of o-quinones catalyzed by dihydrodiol dehydrogenases, members of the aldo-keto reductase (AKR) superfamily (38-40). The widely expressed AKR1A1 can oxidize PAH dihydrodiol metabolites to yield reactive and redox-reactive o-quinones (38, 40). The enzyme is regio- and stereoselective for PAH trans-dihydrodiols. In the case of B[a]P, AKR1A1 preferentially oxidized the trans-7,8-diol to the 7,8-dione (Figure 3.3.7.). The highly reactive quinone can form Michael addition products with DNA resulting in stable and depurinating adducts. Alternatively, the quinone can undergo redox cycling with semiquinone radical and hydroquinone to generate ROS capable of causing oxidative DNA damage as described in Chapter 2.2. for catechol estrogens.

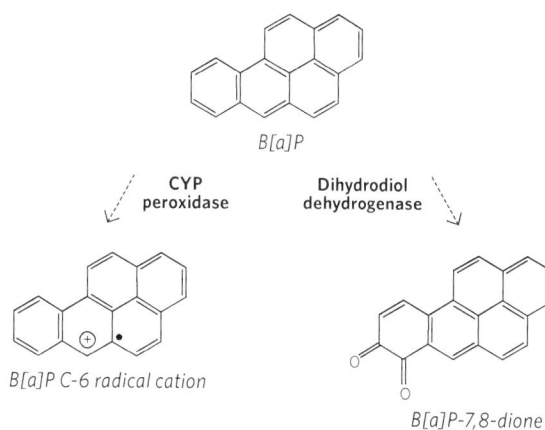

B[a]P

**CYP
peroxidase**   **Dihydrodiol
dehydrogenase**

B[a]P C-6 radical cation   *B[a]P-7,8-dione*

Figure 3.3.7. Oxidative metabolism of B[a]P to C-6 radical cation and 7,8-dione o-quinone

DNA DAMAGE AND REPAIR

The BPDEs are highly unstable and readily react with nucleophiles such as DNA, RNA, protein, or water (13, 14, 41). The BPDE-DNA adduct formation occurs preferentially between the epoxide C10 and the exocyclic amino group of purines, i.e., N6 of deoxyadenosine and N2 of deoxyguanosine (42). As shown in Figure 3.3.6., vicinal diol epoxides such as the BPDEs can be metabolically produced as two enantiomeric pairs of diasteromers, designated as (4)-anti- and (4)-syn-BPDE. Since each of the four isomeric BPDEs can react with the amino group either by cis- or trans-opening of the epoxide ring, a total of eight possible different adducts can be formed from deoxyadenosine and deoxyguanosine, respectively. In principle, both (+)- and

(-)-anti-BPDE react with DNA in vitro to form (+)-trans- and (-)-cis-BPDE-10-N2-dG adducts and (-)-trans- and (+)-cis-BPDE-10-N2-dG adducts, respectively, as shown in Figure 3.3.8. (17). However, >90% of the adduct formed arises from (+)-trans-opening of the epoxide ring by the amino group of deoxyguanosine residues. Similarly, (+)- and (-)-syn-BPDE are theoretically expected to give rise to four adducts, but (-)-cis-opening of the epoxides ring is favored, accounting for ~73% of adducts (42). Thus, BPDEs react extensively with deoxyguanosine via the (+)-anti BPDE isomer and form (+)-trans-BPDE-10-N2-dG and (-)-cis-BPDE-10-N2-dG as the predominant adducts (1, 43, 44).

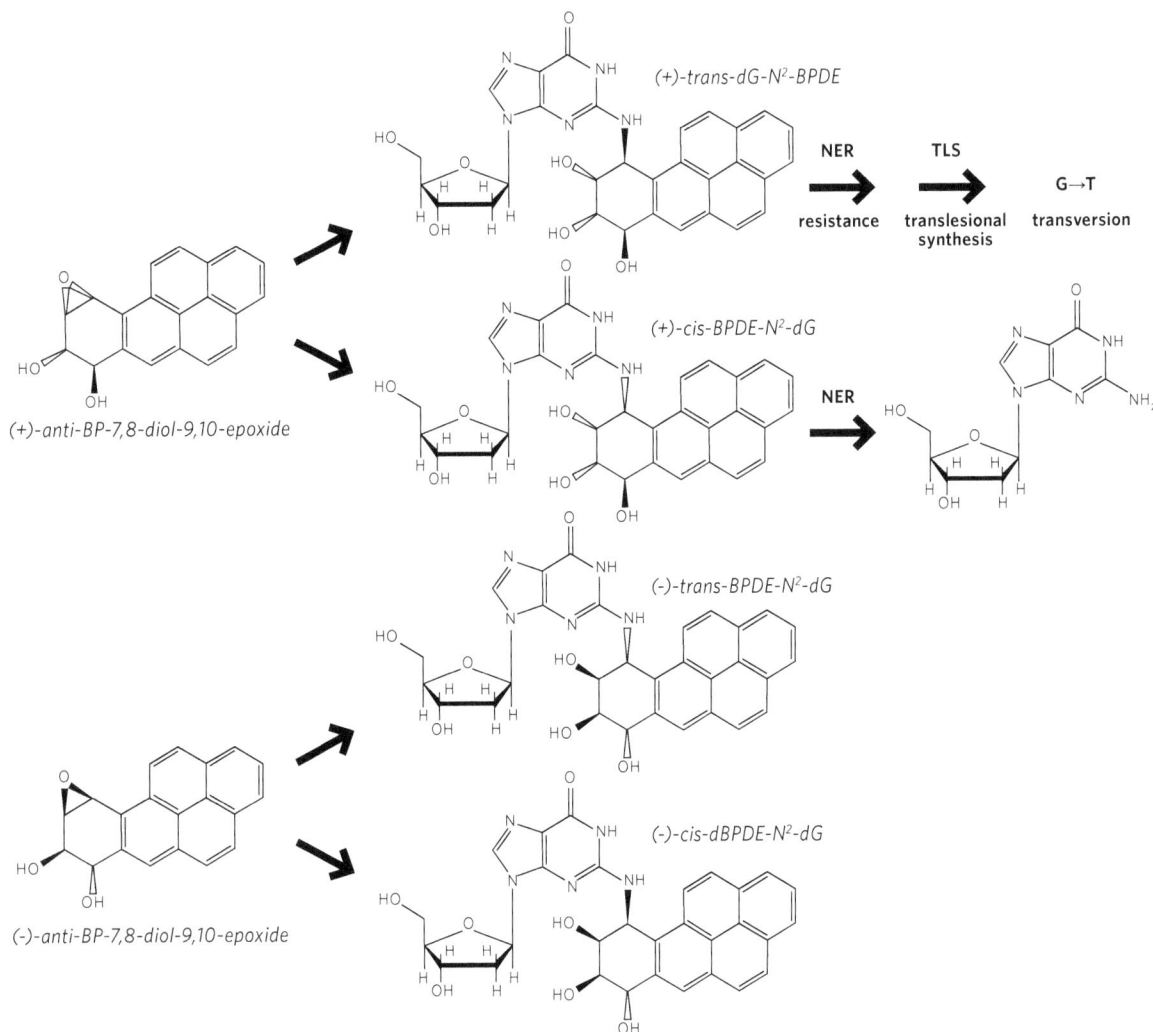

Figure 3.3.8. Formation and repair of stereoisomers of BPDE-10-N2-dG adducts.
Nucleotide excision repair (NER) efficiently recognizes and removes all BPDE-10-N2-dG adducts except (+)-trans-anti-BPDE-10-N2-dG enabling translesional synthesis (TLS) across the adduct and generation of G/T transversions.

In vivo experiments have shown that (+)-*anti*-BPDE induced a complex pattern of mutations, including insertions, deletions, frameshifts, as well as base substitutions, which, for G:C pairs alone included a significant fraction of G:C/T:A, A:T and C:G mutations (45, 46). Most of these mutations derived from the major adduct (+)-*trans*-BPDE-10-N2-dG (44, 46-48). These findings raise two questions: (1) why do the four stereoisomeric BPDE-10-N2-dG adducts arising from (+)-*anti* BPDE differ in their ability to induce mutations, and (2) how can a single adduct, (+)-*trans*-BPDE-10-N2-dG, induce different kinds of mutations? The answers lie in the conformational heterogeneity of PAH-DNA adducts and their recognition by the nucleotide excision repair (NER) machinery (49).

NER is the major mechanism for removing bulky PAH-DNA adducts (50, 51). A key issue is the ability of the NER machinery to recognize and repair certain adducts while failing to repair others. Unfaithful translesion bypass replication of unrepaired adducts can give rise to mutations. The fidelity of the NER recognition process depends on structural factors, particularly the distortion of the DNA double helix in form of diminished Watson-Crick base pairing, DNA bending, and unwinding of the DNA strands. For example, the four isomeric BPDE-10-N2-dG adducts are repaired at different rates because they impose distinctly different structural alterations on DNA. Molecular modeling showed that one of the isomers, (+)-*trans*-BPDE-10-N2-dG, can adopt at least 16 different conformations in double-stranded DNA (52). The B[a]P and dG moieties each can be located in either the major or minor groove. The B[a]P moiety can rotate about the adduct bond while the dG moiety rotates about the glycosidic bond. Thermodynamic and spatial considerations ruled out several conformations (53). NMR analysis showed that (+)-*trans*- and (-)-*trans*-BPDE-10-N2-dG adopt external adduct conformations, which accommodate the B[a]P moiety into the minor groove and retain Watson-Crick hydrogen bonding at all bases (54). In contrast, (+)-*cis*- and (-)-*cis*-BPDE-10-N2-dG adopt internal adduct conformations characterized by B[a]P insertion into the double helix and concomitant disruption of base-pairing. As a consequence, these helix-inserted (+)-*cis* and (-)-*cis* configurations cause displacement of the covalently modified guanine and its cytosine partner into the major and minor groove. As a result, (+)-*cis*- and (-)-*cis*-BPDE-10-N2-dG adducts were excised about 10-fold more efficiently by NER than the corresponding (+)-*trans*- and (-)-*trans*-BPDE-10-N2-dG isomers (54). The structure-mutation relationship of PAH adducts is not just stereospecific, but also sequence-specific. For example, (+)-*trans*-BPDE-10-N2-dG induced >95% G/T transversions in one sequence context (5'-TGC) and ~95% G/A transitions in another (5'-AGA) (44, 46-48). This is due to the profound effect of the 5'-flanking sequence on the adduct-induced conformation (42, 49, 55, 56). For example, replacement of a 5'-flanking dC with dT enhances local flexibility at the modification site and thereby allows several conformations that result in different mutations.

Structural studies of DNA polymerase complexes with PAH-DNA adducts have shown that DNA lesions localized in the solvent-exposed major groove are better tolerated than lesions placed in the minor groove that interact with the polymerase surface. For example, the BPDE-10-N2-dG adduct base-paired with cytosine adopts a conformation that places the B[a]P moiety into the nascent minor groove, resulting in extensive disruption to the interactions between the adducted DNA duplex and the polymerase (57, 58). A high-fidelity polymerase will dissociate from the B[a]P-modified template in preference to replication past the lesion, providing a molecular basis for the blocking effect on replication exerted by B[a]P adducts. By contrast, the error-prone Y family polymerases are capable of replicating past different types of DNA damage and have been shown structurally to accommodate various PAH adducts. There are multiple Y-family DNA polymerases in cells (e.g., polymerases η, κ, and ι in humans) that are capable of extending primers opposite damaged DNA templates in a process termed translesion synthesis (59, 60). The crystal structure of a ternary complex composed of a Y-family polymerase, Dpo4 (a bacterial homolog of human pol κ), a DNA substrate containing a BPDE-10-N6-dA adduct, and an incoming dNTP has been examined (61). The BPDE-10-N6-dA adduct mispaired with dT at the template-primer junction and assumed two different conformations. In one conformation of the ternary complex, B[a]P was placed in the major groove nearly perpendicular to the DNA base pairs. The C10-N6 bond linking B[a]P to the base formed an angle of 47º with the plane of the aromatic hydrocarbon system and the adenine base was shifted ~2Å toward the major groove. The crystal structure offers insight into the mechanism of base mispairing occurring at BPDE adducts. The observed adenine shift results in the juxtaposition of two hydrogen bond acceptors, the N1 of dA and the O4 of its usual base pair partner dT. However, dC and dA, each possessing an exocyclic amino group in the place of O4 of dT, would be better suited than dT to base pair with the dislocated dA. Although DNA polymerase κ preferentially incorporates dT opposite BPDE-10-N6-dA, it does allow incorporation of dC and dA, resulting in A/G and A/T mutations, respectively (62, 63). Other studies have revealed that extrusion of the bulky adduct out of the DNA helix accompanied by template misalignment allows the base 5' of the adduct to serve as template, resulting in a -1 frameshift (64). Thus, the structural analysis of PAH-DNA adducts within DNA polymerases has begun to provide a high-resolution understanding of polymerase blockage and bypass of bulky lesions (65-67).

Analogous metabolic activation pathways, DNA adduct formation and NER processes apply to other PAHs, such as the fjord-region PAHs dibenzo[a,l]pyrene (DBP) and benzo[c]phenanthrene (B[c]Ph), which are among the most carcinogenic PAH known (Figure 3.3.5.) (68, 69). In contrast to the rigid, planar bay-region PAHs, such as B[a]P, the twisted, sterically hindered fjord-region DBP and B[c]Ph diol epoxides differ in their reactivity with DNA bases. For example, BPDEs react extensively with deoxyguanosine via the (+)-anti BPDE isomer and form (+)-trans-BPDE-10-N2-dG and (-)-cis-BPDE-10-N2-dG as the predominant adducts (1, 43, 44). In contrast, DBPDEs preferentially form deoxyadenosine adducts via the (-)-anti- and (+)-syn-DBPDE isomers (70). As expected, PAHs that react preferentially with deoxyguanosine residues in DNA generate mutations of G•C pairs, whereas those that react largely with deoxy-adenosine residues produce mutations preferentially at A•T pairs (42). Moreover, DBP and B[c]Ph adducts cause different degrees of DNA distortion resulting in different repair susceptibilities, providing a molecular basis for the greater mutagenicity and carcinogenicity of fjord-region adducts (51, 68). In summary, the preferential formation of certain PAH stereoisomers during metabolic activation determines the type and level of DNA adducts as well as the efficiency of NER repair, which in turn influence the carcinogenic potency of the PAH compound.

The radical cation pathway of PAH oxidation can also lead to DNA adduct formation. For example, the C6 radical cation of B[a]P can form adducts by reacting with the nucleophilic N7 or C8 of guanine and N3 or N7 of adenine (71-73). These covalent linkages destabilize the N-glycosyl bond to the deoxyribose phosphate backbone of the DNA and generate apurinic sites, giving rise to depurinating PAH adducts such as 8-(benzo[a]pyren-6-yl)guanine (B[a]P-6-C8-Gua) and 7-methylbenz[a]anthracene-12-CH2-N7-adenine, which are produced via the C6 and C12 radical cations of B[a]P and DMBA, respectively (74). While the radical cation pathway leads to depurinating adducts, the diol epoxide pathway causes the formation of stable PAH adducts. There is disagreement which pathway and what type of adduct is principally involved in the carcinogenic activity of PAHs (75). One group reported that incubation of DNA with dibenzo[a,l]pyrene (DBP), horseradish peroxidase, and H_2O_2 led to the formation of adducts at the N7 position (instead of the exocyclic amino group) of purine bases, which destabilize the glycosidic bonds and cause spontaneous depurination, resulting in the generation of apurinic sites (73). However, apurinic sites were not detected in DNA obtained from MCF-7 cells treated with DBP. Instead of depurinating only stable adducts were found, resulting from reaction with the exocyclic amino group (76). Treatment of MCF-7 cells with B[a]P resulted in 5-fold higher levels of stable adducts than apurinic sites (77). These data indicate that depurinating adducts either do not occur in vivo or they are rapidly repaired leaving stable adducts as main cause of mutagenic changes.

The o-quinone pathway via the widely expressed AKR1A1

can also oxidize PAH dihydrodiol metabolites to yield reactive and redox-reactive o-quinones, which can form PAH-DNA adducts as well as oxidative adducts such as 8-oxo-dG described in Chapter 2.3. (38, 39). It is likely that diol epoxides, radical cations, and o-quinones all contribute to PAH-induced carcinogenesis (23). The relative importance of each pathway may depend on the tumor site and the level of expression of the activating enzymes. AKR1A1 was expressed in breast tissue together with CYP1A1 and epoxide hydrolase (38). B[a]P-DNA adducts became more mutagenic when they were irradiated with UVA light (78). It appears that B[a]P-DNA adducts may act as photosensitizers capable of generating ROS in the presence of light (79). Thus, B[a]P-induced mutagenesis may be enhanced by a cooperative effect of covalent B[a]P-DNA and oxidative DNA adducts, such as 8-OH-dG.

Non-genotoxic mechanisms. In addition to causing mutations by DNA misreplication, persistent BPDE-DNA adducts at coding or noncoding sites may interfere with RNA polymerase II-dependent transcription and thereby impair gene expression (80). BPDE-DNA adducts have also been related to changes in promoter methylation in form of hyper- or hypomethylation (81). These mechanisms may explain how BPDE can down- or upregulate the expression of certain genes, e.g., ATF3, E2A, and retinoic acid receptor β (82, 83). B[a]P exposure of HeLa cells has been shown to induce epigenetic activation of mobile elements (i.e., long-interspersed nuclear elements, LINE-1) dispersed throughout the human genome (84).

MOLECULAR BIOLOGY, CELL CULTURE, AND ANIMAL STUDIES

The effect of carcinogens on in vitro growth of normal human mammary epithelial cells (nHMEC) is dependent on the carcinogen concentration (14). For example, B[a]P concentrations above 1.0 µg per ml culture medium (corresponding to 4 µM) completely inhibited nHMEC growth, due to the production of large amounts of BPDE-DNA adducts, which cause S-phase arrest and induction of apoptosis (85, 86). At lower concentrations the cells continued to divide and form BPDE-DNA adducts, predominantly with deoxyguanosine. When cells were exposed to medium containing 0.4 µM B[a]P for 24 h, adducts were detectable as early as 6 h. Adducts persisted for 72 h after replacement with B[a]P-free medium. A comparison of the epithelial cells with fibroblasts showed that the latter cells were 50 – 100 less sensitive to growth inhibition by B[a]P. No adducts were detected in fibroblasts until 96 h after first exposure to B[a]P. These studies were extended to B[a]P metabolism and DNA adduct formation in HMEC derived from 13 reduction mammoplasty specimens, five benign breast lesions, and eight primary breast cancers (87). Exposure to 0.4 µM B[a]

P for 16 h yielded a similar range of metabolites for the three categories of HMEC. However, HMEC from several benign and malignant specimens contained significantly higher levels of BPDE-DNA adducts than HMEC from the reduction mammoplasties. The genotoxicity of PAHs has also been assessed by a DNA repair assay termed unscheduled DNA synthesis (UDS), which is based on [^{3}H]thymidine incorporation into DNA of nondividing cells during excision repair. Cultures of HMEC from reduction mammoplasties of five different women were incubated with PAHs for 24 h in the presence of [^{3}H]thymidine. A positive UDS response was observed with B[a]P but not DMBA (88). These data correlate with in vitro mutagenicity and DNA binding levels in HMEC. In contrast, DMBA was mutagenic in rat MEC, which correlates with the carcinogenic activity of DMBA in the rat mammary gland (89-91). Thus, not surprisingly, there are species-specific differences in mammary mutagenicity and carcinogenicity of PAHs.

An important requirement for the transformation of nHMEC is proliferation in the presence of carcinogen. The proliferating mammary epithelial cell needs to be able to metabolize the carcinogen and to replicate the damaged DNA, which carries the carcinogen-induced mutation to the subsequent cell population (92). Treatment of the spontaneously immortalized human breast epithelial cell line, MCF-10F, with 0.8 μM B[a]P for 24 h resulted in progressive neoplastic transformation. Clones selected from the 8th to 14th passage after B[a]P-treatment showed anchorage-independent growth and invasiveness as well as tumor formation upon injection into T and B cell deficient SCID mice (93). B[a]P has also been shown to decrease BRCA-1 mRNA levels in MCF-7 cells (41). Treatment with BPDE resulted in decreased BRCA-1 protein and increased p53 protein levels. This effect was not seen in ZR75 cells, which contain mutant p53 (Pro152Leu). Thus, the down-regulation of BRCA-1 transcription by BPDE appears to be mediated through a p53-dependent pathway rather than the AhR pathway (41). Treatment of another spontaneously immortalized, nontumorigenic human breast epithelial cell line, MCF-10A, with B[a]P-1,6-dione or B[a]P-3,6-dione produced superoxide and H_2O_2, which led to EGFR activation and increased cell proliferation (94, 95). Thus, B[a]P has genotoxic initiation and nongenotoxic promotion effects and the induction of cell proliferation and DNA synthesis would increase the likelihood of mutations by replication of DNA containing BPDE adducts (42).

It is apparent that PAHs can directly influence cell proliferation through mechanisms other than their primary genotoxic action. In some instances PAHs inhibit cell proliferation while in others they enhance proliferation. This complex effect appears to be mediated through receptors, such as the PAH-induced aryl hydrocarbon receptor (AhR), the estrogen receptor (ERα, ERβ), and the epidermal growth factor receptor (EGFR). The antiprolifera-tive action may involve several antiestrogenic mechanisms including direct interaction of AhR with estrogen responsive genes, competition of AhR and ER for common nuclear coactivators, proteasome-mediated ER degradation, as well as depletion of E_2 through enhanced oxidative metabolism by PAH-induced CYP1A1 and CYP1B1 (96-99). In contrast, several PAHs have also shown estrogenic effects, most likely mediated by hydroxylated PAH metabolites that can activate both ER-dependent reporter genes and ER-regulated endogenous genes (100-104). For example, B[a]P induced proliferation of human breast cancer MCF-7 cells grown in estrogen-free medium at concentrations as low as 100 nM (105). This effect was ER-dependent because it was blocked by the anti-estrogen ICI182780. B[a]P also stimulated the proliferation of nHMEC cultured in limiting amounts of EGF (106).

The principal reason for the stereoselective activation of PAHs resides in the cytochrome P450 enzymes, which catalyze the metabolism of the parent PAHs. While several P450s are capable of metabolizing PAHs, CYP1A1 and CYP1B1 are the most efficient isoforms (107, 108). Human CYP1A1 and CYP1B1 stably expressed in Chinese hamster V79 cells differed in their regio- and stereochemical selectivity of activation of DBP (109). CYP1B1 exclusively generated (-)-anti- and (+)-syn-11,12-DBP-dihydrodiol-13,14-epoxide (DBPDE) and the corresponding (-)-anti- and (+)-syn-DBPDE-DNA adducts. In contrast, CYP1A1 also formed 8,9-, 11,12-, and 13,14-dihydrodiols of DBP, resulting in the production of highly polar DBP-DNA adducts. The human breast cancer cell line MCF-7 has been shown to activate DBP efficiently to several DNA-binding metabolites. MCF-7 cells expressed CYP1A1 only upon induction whereas CYP1B1 is expressed constitutively (110). Thus, the metabolism of DBP is primarily carried out by CYP1B1 in a highly stereoselective manner, i.e., MCF-7 cells exclusively generated (-)-anti- and (+)-syn-DBPDE-DNA adducts (70, 111). The difference in conformation between the (-)-anti- and (+)-syn-DBPDE diastereomers appears to lead to an increased nucleophilic sequestration of the latter by water and cellular proteins. This would explain the observation that treatment of MCF-7 cells with low DBP concentrations generated only (-)-anti-DBPDE-DNA adducts. However, higher doses (≥1 μM) yielded both (-)-anti- and (+)-syn-DBPDE-DNA adducts (109). Above an adduct level of ~15 pmol/mg DNA, both DBPDE isomers induced the expression of the tumor suppressor p53 and one of its target gene products, the cyclin-dependent kinase inhibitor p21[WAF1] (112). These results indicate that the level of adducts rather than the specific structure of the DBPDE-DNA adduct triggers the p53 response. The PAH-DNA adduct level formed may determine whether p53 and p21[WAF1] pathways respond, resulting in cell cycle arrest, or fail to respond and increase the risk of mutation induction by the DNA lesions.

Numerous studies have been performed in animals to

examine the mechanism of PAH carcinogenesis (16, 113). Mice and rats have been the preferred species and skin, lung, and liver the most studied tissues. These investigations revealed that (i) certain isomeric PAH metabolites formed more adducts than other isomers, (ii) both stable and depurinating PAH-DNA adducts were produced, and (iii) the frequency of adduct formation generally correlated with the rate of mutations and ultimately the carcinogenic potency (17, 113-118). These observations also apply to PAH-induced mammary tumors, which include adenomas, adenocarcinomas, and sarcomas. Metabolites of different PAHs differ in mammary tumorigenicity. In general, the twisted fjord-region PAHs exhibit greater tumorigenic activities than the planar bay-region PAHs. For example, DBP was found to be several fold more tumorigenic in the rat mammary model than either B[a]P or DMBA (119). This finding is consistent with the adduct levels detected in mammary DNA following intramammillary exposure to DBP, B[a]P, and DMBA (120). Direct injection into the mammary fat pad of female CD rats of the fjord-region diol epoxide anti-3,4-dihydroxy-epoxy-1,2,3,4-tetrahydrobenzo[c]-phenanthrene induced significantly more adenomas and adenocarcinomas than the bay region *anti*-BPDE (121). A single oral dose of B[a]P (60 mg/kg) administered to Balb/c mice resulted in the formation of several adducts identifiable by ^{32}P postlabeling analysis in DNA extracted from mammary glands (122). The major adduct was (+)-*trans*-dG-N2-BPDE (122, 123). Adduct levels continued to increase for several days, reaching a peak five days after the single injection (120).

The susceptibility of the mammary gland to carcinogens is markedly influenced by its physiological state at the time of exposure (124, 125). Around menarche, the human mammary gland begins to develop and differentiate into defined ducts and lobules. The primary lobules formed at this time are type 1 lobules, which differentiate further into type 2 and type 3 lobules during pregnancy (126). Human nHMEC from type 1 lobules were more sensitive to the transforming effect of DMBA than type 3 lobule cells (127). Virgin rats treated with DMBA developed significantly more mammary tumors than multiparous rats of the same strain. The difference in tumor incidence was attributed to the difference in the extent of mammary tissue differentiation, i.e., mammary glands of virgin rats contain more undifferentiated end buds with a larger number of proliferating epithelial cells than those of parous animals. In the same manner as parity protects rodents from mammary tumorigenesis, full-term pregnancy at an early age decreases breast cancer risk in humans, most likely due to structural changes occurring during pregnancy and lactation (128).

HUMAN STUDIES

The ^{32}P-postlabeling method for the detection of carcinogen-DNA adducts has been utilized in several studies.

In one investigation adducts were analyzed in 5 normal breast tissue samples from women undergoing breast reduction and 26 breast cancers and tumor-adjacent tissues (129). In the mastectomy specimens adduct levels ranged from 1.6 to 10.0 adducts/10^8 nucleotides, with a mean of 4.7 ± 1.9 in tumor tissue, 6.1 4 2.9 in tumor-adjacent tissue, and 5.3 ± 2.4 in tumor and tumor-adjacent tissue combined. Adduct levels in the reduction mammoplasty specimens ranged from 0.43 to 4.41 adducts/10^8 nucleotides, with a mean of 2.3 ± 1.5. Another study of DNA adducts in 87 normal breast tissues adjacent to invasive cancers revealed a putative B[A]P-like adduct in 36 (41%) samples (130, 131). This adduct was absent in all control tissues (29 reduction mammoplasties). The same investigators (132) compared the B[a]P-DNA adduct formation in normal breast tissue of 76 cancer cases and 60 non-cancer (reduction mammoplasty) controls. Minced tissue (1 - 5 g) was cultured at 37ºC in 10 ml of minimal essential medium with 4 µM B[a]P for 24 h. DNA was then extracted and adducts were detected by the ^{32}P-postlabeling assay. The investigators used B[a]P instead of the ultimate carcinogen BPDE to explore the role of carcinogen activation in breast tissue during chemical carcinogenesis. BPDE adduct levels were significantly higher (13.5 ± 2.1/10^8 nucleotides) among cases compared with controls (7.0 ± 0.8/10^8; p = 0.007). Since all tissue samples were exposed to the same concentration of B[a]P under the same experimental conditions, the level of DNA adducts detected in this study reflects intrinsic factors that determine the tissue response to such damage, i.e., carcinogen metabolism and DNA repair capacities. The higher level of DNA damage among cases suggests that case patients had an elevated metabolic activation of B[a]P, deficient detoxification, or reduced DNA repair capacity compared with normal women. Interestingly, family history and CYP1B1 genotype were significant predictors of B[A]P-induced adduct formation in breast tissue. The median adduct levels in both cases and controls were highest for homozygous wild-type CYP1B1 432Val/Val and 453Asn/Asn genotypes and lowest for homozygous variant 432Leu/Leu and 453Ser/Ser genotypes, with intermediate levels recorded for heterozygous genotypes.

Since PAHs have been shown to be mammary carcinogens in animal models and PAH-DNA adducts identified in human breast tissue it is plausible that PAHs affect breast cancer risk. The relation of PAH exposure and breast cancer risk has been investigated in several epidemiologic studies (133). These investigations employed immunological methods (ELISA, immunohistochemistry) utilizing poly- and monoclonal antibodies, which were developed against BPDE-DNA but cross-reacted with varying affinities with other structurally related PAHs (134). An immunohistochemical analysis compared PAH-DNA adducts in tumor and adjacent non-tumor tissue from 100 women with breast cancer and benign breast tissue from 108 control women (135). After controlling for known breast cancer risk factors and two of the major PAH exposure sources (tobacco smoke and diet), PAH-DNA adduct levels were positively associated with the case-control status (OR = 4.43; 95% CI 1.09 – 18.01). Adduct levels in tumor tissue did not vary by stage or tumor size but were positively associated with the percentage of tumor cells expressing estrogen receptor (p = 0.01). Adduct levels in paired tumor and non-tumor tissues of 86 cases were highly correlated (p <0.001), although there was substantial interindividual (27-fold) variability. There was no difference between non-tumor tissue of cases and benign tissue of controls (136). In tumor and non-tumor tissue from cases, the GSTM1-null genotype was associated with increased adduct levels among current alcohol consumers but not among nondrinkers (137).

The population-based Long Island Breast Cancer Study Project was designed to investigate the association between PAH-adducts measured by ELISA in blood mononuclear cells and breast cancer risk. A pooled analysis of 873 cases and 941 controls revealed a positive association between PAH-DNA adducts and breast cancer incidence (OR 1.29; 95% CI 1.05 – 1.58) (138, 139). However, there was no dose-response relationship with increasing quantiles or stratification by source of PAH exposure, i.e., cigarette smoking and consumption of grilled and smoked foods. Since gene-environment interactions may contribute to interindividual differences in suscep-tibility to carcinogens and cancer risk, multiple genes were examined in relation to PAH-DNA adduct levels. An analysis of the interaction with the DNA base-excision repair enzyme XRCC1 showed an increased risk of breast cancer among never smokers with the XRCC1 399Gln allele (OR 1.3; 95% CI 1.0 – 1.7) (140). Further analysis revealed a weak additive interaction between the 399Gln allele and detectable PAH-DNA adducts in never smokers (OR 1.9; 95% CI 1.2 – 3.1). The biological mechanism underlying this observation is unclear but suggests PAH sources other than cigarette smoke. An analysis of five nucleotide excision repair genes (ERCC1, XPA, XPD, XPF, XPG) showed that the presence of at least one variant XPD allele (Asp312Asn in exon 10) was associated with an increase in breast cancer risk (OR 1.25; 95% CI 1.04 – 1.50) (141). The increase associated with 312Asn/Asn homozygosity was even stronger among those with detectable PAH-DNA adduct levels (OR 1.83; 95% CI 1.22 - 2.76 compared to 312Asp/Asp wild-type and nondetectable adducts). Interestingly, another group reported higher levels of PAH-DNA adducts in malignant breast tissue associated with the 312Asn allele (142). Asp312Asn is in linkage disequilibrium with another polymorphism within the XPD gene, Lys751Gln, which was shown to modify the increased risk of breast cancer in smokers (143). A polymorphism in the 3'-untranslated

region of ERCC1 (8092C/A) was not associated with breast cancer risk except in the presence of detectable PAH-DNA adducts (OR 1.92; 95% CI 1.14 – 3.25) (141). There was no association between XPA, XPF, XPG, or XPC genotypes, PAH-DNA adducts, and breast cancer risk (141, 144). An analysis of multiple polymorphisms in the GST gene superfamily (GSTA1, GSTM1, GSTP1, GSTT1) showed no evidence of an interaction with PAH to further increase breast cancer risk (145). There was no association between PAH-DNA adducts and survival among women with breast cancer (146).

Human exposure to PAHs comes from three main sources, cigarette smoke, grilled and fried foods, and environmental pollution, but each of these sources contains other carcinogens as well, e.g., cigarette smoke contains over 100 carcinogens. Thus, it is difficult to link breast cancer to PAH from a specific source. Attempts have been made to measure environmental PAH exposure as potential breast cancer risk factor. One study used total suspended particulates (TSP), a measure of ambient air pollution, as a proxy for PAH exposure (147). In postmenopausal women, exposure to high concentra-tions of TSP (>140 µg/m³) at birth was associated with an increase in breast cancer risk (OR 2.42; 95% CI, 0.97 – 6.09) compared with exposure to low concentrations (<84 µg/m³). However, no association was observed in premenopausal women. Moreover, TSP is a relatively crude measure of air pollution used in the 1960s, which was replaced in 1987 with a more relevant indicator of fine particles that can enter the lower respiratory tract (particulate matter, initially set at <10 µm, currently <2.5 µm). Attempts have also been made to develop a comprehensive analysis of PAH metabolites in urine, which would capture exposure from all sources as well as genetic history and CYP1B1 genotype were significant predictors of B[A]P-induced adduct formation in breast tissue. The median adduct levels in both cases and controls were highest for homozygous wild-type CYP1B1 432Val/Val and 453Asn/Asn genotypes and lowest for homozygous variant 432Leu/Leu and 453Ser/Ser genotypes, with intermediate levels recorded for heterozygous genotypes.

Since PAHs have been shown to be mammary carcinogens in animal models and PAH-DNA adducts identified in human breast tissue it is plausible that PAHs affect breast cancer risk. The relation of PAH exposure and breast cancer risk has been investigated in several epidemiologic studies (133). These inves-tigations employed immunological methods (ELISA, immunohistochemistry) utilizing poly- and monoclonal antibodies, which were developed against BPDE-DNA but cross-reacted with varying affinities with other structurally related PAHs influences on PAH metabolism. These studies revealed that levels of B[a]P metabolites, such as BPDE, were so low that their measurement in urine would not be practical for application in epidemiologic studies (148). However, metabolites of another PAH, phenanthrene (Figure 3.3.5.), were reliably measured in single urine samples by gas chromatography-mass spectrometry. The ratio of *trans, anti*-1,2,3,4-tetrahydroxy-1,2,3,4-tetrahydrophenanthrene (PheT) to phenanthrol (HOPhe), reflecting metabolic activation and detoxification, respectively, was significantly higher in smokers than nonsmokers. Although phenanthrene is not considered a carcinogen, the PheT:HOPhe ratio appears to be characteristic of an individual's ability to metabolically activate or detoxify PAHs (148). Other urinary PAH metabolites are 1-hydroxypyrene and 2-naphthol (149, 150). PAH exposure can also be quantified in healthy individuals by measuring urinary 1-pyrenol and BPDE-DNA adduct levels (151). The depurinated adduct B[a]P-6-N7-Ade was detected in the urine of three of seven cigarette smokers at concentrations of 0.1 – 0.6 fmol adduct/mg creatinine (152). Oxidative PAH metabolites such as BPDE also form covalent adducts with blood proteins, e.g., albumin and hemoglobin, both of which can be used as surrogate biomarkers of DNA damage resulting from the ultimate carcinogen BPDE (153, 154). A nested case-control study in the Shanghai Women's Health Study found no association between two urinary PAH metabolites, 1-hydroxypyrene and 2-naphthol, and breast cancer risk (150).

References

1. Maher VM, Patton JD, Yang JL, Wang YY, Yang LL, Aust AE, et al. *Mutations and homologous recombination induced in mammalian cells by metabolites of benzo[a]pyrene and 1-nitropyrene.* Environ Health Perspect. 1987;76:33-9.

2. Phillips DH. *Polycyclic aromatic hydrocarbons in the diet.* Mutation Res. 1999;443:139-47.

3. Hattemer-Frey HA, Travis CC. *Benzo-a-pyrene: environmental partitioning and human exposure.* Toxicol Ind Health. 1991;7:141-57.

4. Hoffmann D, Hoffmann I. *The changing cigarette, 1950-1995.* J Toxicol Environ Health. 1997;50:307-64.

5. Burbach KM, Poland A, Bradfield CA. *Cloning of the Ah-receptor cDNA reveals a distinctive ligand-activated transcription factor.* Proc Natl Acad Sci USA. 1992;89:8185-9.

6. Perdew GH. *Association of the Ah receptor with the 90-kDa heat shock protein.* J Biol Chem. 1988;263:13802-5.

7. Swanson HI, Bradfield CA. *The AH-receptor: genetics, structure and function.* Pharmacogenetics. 1993;3:213-30.

8. Hoffman EC, Reyes H, Chu FF, Sander F, Conley LH, Brooks BA, et al. *Cloning of a factor required for activity of the Ah (dioxin) receptor.* Science. 1991;252:954-8.

9. Gwinn MR, Keshava C, Olivero OA, Humsi JA, Poirier MC, Weston A. *Transcriptional signatures of normal human mammary epithelial cells in response to benzo[a]pyrene exposure: a comparison of three microarray platforms.* OMICS. 2005;9:334-50.

10. Kress S, Greenlee WF. *Cell-specific regulation of human CYP1A1 and CYP1B1 genes.* Cancer Res. 1997;57:1264-9.

11. Swanson HI, Chan WK, Bradfield CA. *DNA binding specificities and pairing rules of the Ah receptor, ARNT, and SIM proteins.* J Biol Chem. 1995;270:26292-302.

12. Agarwal R, Coffing SL, Baird WM, Kiselyov AS, Harvey RG, Dipple A. *Metabolic activation of benzo[g]chrysene in the human mammary carcinoma cell line MCF-7.* Cancer Res. 1997;57:415-9.

13. MacLeod MC, Selkirk JK. *Physical interactions of isomeric benzo[a]pyrene diol-epoxides with DNA.* Carcinogenesis. 1982;3:287-92.

14. Stampfer MR, Bartholomew JC, Smith HS, Bartley JC. *Metabolism of benzo[a]pyrene by human mammary epithelial cells: toxicity and DNA adduct formation.* Proc Natl Acad Sci USA. 1981;78:6251-5.

15. Szeliga J, Dipple A. *DNA adduct formation by polycyclic aromatic hydrocarbon dihydrodiol epoxides.* Chem Res Toxicology. 1998;11:1-11.

16. Luch A. *Nature and nurture - lessons from chemical carcinogenesis.* Nat Rev Cancer. 2005;5:113-25.

17. Conney AH. *Induction of microsomal enzymes by foreign chemicals and carcinogenesis by polycyclic aromatic hydrocarbons: G.H.A. Clowes Memorial Lecture.* Cancer Res. 1982;42:4875-917.

18. Shimizu Y, Nakatsuru Y, Ichinose M, Takahashi Y, Kume H, Mimura J, et al. *Benzo[a]pyrene carcinogenicity is lost in mice lacking the aryl hydrocarbon receptor.* Proc Natl Acad Sci U S A. 2000;97:779-82.

19. Gelboin HV. *Benzo[a]pyrene metabolism, activation, and carcinogenesis: Role and regulation of mixed-function oxidases and related enzymes.* Physiol Rev. 1980;60:1107-66.

20. Phillips DH. *Fifty years of benzo(a)pyrene.* Nature. 1983;303:468-72.

21. Arlt VM, Hewer A, Sorg BL, Schmeiser HH, Phillips DH, Stiborova M. *3-aminobenzanthrone, a human metabolite of the environmental pollutant 3-nitrobenzanthrone, forms DNA adducts after metabolic activation by human and rat liver microsomes: evidence for activation by cytochrome P450 1A1 and P450 1A2.* Chem Res Toxicol. 2004;17:1092-101.

22. Roberts-Thomson SJ, McManus ME, Tukey RH, Gonzalez FJ, Holder GM. *Metabolism of polycyclic aza-aromatic carcinogens catalyzed by four expressed human cytochromes P450.* Cancer Res. 1995;55:1052-9.

23. Xue W, Warshawsky D. *Metabolic activation of polycyclic and heterocyclic aromatic hydrocarbons and DNA damage: a review.* Toxicol Appl Pharmacol. 2005;206:73-93.

24. Surh YJ, Miller JA. *Roles of electrophilic sulfuric acid ester metabolites in mutagenesis and carcinogenesis by some polynuclear aromatic hydrocarbons.* Chem Biol Interact. 1994;92:351-62.

25. Luch A, Kishiyama S, Seidel A, Doehmer J, Greim H, Baird WM. *The K-region trans-8,9-diol does not significantly contribute as an intermediate in the metabolic activation of dibenzo[a,l]pyrene to DNA-binding metabolites by human cytochrome p450 1A1 or 1B1.* Cancer Res. 1999;59:4603-9.

26. Rinaldi AL, Morse MA, Fields HW, Rothas DA, Pei P, Rodrigo KA, et al. *Curcumin activates the aryl hydrocarbon receptor yet significantly inhibits (-)-benzo(a)pyrene-7R-trans-7,8-dihydrodiol bioactivation in oral squamous cell carcinoma cells and oral mucosa.* Cancer Res. 2002;62:5451-6.

27. Hassett C, Aicher L, Sidhu JS, Omiecinski CJ. *Human microsomal epoxide hydrolase: genetic polymorphism and functional expression in vitro of amino acid variants.* Hum Mol Genetics. 1994;3:421-8.

28. Alexandrov K, Rojas M, Rolando C. *DNA damage by benzo(a)pyrene in human cells is increased by cigarette smoke and decreased by a filter containing rosemary extract, which lowers free radicals.* Cancer Res. 2006;66:11938-45.

29. Bock KW, Raschko FT, Gsdaidmeier H, Seidel A, Oesch F, Grove AD, et al. *Mono- and diglucuronide formation from benzo[a] pyrene and chrysene diphenols by AHH-1 cell-expressed UDP-glucuronosyltransferase UGT1A7.* Biochem Pharmacol. 1999;57:653-6.

30. Nemoto N, Takayama S. *Modification of benzo[a]pyrene metabolism with microsomes by addition of uridine 5'-diphosphoglucuronic acid.* Cancer Res. 1977;37:4125-9.

31. Nemoto N, Takayama S, Gelboin HV. *Enzymatic conversion of benzo[a] pyrene phenols, dihydrodiols and quinones to sulfate conjugates.* Biochem Pharmacol. 1977;26:1825-9.

32. Robertson IGC, Guthenberg C, Mannervik B, Jernstrom B. *Differences in stereoselectivity and catalytic efficiency of three human glutathione transferases in the conjugation of glutathione with 7β,8α-dihydroxy-9α,10α-oxy-7,8,9,10-tetrahydrobenzo(a)pyrene.* Cancer Res. 1986;46:2220-4.

33. Sundberg K, Dreij K, Seidel A, Jernstrom B. *Glutathione conjugation and DNA adduct formation of dibenzo[a,1]pyrene and benzo[a] pyrene diol epoxides in V79 cells stably expressing different human glutathione trasnferases.* Chem Res Toxicol. 2002;15:170-9.

34. Cavalieri EL, Rogan EG. *Central role of radical cations in metabolic activation of polycyclic aromatic hydrocarbons.* Xenobiotica. 1995;25:677-88.

35. Cavalieri EL, Rogan EG,

Cremonesi P, Devanesan PD. *Radical cations as precursors in the metabolic formation of quinones from benzo[a] pyrene and 6-fluorobenzo[a]pyrene. Fluoro substitution as a probe for one-electron oxidation in aromatic substrates.* Biochem Pharmacol. 1988;37:2173-82.

36. Reed MD, Monske ML, Lauer FT, Meserole SP, Born JL, Burchiel SW. *Benzo[a]pyrene diones are produced by photochemical and enzymatic oxidation and induce concentration-dependent decreases in the proliferative state of human pulmonary epithelial cells.* J Toxicol Environ Health. 2003;66:1189-205.

37. Shen AL, Fahl WE, Wrighton SA, Jefcoate CR. *Inhibition of benzo(a)pyrene and benzo(a)pyrene 7,8-dihydrodiol metabolism by benzo(a)pyrene quinones.* Cancer Res. 1979;39:4123-9.

38. Palackal NT, Burczynski ME, Harvey RG, Penning TM. *The ubiquitous aldehyde reductase (AKR1A1) oxidizes proximate carcinogen trans-dihydrodiols to o-quinones: potential role in polycyclic aromatic hydrocarbon activation.* Biochemistry. 2001;40:10901-10.

39. Penning TM, Burczynski ME, Hung CF, McCoull KD, Palackal NT, Tsuruda LS. *Dihydrodiol dehydrogenases and polycyclic aromatic hydrocarbon activation: generation of reactive and redox active o-quinones.* Chem Res Toxicol. 1999;12:1-18.

40. Penning TM, Ohnishi ST, Ohnishi T, Harvey RG. *Generation of reactive oxygen species during the enzymatic oxidation of polycyclic aromatic hydrocarbon trans-dihydrodiols catalyzed by dihydrodiol dehydrogenase.* Chem Res Toxicol. 1996;9:84-92.

41. Jeffy BD, Chirnomas RB, Chen EJ, Gudas JM, Romagnolo F. *Activation of the aromatic hydrocarbon receptor pathway is not sufficient for transcriptional repression of BRCA-1; requirements for metabolism of benzo[a]pyrene to 7r,8t-Dihydroxy-9t,10-epoxy-7,8,9,10-tetrahydrobenzo[a]pyrene.* Cancer Res. 2002;62:113-21.

42. Dipple A, Khan QA, Page JE, Ponten I, Szeliga J. *DNA reactions, mutagenic action and stealth properties of polycyclic aromatic hydrocarbon carcinogens.* Int J Oncol. 1999;14:103-11.

43. Cheng SC, Hilton BD, Roman JM, Dipple A. *DNA adducts form carcinogenic and noncarcinogenic enantiomers of benzo[a]pyrene dihydrodiol epoxide.* Chem Res Toxicol. 1989;2:334-40.

44. Shukla R, Liu T, Geacintov

NE, Loechler EL. *The major, N2-dG adduct of (+)-anti-B[a]PDE shows a dramatically different mutagenic specificity (predominantly, G -->A) in a 5'-CGT-3' sequence context.* Biochemistry. 1997;36:10256-61.

45. Rodriguez H, Loechler EL. *Mutagenesis by the (+)-anti-diol epoxide of benzo[a]pyrene: what controls mutagenic specificity.* Biochemistry. 1993;32:1759-69.

46. Shukla R, Geacintov NE, Loechler EL. *The major, N2-dG adduct of (+)-anti-B[a]PDE induces G-->A mutations in a 5'-AGA-3' sequence context.* Carcinogenesis. 1999;20:261-8.

47. Wei SJ, Chang RL, Merkler KA, Gwynne M, Cui XX, Murthy B, et al. *Dose-dependent mutation profile in the c-Ha-ras proto-oncogene of skin tumors in mice initiated with benzo[a]pyrene.* Carcinogenesis. 1999;20:1689-96.

48. Yang JL, Maher VM, McCormick JJ. *Kinds of mutations formed when a shuttle vector containing adducts of (+/-)-7 beta, 8 alpha-dihydroxy-9 alpha, 10 alpha-epoxy-7, 8, 9, 10-tetrahydrobenzo[a]pyrene replicates in human cells.* Proc Natl Acad Sci USA. 1987;84:3787-91.

49. Cho BP. *Dynamic conformational heterogeneities of carcinogen-DNA adducts and their mutagenic relevance.* J Environ Sci Health C Environ Carcinog Ecotoxicol Rev. 2004;22:57-90.

50. Gillet LC, Scharer OD. *Molecular mechanisms of mammalian global genome nucleotide excision repair.* Chem Rev. 2006;106:253-76.

51. Wu M, Yan S, Patel DJ, Geacintov NE, Broyde S. *Relating repair susceptibility of carcinogen-damaged DNA with structural distortion and thermodynamic stability.* Nucleic Acids Res. 2002;30:3422-32.

52. Kozack RE, Shukla R, Loechler EL. *A hypothesis for what conformation of the major adduct of (+)-anti-B[a]PDE (N2-dG) causes G→T versus G→A mutations based upon a correlation between mutagenesis and molecular modeling results.* Carcinogenesis. 1999;20:95-102.

53. Geacintov NE, Cosman M, Hingerty BE, Amin S, Broyde S, Patel DJ. *NMR solution structures of stereo-isomeric covalent polycyclic aromatic carcinogen-DNA adducts: principles, patterns, and diversity.* Chem Res Toxicol. 1997;10:111-46.

54. Hess MT, Gunz D, Luneva N, Geacintov NE, Naegeli H. *Base pair conformation-dependent excision of

benzo[a]pyrene diol epoxide-guanine adducts by human nucleotide excision repair enzymes.* Mol Cell Biol. 1997;17:7069-76.

55. Ross H, Bigger CAH, Yagi H, Jerina DM, Dipple A. *Sequence specificity in the interaction of the four stereoisomeric benzo[c]phenanthrene dihydrodiol epoxides with the supF gene.* Cancer Res. 1993;53:1273-7.

56. Yan S, Wu M, Buterin T, Naegeli H, Geacintov NE, Broyde S. *Role of base sequence context in conformational equilibria and nucleotide excision repair of benzo[a] pyrene diol epoxide-adenine adducts.* Biochemistry. 2003;42:2339-54.

57. Hsu GW, Huang X, Luneva NP, Geacintov NE, Beese LS. *Structure of a high fidelity DNA polymerase bound to a benzo[a]pyrene adduct that blocks replication.* J Biol Chem. 2005;280:3764-70.

58. Xu P, Oum L, Beese LS, Geacintov NE, Broyde S. *Following an environmental carcinogen N2-dG adduct through replication: elucidating blockage and bypass in a high-fidelity DNA polymerase.* Nucleic Acids Res. 2007;35:4275-88.

59. Pages V, Fuchs RP. *How DNA lesions are turned into mutations within cells?* Oncogene. 2002;21:8957-66.

60. Prakash S, Johnson RE, Prakash L. *Eukaryotic translesion synthesis DNA polymerases: specificity of structure and function.* Annu Rev Biochem. 2005;74:317-53.

61. Ling H, Sayer JM, Plosky BS, Yagi H, Boudsocq F, Woodgate R, et al. *Crystal structure of a benzo[a] pyrene diol epoxide adduct in a ternary complex with a DNA polymerase.* Proc Natl Acad Sci USA. 2004;101:2265-9.

62. Ogi T, Shinkai Y, Tanaka K, Ohmori H. *Polκ protects mammalian cells against the lethal and mutagenic effects of benzo[a]pyrene.* Proc Natl Acad Sci USA. 2002;99:15548-53.

63. Suzuki N, Ohashi E, Kolanovskiy A, Geacintov NE, Grollman AP, Ohmori H, et al. *Translesion synthesis of human DNA polymerase κ on a DNA template containing a single stereoisomer of dG-(+)- or dG-(-)-anti-N2-BPDE (7,8-dihydroxy-anti-9,10-epoxy-7,8,9,10-tetrahydrobenzo[a]pyrene).* Biochemistry. 2002;41:6100-6.

64. Bauer J, Xing G, Yagi H, Sayer JM, Jerina DM, Ling H. *A structural gap in Dpo4 supports mutagenic bypass of a major benzo[a]pyrene dG adduct in DNA through template misalignment.* Proc Natl Acad Sci U S A. 2007;104:14905-10.

65. Broyde S, Wang L, Zhang L, Rechkoblit O, Geacintov NE, Patel DJ. *DNA adduct structure-function relationships: comparing solution with polymerase structures.* Chem Res Toxicol. 2008;21:45-52.

66. Kozack R, Seo KY, Jelinsky SA, Loechler EL. *Toward an understanding of the role of DNA adduct conformation in defining mutagenic mechanism based on studies of the major adduct (formed at N(2)-dG) of the potent environmental carcinogen, benzo[a]pyrene.* Mutation Res. 2000;450:41-59.

67. Seo KY, Jelinsky SA, Loechler EL. *Factors that influence the mutagenic patterns of DNA adducts from chemical carcinogens.* Mutation Res. 2000;463:215-46.

68. Batra VK, Shock DD, Prasad R, Beard WA, Hou EW, Pedersen LC, et al. *Structure of DNA polymerase beta with a benzo[c]phenanthrene diol epoxide-adducted template exhibits mutagenic features.* Proc Natl Acad Sci U S A. 2006;103:17231-6.

69. Buters JT, Mahadeven B, Quintanilla-Martinez L, Gonzalez FJ, Greim H, Baird WM, et al. *Cytochrome P450 1B1 determines susceptibility to dibenzo[a,l]pyrene-induced tumor formation.* Chem Res Toxicol. 2002;15:1127-35.

70. Ralston SL, Seidel A, Luch A, Platt KL, Baird WM. *Stereoselective activation of dibenzo[a,l]pyrene to (-)-anti(11r,12s13s,14r)- and (+)-syn(11s,12r,13s,14r)-11,12-diol-13,14-epoxides which bind extensively to deoxyadenosine residues of DNA in the human mammary carcinoma cell line MCF-7.* Carcinogenesis. 1995;16:2899-907.

71. Cavalieri EL, Rogan EG, Devanesan PD, Cremonesi P, Cerny RL, Gross ML, et al. *Binding of benzo[a]pyrene to DNA by cytochrome P-450 catalyzed one-electron oxidation in rat liver microsomes and nuclei.* Biochemistry. 1990;29:4820-7.

72. Joseph P, Jaiswal AK. *NAD(P)H:quinone oxidoreductase 1 (DT diaphorase) specifically prevents the formation of benzo[a]pyrene quinone-DNA adducts generated by cytochrome P4501A1 and P450 reductase.* Proc Natl Acad Sci USA. 1994;91:8413-7.

73. Rogan EG, Cavalieri EL, Tibbels SR, Cremonesi P, Warner CD, Nagel DL, et al. *Synthesis and identification of benzo[a]pyrene-guanine nucleoside adducts formed by electrochemical oxidation and by horseradish peroxidase catalyzed reaction of benzo[a]pyrene with DNA.* J Am Chem Soc. 1988;110:4023-9.

74. Todorovic R, Ariese F, Devanesan P, Jankowiak R, Small GJ, Rogan EG, et al. *Determination of benzo[a] pyrene- and 7,12-dimethylbenz[a] anthracene-DNA adducts formed in rat mammary glands.* Chem Res Toxicol. 1997;10:941-7.

75. Dowty HV, Xue W, LaDow K, Talaska G, Warshawsky D. *One-electron oxidation is not a major route of metabolic activation and DNA binding for the carcinogen 7H-dibenzo[c,g]carbazole in vitro and in mouse liver and lung.* Carcinogenesis. 2000;21:991-8.

76. Melendez-Colon VJ, Luch A, Seidel A, Baird WM. *Comparison of cytochrome P450- and peroxidase-dependent metabolic activation of the potent carcinogen dibenzo[a,l] pyrene in human cell lines: Formation of stable DNA adducts and absence of a detectable increase in apurinic sites.* Cancer Res. 1999;59:1412-6.

77. Melendez-Colon VJ, Luch A, Seidel A, Baird WM. *Formation of stable DNA adducts and apurinic sites upon metabolic activation of bay and fjord region polycyclic aromatic hydrocarbons in human cell cultures.* Chem Res Toxicol. 2000;13:10-7.

78. Besaratinia A, Pfeifer GP. *Enhancement of mutagenicity of benzo(a)pyrene diol epoxide by a nonmutagenic dose of ultraviolet A radiation.* Cancer Res. 2003;63:8708-16.

79. Kim S-R, Kokubo K, Matsui K, Yamada N, Kanke Y, Fukuoka M, et al. *Light-dependent mutagenesis by benzo[a]pyrene is mediated via oxidative DNA damage.* Environ Mol Mutagen. 2005;46:141-9.

80. Koch KS, Fletcher RG, Grond MP, Inyang AI, Lu XP, Brenner DA, et al. *Inactivation of plasmid reporter gene expression by one benzo(a) pyrene diol-epoxide DNA adduct in adult rat hepatocytes.* Cancer Res. 1993;53:2279-86.

81. Pavanello S, Bollati V, Pesatori AC, Kapka L, Bolognesi C, Bertazzi PA, et al. *Global and gene-specific promoter methylation changes are related to anti-B[a]PDE-DNA adduct levels and influence micronuclei levels in polycyclic aromatic hydrocarbon-exposed individuals.* Int J Cancer. 2009;125:1692-7.

82. Song S, Xu XC. *Effect of benzo[a] pyrene diol epoxide on expression of retinoic acid receptor-β in immortalized esophageal epithelial cells and esophageal cancer cells.* Biochem Biophys Res Commun. 2001;281:872-7.

83. Wang A, Gu J, Judson-Kremer K, Powell KL, Mistry H, Simhambhatla P, et al. *Response of human mammary epitelial cells to DNA damage induced by BPDE: involvement of novel regulatory pathways.* Carcinogenesis. 2003;24:225-34.

84. Stribinskis V, Ramos KS. *Activation of human long interspersed nuclear element 1 retrotransposition by benzo(a)pyrene, an ubiquitous environmental carcinogen.* Cancer Res. 2006;66:2616-20.

85. Chen S, Nguyen N, Tamura K, Karin M, Tukey RH. *The role of the Ah receptor and p38 in benzo[a] pyrene-7,8-dihydrodiol and benzo[a] pyrene-7,8-dihydrodiol-9,10-epoxide-induced apoptosis.* J Biol Chem. 2003;278:19526-33.

86. Khan QA, Dipple A. *Diverse chemical carcinogens fail to induce G1 arrest in MCF-7 cells.* Carcinogenesis. 2000;21:1611-8.

87. Bartley JC, Stampfer MR. *Factors influencing benzo[a]pyrene metabolism in human mammary epithelial cells in culture.* Carcinogenesis. 1985;6:1017-22.

88. Eldridge SR, Gould MN, Butterworth BE. *Genotoxicity of environmental agents in human mammary epithelial cells.* Cancer Res. 1992;52:5617-21.

89. Gould MN, Grau DR, Seidman LA, Moore CJ. *Interspecies comparison of human and rat mammary epithelial cell-mediated mutagenesis by polycyclic aromatic hydrocarbons.* Cancer Res. 1986;46:4942-5.

90. Moore CJ, Eldridge SR, Tricomi WA, Gould MN. *Quantitation of benzo(a)pyrene and 7,12-dimethylbenz(a)anthracene binding to nuclear macromolecules in human and rat mammary epithelial cells.* Cancer Res. 1987;47:2609-13.

91. Moore CJ, Tricomi WA, Gould MN. *Interspecies comparison of polycyclic aromatic hydrocarbon metabolism in human and rat mammary epithelial cells.* Cancer Res. 1986;46:4946-52.

92. Russo J, Russo IH. *Influence of differentiation and cell kinetics on the susceptibility of the rat mammary gland to carcinogenesis.* Cancer Res. 1980;40:2677-87.

93. Calaf G, Russo J. *Transformation of human breast epithelial cells by chemical carcinogens.* Carcinogenesis. 1993;14:483-92.

94. Burdick AD, Davis JW, Liu KJ, Hudson LG, Shi H, Monske ML, et al. *Benzo(a)pyrene quinones increase cell proliferation, generate reactive oxygen species, and transactivate the epidermal growth factor receptor in breast epithelial cells.* Cancer Res. 2003;63:7825-33.

95. Soule HD, Maloney TM, Wolman SR, Peterson WDJ, Brenz R, McGrath CM, et al. *Isolation and characterization of a spontaneously immortalized human breast epithelial cell line, MCF-10.* Cancer Res. 1990;50:6075-86.

96. Arcaro KF, O'Keefe PW, Yang Y, Clayton W, Gierthy JF. *Antiestrogenicity of environmental polycyclic aromatic hydrocarbons in human breast cancer cells.* Toxicology. 1999;133:115-27.

97. Chaloupka K, Krishnan V, Safe S. *Polynuclear aromatic hydrocarbon carcinogens as antiestrogens in MCF-7 human breast cancer cells: role of the Ah receptor.* Carcinogenesis. 1992;13:2233-9.

98. Ebright RH, Wong JR, Chen LB. *Binding of 2-hydroxybenzo(a)pyrene to estrogen receptors in rat cytosol.* Cancer Res. 1986;46:2349-51.

99. Safe S, Wormke M. *Inhibitory aryl hydrocarbon receptor-estrogen receptor α cross-talk and mechanisms of action.* Chem Res Toxicol. 2003;16:807-16.

100. Abdelrahim M, Ariazi E, Kim K, Khan S, Barhoumi R, Burghardt R, et al. *3-Methylcholanthrene and other aryl hydrocarbon receptor agonists directly activate estrogen receptor alpha.* Cancer Res. 2006;66:2459-67.

101. Charles GD, Bartels MJ, Zacharewski TR, Gollapudi BB, Freshour NL, Carney EW. *Activity of benzo[a]pyrene and its hydroxylated metabolites in an estrogen receptor-α reporter gene assay.* Toxicol Sci. 2000;55:320-6.

102. Fertuck KC, Kumar S, Sikka HC, Matthews JB, Zacharewski TR. *Interaction of PAH-related compounds with the α and β isoforms of the estrogen receptor.* Toxicol Lett. 2001;121:167-77.

103. Gozgit JM, Nestor KM, Fasco MJ, Pentecost BT, Arcaro KF. *Differential action of polycyclic aromatic hydrocarbons on endogenous estrogen-responsive genes and on a transfected estrogen-responsive reporter in MCF-7 cells.* Toxicol Appl Pharmacol. 2004;196:58-67.

104. Pearce ST, Liu H, Radhakrishnan I, Abdelrahim M, Safe S, Jordan VC. *Interaction of the aryl hydrocarbon receptor ligand 6-methyl-1,3,8-trichlorodibenzofuran with estrogen receptor α.* Cancer Res. 2004;64:2889-97.

105. Pliskova M, Vondracek J, Vojtesek B, Kozubik A, Machala M. *Deregulation of cell proliferation by polycyclic aromatic hydrocarbons in human breast carcinoma MCF-7 cells reflects both genotoxic and nongenotoxic events.* Toxicol Sci. 2005;83:246-56.

106. Tannheimer SL, Barton SL, Ethier SP, Burchiel SW. *Carcinogenic polycyclic aromatic hydrocarbons increase intracellular ca2+ and cell proliferation in primary human mammary epithelial cells.* Carcinogenesis. 1997;18:1177-82.

107. Shimada T, Fujii-Kuriyama Y. *Metabolic activation of polycyclic aromatic hydrocarbons to carcinogens by cytochromes p450 1A1 and 1B1.* Cancer Sci. 2004;95:1-6.

108. Shimada T, Hayes CL, Yamazaki H, Amin S, Hecht SS, Guengerich FP, et al. *Activation of chemically diverse procarcinogens by human cytochrome P-450 1B1.* Cancer Res. 1996;56:2979-84.

109. Luch A, Coffing SL, Tang YM, Schneider A, Soballa V, Greim H, et al. *Stable expression of human cytochrome p450 1B1 in V79 Chinese hamster cells and metabolically catalyzed DNA adduct formation of dibenzo[a,l]pyrene.* Chem Res Toxicol. 1998;11:686-95.

110. Christou M, Savas U, Schroeder S, Shen X, Thompson T, Gould MN, et al. *Cytochromes CYP1A1 and CYP1B1 in the rat mammary gland: Cell-specific expression and regulation by polycyclic aromatic hydrocarbons and hormones.* Mol Cell Endocrinol. 1995;115:41-50.

111. Ralston SL, Coffing SL, Seidel A, Luch A, Platt KL, Baird WM. *Stereoselective activation of dibenzo[a,l] pyrene and its trans-11,12-dihydrodiol to fjord region 11,12-diol 13,14-epoxides in a human mammary carcinoma MCF-7 cell-mediated V79 cell mutation assay.* Chem Res Toxicol. 1997;10:687-93.

112. Luch A, Kudla K, Seidel A, Doehmer J, Greim H, Baird WM. *The level of DNA modification by (+)-syn-(11s,12r,13s,14r)- and (-)-anti-(11r,12s,13s,14r)-dihydrodiol epoxides of dibenzo[a,l]pyrene determined the effect on the proteins p53 and p21waf1 in the human mammary carcinoma cell line MCF-7.* Carcinogenesis. 1999;20:859-65.

113. Poirier MC. *Chemical-induced DNA damage and human cancer risk.* Nat Rev Cancer. 2004;4:630-7.

114. Chakravarti D, Pelling JC, Cavalieri EL, Rogan EG. *Relating aromatic hydrocarbon-induced DNA adducts and c-H-ras mutations in mouse skin papillomas: the role of apurinic sites.* Proc Natl Acad Sci USA. 1995;92:10422-6.

115. Chen L, Devanesan PD, Higginbotham S, Ariese F, Jankowiak R, Small GJ, et al. *Expanded analysis of benzo[a] pyrene-DNA adducts formed in vitro and in mouse skin: their significance in tumor initiation.* Chem Res Toxicol. 1996;9:897-903.

116. Devanesan PD, RamaKrishna NVS, Todorovic R, Rogan EG, Cavalieri EL, Jeong H, et al. *Identification and quantitation of benzo[a] pyrene-DNA adducts formed by rat liver microsomes in vitro.* Chem Res Toxicol. 1992;5:302-9.

117. Pelling JC, Slaga TJ, DiGiovanni J. *Formation and persistence of DNA, rRNA, and protein adducts in mouse skin exposed to pure optical enantiomers of 7 beta,8 alpha-dihydroxy-9 alpha,10 alpha-epoxy-7,8,9,10-tetrahydrobenzo(a)pyrene in vivo.* Cancer Res. 1984;44:1081-6.

118. Rogan EG, Devanesan PD, RamaKrishna NVS, Higginbotham S, Padmavathi NS, Chapman K, et al. *Identification and quantitation of benzo[a]pyrene-DNA adducts formed in mouse skin.* Chem Res Toxicol. 1993;6:356-63.

119. Cavalieri EL, Higginbotham S, RamaKrishna NVS, Devanesan PD, Todorovic R, Rogan EG, et al. *Comparative dose-response tumorigenicity studies of dibenzo[a,l] pyrene versus 7, 12-dimethylbenz[a] anthracene, benzo[a] and two dibenzo[a,l]pyrene dihydrodiols in mouse skin and rat mammary gland.* Carcinogenesis. 1991;12:1939-44.

120. Arif JM, Smith WA, Gupta RC. *DNA adduct formation and persistence in rat tissues following exposure to the mammary carcinogen dibenzo[a,l]pyrene.* Carcinogenesis. 1999;20:1147-50.

121. Hecht SS, el-Bayoumy K, Rivenson A, Amin S. *Potent mammary carcinogenicity in female CD rats of a fjord region diol-epoxide of benzo[c]phenanthrene compared to a bay region diol-epoxide of benzo[a] pyrene.* Cancer Res. 1994;54:21-4.

122. Walker MP, Jahnke GD, Snedeker SM, Gladen BC, Lucier GW, DiAugustine RP. *32P-post-labeling analysis of the formation and persistence of DNA adducts in mammary glands of parous and nulliparous mice treated with benzo[a]pyrene.* Carcinogenesis. 1992;13:2009-15.

123. Seidman LA, Moore CJ, Gould MN. *32P-postlabeling analysis of DNA adducts in human and rat mammary epithelial cells.* Carcinogenesis. 1988;9:1071-7.

124. Huggins C, Grand LC, Brillantes FP. *Critical significance of breast structure in the induction of mammary cancer in the rat.* Proc Natl Acad Sci U S A. 1959;45:1294-300.

125. Russo J. *Basis of cellular autonomy in the subsceptibility to carcinogenesis.* Toxicol Pathol. 1983;11:149-66.

126. Russo J, Hu YF, Yang X, Russo IH. *Developmental, cellular, and molecular basis of human breast cancer.* J Natl Cancer Inst Monogr. 2000:17-37.

127. Russo J, Reina D, Frederick J, Russo IH. *Expression of phenotypical changes by human breast epithelial cells treated with carcinogens in vitro.* Cancer Res. 1988;48:2837-57.

128. MacMahon B, Cole P, Brown J. *Etiology of human breast cancer: a review.* J Natl Cancer Inst. 1973;50:21-42.

129. Perera FP, Estabrook A, Hewer A, Channing K, Rundle A, Mooney LA, et al. *Carcinogen-DNA adducts in human breast tissue.* Cancer Epidemiol Biomarkers Prev. 1995;4:233-8.

130. Li D, Wang M, Dhingra K, Hittelman WN. *Aromatic DNA adducts in adjacent tissues of breast cancer patients: clues to breast cancer etiology.* Cancer Res. 1996;56:287-93.

131. Li D, Zhang W, Sahin AA, Hittelman WN. *DNA adducts in normal tissue adjacent to breast cancer: A review.* Cancer Detection Prev. 1999;23:454-62.

132. Gorlewska-Roberts K, Green B, Fares M, Ambrosone CB, Kadlubar FF. *Carcinogen-DNA adducts in human breast epithelial cells.* Environ Mol Mutagenesis. 2002;39:184-92.

133. Gammon MD, Santella RM. *PAH, genetic susceptibility and breast cancer risk: an update from the Long Island Breast Cancer Study Project.* Eur J Cancer. 2008;44:636-40.

134. Santella RM. *Immunological methods for detection of carcinogen-DNA damage in humans.* Cancer Epidemiol Biomark Prev. 1999;8:733-9.

135. Rundle A, Tang D, Hibshoosh H, Estabrook A, Schnabel F, Cao W, et al. *The relationship between genetic damage from polycyclic aromatic hydrocarbons in breast tissue and breast cancer.* Carcinogenesis. 2000;21:1281-9.

136. Rundle A, Tang D, Hibshoosh H, Schnabel F, Kelly A, Levine R, et al. *Molecular epidemiologic studies of polycyclic aromatic hydrocarbon-DNA adducts and breast cancer.* Environ Mol Mutagen. 2002;39:201-7.

137. Rundle A, Tang D, Mooney L, Grumet S, Perera F. *The interaction between alcohol consumption and GSTM1 genotype on polycyclic aromatic hydrocarbon-DNA adduct levels in breast tissue.* Cancer Epidemiol Biomarkers Prev. 2003;12:911-4.

138. Gammon MD, Sagiv SK, Eng SM, Shantakumar S, Gaudet MM, Teitelbaum SL, et al. *Polycyclic aromatic hydrocarbon-DNA adducts and breast cancer: a pooled analysis.* Arch Environ Health. 2004;59:640-9.

139. Gammon MD, Santella RM, Neugut AI, Eng SM, Teitelbaum SL, Paykin A, et al. *Environmental toxins and breast cancer on Long Island. I. Polycyclic aromatic hydrocarbon DNA adducts.* Cancer Epidemiol Biomarkers Prev. 2002;11:677-85.

140. Shen J, Gammon MD, Terry MB, Wang L, Wang Q, Zhang F, et al. *Polymorphisms in XRCC1 modify the association between polycyclic aromatic hydrocarbon-DNA adducts, cigarette smoking, dietary antioxidants, and breast cancer risk.* Cancer Epidemiol Biomarkers Prev. 2005;14:336-42.

141. Crew KD, Gammon MD, Terry MB, Zhang FF, Zablotska LB, Agrawal M, et al. *Polymorphisms in nucleotide excision repair genes, polycyclic aromatic hydrocarbon-DNA adducts, and breast cancer risk.* Cancer Epidemiol Biomarkers Prev. 2007;16:2033-41.

142. Tang D, Cho S, Rundle A, Chen S, Phillips D, Zhou J, et al. *Polymorphisms in the DNA repair enzyme XPD are associated with increased levels of PAH-DNA adducts in a case-control study of breast cancer.* Breast Cancer Res Treat. 2002;75:159-66.

143. Terry MB, Gammon MD, Zhang FF, Eng SM, Sagiv SK, Paykin AB, et al. *Polymorphism in the DNA repair gene XPD, polycyclic aromatic hydrocarbon-DNA adducts, cigarette smoking, and breast cancer risk.* Cancer Epidemiol Biomarkers Prev. 2004;13:2053-8.

144. Shen J, Gammon MD, Terry MB, Teitelbaum SL, Eng SM, Neugut AI, et al. *Xeroderma pigmentosum complementation group C genotypes/ diplotypes play no independent or interaction role with polycyclic aromatic hydrocarbons-DNA adducts for breast cancer risk.* Eur J Cancer. 2008;44:710-7.

145. McCarty KM, Santella RM, Steck SE, Cleveland RJ, Ahn J, Ambrosone CB, et al. *PAH-DNA adducts, cigarette smoking, GST polymorphisms, and breast cancer risk.* Environ Health Perspect. 2009;117:552-8.

146. Sagiv SK, Gaudet MM, Eng SM, Abrahamson PE, Shantakumar S, Teitelbaum SL, et al. *Polycyclic aromatic hydrocarbon-DNA adducts and survival among women with breast cancer.* Environ Res. 2009;109:287-91.

147. Bonner MR, Han D, Nie J, Rogerson P, Vena JE, Muti P, et al. *Breast cancer risk and exposure in early life to polycyclic aromatic hydrocarbons using total suspended particulates as a proxy measure.* Cancer Epidemiol Biomarkers Prev. 2005;14:53-60.

148. Hecht SS, Chen M, Yoder A, Jensen J, Hatsukami D, Le C, et al. *Longitudinal study of urinary phenanthrene metabolite ratios: effect of smoking on the diol epoxide pathway.* Cancer Epidemiol Biomarkers Prev. 2005;14:2969-74.

149. Gunier RB, Reynolds P, Hurley SE, Yerabati S, Hertz A, Strickland P, et al. *Estimating exposure to polycyclic aromatic hydrocarbons: a comparison of survey, biological monitoring, and geographic information system-based methods.* Cancer Epidemiol Biomarkers Prev. 2006;15:1376-81.

150. Lee KH, Shu XO, Gao YT, Ji BT, Yang G, Blair A, et al. *Breast cancer and urinary biomarkers of polycyclic aromatic hydrocarbon and oxidative stress in the Shanghai Women's Health Study.* Cancer Epidemiol Biomarkers Prev. 2010;19:877-83.

151. Shuker DE, Farmer PB. *Relevance of urinary DNA adducts as markers of carcinogen exposure.* Chem Res Toxicol. 1992;5:450-60.

152. Casale GP, Singhal M, Bhattacharya S, RamaNathan R, Roberts KP, Barbacci DC, et al. *Detection and quantification of depurinated benzo[a]pyrene-adducted DNA bases in the urine of cigarette smokers and women exposed to household coal smoke.* Chem Res Toxicol. 2001;14:192-201.

153. Scherer G, Frank S, Riedel K, Meger-Kossien I, Renner T. *Biomonitoring of exposure to polycyclic aromatic hydrocarbons of nonoccupationally exposed persons.* Cancer Epidemiol Biomark Prev. 2000;9:373-80.

154. Waidyanatha S, Zheng Y, Serdar B, Rappaport SM. *Albumin adducts of naphthalene metabolites as biomarkers of exposure to polycyclic aromatic hydrocarbons.* Cancer Epidemiol Biomark Prev. 2004;13:117-24.

3.4. DIETARY FACTORS AND LIFESTYLE CHOICES
3.4.1. INTRODUCTION

Incidence rates of breast cancer increase among women who migrate from a country with a low rate to one with a higher rate (1-3). A likely reason for the observed increase is the change in diet associated with the migration (4-6). Obviously, studies on diet and breast cancer risk are difficult to interpret because of the complexity of diets and the attendant challenge of assessing the complexity in dietary interviews or food frequency questionnaires. For example, vegetables are classified into several botanical families: Compositae (e.g., lettuce, spinach), Cruciferae (e.g., broccoli, cabbage), Leguminosae (e.g., beans, peas), Solanaceae (e.g., potatoes, tomatoes), and Umbelliferae (e.g., carrots, celery). Likewise, fruits are classified into several families including Rosaceae (e.g., apples, peaches), Rutaceae (e.g., grapefruits, oranges), and Musaceae (e.g., bananas). Moreover, measurement error affects not only the determination of a single variable (fruit or vegetable intake) but also other dietary variables, such as energy intake, alcohol, smoking, vitamin supplement use, and physical activity (7).

References

1. *SEER Cancer Statistics Review 1973 - 1999.* 2005 http://seer.cancer.gov. csr/1973_1999.

2. McMichael AJ, Giles GG. *Cancer in migrants to Australia: extending the descriptive epidemiological data.* Cancer Res. 1988;48:751-6.

3. Parkin DM, Steinitz R, Khlat M, Kaldor J, Katz L, Young J. *Cancer in Jewish migrants to Israel.* Int J Cancer. 1990;45:614-21.

4. Boyle P, Leake R. *Progress in understanding breast cancer: epidemiological and biological interactions.* Breast Cancer Res Treat. 1988;11:91-112.

5. Doll SR. *The lessons of life: Keynote address to the nutrition and cancer conference.* Cancer Res. 1992;52:2024s-9s.

6. Walker ARP, Walker BF, Stelma S. *Is breast cancer avoidable? Could dietary changes help?* Int J Food Sci Nutr. 1995;46:373-81.

7. Schatzkin A, Kipnis V. *Could exposure assessment problems give us wrong answers to nutrition and cancer questions.* J Natl Cancer Inst. 2004;96:1564-5.

3.4.2. VEGETABLES AND FRUITS

Numerous nutritional studies have examined the association between cancer risk and vegetable consumption. Detailed analysis of the results has shown that consumption of cruciferous vegetables may be associated with a decreased risk of cancer, especially colorectal cancer (1). Cruciferous vegetables belong to the family Cruciferae, which in Western diet consists mostly of the genus Brassica, including cabbage, cauliflower, Brussels sprouts, and broccoli. The protective effect of cruciferous vegetables against cancer appears to be due to their high content of glucosinolates, phytochemicals that are hydrolyzed by plant myrosinase and the gastrointesti-

nal microflora to glucose and an unstable aglycone, the thiohydroxamate-O-sulfonate (Figure 3.4.1.). The latter undergoes spontaneous rearrangement to yield isothiocyanates, thiocyanates, nitriles, and other compounds including indole-3-carbinols (2, 3). These compounds exhibit anticancer activity, which is mediated through several mechanisms. For example, isothiocyanates inhibit phase I enzymes that catalyze the activation of procarcinogens and induce phase II enzymes that detoxify procarcinogens. In addition to the suppression of mutagenic and carcinogenic activity, isothiocyanates have been shown to cause a reduction in systemic oxidative stress and induce apoptosis (4-8). The consumption of even moderate amounts of cruciferous vegetables results in the uptake of tens of milligrams of isothiocyanates, concentrations that have been shown to prevent lung, liver, and esophageal cancer formation in rodents (2, 3, 9).

Isothiocyanates inhibit phase I cytochrome P450 enzymes by competitive and non-competitive mechanisms depending on the CYP isoform (10, 11). In contrast, isothiocyanates induce phase II enzymes, such as glutathione S-transferases (GSTs) (3, 9). Thus, they may enhance the detoxification and excretion of carcinogens and prevent the alteration to DNA. In addition, isothiocyanates are substrates for GSTs undergoing conjugation with glutathione (GSH) to form RNHC(=S)-SG (12, 13). While isothiocyanates inhibit CYPs, indole-3-carbinol and its acid-derived condensation product, indolo[3,2-b]carbazole (ICZ), were shown to bind to the Ah receptor and induce expression of CYP1A1, CYP1A2, and CYP1B1 in both *in vivo* and *in vitro* models (14-16).

Administration of indole-3-carbinol to female Sprague-Dawley rats inhibited mammary tumor formation induced by 7,12-dimethylbenz[a]anthracene (17). Studies in mice with high endogenous rates of mammary tumors have shown that feeding indole-3-carbinol or benzyl isothiocyanate to these animals from an early age resulted in a marked decrease in the incidence of mammary tumors (7, 18). The prevention of mammary carcinogenesis was associated with suppression of cell proliferation, increased apoptosis, and a 5-fold increase in the rate of E_2 hydroxylation at C-2, which suggests metabolic inactivation of endogenous estrogens. Indeed, dietary intake of indole-3-carbinol or Brassica vegetables has been shown to induce C-2 hydroxylation of estrogens in healthy women (19, 20). Therefore, indole-3-carbinole has been proposed as a chemoprotective agent for human breast cancer (21). However, studies in MCF-7 breast cancer cells showed that ICZ increased not only E_2 hydroxylation at C-2 but also at C-4(14). In fact, complete analysis of E_2 metabolites revealed increased rates of hydroxylation at C-2, C-4, C-6α, and C-15α, similar to the profile induced by the environmental pollutant and potent agonist of the Ah receptor, 2,3,7,8-tetrachlorodibenzo-p-dioxin (TCDD) (14). Given the specific pathogenetic role that has emerged for 4-OHE$_2$ and the widespread tissue expression of CYP1B1, the authors suggest that breast cancer chemoprotective studies based on agents that activate the Ah receptor and induce estrogen-metabolizing cytochrome P450 enzymes should be approached with caution. Reanalysis of rodent experiments has shown that indole-3-carbinol treatment resulted in a delay of mammary tumor formation, but did not alter tumor incidence or multiplicity among surviving animals (22).

Epidemiological studies have shown an inverse association between cruciferous vegetable intake and risk of colorectal and other cancers (1, 6). However, the relation between cruciferous vegetable intake and breast cancer risk is uncertain. A comparison of breast cancer rates in European countries suggests that Mediterranean countries have lower rates due to the so-called Mediterranean diet, which is characterized by a high intake of vegetables and fruits in parallel with a high consumption of olive oil and low intake of animal fats. A meta-analysis of 17 studies revealed an inverse association of vegetable intake and breast cancer risk (RR = 0.75; 95% CI 0.66 – 0.85) (23). Fewer studies showed an inverse association with fruit consumption.

A large prospective study of 285,526 women between the ages of 25 and 70 years, participating in the European Prospective Investigation into Cancer and Nutrition (EPIC) found no association with breast cancer risk for either vegetable or fruit intake (24). Participants lived in countries from the north to the south of Europe, including the Mediterranean. Similarly, a pooled analysis of mostly North American cohort studies (7377 breast cancer cases among 351,825 participating women) showed no significant association with total vegetable and fruit consumption (25). A subanalysis of eight botanical groups and 17 specific vegetables and fruits also revealed no associations. On the other hand, a Chinese study found a strong inverse relation of breast cancer risk with fruit and vegetable intake (26). These discrepant findings may be explained by genetic variants of phase I or II enzymes. Women with the GSTP1 codon 105 Val/Val genotype and low cruciferous vegetable intake had a breast cancer risk 1.74-fold (95% CI: 1.13 – 2.67) that of women with the Ile/Ile and Ile/Val genotypes (27). A breast cancer case-control study of Chinese women used urinary isothiocyanate excretion as a biomarker of glucosinolate intake and degradation (4). While there was no association between breast cancer risk and cruciferous vegetable consumption assessed from a food frequency questionnaire, urinary isothiocyanate levels were significantly lower in cases than controls. Although interactions were not statistically significant, trends in the association between urinary isothiocyanates and breast cancer were more consistent with homozygous deletion of GSTM1 and GSTT1, the105 Val/Val genotype of GSTP1 or the 187Pro allele of NADPH quinone oxidoreductase. Another case-control study in premenopausal women also observed an inverse association of cruciferous vegetable consumption with breast cancer risk but found no significant effect of GSTM1 or GSTT1 genotype (28).

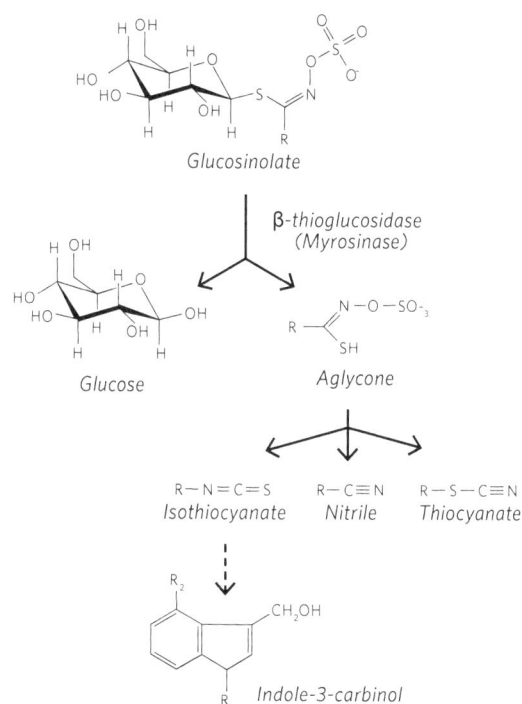

Figure 3.4.1. Formation of isothiocyanates, nitriles, and thiocyanates from glucosinolates by myrosinase-catalyzed degradation.
Isothiocyanates can be converted further to indole-3-carbinol. R = benzyl, phenylethyl, allyl, and other substituents.

References

1. Tijhuis MJ, Wark PA, Arts JMMJG, Visker MHPW, Nagengast FM, Kok FJ, et al. *GSTP1 and GSTA1 polymorphisms interact with cruciferous vegetable intake in colorectal adenoma risk.* Cancer Epidemiol Biomarkers Prev. 2005;14:2943-51.

2. Holst B, Williamson G. *A critical review of the bioavailability of glucosinolates and related compounds.* Nat Prod Rep. 2003;21:425-47.

3. Thornalley PJ. *Isothiocyanates: mechanism of cancer chemopreventive action.* Anti-Cancer Drugs. 2002;13:331-8.

4. Fowke JH, Chung FL, Jin F, Qi D, Cai Q, Conway C, et al. *Urinary isothiocyanate levels, Brassica, and human breast cancer.* Cancer Res. 2003;63:3980-6.

5. Meng Q, Yuan F, Goldberg ID, Rosen EM, Auborn K. *Indole-3-Carbinol is a negative regulator of estrogen receptor-α signaling in human tumor cells.* J Nutr. 2000;130:2927-31.

6. Verhoeven DTH, Goldbohm RA, van Poppe G, Verhagen H, van den Brandt PA. *Epidemiological studies on brassica vegetables and cancer risk.* Cancer Epidemiol Biomarkers Prev. 1996;5:733-48.

7. Warin R, Chambers WH, Potter DM, Singh SV. *Prevention of mammary carcinogenesis in MMTV-neu mice by cruciferous vegetable constituent benzyl isothiocyanate.* Cancer Res. 2009;69:9473-80.

8. Yu R, Mandlekar S, Harvey KJ, Ucker DS, Kong ANT. *Chemopreven-*

tive isothiocyanates induce apoptosis and Caspase-3-like protease activity. Cancer Res. 1998;58:402-8.

9. Hecht SS. *Chemoprevention of cancer by isothiocyanates, modifiers of carcinogen metabolism.* J Nutr. 1999;129:S768-SA74.

10. Maheo K, Morel F, Langouet S, Kramer H, LeFerrec E, Ketterer B, et al. *Inhibition of cytochromes P-450 and induction of glutathione S-transferases by sulforaphane in primary human and rat hepatocytes.* Cancer Res. 1997;57:3649-52.

11. Nakajima M, Yoshida R, Shimada N, Yamazaki H, Yokoi T. *Inhibition and inactivation of human cytochrome P450 isoforms by phenethyl isothiocyanate.* Drug Metab Disp. 2001;29:1110-3.

12. Kolm RH, Danielson UH, Zhang Y, Talalay P, Mannervik B. *Isothiocyanates as substrates for human glutathione transferases: structure--activity studies.* Biochem J. 1995;311:453-9.

13. Meyer DJ, Crease DJ, Ketterer B. *Forward and reverse catalysis and product sequestration by human glutathione S-transferases in the reaction of GSH with dietary aralkyl isothiocyanates.* Biochem J. 1995;306:565-9.

14. Hayes CL, Spink DC, Spink BC, Cao JQ, Walker NJ, Sutter TR. *17b-estradiol hydroxylation catalyzed by human cytochrome P450 1B1.* Proc Natl Acad Sci USA. 1996;93:9776-81.

15. Jellinck PH, Forkert PG, Riddick DS, Okey AB, Michnovicz JJ, Bradlow HL. *Ah receptor binding properties of indole carbinols and induction of hepatic estradiol hydroxylation.* Biochem Pharmacol. 1993;45:1129-36.

16. Tiwari RK, Guo L, Bradlow HL, Telang NT, Osborne MP. *Selective responsiveness of human breast cancer cells to indole-3-carbinol, a chemopreventive agent.* J Natl Cancer Inst. 1994;86:126-31.

17. Wattenberg LW, Loub WD. *Inhibition of polycyclic aromatic hydrocarbon-induced neoplasia by naturally occuring indoles.* Cancer Res. 1978;38:1410-3.

18. Bradlow HL, Michnovicz J, Telang NT, Osborne MP. *Effects of dietary indole-3-carbinol on estradiol metabolism and spontaneous mammary tumors in mice.* Carcinogenesis. 1991;12:1571-4.

19. Bradlow HL, Michnovicz JJ, Halper M, Miller DG, Wong GYC, Osborne MP. *Long-term responses of women to indole-3-carbinol or a high fiber diet.* Cancer Epidemiol Biomarkers Prev. 1994;3:591-5.

20. Fowke JH, Longcope C, Hebert JR. *Brassica vegetable consumption shifts estrogen metabolism in healthy postmenopausal women.* Cancer Epidemiol Biomarkers Prev. 2000;9:773-9.

21. Bradlow HL, Sepkovic DW, Telang NT, Osborne MP. *Indole-3-carbinol. A novel approach to breast cancer prevention.* Ann New York Acad Sci. 1995;768:180-200.

22. Stoner G, Casto B, Ralston S, Roebuck B, Pereira C, Bailey G. *Development of a multi-organ rat model for evaluating chemopreventive agents: efficacy of indole-3-carbinol.* Carcinogenesis. 2002;23:265-72.

23. Gandini S, Merzenich H, Robertson C, Boyle P. *Meta-analysis of studies on breast cancer risk and diet: the role of fruit and vegetable consumption and the intake of associated micronutrients.* Eur J Cancer. 2000;36:636-46.

24. van Gils CH, Peeters PH, Bueno-de-Mesquita H, Boshuizen HC, Lahmann PH, Clavel-Chapelon F, et al. *Consumption of vegetables and fruits and risk of breast cancer.* JAMA. 2005;293:183-93.

25. Smith-Warner SA, Spiegelman D, Yaun SS, Adami HO, Beeson WL, van den Brandt PA, et al. *Intake of fruits and vegetables and risk of breast cancer.* JAMA. 2001;285:769-76.

26. Shannon J, Ray R, Wu C, Nelson Z, Gao DL, et al. *Food and botanical groupings and risk of breast cancer: a case-control study in Shanghai, China.* Cancer Epidemiol Biomarkers Prev. 2005;14:81-90.

27. Lee SA, Fowke JH, Lu W, Ye C, Zheng Y, Cai Q, et al. *Cruciferous vegetables, the GSTP1 Ile105Val genetic polymorphism, and breast cancer risk.* Am J Clin Nutr. 2008;87:753-60.

28. Ambrosone CB, McCann SE, Freudenheim JL, Marshall JR, Zhang Y, Shields P. *Breast cancer risk in premenopausal women is inversely associated with consumption of broccoli, a source of isothiocyanates, but is not modified by GST genotype.* J Nutr. 2004;134:1134-8.

3.4.3. FAT

The observation that breast cancer rates are strongly correlated with national per capita fat consumption has led to the hypothesis that dietary fat may be a causal factor in the etiology of breast cancer (1, 2). However, critical review of these ecological studies revealed the omission of several confounding variables. In developed countries with higher fat intake, the prevalence of breast cancer risk factors, such as low parity, late age at first birth, late age at menarche, increased height, and postmenopausal obesity is higher, thereby predicting higher breast cancer incidence in these countries (3). Overall, epidemiological evidence from case-control and cohort studies designed to examine the dietary fat-human breast cancer relationship has been inconsistent (3-8). In particular, prospective studies in which dietary habits were ascertained before the development of cancer, have not found that dietary fat is related to breast cancer. A 1993 meta-analysis of 23 studies of dietary fat intake and breast cancer risk determined a relative risk of 1.12 (95% CI 1.04 – 1.21). The relative risks separated by case-control and cohort studies yielded 1.21 (95% CI 1.10 – 1.34) and 1.01 (CI 0.90 – 1.13), respectively (9). A 2003 meta-analysis of 45 studies that adjusted for energy intake found that women who consumed the highest levels of total fat had a 13% higher risk of breast cancer than those who consumed the lowest levels (10). The National Institutes of Health-American Association of Retired Persons (NIH-AARP) Health Study of 188,736 post-menopausal women showed an increased risk of 11% (95% CI, 1.00 – 1.24; p trend = 0.017) for women in the highest (40.1%) compared to women in the lowest (20.3% energy from total fat) quintile (11). On the other hand, in a pooled analysis of seven prospective studies in four countries, collectively comprising over 300,000 women among whom nearly 5,000 breast cancers were diagnosed, no association with breast cancer risk was seen between 15 and 50% of energy intake from fat (RR 1.05; 95% CI 0.94 – 1.16) (12).

It has been suggested that both the type and amount of fat (Figure 3.4.2.) are relevant for breast cancer risk (13, 14). Unfortunately, there is no valid biochemical indicator of total fat intake. Several case-control studies have shown a modest increased risk associated with increased intake of various fat subtypes (6). However, prospective cohort studies have provided little support for any association with fat subtypes (6, 15). For example, large prospective studies have detected no correlation between serum cholesterol and triglyceride concentrations and breast cancer rates (8, 16, 17). A prospective study in Norway observed an inverse relation between serum high-density lipoprotein cholesterol (HDL-C) and breast cancer risk in postmenopausal overweight and obese women (18). Such a risk association was not found in the premenopausal age group but low serum HDL-C in premenopausal women was associated with increased levels of free E_2 throughout an entire menstrual cycle (19). Serum HDL-C was also inversely related to serum leptin, insulin, and dehydroepiandrostendione sulfate.

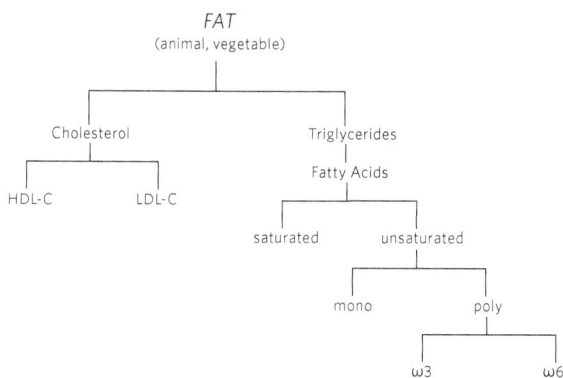

Figure 3.4.2. Schematic overview of lipid fractions found in dietary fats.
Polyunsaturated fatty acids (PUFAs) are classified based on the position of the double bonds. Those with the final carbon-carbon double bond in the n-3 position are classified as omega-3 and those with the double bond in the n-6 position as omega-6 PUFAs. Omega-3 PUFAs include a-linolenic acid (ALA), eicosapentaenoic acid (EPA), and docosahexaenoic acid (DHA), found in fish oils. Omega-6 PUFAs include linoleic acid and arachidonic acid, which are present in vegetable oils.

The intake of specific fatty acids, rather than that of total fat, has been implicated to influence breast cancer risk. A 2003 meta-analysis of epidemiologic studies observed a significant increase in breast cancer risk with high saturated fat intake but no association with monounsaturated and polyunsaturated fatty acids (PUFAs) (10). Evidence from observational epidemiologic studies is conflicting. For example, some studies observed that high intake of monounsaturated fatty acids (e.g., oleic acid) from meat and milk was associated with an increased risk of breast cancer (20, 21). In contrast, diets rich in olive oil, which contains a high level of oleic acid, were shown to reduce breast cancer risk (22). The prospective NIH-AARP Health Study with an average follow-up of 4.4 years showed an increased risk of 12% (95% CI, 1.03 – 1.21) for monounsaturated fat (11). In contrast, the Nurses' Health Study found no evidence for a significant risk association with intake of monounsaturated fat over 14 years of follow-up (23).

Dietary omega-3 PUFAs include α-linolenic acid (ALA), eicosapentaenoic acid (EPA), and docosahexaenoic acid (DHA), which are present at high concentrations in fish and flaxseed oil but also found in nuts, eggs, and meat,

especially from animals fed grass rather than grain. Omega-6 PUFAs include arachidonic acid and linoleic acid, the latter being the most abundant omega-6 PUFA in the American diet, present in most vegetable oils, nuts, and seeds as well as poultry and eggs. Both omega-3 and omega-6 PUFAs are converted to eicosanoids (prostaglandins, thromboxanes, prostacyclins, leukotrienes), which are important in inflammatory and immune system responses. If the rate of synthesis exceeds the rate of metabolism, the excess eicosanoids may have deleterious effects. Since omega-3 PUFAs are converted to eicosanoids at a much slower rate than omega-6 PUFAs, the ω3:ω6 ratio directly affects the level and type of eicosanoids. Low ω3:ω6 ratios have been associated with a variety of health benefits and been claimed to prevent cardiovascular disease and cancer. In view of these putative benefits, numerous epidemiologic studies of omega-3 and omega-6 PUFA intake have been performed to examine the association with breast cancer risk. Overall, these studies have shown inconsistent evidence for a risk association (24-28). A 2006 review of eight prospective studies from six different cohorts assessed breast cancer incidence relative to fish consumption, total and marine omega-3 fatty acid intake, and consumption of specific omega-3 fatty acids, i.e., ALA, EPA, and DHA (29). There was one significant estimate for increased risk, three were for decreased risk, and seven other estimates did not show any significant association (5, 23, 30-35). Importantly, those studies with the largest sample size found no association between omega-3 fatty acid consumption and breast cancer risk. More recent prospective studies have equally failed to demonstrate consistent risk associations with PUFA intake but identified subgroups of women potentially at increased risk (36, 37). For example, intake of heterocyclic amines (HAs) was not associated with breast cancer risk in a cohort of postmenopausal Swedish women, but in individuals with low HA intakes, a significantly increased risk was observed in those with high consumption of omega-6 PUFAs. Fewer studies have examined the association of breast cancer risk with fatty acids in serum, erythrocytes, and adipose tissue. A gas chromatographic analysis of serum phospholipids from 197 pre- and postmenopausal breast cancer patients and matched control women showed no significant difference in the proportion of saturated fatty acids, monounsaturated fatty acids, or omega-3 and omega-6 PUFAs between cases and controls (38). The examination of erythrocyte membranes in Japanese women on a diet rich in fish showed an inverse association of EPA and DHA levels with breast cancer risk (26). The measurement of PUFAs in breast adipose tissue showed an inverse association of the ω3:ω6 ratio with breast cancer risk (39-41). PUFAs undergo dioxygenation, a reaction catalyzed by lipoxygenases, a family of enzymes that includes 5-, 12-, and 15-lipoxygenase. An analysis of polymorphisms in the 5-lipoxygenase gene revealed no association with breast cancer risk (42). However, there was a significant interaction between the -4900 A/G polymorphism and dietary linoleic acid intake. Among women consuming a diet high in linoleic acid (top quartile of intake, >17.4 g/d), carrying the AA genotype was associated with higher breast cancer risk (OR 1.8; 95% CI 1.2 – 2.9) compared with carrying genotypes AG or GG. The functional importance of the -4900 polymorphism is unknown and whether or not it is the causal variant is not clear.

The hypothesis that a low-fat dietary pattern can reduce breast cancer risk has been tested in the Women's Health Initiative Dietary Modification Trial, which included 48,835 postmenopausal women, aged 50 to 29 years (43). The women were randomly assigned to a dietary modification group or a comparison group. The aim of the dietary modification intervention was to reduce the intake of total fat to 20% of total energy and increase the daily consumption of vegetables and fruit and grains. The comparison group was not asked to make dietary changes. After an average of 8.1 years of intervention and follow-up, the incidence of invasive breast cancer was 9% lower in the intervention than the comparison group (95% CI, 0.83 – 1.01; p = 0.09). Eight single nucleotide polymorphisms in intron 2 of the FGFR2 gene, identified as a breast cancer susceptibility locus in genome-wide association studies, were examined in 1676 women who developed breast cancer during the trial follow up (44, 45). Case-only analyses showed that odds ratios for the dietary intervention did not vary significantly with the genotype for any of the eight polymorphisms (46). However, among women whose baseline percent of energy from fat was in the upper quartile there was a significant risk association of the intervention with polymorphism rs3570817 (T/C), ranging from OR 1.06 to 0.53 and 0.62 at 0, 1, and 2 minor alleles (p = 0.03). Thus, women having one or two C alleles of rs3570817 may benefit from reduction from a high-fat to a lower-fat dietary pattern. Another dietary intervention study measured the ω3:ω6 ratio in plasma, breast and gluteal fat samples of 25 women with high-risk localized breast cancer who consumed a low-fat diet and a daily fish oil supplement throughout a three-month period (47). The dietary modulation caused a statistically significant increase in the ω3:ω6 ratio in plasma and breast adipose tissue (47).

To clarify the role of dietary factors in the etiology of breast cancer, much research has been conducted in animals, especially rodents. Numerous studies have examined the effect of fat intake and energy (i.e., caloric) restriction on breast cancer risk. Energy restriction can be achieved in several ways, taking into account that fats (9 kcal/g) have more than twice the caloric content of carbohydrates or

proteins (4 kcal/g). Independent of the type of nutrient restriction, these experiments have consistently shown a positive relationship between energy intake and breast cancer risk, e.g., energy-restricted mice developed 55% less mammary tumors than the control groups (48). Several mechanisms have been proposed to explain how reduced caloric intake may lower carcinogenic risk, including changes in cell cycle regulation, reduced oxidative stress, and decreased angiogenesis. The reduction in caloric intake may influence hormones, cytokines, and growth factors, e.g., decrease insulin and insulin-like growth factors (49, 50). It is not only the amount but also the type of dietary fat that can significantly influence the development and/ or growth of rodent mammary gland tumors and growth of human breast carcinomas in immune deficient mice (51). For example, dietary PUFAs have been shown to modulate mammary tumor formation and proliferation in rodents (52, 53). A meta-analysis of 97 studies in rodents revealed strong tumor-enhancing effects of omega-6 and a nonsignificant protective effect of omega-3 PUFAs (54). It is unclear how much light these results shed on the development of breast cancer in humans. The animal models used to conduct many of these studies do not come close to replicating human exposures (29).

A person on a standard Western diet (40% of calories from fat) usually ingests more calories than a person on a low-fat diet (20% of calories from fat). Thus, fat consumption and energy intake are linked. The association is even more complicated because energy intake is just one component of the overall energy balance, other factors being physical activity and body mass index. Epidemiologic studies investigating the relationship between energy excess and breast cancer risk have been inconsistent. The Prostate, Lung, Colorectal, and Ovarian (PLCO) Cancer Screening Trial observed a positive association between total energy intake and risk of postmenopausal breast cancer (55). However, an extension of the PLCO study using a modified dietary history questionnaire found no difference in risk between women in the highest and lowest quartiles of energy intake (56). Similarly, energy intake was not associated with risk among postmenopausal women in the Breast Cancer Detection Demonstration Project Follow-up Cohort Study nor among pre- and postmenopausal women in the Nurses' Health Study Cohort (7, 23). The Canadian National Breast Screening Study observed an increased risk with high energy intake in premenopausal but not in postmenopausal women (57). Finally, studies of breast cancer risk after caloric restriction in women hospitalized for anorexia nervosa before age 40 or exposed to war-time or post-war famine in early life have been inconclusive (58, 59).

There are several mechanisms by which dietary fats may influence the development of breast cancer. The main mechanisms proposed involve estrogens and PUFAs. The relation between dietary fat intake and circulating estrogen levels is supported by three types of study: epidemiological studies of women in countries with low and high breast cancer rates, epidemiological studies of vegetarian and omnivorous women, and dietary intervention studies. The first type of study compared populations at low risk of developing breast cancer, such as rural Chinese and Japanese women with populations at high risk, such as British and American white women. The Asian women, who consumed 15% or less of their energy as fat, had significantly lower plasma E_2 levels than their Western counterparts on a 40% high-fat diet (60, 61). Postmenopausal women who consume substantial amounts of dietary fat have higher plasma levels of E_1 and E_2 and higher urinary levels of E_3 and total estrogens than vegetarian women (62, 63). The lower plasma and urine estrogen levels in the latter group are the result of enhanced fecal excretion of estrogens (64). Inconsistent associations of dietary fat intake and estrogen concentrations have been observed among premenopausal women (64-66). Dietary intervention studies yielded more consistent results. Reduction of fat intake to 20% of total kilocalories and/or increased dietary fiber consumption resulted in a modest reduction in serum E_2 and E_1 sulfate and an increase in sex hormone-binding globulin (43, 67-71). The absence of changes in serum progesterone and gonadotropins during the dietary intervention suggests that the decrease in serum estrogen levels is due to altered enterohepatic circulation (reduced reabsorption, increased fecal excretion), rather than an effect on the pituitary-ovarian axis.

Based on animal experiments and cell culture studies, there are several ways through which PUFAs may influence carcinogenesis (72). Omega-3 PUFAs inhibit the formation of omega-6 arachidonic acid-derived eicosanoids (e.g., prostaglandin), which are linked to inflammation, tumorigenesis, angiogenesis, cell proliferation and apoptosis induction. In endometriosis-derived stromal cells, the eicosanoid prostaglandin E_2 has also been shown to stimulate the activity of aromatase, which converts androgens to estrogens (73). The lipid peroxidation of PUFAs contributes to the production of free radicals and reactive oxygen species, which can induce oxidative DNA damage via intermediaries, such as malondialdehyde (MDA) (74, 75). The degree of lipid peroxidation depends on the proportion of PUFA in the diet. Animal experiments have shown that the PUFA composition of the diet regulates the fatty acid composition of cell membranes, e.g., hepatic endoplasmic reticulum, and this, in turn, is an important factor controlling the rate and extent of lipid peroxidation *in vitro* and possibly *in vivo* (76). Furthermore, the dietary fatty acid composition can influence the production of MDA-DNA, i.e., M_1G adduct levels, as demonstrated in a study of 59 healthy individuals receiving either a sunflower

oil- (n = 30) or a rapeseed oil-based diet (n = 29). The sunflower oil diet, which is rich in PUFAs, yielded higher concentrations of PUFAs in plasma triglycerides and 3.6-fold higher M_1G levels in leukocyte DNA than the rapeseed oil diet, which is rich in monounsaturated fatty acids (77).

In conclusion, fat intake has been implicated as a risk factor for breast cancer based on ecological studies and animal experiments. However, epidemiological evidence does not support a risk association with either amount or type of dietary fat. One reason may be that fat consumption is linked to energy intake and obesity. Obesity, in turn, is the result of high-energy intake relative to low energy expenditure resulting most often from a sedentary lifestyle. Overall, specific nutritional factors may be less important than the balance of energy intake and expenditure. Weight gain in adulthood has been shown to increase the risk for postmenopausal breast cancer. Avoidance of weight gain, which is best accomplished by regular physical exercise and limited fat intake, is a lifestyle choice that may reduce a woman's lifetime risk of breast cancer (78, 79).

References

1. Armstrong B, Doll R. *Environmental factors and cancer incidence and mortality in different countries, with special reference to dietary practices.* Int J Cancer. 1975;15:617-31.

2. Schatzkin A, Greenwald P, Byar DP, Clifford CK. *The dietary fat-breast cancer hypothesis is alive.* JAMA. 1989;261:3284-7.

3. Hunter DJ. *Role of dietary fat in the causation of breast cancer: Counterpoint.* Cancer Epidemiol Biomarkers Prev. 1999;8:9-13.

4. Byrne C, Rockett H, Holmes MD. *Dietary fat, fat subtypes, and breast cancer risk: lack of an association among postmenopausal women with no history of benign breast disease.* Cancer Epidemiol Biomarkers Prev. 2002;11:261-5.

5. Cho E, Spiegelman D, Hunter DJ, Chen WY, Stampfer MJ, Colditz GA, et al. *Premenopausal fat intake and risk of breast cancer.* J Natl Cancer Inst. 2003;95:1079-85.

6. Howe GR, Friedenreich CM, Jain M, Miller AB. *A cohort study of fat intake and risk of breast cancer.* J Natl Cancer Inst. 1991;83:1035-6.

7. Velie E, Kulldorff M, Schairer C, Block G, Albanes D, Schatzkin A. *Dietary fat, fat subtypes, and breast cancer in postmenopausal women: A prospective cohort study.* J Natl Cancer Inst. 2000;92:833-9.

8. Willett WC, Stampfer MJ, Coldiz GA, Rosner BA, Hennekens CH, Speizer FE. *Dietary fat and the risk of breast cancer.* N Engl J Med. 1987;316:22-8.

9. Boyd NF, Martin LJ, Noffel M, Lockwood GA, Trichler DL. *A meta-analysis of studies of dietary fat and breast cancer risk.* Br J Cancer. 1993;68:627-36.

10. Boyd NF, Stone J, Vogt KN, Connelly BS, Martin LJ, Minkin S. *Dietary fat and breast cancer risk revisited: a meta-analysis of the published literature.* Br J Cancer. 2003;89:1672-85.

11. Thiebaut ACM, Kipnis V, Chang SC, Subar AF, Thompson FE, Rosenberg PS, et al. *Dietary fat and postmenopausal invasive breast cancer in the National Institutes of Health-AARP Diet and Health Study Cohort.* J Natl Cancer Inst. 2007;99:451-62.

12. Hunter DJ, Spiegelman D, Adami H, Beeson L, van Den Brant PA, Folsom AR, et al. *Cohort studies of fat intake and the risk of breast cancer-a pooled analysis.* N Engl J Med. 1996;334:356-561.

13. Cohen LA, Wynder EI. *Do dietary monounsaturated fatty acids play a protective role in carcinogenesis and cardiovascular disease?* Med Hypotheses. 1990;3:83-9.

14. Wynder EL, Cohen LA, Muscat JE, Winters B, Dwyer JT, Blackburn

G. *Breast cancer: weighing the evidence for a promoting role of dietary fat.* J Natl Cancer Inst. 1997;89:766-75.

15. Willett WC, Hunter DJ, Stampfer MJ, Colditz G, Manson JE, Spiegelman D, et al. *Dietary fat and fiber in relation to risk of breast cancer: An 8-year follow-up.* JAMA. 1992;268:2037-44.

16. Gaard M, Tretli S, Urdal P. *Risk of breast cancer in relation to blood lipids: a prospective study of 31,209 Norwegian women.* Cancer Causes Control. 1994;5:501-9.

17. Hiatt RA, Friedman GD, Bawol RD, Ury HK. *Breast cancer and serum cholesterol.* J Natl Cancer Inst. 1992;68:885-9.

18. Furberg AS, Veierod MB, Wilgaard T, Bernstein L. *Serum high-density lipoprotein cholesterol, metabolic profile, and breast cancer risk.* J Natl Cancer Inst. 2004;96:1152-60.

19. Furberg AS, Jasienska G, Bjurstam N, Torjesen PA, Emaus A, Lipson SF, et al. *Metabolic and hormonal profiles: HDL cholesterol as a plausible biomarker of breast cancer risk. The Norwegian EBBA study.* Cancer Epidemiol Biomarkers Prev. 2005;14:33-40.

20. Gaard M, Tretli S, Loken EB. *Dietary fat and the risk of breast cancer: a prospective study of 25,892 Norwegian women.* Int J Cancer. 1995;63:13-7.

21. Gerber M, Richardson S, Crastes de Paulet P, Pujol H, Crastes de Paulet A. *Relationship between vitamin E and polyunsaturated fatty acids in breast cancer. Nutritional and metabolic aspects.* Cancer. 1989;64:2347-53.

22. Martin-Moreno JM, Willett WC, Gorgojo L, Banegas JR, Rodriguez-Artalejo F, Fernandez-Rodriguez JC, et al. *Dietary fat, olive oil intake and breast cancer risk.* Int J Cancer. 1994;58:774-80.

23. Holmes MD, Hunter DJ, Colditz GA, Stampfer MJ, Hankinson SE, Speizer FE, et al. *Association of dietary intake of fat and fatty acids with risk of breast cancer.* JAMA. 1999;281:914-20.

24. Kaizer L, Boyd NF, Kriukov V, Tritchler D. *Fish consumption and breast cancer risk: an ecological study.* Nutr Cancer. 1989;12:61-8.

25. Kohlmeier L. *Biomarkers of fatty acid exposure and breast cancer risk.* Am J Clin Nutr. 1997;66(6 suppl):1548S-56S.

26. Kuriki K, Hirose K, Wakai K, Matsuo K, Ito H, Suzuki T, et al. *Breast cancer risk and erythrocyte compositions of n-3 highly unsaturated fatty acids in Japanese.* Int J Cancer. 2007;121:377-85.

27. McCann SE, Ip C, Ip MM, McGuire MK, Muti P, Edge SB, et al. *Dietary intake of conjugated linoleic acids and risk of premenopausal and postmenopausal breast cancer, Western New York Exposures and Breast cancer study (WEB study).* Cancer Epidemiol Biomarkers Prev. 2004;13:1480-4.

28. Terry PD, Rohan TE, Wolk A. *Intakes of fish and marine fatty acids and the risks of cancers of the breast and prostate and of other hormone-related cancers: a review of the epidemiologic evidence.* Am J Clin Nutr. 2003;77:532-43.

29. MacLean CH, Newberry SJ, Mojica WA, Khanna P, Issa AM, Suttorp MJ, et al. *Effects of Omega-3 fatty acids on cancer risk. A systematic review.* JAMA. 2006;295:403-15.

30. Gago-Dominguez M, Yuan JM, Sun CL, Lee HP, Yu MC. *Opposing effects of dietary n-3 and n-6 fatty acids on mammary carcinogenesis: The Singapore Chinese health study.* Br J Cancer. 2003;89:1686-92.

31. Holmes MD, Colditz GA, Hunter DJ, Hankinson SE, Rosner B, Speizer FE, et al. *Meat, fish and egg intake and risk of breast cancer.* Int J Cancer. 2003;104:221-7.

32. Key TJ, Sharp GB, Appleby PN, Beral V, Goodman MT, Soda M, et al. *Soya foods and breast cancer risk: a prospective study in HIroshima and Nagasaki, Japan.* Br J Cancer. 1999;81:1248-56.

33. Stripp C, Overvad K, Christensen J, Thomsen BL, Olsen A, Moller S, et al. *Fish intake is positively associated with breast cancer incidence rate.* J Nutr. 2003;133:3664-9.

34. Vatten LJ, Solvoll K, Loken E. *Frequency of meat and fish intake and risk of breast cancer in a prospective study of 14,500 Norwegian women.* Int J Cancer. 1990;46:12-5.

35. Voorrips LE, Brants HAM, Kardinaal AFM, Hiddink GJ, van den Brandt PA. *Intake of conjugated linoleic acid, fat and other fatty acids in relation to postmenopausal breast cancer: the Netherlands cohort study on diet and cancer.* Am J Clin Nutr. 2002;76:873-82.

36. Sonestedt E, Ericson U, Gullberg B, Skog K, Olsson H, Wirfalt E. *Do both heterocyclic amines and omega-6 polyunsaturated fatty acids contribute to the incidence of breast cancer in postmenopausal women of the Malmo diet and cancer cohort.* Int J Cancer. 2008;123:1637-43.

37. Thiebaut ACM, Chajes V, Gerber M, Boutron-Ruault MC, Joulin V, Lenoir G, et al. *Dietary intakes of ω-6 and ω-3 polyunsaturated fatty acids and the risk of breast cancer.* Int J Cancer. 2009;124:924-31.

38. Saadatian-Elahi M, Toniolo P, Ferrari P, Goudable J, Akhmedkhanov A, Zeleniuch-Jacquotte A, et al. *Serum fatty acids and risk of breast cancer in a nested case-control study of the New York University women's health study.* Cancer Epidemiol Biomarkers Prev. 2002;11:1353-60.

39. Bougnoux P, Giraudeau B, Couet C. *Diet, cancer, and the lipidome.* Cancer Epidemiol Biomarkers Prev. 2006;15:416-21.

40. Maillard V, Bougnoux P, Ferrari P, Jourdan ML, Pinault M, Lavillonniere F, et al. *N-3 and n-6 fatty acids in breast adipose tissue and relative risk of breast cancer in a case-control study in Tours, France.* Int J Cancer. 2002;98:78-83.

41. Simonsen N, van't Veer P, Strain JJ, Martin-Moreno JM, Huttunen JK, Navajas JF, et al. *Adipose tissue omega-3 and omega-6 fatty acid content and breast cancer in the EURAMIC study. European Community Multicenter Study on antioxidants, myocardial infarction, and breast cancer.* Am J Epidemiol. 1998;147:342-52.

42. Wang J, John EM, Ingles SA. *5-lipoxygenase and 5-lipoxygenase-activating protein gene polymorphisms, dietary linoleic acid, and risk for breast cancer.* Cancer Epidemiol Biomarkers Prev. 2008;17:2748-54.

43. Prentice RL, Caan B, Chlebowski RT, Patterson R, Kuller LH, Ockene JK, et al. *Low-fat dietary pattern and risk of invasive breast cancer. The Women's Health Initiative randomized controlled dietary mdification trial.* JAMA. 2006;295:629-42.

44. Easton DF, Pooley KA, Dunning AM, Pharoah PDP, Thompson D, Ballinger DG, et al. *Genome-wide association study indentifies novel breast cancer susceptibility loci.* Nature. 2007;447:1087-95.

45. Hunter DJ, Kraft P, Jacobs KB, Cox DG, Yeager M, Hankinson SE, et al. *A genome-wide association study identifies alleles in FGFR2 associated with risk of sporadic postmenopausal breast cancer.* Nature Genet. 2007;39:870-4.

46. Prentice RL, Huang Y, Hinds D, Peters U, Cox DR, BeilharzE, et al. *Variation in the FGFR2 gene and the effect of a low-fat dietary pattern on invasive breast cancer.* Cancer Epidemiol Biomarkers Prev. 2010;19:74-9.

47. Bagga D, Capone S, Wang HJ, Heber D, Lill M, Chap L, et al. *Dietary modulation of omega-3/omega-6 polyunsaturated fatty acid ratios in patients with breast cancer.* J Natl Cancer Inst. 1997;89:1123-31.

48. Dirx MJM, Zeegers MPA, Dagnelie PC, van den Bogaard T, van den Brandt PA. *Energy restriction and the risk of spontaneous mammary tumors in mice: a meta-analysis.* Int J Cancer. 2003;106:776-0.

49. Hursting SD, Lavigne JA, Berrigan D, Perkins SN, Barrett JC. *Calorie restriction, aging, and cancer prevention: Mechanisms of action and applicability to humans.* Annu Rev Med. 2003;54:131-52.

50. Thompson HJ, Zhu Z, Jiang W. *Dietary energy restriction in breast cancer prevention.* J Mammary Gland Biol Neoplasia. 2003;8.

51. Welsch CW. *Relationship between dietary fat and experimental mammary tumorigenesis: A review and critique.* Cancer Res. 1992;52:2040s-8s.

52. Hilakivi-Clarke L, Onojafe I, Raygada M, Cho E, Clarke R, Lippman ME. *Breast cancer risk in rats fed a diet high in n-6 polyunsaturated fatty acids during pregnancy.* J Natl Cancer Inst. 1996;88:1821-7.

53. Hilakivi-Clarke L, Stoica A, Raygada M, Martin M. *Consumption of a high-fat diet alters estrogen receptor content, protein kinase C activity, and mammary gland morphology in virgin and pregnant mice and female offspring.* Cancer Res. 1998;58:654-60.

54. Fay MP, Freedman LS, Clifford CK, Midthune DN. *Effect of different types and amounts of fat on the development of mammary tumors in rodents: A review.* Cancer Res. 1997;57:3979-88.

55. Chang SC, Ziegler RG, Dunn B, Stolzenberg-Solomon R, Lacey JV, Huang WY, et al. *Association of energy intake and energy balance with postmenopausal breast cancer in the Prostate, Lung, Colorectal, and Ovarian cancer screening trial.* Cancer Epidemiol Biomarkers Prev. 2006;15:334-41.

56. Sue LY, Schairer C, Ma X, Williams C, Chang SC, Miller AB, et al. *Energy intake and risk of postmenopausal breast cancer: An*

expanded analysis in the prostate, lung, colorectal, and ovarian cancer screening trial (PLCO) cohort. Cancer Epidemiol Biomarkers Prev. 2009;18:2842-50.

57. Silvera SAN, Jain M, Howe G, Miller AB, Rohan TE. *Energy balance and breast cancer risk: a prospective cohort study.* Breast Cancer Res Treat. 2006;97:97-106.

58. Elias SG, Peeters PHM, Grobbee DE, van Noord PA. *Breast cancer risk after caloric restriction during the 1944-1945 Dutch famine.* J Natl Cancer Inst. 2004;96:539-46.

59. Michels KB, Ekbom A. *Caloric restriction and incidence of breast cancer.* JAMA. 2004;291:1226-30.

60. Key TJ, Chen J, Wang DY, Pike MC, Boreham J. *Sex hormones in women in rural China and in Britain.* Br J Cancer. 1990;62:631-6.

61. Shimizu H, Ross RK, Bernstein L, Pike MC, Henderson BE. *Serum oestrogen levels in postmenopausal women: comparison of American whites and Japanese in Japan.* Br J Cancer. 1990;62:451-3.

62. Armstrong BK, Brown JB, Clarke HT, Crooke DK, Hahnel R, Masarei J, et al. *Diet and reproductive hormones: a study of vegetarian and nonvegetarian postmeno-pausal women.* J Natl Cancer Inst. 1981;67:761-7.

63. Goldin BR, Adlercreutz H, Dwyer JT, Swenson L, Warram JH, Gorbach SL. *Effect of diet on excretion of estrogens in postmenoausal women.* Cancer Res. 1981;41:3771-3.

64. Goldin BR, Adlercreutz H, Gorbach SL, Warram JH, Dwyer JT, Swenson L, et al. *Estrogen excretion patterns and plasma levels in vegetarian and omnivorous women.* New Engl J Med. 1982;307:1542-

65. Gray GE, Pike MC, Hirayama T, Tellez J, Gerkins V, Brown JB, et al. *Diet and hormone profiles in teenage girls in four countries at different risk for breast cancer.* Prev Med. 1982;11:108-13.

66. Persky VW, Chatterton RT, Van Horn LV, Grant MD, Langenberg P, Marvin J. *Hormone levels in vegetarian and nonvegetarian teenage girls: Potential implications for breast cancer risk.* Cancer Res. 1992;52:578-83.

67. Bagga D, Ashley JM, Geffrey SP, Wang HJ, Barnard RJ, Korenman S, et al. *Effects of a very low fat, high fiber diet on serum hormones and menstrual functon. Implications for breast cancer prevention.* Cancer. 1995;76:2491-6.

68. Bennett FC, Ingram DM. *Diet and female sex hormone concentra-tions: an intervention study for the type of fat consumed.* Am J Clin Nutr. 1990;52:808-12.

69. Prentice R, Thompson D, Clifford C, Gorbach S, Goldin B, Byar D. *Dietary fat reduction and plasma estradiol concentration in healthy postmenopausal women. The Women's Health Trial Study Group.* J Natl Cancer Inst. 1990;82:129-34.

70. Rose DP, Boyar AP, Cohen C, Strong LE. *Effect of a low-fat diet on hormone levels in women with cystic breast disease. I. Serum steroids and gonadotropins.* J Natl Cancer Inst. 1987;78:623-6.

71. Woods MN, Barnett JB, Spiegelman D, Trail N, Hertzmark E, Longcope C, et al. *Hormone levels during dietary changes in premenopausal African-American women.* J Natl Cancer Inst. 1996;88:1369-74.

72. Larsson SC, Kumlin M, Ingelman-Sundberg M, Wolk A. *Dietary long-chain n-3 fatty acids for the prevention of cancer: a review of potential mechanisms.* Am J Clin Nutr. 2004;79:935-45.

73. Noble LS, Takayama K, Zeitoun KM, Putman JM, Johns DA, Hinshelwood MM, et al. *Protaglandin E2 stimulates aromatase expression in endometriosis-derived stromal cells.* J Clin Endocrinol Metab. 1997;82:600-6.

74. Marnett LJ. *Oxyradicals and DNA damage.* Carcinogenesis. 2000;21:361-70.

75. Welsch CW. *Review of the effects of dietary fat on experimental mammary gland tumorigenesis: role of lipid peroxidation.* Free Radical Biol Med. 1995;18:757-73.

76. Hammer CT, Wills ED. *The role of lipid components of the diet in the regulation of the fatty acid composition of the rat liver endoplasmic reticulum and lipid peroxidation.* Biochem J. 1978;174:585-93.

77. Fang JL, Vaca CE, Valsta LM, Mutanen M. *Determination of DNA adducts of malonaldehyde in humans: effects of dietary fatty acid composition.* Carcinogenesis. 1996;17:1035-40.

78. Bernstein L, Henderson BE, Hanisch R, Sullivan-Halley J, Ross RK. *Physical exercise and reduced risk of breast cancer in young women.* J Natl Cancer Inst. 1994;86:1403-8.

79. Ziegler RG, Hoover RN, Nomura AMY, West DW, Wu AH, Pike MC, et al. *Relative weight, weight change, height, and breast cancer risk in Asian-American women.* J Natl Cancer Inst. 1996;88:650-60.

3.4.4. VITAMINS

Various vitamins have been hypothesized to be determinants of breast cancer, but none have been unequivocally associated with the disease. There are many reasons for this uncertainty, including the complexity of diets, the presence of more than one vitamin in a fruit or vegetable, as well as non-vitamin constituents such as fiber. Moreover, users of vitamin supplements differ from nonusers in numerous ways, including demographic and lifestyle characteristics (1).

VITAMINS A, C, AND E

Retinoids consist of a cyclic end group (the trimethylcyclo-hexenyl ring), a dimethyl-substituted tetraene chain, and a polar hydroxyl (retinol), aldehyde (retinal), or carboxyl (retinoic acid) end group (Figure 3.4.3). The major sources of vitamin A (retinol) in the diet are plant carotenoid pigments, such as β-carotene, which are present in carrots, spinach, and squash. In the mucosa of the small intestine the enzyme carotene-15,15'-dioxygenase cleaves β-carotene yielding two molecules of all-trans retinal, which is reduced

or oxidized to retinol or retinoic acid, respectively (Figure 3.4.3). Other dietary carotenoids with provitamin A activity are α-carotene, β-cryptoxanthin, lycopene, lutein, and zeaxanthin (2). Carotenoids have been shown to have anti-carcinogenic properties through the anti-oxidative action of the polyene chain, which quenches ROS and traps oxidizing agents (2-4). Retinoids including vitamin A play an important role in organogenesis and have been hypothesized to reduce the risk of cancer by regulating cell growth and differentiation (5). The actions of retinoids are mediated by two members of the nuclear receptor superfamily, the retinoic acid receptor (RAR) and retinoid X receptor (RXR), each composed of three subtypes (α, β, and γ). Estradiol induces RARα transcription in ER-positive breast cancer cell lines via half-palindromic ERE and Sp1 motifs in the RARα promoter (6, 7). The RARα, in turn, mediates the inhibitory effect of retinoic acid on the growth of ER-positive cells (8-10). With few exceptions, ER-negative breast cancer cell lines express lower RARα levels and are resistant to the growth-inhibitory effect of retinoic acid (11-14). Transfection experiments with ER-negative MDA-MB-231 breast cancer cells showed that stable ER expression in these cells restored the growth-inhibitory effect of retinoids with or without increased RARα expression (15, 16). In addition, the rate of retinoic acid metabolism appears to differ between retinoic acid-resistant, ER-negative and retinoic acid-sensitive, ER-positive breast cancer cells (17).

Examination of the correlation of RARα and ER expression in breast cancer yielded contradictory results. Two studies determined RARα expression in 135 primary breast cancers by Northern blot or immunohisto-chemistry and found a positive correlation with the ER-positive status (8, 18). Two other studies assessed RARα expression in a total of 103 tumors by *in situ* hybridization or immunohistochemistry and failed to detect a significant correlation with ER status (19, 20). There was also no correlation between RARβ, RARγ, and ER expression.

Most prospective studies of vitamin A/carotenoid intake and breast cancer have not found any associations (21, 22). One study observed no protective association between estimated overall vitamin A intake from evaluated foods but found a protective effect from eating carrots and spinach, which are rich in vitamin A and β-carotene (23). An analysis of over 90,000 premenopausal women in the Nurses' Health Study II also found no association between breast cancer risk and intake of vitamin A or carotenoids (24). However, both compounds may be beneficial for subgroups of women. In the Nurses' Health Study, intake of vitamin A significantly reduced the risk among cigarette smokers and high plasma levels of carotenoids were associated with reduced risk among women with high mammographic density (24, 25).

Vitamin C (ascorbic acid) has a broad spectrum of anti-oxidant activities due to its ability to react with numerous aqueous free radicals and ROS. Vitamin E (α-tocopherol) is the most abundant and effective lipid-soluble anti-oxidant capable of inhibiting lipid peroxidation. Both vitamin C and E have been postulated to have anti-carcinogenic properties through their anti-oxidative actions (3). An analysis of premenopausal women in the Nurses' Health Study II found no association between intake of vitamins C and E and breast cancer risk (24). An analysis of the Swedish Mammographic Screening Cohort showed that consumption of foods high in ascorbic acid may convey protection from breast cancer among overweight women (21). The same study observed no association between intake of vitamin E and breast cancer incidence.

VITAMINS B2, B6, B12, FOLATE AND OTHER NUTRIENTS INVOLVED IN ONE-CARBON METABOLISM

One-carbon metabolism is a network of biochemical reactions that transfer methyl groups ($-CH_3$), regulate DNA synthesis and repair and influence gene expression and integrity (26). The network is complex, involving 19 enzymes or carrier proteins, with various feedback loops and two main cycles, the folate and methionine cycles (27). Several nutrients are involved, including vitamin B2 (riboflavin), vitamin B6 (pyridoxal 5'-phosphate), vitamin B12, folate, methionine, homocysteine, betaine, and choline (Figure 3.4.4.). Key regulatory enzymes are 5,10-methylenetetrahydrofolate reductase (MTHFR), 5-methyltetrahydrofolate-homocysteine methyltransferase (MTR, also

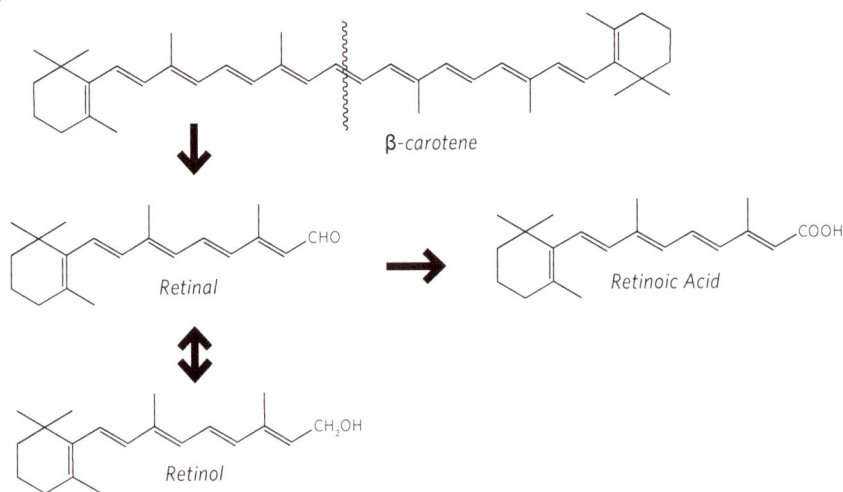

Figure 3.4.3. Structures of β-carotene and the natural retinoids

referred to as methionine synthase), thymidylate synthase (TYMS), cystathionine β-synthase and γ-cystathionase. First, MTHFR reduces 5,10-methylenetetrahydrofolate to 5-methyltetrahydrofolate, the plasma form of folate. Second, MTR, a vitamin B12-dependent enzyme, catalyzes the transfer of a –CH$_3$ group from 5-methyltetrahydrofolate to homocysteine to form methionine, and eventually S-adenosylmethionine (SAM), which is the universal methyl donor for methylation of DNA as well as RNA and protein. Third, by donating the –CH$_3$ group, SAM is converted to S-adenosylhomocysteine (SAH), which, in turn is converted to homocysteine by SAH hydrolase. Fourth, homocysteine is converted to cystathionine to form cysteine via the transsulfuration pathway, which is facilitated by two vitamin B6-dependent enzymes, cystathionine β-synthase (CBS) and γ-cystathionase. Finally, TYMS catalyzes both the conversion of 5,10-methylenetetrahydrofolate to dihydrofolate and of deoxyuridylate to deoxythymidylate, a rate-limiting nucleotide of DNA synthesis. Yet other enzymes are methionine synthase reductase (MTRR), cytosolic serine hydroxymethyltransferase (cSHMT), betaine-homocysteine methyltransferase (BHMT), dihydrofolate reductase (DHFR), and reduced folate carrier 1 (RFC1), the latter being responsible for intestinal absorption of dietary polyglutamyl folate, the predominant form of folate in the diet (28, 29). Thus, the one-carbon metabolic pathway is a network of enzymatic reactions, some of which are redundant. Detailed analysis suggests

that MTHFR and TYMS are the two main rate-limiting enzymes in the pathway (29). Since the imbalance of folate and the B vitamins alters nucleic acid metabolism and disrupts DNA synthesis, repair, and methylation, the deficiency of these nutrients may affect genetics and epigenetics and potentially promote carcinogenesis (30). Thus, investigators have hypothesized an inverse relation between intake of these nutrients and breast cancer risk. However, most epidemiologic studies have not supported this hypothesis.

Most studies of dietary or plasma levels of vitamin B6, methionine, homocysteine, choline, and betaine have not found any association with breast cancer risk (31-38). Homocysteine has been proposed as a risk factor for estrogen-induced hormonal cancer (39). Homocysteine is an immediate precursor for the biosynthesis of SAM and hyperhomocysteinemia results in impaired balance between homocysteine, SAM, and S-adenosylhomocysteine (SAH), with an accumulation of the latter (Figure 3.4.4.) (40-42). SAH is a strong non-competitive inhibitor of catechol-O-methyltransferase (COMT) and thereby decreases the methylation of catechol estrogens, such as 2-OHE$_2$ and 4-OHE$_2$ (43). Thus, hyperhomocysteinemia would indirectly lead to an accumulation of the procarcinogenic 4-OHE$_2$, which causes oxidative DNA damage. While the results of one case-control study supported the hypothesis of an association between hyperhomocysteinemia and increased

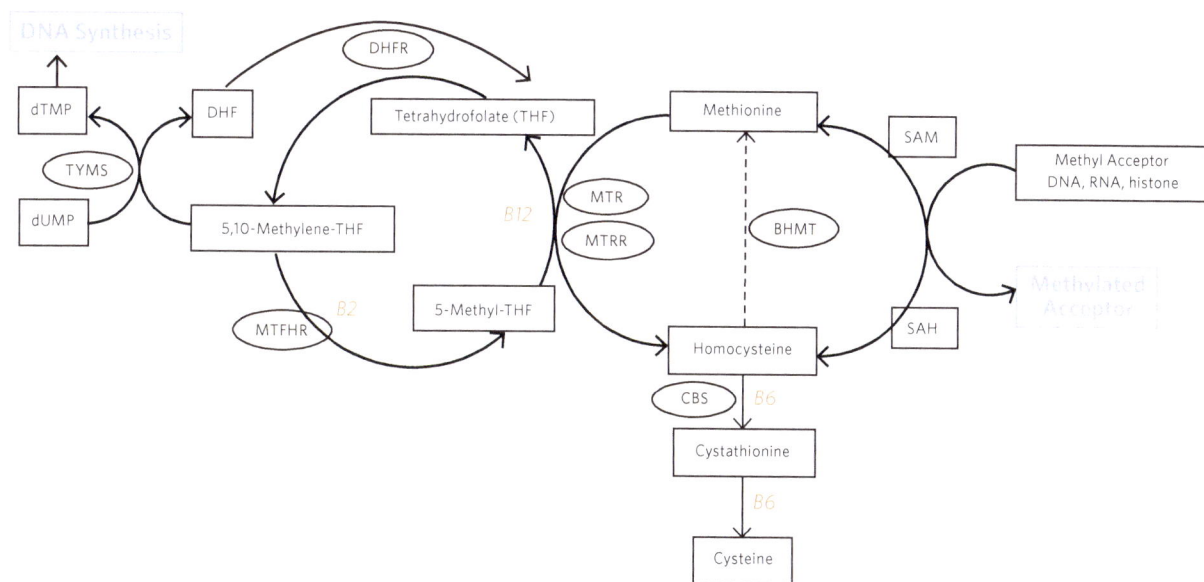

Figure 3.4.4. One-carbon metabolism mediated by folate, vitamin B2, vitamin B6, and vitamin B12.
The reactions are regulated by several enzymes including 5,10-methylenetetrahydrofolate reductase (MTHFR), 5-methyltetrahydrofolate-homocysteine methyltransferase (MTR, also referred to as methionine synthase), thymidylate synthase (TYMS), cystathionine β-synthase (CBS), methionine synthase reductase (MTRR), betaine-homocysteine methyltransferase (BHMT), dihydrofolate reductase (DHFR). SAM, S-adenosylmethionine; SAH, S-adenosylhomocysteine; DHF, dihydrofolate; dUMP, deoxyuridine monophosphate; dTMP, deoxythymidine monophosphate.

Table 3.4.1. Genetic Variants of Enzymes Involved in One-Carbon Metabolism Pathway

Gene	rs#	Nucleotides	Codon (amino acids)	Variant Function
MTHFR	1801133	C677T	Ala222Val	70% reduced
	1801131	A1298C	Glu429Ala	30% reduced
MTR	1805087	A2756G	Asp919Gly	unchanged
TYMS	5'-UTR	28-bp repeat		
	3'-UTR	6-bp del/ins		

bp, base pairs; del, deletion; ins, insertion; MTHFR, 5,10-methylenetetrahydrofolate reductase; MTR, 5-methyltetrahydrofolate-homocysteine methyltransferase; TYMS, thymidylate synthase; UTR, untranslated region

risk of breast cancer (32), two larger studies did not find an association (36, 37). Another study observed an association between breast cancer risk, folate and homocysteine levels, and COMT genotype (44). An increasing number of the variant COMT allele (rs4680; Val108Met) was associated with increased breast cancer risk in women with low levels of folate (p trend = 0.05) or high levels of homocysteine (p trend = 0.02).

Meta-analyses of prospective studies investigating the association between folate intake and breast cancer risk have found no evidence to support any overall risk association (45-47). The detailed analysis of over 90,000 premenopausal women in the Nurses' Health Study II found no association between breast cancer risk and intake of folate or other nutrients involved in one-carbon metabolism (24, 32). In contrast, dietary or plasma levels of vitamin B12 were inversely associated with breast cancer in several studies (32, 36-38).

All enzymes in the one-carbon metabolism pathway possess common genetic variants (Table 3.4.1.) and several have been examined for associations with breast cancer risk. A variant of the MTHFR gene (C677T; rs1801133; Ala222Val) is thermolabile, resulting in loss of activity of approximately 70% (48, 49). Individuals who are homozygous Val/Val have lower levels of genomic DNA methylation and higher levels of formylated tetrahydrofolates in red blood cells than Ala/Ala homozygotes (50). Individuals with the Val/Val genotype might be at increased risk of breast cancer especially at low levels of dietary folate (51, 52). However, a meta-analysis of 17 studies found no difference in breast cancer risk between Val/Val and Ala/Ala homozygotes (46). There was no evidence of an interaction between folate intake and MTFHR genotype on breast cancer risk. Another common MTHFR polymorphism (A1298C; rs1801131; Glu429Ala) was associated with a reduction in enzyme activity of approximately 30% (53). Results of epidemiologic studies of the association of this enzyme variant with breast cancer risk have been inconsistent (29, 52, 54, 55). A common variant of the MTR gene (A2756G; rs1805087; Asp919Gly) was not associated with altered activity and did not influence plasma homocysteine levels (56). Studies of the association

of this enzyme variant with breast cancer risk have been inconsistent (29, 54, 55, 57). Two potentially functional polymorphisms are present in the TYMS gene, a 28-bp tandem repeat in the 5'-untranslated enhanced region and a 6-bp deletion/insertion in the 3'-untranslated region (58, 59). Studies of the association of these polymorphisms with breast cancer risk have been inconsistent (29, 54, 60). Genetic variants of other enzymes involved in the regulation of the one-carbon metabolism (BHMT, cSHMT, DHFR, MTRR, RFC1) were not associated with breast cancer risk (29, 57).

VITAMIN D

The primary source of vitamin D is endogenously synthesized vitamin D3 in the skin. When solar ultraviolet B radiation (wavelength 290 to 315 nm) penetrates the skin, photolysis of cutaneous 7-dehydrocholesterol occurs to form previtamin D3, which is rapidly converted to vitamin D3 (cholecalciferol). Only a minor part (10 – 20%) of the vitamin D body stores comes from dietary sources. The fat-soluble vitamin is chemically described as a secosteroid because one if its four rings is open (Figure 3.4.4.). The vitamin exists in two forms, D2 and D3, which differ structurally in their side chains. Vitamin D2 (ergocalciferol) is manufactured through the ultraviolet irradiation of ergosterol from yeast and plant material. Both are used in over-the-counter vitamin D supplements, which typically contain 10 µg (400 IU) of vitamin D. The form available by prescription in the United States is vitamin D2 (61). The absence of an origin designation, i.e., D2 or D3, implies a mixture of both forms or that the precise origin is unknown. Vitamin D from the skin and diet is metabolized in the liver by vitamin D-25-hydroxylase (CYP27A1) to 25-hydroxyvitamin D, 25(OH)D, the major metabolite in the circulation. In the kidney, 25(OH)D is metabolized by the enzyme 25-hydroxyvitamin D-1α-hydroxylase (CYP27B1) to its active form, 1,25-dihydroxyvitamin D, 1,25-$(OH)_2$D (calcitriol), which interacts with the vitamin D receptor (VDR) to enhance intestinal calcium absorption and bone calcium and phosphorus resorption maintaining calcium and phosphorus levels in the blood (Figure 3.4.5.). The vitamin D and its metabolites are transported in the blood bound to vitamin D-binding protein (DBP, also known as "group-specific component", Gc), an alpha-2 globulin. Adequate plasma levels of 25(OH)D are >30 ng/mL, about 1000 times higher than that of 1,25-$(OH)_2$D, ~30 pg/mL (61). A dose of 2.5 µg (100 IU) of vitamin D supplement raises the plasma 25(OH)D level by ~1 ng/ml (62). 1,25-$(OH)_2$D induces the expression of the enzyme 25-hydroxyvitamin D-24-hydroxylase (CYP24), which catabolizes both 25(OH)D and 1,25-$(OH)_2$D into biologically inactive, water-soluble calcitroic acid.

People living at higher latitudes are at increased risk for colon, prostate, ovarian, breast and other cancers compared with people living at lower latitudes (63). These observations have led to the hypothesis that high levels of vitamin D might reduce the risk of breast cancer. This hypothesis is supported by experimental studies, which have shown anti-carcinogenic properties for both vitamin D and calcium (64). An important piece of evidence is the observation that CYP27B1 is expressed not only in the kidney but also in several non-renal cell lines, including human mammary epithelial cells (65, 66). Thus, mammary epithelial cells can utilize 25(OH)D as substrate to produce the hormonally active 1,25-$(OH)_2$D, which has been shown to bind to the VDR to increase differentiation and decrease proliferation of normal and malignant breast epithelial cells (65, 66). Exposure of MCF-7 cells to 1,25-$(OH)_2$D elevates expression of the cell cycle restricting gene p21, promotes the dephosphorylated form of the retinoblastoma protein and keeps the cell in the G0-G1 stage of the cell cycle (67). 1,25-$(OH)_2$D has also been shown to inhibit breast cancer cell growth by suppressing both the synthesis and biological activity of estrogens. The first effect is brought about by the ability of 1,25-$(OH)_2$D to down-regulate the expression of aromatase in breast cancer cells. This occurs by two mechanisms, a direct repression of aromatase transcription via promoter II through the vitamin D-response elements identified in this promoter and an indirect suppression

by reducing the level of prostaglandins, which are known stimulators of aromatase in breast cancer cells (68). Secondly, 1,25-$(OH)_2$D inhibits estrogen signaling by down-regulating the expression of estrogen receptor α in breast cancer cells (69, 70). In view of these combined biological effects, the vitamin D status as determined by dietary and/or supplemental intakes of vitamin D and circulating levels of 25(OH)D may influence the risk of breast cancer.

Epidemiological studies of dietary and supplemental vitamin D intake in relation to breast cancer risk have yielded inconsistent results (64, 71). For example, pooled data for 980 women showed that the highest vitamin D intake, compared with the lowest, correlated with a 50% lower risk of breast cancer (72). In contrast, the Nurses' Health Study (3172 cases, 88,691 controls) and the Cancer Prevention Study II Nutrition Cohort (2855 cases, 68,567 controls) observed no association between vitamin D intake and breast cancer risk in postmenopausal women (73, 74). However, in premenopausal women the Nurses' Health Study showed an inverse association between vitamin D intake and breast cancer risk (74). Interestingly, the Nurses' Health Study II, which examined dietary vitamin D intake during adolescence found no association with breast cancer risk (75). Similarly, the Women's Lifestyle and Health Cohort Study of over 40,000 women aged 30 to 50 years did not observe any association with dietary vitamin D intake or supplementary multivitamin use (76). There was also no association of breast cancer risk with various solar exposure variables, including sun sensitivity, annual number of sunburns, time spent on sunbathing vacations, or solarium use at any age period of exposure.

Cohort and case-control studies also reported conflicting evidence for an association of plasma levels of vitamin D metabolites with breast cancer risk (71). Both 25(OH)D and 1,25-$(OH)_2$D have been examined, the former

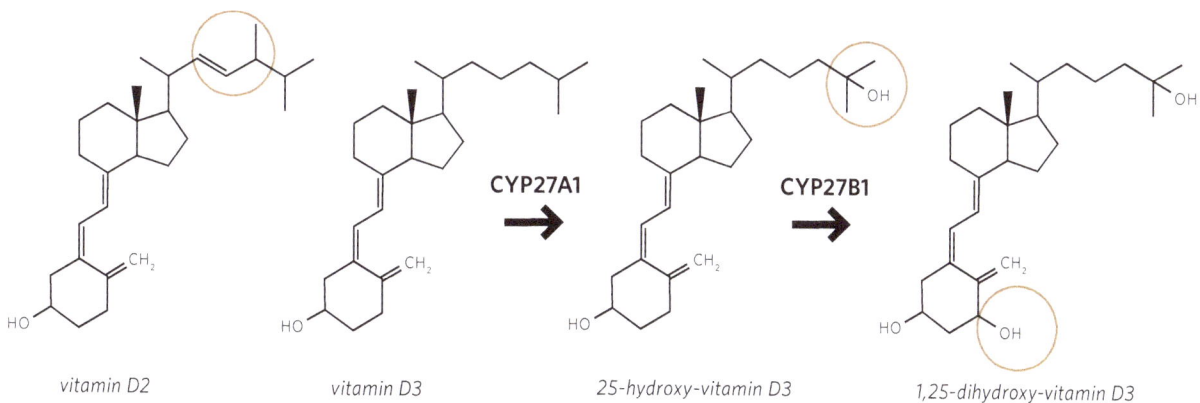

Figure 3.4.5. Structures of vitamin D and key metabolites

vitamin D2   vitamin D3   CYP27A1 →   25-hydroxy-vitamin D3   CYP27B1 →   1,25-dihydroxy-vitamin D3

Table 3.4.2. Genetic Variants of Proteins involved in Vitamin D Transport, Action, and Metabolism

Gene	rs#	Nucleotides	Codon/Region	Variant Allele Frequency	Function Altered
DBP	7041	T/G	Asp416Glu	0.43	25(OH)D binding
	4588	C/A	Lys420Thr	0.24	25(OH)D binding
VDR	1544410	Bsml	3' UTR	African American 0.29 Caucasian 0.41 Hispanic 0.41	VDR mRNA stability
	2228570	Fokl T/C	5' promoter region	African American 0.23 Caucasian 0.41 Hispanic 0.25	VDR transcriptional activity

DBP, vitamin D-binding protein; UTR, untranslated region; VDR, vitamin D receptor

reflecting the vitamin D status from sun exposure, dietary intake, and vitamin supplements combined and the latter representing the active metabolite. In some (77, 78), but not all (79) studies, lower plasma 25(OH)D levels have been observed in women with breast cancer compared with healthy controls. A case-control study of Danish women found lower plasma 25(OH)D levels for cases than controls (80). Compared with the lowest tertile of 25(OH)D levels, the risk of breast cancer was significantly reduced among women in the highest tertile (relative risk, 0.52; 95% CI, 0.32 – 0.85). No association between prediagnostic 25(OH)D and 1,25-(OH)$_2$D levels and risk of developing breast cancer was found in an analysis from the Prostate, Lung, Colorectal, and Ovarian Cancer Screening Trial (81). Although statistically insignificant, an inverse association between plasma 25(OH)D and 1,25-(OH)$_2$D levels and breast cancer risk was observed in a case-control study nested in the Nurses' Health Study (82). Yet another study found a protective effect of 1,25-(OH)$_2$D for breast cancer in Caucasian women but no effect of 25(OH)D (83).

Transport and action of vitamin D are affected by DBP and VDR and polymorphisms in the respective genes (Table 3.4.2.) may modify the anti-carcinogenic effects of vitamin D and thereby could potentially influence breast cancer risk. The vitamin D binding protein consists of three common phenotypic alleles, Gc1s, Gc1f (combined as Gc1) and Gc2, differing by combinations of two non-synonymous single nucleotide polymorphisms, rs4588 and rs7041. Compared with Gc1, Gc2 alleles were associated with lower levels of DBP and plasma 25(OH)D (84). Although homozygote carriers of the Gc2 allele have low plasma 25(OH)D levels, one study showed a significantly reduced risk of breast cancer in postmenopausal women with the Gc2-2 genotype (85). At least 25 polymorphisms of the VDR gene have been identified, many of which occur at high frequency and influence receptor affinity and transcriptional response (86). The rs2228570 polymorphism (T/C transition; Fokl) in exon 2 eliminates the first ATG translation initiation site and allows a second one nine bp downstream to be used resulting in a protein (encoded by the C allele, also known as the F allele) that is three amino acids shorter than the

protein encoded by the T allele (also known as the f allele). The longer protein is less transcriptionally active than the shorter protein (87). A large nested case-control study found a positive association between the ff genotype of Fokl and breast cancer risk (88), whereas four smaller studies found no association with this genotype (64). A pooled analysis of six prospective studies in the National Cancer Institute Breast and Prostate Cancer Cohort Consortium found only a weak association (OR 1.10; 95% CI 0.98 – 1.24 (89). Several polymorphisms in the 3' end of the VDR gene, including Bsml (rs1544410, present in two alleles denoted B and b), Taql, and Apal, occur in strong linkage disequilibrium and are linked with a poly(A) microsatellite repeat that may influence VDR mRNA stability (86). Six of 11 studies have reported higher breast cancer risk associated with the rs1544410 (Bsml) bb genotype (64). However, the pooled analysis of the National Cancer Institute Breast and Prostate Cancer Cohort Consortium observed no association between rs1544410 and breast cancer risk (89). There are several potential explanations for the inconsistencies in findings for these common polymorphisms, such as the small size of many of the individual studies, ethnic variation in allele frequency (Table 3.4.2.), and environmental factors including vitamin D intake (89). In addition to having vitamin D3 25-hydroxylase activity, CYP27A1 also catalyzes the first step in the oxidation of the side chain of various sterol intermediates. Genetic defects in CYP27A1 are the cause of cerebrotendinous xanthomatosis (CTX), a rare sterol storage disorder characterized clinically by progressive neurologic dysfunction, premature atherosclerosis, and cataracts. Genetic defects in CYP27B1 are a cause of vitamin D-dependent rickets type 1 (VDDR-1), an autosomal recessive disease characterized by muscle weakness and early onset of rickets with hypocalcemia (61).

References

1. Slesinski MJ, Subar AF, Kahle LL. *Dietary intake of fat, fiber and other nutrients is related to the use of vitamin and mineral supplements in the United States: The 1992 National Health Interview Survey.* J Nutr. 1996;126:3001-8.

2. Britton G. *Structure and properties of carotenoids in relation to function.* FASEB J. 1995;9:1551-8.

3. Frei B. *Reactive oxygen species and antioxidant vitamins: Mechanisms of action.* Am J Med. 1994;97; suppl 3A:5S - 13S.

4. Nishino H, Tokuda H, Murakoshi M, Satomi Y, Masuda M, Onozuka M, et al. *Cancer prevention by natural carotenoids.* BioFactors. 2000;13:89-94.

5. Chambon P. *A decade of molecular biology of retinoic acid receptors.* FASEB J. 1996;10:940-54.

6. Elgort MG, Zou A, Marschke KB, Allegretto EA. *Estrogen and estrogen receptor antagonists stimulate transcription from the human retinoic acid receptor-a 1 promoter via a novel sequence.* Mol Endocrinol. 1996;10:477-87.

7. Rishi AK, Shao ZM, Baumann RG, Li XS, Sheikh S, Kimura S, et al. *Estradiol regulation of the human retinoic acid receptor a gene in human breast carcinoma cells is mediated via an imperfect half-palindromic estrogen response element and Sp1 motifs.* Cancer Res. 1995;55:4999-5006.

8. Roman SD, Ormandy CJ, Manning DL, Blamey RW, Nicholson RI, Sutherland RL, et al. *Estradiol induction of retinoic acid receptors in human breast cancer cells.* Cancer Res. 1993;53:5940-5.

9. Sheikh MS, Shao ZM, Li XS, Dawson M, Jetten AM, Wu S, et al. *Retinoid-resistant estrogen receptor-negative human breast carcinoma cells transfected with retinoic acid receptor-alpha acquire sensitivity to growth inhibition by retinoids.* J Biol Chem. 1994;269:21440-7.

10. van der Leede BJ, Folkers GE, van den Brink CE, van der Saag PT, van der Burg B. *Retinoic acid receptor alpha 1 isoform is induced by estradiol and confers retinoic acid sensitivity in human breast cancer cells.* Mol Cell Endocrinol. 1995;109:77-86.

11. Fitzgerald P, Teng M, Chandraratna RAS, Heyman RA, Allegretto EA. *Retinoic acid receptor a expression correlates with retinoid-induced growth inhibition of human breast cancer cells regardless of estrogen receptor status.* Cancer Res. 1997;57:2642-50.

12. Rishi AK, Gerald TM, Shao Z, Li X, Baumann RG, Dawson MI. *Regulation of the human retinoic acid receptor a gene in the estrogen receptor-negative human breast carcinoma cell lines SKBR-3 and MDA-MB-4351.* Cancer Res. 1996;56:5246-52.

13. Roman SD, Clarke CL, Hall RE, Alexander IE, Sutherland RL. *Expression and regulation of retinoic acid receptors in human breast cancer cells.* Cancer Res. 1992;52:2236-42.

14. van der Burg B, van der Leede BM, Kwakkenbos-Isbrucker L, Salverda S, de Laat SW, van der Saag PT. *Retinoic acid resistance of estradiol-independent breast cancer cells coincides with diminished retinoic acid receptor function.* Mol Cell Endocrinol. 1993;91:149-57.

15. Rosenauer A, Nervi C, Davison K, Lamph WW, Mader S, Miller WH. *Estrogen receptor expression activates the transcriptional and growth-inhibitory response to retinoids without enhanced retinoic acid receptor a expression.* Cancer Res. 1998;58:5110-6.

16. Sheikh MD, Shao ZM, Chen JC, Hussain A, Jetten AM, Fontana JA. *Estrogen receptor-negative breast cancer cells transfected with the estrogen receptor exhibit increased RAR alpha gene expression and sensitivity to growth inhibition by retinoic acid.* J Cell Biochem. 1993;53:394-404.

17. van der Leede BM, van den Brink CE, Pijnappel WW, Sonneveld E, van der Saag PT, van der Burg B. *Autoinduction of retinoic acid metabolism to polar derivatives with decreased biological activity in retinoic acid-sensitive, but not in retinoic acid-resistant human breast cancer cells.* J Biol Chem. 1997;272:17921-8.

18. Han QX, Allegretto EA, Shao ZM, Kute TE, Ordonez J, Aisner SC, et al. *Elevated expression of retinoic acid receptor-alpha (RAR alpha) in estrogen-receptor positive breast carcinomas as detected by immunohistochemistry.* Diag Mol Pathol. 1997;6:42-8.

19. van der Leede BM, Geertzema J, T.M. V, Decimo D, Lutz Y, van der Saag PT, et al. *Immunohistochemical analysis of retinoic acid receptor-alpha in human breast tumors: retinoic acid receptor-alpha expression correlates with proliferative activity.* Am J Pathol. 1996;148:1905-14.

20. Xu X-c, Sneige N, Liu X, Nandagiri R, Lee JJ, Lukmanji F, et al. *Progressive decrease in nuclear retinoic acid receptor b messenger RNA level during breast carcinogenesis.* Cancer Res. 1997;57:4992-6.

21. Michels KB, Holmberg L, Bergkvist L, Ljung H, Bruce A, Wolk A. *Dietary antioxidant vitamins, retinol, and breast cancer incidence in a cohort of Swedish women.* Int J Cancer. 2001;91:563-7.

22. Terry P, Jain M, Miller AB, Howe GR, Rohan TE. *Dietary carotenoids and risk of breast cancer.* Am J Clin Nutr. 2002;76:883-8.

23. Longnecker MP, Newcomb PA, Mittendorf R, Greenberg ER, Willet WC. *Intake of carrots, spinach, and supplements containing vitamin A in relation to risk of breast cancer.* Cancer Epidemiol Biomarkers Prev. 1997;6:887-92.

24. Cho E, Spiegelman D, Hunter DJ, Chen WY, Zhang SM, Colditz GA, et al. *Premenopausal intakes of vitamins A, C, and E, folate, and carotenoids, and risk of breast cancer.* Cancer Epidemiol Biomarkers Prev. 2003;12:713-20.

25. Tamimi RM, Colditz GA, Hankinson SE. *Circulating carotenoids, mammographic density, and subsequent risk of breast cancer.* Cancer Res. 2009;69:9323-9.

26. Mason JB. B*iomarkers of nutrient exposure and status in one-carbon (methyl) metabolism.* J Nutr. 2003;133:941S-7S.

27. Thomas DC, Conti DV, Baurley J, Nijhout F, Reed M, Ulrich CM. *Use of pathway information in molecular epidemiology.* Hum Genomics. 2009;4:21-42.

28. Leclerc D, Odievre M, Wu Q, Wilson A, Huizenga JJ, Rozen R, et al. *Molecular cloning, expression and physical mapping of the human methionine synthase reductase gene.* Gene. 1999;240:75-88.

29. Xu X, Gammon MD, Zhang H, Wetmur JG, Rao M, Teitelbaum SL, et al. *Polymorphisms of one-carbon-metabolizing genes and risk of breast cancer in a population-based study.* Carcinogenesis. 2007;28:1504-9.

30. Stover PJ. *Physiology of folate and vitamin B12 in health and disease.* Nutrition Rev. 2004;62:S3-S12.

31. Cho E, Holmes M, Hankinson SE, Willet WC. *Nutrients involved in one-carbon metabolism and risk of breast cancer among premenopausal women.* Cancer Epidemiol Biomarkers Prev. 2007;16:2787-90.

32. Chou YC, Lee MS, Wu MH, Shih HL, Yang T, Yu CP, et al. *Plasma homocysteine as a metabolic risk factor for breast cancer: findings from a case-control study in Taiwan.* Breast Cancer Res Treat. 2007;101:199-205.

33. Feigelson HS, Jonas CR, Robertson AS, McCullough ML, Thun MJ, Calle EE. *Alcohol, folate, methionine, and risk of incident breast cancer in the American Cancer Society Cancer Prevention Study II Nutrition Cohort.* Cancer Epidemiol Biomarkers Prev. 2003;12:161-4.

34. Lajous M, Lazcano-Ponce E, Hernandez-Avila M, Willet W, Romieu I. *Folate, vitamin B6, and vitamin B12 intake and the risk of breast cancer among Mexican women.* Cancer Epidemiol Biomarkers Prev. 2006;15:443-8.

35. Rohan TE, Jain MG, Howe GR, Miller AB. *Dietary folate consumption and breast cancer risk.* J Natl Cancer Inst. 2000;92:266-9.

36. Wu K, Helzlsouer KJ, Comstock GW, Hoffman SC, Nadeau MR, Selhub J. *A prospective study on folate, B12, and pyridoxal 5'-phosphate (B6) and breast cancer.* Cancer Epidemiol Biomarkers Prev. 1999;8:209-17.

37. Zhang S, Willet WC, Selhub J, Hunter DJ, Giovannucci EL, Holmes MD, et al. *Plasma folate, vitamin B6, vitamin B12, homocysteine, and risk of breast cancer.* J Natl Cancer Inst. 2003;95:373-80.

38. Lajous M, Romieu I, Sabia S, Boutron-Ruault M, Clavel-Chapelon F. *Folate, vitamin B12 and postmenopausal breast cancer in a prospective study of French women.* Cancer Causes Control. 2006;17:1209-13.

39. Zhu BT. Medical hypothesis: *Hyperhomocysteinemia is a risk factor for estrogen-induced hormonal cancer.* Int J Oncol. 2003;22:499-508.

40. Hoffman DR, Marion DW, Cornatzer WE, Duerre JA. *S-adenosylmethionine and S-adenosylhomocysteine metabolism in isolated rat liver.* J Biol Chem. 1980;255:10822-7.

41. Yi P, Melnyk S, Pogribna M, Pogribny IP, Hine RJ, James SJ. *Increase in plasma homocysteine associated with parallel increases in plasma S-adenosylhomocysteine and lymphocyte DNA hypomethylation.* J Biol Chem. 2000;275:29318-23.

42. Young IS, Woodside JV. *Folate and homocysteine.* Curr Opin Clin Nutr Metab Care. 2000;3:427-32.

43. Zhu BT. *On the mechanism of homocysteine pathophysiology and pathogenesis: a unifying hypothesis.* Histol Histopathol. 2002;17:1283-91.

44. Goodman JE, Lavigne JA, Wu K, Helzlsouer KJ, Strickland PT, Selhub J, et al. *COMT genotype, micronutrients in the folate metabolic pathway and breast cancer risk.* Carcinogenesis. 2001;22:1661-5.

45. Larsson SC, Giovannucci E, Wolk A. *Folate and risk of breast cancer: A meta-analysis.* J Natl Cancer Inst. 2007;99:64-76.

46. Lewis SJ, Harbord RM, Harris R, Smith GD. *Meta-analysis of observational and genetic association studies of folate intakes or levels and breast cancer risk.* J Natl Cancer Inst. 2006;98:1607-22.

47. Zhang S, Hunter DJ, Hankinson SE, Giovannucci EL, Rosner BA,

Colditz GA, et al. *A prospective study of folate intake and the risk of breast cancer.* JAMA. 1999;281:1632-7.

48. Frosst P, Blom HJ, Goyette P, Sheppard CA, Matthews RG, Boers GJH, et al. *A candidate genetic risk factor for vascular disease: a common mutation in methylenetetrahydrofolate reductase.* Nature Genet. 1995;10:111-3.

49. Jacques PF, Bostom AG, Williams RR, Ellison RC, Eckfeldt JH, Rosenberg IH, et al. *Relation between folate status, a common mutation in methylenetetrahydrofolate reductase, and plasma homocysteine concentrations.* Circulation. 1996;93:7-9.

50. Friso S, Choi SW, Girelli D, Mason JB, Dolnikowski GG, Bagley PJ, et al. *A common mutation in the 5,10-methylenetetrahydrofolate reductase gene affects genomic DNA methylation through an interaction with folate status.* Proc Natl Acad Sci. 2002;99:5606-11.

51. Chen J, Gammon MD, Chan W, Palomeque C, Wetmur JG, Kabat GC, et al. *One-carbon metabolism, MTHFR polymorphisms, and risk of breast cancer.* Cancer Res. 2005;65:1606-14.

52. Shrubsole MJ, Gao YT, Cai Q, Shu XO, Dai Q, Hebert JR, et al. *MTHFR polymorphisms, dietary folate intake, and breast cancer risk: Results from the Shanghai breast cancer study.* Cancer Epidemiol Biomarkers Prev. 2004;13:190-6.

53. Weisberg IS, Jacques PF, Selhub J, Bostom AG, Chen Z, Ellison RC, et al. *The 1298A→C polymorphism in methylenetetrahydrofolate reductase (MTHFR): in vitro expression and association with homocysteine.* Atherosclerosis. 2001;156:409-15.

54. Justenhoven C, Hamann U, Pierl CB, Rabstein S, Pesch B, Harth V, et al. *One-carbon metabolism and breast cancer risk: No association of MTHFR, MTR, and TYMS polymorphisms in the GENICA study from Germany.* Cancer Epidemiol Biomarkers Prev. 2005;14:3015-8.

55. Platek ME, Shields PG, Marian C, McCann SE, Bonner MR, Nie J, et al. *Alcohol consumption and genetic variation in methylenetetrahydrofolate reductase and 5-methyltetrahydrofolate-homocysteine methyltransferase in relation to breast cancer risk.* Cancer Epidemiol Biomarkers Prev. 2009;18:2453-9.

56. Klerk M, Lievers KJA, Kluijtmans LAJ, Blom HJ, Den Heijer M, Schouten EG, et al. *The 2756A>G variant in the gene encoding methionine synthase: its relation with plasma homocysteine levels and risk of coronary heart disease in a Dutch case-control study.* Thrombosis Res. 2003;110:87-91.

57. Shrubsole MJ, Gao YT, Cai Q, Shu XO, Dai Q, Jin F, et al. *MTR and MTRR polymorphisms, dietary intake, and breast cancer risk.* Cancer Epidemiol Biomarkers Prev. 2006;15:586-8.

58. Mandola MV, Stoehlmacher J, Zhang W, Groshen S, Yu MC, Iqbal S, et al. *A 6 bp polymorphism in the thymidylate synthase gene causes message instability and is associated with decreased intratumoral TS mRNA levels.* Pharmacogenetics. 2004;14:319-27.

59. Pullarkat ST, Stoehlmacher J, Ghaderi V, Xiong YP, Ingles SA, Sherrod A, et al. *Thymidylate synthase gene polymorphism determines response and toxicity of 5-FU chemotherapy.* Pharmacogenomics J. 2001;1:65-70.

60. Zhai X, Gao J, Hu Z, Tang J, Qin J, Wang S, et al. *Polymorphisms in thymidylate synthase gene and susceptibility to breast cancer in a Chinese population: a case-control analysis.* BMC Cancer. 2006;6:138.

61. Holick MF. *Vitamin D deficiency.* N Engl J Med. 2007;357:266-81.

62. Holick MF. *Vitamin D and sunlight: strategies for cancer prevention and other health benefits.* Clin J Am Soc Nephrol. 2008;3:1548-54.

63. Studzinski GP, Moore DC. *Sunlight -- can it prevent as well as cause cancer?* Cancer Res. 1995;55:4014-22.

64. Cui Y, Rohan TE. *Vitamin D, calcium, and breast cancer risk: A review.* Cancer Epidemiol Biomarkers Prev. 2006;15:1427-37.

65. Kemmis CM, Salvador SM, Smith KM, Welsh J. *Human mammary epithelial cells express CYP27B1 and are growth inhibited by 25-hydroxyvitamin D-3, the major circulating form of vitamin D-3.* J Nutr. 2006;136:887-92.

66. Segersten U, Holm PK, Bjorklund P, Hessman O, Nordgren H, Binderup L, et al. *25-hydroxyvitamin D3 1alpha-hydroxylase expression in breast cancer and use of non-1alpha-hydroxylated vitamin D analogue.* Breast Cancer Res. 2005;7:R980-R6.

67. Jensen SS, Madsen MW, Lukas J, Binderup L, Bartek J. *Inhibitory effects of 1 alpha,25-dihydroxyvitamin D3 on the G1-S phase-controlling machinery.* Mol Endocrinol. 2001;15:1370-80.

68. Krishnan AV, Swami S, Peng L, Wang J, Moreno J, Feldman D. *Tissue-selective regulation of aromatase expression by calcitriol: implications for breast cancer therapy.* Endocrinology. 2010;151:32-42.

69. Stoica A, Saceda M, Fakhro A, Solomon HB, Fenster BD, Martin MB. *Regulation of estrogen receptor-alpha gene expression by 1,25-dihydroxyvitamin D in MCF-7 cells.* J Cell Biochem. 1999;75:640-51.

70. Swami S, Krishnan AV, Feldman D. *1alpha,25-dihydroxyvitamin D3 down-regulates estrogen receptor abundance and suppresses estrogen actions in MCF-7 human breast cancer cells.* Clin Cancer Res. 2000;6:3371-9.

71. Bertone-Johnson ER. *Vitamin D and breast cancer.* Ann Epidemiol. 2009;19:462-7.

72. Garland CF, Garland FC, Gorham FC, Lipkin M, Mohr SB, Newmark H, et al. *The role of vitamin D in cancer prevention.* Am J Public Health. 2006;96:252-61.

73. McCullough ML, Rodriguez C, Diver WR, Spencer Feigelson H, Stevens VL, Thun MJ, et al. *Dairy, calcium, and vitamin D intake and postmenopausal breast cancer risk in the Cancer Prevention Study II Nutrition Cohort.* Cancer Epidemiol Biomarkers Prev. 2005;14:2898-904.

74. Shin MH, Holmes MD, Hankinson SE, Wu K, Colditz GA, Willett WC. *Intake of dairy products, calcium, and vitamin D and risk of breast cancer.* J Natl Cancer Inst. 2002;94:1301-11.

75. Frazier AL, Li L, Cho E, Willett WC, Colditz GA. *Adolescent diet and risk of breast cancer.* Cancer Causes Control. 2004;15:73-82.

76. Kuper H, Yang L, Sandin S, Lof M, Adami H, Weiderpass E. *Prospective study of solar exposure, dietary vitamin D intake, and risk of breast cancer among middle-aged women.* Cancer Epidemiol Biomarkers Prev. 2009;18:2558-61.

77. Abbas S, Linseisen J, Slanger T, Kropp S, Mutschelknauss EJ, Flesch-Janys D, et al. *Serum 25-hydroxyvitamin D and risk of post-menopausal breast cancer -- results of a large case-control study.* Carcinogenesis. 2008;29:93-9.

78. Lowe LC, Guy M, Mansi JL, Peckitt C, Bliss J, Given Wilson R, et al. *Plasma 25-hydroxy vitamin D concentrations, vitamin D receptor genotype and breast cancer risk in a UK Caucasian population.* Eur J Cancer. 2005;41:1164-9.

79. Chlebowski RT, Johnson KC, Kooperberg C, Pettinger M, Wactawski-Wende J, Rohan T, et al. *Calcium plus vitamin D supplementation and the risk of breast cancer.* J Natl Cancer Inst. 2007;100:1581-91.

80. Rejnmark L, Tietze A, Vestergaard P, Buhl L, Lehbrink M, Heickendorff L, et al. *Reduced prediagnostic 25-hydroxyvitamin D levels in women with breast cancer: A nested case-control study.* Cancer Epidemiol Biomarkers Prev. 2009;18:2655-60.

81. Freedman DM, Chang SC, Falk RT, Purdue MP, Huang WY, McCarty CA, et al. *Serum levels of vitamin D metabolites and breast cancer risk in the Prostate, Lung, Colorectal, and Ovarian cancer screening tool.* Cancer Epidemiol Biomarkers Prev. 2008;17:889-94.

82. Bertone-Johnson ER, Chen WY, Holick MF, Hollis BW, Colditz GA, Willett WC, et al. *Plasma 25-hydroxyvitamin D and 1,25-dihydroxyvitamin D and risk of breast cancer.* Cancer Epidemiol Biomarkers Prev. 2005;14:1991-7.

83. Janowsky EC, Lester GE, Weinberg CR, Millikan RC, Schildkraut JM, Garrett PA, et al. *Association between low levels of 1,25-dihydroxyvitamin D and breast cancer risk.* Public Health Nutrition. 1999;2:283-91.

84. Lauridsen AL, Vestergaard P, Hermann AP, Brot C, Heickendorff L, Mosekilde L, et al. *Plasma concentrations of 25-hydroxy-vitamin D and 1,25-dihydroxy-vitamin D are related to the phenotype of Gc (vitamin D-binding protein): a cross-sectional study on 595 early postmenopausal women.* Calcif Tissue Int. 2005;77:15-22.

85. Abbas S, Linseisen J, Slanger T, Kropp S, Mutschelknauss EJ, Flesch-Janys D, et al. *The Gc2 allele of the vitamin D binding protein is associated with a decreased postmenopausal breast cancer risk, independent of the vitamin D status.* Cancer Epidemiol Biomarkers Prev. 2008;17:1339-43.

86. Uitterlinden AG, Fang Y, van Meurs JBJ, van Leeuwen H, Pols HAP. *Vitamin D receptor gene polymorphisms in relation to vitamin D related disease states.* J Steroid Biochem Mol Biol. 2004;89-90:187-93.

87. Arai H, Miyamoto KI, Taketani Y, Yamamoto H, Iemori Y, Morita K, et al. *A vitamin D receptor gene polymorphism in the translation initiation codon: Effect on protein activity and relation to bone mineral density in Japanese women.* J Bone Miner Res. 1997;12:915-21.

88. Chen WY, Bertone-Johnson ER, Hunter DJ, Willett WC, Hankinson SE. *Associations between polymorphisms in the vitamin D receptor and breast cancer risk.* Cancer Epidemiol Biomarkers Prev. 2005;14:2335-9.

89. McKay JD, McCullough ML, Ziegler RG, Kraft P, Saltzman BS, Riboli E, et al. *Vitamin D receptor polymorphisms and breast cancer risk: Results from the National Cancer Institute Breast and Prostate Cancer Cohort Consortium.* Cancer Epidemiol Biomarkers Prev. 2009;18:297-305.

(COMT), UDP-glucuronosyltransferases, and sulfonyl-transferases (13, 14). Interestingly, COMT inactivates tea catechins by O-methylation in a reaction that is faster than that of catechol estrogens and catecholamines (15). EGCG also inhibits methyltransferases, such as COMT and 5-cytosine DNA methyltransferase (DNMT). Thus, EGCG may inhibit COMT-catalyzed methylation of endogenous and exogenous compounds while the inhibition of DNMT may cause reactivation of methylation-silenced genes in cancer cells (13, 16). Moreover, tea polyphenols can induce phase II enzymes, such as glutathione S-transferases, which may enhance the detoxification of carcinogens (17). Finally, physiologic concentrations of EGCG were shown to inhibit telomerase, a key enzyme in cell immortalization (18).

3.4.5. COFFEE AND TEA

Coffee and tea are widely consumed around the world. Green tea is popular in Asia, whereas black tea is favored in the United States and Europe. Both coffee and tea are a source of numerous biochemically active substances. Coffee contains caffeine, phenols (e.g., chlorogenic acid), phytoestrogens (e.g., flavonoids, lignans), and other phytonutrients (e.g., tocopherols). Caffeine (1,3,7-trimethyl-xanthine) is metabolized primarily by CYP1A2 and CYP2E1 to paraxanthine (1,7-dimethylxanthine), theobromine (3,7-dimethylxanthine), and theophylline (1,3-dimethyl-xanthine), which is further metabolized and excreted in the urine (1). Caffeine, theobromine, and xanthine have been shown to have a quenching effect on the production of hydroxyl radicals, as well as on oxidative DNA breakage by hydroxyl radicals (2). In addition to its antioxidant effect, caffeine has anticarcinogenic properties that include antimetastatic effects and the ability to inhibit cell proliferation and enhance apoptosis (3-5). The principal active constituents in green tea are polyphenols, mostly catechins, consisting of a mixture of epicatechin, epigal-locatechin, epicatechin gallate, and epigallocatechin gallate (Figure 3.4.5.). In the processing of black tea, the green tea leaves are dried and crushed upon harvesting to encourage oxidation, which converts the catechins to other polyphenols (e.g., theaflavins, thearubigins; Figure 3.4.6.) that yield the characteristic red-brown color. The catechin content of black tea is one-third that of green tea (6, 7).

Tea polyphenols possess antioxidant and anticarcino-genic activities. Many mechanisms have been proposed for the inhibition of carcinogenesis by tea polyphenols, including the modulation of signal transduction pathways that leads to the inhibition of cell proliferation and trans-formation, induction of apoptosis of preneoplastic and neoplastic cells, as well as inhibition of tumor invasion and angiogenesis (8-11). Epigallocatechin gallate (EGCG) is the major polyphenol in green tea and believed to be the most active anticarcinogenic compound (11). EGCG consists of a meta-5,7-dihydroxyl substituted A ring and trihydroxy phenol structures on both the B and D rings (Figure 3.4.5.). It is a potent antioxidant with a short half-life (about 2 hr in cell culture) (12). After oral absorption, EGCG and other tea catechins undergo extensive methylation, glucuronidation, and sulfation carried out by catechol-O-methyltransferase

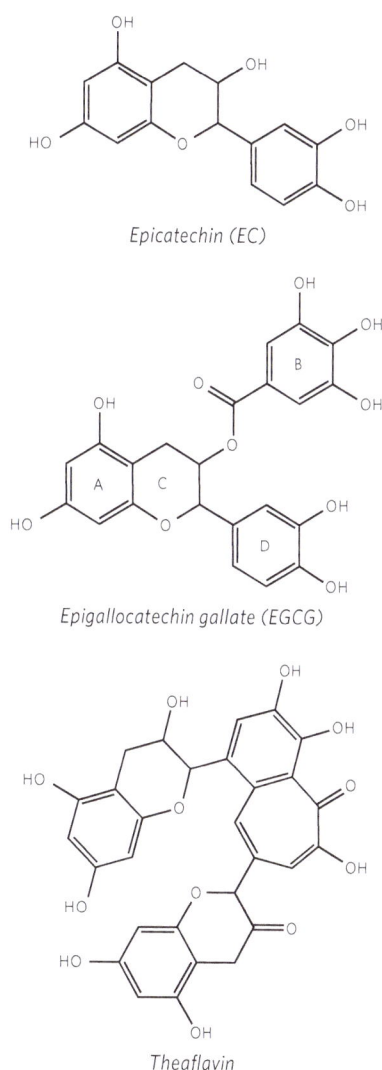

Epicatechin (EC)

Epigallocatechin gallate (EGCG)

Theaflavin

Figure 3.4.6. Chemical structure of polyphenols in green and black tea

The complexity of the chemical constituents found in coffee and tea makes it difficult to isolate any potentially beneficial compound. In addition, the interpretation of epidemiological data is hampered by inter-individual variability in bioavailability and biotransformation of active compounds in both beverages [19]. Some epidemiological studies have shown both beverages to confer protection against breast cancer [20-24]. For example, analysis of the Nurses' Health Study cohort showed a trend for decreased risk associated with increased coffee consumption [25]. However, the majority of studies did not find any effect associated with coffee consumption [26-30]. A meta-analysis of 13 studies examining tea consumption in eight countries revealed differences between green and black tea [31]. There was an inverse association between the consumption of green tea and breast cancer risk (OR 0.78; 95% CI 0.61 – 0.98) [31]. The combined results from eight case-control studies showed a modest inverse association between consumption and risk of breast cancer (OR 0.91; 95% CI 0.84 – 0.98), whereas combined results in five cohort studies showed a modest increase in risk associated with higher levels of black tea intake (OR 1.15; 95% CI 1.02 – 1.31). A study in Asian Americans revealed a reduced risk associated with drinking green tea but not black tea [32]. The reduction in risk was strongest in women who had the low activity COMT alleles, suggesting that these individuals were less efficient in inactivating tea catechins and therefore derive the most benefit from these compounds [33].

References

1. Gu L, Gonzalez FJ, Kalow W, et al. *Biotransformation of caffeine, paraxanthine, theobromine and theophylline by cDNA-expressed human CYP1A2 and CYP2E1.* Pharmacogenetics. 1992;2:73-77.

2. Devasagayam TPA, Kamat JP, Mohan H, et al. *Caffeine as an antioxidant: inhibition of lipid peroxidation induced by reactive oxygen species.* Biochim Biophys Acta. 1996;1282:63-70.

3. Gude RP, Menon LG, Rao SG. *Effect of caffeine, a xanthine derivative, in the inhibition of experimental lung metastasis induced by B16F10 melanoma cells.* J Exp Clin Cancer Res. 2001;20:287-292.

4. Hashimoto T, He Z, Ma W, et al. *Caffeine inhibits cell proliferation by G0/G1 phase arrest in JB6 cells.* Cancer Res. 2004;64:3344-3349.

5. Lu Y, Lou Y, Xie J, et al. *Caffeine and caffeine sodium benzoate have a sunscreen effect, enhance UVB-induced apoptosis, and inhibit UVB-induced skin carcinogenesis in SKH-1 mice.* Carcinogenesis. 2006;28:199-206.

6. Balentine DA, Wiseman SA, Bouwens LCM. *The chemistry of tea flavonoids.* Crit Rev Food Sci Nutr. 1997;37:693-704.

7. Yang CS. *Tea and health.* Nutrition. 1999;15:946-949.

8. Conney AH. *Introduction to drug-metabolizing enzymes: A path to the discovery of human cytochromes P450.* Annu Rev Pharmacol Toxicol. 2003;43:1-30.

9. Kumaraguruparan R, Seshagiri PB, Hara Y, et al. *Chemoprevention of rat mammary carcinogenesis by black tea polyphenols: Modulation of xenobiotic-metabolizing enzymes, oxidative stress, cell proliferation, apoptosis, and angiogenesis.* Mol Carcinogenesis. 2007;46:797-806.

10. Leone M, Zhai D, Sareth S, et al. *Cancer prevention by tea polyphenols is linked to their direct inhibition of antiapoptotic Bcl-2-family proteins.* Cancer Res. 2003;63(23):8118-8121.

11. Yang CS, Maliakal P, Meng X. *Inhibition of carcinogenesis by tea.* Ann Rev Pharmacol Toxicol. 2002;42:25-54.

12. Hong J, Lu H, Meng X, et al. *Stability, cellular uptake, biotransformation, and efflux of tea polyphenol (-)-epigallocatechin-3-gallate in HT-29 human colon adenocarcinoma cells.* Cancer Res. 2002;62:7241-7246.

13. Lu H, Meng X, Yang CS. *Enzymology of methylation of tea catechins and inhibition of catechol-O-methyltransferase by (-)-epigallo-catechin gallate.* Drug Metab Dispos. 2003;31(5):572-579.

14. Lu H, Meng X, Li C, et al. *Glucuronides of tea catechins: enzymology of biosynthesis and biological activities.* Drug Metab Dispos. 2003;31(4):452-4261.

15. Zhu BT. *Catechol-O-Methyltransferase (COMT)-mediated methylation metabolism of endogenous bioactive catechols and modulation by endobiotics and xenobiotics: importance in pathophysiology and pathogenesis.* Curr Drug Metab. 2002;3(3):321-349.

16. Fang MZ, Wang Y, Ai N, et al. *Tea polyphenol (-)-epigallocatechin-3-gallate inhibits DNA methyltransferase and reactivates methylation-silenced genes in cancer cell lines.* Cancer Res. 2003;63:7563-7570.

17. Chow HH, Hakim IA, Vining DR, et al. *Modulation of human glutathione S-transferases by polyphenon E intervention.* Cancer Epidemiol Biomarkers Prev. 2007;16:1662-1666.

18. Naasani I, Oh-Hashi F, Oh-Hara T, et al. *Blocking telomerase by dietary polyphenols is a major mechanism for limiting the growth of human cancer cells in vitro and in vivo.* Cancer Res. 2003;63(4):824-830.

19. Lee MJ, Maliakal P, Chen L, et al. *Pharmacokinetics of tea catechins after ingestion of green tea and (-)-epigallocatechin-3-gallate by humans: formation of different metabolites and individual variability.* Cancer Epidemiol Biomarkers Prev. 2002;11:1025-1032.

20. Baker JA, Beehler GP, C.S. A, et al. *Consumption of coffee, but not black tea, is associated with decreased risk of premenopausal breast cancer.* J Nutr. 2006;136:166-171.

21. Kumar N, Titus-Ernstoff L, Newcomb PA, et al. *Tea consumption and risk of breast cancer.* Cancer Epidemiol Biomarkers Prev. 2009;18:341-345.

22. Lubin F, Ron E, Wax Y, et al. *Coffee and methylxanthines and breast cancer: A case-control study.* J Natl Cancer Inst. 1985;74:569-573.

23. Nkondjock A, Ghadirian P, Kotsopoulos J, et al. *Coffee consumption and breast cancer risk among BRCA1 and BRCA2 mutation carriers.* Int J Cancer 2006;118:103-107.

24. Vatten LJ, Solvoll K, Loken EB. *Coffee consumption and the risk of breast cancer. A prospective study of 14, 593 Norwegian women.* Br J Cancer 1990;62:267-270.

25. Hunter DJ, Manson JE, Stampfer MJ, et al. *A prospective study of caffeine, coffee, tea, and breast cancer.* Am J Epidemiol 1992;136:1000-1001.

26. Folsom AR, McKenzie DR, Bisgard KM, et al. *No association between caffeine intake and post-menopausal breast cancer incidence in the Iowa women's health study.* Am J Epidemiol 1993;138:380-383.

27. McLaughlin CC, Mahoney MC, Nasca PC, et al. *Breast cancer and methylxanthine consumption.* Cancer Causes Control 1992;3:175-178.

28. Michels KB, Holmberg L, Bergkvist L, et al. *Coffee, tea, and caffeine consumption and breast cancer incidence in a cohort of Swedish women.* Ann Epidemiol 2002;12:21-26.

29. Rosenberg L, Miller DR, Helmrich SP, et al. *Breast cancer and the consumption of coffee.* Am J Epidemiol 1985;122:391-399.

30. Stensvold I, Jacobsen BK. *Coffee and cancer: a prospective study of 43,000 Norwegian men and women.* Cancer Causes Control 1994;5:401-408.

31. Sun CL, Yuan JM, Koh WP, et al. *Green tea, black tea and breast cancer risk: a meta-analysis of epidemiological studies.* Carcinogenesis 2006;27:1310-1315.

32. Wu AH, Yu MC, Tseng C, et al. *Green tea and risk of breast cancer in Asian Americans.* Int J Cancer 2003;106:574-579.

33. Wu AH, Tseng CC, Van Den Berg D, et al. *Tea intake, COMT genotype, and breast cancer in Asian-American women.* Cancer Res 2003;63(21):7526-7529.

3.4.6. ALCOHOL

Of the various dietary factors examined, the most consistent observation has been an increase in risk of breast cancer with regular alcohol consumption (1-5). A meta-analysis of 98 studies showed an excess risk of 22% for drinkers versus nondrinkers.(6) Women who drink on average one alcoholic beverage per day (one drink/day corresponds to approximately 15 g alcohol) have a 10 – 30% higher risk of breast cancer than nondrinkers. Risk estimates did not significantly differ by beverage type or menopausal status. There is a dose-response relationship among women who drink moderate to high levels of alcohol. In a pooled analysis from six prospective cohorts, an increment of 10 g/day of alcohol was associated with a 9% increase in breast cancer risk (7). Thus, among women who consume alcohol regularly, reducing alcohol consumption is a potential means to reduce breast cancer risk.

Alcohol and tobacco consumption are closely correlated and published results on their association with breast cancer have not always allowed adequately for confounding between these exposures. The Collaborative Group on Hormonal Factors in Breast Cancer collated, checked, and analyzed centrally over 80% of the relevant information worldwide on tobacco and alcohol consumption and breast cancer (8). Analyses included 58,515 women with invasive breast cancer and 95,067 controls from 53 studies. Relative risks of breast cancer were estimated after stratifying by study, age, parity and, where appropriate, women's age when their first child was born and consumption of tobacco and alcohol. The average consumption of alcohol reported by controls from developed countries was 6.0 g per day, i.e., about half a unit/drink of alcohol per day, and was greater in ever-smokers than never-smokers (8.4 g per day and 5.0 g per day, respectively). Compared with women who reported drinking no alcohol, the relative risk of breast cancer was 1.32 (95% CI 1.19 – 1.45, p < 0.00001) for an intake of 35 – 44 g per day of alcohol and 1.46 (95% CI 1,33 – 1.61; p < 0.00001) for ≥45 g per day alcohol. The relative risk of breast cancer increased by 7.1% (95% CI 5.5 – 8.7%, p < 0.00001) for each additional 10 g per intake of alcohol, i.e., for each extra unit or drink of alcohol consumed on a daily basis. This increase was the same in ever-smokers and never smokers (7.1% per 10 g per day, p < 0.00001, in each group). If the observed relationship for alcohol is causal, these results suggest that about 4% of breast cancers in developed countries are attributable to alcohol. In developing countries, where alcohol consumption among controls averaged only 0.4 g per day, alcohol would have a negligible effect on the incidence of breast cancer.

The relationship between alcohol and breast cancer appears to be causal but the mechanism for this association is not well understood. One potential mechanism is the effect of alcohol on folate and one-carbon metabolism. Alcohol interferes with folate absorption, transport, and metabolism, potentially limiting tissue folate stores (9). Folate deficiency is implicated in carcinogenesis through interference with DNA synthesis and methylation. Indeed, several studies have shown that the excess risk of breast cancer associated with alcohol consumption may be reduced by adequate folate intake (10-13). Another mechanism may be the influence of alcohol intake on estrogen metabolism. Animal experiments have shown that ethanol consumption increases hepatic aromatase activity, which, in turn, could increase the conversion of androgens to estrogens (14). Indeed, several studies observed a positive correlation between alcohol intake in women and both blood and urinary estrogen concentrations (15-19). However, other studies found no correlation or even an inverse association (20-22). Postmenopausal women receiving estrogen replacement therapy experienced a significant and sustained increase in circulating estrogen following ingestion of alcohol (23, 24). Women drinking ≥20 g/day who used HRT had an increased risk of breast cancer (RR 2.24; 95% CI, 1.59 – 3.14) compared to nondrinkers who never used HRT (25). An analysis of postmenopausal women participating in the Women's Health Initiative confirmed the association between alcohol consumption and risk of invasive breast cancer but did not find an association with risk of ductal carcinoma *in situ* suggesting that alcohol may have an effect later in the carcinogenic process (26).

References

1. Feigelson HS, Jonas CR, Robertson AS, et al. *Alcohol, folate, methionine, and risk of incident breast cancer in the American Cancer Society Cancer Prevention Study II Nutrition Cohort.* Cancer Epidemiol Biomarkers Prev. 2003;12:161-164.

2. Gapstur SM, Potter JD, Sellers TA, et al. *Increased risk of breast cancer with alcohol consumption in postmenopausal women.* Am J Epidemiol. 1992;136:1221-1231.

3. Longnecker MP. *Alcoholic beverage consumption in relation to risk of breast cancer: meta-analysis and review.* Cancer Causes Control. 1994;5:73-82.

4. Longnecker MP, Newcomb PA, Mittendorf R, et al. *Risk of breast cancer in relation to lifetime alcohol consumption.* J Natl Cancer Inst. 1995;87:923-929.

5. Swanson CA, Coates RJ, Malone KE, et al. *Alcohol consumption and breast cancer risk among women under age 45 years.* Epidemiology. 1997;8:231-237.

6. Key J, Hodgson S, Omar RZ, et al. *Meta-analysis of studies of alcohol and breast cancer with consideration of the methodological issues.* Cancer Causes Control. 2006;17:759-770.

7. Smith-Warner SA, Spiegelman D, Yaun SS, et al. *Alcohol and breast cancer in women: a pooled analysis of cohort studies.* JAMA. 1998;279:535-540.

8. Hamajima N, Hirose K, Tajima K, et al. *Alcohol, tobacco and breast cancer--collaborative reanalysis of individual data from 53 epidemiological studies, including 58,515 women with breast cancer and 95,067 women without the disease.* Br J Cancer. 2002;87:1234-1245.

9. Platek ME, Shields PG, Marian C, et al. *Alcohol consumption and genetic variation in methylenetetrahydrofolate reductase and 5-methyltetrahydrofolate-homocysteine methyltransferase in relation to breast cancer risk.* Cancer Epidemiol Biomarkers Prev. 2009;18:2453-2459.

10. Larsson SC, Giovannucci E, Wolk A. *Folate and risk of breast cancer: A meta-analysis.* J Natl Cancer Inst. 2007;99:64-76.

11. Rohan TE, Jain MG, Howe GR, et al. *Dietary folate consumption and breast cancer risk.* J Natl Cancer Inst. 2000;92:266-269.

12. Zhang S, Hunter DJ, Hankinson SE, et al. *A prospective study of folate intake and the risk of breast cancer.* JAMA. 1999;281:1632-1637.

13. Zhang S, Willet WC, Selhub J, et al. *Plasma folate, vitamin B6, vitamin B12, homocysteine, and risk of breast cancer.* J Natl Cancer Inst. 2003;95:373-380.

14. Gordon G, Southren AL, Vittek J, et al. *The effect of alcohol ingestion on hepatic aromatase activity and plasma steroid hormones in the rat.* Metabolism. 1979;28:20-24.

15. Hankinson SE, Willett WC, Manson JE, et al. *Alcohol, height and adiposity in relation to estrogen and prolactin levels in postmenopausal women.* J Natl Cancer Inst. 1995;87:1297-1302.

16. Muti P, Trevisan M, Micheli A, et al. *Alcohol consumption and total estradiol in premenopausal women.* Cancer Epidemiol Biomarkers Prev. 1998;7:189-193.

17. Onland-Moret NC, Peeters PHM, van der Schouw YT, et al. *Alcohol and endogenous sex steroid levels in postmenopausal women: A cross-sectional study.* J Clin Endocrinol Metab. 2005;90:1414-1419.

18. Reichman ME, Judd JT, Longcope C, et al. *Effects of alcohol consumption on plasma and urinary hormone concentrations in premenopausal women.* J Natl Cancer Inst. 1993;85:722-727.

19. Verkasalo PK, Thomas HV, Appleby PN, et al. *Circulating levels of sex hormones and their relation to risk factors for breast cancer: a cross-sectional study in 1092 pre- and postmenopausal women (United Kingdom).* Cancer Causes Control. 2001;12:47-59.

20. Cauley JA, Gutai JP, Kuller LH, et al. *The epidemiology of serum sex hormones in postmenopausal women.* Am J Epidemiol. 1989;129:1120-1131.

21. Dorgan JF, Reichman ME, Judd JT, et al. *The relation of reported alcohol ingestion to plasma levels of estrogens and androgens in premenopausal women.* Cancer Causes Control. 1994;5:53-60.

22. Trichopoulos D, Brown J, MacMahon B. *Urine estrogens and breast cancer risk factors among post-menopausal women.* Int J Cancer. 1987;40:721-725.

23. Ginsburg ES, Mello NK, Mendelson JH, et al. *Effects of alcohol ingestion on estrogens in postmenopausal women.* JAMA. 1996;276:1747-1751.

24. McDivit AM, Greendale GA, Stanczyk FZ, et al. *Effects of alcohol and cigarette smoking on change in serum estrone levels in postmenopausal women randomly assigned to fixed doses of conjugated equine estrogens with or without a progestin.* Menopause. 2008;15:382-385.

25. Horn-Ross P, Canchola A, West D, et al. *Patterns of alcohol consumption and breast cancer risk in the California Teachers Study cohort.* Cancer Epidemiol Biomarkers Prev. 2004;13:405-411.

26. Kabat GC, Kim M, Shikany JM, et al. *Alcohol consumption and risk of ductal carcinoma in situ of the breast in a cohort of postmenopausal women.* Cancer Epidemiol Biomarkers Prev. 2010;19:2066-2072.

3.4.7. CIGARETTE SMOKING

Cigarette smoke, a complex chemical mixture containing more than 4000 different compounds, is widely recognized as an important cause of cancer (1). Among the 4000 chemicals, over 100 are carcinogens, mutagens, and tumor promoters, including PAHs, N-nitrosamines, and aromatic amines. Due to the fact that they are lipophilic, smoking-related carcinogens can be stored in breast adipose tissue and then metabolized and activated by human mammary epithelial cells (2). The finding of smoking-specific DNA adducts and p53 gene mutations in the breast tissue of smokers support the biological plausibility of a positive association between cigarette smoking and breast cancer risk (3-8). However, a 2002 review of epidemiological studies has shown positive, inverse, or null associations (2). The overwhelming majority of case-control and cohort studies found no association of breast cancer risk with smoking duration, frequency, pack-years, or age at smoking commencement. For example, a population-based study of 6866 cases and 9529 controls did not observe an influence of smoking on breast cancer risk, even among heavy smokers who began smoking at an early age (9). On the other hand, a cohort study (2552 cases in a cohort of 89,835 women) showed a positive association of smoking for 40 years or longer with breast cancer risk, especially among women who also smoked a packet of cigarettes per day or more (2). The Nurses' Health Study found an increased risk association with smoking for five or more years before a first full-term pregnancy (10).

Nonsmokers frequently are exposed to environmental smoke, referred to as passive or second-hand smoke, e.g., a non-smoking woman's exposure to her husband's smoke. It has been argued that the general lack of an association between active smoking and breast cancer risk makes any association with passive smoking unlikely, given that women who are active smokers are also exposed to their own passive smoke (2, 11). At the same time, the referent (control) group in many studies may have included a large number of women who were exposed to passive smoke thereby introducing a bias toward null. The association between active smoking and breast cancer risk observed in some studies became stronger after removal of passive smokers from the referent group (2, 12). Reviews and large studies have yielded inconsistent results, finding no association between passive smoking and breast cancer risk or increased risk at the highest level of cumulative exposure (2, 10, 13-15). Finally, a 2005 pooled analysis of 19 studies with a thorough assessment of passive exposure concluded that both active and passive smoking were associated with an increased risk of breast cancer, particularly in premenopausal women (16). The pooled risk associated with passive smoking was 1.68 (95% CI 1.33 – 2.12) relative to life-long premenopausal nonsmokers. The pooled risk for pre- and postmenopausal active smokers was 1.46 (95% CI 1.15 – 1.85) compared to women who never were regularly exposed to tobacco smoke.

Tobacco and alcohol consumption are closely correlated and published results on their association with breast cancer have not always allowed adequately for confounding variables between these exposures. The Collaborative Group on Hormonal Factors in Breast Cancer collated, checked, and analyzed centrally over 80% of the relevant information worldwide on tobacco and alcohol consumption and breast cancer (17). Analyses included 58,515 women with invasive breast cancer and 95,067 controls from 53 studies. Relative risks of breast cancer were estimated after stratifying by study, age, parity and, where appropriate, women's age when their first child was born and consumption of tobacco and alcohol. The relationship between smoking and breast cancer was substantially confounded by the effect of alcohol. However, when analyses were restricted to 22,255 women with breast cancer and 40,832 controls who reported drinking no alcohol, smoking was not associated with breast cancer. Compared to never-smokers, the relative risk for ever-smokers was 1.03 (95% CI 0.98 – 1.07) and for current smokers 0.99 (95% CI 0.92 – 1.05).

In spite of biological plausibility, the great majority of epidemiologic studies do not support a role for cigarette smoke in the development of breast cancer. How can we explain the apparent discrepancy? A possible explanation for the discrepant results is the apparent anti-estrogenic effect of cigarette smoking. Clinical studies have shown smoking to be associated with earlier menopause and increased risk of osteoporosis (18). Chemicals in cigarette smoke (e.g., PAHs) are known to induce cytochrome P450 enzymes, which metabolize estrogens, such as E_2 and E_1 (19). Indeed, experimental data demonstrated that PAHs and cigarette smoking dramatically enhanced C-2 hydroxylation of E_2 (20-22). Thus, exposure to smoking-related chemicals might result in decreased levels of parent estrogens in serum and increased levels of estrogen metabolites in the urine. However, circulating levels of E_1 and E_2 as well as sex hormone-binding globulin among current smokers did not differ from those of nonsmokers or former smokers (23-26). There was no difference by stage of menstrual cycle in premenopausal women and no evidence of a dose-response relationship by number of cigarettes smoked. Thus, smoking does not decrease endogenous serum estrogen levels in either pre- or postmenopausal women. One study reported that smoking substantially reduced the levels of urinary E_1 and E_2 during the luteal but not the follicular phase of the menstrual cycle (27). However, other studies found no difference in urinary estrogen levels between premenopausal or postmenopausal smokers and nonsmokers (23, 28-30). The effect of smoking on estrogen levels is clearer in postmenopausal women receiving exogenous estrogens. Smoking significantly reduced serum E_1 and E_2 levels in postmenopausal women on HRT to the point of completely canceling the therapeutic efficacy of estrogen replacement (25, 31, 32). The decrease in serum estrogen levels was seen with oral but not parenteral estrogen replacement, implicating hepatic clearance as mechanism for the reduction in hormone concentration (33, 34).

Another reason for the inconsistency of epidemiological studies may be the omission of potential modifying effects of phase I and II enzymes (cytochrome P450, epoxide hydrolase, glutathione S-transferases, UDP-glucuronsyl-transferases) and their genetic variants. Phase I and II enzymes detoxify and activate the majority of chemicals found in cigarette smoke. Because of the key role played by these enzymes, many epidemiologic studies have examined the association between smoking and breast cancer risk according to variations in the encoding genes (35). Some of the smoking-derived compounds are not only activated to carcinogens by these enzymes but the compounds also induce the expression of the enzymes. For example, B[a]P (or tobacco smoke) is an inducer of CYP1A1 and CYP1B1 and, at the same time, a substrate, being oxidized by the same enzymes to carcinogenic B[a]P-dihydrodiol epoxides (36, 37). Thus, both genetic variants and expression levels of phase I and II enzymes may contribute to the effect of smoking on breast cancer risk. The fact that the very same enzymes are also involved in estrogen metabolism adds another layer of complexity and thereby makes the interpretation even more difficult. For example, members of the CYP1 family function both as aryl hydrocarbon hydroxylases and as E_2 hydroxylases (38, 39).

Since NAT2 plays a key role in the metabolism of aromatic amines, a major class of tobacco smoke carcinogens, Ambrosone examined the association between smoking and breast cancer risk by genotypes for NAT2 (40). For slow acetylators, smoking was associated with increased breast cancer risk in a dose-dependent manner. For example, smoking more than one pack of cigarettes per day for 20 years before the interview was associated with more than a 4-fold increase in risk. A decade after her original report, Ambrosone and her colleagues published a meta-analysis of 13 studies including 4889 premenopausal and 7033 postmenopausal women (41). They also pooled data from 10 of the studies. Both types of analysis confirmed that cigarette smoking was associated with an increase in breast cancer risk among women with NAT2 slow acetylation genotypes. The meta-analysis revealed an increased relative risk of 1.44 with a 95% confidence interval 1.23 – 1.68 for ≥20 pack-years versus never smokers. Because NAT2 slow acetylation genotypes are present in approximately 50% of Caucasians, the increased risk of breast cancer with smoking among this subset of the population would have a large public health effect (42). Combined effects of NAT1 and NAT2 genotypes and smoking on breast cancer risk yielded inconsistent results (43-47).

Studies of CYP1A1 variants found no consistent association between smoking and breast cancer risk according to CYP1A1 genotype (35). Smokers with the CYP1A1 MspI polymorphism had significantly higher level of DNA adducts in normal breast tissue adjacent to cancer than smokers with wild-type CYP1A1 (48). This effect was not seen among non-smokers. A case-only analysis (282 cases) suggested that the CYP1B1 polymorphism Val432Leu may increase breast cancer susceptibility among smokers (49). A pooled analysis of studies examining smoking and breast cancer risk found no interaction with glutathione S-transferase M1, T1, and P1 polymorphisms (50). Similarly, there was no association in African American women (51). Two case-control studies examining a common polymorphism in the sulfotransferase SULT1A1 gene, Arg213His, found no evidence of effect modification (52, 53). Three of four studies examining the superoxide dismutase SOD2 polymorphism Val16Ala observed no interaction of the genotype with cigarette smoking and breast cancer risk (46, 54-56).

Studies of the DNA repair gene XRCC1 have shown inconsistent associations between smoking and breast cancer risk according to XRCC1 genotype (35). An analysis of BRCA1 and BRCA2 mutation carriers by the Hereditary Breast Cancer Clinical Study Group found an increased risk among past but not current smokers with BRCA1 mutations (57). However, other studies observed no increased risk of breast cancer among carriers who had ever smoked (35)

References

1. Hecht SS. *Tobacco carcinogens, their biomarkers and tobacco-induced cancer.* Nature Rev Cancer. 2003;3:733-44.

2. Terry PD, Miller AB, Rohan TE. *Cigarette smoking and breast cancer risk: A long latency period.* Int J Cancer. 2002;100:723-8.

3. Conway K, Edmiston SN, Cui L, Drouin SS, Pang J, He M, et al. *Prevalence and spectrum of p53 mutations associated with smoking in breast cancer.* Cancer Res. 2002;62:1987-95.

4. Denissenko MF, Pao A, Tang M, Pfeifer GP. *Preferential formation of benzo[a]pyrene adducts at lung cancer mutational hotspots in P53.* Science. 1996;274:430-2.

5. Pfeifer GP, Denissenko MF, Olivier M, Tretyakova N, Hecht SS, Hainaut P. *Tobacco smoke carcinogens, DNA damage and p53 mutations in smoking-associated cancers.* Oncogene. 2002;21:7435-51.

6. Perera FP, Estabrook A, Hewer A, Channing K, Rundle A, Mooney LA, et al. *Carcinogen-DNA adducts in human breast tissue.* Cancer Epidemiol Biomarkers Prev. 1995;4:233-8.

7. Santella RM. *Immunological methods for detection of carcinogen-DNA damage in humans.* Cancer Epidemiol Biomarkers Prev. 1999;8:733-9.

8. Santella RM, Gammon MD, Zhang YJ, Motykiewicz G, Young TL, Hayes SC, et al. *Immunohistochemical analysis of polycyclic aromatic hydrocarbon-DNA adducts in breast tumor tissue.* Cancer Lett. 2000;154:143-9.

9. Baron JA, Newcomb PA, Longnecker MP, Mittendorf R, Storer BE, Clapp RW, et al. *Cigarette smoking and breast cancer.* Cancer Epidemiol Biomarkers Prev. 1996;5:399-403.

10. Egan KM, Stampfer MJ, Hunter D, Hankinson S, Rosner BA, Holmes M, et al. *Active and passive smoking in breast cancer: prospective results from the Nurses' Health Study.* Epidemiology. 2002;13:138-45.

11. Khuder SA, Simon VJ, Jr. *Is there an association between passive smoking and breast cancer?* Eur J Epidemiol. 2000;16:1117-21.

12. Wells AJ. *Re: "Breast cancer, cigarette smoking, and passive smoking".* Am J Epidemiol. 1998;147:991-2.

13. Morabia A. *Smoking (active and passive) and breast cancer: epidemiologic evidence up to June 2001.* Environ Mol Mutagen. 2002;39:89-95.

14. Reynolds P, Goldberg D, Hurley S, Nelson DO, Largent J, Henderson KD, et al. *Passive smoking and risk of breast cancer in the California Teachers Study.* Cancer Epidemiol Biomarkers Prev. 2009;18:3389-98.

15. Shrubsole MJ, Gao YT, Dai Q, Shu XO, Ruan ZX, Jin F, et al. *Passive smoking and breast cancer risk among non-smoking Chinese women.* Int J Cancer. 2004;110:605-9.

16. Johnson KC. *Accumulating evidence on passive and active smoking and breast cancer risk.* Int J Cancer. 2005;117:619-28.

17. Hamajima N, Hirose K, Tajima K, Rohan T, Calle EE, Heath CW, Jr., et al. *Alcohol, tobacco and breast cancer--collaborative reanalysis of individual data from 53 epidemiological studies, including 58,515 women with breast cancer and 95,067 women without the disease.* Br J Cancer. 2002;87:1234-45.

18. Baron JA, La Vecchia C, Levi F. *The antiestrogenic effect of cigarette smoking in women.* Am J Obstet Gynecol. 1990;162:502-14.

19. Conney AH. *Introduction to drug-metabolizing enzymes: A path to the discovery of multiple cytochromes P450.* Annu Rev Pharmacol Toxicol. 2003;43:1-30.

20. Michnovicz JJ, Hershcopf RJ, Naganuma H, Bradlow HL, Fishman J. *Increased 2-hydroxylation of estradiol as a possible mechanism for the anti-estrogenic effect of cigarette smoking.* N Engl J Med. 1986;315:1305-9.

21. Michnovicz JJ, Naganuma H, Hershcopf RJ, Bradlow HL, Fishman J. *Increased urinary catechol estrogen excretion in female smokers.* Steroids. 1988;52:69-83.

22. Schneider J, Sassa S, Kappas A. *Metabolism of estradiol in liver cell culture. Differential responses of C-2 and C-16 oxidations to drugs and other chemicals that induce selective species of cytochrome P-450.* J Clin Invest. 1983;72:1420-6.

23. Berta L, Frairia R, Fortunati N, Fazzari A, Gaidano G. *Smoking effects on the hormonal balance of fertile women.* Horm Res. 1992;37:45-8.

24. Cassidenti DL, Pike MC, Vijod AG, Stanczyk FZ, Lobo RA. *A reevaluation of estrogen status in postmenopausal women who smoke.* Am J Obstet Gynecol. 1992;166:1444-8.

25. Jensen J, Christiansen C, Rodbro P. *Cigarette smoking, serum estrogens, and bone loss during hormone-replaceent therapy early after menopause.* N Engl J Med. 1985;313:973-5.

26. Key TJA, Pike MC, Baron JA, Moore JW, Wang DY, Thomas BS, et al. *Cigarette smoking and steroid hormones in women.* J Steroid Biochem Molec Biol. 1991;39:529-34.

27. MacMahon B, Trichopoulos D, Cole P, Brown J. *Cigarette smoking and urinary estrogens.* N Engl J Med. 1982;307:1062-5.

28. Key TJA, Pike MC, Brown JB, Hermon C, Allen DS, Wang DY. *Cigarette smoking and urinary oestrogen excretion in premenopausal and post-menopausal women.* Br J Cancer. 1996;74:1313-6.

29. Trichopoulos D, Brown J, MacMahon B. *Urine estrogens and breast cancer risk factors among post-menopausal women.* Int J Cancer. 1987;40:721-5.

30. Windham GC, Mitchell P, Anderson M, Lasley BL. *Cigarette smoking and effects on hormone function in premenopausal women.* Environ Health Perspect. 2005;113:1285-90.

31. McDivit AM, Greendale GA, Stanczyk FZ, Huang MH. *Effects of alcohol and cigarette smoking on change in serum estrone levels in postmenopausal women randomly assigned to fixed doses of conjugated equine estrogens with or without a progestin.* Menopause. 2008;15:382-5.

32. Mueck AO, Seeger H, Wallwiener D. *Smoking, estradiol metabolism and hormone replacement therapy.* Geburtsh Frauenheilk. 2003;63:213-22.

33. Mueck AO, Seeger H. *Smoking, estradiol metabolism and hormone replacement therapy.* Curr Med Chem Cardiovasc Hematol Agents. 2005;3:45-54.

34. Tanko LB, Christiansen C. *An update on the antiestrogenic effect of smoking: a literature review with implications for researchers and practitioners.* Menopause. 2004;11:104-9.

35. Terry PD, Goodman M. *Is the association between cigarette smoking and breast cancer modified by genotype? A review of epidemiologic studies and meta-analysis.* Cancer Epidemiol Biomarkers Prev. 2006;15:602-11.

36. Port JL, Yamaguchi K, Du B, De Lorenzo M, Chang M, Heerdt PM, et al. *Tobacco smoke induces CYP1B1 in the aerodigestive tract.* Carcinogenesis. 2004;25:2275-81.

37. Spink DC, Katz BH, Hussain MM, Spink BC, Wu SJ, Liu N, et al. *Induction of CYP1A1 and CYP1B1 in T-47D human breast cancer cells by benzo[a]pyrene is diminished by arsenite.* Drug Metab Dispos. 2002;30:262-9.

38. Guengerich FP. *Characterization of human cytochrome P450 enzymes.* FASEB J. 1992;6:745-8.

39. Okey AB. *Enzyme induction in the cytochrome P-450 system.* Pharmacol Therapeut. 1990;45:241-98.

40. Ambrosone CB, Freudenheim JL, Graham S, Marshall JR, Vena JE, Brasure JR, et al. *Cigarette smoking, N-acetyltransferase 2 genetic polymorphisms, and breast cancer risk.* JAMA. 1996;276:1494-501.

41. Ambrosone CB, Kropp S, Yang J, Yao S, Shields PG, Chang-Claude J. *Cigarette Smoking, N-acetyltransferase 2 genotypes, and breast cancer risk: pooled analysis and meta-analysis.* Cancer Epidemiol Biomarkers Prev. 2008;17:15-26.

42. Phillips DH, Garte S. *Smoking and breast cancer: Is there really a link?* Cancer Epidemiol Biomarkers Prev. 2008;17:1-2.

43. Deitz AC, Zheng W, Leff MA, Gross M, Wen W-Q, Doll MA, et al. *N-acetyltransferase-2 genetic polymorphism, well-done meat intake, and breast cancer risk among postmenopausal women.* Cancer Epidemiol Biomarkers Prev. 2000;9:905-10.

44. Delfino RJ, Smith C, West JG, Lin HJ, White E, Liao SY, et al. *Breast cancer, passive and active cigarette smoking and N-acetyltransferase 2 genotype.* Pharmacogenetics. 2000;10:461-9.

45. Millikan RC, Pittman GS, Newman B, Tse CK, Selmin O, Rockhill B, et al. *Cigarette smoking, N-acetyltransferases 1 and 2, and breast cancer risk.* Cancer Epidemiol Biomarkers Prev. 1998;7:371-8.

46. Millikan RC, Player J, Rene de Cotret A, Moorman P, Pittman G, Vannappagari V, et al. *Manganese superoxide dismutase Ala-9Val polymorphism and risk of breast cancer in a population-based case-control study of African Americans and whites.* Breast Cancer Res: BCR. 2004;6:R264-R74.

47. Zheng W, Deitz AC, Campbell DR, Wen W-Q, Cerhan JR, Sellers TA, et al. *N-acetyltransferase 1 genetic polymorphism, cigarette smoking, well-done meat intake, and breast cancer risk.* Cancer Epidemiol Biomarkers Prev. 1999;8:233-9.

48. Firozi PF, Bondy ML, Sahin AA, Chang P, Lukmanji F, Singletary E, et al. *Aromatic DNA adducts and polymorphisms of CYP1A1, NAT2, and GSTM1 in breast center.* Carcinogenesis. 2002;23:301-6.

49. Saintot M, Malaveille C, Hautefeuille A, Gerber M. *Interactions between genetic polymorphism of cytochrome P450-1B1, sulfotransferase 1A1, catechol-o-methyltransferase and tobacco exposure in breast cancer risk.* Int J Cancer. 2003;107:652-7.

50. Vogl FD, Taioli E, Maugard C, Zheng W, Ribeiro Pinto LF, Ambrosone C, et al. *Glutathione S-transferases M1, T1, and P1 and breast cancer: a pooled analysis.* Cancer Epidemiol Biomarkers Prev. 2004;13:1473-9.

51. Van Emburgh BO, Hu JJ, Levine EA, Mosley LJ, Perrier ND, Freimanis RI, et al. *Polymorphisms in CYP1B1, GSTM1, GSTT1 and GSTP1, and susceptibility to breast cancer.* Oncol Reports. 2008;19:1311-21.

52. Lilla C, Risch A, Kropp S, Chang-Claude J. *Sult1A1 genotype, active and passive smoking and breast cancer risk by age 50 in a German case-control study.* Breast Cancer Res: BCR. 2005;7:R229-R37.

53. Sillanpaa P, Kataja V, Eskelinen M, Kosma VM, Uusitupa M, Vainio H, et al. *Sulfotransferase 1A1 genotype as a potential modifer of breast cancer risk among premenopausal women.* Pharmacogenet Genomics. 2005;15:749-52.

54. Gaudet MM, Gammon MD, Santella RM, Britton JA, Teitelbaum SL, Eng SM, et al. *MnSOD Val-9Ala genotype, pro- and anti-oxidant environmental modifiers, and breast cancer among women on Long Island, New York.* Cancer Causes Control. 2005;16:1225-34.

55. Mitrunen K, Sillanpaa P, Kataja V, Eskelinen M, Kosma VM, Benhamou S, et al. *Association between manganese superoxide dismutase (MnSOD) gene polymorphism and breast cancer risk.* Carcinogenesis. 2001;22:827-9.

56. Tamimi RM, Hankinson SE, Spiegelman D, Colditz GA, Hunter DJ. *Manganese superoxide dismutase polymorphism, plasma antioxidants, cigarette smoking, and risk of breast cancer.* Cancer Epidemiol Biomarkers Prev. 2004;13:989-96.

57. Ginsburg O, Ghadirian P, Lubinski J, Cybulski C, Lynch H, Neuhausen S, et al. *Smoking and the risk of breast cancer in BRCA1 and BRCA2 carriers: an update.* Breast Cancer Res Treat. 2009;114:127-35.

CHAPTER 4

Enzymes

The Etiology of Breast Cancer

4.1. INTRODUCTION

Metabolism is the major route of elimination of xenobiotics, including carcinogens and drugs, from the body.

The most common pathways of xenobiotic metabolism are oxidation, reduction, and conjugation. Oxidation and reduction are commonly referred to as phase I metabolism, with cytochromes P450 (CYPs) as main enzymes catalyzing the majority of reactions (1). Conjugation is considered phase II metabolism, which is mediated by multiple enzyme superfamilies, including glutathione S-transferases, N-acetyltransferases, sulfotransferases, and UDP-glucuronosyltransferases (2). Although the primary role of phase I enzymes is the detoxification of xenobiotics, several CYPs can activate procarcinogens to highly reactive ultimate carcinogens (Figure 4.1.). In contrast, phase II enzymes play predominantly a detoxification role by conjugating products with endogenous ligands, such as glutathione, glucuronic acid, and sulfate. In addition to the "classical" phase II enzymes named above, which conjugate xenobiotics with endogenous ligands, there are many other enzymes involved in catalytic protection against electrophile and oxidative damage.

With regard to breast cancer, the study of enzymes is driven by the hypothesis that enzyme-mediated carcinogen metabolism influences the formation of DNA adducts and mutations and thereby mammary carcinogenesis (Figure 4.1.). To assess the role of any enzyme, several questions need to be addressed: (i) Does the enzyme recognize the carcinogen as substrate? (ii) Is the enzyme expressed in normal and malignant mammary epithelial cells? With regard to genetic enzyme variants the questions are: (iii) Does the substitution of a single amino acid alter the rate of protein degradation sufficiently to cause a significant change in protein concentration and activity? (iv) Does the genetic variant differ from the wild-type enzyme in catalytic activity by altering the active site? Genetic variants of an enzyme are potential sources of cancer susceptibility if they differ significantly from the wild-type enzyme in concentration or catalytic activity toward a (pro)carcinogen (3). However, the biological relevance of the observed differences needs to be evaluated in three contexts: (i) the frequency of the high-risk allele in the population; (ii) the likely range of enzyme induction; and (iii) typical exposure levels to the (pro)carcinogen in the general population. For example, dietary exposures to carcinogens, such as B[a]P, are in the nanograms/day range suggesting that cellular concentrations in non-smokers are quite low compared with the K_m of metabolizing enzymes, such as CYP1B1. Consequently, the CYP1B1 genotype is unlikely to play a role in the susceptibility of non-smokers despite differences in K_m between variant and wild-type isoforms (4).

On the other hand, elevated B[a]P exposure levels in smokers have been shown to lead to CYP1B1 induction (up to 40-fold higher mRNA levels), which could create a greater variation in total enzyme activity than that due to small differences in K_m.

References

1. Guengerich FP. *Common and uncommon cytochrome P450 reactions related to metabolism and chemical toxicity.* Chem Res Toxicol. 2001;14:611-50.

2. Jakoby WB, Ziegler DM. *The enzymes of detoxication.* J Biol Chem. 1990;265:20715-8.

3. Weinshilboum R, Wang L. *Pharmacogenetics: Inherited variation in amino acid sequence and altered protein quantity.* Clinl Phamacol Therap. 2004;75:253-8.

4. Mammen JS, Pittman GS, Li Y, Abou-Zahr F, Bejjani BA, Bell DA, et al. *Single amino acid mutations, but not common polymorphisms, decrease the activity of CYP1B1 against (-)benzo[a]pyrene-7R-trans-7,8-dihydrodiol.* Carcinogenesis. 2003;24:1247-55.

DNA DAMAGE

Testosterone

$\downarrow$ CYP19A1

Estradiol

$\downarrow$ CYP1B1

$4\text{-}OHE_2 \longrightarrow E_2\text{-}3,4\text{-}Q$

CAT	rs511895
COMT	rs4680
CYP17A1	rs743572
CYP19A1	rs700519, (TTTA)10
CYP1A1	rs1048943
CYP1B1	rs1056836, 1800440
GSTP1	rs1695
GSTM1	deletion
GSTT1	deletion
HSD17B1	rs605059, 676387
TXN1	rs4135179
TXNRD2	rs5748469, 756661

$$O_2 \xrightarrow[\text{Oxido-reductase}]{\text{P450}} O_2^{\bullet -} \xrightarrow{\text{SOD2}} H_2O_2$$

$H_2O_2 \xrightarrow{\text{Catalase}}$

- H_2O
- $H_2O + O_2$
- $HO^{\bullet}$

$PUFA \longrightarrow$ Malondialdehyde

$PhIP \xrightarrow{CYP1B1} N^2\text{-OH-PhIP}$

$\downarrow$ NAT2

$N^2\text{-OAc-PhIP}$

$B\{a\}P \xrightarrow{CYP1B1}$ Epoxide

$\downarrow$ Epoxide Hydrolase

Diols $\xrightarrow{CYP1B1}$ BPDE

DNA Adduct

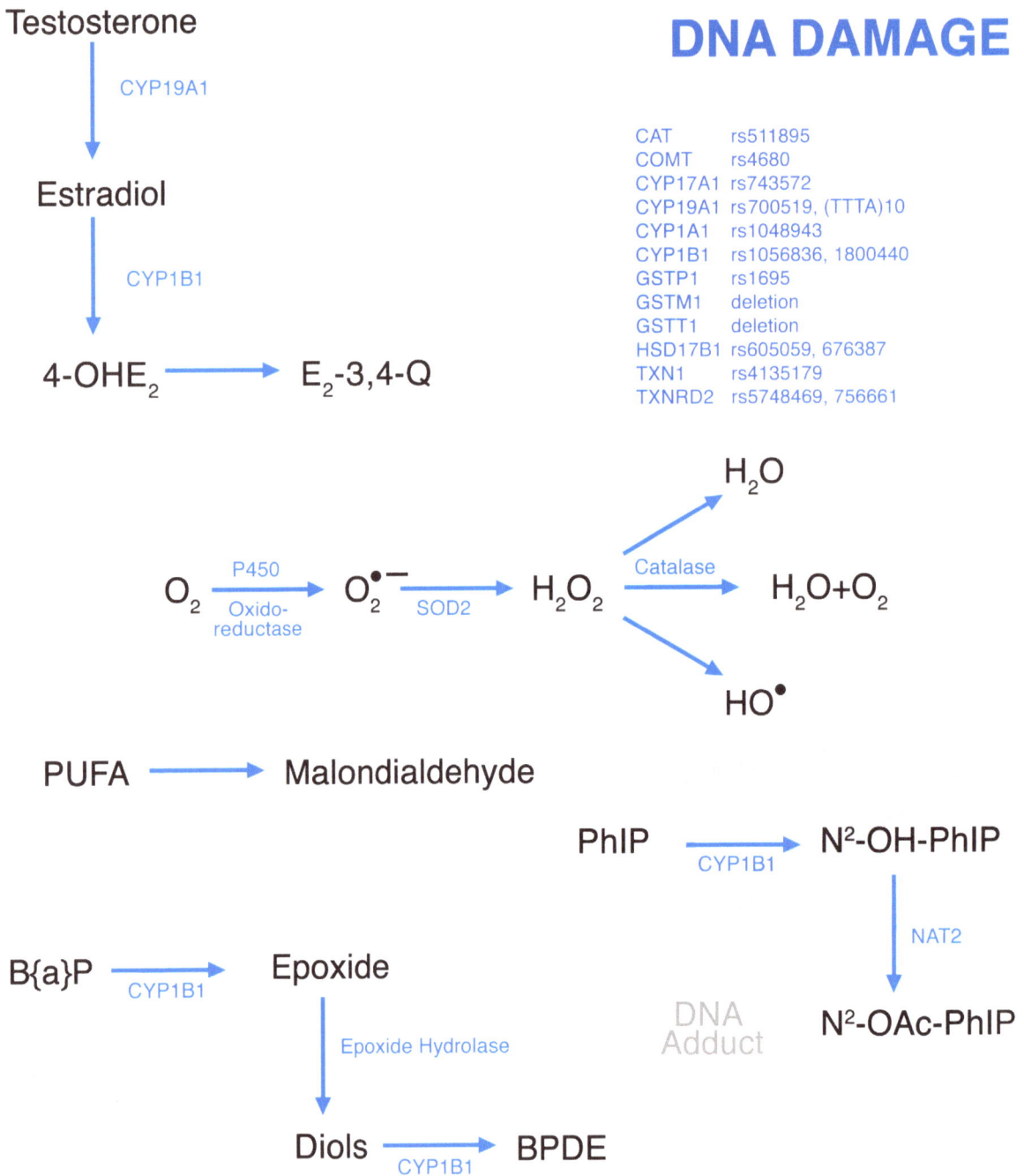

Figure 4.1. Overview of chemical reactions leading to carcinogen formation and DNA damage in the breast.
The majority of the reactions is mediated by enzymes that catalyze the conversion of procarcinogens to carcinogens. The endogenous and exogenous substrates are described in Chapters 2 and 3, respectively. The enzymes are discussed in this Chapter 4. Genome-wide association studies have identified nonsynonymous SNPs and other genetic variants in these enzymes that are linked to breast cancer risk (see inserts with reference SNP ID numbers, rs#). Although each of these polymorphisms has a small effect individually it is likely that the large number of cumulative weak effects contributed by several variants in these pathways plays a major role in mammary carcinogenesis.

4.2. PHASE I OXIDIZING ENZYMES
4.2.1. INTRODUCTION OF CYTOCHROME P450 ENZYMES

Cytochromes P450 (CYPs) constitute a gene superfamily found in all eukaryotes and most prokaryotes that encode heme proteins active as monooxygenase enzymes (1). CYPs catalyze the oxidation of a wide variety of hydrophobic compounds by the two-electron activation of molecular oxygen (2, 3). The electrons are delivered consecutively by NADPH via reductases, the first reducing the heme Fe^{3+} to Fe^{2+}. The ferrous CYP heme iron can then bind O_2. Delivery of the second electron to the ferrous-dioxygen species generates a transient ferryl-oxygen intermediate that cleaves the bound dioxygen. In the productive CYP catalytic cycle, substrate is monooxygenated and a water molecule is formed from the other oxygen atom and two protons (Figure 4.2.1.). In this manner, CYPs catalyze a multitude of reactions, including the hydroxylation of alkanes to alcohols, conversion of alkenes to epoxides, arenes to phenols, sulfides to sulfoxides and sulfones, and the oxidative split of C-N, C-O, C-C, or C-S bonds (3).

CYPs are membrane proteins that can be divided into two classes dependent on their cellular location and the type of reductase system. Class I CYPs, which are found in mitochondria, receive electrons from NADPH via the intermediacy of two proteins, ferredoxin reductase and ferredoxin. Class I CYPs are synthesized with leader sequences that target them to mitochondria. These leader sequences are cleaved upon translocation to the inner mitochondrial membrane where the CYPs bind to the matrix side of the membrane (4). Class II CYPs are present in the endoplasmic reticulum (microsomes), where they receive electrons from NADPH via a single intermediate, P450 oxidoreductase. Class II CYPs are also synthesized with a leader sequence that targets the protein for cotranslational insertion into the endoplasmic reticulum (5). The N-terminal sequence forms a transmembrane domain that anchors the CYP in the microsomal membrane but can be removed without affecting catalytic activity (6, 7). In addition to the N-terminal anchor, the CYP protein associates with the microsomal membrane via a broad, hydrophobic surface formed by noncontiguous portions of the polypeptide chain (8). This interaction places the entrance of the putative substrate access channel in or near the microsomal membrane to facilitate entry and exit of lipophilic substrates and products. The orientation of the protein proximal to the heme cofactor is perpendicular to the plane of the membrane to facilitate interaction with and electron transfer from the P450 oxidoreductase.

All CYPs are proteins of 50–55 kDa with a series of highly conserved helices designated A–L, which are connected by loops and β-sheet structures (8-12). All CYPs contain a cysteine residue as part of a conserved peptide motif Phe-X(6-9)-Cys-X-Gly (where X denotes any amino acid) near the C terminus. The cysteine acts as thiolate ligand to heme, binding octahedral heme iron in the fifth position, and participates in transferring the oxygen atom into the substrate, which is located in the adjacent substrate-binding pocket (13). Comparative analysis of amino acid and coding nucleotide sequences showed conservation of up to six substrate recognition regions along the primary structure in a family-specific manner (14). Interestingly, these regions frequently harbor polymorphisms in form of non-synonymous (amino acid-replacing) codon changes, supporting the idea that evolutionary diversification of CYPs occurred primarily in substrate recognition regions to cope with an increasing number of foreign compounds.

The human genome project has identified 57 CYPs, which are classified by their amino acid similarities and designated by a family number, a subfamily letter, and a number for an individual enzyme within the subfamily, e.g., CYP1A1, 1A2, 1B1 (1, http://drnelson.utmem.edu/cytochromeP450.html). There are 7 class I and 50 class II CYPs, which play essential roles in many cellular reactions, including the synthesis and metabolism of cholesterol, steroids, vitamin D, eicosanoids, biogenic amines, and retinoic acid (13, 15-19). Among the class II CYPs are 15 orphan enzymes whose functions are unknown, suggesting their involvement in additional metabolic pathways. Class II CYPs participate not only in the metabolism of endogenous substrates but also of numerous exogenous chemicals present in the diet and environment, i.e., xenobiotics (13, 20). In addition, they play key roles in the metabolism of many drugs. Collectively, enzymes in the CYP1, CYP2, and CYP3 families are involved in approximately 80% of oxidative drug metabolism and account for almost 50% of the overall elimination of commonly used drugs (21). Polymorphic variants of these enzymes can cause increased or reduced drug metabolism with potential adverse clinical outcome (22). A limited number of CYPs within these families are responsible for the metabolic activation of chemical carcinogens. Specifically, CYP1A1, 1A2, and 1B1 have been shown to be the major enzymes in the metabolism of PAHs, nitro-PAHs and heterocyclic and aryl amines (23).

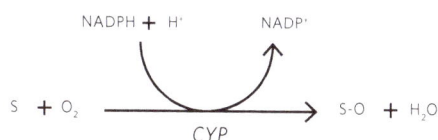

Figure 4.2.1. The CYP catalytic cycle (S = substrate)

In summary, five events are required before a microsomal CYP can fulfill its function: proper anchoring into the microsomal membrane, heme binding, substrate binding, transfer of electrons from P450 oxidoreductase, and O_2 binding. Polymorphisms that affect one or more of these events can be considered to be the molecular basis for altered CYP activity (24). Individual CYPs may have unique substrate specificity, often to a particular region of a molecule, to a particular enantiomer, or to both. However, considerable overlap may also be present. The diversity of CYP-catalyzed reactions arises from structural differences that discriminate between and orient substrates to react with the activated oxygen species formed at the heme. CYPs belonging to families 1 through 3 combine relatively broad metabolic capacities with distinct substrate specificities towards carcinogens. Genomic, phenotypic and clinical information about CYPs has been summarized in http://www.cypalleles.ki.se and http://www.pharmgkb.org (25, 26).

References

1. Nelson DR, Kamataki T, Waxman DJ, Guengerich FP, Estabrook RW, Feyereisen R, et al. *The P450 superfamily: Update on new sequences, gene mapping, accession numbers, early trivial names of enzymes, and nomenclature.* DNA Cell Biol. 1993;12:1-51.

2. Dawson JH. *Probing structure-function relations in heme-containing oxygenases and peroxidases.* Science. 1988;240:433-9.

3. Guengerich FP. *Common and uncommon cytochrome P450 reactions related to metabolism and chemical toxicity.* Chem Res Toxicol. 2001;14:611-50.

4. Ogishima T, Okada Y, Omura T. *Import and processing of the precursor of cytochrome P-450(scc) by bovine adrenal cortex mitochondria.* J Biochem. 1985;98:781-91.

5. Monier S, Van Luc P, Kreibich G, Sabatini DD, Adesnik M. *Signals for the incorporation and orientation of cytochrome P450 in the endoplasmic reticulum membrane.* J Cell Biol. 1988;107:457-70.

6. Larson JR, Coon MJ, Porter TD. *Purification and properties of a shortened form of cytochrome P-450 2E1: Deletion of the NH2-terminal membrane-insertion signal peptide does not alter the catalytic activities.* Proc Natl Acad Sci USA. 1991;88:9141-5.

7. Pernecky SJ, Larson JR, Philpot RM, Coon MJ. *Expression of truncated forms of liver microsomal P450 cytochromes 2B4 and 2E1 in Escherichia coli: Influence of NH2-terminal region on localization in cytosol and membranes.* Proc Natl Acad Sci USA. 1993;90:2651-5.

8. Williams PA, Cosme J, Sridhar V, Johnson EF, McRee DE. *Mammalian microsomal cytochrome P450 monooxygenase: structral adaptations for membrane binding and functional diversity.* Mol Cell. 2000;5:121-31.

9. Graham-Lorence S, Peterson JA. *P450s: Structural sim ilarities and functional differences.* FASEB J. 1996;10:206-14.

10. Hasemann CA, Kurumbail RG, Boddupalli SS, Peterson JA, Deisenhofer J. *Structure and function of cytochromes P450: a comparative analysis of three crystal structures.* Structure. 1995;2:41-62.

11. Poulos TL, Finzel BC, Howard AJ. *High-resolution crystal structure of cytochrome P450cam.* J Mol Biol. 1987;195:687-700.

12. Wester MR, Johnson EF, Marques-Soarres C, Dansette PM, Mansuy D, Stout CD. *Structure of a substrate complex of mammalian cytochrome P450 2C5 at 2.3 Å resolution: Evidence for multiple substrate binding modes.* Biochemistry. 2003;42:6370-9.

13. Nebert DW, Dalton TP. *The role of cytochrome P450 enzymes in endogenous signalling pathways and environmental carcinogenesis.* Nature Rev. 2006;6:947-60.

14. Gotoh O. *Substrate recognition sites in cytochrome P450 family 2 (CYP2) proteins inferred from comparative analyses of amino acid and coding nucleotide sequences.* J Biol Chem. 1992;267:83-90.

15. Bruno RD, Njar VCO. *Targeting cytochrome P450 enzymes: A new approach in anti-cancer drug development.* Bioorg Med Chem. 2007;15:5047-60.

16. Capdevila JH, Harris RC, Falck JR. *Microsomal cytochrome P450 and eicosanoid metabolism.* Cell Mol Life Sci. 2002;59:780-9.

17. Miller WL. *Minireview: regulation of steroidogenesis by electron transfer.* Endocrinology. 2005;146:2544-50.

18. Prosser DE, Jones G. *Enzymes involved in the activation and inactivation of vitamin D.* Trends Biochem Sci. 2004;29:664-73.

19. Yu AM, Idle JR, Byrd LG, Krausz KW, Kupfer A, Gonzalez FJ. *Regeneration of serotonin from 5-methoxytryptamine by polymorphic human CYP2D6.* Pharmacogenetics. 2003;13:173-81.

20. Nebert DW, Russell DW. *Clinical importance of the cytochromes P450.* Lancet. 2002;360:1155-62.

21. Wilkinson GR. *Drug metabolism and variability among patients in drug response.* N Engl J Med. 2005;352:2211-21.

22. Ingelman-Sundberg M, Oscarson M, McLellan RA. *Polymorphic human cytochrome P450 enzymes: an opportunity for individualized drug treatment.* Trends Pharmacol Sci. 1999;20:342-9.

23. Guengerich FP. *Comparisons of catalytic selectivity of cytochrome P450 subfamily enzymes from different species.* Chemico-Biol Interactions. 1997;106:161-82.

24. Yanase T, Simpson ER, Waterman MR. *17α-hydroxylase/17,20-lyase deficiency: From clinical investigation to molecular definition.* Endocrine Rev. 1991;12:91-108.

25. Thorn CF, Klein TE, Altman RB. *PharmGKB: the Pharmacogenomics Knowledge Base.* Methods Mol Biol. 2013;1015:311-20.

26. Zhang T, Zhao M, Pang Y, Zhang W, Angela Liu L, Wei DQ. *Recent progress on bioinformatics, functional genomics, and metabolomics research of cytochrome P450 and its impact on drug discovery.* CurrTopics Medicinal Chem. 2012;12:1346-55.

4.2.2. CYP1A1

The CYP1A1 gene at 15q22-qter is about 7 kb long and contains seven exons (1). The open reading frame starts in the second exon and is 1533 base pairs in length, encoding a protein of 511 amino acids (Figure 4.2.2.). The expression of CYP1A1 is induced by different classes of chemicals including PAHs (e.g., benzo[a]pyrene, 3-methylcholan-threne), halogenated hydrocarbons (e.g., 2,3,7,8-tetra-chlorodibenzo-p-dioxin [TCDD]), and naturally occurring substances like flavones and indole derivatives (2-6). These chemicals induce the expression of CYP1A1 by binding to the aryl hydrocarbon receptor (AhR), a basic helix-loop-helix (bHLH) protein. The AhR is located in the cytoplasm in a multiprotein complex with heat shock protein 90, the X-associated protein 2 (XAP2), and a co-chaperone protein of 23 kDa (7-9). After ligand binding, the AhR complex translocates into the nucleus, dissociates from the protein complex, and binds to a nuclear protein called AhR nuclear translocator protein (ARNT) (10). Formation of the AhR:ARNT heterodimer converts the complex into its high affinity DNA-binding form, which subsequently binds to its specific DNA recognition site, the xenobiotic response element (XRE) located upstream of the CYP1A1 gene, leading to chromatin and nucleosome disruption, increased promoter accessibility, and an increase in transcription of the CYP1A1 gene (7, 9). The AhR is also referred to as the dioxin receptor, since it exhibits high affinity for dioxins, such as TCDD; similarly, the XRE is also called dioxin-responsive element or DRE. All of the XREs are contained in a CpG island located between -178 and -1,712 bp upstream of the CYP1A1 transcription site (9, 11). The AhR-mediated mechanism of gene transcription applies not only to CYP1A1 but also to CYP1A2 and CYP1B1, all of which contain XREs in their promoter regions (12-16). Thus, the AhR and the CYP1 genes that it regulates are central in mediating the genotoxicity of many environmental contaminants. The AhR also regulates the differentiation of T cells in a ligand-specific manner and thereby may participate in the development of autoimmune diseases (17-19).

CYP1A1 is constitutively expressed, albeit at low levels in multiple tissues, including breast, uterus, ovary, prostate, liver, lung, intestine, pancreas, spleen, and thymus (20-22). Immunohistochemical studies revealed CYP1A1 expression in about 90% of breast cancers (23). CYP1A1 levels were elevated in tumors compared with adjacent breast tissue, higher in premenopausal compared with postmenopausal patients, and positively correlated with tumor grade (24). The expression of CYP1A1 is primarily regulated via the AhR and AhR agonists, including polycyclic aromatic hydrocarbons, dioxins, and certain dietary indole carbinols (14, 25). In the human lung, tobacco smoking induces the CYP1A1 enzyme activity up to 100-fold compared with the very low constitutive expression in the lung of non-smokers (4). Female smokers exhibited two-fold higher levels of CYP1A1 expression in nontumor lung tissue than male smokers (26). There are also AhR antagonists, such as the drug salicylamide, which blocks the signal transduction induced by TCDD (27). Cell-type and tissue-specific differences in the expression of CYP1A1 are thought to be due to cell-specific coactivators and corepressors possibly via direct interaction with the AhR:ARNT complexes bound to XREs (13). For example, CYP1A1 is inducible by dioxins (e.g., TCDD) in peripheral blood lymphocytes and by B[a]P in T-47D breast cancer cells (6, 22). A second mechanism of CYP1A1 expression involves the methylation of CpG sites in the CYP1A1 promoter (28). In lung tissue, complete or partial methylation of upstream CpG sites was found in 33% of heavy smokers and in 98% of nonsmokers. The inverse correlation between smoking and CYP1A1 promoter methylation was seen in lung tissue but not in peripheral blood lymphocytes suggesting tissue-specific regulation of methylation. Interestingly, an examination of normal and malignant prostate tissue samples revealed no methylation in the former and hypermethylation in the latter (11).

CYP1A1 metabolizes endogenous and exogenous substrates. The endogenous substrates include steroid hormones and polyunsaturated fatty acids (29-31). CYP1A1 catalyzes E_2 hydroxylation at C-2, C-4, C-6α, and C-15α, but not C-16α (32-34). Kinetic analysis showed that catalytic efficiency for the C-2 hydroxylation was seven-fold higher than the

C-6α and C-15α and 25-fold higher than C-4 activities suggesting that the C-2 hydroxylation of E_2 would be the most physiologically relevant (32). CYP1A1 also catalyzes the subsequent oxidation of 2-OHE_2 to its ortho-quinone, which can react with glutathione, proteins, or DNA (35, 36). The oxidative metabolism of E_1 yields a similar pattern with 2-OHE_1 as main product followed by 4-OHE_1 and detectable levels of 16α-OHE_1 (37). CYP1A1 oxidizes polyunsaturated fatty acids, such as arachidonic acid and eicosapentaenoic acid (29). However, the CYP1A1-mediated reaction differs between these different classes of fatty acids. Arachidonic acid is a (ω – 6) type fatty acid, whereas eicosapentaenoic acid is a (ω - 3) type. With regard to the former, CYP1A1 is a hydroxylase with over 90% of metabolites accounted for by hydroxylation products, such as 19-OH-arachidonic acid. In contrast, with eicosapentaenoic acid as substrate, CYP1A1 is mainly an epoxygenase, resulting in the highly stereoselective formation of 17(R), 18(S)-epoxyeicosatetraenoic acid (29).

CYP1A1 activates many structurally diverse carcinogens, including PAHs, nitro-PAHs, and heterocyclic and aryl amines (3, 38, 39). There is evidence that the carcinogenic potencies of PAHs relate to the potencies of these compounds to induce CYP1A1 and CYP1B1 via the AhR and that the induced CYPs, in turn, activate the PAHs to cause the formation of genotoxic DNA adducts (40). An examination of the relative capacity of each member of the CYP1 enzyme family to activate a variety of carcinogenic chemicals showed that CYP1A1 exhibited unique as well as overlapping substrate specificities with CYP1A2 and CYP1B1 (21). In contrast to CYP1A2, which is expressed predominantly in the liver, both CYP1A1 and 1B1 are expressed in normal breast tissue. The CYP1A1- and 1B1-mediated production of genotoxic metabolites within the breast is a potential cause of mammary carcinogenesis. In fact, DNA adducts derived from CYP1A1- and 1B1-mediated metabolism of B[a]P, PhIP, and ABP were identified in exfoliated ductal epithelial cells in breast milk obtained from healthy, nonsmoking mothers (41). These data show that women are exposed to dietary and environmental xenobiotics, which are converted to genotoxic metabolites capable of reacting with DNA in breast ductal epithelial cells, the cells from which breast cancers arise.

CYP1A1 is not only important for catalyzing endogenous and xenobiotic substrates but also involved in regulating the proliferation of breast cancer cells (42). CYP1A1 knockdown in MCF-7 and MDA-MB-231 cell lines decreased cell proliferation and colony formation, blocked the cell cycle at G0/G1 associated with reduction of cyclin D1, and increased apoptosis associated with reduction of survivin. The inhibitory effect of CYP1A1 silencing on cell proliferation appears to be mediated in part through increased phosphorylation of AMP-activated protein kinase (AMPK), and reduced phosphorylation of AKT, extracellular signal-

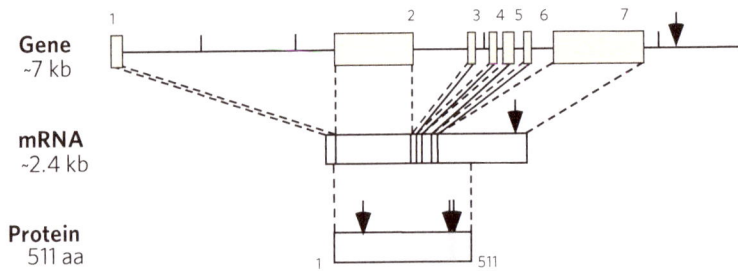

Figure 4.2.2. Overview of CYP1A1 gene, mRNA, and protein.
The CYP1A1 gene at 15q22-qter is about 7 kb long and contains seven exons. The regulatory region (not shown) contains two CpG islands located between -2,813 and -3,567 bp and between -178 and -1,712 bp upstream of the transcription start site. The latter island contains all of the XRE sites that are implicated in the AhR-mediated induction of CYP1A1 expression. The open reading frame starts in the second exon and is 1533 base pairs in length, encoding a protein of 511 amino acids. The arrows indicate the common nonsynonymous polymorphisms Gly45Asp (rs4646422), Thr461Asn (rs1799814), and Ile462Val (rs1048943). The shaded area indicates the heme-binding region. Two polymorphisms are located in the 3'-noncoding region, T3205C (rs4986883) and T3801C (rs4646903), which give rise to MspI restriction sites.

regulated kinases 1 and 2 (ERK1/2), and 70-kDa ribosomal protein S6 kinase (P70S6K).

The human CYP1A1 gene contains over 20 polymorphisms (Figure 4.2.2.)(43-52). The majority are rare or of unknown functional significance, such as a MspI restriction fragment length polymorphism downstream from the polyadenylation sequence (T3801C; rs4646903) and another MspI site in the 3'-noncoding region about 300 bp upstream from the polyadenylation sequence (T3205C; rs4986883), which is present in about 11% of African Americans. Most of the nonsynonymous SNPs are located in either exon 2 or 7, i.e., Gly45Asp (rs4646422), Thr461Asn (rs1799814), and Ile462Val (rs1048943). In spite of their proximity, the latter two SNPs are not linked. The frequencies of these polymorphisms vary between different ethnic groups (51, 53, 54). For example, the variant Gly45Asp is found in about 10% Asians but not in Caucasians or African Americans. Thr461Asn is found in 4% Caucasians and 0.4% African Americans while the variant Ile462Val is present in 5% Caucasians and 23% Asians. Apart from the wild-type (CYP1A1*1), 14 alleles have been described in different populations (55) (http://www.CYPalleles.ki.se/).

The effect of the Ile462Val polymorphism on PAH metabolism has been examined in several studies with mixed results. Urinary metabolites of phenanthrene were determined in smokers, especially r-1,t-2,3,c-4-tetrahydroxy-1,2,3,4-tetrahydro phenanthrene (PheT) as a measure of metabolic activation and phenanthrols (HOPhe) as a measure of detoxification. The ratio PheT/3-HOPhe was significantly increased in individuals heterozygous or homozygous for 462Val (56). The ratio was highest in females and heavy smokers. Other studies showed a range

of effects of the polymorphism on PAH metabolism in lung microsomes or in systems using expressed enzymes (49, 52, 57, 58). Catalytic efficiencies ranged from no difference to three times higher than the wild type in producing mutagenic metabolites. An analysis of B[a]P diolepoxide (BPDE)-DNA adducts in leukocytes of 41 smokers with combinations of genotypes showed significantly higher adduct levels in individuals with the CYP1A1 462Val and GSTM1 null alleles (59). With regard to estrogen metabolism, the 462Val variant displayed significantly higher catalytic activity, especially for the 2-hydroxylation (60, 61). The catalytic efficiencies for 2-OHE_2 and 2-OHE_1 were 5.7- and 12-fold higher, respectively, compared with the wild-type 462Ile enzyme. The Ile462Val polymorphism also affects the metabolism of eicosapentaenoic acid. The catalytic efficiency of the 462Val variant was five times higher for hydroxylation and twice higher for epoxygenation than the wild type 462Ile enzyme (30).

Associations of CYP1A1 polymorphisms and breast cancer risk have been widely reported, but these associations have been inconsistent with some studies reporting increased risk, while others show no effect or even decreased risk (53, 62-70). A meta-analysis found no association of risk with T3801C (rs4646903; 32 studies with 11,909 cases and 16,179 controls), T3205C (rs4986883; 8 studies with 1,378 cases and 1,642 controls), and Thr461Asn (C2453A, rs1799814; 11 studies with 7,189 cases and 8,491 controls) (71). Only the Ile462Val polymorphism (A2455G, rs1048943; 29 studies with 12,257 cases and 20,379 controls) was associated with increased breast cancer risk in Caucasian women carrying the Val/Val genotype (OR 2.185; 95% CI 1.253 – 3.808) but not heterozygous carriers (OR 1.062;

95% CI 0.852 – 1.323). There was no risk association of the Ile462Val polymorphism in Asians or Africans although an earlier meta-analysis observed a reduced risk in east-Asian women with a Val/Val genotype (72). One study reported that the elevated risk of breast cancer associated with postmenopausal estrogen plus progestin hormone therapy was greater among women carrying at least one 462Val allele but two other studies found no interaction (73-75).

The inconsistencies may be due to the fact that there is a striking inter-individual variation in CYP1A1 expression in addition to the genetic variation. For example, CYP1A1 mRNA expression varied ~400-fold between non-tumor breast specimens from breast cancer patients (n = 26) and cancer-free individuals (n = 32) and was independent of CYP1A1 genotype and patient age (76). A second study used quantitative immunoblotting of normal and malignant breast tissues and observed ~150-fold differences in CYP1A1 protein expression between individuals (77). Attempts to explain such high degree of inter-individual variation in CYP1A1 expression have focused primarily on genetic polymorphisms within the CYP1A1 gene with inconsistent results. Since the expression of CYP1A1 is induced via the aryl hydrocarbon receptor (AhR), investigators extended the analysis to the **AhR gene** (78). The AhR-mediated CYP1A1 induction appears to be influenced by the G1721A (Arg554Lys) polymorphism in exon 10 of the AhR gene. The 554Arg residue lies close to the transactivation domain of the AhR protein. Individuals with at least one copy of the variant 1721A allele showed significantly higher levels of induced CYP1A1 activity compared with individuals negative for the polymorphism (p = 0.0001). Interestingly, the Shanghai case-control study showed an association of the A (Lys) allele with decreased risk of breast cancer in premenopausal women (79); the CYP1A1 genotype was not assessed. Levels of 3-methylcholanthrene-induced CYP1A1 activity in lymphocytes also showed a sex difference with women exhibiting significantly lower activity than men (78). Thus, inter-individual variation in levels of induced CYP1A1 activity appears to be associated more with regulatory factors than polymorphisms in the CYP1A1 gene. In general, the effect of smoking-induced increase in CYP1A1 activity appears to override genetic differences in enzyme activity (58).

CYP1A1 polymorphisms have also been examined for joint effects with polymorphisms in other genes, e.g., COMT, ERα, GSTM1 (53, 66, 68). Only the combination of a polymorphism in the 3' noncoding region of the CYP1A1 gene (T6235C; MspI) with variant ERα XbaI and PvuII genotypes was associated with increased risk in a case-control study of 580 Chinese women (68). Finally, somatic mutations in the CYP1A1 gene have

been detected in breast cancers, most likely reflecting mutational selection during tumorigenesis (80).

References

1. Kawajiri K, Watanabe J, Gotah O, Tagashira Y, Sogawa K, Fujii-Kuriyama Y. *Structure and drug inducibility of the human cytochrome P-450c gene.* Eur J Biochem. 1986;159:219-25.

2. Bradfield CA, Bjeldanes LF. *Effect of dietary indole-3 carbinol on intestinal and hepatic monooxygenase, glutathione S-transferase and epoxide hydrolase activities in the rat.* Food Chem Toxicol. 1984;22:977-82.

3. Conney AH. *Induction of drug-metabolizing enzymes: A path to the discovery of multiple cytochromes P450.* Ann Rev Pharmacol Toxicol. 2003;43:1-30.

4. McLemore TL, Adelberg S, Liu MC, McMahon NA, Yu SJ, Hubbard WC, et al. *Expression of CYP1A1 gene in patients with lung cancer: evidence for cigarette smoke-induced gene expression in normal lung tissue and for altered gene regulation in primary pulmonary carcinomas.* J Natl Cancer Inst. 1990;82:1333-9.

5. Nebert DW, Nelson DR, Coon MJ, Estabrook RW, Feyereisen R, Fujii-Kuriyama Y, et al. *The P450 superfamily: update on new sequences, gene mapping, and recommended nomenclature.* DNA Cell Biol. 1991;10:1-14.

6. Spink DC, Katz BH, Hussain MM, Spink BC, Wu SJ, Liu N, et al. *Induction of CYP1A1 and CYP1B1 in T-47D human breast cancer cells by benzo[a]pyrene is diminished by arsenite.* Drug Metab Dispos. 2002;30:262-9.

7. Denison MS, Pandini A, Nagy SR, Baldwin EP, Bonati L. *Ligand binding and activation of the Ah receptor.* Chemico-Biological Interactions. 2002;141:3-24.

8. Perdew GH. *Association of the Ah receptor with the 90-kDa heat shock protein.* J Biol Chem. 1988;263:13802-5.

9. Whitlock JP. *Induction of cytochrome P4501A1.* Ann Rev Pharm Toxicol. 1999;39:103-25.

10. Hoffman EC, Reyes H, Chu FF, Sander F, Conley LH, Brooks BA, et al. *Cloning of a factor required for activity of the Ah (dioxin) receptor.* Science. 1991;252:954-8.

11. Okino ST, Pookot D, Li L-C, Zhao H, Urakami S, Shiina H, et al. *Epigenetic inactivation of the dioxin-responsive cytochrome P4501A1 gene*

in human prostate cancer. Cancer Res. 2006;66:7420-8.

12. Burbach KM, Poland A, Bradfield CA. *Cloning of the Ah-receptor cDNA reveals a distinctive ligand-activated transcription factor.* Proc Natl Acad Sci USA. 1992;89:8185-9.

13. Kress S, Greenlee WF. *Cell-specific regulation of human CYP1A1 and CYP1B1 genes.* Cancer Res. 1997;57:1264-9.

14. Okey AB, Riddick DS, Harper PA. The Ah receptor: *Mediator of the toxicity of 2,3,7,8-tetrachlorodibenzo-p-dioxin (TCDD) and related compounds.* Toxicol Lett. 1994;70:1-22.

15. Swanson HI, Bradfield CA. *The AH-receptor: genetics, structure and function.* Pharmacogenetics. 1993;3:213-30.

16. Swanson HI, Chan WK, Bradfield CA. *DNA binding specificities and pairing rules of the Ah receptor, ARNT, and SIM proteins.* J Biol Chem. 1995;270:26292-302.

17. Quintana FJ, Basso AS, Iglesias AH, Korn T, Farez MF, Bettelli E, et al. *Control of Treg and TH17 cell differentiation by the aryl hydrocarbon receptor.* Nature. 2008;453:65-72.

18. Stevens EA, Bradfield CA. *T cells hand in the balance.* Nature. 2008;453:46-7.

19. Veldhoen M, Hirota K, Westendorf AM, Buer J, Dumoutier L, Renauld J, et al. *The aryl hydrocarbon receptor links TH17-cell-mediated autoimmunity to environmental toxins.* Nature. 2008;453:106-10.

20. Omiecinski CJ, Redlich CA, Costa P. *Induction and developmental expression of cytochrome P4SIA1 messenger RNA in rat and human tissues: detection by the polymerase chain reaction.* Cancer Res. 1990;50:4315-21.

21. Shimada T, Hayes CL, Yamazaki H, Amin S, Hecht SS, Guengerich FP, et al. *Activation of chemically diverse procarcinogens by human cytochrome P-450 1B1.* Cancer Res. 1996;56:2979-84.

22. Vanden Heuvel JP, Clark GC, Thompson CL, McCoy Z, Miller CR, Lucier GW, et al. *CYP1A1 mRNA levels as a human exposure biomarker: use of quantitative polymerase chain reaction to measure CYP1A1* expression in human peripheral blood lymphocytes. Carcinogenesis. 1993;14:2003-6.

23. Murray GI, Patimalla S, Stewart KN, Miller ID, Heys SD. *Profiling the expression of cytochrome P450 in breast cancer.* Histopathology. 2010;57:202-11.

24. Vinothini G, Nagini S. *Correlation of xenobiotic-metabolizing enzymes, oxidative stress and NFkappaB signaling with histological grade and menopausal status in patients with adenocarcinoma of the breast.* Clinica Chim Acta. 2010;411:368-74.

25. Hakkola J, Pasanen M, Pelkonen O, Hukkanen J, Evisalmi S, Anttila S, et al. *Expression of CYP1B1 in human adult and fetal tissues and differential inducibility of CYP1B1 and CYP1A1 by ah receptor ligands in human placenta and cultured cells.* Carcinogenesis. 1997;18:391-7.

26. Mollerup S, Ryberg D, Hewer A, Phillips DH, Haugen A. *Sex differences in lung CYP1A1 expression and DNA adduct levels among lung cancer patients.* Cancer Res. 1999;59:3317-20.

27. MacDonald CJ, Ciolino HP, Yeh GC. *The drug salicylamide is an antagonist of the aryl hydrocarbon receptor that inhibits signal transduction induced by 2,3,7,8-tetrachlorodibenzo-p-dioxin.* Cancer Res. 2004;64:429-34.

28. Anttila S, Hakkola J, Tuominen P, Elovaara E, Husgafvel-Pursiainen K, Karjalainen A, et al. *Methylation of cytochrome P4501A1 promoter in the lung is associated with tobacco smoking.* Cancer Res. 2003;63:8623-8.

29. Schwarz D, Kisselev P, Ericksen SS, Szklarz GD, Chernogolov A, Honeck H, et al. *Arachidonic and eicosapentaenoic acid metabolism by human CYP1A1: highly stereoselective formation of 17(R), 18(S)-epoxyeicosatetraenoic acid.* Biochem Pharmacol. 2004;67:1445-57.

30. Schwarz D, Kisselev P, Chernogolov A, Schunck W, Roots I. *Human CYP1A1 variants lead to differential eicosapentaenoic acid metabolite patterns.* Biochem Biophys Res Commun. 2005;336:779-83.

31. Schwarz D, Kisselev P, Schunck W, Chernogolov A, Boidol W, Cascorbi I, et al. *Allelic variants of human cytochrome P450 1A1 (CYP1A1): effect of T461N and I462V substitutions on steroid hydroxylase specificity.* Pharmacogenetics. 2000;10:519-30.

32. Spink DC, Eugster H, Lincoln DWI, Schuetz JD, Schuetz EG, Johnson JA, et al. *17 beta-estradiol hydroxylation catalyzed by human cytochrome P450 1A1: a comparison of the activities induced by 2,3,7,8-tetrachlorodibenzo-p-dioxin in MCF-7 cells with those from heterologous expression of the cDNA.* Arch Biochem Biophys. 1992;293:342-8.

33. Spink DC, Hayes CL, Young NR, Christou M, Sutter TR, Jefcoate CR, et al. *The effects of 2,3,7,8-tetrachlorodibenzo-p-dioxin on estrogen metabolism in MCF-7 breast cancer cells: evidence for induction of a novel 17 beta-estradiol 4-hydroxylase.* J Steroid Biochem Mol Biol. 1994;51:251-8.

34. Spink DC, Lincoln DWI, Dickerman HW, Gierthy JF. *2,3,7,8-Tetrachlorodibenzo-p-dioxin causes an extensive alteration of 17 beta-estradiol metabolism in MCF-7 breast tumor cells.* Proc Natl Acad Sci USA. 1990;87:6917-21.

35. Roy D, Bernhardt RD, Strobel HW, Liehr JC. *Catalysis of the oxidation of steroid and stilbene estrogens to estrogen quinone metabolites by the beta-naphthoflavone-inducible cytochrome P450 IA family.* Arch Biochem Biophys. 1992;296:450-6.

36. Zhang Y, Gaikwad NW, Olson K, Zahid M, Cavalieri EL, Rogan EG. *Cytochrome P450 isoforms catalyze formation of catechol estrogen quinones that react with DNA.* Metabolism Clin Exper. 2007;56:887-94.

37. Cribb AE, Knight MJ, Dryer D, Guernsey J, Hender K, Tesch M, et al. *Role of polymorphic human cytochrome P450 enzymes in estrone oxidation.* Cancer Epidemiol Biomarkers Prev. 2006;15:551-8.

38. Crofts FG, Sutter TR, Strickland PT. *Metabolism of 2-amino-1-methyl-6-phenylimidazo[4,5-b]pyridine by human cytochrome p4501A1, p4501A2 and p4501B1.* Carcinogenesis. 1998;19:1969-73.

39. Guengerich FP, Shimada T. *Activation of procarcinogens by human cytochrome P450 enzymes.* Mutation Res. 1998;400:201-13.

40. Shimada T, Inoue K, Suzuki Y, Kawai T, Azuma E, Nakajima T, et al. *Arylhydrocarbon receptor-dependent induction of liver and lung cytochromes P450 1A1, 1A2, and 1B1 by polycyclic aromatic hydrocarbons and polychlorinated biphenyls in genetically engineered C57BL/6J mice.* Carcinogenesis. 2002;23:1199-207.

41. Gorlewska-Roberts K, Green B, Fares M, Ambrosone CB, Kadlubar FF. *Carcinogen-DNA adducts in human breast epithelial cells.* Environ Mol Mutagenesis. 2002;39:184-92.

42. Rodriguez M, Potter DA. *CYP1A1 regulates breast cancer proliferation and survival.* Mol Cancer Res : MCR. 2013;11:780-92.

43. Cascorbi I, Brockmoller J, Roots I. *A C4887A polymorphism in Exon 7 of Human CYP1A1: population frequency, mutation linkages, and impact on lung cancer suseptibility.* Cancer Res. 1996;56:4965-9.

44. Chevalier D, Allorge D, Lo-Guidice JM, Cauffiez C, Lhermitte M, Lafitte JJ, et al. *Detection of known and two novel (M331I and R464S) missense mutations in the human CYP1A1 gene in a French Caucasian population.* Hum Mutation. 2001;232:355.

45. Crofts F, Cosma GN, Currie D, Taioli E, Toniolo P, Garte SJ. *A novel CYP1A1 gene polymorphism in African-Americans.* Carcinogenesis. 1993;14:1729-31.

46. Hamada GS, Sugimura H, Suzuki I, Nagura K, Kiyokawa E, Iwase T, et al. *The heme-binding region polymorphism of cytochrome P4501A1 (Cyp1A1), rather than the Rsal polymorphism of IIE1 (CypIIE1), is associated with lung cancer in Rio de Janeiro.* Cancer Epidemiol Biomarkers Prev. 1995;4(/):63-7.

47. Hayashi S, Watanabe J, Nakachi K, Kawajiri K. *Genetic linkage of lung cancer-associated MspI polymorphisms with amino acid replacement in the heme binding region of the human cytochrome P450IA1 gene.* J Biochem. 1991;110:407-11.

48. Kawajiri K, Nakachi K, Imai K, Watanabe J, Hayashi S. *The CYP1A1 gene and cancer susceptibility.* Crit Rev Oncol-Hemat. 1993;14:77-87.

49. Persson I, Johansson I, Ingelman-Sundberg M. *In vitro kinetics of two human CYP1A1 variant enzymes suggested to be associated with interindividual differences in cancer susceptibility.* Biochem Biophys Res Commun. 1997;231:227-30.

50. Saito T, Egashira M, Kiyotani K, Fujieda M, Yamazaki H, Kiyohara C, et al. *Novel nonsynonymous polymorphisms of the CYP1A1 gene in Japanese.* Drug Metab Pharmacokin. 2003;18:218-21.

51. Solus JF, Arietta BJ, Harris JR, Sexton DP, Steward JQ, McMunn C, et al. *Genetic variation in eleven phase I drug metabolism genes in an ethnically diverse population.* Pharmacogenetics. 2004;5:895-931.

52. Zhang Z, Fasco MJ, Huang L, Guengerich FP, Kaminsky LS. *Characterization of purified human recombinant cytochrome P4501A1-Ile 462 and -Val 462 Assessment of a role for the rare allele in carcinogenesis.* Cancer Res. 1996;56:3926-33.

53. Bailey LR, Roodi N, Verrier CS, Yee CJ, Dupont WD, Parl FF. *Breast cancer and CYP1A1, GSTM1, and GSTT1 polymorphisms: Evidence of a lack of association in Caucasians and African Americans.* Cancer Res. 1998;58:65-70.

54. Garte S, Gaspari L, Alexandrie AK, Ambrosone C, Autrup H, Aurup JL, et al. *Metabolic gene polymorphism frequencies in control populations.* Cancer Epidemiol Biomarkers Pre 2001;10:1239-48.

55. Ingelman-Sundberg M, Oscarson M, Daly AK, Garte S, Nebert DW. *Human Cytochrome P-450 (CYP) genes: A web page for the nomenclature of alleles.* Cancer Epidemiol Biomarkers Prev. 2001;10:1307-8.

56. Hecht SS, Carmella SG, Yoder A, Chen M, Li Z, Le C, et al. *Comparison of polymorphisms in genes involved in polycyclic aromatic hydrocarbon metabolism with urinary phenanthrene metabolite ratios in smokers.* Cancer Epidemiol Biomarkers Prev. 2006;15:1805-11.

57. Schwarz D, Kisselev P, Cascorbi I, Schunck W, Roots I. *Differential metabolism of benzo[a]pyrene and benzo[a]pyrene-7,8-dihydrodiol by human CYP1A1 variants.* Carcinogenesis. 2001;22:453-9.

58. Smith GBJ, Harper PA, Wong JMY, Lam MSM, Reid KR, Petsikas D, et al. *Human lung microsomal cytochrome P4501A1 (CYP1A1) activies: Impact of smoking status and CYP1A1, aryl hydrocarbon receptor, and glutathione S-transferase M1 genetic polymorphisms.* Cancer Epidemiol Biomarkers Prev. 2001;10:839-53.

59. Lodovici M, Luceri C, Guglielmi F, Bacci C, Akpan V, Fonnesu ML, et al. *Benzo(a)pyrene diolepoxide (BPDE)-DNA adduct levels in leukocytes of smokers in relation to polymorphism of CYP1A1, GSTM1, GSTP1, GSTT1, and mEH.* Cancer Epidemiol Biomarkers Prev. 2004;13:1342-8.

60. Crooke PS, Ritchie MD, Hachey DL, Dawling S, Roodi N, Parl FF. *Estrogens, enzyme variants, and breast cancer: A risk model.* Cancer Epidemiol Biomarkers Prev. 2006;15:1620-9.

61. Kisselev P, Schunck W-H, Roots I, Schwarz D. *Association of CYP1A1 polymorphisms with differential metabolic activation of 17b-estradiol and estrone.* Cancer Res. 2005;65:2972-8.

63. Masson LF, Sharp L, Cotton SC, Little J. *Cytochrome P-450 1A1 gene polymorphisms and risk of breast cancer: A HuGE review.* Am J Epidemiol. 2005;161:901-15.

64. Mitrunen K, Hirvonen A. *Molecular epidemiology of sporadic breast cancer. The role of polymorphic genes involved in oestrogen biosynthesis and metabolism.* Mutation Res. 2003;544:9-41.

65. Miyoshi Y, Takahashi Y, Egawa C, Noguchi S. *Breast cancer risk associated with CYP1A1 genetic polymorphisms in Japanese women.* Breast J. 2002;8:209-15.

66. Modugno F, Zmuda JM, Potter D, Cai C, Ziv E, Cummings SR, et al. *Estrogen metabolizing polymorphisms and breast cancer risk among older white women.* Breast Cancer Res Treat. 2005;93:261-70.

67. Rebbeck TR, Troxel AB, Walker AH, Panossian S, Gallagher S, DeMichele A, et al. *Pairwise combinations of estrogen metabolism genotypes in postmenopausal breast cancer etiology.* Cancer Epidemiol Biomarkers Prev. 2007;16:444-9.

68. Shen Y, Li D, Wu J, Zhang Z, Gao E. *Joint effects of the CYP1A1 MspI, ERα PvuII, and ERα XbaI polymorphisms on the risk of breast cancer: Results from a population-based case-control study in Shanghai, China.* Cancer Epidemiol Biomarkers Prev. 2006;15:342-7.

69. Taioli E, Trachman J, Chen X, Toniolo P, Garte SJ. *A CYP1A1 restriction fragment length polymorphism is associated with breast cancer in African-American women.* Cancer Res. 1995;55:3757-8.

70. Yao L, Yu X, Yu L. *Lack of significant association between CYP1A1 T3801C polymorphism and breast cancer risk: a meta-analysis involving 25,087 subjects.* Breast Cancer Res Treat. 2010;122:503-7.

71. Sergentanis TN, Economopoulos KP. *Four polymorphisms in cytochrome P450 1A1 (CYP1A1) gene and breast cancer risk: a meta-analysis.* Breast Cancer Res Treat. 2010;122:459-69.

72. Chen C, Huang Y, Li Y, Mao Y, Xie Y. *Cytochrome P450 1A1 (CYP1A1) T3801C and A2455G polymorphisms in breast cancer risk: a meta-analysis.* J Hum Genet. 2007;52:423-35.

73. Diergaarde B, Potter JD, Jupe ER, Manjeshwar S, Shimasaki CD, Pugh TW, et al. *Polymorphisms in genes involved in sex hormone metabolism, estrogen plus progestin hormone therapy use, and risk of postmenopausal breast cancer.* Cancer Epidemiol Biomarkers Prev. 2008;17:1751-9.

74. Rebbeck TR, Troxel AB, Shatalova EG, Blanchard R, Norman S, Bunin G, et al. *Lack of effect modification between estrogen metabolism genotypes and combined hormone replacement therapy in postmenopausal breast cancer risk.* Cancer Epidemiol Biomarkers Prev. 2007;16:1318-20.

75. Marie-Genica Consortium on Genetic Susceptibility for Menopausal Hormone Therapy Related Breast Cancer Risk. *Genetic polymorphisms in phase I and phase II enzymes and breast cancer risk associated with menopausal hormone therapy in postmenopausal women.* Breast Cancer Res Treat. 2010;119:463-74.

76. Goth-Goldstein R, Stampfer MR, Erdmann CA, Russell M. *Interindividual variation in CYP1A1 expression in breast tissue and the role of genetic polymorphism.* Carcinogenesis. 2000;21:2119-22.

77. El-Rayes BF, Ali S, Heilbrun LK, Lababidi S, Bouwman D, Visscher D, et al. *Cytochrome P450 and glutathione transferase expression in human breast cancer.* Clin Cancer Res. 2003;9:1705-9.

78. Smart J, Daly AK. *Variation in induced CYP1A1 levels: relationship to CYP1A1, Ah receptor and GSTM1 polymorphisms.* Pharmacogenetics. 2000;10:11-24.

79. Long J, Egan KM, Dunning L, Shu X, Cai Q, Cai H, et al. *Population-based case-control study of AhR (aryl hydrocarbon receptor) and CYP1A2 polymorphisms and breast cancer risk.* Pharmacogenet Genomics. 2006;16:237-43.

80. Sjoblom T, Jones S, Wood LD, Parsons DW, Lin J, Barber TD, et al. *The consensus coding sequences of human breast and colorectal cancers.* Science. 2006;314:268-74.

4.2.3. CYP1B1

Gene and Protein. The CYP1B1 gene maps to chromosome 2p21 and contains three exons and two introns. The entire coding sequence is contained in exons 2 and 3 (Figure 4.2.3.). The 55-kDa protein contains several conserved regions, including a transmembrane segment, hinge region, meander region, as well as a heme-binding region (1, 2). Analysis of the CYP1B1 amino acid sequence shows extended regions of ≥50% identity to CYP1A1, e.g., the heme-binding region (amino acids 463 to 488, including 470Cys), which forms the recognition motif for binding the axial heme ligand that defines many of the functional and spectral characteristics of the cytochrome P450 superfamily. Putative CYP1B1 substrate recognition sites have been located (3, 4). Several conserved helices are found in the C-terminal half of the molecule (Figure 4.2.3.).

The promoter region of the CYP1B1 gene contains a TATA-like box and two Sp1 binding sites (5, 6). The enhancer region contains eight potential xenobiotic response elements (XREs) (7, 8). The expression of the CYP1B1 gene is regulated by multiple mechanisms at the transcriptional, post-transcriptional, translational, and post-translational levels. One form of transcriptional regulation occurs via binding of the aryl hydrocarbon receptor (AhR) to the XREs, a mechanism responsible for the CYP1B1 induction by cigarette smoking, B[a]P, and dioxin (7-11). Dioxins, such as TCDD, can induce the expression multifold. For example, treatment of epidermal keratinocytes with TCDD increased CYP1B1 mRNA about 100-fold (12). CYP1B1 mRNA expression is also induced by estradiol via ERα (8). A third form of transcriptional regulation occurs by hypomethylation of the promoter region of the CYP1B1 gene, which causes an increase in the amount of mRNA produced (13). MicroRNAs destabilize mRNA and interfere with translation. A specific microRNA, miR-27B, was identified whose concentration was inversely correlated with the expression level of CYP1B1 protein (14). The overexpression of CYP1B1 protein observed in cancer cells may arise because the latter are largely devoid of miR-27B, whereas in normal cells miR-27B inhibits the translation of CYP1B1 mRNA into protein. Finally, a post-translational mechanism involves proteosomal degradation of CYP1B1 protein, targeted through polyubiquination, but not phosphorylation of the protein (15).

Enzymatic Activity. CYP1B1 oxidizes endogenous substrates, e.g., estrogens, retinoids. CYP1B1 is important for the homeostasis of estrogen in extrahepatic tissues such as the breast. The enzyme sequentially oxidizes E_2 to catechol estrogens and estrogen quinones (16-20). Although other CYPs, such as CYP1A2 and CYP3A4, are involved in hepatic and extrahepatic estrogen oxidation, CYP1A1 and CYP1B1 display the highest levels of expression in breast tissue (21-23). In turn, CYP1B1 exceeds CYP1A1 in its catalytic efficiency as E_2 hydroxylase and differs from CYP1A1 in its principal site of catalysis (19, 24, 25). CYP1B1 has its primary activity at the C-4 position of E_2, whereas CYP1A1 has its primary activity at the C-2 position in preference to 4-hydroxylation. Kinetic studies indicate that the E_2 C-4 hydroxylase activity of CYP1B1 has the highest catalytic efficiency and the lowest K_m of all the E_2 hydroxylases reported (19, 26, 27). Thus, CYP1B1 appears to be the main cytochrome P450 enzyme responsible for the C-4 hydroxylation of E_2. The same applies to the oxidation of E_1 to 4-OHE$_1$ (28). Given the carcinogenic potential that has emerged for 4-OHE$_2$ and 4-OHE$_1$, CYP1B1 assumes a special role as the main enzyme producing this catechol estrogen and the corresponding quinones, E_2-3,4-Q and E_1-3,4-Q, which react with glutathione, proteins, or DNA, the latter resulting in predominantly depurinating adducts that can generate mutations (29). CYP1B1 also oxidizes all-*trans*-retinol to all-*trans*-retinal, the rate-limiting step for retinoic acid biosynthesis.(30). CYP1B1 is also involved in the metabolism of the antiestrogen tamoxifen, catalyzing the *trans-cis* isomerization of *trans*-4-hydroxytamoxifen to the weak estrogen agonist *cis*-4-hydroxytamoxifen (31).

CYP1B1 also activates many structurally diverse xenobiotics, including PAHs and their dihydrodiol derivatives, heterocyclic and aryl amines, and nitroaromatic hydrocarbons to genotoxic metabolites (32-36). Investigators compared the relative capacity of each member of the CYP1 enzyme family to activate a variety of carcinogenic chemicals (37). Although CYP1B1 did not produce genotoxic products from classic carcinogens, such as aflatoxin B1, it had the

highest catalytic activity for several procarcinogenic compounds. The involvement of CYP1B1 in the activation of environmental carcinogens is of particular interest with respect to its expression in breast tissue and the storage of these chemicals in mammary adipose tissue. In general, the expression of CYP1B1 in normal breast tissue would enable local production of potentially carcinogenic estrogen metabolites as well as environmental carcinogens, lessening the need to invoke complex pharmacokinetic schemes for the redistribution of more polar and/or conjugated metabolites from the liver.

A number of endogenous and exogenous compounds are capable of inhibiting reactions catalyzed by members of the CYP1 family (38). For example, arachidonic acid, retinoids, and cholecalciferol inhibited the CYP1A1-dependent O-deethylation of 7-ethoxycoumarin (39). Methoxyestrogens were shown to inhibit CYP1B1 as well as CYP1A1 thereby exerting feedback inhibition on the oxidative estrogen metabolism pathway initiated by these two enzymes (40). Several natural and synthetic xenobiotics (e.g., resveratrol, hesperetin, homoeriodictyol, 7-hydroxyfla-vone, rhapontigenin, 2,4,3',5'-tetramethoxystilbene [TMS]) have also been shown to inhibit the enzymatic activity of CYP1 family members (41-43). Hydroxy and/or methoxy substitutions at the 3' and 4' positions in flavanoids conferred selectivity for CYP1 inhibition (42). For example, the flavanone homoeriodictyol selectively inhibited CYP1B1, but not CYP1A1 and CYP1A2. Hesperetin was O-demeth-ylated at the 4' position by both CYP1A1 and CYP1B1 to

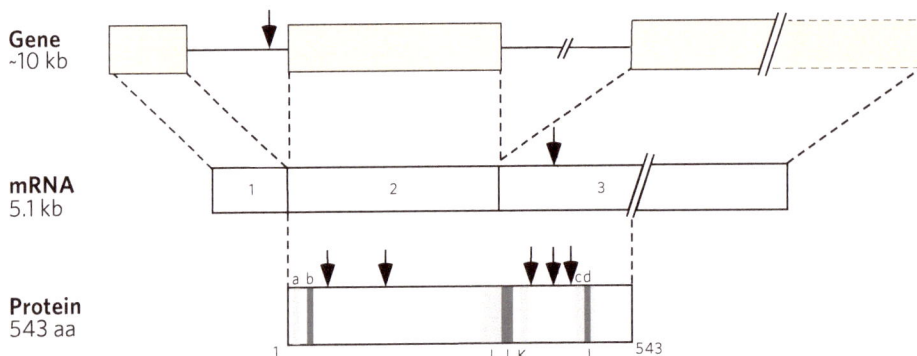

Figure 4.2.3. Overview of CYP1B1 gene, mRNA, and protein.
The CYP1B1 gene at 2p21 is about 10 kb long and contains three exons. The open reading frame starts in the second exon and is 1629 base pairs in length, encoding a protein of 543 amino acids. Conserved regions indicated by shaded boxes include a transmembrane segment (a, amino acids 12–36), hinge region (b, amino acids 51–58), meander region (c, amino acids 438–462), and the heme-binding region (d, amino acids 463–488). Conserved helices named I, J, K, and L are found at amino acids 316–349, 350–364, 379–391, and 471–487, respectively. A variety of mutations have been identified in families with primary congenital glaucoma. Among these are at least 15 missense mutations, found throughout the coding region, representing non-conservative changes in highly conserved regions of the protein. In contrast, the common polymorphisms are present in not well-conserved regions of the protein including five nonsynonymous polymorphisms (arrows): Arg48Gly (rs10012), Ala119Ser (rs1056827), Val432Leu (rs1056836), Ala443Gly (rs4986888), and Asn453Ser (rs1800440). The lengths of the exons are drawn in scale, while those of the introns are not.

Table 4.2.1. CYP1B1 Haplotypes in Caucasians (65, 66)

Allele	Amino acid codon					Frequency (%)
	48	119	432	443	453	
CYP1B1* (wild type)	Arg	Ala	Leu	Ala	Asn	14.7
CYP1B1*2	**Gly**	**Ser**	Leu	Ala	Asn	25.4
CYP1B1*3	Arg	Ala	**Val**	Ala	Asn	38.8
CYP1B1*4	Arg	Ala	Leu	Ala	**Ser**	18.1
CYP1B1*5	**Gly**	Ala	**Val**	Ala	Asn	0.0
CYP1B1*6	**Gly**	**Ser**	**Val**	Ala	Asn	0.4
CYP1B1*7	**Gly**	**Ser**	**Val**	**Gly**	Asn	2.6

the corresponding 4'-hydroxylated flavonoid, eriodictyol, which was then further metabolized by the same enzymes. Glycosides of these flavonoids are major constituents of citrus fruits. Hesperidin (hesperetin-7-rutinoside) is the major flavonoid in orange juice while eriocitrin (eriodictyol-7-rutinoside) is the major flavonoid of lemon and lime juices (44). Considering the possible role of CYP1B1 in activating carcinogens and the apparent selectivity of hesperetin and homoeriodictyol as inhibitors of this enzyme, these dietary compounds may protect against the development of certain cancers. The synthetic TMS, a methoxy derivative of oxyresveratrol, exhibited 50-fold selectivity for CYP1B1 over CYP1A1 and 500-fold selectivity for CYP1B1 over CYP1A2. In particular, TMS strongly inhibited 2- and 4-hydroxylation of E_2 by CYP1B1 with a K_i of 3 nM (41). In addition to its inhibitory action on CYP1B1, TMS induced apoptosis, inhibited microtubule polymerization, and blocked the cell cycle at the G2-M phase, making the drug a potentially beneficial agent for the treatment of hormone-resistant cancer (45, 46). Perillyl alcohol, a dietary monoterpene found in mints, cherries, and cranberries, was shown to inhibit CYP1B1 but not CYP1A1 (47). Several anticancer drugs were also shown to inhibit CYP1B1 activity. Using the ethoxyresorufin O-deethylase assay, paclitaxel, doxorubicin, and tamoxifen yielded K_i values 31.6, 2.6, and 5.0 μM, respectively (48).

Expression. CYP1B1 is widely expressed in extrahepatic tissues. Constitutive expression is observed in steroidogenic glands like the adrenal, ovary, and testis, with inducibility by ACTH and cAMP (12, 49). However, CYP1B1 is not coordinately expressed with steroidogenic enzymes. CYP1B1 is also expressed in steroid-responsive tissue of mesodermal origin, such as the uterus, breast, and prostate. The enzyme is also detected in kidney, lung, and white blood cells (10, 22, 50-52). CYP1B1 is overexpressed in a range of malignant tumors, including breast, prostate, and ovarian cancer (13, 22, 53). Murray and colleagues have performed several immunohistochemical studies of CYP1B1 expression in breast (22, 23). Breast cancer, but not normal breast tissue, revealed CYP1B1 expression (53). Forty-six of 60 (77%) invasive breast cancers showed cytoplasmic staining of tumor cells, ranging from strong

in 10, to moderate in 12, and weak immunoreactivity in 24 cases (21). There was no statistical relationship between the presence of CYP1B1 and the histological type of the tumor, tumor grade, or the presence of lymph node metastasis but a strong positive correlation with ERα and PR expression (23). Interestingly, a comparison of CYP1B1 mRNA levels in breast cancer with adjacent non-tumor tissues showed higher levels in the latter (54). Normal human mammary epithelial cells isolated and cultured from reduction mammoplasty tissue expressed significant levels of CYP1B1 determined by immunoblot analysis (55). The discrepancy between the studies regarding the absence or presence of CYP1B1 protein in normal mammary epithelium may be due to different antibodies or to induction of CYP1B1 as a result of the isolation of the mammary epithelial cells from mammoplasty tissue or their *in vitro* culture over six days (22). Low levels of CYP1B1 expression have also been detected in cultured mammary stromal fibroblasts (56).

Animal Experiments. In animal experiments, the induction of CYP1B1 exhibited sex-dependent expression. E_2 was shown to double constitutive expression of CYP1B1 mRNA and protein in rat mammary gland (57). CYP1B1 mRNA levels were significantly higher in kidney and liver of TCDD-treated female rats than in similarly treated male rats (58). Comparison of TCDD-treated ovari-ectomized and intact female rats showed significantly higher hepatic levels of 8-hydroxydeoxyguanosine, a marker of oxidative DNA damage, in the latter group of animals (59). This result is consistent with the hypothesis that increased metabolism of endogenous estrogens to catechol estrogens by TCDD-induced enzymes may lead to increased DNA damage and hence contribute to TCDD-mediated carcinogenicity in female rats. Cyp1b1-null mice are viable without an observable phenotype (60). Nevertheless, the importance of CYP1B1 in chemical carcinogenesis is well illustrated in this animal model. Embryonic fibroblast cells derived from Cyp1b1-null mice were resistant to 7,12-dimethylbenz[a]anthracene (DMBA)-mediated tumorigenesis and the mice were protected from DMBA-induced malignant lymphomas.

Genetic Variation. Both mutations and polymorphisms have been identified in the CYP1B1 gene (http://www.CYPalleles.ki.se/). Primary congenital glaucoma, an autosomal recessive eye disorder, has been linked to homozygous frameshift and missense mutations in affected Turkish, Saudi Arabian, and Hispanic families (2, 61, 62). At least 15 missense mutations are found throughout the coding region, representing non-conservative changes in highly conserved regions of the protein. These mutations are presumed to result in a severe disruption of CYP1B1 activity, although the precise mechanism by which CYP1B1 inactivation can lead to congenital glaucoma remains unknown (63). Two of the germ line mutations identified in patients with primary congenital glaucoma were also found in women with hepatocellular adenoma, a rare benign liver tumor that has been linked to the use of oral contraceptives (64). Five nonsynonymous polymorphisms exist: Arg48Gly (rs10012), Ala119Ser (rs1056827), Leu432Val (rs1056836), Ala443Gly (rs4986888), and Asn453Ser (rs1800440) (Figure 4.2.3.). There is considerable ethnic variation in the frequency of these polymorphisms. For example, the 432Leu allele is present in about 32% of African-Americans, 55% of Caucasians, and 82% Chinese (67, 68). The 453Ser allele is not found in Chinese (69). Genetic studies have shown absolute linkage of 48Arg to 119Ala and 48Gly to 119Ser in Asian and Caucasian populations (4). The five polymorphisms give rise to seven different haplotypes (Table 4.2.1). It is noteworthy that the sequence deposited as "wild-type" in GenBank is present in a minority of individuals, 14.7% of Caucasians. Additional rare polymorphisms have been identified but the haplotypes have not yet been determined (http://www.CYPalleles.ki.se/).

The polymorphisms listed in Table 4.2.1 are outside conserved regions with the exception of Asn453Ser, which is found within the meander region (Figure 4.2.3.). Several groups have examined the effect of the five polymorphisms on enzyme function (18, 70-75). Although all groups analyzed the 4- and 2-hydroxylation of E_2 by CYP1B1, comparison of the results should take into account differences in expression systems (bacteria, yeast), assay conditions (microsomal membranes, purified proteins), and estrogen metabolite analysis (HPLC, GC/MS). Some groups also provided an incomplete definition of constructs, i.e., only two or three of the five amino acids were listed. Not surprisingly, the results are inconsistent. For example, some investigators observed higher 4-hydroxylation activity for 432Val than 432Leu, whereas others obtained the opposite result, i.e., 432Leu>432Val (4, 18, 70, 72, 75). The effect of the polymorphisms in codons 48, 119, 432, and 453 on PAH metabolism was examined in smokers (77). Urinary metabolites of phenanthrene were examined, especially r-1,t-2,3,c-4-tetrahydroxy-1,2,3,4-tetrahydro-phenanthrene (PheT) as a measure of metabolic activation and phenanthrols (HOPhe) as a measure of detoxification. The PheT/3-HOPhe ratio was significantly lower in individuals heterozygous or homozygous for the Arg48Gly and Ala119Ser polymorphisms compared to the respective homozygous wild types. A gene dosage effect was observed for both variants and the decrease was more pronounced in females than males. In contrast, subjects homozygous for Asn453Ser had significantly higher PheT/3-HOPhe ratios than homozygous wild-type or heterozygous individuals (77). A study of the oxidation of benzo[a]pyrene to benzo[a]pyrene-7,8-diol found higher activity for 432Leu than 432Val (74). However, no difference was seen in the conversion of benzo[a]pyrene-7,8-diol to benzo[a]pyrene-7,8-diol-9,10-epoxides (78). The Ala443Gly polymorphism, observed in a Spanish population with a 2.6% frequency, exhibited a decreased capacity for the conversion of B[a]P to B[a]P-7,8-diol (65). Expression of the CYP1B1.4 allelic variant containing 453Ser in COS-1 cells yielded twofold lower protein levels and enzyme activity compared with the other four allelic CYP1B1 proteins containing 453Asn (15). On balance, it appears that there is at best a two- to threefold difference in catalytic activity between wild-type CYP1B1 and any variant isoform.

Clinical Studies. Since CYP1B1 is the predominant member of the CYP family expressed in breast tissue and plays a key role in the oxidative metabolism of estrogens, multiple epidemiological studies have investigated the possible contribution of CYP1B1 polymorphisms to inter-individual differences in breast cancer susceptibility. The findings from these studies have been inconsistent (79-82). One possible exception is the Leu432Val (rs1056836) polymorphism, the most common in populations of European and African descent. A pooled analysis of nine datasets including 6,842 women (3,391 cases, 3,451 controls) suggested a possible association in Caucasians (for Val/Val and Val/Leu combined) with an odds ratio of 1.5 (95% CI 1.1–2.1) (83). In contrast, a meta-analysis of Arg48Gly (rs10012; 10 studies with 11,321 cases and 13,379 controls), Ala119Ser (rs1056827; 11 studies with 10,715 cases and 11,678 controls), and Asn453Ser (rs1800440; 12 studies with 11,630 cases and 14,053 controls) found no association with breast cancer risk (76). A meta-analysis of Ala119Ser (rs1056827; 4 studies with 3,969 cases and 3,723 controls), Leu432Val (rs1056836; 9 studies with 5,712 cases and 5,107 controls) and Asn453Ser (rs1800440; 3 studies with 2,165 cases and 2,010 controls) found no association with breast cancer risk (69). Several studies have examined the interaction of genes, including CYP1B1, that participate in estrogen biosynthesis and metabolism (82, 84, 85). These gene-gene interaction studies have yielded inconsistent results showing no effect on breast cancer risk, elevated risk or decreased risk. The drawback of any purely genetic assessment is the lack of consideration about functional interactions inherent in complex biological pathways. A pathway-based functional and quantitative approach is necessary to overcome the limitation of gene-gene interactions studies based simply on genotyping. To address these shortcomings we developed a genotypic-phenotypic model for breast cancer risk prediction (86). The model simulates the kinetic effect of genetic variants of the enzymes CYP1A1, CYP1B1, and COMT on the production of the main carcinogenic estrogen metabolite, 4-hydroxyestradiol (4-OHE$_2$), expressed as a 4-OHE$_2$ area under the curve (AUC) metric (4-OHE$_2$-AUC). The model also incorporates phenotypic factors (age, body mass index, hormone replacement therapy, oral contraceptives, family history), which plausibly influence estrogen metabolism and the production of 4-OHE$_2$. We hypothesize that higher 4-OHE$_2$-AUC indicates greater exposure to this carcinogenic metabolite and thereby increases the risk of a woman to develop breast cancer. We applied the model to two independent, population-based breast cancer case-control groups, the German GENICA study and the Nashville Breast Cohort. In the GENICA study, premenopausal women at the 90th percentile of 4-OHE$_2$-AUC among control subjects had a risk of breast cancer that was 2.30 times that of women at the 10th control 4-OHE$_2$-AUC percentile (95% CI 1.7 – 3.2, P = 2.9×10^{-7}). This relative risk was 1.89 (95% CI 1.5 – 2.4, P = 2.2×10^{-8}) in postmenopausal women. In the Nashville Breast Cohort, this relative risk in postmenopausal women was 1.81 (95% CI 1.3 – 2.6, P = 7.6×10^{-4}), which increased to 1.83 (95% CI 1.4 – 2.3, P = 9.5×10^{-7}) when a history of proliferative breast disease was included in the model.

Current smokers carrying the 432Val allele had an increased risk of breast cancer compared to never smokers with the 432Leu/Leu genotype, which was attributed to the increased expression of CYP1B1 by PAHs in smoke and the greater catalytic activity of the Val isoform in producing 4-OHE2 (87). The 432Val allele in combination with obesity (BMI >24 kg/m^2) was also associated with an increased incidence of breast cancer (88). Another study (688 cases, 724 controls) showed an increased risk association with the Asn453Ser (rs1800440) polymorphism but no association with the Leu432Val polymorphism (84). The inconsistent results of these epidemiological studies may be partially explained by differences in environmental exposures, such as diet. For example, the Leu432Val variant was shown to modify the effect of dietary phytoestrogen (flaxseed lignan) on estrogen metabolism (89). Since CYP1B1 is involved in the disposition of exogenous estrogens, investigators have examined whether CYP1B1 variants modify the effect of hormone replacement therapy (HRT; estrogen plus progestin) on postmenopausal breast cancer risk (90). One study found no evidence that Leu432Val or Asn453Ser modify the effect whereas another study observed a significant increase associated with both variants (91, 92). The studies differed in duration of HRT with the first defining

long-term use as ≥3 years and the second as ≥10 years.

Interestingly, two studies observed an association of the 432Val/Val genotype with ERα expression (ERα-positive status) in breast cancer patients whereas two other studies found an association between the 453Ser/Ser genotype and negative ERα status (67, 69, 90, 93). A study of postmenopausal women found that carriers of the 432Leu and 453Ser alleles had modestly higher plasma E$_2$ levels but similar E$_1$ and E$_1$ sulfate levels (93) while another study found no association (94). A large multiethnic study of premenopausal women found no association of the CYP1B1 genotype with serum estrogen levels (95). The Leu432Val polymorphism has also been investigated in relation to other cancers, showing no association with lung cancer, increased risk of ovarian cancer for the 432Leu allele and elevated risk of prostate cancer associated with the 432Val allele (4, 68, 96). Other studies have examined the association of CYP1B1 polymorphisms with endometrial cancer, yielding inconsistent results (97-99).

Proteins such as CYP1B1, which exhibit minimal expression in healthy and overexpression in malignant tissues may be useful targets for immunotherapy. The presence of CYP1B1 in tumors was exploited as a tumor antigen and anti-CYP1B1-specific T cells were shown to kill CYP1B1-expressing tumor cells (50). A phase 1 trial of a CYP1B1 DNA vaccine in 19 patients with advanced cancers (ovary, colon, kidney, prostate, breast) revealed no adverse effects (100). Four of five patients who developed immunity to CYP1B1 and required treatment showed clinically significant responses.

In summary, CYP1B1 has been implicated in the etiology of breast cancer because it is responsible for estrogen metabolism and the formation of oxidative estrogen metabolites capable of forming DNA adducts. The enzyme also plays a role in the metabolism of exogenous carcinogens, such as heterocyclic amines and PAHs. Moreover, CYP1B1 polymorphisms have been implicated in differential breast cancer risk but epidemiological studies have yielded inconsistent results.

References

1. Sissung TM, Price DK, Sparreboom A, Figg WD. *Pharmacogenetics and regulation of human cytochrome P450 1B1: Implications in hormone-mediated tumor metabolism and a novel target for therapeutic intervention.* Mol Cancer Res. 2006;4:135-50.

2. Stoilov I, Akarsu AN, Alozie I, Child A, Barsoum-Hornsy M, Turacli ME, et al. *Sequence analysis and homology modeling suggest primary congenital glaucoma on 2p21 results from mutations disrupting*

either the hinge region or the conserved core structures of cytochrome P450 1B1. Am J Hum Genet. 1998;62:573-84.

3. Gotoh O. *Substrate recognition sites in cytochrome P450 family 2 (CYP2) proteins inferred from comparative analyses of amino acid and coding nucleotide sequences.* J Biol Chem. 1992;267:83-90.

4. Watanabe J, Shimada T, Gillam EM, Ikuta T, Suemasu K, Higashi Y, et al. *Association of CYP1B1 genetic* polymorphism with incidence to breast and lung cancer. Pharmacogenetics. 2000;10:25-33.

5. Tang YM, Wo YP, Stewart J, Hawkins AL, Griffin CA, Sutter TR, et al. *Isolation and characterization of the human cytochrome P450 CYP1B1.* J Biol Chem. 1996;271:28324-30.

6. Wo YP, Stewart J, Greenlee WF. *Functional analysis of the promoter for the human CYP1B1 gene.* J Biol Chem. 1997;272:26702-7.

7. Shehin SE, Stephenson RO, Greenlee WF. *Transcriptional regulation of the human CYP1B1 gene.* J Biol Chem. 2000;275:6770-6.

8. Tsuchiya Y, Nakalima M, Kyo S, Kanaya T, Inoue M, Yokoi T. *Human CYP1B1 is regulated by estradiol via estrogen receptor.* Cancer Res. 2004;64:3119-25.

9. Port JL, Yamaguchi K, Du B, De Lorenzo M, Chang M, Heerdt PM, et al. *Tobacco smoke induces CYP1B1 in the aerodigestive tract.* Carcinogenesis. 2004;25:2275-81.

10. Spencer DL, Masten SA, Lanier KM, Yang X, Grassman JA, Miller CR, et al. *Quantitative analysis of constitutive and 2,3,7,8-tetrachlorodibenzo-p-dioxin-induced cytochrome P450 1B1 expression in human lymphocytes.* Cancer Epidemiol Biomarkers Prev. 1999;8:139-46.

11. Spink DC, Katz BH, Hussain MM, Spink BC, Wu SJ, Liu N, et al. *Induction of CYP1A1 and CYP1B1 in T-47D human breast cancer cells by benzo[a]pyrene is diminished by arsenite.* Drug Metab Dispos. 2002;30:262-9.

12. Sutter TR, Tang YM, Hayes CL, Wo YP, Jabs EW, Li X, et al. *Complete cDNA sequence of a human dioxin-inducible mRNA identifies a new gene subfamily of cytochrome P450 that maps to chromosome 2.* J Biol Chem. 1994;269(May):13092-9.

13. Tokizane T, Shiina H, Igawa M, Enokida H, Urakami S, Kawakami T, et al. *Cytochrome P450 1B1 is overexpressed and regulated by hypomethylation in prostate cancer.* Clinical Cancer Res. 2005;11:5793-801.

14. Tsuchiya Y, Nakajima M, Takagi S, Taniya T, Yokoi T. *MicroRNA regulates the expression of human cytochrome P450 1B1.* Cancer Res. 2006;66:9090-8.

15. Bandiera S, Weidlich S, Harth V, Broede P, Ko Y, Friedberg T. *Proteasomal degradation of human CYP1B1: Effect of the Asn453Ser polymorphism on the post-translational regulation of CYPB1 expression.* Mol Pharmacol. 2005;67:435-43.

16. Belous AR, Hachey DL, Dawling S, Roodi N, Parl FF. *Cytochrome P450 1B1-mediated estrogen metabolism results in estrogen-deoxyribonucleoside adduct formation.* Cancer Res. 2007;67:812-7.

17. Hachey DL, Dawling S, Roodi N, Parl FF. *Sequential action of phase I and II enzymes cytochrome P450 1B1 and glutathione S-transferase P1 in mammary estrogen metabolism.* Cancer Res. 2003;63:8492-9.

18. Hanna IH, Dawling S, Roodi N, Guengerich FP, Parl FF. *Cytochrome P450 1B1 (CYP1B1) pharmacogenetics: association of polymorphisms with functional differences in estrogen hydroxylation activity.* Cancer Res. 2000;60:3440-4.

19. Hayes CL, Spink DC, Spink BC, Cao JQ, Walker NJ, Sutter TR. *17b-estradiol hydroxylation catalyzed by human cytochrome P450 1B1.* Proc Natl Acad Sci USA. 1996;93:9776-81.

20. Jefcoate CR, Liehr JG, Santen RJ, Sutter TR, Yager JD, Yue W, et al. *Tissue-specific synthesis and oxidative metabolism of estrogens.* J Natl Cancer Inst Monogr. 2000;27:95-112.

21. McFadyen MCE, Breeman S, Payne S, Stirk C, Miller ID, Melvin WT, et al. *Immunohistochemical localization of cytochrome P450 CYP1B1 in breast cancer with monoclonal antibodies specific for CYP1B1.* J Histochem Cytochem. 1999;47:1457-64.

22. Murray GI, Melvin WT, Greenlee WF, Burke MD. *Regulation, function, and tissue-specific expression of cytochrome P450 CYP1B1.* Annu Rev Pharmacol Toxicol. 2001;41:297-316.

23. Murray GI, Patimalla S, Stewart KN, Miller ID, Heys SD. *Profiling the expression of cytochrome P450 in breast cancer.* Histopathology. 2010;57:202-11.

24. Spink DC, Eugster H, Lincoln DWI, Schuetz JD, Schuetz EG, Johnson JA, et al. *17 beta-estradiol hydroxylation catalyzed by human cytochrome P450 1A1: a comparison of the activities induced by 2,3,7,8-tetrachlorodibenzo-p-dioxin in MCF-7 cells with those from heterologous expression of the cDNA.* Arch Biochem Biophys. 1992;293:342-8.

25. Spink DC, Hayes CL, Young NR, Christou M, Sutter TR, Jefcoate CR, et al. *The effects of 2,3,7,8-tetrachlorodibenzo-p-dioxin on estrogen metabolism in MCF-7 breast cancer cells: evidence for induction of a novel 17 beta-estradiol 4-hydroxylase.* J Steroid Biochem Mol Biol. 1994;51:251-8.

26. Lee AJ, Cai MX, Thomas PE, Conney AH, Zhu BT. *Characterization of the oxidative metabolites of 17β-estradiol and estrone formed by 15 selectively expressed human cytochrome p450 isoforms.* Endocrinology. 2003;144:3382-98.

27. Waxman DJ, Lapenson DP, Aoyama T, Gelboin HV, Gonzalez FJ, Korzekwa K. *Steroid hormone hydroxylase specificities of eleven cDNA-expressed human cytochrome P450s.* Arch Biochem Biophys. 1991;290:160-6.

28. Cribb AE, Knight MJ, Dryer D, Guernsey J, Hender K, Tesch M, et al. *Role of polymorphic human cytochrome P450 enzymes in estrone oxidation.* Cancer Epidemiol Biomarkers Prev. 2006;15:551-8.

29. Zhang Y, Gaikwad NW, Olson K, Zahid M, Cavalieri EL, Rogan EG. *Cytochrome P450 isoforms catalyze formation of catechol estrogen quinones that react with DNA.* Metabolism Clin Exper. 2007;56:887-94.

30. Chen H, Howald WN, Juchau MR. *Biosynthesis of all-trans-retinoic acid from all-trans-retinol: Catalysis of all-trans-retinol oxidation by human P-450 cytochromes.* Drug Metab Dispos. 2000;28:315-22.

31. Crewe HK, Notley LM, Wunsch RM, Lennard MS, Gillam EMJ. *Metabolism of tamoxifen by recombinant human cytochrome P450 enzymes: Formation of the 4-hydroxy, 4'-hydroxy and n-desmethyl metabolites and isomerization of trans-4-hydroxytamoxifen.* Drug Metab Dispos. 2002;30:869-74.

32. Buters JT, Mahadeven B, Quintanilla-Martinez L, Gonzalez FJ, Greim H, Baird WM, et al. *Cytochrome P450 1B1 determines susceptibility to dibenzo[a,l]pyrene-induced tumor formation.* Chem Res Toxicol. 2002;15:1127-35.

33. Crofts FG, Strickland PT, Hayes CL, Sutter TR. *Metabolism of 2-amino-1-methyl-6-phenylimidazo[4,5-b]pyridine (PhIP) by human cytochrome P4501B1.* Carcinogenesis. 1997;18:1793-8.

34. Crofts FG, Sutter TR, Strickland PT. *Metabolism of 2-amino-1-methyl-6-phenylimidazo[4,5-b]pyridine by human cytochrome p4501A1, p4501A2 and p4501B1.* Carcinogenesis. 1998;19:1969-73.

35. Halberg RB, Larsen MC, Elmergreen TL, Ko AY, Irving AA, Clipson L, et al. *Cyp1b1 exerts opposing effects on intestinal tumorigenesis via exogenous and endogenous substrates.* Cancer Res. 2008;68:7394-402.

36. Savas U, Carstens CP, Jefcoate CR. *Biological oxidations and P450 reactions. Recombinant mouse CYP1B1 expressed in Escherichia coli exhibits selective binding by polycyclic hydrocarbons and metabolism which parallels C3H10T1/2 cell microsomes but differs from human recombinant CYP1B1.* Arch Biochem Biophys. 1997;247:181-92.

37. Shimada T, Hayes CL, Yamazaki H, Amin S, Hecht SS, Guengerich FP, et al. *Activation of chemically diverse procarcinogens by human cytochrome P-450 1B1.* Cancer Res. 1996;56:2979-84.

38. Bruno RD, Njar VCO. *Targeting cytochrome P450 enzymes: A new approach in anti-cancer drug development.* Bioorg Med Chem. 2007;15:5047-60.

39. Yamazaki H, Shimada T. *Effects of arachidonic acid, prostaglandins, retinol, retinoic acid and cholecalciferol on xenobiotic oxidations catalysed by human cytochrome P450 enzymes.* Xenobiotica. 1999;29:231-41.

40. Dawling S, Roodi N, Parl FF. *Methoxyestrogens exert feedback inhibition on cytochrome P450 1A1 and 1B1.* Cancer Res. 2003;63:3127-32.

41. Chun Y-J, Kim S, Kim D, Lee S-K, Guengerich P. *A new selective and potent inhibitor of human cytochrome P450 1B1 and its application to antimutagenesis.* Cancer Res. 2001;61:8164-70.

42. Doostdar H, Burke MD, Mayer RT. *Bioflavonoids: selective substrates and inhibitors for cytochrome P450 CYP1A and CYP1B1.* Toxicology. 2000;144:31-8.

43. Shimada T, Gillam EMJ, Sutter TR, Strickland PT, Guengerich FP, Yamazaki H. *Oxidation of xenobiotics by recombinant human cytochrome P450 1B1.* Drug Metab Dispos. 1997;29:617-22.

44. Mouly PP, Arzouyan CR, Gaydou EM, Estienne JM. *Differentiation of citrus juices by factorial discriminant analysis using liquid chromatography of flavonone glycosides.* J Agric Food Chem. 1994;42:70-9.

45. Chun Y-J, Lee S-K, Kim MY. *Modulation of human cytochrome P450 1B1 expression by 2,4,3',5'-tetramethoxystilbene.* Drug Metab Dispos. 2005;33:1771-6.

46. Park H, Aiyar SE, Fan P, Wang J, Yue W, Okouneva T, et al. *Effects of tetramethylstilbene on hormone-resistant breast cancer cells: Biological and biochemical mechanisms of action.* Cancer Res. 2007;67:5717-26.

The Etiology of Breast Cancer

47. Chan NLS, Wang H, Y. W, Leung HY, Leung LK. *Polycyclic aromatic hydrocarbon-induced CYP1B1 activity is suppressed by perillyl alcohol in MCF-7 cells.* Toxicol Appl Pharmacol. 2006;213:98-104.

48. Rochat B, Morsman JM, Murray GI, Figg WD, McLeod HL. *Human CYP1B1 and anticancer agent metabolism: mechanism for tumor-specific drug activation.* J Pharmacol Exper Ther. 2001;296:537-41.

49. Brake PB, Jefcoate CR. *Regulation of cytochrome P4501B1 in cultured rat adrenocortical cells by cyclic adenosine 3', 5'-monophospate and 2, 3, 7, 8-tetrachlorodiben-zo-p-dioxin.* Endocrinology. 1995;136:5034-41.

50. Maecker B, Sherr DH, Vonderheide RH, von Bergwelt-Baildon MS, Hirano N, Anderson KS, et al. *The shared tumor-associated antigen cytochrome P450 1B1 is recognized by specific cytotoxic T cells.* Blood. 2003;102:3287-94.

51. Spivack SD, Hurteau GJ, Reilly AA, Aldous KM, Ding X, Kaminsky LS. *CYP1B1 expression in human lung.* Drug Metab Dispos. 2001;29:916-22.

52. Toide K, Yamazaki H, Nagashima R, Itoh K, Iwano S, Takahashi Y, et al. *Aryl hydrocarbon hydroxylase represents CYP1B1 and not CYP1A1 in human freshly isolated white cells: Trimodal distribution of Japanese population according to induction of CYP1B1 mRNA by environmental dioxins.* Cancer Epidemiol Biomarkers Prev. 2003;12:219-22.

53. Murray GI, Taylor MC, McFadyen MCE, McKay JA, Greenlee WF, Burke MD, et al. *Tumor-specific expression of cytochrome P450 CYP1B1.* Cancer Res. 1997;57:3026-31.

54. Wen W, Ren Z, Shu XO, Cai Q, Ye C, Gao Y, et al. *Expression of cytochrome P450 1B1 and catechol-o-methyltransferase in breast tissue and their associations with breast cancer risk.* Cancer Epidemiol Biomarkers Prev. 2007;16:917-20.

55. Larsen MC, Angus WGR, Brake PB, Eltom SE, Sukow KA, Jefcoate CR. *Characterization of CYP1B1 and CYP1A1 expression in human mammary epithelial cells: role of the aryl hydrocarbon receptor in polycyclic aromatic hydrocarbon metabolism.* Cancer Res. 1998;58:2366-74.

56. Eltom SE, Larsen MC, Jefcoate CR. *Expression of CYP1B1 but not CYP1A1 by primary cultured human mammary stromal fibroblasts constitutively and in response to dioxin exposure: Role of the Ah receptor.* Carcinogenesis. 1998;19:1437-44.

57. Christou M, Savas U, Schroeder S, Shen X, Thompson T, Gould MN, et al. *Cytochromes CYP1A1 and CYP1B1 in the rat mammary gland: Cell-specific expression and regulation by polycyclic aromatic hydrocarbons and hormones.* Mol Cell Endocrinol. 1995;115:41-50.

58. Walker NJ, Gastel JA, Costa LT, Clark GC, Lucier GW, Sutter TR. *Rat CYP1B1: an adrenal cytochrome P450 that exhibits sex-dependent expression in livers and kidneys of TCDD-treated animals.* Carcinogenesis. 1995;16:1319-27.

59. Tritscher AM, Seacat AM, Yager JD, Groopman JD, Miller BD, Bell D, et al. *Increased oxidative DNA damage in livers of 2,3,7,8-tetrachloro-rodibenzo-p-dioxin treated intact but not ovariectomized rats.* Cancer Lett. 1996;98:219-25.

60. Buters JT, Sakai S, Richter T, Pineau T, Alexander DL, Savas U, et al. *Cytochrome P450 CYP1B1 determines susceptibility to 7, 12-dimethylbenz[a]anthracene-induced lymphomas.* Proc Natl Acad Sci USA. 1999;96:1977-82.

61. Bejjani BA, Lewis RA, Tomey KF, Andersen KL, Dueker DK, Jabak M, et al. *Mutations in CYP1B1, the gene for cytochrome P4501B1, are the predominant cause of primary congenital glaucoma in Saudia Arabia.* Am J Hum Genet. 1998;62:325-33.

62. Stoilov I, Akarsu AN, Sarfarazi M. *Identification of three different truncating mutations in cytochrome P4501B1 (CYP1B1) as the principal cause of primary congenital glaucoma (Buphthalmos) in families linked to the GLC3A locus on chromosome 2p21.* Hum Mol Genet. 1997;6:641-7.

63. Libby RT, Smith RS, Savinova OV, Zabaleta A, Martin JE, Gonzalez FJ, et al. *Modification of ocular defects in mouse developmental glaucoma models by tyrosinase.* Science. 2003;299:1578-81.

64. Jeannot E, Poussin K, Chiche L, Bacq Y, Sturm N, Scoazec JY, et al. *Association of CYP1B1 germ line mutations with hepatocyte nuclear factor 1a-mutated hepatocellular adenoma.* Cancer Res. 2007;67:2611-6.

65. Aklillu E, Ovrebo S, Botnen IV, Otter C, Ingelman-Sundberg M. *Characterization of common CYP1B1 variants with different capacity for benzo[a]pyrene-7, 8-dihydrodiol epoxide formation from benzo[a]pyrene.* Cancer Res. 2005;65:5105-11.

66. Crooke PS, Ritchie MD, Hachey DL, Dawling S, Roodi N, Parl FF.

Estrogens, enzyme variants, and breast cancer: A risk model. Cancer Epidemiol Biomarkers Prev. 2006;15:1620-9.

67. Bailey LR, Roodi N, Dupont WD, Parl FF. *Association of cytochrome P450 1B1 (CYP1B1) polymorphism with steroid receptor status in breast cancer* [Erratum: Cancer Res 1999; 59:1388]. Cancer Res. 1998;58:5038-41.

68. Tang YM, B.L. G, Chen GF, Thompson PA, Lang NP, Shinde A, et al. *Human CYP1B1 Leu432Val gene polymorphism: ethnic distribution in African-Americans, Caucasians and Chinese; oestradiol hydroxylase activity; and distribution in prostrate cancer cases and controls.* Pharmacogenetics. 2000;10:761 *Americans, Caucasians and Chinese; oestradiol hydroxylase activity; and distribution in prostrate cancer cases and controls.* Pharmacogenetics. 2000;10:761-6.

69. Wen W, Cai Q, Shu X-O, Cheng J-R, Parl F, Pierce L, et al. *Cytochrome P450 1b1 and catechol-O-methyltransferase genetic polymorphisms and breast cancer risk in Chinese women: results from the Shanghai Breast Cancer Study and a meta-analysis.* Cancer Epidemiol Biomarkers Prev. 2005;14:329-35.

70. Aklillu E, Oscarson M, Hidestrand M, Leidvik B, Otter C, Ingelman-Sundberg M. *Functional analysis of six different polymorphic CYP1B1 enzyme variants found in an Ethiopian population.* Mol Pharmacol. 2002;61:586-94.

71. Lewis DFV, Gillam EMJ, Everett SA, Shimada T. *Molecular modelling of human CYP1B1 substrate interactions and investigation of allelic variant effects on metabolism.* Chem-Biol Interact. 2003;145:281-95.

72. Li DN, Seidel A, Pritchard MP, Wolf CR, Friedberg T. *Polymorphisms in P450 CYP1B1 affect the conversion of estradiol to the potentially carcinogenic metabolite 4-hydroxyestradiol.* Pharmacogenetics. 2000;10:343-53.

73. McLellan RA, Oscarson M, Hidestrand M, Leidvik B, Jonsson E, Otter C, et al. *Characterization and functional analysis of two common human cytochrome P450 1B1 variants.* Arch Biochem Biophys. 2000;378:175-81.

74. Shimada T, Watanabe J, Inoue K, Guengerich FP, Gillam EMJ. *Specificity of 17β-oestradiol and benzo[a]pyrene oxidation by polymorphic human cytochrome P4501B1 variants substituted at residues 48, 119 and 432.* Xenobiotica. 2001;31:163-76.

75. Shimada T, Watanabe J, Kawajiri K, Sutter TR, Guengerich FP, Gillam EMJ, et al. *Catalytic properties of polymorphic human cytochrome P450 1B1 variants.* Carcinogenesis. 1999;20:1607-13.

76. Economopoulos KP, Sergentanis TN. *Three polymorphisms in cytochrome P450 1B1 (CYP1B1) gene and breast cancer risk: a meta-analysis.* Breast Cancer Res Treat. 2010;122:545-51. PubMed PMID: 20054638.

77. Hecht SS, Carmella SG, Yoder A, Chen M, Li Z, Le C, et al. *Comparison of polymorphisms in genes involved in polycyclic aromatic hydrocarbon metabolism with urinary phenanthrene metabolite ratios in smokers.* Cancer Epidemiol Biomarkers Prev. 2006;15:1805-11.

78. Mammen JS, Pittman GS, Li Y, Abou-Zahr F, Bejjani BA, Bell DA, et al. *Single amino acid mutations, but not common polymorphisms, decrease the activity of CYP1B1 against (-)-benzo[a]pyrene-7R-trans-7,8-dihydrodiol.* Carcinogenesis. 2003;24:1247-55.

79. Huang Y, Trentham-Dietz A, Garcia-Closas M, Newcomb PA, Titus-Ernstoff L, Hampton JM, et al. *Association of CYP1B1 haplotypes and breast cancer risk in Caucasian women.* Cancer Epidemiol Biomarkers Prev. 2009;18:1321-3.

80. Mitrunen K, Hirvonen A. *Molecular epidemiology of sporadic breast cancer. The role of polymorphic genes involved in oestrogen biosynthesis and metabolism.* Mutation Res. 2003;544:9-41.

81. Reding KW, Weiss NS, Chen C, Li CI, Carlson CS, Wilkerson H, et al. *Genetic polymorphisms in the catechol estrogen metabolism pathway and breast cancer risk.* Cancer Epidemiol Biomarkers Prev. 2009;18:1461-7.

82. Delort L, Satih S, Kwiatkowski F, Bignon YJ, Bernard-Gallon DJ. *Evaluation of breast cancer risk in a multigenic model including low penetrance genes involved in xenobiotic and estrogen metabolisms.* Nutrition Cancer. 2010;62:243-51.

83. Paracchini V, Raimondi S, GRam IT, Kang D, Kocabas NA, Kristensen VN, et al. *Meta- and pooled analyses of the cytochrome P-450 1B1 Val432Leu polymorphism and breast cancer: A HuGE-GSEC review.* Am J Epidemiol. 2007;165:115-25.

84. Justenhoven C, Hamann U, Schubert F, Zapatka M, Pierl CB, Rabstein S, et al. *Breast cancer: a candidate gene approach across the estrogen metabolic pathway.* Breast Cancer Res Treat. 2008;108:137-49.

85. Low YL, Li Y, Humphreys K, Thalamuthu A, Li Y, Darabi H, et al. *Multi-variant pathway association analysis reveals the importance of genetic determinants of estrogen metabolism in breast and endometrial cancer susceptibility.* PLoS Genetics. 2010;6:e1001012.

86. Crooke PS, Justenhoven C, Brauch H, Dawling S, Roodi N, Higginbotham KS, et al. *Estrogen metabolism and exposure in a genotypic-phenotypic model for breast cancer risk prediction.* Cancer Epidemiol Biomarkers Prev. 2011;20:1502-15.

87. Saintot M, Malaveille C, Hautefeuille A, Gerber M. *Interactions between genetic polymorphism of cytochrome P450-1B1, sulfotransferase 1A1, catechol-o-methyltransferase and tobacco exposure in breast cancer risk.* Int J Cancer. 2003;107:652-7.

88. Kocabas NA, Sardas S, Cholerton S, Daly AK, Karakaya AE. *Cytochrome P450 CYP1B1 and catechol o-methyltransferase (COMT) genetic polymorphisms and breast cancer susceptibility in a turkish population.* Arch Toxicol. 2002;76:643-9.

89. McCann SE, Wactawski-Wende J, Kufel K, Olson J, Ovando B, Nowell Kadlubar S, et al. *Changes in 2-hydroxyestrone and 16a-hydroxyestrone metabolism with flaxseed consumption: Modification by COMT and CYP1B1 genotype.* Cancer Epidemiol Biomarkers Prev. 2007;16:256-62.

90. Justenhoven C, Pierl CB, Haas S, Fischer HP, Baisch C, Hamann U, et al. *The CYP1B1_1358_GG genotype is associated with estrogen receptor-negative breast cancer.* Breast Cancer Res Treat 2008;111:171-7.

91. Diergaarde B, Potter JD, Jupe ER, Manjeshwar S, Shimasaki CD, Pugh TW, et al. *Polymorphisms in genes involved in sex hormone metabolism, estrogen plus progestin hormone therapy use, and risk of postmenopausal breast cancer.* Cancer Epidemiol Biomarkers Prev. 2008;17:1751-9.

92. Rebbeck TR, Troxel AB, Walker AH, Panossian S, Gallagher S, DeMichele A, et al. *Pairwise combinations of estrogen metabolism genotypes in postmenopausal breast cancer etiology.* Cancer Epidemiol Biomarkers Prev. 2007;16:444-9.

93. De Vivo I, Hankinson SE, Li L, Colditz GA, Huntern DJ. *Association of CP1B1 polymorphisms and breast cancer risk.* Cancer Epidemiol Biomarkers Prev. 2002;11:489-92.

94. Tworoger SS, Chubak J, Aiello EJ, Ulrich CM, Atkinson C, Potter JD, et al. *Association of CYP17, CYP19, CYP1B1, and COMT polymorphisms with serum and urinary sex hormone concentrations in postmenopausal women.* Cancer Epidemiol Biomarkers Prev. 2004;13:94-101.

95. Lurie G, Maskarinec G, Kaaks R, Stanczyk FZ, Marchand LL. *Association of genetic polymorphisms with serum estrogens measured multiple times during a 2-year period in premenopausal women.* Cancer Epidemiol Biomarkers Prev. 2005;14:1521-7.

96. Goodman MT, McDuffie K, Kolonel LN, Terada K, Donlon TA, Wilkens LR, et al. *Case-control study of ovarian cancer and polymorphisms in genes involved in catecholestrogen formation and metabolism.* Cancer Epidemiol Biomarkers Prev. 2001;10:209-16.

97. McGrath M, Hankinson SE, Arbeitman L, Colditz GA, Hunter DJ, De Vivo I. *Cytochrome P450 1B1 and catechol-O-methyltransferase polymorphisms and endometrial cancer susceptibility.* Carcinogenesis. 2004;25:559-65.

98. Rylander-Rudqvist T, Wedren S, Jonasdottir G, Ahlberg S, Weiderpass E, Persson I, et al. *Cytochrome P450 1B1 gene polymorphisms and postmenopausal endometrial cancer risk.* Cancer Epidemiol Biomarkers Prev. 2004;13:1515-20.

99. Sasaki M, Tanaka Y, Kaneuchi M, Sakuragi N, Dahiya R. *CYP1B1 gene polymorphisms have higher risk for endometrial cancer, and positive correlations with estrogen receptor a and estrogen receptor b expressions.* Cancer Res. 2003;63:3913-8.

100. Gribben JG, Ryan DP, Boyajian R, Urban RG, Hedley ML, Beach K, et al. *Unexpected association between induction of immunity to the universal tumor antigen CYP1B1 and response to next therapy.* Clin Cancer Res. 2005;11:4430-6

4.2.4. OTHER CYTOCHROME P450 ENZYMES

Other cytochrome P450 enzymes besides CYP1B1 and CYP1A1 are expressed in breast tissue (1). RT-PCR has been used to detect the mRNA of CYP2C, CYP3A, and CYP2D6 in human breast cancers and in normal breast (2). Breast tissue from reduction mammoplasties expressed mRNA and protein of CYP2A6, CYP2D6, CYP2E1, and CYP4A11 (3). A tissue microarray containing 170 breast cancers of no special type was immunostained for a panel of 21 P450s. The highest percentage of strong immunopositivity in breast cancers was seen for CYP4X1 (50.8%), CYP2S1 (37.5%) and CYP2U1 (32.2%), while CYP2J (98.6%) and CYP3A43 (70.7%) were the P450s that most frequently displayed no immunoreactivity. CYP4V2 (P = 0.01), CYP4X1 (P = 0.01) and CYP4Z1 (P = 0.01) showed correlations with tumor grade. CYP1B1 (P = 0.001), CYP3A5 (P = 0.001) and CYP51 (P = 0.005) showed the most significant correlations with ER status. Correlations with survival were identified for CYP2S1 (P = 0.03), CYP3A4 (P = 0.025), CYP4V2 (P = 0.026) and CYP26A1 (P = 0.03), although none of these P450s was an independent marker of prognosis (4). Genomic, phenotypic and clinical information about these CYPs has been summarized in http://www.cypalleles.ki.se and http://www.pharmgkb.org (5-7).

The **CYP1A2** and CYP1A1 genes are both located at 15q22 in a head-to-head orientation, separated by 23.3 kb (5). The CYP1A2 gene is about 7.8 kb long and contains 7 exons of which exons 2 through 6 encode the 515 amino acids that comprise the 58 kDa protein (8). CYP1A2 is expressed primarily in the liver with constitutive levels far exceeding those of CYP1A1. CYP1A2 plays a key role in the metabolism of estrogens, heterocyclic amines, caffeine, and certain drugs, such as acetaminophen (9-12). Like CYP1A1, CYP1A2 catalyzes C-2 hydroxylation of E_2 at ten times higher rates than C-4 hydroxylation (13, 14). The hydroxylation of E_1 follows the same pattern (10). CYP1A2 also efficiently oxidizes E_2 at C-16α, but is less active in E_1 C-16α hydroxylation (15). CYP1A2 catalyzes the N-hydroxylation of heterocyclic amines, such as 2-amino-1-methyl-6-phenylimidazo[4,5-b]pyridine (PhIP) and 2-amino-3,8-dimethyl imidazo[4,5-f]quinoxaline (MeIQx) (11, 16). Ninety-five percent of caffeine is metabolized by CYP1A2 and caffeine is an inducer of the enzyme (9). The phytochemical sulforaphone, an isothiocyanate found in substantial levels in brassica vegetables, was shown to exert differential effects on CYP1A1 and 1A2. Sulforaphone treatment of rats, at dietary relevant doses, suppressed CYP1A1 activity but induced CYP1A2 expression, although the latter enzyme was not catalytically competent because of bound sulforaphone metabolites (17).

A common polymorphism in intron 1 (-163A/C; rs762551) of the CYP1A2 gene has been associated with decreased enzyme inducibility and enzyme activity, resulting in slower metabolism of caffeine (18). The corresponding CYP1A2*1F genotype has been investigated by several groups in relation to breast cancer risk. Two groups reported a protective effect in the presence of the variant C allele while other groups found no association with breast cancer risk (19-22). In a study of 458 women with breast cancer, moderate to high coffee consumption (≥2 cups per day) in patients with the AA genotype was associated with a later age at diagnosis and ER-negative tumor status compared to low coffee consumption (23). Among healthy premenopausal women, the CC genotype was associated with significantly lower serum E_2 levels compared with the combined AA and AC genotypes (24). Similarly, low CYP1A2 activity was associated with low levels of sex hormone binding globulin (25).

Members of the widely expressed **CYP2 family** have broad substrate specificities and participate in estrogen metabolism (10, 15). For example, CYP2C19 is capable of catalyzing the conversion of E_2 to E_1 and of E_1 to 16α-OHE$_1$ (10, 26). Genetic **CYP2C19** variants result in different metabolizer phenotypes (27). An example is the CYP2C19*17 allele, which results from a SNP in the promoter region (C-806T; rs12248560) and is associated with increased gene expression and an ultrarapid metabolizer phenotype (28). The variant allele was found in 18% of Caucasians and associated with a reduced risk of breast cancer in the GENICA study (1021 cases, 1015 controls: OR 0.77; P = 0.005) (29). A follow-up pooled analysis of four international case-control studies showed a decrease in breast cancer risk for postmeno-pausal women using hormone replacement therapy for 10 years or longer (30). The C-806T (rs12248560) variant is rare in Asians, whereas a C/T (rs4917623) variant is common with a minor allele frequency of 0.38 (31). In a study of Thai women (570 cases, 497 controls) the T allele was associated with increased breast cancer risk for both the heterozygous (OR 1.38; 95% CI 1.04 – 1.83) and homozygous genotypes (OR 1.56; 95% CI 1.06 – 2.30). **CYP2D6** mRNA including splice variants were detected in normal breast tissues and breast cancers (2). Several studies have examined the association of CYP2D6 SNPs with breast cancer susceptibility and reported conflicting results. A large case-control study involving 13,472 women found no evidence of any association between common SNPs and breast cancer risk (32). CYP2D6 is the main CYP isoform catalyzing the hydroxylation of the antiestrogen tamoxifen to the more potent 4-hydroxytamoxifen. Less active genetic variants of CYP2D6 have been implicated in the poor response of ER-positive breast cancers to tamoxifen (33). However, a meta-analysis of 25 studies found no evidence for an association between CYP2D6 genotype and response to tamoxifen treatment (34).

Members of the **CYP3A subfamily**, i.e., CYP3A4, CYP3A5, and CYP3A7, are involved in estrogen hydroxylation and drug clearance, influencing approximately half of all oxidatively metabolized drugs. The CYP3A genes are adjacent to each other at 7q21, but are differentially regulated (35). Substantial inter-individual differences exist in CYP3A expression, exceeding 30-fold in some populations. Sequence diversity in the CYP3A5 promoter is the basis for significant differences in CYP3A5 expression, which contributes at least 50% of total hepatic CYP3A content (36). CYP3A4 contains a common variant in the 5' flanking region, designated CYP3A4*1B. This variant exhibited increased *in vitro* activity compared to the wild-type CYP3A4*1A and showed a striking association with early onset of puberty (37). The expression of CYP3A4 is regulated by the pregnane X receptor and gender. Women have twofold higher hepatic levels of CYP3A4 than men with a corresponding 50% increase in CYP3A-dependent N-dealkylation of verapamil (38). Sex-dependent CYP gene expression due to different underlying mechanisms has also been observed in several rodent P450 enzymes (39-41). Repeated ingestion of green tea catechin resulted in a small reduction in CYP3A4 activity but had no effect on CYP1A2, CYP2C9, and CYP2D6 (42).

Any beneficial effect associated with tea consumption on breast cancer risk does not appear to be due to any clinically insignificant effects of catechins on the disposition of substrates metabolized by CYPs. CYP3A7 is expressed predominantly in fetal life and its expression seems to be silenced shortly after birth; however, some individuals express CYP3A7 into adulthood. CYP3A enzymes are predominantly expressed in the liver, but are also present in intestine, kidney, endometrium, placenta, adrenal, and prostate. CYP3A4 and CYP3A5 mRNAs were detected in 73% (8 of 11) and 82% (9 of 11), respectively, of normal breast tissues, but in only 15% (2 of 13) of breast cancers, and in neither of two fibroadenomas (2, 15). A Japanese-Brazilian case-control study of 873 pairs (403 Japanese, 81 Japanese Brazilians, 389 non-Japanese Brazilians) observed an increased breast cancer risk (OR 1.49; 95% CI 1.10 – 2.04) in Japanese women with a SNP in intron 3 of the CYP3A5 gene (A6986; rs776746) (22).

CYP3A enzymes, together with CYP1A2, catalyze the majority of hepatic estrogen metabolism (10, 43). CYP3A4 and CYP3A5 oxidize both E_2 and E_1 to the corresponding 2- and 4-OH catechols and catalyze the subsequent oxidation to their ortho-quinones, which can react with glutathione, proteins, or DNA (44). In addition, CYP3A4 and CYP3A5 efficiently oxidize both E_2 and E_1 at C-16α, producing 16α-OHE$_2$ and 16α-OHE$_1$, respectively (15, 43, 45, 46). E_1 16α-hydroxylase activity was inhibited by 75% by monoclonal antibodies against CYP3A4/5 and by troleandomycin, a specific CYP3A4/5 inhibitor, indicating that CYP3A4/5 are the principal catalysts of estrogen 16α-hydroxylation. The CYP3A7 isoform distinguished

E_2 and E_1 with >100 times higher $V_{max}:K_m$ ratio for the 16α-hydroxylation of E_1 than E_2. The difference in reaction rates is most likely due to the difference in structure at the C-17 position of E_1 and E_2. The presence of the 17-ketogroup in E_1 appears to be essential for substrate recognition and 16α-hydroxylation by CYP3A7 (46).

The **pregnane X receptor (PXR)**, also called steroid and xenobiotic receptor (SXR), is a key regulator of xenobiotic metabolism (47). The PXR gene at 3q12-q13.3 encodes a 50 kDa nuclear receptor protein with a DNA- and a ligand-binding domain. PXR induces the expression of several hepatic phase I enzymes such as CYP3A4, phase II conjugating enzymes (UGT1A1, GSTs), and phase III drug transporters mediating efflux and uptake (e.g., ATP-binding cassette, ABCB1 protein and organic anion transporting polypeptide-A, OATP-A, respectively) (48, 49). PXR binds to the xenobiotic DNA response elements in the regulatory regions of CYP3A genes as a heterodimer with the retinoid X receptor, RXR. Unlike the steroid, retinoid, and thyroid hormone receptors, which are highly selective for their cognate hormone, PXR has a hydrophobic ligand-binding cavity capable of interacting with diverse compounds (50). Thus, PXR can be activated by a wide variety of structurally divergent agents known to induce CYP3A expression, including antibiotic, anti-inflammatory, anti-lipidemic, anti-depressant, anti-androgen, and anti-cancer drugs (e.g., rifampicin, dexamethasone, spironolactone, lovastatin, St. John's wort, cyproterone acetate, and tamoxifen, respectively) (47). PXR mRNA and protein have been detected in malignant epithelial cells but not in non-neoplastic and stromal cells of breast tumors (48). PXR expression was positively correlated with expression of OATP-A and in ER-positive tumors with the cell proliferation marker, Ki-67. The PXR*1B haplotype was associated with reduced hepatic expression of PXR and its downstream targets, CYP3A4 and ABCB1 (51). This haplotype resulted in lower clearance of the anti-cancer agent doxorubicin, an ABCB1 substrate, in Asian breast cancer patients.

References

1. Suzuki T, Miki Y, Nakamura Y, Moriya Y, Ito K, Ohuchi N, et al. *Sex steroid-producing enzymes in human breast cancer.* Endocrine-Related Cancer. 2005;12:701-20.

2. Huang Z, Fasco MJ, Figge HL, Keyomarsi K, Kaminsky LS. *Expression of cytochromes P450 in human breast tissue and tumors.* Drug Metab Disp. 1996;24:899-905.

3. Hellmold H, Rylander T, Magnusson M, Reihner E, Warner M, Gustafsson JA. *Characterization of cytochrome P450 enzymes in human breast tissue from reduction mammaplasties.* J Clin Endocrin Metabol. 1998;83:886-95.

4. Murray GI, Patimalla S, Stewart KN, Miller ID, Heys SD. *Profiling the expression of cytochrome P450 in breast cancer.* Histopathology. 2010;57:202-11.

5. Jiang Z, Dalton TP, Jin L, Wang B, Tsuneoka Y, Shertzer HG, et al. *Toward the evaluation of function in genetic variability: Characterizing human SNP frequencies and establishing BAC-transgenic mice carrying the human CYP1A1_CYP1A2 locus.* Hum Mutat. 2005;25:196-206.

6. Solus JF, Arietta BJ, Harris JR, Sexton DP, Steward JQ, McMunn C, et al. *Genetic variation in eleven phase I drug metabolism genes in an ethnically diverse population.* Pharmacogenetics. 2004;5:895-931.

7. Thorn CF, Klein TE, Altman RB. *PharmGKB: the Pharmacogenomics Knowledge Base.* Methods Mol Biol. 2013;1015:311-20.

8. Ikeya K, Jaiswal AK, Owens RA, Jones JE, Nebert DW, Kimura S. *Human CYP1A2: sequence, gene structure, comparison with the mouse and rat orthologous gene, and differences in liver 1A2 mRNA expression.* Mol Endocrinol. 1989;3:1399-408.

9. Butler MA, Iwasaki M, Guengerich FP, Kadlubar FF. *Human cytochrome P-450PA (P-450IA2), the phenacetin O-deethylase, is primarily responsible for the hepatic 3-demethylation of caffeine and N-oxidation of carcinogenic arylamines.* Proc Natl Acad Sci USA. 1989;86:7696-700.

10. Cribb AE, Knight MJ, Dryer D, Guernsey J, Hender K, Tesch M, et al. *Role of polymorphic human cytochrome P450 enzymes in estrone oxidation.* Cancer Epidemiol Biomarkers Prev. 2006;15:551-8.

11. Crofts FG, Sutter TR, Strickland PT. *Metabolism of 2-amino-1-methyl-6-phenylimidazo[4,5-b]pyridine by human cytochrome p4501A1, p4501A2 and p4501B1.* Carcinogenesis. 1998;19:1969-73.

12. Guengerich FP. *Characterization of human cytochrome P450 enzymes.* FASEB J. 1992;6:745-8.

13. Aoyama T, Korzekwa K, Nagata K, Gillette J, Gelboin HV, Gonzalez FJ. *Estradiol metabolism by complementary deoxyribonucleic acid-expressed human cytochrome P450s.* Endocrinology. 1990;126:3101-6.

14. Spink DC, Eugster H, Lincoln DWI, Schuetz JD, Schuetz EG, Johnson JA, et al. *17 beta-estradiol hydroxylation catalyzed by human cytochrome P450 1A1: a comparison of the activities induced by 2,3,7,8-tetra-chlorodibenzo-p-dioxin in MCF-7 cells with those from heterologous expression of the cDNA.* Arch Biochem Biophys. 1992;293:342-8.

15. Yamazaki H, Shaw PM, Guengerich FP, Shimada T. *Roles of cytochromes P450 1A2 and 3A4 in the oxidation of estradiol and estrone in human liver microsomes.* Chem Res Toxicol. 1998;11:659-65.

16. Turesky RJ, Constable A, Richoz J, Varga N, Markovic J, Martin MV, et al. *Activation of heterocyclic aromatic amines by rat and human liver microsomes and by purified rat and human cytochrome P450 1A2.* Chemical Res Toxicol. 1998;11:925-36.

17. Yoxall V, Kentish P, Coldham N, Kuhnert N, Sauer MJ, Ioannides C.

Modulation of hepatic cytochromes P450 and phase II enzymes by dietary doses of sulforaphane in rats: Implications for its chemopreventive activity. Int J Cancer. 2005;117:356-62.

18. Sachse C, Brockmoller J, Bauer S, Roots I. *Functional significance of a C → A polymorphism in intron I of the cytochrome P450 CYP1A2 gene tested with caffeine.* Clin Pharmacol. 1999;47:445-9.

19. Kotsopoulos J, Ghadirian P, El-Sohemy A, Lynch HT, Snyder C, Daly M, et al. *The CYP1A2 genotype modifies the association between coffee consumption and breast cancer risk among BRCA1 mutation carriers.* Cancer Epidemiol Biomarkers Prev. 2007;16:912-6.

20. Le Marchand L, Donlon T, Kolonel LN, Henderson BE, Wilkins LR. *Estrogen metabolism-related genes and breast cancer risk: The Multiethnic Cohort Study.* Cancer Epidemiol Biomarkers Prev. 2005;14:1998-2003.

21. Long J, Egan KM, Dunning L, Shu X, Cai Q, Cai H, et al. *Population-based case-control study of AhR (aryl hydrocarbon receptor) and CYP1A2 polymorphisms and breast cancer risk.* Pharmacogenet Genomics. 2006;16:237-43.

22. Shimada N, Iwasaki M, Kasuga Y, Yokoyama S, Onuma H, Nishimura H, et al. *Genetic polymorphisms in estrogen metabolism and breast cancer risk in case-control studies in Japanese, Japanese Brazilians and non-Japanese Brazilians.* J Hum Genet. 2009;54:209-15.

23. Bageman E, Ingvar C, Rose C, Jernstrom H. *Coffee consumption and CYP1A2*1F genotype modify age at breast cancer diagnosis and estrogen receptor status.* Cancer Epidemiol Biomarkers Prev. 2008;17:895-901.

24. Lurie G, Maskarinec G, Kaaks R, Stanczyk FZ, Marchand LL. *Association of genetic polymorphisms with serum estrogens measured multiple times during a 2-year period in premenopausal women.* Cancer Epidemiol Biomarkers Prev. 2005;14:1521-7.

25. Hong CC, Tang BK, Hammond GL, Tritchler D, Yaffe M, Boyd NF. *Cytochrome P450 1A2 (CYP1A2) activity and risk factors for breast cancer: a cross-sectional study.* Breast Cancer Res. 2004;6:R352-R65.

26. Cheng ZN, Shu Y, Liu ZQ, Wang LS, Ou-Yang DS, Zhou HH. *Role of cytochrome P450 in estradiol metabolism in vitro.* Acta Pharmacol Sin. 2000;22:148-54.

27. Klotz U, Schwab M, Treiber G. *CYP2C19 polymorphism and proton pump inhibitors.* Basic Clin Pharmacol Toxicol. 2004;95:2-8.

28. Sim SC, Risinger C, Dahl ML, Aklillu E, Christensen M, Bertilsson L, et al. *A common novel CYP2C19 gene variant causes ultrarapid drug metabolism relevant for the drug response to proton pump inhibitors and antidepressants.* Clin Pharmacol Ther. 2006;79:103-13.

29. Justenhoven C, Hamann U, Pierl CB, Baisch C, Harth V, Rabstein S, et al. *CYP2C19*17 is associated with decreased breast cancer risk.* Breast cancer research and treatment. 2009;115:391-6.

30. Justenhoven C, Obazee O, Winter S, Couch FJ, Olson JE, Hall P, et al. *The postmenopausal hormone replacement therapy-related breast cancer risk is decreased in women carrying the CYP2C19*17 variant.* Breast Cancer Res Treat. 2012;131:347-50.

31. Sangrajrang S, Sato Y, Sakamoto H, Ohnami S, Laird NM, Khuaprema T, et al. *Genetic polymorphisms of estrogen metabolizing enzyme and breast cancer risk in Thai women.* Int J Cancer. 2009;125:837-43.

32. Abraham JE, Maranian MJ, Driver KE, Platte R, Kalmyrzaev B, Baynes C, et al. *CYP2D6 gene variants and their association with breast cancer susceptibility.* Cancer Epidemiol Biomarkers Prev. 2011;20:1255-8.

33. Dezentje VO, Guchelaar HJ, Nortier JWR, Van de Gelde CJH, Gelderblom H. *Clinical implications of CYP2D6 genotyping in tamoxifen reatment for breast cancer.* Clin Cancer Res. 2009;15:15-21.

34. Lum DW, Perel P, Hingorani AD, Holmes MV. *CYP2D6 Genotype and Tamoxifen Response for Breast Cancer: A Systematic Review and Meta-Analysis.* PloS one. 2013;8:e76648.

35. Finta C, Zaphiropoulos PG. *The human cytochrome P450 3A locus. Gene evolution by capture of downstream exons.* Gene. 2000;260:13-23.

36. Kuehl P, Zhang J, Lin Y, Lamba J, Assem M, Schuetz J, et al. *Sequence diversity in CYP3A promoters and characterization of the genetic basis of polymorphic CYP3A5 expression.* Nature Genetics. 2001;27:383-91.

37. Kadlubar FF, Berkowitz GS, Delongchamp RR, Wang C, Green BL, Tang G, et al. *The CYP3A4*1B variant is related to the onset of puberty, a known risk factor for the development of breast cancer.* Cancer Epidemiol Biomarkers Prev. 2003;12:327-31.

38. Wolbold R, Klein K, Burk O, Nussler AK, Neuhaus P, Eichelbaum M, et al. *Sex is a major determinant of CYP3A4 expression in human liver.* Hepatology. 2003;38:978-88.

39. Buggs C, Nasrin N, Mode A, Tollet P, Zhao H, Gustafsson J, et al. *IRE-ABP (Insulin response element-A binding protein), an SRY-like protein, inhibits C/EBPα (CCAAT/Enhancer-binding porteinα)-stimulated expression of the sex-specific cytochrome P450 2C12 gene.* Mol Endocrinol. 1998;12:1294-309.

40. Endo M, Takahashi Y, Sasaki Y, Saito T, Kamataki T. *Novel gender-related regulation of CYP2C12 gene expression in rats.* Mol Endocrinol. 2005;19:1181-90.

41. Wiwi CA, Gupte M, Waxman DJ. *Sexually dimorphic P450 gene expression in liver-specific hepatocyte nuclear factor 4α-deficient mice.* Mol Endocrinol. 2004;18:1975-87.

42. Chow H-HS, Hakim IA, Vining DR, Crowell JA, Cordova CA, Chew WM, et al. *Effects of repeated green tea catechin administration of human cytochrome P450 activity.* Cancer Epidemiol Biomarkers Prev. 2006;15:2473-6.

43. Lee AJ, Cai MX, Thomas PE, Conney AH, Zhu BT. *Characterization of the oxidative metabolites of 17β-estradiol and estrone formed by 15 selectively expressed human cytochrome p450 isoforms.* Endocrinology. 2003;144:3382-98.

44. Zhang Y, Gaikwad NW, Olson K, Zahid M, Cavalieri EL, Rogan EG. *Cytochrome P450 isoforms catalyze formation of catechol estrogen quinones that react with DNA.* Metabolism Clin Exper. 2007;56:887-94.

45. Huang Z, Guengerich FP, Kaminsky LS. *16α-hydroxylation of estrone by human cytochrome p450 3A4/5.* Carcinogenesis. 1998;19:867-72.

46. Lee AJ, Conney AH, Zhu BT. *Human cytochrome p450 3A7 has a distinct high catalytic activity for the 16a-hydroxylation of estrone but not 17b-estradiol.* Cancer Res. 2003;63:6532-6.

47. Kliewer SA, Goodwin B, Willson TM. *The nuclear pregnane X receptor: a key regulator of xenobiotic metabolism.* Endocrine Rev. 2002;23:687-702.

48. Miki Y, Suzuki T, Kitada K, Yabuki N, Shibuya R, Moriya T, et al. *Expression of the steroid and xenobiotic receptor and its possible target gene, organic anion transporting polypeptide-A, in human breast carcinoma.* Cancer Res. 2006;66:535-42.

49. Pascussi JM, Gerbal-Chaloin S, Duret C, Daujat-Chavanieu M, Vilarem MJ, Maurel P. *The tangle of nuclear receptors that controls xenobiotic metabolism and transport: crosstalk and consequences.* Ann Rev Pharmacol Toxicol. 2008;48:1-32.

50. Watkins RE, Wisely GB, Moore LB, Collins JL, Lambert MH, Williams SP, et al. *The human nuclear xenobiotic receptor PXR: structural determinants of directed promiscuity.* Science. 2001;292:2329-33.

51. Sandanaraj E, Lal S, Selvarajan V, Ooi LL, Wong ZW, Wong NS, et al. *PXR pharmacogenetics: association of haplotypes with hepatic CYP3A4 and ABCB1 messenger RNA expression and doxorubicin clearance in Asian breast cancer patients.* Clin Cancer Res. 2008;14:7116-26.

4.3. PHASE II CONJUGATING ENZYMES

4.3.1. CATECHOL-O-METHYLTRANSFERASE

Overview. Catechol-O-methyltransferase (COMT) is an enzyme, which catalyzes the transfer of a methyl group from the methyl donor S-adenosyl-L-methionine (SAM) to one hydroxyl moiety of the catechol ring of a substrate (1) (Figure 4.3.1). Physiological substrates of COMT include the catecholamines (i.e., neurotransmitters epinephrine, norepinephrine, dopamine) and the catechol estrogens (2, 3). The demethylated product of SAM is S-adenosyl-homocysteine (SAH). COMT is inhibited by SAH and homocysteine (4).

Cell fractionation and immunological studies have shown that the enzyme occurs in two distinct forms, in the cytoplasm as a soluble protein (S-COMT) and in association with membranes as a membrane-bound form (MB-COMT) (5-7). The amino acid sequence of S- and MB-COMT is identical, except for an N-terminal extension of 50 hydrophobic amino acids in MB-COMT, which serves as an anchor to the membrane (8, 9). Genetic studies have demonstrated that S- and MB-COMT are encoded by a single gene at 22q11.2. The gene contains six exons, of which exons 1 and 2 are non-coding (Figure 4.3.2.). A proximal promoter gives rise to the 1.3 kb S-COMT mRNA, while a distal promoter gives rise to the 1.5 kb MB-COMT mRNA (10). The MB- and S-COMT proteins occur constitutively, but differ in tissue expression. S-COMT is the predominant form in virtually all tissues, whereas MB-COMT generally accounts for approximately 10% of total enzyme activity (5, 11). For example, S-COMT constitutes >90% of total COMT activity in normal breast tissue as well as MCF-7 breast cancer cells (10).

The COMT gene contains several single nucleotide polymorphisms that have been associated with clinical phenotypes, particularly estrogen-related cancers and a wide spectrum of mental disorders (Table 4.3.1.). Two SNPs in the MB-COMT promoter, C-628T (rs2020917) and intron 1 A701G (rs737865), altered DNA-protein binding patterns and increased transcription (12). Functional studies showed that reporter constructs containing both SNPs increased transcription 2.3-fold compared with the wild type in MDA-468 cells but only 1.3-fold in MCF-7 cells. Two synonymous SNPs, His12/62His (rs4633) and Leu86/136His (rs4818), altered mRNA local stem-loop structures, such that the most stable structure was associated with the lowest protein level and enzymatic activity (13). Two infrequent nonsynonymous SNPs, Ala22Ser (rs6267) and Ala52Thr (rs5031015), were associated with lower catalytic activity and/or lower thermostability (14). A common nonsynonymous SNP in exon 4 (G/A; rs4680) causes a substitution in both the MB-(codon 158) and S-form (codon 108) resulting in Val158Met and Val108Met, respectively (17). The corresponding alleles, COMT*1 and 2, occur with equal frequencies in Caucasians but COMT*1 is twice as common as COMT*2 in African Americans (18) (Table 4.3.1.). S-COMT consists of eight α-helices and seven β-strands (19). Residue 108Val, which is located in the turn between α5 and β3, is not part of the coenzyme (SAM), Mg^{2+} ion, or substrate binding

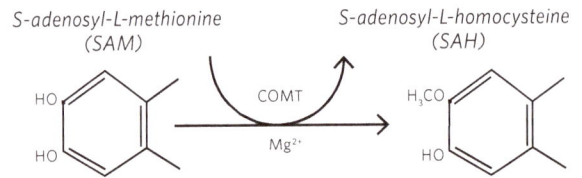

Figure 4.3.1. COMT reaction

sites. Although not directly involved in the methyl transfer reaction, residue 108 appears critically important for overall configuration and stability of the enzyme. The substitution Val108Met renders COMT less thermostable and thereby lowers enzyme efficiency towards all substrates (20). With 3,4-dihydroxybenzoic acid as substrate, COMT activity in red blood cells from individuals with the homozygous Met/Met genotype was about 60% lower than that of homozygous Val/Val individuals, i.e., 0.21 – 0.43 versus 0.55 – 1.03 pmol·min^{-1}·mg^{-1} protein (21). Heterozygotes showed intermediate activity. The COMT*1 allele encodes the high-activity enzyme, and the COMT*2 allele encodes the thermolabile low-activity enzyme. Two infrequent polymorphisms are found at Ala22Ser and Ala52Thr. The former is associated with lower activity and higher thermolability (18).

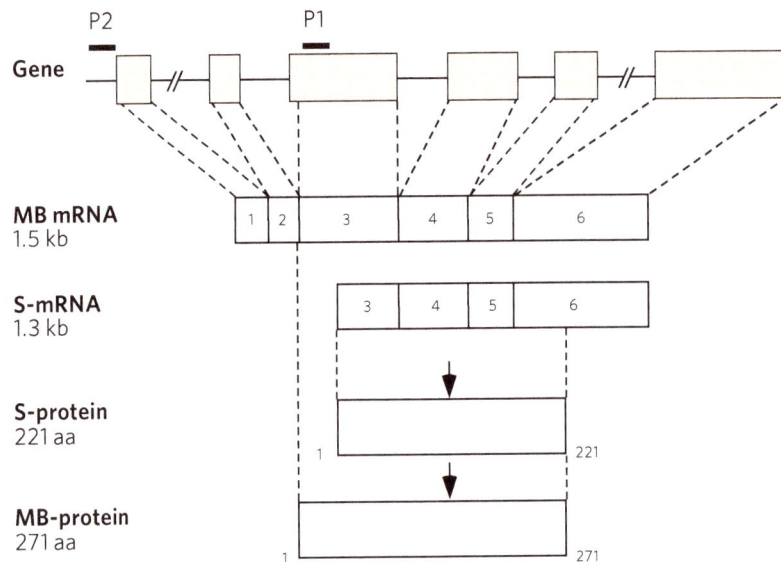

Figure 4.3.2. Overview of COMT gene, mRNA, and protein.
The COMT gene at 22q11.2 is about 11 kb long and contains six exons of which the two most 5' are non-coding. Two promoters, P2 and P1, initiate transcription of membrane-bound (MB) and soluble (S) mRNA, respectively. The 1.5 kb MB-mRNA is translated into the MB form of the enzyme, while the 1.3 kb S-mRNA is translated into the S form. The MB-form of the enzyme is 271 amino acids long, 50 amino acids longer than the S-form. The proximal promoter (P1) is located between the two translation initiation codons. The vertical arrow indicates a common polymorphism (rs4680), which is present in both MB- (codon 158) and S-form (codon 108), as Val158Met and Val108Met, respectively. The lengths of the exons are drawn in scale, while those of the introns are not.

Catechol Estrogen Metabolism. The enzymatic activity of recombinant, purified COMT has been determined for methylation of the catechol estrogen substrates 2-OHE$_2$, 4-OHE$_2$, 2-OHE$_1$, and 4-OHE$_1$ (15, 22, 23). COMT catalyzed the formation of monomethyl ethers at 2-OH, 3-OH, and 4-OH groups. Dimethyl ethers were not observed. COMT interacts differently with the 2-OHE and 4-OHE substrates (15). Methylation of 2-OHE substrates exhibits Michaelis-Menten saturation kinetics and yields two products, i.e., 2-methoxy- and 3-methoxyestrogens. In contrast, the methylation of 4-OHE substrates displays sigmoid saturation kinetics indicating cooperative binding and yields only a single product, i.e., 4-methoxyestrogen. The main structural difference between 2-OH and 4-OH catechol estrogens is the proximity of the 4-OH group to the B ring of the steroid. Clearly, the 2-OH and 3-OH groups in 2-OHE appear to be similar in reactivity, whereas the 3-OH and 4-OH groups in 4-OHE differ in reactivity to the point that in the latter only the 4-OH group becomes methylated.

Expression and Activity in Mammary Epithelial Cells.
Immunohistochemical analysis of benign and malignant breast tissue revealed the presence of COMT in the cytoplasm of all epithelial cells. Immunoreactive COMT was also observed in the nucleus of some benign and malignant epithelial cells (24). There was no correlation between histopathology and number of cells with nuclear COMT, size of foci containing such cells, or intensity of nuclear COMT immunostaining. Staining of both intra- and interlobular stromal cells was always of a much lower intensity than that of epithelial cells in the same tissue sections. Breast cancer cell lines, such as ZR-75 (108Val/Val) and MCF-7 (108Met/Met) have been investigated extensively with regard to COMT catalytic activity, thermal stability, protein concentration, synthesis, and turnover (4, 15-17, 22). The enzyme was less thermostable in MCF-7 than ZR-75 cells but differences in protein turnover (half-life) were influenced by culture conditions. Using the formation of 2-MeOE$_2$ as a measure, the catalytic activity of variant COMT in MCF-7 cells was two- to threefold lower than that of wild-type COMT in ZR-75 cells.

The role of COMT in estrogen-mediated DNA damage was investigated in MCF-10F mammary epithelial cells treated with 4-OHE$_2$ with or without Ro41-0960, a specific inhibitor of COMT (25). MCF-10F cells oxidized 4-OHE$_2$ to the quinones E$_2$(E$_1$)-3,4-Q, which react with DNA to form the depurinating N3Ade and N7Gua adducts. Ro41-0960 blocked the methoxylation of catechol estrogens with concomitant 3- to 4-fold increases in the levels of the depurinating adducts. Thus, low activity of COMT leads to higher levels of depurinating estrogen-DNA adducts that can induce mutations and initiate cancer. The role of COMT on oxidative DNA damage (8-hydroxy-2'-deoxyguanosine) was examined in TCDD-pretreated MCF-7 cells exposed to E$_2$ and Ro41-0960 COMT (26). Administration of the COMT-inhibitor blocked 2-MeOE$_2$ formation and, at the same time, increased 2-OHE$_2$ and 8-oxo-dG levels. In the presence of COMT inhibition, increased oxidative DNA damage was detected in MCF-7 cells exposed to as low as 0.1 μM E$_2$, whereas in the absence of COMT inhibition, no increase in 8-oxo-dG was detected at E$_2$ concentrations ≤10 μM. These results demonstrate that COMT activity is protective against oxidative DNA damage associated with catechol estrogen metabolites. When COMT activity and methoxyestrogens were absent, the authors observed a linear relation between 2-OHE$_2$ plus 4-OHE$_2$ and 8-oxo-dG levels. However, this relationship did not hold under experimental conditions that allowed limited formation of methoxyestrogens, i.e., 8-oxo-dG levels were less than expected for a given concentration of 2-OHE$_2$ plus 4-OHE$_2$ in the presence of 2-MeOE$_2$. These results suggest that 2-MeOE$_2$ may reduce 8-oxo-dG formation.

Epidemiological Studies. Several epidemiological studies have examined the association of COMT genotype with breast cancer risk. The rationale for these studies is as follows. First, by inactivating catechol estrogens, COMT reduces the level of 2-OH and 4-OH estrogen metabolites thereby lowering the potential for mutagenic damage through DNA adduct formation or through superoxide and hydroxy radicals arising from catechol estrogen quinone-semiquinone redox cycling. Since the catalytic

Table 4.3.1. COMT Polymorphisms Associated with Clinical Phenotypes

SNP	rs#	MAF*	Functional Change	Reference
C(-628)T	2020917	0.3	increased transcription	(12)
Intron 1 A(701)G	737865	0.3	increased transcription	(12)
His12/62His	4633	0.5	altered secondary mRNA structure	(13)
Ala22/72Ser	6267	0.016	lower catalytic activity and reduced thermostability	(14)
Ala52/102Thr	5031015	<0.01	reduced thermostability	(14)
Leu86/136Leu	4818	0.4	altered secondary mRNA structure	(13)
Val108/158Met	4680	0.5	reduced thermostability	(4,15,16)

* MAF, minor allele frequency in Caucasians

activity of variant COMT is two- to threefold lower than that of wild type COMT, tissue concentrations of catechol estrogens are expected to differ significantly during the reproductive life of women with the low activity COMT*2 allele compared to women with the high-activity COMT*1 allele. Second, the methoxyestrogen 2-MeOE$_2$, which is produced by COMT from 2-OHE$_2$, has been shown to inhibit the proliferation of breast cancer cells *in vitro* and *in vivo* (27, 28). The inhibitory effect of 2-MeOE$_2$ appears to be due to several mechanisms including the disruption of microtubule function, induction of apoptosis, and inhibition of angiogenesis (29-31). Oral administration of 2-MeOE$_2$ (75 mg/kg) for one month suppressed the growth of human breast cancer in mice by 60% without toxicity (32). Thus, 2-MeOE$_2$ appears to be an endogenous estrogen metabolite that inhibits mammary carcinogenesis and its purported lower level in women with the low-activity COMT*2 allele is expected to increase their risk (33).

In spite of the experimental data, epidemiological evidence does not support an association of low-activity COMT*2 allele with increased breast cancer risk. In larger studies with more than 200 cases, odds ratios were less than unity for two of the three studies in premenopausal women and for all studies in postmenopausal women when low activity COMT*2 homozygotes were compared with high-activity COMT*1 homozygotes (18, 34-37). A meta-analysis of 56 studies including 34,358 breast cancer cases and 45,429 controls found no evidence that the low-activity variant of COMT, as a single factor, leads to increased breast cancer risk (38). Subgroup analysis by menopausal status and ethnicity yielded similar negative results. A comprehensive analysis of ten other single nucleotide polymorphisms in and around the COMT gene locus and two haplotype blocks (one of which included rs4680 Val108Met) did not reveal any association with risk (39). However, two common, functionally significant polymorphisms (rs2020917, rs737865) in the MB-COMT promoter were associated with breast cancer risk reduction in two of three independent studies with a total of 2,327 cases and 2,838 controls (12).

There is evidence for an association between breast cancer risk, COMT genotype, and micronutrients in the folate metabolic pathway (40). These micronutrients (i.e., cysteine, homocysteine, folate, vitamin B12, pyridoxal 5'-phosphate) are known to influence levels of the methyl donor SAM and S-adenosylhomocysteine, a COMT inhibitor generated by the demethylation of SAM (Figure 4.3.1.). High-activity homozygous COMT*1 breast cancer cases had significantly lower levels of homocysteine ($p = 0.05$) and cysteine ($p = 0.04$) and higher levels of pyridoxal 5'-phosphate ($p = 0.02$) than homozygous COMT*1 controls. In contrast, low-activity homozygous COMT*2 cases had higher levels of homocysteine ($p = 0.05$) than low-activity homozygous COMT*2 controls.

An increasing number of COMT*2 alleles was significantly associated with increased breast cancer risk in women with below median levels of folate (p trend = 0.05) or above median levels of homocysteine (p trend = 0.02). No association was seen between vitamin B12, COMT genotype, and breast cancer risk (40). These findings are consistent with a role of certain folate pathway micronutrients in mediating the association between COMT genotype and breast cancer risk. At the same time, these results illustrate the complex interaction of genetic and nutritional factors in breast cancer development. Equally complex is the interaction of the COMT genotype with other risk factors such as mammographic density (41).

References

1. Axelrod J, Tomchick R. *Enzymatic O-methylation of epinephrine and other catechols.* J Biol Chem. 1958;233:702-5.

2. Ball P, Knuppen R. *Catecholoestrogens (2 -and 4-hydroxyoestrogens): chemistry, biogenesis, metabolism, occurrence and physiological signifiicance.* Acta Endocrin Suppl. 1980;232:1-127.

3. Ball P, Knuppen R, Haupt M, Breuer H. *Interactions between estrogens and catecholamines. 3. Studies on the methylation of catechol estrogens, catechol amines and other catechols by the catechol-O-methyltransferases of human liver.* J Clin Endocrin Metab. 1972;34:736-46.

4. Lavigne JA, Helzlsouer KJ, Huang H, Strickland PT, Bell DA, Selmin O, et al. *An association between the allele coding for a low activity variant of catechol-O-methyltransferase and the risk for breast cancer.* Cancer Res. 1997;57:5493-7.

5. Jeffery DR, Roth JA. *Characterization of membrane-bound and soluble catechol-O-methyltransferase from human frontal cortex.* J Neurochem. 1984;42:826-32.

6. Malherbe P, Bertocci B, Caspers P, Zurcher G, Da Prada M. *Expression of functional membrane-bound and soluble catechol-O-methyltransferase in Escherichia coli and a mammalian cell line.* J Neurochem. 1992;58:1782-9.

7. Ulmanen I, Peranen J, Tenhunen J, Tilgmann C, Karhunen T, Panula P, et al. *Expression and intracellular localization of catechol O-methyltransferase in transfected mammalian cells.* Eur J Biochem. 1997;243:452-9.

8. Bertocci B, Miggiano V, Da Prada M, Dembic Z, Lahm HW, Malherbe P. *Human catechol-O-methytransferase: Cloning and expression of the membrane-associated form.* Proc Natl Acad Sci USA. 1991;88:1416-20.

9. Ulmanen I, Lundstrom K. *Cell-free synthesis of rat and human catechol O-methyltransferase.* Eur J Biochem. 1991;202:1013-20.

10. Tenhunen J, Salminen M, Lundstrom K, Kiviluoto T, Savolainen R, Ulmanen I. *Genomic organization of the human catechol O-methyltransferase gene and its expression from two distinct promoters.* Eur J Biochem. 1994;223:1049-59.

11. Grossman MH, Creveling CR, Rybczynski R, Braverman M, Isersky C, Breakefield XO. *Soluble and particulate forms of rat catechol-O-methyltransferase distinguished by gel electrophoresis and immune fixation.* J Neurochem. 1985;44:421-32.

12. Ji Y, Olson J, Zhang J, Hildebrandt M, Wang L, Ingle J, et al. *Breast cancer risk reduction and membrane-bound catechol o-methyltransferase genetic polymorphisms.* Cancer Res. 2008;68:5997-6005.

13. Nackley AG, Shabalina SA, Tchivileva IE, Satterfield K, Korchynskyi O, Makarov SS, et al. *Human catechol-O-methyltransferase haplotypes modulate protein expression by altering mRNA secondary structure.* Science. 2006;314:1930-3.

14. Li Y, Yang X, van Breemen RB, Bolton JL. *Characterization of two new variants of human catechol O-methyltransferase in vitro.* Cancer Lett. 2005;230:81-9.

15. Dawling S, Roodi N, Mernaugh RL, Wang XY, Parl FF. *Catechol-O-methyltransferase (COMT)-mediated metabolism of catechol estrogens: comparison of wild-type and variant COMT isoforms.* Cancer Res. 2001;61:6716-22.

16. Yim DS, Parkb SK, Yoo KS, Chung HH, Kang HL, Ahn SH, et al. *Relationship between the Val158Met polymorphism of catechol O-methyl*

transferase and breast cancer. Pharmacogenetics. 2001;11:279-86.

17. Doyle AE, Yager JD. *Catechol-O-methyltransferase: effects of the val108met polymorphism on protein turnover in human cells.* Biochim Biophys Acta. 2008;1780:27-33.

18. Millikan RC, Pittman GS, Tse CKJ, Duell E, Newman B, Savitz D, et al. *Catechol-O-methyltransferase and breast cancer risk.* Carcinogenesis. 1998;19:1943-7.

19. Vidgren J, Svensson LA, Liljas A. *Crystal structure of catechol O-methyltransferase.* Nature. 1994;368:354-7.

20. Lotta T, Vidgren J, Tilgmann C, Ulmanen I, Melen K, Julkunen I, et al. *Kinetics of human soluble and membrane-bound catechol O-methyltransferase: a revised mechanism and description of the thermolabile variant of the enzyme.* Biochemistry. 1995;34:4202-10.

21. Syvanen AC, Tilgmann C, Rinne J, Ulmanen I. *Genetic polymorphism of catechol-O-methyltransferase (COMT): correlation of genotype with individual variation of S-COMT activity and comparison of the allele frequencies in the normal population and parkinsonian patients in Finland.* Pharmacogenetics. 1997;7:65-71.

22. Goodman JE, Jensen LT, He P, Yager JD. *Characterization of human soluble high and low activity catechol-O-methyltransferase catalyzed catechol estrogen methylation.* Pharmacogenetics. 2002;12:517-28.

23. Lautala P, Ulmanen I, Taskinen J. *Molecular mechanisms controlling the rate and specificity of catechol O-methylation by human soluble catechol O-methyltransferase.* Mol Pharmacol. 2001;59:393-402.

24. Weisz J, Fritz-Wolz G, Gestl S, Clawson GA, Creveling CR, Liehr JG, et al. *Nuclear localization of catechol-o-methyltransferase in neoplastic and nonneoplastic mammary epithelial cells.* Am J Pathol. 2000;156:1841-8.

25. Zahid M, Saeed M, Lu F, Gaikwad N, Rogan E, Cavalieri E. *Inhibition of catechol-O-methyltransferase increases estrogen-DNA adduct formation.* Free Radical Biol Med. 2007;43:1534-40.

26. Lavigne JA, Goodman JE, Fonong T, Odwin S, He P, Roberts DW, et al. *The effects of catechol-o-methyltransferase inhibition on estrogen metabolite and oxidative DNA damage levels in estradiol-treated MCF-7 cells.* Cancer Res. 2001;61:7488-94.

27. Lottering ML, Haag M, Seegers JC. *Effects of 17b-estradiol metabolites on cell cycle events in MCF-7 cells.* Cancer Res. 1992;52:5926-32.

28. Michnovicz JJ, Hershcopf RJ, Naganuma H, Bradlow HL, Fishman J. *Increased 2-hydroxylation of estradiol as a possible mechanism for the anti-estrogenic effect of cigarette smoking.* N England J Med. 1986;315:1305-9.

29. D'Amato RJ, Lin CM, Flynn E, Folkman J, Hamel E. *2-Methoxyestradiol, an endogenous mammalian metabolite, inhibits tubulin polymerization by interacting at the colchicine site.* Proc Natl Acad Sci USA. 1994;91:3964-8.

30. Fotsis T, Zhang Y, Pepper MS, Adlercreutz H, Montesano R, Nawroth PP, et al. *The endogenous oestrogen metabolite 2-methoxyoestradiol inhibits angiogenesis and suppresses tumour growth.* Nature. 1994;368:237-9.

31. Huang P, Feng L, Oldham EA, Keating MJ, Plunkett W. *Superoxide dismutase as a target for the selective killing of cancer cells.* Nature. 2000;407:390-5.

32. Klauber N, Parangi S, Flynn E, Hamel E, D'Amato RJ. *Inhibition of angiogenesis and breast cancer in mice by the microtubule inhibitors 2-methoxyestradiol and taxol.* Cancer Res. 1997;57:81-6.

33. Zhu BT, Conney AH. *Is 2-methoxyestradiol an endogenous estrogen metabolite that inhibits mammary carcinogenesis?* Cancer Res. 1998;58:2269-77.

34. Mitrunen K, Jourenkova N, Kataja V, Eskelinen M, Kosma V-M, Benhamou S, et al. *Polymorphic catechol-O-methyltransferase gene and breast cancer risk.* Cancer Epidemiol Biomarkers Prev. 2001;10:635-40.

35. Reding KW, Weiss NS, Chen C, Li CI, Carlson CS, Wilkerson H, et al. *Genetic polymorphisms in the catechol estrogen metabolism pathway and breast cancer risk.* Cancer Epidemiol Biomarkers Prev. 2009;18:1461-7.

36. Thompson PA, Shields PG, Freudenheim JL, Stone A, Vena JE, Marshall JR, et al. *Genetic polymorphsms in catechol-O-methyltransferase, menopausal status, and breast cancer risk.* Cancer Res. 1998;58:2107-10.

37. Wen W, Cai Q, Shu X-O, Cheng J-R, Parl F, Pierce L, et al. *Cytochrome P450 1b1 and catechol-O-methyltransferase genetic polymorphisms and breast cancer risk in Chinese women: results from the Shanghai Breast Cancer Study and a meta-analysis.* Cancer Epidemiol Biomarkers Prev. 2005;14:329-35.

38. Qin X, Peng Q, Qin A, Chen Z, Lin L, Deng Y, et al. *Association of COMT Val158Met polymorphism and breast cancer risk: an updated meta-analysis.* Diagnostic Pathol. 2012;7:136.

39. Gaudet MM, Chanock S, Lissowska J, Berndt SI, Peplonska B, Brinton LA, et al. *Comprehensive assessment of genetic variation of catechol-o-methyltransferase and breast cancer risk.* Cancer Res. 2006;66:9781-5.

40. Goodman JE, Lavigne JA, Wu K, Helzlsouer KJ, Strickland PT, Selhub J, et al. *COMT genotype, micronutrients in the folate metabolic pathway and breast cancer risk.* Carcinogenesis. 2001;22:1661-5.

41. Hong C-C, Thompson HJ, Jiang C, Hammond GL, Tritchler D, Yaffe M, et al. *Val158Met polymorphism in catechol-O-methyltransferase gene associated with risk factors for breast cancer.* Cancer Epidemiol Biomarkers Prev. 2003;12:838-47.

4.3.2. GLUTATHIONE AND GLUTATHIONE S-TRANSFERASES

GLUTATHIONE

Glutathione (GSH) is a tripeptide composed of γ-glutamyl-cysteine-glycine (1, 2). It is synthesized *de novo* from the amino acids glutamic acid, cysteine, and glycine by the sequential action of two enzymes, γ-glutamylcysteine synthetase (γ-GCS) and GSH synthetase (Figure 4.3.3.). γ-GCS (also known as glutamate cysteine ligase, GCL) is a heterodimeric enzyme that catalyzes the first and rate-limiting step, formation of γ-glutamylcysteine. The enzyme is composed of a catalytically active heavy subunit, γ-GCS-HS (73 kDa) and a regulatory subunit, γ-GCS-LS (30 kDa) (3, 4). γ-GCS is induced by various oxidative and electrophilic compounds via response elements in the promoter regions of the γ-GCS subunit genes, including an antioxidant response element, an electrophilic response element, and an AP-1 binding site (2, 5). The induction is complex and involves additional factors such as the Keap1-Nrf2 pathway (6, 7). Finally, GSH itself regulates the activity of γ-GCS via a negative feedback system. Hence, GSH depletion increases the rate of GSH synthesis (2). GSH is the most abundant nonprotein thiol in mammalian cells with concentrations ranging from 0.1 to 10 mM (8). The tripeptide exists in either reduced (GSH) or oxidized (GSSG) form. The reduction of GSSG to GSH is catalyzed by GSH reductase (GR) in an NADPH-dependent reaction. Cells generally maintain the ratio between oxidized GSSG and reduced GSH as low as 0.1 (9-11). The actual cellular redox state can be measured by quantifying GSSG and GSH and by calculating the half-cell redox potential (E_{hc} expressed in mV) of the GSSG/GSH couple with the Nernst equation (12). In this way, intracellular E_{hc} values ranging from approximately -165 (oxidized) to nearly -260 mV (reduced) have been measured under various conditions in various cell types (12-14). Under physiologic conditions the GSSG/GSH redox change is of a sufficient magnitude to influence the activity of redox sensitive proteins.

GSH serves as a co-factor for several groups of enzymes including glutathione S-transferases (GSTs) and glutathione peroxidases (GPXs). In these enzyme reactions the cysteine thiol acts as a nucleophile donor for numerous exogenous and endogenous electrophilic acceptors. The relative abundance of GSH over GSSG facilitates GSH conjugation by GSTs and GPXs (9-11, 15). In addition to the detoxification of electrophile species, GSH plays a role in other cellular reactions including the reduction of ribonucleotides to deoxyribonucleotides and posttranslational modification of proteins (2). In summary, GSH acts as a storage form of cysteine and is an antioxidant as well as a nucleophile allowing it to suppress carcinogenesis at various levels.

GLUTATHIONE S-TRANSFERASES

Glutathione S-transferases (GSTs) constitute a superfamily of cytosolic, mitochondrial, and microsomal enzymes (16-18). The GSTs catalyze the conjugation of the tripeptide glutathione (GSH) to a wide variety of exogenous and endogenous chemicals with electrophilic functional groups thereby neutralizing their electrophilic sites, and rendering the products more water-soluble (19). They are ubiquitous,

multifunctional enzymes, which protect macromolecules from attack by endogenous reactive electrophiles including α,β-unsaturated aldehydes, quinones, epoxides, and hydroperoxides formed as secondary metabolites during oxidative stress (Figure 4.3.4.). The GSTs also play a key role in the cellular detoxification of electrophilic xenobiotics, such as chemical carcinogens, environmental pollutants, and antitumor agents. In addition to their central role in cellular metabolism, the GSTs participate in other important cellular functions such as signaling and apoptosis (20).

Based on sequence homology and immunological cross-reactivity, human cytosolic GSTs have been grouped into seven families, designated GST Alpha (GSTA), Mu (GSTM), Pi (GSTP), Sigma (GSTS), Omega (GSTO), Theta (GSTT), and Zeta (GSTZ) (21-23). The GSTs have presumably arisen from a single common ancestor and their substrate specificity and diversity have been reshaped by gene duplication, gene recombination, and an accumulation of mutations. Although the promoter regions of the cytosolic GSTs vary between classes they contain one or more of the following response elements and binding sites: antioxidant response element (ARE), xenobiotic response element (XRE), Sp1, AP-1, Maf, Jun, Fos (18). GSTs are induced by many structurally diverse chemicals, including drugs, carcinogens, and dietary compounds, such as tea polyphenols or isothiocyanates derived from cruciferous vegetables (19, 24, 25). Some of the inducing compounds are themselves substrates for the enzyme, suggesting that induction may be an adaptive response mechanism. For example, the pro-oxidant 4-hydroxynonenal can induce GSTs in an apparent adaptation to oxidative stress (26, 27). The transcriptional activation is mediated

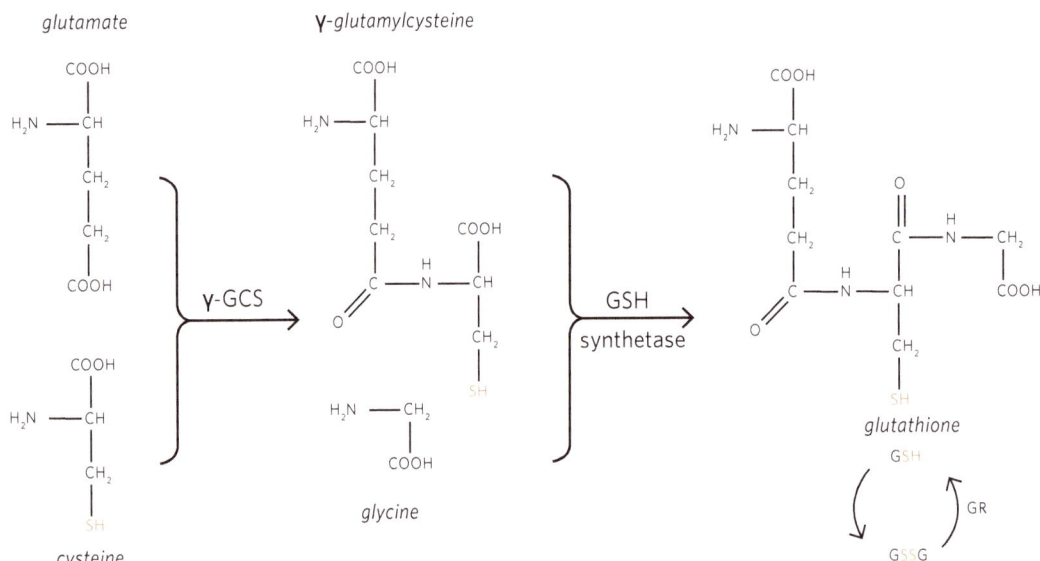

Figure 4.3.3. The synthesis of GSH is accomplished by γ-glutamylcysteine synthetase (γ-GCS) and GSH synthetase. GSH readily oxidizes to glutathione disulfide, GSSG. The latter is reduced enzymatically to GSH by GSH reductase (GR).

The Etiology of Breast Cancer

through the ARE, which is also found in the promoters of other genes inducible by oxidative stress (e.g., NAD(P) H quinone oxidoreductase) (28). Pro-oxidants appear to trigger the Keap1-Nrf2 pathway by modifying Keap1, causing dissociation of Nrf2, which translocates into the nucleus where it heterodimerizes with small Maf proteins and activates ARE-driven transcription (29, 30). Another regulatory pathway involves the pregnane X receptor (31). Certain agents, e.g., tamoxifen, can suppress GST expression (32).

All cytosolic GSTs are active as either homodimers or heterodimers (33). Molecular recognition between the subunits is class specific, which means that heterodimerization occurs only within the same gene class. For example, the A1 and A2 subunits give rise to GSTA1-1, GSTA1-2, and GSTA2-2 (22). The subunits range in size from 24 to 29 kDa. All subunits share a similar topology and two domains, a highly conserved G site for GSH-binding and an H site for hydrophobic substrates (34). The N-terminal domain (residues 1–80) comprises one-third of the protein and forms the G site. It is composed of four β sheets with three flanking α helices, a structural motif common to thioredoxin and other proteins evolved to bind GSH or cysteine. This region contains catalytically essential tyrosine, serine, or cysteine residues that interact with GSH via hydrogen bonds to the cysteinyl carbonyl and sulfur of the thiol group (35-37). The similarity in G-site structure is consistent with similar K_m values for GSH, which range from 0.01 to 0.5 mM for most GSTs. The C-terminal two thirds of the protein (residues 87–210) are all-α helical domain, which together with a loop from the N-terminal domain forms the H site (34). The H-site architecture is much more variable, consistent with significant differences in substrate selectivity among various GST subtypes (37-40). For example, the order of catalytic efficiencies for conjugation of diol epoxides of PAHs like benzo[a]pyrene-7,8-diol-9,10-epoxide (BPDE) was GSTP1 > GSTA1 > GSTM1, whereas the rank order for dibenzo[a]pyrene-11,12-diol-13,14-epoxide

(DBPDE) was GSTA1 > GSTM1 > GSTP1 (41). Thus, GSTs show a high level of specificity toward GSH but the second substrate can vary significantly both between and within the families in spite of their sequence similarity (40, 42).

The class Alpha, Mu, and Pi GSTs are sensitive to product inhibition, and unless the conjugates are eliminated from the cell these transferases will be ineffective at catalyzing detoxification reactions (43). The elimination is accomplished by the so-called multidrug resistance protein 1 (MRP1), which is capable of exporting a variety of GSH-conjugated substrates as well as glucuronidated and sulfated conjugates from the cell (44, 45). MRP1 is part of a family of multidrug resistance proteins involved in drug transport (10).

GSTμ SUBFAMILY

The GSTμ subfamily is encoded by a 100-kb gene cluster at 1p13.3 arranged as 5'-GSTM4-GSTM2-GSTM1-GSTM5-GSTM3-3' (52, 53) (Figure 4.3.5.). Deletion of the GSTM1 gene, GSTM1-0, frequently affects both alleles, resulting in the so-called null genotype, GSTM1-/-. Detailed mapping of the GSTμ gene cluster revealed that the GSTM1 gene is flanked by two almost identical 4.2-kb regions. The GSTM1-0 deletion is caused by a homologous recombination involving the left and right 4.2-kb repeats (52, 53). A polymorphism in the GSTM1 promoter at nucleotide -498C/G (rs412543) was shown to modify binding of the transcription factor AP-2α, resulting in reduced promoter activity and mRNA expression (54). A nonsynonymous polymorphism in exon 7 (Lys172Asn; rs1065411), corresponding to GSTM1*A and GSTM1*B, did not appear to affect the enzyme function (55). Analysis of the other GSTM isoforms shows extensive homologies. For example, exon 8 of the GSTM2 gene and exon 8 of the GSTM1 gene are more than 99% identical over 583 nucleotides and a recombination between these two

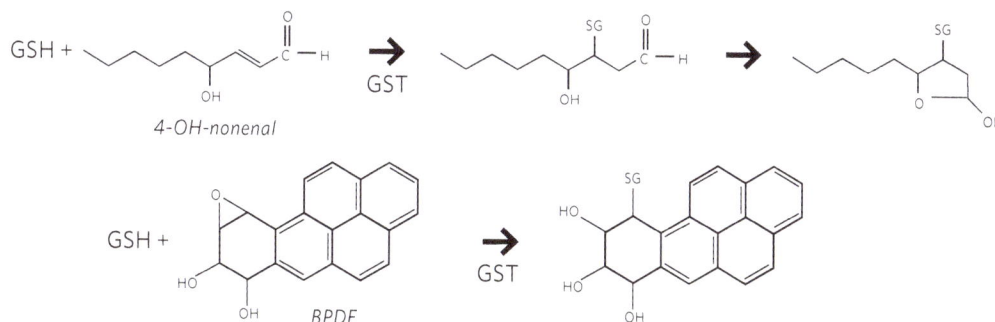

Figure 4.3.4. GSTs catalyze the conjugation of α,β-unsaturated carbonyls, quinones, epoxides, and hydroperoxides with GSH. *An example of an endogenous substrate is the lipid peroxidation product 4-hydroxynonenal, whereas the carcinogen benzo[a]pyrene-7,8-diol-9,10-oxide (BPDE) is an exogenous substrate.*

regions could produce a chimeric GSTM2/GSTM1 gene whose mRNA and protein products would be indistinguishable from those of an unrecombined GSTM2 gene (56). In the GSTM3 gene, the GSTM3*A wild type and GSTM3*B variant allele differ from each other by a deletion of three bp in intron 6, resulting in the generation of a recognition sequence for the YY1 transcription factor in the latter (57). Little is known about the role of GSTM3 in the metabolism of various xenobiotic substrates, except having overlapping substrate specificity with GSTM1 (19). GSTM4 is identical in amino acid sequence to GSTM1 (87%), GSTM2 (83%) and GSTM3 (70%) (58).

GSTπ SUBFAMILY

The single GSTP1 gene at 11q13 is 2.8 kb long and contains seven exons (59-61) (Figure 4.3.6.). The promoter contains multiple SNPs, which give rise to over ten different haplotypes (62, 63). One promoter haplotype was associated with increased mRNA and protein concentrations (62). The open reading frame starts at the 3' end of the first exon and is 630 bp long, encoding a protein of 209 amino acids (most authors exclude the initiator methionine) with a predicted molecular mass of 23,224 Da. GSTP1 undergoes post-translational modification by the Ser/Thr protein kinases, cAMP-dependent protein kinase (PKA) and protein kinase C (PKC), resulting in a more than twofold increase in catalytic activity (64). The phospho-acceptor residues are 42Ser and 184Ser.

Crystallographic analysis has shown that the H-site occupies the same position in three-dimensional space in class μ and π GSTs (65). However, the structure of the H-site differs, forming a hydrophobic cavity in GSTM1, whereas it is approximately half hydrophobic and half hydrophilic in GSTP1. Based on the difference in H-site architecture, one might expect GSTP1 to be more efficient in the detoxification of estrogen metabolites than GSTM1. The steroid ring system would interact with the hydrophobic region of the H-site and the hydroxyl groups of the catechol estrogens with the hydrophilic region, which could stabilize the reaction intermediate in GSTP1 for the GSH conjugation. It was shown that GSTP1 and CYP1B1 are coordinated in sequential reactions of the oxidative estrogen metabolism pathway, i.e., 4-OHE_2 and 2-OHE_2 did not form GSH conjugates in the presence of GSTP1 unless they were first oxidized by CYP1B1 to their corresponding quinones (66). CYP1B1 metabolized E_2 to two products, 4-OHE_2 and 2-OHE_2, and further to $E_2\text{-}3,4\text{-Q}$ and $E_2\text{-}2,3\text{-Q}$, while GSTP1 formed three products, $4\text{-OHE}_2\text{-}2\text{-SG}$, $2\text{-OHE}_2\text{-}4\text{-SG}$, and $2\text{-OHE}_2\text{-}1\text{-SG}$, the last one in smallest amounts. $E_2\text{-}2,3\text{-Q}$ and $E_2\text{-}3,4\text{-Q}$ are products of CYP1B1- and substrates of GSTP1-mediated reactions but also react nonenzymatically with other nucleophiles as indicated by a ten-fold concentration gap between catechol estrogens and GSH-estrogen conjugates. Although both reactions are coordinated qualitatively in terms of product formation and substrate utilization, the quantitative gap would leave room for the accumulation of estrogen quinones and their potential for DNA damage.

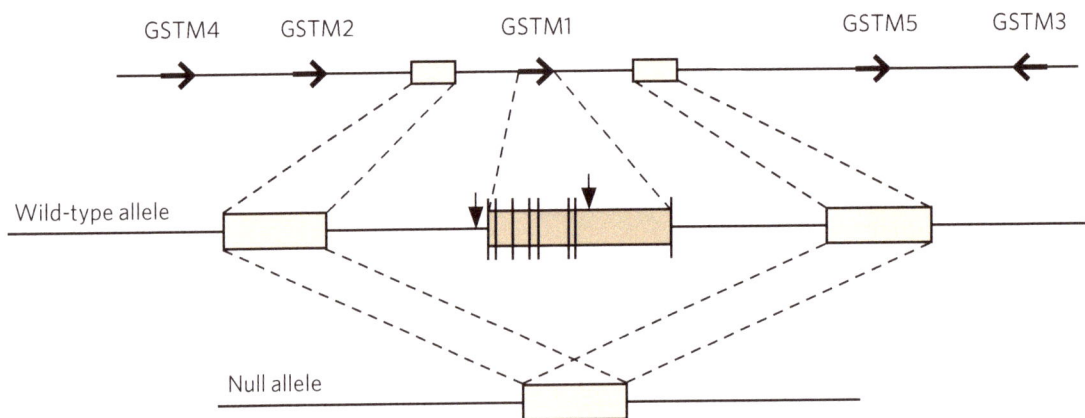

Figure 4.3.5. Diagram of Mu-class GST gene cluster.
The GSTM1 gene is part of the Mu-class GST gene cluster at 1p13.3, which is arranged as 5'-GSTM4-GSTM2-GSTM1-GSTM5-GSTM3-3' (top of diagram). The GSTM1 gene (black box) consists of 8 exons, which range in size from 36 to 112 bp, while the introns vary from 87 to 2,641 bp. GSTM1 is embedded in a region with extensive homologies and flanked by two almost identical 4.2-kb regions (gray boxes). The GSTM1 null allele arises by homologous recombination of the left and right 4.2-kb repeats, which results in a 16-kb deletion containing the entire GSTM1 gene (bottom of diagram). The point of deletion cannot be precisely localized because of the high sequence identity between the repeats. Several SNPs have been identified in the GSTM1 gene, including one at -498 (C/G; rs412543) and another in exon 7 (Lys172Asn; rs1065411; see vertical arrows).

The metabolism of B[a]P presents another example of sequential phase I and II enzyme interaction where mutagenic metabolites produced by CYP1A1 are detoxified by GSTP1-mediated conjugation with GSH. Coexpression of CYP1A1 and GSTP1 in the Chinese hamster lung fibroblast cell line V79MZ showed a reduction in CYP1A1-mediated B[a]P mutagenesis by GSTP1 (67). The reduction was largely due to prevention of the N2-guanine-BPDE adduct, the major mutagenic B[a]P metabolite. However, approximately one-third of the mutations was not prevented suggesting formation by CYP1A1 of a subset of mutagenic metabolites of B[a]P that are not effectively detoxified by GSTP1.

Several non-synonymous polymorphisms have been described in the coding region of the GSTP1 gene (68, 69) (Figure 4.3.6.). The most common, Ile104Val (rs1695), is found in over half of Mexican Americans, one third of Caucasian Americans and 10% of Chinese Americans (69). The next frequent polymorphism, Ala113Val (rs1138272), is present in nearly 10% of Caucasians but only 1 or 2% of African and Mexican Americans. All other non-synonymous polymorphisms occur in less than one percent in any ethnic group: Glu32Val, Asp58Asn, Asp147Tyr, and Arg187Trp (69). With the exception of 147Tyr all other variants showed decreased catalytic activity due to a reduction in enzyme levels, which resulted from an alteration in the rate of protein degradation (69, 70). The amino acid substitution Ile104Val also alters the active site of the enzyme by causing a shift in the side chains of several amino acid residues in the H site, which, in turn, lead to altered dimensions and formation of different hydrogen bonds (65, 71-73). The change in geometry affects thermal stability and accounts for differences in substrate utilization efficiency. For example, 104Ile/Ile was 3- to 4-fold more effective in

catalyzing the relatively small substrate 1-chloro-2,4-di-nitrobenzene (CDNB) than 104Val/Val (71, 74, 75). The converse was true for bulkier substrates, such as benzo[a]pyrene-7,8-diol-9,10-epoxide (BPDE) (74, 76, 77). The substitution in codon 113 did not affect catalytic activity toward CDNB, but exerted a pronounced effect on other substrates. The 104Ile-113Ala variant was most efficient in catalyzing the GSH conjugation of fjord-region diol epoxides of benzo[g]chrysene and benzo[c]phenanthrene, which are nonplanar molecules (78). In contrast, the 104Val-113Val variant was most efficient in the GSH conjugation of bay-region diol epoxides such as BPDE, which is a planar molecule (74, 76). Thus, both amino acids 104 and 113 affect substrate specificity to the point of distinguishing between planar and nonplanar substrates.

GSTP1 is expressed in many tissues including breast where it is the predominant GST (79) (Table 4.3.2.). A CpG island in the GSTP1 promoter was shown by methylation-specific PCR to be unmethylated in normal breast but hypermethylated in approximately one third of primary breast cancers (80, 81). In these tumors, the hypermethylation was associated with loss of GSTP1 expression as demonstrated by immunohistochemistry. A more extended examination of six CpG islands by MALDI mass spectrometry determined that the promoter was methylated in 71% of breast cancers (63). Further analysis revealed that the haplotype structure of a promoter sequence influenced the extent of DNA methylation. The distribution of methylation levels of tumors homozygous for the most frequent GSTP1 promoter haplotype was significantly different from other haplotype combinations. A putative c-Myb response element was identified in one of two minimal promoter haplotypes. *In vitro* analysis

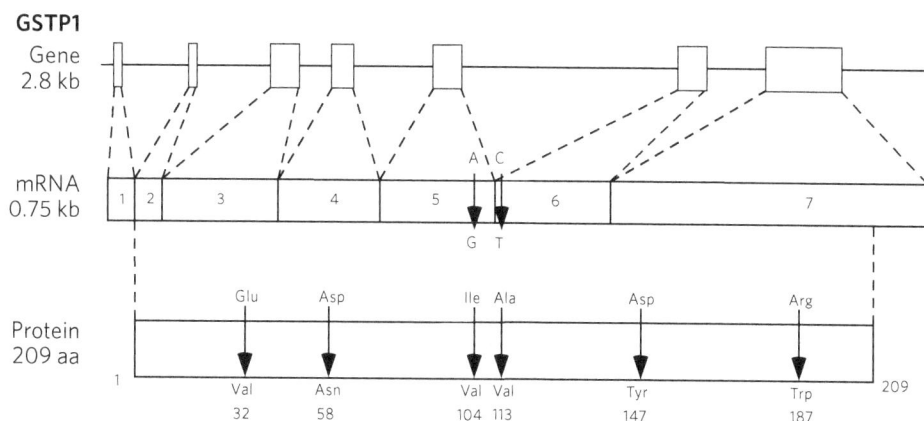

Figure 4.3.6. Overview of GSTP1 gene, mRNA, and protein.
The GSTP1 gene at 11q13 is about 2.8 kb long and contains seven exons. The open reading frame starts at the 3' end of the first exon and is 630 bp long, encoding a protein of 209 amino acids with a relative molecular weight Mr 23,224. The arrows indicate polymorphic sites. The two most common polymorphisms result in amino acid substitutions, Ile104Val (rs1695) and Ala113Val (rs1138272), in exons 5 and 6, respectively.

Table 4.3.2. Immunohistochemical GST Expression Patterns in Normal and Malignant Breast

GST Subtypes	Normal Breast	Breast Cancer	Reference
GSTP1	Ductal epithelium, lymphocytes, macrophages	47% (35/74) 69% (53/77) 75% (141/189)	(88) (80) (89)
GSTM1	Myoepithelial and isolated ductal epithelial cells	42% (32/74) 50% (24/48)	(88) (94)
GSTA1	Ductal epithelium	0% (0/43) 19% (14/74) 43% (22/51)	(87) (88) (94)
GSTO1	Myoepithelium	Not done	(97)

showed that binding of c-Myb to its response element was weakened by the sequence variation, leading to a change in the expression from the GSTP1 promoter. These results suggest that GSTP1 promoter haplotypes affect DNA methylation levels and promoter activity in breast carcinomas (63). Other studies implicate the methyl-CpG-binding domain (MBD) protein MBD2 in CpG island hypermethylation and GSTP1 silencing in breast cancer cells (82). Several studies reported an inverse correlation between GSTP1 and estrogen receptor expression, whereas other studies did not find an association with ER status (80, 83-87). There was a significant association between ER status, promoter haplotype and DNA methylation (63). Studies of GSTP1 expression levels in relation to prognosis have yielded inconsistent results (87-90). CpG island hypermethylation was found to be associated with longer relapse-free and overall survival in breast cancer in one study and shorter survival in another (63, 91). Two studies investigated the degree of GSTP1 hypermethylation in precursor lesions with one showing a progression from normal breast tissue to ductal hyperplasia and *in situ* and invasive carcinoma while the other detected no difference between normal and hyperplastic epithelium but a higher degree in both *in situ* and invasive lesions (92, 93).

Besides GSTP1, at least two other members of the GST family are expressed in breast tissue, namely GSTA1 and GSTM1 (79, 88, 94, 95). However, GSTM1 and GSTA1 are not consistently expressed in breast tissue. About 50% Caucasian and 25% African American women possess the GSTM1 null genotype and therefore lack GSTM1 expression in all tissues including breast (96). On the other hand, GSTA1 was either completely absent or present in the lowest percentage of breast cancer cases examined by immunohistochemistry (87, 88, 94). Thus, GSTP1 is the predominant GST catalyzing the GSH conjugation of estrogen quinones and PAH diol epoxides in breast tissue with variable contributions by GSTM1 and GSTA1.

GSTθ SUBFAMILY

The GSTθ subfamily consists of two genes, GSTT1 and GSTT2, which are located at 22q11.2 and separated by about 50 kb (98, 99) (Figure 4.3.7.). Both genes have five exons with identical intron/exon boundaries but share only 55% amino acid identity. About 20% of Caucasians are homozygous for a GSTT1 null allele, GSTT1-0. The GSTT1 null genotype is observed in 47% Asians (96). The deletion of the GSTT1 gene does not include GSTT2 (98). Analysis of a 119-kb section containing the GSTT1 and GSTT2 genes revealed two identical 403-bp repeats, which were identified as deletion/junction regions of the GSTT1 null allele (100). Like GSTM1*0, the GSTT1*0 deletion appears to be caused by a homologous recombination event involving the left and right 403-bp repeats. The recombination results in a ~54-kb deletion containing the entire GSTT1 gene. A non-synonymous polymorphism, Thr104Pro, results in a conformational change of an α helix that decreases the enzyme activity and mimics the null phenotype (101).

CLINICAL STUDIES

The majority of polymorphisms affecting genes involved in carcinogen metabolism are single nucleotide polymorphisms. Deletions are less common and the complete absence of a gene in form of a null allele is rare. It is for this reason that the GSTM1 and GSTT1 -/- genotypes have attracted so much attention and become the focus of numerous studies in molecular epidemiology. The underlying hypothesis of these investigations is that normal or increased GST enzyme activity may protect susceptible tissues from somatic DNA mutations by facilitating the detoxification of electrophilic carcinogens. In contrast, homozygous deletions of GSTM1 or GSTT1 are expected to have an impaired ability to metabolically eliminate carcinogenic compounds and may therefore place GSTM1 -/- or GSTT1 -/- individuals at increased cancer risk. Indeed, the first part of the assumption is supported by the observation of a strong association between GSTM1 deletion and increased PAH-DNA adduct levels in breast cancer tissue (102). However, the second part of the hypothesis was not supported as the same study showed no evidence of an association between GSTM1 genotype and breast cancer risk (102).

A meta-analysis of 30 studies involving over 10,000 individuals identified the GSTM1 null genotype in 53% Caucasians (with a 42 to 60% range for individual studies) (96). The frequency of the GSTM1 null genotype was similar in Asians but lower in African-Americans, 27% (16–36%). A search of the literature published from 1993 to 2003 listed over 500 studies of the GSTM1 genotype in relation to lung, breast, colon, brain, and various other types of cancer. These studies have in common PCR-based genotyping using an assay designed to identify the wild-type allele of GSTM1 (103, 104). In this assay, the absence of a PCR product (273 bp) indicates the GSTM1 -/- genotype and study participants were categorized as either wild-type or null 'genotypes'. This analytical approach has one basic flaw in that it does not positively identify the null allele and, therefore, cannot distinguish homozygous +/+ from heterozygous +/- individuals (105). Assuming that the presence of 2, 1, or 0 GSTM1 alleles is associated with a gene dosage effect resulting in high-, low-, or non-GSTM1 conjugator phenotypes, this approach oversimplifies phenotypes as all or none (106). Not surprisingly, the large number of studies utilizing this approach has yielded confusing data, which resulted in inconsistent or contradictory

publications on the association of the GSTM1 'genotype' with various malignancies (107-109). We analyzed the GSTM gene cluster and developed a long-range PCR assay (14 kb product) to allow positive identification of the null allele. In combination with the identification of the wild-type allele, we could perform true GSTM1 genotyping and define the +/+, +/-, and -/- genotypes. Thus, we could determine the frequency of the GSTM1 wild-type and null alleles in the Caucasian and African-American populations. We found that the GSTM1 wild-type allele is nearly twice as common in African-American (0.407) than in Caucasian (0.225) women (110). Several other studies have used long-range or real-time PCR to positively identify the null allele and determine the frequency of +/+, +/-, and -/- genotypes (54, 106, 111-113).

Application of true GSTM1 genotyping to a breast cancer case-control study revealed that the relative risk of breast cancer for Caucasian women with the +/+ genotype compared to women with the -/- genotype was 2.82 (95% CI 1.45 – 5.49; p = 0.002) (110). The association between the GSTM1 +/+ genotype and elevated breast cancer risk was unexpected and requires an explanation, which is speculative at this time, ranging from linkage of GSTM1 with other genes to the substrate GSH. With regard to GSH, mammalian cells have evolved protective mechanisms such as GSH conjugation to minimize injurious events that result from toxic chemicals and normal oxidative products of cellular metabolism. GSH depletion to about 20–30% of total glutathione levels

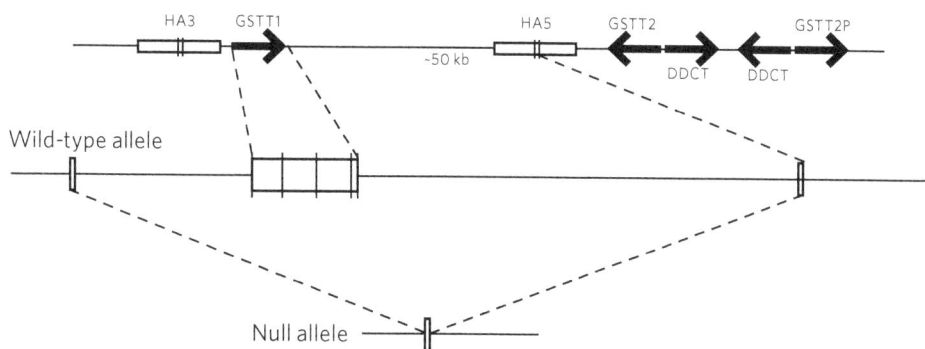

Figure 4.3.7. Diagram of Theta-class GST gene cluster.
The GSTT1 gene is part of the Theta-class GST gene cluster at 22q11.2 (top of diagram). GSTT1 and GSTT2 are separated by approximately 50 kb. GSTT2 lies head-to-head with a gene encoding the D-dopachrome tautomerase (DDCT). The GSTT2 and DDCT genes have been duplicated in an inverted repeat. The duplicated GSTT2 is a pseudogene (named GSTT2P) because an abnormal exon2/intron 2 splice site causes a premature translation stop. The GSTT1 gene (black box) consists of five exons, which range in size from 88 to 195 bp, while the introns vary from 205 to 2,363 bp. The GSTT1 gene is embedded in a region with extensive homologies and flanked by two 18 kb regions, HA3 and HA5 (gray boxes), which are more than 90% homologous. In their central portions HA3 and HA5 share a 403-bp sequence with 100% identity. The GSTT1 null allele arises by homologous recombination of the left and right 403-bp repeats, which results in a 54-kb deletion containing the entire GSTT1 gene (bottom of diagram). The point of deletion cannot be precisely localized because of the sequence identity between the 403-bp repeats.

can impair the conjugation defense against the toxic actions of such compounds and become detrimental to cellular processes (15). Thus, the combined conjugation activities of all GSTs may lead to GSH depletion and thereby become counterproductive. Instead of protecting, the GSTs collectively may expose the cell to injurious effects, such as oxidative DNA damage and associated mutagenic lesions. Alternatively, GSTs can convert several classes of compounds, via conjugation with GSH, into cytotoxic, genotoxic, or mutagenic metabolites (114). Although conjecture, this may explain the high frequency of the GSTM1 and GSTT1 -/- genotypes in the general population. Finally, one should consider other cellular functions of GSTM1, such as the suppression of apoptosis signal-regulating kinase 1 (ASK1) (115). Overexpression of GSTM1 was shown to repress ASK1-dependent apoptotic cell death and, by inference, one would expect the GSTM1 null genotype and the associated lack of GSTM1 to lead to increased apoptosis (116). A polymorphism in the GSTM1 promoter at nucleotide -498C/G (rs412543) was shown to modify binding of the transcription factor AP-2α, resulting in reduced promoter activity and mRNA expression (54). Depending on the presence of the -498C nucleotide and the GSTM1 + allele, all GSTM1 genotypes could be classified according to expression levels and breast cancer risk. Based on these results, the authors suggested a U-shaped association of GSTM1 with breast cancer risk in which both null activity (-/-) and overactivity (+/+) are disadvantageous (54).

Similar to GSTM1, the majority of epidemiological studies of the GSTT1 genotype employed a PCR-based assay that identified the wild-type allele (117, 118). A meta-analysis of 48 studies (17,254 cases, 21,163 controls) using this analytical approach showed an increased breast cancer risk associated with the GSTT1 null genotype (OR 1.138; 95% CI 1.051 – 1.232) (119). When stratified by ethnicity, an increased risk was found for Caucasian women (OR 1.185; 95% CI 1.075 – 1.306) but not Asian or African women. The absence of a PCR product signified the GSTT1 -/- genotype. Hence, this assay also did not positively identify the null allele and therefore could not distinguish homozygous +/+ from heterozygous +/- individuals. This distinction is important because of a gene-dosage effect. Determination of GSTT1 enzyme activity in erythrocytes showed a trimodal phenotypic distribution of high-, intermediate-, and null activity corresponding to the +/+, +/-, and -/- genotypes, indicating a gene-dosage effect (100, 120). The development of long-range and real-time PCR assays allowed the positive identification of the null allele (100, 106, 111-113). In Caucasians, the frequencies of the wild-type and null alleles were approximately 0.6 and 0.4, respectively. It is unknown whether the GSTT1 gene is expressed in breast tissue.

Several epidemiological studies have examined the association of the GSTP1 polymorphism Ile104Val (rs1695) with breast cancer risk yielding inconclusive results (25, 121-130). A meta-analysis of 30 studies (15,901 cases, 18,757 controls) found no association with breast cancer risk in the overall population (131). However, a subgroup analysis by ethnicity showed an increased risk among Asian women carrying the 104Val allele (OR 1.42; 95% CI 1.20 – 1.69).

Since GSTs have overlapping substrate specificities, deficiency of an individual GST isoenzyme may be compensated by other isoforms. Therefore, simultaneous determination of all GST genotypes appears to be a prerequisite for reliable interpretation of the role of the GST family in breast cancer. Several epidemiological studies have examined the relation between breast cancer risk and genotypes of two or more GST subtypes (107, 126). Overall, no clear pattern has emerged. Individual studies of two GSTs observed associations that were not confirmed by others (121, 123, 130, 132-138). Simultaneous analysis of three GSTs did not clarify risk associations but rather led to more contradictory results. For example, one study reported significantly increased risk for women with GSTM1 null and GSTT1 null genotypes together with the GSTP1 (104Val/Val) genotype, whereas another did not observe any association with this genotype combination (124, 126). A third study found the lowest risk for women simultaneously carrying the GSTM1 null, GSTT1 null, and GSTP1 (104Val/Val or Ile/Val) genotypes, whereas another noted an increased risk in premenopausal women lacking the GSTM1 and GSTT1 genes and carrying the GSTP1 (104Ile/Ile) genotype (125, 126). A fourth study found no association with GSTM1 and GSTT1 but a decreased risk for women with a homozygous 104Val/Val genotype (139). Yet another study of 500 breast cancer patients and 395 controls found no increase in risk associated with any combination of GSTM1, GSTP1, and GSTT1 genotypes (140). A similar inconsistency exists in reports relating breast cancer survival to GSTM1, GSTP1, and GSTT1 polymorphisms (141-144). A meta-analysis of 10 studies showed no risk association of GSTM1, GSTP1, and GSTT1 genotypes with breast cancer risk except in Asian women who exhibited an increased risk association with the GSTP1 (104Val/Val) genotype (121). A pooled analysis of studies examining polymorphisms in GSTM1, GSTP1, and GSTT1 found no association with breast cancer risk or an interaction with smoking (145). Similarly, there was no association in African American women (129). A population-based study (3155 cases, 5496 controls) of GSTM1, GSTM3, GSTP1, and GSTT1 showed an increased postmenopausal breast cancer risk associated with hormone replacement therapy for carriers of the functional GSTT1 allele (127). Compared to noncarriers the risk increased 4% per year of HRT use (OR 1.04; 95% CI 1.03 – 1.05; P = 0.0001). An analysis of nine SNPs located in the infrequently studied GSTA2, GSTM2, GSTO1, GSTO2, and GSTZ1 genes showed no evidence for an association with breast cancer risk in a population of 1021 cases and 1015 controls (146).

Some of the discrepant findings may be explained by dietary factors. Women with the GSTP1 Val/Val genotype and low cruciferous vegetable intake had a breast cancer risk 1.74-fold (95% CI: 1.13 – 2.67) that of women with the Ile/Ile and Ile/Val genotype (25). The interaction of the three GSTs with other genes has also been investigated, e.g., CYP1B1, epoxide hydrolase, showing associations with breast cancer risk (128, 129). However, the authors caution that further studies are required to rule out statistical artifacts. The contradictory results of these studies may be attributed to differences in the study populations and their exposure to environmental or dietary factors. On the other hand, none of the preceding studies truly genotyped GSTM1 and GSTT1. They only identified -/- homozygosity and thereby oversimplified the phenotype as all or none. The positive identification of wild-type and null alleles with the clear definition of +/+, +/-, and -/- genotypes and unambiguous assignment of high, low, and none conjugator phenotypes might help in resolving apparent inconsistencies in the epidemiological GST literature. GSTs are expressed in large amounts (between 5 and 100 μM or ~1% of total cell protein) in most tissues (19). Several points can be made regarding GST expression. (i) The GST subtype composition varies among tissues resulting in tissue-specific detoxifying capabilities. (ii) The level of GST expression among individual tissues varies considerably as a result of the high degree of inducibility by a variety of agents. The molecular mechanism of induction involves several response elements, including the xenobiotic- and antioxidant-responsive elements (19). At least 100 chemicals have been shown to induce GSTs (19). A significant number of these chemical inducers occur naturally and, as they are found as non-nutrient components in vegetables and citrus fruits, it is apparent that humans are likely to be exposed regularly to such compounds. Since diets vary, as much as 10-fold inter-individual variation in total GST activity has been reported in certain organs. For example, the total GST content in 43 samples of normal human pancreas was shown to vary 7-fold and levels of expression among GSTA, GSTM, and GSTP isozymes varied 6- to 30-fold (147). (iii) Expression of GSTs also appears to undergo sex- and age-specific regulations. (iv) The level of GST expression is generally higher in tumor than in normal tissues. Analysis of total GST activity in matched normal and malignant breast tissue from 30 patients revealed 5-fold higher GST levels in the cancer tissue than in normal breast tissue from the same patient (79, 148). Increased expression of GSTM3 was identified as a predictor of short survival of breast cancer patients (149). (v) The phenotypic variability may be greater than the effect of genetic polymorphisms. GSTP1 illustrates this point. The difference in enzymatic activity between recombinant wild-type GSTP1 104Ile/Ile and variant GSTP1 104Val/Val is only three- to four-fold (71), whereas the level of inter-individual GSTP1 expression may vary more than five-fold (147). In multidrug-resistant MCF-7 cells (AdrR MCF-7) initially selected for resistance to doxorubicin (former generic name, adriamycin), the level of GSTP1 expression was 40-fold higher than in the parental MCF-7 cell line (86). The overexpression was not the result of GSTP1 gene amplification. This means that the phenotypic expression of GSTs in a single organ or cell type may exhibit greater variation in enzyme activity than that predicted from genetic polymorphisms. The wide variability of phenotypic GST expression can therefore obscure the effect of genetic polymorphisms on individual cancer susceptibility (150).

References

1. Sies H. *Glutathione and its role in cellular functions.* Free Radical Biol Med. 1999;27:916-21.

2. Townsend DM, Tew KD, Tapiero H. *The importance of glutathione in human disease.* Biomed Pharmacother. 2003;57:145-55.

3. Sierra-Rivera E, Dasouki M, Summar ML, Krishnamani MR, Meredith M, Rao PN, et al. *Assignment of the human gene (GLCLR) that encodes the regulatory subunit of γ-glutamyl-cysteine synthetase to chromosome 1p21.* Cytogenet Cell Genet. 1996;72:252-4.

4. Sierra-Rivera E, Summar ML, Dasouki M, Krishnamani MR, Phillips III JA, Freeman ML. *Assignment of the gene (GLCLC) that encodes the heavy subunit of gamma-glutamylcysteine synthetase to human chromosome 6.* Cytogenet Cell Genet. 1995;70:278-9.

5. Moinova HR, Mulcahy RT. *An electrophile responsive element (EpRE) regulates beta-naphthoflavone induction of the human gamma-glutamylcysteine synthetase regulatory subunit gene.* J Biol Chem. 1998;273:14683-9.

6. Sekhar KR, Crooks PA, Sonar VN, Friedman DB, Chan JY, Meredith MJ, et al. *NADPH oxidase activity is essential for Keap1/Nrf2-mediated induction of GCLC in response to 2-indol-3-yl-methyleneiquinuclidin-3-ols.* Cancer Res. 2003;63:5636-45.

7. Wild AC, Moinova HR, Mulcahy RT. *Regulation of gamma-glutamycysteine synthetase subunit gene expression by the transcription factor Nrf2.* J Biol Chem. 1999;274:33627-36.

8. Manoharan TH, Gulick AM, Puchalski RB, Servais AL, Fahl WE. *Structural studies on human glutathione S-transferase p.* J Biol Chem. 1992;267:18940-5.

9. Boyland E, Chasseaud LF. *The role of glutathione and glutathione S-transferases in mercapturic acid biosynthesis.* Adv Enzymol Rel Areas Mol Biol. 1969;32:173-219.

10. Hayes JD, McLellan LI. *Glutathione and glutathione-dependent enzymes represent a co-ordinately regulated defence against oxidative stress.* Free Rad Res. 1999;31:273-300.

11. Taylor GW, Donnelly LE, Murray S, Rendell NB. *Excursions in biomedical mass spectrometry.* Br J Clin Pharmacol. 1996;42:119-26.

12. Schafer FQ, Buettner GR. *Redox environment of the cell as viewed through the redox state of the glutathione disulfide/glutathione couple.* Free Radical Biol Med.2001;30:1191-212.

13. Conour JE, Graham WV, Gaskins HR. *A combined in vitro/bioinformatic investigation of redox regulatory mechanisms governing cell cycle progression.* Physiol Genomics. 2004;18:196-205.

14. Kirlin WG, Cai J, Thompson SA, Diaz D, Kavanagh TJ, Jones DP. *Glutathione redox potential in response to differentiation and enzyme inducers.* Free Radical Biol Med. 1999;27:1208-18.

15. Reed DJ. *Glutathione: toxicological implications.* Annu Rev Pharmacol Toxicol. 1990;30:603-31.

16. Hayes JD, Flanagan JU, Jowsey

IR. Glutathione transferases. Annu Rev Pharmacol Toxicol. 2005;45:51-88.

17. Strange RC, Spiteri MA, Ramachandran S, Fryer A. Glutathione-S-transferase family of enzymes. Mutat Res. 2001;482:21-6.

18. Townsend DM, Tew KD. Cancer drugs, genetic variation and the glutathione-S-transferase gene family. Am J Pharmacogenomics. 2003;3:157-72.

19. Hayes JD, Pulford DJ. The glutathione S-transferase supergene family: Regulation of GST and the contribution of the isoenzymes to cancer chemoprotection and drug resistance. Crit Rev Biochem Mol Biol. 1995;30:445-600.

20. Laborde E. Glutathione transferases as mediators of signaling pathways involved in cell proliferation and cell death. Cell Death Differ. 2010;17:1373-80.

21. Board PG, Coggan M, Chelvanayagam G, Easteal S, Jermiin LS, Schulte GK, et al. Identification, characterization and crystal structure of the omega class glutathione transferases. J Biol Chem. 2000;275:24798-806.

22. Mannervik B, Awasthi YC, Board PG, Hayes JD, Di Ilio C, Ketterer B, et al. Nomenclature for human glutathione transferases. Biochem J. 1992;282:305-8.

23. Pemble SE, Taylor JB. An evolutionary perspective on glutathione transferases inferred from class-theta glutathione transferase cDNA sequences. Biochem J. 1992;287:957-63.

24. Chow HH, Hakim IA, Vining DR, Crowell JA, Tome ME, Ranger-Moore J, et al. Modulation of human glutathione S-transferases by polyphenon E intervention. Cancer Epidemiol Biomarkers Prev. 2007;16:1662-6.

25. Lee SA, Fowke JH, Lu W, Ye C, Zheng Y, Cai Q, et al. Cruciferous vegetables, the GSTP1 Ile105Val genetic polymorphism, and breast cancer risk. Am J Clin Nutr. 2008;87:753-60.

26. Flier J, Van Muiswinkel FL, Jongenelen CAM, Drukarch B. The neuroprotective antioxidant α-lipoic acid induces detoxication enzymes in cultured astroglial cells. Free Rad Res. 2002;36:695-9.

27. Fukuda A, Nakamura Y, Ohigashi H, Osawa T, Uchida K. Cellular response to the redox active lipid peroxidation products: induction of glutathione s-transferase P by 4-hydroxy-2-nonenal.

Biochem Biophys Res Commun. 1997;236:505-9.

28. Ross D, Kepa JK, Winski SL, Beall HD, Siegel D. NAD(P) H:quinone oxidoreductase 1 (NQO1):chemoprotection, bioactivation, gene regulation and genetic polymorphisms. Chem Biol Interact 2000;129:77-97.

29. Itoh K, Wakabayashi N, Katoh Y, Ishii T, Igarashi K, Engel JD, et al. Keap1 represses nuclear activation of antioxidant responsive elements by Nrf2 through binding to the amino-terminal Neh2 domain. Genes Develop. 1999;13:76-86.

30. Wakabayashi N, Dinkova-Kostova AT, Holtzclaw WD, Kang MI, Kobayashi A, Yamamoto M, et al. Protection against electrophile and oxidant stress by induction of the phase 2 response: Fate of cysteines of the Keap1 sensor modified by inducers. Proc Natl Acad Sci USA. 2004;101:2040-5.

31. Gong H, Singh SV, Singh SP, Mu Y, Lee JH, Saini SP, et al. Orphan nuclear receptor pregnane X receptor sensitizes oxidative stress responses in transgenic mice and cancerous cells. Mol Endocrinol. 2006;20:279-90.

32. Nuwaysir EF, Daggett DA, Jordan VC, Pitot HC. Phase II enzyme expression in rat liver in response to the antiestrogen tamoxifen. Cancer Res. 1996;56:3704-10.

33. Singh SV, Leal T, Ansari GA, Awasthi YC. Purification and characterization of glutathione S-transferases of human kidney. Biochem J. 1987;246:179-86.

34. Armstrong RN. Structure, catalytic mechanism, and evolution of the glutathione transferases. Chem Res Toxicol. 1997;10:2-18.

35. Ji X, E.C. vR, Johnson WW, Tomarev SI, Piatigorsky J, Armstrong RN, et al. Three-dimensional structure, catalytic properties, and evolution of a sigma class glutathione transferase from squid, a progenitor of the lens S-crystallins of cephalopods. Biochemistry. 1995;34:5317-28.

36. Lim K, Ho JX, Keeling K, Gilliland GL, Ji X, Rucker F, et al. Three-demensional structure of Schistosoma japonicum glutathione S-transferase fused with a six-amino acid conserved neutralizing epitope of gp41 from HIV. Protein Sci. 1994;3:2233-44.

37. Sinning I, Kleywegt GJ, Cowan SW, Reinemer P, Dirr HW, Huber R, et al. Structure determination and refinement of human alpha class glutathione transferase A1-1, and a

comparison with the Mu and Pi class enzymes. J Mol Biol. 1993;232:192-212.

38. Berhane K, Widersten M, Engstrom A, Kozarich JW, Mannervik B. Detoxification of base propenals and other a,b-unsaturated aldehyde products of radical reactions and lipid peroxidation by human glutathione transferases. Proc Natl Acad Sci USA. 1994;91:1480-4.

39. Reinemer P, Dirr HW, Ladenstein R, Huber R, Lo Bello M, Federici G, et al. Three-dimensional structure of class pi glutathione S-transferase from human placenta in complex with S-hexylglutathione at 2.8 A resolution. J Mol Biol. 1992;227:214-26.

40. Salinas AE, Wong MG. Glutathione S-transferases - A review. Curr Medicinal Chem. 1999;6:279-309.

41. Sundberg K, Dreij K, Seidel A, Jernstrom B. Glutathione conjugation and DNA adduct formation of dibenzo[a,1]pyrene and benzo[a] pyrene diol epoxides in V79 cells stably expressing different human glutathione trasnferases. Chem Res Toxicol. 2002;15:170-9.

42. Pal A, Seidel A, Xia H, Hu X, Srivastava SK, Oesch F, et al. Specificity of murine glutathione S-transferase isozymes in the glutathione conjugation of (-)-anti- and (+)-syn-stereoisomers of benzo[g] chrysene 11,12-diol 13,14-epoxide. Carcinogenesis. 1999;20:1997-2001.

43. Meyer DJ. Significance of an unusually low Km for glutathione in glutathione transferases of the alpha, mu, pi classes. Xenobiotica. 1993;23:823-34.

44. Jedlitschky G, Leier I, Buchholz U, Barnouin K, Kurz G, Keppler D. Transport of glutathione, glucuronate, and sulfate conjugates by the MRP gene-encoded conjugate export pump. Cancer Res. 1996;56:988-94.

45. Morrow CS, Diah S, Smitherman PK, Schneider E, Townsend AJ. Multidrug resistance protein and glutathione S-transferase P1-1 act in synergy to confer protection from 4-nitroquinoline 1-oxide toxciity. Carcinogenesis. 1998;19:109-15.

46. Morel F, Rauch C, Coles B, Le Ferrec E, Guillouzo A. The human glutathione transferase alpha locus: genomic organization of the gene cluster and functional characterization of the genetic polymorphism in the hGSTA1 promoter. Pharmacogenetics. 2002;12:277-86.

47. Coles BF, Morel F, Rauch C, Huber WH, Yang M, Teitel CH, et al. Effect of polymorphism in the

human glutathione S-transferase A1 promoter on hepatic GSTA1 and GSTA2 expression. Pharmacogenetics. 2001;11:663-9.

48. Tijhuis MJ, Wark PA, Arts JMMJG, Visker MHPW, Nagengast FM, Kok FJ, et al. GSTP1 and GSTA1 polymorphisms interact with cruciferous vegetable intake in colorectal adenoma risk. Cancer Epidemiol Biomarkers Prev. 2005;14:2943-51.

49. Ning B, Wang C, Morel F, Nowell S, Ratnasinghe DL, Carter W, et al. Human glutathione S-transferase A2 polymorphisms: variant expression, distribution in prostate cancer cases/controls and a novel form. Pharmacogenetics. 2004;14:35-44.

50. Tetlow N, Board PG. Functional polymorphism of human glutathione transferase A2. Pharmacogenetics. 2004;14:111-6.

51. Lin D, Meyer DJ, Ketterer B, Lang NP, Kadlubar FF. Effects of human and rat glutathione S-transferases on the covalent DNA binding of the N-acetoxy derivatives of heterocyclic amine carcinogens in vitro: a possible mechanism of organ specificity in their carcinogenesis. Cancer Res. 1994;54:4920-6.

52. Pearson WR, Vorachek WR, Xu SJ, Berger R, Hart J, Vannais D, et al. Identification of class-mu glutathione transferase genes GSTM1-GSTM5 on human chromosome 1p13. Am J Hum Genet. 1993;53:220-33.

53. Xu SJ, Wang YP, Roe B, Pearson WR. Characterization of the human class mu glutathione S-transferase gene cluster and the GSTM1 deletion. J Biol Chem. 1998;273:3517-27.

54. Yu K, Di G, Fan L, Wu J, Hu Z, Shen Z, et al. A functional polymorphism in the promoter region of GSTM1 implies a complex role for GSTM1 in breast cancer. FASEB J. 2009;23:2274-87.

55. Widersten M, Pearson WR, Engstrom A, Mannervik B. Heterologous expression of the allelic variant mu-class glutathione transferases mu and psi. Biochem J. 1991;276:519-24.

56. Vorachek WR, Pearson WR, Rule GS. Cloning, expression, and characterization of a class-mu glutathione transferase from human muscle, the product of the GST4 locus. Proc Natl Acad Sci USA. 1991;88:4443-7.

57. Inskip A, Elexperu-Camiruaga J, Buxton N, Dias PS, MacIntosh J, Campbell MD, et al. Identification of polymorphism at the glutathione S-transferase, GSTM3 locus: evidence

for linkage with GSTM1*A. Biochem J. 1995;312:713-6.

58. Comstock KE, Johnson KJ, Rifenbery D, Henner WD. *Isolation and analysis of the gene and cDNA for human Mu class glutathione S-transferease, GSTM4.* J Biol Chem. 1993;268:16958-65.

59. Cowell IG, Dixon KH, Pemble SE, Ketterer B, Taylor JB. *The structure of the human glutathione S-transference pi gene.* Biochem J. 1988;255:79-83.

60. Kano T, Sakai M, Muramatsu M. *Structure and expression of a human class pi glutathione S-transferase messenger RNA.* Cancer Res. 1987;47:5626-30.

61. Morrow CS, Cowan KH, Goldsmith ME. *Structure of the human genomic glutathione S-trans-ferase-pi gene.* Gene. 1989;75:3-11.

62. Cauchi S, Han W, Kumar SV, Spivack SD. *Haplotype-environment interactions that regulate the human glutathione s-transferase P1 promoter.* Cancer Res. 2006;66:6439-48.

63. Ronneberg JA, Tost J, Solvang HK, Alnaes GI, Johansen FE, Brendeford EM, et al. *GSTP1 promoter haplotypes affect DNA methylation levels and promoter activity in breast carcinomas.* Cancer Res. 2008;68:5562-71.

64. Lo HW, Antoun GR, Ali-Osman F. *The human glutathione S-transferase P1 protein is phos-phorylated and its metabolic function enhanced by the Ser/Thr protein kinases, cAMP-dependent protein kinase and protein kinase C, in glioblastoma cells.* Cancer Res. 2004;64:9131-8.

65. Ji X, Tordova M, O'Donnell R, Parsons JF, Hayden JB, Gilliland GL, et al. *Structure and function of the xenobiotic substrate-binding site and location of a potential non-substrate-binding site in a class pglutathione S-transferase.* Biochemistry. 1997;36:9690-702.

66. Hachey DL, Dawling S, Roodi N, Parl FF. *Sequential action of phase I and II enzymes cytochrome P450 1B1 and glutathione S-transferase P1 in mammary estrogen metabolism.* Cancer Res. 2003;63:8492 - 9.

67. Kushman ME, Kabler SL, Fleming MH, Ravoori S, Gupta RC, Doehmer J, et al. *Expression of human glutathione S-transferase P1 confers resistance to benzo[a] pyrene or benzo[a]pyrene-7,8-di-hydrodiol mutagenesis, macromo-lecular alkylation and formation of stable N2-Gua-BPDE adducts in stably transfected V79MZ cells co-expressing hCYP1A1.* Carcinogen-esis. 2007;28:207-14.

68. Board PG, Webb GC, Coggan M. *Isolation of a cDNA clone and localization of the human glutathione S-transferase 3 genes to chromosome bands 11q13 and 12q13-14.* Ann Hum Genet. 1989;53:205-13.

69. Moyer AM, Salavaggione OE, Wu TY, Moon I, Eckloff BW, Hildebrandt AT, et al. *Glutathione S-transferase P1: gene sequence variation and functional genomic studies.* Cancer Res. 2008;68:4791-801.

70. Weinshilboum R, Wang L. *Pharmacogenetics: Inherited variation in amino acid sequence and altered protein quantity.* Clin Phamacol Therap. 2004;75:253-8.

71. Ali-Osman F, Akande O, Antoun G, Mao J, Buolamwini J. *Molecular cloning, characteriza-tion, and expression in Escherichia coli of full-length cDNAs of three human glutathione S-transferase Pi gene variants.* J Biol Chem. 1997;272:10004-12.

72. Ji X, Blaszczyk J, Xiao B, O'Donnell R, Hu X, Herzog C, et al. *Structure and function of residue 104 and water molecules in the xenobiotic substrate-binding site in human glutathione S-transferase P1-1.* Biochemistry. 1999;38:10231-8.

73. Johansson AS, Stenberg G, Widersten M, Mannervik B. *Structure-activity relationships and thermal stability of human glutathione transferase P1-1 governed by the H-site residue 105.* J Mol Biol. 1998;278:687-98.

74. Hu X, O'Donnell R, Srivastava SK, Xia H, Zimniak P, Nanduri B, et al. *Active site architecture of polymorphic forms of human glutathione S-transferase P1-1 accounts for their enantioselectiv-ity and disparate activity in the glutathione conjugation of 7beta, 8alpha-dihydroxy-9alpha, 10alpha-oxy-7,8,9,10-tetrahydrobenzo(a) pyrene.* Biochem Biophys Res Comm. 1997;235:424-8.

75. Zimniak P, Nanduri B, Pikula S, Bandorowicz-Pinkula J, Singhal SS, Srivastava SK, et al. *Naturally occurring human glutathione S-transferase GSTP1-1 isoforms with isoleucine and valine in position 104 differ in enzymic properties.* Eur J Biochem. 1994;224:893-9.

76. Hu X, Xia H, Srivastava SK, Herzog C, Awasthi YC, Ji X, et al. *Activity of four allelic forms of glutathione S-transferase hGSTP1-1 for diol epoxides of polycyclic aromatic hydrocarbons.* Biochem Biophys Res Commun. 1997;238:397-402.

77. Sundberg K, Johansson AS, Stenberg G, Widersten M, Seidel

A, Mannervik B, et al. *Differences in the catalytic efficiencies of allelic variants of glutathione transferase P1-1 towards carcinogenic diol epoxides of polycyclic aromatic hydrocarbons.* Carcinogenesis. 1998;19:433-6.

78. Hu X, Xia H, Srivastava SK, Pal A, Wasthi YC, Zimniak P, et al. *Catalytic efficiencies of allelic variants of human glutathione S-transferase P1-1 toward carcinogenic anti-diol epoxides of benzo[c]phenanthrene and benzo[g]chrysene.* Cancer Res. 1998;58:5340-3.

79. Kelley MK, Engqvist-Goldstein A, Montali JA, Wheatley JB, Schmidt J, D.E., Kauvar LM. *Variability of glutathione S-transferase isoenzyme patterns in matched normal and cancer human breast tissue.* Biochem J. 1994;304:843-8.

80. Esteller M, Corn PG, Urena JM, Gabrielson E, S.B. B, Herman JG. *Inactivation of glutathione S-transferase P1 gene by promotor hypermethylation in human neoplasia.* Cancer Res. 1998;58:4515-8.

81. Jhaveri MS, Morrow CS. *Methylation-mediated regulation of the glutathione S-transferase P1 gene in human breast cancer cells.* Gene. 1998;210:1-7.

82. Lin X, Nelson WG. *Methyl-CpG-binding domain protein-2 mediates transcriptional repression associated with hypermethylated GSTP1 CpG islands in MCF-7 breast cancer cells.* Cancer Res. 2003;63:498-504.

83. Frierson HE, Gaffey MJ, Meredith SD, Boyd JC, Williams ME. *Immunohistochemical staining and Southern blot hybridization for glutathione S-transferase pi in mammary infiltratiing ductal carcinoma.* Mod Pathol. 1995;8:643-7.

84. Molina R, Oesterreich S, Zhou JL, Tandon AK, Clark GM, Allred DC, et al. *Glutathione transferase GST pi in breast tumors evaluated by three techniques.* Dis Markers. 1993;11:71-82.

85. Morrow CS, Chiu J, Cowan KH. *Posttranscriptional control of glutathione S-transferase pi gene expression in human breast cancer cells.* J Biol Chem. 1992;267:10544-50.

86. Moscow JA, Townsend AJ, Goldsmith ME, Whang-Peng J, Vickers PJ, Poisson R, et al. *Isolation of the human anionic glutathione S-transferase cDNA and the relation of its gene expression to estrogen-receptor content in primary breast cancer.* Proc Natl Acad Sci USA. 1988;85:6518-22.

87. Shea TC, Claflin G, Comstock KE, Sanderson BJ, Burstein NA,

Keenan EJ, et al. *Glutathione transferase activity and isoenzyme composition in primary human beast cancers.* Cancer Res. 1990;50:6848-53.

88. Cairns J, Wright C, Cattan AR, Hall AG, Cantwell BJ, Harris AL, et al. *Immunohistochemi-cal demonstration of glutathione S-transferases in primary human breast carcinomas.* J Pathol. 1992;166:19-25.

89. Gilbert L, Elwood LJ, Merino M, Masood S, Barnes R, Steinberg SM, et al. *A pilot study of pi-class glutathione S-transferase expression in breast cancer: correlation with estrogen receptor expression and prognosis in node-negative breast cancer.* J Clin Oncol. 1993;11:49-58.

90. Silvestrini R, Veneroni S, Benini W, Daidone MG, Luisi A, Leutner M, et al. *Expression of p53, glutathione S-transferase-pi, and Bcl-2 proteins and benefit from adjuvant radiotherapy in breast cancer.* J Natl Cancer Inst. 1997;89:639-45.

91. Arai T, Miyoshi Y, Kim SJ, Taguchi T, Tamaki Y, Noguchi S. *Association of GSTP1 CpG islands hypermethylation with poor prognosis in human breast cancers.* Breast Cancer Res Treat. 2006;100:169-76.

92. Lee JS. *GSTP1 promoter hyper-methylation is an early event in breast carcinogenesis.* Virchows Arch. 2007;450:637-42.

93. Pasquali L, Bedeir A, Ringquist S, Styche A, Bhargava R, Trucco G. *Quantification of CpG island methylation in progressive breast lesions from normal to invasive carcinoma.* Cancer Lett. 2007;257:136-44.

94. Alpert LC, Schecter RL, Berry DA, Melnychuk D, Peters WP, Caruso JA, et al. *Relation of glutathione S-transferase aandm isoforms to response to therapy in human breast cancer.* Clin Cancer Res. 1997;3:661-7.

95. Forrester LM, Hayes JD, Millis R, Barnes D, Harris AL, Schlager JJ, et al. *Expression of glutathione S-transferases and cytochrome P450 in normal and tumor breast tissue.* Carcinogenesis. 1990;11:2163-70.

96. Garte S, Gaspari L, Alexandrie AK, Ambrosone C, Autrup H, Aurup JL, et al. *Metabolic gene polymorphism frequencies in control populations.* Cancer Epidemiol Biomark Prev. 2001;10:1239-48.

97. Yin ZL, Dahlstrom JE, Le Couteur DG, Board P. *Immuno-histochemistry of omega class glutathione S-transferase in human tissues.* J Histochem Cytochem.

2001;49:983-7.

98. Coggan M, Whitbread L, Whittington A, Board P. *Structure and organization of the human theta-class glutathione S-transferase and D-dopachrome tautomerase gene complex.* Biochem J. 1998;334:617-23.

99. Whittington A, Vichai V, Webb G, Baker R, Pearson W, Board P. *Gene structure, expression and chromosomal localization of murine theta class glutathione transferase mGSTT1-1.* Biochem J. 1999;337:141-51.

100. Sprenger R, Schlagenhaufer R, Kerb R, Bruhn C, Brockmoller J, Roots T, et al. *Characterization of the glutathione S-transferase GSTT1 deletion: discrimination of all genotypes by polymerase chain reaction indicates a trimodular genotype-phenotype correlation.* Pharmacogenetics. 2000;10:557-65.

101. Alexandrie AK, Rannug A, Juronen E, Tasa G, Warholm M. *Detection and characterization of a novel functional polymorphism in the GSTT1 gene.* Pharmacogenetics. 2002;12:613-9.

102. Rundle A, Tang D, Zhou J, Cho S, Perera F. *The association between glutathione S-transferase M1 genotype and polycyclic aromatic hydrocarbon-DNA adducts in breast tissue.* Cancer Epidemiol Biomarkers Prev. 2000;9:1079-85.

103. Bell DA, Taylor JA, Paulson DF, Robertson CN, Mohler JL, Lucier GW. *Genetic risk and carcinogen exposure: a common defect of the carcinogen-metabolism gene glutathione S-transferase M1 (GSTM1) that increases susceptibility to bladder cancer.* J Natl Cancer Inst. 1993;85:1159-64.

104. Seidegard J, Vorachek WR, Pero RW, Pearson WR. *Hereditary differences in the expression of the human glutathione transferase active on trans-stilbene oxide are due to a gene deletion.* Proc Natl Acad Sci USA. 1988;85:7293-7.

105. Parl FF. *Glutathione S-transferase genotypes and cancer risk.* Cancer Lett. 2005;221:123-9.

106. Moore LE, Huang W, Chatterjee N, Gunter M, Chanock S, Yeager M, et al. *GSTM1, GSTT1, and GSTP1 polymorphisms and risk of advanced colorectal adenoma.* Cancer Epidemiol Biomarkers Prev. 2005;14:1823-7.

107. Dunning AM, Healey CS, Pharoah PDP, Teare MD, Ponder BAJ, Easton DF. *A systematic review of genetic polymorphisms and breast cancer risk.* Cancer Epidemiol Biomark Prev. 1999;8:843-54.

108. Geisler SA, Olshan AF. *GSTM1, GSTT1, and the risk of squamous cell carcinoma of the head and neck: a mini-HuGE review.* Am J Epidemiol. 2001;154:95-105.

109. Rebbeck TR. *Molecular epidemiology of the human glutathione S-transferase genotypes GSTM1 and GSTT1 in cancer susceptibility.* Cancer Epidemiol Biomark Prev. 1997;6:733-43.

110. Roodi N, Dupont WD, Moore JH, Parl FF. *Association of homozygous wild-type glutathione S-transferase M1 genotype with increased breast cancer risk.* Cancer Res. 2004;64:1233-6.

111. Bediaga NG, Alfonso-Sanchez MA, de Ronobales M, Rocandio AM. *GSTT1 and GSTM1 gene copy number analysis in paraffin-embedded tissue using quantitative real-time PCR.* Anal Biochem. 2008;278:221-3.

112. Covault J, Abreu C, Kranzler H, Oncken C. *Quantitative real-time PCR for gene dosage determinations in microdeletion genotypes.* BioTechniques. 2003;35:594-8.

113. Girault I, Lidereau R, Bieche I. *Trimodal GSTT1 and GSTM1 genotyping assay by real-time PCR.* Int J Biol Markers. 2005;20:81-6.

114. Monks TJ, Anders MW, Dekant W, Stevens JL, Lau SS, van Bladeren PJ. *Glutathione conjugate mediated toxicities.* Toxicol Applied Pharmacol. 1990;106:1-19.

115. Dorion S, Lambert H, Landry J. *Activation of the p38 signaling pathway by heat shock involves the dissociation of the glutathione S-transferase Mu from Ask1.* J Biol Chem. 2002;277:30792-7.

116. Cho SG, Lee YH, Park HS, Ryoo K, Kang KW, Park J, et al. *Glutathione S-transferase mu modulates the stress-activated signals by suppressing Apoptosis Signal-regulating Kinase 1.* J Biol Chem. 2001;276:12749-55.

117. Pemble S, Schroeder KR, Spencer SR, Meyer DJ, Hallier E, Bolt HM, et al. *Human glutathione S-transferase theta (GSTT1): cDNA cloning and the characterization of a genetic polymorphism.* Biochem J. 1994;300:271-6.

118. Wiencke JK, Pemble S, Ketterer B, Kelsey KT. *Gene deletion of glutathione S-transferase theta: Correlation with induced genetic damage and potential role in endogenous mutagenesis.* Cancer Epidemiol Biomarkers Prev. 1995;4:253-9.

119. Chen XX, Zhao RP, Qiu LX, Yuan H, Mao C, Hu XC, et al. *Glutathione S-transferase T1 polymorphism is associated with breast cancer susceptibility.* Cytokine. 2011;56:477-80.

120. Bruhn C, Brockmoller J, Kerb R, Roots I, Borchert HH. *Concordance between enzyme activity and genotype of glutathione S-transferase theta (GSTT1).* Biochem Pharmacol. 1998;59:1189-93.

121. Egan KM, Cai Q, Shu XO, Jin F, Zhu TL, Dai Q, et al. *Genetic polymorphisms in GSTM1, GSTP1, and GSTT1 and the risk for breast cancer: results from the Shanghai breast cancer study and meta-analysis.* Cancer Epidemiol Biomarkers Prev. 2004;13:197-204.

122. Gudmundsdottir K, Tryggvadottir L, Eyfjord JE. *GSTM1, GSTTI, and GSTP1 genotypes in relation to breast cancer risk and frequency of mutations in the p53 gene.* Cancer Epidemiol Biomarkers Prev. 2001;10:1169-73.

123. Harries LW, Stubbins MJ, Forman D, Howard GC, Wolf CR. *Identification of genetic polymorphisms at the glutathione S-transferase Pi locus and association with susceptibility to bladder, testicular and prostate cancer.* Carcinogenesis. 1997;18:641-4.

124. Helzlsouer KJ, Selmin O, Huang H, Strickland PT, Hoffman S, Alberg AJ, et al. *Association between glutathione S-transferase M1, P1, and T1 genetic polymorphisms and development of breast cancer.* J Natl Cancer Inst. 1998;90:512-8.

125. Millikan R, Pittman G, Tse CK, Savitz DA, Newman B, Bell D. *Glutathione S-transferases M1, T1 and P1 and breast cancer.* Cancer Epidemiol Biomark Prev. 2000;9:567-73.

126. Mitrunen K, Jourenkova N, Kataja V, Eskelinen M, Kosma VM, Benhamou S, et al. *Glutathione S-transferase M1, M3, P1, and T1 genetic polymorphisms and susceptibility to breast cancer.* Cancer Epidemiol Biomark Prev. 2001;10:229-36.

127. Marie-Genica Consortium on Genetic Susceptibility for Menopausal Hormone Therapy Related Breast Cancer Risk. *Genetic polymorphisms in phase I and phase II enzymes and breast cancer risk associated with menopausal hormone therapy in postmenopausal women.* Breast Cancer Res Treat. 2010;119:463-74.

128. Spurdle AB, Chang JH, Byrnes GB, Chen X, Dite GS, McCredie MR, et al. *A systematic approach to analysing gene-gene interactions: polymorphisms at the microsomal epoxide hydrolase EPHX and glutathione S-transferase GSTM1, GSTT1, and GSTP1 loci and breast cancer risk.* Cancer Epidemiol Biomarkers Prev. 2007;16:769-74.

129. Van Emburgh BO, Hu JJ, Levine EA, Mosley LJ, Perrier ND, Freimanis RI, et al. *Polymorphisms in CYP1B1, GSTM1, GSTT1 and GSTP1, and susceptibility to breast cancer.* Oncol Reports. 2008;19:1311-21.

130. Zhao M, Lewis R, Gustafson DR, Wen WQ, Cerhan JR, Zheng W. *No apparent association of GSTP1 A313 G polymorphism with breast cancer risk among postmenopausal Iowa women.* Cancer Epidemiol Biomark Prev. 2001;10:1301-2.

131. Lu S, Wang Z, Cui D, Liu H, Hao X. *Glutathione S-transferase P1 Ile105Val polymorphism and breast cancer risk: a meta-analysis involving 34,658 subjects.* Breast Cancer Res Treat. 2011;125:253-9.

132. Ambrosone CB, Freudenheim JL, Graham S, Marshall JR, Vena JE, Brasure JR, et al. *Cytochrome P4501A1 and glutathione S-transferase (M1) genetic polymorphisms and postmenopausal breast cancer risk.* Cancer Res. 1995;55:3483- 5.

133. Bailey LR, Roodi N, Verrier CS, Yee CJ, Dupont WD, Parl FF. *Breast cancer and CYP1A1, GSTM1, and GSTT1 polymorphisms: Evidence of a lack of association in Caucasians and African Americans.* Cancer Res. 1998;58:65-70.

134. Charrier J, Maugard CM, Le Mevel B, Bignon YJ. *Allelotype influence at glutathione S-transferase M1 locus on breast cancer susceptibility.* Br J Cancer. 1999;79:346-53.

135. Coughlin SS, Piper M. *Genetic polymorphisms and risk of breast cancer.* Cancer Epidemiol Biomark Prev. 1999;8:1023-32.

136. Garcia-Closas M, Kelsey KT, Hankinson SE, Spiegelman D, Springer K, Willett WC, et al. *Glutathione S-transferase mu and theta polymorphisms and breast cancer susceptibility.* J Natl Cancer Inst. 1999;91:1960-4.

137. Kelsey KT, Hankinson SE, Colditz GA, Springer K, Garcia-Closas M, Spiegelman D, et al. *Glutathione S-transferase classmdeletion polymorphism and breast cancer: Results from prevalent versus incident cases.* Cancer Epidemiol Biomark Prev. 1997;6:511-5.

138. Park SK, Yoo KY, Lee SJ, Kim SU, Ahn SH, Noh DY, et al.

Alcohol consumption, glutathione S-transferase M1 and T1 genetic polymorphisms and breast cancer risk. Pharmacogenetics. 2000;10:301-9.

139. Reding KW, Weiss NS, Chen C, Li CI, Carlson CS, Wilkerson H, et al. *Genetic polymorphisms in the catechol estrogen metabolism pathway and breast cancer risk.* Cancer Epidemiol Biomarkers Prev. 2009;18.

140. Gudmundsdottir K, Tryggvadottir L, Eyfjord JE. *GSTM1, GSTT1, and GSTP1 genotypes in relation to breast cancer risk and frequency of mutations in the p53 gene.* Cancer Epidemiol Biomark Prev. 2001;10:1169-73.

141. Ambrosone CB, Sweeney C, Coles BF, Thompson PA, McClure GY, Korourian S, et al. *Polymorphisms in glutathione S-transferases (GSTM1 and GSTT1) and survival after treatment for breast cancer.* Cancer Res. 2001;61:7130-5.

142. Goode EL, Dunning AM, Kuschel B, Healey CS, Day NE, Ponder BAJ, et al. *Effect of germ-line genetic variation on breast cancer survival in a population-based study.* Cancer Res. 2002;62:3052-7.

143. Lizard-Nacol S, Coudert B, Colosetti P, Riedinger JM, Fargeot P, Brunet-Lecomte P. *Glutathione S-transferase M1 null genotype: lack of association with tumour characteristics and survival in advanced breast cancer.* Breast Cancer Res: BCR. 1999;1:81-7.

144. Sweeney C, McClure GY, Fares MY, Stone A, Coles BF, Thompson PA, et al. *Association between survival after treatment for breast cancer and glutathione S-trasferase P1*

Ile105Val polymorphism. Cancer Res. 2000;60:5621-4.

145. Vogl FD, Taioli E, Maugard C, Zheng W, Ribeiro Pinto LF, Ambrosone C, et al. *Glutathione S-transferases M1, T1, and P1 andf breast cancer: a pooled analysis.* Cancer Epidemiol Biomarkers Prev. 2004;13:1473-9.

146. Andonova IE, Justenhoven C, Winter S, Hamann U, Baisch C, Rabstein S, et al. *No evidence for glutathione S-transferases GSTA2, GSTM2, GSTO1, GSTO2, and GSTZ1 in breast cancer risk.* Breast Cancer Res Treat. 2010;121:497-502.

147. Coles BF, Anderson KE, Doerge DR, Churchwell ML, Lang NP, Kadlubar FF. *Quantitative analysis of interindividual variation of glutathione S-transferase expression in human pancreas and the ambiguity of correlating genotype with phenotype.* Cancer Res. 2000;60:573-9.

148. Albin N, Massaad L, Toussaint C, Mathieu M, Morizet J, Parise O, et al. *Main drug-metabolizing enzyme systems in human breast tumors and peritumoral tissues.* Cancer Res. 1993;53:3541-6.

149. Glinksky GV, Higashiyama T, Glinskii AB. *Classification of human breast cancer using gene expression profiling as a component of the survival predictor algorithm.* Clin Cancer Res. 2004;10:2272-83.

150. Coles BF, Kadlubar FF. *Detoxification of electrophilic compounds by glutathione S-transferase catalysis: determinants of individual response to chemical carcinogens and chemotherapeutic drugs?* BioFactors 2003;17:115-30.

4.3.3. N-ACETYLTRANSFERASES
4.3.3.1. INTRODUCTION

The arylamine N-acetyltransferases (NATs) are cytoplasmic enzymes, which catalyze the transfer of an acetyl group from endogenous acetyl coenzyme A (acetylCoA) to environmental acceptor amines bearing primary amino or hydrazine functional groups (1). NATs recognize a wide range of substrates including xenobiotics, such as aromatic (aryl) and heterocyclic amines as well as hydrazine drugs. An endogenous target is the catabolism of folic acid and the N-acetylation of a folate metabolite (2). In addition to N-acetylation, NATs are capable of catalyzing the O-acetylation of hydroxylamine and hydroxamic acid metabolites (3-6) (Figure 4.3.8.). This is pertinent to heterocyclic amines, which do not undergo

significant N-acetylation, but, instead, after cytochrome P450-mediated N-hydroxylation undergo O-acetylation catalyzed by NAT. Furthermore, NATs can catalyze N,O-acetyltransfer reactions without the involvement of acetylCoA. For example, NAT can mediate the transfer of an acetyl group from the N to the O of acetylhydroxamates to generate N-acetoxyesters (1). The N-acetoxyarylamine metabolites formed are highly unstable and undergo spontaneous hydrolysis to arylnitrenium ions that bind to nucleophiles, such as DNA to produce covalent arylamine-DNA adducts (Figure 4.3.8.). In humans, acetylation is a major route of biotransformation for many heterocyclic and aromatic amines present in the diet, cigarette smoke, and the environment (7-9). While the N-acetylation leads to innocuous acetamides, the O-acetylation and N,O-acetyl transfer of hydroxylamine and hydroxamic acid metabolites results in the formation of reactive, protein- and DNA-binding acetoxy esters (5, 10). Human and animal studies have clearly shown the important role of NATs in carcinogen bioactivation and DNA adduct formation that can lead to mutations and thereby influence individual cancer susceptibility.

AcetylCoA carries the acetyl group in an activated state, attached to CoA by a relatively weak thioester linkage. The hydrolysis of the high-energy thioester bond is thermodynamically more favorable than that of an oxygen ester. All acetylCoA-dependent enzymes exploit the fact that the loosely bound acetyl group has a high transfer potential. The transfer can occur either in a single step directly to a substrate or involves a two-step substituted-enzyme ("ping-pong") process as in the case of NAT1 and NAT2. Initially, the active site cysteine of the enzyme (residue 68 in both NATs) accepts the acetyl group from acetylCoA, forming a thiol ester. The second step involves the binding of substrate to the acetylated enzyme, transfer of the acetyl group to the substrate, and release of the acetylated product. Structural studies of NATs have revealed three domains, the first predominantly alpha helical, the second a beta barrel, and the third in form of an alpha-beta lid (1). A conserved cysteine protease-like catalytic triad (68Cys, 107His, 122Asp) confirms the fundamental role in catalysis of the active site cysteine residue, 68Cys (11, 12). The acetylCoA molecule is bound in a cleft on the concave surface of the protein with hydrogen bonding to the pyrophosphate portion of acetylCoA (13-15). Hydrogen peroxide and S-nitrosothiols induce reversible inhibition of NAT1 through direct interaction with its catalytic 68Cys (16). Peroxynitrite leads to irreversible inhibition of the enzyme.

NAT1 and NAT2 share 87% nucleotide and 81% amino acid identity (1). This translates into 55 different amino acids (19%) of the 290-amino-acid protein products, of which only 28 (10%) are non-conservative changes. Both enzymes catalyze the same acetylation reaction but differ in substrate specificity and tissue expression. For example,

NAT1 preferentially metabolizes p-aminobenzoate and p-aminosalicylate, whereas sulfamethazine, procainamide, isoniazid, and caffeine are model substrates for NAT2 (Figure 4.3.9.). A central region (amino acids 112–210) on each of the NAT proteins, which is distinct from the active site 68Cys residue in the linear amino acid sequence, was shown to impart the respective NAT1-type or NAT2-catalytic specificity to these proteins (17). Within this central region, a highly conserved 42-amino acid segment (amino acids 107–148) differs at only three amino acid positions between the two proteins: 125Phe, 127Arg, 129Tyr in NAT1 and 125Ser, 127Ser, 129Ser in NAT2 (18). The effect of having two bulky groups (127Arg, 129Tyr) protruding into the substrate-binding site makes the binding pocket of NAT1 smaller than that of NAT2, which is lined by the serine residues (15).

NAT1 expression is widely distributed throughout the human body, whereas NAT2 expression is more limited (19). The tissue-specific expression and the substrate selectivity are important in understanding the roles of NAT isozymes in carcinogenesis. For example, the activation of the suspected human carcinogen 3-nitrobenzanthrone (3-NBA) involves hepatic NAT2 but not NAT1 to form 3-NBA-DNA adducts (20). NAT1 is expressed and active in normal and malignant breast tissue (21-23). Immunohistochemical staining also identified NAT2 protein expression in human mammary epithelial cells but NAT2-specific sulfamethazine acetylation activity was not detectable in mammary cytosol (24). NAT1 mRNA transcript levels were two- to threefold higher than NAT2 mRNA transcripts. Nat1/2 (-/-) double-knockout mice were normal, viable, and fertile (25).

There are over 50 SNPs in the NAT1 and NAT2 genes of which many are associated with decreased or increased rates of acetylation (9). Thus, the acetylation polymorphism is one of the most common inherited variations in the biotransformation of drugs and chemicals. Its association with drug toxicity and cancer risk has made it one of the oldest examples of a pharmacogenetic condition (26).

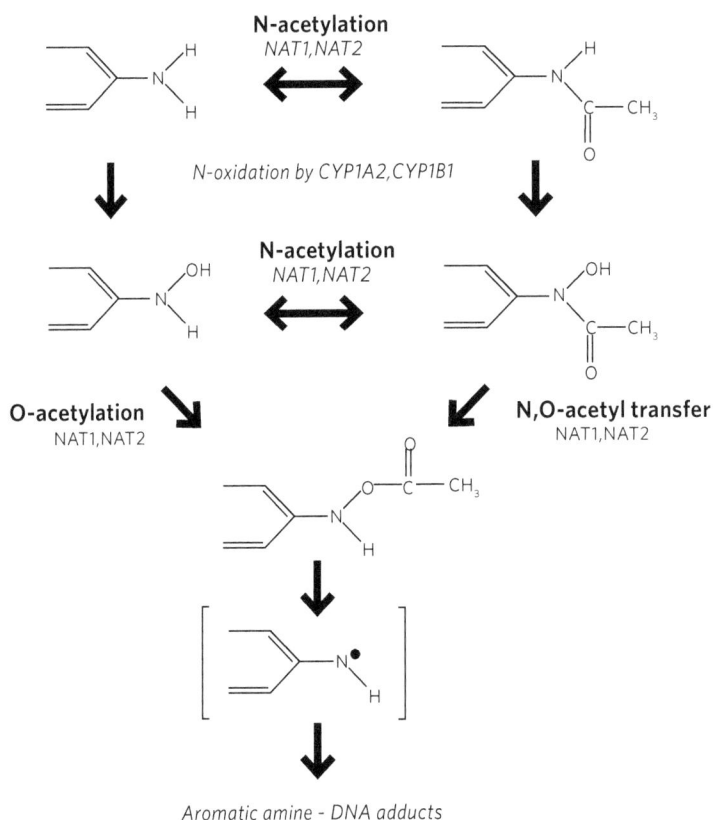

Figure 4.3.8. Role of N-acetyltransferases (NATs) in the metabolism of aromatic and heterocyclic amines.
NATs catalyze the hydrolysis of acetylCoA and the N-acetylation of aromatic and heterocyclic amines. They also catalyze the O-acetylation of arylhydroxylamines. Furthermore, NATs can catalyze intramolecular N,O-acetyl transfer reactions without the involvement of acetylCoA. The N-acetoxyarylamine metabolites formed are highly unstable and undergo spontaneous hydrolysis to arylnitrenium ions that bind to nucleophiles, such as DNA, to produce covalent arylamine-DNA adducts.

More than 50% of Caucasians are of the "slow acetylator" phenotype and are less efficient than "rapid acetylators" in the metabolism of numerous drugs and environmental chemicals containing primary aromatic amines (27). In contrast, only 10% of Asians are slow acetylators. Individuals with the 'slow acetylator' phenotype are at higher risk to develop bladder cancer, whereas rapid acetylators are at higher risk for colorectal cancer (7, 28-33). The majority of the over 50 SNPs observed in NAT1 and NAT2 do not involve the active site and cofactor binding regions but appear to be distributed around the periphery of the proteins, affecting their structural integrity and half-life rather than catalytic activity (1, 15, 34). Thus, these SNPs account for alterations in amino acids, which result in unstable proteins and therefore lower enzyme concentrations in slow acetylators (35, 36). Because most NAT1 and NAT2 alleles have multiple nucleotide substitutions, it is important that the correct phase be determined to identify their location on one or the other of the homologous chromosomes in the diploid genome (32). Genotype and haplotype misclassifications may have introduced substantial bias in some of the earlier epidemiological investigations, affecting the interpretation of gene-environment interactions.

4.3.3.2. N-ACETYLTRANSFERASE 1

The NAT1 gene at 8p22 consists of an intronless open reading frame of 870 bp, which is translated into the 290 amino acids that constitute the 33 kDa protein (2, 37). The NAT1 gene has several alternatively spliced non-coding exons up to 55 kb upstream of the single open reading frame, which results in a highly complex regulation of expression (1, 38). A promoter located 11.8 kb upstream of the coding exon directs expression of a breast-specific transcript (39, 40). NAT1 is induced by androgens via an androgen-responsive element located 745 bases upstream of the coding exon (41). NAT1 expression is also regulated by post-translational and environmental factors (42). The half-life of NAT1 is ~22 h in human cells (35).

NAT1 plays a major role in the bioactivation of the aromatic amine, 4-aminobiphenyl (ABP) to a DNA-binding species (43). Following absorption of the xenobiotic, hepatic CYP1A2 catalyzes the N-hydroxylation of ABP to the N-hydroxylamine, N-OH-ABP, which enters the circulation und undergoes renal filtration into the urinary bladder lumen (44). Upon reabsorption into the urinary bladder epithelium, N-OH-ABP is converted by NAT1 to N-acetoxy-ABP ester, which is capable of reacting with DNA, forming adducts such as N-(deoxyguanosine-8-yl)-4-aminobiphenyl (dG-ABP) and dA-ABP. In the human breast cell line MCF7, N-OH-ABP was similarly converted by NAT1 to the N-acetoxy ester, which resulted in the formation of dG-ABP as the primary adduct and dA-ABP as minor adduct (6).

Immunohistochemical analysis showed the expression of NAT1 in human mammary epithelial cells (21). N-acetylation of p-aminosalicylate, an activity specific to NAT1, but not acetylation of sulfamethazine, an activity specific to NAT2, was detected in mammary gland cytosols from 10 women and lysates from a primary culture of human mammary epithelial cells (21). An investigation of the methylation status of the NAT1 promoter region showed significantly less methylated CpG islands in breast cancer tissue than in normal and benign tissues (45). The expression of NAT1 at the mRNA and protein levels was increased in invasive ductal and lobular breast carcinomas when compared with normal breast tissue (22). Microarray analysis revealed higher

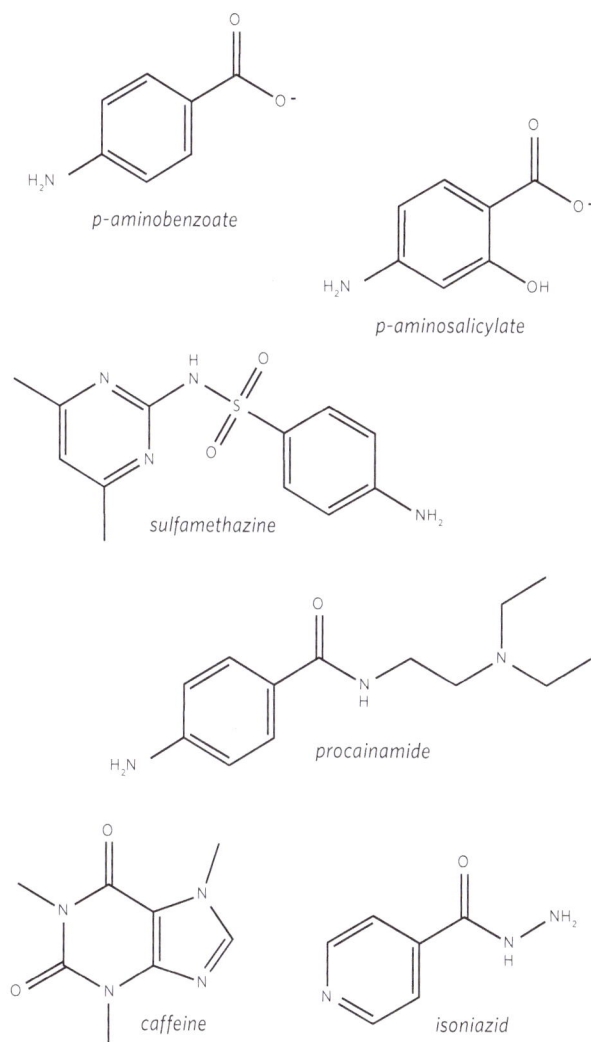

Figure 4.3.9. N-acetyltransferase substrates include p-aminobenzoate and p-aminosalicylate for NAT1 and sulfamethazine, procainamide, caffeine, and isoniazid for NAT2.

Table 4.3.3. Human NAT1 alleles (9, 48, 49, 51, 52)

Allele	Phenotype	Nucleotide change(s)	Amino acid change(s)
NAT1*4	normal	none	none
NAT1*3	normal	C1095A	none
NAT1*5	normal	G350, 351C, G497-499C, A884G, Δ976, Δ1105	Arg117Thr, Arg166Thr, Glu167Gln
NAT1*10	rapid?	T1088A, C1095A	none
NAT1*11A	normal	C-344T, A-40T, G445A, G459A, T640G, Δ1105, C1095A	Val149Ile, Ser214Ala
NAT1*11B	normal	C-344T, A-40T, G445A, G459A, T640G, Δ1105	Val149Ile, Ser214Ala
NAT1*11C	normal	C-344T, A-40T, G459A, T640G, Δ1105	Ser214Ala
NAT1*14A	slow	G560A, T1088A, C1095A	Arg187Gln
NAT1*14B	slow	G560A	Arg187Gln
NAT1*15	slow	C559A	Arg187Stop
NAT1*16	slow	AAA insertion after 1091, C1095A	none
NAT1*17	slow	C190T	Arg64Trp
NAT1*18A	unknown	Δ3(1064-1087), T1088A, C1095A	none
NAT1*18B	unknown	Δ3(1064-1087)	none
NAT1*19	slow	C97T	Arg33Stop
NAT1*20	unknown	T402C	none
NAT1*21	rapid	A613G	Met205Val
NAT1*22	slow	A752T	Asp251Val
NAT1*23	unknown	T777C	none
NAT1*24	rapid	G781A	Glu261Lys
NAT1*25	rapid	A787G	Ile263Val
NAT1*26A	unknown	TAA insertion (1066-1091), C1095A	none
NAT1*26B	unknown	TAA insertion (1066-1091)	none
NAT1*27	unknown	T21G, T777C	none
NAT1*28	unknown	TAATAA deletion (1085-1090)	none
NAT1*29	unknown	T1088A, C1095A, Δ1025	none

NAT1 expression in luminal than basal-like cancers (46, 47). NAT1 expression clustered with ER expression and, in fact, several microarray studies indicate that NAT1 falls within the top three most expressed genes in ER-positive breast tumors (38). Immunohistochemical analysis confirmed a strong association of NAT1 staining with the ER-positive phenotype. Among ER-positive breast cancer cell lines, ZR-75 showed by far the highest NAT1 activity of 202±28 nmol/min/mg protein (38). Overexpression of NAT1 in the human mammary luminal epithelial cell line HB4a conveyed enhanced growth and etoposide resistance relative to control cells (22). Exposure of MCF-7 cells to physiologic concentrations of peroxynitrite resulted in the irreversible inactivation of NAT1 due to oxidative modification of the active site cysteine (23).

At least 26 different alleles have been identified in the NAT1 locus, resulting from the existence of multiple single-nucleotide polymorphisms (up to five SNPs per allele; Table 4.3.3.). A consensus allele nomenclature was published in 1995 and subsequently updated (32, 48, 49). A detailed description of NAT1 alleles can be found at www.louisville.edu/medschool/pharmacology/NAT.html. NAT1*4 is defined as the reference NAT1 allele because it is associated with high activity and the most frequent

allele in the original population studied (Caucasian). Several NAT1 variants undergo more rapid ubiquination and proteasomal degradation than the wild-type NAT1*4 protein resulting in half-lives of less than 4 h and lower enzyme concentrations in slow acetylators (35, 36). The effect of each SNP is similar on N-acetylation and O-acetylation (9). The examination of recombinant NAT1 variants possessing single SNPs in the open reading frame revealed reduced protein levels and catalytic activities to levels below detection for C97T (Arg33stop), C190T (Arg64Try), C559T (Arg187stop), and A752T (Asp251Val) (50). G560A (Arg187Gln) substantially reduced NAT1 protein level and catalytic activity and increased substrate K_m. In contrast, the polymorphisms G445A (Val149Ile), G459A (synonymous) and T640G (Ser214Ala) haplotypes significantly increased NAT1 protein level and catalytic activity. On the other hand, neither T21G (synonymous), T402C (synonymous), A613G (Met205Val), T777C (synonymous), G781A (Glu261Lys) nor A787G (Ile263Val) significantly affected K_m, catalytic activity, mRNA or protein levels compared to reference NAT1*4 (50). Since the structure-function information was derived from a recombinant expression system, more data from human tissues are needed to investigate the role of tissue-specific regulatory factors.

NAT1 was expressed at high levels in tissue cytosol prepared from human bladder samples, whereas NAT2 activity was not detectable (28). NAT1 activity in the bladder of individuals with the heterozygous NAT1*10 allele was twofold higher than in subjects homozygous for the wild-type NAT1*4 allele. Likewise, DNA adduct levels in the mucosa of the urinary bladder were found to be twofold higher in individuals with the heterozygous NAT1*10 allele (3.5 ± 2.1 adducts/10^8 dNp) as compared to the homozygous NAT1*4 allele (1.8 ± 1.9 adducts/10^8 dNp). In contrast, there was no association between NAT1 genotypes and DNA adduct levels in normal breast tissue samples from 42 women undergoing mastectomy for breast cancer or reduction mammoplasty (53).

Epidemiological studies have not shown any consistent association of NAT1 genotypes with breast cancer risk (32, 54). However, environmental factors, such as smoking and red meat consumption, may have a modifying effect leading to elevated risk of breast cancer in smokers who consistently consumed well-done red meat and possessed the NAT1*11 allele (55, 56). Two studies have identified the NAT1 protein expression level in tumor samples as a predictor of good outcome in breast cancer patients (57, 58).

4.3.3.3. N-ACETYLTRANSFERASE 2

The NAT2 gene, like NAT1, is located at 8p22 and consists of an intronless open reading frame of 870 bp (37). Both NAT genes are separated by only ~177 kb and share 87% nucleotide identity. However, NAT2 contains a single non-coding exon at -8.7 kb compared to multiple distant non-coding exons in NAT1 (1). The expression of NAT2 exhibits sexual dimorphism (59). The gender-specific expression is mediated via an androgen-responsive element in the NAT2 promoter. Over 25 different alleles have been identified in the NAT2 locus, resulting from the existence of multiple single-nucleotide polymorphisms (Table 4.3.4.). A detailed description of NAT2 alleles can be found at www.louisville.edu/medschool/pharmacology/NAT.html (32, 48, 49). NAT2*4 is defined as the reference NAT2 allele because it is associated with high activity and the most frequent allele in the original population studied (Japanese). Each of the variant alleles comprises between one and four nucleotide substitutions, of which 13 are located in the protein-coding region of the gene. Four of these are silent, whereas nine lead to a change in the encoded amino acids: Arg64Trp, Arg64Gln, Ile114Thr, Gln145Pro, Glu167Lys, Arg197Gln, Lys268Arg, Lys282Thr, and Gly286Glu. The first eight wild-type amino acids are conserved in multiple species. The relationship between polymorphic substitutions and the activity of the variant enzymes have

been investigated in eukaryotic and prokaryotic expression systems (60). Compared to the reference NAT2*4, most variants have a slow acetylator phenotype (Table 4.3.4.). These variants are associated with lower NAT2 protein production or stability, which result in low levels of immuno-reactive NAT2 protein and account for the slow acetylator phenotype (61, 62). Only the Lys268Arg substitution (alleles NAT2*12A–C) and the silent C282T variant (allele NAT2*13) are associated with rapid acetylation (60). The effect of each SNP is similar on N-acetylation and O-acetylation (9). There are major ethnic differences in NAT2* allele distribution. For example, the Gly286Glu substitution (alleles NAT2*7A and NAT2*7B) is frequent in Asians (12%) but rare in Caucasians and Africans (1–2%) (27, 63). In contrast, the Arg64Gln substitution (alleles NAT*14A – G) is more common in Africans (7–19%) than in Asians and Caucasians (1%) (64). The slow acetylator phenotype is not homogeneous but rather consists of multiple phenotypes dependent upon the inheritance of specific SNPs and alleles.

Several case-control studies have investigated the association of NAT2 genotype and acetylator phenotype with breast cancer risk yielding conflicting results (55, 65-69). Since NAT2 plays a key role in the metabolism of aromatic amines, a major class of tobacco smoke carcinogens, Ambrosone examined the association between smoking and breast cancer risk by genotypes for NAT2 (65). For slow acetylators, smoking was associated with increased breast cancer risk in a dose-dependent manner. For example, smoking more than one pack of cigarettes per day for 20 years before the interview was associated with more than a 4-fold increase in risk. Because NAT2 slow acetylation genotypes are present in approximately 50% of Caucasians, the increased risk of breast cancer with smoking among this subset of the population would have a large public health effect (70). Consequently, several studies were conducted during the following decade to investigate associations between smoking, NAT2 genotypes, and breast cancer risk. In 2008, Ambrosone and her colleagues published a meta-analysis of 13 studies including 4,889 premenopausal and 7,033 postmenopausal women (66). They also pooled data from 10 of the studies. Both types of analysis confirmed that cigarette smoking was associated with an increase in breast cancer risk among women with NAT2 slow acetylation genotypes. The meta-analysis revealed an increased relative risk of 1.44 (95% CI 1.23–1.68) for ≥20 pack-years versus never smokers. Higher levels of DNA adducts were detected by ^{32}P-postlabeling analysis in normal mammary tissue from women genotyped as slow acetylators for NAT2 than in tissue from rapid acetylators (53). In contrast to the interaction between smoking and slow acetylation, one study showed an association of NAT2 rapid acetylation with increased breast cancer risk in women consuming well-done meat (32).

Table 4.3.4. Human NAT2 alleles (9, 48, 49, 51, 52)

Allele	Phenotype	Nucleotide change(s)	Amino acid change(s)
NAT2*4	rapid	none	none
NAT2*5A	slow	T341C, C481T	Ile114Thr
NAT2*5B	slow	T341C, C481T, A803G	Ile114Thr, Lys268Arg
NAT2*5C	slow	T341C, A803G	Ile114Thr, Lys268Arg
NAT2*5D	slow	T341C	Ile114Thr
NAT2*5E	slow	T341C, G590A	Ile114Thr, Arg197Gln
NAT2*5F	slow	T341C, C481T, C759T, A803G	Ile114Thr, Lys268Arg
NAT2*6A	slow	C282T, G590A	Arg197Gln
NAT2*6B	slow	G590A	Arg197Gln
NAT2*6C	slow	C282T, G590A, A803G	Arg197Gln, Lys268Arg
NAT2*6D	slow	T111C, C282T, G590	Arg197Gln
NAT2*7A	slow	G857A	Gly286Glu
NAT2*7B	slow	C282T, G857A	Gly286Glu
NAT2*10	unknown	G499A	Glu167Lys
NAT2*11	unknown	C481T	none
NAT2*12A	rapid	A803G	Lys268Arg
NAT2*12B	rapid	C282T, A803G	Lys268Arg
NAT2*12C	rapid	C481T, A803G	Lys268Arg
NAT2*13	rapid	C282T	none
NAT2*14A	slow	G191A	Arg64Gln
NAT2*14B	slow	G191A, C282T	Arg64Gln
NAT2*14C	slow	G191A, T341C, C481T, A803G	Arg64Gln, Ile114Thr, Lys268Arg
NAT2*14D	slow	G191A, C282T, G590A	Arg64Gln, Arg197Gln
NAT2*14E	slow	G191A, A803G	Arg64Gln, Lys268Arg
NAT2*14F	slow	G191A, T341C, A803G	Arg64Gln, Ile114Thr, Lys268Arg
NAT2*14G	slow	G191A, C282T, A803G	Arg64Gln, Lys268Arg
NAT2*17	slow	A434C	Gln145Pro
NAT2*18	unknown	A845C	Lys282Thr
NAT2*19	slow	C190T	Arg64Trp

References

1. Sim E, Lack N, Wang C, Long H, Westwood I, Fullam E, et al. *Arylamine N-acetyltransferases: Structural and functional implications of polymorphisms.* Toxicology. 2008;254:170-83.

2. Minchin RF, Hanna PE, Dupret J, Wagner CR, Rodrigues-Lima F, Butcher NJ. *Arylamine N-acetyltransferase I.* Int J Biochem Cell Biol. 2007;39:1999-2005.

3. Land SJ, Zukowski K, Lee M, Debiec-Rychter M, King CM, Wang CY. *Metabolism of aromatic amines: relationships of N-acetylation, O-acetylation, N,O-acetyltransfer and deacetylation in human liver and urinary bladder.* Carcinogenesis. 1989;10:727-31.

4. Minchin RF, Reeves PT, Teitel CH, McManus ME, Mojarrabi B, Ilett KF, et al. *N- and o-acetylation of aromatic and heterocyclic amine carcinogens by human monomorphic and polymorphic acetyltransferases expressed in COS-1 cells.* Biochem Biophys Res Commun. 1992;185:839-44.

5. Hein DW, Doll MA, Rustan TD, Gray K, Feng Y, Ferguson RJ, et al. *Metabolic activation and deactivation of arylamine carcinogens by recombiant human NAT1 and polymorphic NAT2 acetyltransferases.* Carcinogenesis. 1993;14:1633-8.

6. Swaminathan S, Frederickson SM, Hatcher JF. *Metabolic activation of n-hydroxy-4-acetylaminobiphenyl by cultured human breast epithelial cell line MCF 10A.* Carcinogenesis. 1994;15:611-7.

7. Hayes RB, Bi W, Rothman N, Broly F, Caporaso N, Feng P, et al. *N-acetylation phenotype and genotype and risk of bladder cancer in benzidine-exposed workers.* Carcinogenesis. 1993;14(4):675-8.

8. Vineis P, Bartsch H, Caporaso N, Harrington AM, Kadlubar FF, Landi MT, et al. *Genetically based N-acetyltransferase metabolic polymorphism and low-level environmental exposure to carcinogens.* Nature. 1994;369:154-6.

9. Hein DW. *Molecular genetics and function of NAT1 and NAT2: role in aromatic amine metabolism and carcinogenesis.* Mutation Res. 2002;506-507:65-77.

10. King CM, Traub NR, Lortz ZM, Thissen MR. *Metabolic activation of arylhydroxamic acids by N-O-acyltransferase of rat mammary gland.* Cancer Res. 1979;39:3369-72.

11. Sinclair JC, Sandy J, Delgoda R, Sim E, Noble MEM. *Structure of arylamine N-acetyltransferase reveals a catalytic triad.* Nature Struct Biol. 2000;7:560-4.

12. Pompeo F, Brooke E, Kawamura A, Mushtaq A, Sim E. *The pharmacogentics of NAT: structural aspects.* Pharmacogenomics. 2002;3:19-30.

13. Dutnall RN, Tafrov ST, Sternglanz R, Ramakrishnan V. *Structure of the histone acetyltransferase Hat1: A paradigm for the GCN5-related N-acetyltransferase superfamily.* Cell. 1998;94:427-38.

14. Wolf E, Vassilev A, Makino Y, Sali A, Nakatani Y, Burley SK. *Crystal structure of a GCN5-related N-acetyltransferase: Serratia marcescens aminoglycoside 3-N-acetyltransferase.* Cell. 1998;94:439-49.

15. Wu H, Dombrovsky L, Tempel W, Martin F, Loppnau P, Goodfellow GH, et al. *Structural basis of substrate-binding specificity of human arylamine N-acetyltransferases.* J Biol Chem. 2007;282:30189-97.

16. Dupret JM, Rodrigues-Lima F. *Structure and regulation of the drug-metabolizing enzymes arylamine N-acetyltransferases.* Curr Medicinal Chem. 2005;12:311-8.

17. Dupret JM, Goodfellow GH, Janezic SA, Grant DM. *Structure-function studies of human arylamine N-acetyltransferases NAT1 and NAT2.* J Biol Chem. 1994;269(43):26830-5.

18. Goodfellow GH, Dupret JM, Grant DM. *Identification of amino acids imparting acceptor substrate selectivity to human arylamine acetyltransferase NAT1 and NAT2.* Biochem J. 2000;348:159-66.

19. Windmill KF, Gaedigk A, Hall PD, Samaratunga H, Grant DM, McManus ME. *Localization of N-acetyltransferases NAT1 and NAT2 in human tissues.* Toxicol Sci. 2000;54:19-29.

20. Arlt VM, Stiborova M, Henderson CJ, Osborne MR, Bieler CA, Frei E, et al. *Environmental pollutant and potent mutagen 3-nitrobensanthrone forms DNA adducts after reduction by NAD(P)H:Quinone oxidoreductase and conjugation by acetyltransferases and sulfotransferases in human hepatic cytosols.* Cancer Res. 2005;65(7):2644-52.

21. Sadrieh N, Davis CD, Snyderwine EG. *N-acetyltransferase expression and metabolic activation of the food-derived heterocyclic amines in the human mammary gland.* Cancer Res. 1996;56:2683-7.

22. Adam PJ, Berry J, Loader JA, Tyson KL, Craggs G, Smith P, et al. *Arylamine n-acetyltransferase-1 is highly expressed in breast cancers and conveys enhanced growth and resistance to etoposide in vitro.* Mol Cancer Res. 2003;1:826-35.

23. Dairou J, Atmane N, Rodrigues-Lima F, Dupret JM. *Peroxynitrite irreversibly inactivates the human xenobiotic-metabolizing enzyme arylamine N-acetyltransferase 1 (NAT1) in human breast cancer cells.* J Biol Chem. 2004;279(9):7708-14.

24. Williams JA, Stone EM, Fakis G, Johnson N, Cordell JA, Meinl W, et al. *N-acetyltransferases, sulfotransferases and heterocyclic amine activation in the breast.* Pharmacogenetics. 2001;11:373-88.

25. Sugamori KS, Wong S, Gaedigk A, Yu V, Abramovici H, Rozmahel R, et al. *Generation and functional characterization of arylamine N-acetyltransferase NAT1/NAT2 double-knockout mice.* Mol Pharmacol. 2003;64:170-9.

26. Evans DAP. *N-acetyltransferase.* Pharmac Ther. 1989;42:157-208.

27. Blum M, Demierre A, Grant DM, Heim M, Meyer UA. *Molecular mechanism of slow aceetylation of drugs and carcinogens in humans.* Proc Natl Acad Sci USA. 1991;88:5237-41.

28. Badawi AF, Hirvonen A, Bell DA, Lang NP, Kadlubar FF. *Role of aromatic amine acetyltransferases, NAT1 and NAT2, in carcinogen-DNA adduct formation in the human urinary bladder.* Cancer Res. 1995;55:5230-7.

29. Bell DA, Badawi AF, Lang NP, Ilett KF, Kadlubar FF, Hirvonen A. *Polymorphism in the N-acetyltransferase 1 (NAT1) polyadenylation signal: Association of NAT1*10 allele with higher N-acetylation activity in bladder and colon tissue.* Cancer Res. 1995;55:5226-9.

30. Bell DA, Stephens EA, Castranio T, Umbach DM, Watson M, Deakin M, et al. *Polyadenylation polymorphism in the acetyltransferase 1 gene (NAT1) increases risk of colorectal cancer.* Cancer Res. 1995;55:3537-42.

31. Brockmoller J, Cascorbi I, Kerb R, Roots I. *Combined analysis of inherited polymorphisms in arylamine N-acetyltransferase 2, glutathione S-transferases M1 and T1, microsomal epoxide hydrolase, and cytochrome P450 enzymes as modulators of bladder cancer risk.* Cancer Res. 1996;56:3915-25.

32. Hein DW, Doll MA, Fretland AJ, Leff MA, Webb SJ, Xiao GH, et al. *Molecular genetics and epidemiology of the NAT1 and NAT2 acetylation polymorphisms.* Cancer Epidemiol Biomarkers Prev. 2000;9:29-42.

33. Lilla C, Verla-Tebit E, Risch A, Jager B, Hoffmeister M, Brenner H, et al. *Effect of NAT1 and NAT2 genetic polymorphisms on colorectal cancer risk associated with exposure to tobacco smoke and meat consumption.* Cancer Epidemiol Biomarkers Prev. 2006;15:99-107.

34. Walraven JM, Trent JO, Hein DW. *Structure-function analyses of single nucleotide polymorphisms in human N-acetyltransferase I.* Drug Metab Rev. 2008;40:169-84.

35. Butcher NJ, Arulpragasam A, Minchin RF. *Proteasomal degradation of N-acetyltransferase 1 is prevented by acetylation of the active site cysteine.* J Biol Chem. 2004;279:22131-7.

36. Liu F, Zhang N, Zhou X, Hanna PE, Wagner CR, Koepp DM, et al. *Arylamine N-acetyltransferase aggregation and constitutive ubiquitylation.* J Mol Biol. 2006;361:482-92.

37. Grant DM. *Molecular genetics of the N-acetyltransferases.* Pharmacogenetics. 1993;3:45-50.

38. Wakefield L, Robinson J, Long H, Ibbitt JC, Cooke S, Hurst HC, et al. *Arylamine N-acetyltransferase I expression in breast cancer cell lines: A potential marker in estrogen receptor-positive tumors.* Genes Chromosomes Cancer. 2008;47:118-26.

39. Husain A, Barker DF, States JC, Doll MA, Hein DW. *Identification of the major promoter and non-coding exons of the human arylamine N-acetyltransferase 1 gene (NAT1).* Pharmacogenetics. 2004;14:397-406.

40. Butcher NJ, Arulpragasam A, Li Goh H, Davey T, Minchin RF. *Genomic organization of human arylamine N-acetyltransferase type I reveals alternative promoters that generate different 5'-UTR splice variants with altered translational activities.* Biochem J. 2005;387:119-27.

41. Butcher NJ, Tetlow NL, Cheung C, Broadhurst GM, Minchin RF. *Induction of human arylamine N-acetyltransferase type I by androgens in human prostate cancer cells.* Cancer Res. 2007;67:85-92.

42. Butcher NJ, Minchin RF. *Arylamine N-acetyltransferase 1: a novel drug target in cancer development.* Pharmacol Rev. 2012;64:147-65.

43. Culp SJ, Roberts DW, Talaska G, Lang NP, Fu PP, Lay Jr JO, et al. *Immunochemical, 32P-postlabeling, and GC/MS detection of 4-aminobiphenyl-DNA adducts in human peripheral lung in relation to metabolic activation pathways involving pulmonary N-oxidation, conjugation, and peroxidation.* Mutation Res. 1997;378:97-112.

44. Kaderbhai MA, Ugochukwu CC, Lamb DC, Kelly SL. *Targeting of active human cytochrome P4501A1 (CYP1A1) to the periplasmic space of Escherichia coli.* Biochem Biophys Res Commun. 2000;279:803-7.

45. Kim SJ, Kang HS, Chang HL, Jung YC, Sim HB, Lee KS, et al. *Promoter hypomethylation of the N-acetyltransferase 1 gene in breast cancer.* Oncology Reports. 2008;19:663-8.

46. Farmer P, Bonnefoi H, Becette V, Tubiana-Hulin M, Fumoleau P, Larsimont D, et al. *Identification of molecular apocrine breast tumours by microarray analysis.* Oncogene. 2005;24:4660-71.

47. Sorlie T, Perou CM, Tibshirani R, Aas T, Geisler S, Johnsen H, et al. *Gene expression patterns of breast carcinomas distinguish tumor subclasses with clinical implications.* Proc Natl Acad Sci U S A. 2001;98:10869-74.

48. Vatsis KP, Weber WW, Bell DA, Dupret JM, Price Evans DA, Grant DM, et al. *Nomenclature for N-acetyltransferases.* Pharmacogenetics. 1995;5:1-17.

49. Hein DW, Boukouvala S, Grant DM, Minchin RF, Sim E. *Changes in consensus arylamine N-acetyltransferase gene nomenclature.* Pharmacogenetics Genomics. 2008;18:367-8.

50. Zhu Y, Hein DW. *Functional effects of single nucleotide polymorphisms in the coding region of human N-acetyltransferase 1.* Pharmacogenomics J. 2008;8:339-48.

51. Butcher NJ, Boukouvala S, Sim E, Minchin RF. *Pharmacogenetics of the arylamine N-acetyltransferases.* Pharmacogenomics J. 2002;2:30-42.

52. Hein DW, Grant DM, Sim E. *Update on consensus arylamine B-acetyltransferase gene nomenclature.* Pharmacogenetics. 2000;10:291-2.

53. Pfau W, Stone EM, Brockstedt U, Carmichael PL, Marquardt H, Phillips DH. *DNA adducts in human breast tissue: Association with N-acetyltransferase-2 (NAT2) and NAT1 genotypes.* Cancer Epidemiol Biomarkers Prev. 1998;7:1019-25.

54. Millikan RC. *NAT1*10 and NAT1*11 polymorphisms and breast cancer risk.* Cancer Epidemiol Biomarkers Prev. 2000;9:217-9.

55. Millikan RC, Pittman GS, Newman B, Tse CK, Selmin O, Rockhill B, et al. *Cigarette smoking, N-acetyltransferases 1 and 2, and breast cancer risk.* Cancer Epidemiol Biomarkers Prev. 1998;7:371-8.

56. Zheng W, Deitz AC, Campbell DR, Wen W-Q, Cerhan JR, Sellers TA, et al. *N-acetyltransferase 1 genetic polymorphism, cigarette smoking, well-done meat intake, and breast cancer risk.* Cancer Epidemiol Biomarkers Prev. 1999;8:233-9.

57. Dolled-Filhart M, Ryden L, Cregger M, Jirstrom K, Harigopal M, Camp RL, et al. *Classification of breast cancer using genetic algorithms and tissue microarrays.* Clin Cancer Res. 2006;12:6459-68.

58. Ring BZ, Seitz RS, Beck R, Shasteen WJ, Tarr SM, Cheang MCU, et al. *Novel prognostic immunohistochemical biomarker panel for estrogen receptor-positive breast cancer.* J Clin Oncol. 2006;24:3039-47.

59. Estrada L, Kanelakis KC, Levy GN, Weber WW. *Tissue- and gender-specific expression of N-acetyltransferase 2 (NAT2*) during development of the outbred mouse strain CD-1.* Drug Metabol Dispos. 2000;28:139-46.

60. Hein DW, Doll MA, Rustan TD, Ferguson RJ. *Metabolic activation of N-hydroxyarylamines and N-hydroxyarylamides by 16* recombinant human NAT2 allozymes: Effects of 7 specific NAT2 nucleic acid substitutions. Cancer Res. 1995;55:3531-6.

61. Hein DW, Ferguson RJ, Doll MA, Rustan TD, Gray K. *Molecular genetics of human polymorphic N-acetyltransferase: enzymatic analysis of 15 recombinant wild-type, mutant, and chimeric NAT2 allozymes.* Hum Mol Genet. 1994;3:729-34.

62. Fretland AJ, Leff MA, Doll MA, Hein DW. *Functional characterization of human N-acetyltransferase 2 (NAT2) single nucleotide polymorphisms.* Pharmacogenetics. 2001;11:207-15.

63. Lin HJ, Han CY, Lin BK, Hardy S. *Slow acetylator mutations in the human polymorphic N-acetyltransferase gene in 786 Asians, Blacks, Hispanics, and Whites: Application to metabolic epidemiology.* Am J Hum Genet. 1993;52:827-34.

64. Bell DA, Taylor JA, Butler MA, Stephens EA, Wiest J, Brubaker LH, et al. *Genotype/phenotype discordance for human arylamine N-acetyltransferase (NAT2) reveals a new slow-acetylator allele common in African-Americans.* Carcinogenesis. 1993;14:1689-92.

65. Ambrosone CB, Freudenheim JL, Graham S, Marshall JR, Vena JE,

Brasure JR, et al. *Cigarette smoking, N-acetyltransferase 2 genetic polymorphisms, and breast cancer risk.* JAMA. 1996;276:1494-501.

66. Ambrosone CB, Kropp S, Yang J, Yao S, Shields PG, Chang-Claude J. *Cigarette Smoking, N-acetyl-transferase 2 genotypes, and breast cancer risk: pooled analysis and meta-analysis.* Cancer Epidemiol Biomarkers Prev. 2008;17:15-26.

67. Firozi PF, Bondy ML, Sahin AA, Chang P, Lukmanji F, Singletary E, et al. *Aromatic DNA adducts and polymorphisms of CYP1A1, NAT2, and GSTM1 in breast center.* Carcinogenesis. 2002;23:301-6.

68. Hunter DJ, Hankinson SE, Hough H, Gertig DM, Garcia-Closas M, Spiegelman D, et al. *A prospective study of NAT2 acetylation genotype, cigarette smoking, and risk of breast cancer.* Carcinogenesis. 1997;18:2127-32.

69. Ilett KF, Detchon P, Ingram DM, Castleden WM. *Acetylation phenotype is not associated with breast cancer.* Cancer Res. 1990;50:6649-51.

70. Phillips DH, Garte S. Smoking and breast cancer: *Is there really a link?* Cancer Epidemiol Biomarkers Prev. 2008;17.

4.3.4. SULFOTRANSFERASES

Sulfotransferases transfer the sulfonyl (sulfur trioxide) moiety from the cofactor 3'-phosphoadenosine-5'-phosphosulfate (PAPS) to hydroxyl, amino, sulfhydryl, and N-oxide groups of numerous endogenous and exogenous substrates (1, 2) (Figure 4.3.10.). Depending on the author, this reaction has been termed sulfonation (transfer of a sulfonate group) or sulfation (formation of a sulfated product). In the mammalian organism, two classes of sulfotransferases can be distinguished (3). One class metabolizes macromolecular endogenous structures and comprises mainly membrane-bound forms localized in the Golgi apparatus. No xenobiotic-metabolizing activities have been reported for these forms. The other class of enzymes is soluble (usually cytosolic) and metabolizes numerous xenobiotics and small endogenous compounds, such as steroid, catecholamine, and thyroid hormones. All cytosolic sulfotransferases studied are members of a single sulfo-transferase (SULT) superfamily consisting of at least 11 different enzymes, which are classified into three families in humans, the phenol SULTs (SULT1), the hydroxysteroids SULTs (SULT2), and a brain-specific SULT4A1 based on amino acid sequence identity (4-7). Structural analyses of

SULT1A1 and 1E1 revealed an extended hydrophobic binding site, which is consistent with binding of both small and large phenolic substrates (8, 9). Two conserved residues (47Lys and 137Ser in SULT1A1) play key roles in the interaction with PAPS and in the sulfonyl transfer reaction (10). In parallel with the sulfonation reaction, which produces stable and long half-life products, SULTs can also generate reactive byproducts that can bind to nucleic acids and cause DNA damage (Figure 4.3.10.) (11). PAPS is synthesized by two isoforms of PAPS synthase, PAPSS1 and PAPSS2 (2).

The sulfonation of steroids generally has an inactivating effect by increasing the solubility of these compounds and thereby enhancing urinary excretion. The importance of SULTs for estrogen conjugation is demonstrated by the observation that the majority of circulating estrogen in humans is sulfate conjugated (12). Estrone sulfate, the main form of estrogen found in the circulation, is converted back into estrone by steroid sulfatase in target tissues (see Chapter 2). At the cellular level, it has been shown that addition of the charged sulfate group to E_2 prevents its binding to the estrogen receptor and thereby ameliorates the mitogenic action of E_2. In addition, SULTs are known to play critical roles in the detoxification and elimination of numerous xenobiotics. However, they can also metabolically activate a large number of procarcinogens, including N-hydroxylated derivatives of heterocyclic and aromatic amines as well as benzylic alcohols derived from polycyclic aromatic hydrocarbons (11, 13-16). Incubation of mammary epithelial cells with [^{3}H]N-OH-PhIP in the presence of PAPS revealed the formation of DNA adducts, indicating that SULTs play a role in carcinogen bioactivation in human breast (17).

Members of the SULT1A subfamily, namely SULT1A1, 1A2, and 1A3, are encoded by neighboring genes located at 16p11.2-12.1 (18). They are highly homologous in their amino acid sequence but differ in their substrate specificity. The **SULT1A1** gene is about 4.4 kb long and contains eight exons (Figure 4.3.11.). Exon 1 exists in two alternative forms, 1A and 1B. The open reading frame starts in the second exon and is 888 base pairs in length, encoding a protein of 296 amino acids with a molecular mass of 35 kDa. The enzyme exists in the native state as a 70-kDa homodimer, which exhibits strong substrate inhibition (9). SULT1A1 sulfonates a wide assortment of molecules, including simple uncharged substituted phenols, flavonoids, aromatic hydroxylamines, aromatic heterocyclic hydroxylamines, polycyclic aromatic compounds, iodothyronines, and estrogens. The dietary polyphenols quercetin and resveratrol are substrates of SULT1A1, which inhibit the sulfonation of E_2 in a competitive manner (19). Thus, the substrate-binding site can accept small flat aromatic compounds, larger L-shaped aromatics, and extended planar aromatic or aliphatic ring systems (9). The SULT1A1 gene contains polymorphisms that are associated with decreased enzyme activity and thermal

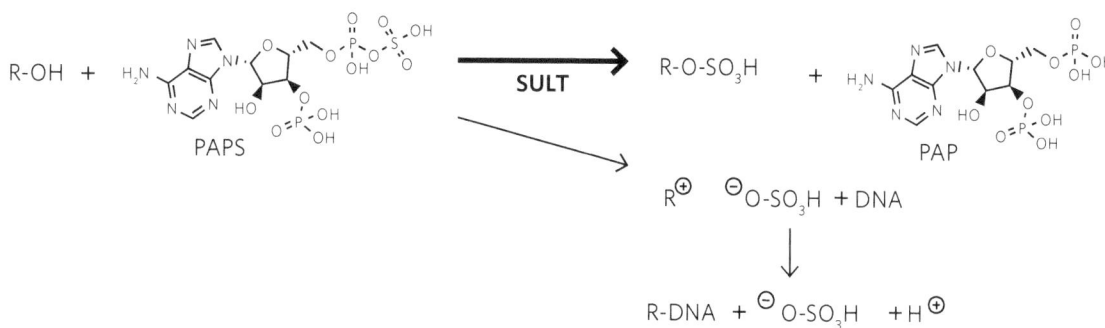

Figure 4.3.10. Sulfotransferases transfer the sulfonyl moiety from the cofactor 3'-phosphoadenosine-5'-phosphosulfate (PAPS) to nucleophilic groups of numerous endogenous and exogenous substrates.
In certain classes of compounds, e.g., aromatic hydroxylamines, the sulfate group is electron-withdrawing and may be cleaved off, leading to an electrophilic cation that can bind to nucleic acids and cause DNA damage (11).

stability (9, 20, 21). Two polymorphic sites have been described, Arg213His (rs9282861) and Met223Val, resulting in three alleles, SULT1A1*1 (213Arg, 223Met), SULT1A1*2 (213His, 223Met), and SULT1A1*3 (213His, 223Val) (21-24). Allele frequencies for SULT1A1*1, *2, and *3 were 66%, 33%, and 1% for Caucasians, 48%, 29%, and 23% for African-Americans, and 91%, 8%, and 0.1%, respectively, for Asians (21). SULT1A1*2 was shown to have lower thermal stability, a shorter half-life, and ~10-fold lower activity for p-nitrophenol and 2-MeOE$_2$ conjugation than SULT1A1*1 (20, 25). Several single-nucleotide polymorphisms in the promoter region are in linkage disequilibrium with the Arg213His polymorphism (26).

The **SULT1E1** gene at 4q13.1 is about 20 kb long and contains eight exons (Figure 4.3.12.). The open reading frame starts in the second exon and is 882 base pairs in length, encoding a protein of 294 amino acids with a molecular mass of 35 kDa. Similar to SULT1A1, the SULT1E1 enzyme exists in the native state as a 70-kDa homodimer, which exhibits substrate inhibition. This is not surprising since both SULT isoforms share ~60% amino acid identity. The SULT1E1 gene

contains polymorphisms that are associated with decreased enzyme activity and thermal stability (27). Three SULT1E1 polymorphisms cause amino acid substitutions Asp22Tyr, Ala32Val, and Pro253His (27). Kinetic studies with E$_2$ and recombinant SULT1E1 variant 22Tyr revealed an increase in apparent K$_m$, resulting in 40-fold lower activity compared to wild type enzyme, consistent with the location of residue 22 at the entrance of the substrate-binding pocket (10). Expression studies in COS-1 cells revealed a drastic decrease in immunoreactive protein for variant 22Tyr (only 10% compared to 100% for the wild type 22Asp) due to decreased thermal stability or increased proteasome-mediated degradation. The Ala32Val substitution was associated with a less severe effect on kinetic properties and immunoreactive protein level, whereas the Pro253His variant behaved like wild type SULT1E1. The striking decrease in enzyme activity and concentration observed for Ala32Val and especially Asp22Tyr are expected to have considerable impact on mammary estrogen metabolism. However, the allele frequency of these SULT1E1 variants is <1% (27), much lower than the variant SULT1A1 allele frequency and raising the question whether they are indeed polymorphisms or mutations.

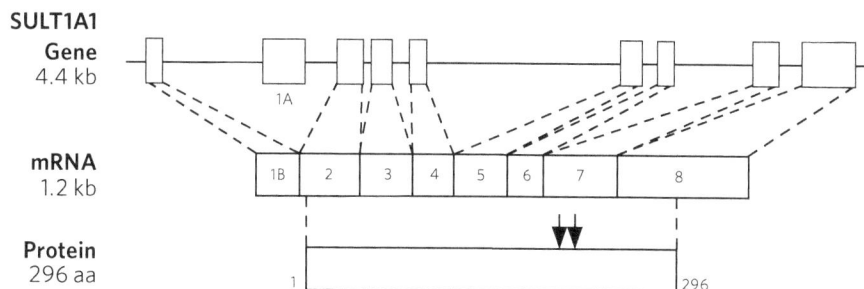

Figure 4.3.11. Overview of SULT1A1 gene, mRNA, and protein.
The SULT1A1 gene at 16p11.2-12.1 is about 4.4 kb long and contains eight exons. Exon 1 exists in two alternative forms, 1A and 1B. The open reading frame starts in the second exon and is 888 base pairs in length, encoding a protein of 296 amino acids with a molecular mass of 35 kDa. The arrows indicate polymorphisms resulting in amino acid substitutions Arg213His (rs9282861) and Met223Val. The enzyme exists in the native state as a 70-kDa homodimer which exhibits strong substrate inhibition.

SULT2A1 is the isoform responsible for the sulfonation of DHEA in the fetal adrenal gland and the reticular layer of the adult adrenal cortex. The SULT2A1 gene contains polymorphisms that are associated with decreased enzyme activity and thermal stability (28). Two polymorphisms, Ala63Pro and Ala261Thr, are present in 5 – 10% of African Americans (26).

Growing recognition of the carcinogenic potential of catechol estrogens has led to increased interest in the role of SULTs in intracellular estrogen metabolism (29). These studies are driven by the hypothesis that SULT-mediated estrogen conjugation reduces catechol estrogen levels and thereby decreases breast cancer risk. The identification of new SULT isoforms during the past few years has shown that earlier tissue studies frequently encompassed unrecognized isoforms, clouding the issue of SULT specificity in estrogen conjugation (4). A comprehensive study of ten recombinant SULT isoforms showed that seven (1A1, 1A2, 1A3, 1E1, 2A1, 2B1a, 2B1b) catalyzed the sulfate conjugation of catechol estrogens, whereas three (1B1, 1C1, 4A1) did not (30). Although seven SULT isoforms were shown to be capable of conjugating estrogens, they differ significantly in substrate affinity. There is consensus among investigators that only SULT1E1 can conjugate E_2, 2-OHE$_2$, and 4-OHE$_2$ at nanomolar concentrations, in contrast to micromolar concentrations observed for SULT1A1, 1A2, 1A3, and 2A1 (30-35). There is disagreement with respect to the sulfation of methoxyestrogens. One study observed SULT1E1-mediated conjugation of nanomolar concentrations of 2-MeOE$_2$ (36). This does not match the results of another study, which identified the presence of 2-MeOE$_2$-3S in MCF-7 culture media by electrospray ion-trap mass spectrometry (37). About 96% of 2-MeOE$_2$ was conjugated, but only 27% of 4-MeOE$_2$, apparently by the combined action of COMT and SULT. To determine the identity of the SULT isoform, the authors examined the ability of SULT1A1, 1A2, 1A3, 1E1, and 2A1 to conjugate 2-MeOE$_2$ and 4-MeOE$_2$. Surprisingly, only SULT1A1 was capable of conjugating the methoxyestrogens; SULT1E1 was inactive. Moreover, SULT1A1 conjugated 2-MeOE$_2$ at nanomolar levels instead of micromolar concentrations for E_2, 2-OHE$_2$, and 4-OHE$_2$, suggesting that 2-MeOE$_2$ may be a physiological substrate of SULT1A1. In summary, SULT1E1 is the principal isoform for the sulfonation of E_2, 2-OHE$_2$, and 4-OHE$_2$, but uncertainty exists whether SULT1A1 or 1E1 is responsible for the sulfonation of methoxyestrogens. The identification of these key enzymes represents progress in our understanding of estrogen conjugation but, at the same time, raises questions about the integration of sulfation reactions into the overall pathway of E_2 metabolism in the context of CYP1B1, CYP1A1, and COMT activity. While the kinetics of single SULT reactions have been determined, it is obvious that key estrogen metabolites are substrates for two or even three enzymes (Table 4.3.5.). For example, 4-OHE$_2$ serves as substrate for SULT1E1 as well as COMT and CYP1B1. The CYP1B1-mediated production of E_2-3,4-Q leads to DNA adduct formation, whereas the reactions mediated by COMT and SULT1E1 result in noncarcinogenic products, 4-MeOE$_2$ and 4-OHE$_2$-3S, respectively (38).

There are conflicting results published with regard to which SULT isoforms are expressed in normal breast tissue. Immunocytochemical studies have shown that SULT1E1 is the principal isoform in normal mammary epithelial cells derived from reduction mammoplasties, the non-tumor derived cell line 184A1, and epithelial cells in normal breast tissues (33, 37, 39). Other isoforms, such as SULT1A1, were not detectable immunologically in normal mammary epithelium, although reverse transcription-PCR revealed SULT1A2 and 1A3 mRNA in 184A1 cells (37). In contrast, Western blot analysis of mammary cytosolic protein prepared directly from breast tissue samples showed detectable levels of SULT1A1

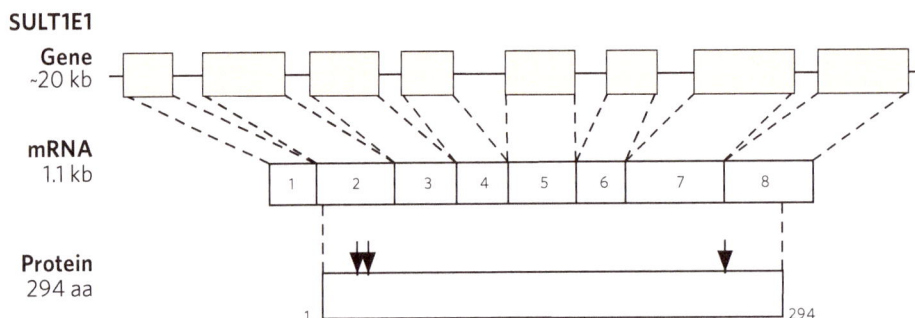

Figure 4.3.12. Overview of SULT1E1 gene, mRNA, and protein.
The SULT1E1 gene at 4q13.1 is about 20 kb long and contains eight exons. The open reading frame starts in the second exon and is 882 base pairs in length, encoding a protein of 294 amino acids with a molecular mass of 35 kDa. The arrows indicate polymorphisms resulting in amino acid substitutions Asp22Tyr, Ala32Val, and Pro253His. The lengths of the exons are drawn in scale, while those of the introns are not. The enzyme exists in the native state as a 70 kDa homodimer which exhibits strong substrate inhibition.

Table 4.3.5. Estrogen Substrates for SULT1E1, SULT1A1, CYP1B1, CYP1A1, and COMT

Substrate	SULT		Competing Enzyme	
	Isoform	Product	Enzyme	Product
E_2	1E1	E_2-3S	CYP1A1, 1B1	2-OHE_2, 4-OHE_2
2-OHE_2	1E1	2-OHE_2-3S	CYP1A1, 1B1	E_2-2,3-Q
			COMT	2-$MeOE_2$, 2-OH-3-$MeOE_2$
4-OHE_2	1E1	4-OHE_2-3S	CYP1B1	E_2-3,4-Q
			COMT	4-$MeOE_2$
2-$MeOE_2$	1A1	2-$MeOE_2$-3S	CYP1A1, 1B1	2-OHE_2

and SULT1A3 but not SULT1E1 (40). The latter findings raise questions about the effect of culture conditions on SULT expression in mammary epithelial cells. The SULT expression pattern was almost opposite in breast cancer cell lines and tissues. Virtually all malignant cell lines express one or more members of the SULT1A subfamily. For example, SULT1A1 protein and mRNA levels were particularly high in BT-20, MCF-7, T47D, and ZR-75 cells. In contrast, SULT1E1 was present in trace amounts or undetectable in most malignant cell lines (33, 37, 41). RT-PCR analysis of 42 normal and malignant breast tissue samples revealed expression of SULT1A1, 1A2, 1A3, 1B1, 1C1, 1E1, 2A1, and 2B1 mRNA (42). Unspliced SULT1A2, containing the complete intron between exons 7 and 8, was found in 4 of 16 normal but none of 26 malignant specimens. SULT1E1 was detected by immunocytochemistry in 50 of 113 (44.2%) invasive ductal carcinomas (39). A subgroup analysis of 35 cases showed a significant correlation (p < 0.01) of the semiquantitative immunohistochemical SULT1E1 score with mRNA levels and enzymatic activity. Interestingly, women with SULT1E1-positive tumors had a better prognosis (longer disease-free interval [p = 0.0044] and overall survival [p = 0.0026]) than their SULT1E1-negative counterparts. Both the expression of SULT1E1 in normal mammary epithelium and the poor clinical outcome of SULT1E1-negative breast cancers support the view that SULT1E1-mediated conjugation is important in limiting long-term mammary exposure to carcinogenic catechol estrogens.

Clinical studies have assessed the influence of the SULT1A1 SNP Arg213His (rs9282861). Premenopausal women carrying the 213His allele had 11% lower mean follicular phase concentrations of circulating E_1 (p = 0.02) compared to the 213Arg wildtype and the 213His/His genotype was associated with 16% lower breast density (p = 0.001) compared to the 213Arg/Arg genotype (43, 44). Epidemiological studies of the SULT1A1*1, 1*2, and 1*3 genotypes have shown inconsistent associations with breast cancer risk, reporting either a lack of association or an increased risk associated with the SULT1A1*2 genotype

(45-51). However, other risk factors such as parity and body mass index may modify the effect of SULT1A1 variants (52, 53).

Clinical studies have also assessed the influence of SULT1E1 polymorphisms including two SNPs located in intron 1, 69A/G (rs3775768; MAF 27%) and -73G/C (rs4149530; MAF 22%) (44). Circulating E_1 and E_2 (p = 0.05) concentrations in premenopausal women were significantly lower with the A-C than the A-G haplotype. Another study observed a moderately decreased breast cancer risk associated with a polymorphism in the 3' flanking region, 959G/A (rs3786599; odds ratio 0.8; 95% CI 0.70-1.00) (50).

References

1. Weinshilboum R, Otterness D. *Sulfotransferase Enzymes*. Handbook Exp Pharmacol. 1994;112:45-78.

2. Strott CA. *Sulfonation and molecular action*. Endocrine Rev. 2002;23:703-32.

3. Glatt H, Boeing H, Engelke CEH, Ma L, Kuhlow A, Pabel U, et al. *Human cytosolic sulphotransferases: genetics, characteristics, toxicological aspects*. Mutation Res. 2001;482:27-40.

4. Falany CN, Xie X, Wang J, Ferrer J, Falany JL. *Molecular cloning and expression of novel sulphotransferase-like cDNAs from human and rat brain*. Biochem J. 2000;346:857-64.

5. Glatt H, Engelke CEH, Pabel U, Teubner W, Jones AL, Coughtrie MW, et al. *Sulfotransferases: genetics and role in toxicology*. Toxicol Lett. 2000;112-113:341-8.

6. Liyou NE, Buller KM, Tresillian MJ, Elvin CM, Scott HL, Dodd PR, et al. *Localization of a brain sulfotransferase, SULT4A1, in the human and rat brain: an immunohistochemical study*. J Histochem Cytochem. 2003;51:1655-64.

7. Blanchard RL, Freimuth RR, Buck J, Weinshilboum RM, Coughtrie MWH. *A proposed nomenclature system for the cytosolic sulfotransferase (SULT) superfamily*. Phamacogenetics. 2004;14:199-211.

8. Kakuta Y, Pedersen LG, Carter CW, Negishi M, Pedersen LC. *Crystal structure of estrogen sulphotransferase*. Nature Struct Biol. 1997;4:904-8.

9. Gamage NU, Duggleby RG, Barnett AC, Tresilliant M, Latham CF, Liyou NE, et al. *Structure of a human carcinogen-converting enzyme, SULT1A1*. J Biol Chem. 2003;278:7655-62.

10. Pedersen LC, Petrotchenko E, Shevtsov S, Negishi M. *Crystal structure of the human estrogen sulfotransferase-PAPS complex*. J Biol Chem. 2002;277:17928-32.

11. Glatt H. *Bioactivation of mutagens via sulfation*. FASEB J. 1997;11:314-21.

12. Pasqualini JR, Chetrite G, Blacker C, Feinstein M-C, Delalonde L, Talbi M, et al. *Concentrations of estrone, estradiol, and estrone sulfate and evaluation of sulfatase*

and aromatase activities in pre- and postmenopausal breast cancer patients. J Clin Endocrinol Metabol. 1996;81:1460-4.

13. Surh Y, Kwon H, Tannenbaum SR. *Sulfotransferase-mediated activation of 4-hydroxy- and 3,4-dihydroxy-3,4-dihydrocyclopenta[c,d] pyrene, major metabolites of cyclopenta[c,d]pyrene.* Cancer Res. 1993;53:1017-22.

14. Chou HC, Lang NP, Kadlubar FF. *Metabolic activation of the N-hydroxy derivative of the carcinogen 4-aminobiphenyl by human tissue sulfotransferases.* Carcinogenesis. 1995;16:413-7.

15. Dubuisson JG, Gaubautz JW. *Bioactivation of the proximal food mutagen 2-hydroxyamino-1-methyl-6-phenylimidazo[4,5-b]pyridine (N-OH-PhIP) to DNA-binding species by human mammary gland enzymes.* Nutrition. 1998;14:683-6.

16. Glatt H, Bartsch I, Christoph S, Coughtrie MWH, Falany CN, Hagen M, et al. *Sulfotransferase-mediated activation of mutagens studied using heterologous expression systems.* Chem-Biol Interact. 1998;109:195-219.

17. Lewis AJ, Walle UK, King RS, Kadlubar FF, Falany CN, Walle T. *Bioactivation of the cooked food mutagen N-hydroxy-2-amino-1-methyl-6-phenylimidazo[4,5-b]pyridine by estrogen sulfotransferase in cultured human mammary epithelial cells.* Carcinogenesis. 1998;19:2049-53.

18. Dooley TP, Huang Z. *Genomic organization and DNA sequences of two human phenol sulfotransferase genes (STP1 and STP2) on the short arm of chromosome 16.* Biochem Biophys Res Commun. 1996;228:131-40.

19. Otake Y, Nolan AL, Walle UK, Walle T. *Quercetin and resveratrol potently reduce estrogen sulfotransferase activity in normal human mammary epithelial cells.* J Steroid Biochem Mol Biol. 2000;73:265-70.

20. Raftogianis RB, Wood TC, Otterness DM, Van Loon JA, Weinshilboum RM. *Phenol sulfotransferase pharmacogenetics in humans: association of common SULT1A1 alleles with TS PST phenotype.* Biochem Biophys Res Comm. 1997;239:298-304.

21. Carlini EJ, Raftogianis RB, Wood TC, Jin F, Zheng W, Rebbeck TR, et al. *Sulfation pharmacogenetics: SULT1A1 and SULT1A2 allele frequencies in Caucasian, Chinese and African-American subjects.* Pharmacogenetics. 2001;11:57-68.

22. Raftogianis RB, Wood TC, Weinshilboum RM. *Human phenol sulfotransferases SULT1A2 and SULT1A1.* Biochem Pharmacol. 1999;58:605-16.

23. Ozawa S, Shimizu M, Katoh T, Miyajima A, Ohno Y, Matsumoto Y, et al. *Sulfating-activity and stability of cDNA-expressed allozymes of human phenol sulfotransferase, ST1A3*1 ((213)Arg) and ST1A3*2 ((213) His), both of which exist in Japanese as well as Caucasians.* J Biochem. 1999;126:271-7.

24. Coughtrie MW, Gilissen RA, Shek B, Strange RC, Fryer AA, Jones PW, et al. *Phenol sulphotransferase SULT1A1 polymorphism: molecular diagnosis and allele frequencies in Caucasian and African populations.* Biochem J. 1999;337:45-9.

25. Nagar S, Walther S, Blanchard RL. *Sulfotransferase (SULT) 1A1 polymorphic variants *1, *2, and *3 are associated with altered enzymatic activity, cellular phenotype, and protein degradation.* Mol Pharmacol. 2006;69:2084-92.

26. Nowell S, Falany CN. *Pharmacogenetics of human cytosolic sulfotransferases.* Oncogene. 2006;25:1673-8.

27. Adjei AA, Thomae BA, Prondzinski JL, Eckloff BW, Wieben ED, Weinshilboum RM. *Human estrogen sulfotransferase (SULT1E1) pharmacogenomics: gene resequencing and functional genomics.* Br J Pharmacol. 2003;139:1373-82.

28. Thomae BA, Eckloff BW, Freimuth RR, Weiben ED, Weinshilboum RM. *Human sulfotransferase SULT2A1 pharmacogenetics: genotype-to-phenotype studies.* Pharmacogenomics J. 2002;2:48-56.

29. Raftogianis R, Creveling C, Weinshilboum R, Weisz J. *Estrogen metabolism by conjugation.* J Natl Cancer Inst Monogr. 2000;27:113-24.

30. Adjei AA, Weinshilboum RM. *Catecholestrogen sulfation: possible role in carcinogenesis.* Biochem Biophys Res Comm. 2002;292:402-8.

31. Falany CN, Wheeler J, Oh TS, Falany JL. *Steroid sulfation by expressed human cytosolic sulfotransferases.* J Steroid Biochem Mol Biol. 1994;48:369-75.

32. Lee YC, Komatsu K, Driscoll WJ, Strott CA. *Structural and functional characterization of estrogen sulfotransferase isoforms: distinct catalytic and high affinity binding activities.* Mol Endocrinol. 1994;8:1627-35.

33. Falany JL, Falany CN. *Expression of cytosolic sulfotransferases in normal mammary epithelial cells and breast cancer cell lines.* Cancer Res. 1996;56(April):1551-5.

34. Zhang H, Varlamova O, Vargas FM, Falany CN, Leyh TS, Varmalova O. *Sulfuryl transfer: the catalytic mechanism of human estrogen sulfotransferase.* J Biol Chem. 1998;273:10888-92.

35. Faucher F, Lacoste L, Dufort I, Luu-The V. *High metabolization of catecholestrogens by type 1 estrogen sulfotransferase (hEST1).* J Steroid Biochem Mol Biol. 2001;77:83-6.

36. Adjei AA. *2-Methoxyestradiol (2-ME2) sulfation: possible metabolic pathway.* Clin Pharm Ther. 2001;69:75.

37. Spink BC, Katz BH, Hussain MM, Pang S, Connor SP, Aldous KM, et al. *SULT1A1 catalyzes 2-methoxyestradiol sulfonation in MCF-7 breast cancer cells.* Carcinogenesis. 2000;21:1947-57.

38. Hui Y, Yasuda S, Liu MY, Wu YY, Liu MC. *On the sulfation and methylation of catecholestrogens in human mammary epithelial cells and breast cancer cells.* Biol Pharmaceutical Bull. 2008;31:769-73.

39. Suzuki T, Nakata T, Miki Y, Kaneko C, Moriya T, Ishida T, et al. *Estrogen sulfotransferase and steroid sulfatase in human breast carcinoma.* Cancer Res. 2003;63:2762-70.

40. Williams JA, Stone EM, Fakis G, Johnson N, Cordell JA, Meinl W, et al. *N-acetyltransferases, sulfotransferases and heterocyclic amine activation in the breast.* Pharmacogenetics. 2001;11:373-88.

41. Falany JL, Macrina N, Falany CN. *Regulation of MCF-7 breast cancer cell growth by beta-estradiol sulfation.* Breast Cancer Res Treat. 2002;74:167-76.

42. Aust S, Obrist P, Klimpfinger M, Tucek G, Jager W, Thalhammer T. *Altered expression of the hormone- and xenobiotic-metabolizing sulfotransferase enzymes 1A2 and 1C1 in malignant breast tissue.* Int J Oncol. 2005;26:1079-85.

43. Yong M, Schwartz SM, Atkinson C, Makar KW, Thomas SS, Newton KM, et al. *Associations between polymorphisms in glucuronidation and sulfation enzymes and mammographic breast density in premenopausal women in the United States.* Cancer Epidemiol Biomarkers Prev 2010;19:537-46.

44. Yong M, Schwartz SM, Atkinson C, Makar KW, Thomas SS, Stanczyk FZ, et al. *Associations between polymorphisms in glucuronidation and sulfation enzymes and sex steroid concentrations in premenopausal women in the United States.* J Steroid Biochem Mol Biol. 2011;124:10-8.

45. Seth P, Lunetta KL, Bell DW, Gray H, Nasser SM, Rhei E, et al. *Phenol sulfotransferases: hormonal regulation, polymorphism, and age of onset of breast cancer.* Cancer Res. 2000;60:6859-63.

46. Zheng W, Xie D, Cerhan JR, Sellers TA, Wen W, Folsom AR. *Sulfotransferase 1A1 polymorphism, endogenous estrogen exposure, well-done meat intake, and breast cancer risk.* Cancer Epidemiol Biomarkers Prev. 2001;10:89-94.

47. Tang D, Rundle A, Mooney L, Cho S, Schnabel F, Estabrook A, et al. *Sulfotransferase 1A1 (SULT1A1) polymorphism, PAH-DNA adduct levels in breast tissue and breast cancer risk in a case-control study.* Breast Cancer Res Treat. 2003;78:217-22.

48. Han D, Zhou X, Hu M, Wang C, Xie W, Tan X, et al. *Sulfotransferase 1A1 (SULT1A1) polymorphism and breast cancer risk in Chinese women.* Toxicol Lett. 2004;150:167-77.

49. Sparks R, Ulrich CM, Bigler J, Tworoger SS, Yasui Y, Rajan KB, et al. *UDP-glucuronosyltransferase and sulfotransferase polymorphisms, sex hormone concentrations, and tumor receptor status in breast cancer patients.* Breast Cancer Res. 2004;6:R488-R98.

50. Choi J, Lee K, Park SK, Noh D, Ahn S, Chung H, et al. *Genetic polymorphisms of SULT1A1 and SULT1E1 and the risk and survival of breast cancer.* Cancer Epidemiol Biomarkers Prev. 2005;14:1090-5.

51. Le Marchand L, Donlon T, Kolonel LN, Henderson BE, Wilkins LR. *Estrogen metabolism-related genes and breast cancer risk: The Multiethnic Cohort Study.* Cancer Epidemiol Biomarkers Prev. 2005;14:1998-2003.

52. Sillanpaa P, Kataja V, Eskelinen M, Kosma VM, Uusitupa M, Vainio H, et al. *Sulfotransferase 1A1 genotype as a potential modifier of breast cancer risk among premenopausal women.* Pharmacogenet Genomics. 2005;15:749-52.

53. Yang G, Gao YT, Cai QY, Shu XO, Cheng JR, Zheng W. *Modifying effects of sulfotransferase 1A1 gene polymorphism on the association of breast cancer risk with body mass index or endogenous steroid hormones.* Breast Cancer Res Treat. 2005;94:63-70.

The Etiology of Breast Cancer

4.3.5. UDP-GLUCURONOSYLTRANSFERASES

The uridine diphosphate (UDP)-glucuronosyltransferase (UGT) superfamily currently consists of 16 functional genes, which are organized into two families of enzymes, UGT1 and UGT2 (1-4). UGTs are microsomal enzymes that catalyze the transfer of the polar D-glucuronic acid moiety from UDP-glucuronic acid to a wide variety of exogenous and endogenous lipophilic compounds, including drugs, xenobiotics, steroids, bile acids, and bilirubin (5-7). The conjugation reaction typically involves a nucleophilic functional group of the lipophilic acceptor compound (aglycone), such as a hydroxyl group, resulting in the formation of a more water-soluble β-D-glucuronide (Figure 4.3.13.).

The **UGT1 gene** locus spans over 500 kb on chromosome 2q37 (8). The 5′ region of the UGT1A complex contains 13 unique exons 1, which are arranged in a tandem array with each having its own proximal TATA box element. Each first exon can be spliced and joined with common exons 2 through 5, resulting in proteins with unique amino termini preceding a shared carboxyl terminus of 245 amino acids. The unique amino termini specify acceptor-substrate selection and the common carboxyl terminus apparently specifies the interaction with the common donor substrate, UDP-glucuronic acid. In this nomenclature system, each first exon (2nd, 11th – 13th are pseudogenes) is regarded as a distinct gene, i.e., UGT1A1, 1A3, 1A4, 1A5, 1A6, 1A7, 1A8, 1A9, and 1A10 (9). Each UGT1A isoform is under the control of individual promoter sequences and isoform-specific regulation and tissue expression have been observed (6, 10). Constitutive expression is controlled by hepatocyte nuclear factor 1, CAAT-enhancer binding protein, and other factors. Hormones, drugs, and xenobiotics induce transcription via members of the nuclear receptor superfamily, such as the pregnane X and androstane receptors, and the arylhydrocarbon receptor (6). In contrast, gene products of the **UGT2 family** are transcribed from unique genes clustered on chromosome 4 and therefore exhibit differences in amino acid sequence throughout the entire polypeptide chain (11, 12). Family 2 is divided into two subfamilies, UGT2A, which is only expressed in olfactory cells and UGT2B, which is expressed in hepatic and extra-hepatic cells. The UGT2B subfamily is composed of 11 members, of which five are pseudogenes. Five of the remaining genes, UGT2B4, 2B7, 2B11, 2B15, and 2B17, encode enzymes involved in steroid metabolism, while 2B10 encodes an orphan enzyme without known endogenous substrates (4, 13-15). Estrogens up-regulate the expression and activity of UGT2B15 in ER-positive human breast cancer cells (16). Based on substrate specificity, conjugation efficiency, and level of expression in breast epithelial cells UGT2B7 is the predominant enzyme of the UGT2 family involved in mammary estrogen metabolism.

The elimination of many hydrophobic xenobiotics and/or metabolites is facilitated by their UGT-mediated conjugation with hydrophilic UDP-glucuronic acid (1, 6). For example, hydroxylated metabolites of the polycyclic aromatic hydrocarbon benzo[a]pyrene are eliminated substantially via glucuronidation, which may prevent subsequent bioactivation by cytochrome P450s or peroxidases to an ultimate reactive intermediate capable of forming DNA adducts and initiating carcinogenesis (17). With regard to estrogens, glucuronidation of E_2 can occur at either the C-3 or C-17β hydroxyl group, abolishing its affinity for the estrogen receptor due to steric hindrance (18). By increasing polarity, glucuronidation of estrogens also facilitates partitioning of the lipophilic steroids into the aqueous compartment. The resulting glucuronidated metabolites constitute the most abundant circulating estrogen conjugates. They are more hydrophilic and can be excreted in bile and urine. The kidney of Syrian hamsters was shown to contain lysosomal glucuronidase activity capable of deconjugating E_2- and E_1-3β-glucuronides (19). The deglucuronidation is an important source of parent estrogens and catechol estrogens facilitating the development of estrogen-induced renal cancer in hamsters.

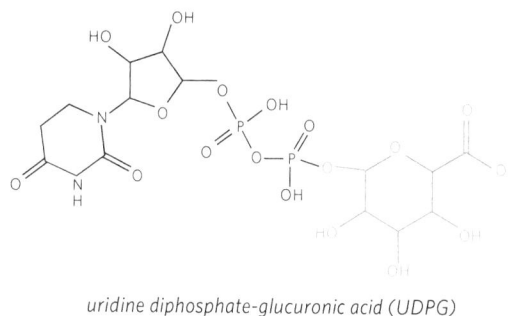

uridine diphosphate-glucuronic acid (UDPG)

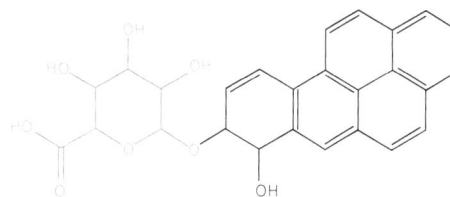

B[a]P-7,8-diol-8-O-glucuronide

Figure 4.3.13. UGTs catalyze the conjugation of D-glucuronic acid from UDP-glucuronic acid to a wide variety of exogenous and endogenous lipophilic compounds making them more water-soluble.
An example is B[a]P-7,8-diol, which is converted to B[a]P-7,8-diol-8-O-glucuronide.

Several UGT isoforms are capable of estrogen conjugation, i.e., UGT1A1, 1A3, 1A7, 1A8, 1A9, 1A10, 2B4, 2B7, 2B11, and 2B15 (1, 3, 13, 15, 20-31). Of the UGTs listed, UGT1A1, 1A3, 1A8, 1A9, and 2B7 have the highest activity toward estrogens. The parent hormones, E_2 and E_1, are recognized as substrates but individual isoforms display distinct differences in substrate specificity and conjugation efficiency. Comparison of UGT1A1 and 2B7 showed regioselective conjugation of E_2, i.e., UGT1A1 only conjugated the C-3 hydroxyl group of the A-ring, whereas UGT2B7 conjugated the 17β-hydroxyl in the D-ring, yielding E_2-3Glu and E_2-17βGlu, respectively (28, 32). UGTs recognize not only E_2 and E_1 as substrates but also their respective catechols. In fact, several isoforms were more active toward the catechol estrogens than the parent hormones, including UGT1A1, 1A9, and 2B7 (15, 27, 31, 33). On the other hand, comparison of catechol estrogen substrates revealed that UGT1A1 and 1A3 were more active toward 2-OHE_2, while UGT1A9 and 2B7 conjugated 4-OHE_2 more efficiently (25, 27, 33). Although the catechols derived from E_2 and E_1 are generally metabolized with similar efficiencies, UGT2B7 is a notable exception, which displays a 7- to 25-fold higher activity toward 4-OHE_1 than 4-OHE_2 (15, 25, 31, 34). Catechol estrogens are potentially conjugated at the C2-, 3-, 4-, or 17-hydroxyl positions. In the case of 4-OHE_2 and 4-OHE_1, glucuronidation occurs predominantly at the C3- and 4-hydroxyl positions with the latter as the preferred site (31, 34). Finally, conjugation of 2-OHE_1/E_2 and 2- and 4-$MeOE_1$/E_2 was selective at position C3, mostly catalyzed by UGT1A1 and 1A8 (31).

With regard to breast cancer, the study of UGTs is driven by the hypothesis that UGT-mediated estrogen conjugation increases clearance of parent hormones and catechol estrogen metabolites and thereby decreases breast cancer risk (35). In most studies the examination of UGT expression in breast tissue was limited to transcript detection. Relevant to the isoforms with the highest activity toward estrogen conjugation, it is noteworthy that UGT1A9 mRNA was detectable in breast tissue whereas UGT1A1 mRNA was not detected (27, 30). UGT2B7 transcript was present in normal mammary tissue, but not in T47D and ZR-75 breast cancer cells (15). Immunostaining of UGT1A8/1A9 and UGT2B7 was observed in the cytoplasm of ductal epithelial cells of normal breast tissue obtained from postmenopausal women (34). A detailed immuno-histochemical study demonstrated expression of UGT2B7 protein in normal mammary epithelium obtained from either reduction mammoplasties or tissue distant from invasive cancer in mastectomy specimens (36). In contrast, UGT2B7 protein expression was significantly reduced in malignant cells. The observed difference of UGT2B7 expression between benign and malignant cells is consistent with the hypothesis that UGT-mediated conjugation of catechol estrogens prevents the formation of potentially carcinogenic estrogen quinones. Based on efficiency of estrogen conjugation and expression in breast tissue, it appears that UGT1A8, 1A9, and 2B7 are the predominant isoforms to be considered in mammary estrogen metabolism.

Numerous polymorphisms have been described in human UGT genes, including several that are associated with altered catalytic activity (Table 4.3.6.) (29, 37-39). Functional significance has been convincingly demonstrated for a polymorphism in a TA repeat element in the TATA box region of the **UGT1A1** promoter, $A(TA)_{5-8}TAA$ (rs8175347). The most common allele, UGT1A1*1, contains six TA repeats, whereas the most common variant consists of seven TA repeats, designated UGT1A1*28. Two less frequent alleles, $A(TA)_5TAA$ and $T(TA)_8TAA$ are referred to as UGT1A1*36 and UGT1A1*37, respectively (4). In vitro investigations have shown that the length of the TA-repeat influences UGT1A1 transcription, i.e., UGT1A1 expression decreases with increasing number of repeats. Thus, the presence of seven or eight repeats leads to a decrease in UGT1A1 expression, which results in impaired glucuronidation of bilirubin in Gilbert's syndrome (40, 41). The UGT1A1*28 allele also resulted in decreased formation of the glucuronide conjugate of the B[a]P metabolite, B[a]P-7,8-dihydrodiol, in liver microsomes (42). The **UGT1A8** gene contains the nonsynonymous polymorphisms Ala173Gly and Cys277Tyr (rs17863762). The wild-type allele UGT1A8*1 is present in 75% of Caucasians, whereas the variant alleles UGT1A8*2 and *3 are seen in 23.8% and 1.2%, respectively (34, 38). The 277Cys in the substrate-binding region is highly conserved and the tyrosine substitution resulted in marked reduction of UGT1A8 activity at positions C3 and 4 of both 4-OHE_1 and 4-OHE_2 (34). The **UGT1A9** gene contains the nonsynonymous polymorphisms Cys3Tyr and Met33Thr. The wild-type and variant alleles UGT1A9*1 and *3 are present in 97.8% and 2.2%, respectively, of Caucasians. The UGT1A9*2 allele, corresponding to Cys3Tyr, is present in 2.5% of African-Americans (34). The Met33Thr substitution was associated with a substrate-dependent reduction in activity (34, 43). Fifty percent of Caucasians possess a polymorphism in the **UGT2B7** gene at His268Tyr (rs7439366) (44). The polymorphism was associated with a substrate-specific effect, showing no change in the in vitro glucuronidation of androsterone, reduced activity towards E_2, and increased catalytic efficiency with regard to 4-OHE_1 and 4-OHE_2 glucuronidation (28, 34, 45). Liver microsomes from subjects with the homozygous variant 268Tyr/Tyr genotype exhibited significantly reduced glucuronidation of the major metabolite of a potent tobacco-derived nitrosamine procarcinogen, 4-(methylnitrosamino)-1-(3-pyridyl)-1-butanol (46). About 5% of Caucasians also possess a functional variation in the UGT2B7 promoter at position -79 G/A, referred to as haplotype III, which is in linkage disequilibrium with the codon 268 variation (47). This promoter polymorphism was associated with decreased transcription in hepatoma and colon cells and resulted in reduced enzymatic activity (34,

Table 4.3.6. Functional UGT Polymorphisms

Gene	SNP (rs#)	Old designation	MAF (%)	Function
UGT1A1	A(TA)$_{5-8}$TAA (rs8175347)	(TA)$_5$ = UGT1A1*36		UGT1A1 expression decreases with increasing number of repeats
		(TA)$_6$ = UGT1A1*1		
		(TA)$_7$ = UGT1A1*28		
		(TA)$_8$ = UGT1A1*37		
UGT1A8	Ala173Gly	UGT1A8*2	23.8	reduced activity
	Cys277Tyr (rs17863762)	UGT1A8*3	1.2	
UGT1A9	Cys3Tyr	UGT1A9*2	2.5	reduced activity
	Met33Thr	UGT1A9*3	2.2	substrate-specific effect
UGT2B7	His268Tyr (rs7439366)		50	
UGT2B15	Asp85Tyr (rs1902023)			increased activity

MAF, minor allele frequency

47). The **UGT2B15** gene contains a polymorphism, Asp85Tyr (rs1902023), which results in increased activity toward dihydrotestosterone (37).

Premenopausal women with the UGT1A1 (TA$_7$/TA$_7$) genotype had 25% lower mean follicular phase concentrations of circulating E$_2$ compared to the wildtype (TA$_6$/TA$_6$; p = 0.02) whereas postmenopausal women showed no difference (48, 49). Postmenopausal women with the UGT2B15 85Asp/Tyr and Tyr/Tyr genotypes had significantly higher concentrations of circulating E$_2$ than women with the wild-type Asp/Asp genotype (50). The UGT1A1 A(TA)$_{5-8}$TAA polymorphism was associated with a marginal effect (p = 0.06) on breast cancer risk of premenopausal but not postmenopausal African-American women (51). Similarly, the variant UGT1A1*28 allele was associated with a higher risk of breast cancer in Chinese women younger than 40 years but not among older women (52). However, the Nurses' Health Study showed no risk association in pre- or postmenopausal Caucasian women (49).

References

1. King CD, Rios GR, Green MD, Tephly TR. UDP-glucuronosyltrans-ferases. Current Drug Metabol. 2000;1:143-61.

2. Mackenzie PI, Owens IS, Burchell B, Bock KW, Bairoch A, Belanger A, et al. The UDP glycoyltransferase gene superfamily: recommended nomenclature update based on evolutionary divergence. Pharmacogenetics. 1997;7:255-69.

3. Tukey RH, Strassburg CP. Human UDP-glucuronosyltransferases: metabolism, expression, and disease. Ann Rev Pharmacol Toxicol. 2000;40:581-616.

4. Guillemette C, Belanger A, Lepine J. Metabolic inactivation of estrogens in breast tissue by UDP-glucuronosyltransferase enzymes: an overview. Breast Cancer Res. 2004;6:246-54.

5. Mackenzie PI, Mojarrabi B, Meech R, Hansen A. Steroid UDP glucuronosyltransferases: characterization and regulation. J Endocrinol. 1996;150:S79-S86.

6. Mackenzie PI, Gregory PA, Gardner-Stephen DA, Lewinsky RH, Jorgensen BR, Nishiyama T, et al. Regulation of UDP glucuronosyltransferase genes. Current Drug Metabol. 2003;4:249-57.

7. Lazarus P, Blevins-Primeau AS, Zheng Y, Sun D. Potential role of UGT pharmogenetics in cancer treatment and prevention. Ann NY Acad Sci. 2009;1155:99-111.

8. Owens IS, Ritter JK. Gene structure at the human UGT1 locus creates diversity in isozyme structure, substrate specificity, and regulation. Progr Nucl Acid Res Mol Biol. 1995;51:305-38.

9. Gong Q, Cho JW, Huang T, Potter C, Gholami N, Basu NK, et al. Thirteen UDPglucuronosyltransferase genes are encoded at the human UGT1 gene complex locus. Pharmacogenetics. 2001;11:357-68.

10. Fisher MB, Paines MF, Strelevitz TJ, Wrighton SA. The role of hepatic and extrahepatic UDP-glucuronosyltransferases in human drug metabolism. Drug Metabol Rev. 2001;33:273-97.

11. Carrier J-S, Turgeon D, Journault K, Hum DW, Belanger A. Isolation and characterization of the human UGT2B7 gene. Biochem Biophys Res Comm. 2000;272:616-21.

12. Turgeon D, Carrier J-S, Levesque E, Beatty BG, Belanger A, Hum DW. Isolation and characterization of the human UGT2B15 gene, localized within a cluster of UGT2B genes and pseudogenes on chromosome 4. J Mol Biol. 2000;295:489-504.

13. Jin CJ, Miners JO, Lillywhite KJ, Mackenzie PI. cDNA cloning and expression of two new members of the human liver UDP-glucuronosyltransferase 2B subfamily. Biochem Biophys Res Comm. 1993;194:496-503.

14. Belanger A, Hum DW, Beaulieu M, Levesque E, Guillemette C, Tchernof A, et al. Characterization and regulation of UDP-glucuronosyltransferases in steroid target tissues. J Steroid Biochem Mol Biol. 1998;65:301-10.

15. Turgeon D, Carrier J-S, Levesque E, Hum DW, Belanger A. Relative enzymatic activity, protein stability, and tissue distribution of human steroid-metabolizing UGT2B subfamily members. Endocrinology. 2001;142:778-87.

16. Harrington WR, Sengupta S, Katzenellenbogen BS. Estrogen regulation of the glucuronidation enzyme UGT2B15 in estrogen receptor-positive breast cancer cells. Endocrinology. 2006;147:3843-50.

17. Vienneau DS, DeBoni U, Wells PG. Potential genoprotective role for UDP-glucuronosyltransferases in chemical carcinogenesis: Initiation of micronuclei by benzo(a)pyrene and benzo(e)pyrene in UDP-glucuronosyltransferase-deficient cultured rat skin fibroblasts. Cancer Res. 1995;55:1045-51.

18. Roy AK. Regulation of steroid hormone action in target cells by specific hormone-inactivating enzymes. Proc Soc Exp Biol Med. 1992;199:265-72.

19. Zhu BT, Evaristus EN, Antoniak SK, Sarabia SF, Ricci MJ, Liehr JG. Metabolic deglucuronidation and demethylation of estrogen conjugates as a source of parent estrogens and catecholestrogen metabolites in Syrian hamster kidney, a target organ of estrogen-induced tumorigenesis. Toxicol Appl Pharmacol. 1996;136:186-93.

20. Ritter JK, Sheen YY, Owens IS. *Cloning and expression of human liver UDP-glucuronosyltransferase in COS-1 cells. 3,4-catechol estrogens and estriol as primary substrates.* J Biol Chem. 1990;265:7900-6.

21. Ritter JK, Chen F, Sheen YY, Lubet RA, Owens IS. *Two human liver cDNAs encode UDP-glucuronosyltransferases with 2 log differences in activity toward parallel substrates including hyodeoxycholic acid and certain estrogen derivatives.* Biochemistry. 1992;31:3409-14.

22. King CD, Green MD, Rios GR, Coffman BL, Owens IS, Bishop WP, et al. *The glucuronidation of exogenous and endogenous compounds by stably expressed rat and human UDP-glucuronosyltransferase 1.1.* Arch Biochem Biophys. 1996;332:92-100.

23. Mojarrabi B, Butler R, Mackenzie PI. *cDNA cloning and characterization of the human UDP glucuronosyltransferase, UGT1A3.* Biochem Biophys Res Commun. 1996;225:785-90.

24. Cheng Z, Radominska-Pandya A, Tephly TR. *Cloning and expression of human UDP-glucuronosyltransferase (UGT) 1A8.* Arch Biochem Biophys. 1998;356:301-5.

25. Cheng Z, Rios GR, King CD, Coffman BL, Green MD, Mojarrabi B, et al. *Glucuronidation of catechol estrogens by expressed human UDP-glucuronosyltransferases (UHTs) 1A1, 1A3, and 2B7.* Toxicol Sci. 1998;45:52-7.

26. Strassburg CP, Manns MP, Tukey RH. *Expression of the UDP-gucuronosyltransferase 1A locus in human colon. Identification and characterization of the novel extrahepatic UGT1A8.* J Biol Chem. 1998;273:8719-26.

27. Albert C, Vallee M, Guillaume B, Belanger A, Hum DW. *The monkey and human uridine diphosphate-glucuronosyltransferase UGT1A9, expressed in steroid target tissues, are estrogen-conjugating enzymes.* Endocrinology. 1999;140:3292-302.

28. Gall WE, Zawada G, Mojarrabi B, Tephly TR, Green MD, Coffman BL, et al. *Differential glucuronidation of bile acids, androgens and estrogens by human UGT1A3 and 2B7.* J Steroid Biochem Mol Biol. 1999;70:101-8.

29. Levesque E, Beaulieu M, Hum DW, Belanger A. *Characterization and substrate specificity of UGT2B4 (E458): a UDP-glucuronosyltransferase encoded by a polymorphic gene.* Pharmacogenetics. 1999;9:207-16.

30. Vallee M, Albert C, Beaudry G, Hum DW, Belanger A. *Isolation and characterization of the monkey UDP-glucuronosyltransferase cDNA clone monUGT1A01 active on bilirubin and estrogens.* J Steroid Biochem Mol Biol. 2001;77:239-49.

31. Lepine J, Bernard O, Plante M, Tetu B, Pelletier G, Labrie F, et al. *Specificity and regioselectivity of the conjugation of estradiol, estrone, and their catecholestrogen and methoxyestrogen metabolites by human uridine diphospho-glucuronosyltransferases expressed in endometrium.* J Clin Endocrinol Metab. 2004;89:5222-32.

32. Senafi SB, Clarke DJ, Burchell B. *Investigation of the substrate specificity of a cloned expressed human bilirubin UDP-glucuronosyltransferase: UDP-sugar specificity and involvement in steroid and xenobiotic glucuronidation.* Biochem J. 1994;303:233-40.

33. Duguay Y, McGrath M, Lepine J, Gagne J, Hankinson SE, Colditz GA, et al. *The functional UGT1A1 promoter polymorphism decreases endometrial cancer risk.* Cancer Res. 2004;64:1202-7.

34. Thibaudeau J, Lepine J, Tojcic J, Duguay Y, Pelletier G, Plante M, et al. *Characterization of common UGT1A8, UGT1A9, and UGT2B7 variants with different capacities to inactivate mutagenic 4-hydroxylated metabolites of estradiol and estrone.* Cancer Res. 2006;66:125-33.

35. Raftogianis R, Creveling C, Weinshilboum R, Weisz J. *Estrogen metabolism by conjugation.* J Natl Cancer Inst Monogr. 2000;27:113-24.

36. Gestl SA, Green MD, Shearer DA, Frauenhoffer E, Tephly TR, Weisz J. *Expression of UGT2B7, a UDP-glucuronosyltransferase implicated in the metabolism of 4-hydroxyestrone and all-trans retinoic acid, in normal human breast parenchyma and in invasive and in situ breast cancers.* Am J Pathol. 2002;160:1467-79.

37. Levesque E, Beaulieu M, Green MD, Tephly TR, Belanger A, Hum DW. *Isolation and characterization of UGT2B15(Y85): a UDP-glucuronosyltransferase encoded by a polymorphic gene.* Pharmacogenetics. 1997;7:317-25.

38. Huang YH, Galijatovic A, Nguyen N, Geske D, Beaton D, Green J, et al. *Identification and functional characterization of UDP-glucuronosyltransferases UGT1A8*1, UGT1A8*2 and UGT1A8*3.* Pharmacogenetics. 2002;12:287-97.

39. Miners JO, McKinnon RA, Mackenzie PI. *Genetic polymorphisms of UDP-glucurono-syltrans ferases and their functional significance.* Toxicology. 2002;181-182:453-6.

40. Bosma PJ, Chowdhury JR, Bakker C, Gantla S, De Boer A, Oostra BA, et al. *The genetic basis of the reduced expression of bilirubin UDP-glucuronosyltransferase 1 in Gilbert's syndrome.* N Engl J Med. 1995;333:1171-5.

41. Beutler E, Gelbart T, Demina A. *Racial variability in the UDP-glucoronosyltransferase 1 (UGT1A1) promoter: a balanced polymorphism for regulation of bilirubin metabolism.* Proc Natl Acad Sci USA. 1998;95:8170-4.

42. Fang J, Lazarus P. *Correlation between the UDP-Glucuronosyl-transferase (UGT1A1) TATAA box polymorphism and carcinogen detoxification phenotype: significantly decreased glucuronidating activity against Benzo(a)pyrene-7,8-dihy-drodiol(-) in liver microsomes from subjects with the UGT1A1*28 variant.* Cancer Epidemiol Biomarkers Prev. 2004;13:102-9.

43. Bernard O, Guillemette C. *The main role of UGT1A9 in the hepatic metabolism of mycophenolic acid and the effects of naturally occurring variants.* Drug Metab Dispos. 2004;32:775-8.

44. Bhasker CR, McKinnon W, Stone A, Lo AC, Kubota T, Ishizaki T, et al. *Genetic polymorphism of UDP-glucuronosyltransferase 2B7 (UGT2B7) at amino acid 268: ethnic diversity of alleles and potential clinical significance.* Pharmacogenetics. 2000;10:679-85.

45. Coffman BL, King CD, Rios GR, Tephly TR. *The glucuronidation of opioids, other xenobiotics, and androgens by human UGT2B7Y(268) and UGT2B7H(268).* Drug Metab Dispos. 1998;26:73-7.

46. Wiener D, Fang J, Dossett N, Lazarus P. *Correlation between UDP-glucuronosyltransferase genotypes and 4-(methylnitrosamino)-1-(3-pyridyl)-1-butanone glucuronidation phenotype in human liver microsomes.* Cancer Res. 2004;64:1190-6.

47. Duguay Y, Baar C, Skorpen F, Guillemette C. *A novel functional polymorphism in the uridine diphosphate-glucuronosyltransferase 2B7 promoter with significant impact on promoter activity.* Clin Pharmacol Ther. 2004;75:223-33.

48. Yong M, Schwartz SM, Atkinson C, Makar KW, Thomas SS, Stanczyk FZ, et al. *Associations between polymorphisms in glucuronidation and sulfation enzymes and sex steroid concentrations in premenopausal women in the United States.* J Steroid Biochem Mol Biol. 2011;124:10-8.

49. Guillemette C, De Vivo I, Hankinson SE, Haiman CA, Spiegelman D, Housman DE, et al. *Association of genetic polymorphisms in UGT1A1 with breast cancer and plasma hormone levels.* Cancer Epidemiol Biomark Prev. 2001;10:711-4.

50. Sparks R, Ulrich CM, Bigler J, Tworoger SS, Yasui Y, Rajan KB, et al. *UDP-glucuronosyltransferase and sulfotransferase polymorphisms, sex hormone concentrations, and tumor receptor status in breast cancer patients.* Breast Cancer Res. 2004;6:R488-R98.

51. Guillemette C, Millikan RC, Newman B, Housman DE. *Genetic polymorphisms in uridine diphospho-glucuronosyltransferase 1A1 and association with breast cancer among African Americans.* Cancer Res. 2000;60:950-6.

52. Adegoke OJ, Shu XO, Gao Y, Cai Q, Breyer J, Smith J, et al. *Genetic polymorphisms in uridine diphospho-glucuronosyltransferase 1A1 (UGT1A1) and risk of breast cancer.* Breast Cancer Res Treat. 2004;85:239-45.

The Etiology of Breast Cancer

4.4. ANTI-OXIDANT ENZYMES
4.4.1. INTRODUCTION

In addition to the "classical" phase II enzymes, such as glutathione transferases and UDP-glucuronsyltransferases which conjugate xenobiotics with endogenous ligands, there are many other enzymes involved in catalytic protection against electrophile and oxidative damage (1). These include anti-oxidant enzymes, such as superoxide dismutase, catalase, glutathione peroxidase, glutathione reductase, thioredoxin, and thioredoxin reductase, as well as NAD(P)H:quinone oxidoreductase 1 (NQO1), and epoxide hydrolase. Several of these enzymes are regulated via anti-oxidant response elements (ARE) that are targets of the Nrf2-Keap1 pathway, composed of the transcription factor Nrf2 (nuclear factor erythroid 2-related factor 2) and the repressor protein Keap1 (Kelch-like ECH-associated protein 1) (2-4).

Reactive oxygen species (ROS) play physiological roles in several cell processes including signal transduction, cell cycle progression, cell proliferation and apoptosis (5-10). However, when the ROS production is increased during oxidative stress, damage to cellular macromolecules such as DNA, proteins, and lipids may ensue contributing to the pathogenesis of cancer and various other diseases (11-13). There are few direct measurements of ROS concentrations in normal or malignant breast tissues with one reporting elevated levels in breast cancer (14, 15). All cells possess two defense systems to protect against ROS and oxidative damage: anti-oxidant enzymes and reducing substances such as glutathione, ascorbate, α-tocopherol, α-lipoic acid, melatonin, and others (12, 16-18) (Table 4.4.1.).

Although antioxidant enzymes are critically important in limiting ROS-mediated damage to cellular macromolecules, they are clearly not 100% effective at performing this task since DNA oxidation products can be detected in blood and urine under physiological conditions. In the following, I will focus on antioxidant enzymes with an emphasis on studies relevant to breast cancer (Figure 4.4.1).

Table 4.4.1. Anti-oxidant Enzymes and Reducing Substances

I. Anti-oxidant enzymes	II. Reducing substances
Superoxide dismutase	glutathione
Catalase	ascorbic acid (vitamin C)
Glutathione peroxidases	α-tocopherol (vitamin E)
Thioredoxin system	α-lipoic acid
Peroxiredoxins	melatonin

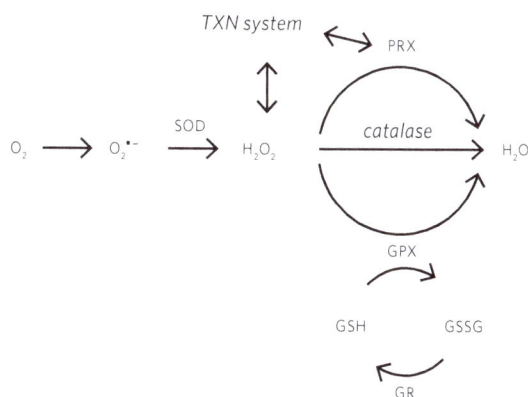

Figure 4.4.1. Overview of antioxidant enzymes and their roles in minimizing ROS levels.
SOD, superoxide dismutase, GPX, glutathione peroxidase, TXN, thioredoxin, PRX, peroxiredoxin, GR, glutathione reductase

References

1. Talalay P, Dinkova-Kostova AT, Holtzclaw WD. *Importance of phase 2 gene regulation in protection against electrophile and reactive oxygen toxicity and carcinogenesis.* Adv Enzyme Regul. 2003;43:121-34.

2. Hayes JD, McMahon M. *NRF2 and KEAP1 mutations: permanent activation of an adaptive response in cancer.* Trends Biochem Sci. 2009;34:176-88.

3. Itoh K, Wakabayashi N, Katoh Y, Ishii T, Igarashi K, Engel JD, et al. *Keap1 represses nuclear activation of antioxidant responsive elements by Nrf2 through binding to the amino-terminal Neh2 domain.* Genes Develop. 1999;13:76-86.

4. Wakabayashi N, Dinkova-Kostova AT, Holtzclaw WD, Kang MI, Kobayashi A, Yamamoto M, et al. *Protection against electrophile and oxidant stress by induction of the phase 2 response: Fate of cysteines of the Keap1 sensor modified by inducers.* Proc Natl Acad Sci USA. 2004;101:2040-5.

5. Burdon RH, Rice-Evans C. *Free radicals and the regulation of mammalian cell proliferation.* Free Rad Res Commun. 1989;6:345-58.

6. Finkel T. *Oxygen radicals and signaling.* Curr Opinion Cell Biol. 1998;10:248-53.

7. Thannickal VJ, Fanburg BL. *Reactive oxygen species in cell signaling.* Am J Physiol Lung Cell Mol Physiol. 2000;279:L1005-L28.

8. Ueda S, Masutani H, Nakamura H, Tanaka T, Ueno M, Yodoi J. *Redox control of cell death.* Antiox Redox Signal. 2002;4:405-14.

9. Conour JE, Graham WV, Gaskins HR. *A combined in vitro/bioinformatic investigation of redox regulatory mechanisms governing cell cycle progression.* Physiol Genomics. 2004;18:196-205.

10. Menon SG, Goswami PC. *A redox cycle within the cell cycle: ring in the old with the new.* Oncogene. 2007;26:1101-9.

11. Shigenaga MK, Hagen TM, Ames BN. *Oxidative damage and mitochondrial decay in aging.* Proc Natl Acad Sci USA. 1994;91:10771-8.

12. Ambrosone CB. *Oxidants and antioxidants in breast cancer.* Antiox Redox Signal. 2000;2:903-17.

13. Laurent A, Nicco C, Chereau C, Goulvestre C, Alexandre J, Alves A, et al. *Controlling tumor growth by modulating endogenous production of reactive oxygen species.* Cancer Res. 2005;65:948-56.

14. Wright RM, McManaman JL, Repine JE. *Alcohol-induced breast cancer: A proposed mechanism.* Free Radic Biol Med. 1999;26:348-54.

15. Haklar G, Sayin-Ozveri E, Yuksel M, Aktan AO, Yalcin AS. *Different kinds of reactive oxygen and nitrogen species were detected in colon and breast tumors.* Cancer Lett. 2001;165:219-24.

16. Moini H, Packer L, Saris NL. *Antioxidant and prooxidant activities of a-Lipoic acid and dihydrolipoic acid.* Toxicol Appl Pharmacol. 2002;182:84-90.

17. Reiter RJ, Tan D, Osuna C, Gitto E. *Actions of melatonin in the reduction of oxidative stress.* J Biomed Sci. 2000;7:444-58.

18. Kalpakcioglu B, Senel K. *The interrelation of glutathione reductase, catalase, glutathione peroxidase, superoxide dismutase, and glucose-6-phosphate in the pathogenesis of rheumatoid arthritis.* Clin Rheumatol. 2008;27:141-5.

4.4.2. SUPEROXIDE DISMUTASES

The superoxide dismutases (SODs) are ubiquitous components of the cellular antioxidant system that catalyze the dismutation or disproportionation of superoxide anion to oxygen and hydrogen peroxide (Figure 4.4.2.) (1-3).

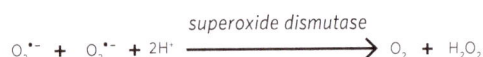

$$O_2^{\bullet-} + O_2^{\bullet-} + 2H^+ \xrightarrow{\text{superoxide dismutase}} O_2 + H_2O_2$$

Figure 4.4.2. SOD-mediated reaction

SODs require a redox active transition metal in the active site to accomplish the catalytic breakdown of superoxide anion. The nature of the superoxide anion is such that it may act either as a reductant or as an oxidant. In the SOD-mediated reaction, the metal co-factors catalyze both a one-electron oxidation and a one-electron reduction of separate superoxide anions to give the overall disproportionation reaction (Figure 4.4.3.). The SODs are catalytically efficient because the redox potential of the metal site (+200 to +400 mV) lies between that of oxygen/superoxide (-160 mV) and superoxide/hydrogen peroxide (+890 mV) (3, 4).

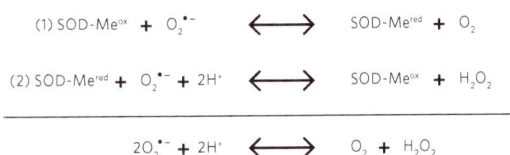

$$(1)\ \text{SOD-Me}^{ox} + O_2^{\bullet-} \longleftrightarrow \text{SOD-Me}^{red} + O_2$$
$$(2)\ \text{SOD-Me}^{red} + O_2^{\bullet-} + 2H^+ \longleftrightarrow \text{SOD-Me}^{ox} + H_2O_2$$
$$\overline{\qquad 2O_2^{\bullet-} + 2H^+ \longleftrightarrow O_2 + H_2O_2 \qquad}$$

Figure 4.4.3. Steps in SOD-mediated reaction

SODs are present in all aerobic organisms and employ various transition metals to carry out the disproportionation of superoxide anion (5, 6). In humans, cytosolic SOD1 contains Cu/Zn and mitochondrial SOD2 contains Mn. Both enzymes are expressed in all tissues. In contrast, SOD3, an extracellular enzyme, is expressed selectively in blood vessels, lung, kidney, and uterus (7). Like SOD1, the SOD3 protein binds Cu and Zn; SOD3 is approximately 50% homologous to SOD1 but shows minimal homology with SOD2.

SOD1 (Cu/ZnSOD)

Gene and Protein. The SOD1 gene at 21q22 encodes the homodimeric 32 kDa copper, zinc enzyme (8, 9) (Figure 4.4.4.). SOD1 is present in the cytoplasm of all cells comprising about 1% of total cellular protein. It is thus a major cytosolic protein and represents the predominant SOD in all tissues, comprising 70–80% of total SOD activity. Despite its constitutive expression, SOD1 mRNA levels can vary in response to regulation by transcription factors Sp1, Egr-1 and Wilms' tumor WT1 via binding sites in the proximal SOD1 promoter (6, 10).

Each SOD1 subunit binds one copper and one zinc ion. The copper is bound to four histidine residues 46, 48, 63, and 120 while zinc is linked to residues 63His, 71His, 80His, and 83Asp. Residue 63His binds both metal ions simultaneously via the two imidazole nitrogen atoms, placing the Cu^{2+} and Zn^{2+} at a distance of about 6 Å (5, 11). The positively charged guanidinium group of the conserved 143Arg stabilizes the superoxide anion substrate opposite the catalytic Cu^{2+} (5, 6). Copper is directed to SOD1 by the concerted action of cell surface and intracellular copper transporters and finally a copper chaperone known as CCS. CCS can insert copper into newly synthesized SOD1 in the presence of oxygen, providing a means of regulating SOD1 activity in response to oxygen status (11). The post-translational activation of the SOD1 enzyme occurs in a stepwise fashion: (i) incorporation of Zn^{2+}, (ii) formation of a heterodimeric complex between CCS and SOD1, (iii) translocation of Cu^+ from CCS to SOD1, followed by oxidation to Cu^{2+} with binding to 63His and release of CCS, and (iv) intra-molecular disulfide formation between cysteine residues 57 and 146 followed by dimerization (3, 11).

Clinical Studies. Patients with Down syndrome (trisomy 21) have three copies of the SOD1 gene. Overexpression of SOD1 has been shown experimentally to cause advanced cellular senescence and linked to the premature aging observed in Down syndrome (12). Amyotrophic lateral sclerosis (ALS) is an adult onset neurodegenerative disorder characterized by the loss of motor neurons, culminating in muscle wasting and death from respiratory failure. About 90% of ALS cases occur sporadically and of the remaining 10% familial ALS approximately 25% are caused by over 100 different missense mutations of the SOD1 gene (6, 13-16). These mutations appear to result in a gain of function rather than a deficiency of SOD1 activity (17). Alternatively, the mutant SOD1 protein can undergo oligomerization and form insoluble aggregates that initiate cell death (3). The degenerative process in the motor neurons appears to be caused by diffusible factors originating in adjacent glial cells (18, 19).

Erythrocytes are particularly vulnerable to oxidative damage due to continuous exposure to high oxygen tension as well as the presence of large amounts of iron, a potent catalyst for ROS production (20). For example, the autoxidation of oxyhemoglobin has been shown to generate superoxide anion through the transfer of electrons from Fe^{2+} to O_2 (21). However, the normal red cell is resistant to oxidative damage because it is endowed with several antioxidant enzymes: SOD, catalase, glutathione peroxidase. Cancer and other conditions associated with oxidative stress cause not only oxidative tissue damage but via unknown mechanisms may also influence the activity of erythrocyte antioxidant enzymes. With regard to breast cancer, several studies have shown significantly lower levels of SOD, catalase, and glutathione peroxidase activity in red cells of breast cancer patients compared with control women (22-24).

SOD2 (MnSOD)

Gene and Protein. The mitochondrial SOD2 enzyme is encoded by a nuclear gene at 6q25 (25, 26) (Figure 4.4.5.). SOD2 translation takes place in ribosomes juxtaposed to mitochondria. The protein is synthesized as a precursor polypeptide composed of 222 amino acids with a cleavable, N-terminal mitochondrial targeting sequence (MTS) (27, 28). The 24-amino acid long MTS drives the mitochondrial import of the SOD2 precursor through the translocase of the outer membrane (TOM) and the translocase of the inner membrane (TIM) (29, 30). Inside the mitochondrial matrix, the SOD2-MTS is cleaved by mitochondrial processing peptidase (MPP). As the protein enters the mitochondria, SOD2 remains sufficiently unfolded to allow Mn insertion, which is facilitated by Mtm1p, a member of the mitochondrial carrier family of transporters in the inner membrane (31). Following Mn binding and complete entry into the mitochondrial matrix, four monomers of the mature kDa SOD2 protein assemble into the active 96 kDa

homotetramer (32, 33). Overall, translation, mitochondrial import, Mn insertion, and tetramerization are tightly coupled (3, 34). Each SOD2 monomer contains one Mn atom. The metal is buried well within the protein interior and coordinated in a trigonal bipyramidal geometry to three histidines, one aspartate, and a solvent water molecule (27, 35).

SOD2 can be induced by its substrate, superoxide anion, and by ionizing and ultraviolet irradiation (36-38). It is also induced by proinflammatory mediators like interleukin 1, tumor necrosis factor α, and interferon γ (39). SOD2 activity changes during the cell cycle, increasing during G1 and decreasing during S phase, while superoxide levels decrease and increase, respectively (40). In MCF-7 cells SOD2 regulates the expression of hypoxia-inducible factor-1α (HIF-1α) by modulating the steady-state level of superoxide (41). Mitochondrial SOD2 plays an essential role in oxidative stress protection. Complete loss of the SOD2 gene results in neonatal lethality in mice and even haploinsufficiency can be detrimental (42, 43). Life-long reduction in SOD2 activity results in increased DNA damage in form of elevated levels of 8-oxo-2'-deoxyguanosine in all tissues and a higher incidence of cancer (44).

A polymorphism in the MTS, Ala16Val (rs4880; previously rs1799725), modulates the efficiency of mitochondrial SOD2 import (29). The partial α-helical structure of Ala-MTS allows efficient SOD2 import while the β-sheet configuration of the Val-variant causes partial arrest of the precursor within the inner membrane and decreased formation of the SOD2 tetramer in the mitochondrial matrix (45). The result is a 30% higher activity of Ala-SOD2 than the variant Val-SOD2. Since the substitution in codon 16 is located at position –9 of the mature SOD2 protein, it is also referred to as the –9Ala/Val dimorphism. Each allele is present in about 50% of Caucasians whereas the Val

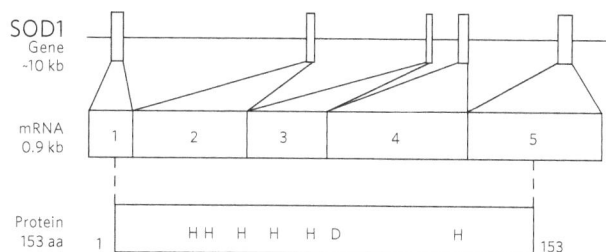

Figure 4.4.4. Overview of SOD1 gene, mRNA, and protein.
The SOD1 gene at 21q22 is 11 kb long and contains five exons arranged in a 0.9 kb mRNA transcript. The protein consists of 153 amino acids with a molecular weight of 15.6 kDa. The metals copper and zinc are bound to histidine (H) and aspartic acid (D) residues. The coordination sphere of Cu^{2+} is formed by four histidine residues (46, 48, 63, 120) and one water molecule. The Zn^{2+} is coordinated in an approximately tetrahedral geometry by 63His, 71His, 80His, and 83Asp. The positively charged guanidinium group of the conserved 143Arg stabilizes the superoxide anion substrate opposite the catalytic Cu^{2+}. The active SOD1 enzyme is a non-covalently linked 32 kDa homodimer. There is no evidence of common nonsynonymous SOD1 polymorphisms.

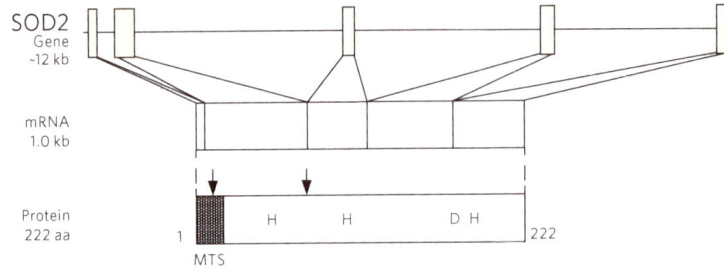

Figure 4.4.5. Overview of SOD2 gene, mRNA, and protein.
The SOD2 gene at 6q25 is 12 kb long and contains five exons arranged in a 1 kb mRNA transcript. The cDNA consists of 95 bp of 5' untranslated sequence, 666 bp of coding sequence, and 216 bp of 3' untranslated sequence. The amino acid sequence is 222 residues in length. The first 24 constitute the leader peptide for mitochondrial targeting (MTS) that is not present in the mature 23 kDa protein. The Mn atom is bound to amino acids 52His, 107His, 194Asp, and 198His. The active SOD2 enzyme is a non-covalently linked homotetramer. The arrows indicate polymorphisms at Ala16(-9)Val and Ile58Thr.

allele occurs in about 80% of Japanese (29, 46). Another polymorphism, Ile58Thr (Ile82Thr; rs1141718), destabilizes the tetrameric interface resulting in a shortened SOD2 half-life (33); this polymorphism is rare (<0.05%) in the population (47).

Clinical Studies. An immunohistochemical study of 259 invasive breast cancers revealed SOD2 expression in only 44% (48). Another study observed less frequent SOD2 expression in tumor cells of invasive breast cancers than in *in situ* or non-neoplastic breast epithelial cells (49). Several studies of the -9Ala/Val (rs4880) polymorphism have been reported to modulate the risk of diverse cancers, neurodegenerative disorders, alcoholic liver disease, and cardiomyopathies (29, 45, 46). Three studies have reported a positive association of the –9Ala/Val SNP with breast cancer risk (46, 50, 51). A multicenter study involving 8732 cases and 8458 controls observed a borderline association, p = 0.056 (52). Most studies have found no association with breast cancer risk (47, 53-57). Two large studies examined the association of breast cancer risk with common SNPs in 10 genes encoding enzymes in the antioxidant defence system including SOD1 and SOD2. A case-control study (4474 cases, 4580 controls) examined 54 SNPs that tag all known variants in the 10 genes including SOD2 –9Ala/Val (58). There was no evidence for an association of this or any other SOD2 or SOD1 polymorphism with breast cancer risk. A multilocus study of the same population limited to 2271 cases and 2280 controls examined gene-gene interactions between 52 SNPs in the 10 genes (59). The statistical analysis by three different methods (unconditional logistic regression, multifactor dimensionality reduction, and hierarchical cluster analysis) found no evidence for an association of the SOD2 –9Ala/Val or any other variant in the other genes with breast cancer risk.

SOD3

Gene and Protein. The SOD3 gene at 4q21 is approximately 5.9 kb long and contains 3 exons arranged in a 4.2 kb mRNA (Figure 4.4.6.). The 720 bp coding region is located entirely within exon 3 (60). The 240 amino acid propeptide includes an 18 amino acid signal peptide, which is removed to yield the mature 222 amino acid protein (61).

The carboxy-terminal of SOD3 contains a cluster of nine amino acids with a positive charge (six arginines, three lysines). This region shows strong affinity to heparin and heparan sulfate, which accounts for the location of SOD3 on cell surfaces and in the extracellular matrix. The active SOD3 exists as a 135 kDa homotetrameric copper-zinc containing glycoprotein and functions as the predominant superoxide scavenger in extracellular space (62).

In contrast to SOD1 and SOD2, the expression of SOD3 is restricted to very few cell types: vascular smooth muscle cells, alveolar type II cells, proximal renal tubular cells, and cultured endothelial and fibroblast cell lines (6, 7). Nitric oxide, an essential regulator of vascular tone, can react with superoxide anion to form the toxic peroxynitrite anion (Figure 4.4.7.). The latter can break down to produce the hydroxyl radical. By inactivating superoxide anion, SOD3 is likely to play an important role in regulating nitric oxide-induced signaling events in the vascular wall.

$$O_2^{\bullet-} + NO^{\bullet} \xrightarrow{K\text{-}10^{10}} ONOO^- \longrightarrow {}^{\bullet}OH + NO_2$$

Figure 4.4.7. SOD3-mediated reaction with superoxide anion and nitric oxide

The SOD3 gene contains several non-synonymous SNPs. The substitution Arg213Gly (rs1799895) is observed in 3 to 6% of Caucasians and Asians and associated with a 10-fold increase in plasma SOD3 concentration (63-68). Residue

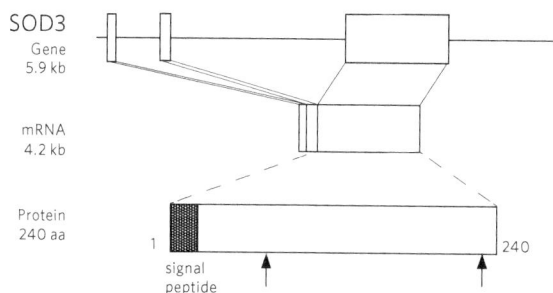

Figure 4.4.6. Overview of SOD3 gene, mRNA, and protein.
The SOD3 gene at 4q21 is 5.9 kb long and contains three exons arranged in a 4.2 kb mRNA transcript. The translated protein consists of 240 residues from which a signal peptide of 18 amino acids is cleaved to form the mature 222 amino acid SOD3. The carboxy-terminal segment contains nine positively charged amino acids that confer high affinity for heparin binding. The active SOD3 enzyme is a 135 kDa homotetramer. The arrows indicate nonsynonymous polymorphisms: Ala58Thr (rs2536512) and Arg213Gly (rs1799895).

213 is located in the center of the carboxy-terminal cluster of positively charged amino acids, which defines the heparin-binding domain, and the Arg213Gly substitution impaired the affinity for heparin and endothelial cell surfaces. The Ala58Thr (rs2536512) polymorphism showed a positive association of the 58Thr allele with ER-positive breast cancer (69).

Clinical Studies. SOD3 is secreted into the extracellular environment and present in plasma, cerebrospinal fluid, lymph, and ascites (70). Plasma levels are highest in childhood, decreasing to adult levels around age 20 (71). SOD3 mRNA levels are higher in ER-negative than ER-positive breast cancers (64). The growth rate of MDA-MB231 and MDA-MB435 breast cancer cells was only moderately reduced by overexpressing full-length SOD3, but dramatically by SOD3 lacking the heparin-binding domain and concomitant heparin treatment (72).

References

1. McCord JM, Fridovich I. *Superoxide dismutase. An enzymic function for erythrocuprein (hemocuprein).* J Biol Chem. 1969;244:6049-55.

2. Fridovich I. *Reflections of a fortunate biochemist.* J Biol Chem. 2001;276:28629-36.

3. Culotta VC, Yang M, O'Halloran TV. *Activation of superoxide dismutases: Putting the metal to the pedal.* Biochim Biophys Acta. 2006;1763:747-58.

4. Vance CK, Miller A-F. *A simple proposal that can explain the inactivity of metal-substituted superoxide dismutases.* J Am Chem Soc. 1998;120:461-7.

5. Bordo D, Djinovic K, Bolognesi M. *Conserved patterns in the Cu,Zn superoxide dismutase family.* J Mol Biol. 1994;238:366-86.

6. Zelko IN, Mariani TJ, Folz RJ. *Superoxide dismutase multigene family: a comparison of the CuZn-SOD (SOD1), Mn-SOD (SOD2), and EC-SOD (SOD3) gene structures, evolution, and expression.* Free Radic Biol Med. 2002;33:337-49.

7. Fattman CL, Schaefer LM, Oury TD. *Extracellular superoxide dismutase in biology and medicine.* Free Radic Biol Med. 2003;35:236-56.

8. Levanon D, Lieman-Hurwitz J, Dafni N, Wigderson M, Sherman L, Bernstein Y, et al. *Architecture and anatomy of the chromosomal locus in human chromosome 21 encoding the Cu/Zn superoxide dismutase.* EMBO J. 1985;4:77-84.

9. Danciger E, Dafni N, Bernstein Y, Laver-Rudich Z, Neer A, Groner Y. *Human Cu/Zn superoxide dismutase gene family: molecular structure and characterization of four Cu/Zn superoxide dismutase-related pseudogenes.* Proc Natl Acad Sci U S A. 1986;83:3619-23.

10. Minc E, de Coppet P, Masson P, Thiery L, Dutertre S, Amor-Gueret M, et al. *The human copper-zinc superoxide dismutase gene (SOD1) proximal promoter is regulated by Sp1, Egr-1, and WT1 via non-canonical binding sites.* J Biol Chem. 1999;274:503-9.

11. Furukawa Y, Torres AS, O'Halloran TV. *Oxygen-induced maturation of SOD1: a key role for disulfide formation by the copper chaperone CCS.* EMBO J. 2004;23:2872-81.

12. de Haan JB, Cristiano F, Iannello R, Bladier C, Kelner MJ, Kola I. *Elevation in the ratio of Cu/Zn-superoxide dismutase to glutathione peroxidase activity induces features of cellular senescence and this effect is mediated by hydrogen peroxide.* Hum Mol Genet. 1996;5:283-92.

13. Rosen DR, Siddique T, Patterson D, Figlewicz DA, Sapp P, Hentati A, et al. *Mutations in Cu/Zn superoxide dismutase gene are associated with familial amyotrophic lateral sclerosis.* Nature. 1993;362:59-62.

14. Esteban J, Rosen DR, Bowling AC, Sapp P, McKenna-Yasek D, O'Regan JP, et al. *Identification of two novel mutations and a new polymorphism in the gene for Cu/Zn superoxide dismutase in patients with amyotrophic lateral sclerosis.* Hum Mol Genet. 1994;3:997-8.

15. Gurney ME, Pu H, Chiu AY, Dal Canto MC, Polchow CY, Alexander DD, et al. *Motor neuron degeneration in mice that express a human Cu,Zn superoxide dismutase mutation.* Science. 1994;264:1772-5.

16. Valentine JS, Doucette PA, Zittin Potter S. *Copper-zinc superoxide dismutase and amyotrophic lateral sclerosis.* Annu Rev Biochem. 2005;74:563-93.

17. Yim MB, Kang JH, Yim HS, Kwak HS, Chock PB, Stadtman ER. *A gain-of-function of an amyotrophic lateral sclerosis-associated Cu,Zn-superoxide dismutase mutant: An enhancement of free radical formation due to a decrease in Km for hydrogen peroxide.* Proc Natl Acad Sci USA. 1996;93:5709-14.

18. Di Giorgio FP, Carrasco MA, Siao MC, Maniatis T, Eggan K. *Non-cell autonomous effect of glia on motor neurons in an embryonic stem cell-based ALS model.* Nat Neurosci. 2007;10:608-14.

19. Nagai M, Re DB, Nagata T, Chalazonitis A, Jessell TM, Wichterle H, et al. *Astrocytes expressing ALS-linked mutated SOD1 release factors selectively toxic to motor neurons.* Nat Neurosci. 2007;10:615-22.

20. Scott MD, Lubin BH, Zuo L, Kuypers FA. *Erythrocyte defense against hydrogen peroxide: preeminent importance of catalase.* J Lab Clin Med. 1991:7-16.

21. Misra HP, Fridovich I. *The generation of superoxide radical during the autoxidation of hemoglobin.* J Biol Chem. 1972;247:6960-2.

22. Kumar K, Thangaraju M, Sachdanandam P. *Changes observed in antioxidant system in the blood of post-menopausal women with breast cancer.* Biochem Int. 1991;25:371-80.

23. Abiaka C, Al-Awadi F, Al-Sayer H, Gulshan S, Behbehani A, Farghally M. *Activities of erythrocyte antioxidant enzymes in cancer patients.* J Clin Lab Anal. 2002;16:167-71.

24. Kumaraguruparan R, Subapriya R, Kabalimoorthy J, Nagini S. *Antioxidant profile in the circulation of patients with fibroadenoma and adenocarcinoma of the breast.* Clin Biochem. 2002;35:275-9.

25. Beck Y, Oren R, Amit B, Levanon A, Gorecki M, Hartman JR. *Human Mn superoxide dismutase cDNA sequence.* Nucleic Acids Res. 1987;15:9076.

26. Church SL, Grant JW, Meese EU, Trent JM. *Sublocalization of the gene encoding manganese superoxide dismutase (MnSOD/SOD2) to 6q25 by fluorescence in situ hybridization and somatic cell hybrid mapping.* Genomics. 1992;14:823-5.

27. Wispe JR, Clark JC, Burhans MS, Kropp KE, Korfhagen TR, Whitsett JA. *Synthesis and processing of the precursor for human mangano-super-oxide dismutase.* Biochim Biophys Acta. 1989;994:30-6.

28. Wan XS, Devalaraja MN, St Clair DK. *Molecular structure and organization of the human manganese superoxide dismutase gene.* DNA Cell Biol. 1994;13:1127-36.

29. Shimoda-Matsubayashi S, Matsumine H, Kobayashi T, Nakagawa-Hattori Y, Shimizu Y, Mizuno Y. *Structural dimorphism in the mitochondrial targeting sequence in the human manganese superoxide dismutase gene. A predictive evidence for conformational change to influence mitochondrial transport and a study of allelic association in Parkinson's disease.* Biochem Biophys Res Commun. 1996;226:561-5.

30. Pfanner N, Geissler A. *Versatility of the mitochondrial protein import machinery.* Nat Rev Mol Cell Biol. 2001;2:339-49.

31. Luk E, Carroll M, Baker M, Culotta VC. *Manganese activation of superoxide dismutase 2 in Saccharomyces cerevisiae requires MTM1, a member of the mitochondrial carrier family.* Proc Natl Acad Sci USA. 2003;100:10353-7.

32. Matsuda Y, Higashiyama S, Kijima Y, Suzuki K, Kawano K, Akiyama M, et al. *Human liver manganese superoxide dismutase. Purification and crystallization, subunit association and sulfhydryl reactivity.* Eur J Biochem. 1990;194:713-20.

33. Borgstahl GE, Parge HE,

Hickey MJ, Johnson MJ, Boissinot M, Hallewell RA, et al. *Human mitochondrial manganese superoxide dismutase polymorphic variant Ile58Thr reduces activity by destabilizing the tetrameric interface.* Biochemistry. 1996;35:4287-97.

34. Luk E, Yang M, Jensen LT, Bourbonnais Y, Culotta VC. *Manganese activation of superoxide dismutase 2 in the mitochondria of Saccharomyces cerevisiae.* J Biol Chem. 2005;280:22715-20.

35. Wintjens R, Noel C, May AC, Gerbod D, Dufernez F, Capron M, et al. *Specificity and phenetic relationships of iron- and manganese-containing superoxide dismutases on the basis of structure and sequence comparisons.* J Biol Chem. 2004;279:9248-54.

36. Poswig A, Wenk J, Brenneisen P, Wlasschek M, Hommel C, Quel G, et al. *Adaptive antioxidant response of manganese-superoxide dismutase following repetitive UVA irradiation.* J Invest Dermatol. 1999;112:13-8.

37. Liu R, Buettner GR, Oberley LW. *Oxygen free radicals mediate the induction of manganese superoxide dismutase gene expression by TNF-alpha.* Free Radic Biol Med. 2000;28:1197-205.

38. Menon SG, Sarsour EH, Kalen AL, Venkataraman S, Hitchler MJ, Domann FE, et al. *Superoxide signaling mediates N-acetyl-L-cysteine-induced G1 arrest: Regulatory role of cyclin D1 and manganese superoxide dismutase.* Cancer Res. 2007;67:6392-9.

39. Hirose K, Longo DL, Oppenheim JJ, Matsushima K. *Overexpression of mitochondrial manganese superoxide dismutase promotes thesurvival of tumor cells exposed to interleukin-1, tumor necrosis factor, selected anticancer drugs, and ionizing radiation.* FASEB J. 1993;7:361-8.

40. Sarsour EH, Kalen AL, Xiao Z, Veenstra TD, Chaudhuri L, Venkataraman S, et al. *Manganese superoxide dismutase regulates a metabolic switch during the mammalian cell cycle.* Cancer Res. 2012;72:3807-16.

41. Kaewpila S, Venkataraman S, Buettner GR, Oberley LW. *Manganese superoxide dismutase modulates hypoxia-inducible factor-1 alpha induction via superoxide.* Cancer Res. 2008;68:2781-8.

42. Lebovitz RM, Zhang H, Vogel H, Cartwright J, Jr., Dionne L, Lu N, et al. *Neurodegeneration, myocardial injury, and perinatal death in mitochondrial superoxide dismutase-deficient mice.* Proc Natl Acad Sci

USA. 1996;93:9782-7.

43. Strassburger M, Bloch W, Sulyok S, Schuller J, Keist AF, Schmidt A, et al. *Heterozygous deficiency of manganese superoxide dismutase results in severe lipid peroxidation and spontaneous apoptosis in murine myocardium in vivo.* Free Radic Biol Med. 2005;38:1458-70.

44. Van Remmen H, Ikeno Y, Hamilton M, Pahlavani M, Wolf N, Thorpe SR, et al. *Life-long reduction in MnSOD activity results in increased DNA damage and higher incidence of cancer but does not accelerate aging.* Physiol Genomics. 2003;16:29-37.

45. Sutton A, Khoury H, Prip-Buus C, Cepanec C, Pessayre D, Degoul F. *The Ala16Val genetic dimorphism modulates the import of human manganese superoxide dismutase into rat liver mitochondria.* Pharmacogenetics. 2003;13:145-57.

46. Ambrosone CB, Freudenheim JL, Thompson PA, Bowman E, Vena JE, Marshall JR, et al. *Manganese superoxide dismutase (MnSOD) genetic polymorphisms, dietary antioxidants, and risk of breast cancer.* Cancer Res. 1999;59:602-6.

47. Cai Q, Shu XO, Wen W, Cheng JR, Dai Q, Gao YT, et al. *Genetic polymorphism in the manganese superoxide dismutase gene, antioxidant intake, and breast cancer risk: results from the Shanghai breast cancer study.* Breast cancer research : BCR. 2004;6:R647-55.

48. Karihtala P, Kinnula VL, Soini Y. *Antioxidative response for nitric oxide production in breast carcinoma.* Oncol Rep. 2004;12:755-9.

49. Soini Y, Vakkala M, Kahlos K, Paakko P, Kinnula V. *MnSOD expression is less frequent in tumour cells of invasive breast carcinomas than in in situ carcinomas or non-neoplastic breast epithelial cells.* J Pathol. 2001;195:156-62.

50 Mitrunen K, Sillanpaa P, Kataja V, Eskelinen M, Kosma VM, Benhamou S, et al. *Association between manganese superoxide dismutase (MnSOD) gene polymorphism and breast cancer risk.* Carcinogenesis. 2001;22:827-9.

51. Bergman M, Ahnstrom M, Palmeback Wegman P, Wingren S. *Polymorphism in the manganese superoxide dismutase (MnSOD) gene and risk of breast cancer in young women.* J Cancer Res Clin Oncol. 2005;131:439-44.

52. Pharoah P. *Commonly studied single-nucleotide polymorphisms and breast cancer: results from the Breast*

Cancer Association Consortium. J Natl Cancer Inst. 2006;98:1382-96.

53. Egan KM, Thompson PA, Titus-Ernstoff L, Moore JH, Ambrosone CB. *MnSOD polymorphism and breast cancer in a population-based case-control study.* Cancer Lett. 2003;199:27-33.

54. Knight JA, Onay UV, Wells S, Li H, Shi EJ, Andrulis IL, et al. *Genetic variants of GPX1 and SOD2 and breast cancer risk at the Ontario site of the breast cancer family registry.* Cancer Epidemiol Biomarkers Prev. 2004;13:146-9.

55. Millikan RC, Player J, Rene de Cotret A, Moorman P, Pittman G, Vannappagari V, et al. *Manganese superoxide dismutase Ala-9Val polymorphism and risk of breast cancer in a population-based case-control study of African Americans and whites.* Breast Cancer Res: BCR. 2004;6:R264-R74.

56. Tamimi RM, Hankinson SE, Spiegelman D, Colditz GA, Hunter DJ. *Manganese superoxide dismutase polymorphism, plasma antioxidants, cigarette smoking, and risk of breast cancer.* Cancer Epidemiol Biomarkers Prev. 2004;13:989-96.

57. Gaudet MM, Gammon MD, Santella RM, Britton JA, Teitelbaum SL, Eng SM, et al. *MnSOD Val-9Ala genotype, pro- and anti-oxidant environmental modifiers, and breast cancer among women on Long Island, New York.* Cancer Causes Control. 2005;16:1225-34.

58. Cebrian A, Pharoah PD, Ahmed S, Smith PL, Luccarini C, Luben R, et al. *Tagging single-nucleotide polymorphisms in antioxidant defense enzymes and susceptibility to breast cancer.* Cancer Res. 2006;66:1225-33.

59. Oestergaard MZ, Tyrer J, Cebrian A, Shah M, Dunning AM, Ponder BA, et al. *Interactions between genes involved in the antioxidant defence system and breast cancer risk.* Br J Cancer. 2006;95:525-31.

60. Folz RJ, Crapo JD. *Extracellular superoxide dismutase (SOD3): tissue-specific expression, genomic characterization, and computer-assisted sequence analysis of the human EC SOD gene.* Genomics. 1994;22:162-71.

61. Hjalmarsson K, Marklund SL, Engstrom A, Edlund T. *Isolation and sequence of complementary DNA encoding human extracellular superoxide dismutase.* Proc Natl Acad Sci USA. 1987;84:6340-4.

62. Marklund SL. *Human copper-containing superoxide dismutase of* *high molecular weight.* Proc Natl Acad Sci USA. 1982;79:7634-8.

63. Folz RJ, Peno-Green L, Crapo JD. *Identification of a homozygous missense mutation (Arg to Gly) in the critical binding region of the human EC-SOD gene (SOD3) and its association with dramatically increased serum enzyme levels.* Hum Mol Genet. 1994;3:2251-4.

64. Gruvberger S, Ringner M, Chen Y, Panavally S, Saal LH, Borg A, et al. *Estrogen receptor status in breast cancer is associated with remarkably distinct gene expression patterns.* Cancer Res. 2001;61:5979-84.

65. Marklund SL, Nilsson P, Israelsson K, Schampi I, Peltonen M, Asplund K. *Two variants of extracellular-superoxide dismutase: relationship to cardiovascular risk factors in an unselected middle-aged population.* J Intern Med. 1997;242:5-14.

66. Sandstrom J, Nilsson P, Karlsson K, Marklund SL. *10-fold increase in human plasma extracellular superoxide dismutase content caused by a mutation in heparin-binding domain.* J Biol Chem. 1994;269:19163-6.

67. Yamada H, Yamada Y, Adachi T, Goto H, Ogasawara N, Futenma A, et al. *Molecular analysis of extra-cellular-superoxide dismutase gene associated with high level in serum.* Jpn J Hum Genet. 1995;40:177-84.

68. Yamada H, Yamada Y, Adachi T, Goto H, Ogasawara N, Futenma A, et al. *Polymorphism of extracellular superoxide dismutase (EC-SOD) gene: relation to the mutation responsible for high EC-SOD level in serum.* Jpn J Hum Genet. 1997;42:353-6.

69. Hubackova M, Vaclavikova R, Ehrlichova M, Mrhalova M, Kodet R, Kubackova K, et al. *Association of superoxide dismutases and NAD(P)H quinone oxidoreductases with prognosis of patients with breast carcinomas.* Int J Cancer. 2012;130:338-48.

70. Marklund SL, Holme E, Hellner L. *Superoxide dismutase in extracellular fluids.* Clin Chim Acta. 1982;126:41-51.

71. Adachi T, Wang J, Wang XL. *Age-related change of plasma extracellular-superoxide dismutase.* Clin Chim Acta. 2000;290:169-78.

72. Teoh ML, Fitzgerald MP, Oberley LW, Domann FE. *Overexpression of extracellular superoxide dismutase attenuates heparanase expression and inhibits breast carcinoma cell growth and invasion.* Cancer Res. 2009;69:6355-63.

4.4.3. CATALASE

Catalase is a ubiquitous enzyme that reduces H_2O_2 to H_2O in all organisms (1, 2). In contrast to peroxidases, catalase uses only H_2O_2 as a substrate and functions when the H_2O_2 concentration is above physiological levels (>10^{-6} M) as can happen in oxidative bursts characteristic of stress responses (Figure 4.4.8.). By converting H_2O_2 to water and molecular oxygen catalase protects cells from oxidative stress.

$$2H_2O_2 \xrightarrow{\text{catalase}} 2H_2O + O_2$$

Figure 4.4.8. Catalase-mediated reaction

Catalase is a homotetramer with each subunit containing one heme group and binding one NADPH (3). Each 60 kDa-monomer is composed of 526 amino acid residues and the four subunits are arranged in a tetrahedrally symmetric ellipsoid tetramer (4). The heme iron is in the ferric state and the high-spin Fe^{3+} reacts rapidly with H_2O_2; 1 mol of catalase can dispose of over 40,000 mol H_2O_2 per second at 0ºC. The reaction is first order and depends strictly on the concentration of H_2O_2 (1).

The catalase gene at 11p13 is 34 kb long and contains 13 exons (5) (Figure 4.4.9.). There are no common polymorphisms in the coding region of the human catalase gene (6). However, a C/T polymorphism in the promoter region (rs1001179; –262 bp from the transcription site) is common with the variant T allele present in 28% of a Swedish population (6). Reporter constructs showed higher transcriptional activity of the T variant. Catalase levels determined by Western immunoblot were significantly higher in red blood cells from donors containing the T allele compared to donors homozygous for the C allele. Surprisingly, enzyme activity studies of erythrocytes yielded opposite results to the Western data. Higher catalase activity levels were associated with the CC than the CT or TT genotypes (7). The differences in catalase activity were affected by dietary factors, being most pronounced among those individuals in the highest tertiles of consumption of fruits and vegetables (8). The association between catalase genotype and enzyme activity was only observed among Caucasians (p < 0.0001). There was no association among African Americans (p = 0.91) suggesting a complex relationship between genotype and phenotypic activity.

Clinical Studies. The highest levels of catalase are found in the liver, kidney, and erythrocytes. In tissues such as liver and breast, catalase is present predominantly in peroxisomes, whereas in mature erythrocytes catalase is found free in the cytosol. Using H_2O_2 as substrate, catalase activity was significantly lower in homogenates of 36 breast cancers than in the adjacent normal tissue (9). Estradiol treatment of normal human breast epithelial cells in culture resulted in decreased catalase activity (10). In breast cancer cells E_2 treatment caused a decrease in enzyme activity in ER-positive but not ER-negative cell lines (11). An immunohistochemical study of 357 invasive breast cancers revealed catalase expression in only 51% (12).

Rare mutations of the catalase gene are associated with acatalasemia, an autosomal recessive trait characterized by erythrocyte catalase levels 0.2–4% of normal (13). Affected individuals did not show any signs of tissue damage with the exception of occasional oral lesions. Mutant CH3 mice lacking the catalase gene showed an increased incidence of spontaneous mammary cancers (14). Vitamin E supplementation prevented tumor formation in half of the animals.

Several studies have examined the association of the CAT -262 C/T (rs1001179) polymorphism with breast cancer risk. The Long Island Breast Cancer Study Project (1008 cases, 1056 controls) found a 17% reduction in breast cancer risk for the -262 CC genotype) compared with having at least one variant T allele (OR 0.83; 95% CI 0.69–1.00) (7). This association was more pronounced in women with high fruit intake who did not use vitamin

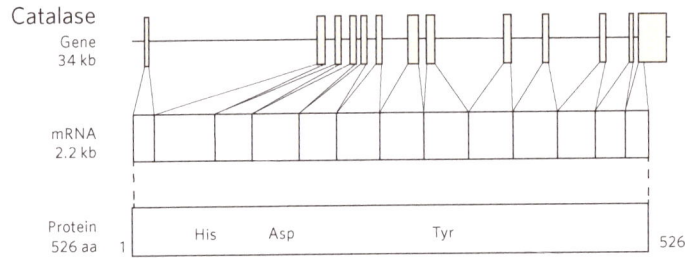

Figure 4.4.9. Overview of catalase gene, mRNA, and protein.
The CAT gene at 11p13 is 34 kb long and split into 13 exons. The introns range in size from 400 bp to 10.5 kb, the latter separating exons 1 and 2. The promoter is GC-rich, lacks a TATA box, and has several putative Sp1 binding sites. The 2.4 kb mRNA is translated into a 60 kDa protein consisting of 526 amino acids. The enzyme is a 240 kDa homotetramer with each subunit containing one heme group held in place by amino acid residues 74His, 147Asp, and 357Tyr.

supplements (OR 0.59; 95% CI 0.38–0.89). A study of the Western New York Exposures and Breast Cancer cohort (WEB; 616 cases, 1,082 controls) found no association of the CAT genotype with breast cancer risk but observed an effect modification for hormone replacement therapy (15). Ever-use of hormone replacement therapy by itself was associated with increased risk (OR 1.39; 95% CI 1.11 – 1.75). The increase was more pronounced among women with variant CT or TT genotypes (OR 1.88; 95% CI 1.29 - 2.75) than among those with CC (OR 1.15; 95% CI 0.86 - 1.54). Similarly, risk associated with ≥5 years of HRT use was greater among those with at least one variant T allele (OR 2.32; 95% CI 1.50 – 3.59). Increased risk was limited to ER-positive tumors. A study from the East Anglian region of the UK (4474 cases, 4580 controls) found no evidence for an association of the -262 polymorphism with breast cancer risk (16). However, another polymorphism (27168G/A; rs511895) showed a borderline association (p = 0.06). A follow-up analysis using a multilocus approach with polymorphisms in other anti-oxidant enzymes found no evidence of an association of the CAT genotype with breast cancer risk (17).

References

1. Chance B, Sies H, Boveris A. *Hydroperoxide metabolism in mammalian organs.* Physiol Rev. 1979;59:527-605.

2. Eaton JW, Ma M. *Acatalasemia.* In: Scriver CR, Beaudet AL, Sly WS, Valle D, editors. The Metabolic and Molecular Bases of Inherited Disease. New York: McGraw-Hill; 1995. p. 2371-83.

3. Kirkman HN, Gaetani GF. Catalase: *A tetrameric enzyme with four tightly bound molecules of NADPH.* Proc Natl Acad Sci USA. 1984;81:4343-7.

4. Reid TJ, Murthy MR, Sicignano A, Tanaka N, Musick WD, Rossman MG. *Structure and heme environment of beef liver catalase at 2.5 Å resolution.* Proc Natl Acad Sci USA. 1981;78:4767-71.

5. Quan F, Korneluk RG, Tropak MB, Gravel RA. *Isolation and characterization of the human catalase gene.* Nucl Acids Res. 1986;14:5321-35.

6. Forsberg L, Lyrenas L, De Faire U, Morgenstern R. *A common functional C-T substitution polymorphism in the promoter region of the human catalase gene influences transcription factor binding, reporter gene transcription and is correlated to blood catalase levels.* Free Radic Biol Med. 2001;30:500-5.

7. Ahn J, Gammon MD, Santella RM, Gaudet MM, Britton JA, Teitelbaum SL, et al. *Associations between breast cancer risk and the catalase genotype, fruit and vegetable consumption, and supplement use.* Am J Epidemiol. 2005;162:943-52.

8. Ahn J, Nowell S, McCann SE, Yu J, Carter L, Lang NP, et al. *Associations between catalase phenotype and genotype: Modification by epidemiologic factors.* Cancer Epidemiol Biomarkers Prev. 2006;15:1217-22.

9. Bouhtoury FE, Keller JM, Colin S, Parache RM, Dauca M. *Peroxisomal enzymes in normal and tumoral human breast.* J Pathol. 1992;166:27-35.

10. Dabrosin C, Hammar M, Ollinger K. *Impact of oestradiol and progesterone on antioxidant activity in normal human breast epithelial cells in culture.* Free radical research. 1998;28:241-9.

11. Mobley JA, Brueggemeier RW. *Estrogen receptor-mediated regulation of oxidative stress and DNA damage in breast cancer.* Carcinogenesis. 2004;25:3-9.

12. Karihtala P, Kinnula VL, Soini Y. *Antioxidative response for nitric oxide production in breast carcinoma.* Oncol Rep. 2004;12:755-9.

13. Ogata M. *Acatalasemia.* Hum Genet. 1991;86:331-40.

14. Ishii K, Zhen LX, Wang D, Funamori Y, Ogawa K, Taketa K. *Prevention of mammary tumorigenesis in acatalasemic mice by vitamin E supplementation.* Jpn J Cancer Res. 1996;87:680-4.

15. Quick SK, Shields PG, Nie J, Platek ME, McCann SE, Hutson AD, et al. *Effect modification by catalase genotype suggests a role for oxidative stress in the association of hormone replacement therapy with postmenopausal breast cancer risk.* Cancer Epidemiol Biomarkers Prev. 2008;17:1082-7.

16. Cebrian A, Pharoah PD, Ahmed S, Smith PL, Luccarini C, Luben R, et al. *Tagging single-nucleotide polymorphisms in antioxidant defense enzymes and susceptibility to breast cancer.* Cancer Res. 2006;66:1225-33.

17. Oestergaard MZ, Tyrer J, Cebrian A, Shah M, Dunning AM, Ponder BA, et al. *Interactions between genes involved in the antioxidant defence system and breast cancer risk.* Br J Cancer. 2006;95:525-31.

4.4.4. GLUTATHIONE PERIOXIDASES

Glutathione peroxidases (GPXs) catalyze the reduction of H_2O_2 as well as organic hydroperoxides (ROOH) to water and alcohols, respectively, by oxidizing glutathione (GSH) to glutathione disulfide (GSSG) as shown in Figure 4.4.10. (1-3). In contrast to catalase, GPXs respond to small variations in H_2O_2 or ROOH concentration. There are four GPXs all of which contain selenium in their active sites. There are also selenium-independent GPX activities in mammalian cells, in particular the glutathione-S-transferases, whose action is directed mainly towards organic hydroperoxides rather than against H_2O_2 (4). Organic hydroperoxides are physiological (e.g., fatty acid, cholesterol, and phospholipid hydroperoxides) or synthetic (e.g., tert-butyl and cumene hydroperoxide). GPXs have other functions besides their peroxidase activity. They can serve as modulators of inflammatory and innate immune responses and are involved as intermediates in signal transduction pathways (5, 6).

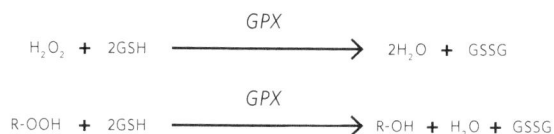

Figure 4.4.10. GPX-mediated reactions

The GPX family of proteins is divided into four members (GPX1 to GPX4) based on their primary sequence, substrate specificity, and tissue expression (2) (Table 4.4.2.). Each GPX gene has been mapped to a different chromosome (14, 15). Gene sequence comparisons and dendrogram analyses suggest that the members of the GPX family are derived from an ancestral gene through duplication and intron-exon shuffling (11). The gene duplication events were followed by the establishment of strict tissue-specific controls of GPX gene expression. Although the specificity for the hydroperoxide substrates is generally low, there are distinct differences between the GPX isomers, which may be explained by modified accessibility to the respective catalytic sites (11).

The trace metal selenium is a catalytically active constituent of several enzymes, serving as electron donor in GPXs,

thioredoxin reductases, and iodothyronine deiodinases (16-18). All these enzymes contain selenium in form of a single selenocysteine residue, which becomes incorporated in an unusual manner during protein synthesis. Ordinarily, the genetic code TGA serves as one of three translation termination signals or stop codon, but it can be recoded to direct the incorporation of selenocysteine. This occurs via a complex mechanism that involves several factors (19-23). (i) TGA appears to be the universal codon for selenocysteine, which may be considered as the 21st amino acid. (ii) The UGA codon in the corresponding mRNA is recognized by the anticodon of a specific tRNA(ser)sec carrying the selenocystine residue. Unlike the usual protein synthesis, the tRNA is not charged directly with selenocysteine but first with serine. Seryl-tTNA is transformed into selenocysteinyl-tRNA(ser)sec by means of selenophosphate, which is synthesized from H_2Se and ATP. (iii) Finally, translation of selenoprotein mRNA depends upon a 60-to-100-nucleotide sequence element in the 3'-untranslated region. This selenocysteine insertion sequence (SECIS) forms a stem-loop secondary structure that is recognized by specific translation factors, such as the SECIS-binding protein 2, SBP2 (23-26). Overall, selenocysteine incorporation appears to be an inefficient process. Aminoglycosides decrease activity by interfering with selenocysteine incorporation (27).

Structural analysis of GPX1 showed that the selenium in form of selenocysteine (residue 47) is located in the interior of the protein, forming hydrogen bridges with glutamine (residue 82) and tryptophan (residue 160) (28). This so-called catalytic triad of selenocysteine, glutamine, and tryptophan is common to all GPXs (29). In the catalytic cycle, the selenium shuttles between the selenol anion (GPX-Se⁻) and the selenic acid (GPX-SeOH) (Figure 4.4.11.). In the ground state of the enzyme, the selenium is in its reduced form facilitating nucleophilic

Table 4.4.2. Overview of glutathione peroxidases (1, 2, 6-13)

	GPX1	GPX2	GPX3	GPX4
General GPX name	cellular	gastrointestinal	plasma	phospholipid hydroperoxide
Tissue expression	ubiquitous	gastrointestinal	kidney	brain, testis
Cellular location	cytosol	cytosol	extracellular	membrane-bound, mitochondria, nuclei
Chromosome locus	3p21.3	14q24.1	5q33.1	19p13.3
KDa	21	22	22.5	19
Exons	2	2	5	8
Polymerization	tetramer	tetramer	tetramer	monomer
Polymorphisms	GCG repeat: 5-7 Ala -1040 G/A; rs3448 Pro198(200)Leu;rs1050450			3' SECIS region

SECIS, selenocysteine insertion sequence.

attack on the peroxide. Upon collision with the peroxide, the selenium changes its oxidation state from –2 to 0 (11). It is regenerated to the ground state by reducing two molecules of GSH (29, 30). In GPX1, the catalytic triad is surrounded by four arginines to position the electron donor GSH. The correct orientation is achieved by electrostatic interaction of the arginine residues with each of the three amino acids in GSH, i.e., glycine, cysteine, and γ-glutamine.

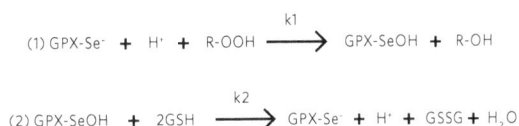

$$(1)\ GPX\text{-}Se^- \ +\ H^\cdot \ +\ R\text{-}OOH \ \xrightarrow{k1}\ GPX\text{-}SeOH \ +\ R\text{-}OH$$

$$(2)\ GPX\text{-}SeOH \ +\ 2GSH \ \xrightarrow{k2}\ GPX\text{-}Se^- \ +\ H^\cdot \ +\ GSSG + H_2O$$

Figure 4.4.11. Steps in GPX-mediated reaction

GPX1, the predominant GPX isoform, is expressed in all cells. Selenium supplementation is known to increase erythrocyte GPX1 activity in populations with a low intake of the trace element (<40 μg/day) (31). Intake of fruit and vegetables was also shown to significantly increase GPX1 activity in red cells (32). GPX1 can catalyze the removal of inorganic and lipid hydroperoxides. Interestingly, GPX1-/- mice developed normally but became more sensitive to oxidative stress (33, 34).

Exon 1 contains a variable number of GCG triplet nucleotides encoding alanine in a polyalanine tract (35). A study of two different normal populations revealed three GPX1 alleles with five, six or seven alanines in the polyalanine sequence. Each allele occurred frequently and about 70% of genotypes were heterozygous. In Caucasians, the GPX1*Ala6 allele was linked to three substitutions, Pro198Leu, +2 C/T (rs1987628), and –592A/G (35). The Pro198Leu SNP (rs1050450; now designated Pro200Leu) is in complete linkage disequilibrium with rs1987628 (13). Chinese do not possess the Pro198Leu SNP, consistent with the finding that Asians have fewer GPX1 haplotypes than African Americans, Caucasians, and Hispanics (8, 36). Three larger studies showed lower erythrocyte GPX1 activity for the 198Leu allele, whereas two smaller studies found no association (13, 37, 38). Expression studies have shown that GPX1 activity in cells with the Leu allele was less responsive to increasing selenium supplementation compared with the Pro allele (39). It remained undetermined how the proline versus leucine amino acid substitution could influence selenium stimulation of GPX1 activity.

GPX2 is expressed predominantly in gastrointestinal epithelial cells with similar substrate specificities as GPX1 (7). Double knockout mice deficient in GPX1 and GPX2 developed ileocolitis followed by ileocolonic tumors at six to nine months of age (40). The tumor DNA showed a threefold increase in deletions at mononucleotide repeats, a signature mutation associated with oxidative stress.

GPX3 is the only extracellular GPX. The enzyme is secreted into the circulation and capable of reducing lipid hydroperoxides, such as free fatty acid hydroperoxides (9, 10, 41). It is preferentially expressed in the kidney and at lower levels in a wide array of tissues (42). Hypoxia increases GPX3 expression by stimulating transcription via the hypoxia-inducible factor-1 (HIF-1) binding site in the GPX3 promoter (43, 44). The most abundant Se-containing protein in the circulation is selenoprotein P, which contains ten seleno-cysteine residues (18, 23, 30). Selenoprotein P functions as a Se delivery molecule and perhaps as an extracellular antioxidant with GPX-like activity.

The **GPX4** gene at 19p13.3 is composed of eight exons. Distinct promoters direct transcription into mRNAs of different lengths, which lead to three GPX4 isoforms differing in their N-terminal extensions and cellular locations: cytosolic, mitochondrial, and nuclear (45). A common T/C polymorphism has been identified in the 3' SECIS, which potentially affects the translational incorporation of selenocysteine and GPX4 function (12). In contrast to the other GPX family members, GPX4 is a monomer. The monomeric nature may make the active site more freely accessible, which would explain the ability of GPX4 to attack more complex hydroperoxy lipids including those integrated into lipid layers (45, 46). In a concerted action with vitamin E, GPX4 is capable of breaking the chain reactions that lead to oxidative lipid destruction (11). Specifically, vitamin E reduces lipid peroxy radicals to hydroperoxides, which are further reduced by GPX4 with the consumption of GSH. This prevents decomposition of the hydroperoxides into alkoxy radicals and reinitiation of free radical chains. Since GPX4 preferentially catalyzes the peroxidation of phospholipid hydroperoxides, its principal function appears to be the protection of cell membranes from oxidative damage. The highest level of GPX expression occurs in brain and testis where the enzyme is essential in sperm maturation (1, 47). Interestingly, a fifth enzyme, GPX5, is expressed only in the epididymis. GPX5 does not contain selenium (6).

GPX Expression in Breast Tissue and Cell Lines. Using H_2O_2 or cumene hydroperoxide as substrates, total GPX activity was twofold higher in cytosol preparations from breast cancers than normal tissues (48, 49). Due to the co-expression of several isoforms in breast tissue it has been difficult to demonstrate the importance of each individual GPX. **GPX1** expression is generally present in ER-negative but absent in ER-positive breast cancer cell lines (50). However, high levels were observed in ER-positive MDA-MB-134 cells. MCF-7 cells, which lack GPX1 expression, were engineered to exclusively express GPX1 allelic variants representing a combination of 198Leu or 198Pro with Ala5 or Ala7 repeats (51). Selenium addition induced a significantly higher GPX1 activity for the 198Leu/Ala5 construct than for the 198Pro/Ala5, 198Leu/Ala7 or 198Pro/Ala7 variants suggesting that intrinsic and extrinsic factors cooperate to determine the enzyme activity level. The expression of **GPX2** was commonly elevated in mammary carcinomas of Hras128 rats, which had been induced by three different carcinogens: N-Methyl-N-nitro-surea (MNU), 7,12-dimethyl benz[a]anthracene (DMBA), and 2-amino-1-methyl-6-phenylimidazo[4,5-b]pyridine (PhIP) (52). Forced suppression of GPX2 expression by siRNA resulted in growth inhibition of both rat and human breast cancer cell lines. In MCF-7 cells, retinoic acid treatment induced expression of the GPX2 but not the GPX1 gene (53). In human breast tissue, GPX2 mRNA was detected in normal and malignant epithelial cells (50, 53). There was an inverse correlation between histologic grade and intensity of immunohistochemical GPX2 staining (52). **GPX3** activity has been detected in human milk consistent with the presence of GPX3 mRNA in normal breast tissue (54, 55). GPX3 mRNA was also found in MDA-MB-231, SK-BR-3, and T47D but not MCF-7 cell lines. **GPX4** activity was detectable at low levels in ER-negative and ER-positive cell lines (46, 50).

Clinical Studies. Epidemiological studies and nutritional trials have shown that dietary selenium supplementation can prevent lung, colon, and prostate cancer but there is no convincing evidence for a reduction in the incidence of breast cancer (56). Nevertheless, selenoenzymes such as GPXs and thioredoxin reductases (Section 4.4.5.) may play a role in mammary carcinogenesis. GPX activity has been measured in erythrocytes and shown to be lower in women with breast cancer compared with healthy controls (57, 58). When subdivided into quartiles on the basis of red cell GPX activity, a prospective cohort study of premenopausal women (377 cases, 377 controls) also showed that lower GPX activity was associated with higher risk of breast cancer (38).

Cytogenetic studies have shown that the allele loss of chromosome 3p is a common event in the pathogenesis of breast cancer. The 3p21.3 region containing the GPX1 locus had the highest frequency (36%) of loss of heterozygosity (LOH) in preneoplastic breast epithelium (59). For unknown reasons the LOH in breast cancer was associated with an overrepresentation of the variant 198Leu allele compared with the wild-type 198Pro allele (OR 1.91; 95% CI 1.02–3.58; p < 0.05) (39). Whether this reflects higher risk for developing breast cancer by carrying the variant allele or LOH during tumor development could not be determined. A case-control study did not examine the 3p LOH but the Pro198Leu (rs1050450) polymorphism and number of alanine repeats (60). There was no association between breast cancer risk and the SNP in codon 198. However,

premenopausal women with the GPX1*Ala5 repeat allele exhibited an increased risk (OR 1.55; 95% CI 1.04–2.30). A study of postmenopausal women (377 cases, 377 controls) examined only the Pro198Leu polymorphism and found that carriers of the Leu allele were at 1.43-fold higher risk of breast cancer compared with non-carriers (38). However, no risk association was observed in either the Long Island Breast Cancer Study Project (1038 cases, 1088 controls) or the Nurses' Health Study (1293 cases, 1695 controls) (61, 62). However, The Nurses' Health Study found an interaction between the GPX1 Pro198Leu and SOD2 Ala16Val (rs1799725) polymorphisms (63). While neither allele alone showed any change in breast cancer risk, an increase in risk (OR 1.87, 95% CI 1.09 – 3.19) was observed in individuals who carry both the GPX1 198Leu/Leu and SOD2 16 Ala/Ala genotypes. A large UK study (4474 cases, 4580 controls) also found no evidence for an association of the Pro198Leu SNP or another GPX1 polymorphism, -1040G/A (rs3448), with breast cancer risk (64, 65). Genetic variants of GPX4 have been identified, such as the T/C polymorphism in the 3' SECIS, but their functional significance for breast cancer is unknown (12, 47).

References

1. Brigelius-Flohe R. *Tissue-specific functions of individual glutathione peroxidases.* Free Radic Biol Med. 1999;27:951-65.

2. Arthur JR. *The glutathione peroxidases.* Cell Mol Life Sci. 2000;57:1825-35.

3. Townsend DM, Tew KD, Tapiero H. *The importance of glutathione in human disease.* Biomed Pharmacother. 2003;57:145-55.

4. Hayes JD, McLellan LI. *Glutathione and glutathione-dependent enzymes represent a co-ordinately regulated defence against oxidative stress.* Free Rad Res. 1999;31:273-300.

5. Chu FF, Esworthy RS, Doroshow JH. *Role of se-dependent glutathoine peroxidases in gastrointestinal inflammation and cancer.* Free Radic Biol Med. 2004;36:1481-95.

6. Drevet JR. *The antioxidant glutathione peroxidase family and spermatozoa: a complex story.* Mol Cell Endocrinol. 2006;250:70-9.

7. Chu FF, Doroshow JH, Esworthy RS. *Expression, characterization, and tissue distribution of a new cellular selenium-dependent glutathione peroxidase, GSHPx-GI.* J Biol Chem. 1993;268:2571-6.

8. Foster CB, Aswath K, Chanock SJ, McKay HF, Peters U. *Polymorphism analysis of six selenoprotein genes:* support for a selective sweep at the glutathione peroxidase I locus (3p21) in asian populations. BMC Genetics. 2006;7.

9. Maddipati KR, Marnett LJ. *Characterization of the major hydroperoxide-reducing activity of human plasma.* J Biol Chem. 1987;262:17398-403.

10. Takahashi K, Avissar N, Whitin J, Cohen H. *Purification and characterization of human plasma glutathione peroxidase: a selenoglycoprotein distinct from the known cellular enzyme.* Arch Biochem Biophys. 1987;256:677-86.

11. Ursini F, Maiorino M, Brigelius-Flohé R, Aumann KD, Roveri A, Schomburg D, et al. *Diversity of glutathione peroxidases.* Methods Enzymol. 1995;252:38-53.

12. Villette S, Kyle JA, Brown KM, Pickard K, Milne JS, Nicol F, et al. *A novel single nucleotide polymorphism in the 3' untranslated region of human glutathione peroxidase 4 influences lipoxygenase metabolism.* Blood Cells Mol Dis. 2002;29:174-8.

13. Takata Y, King IB, Lampe JW, Burk RF, Hill KE, Santella RM, et al. *Genetic variation in GPX1 is associated with GPX1 activity in a comprehensive analysis of genetic variations in selenoenzyme genes and their activity and oxidative stress in humans.* J Nutrition. 2012;142:419-26.

14. Chu FF. *The human glutathione peroxidase genes GPX2, GPX3, and GPX4 map to chormosomes 14, 5, and 19, respectively.* Cytogenet Cell Genet. 1994;66:96-8.

15. Chu FF, De Silva HAR, Esworthy RS, Boteva KK, Walters CE, Roses A, et al. *Polymorphism and chromosomal localization of the GI-form of human glutathione peroxidase (GPX2) on 14q24.1 by in situ hybridization.* Genomics. 1996;32:272-6.

16. Stadtman TC. *Selenium biochemistry: mammalian selenoenzymes.* Ann NY Acad Sci. 2000;899:399-402.

17. Arteel GE, Sies H. *The biochemistry of selenium and the glutathione system.* Environ Toxicol Pharmacol. 2001;10:153-8.

18. Behne D, Kyriakopoulos A. *Mammalian selenium-containing proteins.* Annu Rev Nutr. 2001;21:453-73.

19. Chambers I, Frampton J, Goldfarb P, Affara N, McBain W, Harrison PR. *The structure of the mouse glutathione peroxidase gene: the selenocysteine in the active site is encoded by the 'termination' codon, TGA.* EMBO J. 1986;5:1221-7.

20. Berry MJ, Banu L, Harney JW, Larsen PR. *Functional characterization of the eukaryotic SECIS elements which direct selenocysteine insertion at UGA codons.* EMBO J. 1993;12:3315-22.

21. Shen Q, Chu FF, Newburger PE. *Sequences in the 3'-untranslated region of the human cellular glutathione peroxidase gene are necessary and sufficient for selenocysteine incorporation at the UGA codon.* J Biol Chem. 1993;268:11463-9.

22. Nasim MT, Jaenecke S, Belduz A, Kollmus H, Flohé L, McCarthy JEG. *Eukaryotic selenocysteine incorporation follows a nonprocessive mechanism that competes with translational termination.* J Biol Chem. 2000;275:14846-52.

23. Copeland PR. *Regulation of gene expression by stop codon recoding: selenocysteine.* Gene. 2003;312:17-25.

24. Shen Q, Wu R, Leonard JL, Newburger PE. *Identification and molecular cloning of a human selenocysteine insertion sequence-binding protein.* J Biol Chem. 1998;273:5443-6.

25. Lescure A, Gautheret D, Carbon P, Krol A. *Novel selenoproteins identified in silico and in vivo by using a conserved RNA structural motif.* J Biol Chem. 1999;274:38147-54.

26. Copeland PR, Fletcher JE, Carlson BA, Hatfield DL, Driscoll DM. *A novel RNA binding protein, SBP2, is required for the translation of mammalian selenoprotein mRNAs.* EMBO J. 2000;19:306-14.

27. Handy DE, Hang G, Scolaro J, Metes N, Razaq N, Yang Y, et al. *Aminoglycosides decrease glutathione peroxidase-1 activity by interfering with selenocysteine incorporation.* J Biol Chem. 2006;281:3382-8.

28. Epp O, Ladenstein R, Wendel A. *The refined structure of the selenoenzyme glutathione peroxidase at 0.2-nm resolution.* Eur J Biochem. 1983;133:51-69.

29. Maiorino M, Aumann KD, Brigelius-Flohe R, Doria D, van den Heuvel J, McCarthy J, et al. *Probing the presumed catalytic triad of selenium-containing peroxidases by mutational analysis of phospholipid hydroperoxide glutathione peroxidase (PHGPx).* Biol Chem Hoppe Seyler. 1995;376:651-60.

30. Takebe G, Yarimizu J, Saito Y, Hayashi T, Nakamura H, Yodoi J, et al. *A comparative study on the hydroperoxide and thiol specificity of the glutathione peroxidase family and selenoprotein P.* J Biol Chem. 2002;277:41254-8.

31. Neve J. *Human selenium supplementation as assessed by changes in blood selenium concentration and glutathione peroxidase activity.* J Trace Elements Med Biol. 1995;9:65-73.

32. Dragsted LO, Pederson A, Hermetter A, Basu S, Hansen M, Haren GR, et al. *The 6-a-day study: effects of fruit and vegetables on markers of oxidative stress and antioxidative defense in healthy nonsmokers.* Am J Clin Nutr. 2004;79:1060-72.

33. Ho YS, Magnenat JL, Bronson RT, Cao J, Gargano M, Sugawara M, et al. *Mice deficient in cellular glutathione peroxidase develop normally and show no increased sensitivity to hyperoxia.* J Biol Chem. 1997;272:16644-51.

34. de Haan JB, Bladier C, Griffiths P, Kelner M, O'Shea RD, Cheung NS, et al. *Mice with a homozygous null mutation for the most abundant glutathione peroxidase, Gpx1, show increased susceptibility to the oxidative stress-inducing agents paraquat and hydrogen peroxide.* J Biol Chem. 1998;273:22528-36.

35. Moscow JA, Schmidt L, Ingram DT, Gnarra J, Johnson B, Cowan KH. *Loss of heterozygosity of the human cytosolic glutathione peroxidase I gene in lung cancer.* Carcinogenesis.

1994;15:2769-73.

36. Ratnasinghe D, Tangrea JA, Andersen MR, Barrett MJ, Virtamo J, Taylor PR, et al. *Glutathione peroxidase codon 198 polymorphism variant increases lung cancer risk.* Cancer Res. 2000;60:6381-3.

37. Forsberg L, de Faire U, Marklund SL, Andersson PM, Stegmayr B, Morgenstern R. *Phenotype determination of a common pro-leu polymorphism in human glutathione peroxidase 1.* Blood Cells Mol Dis. 2000;26:423-6.

38. Ravn-Haren G, Olsen A, Tjønneland A, Dragsted LO, Nexø BA, Wallin H, et al. *Associations between GPX1 Pro198Leu polymorphism, erythrocyte GPA activity, alcohol consumption and breast cancer risk in a prospective cohort study.* Carcinogenesis. 2006;27:820-5.

39. Hu YJ, Diamond AM. *Role of glutathione peroxidase 1 in breast cancer: loss of heterozygosity and allelic differences in the response to selenium.* Cancer Res. 2003;63:3347-51.

40. Lee DH, Esworthy RS, Chu C, Pfeifer GP, Chu FF. *Mutation accumulation in the intestine and colon of mice deficient in two intracellular glutathione peroxidases.* Cancer Res.2006;66:9845-51.

41. Esworthy RS, Chu FF, Geiger P, Girotti AW, Doroshow JH. *Reactivity of plasma glutathione peroxidase with hydroperoxide substrates and glutathione.* Arch Biochem Biophys. 1993;307:29-34.

42. Maser RL, Magenheimer BS, Calvet JP. *Mouse plasma glutathione peroxidase.* J Biol Chem. 1994;269:27066-73.

43. Bierl C, Voetsch B, Jin RC, Handy DE, Loscalzo J. *Determinants of human plasma glutathione peroxidase (GPx-3) expression.* J Biol Chem. 2004;279:26839-45.

44. Saito Y, Hayashi T, Tanaka A, Watanabe Y, Suzuki M, Saito E, et al. *Selenoprotein P in human plasma as an extracellular phospholipid hydroperoxide glutathione peroxidase.* J Biol Chem. 1999;274:2866-71.

45. Maiorino M, Scapin M, Ursini F, Biasolo M, Bosello V, Flohé L. *Distinct promoters determine alternative transcription of gpx-4 into phospholipid-hydroperoxide glutathione peroxidase variants.* J Biol Chem. 2003;278:34286-90.

46. Maiorino M, Chu FF, Ursini F, Davies KJA, Doroshow JH, Esworthy RS. *Phospholipid*

hydroperoxide glutathione peroxidase is the 18-kDa selenoprotein expressed in human tumor cell lines. J Biol Chem. 1991;266:7728-32.

47. Maiorino M, Bosello V, Ursini F, Foresta C, Garolla A, Scapin M, et al. *Genetic variations of gpx-4 and male infertility in humans.* Biol Reprod. 2003;68:1134-41.

48. Di Ilio C, Sacchetta P, Del Boccio G, La Rovere G, Federici G. *Glutathione peroxidase, glutathione s-transferase and glutathione reductase activities in normal and neoplastic human breast tissue.* Cancer Lett. 1985;29:37-42.

49. Howie AF, Forrester LM, Glancey MJ, Schlager JJ, Powis G, Beckett GJ, et al. *Glutathione S-transferase and glutathione peroxidase expression in normal tumour human tissues.* Carcinogenesis. 1990;11:451-8.

50. Esworthy RS, Baker MA, Chu FF. *Expression of selenium-dependent glutathione peroxidase in human breast tumor cell lines.* Cancer Res. 1995;55:957-62.

51. Zhuo P, Goldberg M, Herman L, Lee BS, Wang H, Brown RL, et al. *Molecular consequences of genetic variations in the glutathione peroxidase 1 selenoenzyme.* Cancer Res. 2009;69:8183-90.

52. Naiki-Ito A, Asamoto M, Hokaiwado N, Takahashi S, Yamashita H, Tsuda H, et al. *Gpx2 is an overexpressed gene in rat breast cancers induced by three different chemical carcinogens.* Cancer Res.200;67:11353-8.

53. Chu FF, Esworthy RS, Lee L, Wilczynski S. *Retinoic acid induces Gpx2 gene expression in MCF-7 human breast cancer cells.* J Nutrition. 1999;129:1846-54.

54. Avissar N, Slemmon JR, Palmer IS, Cohen HJ. *Partial sequence of human plasma glutathione peroxidase and immunologic identification of milk glutathione peroxidase as the plasma enzyme.* J Nutrition. 1991;121:1243-9.

55. Chu FF, Esworthy RS, Doroshow JH, Doan K, Liu XF. *Expression of plasma glutathione peroxidase in human liver in addition to kidney, heart, lung, and breast in humans and rodents.* Blood. 1992;79:3233-8.

56. Rayman MP. *Selenium in cancer prevention: a review of the evidence and mechanism of action.* Proc of Nutrition Society. 2005;64:527-42.

57. Abiaka C, Al-Awadi F, Al-Sayer H, Gulshan S, Behbehani A, Farghally M. *Activities of erythrocyte antioxidant enzymes* in cancer patients. J Clin Lab Anal. 2002;16:167-71.

58. Kumaraguruparan R, Subapriya R, Kabalimoorthy J, Nagini S. *Antioxidant profile in the circulation of patients with fibroadenoma and adenocarcinoma of the breast.* Clin Biochem. 2002;35:275-9.

59. Maitra A, Wistuba II, Washington C, Virmani AK, Ashfaq R, Milchgrub S, et al. *High-resolution chromosome 3p allelotyping of breast carcinomas and precursor lesions demonstrates frequent loss of heterozygosity and a discontinuous pattern of allele loss.* Am J Pathol. 2001;159:119-30.

60. Knight JA, Onay UV, Wells S, Li H, Shi EJ, Andrulis IL, et al. *Genetic variants of GPX1 and SOD2 and breast cancer risk at the Ontario site of the breast cancer family registry.* Cancer Epidemiol Biomarkers Prev. 2004;13:146-9.

61. Ahn J, Gammon MD, Santella RM, Gaudet MM, Britton JA, Teitelbaum SL, et al. *No association between glutathione peroxidase*

Pro198Leu polymorphism and breast cancer risk. Cancer Epiderniol Biomarkers Prev. 2005;14:2459-61.

62. Cox DG, Hankinson SE, Kraft P, Hunter DJ. *No association between GPX1 Pro198Leu and breast cancer risk.* Cancer Epidemiol Biomarkers Prev. 2004;13:1821-2.

63. Cox DG, Tamimi RM, Hunter DJ. *Gene x Gene interaction between MnSOD and GPX-1 and breast cancer risk: a nested case-control study.* BMC Cancer. 2006;6:217.

64. Cebrian A, Pharoah PD, Ahmed S, Smith PL, Luccarini C, Luben R, et al. *Tagging single-nucleotide polymorphisms in antioxidant defense enzymes and susceptibility to breast cancer.* Cancer Res. 2006;66:1225-33.

65. Oestergaard MZ, Tyrer J, Cebrian A, Shah M, Dunning AM, Ponder BA, et al. *Interactions between genes involved in the antioxidant defence system and breast cancer risk.* Br J Cancer. 2006;95:525-31.

cyclic reduction-oxidation of the S-S group.

The two main intracellular thiols, GSH and TXN, exist in equilibria of oxidized and reduced forms, i.e., GSSG $\leftrightarrow$ GSH and TXN-S$_2$ $\leftrightarrow$ TXN-(SH)$_2$. Two structurally related enzymes, GSH reductase and TXN reductase use NADPH as a source of reducing equivalents to ensure that most GSH and TXN are available in reduced form. Thus, the intracellular milieu is normally maintained under strong reducing conditions, primarily by the redox-buffering capacity of GSH and TXN, to counteract oxidative stress and ROS formation. In addition to their antioxidant functions, GSH and TXN participate in cell signaling processes (9). For example, TXN induces hypoxia inducible factor 1 (HIF-1) and vascular endothelial growth factor (VEGF) (1). There is also evidence of crosstalk between TXN and GSH (10). GSH can attach to TXN and form mixed GSH-TXN disulfides abolishing TXN activity. TXN has also been shown to inhibit apoptosis via apoptosis signal regulating kinase 1 (1, 11).

Thioredoxin reductases (TXNRDs) are a family of homodimeric flavoenzymes that catalyze the NADPH-dependent reduction of oxidized thioredoxin (12-14). Each subunit contains separate FAD- and NADPH-binding domains and a conserved active site -Cys-Val-Asn-Val-Gly-Cys- (3, 15). Like glutathione peroxidases, TXNRDs also contain selenocysteine, which is indispensable for their enzymatic activity (5). The human TXNRDs exhibit unusually broad substrate specificity. In addition to thioredoxin, TXNRDs are able to use natural killer cell lysine, lipid hydroperoxides, menadione (vitamin K), 5,5'-dithiobis-2-nitrobenzoic acid (DTNB), and dehydroascorbate as substrates (2, 12, 16). This broad substrate specificity has been attributed to the location of selenocysteine as penultimate amino acid residue (496SeCys) of the flexible C-terminus (17). The catalytic mechanism of TXNRDs involves the flow of electrons from NADPH to FAD, from FAD to the active site disulfide, and then to the C-terminal SeCys (Figure 4.4.12.) (17, 18). The reduced SeCys moves away to transfer the electrons to a substrate, such as TXN, with reduction of disulfide bonds.

4.4.5. THIOREDOXIN SYSTEM

The thioredoxin system consists of two proteins, the redox protein thioredoxin (TXN) and the enzyme TXN reductase (TXNRD; Table 4.4.3.). TXNRD reduces and activates TXN, which in turn reduces oxidized cysteine residues on cellular proteins and binds ROS thereby protecting cells against oxidative stress (1, 2).

Thioredoxins (TXNs) are a family of small (10–12 kDa) ubiquitous proteins with two redox-active cysteine residues in a conserved amino acid sequence: -Cys-Gly-Pro-Cys (4, 8). TXNs exist either in reduced form with a dithiol, TXN-(SH)$_2$, or in oxidized form, TXN-S$_2$, with an intramolecular disulfide bridge. The name thioredoxin reflects the

Table 4.4.3. Overview of TXN System (2-7)

Name	TXN1	TXN2	TXNRD1	TXNRD2
Cellular location	cytosol	mitochondria	cytosol	mitochondria
Chromosome	9q31	22q11	12q23-24	22q11.2
Amino acids	104	106	497	495
Redox-active site	-**Cys**-Gly-Pro-**Cys**-		-**Cys**-Val-Asn-Val-Gly-**Cys**-	
Polymorphisms	rs4135208	rs2267337		Ala66Ser (rs574869)
	rs2301241	rs2281082		g23524 (rs756661)
		rs4821494		

TXNRD1 is a cytosol enzyme composed of two identical 56 kDa subunits (3, 5). TXNRD2 is a mitochondrial enzyme, which is 54% identical to TXNRD1. TXNRD2 differs from TXNRD1 by an N-terminal mitochondrial leader sequence (15). The TXNRD2 gene has been mapped to chromosome 22q11.2, which also contains the mitochondrial TXN gene.

$$TXN\text{-}S_2 \ + \ NADPH \ + \ H^+ \ \xrightarrow{\ TXNRD\ } \ TXN\text{-}(SH)_2 \ + \ NADP^+$$

Figure 4.4.12. TXNRD-mediated reaction

Thioredoxin interacting protein (TXNIP). In contrast to TXNRD, which supports TXN function, TXNIP (also called vitamin D3-upregulated protein 1, VDUP1) binds to and inhibits the reduced form of TXN (19).

The TXN system is a major cellular reductant involved in many thiol-dependent reactions. For example, TXN serves as hydrogen donor for ribonucleotide reductase, an essential enzyme that forms the deoxyribonucleotide precursors of DNA from the corresponding ribonucleotides (Figure 4.4.13.). In general, the TXN system can function as a protein disulfide oxidoreductase to catalyze dithiol-disulfide exchange reactions. TXN is located predominantly in the cytoplasm but translocates into the nucleus in response to environmental stress (20). In the nucleus, TXN can undergo dithiol-disulfide exchange with transcription factors, affect their DNA binding and thereby regulate gene transcription (21). TXNRD1 has been shown to reduce H_2O_2 and lipid hydroperoxides in the presence of NADPH but without TXN, while TXN by itself can scavenge H_2O_2 (16, 22).

Clinical Studies. An immunohistochemical study revealed overexpression of TXN and TXNRD in breast cancer relative to normal breast tissue (23). Both TXN and TXNRD were present in cytoplasm and nucleus. Over-

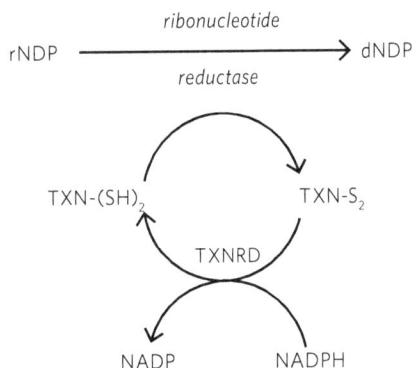

Figure 4.4.13. The thioredoxin system, composed of thioredoxin (TXN) and thioredoxin reductase (TXNRD), serves as hydrogen donor for ribonucleotide reductase, a key enzyme in DNA synthesis.

expression was observed in 30–40% of lobular cancers and 60–70% of ductal carcinomas. The expression level of TXN1 correlated with the tumor grade (24). HER2-positive tumors expressed significantly higher levels of TXNRD1 than HER2-negative cancers (25). Breast cancer cell lines also exhibited TXN and TXNRD expression and MCF-7 cells were shown to secrete TXN into the culture medium (26). Induction of HER2 in MCF-7 cells caused an increase in TXNRD1 and a decrease in TXNIP (25). Transfection of a dominant negative redox-inactive thioredoxin system TXN mutant prevented *in vivo* growth of MCF-7 cells (26). Each of the TXN and TXNRD genes contain polymorphisms (Table 4.4.3.). A population-based UK study called SEARCH (4474 cases, 4580 controls) found borderline evidence for an association of TXN1 SNP t2715c (rs4135179) and TXNRD2 SNPs g23524a (rs756661) and Ala66Ser (rs5748469) with breast cancer risk (7). However, a follow-up analysis using a multi-locus approach with polymorphisms in other anti-oxidant enzymes found no evidence (27). A German study named MARIE (2470 cases, 4715 controls) observed an increased risk of postmenopausal breast cancer for cytosolic TXN1 rs2301241 (OR_{trend} 1.08; 95% CI 1.01 – 1.16) and a decreased risk for mitochondrial TXN2 rs2281082 (OR_{trend} 0.86; 95% CI 0.78 – 0.94) (6). The combination of SEARCH and MARIE data with results from the Nurses' Health Study (CGEMS data) and the Swedish SASBAC study confirmed the increased risk associated with TXN1 rs2301241 (OR_{trend} 1.05; 95% CI 1.00 – 1.10) (6, 28, 29).

References

1. Arner ES, Holmgren A. *The thioredoxin system in cancer.* Semin Cancer Biol 2006;16:420-6.

2. Nordberg J, Arner ES. *Reactive oxygen species, antioxidants, and the mammalian thioredoxin system.* Free Radic Biol Med. 2001;31:1287-312.

3. Gasdaska PY, Gasdaska JR, Cochran S, Powis G. *Cloning and sequencing of a human thioredoxin reductase.* FEBS Lett. 1995;373:5-9.

4. Spyrou G, Enmark E, Miranda-Vizuete A, Gustafsson JÅ. *Cloning and expression of a novel mammalian thioredoxin.* J Biol Chem. 1997;272:2936-41.

5. Tamura T, Stadtman TC. *A new selenoprotein from human lung adenocarcinoma cells: Purification, properties, and thioredoxin reductase activity.* Proc Natl Acad Sci USA. 1996;93:1006-11.

6. Seibold P, Hein R, Schmezer P, Hall P, Liu J, Dahmen N, et al. *Polymorphisms in oxidative stress-related genes and postmenopausal breast cancer risk.* Int J Cancer. 2011;129:1467-76.

7. Cebrian A, Pharoah PD, Ahmed S, Smith PL, Luccarini C, Luben R, et al. *Tagging single-nucleotide polymorphisms in antioxidant defense enzymes and susceptibility to breast cancer.* Cancer Res. 2006;66:1225-33.

8. Holmgren A. *Thioredoxin.* Annu Rev Biochem. 1985;54:237-71.

9. Thannickal VJ, Fanburg BL. *Reactive oxygen species in cell signaling.* Am J Physiol Lung Cell Mol Physiol. 2000;279:L1005-L28.

10. Casagrande S, Bonetto V, Fratelli M, Gianazza E, Eberini I, Massignan T, et al. *Glutathionylation of human thioredoxin: A possible crosstalk between the glutathione and thioredoxin systems.* Biochemistry. 2002;99:9745-9.

11. Ueda S, Masutani H, Nakamura H, Tanaka T, Ueno M, Yodoi J. *Redox control of cell death.* Antiox Redox Signal. 2002;4:405-14.

12. Mustacich D, Powis G. *Thioredoxin reductase.* Biochem J. 2000;346:1-8.

13. Stadtman TC. *Selenium*

biochemistry: mammalian selenoenzymes. Ann NY Acad Sci. 2000;899:399-402.

14. Behne D, Kyriakopoulos A. *Mammalian selenium-containing proteins.* Annu Rev Nutr. 2001;21:453-73.

15. Miranda-Vizuete A, Damdimopoulos AE, Pedrajas JR, J. A. G, Spyrou G. *Human mitochondrial thioredoxin reductase cDNA cloning, expression and genomic organization.* Eur J Biochem. 1999;261:405-12.

16. Bjornstedt M, Hamberg M, Kumar S, Xue J, Holmgren A. *Human thioredoxin reductase directly reduces lipid hydroperoxides by NADPH and selenocystine strongly stimulates the reaction via catalytically generated selenols.* J Biol Chem. 1995;270:11761-4.

17. Gromer S, Wissing J, Behne D, Ashman K, Schirmer RH, Flohe L, et al. *A hypothesis on the catalytic mechanism of the selenoenzyme thioredoxin reductase.* Biochem J. 1998;332:591-2.

18. Zhong L, Arner ES, Holmgren A. *Structure and mechanism of mammalian thioredoxin reductase: The active site is a redox-active selenolthiol/selenenylsulfide formed from the conserved cysteine-seleno-cysteine sequence.* Biochemistry. 2000;97:5854-9.

19. Kim SY, Suh HW, Chung JW, Yoon SR, Choi I. *Diverse functions of VDUP1 in cell proliferation, differentiation, and diseases.* Cell Mol Immunol. 2007;4:345-51.

20. Wei SJ, Botero A, Hirota K, Bradbury CM, Markovina S, Laszlo A, et al. *Thioredoxin nuclear translocation and interaction with redox factor-1 activates the activator protein-1 transcription factor in response to ionizing radiation.* Cancer Res. 2000;60:6688-95.

21. Arrigo A-P. *Gene expression and the thiol redox state.* Free Radic Biol Med. 1999;27:936-44.

22. Spector A, Yan GZ, Huang RC, McDermott MJ, Gascoyne PR, Pigiet V. The effect of H2O2 upon thioredoxin-enriched lens epithelial cells J Biol Chem. 1988;263:4984-90.

23. Lincoln DT, Ali Emadi EM, Tonissen KF, Clarke FM. *The thioredoxin-thioredoxin reductase system: Over-expression in human cancer.* AntiCancer Res. 2003;23:2425-33.

24. Cha MK, Suh KH, Kim IH. *Overexpression of peroxiredoxin I and thioredoxin1 in human breast carcinoma.* J Exp Clin Cancer Res : CR. 2009;28:93.

25. Cadenas C, Franckenstein D, Schmidt M, Gehrmann M, Hermes M, Geppert B, et al. *Role of thioredoxin reductase 1 and thioredoxin interacting protein in prognosis of breast cancer.* Breast Cancer Res: BCR. 2010;12:R44.

26. Berggren M, Gallegos A, Gasdaska JR, Gasdaska PY, Warneke J, Powis G. *Thioredoxin and thioredoxin reductase gene expression in human tumors and cell lines, and the effects of serum stimulation and hypoxia.* Anticancer Res. 1996;16:3459-66.

27. Oestergaard MZ, Tyrer J, Cebrian A, Shah M, Dunning AM, Ponder BA, et al. *Interactions between genes involved in the antioxidant defence system and breast cancer risk.* Br J Cancer. 2006;95:525-31.

28. Einarsdottir K, Darabi H, Li Y, Low YL, Li YQ, Bonnard C, et al. *ESR1 and EGF genetic variation in relation to breast cancer risk and survival.* Breast Cancer Res: BCR. 2008;10:R15.

29. Hunter DJ, Kraft P, Jacobs KB, Cox DG, Yeager M, Hankinson SE, et al. *A genome-wide association study identifies alleles in FGFR2 associated with risk of sporadic post-menopausal breast cancer.* Nature Genet. 2007;39:870-4

4.4.6. PEROXIREDOXINS

The human peroxiredoxin (PRX) family consists of six distinct proteins located in various subcellular compartments, including mitochondria and peroxisomes, where oxidative stress is most evident (1-3) (Table 4.4.4.). The PRX family differs from other peroxidases, such as catalase and the GPX family. Instead of having heme or selenocysteine, the PRXs have two cysteine residues as their active site, except PRX6, which contains only one conserved cysteine (4). A Cys residue in the N-terminal portion of the protein is the primary reaction site. It is oxidized by H_2O_2 or alkyl hydroperoxides to sulfenic acid, Cys-SOH, which reacts with a conserved C-terminal Cys-SH of another subunit to form an intermolecular disulfide bond (5, 6). Thus, when reducing peroxides, PRXs themselves are oxidized and therefore act both as enzyme and co-substrate (7). In the next step, the PRXs are reduced by thioredoxin, except PRX6 for which the electron donor is not known. Since thioredoxin is the electron donor for most PRXs, they are also referred to as thioredoxin peroxidases (3).

Together with catalase and GPXs, PRX enzymes play an important role in eliminating peroxides generated during metabolism. The catalytic efficiency of PRXs indicated by the k_{cat}/K_m is significantly lower than that of catalase or GPXs. However, catalase is mainly in peroxisomes and GPX, which is mainly in cytosol, exists in low amounts in most tissues. In contrast, several of the PRX isoforms are present in cytosol in relative abundance (0.2-0.4% of total soluble protein) (8). Like the other enzymes, PRXs regulate H_2O_2 concentration to influence various biological processes, such as the stimulation of cell proliferation and inhibition of apoptosis. For example, overexpression of PRX in cultured cells eliminated the H_2O_2 generated in response to growth factors (4, 9). Furthermore, PRX overexpression inhibited both the activation of nuclear factor κB and apoptosis elicited by H_2O_2 or tumor necrosis factor α.

Clinical Studies. Western immunoblots showed overexpression of PRX1-3 in 24 invasive breast cancers relative to normal breast tissue (10). An immunohistochemical analysis of PRX1-6 in over 400 invasive breast cancers also showed overexpression of all isoforms, especially PRX3-5, compared to non-neoplastic breast tissue (11). The increased expression of the mitochondrial and peroxisomal isoforms was associated with the presence of estrogen (PRX3) and progesterone (PRX3, 4) receptors. PRX5 expression correlated with larger tumor size, positive lymph node status, and shorter survival, whereas PRX3 and 4 were related to better prognosis. Interestingly, there was no correlation with cell proliferation measured by Ki-67 and PRX expression. These data suggest that the overexpression of PRXs is in response to the increased production of ROS in cancer tissue rather than the anti-apoptotic and proliferative effect of these proteins (11). PRX2 and 3 were found to be overexpressed in metastatic human breast cancer cells elicited in nude mice (12). PRX2 was upregulated in radiation-resistant MCF+FIR3 human breast cancer cells but not in MCF+FIS4 cells (8). The resistance of MCF-FIR3 cells to ionizing radiation was partially reversed by silencing PRX2 expression through PRX2 siRNA treatment suggesting that PRX2 is involved in the cellular response to radiation.

Table 4.4.4. Overview of peroxiredoxins (1, 2, 4)

Name	PRX1	PRX2	PRX3	PRX4	PRX5	PRX6
Cellular location	cytosol	cytosol	mitochondria	cytosol, plasma	peroxisome	cytosol, plasma
Subunit structure	dimer	dimer	dimer	dimer	monomer	monomer
Active site	2 Cys	2 Cys	2 Cys	2 Cys	2 Cys	1 Cys
Electron donor	TXN GSH	TXN	TXN	TXN GSH	TXN	unknown

References

1. Chae HZ, Robison K, Poole LB, Church G, Storz G, Rhee SG. *Cloning and sequencing of thiol-specific antioxidant from mammalian brain: Alkyl hydroperoxide reductase and thiol-specific antioxidant define a large family of antioxidant enzymes.* Proc Natl Acad Sci USA. 1994;91:7017-21.

2. Fujii J, Ikeda Y. *Advances in our understanding of peroxiredoxin, a multifunctional, mammalian redox protein.* Redox Rep. 2002;7:123-30.

3. Rhee SG, Chae HZ, Kim K. *Peroxiredoxins: a historical overview and speculative preview of novel mechanisms and emerging concepts in cell signaling.* Free Radical Biol Med. 2005;38:1543-52.

4. Kang SW, Baines IC, Rhee SG. *Characterization of a mammalian peroxiredoxin that contains one conserved Cysteine.* J Biol Chem. 1998;273:6303-11.

5. Verdoucq L, Vignols F, Jacquot JP, Chartier Y, Meyer Y. *In vivo characterization of a thioredoxin h target protein defines a new peroxiredoxin family.* J Biol Chem. 1999;274:19714-22.

6. Declercq JP, Evrard C, Clippe A, Vander Stricht D, Bernard A, Knoops B. *Crystal structure of human peroxiredoxin 5, a novel type of mammalian peroxiredoxin at 1.5 Å resolution.* J Mol Biol. 2001;311:751-9.

7. Rabilloud T, Heller M, Gasnier F, Luche S, Rey C, Aebersold R, et al. *Proteomics analysis of cellular response to oxidative stress. Evidence for in vivo overoxidation of peroxiredoxins at their active site.* J Biol Chem. 2002;277:19396-401.

8. Wang T, Tamae D, LeBon T, Shively JE, Yen Y, Li JJ. *The role of peroxiredoxin II in radiation-resistant MCF-7 breast cancer cells.* Cancer Res. 2005;65:10338-46.

9. Kang SW, Chae HZ, Seo MS, Kim K, Baines IC, Rhee SG. *Mammalian peroxiredoxin isoforms can reduce hydrogen peroxide generated in response to growth factors and tumor necrosis factor-a.* J Biol Chem. 1998;273:6297-302.

10. Noh DY, Ahn SJ, Lee RA, Kim SW, Park IA, Chae HZ. *Overexpression of peroxiredoxin in human breast cancer.* AntiCancer Res. 2001;21:2085-90.

11. Karihtala P, Mantyniemi A, Kang SW, Kinnula VL, Soini Y. *Peroxiredoxins in breast carcinoma.* Clin Cancer Res. 2003;9:3418-24.

12. España L, Martín B, Aragüés R, Chiva C, Oliva B, Andreu D, et al. *Bcl-xL -mediated changes in metabolic pathways of breast cancercells. From survival in the blood stream to organ specific metastasis.* Am J Pathol. 2005;167:1125-37.

to H_2O by catalase, GPX, and PRX. The balance between the first and second step antioxidant enzymes may be critical in protecting cells against ROS. Since SOD removes superoxide anions, it is considered a detoxifying enzyme. On the one hand, too little SOD relative to catalase/GPX/ PRX could lead to accumulation of superoxide anion, which are toxic to macromolecules. On the other hand, too much SOD relative to catalase/GPX/PRX could lead to increased production of the H_2O_2 intermediate, which can generate the even more toxic OH˙ via the Fenton reaction. These are theoretical considerations because there are insufficient data available at the present time to gauge the relative concentrations of all antioxidant enzymes in breast tissue. The GPX and PRX families were discovered much later than catalase and SODs and the sheer number of enzyme isoforms makes it difficult to reliably assess their expression in normal and malignant epithelium. The up/down arrows in Table 4.4.5. reflect results of studies of over/underexpression of individual enzymes in breast cancer relative to normal breast tissue. However, the overall balance and capability of the antioxidant enzyme system in breast tissue remain unknown, largely a reflection of its complexity. There are additional enzymes, such as NAD(P)H:quinone oxidoreductase, lactoperoxidase, myeloperoxidase, and NADPH oxidase (see Section 4.5.) that indirectly affect the antioxidant pathway and further increase the biological complexity.

Nonsynonymous polymorphisms have been described in virtually all genes encoding antioxidant enzymes, some with suspected or proven alteration in enzyme activity. For example, the Ala16(-9)Val polymorphism (rs4880) in SOD2 is associated with 30% higher activity of the wild-type Ala enzyme than the variant Val isoform. Thus –9Ala/Ala may mean too much SOD2 activity in certain conditions whereas –9Val/Val may not be enough in other circumstances. However, the majority of epidemiological studies did not find an association of the -9Ala/Val polymorphism with breast cancer risk.

Since the antioxidant defense system consists of multiple enzymes, several studies have examined the association of breast cancer risk with SNPs in combination of genes including SOD1, SOD2, catalase, GPX1, GPX4, TXN1, TXNRD1, TXN2, TXNRD2, and glutathione reductase.

4.4.7. SUMMARY OF ANTI-OXIDANT ENZYMES

Like other cells, mammary epithelium generates toxic ROS during oxidative metabolism. A complex antioxidant system composed of enzymatic and non-enzymatic factors protects against the oxidant injury. There are six different families of antioxidant enzymes with a total of 16 isoforms (Table 4.4.5.). The enzymatic antioxidant pathway consists of essentially two steps: dismutation of superoxide anion to H_2O_2 by SODs and conversion of H_2O_2

Table 4.4.5. Overview of Anti-Oxidant Enzymes

Enzymes	Isoforms	Active Sites	Polymorphisms	Over/underexpression in breast cancer relative to normal breast tissue
Superoxide Dismutase	1 - 3	Cu/Zn, Mn	SOD2 Ala16(-9)Val (rs4880) SOD2 Ile58(82)Thr (rs1141718) SOD3 Ala58Thr (rs2536512) SOD3 Arg213Gly (rs1799895)	↓
Catalase	1	heme	-262 C/ T (rs1001179)	↓
Glutathione Peroxidase	1 - 4	selenium	GPX1 GCG repeat (5–7 Ala)	↑
			GPX1 -1040 G/A; (rs3448) GPX1 Pro198(200)Leu (rs1050450)	↑
Thioredoxin System	TXN1 + 2	cysteine	TXN1 t2715c (rs4135179)	↑
	TXNRD 1 + 2	selenium	TXNRD2 g23524 TXNRD2 Ala66Ser (rs5748469)	↑
Peroxiredoxin	1 - 6	cysteine	not defined	↑

A UK study called SEARCH (4474 cases, 4580 controls) examined 54 SNPs that tag all known variants in these genes including SOD2 –9Ala/Val (1). There was only borderline evidence for an association of any polymorphism with breast cancer risk. A multilocus study of the same population limited to 2271 cases and 2280 controls examined gene-gene interactions between 52 SNPs in the 10 genes (2). The statistical analysis by three different methods (unconditional logistic regression, multifactor dimensionality reduction, and hierarchical cluster analysis) found no evidence for an association of the SOD2 –9Ala/Val or any other variant in the other genes with breast cancer risk. A German study called MARIE (2470 cases, 4715 controls) observed an increased risk of postmenopausal breast cancer for cytosolic TXN1 rs2301241 (OR_{trend} 1.08; 95% CI 1.01 – 1.16) and a decreased risk for mitochondrial TXN2 rs2281082 (OR_{trend} 0.86; 95% CI 0.78 – 0.94) (3). The combination of SEARCH and MARIE data with results from the Nurses' Health Study (CGEMS data) and the Swedish SASBAC study confirmed the increased risk associated with TXN1 rs2301241 (OR_{trend} 1.05; 95% CI 1.00 – 1.10) (3-5). This result does not rule out a significant role of the other antioxidant enzymes in mammary carcinogenesis. Their level of expression in breast tissue and the concentration of reducing factors (GSH, ascorbate, α-tocopherol, α-lipoic acid, melatonin, and others) may be as important as the enzyme genotypes.

References

1. Cebrian A, Pharoah PD, Ahmed S, Smith PL, Luccarini C, Luben R, et al. *Tagging single-nucleotide polymorphisms in antioxidant defense enzymes and susceptibility to breast cancer.* Cancer Res. 2006;66:1225-33.

2. Oestergaard MZ, Tyrer J, Cebrian A, Shah M, Dunning AM, Ponder BA, et al. *Interactions between genes involved in the antioxidant defence system and breast cancer risk.* Br J Cancer. 2006;95:525-31.

3. Seibold P, Hein R, Schmezer P, Hall P, Liu J, Dahmen N, et al. *Polymorphisms in oxidative stress-related genes and postmenopausal breast cancer risk.* Int J Cancer 2011;129:1467-76.

4. Einarsdottir K, Darabi H, Li Y, Low YL, Li YQ, Bonnard C, et al. *ESR1 and EGF genetic variation in relation to breast cancer risk and survival.* Breast Cancer Res : BCR. 2008;10:R15.

5. Hunter DJ, Kraft P, Jacobs KB, Cox DG, Yeager M, Hankinson SE, et al. *A genome-wide association study identifies alleles in FGFR2 associated with risk of sporadic postmenopausal breast cancer.* Nature Genet. 2007;39:870-4.

4.5. OTHER PHASE II ENZYMES

4.5.1. P450 OXIDOREDUCTASE

Overview. P450 Oxidoreductase (POR) is the electron donor protein for several oxygenase enzymes located in the endoplasmic reticulum (microsomes) and nuclear envelope of most eukaryotic cells (1-3). The oxygenase enzymes include all 50 microsomal CYPs, such as the steroidogenic CYP17 (17α-hydroxylase, 17,20 lyase), CYP21 (21-hydroxylase), and CYP19A1 (aromatase). POR belongs to a family of dual flavin reductases that contain both flavin adenine dinucleotide (FAD) and flavin mononucleotide (FMN) (4). POR functions as an internal electron transport chain, transferring electrons from NADPH to FAD to FMN and finally to cytochrome P450 (5). POR is an obligatory partner for all microsomal CYPs. The concentration of POR in the microsomal membrane is generally lower than the CYP concentration. Based on these factors, POR is often the rate-limiting component of CYP-mediated enzyme reactions (6).

Gene and Protein. The POR gene at 7q11.2 is over 50 kb long and composed of 16 exons (Figure 4.5.1.). The gene

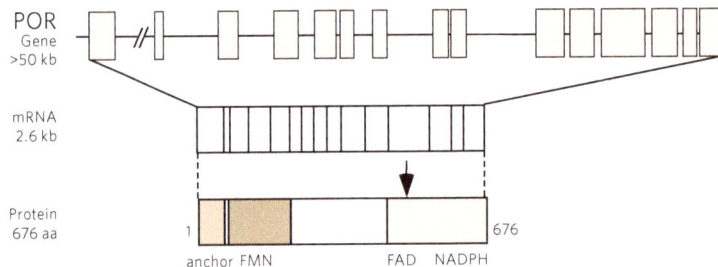

Figure 4.5.1. Overview of POR gene, mRNA, and protein.
The POR gene at 7q11.2 is over 50 kb long and contains 16 exons transcribed from a TATA-less promoter. The untranslated first exon resides >20 kb upstream of the coding region, which starts with exon 2. The lengths of the exons are drawn in scale, while those of the introns are not. Structural and functional domains of the protein are indicated. A non-synonymous SNP Ala503Val (rs1057868; arrow) is associated with reduced catalytic activity.

has an extended first intron, followed by exons 2 – 16, which encode the 78 kDa protein consisting of 676 amino acids (7-9). A comprehensive analysis of the POR gene sequence in 842 individuals from four ethnic groups (African, Caucasian, Chinese, and Mexican Americans) identified 108 noncoding SNPs, 32 coding SNPs, 15 of which were non-synonymous and 8 insertions/deletions (10). Most of these SNPs were rare, but 36 noncoding and 7 coding SNPs (one non-synonymous Ala503Val) were found at allele frequencies of ≥1% in at least one ethnic group. Ala503Val (rs1057868) was the most common, occurring on 27.9% of all alleles with ethnic predilection ranging from 19.1% in African Americans to 36.7% in Chinese Americans. The catalytic activity of the 503Val variant was 56 – 67% of the 503Ala wild type.

The organization of the POR gene reveals a general correspondence between exons and structural domains of the protein (11). The N-terminal membrane-anchoring domain has a type I signal-anchor sequence, which shows amino-terminus-lumen and carboxy-terminus-cytoplasm topology (12). Sequence comparisons and crystallographic studies revealed four additional domains from N- to C-terminus: the FMN-binding domain, the connecting domain, and the FAD- and NADPH-binding domains (13, 14). During the internal electron transfer, POR cycles between one- and three-electron reduced states resulting from coordinated changes in the oxidation-reduction states of FMN and FAD (5, 15-18) (Figure 4.5.2.). The reaction starts with binding of NADPH to POR. NADPH gives up a pair of electrons, which are accepted by FAD. Electron receipt elicits a conformational change in the flexible connecting domain of POR, permitting the isoalloxazine rings of the FAD and FMN moieties to come closer together so that the electrons can pass from FAD to FMN. Upon receipt of electrons by FMN, the hinge region flexes once more, permitting the FMN domain to get closer to the heme-binding domain of CYP. The surface charge of the FMN domain of POR is negative due to a cluster of aspartate residues, whereas the redox partner binding site of CYP has a positive surface charge produced by lysine and arginine residues (13, 19-23). FMN

then donates the electrons one at a time to the heme group in CYP, associated with another conformational change that returns POR to its original orientation. The electrons travel about 18 Å from the FMN moiety to reach the heme iron in CYP (24). The molecular pathways for this electron flow vary between CYPs depending on their individual structures. The hydrophobic N-terminal domain is not only responsible for anchoring POR in the microsomal membrane but also for the binding to cytochromes P450 (11, 25). It is remarkable that POR is the sole enzyme responsible for the electron transfer from NADPH to all known microsomal CYPs. While CYP genes exhibit tissue-specific expression and their promoters contain either a TATA or CCAAT box, the POR gene is expressed in virtually every tissue and, like many ubiquitous enzymes, has a GC-rich TATA- and CCAAT-less promoter, characteristic of housekeeping genes (26). As a housekeeping gene, POR shows little developmental, tissue-specific, or inter-individual variation in its expression (27). POR is expressed as early as the two-cell stage of embryonic development (28). POR gene disruption in mice showed that loss of POR does not block early embryonic development but is essential for progression past mid-gestation (29). Early embryogenesis in POR -/- mice was not affected but neural tube defects and cardiac abnormalities were observed by day 10.5 of gestation, leading to lethality by day 13.5.

Clinical Studies. POR deficiency has been identified in humans. Recessive POR mutations cause disordered steroidogenesis associated with ambiguous genitalia and/or the Antley-Bixler syndrome, a skeletal malformation featuring craniosynostosis that can also result from mutations in fibroblast growth factor receptor 2 (2, 30). A study of the Multiethnic Cohort (1,615 cases, 1,962 controls) sequenced the POR gene and found no association of the common SNP Ala503Val (rs1057868) with breast cancer risk (31). However, a rare silent exon 1 SNP, Gly5Gly (A15G), which may be associated with altered POR RNA splicing, conferred a modestly increased risk for breast cancer in African-American women (OR 1.58; 95% CI 1.04 - 2.41; P = 0.03).

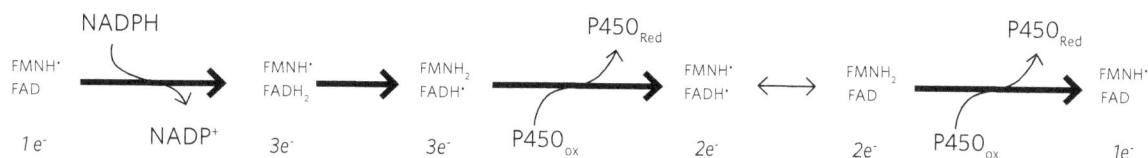

Figure 4.5.2. Electron transfer from NADPH via POR to cytochrome P450.
The FAD/FMN-containing POR cycles between a one- and three-electron reduced state with FADH• and FMNH• semiquinone intermediates. The FAD/FMN interflavin electron transfer is faster than the rate of electrons entering POR from NADPH to FAD. The only species capable of transferring electrons to the oxidized CYP (P450ox) is $FMNH_2$.

References

1. Kasper CB. *Biochemical distinctions between the nuclear and microsomal membranes from rat hepatocytes.* J Biol Chem. 1971;246:577-81.

2. Miller WL, Huang N, Pandey AV, Fluck CE, Agrawal V. *P450 oxidoreductase deficiency: a new disorder of steroidogenesis.* Ann N Y Acad Sci. 2005;1061:100-8.

3. Wang J, Ortiz de Montellano PR. *The binding sites on human heme oxygenase-1 for cytochrome P450 reductase and biliverdin reductase.* J Biol Chem. 2003;278:20069-76.

4. Dohr O, Paine MJI, Freidberg T, Roberts GCK, Wolf CR. *Engineering of a functional human NADH-dependent cytochrome P450 system.* Proc Natl Acad Sci USA. 2001;98:81-6.

5. Vermilion JL, Ballou DP, Massey V, Coon MJ. *Separate roles for FMN and FAD catalysis by liver microsomal NADPH-cytochrome P-450 reductase.* J Biol Chem. 1981;256:266-77.

6. Cawley GF, Batie CJ, Backes WL. *Substrate-dependent competition of different P450 isozymes for limiting NADPH-cytochrome P450 reductase.* Biochemistry. 1995;34:1244-7.

7. Haniu M, McManus ME, Birkett DJ, Lee TD, Shively JE. *Structural and functional analysis of NADPH-cytochrome P-450 reductase from human liver: complete sequence of human enzyme and NADPH-binding sites.* Biochemistry. 1989; 28:8639-45.

8. Shephard EA, Phillips IR, Santisteban I, West LF, Palmer CNA, Ashworth A, et al. *Isolation of a human cytochrome P-450 reductase cDNA clone and localization of the corresponding gene to chromsome 7q11.2.* Ann Hum Genet. 1989;53:291-301.

9. Yamano S, Aoyama T, McBride OW, Hardwick JP, Gelboin HV, Gonzalez FJ. *Human NADPH-P450 oxidoreductase: complementary DNA cloning, sequence and vaccinia virus-mediated expression and localization of the CYPOR gene to chromosome 7.* Mol Pharmacol. 1989;35:83-8.

10. Huang N, Agrawal V, Giacomini KM, Miller WL. *Genetics of P450 oxidoreductase: sequence variation in 842 individuals of four ethnicities and activities of 15 missense mutations.* Proc Natl Acad Sci USA. 2008;105:1733-8.

11. Porter TD, Beck TW, Kasper CB. *NADPH-cytochrome P-450 oxidoreductase gene organization correlates with structural domains of the protein.* Biochemistry. 1990;29:9814-8.

12. Kida Y, Ohgiya S, Mihara K, Sakaguchi M. *Membrane topology of NADPH-cytochrome P450 reductase on the endoplasmic reticulum.* Arch Biochem Biophys. 1998;351:175-9.

13. Wang M, Roberts DL, Paschke R, Shea TM, Masters BS, Kim J-JP. *Three-dimensional structure of NADPH-cytochrome P450 reductase: prototype for FMN- and FAD-containing enzymes.* Proc Natl Acad Sci USA. 1997;94:8411-6.

14. Zhao Q, Modi S, Smith G, Paine M, McDonaugh PD, Wolf CR, et al. *Crystal structure of the FMN-binding domain of human cytochrome P450 reductase at 1.93 A resolution.* Protein Sci. 1999;8:298-306.

15. Oprian DD, Coon MJ. *Oxidation-reduction states of FMN and FAD in NADPH-cytochrome P-450 reductase during reduction by NADPH.* J Biol Chem. 1982;257:8935-44.

16. Munro AW, Noble MA, Robledo L, Daff SN, Chapman SK. *Determination of the redox properties of human NADPH-cytochrome P450 reductase.* Biochemistry. 2001;40:1956-63.

17. Gutierrez A, Munro AW, Grunau A, Wolf CR, Scrutton NS, Roberts GCK. *Interflavin electron transfer in human cytochrome P450 reductase is enhanced by coenzyme binding.* Eur J Biochem. 2003;270:2612-21.

18. Daff S. *An appraisal of multiple NADPH binding-site models proposed for cytochrome P450 reductase, NO synthase, and related Diflavin Reductase systems.* Biochemistry. 2004;43:3929-32.

19. Hasemann CA, Kurumbail RG, Boddupalli SS, Peterson JA, Deisenhofer J. *Structure and function of cytochromes P450: a comparative analysis of three crystal structures.* Structure. 1995;2:41-62.

20. Shen AL, Kasper CB. *Role of acidic residues in the interaction of NADPH-cytochrome P450 oxidoreductase with cytochrome P450 and cytochrome c.* J Biol Chem. 1995;270:27475-80.

21. Estabrook RW, Shet MS, Fisher CW, Jenkins CM, Waterman MR. *The interaction of NADPH-P450 reductase with P450: an electrochemical study of the role of the flavin mononucleotide-binding domain.* Arch Biochem Biophys. 1996;333:308-15.

22. Kondo S, Sakaki T, Ohkawa H, Inouye K. *Electrostatic interaction between cytochrome P450 and NADPH-P450 reductase: comparison of mixed and fused systems consisting of rat cytochrome P450 1A1 and yeast NADPH-P450 reductase.* Biochem Biophys Res Commun. 1999;257:273.

23. Davydov DR, Kariakin AA, Petushkova NA, Peterson JA. *Association of cytochromes P450 with their reductases: opposite sign of the electrostatic interactions in P450BM-3 as compared with the microsomal 2B4 system.* Biochemistry. 2000;39:6489-97.

24. Sevrioukova IF, Li H, Zhang H, Peterson JA, Poulos TL. *Structure of a cytochrome P450-redox partner electron-transfer complex.* Proc Natl Acad Sci USA. 1999;96:1863-8.

25. Black SD, French JS, Williams Jr. CJ, Coon MJ. *Role of a hydrophobic polypeptide in the N-terminal region of NADPH-cytochrome P-450 reductase in complex formation with P-450LM.* Biochem Biophys Res Commun. 1979;91:1528-35.

26. O'Leary KA, Kasper CB. *Molecular basis for cell-specific regulation of the NADPH-cytochrome P450 oxidoreductase gene.* Arch Biochem Biophys. 2000;379:97-108.

27. Shephard EA, Palmer CNA, Segall HJ, Phillips IR. *Quantification of cytochrome P450 reductase gene expression in human tissues.* Arch Biochem Biophys. 1992;294:168-72.

28. Stromstedt M, Keeney DS, Waterman MR, Paria BC, Conley AJ, Dey SK. *Preimplantation mouse blastocytes fail to express CYP genes required for estrogen biosynthesis.* Mol Reprod Develop. 1996;43:428-36.

29. Shen AL, O'Leary KA, Kasper CB. *Association of multiple developmental defects and embryonic lethality with loss of microsomal NADPH-cytochrome P450 oxidoreductase.* J Biol Chem. 2002;277:6536-41.

30. Marohnic CC, Panda SP, Martasek P, Masters BS. *Diminished FAD binding in the Y459H and V492E Antley-Bixler syndrome mutants of human cytochrome P450 reductase.* J Biol Chem. 2006;281:35975-82.

31. Haiman CA, Setiawan VW, Xia LY, Le Marchand L, Ingles SA, Ursin G, et al. *A variant in the cytochrome p450 oxidoreductase gene is associated with breast cancer risk in African Americans.* Cancer Res. 2007;67:3565-8

4.5.2. NAD(P)H:QUINONE OXIDOREDUCTASE

Overview. NAD(P)H:quinone oxidoreductase 1 (NQO1), is a phase II enzyme, formerly known as DT-diaphorase. NQO1 is a cytosolic, homodimeric FAD-enzyme that uses NADH or NADPH to reduce quinones in a two-electron process to hydroquinones (1-3). The reduction of quinones to hydroquinones bypasses the formation of semiquinones and thereby prevents redox cycling and the production of ROS. Thus, the obligate two-electron reductase activity of NQO1 is the basis for its postulated role as a defense mechanism against the carcinogenic effects of quinone xenobiotics. The hydroquinones produced by NQO1 are less toxic and readily excreted upon conjugation. However, NQO1 can also use nitro compounds as electron acceptors, reducing them to nitrosamines, some of which are potent carcinogens (2).

Gene and Protein. The NQO1 gene on chromosome 16q22 is organized in 6 exons (4, 5) (Figure 4.5.3.). The sixth exon contains four polyadenylation sites, which give rise to four mRNA transcripts ranging from 1.2 to 2.7 kb (6). The gene encodes 273 amino acids in a 31 kDa protein that forms a homodimer located in the cytosol.

Each subunit noncovalently binds one molecule of FAD and uses NADH or NADPH to reduce quinones in a two-electron process to hydroquinones. The electron donors NADH and NADPH have similar V_{max}, but the k_m for NADH is greater than that for NADPH (1). Analysis of the NQO1 crystal structure revealed extensive overlap between the NAD(P)H and quinone binding sites (7). This provides the molecular basis for the observed ping-pong mechanism: NAD(P)H binds to NQO1, reduces FAD to $FADH_2$, and is then released, allowing the quinone substrate to bind the enzyme and to be reduced by $FADH_2$ to the hydroquinone. Thus, NAD(P)H cycles in and out of the enzyme before the quinone substrate can bind while FAD remains bound during catalytic cycling via hydrogen bonds and van der Waals interactions (8). NQO1 can also scavenge superoxide anion $O_2^{\cdot-}$, which reacts with $FADH_2$ (Figure 4.5.4.). Since $O_2^{\cdot-}$ can be generated by autoxidation of hydroquinones, removal of $O_2^{\cdot-}$ by NQO1 would prevent the acceleration of hydroquinone autoxidation, preserve the hydroquinone, and facilitate its conjugation and excretion (9).

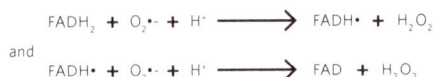

$$FADH_2 + O_2^{\cdot-} + H^+ \longrightarrow FADH\cdot + H_2O_2$$

and

$$FADH\cdot + O_2^{\cdot-} + H^+ \longrightarrow FAD + H_2O_2$$

Figure 4.5.4. Enzymatic reaction of FADH2 with $O_2^{\cdot-}$

There are several transcriptional response elements in the promoter region of the NQO1 gene that mediate the high inducibility of the enzyme, such as the antioxidant response element and the xenobiotic response element, involving the aromatic hydrocarbon (Ah) receptor (2, 4, 10). Many dietary compounds including isothiocyanates, such as sulforaphane in cruciferous vegetables (e.g., broccoli, cauliflower) can increase NQO1 levels (11-13). Ionizing radiation, heat shock, and various chemicals have been shown to induce NQO1 expression, e.g., azo dyes, dioxins, diphenyls, dithiolethiones, α-lipoic acid, isothiocyanates, and polycyclic aromatic hydrocarbons (3, 4, 14-16). Exposure to different inducing agents may explain the overlap in NQO1 activity levels between homozygous (187Pro/Pro) and heterozygous (187Pro/Ser) individuals. Interestingly, NQO1 is suppressed by estrogen receptor agonists, such as estradiol, whereas antiestrogens, such as tamoxifen, induce NQO1 expression (10, 17, 18).

Inactivation of $E_2(E_1)$-3,4-Q can occur in two ways (i) by reaction with glutathione (GSH), which is catalyzed by GST, and (ii) by reduction to 4-$OHE_2(1)$, which is catalyzed by NQO1. By reducing $E_2(E_1)$-3,4-Q levels, both GST and

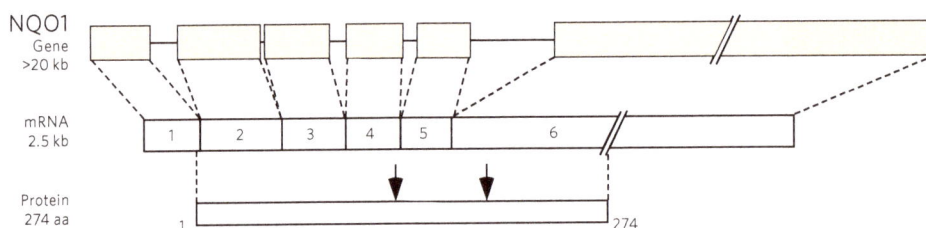

Figure 4.5.3. Overview of NQO1 gene, mRNA, and protein.
The NQO1 gene at 16q22 is about 20 kb long and contains six exons. The sixth exon contains four polyadenylation sites, which give rise to four mRNA transcripts ranging from 1.2 to 2.7 kb. The open reading frame starts at the end of the first exon, which codes for two amino acids including the initiating methionine and one G for the first codon of the second exon. The sixth exon is the largest among the exons, consisting of 1833 bp of which the first 305 are translated. The arrows indicate the polymorphisms T/C in exon 4 (Arg139Trp; rs1131341) and C/T in exon 6 (Pro187Ser; rs1800566). The lengths of the exons are drawn in scale, while those of the introns are not.

NQO1 prevent the formation of estrogen-DNA adduct formation. NQO1 displays substrate specificity towards estrogen quinones. The enzyme recognizes the quinone of the synthetic estrogen, diethylstilbestrol (DES), converting DES-Q to its hydroxyquinone, 2-OH-DES (17). NQO1 also recognizes the o-quinone of the equine estrogen equilenin, EN-3,4-Q, reducing the latter to 4-hydroxyequilenin, 4-OH-EN (19). However, NQO1 does not recognize the o-quinone of estrone, E_1-3,4-Q, as substrate (20). The structural difference between E_1/E_2 and equine estrogens is the aromatic ring of the latter, which influences the reactivity of their respective metabolites. For example, 4-hydroxyequilenin and 4-hydroxyequilin readily autoxidize to quinones, whereas 4-OHE$_1$ and 4-OHE$_2$ do not, i.e., they are more stable.

Many other quinones derived from endogenous and exogenous compounds (e.g. benzo- and naphthoqui-nones) can serve as substrates for NQO1. *In vitro* studies of NQO1 generally utilize 2-methyl-1,4-naphthoquinone (menadione, vitamin K3) or 2,6-dichlorophenol-indophenol (DCPIP) as electron acceptors (21-24). In addition to cytosolic NQO1, quinones can be reduced by several other NAD(P)H-oxidizing flavoenzymes, e.g., mitochondrial NADH dehydrogenase, microsomal NADPH-cytochrome P450 reductase, and microsomal NADH-cytochrome b5 reductase. The mitochondrial and microsomal enzymes reduce quinones by one electron to semiquinones, which are readily reoxidized by molecular oxygen with the formation of superoxide anion radical and other ROS (see Chapter 2.3). NQO1 is unique as a two-electron transferring quinone reductase (25, 26) (Figure 4.5.5.). The NQO1-mediated reduction of quinones to hydroquinones bypasses the formation of semiquinones and thereby prevents redox cycling and the production of ROS. Thus, the obligate two-electron reductase activity of NQO1 is the basis for its postulated role as a defense mechanism against the carcinogenic effects of quinone xenobiotics. For example, NQO1 prevented the formation of benzo[a]pyrene quinone-DNA adducts by CYP1A1 and P450 reductase (27). The hydroquinones produced by NQO1 are less toxic and

readily excreted upon conjugation. It should be mentioned that NQO1 is not always a detoxicating enzyme. Besides quinones, NQO1 can use nitro compounds as electron acceptors, reducing them to nitrosamines, some of which are potent carcinogens (2). For example, NQO1 catalyzes the reductive activation of nitro-polycyclic aromatic hydrocarbons, such as 3-nitrobenzanthrone, a potent mutagen and suspected human carcinogen identified in diesel exhaust and air pollution (28).

In addition to its role in the detoxification of quinones, NQO1 binds and stabilizes several short-lived proteins including the tumor suppressors p53 and p73 and the enzyme ornithine decarboxylase (29-31). The anticoagulant dicoumarol is the strongest known inhibitor of NQO1 (k$_i$ 10^{-8} to 10^{-10} M). Dicoumarol competes with NAD(P)H for binding to NQO1 and prevents the electron transfer to FAD (1). Dicoumarol also disrupts the binding of NQO1 to p53, p73, and ornithine decarboxylase by inducing conformational changes in the positions of 128Tyr and 232Phe on the surface of NQO1 (32). A less potent inhibitor is the natural phenolic compound curcumin found in the spice turmeric that gives the yellow color and flavor to curry. Curcumin inhibited NQO1 activity by competing with NAD(P)H binding, dissociated NQO1-p53 complexes, and induced p53 degradation (33).

The NQO1 gene contains a common nonsynonymous SNP (minor allele frequency 0.24) at nucleotide position 609 C/T in exon 6, resulting in Pro187Ser (rs1800566) (34-36). The proline at position 187 is located at the end of a beta sheet, adjacent to an exposed loop but distant from the substrate binding sites (7). Thus, the Pro187Ser substitution does not affect the binding of NAD(P)H or quinone substrates. Transfection studies showed that the variant allele results in a tenfold reduction of enzyme activity and a two- to threefold reduction in protein levels compared with wild-type transfectants (37). These findings indicate that the substitution results in a variant NQO1 protein that has minimal catalytic activity and is rapidly degraded by the ubiquitin/proteasomal system (38). The half-life of the 187Ser variant protein was 1.2 h compared to over 18 h for wild-type NQO1. Compared with cell lines and tissues possessing the wild type (C/C) genotype, the homozygous variant (T/T) has only 2 to 4% of the quinone reductase activity, whereas the heterozygous variant (C/T) has a 2-fold decrease in enzyme activity (37, 39). Although heterozygous cell lines have on average half the activity of wild-type cells, the ranges are overlapping. Another polymorphism has been observed at nucleotide 464 T/C, which results in an Arg139Trp substitution with diminished catalytic activity and decreased protein stability (40).

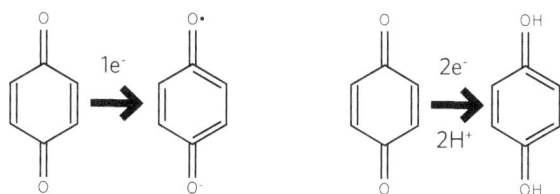

Figure 4.5.5. Enzymatic reduction reactions of a model quinone showing a one-electron reduction to the semiquinone (left) and a two-electron reduction to the hydroquinone (right).

The **NQO2** gene at 6p25 encodes a cytosolic quinone reductase, NQO2, which has 49% amino acid identity to NQO1 (41, 42). However, the NQO2 isoenzyme uses dihydronicotinamide riboside as electron donor instead of NAD(P)H (43). The crystal structure of NQO2 contains a metal binding site for copper or zinc that is not present in NQO1 (44). NQO1 and NQO2 have overlapping substrate specificities and the deficiency of either enzyme in NQO1-/- and NQO2-/- mice resulted in increased susceptibility to benzo[a]pyrene- and 7,12,-dimethylbenz[a] anthracene-induced skin carcinogenesis, presumably due to the inability to reduce the respective quinone metabolites (45). The NQO2 promoter sequence contains a 29 bp-insertion/deletion and a noncoding SNP, A+237C (rs2071002). The Ins-29 allele introduced transcriptional repressor Sp3 binding sites whereas +237A abolished a transcriptional activator Sp1 site (46).

Clinical Studies. NQO1 is expressed in many tissues, most abundantly in stomach and kidney. The enzyme is overexpressed in cancer of the lung, colon, pancreas, and breast relative to surrounding normal tissue (47, 48). In cancer cells, NQO1 is not only present in the cytosol but also detected in the nucleus by confocal and immunoelectron microscopy (49). Expression levels of three different reductase enzymes (NQO1, NADPH-cytochrome P450 reductase, NADH-cytochrome b5 reductase) in 69 human cancer cell lines varied among cell lines derived from the same tissue as well as between lines derived from different tissues (50). In general, NQO1 activity levels were higher than those observed for the other two reductases across the entire cell line panel. The human breast cancer cell lines MDA-MB-231 and MDA-MB-468 carry the homozygous 187Ser/Ser genotype and exhibit little or no NQO1 activity (15, 39).

The variant allele 464C (139Trp) occurs with a frequency of 5% in Caucasians and 4% in Asians (51). The frequency of the 609T (187Ser) allele varies more across ethnic groups, ranging from 25% in Caucasians to 44% in Asians (52). The observed frequency of homozygous 187Ser/Ser was 4.4% in Caucasians, 5.2% in African-Americans, and 20.3% in Asians. Individuals heterozygous or homozygous for the 187Ser allele have an increased risk of developing benzene poisoning and leukemias, specifically acute myelogenous leukemia (53-55). Meta-analysis of bladder and colorectal cancer showed a significant association with the variant 187Ser allele in Caucasians (5).

Several studies have examined the association of the variant 187Ser allele with breast cancer risk yielding inconsistent results. An American study (346 cases, 235 controls) and a Japanese study (237 cases, 640 controls) found no association (56, 57). In contrast, a European study of two separate populations involving a total of 1063 women (655 cases, 408 controls) observed a significant difference of genotype and allele frequencies (OR = 1.46; 95% CI 1.16 – 1.85; p = 0.001) (58). The

Shanghai Breast Cancer Study (1039 cases, 1121 controls) showed no association with the NQO1 genotype (59). However, in pre- and postmenopausal Chinese women with a history of oral contraceptive usage the wild-type 187Pro allele was associated with a lower breast cancer risk. A nested case-control study of postmenopausal women (505 cases, 502 controls) from the American Cancer Society Prevention II Nutrition Cohort found no association of the Pro187Ser SNP with breast cancer risk nor for variants in three other genes involved in iron-related oxidative stress pathways, nitric oxide synthase (NOS3), heme oxygenase (HO), and nuclear factor erythroid 2-related factor 2 (Nrf2) (60). When examined In relation to supplemental iron intake, carriage of three or more high-risk alleles resulted in a greater than twofold increased risk compared with women with no high-risk alleles (OR 2.39; 95% CI 1.09 – 5.26).

A Chinese hospital-based study (893 cases, 711 controls) examined the association of NQO2 polymorphisms with breast cancer risk (46). The 29-bp deletion allele and the +237C (rs2071002) allele were both significantly associated with reduced risk, notably for tumors with wild-type p53. The results were validated in an independent population set (403 familial/early onset cases, 1039 community-based controls).

Summary. NQO1 reduces quinones to hydroquinones and thereby bypasses the formation of semiquinones and prevents redox cycling. The reduced production of ROS and presumably oxidative DNA adducts is the basis for the postulated role of NQO1 as a defensive mechanism against the carcinogenic effect of quinones. A variant allele (Pro187Ser; rs1800566) results in a virtual null-type with absence of enzyme activity. About 4% Caucasians and 20% Asians are homozygous 187Ser/Ser. Epidemiological studies of the association of this variant allele with breast cancer risk have yielded inconsistent results.

References

1. Ernster L. *DT diaphorase: a historical review.* Chemica Scripta. 1987;27A:1-13.

2. Ross D, Kepa JK, Winski SL, Beall HD, Siegel D. *NAD(P)H:quinone oxidoreductase 1 (NQO1):chemoprotection, bioactivation, gene regulation and genetic polymorphisms.* Chem Biol Interact 2000;129:77-97.

3. Begleiter A, Fourie J. *Induction of NQO1 in cancer cells.* Methods Enzymol. 2004;382:320-51.

4. Jaiswal AK. *Human NAD(P)H:quinone oxidoreductase (NQO1) gene structure and induction by dioxin.* Biochemistry. 1991;30:10647-53.

5. Chao C, Zhang Z-F, Berthiller J,

Boffetta P, Hashibe M. *NAD(P)H:quinone oxidoreductase 1 (NQO1) pro187ser polymorphism and the risk of lung, bladder, and colorectal cancers: a meta-analysis.* Cancer Epidemiol Biomarkers Prev. 2006;15:979-87.

6. Gasdaska PY, Fisher H, Powis G. *An alternatively spliced form of NQO1 (DT-diaphorase) messenger RNA lacking the putative quinone substrate binding site is present in human normal and tumor tissues.* Cancer Res. 1995;55:2542-7.

7. Li R, Bianchet MA, Talalay P, Amzel LM. *The three-dimensional structure of NAD(P)H:quinone reductase, a flavoprotein involved in cancer chemoprotection and chemotherapy: mechanism of the*

two-electron reduction. Proc Natl Acad Sci USA. 1995;92:8846-50.

8. Faig M, Bianchet MA, Talalay P, Chen S, Winski S, Ross D, et al. *Structures of recombinant human and mouse NAD(P)H:quinone oxidoreductases: species comparison and structural changes with substrate binding and release.* Proc Natl Acad Sci USA. 2000;97:3177-82.

9. Siegel D, Gustafson DL, Dehn DL, Han JY, Boonchoong P, Berliner LJ, et al. *NAD(P)H:quinone oxidoreductase 1: role as a superoxide scavenger.* Mol Pharmacol. 2004;65:1238-47.

10. Montano MM, Katzenellenbogen BS. *The quinone reductase gene: A unique estrogen receptor-regulated gene that is activated by antiestrogens.* Proc Natl Acad Sci USA. 1997;94:2581-6.

11. Sreerama L, Hedge MW, Sladek NE. *Identification of a class 3 aldehyde dehydrogenase in human saliva and increased levels of this enzyme, glutathione S-transferases, and DT-diaphorase in the saliva of subjects who continually ingest large quantities of coffee or broccoli.* Clin Cancer Res. 1995;1:1153-63.

12. Yoxall V, Kentish P, Coldham N, Kuhnert N, Sauer MJ, Ioannides C. *Modulation of hepatic cytochromes P450 and phase II enzymes by dietary doses of sulforaphane in rats: Implications for its chemopreventive activity.* Int J Cancer. 2005;117:356-62.

13. Zhang Y, Talalay P, Cho C-G, Posner GH. *A major inducer of anticarcinogenic protective enzymes from broccoli: isolation and elucidation of structure.* Proc Natl Acad Sci USA. 1992;89:2399-403.

14. Benson AM, Hunkeler MJ, Talalay P. *Increase of NAD(P)H:quinone reductase by dietary antioxidants: possible role in protection against carcinogenesis and toxicity.* Proc Natl Acad Sci USA. 1980;77:5216-20.

15. Begleiter A, Leith MK, Curphey TJ, Doherty GP. *Induction of DT-diaphorase in cancer chemoprevention and chemotherapy.* Oncology Res. 1997;9:371-82.

16. Park HJ, Choi EK, Choi J, Ahn K-J, Kim EJ, Ji I-M, et al. *Heat-induced up-regulation of NAD(P)H:quinone oxidoreductase potentiates anticancer effects of β-lapachone.* Clin Cancer Res. 2005;11:8866-71.

17. Roy D, Liehr JG. *Temporary decrease in renal quinone reductase activity induced by chronic administration of estradiol to male Syrian hamsters.* J Biol Chem. 1988;263:3646-51.

18. Bianco NR, Perry G, Smith MA, Templeton DJ, Montano MM. *Functional implications of antiestrogen induction of quinone reductase: inhibition of estrogen-induced deoxyribonucleic acid damage.* Mol Endocrinol. 2003;17:1344-55.

19. Shen L, Pisha E, Huang Z, Pezzuto JM, Krol E, Alam Z, et al. *Bioreductive activation of catechol estrogen-ortho-quinones: aromatization of the B ring in 4-hydroxyequilenin markedly alters quinoid formation and reactivity.* Carcinogenesis. 1997;18:1096-101.

20. Nutter LM, Wu YY, Ngo EO, Sierra EE, Gutierrez PL, Abul-Hajj YJ. *An o-quinone form of estrogen produces free radicals in human breast cancer cells: correlation with DNA damage.* Chem Res Toxicol. 1994;7:23-8.

21. Prochaska HJ, Talalay P, Sies H. *Direct protective effect of NAD(P)H:quinone reductase against menadione-induced chemiluminescence of postmitochondrial fractions of mouse liver.* J Biol Chem. 1987;262:1931-4.

22. Ma Q, Wang R, Yang CS, Lu AYH. *Expression of mammalian DT-diaphorase in Escherichia coli: purification and characterization of the expresed protein.* Arch Biochem Biophys. 1990;283:311-7.

23. Chen HH, Ma JX, Forrest GL, Deng PS-K, Martino PA, Lee TD, et al. *Expression of rat liver NAD(P)H:quinone-acceptor oxidoreductase in Escherichia coli and mutagenesis in vitro at Arg-177.* Biochem J. 1992;284:855-60.

24. Chen S, Knox R, Wu K, Deng PS-K, Zhou D, Bianchet MA, et al. *Molecular basis of the catalytic differences among DT-diaphorase of human, rat, and mouse.* J Biol Chem. 1997;272:1437-9.

25. Giulivi C, Cadenas E. *One- and two-electron reduction of 2-methyl-1,4-naphthoquinone bioreductive alkylating agents: kinetic studies, free-radical production, thiol oxidation and DNA-strand-break formation.* Biochem J. 1994;301:21-30.

26. Workman P. *Enzyme-directed bioreductive drug development revisited: a commentary on recent progress and future prospects with emphasis on quinone anticancer agents and quinone metabolizing enzymes, particularly DT-diaphorase.* Oncol Res. 1994;6:461-75.

27. Joseph P, Jaiswal AK. *NAD(P)H:quinone oxidoreductase 1 (DT diaphorase) specifically prevents the formation of benzo[a]pyrene quinone-DNA adducts generated by cytochrome P4501A1 and P450 reductase.* Proc Natl Acad Sci USA.

1994;91:8413-7.

28. Arlt VM, Stiborova M, Henderson CJ, Osborne MR, Bieler CA, Frei E, et al. *Environmental pollutant and potent mutagen 3-nitrobensanthrone forms DNA adducts after reduction by NAD(P)H:Quinone oxidoreductase and conjugation by acetyltransferases and sulfotransferases in human hepatic cytosols.* Cancer Res. 2005;65:2644-52.

29. Asher G, Lotem J, Cohen B, Sachs L, Shaul Y. *Regulation of p53 stability and p53-dependent apoptosis by NADH quinone oxidoreductase 1.* Proc Natl Acad Sci USA. 2001;98:1188-93.

30. Anwar A, Dehn D, Siegel D, Kepa JK, Tang LJ, Pietenpol JA, et al. *Interaction of human NAD(P)H:quinone oxidoreductase 1 (NQO1) with the tumor suppressor protein p53 in cells and cell-free systems.* J Biol Chem. 2003;278:10368-73.

31. Asher G, Bercovich Z, Tsvetkov P, Shaul Y, Kahana C. *20S proteasomal degradation of ornithine decarboxylase is regulated by NQO1.* Mol Cell. 2005;17:645-55.

32. Asher G, Dym O, Tsvetkov P, Adler J, Shaul Y. *The crystal structure NAD(P)H quinone oxidoreductase 1 in complex with its potent inhibitor dicoumarol.* Biochemistry. 2006;45:6372-8.

33. Tsvetkov P, Asher G, Reiss V, Shaul Y, Sachs L, Lotem J. *Inhibition of NAD(P)H:quinone oxidoreductase 1 activity and induction of p53 degradation by the natural phenolic compound curcumin.* Proc Natl Acad Sci USA. 2005;102:5535-40.

34. Traver RD, Horikoshi T, Danenberg KD, Stadlbauer THW, Danenberg PV, Ross D, et al. *NAD(P)H:quinone oxidoreductase gene expression in human colon carcinoma cells: characterization of a mutation which modulates DT-diaphorase activity and mitomycin sensitivity.* Cancer Res. 1992;52:797-802.

35. Kuehl BL, Paterson JWE, Peacock JW, Paterson MC, Rauth AM. *Presence of a heterozygous substitution and its relationship to DT-diaphorase activity.* Br J Cancer. 1995;72:555-61.

36. Traver RD, Siegel D, Beall HD, Phillips RM, Gibson NW, Franklin WA, et al. *Characterization of a polymorphism in NAD(P)H: quinone oxidoreductase (DT-diaphorase).* Br J Cancer. 1997;75:69-75.

37. Misra V, Klamut HJ, Rauth AM. *Transfection of COS-1 cells with DT-diaphorase cDNA: role of a base change at position 609.* Br J Cancer.

1998;77:1236-40.

38. Siegel D, Anwar A, Winski SL, Kepa JK, Zolman KL, Ross D. *Rapid polyubiquitination and proteasomal degradation of a mutant form of NAD(P)H:quinone oxidoreductase 1.* Mol Pharmacol. 2001;59:263-8.

39. Siegel D, McGuinness SM, Winski SL, Ross D. *Genotype-phenotype relationships in studies of a polymorphism in NAD(P)H:quinone oxidoreductase 1.* Pharmacogenetics. 1999;9:113-21.

40. Pan SS, Forrest GL, Akman SA, Hu LT. *NAD(P)H:quinone oxidoreductase expression and mitomycin C resistance developed by human colon cancer HCT 116 cells.* Cancer Res. 1995;55:330-5.

41. Jaiswal AK. *Human NAD(P)H:quinone oxidoreductase 2.* J Biol Chem. 1994;269:14502-8.

42. Jaiswal AK, Bell DW, Radjendirane V, Testa JR. *Localization of human NQO1 gene to chromosome 16q22 and NQO2-6p25 and associated polymorphisms.* Pharmacogenetics. 1999;9:413-8.

43. Wu K, Knox R, Sun XZ, Joseph P, Jaiswal AK, Zhang D, et al. *Catalytic properties of NAD(P)H:quinone oxidoreductase-2 (NQO2), a dihydronicotinamide riboside dependent oxidoreductase.* Arch Biochem Biophys. 1997;347:221-8.

44. Foster CE, Bianchet MA, Talalay P, Zhao Q, Amzel LM. *Crystal structure of human quinone reductase type 2, a metalloflavoprotein.* Biochemistry. 1999;38:9881-6.

45. Iskander K, Paquet M, Brayton C, Jaiswal AK. *Deficiency of NRH:quinone oxidoreductase 2 increases susceptibility to 7,12-dimethylbenz(a)anthracene and benzo(a)pyrene-induced skin carcinogenesis.* Cancer Res. 2004;64:5925-8.

46. Yu KD, Di GH, Yuan WT, Fan L, Wu J, Hu Z, et al. *Functional polymorphisms, altered gene expression and genetic association link NRH:quinone oxidoreductase 2 to breast cancer with wild-type p53.* Human molecular genetics. 2009;18:2502-17.

47. Cullen JJ, Hinkhouse MM, Grady M, Gaut AW, Liu J, Zhang YP, et al. *Dicumarol inhibition of NADPH:quinone oxidoreductase induces growth inhibition of pancreatic cancer via a superoxide-mediated mechanism.* Cancer Res. 2003;63:5513-20.

48. Schlager JJ, Powis G. *Cytosolic NAD(P)H:(quinone-acceptor) oxidoreductase in human normal* and tumor tissue: effects of cigarette smoking and alcohol. Int J Cancer. 1990;45:403-9.

49. Winski SL, Koutalos Y, Bentley DL, Ross D. *Subcellular localization of NAD(P)H:quinone oxidoreductase 1 in human cancer cells.* Cancer Res. 2002;62:1420-4.

50. Fitzsimmons SA, Workman P, Grever M, Paull K, Camalier R, Lewis AD. *Reductase enzyme expression across the National Cancer Institute tumor cell line panel: correlation with sensitivity to mitomycin C and EO9.* J Natl Cancer Inst. 1996;88:259-69.

51. Gaedigk A, Tyndale RF, Jurima-Romet M, Sellers EM, Grant DM, Leeder JS. *NAD(P)H:quinone oxidoreductase: polymorphisms and allele frequencies in Caucasian, Chinese and Canadian native indian and Inuit populations.* Pharmacogenetics. 1998;8:305-13.

52. Kelsey KT, Ross D, Traver RD, Christiani DC, Zuo Z-F, Spitz MR, et al. *Ethnic variation in the prevalence of a common NAD(P)H quinone oxidoreductase polymorphism and its implications for anti-cancer chemotherapy.* Br J Cancer. 1997;76:852-4.

53. Rothman N, Smith MT, Hayes RB, Traver RD, Hoener B-A, Campleman S, et al. *Benzene poisoning, a risk factor for hematological malignancy, is associated with the NQO1 609C → T mutation and rapid fractional excretion of chlorzoxazone.* Cancer Res. 1997;57:2839-42.

54. Smith MT, Wang Y, Kane E, Rollinson S, Wiemels JL, Roman E, et al. *Low NAD(P)H:quinone oxidoreductase 1 activity is associated with increased risk of acute leukemia in*

adults. Blood. 2001;97:1422-6.

55. Bauer AK, Faiola B, Abernethy DJ, Marchan R, Pluta LJ, Wong VA, et al. *Genetic susceptibility to benzene-induced toxicity: role of NADPH: quinone oxidoreductase-1.* Cancer Res. 2003;63:929-35.

56. Siegelmann-Danieli N, Buetow KH. *Significance of genetic variation at the glutathione S-transferase M1 and NAD(P)H:quinone oxidoreductase 1 detoxification genes in breast cancer development.* Oncology 2002;62:39-45.

57. Hamajima N, Matsuo K, Iwata H, Shinoda M, Yamamura Y, Kato T, et al. *NAD(P)H: quinone oxidoreductase 1 (NQO1) C609T polymorphism and the risk of eight cancers for Japanese.* Int J Clin Oncol. 2002;7:103-8.

58. Menzel H-J, Sarmanova J, Soucek P, Berberich R, Grunewald K, Haun M, et al. *Association of NQO1 polymorphism with spontaneous breast cancer in two independent populations.* Br J Cancer. 2004;90:1989-94.

59. Fowke JH, Shu X-O, Dai Q, Jin F, Cai Q, Gao Y-T, et al. *Oral contraceptive use and breast cancer risk: modification by NAD(P)H:quinone oxoreductase (NQO1) genetic polymorphisms.* Cancer Epidemiol Biomarkers Prev. 2004;13:1308-15.

60. Hong CC, Ambrosone CB, Ahn J, Choi JY, McCullough ML, Stevens VL, et al. Genetic variability in iron-related oxidative stress pathways (Nrf2, NQO1, NOS3, and HO-1), iron intake, and risk of postmenopausal breast cancer. Cancer Epidemiol Biomarkers Prev. 2007;16:1784-94.

dihydrodiol-epoxides that form covalent adducts with DNA. Thus, EPHX1 is important for its dual functional role in detoxification as well as bioactivation.

The EPHX1 gene is located at 1q42.1 and contains nine exons that encode a 53 kDa protein composed of 455 amino acids (2, 3). The coding region contains eight non-synonymous polymorphisms (4). The two most common polymorphisms are Tyr113His (rs1051740) in exon 3 and His139Arg (rs2234922) in exon 4 (5). Functional analysis of recombinant wild-type and variant enzymes showed that the 113Tyr/139His combination was twice as active as the other alleles (4). However, no significant differences were evident in the reaction rates when enzyme activities were analyzed in human liver microsomal fractions from individuals with variant genotypes in codons 113 and 139 (4). These results suggest that structural differences encoded by the Tyr113His and His139Arg variant alleles do not exert a major effect on *in vivo* EPHX1 activity. Other studies suggest that differences in EPHX1 activity are more likely to result from variation in protein stability than from changes in catalytic activity although differences between calculated half-lives were not significant: 113Tyr/139His 15.2 hr, 113His/139His 10.7 hr, 113Tyr/139Arg 16.9 hr, 113His/139Arg 16.0 hr (6, 7). Additional polymorphisms in the 5' flanking region of the EPHX1 gene may also contribute to the range of functional enzyme expression existing in human populations (8). EPHX1 is expressed in most organs including normal and malignant breast tissues (9)

A population-based study (47,089 individuals from the Danish general population) found no association of the EPHX1 genotype (Tyr113His, rs1051740; His139Arg, rs2234922) with breast cancer risk regardless of smoking history (10). Similarly, several case-control studies found no association between EPHX1 genotype and breast cancer risk stratified by both menopausal and smoking status (11, 12). One study (1,246 cases, 664 controls) observed a decrease in risk associated with 113His/His homozygosity (odds ratio 0.60; 95% CI 0.43 – 0.84; p = 0.05). Interestingly, the risk was increased in women with GSTM1 null and GSTT1 null genotypes who were either heterozygous 113Tyr/His (OR 2.02; 95% CI 1.19 – 3.45; p = 0.009) or homozygous 113His/His (OR 3.54; 95% CI 1.29 – 9.72; p = 0.14) (13).

4.5.3. EPOXIDE HYDROLASE

Epoxides are organic three-membered oxygen compounds that arise from oxidative metabolism of endogenous and exogenous substrates via chemical and enzymatic oxidation processes. The resultant epoxides are highly reactive and may be mutagenic and carcinogenic. The microsomal enzyme epoxide hydrolase (EPHX1) catalyzes the hydrolysis of epoxides to *trans*-dihydrodiols and is responsible for the detoxification of a wide variety of genotoxic epoxides (1). On the other hand, EPHX1 produces the bay-region diol-epoxides of carcinogenic polyaromatic hydrocarbons, such as benzo[a]pyrene, which are further activated by cytochrome P450 enzymes to generate highly reactive

References

1. Fretland AJ, Omiecinski CJ. *Epoxide hydrolases: biochemistry and molecular biology.* Chem Biol Interact. 2000;129:41-59.

2. Skoda RC, Demierre A, McBride OW, Gonzalez FJ, Meyer UA. *Human microsomal xenobiotic epoxide hydrolase.* J Biol Chem. 1988;263:1549-54.

3. Hartsfield Jr. JK, Sutcliffe MJ, Everett ET, Hassett C, Omiecinski CJ, Saari JA. *Assignment of microsomal epoxide hydrolase (EPHX1) to human chromosome 1q42.1 by in situ hybridization.* Cytogenet Cell Genet. 1998;83:44-5.

4. Hosagrahara VP, Rettie AE, Hassett C, Omiecinski CJ. *Functional analysis of human microsomal epoxide hydrolase genetic variants.* Chem Biol Interact. 2004;150:149-59.

5. Hassett C, Robinson KB, Beck

NB, Omiecinski CJ. *The human microsomal epoxide hydrolase gene (EPXH1): complete nucleotide sequence and structural characterization.* Genomics. 1994;23:433-42.

6. Laurenzana EM, Hasett C, Omiecinski CJ. *Post-transcriptional regulation of human microsomal epoxide hydrolase.* Pharmacogenetics. 1998;8:157-67.

7. Omiecinski CJ, Hassett C, Vinayak H. *Epoxide hydrolase - polymorphism and role in toxicology.* Toxicol Lett. 2000;112-113:365-70.

8. Raaka S, Hassett C, Omiecinski CJ. *Human microsomal epoxide hydrolase: 5'-flanking region genetic polymorphisms.* Carcinogenesis. 1998;19:387-93.

9. Coller JK, Fritz P, Zanger UM, Siegle I, Eichelbaum M, Kroemer HK, et al. *Distribution of microsomal epoxide hydrolase in humans: An immunohistochemical study in normal tissues, and benign and malignant tumours.* Histochem J. 2001;33:329-36.

10. Lee J, Dahl M, Nordestgaard BG. *Genetically lowered microsomal epoxide hydrolase activity and tobacco-related cancer in 47,000 individuals.* Cancer Epidemiol Biomarkers Prev. 2011;20:1673-82.

11. de Assis S, Ambrosone CB, Wustrack S, Krishnan S, Freudenheim JL, Shields PG. *Microsomal epoxide hydrolase variants are not associated with risk of breast cancer.* Cancer Epidemiol Biomarkers Prev. 2002;11:1697-8.

12. Justenhoven C, Hamann U, Schubert F, Zapatka M, Pierl CB, Rabstein S, et al. *Breast cancer: a candidate gene approach across the estrogen metabolic pathway.* Breast Cancer Res Treat. 2008;108:137-49.

13. Spurdle AB, Chang JH, Byrnes GB, Chen X, Dite GS, McCredie MR, et al. *A systematic approach to analysing gene-gene interactions: polymorphisms at the microsomal epoxide hydrolase EPHX and glutathione S-transferase GSTM1, GSTT1, and GSTP1 loci and breast cancer risk.* Cancer Epidemiol Biomarkers Prev. 2007;16:769-74.

4.5.4. MYELOPEROXIDASE, LACTOPEROXIDASE, AND NADPH OXIDASE

Human milk harbors three heme-containing enzymes: lactoperoxidase (LPO), myeloperoxidase (MPO), and NADPH oxidase (NOX) (1-4). LPO is secreted by mammary ductal epithelial cells into the breast ducts and hence into milk. Milk also contains abundant neutrophils, which are rich in MPO and NOX located in phagolysosomes (2). Milk that is stored in the mammary ducts is potentially an ideal medium for bacterial growth. All three enzymes protect milk against the growth of bacteria by catalyzing the production of reactive oxygen species (ROS) and other oxidants that destroy microorganisms (1, 3). NOX catalyzes the NADPH-dependent reduction of oxygen to form superoxide, which can react with itself to form H_2O_2 that serves as substrate for LPO and MPO. MPO produces hypochlorous acid (HOCl, the reactive agent in bleach) from H_2O_2 and chloride anion (Cl⁻) during the neutrophil's respiratory burst.

Lactoperoxidase. The LPO gene is arranged tail-to-tail with the MPO gene at 17q12-24 (4). The LPO gene spans 28 kb, contains 12 exons, and encodes a 80 kDa protein of 712 amino acids.

Partially purified LPO from human milk in the presence of H_2O_2 and calf thymus DNA was shown to oxidatively metabolize aromatic and heterocyclic amines to DNA-binding intermediates. The order of DNA-adduct formation was benzidine > ABP > IQ > MeIQx > PhIP (5). The results from this *in vitro* study are supported by an analysis of DNA-carcinogen adducts present in DNA isolated from epithelial cells from human breast milk. Adducts derived from three main groups of carcinogens, namely heterocyclic amines (PhIP), aromatic amines (ABP), and polycyclic aromatic hydrocarbons (benzo[a] pyrene) were identified (6).

Myeloperoxidase. The MPO gene is organized in a similar manner as the LPO gene and encodes a protein of 719 amino acids, of which 51% are identical to those of LPO (4). The initial translation product of the MPO gene is an ~80 kDa protein, which undergoes proteolytic removal of the 41 amino acid signal peptide, followed by N-glycosylation with the incorporation of a mannose-rich side chain to generate the ~90 kDa apoproMPO (2). The covalent addition of heme converts the enzymatically inactive apoproMPO to the active proMPO, which undergoes two sequential proteolytic reactions (7). The first reaction, removal of the N-terminal 125 amino acid proregion results in a 72 – 75 kDa protein, which undergoes cleavage to generate the 467 amino acid heavy subunit (57 kDa) and the 112 amino acid light subunit (12 kDa) that associate as a heavy-light protomer. Mature MPO has a mass of ~150 kDa and consists of a pair of heavy-light protomers whose heavy subunits are linked by a disulfide bond along their long axis (2).

An Alu sequence in the MPO promoter contains a polymorphic site at -463 G/A of a composite Sp1-thyroid hormone-retinoic acid response element (8). The G allele strongly binds the Sp1 transcription factor to increase MPO expression whereas the A allele leads to lower transcription rates. The frequency of the -463 A allele is about 20% in Caucasians (9). The -463 A allele was associated with reduced B[a]P diol epoxide DNA adduct levels in the skin of coal tar ointment-treated patients with atopic dermatitis (10). In contrast, the variant allele was associated with significantly higher DNA adduct levels in breast tissue from reduction mammoplasties and in normal tissue adjacent to breast cancer (11). However, the ^{32}P-postlabeling method did not allow adduct identification in breast tissues whereas the identity of the BPDE DNA adducts in skin was determined by HPLC. In epidemiological studies, the -463 A allele has been associated with decreased risk of lung cancer by some investigators but not by others (9, 12). A study of 1,011 breast cancer cases and 1,067 controls showed that having at least one A allele was associated with an overall 13% reduction in breast cancer risk (13). The risk reduction was greatest among women with higher consumption of fruits and vegetables (odds ratio 0.75; 95% CI 0.58 – 0.97).

In summary, these findings indicate that women are exposed to several classes of dietary and environmental carcinogens, which are activated by locally expressed enzymes like LPO and MPO and form DNA adducts in ductal epithelial cells, the cells from which most breast cancers arise (1, 14).

NADPH Oxidase. The NOX family consists of five trans-membranous enzymes, NOX1 – NOX5, located in the membranes of phagolysosomes and plasma membranes of selected cells, such as vascular smooth muscle, colon and mammary epithelium (3). NOX is inactive in resting neutrophils but is activated by exposure to microorganisms or inflammatory mediators, resulting in the non-mitochondrial production of ROS. The NOX of phagocytes (Phox; mainly neutrophils and phagocytes) is a complex of six proteins consisting of the catalytic subunit gp91phox (otherwise known as NOX2), the regulatory subunits p22phox, p47phox, p40phox, p67phox and p67phox and the small GTPase RAC (15).

Mutations in the NOX subunit genes cause chronic granulomatous diseases, such as X-linked chronic granulomatous disease (CGD). CGD cells have a low capacity for phagocytosis and persistent bacterial infections occur resulting in the formation of granulomas (16). NOX also plays a major role in atherosclerosis and hence SNPs in NOX subunit genes have been examined in cardiovascular disease (17, 18). A nonsynonymous SNP (Tyr72His; rs4673) in the CYBA gene, which encodes p22phox was associated with increased risk of non-Hodgkin lymphoma (19).

A study of SNPs in oxidative stress-related genes and postmenopausal breast cancer risk (2476 cases, 4748 controls) observed a decreased risk of breast cancer for an intronic SNP (rs3794624) of the CYBA gene (OR_{trend} 0.91; 95% CI 0.85 – 0.98) (20).

References

1. Josephy PD. *The role of peroxidase-catalyzed activation of aromatic amines in breast cancer.* Mutagenesis. 1996;11:3-7.

2. Klebanoff SJ. *Myeloperoxidase: friend and foe.* J Leukoc Biol. 2005;77:598-625.

3. Lambeth JD. *NOX enzymes and the biology of reactive oxygen.* Nat Rev Immunol. 2004;4:181-9.

4. Ueda T, Sakamaki K, Kuroki T, Yano I, Nagata S. *Molecular cloning and characterization of the chromosomal gene for human lactoperoxidase.* Eur J Biochem. 1997;243:32-41.

5. Gorlewska-Roberts KM, Teitel CH, Lay Jr. JO, Roberts DW, Kadlubar FF. *Lactoperoxidase-catalysed activation of carcinogenic aromatic and heterocyclic amines.* Chem Res Toxicol. 2004;17:1659-66.

6. Gorlewska-Roberts K, Green B, Fares M, Ambrosone CB, Kadlubar FF. *Carcinogen-DNA adducts in human breast epithelial cells.* Environ Mol Mutagenesis. 2002;39:184-92.

7. Pinnix IB, Guzman GS, Bonkovsky HL, Zaki SR, Kinkade J, J.M. *The post-translational processing of myeloperoxidase is regulated by the availability of heme.* Arch Biochem Biophys. 1994;312:447-58.

8. Piedrafita FJ, Molander RB, Vansant G, Orlova EA, Pfahl M, Reynolds WF. *An Alu element in the myeloperoxidase promoter contains a composite SP1-Thyroid Hormone-Retinoic Acid response element.* J Biol Chem. 1996;271:14412-20.

9. Feyler A, Voho A, Bouchardy C, Kuokkane K, Dayer P, Hirvonen A, et al. *Point: Myeloperoxidase -463G- A Polymorphism and Lung Cancer Risk 1.* Cancer Epidemiol Biomarkers Prev. 2002;11:1550-4.

10. Rojas M, Godschalk R, Alexandrov K, Cascorbi I, Kriek E, Ostertag J, et al. *Myeloperoxidase -463A variant reduces benzo(a)pyrene diol epoxide DNA adducts in skin of coal tar patients.* Carcinogenesis. 2001;22:1015-8.

11. Brockstedt U, Krajinovic M, Richer C, Mathonnet G, Sinnett D, Wolfgang P, et al. *Analyses of bulky DNA adduct levels in human breast tissue and genetic polymorphisms of cytochromes P450 (CYPs),myeloperoxidase (MPO), quinone oxidoreductase (NQO1), and glutathione S-transferases (GSTs).* Mutation Res. 2002;516:41-7.

12 Xu L, Liu G, Miller DP, Zhou W, Lynch TJ, Wain JC, et al. *Counterpoint: The myeloperoxidase -463G-A polymorphism does not decrease lung cancer suscptibility in Caucasians.* Cancer Epidemiol Biomarkers Prev. 2002;11:1555-9.

13. Ahn J, Gammon MD, Santalla RM, Gaudet MM, Britton JA, Teitelbaum SL, et al. *Myeloperoxidase genotype, fruit and vegetable consumption, and breast cancer risk.* Cancer Res. 2004;64:7634-9.

14. Williams JA. *Single nucleotide polymorphisms, metobolic activation and environmental carcinogenesis: why molecular epidemiologists should think about enzyme expression.* Carcinogenesis. 2001;22:209-14.

15. Vignais PV. *The superoxide-generating NADPH oxidase: structural aspects and activation mechanism.* Cell Mol Life Sci. 2002;59:1428-59.

16. Song E, Jaishankar GB, Saleh H, Jithpratuck W, Sahni R, Krishnaswamy G. *Chronic granulomatous disease: a review of the infectious and inflammatory complications.* Clin Mol Allergy. 2011;9:10.

17. Curtiss LK. *Reversing atherosclerosis?* N Engl J Med. 2009;360:1144-6.

18. San Jose G, Fortuno A, Beloqui O, Diez J, Zalba G. *NADPH oxidase CYBA polymorphisms, oxidative stress and cardiovascular diseases.* Clin Sci (Lond). 2008;114:173-82.

19. Lan Q, Zheng T, Shen M, Zhang Y, Wang SS, Zahm SH, et al. *Genetic polymorphisms in the oxidative stress pathway and susceptibility to non-Hodgkin lymphoma.* Hum Genet. 2007;121:161-8.

20. Seibold P, Hein R, Schmezer P, Hall P, Liu J, Dahmen N, et al. *Polymorphisms in oxidative stress-related genes and postmenopausal breast cancer risk.* Int J Cancer. 2011;129:1467-76.

CHAPTER 5

DNA Repair

5.1. INTRODUCTION

Each of the ~10^{13} cells in the human body receives tens of thousands of DNA lesions per day (1).

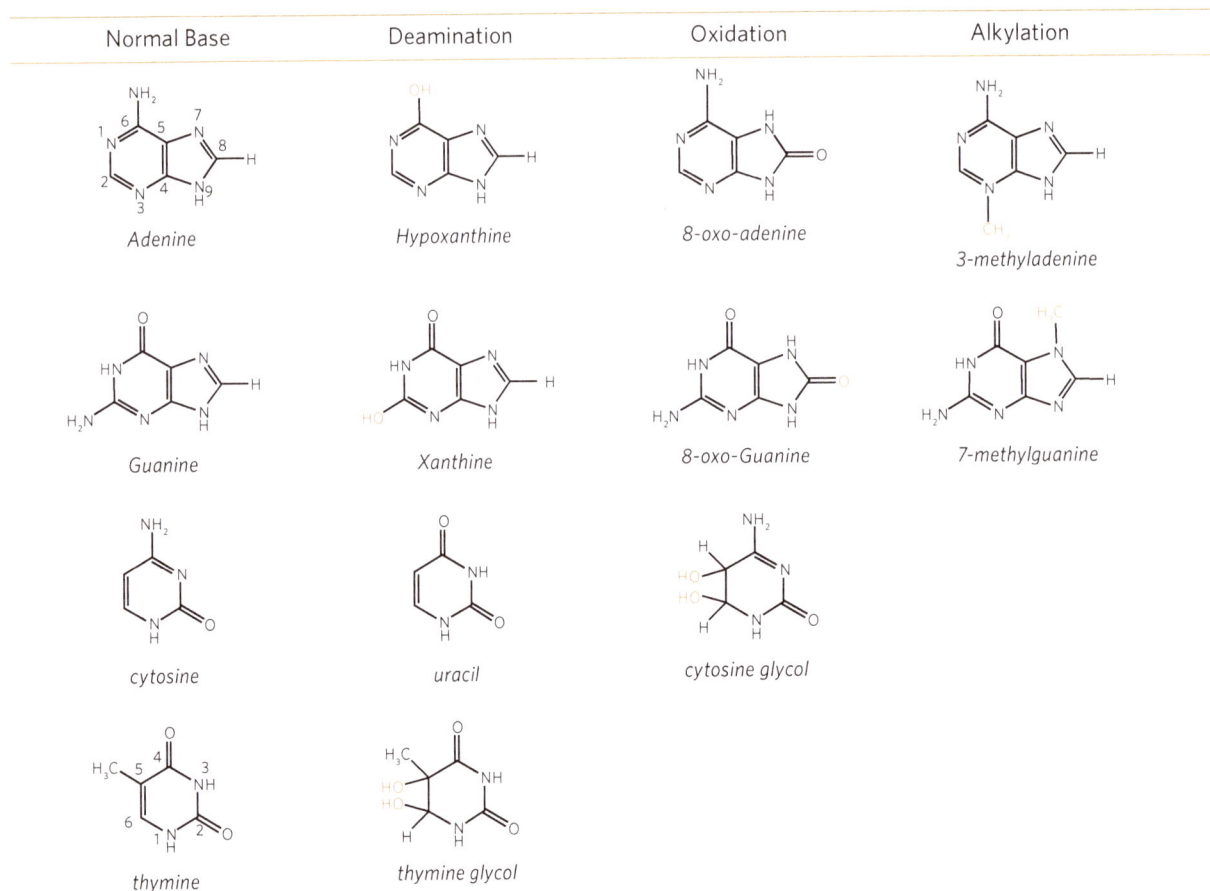

Normal Base	Deamination	Oxidation	Alkylation
Adenine	Hypoxanthine	8-oxo-adenine	3-methyladenine
Guanine	Xanthine	8-oxo-Guanine	7-methylguanine
cytosine	uracil	cytosine glycol	
thymine	thymine glycol		

Figure 5.1.1. Overview of common spontaneous base modifications generated by deamination, oxidation, and alkylation (methylation)

Some alterations arise spontaneously, such as base mismatches or strand breaks occurring during replication. Taking into account that a normal diploid human cell contains about 5 x 10^9 base pairs (bp), DNA mismatches arising during replication are rare events with an estimated rate lower than 1 x 10^{-9}, i.e., less than one error for every billion bp (2). A more common type of DNA alteration is the continuous loss of bases through spontaneous hydrolysis of the base N-glycosyl bond, which generates abasic or apurinic/apyrimidinic (AP) sites at an estimated rate of 2,000 to 10,000 AP sites per cell per day (3, 4). Guanine and adenine are liberated from DNA at similar rates, whereas cytosine and thymine are lost at approximately 5% of the rate of purines (1). Thus, depurination accounts for the majority of AP sites. In addition to the intrinsic lability of the N-glycosyl bond, DNA base residues are susceptible to hydrolytic deamination. Cytosine is the main base

undergoing spontaneous deamination to yield uracil, while its homologue, 5-methylcytosine, yields thymine (Figure 5.1.1.) (1, 5). By comparison with the hydrolytic conversion of cytosine to uracil, the deamination of purines is a minor reaction. For example, adenine is converted to hypoxanthine in DNA at 2–3% of the rate of cytosine deamination (6). The rate of deamination of guanine to xanthine is similar. As described in Chapter 2.3. (Reactive Oxygen Species, ROS), DNA is also subject to oxidation by ROS although the nucleus is a poorly oxygenated cellular compartment, which lacks detectable oxygen metabolism (7). Moreover, histones quench the generation of oxygen radical-inflicted DNA lesions (8). Nevertheless, oxidation of DNA occurs, the most important reaction converting guanine residues to 8-oxo-guanine (Figure 5.1.1.). The oxidation of guanine to 8-oxo-guanine and the deamination of cytosine to uracil are the two major reactions causing spontaneous DNA damage, each taking place at an estimated rate of 400 to 1000 times per day in a human cell (1, 5, 9). Oxidation also affects pyrimidine residues leading to ring-saturation in form of pyrimidine hydrates such as thymine and cytosine glycols (Figure 5.1.1.). These derivatives have lost the 5,6 double bond and, as a consequence, their planar ring structure (10). In addition to oxygen, cells contain several other small reactive molecules capable of damaging DNA. The most important of these is S-adenosylmethione (SAM), which is an efficient methyl group donor used as cofactor in most cellular trans-methylation reactions. As a weak alkylating agent, SAM can cause nonenzymatic methylation of proteins and DNA. The main DNA targets are the ring nitrogens of purine residues, leading to the formation of 3-methyladenine and 7-methylguanine (Figure 5.1.1.) (11, 12). At an intracellular SAM concentration of 4×10^{-5} M, nonenzymatic methylation of DNA generated about 600 3-methyladenine residues per day in a human cell (12).

Historically, studies of chemical carcinogenesis have focused on exogenous or xenobiotic compounds, but there is growing evidence that endogenously generated compounds may contribute to the carcinogenic process (1, 13-15). In Chapter 2, I showed how several types of DNA adducts are generated by two classes of endogenous compounds, namely catechol estrogens, which are formed during the oxidative metabolism of estrogens, and malon-dialdehyde (MDA), which is formed enzymatically as a byproduct of the cyclooxygenase pathway of arachidonic acid metabolism, as well as nonenzymatically during the peroxidation of PUFAs (16, 17). In principle, any endogenous DNA alteration occurring 'spontaneously' by a physiological mechanism is potentially mutagenic. For example, AP sites are devoid of bases and therefore non-instructional for DNA polymerases. Consequently, AP sites may give rise to base substitutions and cause a mutation unless the AP site is repaired beforehand (18).

Chapter 3 documented how exogenous carcinogens cause the formation of DNA lesions. Women are regularly exposed to dietary and environmental carcinogens including heterocyclic amines (e.g., 2-amino-1-methyl-6-phenylimidazo[4,5-b]pyridine, PhIP), aromatic amines (e.g., 4-aminobiphenyl, ABP), and polycyclic aromatic hydrocarbons (e.g., benzo[a]pyrene, B[a]P), which have been shown to react with DNA in breast ductal epithelial cells, the cells from which breast cancers arise (19). DNA isolated from exfoliated ductal epithelial cells in breast milk obtained from healthy, nonsmoking mothers contained detectable levels of PhIP adducts (mean value of 4.7 adducts/10^7 nucleotides), ABP adducts (mean 4.7 adducts/10^7 nucleotides), and B[a]P adducts (mean 1.9 adducts/10^7 nucleotides). PhIP and B[a]P also induced DNA single-strand breaks in viable exfoliated cells isolated from breast milk (20). Figure 5.1.2. summarizes the DNA lesions generated by these endogenous and exogenous compounds.

The fate of a cell exposed to a carcinogen depends on the extent of DNA damage (14, 21-23). Cells that incur severe DNA damage undergo apoptosis and therefore are no longer candidates for malignant transformation. When less severe DNA damage is induced, the cellular response allows repair of the damage. However, if the damage fails to be repaired, polymerases may insert an incorrect base opposite a given lesion and the mutagenic lesions may be propagated and lead to tumor initiation by activating oncogenes or inactivating tumor suppressor genes (24-26). Thus, the ultimate fate, i.e., apoptosis or carcinogenesis, depends on the extent of DNA damage and the ability to repair DNA. To combat threats posed by DNA damage, cells have evolved mechanisms – collectively termed the DNA-damage response – to detect DNA lesions, signal their presence and promote their repair (22). The wide diversity of DNA lesions necessitates multiple DNA repair mechanisms. Five overlapping DNA repair pathways utilize either excision (i - iii) or double-strand break (iv and v) repair mechanisms (27, 28).

    i. Base excision repair
    ii. Nucleotide excision repair
    iii. Mismatch repair
    iv. Homologous recombination repair
    v. Nonhomologous end-joining repair

Base excision repair (BER). Structural studies of DNA adducts have revealed a spectrum of lesions ranging from nondistorting ones in which the ligand fits into the major groove with minimal perturbation of the double helix to denaturing lesions in which the ligand is intercalated into the helix with local unwinding and base displacement (29). The nondistorting lesions are typically substrates for BER, which functions as the main guardian against damage due to cellular metabolism. These small chemical alterations of bases resulting from oxidation by reactive oxygen species, methylation and deamination may impede transcription and replication, but frequently result in miscoding. Thus, BER protects cells from deleterious effects secondary to modified or missing bases and thereby is particularly relevant for preventing mutagenesis.

Nucleotide excision repair (NER) deals with a wide class of helix-distorting lesions that interfere with base pairing and generally obstruct transcription and normal replication. The bulky lesions are excised as oligonucleotides of about 30 bp in length by a multiprotein complex (30). While BER is mostly concerned with damage of endogenous origin, most NER lesions arise from exogenous sources.

Mismatch repair (MMR) corrects DNA replication errors and removes nucleotides mispaired by DNA polymerases. MMR also removes insertion/deletion loops (ranging from one to ten or more bases) that result from slippage during replication of repetitive sequences or during recombination. The hallmark of a defective MMR system is microsatellite

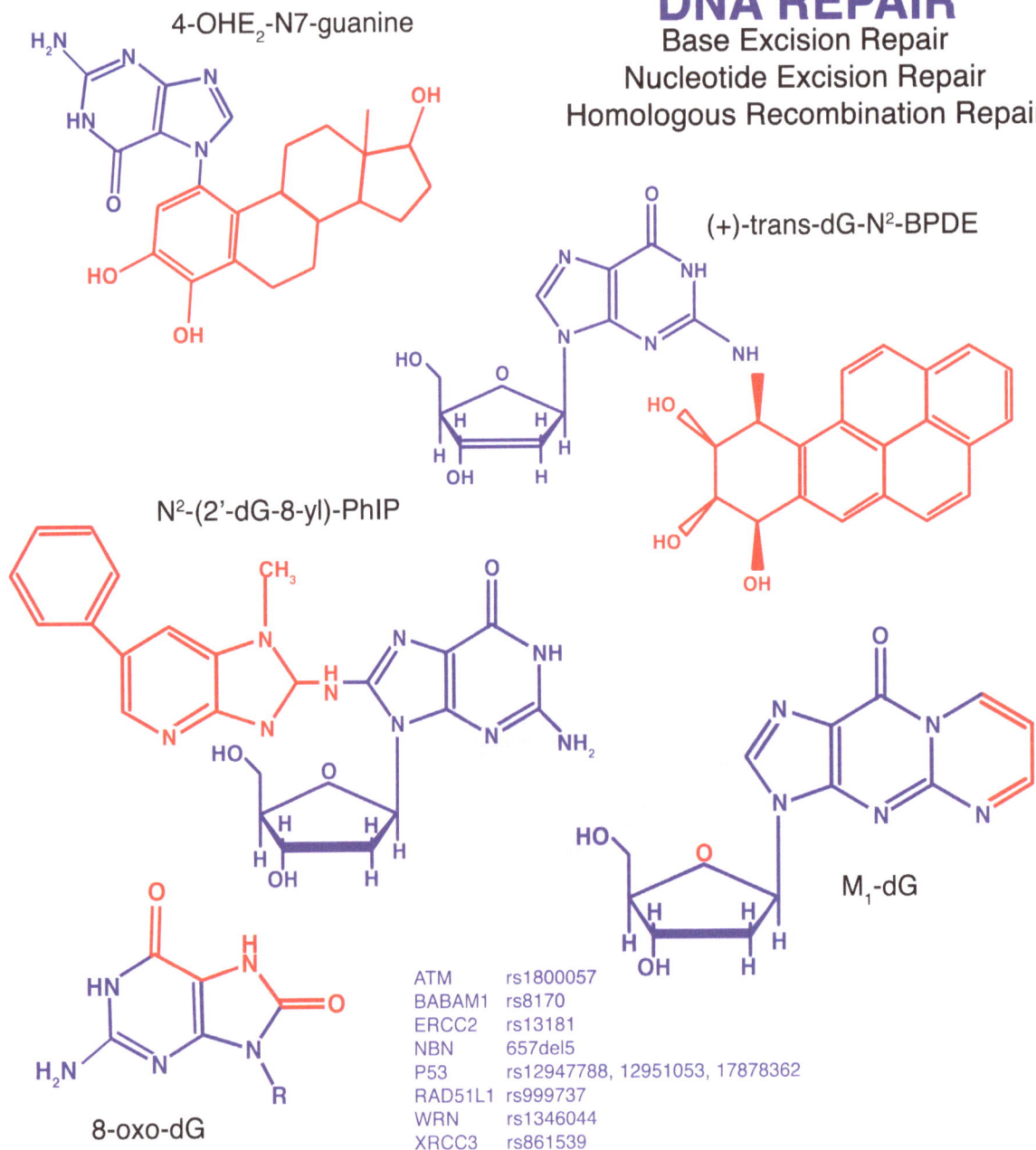

Figure 5.1.2. Overview of DNA adducts generated by endogenous and exogenous carcinogens and found in benign or malignant human mammary epithelial cells.

A causal relation to breast cancer is difficult to establish because DNA repair mechanisms are extremely efficient erasing these adducts almost immediately after they are formed. The insert lists DNA repair proteins and enzymes with genetic variants implicated in breast cancer risk by genome-wide association studies. Details are discussed in subsequent sections.

instability (MSI), the consequence of non-repaired, slippage-induced strand misalignments that occur during DNA replication. MMR defects dramatically increase mutation rates and thereby accelerate oncogenesis.

Homologous recombination repair (HRR) and **nonhomologous end-joining repair (NHEJR).** Chromosomal breaks in form of single-strand breaks (SSBs) or double-strand DNA breaks (DSBs) can occur spontaneously as a result of stalled replication forks during S phase and after exposure to DNA-damaging agents, such as ionizing radiation or X-rays, free radicals, and numerous chemicals (27, 28, 31). When DSBs occur, the complementary copies of DNA are severed in the same place and therefore cannot act as templates for each other. If unrepaired or repaired inappropriately, DSBs can lead to chromosome loss, deletions, duplications, or translocations. Cells use two mechanisms to accomplish DSB repair: HRR and NHEJR. HRR takes place after replication and uses the intact copy on the sister chromatid as template to properly align and seal the broken ends in an error-free manner. Otherwise cells rely on NHEJR, which simply links ends of a DSB together without any template. Although NHEJR is designed to repair, the end-joining is sometimes associated with gain or loss of a few nucleotides if internal microhomologies are used for annealing before sealing. Thus, both HRR and NHEJR are mechanisms to repair DSBs but NHEJR itself is error-prone and may thereby enhance mutagenesis.

References

1. Lindahl T. *Instability and decay of the primary structure of DNA.* Nature. 1993;362:709-15.

2. McCulloch SD, Kunkel TA. *The fidelity of DNA synthesis by eukaryotic replicative and translesion synthesis polymerases.* Cell Res. 2008;18:148-61.

3. Atamna H, Cheung I, Ames BN. *A method for detecting abasic sites in living cells: age-dependent changes in base excision repair.* Proc Natl Acad Sci USA. 2000;97:686-91.

4. Lindahl T, Nyberg B. *Rate of depurination of native deoxyribonucleic acid.* Biochemistry. 1972;11:3610-8.

5. Stivers JT, Jiang YL. *A mechanistic perspective on the chemistry of DNA repair glycosylases.* Chem Rev. 2003;103:2729-59.

6. Karran P, Lindahl T. *Hypoxanthine in deoxyribonucleic acid: Generation by heat-induced hydrolysis of adenine residues and release in free form by a deoxyribonucleic acid glycosylase from calf thymus.* Biochemistry. 1980;19:6005-11.

7. Joenje H. *Genetic toxicology of oxygen.* Mutation Res. 1989;219:193-208.

8. Ljungman M, Hanawalt PC. *Efficient protection against oxidative DNA damage in chromatin.* Mol Carcinogenesis. 1992;5:264-9.

9. Kunkel TA. *The high cost of living.* Trends Genet. 1999;15:93-4.

10. Breimer LH. *Molecular mechanisms of oxygen radical carcinogenesis and mutagenesis: The role of DNA base damage.* Mol Carcinogenesis. 1990;3:188-97.

11. Barrows LR, Magee PN. *Nonenzymatic methylation of DNA by S-adenosylmethionine in vitro.* Carcinogenesis. 1982;3:349-51.

12. Rydberg B, Lindahl T. *Nonenzymatic methylation of DNA by the intracellular methyl group donor S-adenosyl-L-methionine is a potentially mutagenic reaction.* EMBO J. 1982;1:211-6.

13. Ames BN, Gold LS. *Endogenous mutagens and the causes of aging and cancer.* Mutat Res. 1991;250:3-16.

14. Loeb LA, Harris CC. *Advances in chemical carcinogenesis: a historical review and prospective.* Cancer Res. 2008;68:6863-72.

15. Marnett LJ. *Oxyradicals and DNA damage.* Carcinogenesis. 2000;21:361-70.

16. Esterbauer H, Schaur RJ, Zollner H. *Chemistry and biochemistry of 4-hydroxynonenal, malonaldehyde and related aldehydes.* Free Radical Biol Med. 1991;11:81-128.

17. Janero DR. *Malondialdehyde and thiobarbituric acid-reactivity as diagnostic indices of lipid peroxidation and peroxidative tissue injury.* Free Radical Biol Med. 1990;9:515-40.

18. Loeb LA, Preston BD. *Mutagenesis by apurinic/apyrimidinic sites.* Ann Rev Genetics. 1986;20:201-30.

19. Gorlewska-Roberts K, Green B, Fares M, Ambrosone CB, Kadlubar FF. *Carcinogen-DNA adducts in human breast epithelial cells.* Environ Mol Mutagenesis. 2002;39:184-92.

20. Martin FL, Cole KJ, Williams JA, Millar BC, Harvey D, Weaver G, et al. *Activation of genotoxins to DNA-damaging species in exfoliated breast milk cells.* Mutation Res. 2000;470:115-24.

21. Harper JW, Elledge SJ. *The DNA damage response: ten years after.* Mol Cell. 2007;28:739-45.

22. Jackson SP, Bartek J. *The DNA-damage response in human biology and disease.* Nature. 2009;461:1071-8.

23. Thompson CB. *Apoptosis in the pathogenesis and treatment of disease.* Science. 1995;267:1456-62.

24. El-Bayoumy K. *Environmental carcinogens that may be involved in human breast cancer etiology.* Chem Res Toxicol. 1992;5:585-90.

25. Hiraku Y, Kawanishi S. *Oxidative DNA damage and apoptosis induced by benzene metabolites.* Cancer Res. 1996;56:5172-8.

26. Hollstein M, Sidransky D, Vogelstein B, Harris CC. *p53 mutations in human cancers.* Science. 1991;253:49-53.

27. Hoeijmakers JH. *DNA damage, aging, and cancer.* N Engl J Med. 2009;361:1475-85.

28. Hoeijmakers JHJ. *Genome maintenance mechanisms for preventing cancer.* Nature. 2001;411:366-74.

29. Patel DJ, Mao B, Gu Z, Hingerty BE, Gorin A, Basu AK, et al. *Nuclear magnetic resonance solution structures of covalent aromatic amine-DNA adducts and their mutagenic relevance.* Chem Res Toxicol. 1998;11:391-407.

30. de Laat WL, Jaspers NG, Hoeijmakers JH. *Molecular mechanism of nucleotide excision repair.* Genes Dev. 1999;13:768-85.

31. Marians KJ. *Replication and recombination intersect.* Curr Opin Genet Dev. 2000;10:151-6.

5.2. BASE EXCISION REPAIR

5.2.1. INTRODUCTION

DNA base excision repair (BER) counteracts the mutagenic effects of various kinds of base alterations that do not significantly distort the secondary structure of the double helix. This includes spontaneous base modifications generated by deamination, oxidation, and alkylation (Figure 5.2.1.). The BER pathway is initiated by a lesion-specific DNA glycosylase, which recognizes and removes the damaged base (1-4). Table 5.2.1. summarizes the most common base modifications of endogenous origin and their glycosylases.

The DNA glycosylases flip the suspected base out of the helix by DNA backbone compression to accommodate it in an internal cavity of the protein (4, 10-13). Inside the enzyme, the damaged base is removed from the DNA by cleaving the N-C1' glycosylic bond between the base and the sugar-phosphate backbone. The enzymatic removal of the bases leaves apurinic and apyrimidinic (AP or abasic) sites in the DNA backbone. AP sites also arise spontaneously in DNA. In general, nucleic acids will undergo spontaneous decomposition in solution. RNA is particularly vulnerable because of the 2'-hydroxyl group of ribose, which renders the phosphodiester bond of RNA molecules susceptible to hydrolysis, especially in the presence of divalent cations, such as Mg^{2+} and Ca^{2+} (14). Reduction of the ribose moiety to the unusual sugar, deoxyribose, is associated with greatly increased chemical stability of the phosphodiester bond in DNA. However, removal of the sugar 2'-OH group makes the N-glycosyl bond more labile, leading to release of the base under physiological conditions (2, 15, 16). Thus, RNA is characterized by instability of the phosphodiester bonds and DNA by instability of the base-sugar linkage. Guanine and adenine are liberated from DNA at similar rates, whereas cytosine and thymine are lost at approximately 5% of the rate of purines (17). Thus, depurination accounts for the majority of AP sites. Due to the intrinsic lability of the DNA glycosyl bonds, AP sites are generated spontaneously at an estimated rate of 2,000 to 10,000 AP sites per cell per day (2, 5). The number of AP sites differs between tissues, being greatest in brain, followed by colon and heart, and then liver, lung, and kidney (18). In summary, AP sites are a common intermediate of the BER pathway, which arise as a consequence of removal of altered bases by DNA glycosylases or as spontaneous detachment of normal bases from the deoxyribose-phosphate backbone. Because of the base loss, AP sites are non-instructional for DNA polymerases. Consequently, AP sites may give rise to base substitutions and are potentially mutagenic (16). Translesional synthesis on DNA templates containing a single AP site shows preferential incorporation of dAMP opposite the lesion, a phenomenon known as the "A rule" (19).

Table 5.2.1. DNA Base Modifications and corresponding DNA Glycosylases (2, 4-9)

Base Modification		DNA Glycosylase	
Name (Frequency per cell per day)	Gene (locus)	Name (Number of molecules per cell)	Polymorphism
AP site (2,000 - 10,000)	APE1 (14q11)	Apurinic/Apyrimidinic endonuclease 1, APE1 (300,000 - 7,000,000)	Asp148Glu Leu104Arg Glu126Asp Arg237Ala Asp283Gly
8-oxoG, FapyG (1000)	OGG1 (3p26)	8-oxoG-DNA Glycosylase 1; OGG1 (123,000 ± 22,000)	Ser326Cys
Uracil in ss/ds DNA (400)	UDG (12q13)	Uracil DNA glycosylase, UDG or Uracil-N-glycosylase, UNG (178,000 ± 20,000)	
7-methylguanine; 7-MeG 3-methyladenine; 3-MeA (600 – 4000)	AAG	7-methylguanine DNA glycosylase, 3-methyladenine DNA glycosylase, MAG	
1-methyladenine (1-MeA) 3-methylcytosine (3-MeC)	ABH2 (12q24.1) ABH3 (11q11)		
Uracil in ss/ds DNA	SMUG1	Single-strand Selective Monofunctional Uracil DNA Glycosylase, SMUG1	
U:G, T:G	TDG	Thymine DNA glycoslyase, TDG	
U:G, T:G in CpG sequences	MBD4	Methyl-CpG binding domain 4, MBD4	
A:8-oxoG	MUTYH		
5-guanidinohydantoin (Gh), spiroiminohydantoin (Sp)	NEIL		
Thymine glycol, oxidized pyrimidines	NTH1		

The main enzymes that initiate repair of AP sites are AP endonucleases, which recognize the lesion and hydrolyze the phosphodiester immediately 5' to the abasic site, resulting in a single-strand break with normal 3'-OH and baseless 5'-deoxyribose phosphate termini (20). The 5'-terminal deoxyribose-phosphate group created by the incision must be removed prior to re-ligation. This process is achieved differently by two sub-pathways of BER, "short-patch" and "long-patch", depending upon the size of the repair gap (Figure 5.2.1.). In short-patch BER, polymerase β (Pol β) attaches a single nucleotide to the newly generated 3'-OH and thereby displaces the 5' baseless sugar-phosphate, which it subsequently removes by deploying its inherent AP-lyase activity (21). DNA ligase III seals the remaining nick to restore the original DNA sequence. The protein XRCC1, which has no catalytic activity, serves as a scaffold by interacting with both Pol β and DNA ligase III (22). Thus, short-patch BER is a highly coordinated pathway with functional interactions between DNA glycosylases and APE1, APE1 and Pol β, and Pol β, DNA ligase III, and XRCC1 (23-27). A variant of the short-patch pathway involves a bifunctional DNA glycosylase/AP-lyase. The other components, APE1, Pol β, and DNA ligase III are still required but Pol β does not participate in the removal of the baseless sugar-phosphate. In long-patch BER, an oligonucleotide containing two to six nucleotides is inserted by polymerases δ, ε, or β, supported by two essential proteins, replication factor C (RFC) and proliferating cell nuclear antigen (PCNA) (28, 29). The short oligonucleotide overhang is excised by the flap structure-specific endonuclease 1 (FEN-1) and the nick then sealed by DNA ligase I (30-36). Short-patch BER is the predominant

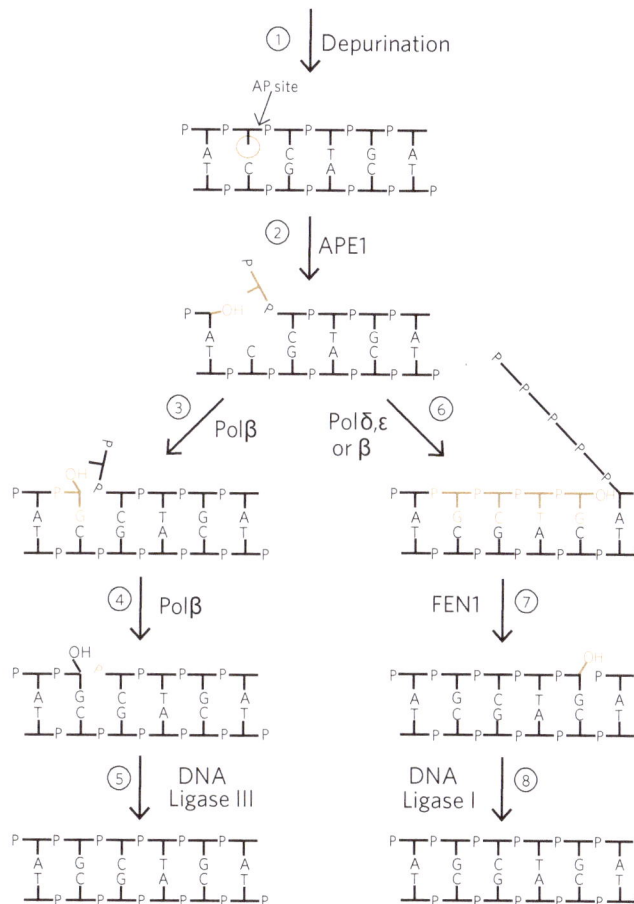

Figure 5.2.1. Overview of BER pathway.
The initial reaction is (1) a depurination by hydrolysis of the N-glycosylic bond between base and sugar, leaving an apurinic (AP) site in the double-stranded DNA. (2) AP endonuclease APE1 cleaves 5' to the AP site, creating a single-strand break. The bottom sequence of reactions shows the BER pathway with its two branches. In the predominant short-patch pathway (left) (3) DNA polymerase β (Pol β) attaches dGTP to the newly generated 3'-OH and thereby displaces the 5' baseless sugar-phosphate, which (4) Pol β subsequently removes by deploying its inherent AP-lyase activity. (5) DNA ligase III seals the remaining nick to restore the original DNA sequence. In the alternate long-patch pathway (right), (6) DNA polymerases δ, ε, or β insert an oligonucleotide containing two to six nucleotides. (7) The oligonucleotide overhang is excised by the endonuclease FEN1 and (8) the nick sealed by DNA ligase 1.

pathway and long-patch BER serves as back-up pathway to remove, for example, modified abasic sites that are resistant to the AP-lyase activity of Pol β (37, 38). The ratio of BER proteins at the site of repair can also influence the choice between short- and long-patch pathways (39). Treatment of mice with the peroxisome proliferator, WY-14,643, a carcinogen causing oxidative stress, resulted in the induction of genes specific for the long-patch but not the short-patch BER (40). Mouse knockout studies revealed that deficiencies in DNA glycosylases are associated with normal development whereas deficiencies in APE1, DNA polymerase β, ligase I, FEN1, and XRCC1 are associated with an embryonic lethal phenotype (41).

In summary, the BER pathway repairs damage to DNA that arises spontaneously as a result of deamination, oxidation, and alkylation events. The BER is also responsible for repairing small, non-helix distorting lesions that may be induced by chemical carcinogens and for the repair of AP sites, which arise spontaneously or as intermediates in the DNA repair process. In line with the predominance of these types of DNA damage, it is estimated that the BER pathway repairs up to 1,000,000 nucleotides/cell/day (42). This is accomplished by the well-coordinated interaction of enzymes (DNA glycosylases, AP endonucleases, DNA polymerases, ligases) and structural proteins (RFC, PCNA, XRCC1). Most of the genes encoding these BER components display nonsynonymous polymorphisms and other variants, none of which has shown a consistent association with breast cancer risk.

References

1. Friedberg EC. *Inroads into base excision repair II. The discovery of DNA glycosylases.* DNA Repair. 2004;3:1531-6.

2. Lindahl T, Nyberg B. *Rate of depurination of native deoxyribonucleic acid.* Biochemistry. 1972;11:3610-8.

3. Scharer OD, Jiricny J. *Recent progress in the biology, chemistry and structural biology of DNA glycosylases.* BioEssays. 2001;23:270-81.

4. Stivers JT, Jiang YL. *A mechanistic perspective on the chemistry of DNA repair glycosylases.* Chem Rev. 2003;103:2729-59.

5. Atamna H, Cheung I, Ames BN. *A method for detecting abasic sites in living cells: age-dependent changes in base excision repair.* Proc Natl Acad Sci USA. 2000;97:686-91.

6. Cappelli E, Hazra T, Hill JW, Slupphaug G, Bogliolo M, Frosina G. *Rates of base excision reapir are not solely dependent on levels of initiating enzymes.* Carcinogenesis. 2001;22:387-93.

7. Chen DS, Herman T, Demple B. *Two distinct human DNA diesterases that hydrolyze 3'-blocking deoxyribose fragments from oxidized DNA.* Nucleic Acids Res. 1991;19:5907-14.

8. Kunkel TA. *The high cost of living.* Trends Genet. 1999;15:93-4.

9. Rydberg B, Lindahl T. *Nonenzymatic methylation of DNA by the intracellular methyl group donor S-adenosyl-L-methionine is a potentially mutagenic reaction.* EMBO J. 1982;1:211-6.

10. Barnes DE, Lindahl T. *Repair and genetic consequences of endogenous DNA base damage in mammalian cells.* Annu Rev Genet. 2004;38:445-76.

11. David SS, O'Shea VL, Kundu S. *Base-excision repair of oxidative DNA damage.* Nature. 2007;447:941-50.

12. Kunkel TA, Wilson SH. *Push and pull of base flipping.* Nature. 1996;384:25-6.

13. Slupphaug G, Mol CD, Kavli B, Arvai AS, Krokan HE, Tainer JA. *A nucleotide-flipping mechanism from the structure of human uracil-DNA glycosylase bound to DNA.* Nature. 1996;384:87-92.

14. Lindahl T. *Irreversible heat inactivation of transfer ribonucleic acids.* J Biol Chem. 1967;242:1970-3.

15. Lindahl T, Andersson A. *Rate of chain breakage at apurinic sites in double-stranded deoxyribonucleic acid.* Biochemistry. 1972;11:3618-23.

16. Loeb LA, Preston BD. *Mutagenesis by apurinic/apyrimidinic sites.* Ann Rev Genetics. 1986;20:201-30.

17. Lindahl T. *Instability and decay of the primary structure of DNA.* Nature. 1993;362:709-15.

18. Nakamura J, Swenberg JA. *Endogenous apurinic/apyrimidinic sites in genomic DNA of mammalian tissues.* Cancer Res. 1999;59:2522-6.

19. Shibutani S, Takeshita M, Grollman AP. *Translesional synthesis on DNA templates containing a single abasic site.* J Biol Chem. 1997;272:13916-22.

20. Kelley MR, Kow YW, Wilson DM. *Disparity between DNA base excision repair in yeast and mammals: Translational implications.* Cancer Res. 2003;63:549-54.

21. Matsumoto Y, Kim K. *Excision of deoxyribose phosphate residues by DNA polymerase β during DNA repair.* Science. 1995;269:699-702.

22. Kubota Y, Nash RA, Klungland A, Schar P, Barnes DE, Lindahl T. *Reconstitution of DNA base excision-repair with purified human proteins: interaction between DNA polymerase β and the XRCC1 protein.* EMBO J. 1996;15:6662-70.

23. Bennett RA, Wilson DM, Wong D, Demple B. *Interaction of human apurinic endonuclease and DNA polymerase β in the base excision repair pathway.* Proc Natl Acad Sci USA. 1997;94:7166-9.

24. Mol CD, Izumi T, Mitra S, Tainer JA. *DNA-bound structures and mutants reveal abasic DNA binding by APE1 DNA repair and coordination.* Nature. 2000;403:451-6.

25. Srivastava DK, Vande Berg BJ, Prasad R, Molina JT, Beard WA, Tomkinson AE, et al. *Mammalian abasic site base excision repair. Identification of the reaction sequence and rate-determining steps.* J Biol Chem. 1998;273:21203-9.

26. Waters TR, Gallinari P, Jiricny J, Swann PF. *Human thymine DNA glycosylase binds to apurinic sites in DNA but is displaced by human apurinic endonuclease 1.* J Biol Chem. 1999;274:67-74.

27. Wilson SH, Kunkel TA. *Passing the baton in base excision repair.* Nature Struct Biol. 2000;7:176-8.

28. Jonsson ZO, Hubscher U. *Proliferating cell nuclear antigen: more than a clamp for DNA polymerases.* BioEssays. 1997;19:967-75.

29. Mossi R, Hubscher U. *Clamping down on clamps and clamp loaders. The eukaryotic replication factor C.* Eur J Biochem. 1998;254:209-16.

30. Klungland A, Lindahl T. *Second pathway for completion of human DNA base excision-repair: reconstitution with purified proteins and requirement for DNase IV (FEN1).* EMBO J. 1997;16:3341-8.

31. Lieber MR. *The FEN-1 family of structure-specific nucleases in eukaryotic DNA replication, recombination and repair.* BioEssays. 1997;19:233-40.

32. Liu Y, Kao HI, Bambara RA. *Flap endonuclease1: A central component of DNA metabolism.* Annu Rev Biochem. 2004;73:589-615.

33. Matsumoto Y, Kim K, Hurwitz J, Gary R, Levin DS, Tomkinson AE, et al. *Reconstitution of proliferating cell nuclear antigen-dependent repair of apurinic/apyrimidinic sites with purified human proteins.* J Biol Chem. 1999;274:33703-8.

34. Pascucci B, Stucki M, Jonsson ZO, Dogliotti E, Hubscher U. *Long patch base excision repair with purified human proteins.* J Biol Chem. 1999;274:33696-702.

35. Prasad R, Singhal RK, Srivastava DK, Molina JT, Tomkinson AE, Wilson SH. *Specific interaction of DNA Polymerase β and DNA Ligase I in a multiprotein base excision repair complex from bovine testis.* J Biol Chem. 1996;271:16000-7.

36. Shen B, Singh P, Liu R, Qiu J, Zheng L, Finger LD, et al. *Multiple but dissectible functions of FEN-1 nucleases in nucleic acid processing, genome stability and diseases.* BioEssays. 2005;27:717-29.

37. Fortini P, Parlanti E, Sidorkina OM, Laval J, Dogliotti E. *The type of DNA glycosylase determines the base excision repair pathway in mammalian cells.* J Biol Chem. 1999;274:15230-6.

38. Fortini P, Pascucci B, Parlanti E, D'Errico M, Simonelli V, Dogliotti E. *8-Oxoguanine DNA damage: at the crossroad of alternative repair pathways.* Mutation Res. 2003;531:127-39.

39. Sukhanova MV, Khodyreva SN, Lebedeva NA, Prasad R, Wilson SH, Lavrik OI. *Human base excision repair enzymes apurinic/apyrimidinic endonuclease1 (APE1), DNA polymerase β and poly(ADP-ribose) polymerase 1: interplay between strand-displacement DNA synthesis and proofreading exonuclease activity.* Nucleic Acids Res. 2005;33:1222-9.

40. Rusyn I, Asakura S, Pachkowski BF, Bradford BU, Denissenko MF, Peters JM, et al. *Expression of base excision DNA repair genes is a sensitive biomarker for in vivo detection of chemical-induced chronic oxidative stress: Identification of the molecular source of radicals responsible for DNA damage by peroxisome proliferators.* Cancer Res. 2004;64:1050-7.

41. Wilson DM, Thompson LH. *Life without DNA repair.* Proc Natl Acad Sci USA. 1997;94:12754-7.

42. Holmquist GP. *Endogenous lesions, S-phase-independent spontaneous mutations, and evolutionary strategies for base excision repair.* Mutation Res. 1998;400:59-68.

5.2.2. DNA GLYCOSYLASES

The enzymes central to the BER pathway are the DNA glycosylases, which recognize and remove specific modified bases. The low frequency of damaged bases requires an extremely fast process of interrogating millions of base pairs of undamaged DNA. Because DNA glycosylases consume no biochemical energy during the search for damage, they must rely solely on thermally-driven translocation along vast expanses of the genome. This rapid search process was visualized by using single-molecule detection to track the real-time movement of a DNA glycosylase, 8-oxoG-DNA glycosylase (OGG1), along a normal DNA duplex (1). OGG1 was found to slide along DNA as a consequence of Brownian motion with a diffusion constant approaching the theoretical upper limit for one-dimensional diffusion, indicating that OGG1 checks millions of base pairs per second. Since each damaged base contains a relatively minor structural change from the normal base, the process also requires perfect geometric alignment of the damaged base within the active site pocket. Thus, DNA glycosylases must combine speed and a high level of quality control, which is accomplished in a multi-step process that includes enzyme-initiated DNA bending, disruption of the base pair, extrusion of the damaged nucleotide from the interior of the DNA helix, and placement of the target base into an extrahelical base-specific pocket of the enzyme (2-4). This DNA glycosylase-driven process is often referred to as 'base flipping' but might be described more accurately as 'nucleotide flipping' because the entire nucleotide is rotated out of the helix to accommodate the base in the base-specific pocket (5).

At least ten different human DNA glycosylases have been cloned to date, each with unique substrate specificities (Table 5.2.1.). Extensive reviews have been published that provide detailed discussions of DNA glycosylases (4-8). Here, I present a limited overview of several of these

enzymes all of which have in common that they remove modified bases in a free form and thereby leave an abasic or AP site in the DNA. DNA glycosylases can be classified based on structural and functional characteristics. The latter classification distinguishes monofunctional and bifunctional glycosylases (8, 9). For bifunctional DNA glycosylases, the AP site is acted upon by an AP lyase activity inherent in the glycosylase itself. For monofunctional DNA glycosylases, the enzyme will protect the AP site until acted upon by an AP endonuclease. In both cases, a strand break is formed that needs further processing by other proteins (lyases and/or nucleases) in order to remove the sugar-phosphate residue remaining at the 3'- or 5'-end, respectively. Uracil DNA glycosylase (UDG) is an example of a monofunctional enzyme whereas OGG1 represents a bifunctional enzyme. Structural studies have shown that DNA glycosylases fall into two families (5, 8). The first family includes UDG as a prototype, which contains a central four-stranded, parallel twisted β-sheet flanked by α-helices. This motif is involved in pyrimidine binding and glycosidic bond hydrolysis. The second structural family includes OGG1, which contains a helix-hairpin-helix (HhH) motif that binds DNA in a non-sequence specific manner via the formation of hydrogen bonds between protein backbone nitrogens and DNA phosphate groups (10). The core fold of the HhH motif is characterized by two α-helical N- and C-terminal domains. The N-terminal motif typically has four α-helices whereas the C-terminal domain has six or seven α-helices. The hairpin loop contains a conserved sequence and an aspartate ~20 residues away that is essential for the catalytic activity.

8-oxo-7,8-dihydroguanine (8-oxoG) and other Guanine Oxidation Products

As described in Chapter 2.3. (Reactive Oxygen Species), 8-oxo-7,8-dihydroguanine (8-oxoG or GO), an oxidatively damaged form of guanine, is an important mutagenic DNA lesion because it can form a stable base pair with adenine as well as with cytosine. Thus, DNA polymerases can mispair 8-oxoG with adenine during replication, generating G:C/T:A transversions. To prevent 8-oxoG-induced mutagenesis, cells possess a mechanism called the GO system, which consists of three components, i.e., MutT, MutM, and MutY in *E.coli*, and their human homologs MTH, 8-oxoG-DNA glycosylase 1 (OGG1), and MUTYH. The MTH protein is a phosphatase that specifically converts 8-oxo-dGTP in the nucleotide pool into 8-oxo-dGMP to keep the oxidized nucleotides from being incorporated during replication (11, 12). Gene knockout mice deficient in MTH are viable but develop a greater number of tumors in lung, liver, and stomach than wild-type mice (13). Both MUTYH and OGG1 are 8-oxoG-specific DNA glycosylases. The BER enzyme OGG1 removes 8-oxoG from the 8-oxoG:C base pair so that subsequent processing by other enzymes in the BER pathway can restore the G:C base pair. However, if this does not occur, and replication takes place, then MUTYH intercepts the resultant 8-oxoG:A base pair and excises the adenine residues incorporated inappropriately by DNA polymerases opposite 8-oxoG. Subsequent processing of the AP site via the BER pathway provides an opportunity to create an 8-oxo-G:C substrate for OGG1 and elimination of 8-oxoG (2, 6, 14-16). The oxidation of guanine leads not only to 8-oxoG but also to the formation of several other products including 2,6-diamino-4-hydroxy-5-formamido-pyrimidine (FapyG) and two hydantoins, spiroiminodihydan-toin (Sp) and 5-guanidinohydantoin (Gh) (17) (see Chapter 2.3.3., Figure 2.3.5.). Sp can be generated by oxidation of G and 8-oxoG by a large number of oxidants and, depending on the conditions, Gh is formed in addition to Sp (18). When the substrate is a nucleoside or single-stranded DNA, the main product is Sp, whereas Gh predominates in double-stranded DNA (19). In *E. coli*-based mutagenesis assays, 8-oxoG was mildly mutagenic (3%) whereas both hydantoin lesions were greater than 95% mutagenic, mediating G-to-C and/or G-to-T transversions (20, 21).

8-oxoG-DNA Glycosylase 1 (OGG1)

In all organisms, 8-oxoG is primarily repaired by the BER pathway, which is initiated in mammalian cells by the action of OGG1. OGG1 recognizes 8-oxodG:dC duplexes as substrate and catalyzes the hydrolysis of the N-glycosyl bond linking 8-oxoG to the sugar phosphate backbone, generating an AP site. OGG1 is also endowed with an AP lyase activity that can incise the phosphodiester bond immediately 3' of the AP site, yielding a 3'-terminal sugar phosphate (22, 23). However, in the presence of the abundant cellular AP endonuclease APE1, AP sites resulting from the removal of 8-oxoG are primarily processed by APE1 (24). OGG1 does not recognize the hydantoins Sp or Gh as substrates, but can catalyze the excision of FapyG from DNA, although less efficiently than the removal of 8-oxoG (25, 26).

The OGG1 gene at 3p26 spans ~16.7 kb and encodes 8 exons (22). Alternative splicing results in the formation of several splice variants that have been classified into two major types (27, 28). Type 1 splice variants end with exon 7, whereas type 2 variants end with exon 8. All variants share the same N-terminus with a common mitochondrial-targeting signal but each possesses a unique C-terminus. OGG1 type 1a encodes a 36-kDa protein containing 345 amino acids with a nuclear localization signal at its C-terminus. The nuclear type 1a protein is the most abundant OGG1 isoform in normal and malignant cells (28, 29). The N-terminal region of OGG1 contributes to the 8-oxoG-binding pocket and two HhH-GPD domains (a helix-hairpin-helix structural element followed by a Gly/Pro-rich loop and a conserved Asp residue) provide both the catalytic and DNA-binding functions of OGG1 (9, 30). OGG1 searches for 8-oxoG by intrahelical interrogation of normal base pairs. Residue 114Phe serves as a sensor of the stability and/or deformability of base pairs. The side chain

of 114Phe is wedged into the helix from the minor groove resulting in severe bending of the DNA and buckling of the base pair (31). The 8-oxoG:C base pair is less stable than G:C or A:T in withstanding the insertion of 114Phe. The identified 8-oxoG lesion is extruded from the DNA helix and inserted into an extrahelical pocket composed of amino acid residues 319Phe, 315Gln, 42Gly, and 253Cys. In this active site pocket, 319Phe and 253Cys sandwich 8-oxoG, while 315Gln amide-NH_2 and a water molecule tightly bound to the residue cooperate to recognize O6 of the oxoG, and the side-chain carbonyl of 315Gln and a second tightly bound water form hydrogen bonds with N1, NH_2, and O6 of 8-oxoG (30). In addition, 268Asp is required to deprotonate 249Lys and 270His is needed to protonate 249Lys. The catalytically active residue 249Lys forms a transient covalent imino enzyme DNA intermediate that displaces the 8-oxoG base and promotes conjugate elimination of the 3'-phosphodies-ter through Schiff base chemistry (26, 32).

Gene knockout mice defective in OGG1 are viable but accumulate higher levels of 8-oxoG lesions compared with wild-type controls and exhibit elevated spontaneous mutations, especially when exposed to higher levels of oxidative stress (33, 34). However, OGG1-/- mice showed no increase in tumorigenesis, most likely because of the protective effect of MYH. On the other hand, double knockout mice defective in both OGG1 and MYH developed lung tumors associated with an age-dependent accumulation of 8-oxoG lesions and mutations in codon 12 of the K-ras oncogene (35, 36). The human breast cancer cell line, HCC1937, is deficient in OGG1 protein expression and contains elevated levels of 8-oxoG (37).

Mutations in the OGG1 gene have been identified in human lung, kidney, and colon tumors (25, 38, 39). OGG1 plays a role in neurodegenerative disorders (40). Four of 14 patients with Alzheimer's disease contained OGG1 mutations in their brain lesions, specifically non-synon-ymous mutations Ala53Thr, Ala288Val, and single base deletions (796C), which alter the C-terminus of OGG1 (41). At least 20 validated sequence variants have been identified in the OGG1 gene. A non-synonymous polymorphism occurs in exon 7 at codon Ser326Cys with a variant allele frequency of about 20% in Caucasians and 40% in Asians (42). Functional investigations of enzyme activity associated with the 326Ser and Cys alleles found no difference in two studies and higher activity for the 326Ser wild type in another (43-45). Seven epidemiological studies showed an increased risk of lung cancer associated with the 326Cys/Cys genotype (46). However, three studies of breast cancer found no risk association with the Ser326Cys polymorphism (47-49). Nine additional OGG1 polymorphisms are located in the 3' non-coding region, of which two, 7143A/G and 11657A/G, were found to be associated with increased prostate cancer risk (50). However, none of the polymorphisms, haplotypes, or diplotypes were associated with breast cancer risk (48).

MUTYH

The 8-oxoG residues that escape OGG1-initiated repair can generate misincorporation of an adenine during replication, thereby producing 8-oxoG/A mismatches. In mammalian cells, the removal of the mispaired adenine is catalyzed by MUTYH, the human homolog of the E.coli MutY enzyme (45, 51, 52). MUTYH relies heavily on the initial recognition of 8-oxoG to locate adenine bases for excision (53). The MUTYH gene at 1p34.3-32.1 is 11 kb long and encodes 16 exons, which undergo alternative splicing. The main 52 kDa MUTYH protein consists of 535 amino acids and contains binding sites for other BER proteins, such as proliferating cell nuclear antigen (PCNA) at the C terminus and trimeric replication protein A (RPA) at the N terminus (54). Binding of the AP endonuclease APE1 at a region around residue 300 enhances the formation of MUTYH-DNA complexes and stimulates the catalytic activity of MUTYH, which resides in the N-terminal region (55). MUTYH also interacts with the mismatch repair enzymes MSH2/MSH6 resulting in increased DNA binding and glycosylase activity (56). Both MUTYH and MSH6 bind PCNA, which may act as coordinator of two different repair pathways, MUTYH-dependent BER and MMR that contribute to the removal of the adenine misincorporated opposite 8-oxoG.

Hereditary colorectal cancer is caused by several syndromes, including MUTYH-associated polyposis (MAP), an autosomal recessive disorder characterized by mutations in both alleles of the MUTYH gene (2, 57, 58). 59, 60). Several founder mutations were found to be associated with a decrease in MUTYH expression and/or activity (61). The reduced capacity of the MUTYH variants to recognize and repair 8-oxoG:A mismatches leads to increased numbers of G-to-T transversions in the APC gene and eventually results in inactivation of the APC protein (62). A population-based study of Jewish women of North African origin (389 breast cancer cases, 541 controls) examined two MUTYH variants, Tyr179Cys and Gly396Asp, and observed an increased risk association with the latter (odds ratio 1.86; 95% CI 1.02 – 3.39; P = 0.039) (57). However, a Dutch study (1469 sporadic and 471 familial non-BRCA1/2 breast cancer cases, 1666 controls) found no association with Gly396Asp, Tyr179Cys or three other variants, Arg309Cys, Pro405Leu, and Ser515Phe (58).

NEIL and NTH1

The mammalian homologs of the *E. coli* DNA glycosylase Nei were designated Nei-like or NEIL (63-65). NEIL1 and NEIL2 specifically excise FapyA, FapyG, and the hydantoins Sp and Gh, but not 8-oxoG (66, 67). Both Sp and Gh contain a tetrahedral, sp3, carbon within the normally planar nucleic acid base ring structure. This conformational change would be expected to disrupt normal base pair stacking in duplex DNA and thereby distort the DNA helix. This distortion may play a role in the ability of NEIL to recognize and cleave the hydantoin lesions within DNA. NEIL1 stimulates OGG1 activity, evidence for functional collaboration between two DNA glycosylases (68). Interestingly, the number of NEIL1 molecules in nuclear extract of HeLa cells was about twice that of OGG1, suggesting that NEIL1 activation of OGG1 could be physiologically relevant. **Human Endonuclease III Homolog I (hNTH1**, a homolog of E. coli endonuclease III) excises oxidatively damaged pyrimidines, such as the ring-saturated thymine glycol and 5-hydroxycytosine (69-71) (Figure 5.1.1.).

Uracil DNA glycosylases

Uracil arises in DNA from spontaneous deamination of cytosine due to the inherent instability of this base (Figure 5.2.2.) (72). Because cytosine normally pairs with guanine, cytosine deamination to uracil generates a G:U mispair. If the uracil is not corrected by BER before the next round of replication, then adenine will be incorporated opposite the uracil, ultimately yielding a G:C to A:T transition (73). At least four different uracil DNA glycosylase activities have been identified in mammalian cells, removing uracil and generating an AP site. These enzymes vary in substrate specificity and localization in the cell: UDG, SMUG1, TDG, and MBD4.

Uracil DNA glycosylase (UDG), also known as uracil-N-glycosylase (UNG), removes uracil from DNA. The UDG gene at 12q13.11-13.3 consists of seven exons and encodes both the nuclear and mitochondrial forms of the enzyme, referred to as UDG2 and UDG1, respectively (74). The two forms are generated by the use of two promoters and alternative splicing (75). Nuclear UDG2 is a 32-kDa protein composed of 269 amino acids, whereas the mitochondrial UDG1 has additional N-terminal residues, which are responsible for mitochondrial import. UDG2 is the major species in most tissues, representing about 70% of total UDG activity (74). Both UDG2 and UDG1 are expressed in proliferating tissues whereas UDG1 is the predominant form in tissues with low or no proliferation (76).

Yeast mutants in the UDG gene exhibit a 20-fold increase in the spontaneous mutation rate (77). UDG is extraordinarily specific, being able to discriminate between uracil, which is excised, and thymine, which is a natural component of DNA and differs from uracil by a single methyl group. The structural basis for the specificity of the enzyme is provided by sequence-conserved residues that form a positively charged, active site groove the width of duplex DNA (78). Uracil binds at the base of the groove within a rigid preformed pocket that confers selectivity over other bases by shape complementarity. Residues 272Leu and 268His play important roles. Spontaneous thermally-induced opening of base pairs allows full insertion of the leucine side chain into the DNA base stack only when a uracil has

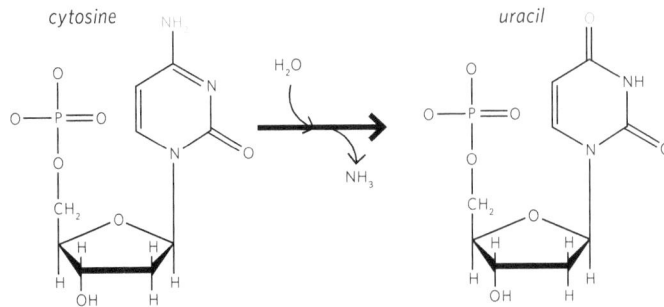

Figure 5.2.2. Cytosine deamination yields uracil

been detected (79, 80). The insertion leads to compression of the DNA backbone flanking the uracil and flipping out of the damaged base from the major groove (3). This extrahelical location allows the nearby histidine to move into catalytic position to carry out the glycosylic bond cleavage. Single-strand **Selective Monofunctional Uracil DNA Glycosylase (SMUG1)** serves as a backup enzyme for UDG. SMUG1 also excises 5-hydroxyuracil, 5-hydroxymethyluracil and 5-formyluracil bearing an oxidized group at ring C5 (81).

In mammalian DNA, 2 – 7% of the total cytosine is methylated, 5-methylcytosine, mostly in the sequence MeCpG. Spontaneous deamination of 5-methylcytosine yields thymine and generates G:T mispairs in DNA. The repair of these G:T mismatches in initiated by **Thymine DNA glycosylase (TDG)**, which excises the mismatched thymine (82). The enzyme can also remove uracil from G:U mispairs and thereby serves as a backup enzyme for UDG (83). The same reactions are performed by the repair enzyme **Methyl-CpG binding domain 4 (MBD4)**, which contains two DNA binding domains, a G:T mismatch-specific DNA glycosylase and a methyl-CpG binding domain. This apparent fusion of functions results in an enzyme that recognizes mismatched G:T residues at methylated CpG sites but will also excise uracil from G:U mispairs (84). MBD4 interacts with the mismatch repair protein MLH1 (85). About 20% of hereditary non-polyposis colorectal cancers that carried MLH1 mutations and displayed microsatellite instability also harbored mutations in MBD4 (86). Knockout mice deficient in MBD4 develop normally and do not show increased cancer susceptibility. However, the animals acquire more C:G/T:A transition mutations at CpG sites in epithelial cells of the small intestine (87).

3-methyladenine DNA glycosylase (MAG), alternatively named alkyladenine glycosylase (AAG) or N-methylpurine glycosylase (MPG), removes a diverse group of damaged bases from DNA, including cytotoxic and mutagenic alkylation adducts of purines (88). MAG interacts with the NER protein, HR23, the human homolog of the *S. cerevisiae* RAD 23 protein (89). The interaction results in greater binding affinity of the MAG-HR23 complex to damaged

DNA and increased catalytic activity of MAG. MAG was also shown to interact with estrogen receptor α, altering the action of both proteins (90). ERα increased the binding of MAG to damaged DNA and increased the catalytic removal of modified bases by MAG. In turn, MAG enhanced the binding of ERα to the estrogen response element and reduced estrogen-mediated transcription.

ABH2 and **ABH3** belong to a family of DNA repair enzymes that catalyzes the oxidative demethylation of 1-methyladenine and 3-methylcytosine (91, 92). The ABH2 gene at 12q24.1 spans 5.3 kb and contains four exons while the ABH3 gene is located at 11q11 and spans 39.4 kb with 10 exons (93). ABH2 also removes etheno DNA lesions, such as 1,N6-ethenoadenine (εA), a mutagenic adduct formed by lipid peroxidation (94). ABH2 and ABH3 are dioxygenases that require Fe(II) and α-ketoglutarate as essential cofactors to catalyze the reaction (95). Both enzymes remove 1-methyladenine and 3-methylcytosine from methylated polynucleotides in an α-ketoglutarate dependent reaction, and act by direct damage reversal with the regeneration of the unsubstituted bases.

References

1. Blainey PC, van Oijen AM, Banerjee A, Verdine GL, Xie XS. *A base-excision DNA-repair protein finds intrahelical lesion bases by fast sliding in contact with DNA.* Proc Natl Acad Sci USA. 2006;103:5752-7.

2. David SS, O'Shea VL, Kundu S. *Base-excision repair of oxidative DNA damage.* Nature. 2007;447:941-50.

3. Slupphaug G, Mol CD, Kavli B, Arvai AS, Krokan HE, Tainer JA. *A nucleotide-flipping mechanism from the structure of human uracil-DNA glycosylase bound to DNA.* Nature. 1996;384:87-92.

4. Stivers JT, Jiang YL. *A mechanistic perspective on the chemistry of DNA repair glycosylases.* Chem Rev. 2003;103:2729-59.

5. Huffman JL, Sundheim O, Tainer

JA. *DNA base damage recognition and removal: New twists and grooves.* Mutation Res. 2005;577:55-76.

6. Barnes DE, Lindahl T. *Repair and genetic consequences of endogenous DNA base damage in mammalian cells.* Annu Rev Genet. 2004;38:445-76.

7. Nilsen H, Krokan HE. *Base excision repair in a network of defence and tolerance.* Carcinogenesis. 2001;22:987-98.

8. Scharer OD, Jiricny J. *Recent progress in the biology, chemistry and structural biology of DNA glycosylases.* BioEssays. 2001;23:270-81.

9. Labahn J, Scharer OD, Long A, Ezaz-Nikpay K, Verdine GL, Ellenberger TE. *Structural basis* for the excision repair of alkylation-damaged DNA. Cell. 1996;86:321-9.

10. Doherty AJ, Serpell LC, Ponting CP. *The helix-hairpin-helix DNA-binding motif: a structural basis for non-sequence-specific recognition of DNA.* Nucleic Acids Res. 1996;24:2488-97.

11. Fujikawa K, Kamiya H, Yakushiji H, Fujii Y, Nakabeppu Y, Kasai H. *The oxidized forms of dATP are substrates for the human MutT homologue, the hMTH1 protein.* J Biol Chem. 1999;274:18201-5.

12. Sakumi K, Furuichi M, Tsuzuki T, Kakuma T, Kawabata SI, Maki H, et al. *Cloning and expression of cDNA for a human enzyme that hydrolyzes 8-Oxo-dGTP, a mutagenic substrate for DNA synthesis.* J Biol Chem. 1993;268:23524-30.

13. Tsuzuki T, Egashira A, Igarashi H, Iwakuma T, Nakatsuru Y, Tominaga Y, et al. *Spontaneous tumorigenesis in mice defective in the MTH1 gene encoding 8-oxo-dGTPase.* Proc Natl Acad Sci USA. 2001;98:11456-61.

14. Christmann M, Tomicic MJ, Roos WP, Kaina B. *Mechanisms of human DNA repair: an update.* Toxicology. 2003;193:3-34.

15. Lindahl T, Wood RD. *Quality control by DNA repair.* Science. 1999;286:1897-905.

16. Wood RD, Mitchell M, Sgouros J, Lindahl T. *Human DNA repair genes.* Science. 2001;291:1284-9.

17. Neeley WL, Essigmann JM. *Mechanisms of formation, genotoxicity, and mutation of guanine oxidation products.* Chem Res Toxicol. 2006;19:491-505.

18. Luo W, Muller JG, Rachlin EM, Burrows CJ. *Characterization of hydantoin products from one-electron oxidation of 8-oxo-7,8-dihydroguanosine in a nucleoside model.* Chem Res Toxicol. 2001;14:927-38.

19. Burrows CJ, Muller JG, Kornyushyna O, Luo W, Duarte V, Leipold MD, et al. *Structure and potential mutagenicity of new hydantoin products from guanosine and 8-Oxo-7,8-Dihydroguanine oxidation by transition metals.* Environ Health Perspect. 2002;110:713-7.

20. Delaney S, Neeley WL, Delaney JC, Essigmann JM. *The substrate specificity of MutY for hyperoxidized guanine lesions in vivo.* Biochemistry. 2007;46:1448-55.

21. Henderson PT, Delaney JC, Muller JG, Neeley WL, Tannenbaum SR, Burrows CJ, et al. *The hydantoin lesions formed from oxidation of 7,8-dihydro-8-oxoguanine are potent sources of replication errors in vivo.* Biochemistry. 2003;42:9257-62.

22. Radicella JP, Dherin C, Desmaze C, Fox MS, Boiteux S. *Cloning and characterization of hOGG1, a human homolog of the OGG1 gene of Saccharomyces cerevisiae.* Proc Natl Acad Sci USA. 1997;94:8010-5.

23. Rosenquist TA, Zharkov DO, Grollman AP. *Cloning and characterization of a mammalian 8-oxoguanine DNA glycosylase.* Proc Natl Acad Sci USA. 1997;94:7429-34.

24. Fortini P, Pascucci B, Parlanti E, D'Errico M, Simonelli V, Dogliotti E. *8-Oxoguanine DNA damage: at the crossroad of alternative repair pathways.* Mutation Res. 2003;531:127-39.

25. Audebert M, Chevillard S, Levalois C, Gyapay G, Vieillefond A, Klijanienko J, et al. *Alterations of the DNA repair gene OGG1 in human clear cell carcinomas of the kidney.* Cancer Res. 2000;60:4740-4.

26. Nash HM, Lu R, Lane WS, Verdine GL. *The critical active-site amine of the human 8-oxoguanine DNA glycosylase, hOgg1: direct identification, ablation and chemical reconstitution.* Chem Biol. 1997;4:693-702.

27. Nishioka K, Ohtsubo T, Oda H, Fujiwara T, Kang D, Sugimachi K, et al. *Expression and differential intracellular localization of two major forms of human 8-oxoguanine DNA glycosylase encoded by alternatively spliced OGG1 mRNAs.* Mol Biol Cell. 1999;10:1637-52.

28. Shinmura K, Kohno T, Takeuchi-Sasaki M, Maeda M, Segawa T, Kamo T, et al. *Expression of the OGG1-type 1a (nuclear form) protein in cancerous and non-cancerous human cells.* Int J Oncol. 2000;16:701-7.

29. Mambo E, Nyaga SG, Bohr VA, Evans MK. *Defective repair of 8-hydroxyguanine in mitochondria of MCF-7 and MDA-MB-468 human breast cancer cell lines.* Cancer Res. 2002;62:1349-55.

30. Bruner SD, Norman DP, Verdine GL. *Structural basis for recognition and repair of the endogenous mutagen 8-oxoguanine in DNA.* Nature. 2000;403:859-66.

31. Banerjee A, Santos WL, Verdine GL. *Structure of a DNA glycosylase searching for lesions.* Science. 2006;311:1153-7.

32. Zharkov DO, Rosenquist TA, Gerchman SE, Grollman AP. *Substrate specificity and reaction mechanism of murine 8-oxoguanine-DNA glycosylase.* J Biol Chem. 2000;275:28607-17.

33. Arai T, Kelly VP, Komoro K, Minowa O, Noda T, Nishimura S. *Cell proliferation in Liver of Mmh/Ogg1-deficient mice enhances mutation frequency because of the presence of 8-hydroxyguanine in DNA.* Cancer Res. 2003;63:4287-92.

34. Klungland A, Rosewell I, Hollenbach S, Larsen E, Daly G, Epe B, et al. *Accumulation of premutagenic DNA lesions in mice defective in removal of oxidative base damage.* Proc Natl Acad Sci USA. 1999;96:13300-5.

35. Russo MT, De Luca G, Degan P, Parlanti E, Dogliotti E, Barnes DE, et al. *Accumulation of the oxidative base lesion 8-hydroxyguanine in DNA of tumor-prone mice defective in both the Myh and Ogg1 DNA glycosylases.* Cancer Res. 2004;64:4411-4.

36. Xie Y, Yang H, Cunanan C, Okamoto K, Shibata D, Pan J, et al. *Deficiencies in mouse Myh and Ogg1 result in tumor predisposition and G to T mutations in codon 12 of the K-Ras oncogene in lung tumors.* Cancer Res. 2004;64:3096-102.

37. Nyaga SG, Lohani A, Jaruga P, Trzeciak AR, Dizdaroglu M, Evans MK. *Reduced repair of 8-hydroxyguanine in the human breast cancer cell line, HCCI997.* BMC Cancer. 2006;6:297.

38. Chevillard S, Radicella JP, Levalois C, Lebeau J, Poupon MF, Oudard S, et al. *Mutations in OGG1, a gene involved in the repair of oxidative DNA damage, are found in human lung and kidney tumours.* Oncogene. 1998;16:3083-6.

39. Park YJ, Choi EY, Choi JY, Park JG, You HJ, Chung MH. *Genetic changes of hOGG1 and the activity of OH8Gua glycosylase in colon cancer.* Eur J Cancer. 2001;37:340-6.

40. Kovtun IV, Liu Y, Bjoras M, Klungland A, Wilson SH, McMurray CT. *OGG1 initiates age-dependent CAG trinucleotide expansion in somatic cells.* Nature. 2007;447:447-52.

41. Mao G, Pan X, Zhu BB, Zhang Y, Yuan F, Huang J, et al. *Identification and characterization of OGG1 mutations in patients with Alzheimer's disease.* Nucleic Acids Res. 2007;35:2759-66.

42. Goode EL, Ulrich CM, Potter JD. *Polymorphisms in DNA repair genes and associations with cancer risk.* Cancer Epidemiol Biomark Prev. 2002;11:1513-30.

43. Dherin C, Radicella JP, Dizdaroglu M, Boiteux S. *Excision of oxidatively damaged DNA bases by the human α-hOgg1 protein and the polymorphic α-hOgg1(Ser326Cys) protein which is frequently found in human populations.* Nucleic Acids Res. 1999;27:4001-7.

44. Weiss JM, Goode EL, Ladiges WC, Ulrich CM. *Polymorphic variation in hOGG1 and risk of cancer: A review of the functional and epidemiologic literature.* Mol Carcinogenesis. 2005;42:127-41.

45. Yamane A, Shinmura K, Sunaga N, Saitoh T, Yamaguchi S, Shinmura Y, et al. *Suppressive activities of OGG1 and MYH proteins against G:C to T:A mutations caused by 8-hydroxyguanine but not by benzo[a]pyrene diol epoxide in human cells in vivo.* Carcinogenesis. 2003;24:1031-7.

46. Hung RJ, Hall J, Brennan P, Boffetta P. *Genetic polymorphisms in the base excision repair pathway and cancer risk: a HuGE review.* Am J Epidemiol. 2005;162:925-42.

47. Choi JY, Hamajima N, Tajima K, Yoo KY, Yoon KS, Park SK, et al. *hOGG1 Ser326Cys polymorphism and breast cancer risk among Asian women.* Breast Cancer Res Treat. 2003;79:59-62.

48. Rossner P, Terry MB, Gammon MD, Zhang FF, Teitelbaum SL, Eng SM, et al. *OGG1 polymorphisms and breast cancer risk.* Cancer Epidemiol Biomarkers Prev. 2006;15:811-5.

49. Vogel U, Nexo BA, Olsen A, Thomsen B, Jacobsen NR, Wallin H, et al. *No association between OGG1 Ser326Cys polymorphism and breast cancer risk.* Cancer Epidemiol Biomarkers Prev. 2003;12:170-1.

50. Xu J, Zheng SL, Turner A, Isaacs SD, Wiley KE, Hawkins GA, et al. *Associations between hOGG1 sequence variants and prostate cancer susceptibility.* Cancer Res. 2002;62:2253-7.

51. Slupska MM, Baikalov C, Luther WM, Chiang JH, Wei YF, Miller JH. *Cloning and sequencing a human homolog (hMYH) of the Escherichia coli mutY gene whose function is required for the repair of oxidative DNA damage.* J Bacteriol. 1996;178:3885-92.

52. Zharkov DO, Grollman AP. *MutY DNA glycosylase: Base release and intermediate complex formation.* Biochemistry. 1998;37:12384-94.

53. Fromme JC, Banerjee A, Huang SJ, Verdine GL. *Structural basis for removal of adenine mispaired with 8-oxoguanine by MutY adenine DNA glycosylase.* Nature. 2004;427:652-6.

54. Parker A, Gu Y, Mahoney W, Lee SH, Singh KK, Lu AL. *Human* homolog of the MutY repair protein (hMYH) physically interacts with proteins involved in long patch DNA base excision repair. J Biol Chem. 2001;276:5547-55.

55. Yang H, Clendenin WM, Wong D, Demple B, Slupska MM, Chiang JH, et al. *Enhanced activity of adenine-DNA glycosylase (Myh) by apurinic/apyrimidinic endonuclease (Ape1) in mammalian base excision repair of an A/GO mismatch.* Nucleic Acids Res. 2001;29:743-52.

56. Gu Y, Parker A, Wilson TM, Bai H, Chang DY, Lu AL. *Human MutY homolog, a DNA glycosylase involved in base excision repair, physically and functionally interacts with mismatch repair proteins human MutS homolog 2/human MutS homolog 6.* J Biol Chem. 2002;277:11135-42.

57. Rennert G, Lejbkowicz F, Cohen I, Pinchev M, et al. *MutY mutation carriers have increased breast cancer risk.* Cancer. 2012;118:1989-93.

58. Out AA, Wasielewski M, Huijts PEA, van Minderhout IJHM, et al. *MUTYH gene variants and breast cancer in a Dutch case-control study.* Breast Cancer Res Treat. 2012;134:219-27.

59. Chmiel NH, Livingston AL, David SS. *Insight into the functional consequences of inherited variants of the hMYH adenine glycosylase associated with colorectal cancer: Complementation assays with hMYH variants and pre-steady-state kinetics of the corresponding mutated E. coli enzymes.* J Mol Biol. 2003;327:431-43.

60. Sieber OM, Lipton L, Crabtree M, Heinimann K, Fidalgo P, Phillips RK, et al. *Multiple colorectal adenomas, classic adenomatous polyposis, and germ-line mutations in MYH.* N Engl J Med. 2003;348:791-9.

61. Yamaguchi S, Shinmura K, Saitoh T, Takenoshita S, Kuwano H, Yokota J. *A single nucleotide polymorphism at the splice donor site of the human MYH base excision repair gene results in reduced translation efficiency of its transcripts.* Genes to Cells. 2002;7:461-74.

62. Cheadle JP, Sampson JR. *MUTYH-associated polyposis. From defect in base excision repair to clinical genetic testing.* DNA Repair. 2007;6:274-9.

63. Bandaru V, Sunkara S, Wallace SS, Bond JP. *A novel human DNA glycosylase that removes oxidative DNA damage and is homologous to Escherichia coli endonuclease VIII.* DNA Repair. 2002;1:517-29.

64. Hazra TK, Izumi T, Boldogh I, Imhoff B, Kow YW, Jaruga P, et al.

Identification and characterization of a human DNA glycosylase for repair of modified bases in oxidatively damaged DNA. Proc Natl Acad Sci USA. 2002;99:3523-8.

65. Morland I, Rolseth V, Luna L, Rognes T, Bjoras M, Seeberg E. *Human DNA glycosylases of the bacterial Fpg/MutM superfamily: an alternative pathway for the repair of 8-oxoguanine and other oxidation products in DNA.* Nucl Acids Res. 2002;30:4926-36.

66. Hailer MK, Slade PG, Martin BD, Rosenquist TA, Sugden KD. *Recognition of the oxidized lesions spiroiminodihydantoin and guanidino-hydantoin in DNA by the mammalian base excision repair glycosylases NEIL1 and NEIL2.* DNA Repair. 2005;4:41-50.

67. Jaruga P, Birincioglu M, Rosenquist TA, Dizdaroglu M. *Mouse NEIL1 protein is specific for excision of 2,6-diamino-4-hydroxy-5-formamidopyrimidine and 4,6-diamino-5-formamidopyrimidine from oxidatively damaged DNA.* Biochemistry. 2004;43:15909-14.

68. Mokkapati SK, Wiederhold L, Hazra TK, Mitra S. *Stimulation of DNA glycosylase activity of OGG1 by NEIL1: Functional collaboration between two human DNA glycosylases.* Biochemistry. 2004;43:11596-604.

69. Aspinwall R, Rothwell DG, Roldan-Arjona T, Anselmino C, Ward CJ, Cheadle JP, et al. *Cloning and characterization of a functional human homolog of Escherichia coli endonuclease III.* Proc Natl Acad Sci USA. 1997;94:109-14.

70. Eide L, Luna L, E.C. G, Henderson PT, Essigmann JM, Demple B, et al. *Human endonuclease III acts preferentially on DNA damage opposite guanine residues in DNA.* Biochemistry. 2001;40:6653-9.

71. Ikeda S, Biswas T, Roy R, Izumi T, Boldogh I, Kurosky A, et al. *Purification and characterization of human NTH1, a homolog of Escherichia coli endonuclease III.* J Biol Chem. 1998;273:21585-93.

72. Lindahl T. *Instability and decay of the primary structure of DNA.* Nature. 1993;362:709-15.

73. Kunkel TA, Wilson SH. *Push and pull of base flipping.* Nature. 1996;384:25-6.

74. Slupphaug G, Markussen FH, Olsen LC, Aasland R, Aarsaether N, Bakke O, et al. *Nuclear and mitochondrial forms of human uracil-DNA glycosylase are encoded by the same gene.* Nucleic Acids Res.

1993;21:2579-84.

75. Nilsen H, Otterlei M, Haug T, Solum K, Nagelhus TA, Skorpen F, et al. *Nuclear and mitochondrial uracil-DNA glycosylases are generated by alternative splicing and transcription from different positions in the UNG gene.* Nucleic Acids Res. 1997;25:750-5.

76. Haug T, Skorpen F, Aas PA, Malm V, Skjelbred C, Krokan HE. *Regulation of expression of nuclear and mitochondrial forms of human uracil-DNA glycosylase.* Nucleic Acids Res. 1998;26:1449-57.

77. Impellizzeri KJ, Anderson B, Burgers PM. *The spectrum of spontaneous mutations in a Saccharomyces cerevisiae uracil-DNA-glycosylase mutant limits the function of this enzyme to cytosine deamination repair.* J Bacteriol. 1991;173:6807-10.

78. Mol CD, Arvai AS, Slupphaug G, Kavli B, Alseth I, Krokan HE, et al. *Crystal structure and mutational analysis of human uracil-DNA glycosylase: Structural basis for specificity and catalysis.* Cell. 1995;80:869-78.

79. Mol CD, Arvai AS, Sanderson RJ, Slupphaug G, Kavli B, Krokan HE, et al. *Crystal structure of human uracil-DNA glycosylase in complex with a protein inhibitor: Protein mimicry of DNA.* Cell. 1995;82:701-8.

80. Parker JB, Bianchet MA, Krosky DJ, Friedman JI, Amzel LM, Stivers JT. *Enzymatic capture of an extrahelical thymine in the search for uracil in DNA.* Nature. 2007;449:433-7.

81. Matsubara M, Tanaka T, Terato H, Ohmae E, Izumi S, Katayanagi K, et al. *Mutational analysis of the damage-recognition and catalytic mechanism of human SMUG1 DNA glycosylase.* Nucleic Acids Res. 2004;32:5291-302.

82. Waters TR, Gallinari P, Jiricny J, Swann PF. *Human thymine DNA glycosylase binds to apurinic sites in DNA but is displaced by human apurinic endonuclease 1.* J Biol Chem. 1999;274:67-74.

83. Neddermann P, Jiricny J. *Efficient removal of uracil from G•U mispairs by the mismatch-specific thymine DNA glycosylase from HeLa cells.* Proc Natl Acad Sci USA. 1994;91:1642-6.

84. Petronzelli F, Riccio A, Markham GD, Seeholzer SH, Genuardi M, Karbowski M, et al. *Investigation of the substrate spectrum of the human mismatch-specific DNA N-Glycosylase MED1 (MBD4): Fundamental role of the catalytic domain.* J Cell Physiol. 2000;185:473-80.

85. Riccio A, Aaltonen LA, Godwin AK, Loukola A, Percesepe A, Salovaara R, et al. *The DNA repair gene MBD4 (MED1) is mutated in human carcinomas with microsatellite instability.* Nature Genet. 1999;23:266-8.

86. Bellacosa A, Cicchillitti L, Schepis F, Riccio A, Yeung AT, Matsumoto Y, et al. *MED1, a novel human methyl-CpG-binding endonuclease, interacts with DNA mismatch repair protein MLH1.* Proc Natl Acad Sci USA. 1999;96:3969-74.

87. Wong E, Yang K, Kuraguchi M, Werling U, Avdievich E, Fan K, et al. *Mbd4 inactivation increases C→T transition mutations and promotes gastrointestinal tumor formation.* Proc Natl Acad Sci USA. 2002;99:14937-42.

88. Lau AY, Scharer OD, Samson L, Verdine GL, Ellenberger T. *Crystal structure of a human alkylbase-DNA repair enzyme complexed to DNA: Mechanisms for nucleotide flipping and base excision.* Cell. 1998;95:249-58.

89. Miao F, Bouziane M, Dammann R, Masutani C, Hanaoka F, Pfeifer G, et al. *3-methyladenine-DNA glycosylase (MPG protein) interacts with human RAD23 proteins.* J Biol Chem. 2000;275:28433-8.

90. Likhite VS, Cass EI, Anderson SD, Yates JR, Nardulli AM. *Interaction of estrogen receptor α with 3-methyladenine DNA glycosylase modulates transcription and DNA repair.* J Biol Chem. 2004;279:16875-82.

91. Duncan T, Trewick SC, Koivisto P, Bates PA, Lindahl T, Sedgwick B. *Reversal of DNA alkylation damage by two human dioxygenases.* Proc Natl Acad Sci USA. 2002;99:16660-5.

92. Singer B, Hang B. *What structural features determine repair enzyme specificity and mechanism in chemically modified DNA?* Chem Res Toxicol. 1997;10:713-32.

93. Aas PA, Otterlei M, Falnes PO, Vagbo CB, Skorpen F, Akbari M, et al. *Human and bacterial oxidative demethylases repair alkylation damage in both RNA and DNA.* Nature. 2003;421:859-63.

94. Ringvoll J, Moen MN, Nordstrand LM, Meira LB, Pang B, Bekkelund A, et al. *AlkB homologue 2-mediated repair of ethenoadenine lesions in mammalian DNA.* Cancer Res. 2008;68:4142-9.

95. Yang CG, Yi C, Duguid EM, Sullivan CT, Jian X, Rice PA, et al. *Crystal structures of DNA/RNA repair enzymes AlkB and ABH2 bound to dsDNA.* Nature. 2008;452:961-5.

5.2.3. APURINIC/APYRIMIDINIC (AP) ENDONUCLEASES

AP sites are probably the most common lesions in cellular DNA. AP endonucleases recognize AP sites in the genome and initiate their repair by hydrolyzing the phosphodiester on the 5'-side of the abasic residues (1). AP endonucleases, such as the major human endonuclease, APE1, are quite abundant in most cells ranging from 300,000 to 7,000,000 APE1 molecules per cell in fibroblasts and HeLa cells, respectively (2-4). The abundance of AP endonucleases is consistent with the observation that most endogenous AP sites in mammalian cells are recovered in a form already cleaved on the 5' side of the lesion (5).

Apurinic/Apyrimidinic endonuclease 1 (APE1; also called Ref 1, Hap1, or Apex). The APE1 gene at 14q11.2 - 12 is only 2.6 kb in size due to unusually short introns separating five exons (6-8). The promoter region contains a CCAAT box, but no other regulatory site or a TATA box, consistent with the constitutive expression of the APE gene. The first exon is non-coding and exons 2 – 5 are translated into a 37-kDa protein composed of 318 amino acids. The DNA repair function is located in the C-terminal residues 61 – 318. Biochemical and structural studies have elucidated the repair function carried out by APE1 (4, 9-12). The protein first displaces bound DNA glycosylases, then uses a rigid, positively charged surface to bend the DNA helix ~35º to flip-out the AP site into its active site pocket. The APE1-induced DNA distortion at the AP site is followed by cleavage of the lesion, i.e., incision of the phosphodiester backbone 5' to the AP site, leaving 3'-hydroxyl and 5'-deoxyribose phosphate termini. The APE1-DNA complex is then recognized by DNA polymerase β, which displaces APE1 and bends the DNA helix further to 82º to carry out its own catalytic action (13, 14). The APE1-cleaved DNA intermediate is also recognized by poly(ADP-ribose) polymerase-1 (PARP-1), which appears to compete with APE1 binding (15).

APE1 plays an additional, unrelated role as a redox regulator of transcription factors. N-terminal residues 1 – 127 including the redox active residue cytosine 65 stimulate the DNA-binding activity of AP-1 transcription factors (i.e., Fos and Jun) through reduction of conserved cysteine residues in the DNA-binding domain of each protein (16, 17). APE1 is also able to stimulate the DNA-binding activity of other classes of redox-regulated transcription factors including p53 (18, 19). Thus, APE1 connects DNA repair and gene expression processes through endonuclease and redox activities encoded by distinct regions of the protein. Not surprisingly, APE1 is essential for viability as judged by the embryonic-lethal phenotype of mice defective in APE1 (20). Exposure of fibroblasts and HeLa cells to ROS resulted in increased APE1 expression and translocation of the enzyme to the nucleus (21). The ROS-induced increase in APE1 activity leads to enhanced BER, which may explain the increasing cytotoxic resistance observed in cells treated with ROS generators, such as H_2O_2 and bleomycin.

The APE1 gene contains over 20 polymorphic sites, which have been assessed functionally as recombinant proteins or based on structural predictions (10, 22-24). The most common substitution, Asp148Glu, observed in 38% of Caucasians, does not alter enzyme activity. There was no association of the Asp148Glu polymorphism with either lung or upper gastrointestinal cancer risk (25). The frequencies of all other variant alleles are less than 4%. Several of these less common variants are associated with substantially reduced activity. For example, Leu104Arg, Glu126Asp, and Arg237Ala exhibit 50% and Asp283Gly 90% reduction in repair capacity (22). The variant alleles in codons 104, 126, and 283 have been detected in patients with sporadic amyotrophic lateral sclerosis (26-28).

References

1. Barzilay G, Hickson ID. *Structure and function of apurinic/apyrimidinic endonucleases.* BioEssays. 1995;17:713-9.

2. Cappelli E, Hazra T, Hill JW, Slupphaug G, Bogliolo M, Frosina G. *Rates of base excision reapir are not solely dependent on levels of initiating enzymes.* Carcinogenesis. 2001;22:387-93.

3. Chen DS, Herman T, Demple B. *Two distinct human DNA diesterases that hydrolyze 3'-blocking deoxyribose fragments from oxidized DNA.* Nucleic Acids Res. 1991;19:5907-14.

4. Mol CD, Izumi T, Mitra S, Tainer JA. *DNA-bound structures and mutants reveal abasic DNA binding by APE1 DNA repair and coordination.* Nature. 2000;403:451-6.

5. Nakamura J, Swenberg JA. *Endogenous apurinic/apyrimidinic sites in genomic DNA of mammalian tissues.* Cancer Res. 1999;59:2522-6.

6. Demple B, Herman T, Chen DS. *Cloning and expression of APE, the cDNA encoding the major human apurinic endonuclease: Definition of a family of DNA repair enzymes.* Proc Natl Acad Sci USA. 1991;88:11450-4.

7. Harrison L, Ascione G, Menninger JC, Ward DC, Demple B. *Human apurinic endonuclease gene (APE): structure and genomic mapping (chromosome 14q11.2-12).* Hum Mol Genet. 1992;1:677-80.

8. Robson CN, Hochhauser D, Craig R, Rack K, Buckle VJ, Hickson ID. *Structure of the human DNA repair gene HAP1 and its localisation to chromosome 14q 11.2-12.* Nucleic Acids Res. 1992;20:4417-21.

9. Gorman MA, Morera S, Rothwell DG, de La Fortelle E, Mol CD, A. TJ, et al. *The crystal structure of the human DNA repair endonuclease HAP1 suggests the recognition of extra-helical deoxyribose at DNA abasic sites.* EMBO J. 1997;16:6548-58.

10. Masuda Y, Bennett RA, Demple B. *Dynamics of the interaction of human apurinic endonuclease (Ape1) with its substrate and product.* J Biol Chem. 1998;273:30352-9.

11. Waters TR, Gallinari P, Jiricny J, Swann PF. *Human thymine DNA glycosylase binds to apurinic sites in DNA but is displaced by human apurinic endonuclease 1.* J Biol Chem. 1999;274:67-74.

12. Wilson DM, Takeshita M, Demple B. *Abasic site binding by the human apurinic endonuclease, Ape, and determination of the DNA contact sites.* Nucl Acids Res. 1997;25:933-9.

13. Bennett RA, Wilson DM, Wong D, Demple B. *Interaction of human apurinic endonuclease and DNA polymerase β in the base excision repair pathway.* Proc Natl Acad Sci USA. 1997;94:7166-9.

14. Wilson SH, Kunkel TA. *Passing the baton in base excision repair.* Nature Struct Biol. 2000;7:176-8.

15. Cistulli C, Lavrik OI, Prasad R, Hou E, Wilson SH. *AP endonuclease and poly(ADP-ribose) polymerase-1 interact with the same base excision* *repair intermediate.* DNA Repair. 2004;3:581-91.

16. Walker LJ, Robson CN, Black E, Gillespie D, Hickson ID. *Identification of residues in the human DNA repair enzyme HAP1 (Ref-1) that are essential for redox regulation in Jun DNA binding.* Mol Cell Biol. 1993;13:5370-6.

17. Xanthoudakis S, Miao GG, Curran T. *The redox and DNA-repair activities of Ref-1 are encoded by nonoverlapping domains.* Proc Natl Acad Sci USA. 1994;91:23-7.

18. Hainaut P, Milner J. *Redox modulation of p53 conformation and sequence-specific DNA binding in vitro.* Cancer Res. 1993;53:4469-73.

19. Jayaraman L, Murthy KG, Zhu C, Curran T, Xanthoudakis S, Prives C. *Identification of redox/repair protein Ref-1 as a potent activator of p53.* Genes Dev. 1997;11:558-70.

20. Xanthoudakis S, Smeyne RJ, Wallace JD, Curran T. *The redox/DNA repair protein, Ref-1, is essential for early embryonic development in mice.* Proc Natl Acad Sci USA. 1996;93:8919-23.

21. Ramana CV, Boldogh I, Izumi T, Mitra S. *Activation of apurinic/apyrimidinic endonuclease in human cells by reactive oxygen species and its correlation with their adaptive response to genotoxicity of free radicals.* Proc Natl Acad Sci USA. 1998;95:5061-6.

22. Hadi MZ, Coleman MA, Fidelis K, Mohrenweiser HW, Wilson DM. *Functional characterization of Ape1 variants identified in the human population.* Nucleic Acids Res. 2000;28:3871-9.

23. Mohrenweiser HW, Xi T, Vazquez-Matias J, Jones IM. *Identification of 127 amino acid substitution variants in screening 37 DNA repair genes in humans.* Cancer Epidemiol Biomark Prev. 2002;11:1054-64.

24. Xi T, Jones IM, Mohrenweiser HW. *Many amino acid substitution variants identified in DNA repair genes during human population screenings are predicted to impact protein function.* Genomics. 2004;83:970-9.

25. Hung RJ, Hall J, Brennan P, Boffetta P. *Genetic polymorphisms in the base excision repair pathway and cancer risk: a HuGE review.* Am J Epidemiol. 2005;162:925-42.

26. Fishel ML, Vasko MR, Kelley MR. *DNA repair in neurons: So if they don't divide what's to repair?* Mutat Res. 2007;614:24-36.

27. Hayward C, Colville S, Swingler RJ, Brock DJ. *Molecular genetic analysis of the APEX nuclease gene in amyotrophic lateral sclerosis.* Neurology. 1999;52:1899-901.

28. Schymick JC, Talbot K, Traynor BJ. *Genetics of sporadic amyotrophic lateral sclerosis.* Hum Mol Genet. 2007;16:R233-R42.

5.2.4. BER SCAFFOLD PROTEINS

Replication Factor C (RFC)

DNA replication and repair require the concerted action of many enzymes. The interaction of these enzymes with DNA and their optimal catalytic action is dependent on the presence of several accessory proteins. Among the main accessory proteins are replication factor C (RFC) and proliferating cell nuclear antigen (PCNA). RFC functions as a DNA "clamp loader" because it can bind to a template-primer junction and, in the presence of ATP, load the PCNA clamp onto DNA, thereby permitting the recruitment of DNA polymerases δ and ε to the site of DNA synthesis (1, 2). RFC is a heteropentameric complex composed of one large (140 kDa) and four small (36, 37, 38, 40 kDa) subunits, each encoded by a separate gene (3). Different functional regions have been identified in the subunits, such as DNA- and PCNA-binding regions between residues 369 – 480 and 481 – 728, respectively, of the large RFC 140 subunit (4). The C-terminal regions of all five subunits are required for assembly of the RFC complex and the N-termini of the four small subunits are important for DNA replication activity (5, 6). RFC 140 interacts with several other proteins, such as BRCA1 and ATM, in a large protein complex termed the BRCA1-associated genome surveillance complex (BASC), which has been postulated to act as sensor for DNA damage (7). Moreover, RFC 140 can bind to retinoblastoma protein (Rb) and histone deacetylase 1 (HDAC1), suggesting a role in gene expression in addition to its primary function as a clamp loader (8, 9).

Proliferating Cell Nuclear Antigen (PCNA)

The PCNA gene at 20pter-p12 contains six exons, which encode a 29-kDa protein composed of 261 amino acids (10). PCNA is an essential component of cell cycle regulation and DNA replication and repair (2, 11-13). These diverse functions are based on the interaction of PCNA with multiple proteins and DNA via a conserved PCNA-interacting protein (PIP) box motif. The minimal consensus PIP box is defined by Glu-Xxx-Xxx-Hhh-Xxx-Xxx-Aaa-Aaa, in which Hhh are residues with moderately hydrophobic side chains (e.g., Leu, Ile, Met), Aaa residues with highly hydrophobic, aromatic side chains (e.g., Phe, Tyr), and Xxx any residue (13). One of the proteins bound to the PIP box is the cell cycle inhibitor, p21 (14-16). The interaction between PCNA and p21 blocks the ability of PCNA to activate pol δ, the principal replicative DNA polymerase (17, 18). PCNA forms a trimeric ring with three-fold symmetry perpendicular to the ring plane and a central hole, which allows the protein to freely slide on double-stranded DNA (2, 19). Thus, PCNA acts as a "sliding clamp" molecular adaptor that binds several DNA-editing enzymes enabling them to participate in DNA replication and repair, e.g., pol δ/ε, pol β, ligase 1, and FEN1 (1, 12, 13, 20-23). PCNA facilitates excision during long-patch BER through its interaction with FEN1 and also alters the rate constant for polymerase-DNA dissociation, resulting in a lower K_m of the DNA polymerases for dNTP substrates (24). Other partners of PCNA include XP-G endonuclease, active in NER, and the mismatch repair proteins MSH2 and MLH1 (11, 12, 25).

Sequence analysis of the PCNA gene in 60 healthy Northern Europeans revealed seven common intronic SNPs and two rare synonymous exonic SNPs (26). Six of the seven intronic SNPs always co-segregated with an allele frequency of 9%. The frequency of the linked intronic SNPs was significantly lower (3%) in a breast cancer population of 118 women. The expression of PCNA in breast cancer has been correlated with mitotic rate and S-phase fraction (27, 28). Overexpression was correlated with poor histological differentiation and ER-negative status but not with prognosis (29).

Poly(ADP-ribose) Polymerase (PARP)

PARP-1 and –2 are members of the poly(ADP-ribose) polymerase family that catalyzes the successive covalent addition of homopolymers of adenosine diphosphate ribose units from nicotinamide adenine dinucleotide (NAD+) to certain nuclear acceptor proteins (30, 31). In this process, PARP-1 and –2 act as catalytic dimers, forming homo- and heterodimers to attach ADP-ribose units to each other as well as proteins involved in DNA metabolism (topoisomerases, DNA replication factors) and chromatin architecture (histones H1, H2B, lamin B). Both PARP-1 and –2 are zinc finger proteins that do not bind specific sequences in the DNA but rather detect single- and double-strand DNA breaks. The zinc fingers confer upon PARP the ability to recognize and rapidly bind DNA strandbreaks, including those arising during BER (32). Following DNA binding, PARP catalyzes the transfer of ADP ribose resulting in the addition and synthesis of branched anionic ADP-ribose homopolymers, which facilitate dissociation of the modified PARP from DNA, allowing other enzymes to access and repair the DNA strand break (30, 31, 33). PARP-1 and –2 also contain a domain defined by distinct hydrophobic clusters of amino acids that was originally detected in the C-terminal region of BRCA-1 and therefore named BRCT domain (BRCA-1 C-terminus) (34). The BRCT domain occurs as an autonomous folding unit of ~95 amino acids in proteins involved in DNA repair, recombination, and cell cycle control (35-37). The BRCT domain of PARP-1 interacts with the scaffold protein X-ray repair cross complementing 1 (XRCC1) (38, 39). The interaction with XRCC1 downregulates PARP-1 activity. PARP-1 also interact with two more BER proteins, DNA pol β and ligase III (33, 40-42). Treatment of cells with certain chemical or physical DNA-damaging agents, including ROS, alkylating agents, and γ- or UV radiation induces a dose-dependent stimulation of PARP synthesis (43). Evidence for the involvement of PARP in BER was provided by the observation that PARP-1 -/- mice treated with the alkylating agent N-methyl-N-nitrosurea or γ-irradiation showed extensive DNA damage and genomic instability (44-47). Furthermore, reconstitution of BER using purified proteins

showed that PARP-1 stimulates two of the key steps in long-patch BER, namely strand displacement synthesis by pol β and 5′-flap cleavage by FEN1 (48).

The PARP-1 gene at 1q41-42 is 43 kb long and divided in 23 exons encoding a 113 kDa protein. PARP-1 is constitutively expressed at a basal level and its catalytic activity is strongly stimulated in response to single- or double-stranded breaks (32). PARP-1 has a modular organization (A – F) with the N-terminal DNA-binding domain encompassing two zinc finger motifs, A and B (residues 21 – 56 and 125 – 162, respectively), and a bipartite nuclear localization signal in domain C (Figure 5.2.3.) (49). The central region of the protein contains the BRCT domain, the auto-poly(ADP-ribosylation) sites, and the dimerization domain E. The NAD+ binding site and the catalytic domain are located in the C-terminal region. Five non-synonymous polymorphisms have been identified involving codons Ala188Thr, Val334Ile, Ser383Tyr, Val761Ala, Lys940Arg (50). The variant allele frequency is 18% for 761Ala but less than 2% for the other variants.

One study reported a marginally lower poly(ADP-ribosyl)ation activity in peripheral blood lymphocytes of breast cancer patients compared with age-matched control women (51). Functional analysis of peripheral lymphocytes from 354 cancer-free subjects showed a significantly lower PARP-1 catalytic activity in response to H_2O_2 associated with the 761Ala/Ala than Val/Ala or Val/Val genotypes (52). Another study did not observe an association between the Ala/Ala genotypes and enzyme activity (53). The environmental genotoxic agent 2,3,7,8-tetrachlorobenzo-p-dioxin (TCDD) induced oxidative stress, DNA strand breaks, and PARP-1 in MCF-7 and MDA-MB-231 breast cancer cells (43). The increased levels and activation of PARP-1 were associated with decreased intracellular NAD(P)H and NAD+ levels.

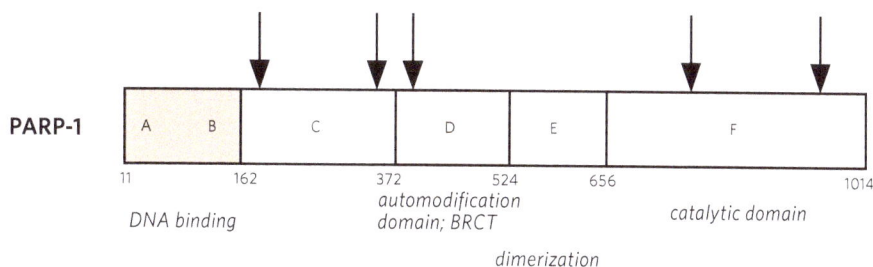

Figure 5.2.3. Diagram of PARP-1 showing functional domains, conserved sequences, and non-synonymous polymorphisms. *The N-terminal DNA-binding domain contains two zinc finger motifs, A and B at residues 21 – 56 and 125 – 162, respectively. Two nuclear localization signals of about ten amino acids have been identified in positions 225 and 350. The central region contains the auto-modification sites and a BRCT domain, which interacts with the repair protein XRCC1. The dimerization domain E is involved in the formation of PARP-1 and -2 homo- and heterodimers. The C-terminal region harbors the NAD+ binding site and the catalytic domain, which is involved in the nick-dependent poly(ADP-ribose) synthesis. The arrows indicate non-synonymous polymorphisms involving codons Ala188Thr, Val334Ile, Ser383Tyr, Val761Ala, and Lys940Arg.*

X-ray Repair Cross Complementing 1 (XRCC1)

The XRCC1 gene at 19q13.2-13.3 consists of 17 exons and encodes a 70-kDa protein composed of 633 amino acids (Figure 5.2.4.) (54, 55). XRCC1 has no known catalytic activity but serves as a scaffold at the site of DNA damage interacting with several other proteins during BER and single-strand break repair (38, 56). The N-terminal domain of XRCC1 interacts with pol β (57). Structural analysis of the N-terminal domain revealed simultaneous contacts with single-strand gap DNA and the palm-thumb of pol β (58, 59). A nuclear localization domain is present in residues 271 – 276. XRCC1 possesses two BRCA-1 carboxyl-terminal (BRCT) domains, denoted BRCT I and II, that are located centrally (residues 319 – 402) and at the C-terminus (residues 538 – 633) of the protein (34, 35). BRCT I binds PARP-1 whereas BRCT II is responsible for binding and stabilizing DNA ligase IIIα (37, 38, 60-62). It is likely that the interaction with PARP-1 serves to recruit the XRCC1 protein complex to sites of single-strand breakage, since XRCC1 preferentially binds the activated form of PARP-1 that arises once the latter protein has bound to a single-strand break. This ability to discriminate between active and inactive PARP-1 most likely reflects the ability of XRCC1 to bind poly(ADP-ribose), the polymeric product of PARP-1 activity (38). Furthermore, XRCC1 promotes the efficiency of the repair process by bringing together DNA polymerase β and ligase IIII, which do not interact directly (57, 60). Finally, XRCC1 binds and stimulates polynucleotide kinase at damaged DNA termini and thereby accelerates the overall repair reaction (63). XRCC1 also interacts with APE1, OGG1, and PCNA (56). In view of the interaction with multiple proteins, it is not surprising that XRCC1 is essential for viability as judged by the embryonic-lethal phenotype of mice defective in XRCC1 (64).

The XRCC1 gene contains over 60 single nucleotide polymorphisms of which approximately 30 are located in exons or promoter regions. Nine non-synonymous SNPs involve codons Val72Ala, Pro161Leu, Phe173Leu, Arg194Trp, Arg280His, Pro309Ser, Arg399Gln, Arg560Trp, and Tyr576Ser (50). Most of the variant alleles occur in less than 4% of the population except 194Trp and 399Gln, which are found in 5 - 7% and 34 - 36%, respectively, of Caucasians

(65). The frequency of these alleles varies between ethnic groups, i.e., 194Trp occurs in 34% Asian and 399Gln in 14% African-American women.

Residue 72Val is located in a β-strand contacting pol β (59). This amino acid is conserved in hamster and mouse but polymorphic in humans, Val72Ala. The Arg194Trp polymorphism has been shown to be associated with lower bleomycin and benzo(a)pyrene diol epoxide sensitivity *in vitro* (66). The Arg280His polymorphism is located in the PCNA binding region and was suggested in a small study to be associated with higher bleomycin sensitivity (56, 67). The Arg399Gln polymorphism is located in the BRCT I domain. Interestingly, epidemiological studies have reported an overrepresentation of one allele over the other among groups of individuals with a variety of malignancies (66). However, there is no consensus among these studies as to which of the alleles is detrimental. One possible explanation for the lack of consensus is the location of several other DNA repair genes within a 2 Mb region of chromosome 19q13.2-13.3, such as DNA ligase I, ERCC1, ERCC2, and polynucleotide kinase, which could harbor polymorphisms that contribute to the apparent impact of this common XRCC1 polymorphism (54, 68). Several *in vivo* studies have examined the functional effect of the Arg399Gln polymorphism. Lymphocytes cultured in the presence of bleomycin acquired significantly more chromosome breaks per cell in homozygous 399Gln/Gln than Arg/Gln or Arg/Arg individuals (66). The levels of white blood cell and placental DNA ^{32}P, polyphenol, and aflatoxin B1 adducts were higher in Gln/Gln than Arg/Arg homozygotes (69-71). However, an *in vitro* comparison of the Arg and Gln alleles revealed no difference in DNA single-strand break repair or cell survival after DNA alkylation (68). The polymorphic residues Arg560Trp and Tyr576Ser are located in conserved sequences of the BRCT domain (37). X-ray crystallography revealed a salt bridge between 560Arg and 572Glu, which could not be formed with a 560Trp substitution.

Several epidemiological studies have examined the association between breast cancer risk and XRCC1 polymorphisms, especially the common Arg194Trp and Arg399Gln variants. The interpretation is complicated

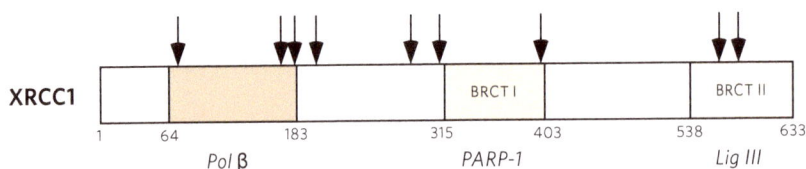

Figure 5.2.4. Diagram of XRCC1 showing functional domains, conserved sequences, and non-synonymous polymorphisms. *XRCC1 interacts with several other BER proteins, i.e., pol β in the N-terminal domain, PARP-1 in the mid-molecule BRCT I domain, and DNA ligase IIIα in the C-terminal BRCT II domain. The arrows indicate non-synonymous polymorphisms involving codons Val72Ala, Pro161Leu, Phe173Leu, Arg194Trp, Arg280His, Pro309Ser, Arg399Gln, Arg560Trp, and Tyr576Ser.*

by the fact that both polymorphisms show marked differences in minor allele frequency (MAF) between different ethnic groups and populations. The range in MAFs is 0.05 – 0.37 for 194Trp and 0.13 – 0.39 for 399Gln (72). Four large studies (Nurses' Health Study, Shanghai Breast Cancer Study, Long Island Breast Cancer Study Project, HuGE review) found no association of either of the two polymorphisms with breast cancer risk (65, 73-75). The population-based Carolina Breast Cancer Study found a positive association between cancer risk and codon 399Gln/Gln or Arg/Gln genotypes compared with Arg/Arg among African-American women (253 cases, 266 controls; OR 1.7; 95%CI 1.1 – 2.4) (76). However, there was no significant association in Caucasian women (386 cases, 381 controls; OR 1.0; 95% CI 0.8 – 1.4). Another study observed the combination of the 399Gln with the 280His allele more frequently in cases than in controls (OR 2.54; 95% CI 1.04 – 6.22) (77). Few studies have examined the interaction of XRCC1 polymorphisms with cigarette smoking and dietary factors on breast cancer risk. The Long Island Breast Cancer Study Project showed a decreased risk of breast cancer in women with at least one 194Trp allele and high intake of fruits and vegetables (OR 0.58; 95% CI 0.38 – 0.89) (65). The same study indicated an increased risk of breast cancer among never smokers with the 399Gln allele (OR 1.3; 95% CI 1.0 – 1.7). Further analysis revealed a weak additive interaction between the 399Gln allele and detectable PAH-DNA adducts in never smokers (OR 1.9; 95% CI 1.2 – 3.1). An extended analysis of the Carolina Breast Cancer Study (2,077 cases with 786 African American and 1,281 Caucasian women; 1,818 controls with 681 African American and 1,137 Caucasian women) showed a positive association between breast cancer risk and dose of active smoking for participants with XRCC1 codon 194Arg/Arg (P_{trend} = 0.046), 399Arg/Arg (Ptrend = 0.012), and 280His/His or His/Arg (P_{trend} = 0.047) genotypes (78).

References

1. Indiani C, O'Donnell M. *The replication clamp-loading machine at work in the three domains of life.* Nature Rev Mol Cell Biol. 2006;7:751-61.

2. Kelman Z. *PCNA: structure, functions and interactions.* Oncogene. 1997;14:629-40.

3. Mossi R, Hubscher U. Clamping down on clamps and clamp loaders. The eukaryotic replication factor. C Eur J Biochem. 1998;254:209-16.

4. Fotedar R, Mossi R, Fitzgerald P, Rousselle T, Maga G, Brickner H, et al. A conserved domain of the large subunit of replication factor C binds PCNA and acts like a dominant negative inhibitor of DNA replication in mammalian cells. EMBO J. 1996;15:4423-33.

5. Uhlmann F, Cai J, Gibbs E, O'Donnell M, Hurwitz J. *Deletion analysis of the large subunit p140 in human replication factor C reveals regions required for complex formation and replication activities.* J Biol Chem. 1997;272:10058-64.

6. Uhlmann F, Gibbs E, Cai J, O'Donnell M, Hurwitz J. *Identification of regions within the four small subunits of human replication factor C required for complex formation and DNA replication.* J Biol Chem. 1997;272:10065-71.

7. Wang Y, Cortez D, Yazdi P, Neff N, Elledge SJ, Qin J. *BASC, a super complex of BRCA1-associated proteins involved in the recognition and repair of aberrant DNA structures.* Genes Dev. 2000;14:927-39.

8. Anderson LA, Perkins ND. *The large subunit of replication factor C interacts with the histone deacetylase, HDAC1.* J Biol Chem. 2002;277:29550-4.

9. Pennaneach V, Salles-Passador I, Munshi A, Brickner H, Regazzoni K, Dick F, et al. *The large subunit of replication factor C promotes cell survival after DNA damage in an LxCxE Motif- and Rb-dependent manner.* Mol Cell. 2001;7:715-27.

10. Travali S, Ku DH, Rizzo MG, Ottavio L, Baserga R, Calabretta B. *Structure of the human gene for the proliferating cell nuclear antigen.* J Biol Chem. 1989;264:7466-72.

11. Jonsson ZO, Hubscher U. *Proliferating cell nuclear antigen: more than a clamp for DNA polymerases.* BioEssays. 1997;19:967-75.

12. Maga G, Hubscher U. *Proliferating cell nuclear antigen (PCNA): a dancer with many partners.* J Cell Sci. 2003;116:3051-60.

13. Warbrick E. *PCNA binding through a conserved motif.* BioEssays. 1998;20:195-9.

14. Chen J, Jackson PK, Kirschner MW, Dutta A. *Separate domains of p21 involved in the inhibition of cdk kinase and PCNA.* Nature. 1995;374:386-8.

15. Luo Y, Hurwitz J, Massague J. *Cell-cycle inhibition by independent CDK and PCNA binding domains in p21Cip1.* Nature. 1995;375:159-61.

16. Xiong Y, Zhang H, Beach D. *D-type cyclins associate with multiple protein kinases and the DNA replication and repair factor PCNA.* Cell. 1992;71:505-14.

17. Ducoux M, Urbach S, Baldacci G, Hubscher U, Koundrioukoff S, Christensen J, et al. *Mediation of proliferating cell nuclear antigen (PCNA)-dependent DNA replication through a conserved p21Cip1-like PCNA-binding motif present in the third subunit of human DNA polymerase δ.* J Biol Chem. 2001;276:49258-66.

18. Waga S, Hannon GJ, Beach D, Stillman B. *The p21 inhibitor of cyclin-dependent kinases controls DNA replication by interaction with PCNA.* Nature. 1994;369:574-8.

19. Krishna TS, Kong XP, Gary S, Burgers PM, Kuriyan J. *Crystal structure of the eukaryotic DNA polymerase processivity factor PCNA.* Cell. 1994;79:1233-43.

20. Kedar PS, Kim SJ, Robertson A, Hou E, Prasad R, Horton JK, et al. *Direct interaction between mammalian DNA polymerase β and proliferating cell nuclear antigen.* J Biol Chem. 2002;277:31115-23.

21. Levin DS, Bai W, Yao N, O'Donnell M, Tomkinson AE. *An interaction between DNA ligase I and proliferating cell nuclear antigen: Implications for Okazaki fragment synthesis and joining.* Proc Natl Acad Sci USA. 1997;94:12863-8.

22. Sakurai S, Kitano K, Yamaguchi H, Hamada K, Okada K, Fukuda K, et al. *Structural basis for recruitment of human flap endonuclease 1 to PCNA.* EMBO J. 2005;24:683-93.

23. Zhang P, Mo JY, Perez A, Leon A, Liu L, Mazloum N, et al. *Direct interaction of proliferating cell nuclear antigen with the p125 catalytic subunit of mammalian DNA polymerase δ.* J Biol Chem. 1999;274:26647-53.

24. Gary R, Kim K, Cornelius HL, Park MS, Matsumoto Y. *Proliferating cell nuclear antigen facilitates excision in long-patch base excision repair.* J Biol Chem. 1999;274:4354-63.

25. Umar A, Buermeyer AB, Simon JA, Thomas DC, Clark AB, Liskay RM, et al. *Requirement for PCNA in DNA mismatch repair at a step preceding DNA resynthesis.* Cell. 1996;87:65-73.

26. Ma X, Jin Q, Forsti A, Hemminki K, Kumar R. *Single nucleotide polymorphism analyses of the human proliferating cell nuclear antigen (PCNA) and flap endonuclease (FEN1) genes.* Int J Cancer. 2000;88:938-42.

27. Keshgegian AA, Cnaan A. *Proliferation markers in breast carcinoma. Mitotic figure count, S-phase fraction, proliferating cell nuclear antigen, Ki-67 and MIB-1.* Am J Clin Pathol. 1995;104:42-9.

28. van Dierendonck JH, Wijsman JH, Keijzer R, van de Velde CJ, Cornelisse CJ. *Cell-cycle-related staining patterns of anti-proliferating cell nuclear antigen monoclonal antibodies.* Am J Pathol. 1991;138:1165-72.

29. Haerslev T, Jacobsen GK, Zedeler K. *Correlation of growth fraction by Ki-67 and proliferating cell nuclear antigen (PCNA) immunohis-tochemistry with histopathological parameters and prognosis in primary breast carcinomas.* Breast Cancer Res Treat. 1996;37:101-13.

30. Lautier D, Lagueux J, Thibodeau J, Menard L, Poirier GG. *Molecular and biochemical features of poly(ADP-ribose) metabolism.* Mol Cell Biol. 1993;122:171-93.

31. Oei SL, Griesenbeck J,

Schweiger M. *The role of poly(ADP-ribosyl)ation.* Rev Physiol Biochem Pharmacol. 1997;131:128-73.

32. Lindahl T, Satoh MS, Poirier GG, Klungland A. *Post-translational modification of poly(ADP-ribose) polymerase induced by DNA strand breaks.* Trends Biochem Sci. 1995;20:405-11.

33. Oei SL, Ziegler M. *ATP for the DNA ligation step in base excision repair is generated from poly(ADP-ribose).* J Biol Chem. 2000;275:23234-9.

34. Koonin EV, Altschul SF, Bork P. *Functional motifs.* Nature Genet. 1996;13:266-8.

35. Callebaut I, Mornon JP. *From BRCA1 to RAP1: a widespread BRCT module closely associated with DNA repair.* FEBS Lett. 1997;400:25-30.

36. Yu X, Chini CC, He M, Mer G, Chen J. *The BRCT domain is a phospho-protein binding domain.* Science. 2003;302:639-42.

37. Zhang X, Morera S, Bates PA, Whitehead PC, Coffer AI, Hainbucher K, et al. *Structure of an XRCC1 BRCT domain: a new protein-protein interaction module.* EMBO J. 1998;17:6404-11.

38. Masson M, Niedergang C, Schreiber V, Muller S, Menissier-De Murcia J, De Murcia G. *XRCC1 is specifically associated with poly(ADP-ribose) polymerase and negatively regulates its activity following DNA damage.* Mol Cell Biol. 1998;18:3563-71.

39. Schreiber V, Ame JC, Dolle P, Schultz I, Rinaldi B, Fraulob V, et al. *Poly(ADP-ribose) polymerase-2 (PARP-2) is required for efficient base excision DNA repair in association with PARP-1 and XRCC1.* J Biol Chem. 2002;277:23028-36.

40. Caldecott KW, Aoufouchi S, Johnson P, Shall S. *XRCC1 polypeptide interacts with DNA polymerase β and possibly poly(ADP-ribose) polymerase, and DNA ligase III is a novel molecular 'nick-sensor' in vitro.* Nucleic Acids Res. 1996;24:4387-94.

41. Leppard JB, Dong Z, Mackey ZB, Tomkinson AE. *Physical and functional interaction between DNA ligase IIIα and poly(ADP-Ribose) polymerase 1 in DNA single-strand break repair.* Mol Cell Biol. 2003;23:5919-27.

42. Mackey ZB, Niedergang C, de Murcia JM, Leppard J, Au K, Chen J, et al. *DNA ligase III is recruited to DNA strand breaks by a zinc finger motif homologous to that of poly(ADP-ribose) polymerase.* J Biol Chem. 1999;274:21679-87.

43. Lin PH, Lin CH, Huang CC, Chuang MC, Lin P. *2,3,7,8-Tetra-chlorodibenzo-p-dioxin (TCDD) induces oxidative stress, DNA strand breaks, and poly(ADP-ribose) polymerase-1 activation in human breast carcinoma cell lines.* Toxicol Lett. 2007;172:146-58

44. Beneke R, Geisen C, Zevnik B, Bauch T, Muller WU, Kupper JH, et al. *DNA excision repair and DNA damage-induced apoptosis are linked to poly(ADP-Ribosyl)ation but have different requirements for p53.* Mol Cell Biol. 2000;20:6695-703.

45. Masutani M, Nozaki T, Nakamoto K, Nakagama H, Suzuki H, Kusuoka O, et al. *The response of Parp knockout mice against DNA damaging agents.* Mutat Res. 2000;462:159-66.

46. Menissier-De Murcia J, Niedergang C, Trucco C, Ricoul M, Dutrillaux B, Marks M, et al. *Requirement of poly(ADP-ribose) polymerase in recovery from DNA damage in mice and in cells.* Proc Natl Acad Sci USA. 1997;94:7303-7.

47. Trucco C, Oliver FJ, De Murcia G, Menissier-De Murcia J. *DNA repair defect in poly(ADP-ribose) polymerase-deficient cell lines.* Nucleic Acids Res. 1998;26:2644-9.

48. Prasad R, Lavrik OI, Kim SJ, Kedar P, Yang XP, Vande Berg BJ, et al. *DNA polymerase β-mediated long patch base excision repair.* J Biol Chem. 2001;276:32411-4.

49. Schreiber V, Molinete M, Boeuf H, De Murcia G, Menissier-De Murcia J. *The human poly(ADP-ribose) polymerase nuclear localization signal is a bipartite element functionally separate from DNA binding and catalytic activity.* EMBO J. 1992;11:3263-9.

50. Mohrenweiser HW, Xi T, Vazquez-Matias J, Jones IM. *Identification of 127 amino acid substitution variants in screening 37 DNA repair genes in humans.* Cancer Epidemiol Biomarkers Prev. 2002;11:1054-64.

51. Hu JJ, Roush GC, Dubin N, Berwick M, Roses DF, Harris MN. *Poly(ADP-ribose) polymerase in human breast cancer: a case-control analysis.* Pharmacogenetics. 1997;7:309-16.

52. Lockett KL, Hall MC, Xu J, Zheng SL, Berwick M, Chuang SC, et al. *The ADPRT V762A genetic variant contributes to prostate cancer susceptibility and deficient enzyme function.* Cancer Res. 2004;64:6344-8.

53. Cottet F, Blanche H, Verasdonck P, Le Gall I, Schachter F, Burkle A, et al. *New polymorphisms in the human poly(ADP-ribose) polymerase-1 coding sequence: lack of association with longevity or with increased cellular poly(ADP-ribosyl)ation capacity.* J Mol Med. 2000;78:431-40.

54. Lamerdin JE, Montgomery MA, Stilwagen SA, Scheidecker LK, Tebbs RS, Brookman KW, et al. *Genomic sequence comparison of the human and mouse XRCC1 DNA repair gene regions.* Genomics. 1995;25:547-54.

55. Thompson LH, Brookman KW, Jones NJ, Allen SA, Carrano AV. *Molecular cloning of the human XRCC1 gene, which corrects defective DNA strand break repair and sister chromatid exchange.* Mol Cell Biol. 1990;10:6160-71.

56. Fan J, Otterlei M, Wong HK, Tomkinson AE, Wilson DM. *XRCC1 co-localizes and physically interacts with PCNA.* Nucleic Acids Res. 2004;32:2193-201.

57. Kubota Y, Nash RA, Klungland A, Schar P, Barnes DE, Lindahl T. *Reconstitution of DNA base excision-repair with purified human proteins: interaction between DNA polymerase β and the XRCC1 protein.* EMBO J. 1996;15:6662-70.

58. Gryk MR, Marintchev A, Maciejewski MW, Robertson A, Wilson SH, Mullen GP. *Mapping of the interaction interface of DNA polymerase β with XRCC1.* Structure. 2002;10:1709-20.

59. Marintchev A, Mullen MA, Maciejewski MW, Pan B, Gryk MR, Mullen GP. *Solution structure of the single-strand break repair protein XRCC1 N-terminal domain.* Nature Struct Biol. 1999;6:884-93.

60. Cappelli E, Taylor R, Cevasco M, Abbondandolo A, Caldecott K, Frosina G. *Involvement of XRCC1 and DNA ligase III gene products in DNA base excision repair.* J Biol Chem. 1997;272:23970-5.

61. Krishnan VV, Thornton KH, Thelen MP, Cosman M. *Solution structure and backbone dynamics of the human DNA ligase III5.2.5. DNA Polymerases BRCT domain.* Biochemistry. 2001;40:13158-66.

62. Nash RA, Caldecott KW, Barnes DE, Lindahl T. *XRCC1 protein interacts with one of two distinct forms of DNA ligase III.* Biochemistry. 1997;36:5207-11.

63. Whitehouse CJ, Taylor RM, Thistlethwaite A, Zhang H, Karimi-Busheri F, Lasko DD, et al. *XRCC1 stimulates human polynucleotide kinase activity at damaged DNA termini and accelerates DNA single-strand break repair.* Cell. 2001;104:107-17.

64. Tebbs RS, Flannery ML, Meneses JJ, Hartmann A, Tucker JD, Thompson LH, et al. *Requirement for the Xrcc1 DNA base excision repair gene during early mouse development.* Dev Biol. 1999;208:513-29.

65. Shen B, Singh P, Liu R, Qiu J, Zheng L, Finger LD, et al. *Multiple but dissectible functions of FEN-1 nucleases in nucleic acid processing, genome stability and diseases.* BioEssays. 2005;27:717-29.

66. Wang Y, Spitz MR, Zhu Y, Dong Q, Shete S, Wu X. *From genotype to phenotype: correlating XRCC1 polymorphisms with mutagen sensitivity.* DNA Repair. 2003;2:901-8.

67. Tuimala J, Szekely G, Gundy S, Hirvonen A, Norppa H. *Genetic polymorphisms of DNA repair and xenobiotic-metabolizing enzymes: role in mutagen sensitivity.* Carcinogenesis. 2002;23:1003-8.

68. Taylor RM, Thistlethwaite A, Caldecott KW. *Central role for the XRCC1 BRCT I domain in mammalian DNA single-strand break repair.* Mol Cell Biol. 2002;22:2556-63.

69. Duell EJ, Wiencke JK, Cheng TJ, Varkonyi A, Zuo ZF, Ashok TD, et al. *Polymorphisms in the DNA repair genes XRCC1 and ERCC2 and biomarkers of DNA damage in human blood mononuclear cells.* Carcinogenesis. 2000;21:965-71.

70. Lunn RM, Langlois RG, Hsieh LL, Thompson CL, Bell DA. *XRCC1 polymorphisms: Effects on aflatoxin B1-DNA adducts and glycophorin A variant frequency.* Cancer Res. 1999;59:2557-61.

71. Matullo G, Palli D, Peluso M, Guarrera S, Carturan S, Celentano E, et al. *XRCC1, XRCC3, XPD gene polymorphisms, smoking and 32P-DNA adducts in a sample of healthy subjects.* Carcinogenesis. 2001;22:1437-45.

72. Hodgson ME, Poole C, Olshan AF, North KE, Zeng D, Millikan RC. *Smoking and selected DNA repair gene polymorphisms in controls: systematic review and meta-analysis.* Cancer Epidemiol Biomarkers Prev. 2010;19:3055-86.

73. Han J, Hankinson SE, De Vivo I, Spiegelman D, Tamimi RM, Mohrenweiser HW, et al. *A prospective study of XRCC1 haplotypes and their interaction with plasma carotenoids on breast cancer risk.* Cancer Res. 2003;63:8536-41.

74. Hung RJ, Hall J, Brennan P, Boffetta P. *Genetic polymorphisms* *in the base excision repair pathway and cancer risk: a HuGE review.* Am J Epidemiol. 2005;162:925-42.

75. Shu XO, Cai Q, Gao YT, Wen W, Jin F, Zheng W. *A population-based case-control study of the Arg399Gln polymorphism in DNA repair gene XRCC1 and risk of breast cancer.* Cancer Epidemiol Biomarkers Prev. 2003;12:1462-7.

76. Duell EJ, Millikan RC, Pittman GS, Winkel S, Lunn RM, Tse CJ, et al. *Polymorphisms in the DNA repair gene XRCC1 and breast cancer.* Cancer Epidemiol Biomarkers Prev. 2001;10:217-22.

77. Moullan N, Cox DG, Angele S, Romestaing P, Gerard JP, Hall J. *Polymorphisms in the DNA repair gene XRCC1, breast cancer risk, and response to radiotherapy.* Cancer Epidemiol Biomarkers Prev. 2003;12:1168-74.

78. Pachkowski BF, Winkel S, Kubota Y, Swenberg JA, Millikan RC, Nakamura J. *XRCC1 genotype and breast cancer: Functional studies and epidemiologic data show interactions between XRCC1 codon 280 his and smoking.* Cancer Res. 2006;66:2860-8.

transferase subdomain contains three aspartates (residues 190, 192, and 256) that coordinate two divalent metal cations, Mg^{2+}, which assist in the enzymatic reaction. Altogether, the 31-kDa domain covers about six nucleotides in the template strand and on the downstream side of the gap about six nucleotides are covered by the 8-kDa domain. The physical association of the lyase and polymerase activities in a single enzyme enhances the efficiency of repair. *In vitro* studies demonstrated that the lyase activity was the rate-limiting step for BER in the presence of high concentrations of BER enzymes (11). These studies have also revealed direct interaction of pol β with ligase I within the multiprotein BER complex (12). The interaction involves regions close to the N-terminus of both proteins (13). Other structural studies have shown the interaction of the palm-thumb region of pol β with the scaffold protein XRCC1 (14-16).

5.2.5. DNA POLYMERASES

DNA polymerase β (Pol β). The pol β gene at 8p11 – p12 is 33 kb long and composed of 14 exons encoding a 39-kDa protein consisting of 335 amino acids (1-5). While the majority of DNA polymerases play a role in semi-conservative DNA replication, pol β catalyzes DNA synthesis during BER (6). Pol β fills the gap in a processive rather than distributive manner. In addition to filling the gap created by APE1, pol β catalyzes release of the 5' terminal deoxyribose phosphate (dRP) residue from the incised AP site. The catalytic domain for this so-called AP or dRP lyase activity resides with an N-terminal 8-kDa fragment of 75 amino acid residues, which comprise a distinct structural domain of the enzyme (Figure 5.2.5.) (7). The crystal structure of a complex of the 8-kDa domain with DNA substrates identified key residues 34His, 35Lys, 39Tyr, and 72Lys involved in DNA binding and lyase reaction (8, 9). The DNA polymerase activity resides in the C-terminal 31-kDa portion of approximately 260 amino acid residues. The latter segment is subdivided into "fingers", "palm", and "thumb" subdomains to "grasp" DNA, associated with double-stranded DNA binding, nucleotidyl transferase, and dNTP selection functions, respectively (1, 5, 9, 10). The nucleotidyl

Non-synonymous polymorphisms are found in codons Gln8Arg, Arg137Gln, and Pro242Arg with frequencies of the variant alleles below 2% (17, 18). Site-directed mutagenesis close to the polymorphic codon Pro242Arg has identified several residues (i.e., 246Asp, 249Glu, 253Arg) that do not interact with substrates but influence catalytic activity and/or fidelity when altered (1). For example, a Asp246Val mutant exhibited decreased discrimination by enhancing the efficiency of incorrect nucleotide insertion (19). Splice variants have been detected in normal and malignant tumor tissues, including fibroadenomas and invasive breast cancers (2, 20, 21). Loss of exon 2 is observed most frequently, creating an early stop codon in exon 3 that would be predicted to produce a truncated protein of only 26 amino acids. This oligopeptide would include the first 20 amino acids of the single-stranded DNA binding AP lyase domain of pol β (5). In addition to the intact gene, fibroadenomas and breast cancers may contain a copy with an 87 bp deletion encoding amino acid residues 208 – 236 in the palm region of the protein (20). This truncated variant does not disrupt the DNA binding or gap filling synthesis activity of wild-type pol β. However, the truncated pol β formed a complex with XRCC1 that inhibited both functions of the wild-type enzyme (22).

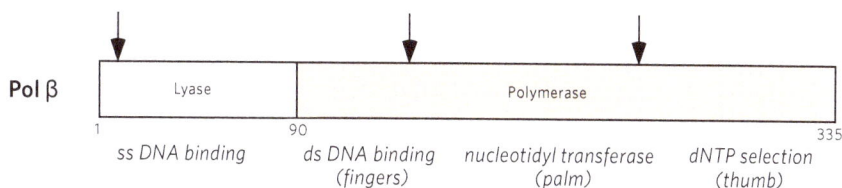

Figure 5.2.5. Diagram of pol β showing functional domains, conserved sequences, and non-synonymous polymorphisms. *The N-terminal 8-kDa lyase domain binds single-stranded DNA and catalyzes release of the 5' terminal deoxyribose phosphate (dRP) residue from the incised AP site. The C-terminal 31-kDa polymerase domain is subdivided into "fingers", "palm", and "thumb" subdomains, associated with double-stranded DNA binding, nucleotidyl transferase, and dNTP selection functions, respectively. The arrows indicate three non-synonymous polymorphisms, Gln8Arg, Arg137Gln, and Pro242Arg.*

Pol β is essential for viability as shown by the embryonic-lethal phenotype of mice lacking both alleles of the pol β gene (23). Heterozygous pol β +/- mice survive but exhibit an increased mutational response to carcinogens (24). The expression level of pol β is independent of cell cycle stage but was shown to be variable in cell lines and tumor tissue. In cell lines from normal breast or colon, the level of pol β was ~1 ng/ml cell extract, whereas in all of the breast and colon carcinoma cell lines tested, a higher level of pol β was observed (25). Invasive breast cancers exhibited a wide range of pol β expression; one tumor had a much higher level of pol β (286 ng/ml extract) than adjacent normal breast tissue, whereas another tumor had the same level of pol β as adjacent normal tissue. In contrast, all renal cell carcinomas had slightly lower pol β levels than adjacent normal tissue. Thus, pol β is upregulated in some types of cancer and cell lines but not in others (25).

DNA polymerases δ and ε (Pol δ and ε). DNA replication requires the participation of pol α, δ, and ε. Among these, pol α is the only enzyme that can start DNA synthesis *de novo*. It can synthesize short primers to initiate leading strand synthesis at the replication origin and Okazaki fragments on the lagging strand. Both pol δ and ε can elongate primers synthesized by pol α and appear to have overlapping or complementary functions (26). While pol β is a single 39-kDa protein, pol δ and ε are complex proteins composed of multiple heteromeric subunits (26, 27). For example, pol δ contains a catalytic subunit of 125 kDa, and two smaller subunits of 66 – 69 and 48 – 50 kDa (28).

Flap Endonuclease 1 (FEN1)

The FEN1 gene at 11q12 encodes a 43-kDa protein composed of 380 amino acids (29, 30). FEN1 is a multifunctional, structure-specific nuclease involved in DNA replication and DNA repair (long-patch BER, homologous recombination, nonhomologous end-joining repair), as well as resolution of di- and trinucleotide repeat secondary structures and apoptotic DNA fragmentation (29-33). In DNA replication, FEN1 processes the 5' ends of Okazaki fragments in lagging-strand DNA synthesis. In long-patch BER, FEN1 is essential for cleaving the 5' overhanging single-stranded DNA flaps at the single strand-double strand junction (34, 35).

FEN1 is a member of the XPG/RAD2 nuclease family based on sequence similarities (30). Crystallographic, genetic, and biochemical studies have revealed the conservation of key structures between FEN1 homologues of distantly related organisms (35-38). One key element is a flexible loop in the central portion of the molecule, which contains several positively charged amino acids lining the inner surface of a helical motif. The flexible loop appears to form a hole (8 x 25 Å) that enables FEN1 to thread over the single-stranded DNA to reach the branch point for flap cleavage at the active site

of the enzyme. The active site is formed by two clusters of conserved acidic residues, which surround two tightly bound Mg^{2+}. The C-terminal region, which is rich in lysine residues, is involved in DNA binding and interacts with PCNA. The interaction with PCNA stabilizes FEN1 at the branch point and thereby stimulates the enzyme activity up to 50-fold (35, 38). The C-terminal region also contains a nuclear localization signal (39). The nuclear import of FEN1 is cell cycle dependent and inducible by treatment of cells with DNA damaging agents. In addition to PCNA, FEN1 also interacts with APE1. The interaction between APE1 and FEN1 provides coordinated loading of the proteins onto the substrate, thus passing the substrate from one enzyme to another (40).

The human FEN1 gene consists of a single exon. Sequence analysis of this exon in 132 healthy individuals of African, Asian, or Caucasian origin revealed no polymorphism, suggesting that the coding region of FEN1 is highly conserved compared with other DNA repair genes (18, 41). The importance of the FEN1 gene has been demonstrated in mouse models. Mutation of a single amino acid in the nuclease domain was associated with an elevated incidence of B-cell lymphomas while the mutation of two amino acids in the PCNA binding domain led to extensive apoptosis of the forebrain (42). Complete loss of both FEN1 alleles leads to embryonic lethality whereas animals with one intact allele appear to be free of disease (43). However, FEN1 heterozygous knockout mice that were also heterozygous for the adenomatous polyposis coli (APC) gene developed adenocarcinomas suggesting that FEN1 is a tumor suppressor gene (44). Interestingly, APC was shown to interact with both FEN1 and pol β and block long-patch BER (45). The transformation of pre-malignant breast epithelial cells, MCF-10A, by treatment with benzo[a] pyrene and cigarette smoke condensate was associated with increased levels of APC and impaired long-patch BER (46). FEN1 has also been shown to participate in physical and functional interactions with the Werner syndrome protein WRN, which is a member of the RecQ helicase family (47). Werner syndrome is an inherited premature aging disorder characterized by chromosomal instability. Finally, FEN1 was shown to interact with ERα, to enhance the interaction of ERα with estrogen response element-containing DNA, and to modulate the expression of estrogen-responsive genes, such as pS2 and progesterone receptor (48). The identification of FEN1 as a regulator of ERα function suggests a link between estrogen-responsive gene expression and DNA repair. Interestingly, E_2 was shown to regulate the expression of FEN1 in epithelial and myometrial cells of the mouse uterus (48). In summary, FEN1 is a multifunctional and structure-specific nuclease involved in several nucleic acid processing pathways guarding genomic integrity. The enzyme can carry out multiple reactions by interacting with different proteins, e.g., PCNA, RFC, APE1, pol β, APC, WRN, and ERα.

References

1. Beard WA, Wilson SH. *Structure and mechanism of DNA polymerase β*. Chem Rev. 2006;106:361-82.

2. Chyan YJ, Ackerman S, Shepherd NS, McBride OW, Widen SG, Wilson SH, et al. *The human DNA polymerase β gene structure. Evidence of alternative splicing in gene expression*. Nucleic Acids Res. 1994;22:2719-25.

3. McBride OW, Zmudzka BZ, Wilson SH. *Chromosomal location of the human gene for DNA polymerase β*. Proc Natl Acad Sci USA. 1987;84:503-7.

4. Widen SG, Kedar P, Wilson SH. *Human β-polymerase gene. Structure of the 5'-flanking region and active promoter*. J Biol Chem. 1988;263:16992-8.

5. Wilson SH. *Mammalian base excision repair and DNA polymerase beta*. Mutat Res. 1998;407:203-15.

6. Sobol RW, Horton JK, Kuhn R, Gu H, Singhal RK, Prasad R, et al. *Requirement of mammalian DNA polymerase-β in base-excision repair*. Nature. 1996;379:183-6.

7. Matsumoto Y, Kim K. *Excision of deoxyribose phosphate residues by DNA polymerase β during DNA repair*. Science. 1995;269:699-702.

8. Maciejewski MW, Liu D, Prasad R, Wilson SH, Mullen GP. *Backbone dynamics and refined solution structure of the N-terminal domain of DNA polymerase β. Correlation with DNA binding and dRP lyase activity*. J Mol Biol. 2000;296:229-53.

9. Sawaya MR, Prasad R, Wilson SH, Kraut J, Pelletier H. *Crystal structures of human DNA polymerase β complexed with gapped and nicked DNA: Evidence for an induced fit mechanism*. Biochemistry. 1997;36:11205-15.

10. Pelletier H, Sawaya MR, Kumar A, Wilson SH, Kraut J. *Structures of ternary complexes of rat DNA polymerase β, a DNA template-primer, and ddCTP*. Science. 1994;264:1891-903.

11. Srivastava DK, Vande Berg BJ, Prasad R, Molina JT, Beard WA, Tomkinson AE, et al. *Mammalian abasic site base excision repair. Identification of the reaction sequence and rate-determining steps*. J Biol Chem. 1998;273:21203-9.

12. Prasad R, Singhal RK, Srivastava DK, Molina JT, Tomkinson AE, Wilson SH. *Specific interaction of DNA Polymerase β and DNA Ligase I in a multiprotein base excision repair complex from bovine testis*. J Biol Chem. 1996;271:16000-7.

13. Dimitriadis EK, Prasad R, Vaske MK, Chen L, Tomkinson AE, Lewis MS, et al. *Thermodynamics of human DNA ligase I trimerization and association with DNA polymerase β*. J Biol Chem. 1998;273:20540-50.

14. Gryk MR, Marintchev A, Maciejewski MW, Robertson A, Wilson SH, Mullen GP. *Mapping of the interaction interface of DNA polymerase β with XRCC1*. Structure. 2002;10:1709-20.

15. Kubota Y, Nash RA, Klungland A, Schar P, Barnes DE, Lindahl T. *Reconstitution of DNA base excision-repair with purified human proteins: interaction between DNA polymerase β and the XRCC1 protein*. EMBO J. 1996;15:6662-70.

16. Marintchev A, Mullen MA, Maciejewsik MW, Pan B, Gryk MR, Mullen GP. *Solution structure of the single-strand break repair protein XRCC1 N-terminal domain*. Nature Struct Biol. 1999;6:884-93.

17. Eydmann ME, Knowles MA. *Mutation analysis of 8p genes POLB and PPP2CB in bladder cancer*. Cancer Genet Cytogenet. 1997;93:167-71.

18. Mohrenweiser HW, Xi T, Vazquez-Matias J, Jones IM. *Identification of 127 amino acid substitution variants in screening 37 DNA repair genes in humans*. Cancer Epidemiol Biomark Prev. 2002;11:1054-64.

19. Dalal S, Kosa JL, Sweasy JB. *The D246V mutant of DNA polymerase β misincorporates nucleotides*. J Biol Chem. 2004;279:577 - 84.

20. Bhattacharyya N, Chen HC, Grundfest-Broniatowski S. *Alteration of hMSH2 and DNA polymerase β genes in breast carcinomas and fibroadenomas*. Biochem Biophys Res Comm. 1999;259:429-35.

21. Thompson TE, Rogan PK, Risinger JI, Taylor JA. *Splice variants but not mutations of DNA polymerase β are common in bladder cancer*. Cancer Res. 2002;62:3251-6.

22. Bhattacharyya N, Banerjee S. *A novel role of XRCC1 in the functions of a DNA polymerase β variant*. Biochemistry. 2001;40:9005-13.

23. Gu H, Marth JD, Orban PC, Mossmann H, Rajewsky K. *Deletion of a DNA polymerase β gene segment in T cells using cell type-specific gene targeting*. Science. 1994;265:103-6.

24. Cabelof DC, Guo ZM, Raffoul JJ, Sobol RW, Wilson SH, Richardson A, et al. *Base excision repair deficiency caused by polymerase β haploinsufficiency: Accelerated DNA damage and increased mutational response to carcinogens*. Cancer Res. 2003;63:5799-807.

25. Srivastava DK, Husain I, Arteaga CL, Wilson SH. *DNA polymerase β expression differences in selected human tumors and cell lines*. Carcinogenesis. 1999;20:1049-54.

26. Waga S, Stillman B. *The DNA replication fork in eukaryotic cells*. Annu Rev Biochem. 1998;67:721-51.

27. Dua R, Levy DL, Li CM, Snow PM, Campbell JL. *In vivo reconstitution of Saccharomyces cerevisiae DNA polymerase e in insect cells. Purification and characterization*. J Biol Chem. 2002;277:7889-96.

28. Ducoux M, Urbach S, Baldacci G, Hubscher U, Koundrioukoff S, Christensen J, et al. *Mediation of proliferating cell nuclear antigen (PCNA)-dependent DNA replication through a conserved p21Cip1-like PCNA-binding motif present in the third subunit of human DNA polymerase δ*. J Biol Chem. 2001;276:49258-66.

29. Liu Y, Kao HI, Bambara RA. *Flap endonuclease1: A central component of DNA metabolism*. Annu Rev Biochem. 2004;73:589-615.

30. Shen B, Singh P, Liu R, Qiu J, Zheng L, Finger LD, et al. *Multiple but dissectible functions of FEN-1 nucleases in nucleic acid processing, genome stability and diseases*. BioEssays. 2005;27:717-29.

31. Hoeijmakers JHJ. *Genome maintenance mechanisms for preventing cancer*. Nature. 2001;411:366-74.

32. Liu R, Qiu J, Finger LD, Zheng L, Shen B. *The DNA-protein interaction modes of FEN-1 with gap substrates and their implication in preventing duplication mutations*. Nucleic Acids Res. 2006;34:1772-84.

33. Parrish JZ, Yang C, Shen B, Xue D. *CRN-1, a caenorhabditis elegans FEN-1 homologue, cooperates with CPS-6/EndoG to promote apoptotic DNA degradation*. EMBO J. 2003;22:3451-60.

34. Klungland A, Lindahl T. *Second pathway for completion of human DNA base excision-repair: reconstitution with purified proteins and requirement for DNase IV (FEN1)*. EMBO J. 1997;16:3341-8.

35. Lieber MR. *The FEN-1 family of structure-specific nucleases in eukaryotic DNA replication, recombination and repair*. BioEssays. 1997;19:233-40.

36. Hosfield DJ, Mol CD, Shen B, Tainer JA. *Structure of the DNA repair and replication endonuclease and exonuclease FEN-1: coupling DNA and PCNA binding to FEN-1 activity*. Cell. 1998;95:135-46.

37. Hwang KY, Baek K, Kim HY, Cho Y. *The crystal structure of flap endonuclease-1 from Methanococcus jannaschii*. Nature Struct Biol. 1998;5:707-13.

38. Sakurai S, Kitano K, Yamaguchi H, Hamada K, Okada K, Fukuda K, et al. *Structural basis for recruitment of human flap endonuclease 1 to PCNA*. EMBO J. 2005;24:683-93.

39. Qiu J, Li X, Frank G, Shen B. *Cell cycle-dependent and DNA damage-inducible nuclear localization of FEN-1 nuclease is consistent with its dual functions in DNA replication and repair*. J Biol Chem. 2001;276:4901-8.

40. Dianova II, Bohr VA, Dianov GL. *Interaction of human AP endonuclease 1 with flap endonuclease 1 and proliferating cell nuclear antigen involved in long-patch base excision repair*. Biochemistry. 2001;40:12639-44.

41. Ma X, Jin Q, Forsti A, Hemminki K, Kumar R. *Single nucleotide polymorphism analyses of the human proliferating cell nuclear antigen (PCNA) and flap endonuclease (FEN1) genes*. Int J Cancer. 2000;88:938-42.

42. Larsen E, Kleppa L, Meza TJ, Meza-Zepeda LA, Rada C, Castellanos CG, et al. *Early-onset lymphoma and extensive embryonic apoptosis in two domain-specific Fen1 mice mutants*. Cancer Res. 2008;68:4571-9.

43. Kucherlapati M, Yang K, Kuraguchi M, Zhao J, Lia M, Heyer J, et al. *Haploinsufficiency of flap endonuclease (Fen1) leads to rapid tumor progression*. Proc Natl Acad Sci USA. 2002;99:9924-9.

44. Henneke G, Friedrich-Heineken E, Hubscher U. *Flap endonuclease 1: a novel tumour suppressor protein*. Trends Biochem Sci. 2003;28:384-90.

45. Jaiswal AS, Balusu R, Armas ML, Kundu CN, Narayan S. *Mechanism of adenomatous polyposis coli (APC)-mediated blockage of long-patch base excision repair*. Biochemistry. 2006;45:15903-14.

46. Kundu CN, Balusu R, Jaiswal AS, Gairola CG, Narayan S. *Cigarette smoke condensate-induced level of* *adenomatous polyposis coli blocks long-patch base excision repair in breast epithelial cells.* Oncogene. 2007;26:1428-38.

47. Brosh RM, von Kobbe C, Sommers JA, Karmakar P, Opresko PL, Piotrowski J, et al. *Werner syndrome protein interacts with human flap endonuclease 1 and stimulates its cleavage activity.* EMBO J. 2001;20:5791-801.

48. Schultz-Norton JR, Walt KA, Ziegler YS, McLeod IX, Yates JR, Raetzman LT, et al. *The deoxyribonucleic acid repair protein flap endonuclease-1 modulates estrogen-responsive gene expression.* Mol Endocrinol. 2007;21:1569-80.

5.2.6. DNA LIGASES

DNA ligases catalyze the joining of strand breaks in the phosphodiester backbone of DNA and play essential roles in DNA replication, recombination, and repair (1-4). Several DNA ligases are required for these diverse roles. Three mammalian genes encoding DNA ligases have been isolated: ligase I, III, and IV; ligase II appears to be a proteolysis fragment of ligase III. The enzymes utilize ATP as cofactor and form a covalent ligase-AMP complex in the first step of the ligation reaction. The AMP moiety is attached to an active site lysine residue within the conserved sequence Lys-Xxx-Asp-Gly-Xxx-Arg of the catalytic core (3, 5). In the later steps, the AMP group is transferred to the 5'-phosphate terminus of a nick in a duplex DNA molecule, generating a covalent DNA-AMP complex and then the non-adenylated enzyme catalyzes the formation of a phosphodiester bond with the concomitant release of AMP. Although all three ligases use the same basic reaction mechanism, they have distinct functions, interact with different protein partners, and are not inter-changeable (2, 3). Laser microirradiation of human HeLa or HEK 293IIT cells introduced DNA lesions, which were repaired *in vivo* by DNA ligases. Time-lapse microscopy of fluorescently tagged proteins showed that ligase III accumulated at irradiated sites before ligase I, with only a faint accumulation of ligase IV (6). The differential recruitment of ligase III to the repair site was mediated by interaction with XRCC1, while ligase I was recruited via interaction with PCNA, emphasizing that these enzymes carry out distinct functions together with their respective partner proteins.

DNA Ligase III

The DNA ligase III gene at 17q11.2-12 encodes two nuclear proteins denoted Lig-IIIα and Lig-IIIβ, which differ at their C-termini as a consequence of tissue- and cell type-specific splicing (7, 8). The 103-kDa Lig-IIIα (922 amino acids) is expressed in both somatic and germ cells and contains a C-terminal BRCT motif (9, 10). In the case of Lig-IIIα, the BRCT domain binds XRCC1, which is required for intracellular stability of Lig-IIIα and consequently for efficient DNA ligation following the excision of damaged DNA bases (11, 12). In contrast, the 96-kDa Lig-IIIβ (862 amino acids) is only expressed in male meiotic cells, lacks the C-terminal BRCT domain, and consequently does not bind XRCC1 (7). Both ligase III isoforms contain a 38-residue zinc finger motif near the N-terminus (amino acids 18 – 55), which is homologous to two zinc fingers of PARP (Figure 5.2.6.) (13, 14). Several studies have demonstrated the interaction between ligase III and PARP1 and PARP2 (15-17). The zinc finger of ligase III appears to increase DNA joining in the presence of PARP, enables the ligase to bind DNA duplexes harboring a variety of secondary structures and rejoin adjacent strand breaks (18). Both ligase III isoforms contain the active site motif, Lys-Xxx-Asp-Gly-Xxx-Arg, as amino acids 421 – 426. A conserved peptide (residues 712 – 727, including Arg-Phe-Pro-Arg) is required for the transfer of the AMP group from the enzyme to the 5'-phosphate terminus at the DNA nick (16). In addition to playing a role in BER, DNA ligase III participates in a backup pathway of nonhomologous end-joining repair (19). Mice homozygous for ligase III inactivation showed embryonic lethality (20).

The ligase III gene contains three non-synonymous polymorphisms in the C-terminal region involving codons Arg780His, Lys811Thr, and Pro899Ser with variant allele frequencies less than 4% (21). Residue 899Pro is a

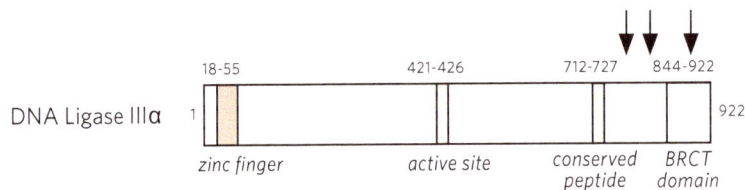

Figure 5.2.6. Diagram of Ligase IIIα showing functional domains, conserved sequences, and non-synonymous polymorphisms. *The C-terminus (residues 18 – 55) contains a zinc finger motif homologous with the zinc fingers of PARP that interact with DNA strand breaks. The active site motif, Lys-Xxx-Asp-Gly-Xxx-Arg, occupies residues 421 – 426. A conserved peptide (residues 712 – 727) plays a role in the AMP transfer reaction that produces the adenylate intermediate at the nicked DNA. The C-terminal end (residues 844 – 922) is homologous to the BRCT domain, which binds XRCC1 to stabilize the enzyme for efficient DNA ligation. The arrows indicate three non-synonymous polymorphisms, which are located in the C-terminal region involving codons Arg780His, Lys811Thr, and Pro899Ser with variant allele frequencies less than 4%.*

DNA Ligase I

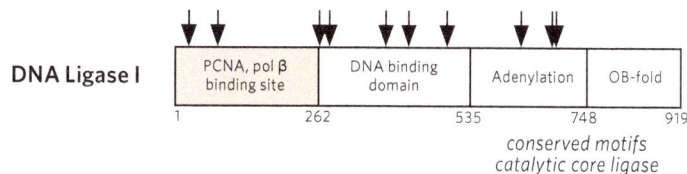

PCNA, pol β binding site	DNA binding domain	Adenylation	OB-fold

1   262   535   748   919

conserved motifs
catalytic core ligase

Figure 5.2.7. Diagram of Ligase I showing functional domains, conserved sequences, and non-synonymous polymorphisms. *The N-terminal region (residues 1 – 262) interacts with PCNA and pol β and also contains the nuclear localization signal. The mid-molecule region (residues 262 – 535) interacts with DNA. The C-terminal region contains the catalytic domain including several conserved motifs as well as an OB-fold domain. The arrows indicate ten non-synonymous polymorphisms involving codons Ala24Val, Arg62Trp, Gly249Glu, Asn267Ser, Val369Ile, Arg409His, Met480Val, Thr614Ile, Glu673Asp, and Arg677Leu with variant allele frequencies less than 3%.*

non-conserved amino acid located in the BRCT domain of Lig-IIIα. It is one of six prolines in the BRCT domain, all of which have peptide bonds in the *trans* conformation (9, 14, 22).

DNA Ligase I

DNA ligase 1 constitutes the major ligase activity in proliferating cells. The gene at 19q13.2-13.3 encodes a 125-kDa protein with the catalytic activity residing in the C-terminal part (23, 24). Residue 568Lys forms a covalent enzyme-AMP adduct during step 1 of the ligation reaction (25). A crystal structure of ligase I (residues 233 – 919) complexed with nicked DNA substrate revealed that three domains of the enzyme, namely the DNA binding domain (residues 262 – 535), the adenylation domain (residues 535 – 746), and OB-fold (residues 746 – 919) completely encircle and partially unwind the DNA to position the catalytic core on the nick (Figure 5.2.7.) (25). Key residues identified in the DNA binding pocket were 448Gly, 451Arg, and 455Ala (26). The N-termini of ligase I and the large subunit of RFC contain a homologous sequence of 20 amino acids that interact with PCNA (27). The interaction with PCNA directs the end-joining activity of ligase I to DNA replication sites. The N-terminal region of ligase I also contains nuclear localization signals (28). Residue 66Ser is phosphorylated in a cell cycle-dependent manner (29). Dephosphorylation of the enzyme in early G1 is linked to nuclear localization. The N-terminal region of ligase I also interacts with pol β (30, 31). Both the noncatalytic N-terminal and the catalytic C-terminal domains of ligase I also interact directly with replication factor C, specifically the large RFC subunit p140 and the p36 and p38 subunits of RFC (32). The interaction of ligase I with PCNA is critical for long-patch BER (33).

Loss of ligase I function leads to defective long-patch BER (33). Ten non-synonymous polymorphisms have been identified in the ligase I gene (21). The following codons are affected: Ala24Val, Arg62Trp, Gly249Glu, Asn267Ser, Val369Ile, Arg409His, Met480Val, Thr614Ile, Glu673Asp, and Arg677Leu. Codon 24 is located in the N-terminal PCNA and pol β binding domain of the ligase I protein. Although none of the 10 amino acid substitution variants exist at a frequency of over 2%, the total variant allele frequency for the ligase I gene is 11%. Residue 267Asn is

part of a peptide sequence homologous to bovine ligase I, Tyr-Asn-Pro-Ala (23). Residue 673Glu is located in a peptide motif conserved between ligases and homologous to residue 313Glu in DNA ligase IV (14). An A/C polymorphism in exon 6 does not cause an amino acid change and was not associated with the risk of lung cancer (34).

References

1. Cao W. *DNA ligases: Structure, function and mechanism.* Curr Org Chem. 2002;6:827-39.

2. Martin IV, MacNeill SA. *ATP-dependent DNA ligases.* Genome Biol. 2002;3:3005.1-.7.

3. Timson DJ, Singleton MR, Wigley DB. *DNA ligases in the repair and replication of DNA.* Mutat Res. 2000;460:301-18.

4. Tomkinson AE, Vijayakumar S, Pascal JM, Ellenberger T. *DNA ligases: Structure, reaction mechanism, and function.* Chem Rev. 2006;106:687-99.

5. Tomkinson AE, Mackey ZB. *Structure and function of mammalian DNA ligases.* Mutat Res. 1998;407:1-9.

6. Mortusewicz O, Rothbauer U, Cardoso MC, Leonhardt H. *Differential recruitment of DNA ligase I and III to DNA repair sites.* Nucleic Acids Res. 2006;34:3523-32.

7. Mackey ZB, Ramos W, Levin DS, Walter CA, McCarrey JR, Tomkinson AE. *An alternative splicing event which occurs in mouse pachytene spermatocytes generates a form of DNA ligase III with distinct biochemical properties that may function in meiotic recombination.* Mol Cell Biol. 1997;17.

8. Nash RA, Caldecott KW, Barnes DE, Lindahl T. *XRCC1 protein interacts with one of two distinct forms of DNA ligase III.* Biochemistry. 1997;36:5207-11.

9. Callebaut I, Mornon JP. *From BRCA1 to RAP1: a widespread BRCT module closely associated with DNA repair.* FEBS Lett. 1997;400:25-30.

10. Koonin EV, Altschul SF, Bork P. *Functional motifs.* Nature Genet. 1996;13:266-8.

11. Cappelli E, Taylor R, Cevasco M, Abbondandolo A, Caldecott K, Frosina G. *Involvement of XRCC1 and DNA ligase III gene products in DNA base excision repair.* J Biol Chem. 1997;272:23970-5.

12. Kubota Y, Nash RA, Klungland A, Schar P, Barnes DE, Lindahl T. *Reconstitution of DNA base excision-repair with purified human proteins: interaction between DNA polymerase β and the XRCC1 protein.* EMBO J. 1996;15:6662-70.

13. Caldecott KW, Aoufouchi S, Johnson P, Shall S. *XRCC1 polypeptide interacts with DNA polymerase β and possibly poly(ADP-ribose) polymerase, and DNA ligase III is a novel molecular 'nick-sensor' in vitro.* Nucleic Acids Res. 1996;24:4387-94.

14. Wei YF, Robins P, Carter K, Caldecott K, Pappin DJ, Yu GL, et al. *Molecular cloning and expression of human cDNAs encoding a novel DNA ligase IV and DNA ligase III, an enzyme active in DNA repair and recombination.* Mol Cell Biol. 1995;15:3206-16.

15. Leppard JB, Dong Z, Mackey ZB, Tomkinson AE. *Physical and functional interaction between DNA ligase IIIα and poly(ADP-Ribose) polymerase 1 in DNA single-strand break repair.* Mol Cell Biol. 2003;23:5919-27.

16. Mackey ZB, Niedergang C, de Murcia JM, Leppard J, Au K, Chen J, et al. *DNA ligase III is recruited to DNA strand breaks by a zinc finger motif homologous to that of poly(ADP-ribose) polymerase.* J Biol Chem. 1999;274:21679-87.

17. Schreiber V, Ame JC, Dolle P, Schultz I, Rinaldi B, Fraulob V, et al. *Poly(ADP-ribose) polymerase-2 (PARP-2) is required for efficient base excision DNA repair in association with PARP-1 and XRCC1.* J Biol Chem. 2002;277:23028-36.

18. Taylor RM, Whitehouse CJ, Caldecott KW. *The DNA ligase III zinc finger stimulates binding to DNA secondary structure and pormotes end joining.* Nucleic Acids Res. 2000;28:3558-63.

19. Wang H, Rosidi B, Perrault R, Wang M, Zhang L, Windhofer F, et al. *DNA ligase III as a candidate component of backup pathways of nonhomologous end joining.* Cancer Res. 2005;65:4020-30.

20. Puebla-Osorio N, Lacey DB, Alt FW, Zhu C. *Early embryonic lethality due to targeted inactivation of DNA ligase III.* Mol Cell Biol. 2006;26:3935-41.

21. Mohrenweiser HW, Xi T, Vazquez-Matias J, Jones IM. *Identification of 127 amino acid substitution variants in screening 37 DNA repair genes in humans.* Cancer Epidemiol Biomarkers Prev. 2002;11:1054-64.

22. Krishnan VV, Thornton KH, Thelen MP, Cosman M. *Solution structure and backbone dynamics of the human DNA ligase IIIα BRCT domain.* Biochemistry. 2001;40:13158-66.

23. Barnes DE, Johnston LH, Kodama KI, Tomkinson AE, Lasko DD, Lindahl T. *Human DNA ligase I cDNA: Cloning and functional expression in Saccharomyces cerevisiae.* Proc Natl Acad Sci USA. 1990;87:6679-83.

24. Tomkinson AE, Lasko DD, Daly G, Lindahl T. *Mammalian DNA ligases. Catalytic domain and size of DNA ligase I.* J Biol Chem. 1990;265:12611-7.

25. Pascal JM, O'Brien PJ, Tomkinson AE, Ellenberger T. *Human DNA ligase I completely encircles and partially unwinds nicked DNA.* Nature. 2004;432:473-8.

26. Chen X, Zhong S, Zhu X, Dziegielewska B, Ellenberger T, Wilson GM, et al. *Rational design of human DNA ligase inhibitors that target cellular DNA replication and repair.* Cancer Res. 2008;68:3169-77.

27. Montecucco A, Rossi R, Levin DS, Gary R, Park MS, Motycka TA, et al. *DNA ligase I is recruited to sites of DNA replication by an interaction with proliferating cell nuclear antigen: identification of a common targeting mechanism for the assembly of replication factories.* EMBO J. 1998;17:3786-95.

28. Cardoso MC, Joseph C, Rahn HP, Reusch R, Nadal-Ginard B, Leonhardt H. *Mapping and use of a sequence that targets DNA ligase I to sites of DNA replication in vivo.* J Cell Biol. 1997;139:579-87.

29. Rossi R, Villa A, Negri C, Scovassi I, Ciarrocchi G, Biamonti G, et al. *The replication factory targeting sequence/PCNA-binding site is required in G1 to control the phosphorylation status of DNA ligase I.* EMBO J. 1999;18:5745-54.

30. Dimitriadis EK, Prasad R, Vaske MK, Chen L, Tomkinson AE, Lewis MS, et al. *Thermodynamics of human DNA ligase I trimerization and association with DNA polymerase β.* J Biol Chem. 1998;273:20540-50.

31. Prasad R, Singhal RK, Srivastava DK, Molina JT, Tomkinson AE, Wilson SH. *Specific interaction of DNA Polymerase β and DNA Ligase I in a multiprotein base excision repair complex from bovine testis.* J Biol Chem. 1996;271:16000-7.

32. Levin DS, Vijayakumar S, Liu X, Bermudez VP, Hurwitz J, Tomkinson AE. *A conserved interaction between the replicative clamp loader and DNA ligase in eukaryotes.* J Biol Chem. 2004;279:55196-201.

33. Levin DS, McKenna AE, Motycka TA, Matsumoto Y, Tomkinson AE. *Interaction between PCNA and DNA ligase I is critical for joining of Okazaki fragments and long-patch base-excision repair.* Curr Biol. 2000;10.

34. Sobti RC, Kaur P, Kaur S, Janmeja AK, Jindal SK, Kishan J, et al. *No association of DNA ligase-I polymorphism with the risk of lung cancer in North-Indian population.* DNA Cell Biol. 2006;25.

5.3. NUCLEOTIDE EXCISION REPAIR

5.3.1. INTRODUCTION

Nucleotide excision repair (NER) corrects a wide variety of helix-distorting DNA lesions that interfere with base pairing and generally obstruct transcription and normal replication (1-5). Bulky PAH-DNA adducts and UV-induced thymine dimers are examples of these chemically and structurally diverse lesions. Recognition of the distorted DNA structure leads to the excision of a single-stranded segment of about 30 bp in length that includes the lesion. The removal creates a single-strand gap in the DNA, which is filled in by DNA polymerase using the undamaged strand as template. NER can be divided into two sub-pathways, "global-genome NER" and "transcription-coupled NER", which differ only in their recognition of the helix-distorting DNA damage. Global-genome NER surveys the entire genome for distorting injury and is able to eliminate lesions at any moment in the cell cycle. Transcription-coupled NER focuses on damage in active genes and ensures the efficient clearance of lesions on the transcribed strand that block elongating RNA polymerases. Figure 5.3.1. summarizes global-genome NER in non-transcribed regions, which make up the bulk of DNA.

Several rare, autosomal recessive disorders are associated with inborn errors in NER resulting in cancer, complex developmental abnormalities, and neurodegenerative disorders (3, 6). These include xeroderma pigmentosum, Cockayne syndrome, and trichothiodystropy, all characterized by exquisite sun sensitivity. The prototype NER disorder, xeroderma pigmentosum (XP), exhibits a dramatic 1000-fold higher incidence of sun-induced skin cancer than normal individuals. XP arises from mutations in one of seven genes, XPA – XPG, all of which encode key proteins involved in NER. Many XP skin tumors also contain mutations in the p53 gene, over half of which are CC/TT tandem mutations. Such mutations are found rarely in internal cancers in the general population and are a "signature" for mutations produced by UV light. Cockayne syndrome (CS) is a multisystem disorder characterized by cachectic dwarfism, mental retardation, premature aging, retinopathy, and sun sensitivity. CS is caused by mutations in the CSA or CSB genes, which encode key proteins involved in transcription-coupled NER. CS is not prone to cancer development, which may be explained by the observation that CS cells undergo cytotoxic rather than mutagenic lesions with the lesion-induced apoptosis providing protection against tumorigenesis (4). In trichothiodystrophy (TTD), sulfur-deficient brittle hair is accompanied by scaly skin (ichthyosis), mental and physical retardation, abnormal facies, and in about one-half the cases, sun sensitivity. The features of TTD can result from mutations in one of three genes (XPD, XPB, or TTDA), all of whose products are subunits of transcription factor II H (TFIIH), an essential component of NER as well

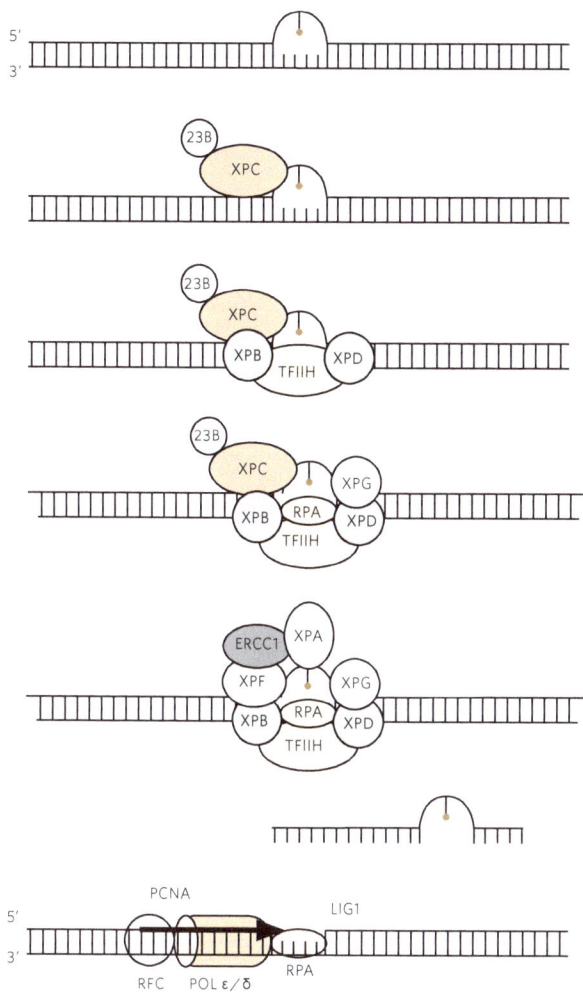

Figure 5.3.1. Overview of global-genome NER pathway.
The helix-distorting DNA lesion is recognized by the protein dimer XPC-hHR23B. Next, the multi-subunit transcription factor TFIIH is recruited, which includes two DNA helicases with opposite polarities, XPB and XPD. The helicase activities of XPB and XPD unwind the DNA around the lesion allowing the binding of RPA to the single-stranded DNA, which stabilizes the open complex and creates specific sites for cleavage on both sides of the lesion. The endonuclease XPG is recruited at the 3' side and the XPC-hHR23B duplex is replaced by XPA, which is believed to verify the presence of a lesion. The endonuclease dimer ERCC1-XPF is recruited at the 5' side of the lesion and dual incisions are made in the DNA strand flanking the lesion. After a 24- to 32-residue oligonucleotide is released, the gap is filled in by PCNA-dependent DNA polymerases ε or δ and sealed by DNA ligase 1.

as transcription initiation. Despite severe defects in NER, TTD is also not predisposed to cancer, demonstrating that genotype/phenotype relations are not always straightforward. The mutations not only compromise NER but also affect transcription, causing developmental delay and reduced expression of matrix proteins in hair and skin.

References

1. De Laat WI, Jaspers NG, Hoeijmakers JH. *Molecular mechanism of nucleotide excision repair.* Genes Dev. 1999;13:768-85.

2. Gillet LC, Scharer OD. *Molecular mechanisms of mammalian global genome nucleotide excision repair.* Chem Rev. 2006;106:253-76.

3. Hoeijmakers JH. Genome maintenance mechanisms for preventing cancer. Nature. 2001;411:366-74.

4. Mitchell JR, Hoeijmakers JH, Niedernhofer LJ. *Divide and conquer:*

Nucleotide excision repair battles cancer and ageing. Curr Opinion Cell Biol. 2003;15:232-40.

5. Wu M, Yan S, Patel DJ, Geacintov NE, Broyde S. *Relating repair susceptibility of carcinogen-damaged DNA with structural distortion and thermodynamic stability.* Nucleic Acids Res. 2002;30:3422-32.

6. Cleaver JE. *Cancer in Xeroderma Pigmentosum and related disorders of DNA Repair.* Nat Rev Cancer. 2005;5:564-73.

5.3.2. NER GENES AND PROTEINS

The products of over 20 genes are involved in the multi-step global-genome NER pathway consisting of DNA damage recognition, lesion demarcation, dual incision, release of damaged oligonucleotide, and gap-filling DNA synthesis (Table 5.3.1.). The pathway begins with detection of the damaged DNA site by XPC, which exists in a complex with hHR23B, a multifunctional protein that protects XPC from ubiquitin-mediated proteasome degradation (1, 2). XPC recognizes the disordered DNA structure irrespective of the chemical nature of the lesion, which explains the versatility of global-genome NER in dealing with chemically and structurally diverse lesions (3). Certain types of lesions (e.g., UV photoproducts) also require recruitment of the XPE heterodimer, consisting of damage-specific DNA binding proteins 1 (DDB1 or p127) and 2 (DDB2 or p48), the latter encoded by the XPE gene and linked to ubiquitin-mediated proteasome degradation (4, 5). XPC-hHR23B binding causes partial opening of the DNA duplex around the lesion, which leads to the recruitment of the transcription factor TFIIH complex. Thus, separate regions of the 125 kDa XPC protein are involved in the binding to DNA, hHR23B, and TFIIH (4). TFIIH is a ten-subunit protein complex with multiple enzymatic functions, including bidirectional helicase activity conferred by the XPB and XPD subunits. The ATP-dependent helicase activities of XPB and XPD cause local unwinding of the DNA around the lesion, which permits binding of replication protein A (RPA) to the single-stranded DNA leading to stabilization of the open complex and creation of specific sites for cutting on the 3' and 5' sides of the lesion. Then XPG, a member of the FEN1 family of structure-specific endonucleases, is recruited at the 3' side and the XPC-hHR23B duplex is replaced by XPA, which is believed to verify the presence of a lesion. Finally, a heterodimer endonuclease, ERCC1 (excision repair cross complementation group 1)-XPF, is recruited at the 5' side of the lesion. Incisions are made in the phosphodiester bonds

6 ± 3 bases 3' of the lesion by XPG and 20 ± 5 bases 5' of the lesion by ERCC1-XPF (6). Following incision, the gap of 24 to 32 nucleotides is filled in by DNA polymerases ε or δ with the assistance of two accessory factors, proliferating cell nuclear antigen (PCNA) and replication factor C (RFC). The repaired DNA is joined to the preexisting DNA strand by DNA ligase 1. Considering that several of the NER factors consist of subunits (e.g., 3 in RPA and PCNA, 5 in RFC), a total of about 30 polypeptides participate in the pathway in a sequential order assembly (4, 7, 8). Transcription-coupled NER requires all of the factors involved in global-genome NER except XPC. The site of damage is recognized when the lesion causes RNA polymerase II to stall during transcription. The blocked transcription triggers the recruitment of CSA and CSB, which displace the stalled polymerase, followed by recruitment of TFIIH. The subsequent steps of transcription-coupled and global-genome NER are identical (Figure 5.3.1.).

What is the importance of NER in the pathogenesis of non-skin cancers, in particular breast cancer? Overall, there is no evidence from experimental studies of individual NER pathway factors for any consistent association with breast cancer. Mouse knockout strains which reflect the human NER defects display increased skin cancer from UV light or severe developmental disorders (9). For example, Xpa -/- knockout mice, which are completely deficient in NER activity, developed tumors of the skin and liver. After treatment with PAHs, they also developed lymphomas but not mammary tumors (10). Xpc -/- knockout mice showed a 30-fold increased spontaneous mutation rate in the Hprt gene in T lymphocytes but no increased tumor incidence or premature ageing (11). ERCC1 -/- knockout mice display premature ageing with growth retardation, neurodegeneration, as well as renal and hepatic failure (12). These data suggest that lack of ERCC1 results in cytotoxic lesions and ageing rather than mutagenic lesions and cancer (13). Overall, the absence of mammary tumors in these animal models indicates that spontaneous or carcinogen-induced DNA damage capable of initiating breast cancer was corrected by pathways other than NER.

Investigations of breast cancer risk have focused on two NER components, XPD and ERCC1/XPF. **XPD.** Clinical evidence indicates that mutations in XPD helicase cause three distinct phenotypes: XP, CS, and TTD. The question how point mutations in adjacent residues in a single enzyme can give rise to such different disease phenotypes has been answered by structural studies. The helicase activity of XPD is dependent on a small iron-sulfur (FeS) binding domain near the N terminus. This cluster is liganded by four cysteine residues, three of which are absolutely conserved. Mutations of any of the three cysteines result in disruption of the FeS cluster and loss of helicase activity (17, 18). Other mutations have been mapped to the ATP-binding domain and the DNA-binding channel of XPD providing a molecular understanding for the wide spectrum of clinical phenotypes associated with XPD deficiency. The FeS domain is also present in a related DNA helicase, FancJ (alias Bach1 or Brip1) that interacts with the highly conserved, C-terminal BRCT repeats of BRCA1 (19, 20).

Table 5.3.1. NER Genes and Proteins (9, 14-16)

Protein	Function	Chromosome	Polymorphism
XPA	damage verification	9p34.1	5'-untranslated region -4G/A
XPB	helicase, 3'-5'	2q21	
XPC	damage binding	3p25.1	Ala499Val (rs2228000 C/T) in exon 9
			Lys939Gln (rs2228001 A/C) in exon 16
HR23B	XPC protection	9q31.2	Ala249Val (rs1805329 C/T) in exon 7
XPD	helicase, 5'-3'	19q13.2	Asp312Asn (rs1799793 G/A) in exon 10 in linkage disequilibrium with
(ERCC2)			Lys751Gln (rs13181 A/C) in exon 23
XPE	damage binding	11p11-12	
XPF	5' nuclease	16p13.3	Arg415Gln (rs1800067 G/A) in exon 8
(ERCC4)			G807C (rs744154) in intron 1
ERCC1	XPF stabilization	19q13.2-3	Asn118Asn (rs11615 C/T) in exon 4
			3'-untranslated C8092A (rs3212986)
XPG	3' nuclease	13q32-33	Asp1104His (rs17655 G/C) in exon 15
CSA	ubiquitylation	5q12.1	
CSB	DNA-dependent ATPase	10q11-21	Arg1213Gly (rs2228527)
			Arg1230Pro (rs4253211 G/C) in exon 18

Interestingly, two women with early-onset breast cancer and no abnormalities in either BRCA1 or BRCA2 carried germline FancJ mutations affecting helicase activity, suggesting that abnormal FancJ function contributes to breast cancer development.

The XPD gene contains two common non-synonymous polymorphisms in exons 10 and 23: Asp312Asn (rs1799793 G/A) and Lys751Gln (rs1052559 A/C), respectively, which are in linkage disequilibrium (21, 22). The Asp312Asn polymorphism shows a marked difference in the frequency of the minor 312Asn allele, which ranged from 0.06 to 0.44 between different ethnic groups and populations (23). One study observed higher levels of PAH-DNA adducts in malignant breast tissue associated with the variant 312Asn allele (24). An epidemiologic study (688 cases, 724 controls) observed a highly significant association between the Asp312Asn polymorphism and breast cancer risk with homozygous wild-type 312Asp/Asp women having a twofold increase in risk (25). However, the Long Island Breast Cancer Project (1053 cases, 1102 controls) found an effect in the opposite direction. The presence of at least one variant 312Asn allele was associated with an increase in breast cancer risk (OR 1.25; 95% CI 1.04 – 1.50) (26). The increase associated with 312Asn/Asn homozygosity was even stronger in a subgroup of women with detectable PAH-DNA adduct levels (OR 1.83; 95% CI 1.22 – 2.76 compared to 312Asp/Asp wild-type and nondetectable adducts). An analysis of DNA adduct levels (determined by ^{32}P-postlabelling assay) in white blood cells of 308 healthy individuals showed significantly higher adduct levels in never smokers with 751Gln/Gln compared to the 751Lys/Lys genotype but no association was observed in current or former smokers (27). Another study observed that 751Gln/Gln compared to the 751Lys/Lys genotype was associated with an increased risk of breast cancer in current smokers and in women with white blood cell PAH-DNA adduct levels above the median (22). In the largest analysis, comprised of a combination of three case-control studies (in total 3634 cases, 3340 controls), neither the Asp312Asn nor the Lys751Gln polymorphism showed a significant association with breast cancer risk (28). Similarly, a meta-analysis of 10 breast cancer studies found no risk association with either polymorphism (29). The Long Island Breast Cancer Project found no association between XPA, XPC, XPF, and XPG genotypes, PAH-DNA adducts, and breast cancer risk (26, 30).

ERCC1/XPF. Structural analysis of the ERCC1/XPF heterodimer indicates a complementary role for the two proteins in NER. ERCC1 is responsible for single-stranded DNA binding whereas XPF possesses nuclease activity (31, 32). There are no known mutations in the ERCC1 gene associated with any of the NER-deficiency syndromes, but partially inactivating mutations in its binding partner

XPF give rise to a mild XP phenotype characterized by slight photosensitivity and late-onset skin cancer (33). In addition to NER, the ERCC1/XPF complex is also involved in double strand break and interstrand crosslink repair by cutting DNA overhangs around a lesion (34). Moreover, ERCC1/XPF plays a role in telomere maintenance by degrading 3' G-rich overhangs (35). Thus, ERCC1/XPF plays several important roles in genome maintenance, which makes it difficult to elucidate the importance of each function in isolation. Although ERCC1/XPF is essential for genome maintenance, high ERCC1/XPF mRNA and protein levels can be counterproductive in cancer patients receiving chemotherapy, often leading to chemoresistance and poor outcome (34). Experimental and clinical evidence implicates ERCC1 as critically important for the removal of platinum-DNA adducts from tumor DNA introduced by cisplatin chemotherapy (36). ERCC1 protein expression levels in non-small cell lung carcinoma (NSCLC) correlated with clinical sensitivity to cisplatin therapy (37). Patients with ERCC1-positive NSCLC tumors did not benefit from adjuvant platinum chemotherapy whereas patients with ERCC1-negative tumors experienced prolonged survival.

The ERCC1 gene contains two polymorphisms, a synonymous Asn118Asn (rs11615 C/T) in exon 4 and C8092A (rs3212986) in the 3'-untranslated region (Table 5.3.1.). The ERCC1 gene at 19q13.2-3 is positioned in an anti-sense orientation to and overlapping with the ASE-1 gene, which encodes a nucleolar protein. The C/A (rs3212986) polymorphism results in a Gln504Lys substitution in the ASE-1 gene (21, 38). Although neither of the two polymorphisms alters the ERCC1 amino acid sequence, the mRNA stability and protein expression may be impaired. PAH-DNA adduct levels in blood lymphocytes of 707 healthy individuals were significantly higher than the median value for the rs3212986 A/A genotype compared with the homozygous C/C wild-type (39). The rs11615 C/T polymorphism modified the association between cigarette smoking and lung cancer risk but no association was detected with smoking, alcohol consumption and risk of colorectal cancer (40, 41). The XPF gene contains an Arg415Gln polymorphism (Table 5.3.1.). One breast cancer study (336 cases, 416 controls) observed the homozygous 415Gln/Gln genotype in seven cases but only one control (p = 0.05) (42). A polymorphism in XPF intron 1, G807C (rs744154) was associated with breast cancer risk in one study but a much larger study (over 30,000 cases and 30,000 controls) found no evidence of an association (43, 44).

Cell-cycle checkpoints delay cell proliferation to repair damaged DNA before replication or cell division proceed. Replicative DNA polymerases, such as Pols ε or δ, play a crucial role by passing genetic information in a stable, accurate fashion to succeeding progeny. They synthesize

DNA with a high degree of accuracy and are blocked by lesions that significantly distort the geometry of DNA (45). Thus, the high-fidelity Pols ε or δ will dissociate from the B[a]P-modified template in preference to replication past the lesion, providing a molecular basis for the blocking effect on replication exerted by B[a]P adducts. At the same time, it has become apparent that cells can tolerate much damage in their genomes without removing it. This tolerance is achieved by another class of DNA polymerases, called Y family polymerases, which are capable of replicating past different types of DNA damage including various PAH adducts. There are several Y-family DNA polymerases in cells (e.g., Pol η, ι, and κ in humans) that are capable of extending primers opposite damaged DNA templates in a process termed translesion synthesis (46, 47). Crystallographic data have shown that DNA polymerases of the Y-family adopt the conventional right-handed grip on the DNA by the palm, fingers, and thumb domains similar to the structure found in high-fidelity DNA polymerases. However, shorter 'fingers' and 'thumb' provide a wider open active cleft capable of accommodating various distortions more easily (48-51). Despite the overall similarity of structural features among the Y-family polymerases, there is a high degree of specificity in their lesion bypass properties (47). Thus, Pol η consistently bypasses UV-induced thymine dimers by correctly inserting two dAMPs opposite the lesions (52). Pol ι is unable to replicate through a thymine dimer but can proficiently incorporate nucleotides opposite N2-adducted guanines, such as trans-4-hydroxy-2-nonenal-deoxyguanosine, that results from lipid peroxidation (53). Pol κ, on the other hand, is capable of bypassing all stereoisomers of anti-BPDE-N2-dG with dCMP insertion opposite the lesion at least three times more frequently than other incorporations (54). In contrast, Pol η incorrectly inserts dAMP opposite BPDE-N2-dG and thereby can introduce mutations (49). Compared to high-fidelity polymerases, Y-family polymerases also exhibit a significant error rate (as high as 1%) on undamaged DNA templates (55). Thus, Y-family polymerases are lesion-tolerant but at the same time error-prone and mutagenic.

About 20% of XP patients have a normal NER system but are deficient in Pol η. This subset of XP patients is classified as XP variant (XP-V) (9, 56, 57). At first sight, it seems paradoxical that the inherited deficiency of an error-prone DNA polymerase actually increases the mutation rate and causes XP. However, in the absence of Pol η, TT dimers are replicated less efficiently and with increased error rate by other DNA polymerases, leading to the high incidence of skin cancer. Pol η -/- knockout mice showed enhanced sensitivity to UV damage but exhibited no increased mutagenesis in internal organs, such as liver, spleen or small intestine (58). Pol ι -/- knockout mice have a normal phenotype but human fibroblasts made deficient in Pol ι showed enhanced sensitivity to oxidative damage (59, 60). Two breast cancer cell lines, MCF-7 and MDA-MB-468, exhibited overexpression of Pol ι compared to normal mammary epithelial and MCF-10A cells (61). UV treatment of the malignant cells resulted in a higher rate of mutagenesis compared to the normal cells.

The antiestrogen tamoxifen is widely used for the treatment or prevention of breast cancer. However, the drug is also carcinogenic in human uterus and rat liver, highlighting the complexity of its actions. The analysis of endometrial DNA samples of women receiving tamoxifen revealed the presence of dG-N2-tamoxifen adducts at a frequency of 1.5 – 13.1 adducts/10^8 nucleotides in six patients, whereas no tamoxifen adducts were detected in the other seven women in the study group (62). *In vitro* studies demonstrated that the reconstituted NER repairs dG-N2-tamoxifen adducts with poor efficiency suggesting that individual variation in repair capacity may play a role in the development of tamoxifen-induced endometrial cancer (63). Tamoxifen treatment of chicken B lymphocytes (DT40), which are deficient in translesion DNA synthesis enzymes Rad18, Rev3, and Polk, resulted in extensive chromosomal breaks (64). The mutant cells also exhibited increased sensitivity to 4-OHE$_2$ compared with wild-type cells, whereas sensitivity to E$_2$ or 2-OHE$_2$ did not differ between the two cell types. These results suggest that error-prone DNA synthesis contributes to mutagenic lesions induced by tamoxifen and 4-OHE$_2$.

References

1. Ortolan TG, Chen L, Tongaonkar P, Madura K. *Rad23 stabilizes Rad4 from degradation by the Ub/proteasome pathway.* Nucleic Acids Res. 2004;32:6490-500.

2. Sugasawa K, Ng JM, Masutani C, Iwai S, van der Spek PJ, Eker AP, et al. *Xeroderma pigmentosum group C protein complex is the initiator of global genome nucleotide excision repair.* Mol Cell. 1998;2:223-32.

3. Min JH, Pavletich NP. *Recognition of DNA damage by the Rad4 nucleotide excision repair protein.* Nature. 2007;449:570-5.

4. Dip R, Camenisch U, Naegeli H. *Mechanisms of DNA damage recognition and strand discrimination in human nucleotide excision repair.* DNA Repair (Amst). 2004;3:1409-23.

5. Groisman R, Polanowska J, Kuraoka I, Sawada J, Saijo M, Drapkin R, et al. *The ubiquitin ligase activity in the DDB2 and CSA complexes is differentially regulated by the COP9 signalosome in response to DNA damage.* Cell. 2003;113:357-67.

6. Huang JC, Svoboda DL, Reardon JT, Sancar A. *Human nucleotide excision nuclease removes thymine dimers from DNA by incising the 22nd phosphodiester bond 5' and the* 6th phosphodiester bond 3' to the photodimer. Proc Natl Acad Sci U S A. 1992;89:3664-8.

7. Aboussekhra A, Biggerstaff M, Shivji MK, Vilpo JA, Moncollin V, Podust VN, et al. *Mammalian DNA nucleotide excision repair reconstituted with purified protein components.* Cell. 1995;80:859-68.

8. Kesseler KJ, Kaufmann WK, Reardon JT, Elston TC, Sancar A. *A mathematical model for human nucleotide excision repair: damage recognition by random order assembly and kinetic proofreading.* J Theor Biol. 2007;249:361-75.

9. Cleaver JE. *Cancer in xeroderma pigmentosum and related disorders of DNA repair.* Nat Rev Cancer. 2005;5:564-73.

10. de Vries A, van Oostrom CT, Dortant PM, Beems RB, van Kreijl CF, Capel PJ, et al. *Spontaneous liver tumors and benzo[a]pyrene-induced lymphomas in XPA-deficient mice.* Mol Carcinogenesis. 1997;19:46-53.

11. Wijnhoven SW, Kool HJ, Mullenders LH, van Zeeland AA, Friedberg EC, van der Horst GT, et al. *Age-dependent spontaneous mutagenesis in Xpc mice defective in nucleotide excision repair.* Oncogene. 2000;19:5034-7.

12. Weeda G, Donker I, de Wit J, Morreau H, Janssens R, Vissers CJ, et al. *Disruption of mouse ERCC1 results in a novel repair syndrome with growth failure, nuclear abnormalities and senescence.* Curr Biol. 1997;7:427-39.

13. Mitchell JR, Hoeijmakers JH, Niedernhofer LJ. *Divide and conquer: nucleotide excision repair battles cancer and ageing.* Curr Opinion Cell Biol. 2003;15:232-40.

14. Goode EL, Ulrich CM, Potter JD. *Polymorphisms in DNA repair genes and associations with cancer risk.* Cancer Epidemiol Biomarkers Prev. 2002;11:1513-30.

15. Millikan RC, Hummer A, Begg C, Player J, de Cotret AR, Winkel S, et al. Polymorphisms in nucleotide excision repair genes and risk of multiple primary melanoma: the Genes Environment and Melanoma Study. Carcinogenesis. 2006;27:610-8.

16. Mohrenweiser HW, Xi T, Vazquez-Matias J, Jones IM. *Identification of 127 amino acid substitution variants in screening 37 DNA repair genes in humans.* Cancer Epidemiol Biomarkers Prev. 2002;11:1054-64.

17. Fan L, Fuss JO, Cheng QJ, Arvai AS, Hammel M, Roberts VA, et al. XPD helicase structures and activities: insights into the cancer and aging phenotypes from XPD mutations. Cell. 2008;133:789-800.

18. Liu H, Rudolf J, Johnson KA, McMahon SA, Oke M, Carter L, et al. *Structure of the DNA repair helicase XPD.* Cell. 2008;133:801-12.

19. Cantor S, Drapkin R, Zhang F, Lin Y, Han J, Pamidi S, et al. *The BRCA1-associated protein BACH1 is a DNA helicase targeted by clinically relevant inactivating mutations.* Proc Natl Acad Sci U S A. 2004;101:2357-62.

20. Levran O, Attwooll C, Henry RT, Milton KL, Neveling K, Rio P, et al. *The BRCA1-interacting helicase BRIP1 is deficient in Fanconi anemia.* Nature Genet. 2005;37:931-3.

21. Shen MR, Jones IM, Mohrenweiser H. Nonconservative amino acid substitution variants exist at polymorphic frequency in DNA repair genes in healthy humans. Cancer Res. 1998;58:604-8.

22. Terry MB, Gammon MD, Zhang FF, Eng SM, Sagiv SK, Paykin AB, et al. *Polymorphism in the DNA repair gene XPD, polycyclic aromatic hydrocarbon-DNA adducts, cigarette smoking, and breast cancer risk.* Cancer Epidemiol Biomarkers Prev. 2004;13:2053-8.

23. Hodgson ME, Poole C, Olshan AF, North KE, Zeng D, Millikan RC. Smoking and selected DNA repair gene polymorphisms in controls: systematic review and meta-analysis. Cancer Epidemiol Biomarkers Prev. 2010;19:3055-86.

24. Tang D, Cho S, Rundle A, Chen S, Phillips D, Zhou J, et al. *Polymorphisms in the DNA repair enzyme XPD are associated with increased levels of PAH-DNA adducts in a case-control study of breast cancer.* Breast Cancer Res Treat. 2002;75:159-66.

25. Justenhoven C, Hamann U, Pesch B, Harth V, Rabstein S, Baisch C, et al. ERCC2 genotypes and a corresponding haplotype are linked with breast cancer risk in a German population. Cancer Epidemiol Biomarkers Prev. 2004;13:2059-64.

26. Crew KD, Gammon MD, Terry MB, Zhang FF, Zablotska LB, Agrawal M, et al. *Polymorphisms in nucleotide excision repair genes, polycyclic aromatic hydrocarbon-DNA adducts, and breast cancer risk.* Cancer Epidemiol Biomarkers Prev. 2007;16:2033-41.

27. Matullo G, Palli D, Peluso M, Guarrera S, Carturan S, Celentano E, et al. *XRCC1, XRCC3, XPD gene polymorphisms, smoking and 32P-DNA adducts in a sample of healthy subjects.* Carcinogenesis. 2001;22:1437-45.

28. Kuschel B, Chenevix-Trench G, Spurdle AB, Chen X, Hopper JL, Giles GG, et al. *Common polymorphisms in ERCC2 (Xeroderma pigmentosum D) are not associated with breast cancer risk.* Cancer Epidemiol Biomarkers Prev. 2005;14:1828-31.

29. Wang F, Chang D, Hu FL, Sui H, Han B, Li DD, et al. *DNA repair gene XPD polymorphisms and cancer risk: a meta-analysis based on 56 case-control studies.* Cancer Epidemiol Biomarkers Prev. 2008;17:507-17.

30. Shen J, Gammon MD, Terry MB, Teitelbaum SL, Eng SM, Neugut AI, et al. Xeroderma pigmentosum complementation group C genotypes/diplotypes play no independent or interaction role with polycyclic aromatic hydrocarbons-DNA adducts for breast cancer risk. Eur J Cancer. 2008;44:710-7.

31. Orelli B, McClendon TB, Tsodikov OV, Ellenberger T, Niedernhofer LJ, Scharer OD. *The XPA-binding domain of ERCC1 is required for nucleotide excision repair but not other DNA repair pathways.* J Biol Chem. 2010;285:3705-12.

32. Tripsianes K, Folkers G, Ab E, Das D, Odijk H, Jaspers NG, et al. The structure of the human ERCC1/XPF interaction domains reveals a complementary role for the two proteins in nucleotide excision repair. Structure. 2005;13:1849-58.

33. Sijbers AM, van Voorst Vader PC, Snoek JW, Raams A, Jaspers NG, Kleijer WJ. *Homozygous R788W point mutation in the XPF gene of a patient with xeroderma pigmentosum and late-onset neurologic disease.* J Invest Dermatol. 1998;110:832-6.

34. Kirschner K, Melton DW. *Multiple roles of the ERCC1-XPF endonuclease in DNA repair and resistance to anticancer drugs.* Anticancer Res. 2010;30:3223-32.

35. Zhu XD, Niedernhofer L, Kuster B, Mann M, Hoeijmakers JH, de Lange T. ERCC1/XPF removes the 3' overhang from uncapped telomeres and represses formation of telomeric DNA-containing double minute chromosomes. Mol Cell. 2003;12:1489-98.

36. Reed E. *ERCC1 measurements in clinical oncology.* N Engl J Med. 2006;355:1054-5.

37. Olaussen KA, Dunant A, Fouret P, Brambilla E, Andre F, Haddad V, et al. *DNA repair by ERCC1 in non-small-cell lung cancer and cisplatin-based adjuvant chemotherapy.* N Engl J Med. 2006;355:983-91.

38. van Duin M, van Den Tol J, Hoeijmakers JH, Bootsma D, Rupp IP, Reynolds P, et al. *Conserved pattern of antisense overlapping transcription in the homologous human ERCC-1 and yeast RAD10 DNA repair gene regions.* Mol Cell Biol. 1989;9:1794-8.

39. Zhao H, Wang LE, Li D, Chamberlain RM, Sturgis EM, Wei Q. *Genotypes and haplotypes of ERCC1 and ERCC2/XPD genes predict levels of benzo[a]pyrene diol epoxide-induced DNA adducts in cultured primary lymphocytes from healthy individuals: a genotype-phenotype correlation analysis.* Carcinogenesis. 2008;29:1560-6.

40. Hansen RD, Sorensen M, Tjonneland A, Overvad K, Wallin H, Raaschou-Nielsen O, et al. *A haplotype of polymorphisms in ASE-1, RAI and ERCC1 and the effects of tobacco smoking and alcohol consumption on risk of colorectal cancer: a Danish prospective case-cohort study.* BMC Cancer. 2008;8:54.

41. Zhou W, Liu G, Park S, Wang Z, Wain JC, Lynch TJ, et al. *Gene-smoking interaction associations for the ERCC1 polymorphisms in the risk of lung cancer.* Cancer Epidemiol Biomarkers Prev. 2005;14:491-6.

42. Smith TR, Levine EA, Freimanis RI, Akman SA, Allen GO, Hoang KN, et al. *Polygenic model of DNA repair genetic polymorphisms in human breast cancer risk.* Carcinogenesis. 2008;29:2132-8.

43. Gaudet MM, Milne RL, Cox A, Camp NJ, Goode EL, Humphreys MK, et al. *Five polymorphisms and breast cancer risk: results from the Breast Cancer Association Consortium.* Cancer Epidemiol Biomarkers Prev. 2009;18:1610-6.

44. Milne RL, Ribas G, Gonzalez-Neira A, Fagerholm R, Salas A, Gonzalez E, et al. *ERCC4 associated with breast cancer risk: a two-stage case-control study using high-throughput genotyping.* Cancer Res. 2006;66:9420-7.

45. Federley RG, Romano LJ. *DNA polymerase: structural homology, conformational dynamics, and the effects of carcinogenic DNA adducts.* J Nucleic Acids. 2010;2010.

46. Pages V, Fuchs RP. *How DNA lesions are turned into mutations within cells?* Oncogene. 2002;21:8957-66.

47. Prakash S, Johnson RE, Prakash L. *Eukaryotic translesion synthesis DNA polymerases: specificity of structure and function.* Annu Rev Biochem. 2005;74:317-53.

48. Biertumpfel C, Zhao Y, Kondo Y, Ramon-Maiques S, Gregory M, Lee JY, et al. *Structure and mechanism of human DNA polymerase eta.* Nature. 2010;465:1044-8.

49. Chandani S, Jacobs C, Loechler EL. *Architecture of y-family DNA polymerases relevant to translesion DNA synthesis as revealed in structural and molecular modeling studies.* J Nucleic Acids. 2010;2010.

50. Friedberg EC, Lehmann AR, Fuchs RP. *Trading places: how do DNA polymerases switch during translesion DNA synthesis?* Mol Cell. 2005;18:499-505.

51. Lone S, Townson SA, Uljon SN, Johnson RE, Brahma A, Nair DT, et al. *Human DNA polymerase kappa encircles DNA: implications for mismatch extension and lesion bypass.* Mol Cell. 2007;25:601-14.

52. Matsuda T, Bebenek K, Masutani C, Hanaoka F, Kunkel TA. *Low fidelity DNA synthesis by human DNA polymerase-eta.* Nature. 2000;404:1011-3.

53. Wolfle WT, Johnson RE, Minko IG, Lloyd RS, Prakash S,

Prakash L. *Replication past a trans-4-hydroxynonenal minor-groove adduct by the sequential action of human DNA polymerases iota and kappa.* Mol Cell Biol. 2006;26:381-6.

54. Suzuki N, Ohashi E, Kolanovskiy A, Geacintov NE, Grollman AP, Ohmori H, et al. *Translesion synthesis of human DNA polymerase κ on a DNA template containing a single stereoisomer of dG-(+)- or dG-(-)-anti-N2-BPDE (7,8-dihydroxy-anti-9,10-epoxy-7,8,9,10-tetrahydrobenzo[a]pyrene).* Biochemistry. 2002;41:6100-6.

55. McCulloch SD, Kunkel TA. *The fidelity of DNA synthesis by eukaryotic replicative and translesion synthesis polymerases.* Cell Res. 2008;18:148-61.

56. Inui H, Oh KS, Nadem C, Ueda T, Khan SG, Metin A, et al. *Xeroderma pigmentosum-variant patients from America, Europe, and Asia.* J Invest Dermatol. 2008;128:2055-68.

57. Yuasa M, Masutani C, Eki T, Hanaoka F. *Genomic structure, chromosomal localization and identification of mutations in the xeroderma pigmentosum variant (XPV) gene.* Oncogene. 2000;19:4721-8.

58. Busuttil RA, Lin Q, Stambrook PJ, Kucherlapati R, Vijg J. *Mutation frequencies and spectra in DNA polymerase eta-deficient mice.* Cancer Res. 2008;68:2081-4.

59. McDonald JP, Frank EG, Plosky BS, Rogozin IB, Masutani C, Hanaoka F, et al. *129-derived strains of mice are deficient in DNA polymerase iota and have normal immunoglobulin hypermutation.* J Exp Med. 2003;198:635-43.

60. Petta TB, Nakajima S, Zlatanou A, Despras E, Couve-Privat S, Ishchenko A, et al. *Human DNA polymerase iota protects cells against oxidative stress.* EMBO J. 2008;27:2883-95.

61. Yang J, Chen Z, Liu Y, Hickey RJ, Malkas LH. *Altered DNA polymerase iota expression in breast cancer cells leads to a reduction in DNA replication fidelity and a higher rate of mutagenesis.* Cancer Res. 2004;64:5597-607.

62. Shibutani S, Suzuki N, Terashima I, Sugarman SM, Grollman AP, Pearl ML. *Tamoxifen-DNA adducts detected in the endometrium of women treated with tamoxifen.* Chem Res Toxicol. 1999;12:646-53.

63. Shibutani S, Reardon JT, Suzuki N, Sancar A. *Excision of tamoxifen-DNA adducts by the human nucleotide excision repair system.* Cancer Res. 2000;60:2607-10.

64. Mizutani A, Okada T, Shibutani S, Sonoda E, Hochegger H, Nishigori C, et al. *Extensive chromosomal breaks are induced by tamoxifen and estrogen in DNA repair-deficient cells.* Cancer Res. 2004;64:3144-7.

Table 5.4.1. MMR Genes and Proteins (24)

Protein	Chromosome	Polymorphisms
MSH2	2p21	Asn127Ser (rs17217723 A/G) Gly322Asp (rs4987188 A/G)
MSH3	2p16	Arg940Gln (rs184967 A/G) Thr1036Ala (rs26279 A/G)
MSH6	7p22	Gly39Glu (rs1042821 C/T)
MLH1	3p21	Ile219Val (rs1799977 G/A)
PMS2	7p22	Glu41Lys (rs34506829 A/G)
PMS1	2q31-33	Pro277Ser (rs1805321 C/T)

nucleotides, so-called insertion/deletion loops (IDLs). Replication-associated transactions, such as mispairs and IDLs that have escaped the proofreading exonuclease, become substrates for MMR whose task is to restore the information contained in the template strand.

There are three types of excision repair, namely BER, NER, and MMR. MMR differs from all other DNA repair pathways in that mismatches and IDLs are composed of unmodified nucleotides and exist as detectable moieties in DNA solely in the double-stranded form. Thus, successful restoration of the original DNA sequence by MMR requires four steps: (i) recognition of base-base mismatches and IDLs, (ii) distinction of the parent strand from the wrong (newly synthesized) daughter strand, (iii) degradation past the mismatch, and (iv) resynthesis of the excised tract (3, 5, 6). The best-studied MMR system with respect to both genetics and biochemistry is the mutSLH pathway of E. coli (4, 7). The human homologs of the E.coli proteins are listed in Table 5.4.1. (8-12). Human MutS homologs (hMSH) 2 and 6 form heterodimers, referred to as hMutSα, recognize mismatches and single-base loops, whereas hMSH2/3 dimers (hMutSβ) recognize IDLs. Heterodimeric complexes of the hMutL-like proteins hMLH1/hPMS2 (hMutLα) and hMLH1/hPMS1 (hMutLβ) interact with MSH complexes and nearby replication factors, the latter permitting strand discrimination. Several proteins are involved in the excision of the new strand past the mismatch and resynthesis steps, including polδ/ε, RFC, RPA, PCNA, exonuclease I, and endonuclease FEN1 (3, 13). MMR components also interact with NER and recombination proteins as well as BRCA1 (6, 14).

Bacteria and eukaryotes deficient in MMR are prone to mismatch and recombination errors, resulting in up to 1000-fold increased rates of spontaneous mutations including destabilization of microsatellites (so-called microsatellite instability, MSI, with the replication error-positive RER+ or mutator phenotype) and promiscuity in inter- and intragenomic recombination (the recombinator phenotype) (10, 15-18). Microsatellites (also known as

5.4. MISMATCH REPAIR

DNA replication is a complex process, whose fidelity is estimated to be in the range of one error per 10^{10} nucleotides synthesized, i.e., less than one error for every billion base pairs copied (1, 2). This extraordinary fidelity depends on three factors: DNA polymerases, exonucleolytic proofreading, and mismatch repair (MMR) (3-5). Replicative DNA polymerases are extremely precise enzymes, which can duplicate the sequence of a given genome within hours with an error rate of ~10^{-5}. All replicative polymerases possess proofreading 3' → 5' exonuclease activity that adds two orders of magnitude to the fidelity of the replication process (1). In rare cases, a mispair eludes the proofreading, especially the G/T wobble pair, which is stabilized by two hydrogen bonds and causes only a slight distortion of the double helix. A similar situation can arise when the primer and template strands slip with respect to one another. Such events occur relatively frequently in runs of repeated mono- or dinucleotides (ranging from one to ten or more bases), where they give rise to loops containing extrahelical short tandem repeats or simple sequence repeats) are sequences of DNA comprising multiple copies of a repeat unit of 1 – 6 base pairs. They are common, polymorphic, and distributed widely throughout the genome. Therefore, MSI is observed in nucleotide repeats of non-coding as well as coding regions of genes such as transforming growth factor β1 receptor II (TGFβ1RII), insulin-like growth factor II receptor (IGFIIR), BCL-2-associated X (BAX), retinoblastoma-protein-interacting zinc finger (RIZ), and transcription factor 4 (TCF 4) (19-23). The genes with coding microsatellites accumulate frame-shift mutations and thereby lose function.

Hereditary colorectal cancer is caused by several syndromes, including hereditary non-polyposis colorectal cancer (HNPCC), familial adenomatous polyposis (FAP), and MUTYH-associated polyposis (MAP) (20, 25-28). Approximately half of all patients with HNPCC possess germline mutations in MSH2 and MLH1. Defects in MSH6 cause late-onset atypical HNPCC. The HNPCC syndrome predisposes not only to colorectal cancer but also to extracolonic tumors such as carcinomas of the endometrium and ovary. The reason why these MMR defects cause predominantly cancers of the colon, endometrium and ovary is unclear. The mutations in MSH2 and MLH1 result in genomic instability, which is apparent as MSI. MSI also occurs in sporadic colorectal cancers (about 15%) and has been detected in a variety of other cancer types (29-31). Investigations of breast cancers have shown that MSI occurs in less than 5% of primary tumors (19, 32, 33).

Two mechanisms by which MMR genes can be inactivated during carcinogenesis have been proposed: mutations in MMR genes and epigenetic silencing of MMR genes by promoter hypermethylation (20, 34). Genetic alterations of MSH2 and MLH1 and epigenetic modification (hyper-methylation) of the MLH1 promoter have been shown to be associated with breast cancers displaying MSI (35). The hypermethylation, in turn, was associated with reduced expression of MLH1 protein in 26 of 83 (31%) primary breast cancers (36). Nine of 20 breast cancers contained a nonsense mutation in nucleotide 1862 of the PMS2 gene, which resulted in a premature stop in codon 613 (37). Western immunoblot revealed two bands at 100 kDa, corresponding to the wild-type form, and at 75 kDa, corresponding to the truncated protein. The truncated form was also found in two samples of normal appearing tissue adjacent to their corresponding cancerous lesions. Three of 11 breast cancers contained a missense mutation in the MSH2 gene at codon Asp475Gly (38). Another tumor contained an in-frame 1476 bp deletion.

A study of MMR deficiency in ten human cancer cell lines (colon, endometrium, ovary, and prostate) revealed mutations in the MLH1 (five), MSH2 (four), MSH3 (one), and PMS2 (one) genes (39). A prostate cancer cell line, DU145, was mutated in both MLH1 and PMS2. A wild-type copy

corresponding to the mutant gene was not detected in any of these lines, consistent with the hypothesis that defective repair, MSI, and possibly tumor formation all require loss of function of both alleles of a particular MMR gene. In two breast cancer cell lines, CAL51 and MT-3, MSI was detected within non-coding and within coding repeat sequences of genes mutated in HNPCC, such as TGFβ2, IGF2R, and BAX (21, 23). Analysis of MMR genes revealed MLH1 promoter hypermethylation in CAL51 cells and a hemizygous missense mutation, Tyr117Met, in MT-3 cells.

Each of the MMR genes contains multiple synonymous and non-synonymous polymorphisms, e.g., 43 in MSH6 (20, 24, 37, 40). Most of these polymorphisms occur in frequencies below 5%, except for MSH3 Arg94Gln (10%), MSH3 Thr1036Ala (30%), MSH6 Gly39Glu (24%), and MLH1 Ile219Val (12%). A case-control study of 300 women showed a strong association of the MSH2 Gly322Asp polymorphism with breast cancer risk (41). The odds ratio for the Gly/Gly genotype was 8.39 (95% CI 1.44 – 48.8). In contrast, the Asn127Ser polymorphism showed no association.

In summary, deficiency of MMR proteins plays a causative role in the HNPCC syndrome, which includes colorectal, endometrial, and ovarian cancers. There is no consistent association of MMR defects with breast cancer risk. MSI is present in less than 5% of primary breast cancers.

References

1. Kunkel TA. *DNA replication fidelity.* J Biol Chem. 1992;267:18251-4.

2. McCulloch SD, Kunkel TA. *The fidelity of DNA synthesis by eukaryotic replicative and translesion synthesis polymerases.* Cell Res. 2008;18:148-61.

3. Iyer RR, Pluciennik A, Burdett V, Modrich PL. *DNA mismatch repair: Functions and mechanisms.* Chem Rev. 2006;106:302-23.

4. Jiricny J. *Replication errors: cha(lle)nging the genome.* EMBO J. 1998;17:6427-36.

5. Jiricny J. *The multifaceted mismatch-repair system.* Nature Rev. 2006;7:335-46.

6. Hoeijmakers JHJ. *Genome maintenance mechanisms for preventing cancer.* Nature. 2001;411:366-74.

7. Modrich P. *Mismatch repair, genetic stability, and cancer.* Science. 1994;266:1959-60.

8. Bronner CE, Baker SM, Morrison PT, Warren G, Smith LG, Lescoe MK, et al. *Mutation in the DNA*

mismatch repair gene homologue hMLH1 is associated with hereditary non-polyposis colon cancer. Nature. 1994;368:258-61.

9. Edelmann W, Yang K, Umar A, Heyer J, Lau K, Fan K, et al. *Mutation in the mismatch repair gene Msh6 causes cancer susceptibility.* Cell. 1997;91:467-77.

10. Fishel R, Lescoe MK, Rao MR, Copeland NG, Jenkins NA, Garber J, et al. *The human mutator gene homolog MSH2 and its association with hereditary nonpolyposis colon cancer.* Cell. 1993;75:1027-38.

11. Nicolaides NC, Papadopoulos N, Liu B, Wei YF, Carter KC, Ruben SM, et al. *Mutations of two PMS homologues in hereditary nonpolyposis colon cancer.* Nature. 1994;371:75-80.

12. Papadopoulos N, Nicolaides NC, Liu B, Parsons R, Lengauer C, Palombo F, et al. *Mutations of GTBP in genetically unstable cells.* Science. 1995;268:1915-7.

13. Umar A, Buermeyer AB, Simon JA, Thomas DC, Clark AB, Liskay RM, et al. *Requirement for PCNA* in DNA mismatch repair at a step preceding DNA resynthesis. Cell. 1996;87:65-73.

14. Wang Y, Cortez D, Yazdi P, Neff N, Elledge SJ, Qin J. BASC, a super complex of BRCA1-associated proteins involved in the recognition and repair of aberrant DNA structures. Genes Dev. 2000;14:927-39.

15. Aaltonen LA, Peltomaki P, Leach FS, Sistonen P, Pylkkanen L, Mecklin J, et al. Clues to the pathogenesis of familial colorectal cancer. Science. 1993;260:812-6.

16. Loeb LA. Microsatellite instability: Marker of a mutator phenotype in cancer. Cancer Res. 1994;54:5059-63.

17. Loeb LA. A mutator phenotype in cancer. Cancer Res. 2001;61:3230-9.

18. Strand M, Prolla TA, Liskay RM, Petes TD. Destabilization of tracts of simple repetitive DNA in yeast by mutations affecting DNA mismatch repair. Nature. 1993;365:274-6.

19. Anbazhagan R, Fujii H, Gabrielson E. Microsatellite instability is uncommon in breast cancer. Clin Cancer Res. 1999;5:839-44.

20. Lynch HT, de la Chapelle A. Hereditary colorectal cancer. N Engl J Med. 2003;348:919-32.

21. Markowitz S, Wang J, Myeroff L, Parsons R, Sun L, Lutterbaugh

J, et al. Inactivation of the type II TGF-b receptor in colon cancer cells with micrsatellite instability. Science. 1995;268:1336-8.

22. Rampino N, Yamamoto H, Ionov Y, Li Y, Sawai H, Reed JC, et al. Somatic frameshift mutations in the BAX gene in colon cancers of the microsatellite mutator phenotype. Science. 1997;275:967-9.

23. Seitz S, Waβmuth P, Plaschke J, Schackert HK, Karsten U, Santibanez-Koref MF, et al. Identification of microsatellite instability and mismatch repair gene mutations in breast cancer cell lines. Genes Chromosomes Cancer. 2003;37:29-35.

24. Doss CG, Sethumadhavan R. Investigation on the role of nsSNPs in HNPCC genes--a bioinformatics approach. J Biomed Sci. 2009;16:42.

25. Clendenning M, Baze ME, Sun S, Walsh K, Liyanarachchi S, Fix D, et al. Origins and prevalence of the American founder mutation of MSH2. Cancer Res. 2008;68:2145-53.

26. Jo WS, Chung DC. Genetics of hereditary colorectal cancer. Semin Oncol. 2005;32:11-23.

27. Markowitz SD, Bertagnolli MM. Molecular origins of cancer: Molecular basis of colorectal cancer. N Engl J Med. 2009;361:2449-60.

28. Strate LL, Syngal S. Hereditary colorectal cancer syndromes. Cancer

Causes Control. 2005;16:201-13.

29. Ionov Y, Peinado MA, Malkhosyan S, Shibata D, Perucho M. Ubiquitous somatic mutations in simple repeated sequences reveal a new mechanism for colonic carcinogenesis. Nature. 1993;363:558-61.

30. Peltomaki P, Lothe RA, Aaltonen LA, Pylkkanen L, Nystrom-Lahti M, Seruca R, et al. Microsatellite instability is associated with tumors that characterize the hereditary non-polyposis colorectal carcinoma syndrome. Cancer Res. 1993;53:5853-5.

31. Thibodeau SN, Gren G, Schaid D. Microsatellite instability in cancer of the proximal colon. Science. 1993;260:816-9.

32. Dillon EK, de Boer WB, Papadimitriou JM, Turbett GR. Microsatellite instability and loss of heterozygosity in mammary carcinoma and its probable precursors. Br J Cancer. 1997;76:156-62.

33. Yee CJ, Roodi N, Verrier CS, Parl FF. Microsatellite instability and loss of heterozygosity in breast cancer. Cancer Res. 1994;54:1641-4.

34. Baylin SB, Herman JG. DNA hypermethylation in tumorigenesis. Epigenetics joins genetics. Trends Genet. 2000;16:168-74.

35. Murata H, Khattar NH, Kang Y, Gu L, Li GM. Genetic and epigenetic modification of mismatch repair genes

hMSH2 and hMLH1 in sporadic breast cancer with microsatellite instability. Oncogene. 2002;21:5696-703.

36. Murata H, Khattar NH, Gu L, Li GM. Roles of mismatch repair proteins hMSH2 and hMLH1 in the development of sporadic breast cancer. Cancer Lett. 2005;223:143-50.

37. Balogh GA, Heulings RC, Russo J. The mismatch repair gene hPMS2 is mutated in primary breast cancer. Int J Mol Med. 2006;18:853-7.

38. Bhattacharyya N, Chen HC, Grundfest-Broniatowski S. Alteration of hMSH2 and DNA polymerase β genes in breast carcinoma and fibroadenomas. Biochem Biophys Res Comm. 1999;259:429-35.

39. Boyer JD, Umar A, Risinger JI, Lipford JR, Kane M, Yin S, et al. Microsatellite instability, mismatch repair deficiency, and genetic defects in human cancer cell lines. Cancer Res. 1995;55:6063-70.

40. Mohrenweiser HW, Xi T, Vazquez-Matias J, Jones IM. Identification of 127 amino acid substitution variants in screening 37 DNA repair genes in humans. Cancer Epidemiol Biomarkers Prev. 2002;11:1054-64.

41. Poplawski T, Zadrozny M, Kolacinska A, Rykala J, Morawiec Z, Blasiak J. Polymorphisms of the DNA mismatch repair gene HMSH2 in breast cancer occurrence and progression. Breast Cancer Res Treat. 2005;94:199-204.

5.5. HOMOLOGOUS RECOMBINATION REPAIR
5.5.1. INTRODUCTION

Double-strand DNA breaks (DSBs) arise from ionizing radiation or X-rays, free radicals, chemicals and during replication of single-strand breaks (SSBs) (1-4). When DSBs occur, the complementary copies of DNA are severed in the same place and therefore cannot act as templates for each other. If unrepaired or repaired inappropriately, DNA sequences can be joined to sequences originally distant in the genome resulting in genome rearrangement. Thus, DSBs can lead to chromosome loss, deletions, duplications, or translocations. Cells rely on two mechanisms to accomplish DSB repair: non-homologous end-joining repair (NHEJR) in G1 and early S phase and homologous recombination repair (HRR) in late S and G2, when the intact copy on the sister chromatid can be used as template to properly align and seal the broken ends in an error-free manner. HRR is also involved in repairing SSBs and in bypassing interstrand cross-links.

The HRR pathway is initiated by DSB detection, which triggers a complex cascade of reactions involving the interaction of DNA and chromatin with multiple proteins that undergo phosphorylation, ubiquitylation, sumoylation, and acetylation in a process aimed at halting the cell-cycle machinery and allowing repair (5-12). In comparison to other DNA damage repair pathways, HRR is far more complex than BER, NER, MRR, or NHEJR based on the number of proteins participating in multiple levels of interaction including cell-cycle checkpoint activation. For example, the protein kinase ATM (ataxia-teleangiectasia mutated) mediates a two-step response to DSBs. In the rapid response within minutes after DNA damage, the local chromatin structure is altered and ATM phosphorylates the variant histone H2AX (13). Activated ATM also phosphorylates checkpoint kinase 2 (CHK2), which phosphorylates CDC25A, targeting it for ubiquitylation and degradation.

Therefore, phosphorylated CDK2-cyclin accumulates and progression through the cell cycle is blocked. In the delayed response, ATM phosphorylates the inhibitor of p53, MDM2, and p53, which is also phosphorylated by CHK2 (5). The resulting activation and stabilization of p53 leads to an increased expression of Cdk inhibitor p21, which further helps to keep Cdk activity low and to maintain G1/S cell cycle arrest. Furthermore, the ATM-mediated signaling cascade phosphorylates several other substrates, including ATM itself, BRCA1, NBS1, SMC1, SSB1, and 53BP1, reinforcing the activation of DSB repair, cell cycle checkpoints or apoptosis (14-19).

Several models of HRR have been proposed, one of which is shown in Figure 5.5.1. The HRR pathway consists of several steps beginning with DSB recognition and ending with ligation and reconstitution of intact DNA. Several key players act as preformed complexes in the HRR pathway, most notably the MRN complex, which is composed of a single NBS1 molecule and two dimers of the MRE11 nuclease and the RAD50 DNA binding protein. Other proteins, e.g., BRCA1, set in motion the sequential recruitment of several other proteins. To add to the complexity of HRR, the MRN complex, ATM, BRCA1, BRCA2, and other proteins are involved in more than one step. For example, the MRN complex participates in the initial recognition of DSBs as well as the resection of DNA ends and the following recombination (20). BRCA1 participates in damage recognition, end processing, and DNA strand exchange.

DSB recognition. HRR is initiated when the trimeric MRN (MRE11, RAD50, NBS1) protein complex recognizes and binds to a DSB followed by recruitment of ATM, which phosphorylates multiple targets (20-22). One of the targets is the histone H2AX, one of three H2A subfamily members involved in packaging DNA into nucleosomes. Phosphorylation of H2AX at 139Ser occurs following DNA double-strand but not single-strand break induction and is

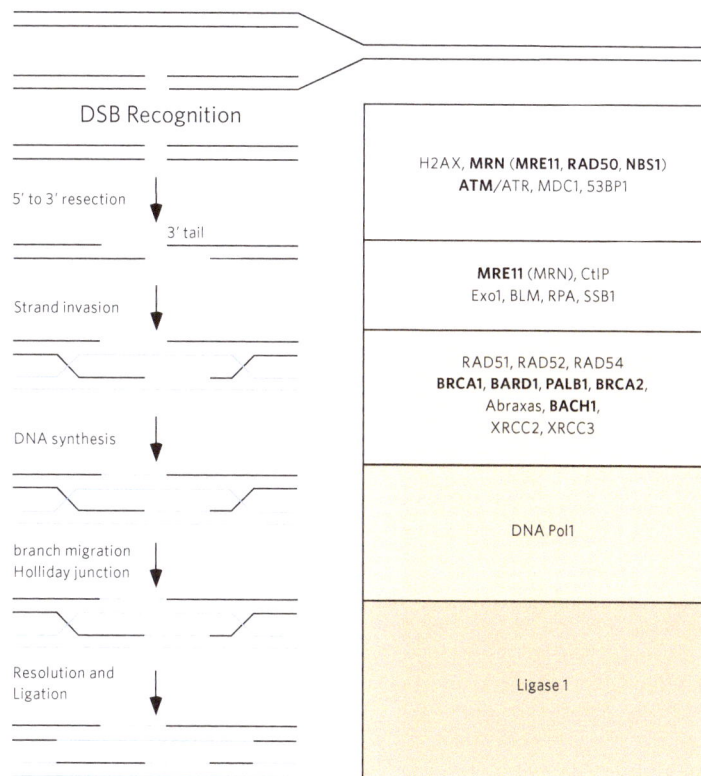

Figure 5.5.1. Schematic overview of HRR pathway and listing of participating proteins.
Following recognition of the DSB, the broken ends are resected on the 5' strands to produce 3' single-stranded DNA tails. The 3' end invades the homologous DNA of the intact sister chromatid, displaces the identical strand of the sister duplex and forms a displacement loop (D-loop) allowing recombination with the complementary strand. The latter provides a template to permit faithful DNA replication to form an intermediate containing Holliday junctions, which are resolved and ligated resulting in either crossover or non-crossover gene conversion products. Only one of several possible recombination products is shown. The cell cycle is arrested in response to DSBs either at the G1/S or G2/M checkpoint. These arrests are released once the DSBs have been repaired in the HRR pathway. Each protein is listed only once but may be involved in more than one step in the pathway. For example, the MRN complex participates in DSB detection, end resection, and recombination. Proteins highlighted in bold are affected by germ-line mutations and associated with familial breast cancer.

one of the earliest damage responses, reaching maximum levels within minutes after exposure to ionizing radiation (23-26). The phosphorylated form of H2AX, termed γH2AX, sets in motion the recruitment of several proteins starting with MDC1 (for mediator of DNA damage checkpoint protein 1) and followed by RNF8, an E3 ubiquitin ligase, which interacts with the E2 ubiquitin-conjugating enzyme UBC13 to ubiquinate γH2AX (25, 27-29). In turn, the ubiquitylation of γH2AX allows the attachment of ubiquitylated receptor-associated protein 80 (RAP80), which recruits the phosphoprotein Abraxas that finally binds BRCA1 via its breast cancer carboxy-terminal (BRCT) repeats (30, 31). In the process, ATM also phosphorylates two BRCA1 residues, 1423Ser and 1524Ser, as well as the BARD1 residue 714Thr, which are important for the DNA damage function of the BRCA1-BARD1 complex (32, 33). To complicate matters, BRCA1 is endowed with E3 ubiquitin ligase activity in its RING domain, which is even more pronounced in the BRCA1-BARD1 complex, allowing ubiquitylation of other proteins as well possible auto-ubiquitylation of BRCA1-BARD1 heterodimers (34). Finally, BRCA1 is covalently modified by attachment of small ubiquitin-like modifier (SUMO) proteins to lysine residues such as 119Lys (35). Thus, BRCA1 plays a key role in the DNA damage response pathways by interacting with multiple proteins with modification reactions of phosphorylation, ubiquitylation, sumoylation, and acetylation (11, 35-39). Moreover, BRCA1 appears to influence the choice of repair pathway by promoting error-free HRR over error-prone NHEJR (40). This occurs through at least two mechanisms involving 53BP1 (p53-binding protein 1) and CtIP (interactor for CtBP or RBBP8). 53BP1 co-localizes at DSBs with the MRN complex, γH2AX, MDC1, and BRCA1 (41). The interaction with the MRN complex occurs via the BRCT repeats in 53BP1 and is accompanied by ATM-mediated phosphorylation of both 53BP1 and NBS1 (18, 42). In BRCA1-deficient cells, 53BP1 inhibits HRR by blocking resection of DNA breaks leading to error-prone repair by NHEJR (43, 44). In wild-type cells, BRCA1 overcomes the inhibitory effect of 53BP1 by presently unknown mechanisms and thereby promotes active DSB resection and accurate repair by HRR. The other process involves another binding partner of BRCA1, CtIP (yeast homolog Sae2), which undergoes phosphorylation on residue 327Ser and recruits BRCA1 to shift the balance of DSB repair from NHEJR to HRR (45). CtIP is also acetylated by histone acetyltransferases resulting in its degradation by autophagy (46, 47). Instead of Abraxas or CtIP, BRCA1 can interact with BACH1 (for BRCA1-associated C-terminal helicase) and is then joined by RAD51 in HRR or by RAD50-MRE11-NBS1 in NHEJR (48-51). Thus, after DSB detection, a complex cascade of reactions is triggered aimed at halting the cell-cycle machinery and recruiting multiple repair factors, which accumulate in the nucleus in discrete dots or "foci" (8). Among these factors are also RAD54, a DNA-dependent ATPase, and RAD52, which co-localizes with RAD50 and binds directly to DSBs, protecting them from exonuclease activity (4, 52, 53).

End resection. Nucleolytic degradation of the DSB ends in the 5' to 3' direction producing 3' single-stranded DNA tails, which serve as templates for the subsequent recombination with sister chromatids (54). The resection is carried out in two steps involving four nucleases. In the first step, the nuclease activity of the MRN complex is stimulated by interaction of MRE11 with CtIP resulting in the removal of a short oligonucleotide to form an early intermediate (55, 56). Second, Exo1 and/or Sgs1 (human homolog is BLM, the Bloom syndrome helicase) rapidly process this intermediate to generate extensive tracts of ssDNA (55, 57). The ssDNA generated by the nucleases then serves as substrate for binding replication protein A (RPA) and single-stranded binding protein SSB1, which facilitate RAD51 nucleofilament formation and strand invasion in the next step (4, 19). The transition from DSB to ssDNA is also required for activation of the ATR-mediated checkpoint response (58). Moreover, CtIP interacts not only with MRE11 but also with BRCA1 (59).

Strand invasion and homologous recombination of sister chromatid with branch migration, resolution of Holliday junctions, and ligation. The central protein of HRR is the recombinase RAD51, a relatively small 38-kDa protein (4). It is functional as a long helical polymer, made of hundreds of monomers, that wraps around the ssDNA generated at the site of a DSB to form a nucleoprotein filament. RAD51 interacts with an undamaged DNA molecule and, when a homologous region has been located, RAD51 catalyzes strand-exchange in an ATP-dependent reaction in which the damaged molecule invades the other DNA duplex, displacing one strand as a D-loop (7, 60). The 3' tail of the damaged DNA molecule is then extended by DNA polymerase that copies information from the undamaged partner, and the ends are ligated by DNA ligase I. Finally, after migration, the DNA crossovers (Holliday junctions) are resolved by cleavage and ligation to yield two intact DNA molecules (4). These events are influenced by other proteins, such as the RAD51-related proteins XRCC2 and XRCC3, RAD52, and especially BRCA2, which binds up to six RAD51 molecules in the presence of DSS1, a 70-amino acid polypeptide (8, 61, 62). BRCA2 enhances the binding of RAD51 to ssDNA, enabling RAD51 to displace RPA from ssDNA. Additionally, BRCA2 stabilizes RAD51-ssDNA filaments by blocking ATP hydrolysis (61, 63). Thus, BRCA2 promotes specific and stable assembly of RAD51 on ssDNA and thereby enables RAD51-mediated strand exchange between homologous DNA sequences (63-65). An important partner in localizing BRCA2 and RAD51 at the DNA damage site is PALB2 (for partner and localizer of BRCA2), which contains a C-terminal β-propeller domain that binds to the N-terminal third of BRCA2 (66). At its N-terminus PALB2 also contains a coiled-coil region, which interacts with a coiled-coil region upstream of the BRCT repeats in BRCA1, thereby serving as an important link between BRCA2 and BRCA1 (67-69).

References

1. Hoeijmakers JH. *DNA damage, aging, and cancer.* N Engl J Med. 2009;361:1475-85.

2. Hoeijmakers JHJ. *Genome maintenance mechanisms for preventing cancer.* Nature. 2001;411:366-74.

3. Marians KJ. *Replication and recombination intersect.* Curr Opinion Genet Dev. 2000;10:151-6.

4. West SC. *Molecular views of recombination proteins and their control.* Nature Rev Mol Cell Biol. 2003;4:435-45.

5. Bartkova J, Horejsi Z, Koed K, Kramer A, Tort F, Zieger K, et al. *DNA damage response as a candidate anti-cancer barrier in early human tumorigenesis.* Nature. 2005;434:864-70.

6. Connelly JC, Leach DR. *Tethering on the brink: the evolutionarily conserved Mre11-Rad50 complex.* Trends Biochem Sci. 2002;27:410-8.

7. Ishino Y, Nishino T, Morikawa K. *Mechanisms of maintaining genetic stability by homologous recombination.* Chem Rev. 2006;106:324-39.

8. Khanna KK, Jackson SP. *DNA double-strand breaks: signaling, repair and the cancer connection.* Nature Genet. 2001;27:247-54.

9. Moynahan ME, Jasin M. Mitotic homologous recombination maintains genomic stability and suppresses tumorigenesis. Nature Rev Mol Cell Biol. 2010;11:196-207.

10. Oberle C, Blattner C. Regulation of the DNA Damage Response to DSBs by Post-Translational Modifications. Curr Genomics. 2010;11:184-98.

11. Potenski CJ, Klein HL. Molecular biology: The expanding arena of DNA repair. Nature. 2011;471:48-9.

12. Rossetto D, Truman AW, Kron SJ, Cote J. *Epigenetic modifications in double-strand break DNA damage signaling and repair.* Clin Cancer Res. 2010;16:4543-52.

13. Ayoub N, Jeyasekharan AD, Bernal JA, Venkitaraman AR. *HP1-beta mobilization promotes chromatin changes that initiate the DNA damage response.* Nature. 2008;453:682-6.

14. Barzilai A, Rotman G, Shiloh Y. *ATM deficiency and oxidative stress: a new dimension of defective response to DNA damage.* DNA Repair. 2002;1:3-25.

15. Kim ST, Lim DS, Canman CE, Kastan MB. *Substrate specificities and identification of putative substrates of ATM kinase family members.* J Biol Chem. 1999;274:37538-43.

16. Kitagawa R, Bakkenist CJ, McKinnon PJ, Kastan MB. *Phosphorylation of SMC1 is a critical downstream event in the ATM-NBS1-BRCA1 pathway.* Genes Dev. 2004;18:1423-38.

17. Kurz EU, Lees-Miller SP. *DNA damage-induced activation of ATM and ATM-dependent signaling pathways.* DNA Repair (Amst). 2004;3:889-900.

18. Matsuoka S, Ballif BA, Smogorzewska A, McDonald ER, 3rd, Hurov KE, Luo J, et al. *ATM and ATR substrate analysis reveals extensive protein networks responsive to DNA damage.* Science. 2007;316:1160-6.

19. Richard DJ, Bolderson E, Cubeddu L, Wadsworth RI, Savage K, Sharma GG, et al. *Single-stranded DNA-binding protein hSSB1 is critical for genomic stability.* Nature. 2008;453:677-81.

20. Yuan J, Chen J. *MRE11-RAD50-NBS1 complex dictates DNA repair independent of H2AX.* J Biol Chem. 2010;285:1097-104.

21. Lee JH, Paull TT. *ATM activation by DNA double-strand breaks through the Mre11-Rad50-Nbs1 complex.* Science. 2005;308:551-4.

22. Lee JH, Paull TT. *Activation and regulation of ATM kinase activity in response to DNA double-strand breaks.* Oncogene. 2007;26:7741-8.

23. Fernandez-Capetillo O, Lee A, Nussenzweig M, Nussenzweig A. *H2AX: the histone guardian of the genome.* DNA Repair (Amst). 2004;3:959-67.

24. Mahadevaiah SK, Turner JM, Baudat F, Rogakou EP, de Boer P, Blanco-Rodriguez J, et al. *Recombinational DNA double-strand breaks in mice precede synapsis.* Nature Genet. 2001;27:271-6.

25. Paull TT, Rogakou EP, Yamazaki V, Kirchgessner CU, Gellert M, Bonner WM. *A critical role for histone H2AX in recruitment of repair factors to nuclear foci after DNA damage.* Curr Biol. 2000;10:886-95.

26. Rogakou EP, Pilch DR, Orr AH, Ivanova VS, Bonner WM. *DNA double-stranded breaks induce histone H2AX phosphorylation on serine 139.* J Biol Chem. 1998;273:5858-68.

27. Goldberg M, Stucki M, Falck J, D'Amours D, Rahman D, Pappin D, et al. *MDC1 is required for the intra-S-phase DNA damage checkpoint.* Nature. 2003;421:952-6.

28. Stewart GS, Wang B, Bignell CR, Taylor AM, Elledge SJ. *MDC1 is a mediator of the mammalian DNA damage checkpoint.* Nature. 2003;421:961-6.

29. Wang B, Elledge SJ. *Ubc13/Rnf8 ubiquitin ligases control foci formation of the Rap80/Abraxas/Brca1/Brcc36 complex in response to DNA damage.* Proc Natl Acad Sci U S A. 2007;104:20759-63.

30. Kim H, Chen J, Yu X. *Ubiquitin-binding protein RAP80 mediates BRCA1-dependent DNA damage response.* Science. 2007;316:1202-5.

31. Wang B, Matsuoka S, Ballif BA, Zhang D, Smogorzewska A, Gygi SP, et al. *Abraxas and RAP80 form a BRCA1 protein complex required for the DNA damage response.* Science. 2007;316:1194-8.

32. Cortez D, Wang Y, Qin J, Elledge SJ. *Requirement of ATM-dependent phosphorylation of brca1 in the DNA damage response to double-strand breaks.* Science. 1999;286:1162-6.

33. Kim HS, Li H, Cevher M, Parmelee A, Fonseca D, Kleiman FE, et al. *DNA damage-induced BARD1 phosphorylation is critical for the inhibition of messenger RNA processing by BRCA1/BARD1 complex.* Cancer Res. 2006;66:4561-5.

34. Simons AM, Horwitz AA, Starita LM, Griffin K, Williams RS, Glover JN, et al. *BRCA1 DNA-binding activity is stimulated by BARD1.* Cancer Res. 2006;66:2012-8.

35. Morris JR, Boutell C, Keppler M, Densham R, Weekes D, Alamshah A, et al. *The SUMO modification pathway is involved in the BRCA1 response to genotoxic stress.* Nature. 2009;462:886-90.

36. Bergink S, Jentsch S. *Principles of ubiquitin and SUMO modifications in DNA repair.* Nature. 2009;458:461-7.

37. Foulkes WD. *Traffic control for BRCA1.* N Engl J Med. 2010;362:755-6.

38. Galanty Y, Belotserkovskaya R, Coates J, Polo S, Miller KM, Jackson SP. *Mammalian SUMO E3-ligases PIAS1 and PIAS4 promote responses to DNA double-strand breaks.* Nature. 2009;462:935-9.

39. Huen MS, Sy SM, Chen J. *BRCA1 and its toolbox for the maintenance of genome integrity.* Nature Rev Mol Cell Biol. 2010;11:138-48.

40. Boulton SJ. DNA repair: *Decision at the break point.* Nature. 2010;465:301-2.

41. Ward IM, Minn K, Jorda KG, Chen J. *Accumulation of checkpoint protein 53BP1 at DNA breaks involves its binding to phosphorylated histone H2AX.* J Biol Chem. 2003;278:19579-82.

42. Lee JH, Goodarzi AA, Jeggo PA, Paull TT. *53BP1 promotes ATM activity through direct interactions with the MRN complex.* EMBO J. 2010;29:574-85.

43. Bouwman P, Aly A, Escandell JM, Pieterse M, Bartkova J, van der Gulden H, et al. *53BP1 loss rescues BRCA1 deficiency and is associated with triple-negative and BRCA-mutated breast cancers.* Nature Struct Mol Biol. 2010;17:688-95.

44. Bunting SF, Callen E, Wong N, Chen HT, Polato F, Gunn A, et al. *53BP1 inhibits homologous recombination in Brca1-deficient cells by blocking resection of DNA breaks.* Cell. 2010;141:243-54.

45. Yun MH, Hiom K. *CtIP-BRCA1 modulates the choice of DNA double-strand-break repair pathway throughout the cell cycle.* Nature. 2009;459:460-3.

46. Kaidi A, Weinert BT, Choudhary C, Jackson SP. *Human SIRT6 promotes DNA end resection through CtIP deacetylation.* Science. 2010;329:1348-53.

47. Robert T, Vanoli F, Chiolo I, Shubassi G, Bernstein KA, Rothstein R, et al. *HDACs link the DNA damage response, processing of double-strand breaks and autophagy.* Nature. 2011;471:74-9.

48. Cantor S, Drapkin R, Zhang F, Lin Y, Han J, Pamidi S, et al. *The BRCA1-associated protein BACH1 is a DNA helicase targeted by clinically relevant inactivating mutations.* Proc Natl Acad Sci U S A. 2004;101:2357-62.

49. Cousineau I, Abaji C, Belmaaza A. *BRCA1 regulates RAD51 function in response to DNA damage and suppresses spontaneous sister chromatid replication slippage: implications for sister chromatid cohesion, genome stability, and carcinogenesis.* Cancer Res. 2005;65:11384-91.

50. Scully R, Chen J, Ochs RL, Keegan K, Hoekstra M, Feunteun J, et al. *Dynamic changes of BRCA1 subnuclear location and phosphorylation state are initiated by DNA damage.* Cell. 1997;90:425-35.

51. Zhong Q, Chen CF, Li S, Chen Y, Wang CC, Xiao J, et al. *Association of BRCA1 with the hRad50-hMre11-p95 complex and the DNA damage response.* Science. 1999;285:747-50.

52. Dronkert ML, Beverloo HB, Johnson RD, Hoeijmakers JH, Jasin M, Kanaar R. *Mouse RAD54 affects DNA double-strand break repair and sister chromatid exchange.* Mol Cell Biol. 2000;20:3147-56.

53. Haber JE. DNA repair. *Gatekeepers of recombination.* Nature. 1999;398:665, 7.

54. Klein HL. *Molecular biology: DNA endgames.* Nature. 2008;455:740-1.

55. Mimitou EP, Symington LS. *Sae2, Exo1 and Sgs1 collaborate in DNA double-strand break processing.* Nature. 2008;455:770-4.

56. Sartori AA, Lukas C, Coates J, Mistrik M, Fu S, Bartek J, et al. *Human CtIP promotes DNA end resection.* Nature. 2007;450:509-14.

57. Mimitou EP, Symington LS. *DNA end resection--unraveling the tail.* DNA Repair (Amst). 2011;10:344-8.

58. Zou L, Elledge SJ. *Sensing DNA damage through ATRIP recognition of RPA-ssDNA complexes.* Science. 2003;300:1542-8.

59. Yu X, Chen J. *DNA damage-induced cell cycle checkpoint control requires CtIP, a phosphorylation-dependent binding partner of BRCA1 C-terminal domains.* Mol Cell Biol. 2004;24:9478-86.

60. Kowalczykowski SC. *Structural biology: snapshots of DNA repair.* Nature. 2008;453:463-6.

61. Liu J, Doty T, Gibson B, Heyer WD. *Human BRCA2 protein promotes RAD51 filament formation on RPA-covered single-stranded DNA.* Nature Struct Mol Biol. 2010;17:1260-2.

62. Wray J, Liu J, Nickoloff JA, Shen Z. *Distinct RAD51 associations with RAD52 and BCCIP in response to DNA damage and replication stress.* Cancer Res. 2008;68:2699-707.

63. Jensen RB, Carreira A, Kowalczykowski SC. *Purified human BRCA2 stimulates RAD51-mediated recombination.* Nature. 2010;467:678-83.

64. Thorslund T, McIlwraith MJ, Compton SA, Lekomtsev S, Petronczki M, Griffith JD, et al. *The breast cancer tumor suppressor BRCA2 promotes the specific targeting of RAD51 to single-stranded DNA.* Nature Struct Mol Biol. 2010;17:1263-5.

65. Zou L. DNA repair: *A protein giant in its entirety.* Nature. 2010;467:667-8.

66. Xia B, Sheng Q, Nakanishi K, Ohashi A, Wu J, Christ N, et al. *Control of BRCA2 cellular and clinical functions by a nuclear partner, PALB2.* Mol Cell. 2006;22:719-29.

67. Sy SM, Huen MS, Chen J. *PALB2 is an integral component of the BRCA complex required for homologous recombination repair.* Proc Natl Acad Sci U S A. 2009;106:7155-60.

68. Zhang F, Fan Q, Ren K, Andreassen PR. *PALB2 functionally connects the breast cancer susceptibility proteins BRCA1 and BRCA2.* Mol Cancer Res. 2009;7:1110-8.

69. Zhang F, Ma J, Wu J, Ye L, Cai H, Xia B, et al. *PALB2 links BRCA1 and BRCA2 in the DNA-damage response.* Curr Biol. 2009;19:524-9.

5.5.2. GERM-LINE MUTATIONS AND POLYMORPHISMS

The BRCA proteins play key roles in HRR. BRCA1 can recruit PALB2, which in turn organizes BRCA2 and RAD51 (1-3). Abolishing the interaction of PALB2 with either BRCA2 or BRCA1 has been shown to impair DSB repair by homologous recombination (1-3). These results highlight the physical and functional linkage of all three proteins in maintaining genomic stability and suggest that impaired HRR is the major cause for genomic instability and tumorigenesis observed in patients carrying BRCA1, BRCA2, or PALB2 germ-line mutations

(4-7). BRCA1, BRCA2, ATM, PALB2 and other genes and their protein products are discussed in greater detail in a separate chapter (see Chapter 8: Familial Risk Factors). Interestingly, germ-line mutations in other HRR genes that were known to be associated with malignancies other than breast or ovarian cancers have also been linked to familial breast cancer. For example, NBS1 and MRE11 gene mutations cause the Nijmegen breakage syndrome (NBS) and ataxia teleangiectasia-like disorder (ATLD), respectively, which are associated with multisystem defects including immunodeficiency and increased lymphoid malignancies (8, 9). However, sequencing of the MRE11, RAD50, and NBS1 genes in patients from non-BRCA1/2 breast cancer families has identified germ-line mutations in each of the three genes encoding the MRN protein complex (10-12). A germ-line frameshift mutation was identified in the DNA polymerase theta gene (PolQ c.3605delT) in a patient from a high-risk breast cancer family without BRCA1 or BRCA2 mutations (13). In summary, these results indicate that the HRR is an essential anti-cancer barrier and germ-line mutations in HRR-related genes collectively account for the majority of familial breast cancers (4).

Several studies have examined the association of polymorphisms in HRR-related genes with breast cancer risk. A 2002 study of the NBS1, RAD51, RAD52, XRCC2, and XRCC3 genes in 2205 cases and 1826 controls identified eight haplotypes of the XRCC3 gene and observed an association of a rare haplotype (0.3%) with increased breast cancer risk (14). Moreover, homozygous carriers of a polymorphic XRCC3 variant (rs861539; Thr241Met) also showed an increased risk. A 2006 meta-analysis of 10,979 cases and 10,423 controls provided support for an association of the 241Met/Met genotype with increased risk (odds ratio 1.16; 95% CI 1.04 – 1.30; p = 0.009) (15). However, the Breast Cancer Association Consortium (BCAC) found no risk association with the Thr241Met polymorphism by comparing 12,365 cases and 13,138 controls (16). Similarly, a 2008 study of 4,470 cases and 4,560 controls observed no association with XRCC3 rs861539 (17). The 2002 study also observed an increased risk association for a polymorphic variant in the XRCC2 gene (rs3218536; Arg188His) (14). However, neither the 2006 meta-analysis nor the BCAC corroborated such association and the 2008 study observed the opposite effect, a protective association of the rare 188His allele (15-17). No risk associations or inconsistent results have been obtained for SNPs in other genes, whose protein products participate in the HRR pathway: ATR, BARD1, BRIP1 (BACH1), CHEK1, MRE11A, NBS1, RAD50, RAD51, RAD51C, RAD52, and RAD54 (14, 15, 17). One study examined 15 SNPs evenly distributed through the genes of the MRN complex. A polymorphism in an NBS1 intron (rs1805790; A/G) showed a borderline significant association with increased risk of breast cancer and a combination of high-risk MRN genotypes was associated with a significant trend toward increased risk (18). The same group of investigators used the same approach for the BLM and RAD51 genes and found that certain SNP combinations (all located in introns) were associated with increased risk (19).

References

1. Sy SM, Huen MS, Chen J. *PALB2 is an integral component of the BRCA complex required for homologous recombination repair.* Proc Natl Acad Sci U S A. 2009;106:7155-60.

2. Zhang F, Fan Q, Ren K, Andreassen PR. *PALB2 functionally connects the breast cancer susceptibility proteins BRCA1 and BRCA2.* Mol Cancer Res. 2009;7:1110-8.

3. Zhang F, Ma J, Wu J, Ye L, Cai H, Xia B, et al. *PALB2 links BRCA1 and BRCA2 in the DNA-damage response.* Curr Biol. 2009;19:524-9.

4. Bartkova J, Horejsi Z, Koed K, Kramer A, Tort F, Zieger K, et al. *DNA damage response as a candidate anti-cancer barrier in early human tumorigenesis.* Nature. 2005;434:864-70.

5. Huen MS, Sy SM, Chen J. *BRCA1 and its toolbox for the maintenance of genome integrity.* Nature Rev Mol Cell Biol. 2010;11:138-48.

6. Moynahan ME, Jasin M. *Mitotic homologous recombination maintains genomic stability and suppresses tumorigenesis.* Nature Rev Mol Cell Biol. 2010;11:196-207.

7. Rahman N, Seal S, Thompson D, Kelly P, Renwick A, Elliott A, et al. *PALB2, which encodes a BRCA2-interacting protein, is a breast cancer susceptibility gene.* Nature Genet 2007;39:165-7.

8. Lee JH, Xu B, Lee CH, Ahn JY, Song MS, Lee H, et al. *Distinct functions of Nijmegen breakage syndrome in ataxia telangiectasia mutated-dependent responses to DNA damage.* Mol Cancer Res. 2003;1:674-81.

9. Varon R, Vissinga C, Platzer M, Cerosaletti KM, Chrzanowska KH, Saar K, et al. *Nibrin, a novel DNA double-strand break repair protein, is mutated in Nijmegen breakage syndrome.* Cell. 1998;93:467-76.

10. Bartkova J, Tommiska J, Oplustilova L, Aaltonen K, Tamminen A, Heikkinen T, et al. *Aberrations of the MRE11-RAD50-NBS1 DNA damage sensor complex in human breast cancer: MRE11 as a candidate familial cancer-predisposing gene.* Mol Oncol. 2008;2:296-316.

11. Heikkinen K, Rapakko K, Karppinen SM, Erkko H, Knuutila S, Lundan T, et al. *RAD50 and NBS1 are breast cancer susceptibility genes associated with genomic instability.* Carcinogenesis. 2006;27:1593-9.

12. Tommiska J, Seal S, Renwick A, Barfoot R, Baskcomb L, Jayatilake H, et al. *Evaluation of RAD50 in familial breast cancer predisposition.* Int J Cancer. 2006;118:2911-6.

13. Wang X, Szabo C, Qian C, Amadio PG, Thibodeau SN, Cerhan JR, et al. *Mutational analysis of thirty-two double-strand DNA break repair genes in breast and pancreatic cancers.* Cancer Res. 2008;68:971-5.

14. Kuschel B, Auranen A, McBride S, Novik KL, Antoniou A, Lipscombe JM, et al. *Variants in DNA double-strand break repair genes and breast cancer susceptibility.* Hum Mol Genet. 2002;11:1399-407.

15. Garcia-Closas M, Egan KM, Newcomb PA, Brinton LA, Titus-Ernstoff L, Chanock S, et al. *Polymorphisms in DNA double-strand break repair genes and risk of breast cancer: two population-based studies in USA and Poland, and meta-analyses.* Hum Genet. 2006;119:376-88.

16. *Commonly studied single-nucleotide polymorphisms and breast cancer: Results from the Breast Cancer Association Consortium.* J Natl Cancer Inst. 2006;98:1382-96.

17. Pooley KA, Baynes C, Driver KE, Tyrer J, Azzato EM, Pharoah PD, et al. *Common single-nucleotide polymorphisms in DNA double-strand break repair genes and breast cancer risk.* Cancer Epidemiol Biomarkers Prev. 2008;17:3482-9.

18. Hsu HM, Wang HC, Chen ST, Hsu GC, Shen CY, Yu JC. *Breast cancer risk is associated with the genes encoding the DNA double-strand break repair Mre11/Rad50/Nbs1 complex.* Cancer Epidemiol Biomarkers Prev. 2007;16:2024-32.

19. Ding SL, Yu JC, Chen ST, Hsu GC, Kuo SJ, Lin YH, et al. *Genetic variants of BLM interact with RAD51 to increase breast cancer susceptibility.* Carcinogenesis. 2009;30:43-9.

5.6. NONHOMOLOGOUS END-JOINING REPAIR (NHEJR)

As the name implies, NHEJR involves the religation of DSBs with little or no homology. Without a template, the end-joining is sometimes associated with gain or loss of a few nucleotides if internal microhomologies are used for annealing before sealing. Thus, NHEJR is designed to repair DNA damage but the process itself is error-prone and may thereby enhance mutagenesis. NHEJR is the repair pathway of choice of non-dividing and of mitotic cells during G1 and early S phase of the cell cycle. Since most cells are in a quiescent state (G0), NHEJR is considered the major DSB repair pathway for the human genome (1-3). NHEJR is also an essential physiological pathway for the completion of V(D)J recombination, the programmed DNA rearrangement that assembles the antigen receptors of B and T lymphocytes.

The NHEJR pathway consists of four consecutive steps: (i) DSB recognition, (ii) recruitment of repair factors, (iii) processing of DNA ends, and (iv) sealing of DNA break. These steps are mediated by four core proteins: Ku70/80, the catalytic subunit of the DNA-dependent protein kinase (DNA-PKcs), and the XRCC4/DNA ligase IV heterodimer. Central to NHEJR is the Ku protein, a heterodimer composed of two subunits called Ku70 (69 kDa) and Ku80 (83 kDa). Ku70/80 binds with high avidity to DNA termini and interacts with DNA-PKcs to form the DNA-PK holoenzyme (1). In case one of the termini has lost its base, Ku70/80 has 5'-deoxyribose-5-phosphate (5'-dRP) lyase activity allowing the excision of the AP (apurinic/apyrimidinic) site prior to joining of the broken ends (4). DNA-PKcs is a 465-kDa protein, whose C-terminal region has homology to a family of phosphatidylinositol-3 kinase-related kinases (PIKKs), which includes ATM, ATR (ATM- and RAD3-related), and mTOR (mammalian target of rapamycin) (5-7). DNA-PKcs, ATM, and ATR participate in DNA damage response pathways and phosphorylate multiple downstream substrates at serines or threonines followed by glutamine, i.e., SQ or TQ motifs (8-12). DNA-PKcs has affinity for DNA ends and its activation appears to be triggered by simultaneous interaction with both ssDNA ends derived from a DSB (13). One of the phosphorylation targets of DNA-PKcs is XRCC4 (for X-ray repair complementing group 4 protein), a nuclear phospho-protein, which forms a tight complex with DNA ligase IV (1, 14). The DNA ligase IV gene at 13q33-34 encodes a 96-kDa protein (15-17). The N-terminal half contains the catalytic core while the C-terminal half contains two BRCT domains, which bind XRCC4. The stability of XRCC4 is affected by post-translational modification in form of ubiquitylation and sumoylation (18). A backup pathway involves DNA ligase III (19). The MRN (MRE11, RAD50, NBS1) complex, which plays a key role in HRR, may also become involved in NHEJR, particularly if the DNA ends require processing before ligation (20). Similarly, ATM and H2AX appear to play a role in conjunction with the XRCC4-like factor, XLF (21).

Both HRR and NHEJR are DSB repair pathways but the former is mostly error-free whereas the latter is error-prone and may thereby enhance mutagenesis. Consequently, the relative contribution of these two competing pathways to DSB repair is important for maintaining genomic integrity. Several mechanisms are in place to direct DSBs to either HRR or NHEJR. One mechanism involves 53BP1 (p53-binding protein 1) and BRCA1 (22). In BRCA1-deficient cells, 53BP1 inhibits HRR by blocking resection of DNA breaks leading to error-prone repair by NHEJR (23, 24). The other process involves another binding partner of BRCA1, CtIP (yeast homolog Sae2), which undergoes phosphory-lation on residue 327Ser and recruits BRCA1 to shift the balance of DSB repair from NHEJR to HRR (25).

Several studies have examined the association of polymorphisms in NHEJR-related genes with breast cancer risk. A 2002 study of the KU70, Ku80, and ligase IV genes in 2205 cases and 1826 controls observed an association of a silent polymorphism in the latter gene (rs1805386; Asp568Asp) with decreased breast cancer risk (26). However, a 2006 meta-analysis of 10,979 cases and 10,423 controls and the Breast Cancer Association Consortium (BCAC; 8,933 cases, 9,874 controls) observed no association of ligase IV polymorphism rs1805386 with risk (27, 28). A study of Chinese women (254 cases, 379 controls) found an increased risk of breast cancer associated with non-synonymous polymorphisms in Ku70 (rs2267437; Cys61Gly) and XRCC4 (rs2075685; Thr1394Gly) (29). Polymorphisms in Ku80 (rs3835; G69506A), DNA-PKcs (rs2231178; C55966T), and ligase IV (rs1805388; C4062T) had no effect individually but in combination with the former two constituted a high-risk profile for developing breast cancer. A subsequent case-control study (469 cases, 740 controls) examined the risk association of the five-gene high-risk profile with the BRCA1 Glu1038Gly polymorphism. Women with at least one variant BRCA1 allele and high-risk genotypes of the NHEJR genes had a significantly increased risk (30). A study of American women who had a strong family history of breast cancer but were negative for BRCA1/2 mutations revealed significant associations with two tagging SNP haplotypes in the XRCC4 gene. One 2-locus haplotype was associated with reduced risk, the other with increased risk (31). The latter was in strong linkage disequilibrium with the XRCC4 SNP rs2075685 (Thr1394Gly), supporting the results of the Chinese study.

References

1. Jackson SP. *Sensing and repairing DNA double-strand breaks.* Carcinogenesis. 2002;23:687-96.

2. Oberle C, Blattner C. *Regulation of the DNA Damage Response to DSBs by Post-Translational Modifications.* Curr Genomics. 2010;11:184-98.

3. Valerie K, Povirk LF. *Regulation and mechanisms of mammalian double-strand break repair.* Oncogene. 2003;22:5792-812.

4. Roberts SA, Strande N, Burkhalter MD, Strom C, Havener JM, Hasty P, et al. *Ku is a 5'-dRP/AP lyase that excises nucleotide damage near broken ends.* Nature. 2010;464:1214-7.

5. Keith CT, Schreiber SL. *PIK-related kinases: DNA repair, recombination, and cell cycle checkpoints.* Science. 1995;270:50-1.

6. Lempiainen H, Halazonetis TD. *Emerging common themes in regulation of PIKKs and PI3Ks.* EMBO J. 2009;28:3067-73.

7. Perry J, Kleckner N. *The ATRs, ATMs, and TORs are giant HEAT repeat proteins.* Cell. 2003;112:151-5.

8. Bao S, Tibbetts RS, Brumbaugh KM, Fang Y, Richardson DA, Ali A, et al. *ATR/ATM-mediated phosphory-lation of human Rad17 is required for genotoxic stress responses.* Nature. 2001;411:969-74.

9. Chen J. *Ataxia telangiectasia-related protein is involved in the phosphorylation of BRCA1 following deoxyribonucleic acid damage.* Cancer Res. 2000;60:5037-9.

10. Gatei M, Zhou BB, Hobson K, Scott S, Young D, Khanna KK. *Ataxia telangiectasia mutated (ATM) kinase and ATM and Rad3 related kinase mediate phosphorylation of Brca1 at distinct and overlapping sites. In vivo assessment using phospho-specific antibodies.* J Biol Chem. 2001;276:17276-80.

11. Kim ST, Lim DS, Canman CE, Kastan MB. *Substrate specificities and identification of putative substrates of ATM kinase family members.* J Biol Chem. 1999;274:37538-43.

12. Matsuoka S, Ballif BA, Smogorzewska A, McDonald ER, 3rd, Hurov KE, Luo J, et al. *ATM and ATR substrate analysis reveals extensive protein networks responsive to DNA damage.* Science. 2007;316:1160-6.

13. Martensson S, Hammarsten O. *DNA-dependent protein kinase catalytic subunit. Structural requirements for kinase activation by DNA ends.* J Biol Chem. 2002;277:3020-9.

14. Grawunder U, Wilm M, Wu X, Kulesza P, Wilson TE, Mann M, et al. *Activity of DNA ligase IV stimulated by complex formation with XRCC4 protein in mammalian cells.* Nature. 1997;388:492-4.

15. Timson DJ, Singleton MR, Wigley DB. *DNA ligases in the repair and replication of DNA.* Mutat Res. 2000;460:301-18.

16. Tomkinson AE, Mackey ZB. *Structure and function of mammalian DNA ligases.* Mutat Res. 1998;407:1-9.

17. Wei YF, Robins P, Carter K, Caldecott K, Pappin DJ, Yu GL, et al. *Molecular cloning and expression of human cDNAs encoding a novel DNA ligase IV and DNA ligase III, an enzyme active in DNA repair and recombination.* Mol Cell Biol. 1995;15:3206-16.

18. Yurchenko V, Xue Z, Sadofsky MJ. *SUMO modification of human XRCC4 regulates its localization and function in DNA double-strand break repair.* Mol Cell Biol. 2006;26:1786-94.

19. Wang H, Rosidi B, Perrault R, Wang M, Zhang L, Windhofer F, et al. *DNA ligase III as a candidate component of backup pathways of nonhomologous end joining.* Cancer Res. 2005;65:4020-30.

20. Connelly JC, Leach DR. *Tethering on the brink: the evolution-arily conserved Mre11-Rad50 complex.* Trends Biochem Sci. 2002;27:410-8.

21. Zha S, Guo C, Boboila C, Oksenych V, Cheng HL, Zhang Y, et al. *ATM damage response and XLF repair factor are functionally redundant in joining DNA breaks.* Nature. 2011;469:250-4.

22. Boulton SJ. *DNA repair: Decision at the break point.* Nature. 2010;465:301-2.

23. Bouwman P, Aly A, Escandell JM, Pieterse M, Bartkova J, van der Gulden H, et al. *53BP1 loss rescues BRCA1 deficiency and is associated with triple-negative and BRCA-mutated breast cancers.* Nature Struct Mol Biol. 2010;17:688-95.

24. Bunting SF, Callen E, Wong N, Chen HT, Polato F, Gunn A, et al. *53BP1 inhibits homologous recombination in Brca1-deficient cells by blocking resection of DNA breaks.* Cell. 2010;141:243-54.

25. Yun MH, Hiom K. *CtIP-BRCA1 modulates the choice of DNA double-strand-break repair pathway throughout the cell cycle.* Nature. 2009;459:460-3.

26. Kuschel B, Auranen A, McBride S, Novik KL, Antoniou A, Lipscombe JM, et al. *Variants in DNA double-strand break repair genes and breast cancer susceptibility.* Hum Mol Genet. 2002;11:1399-407.

27. *Commonly studied single-nucleotide polymorphisms and breast cancer: results from the Breast Cancer Association Consortium.* J Natl Cancer Inst. 2006;98:1382-96.

28. Garcia-Closas M, Egan KM, Newcomb PA, Brinton LA, Titus-Ernstoff L, Chanock S, et al. *Polymorphisms in DNA double-strand break repair genes and risk of breast cancer: two population-based studies in USA and Poland, and meta-analyses.* Hum Genet. 2006;119:376-88.

29. Fu YP, Yu JC, Cheng TC, Lou MA, Hsu GC, Wu CY, et al. *Breast cancer risk associated with genotypic polymorphism of the nonhomologous end-joining genes: a multigenic study on cancer susceptibility.* Cancer Res. 2003;63:2440-6.

30. Bau DT, Fu YP, Chen ST, Cheng TC, Yu JC, Wu PE, et al. *Breast cancer risk and the DNA double-strand break end-joining capacity of nonhomologous end-joining genes are affected by BRCA1.* Cancer Res. 2004;64:5013-9.

31. Allen-Brady K, Cannon-Albright LA, Neuhausen SL, Camp NJ. *A role for XRCC4 in age at diagnosis and breast cancer risk.* Cancer Epidemiol Biomarkers Prev. 2006;15:1306-10.

CHAPTER 6

Cell Proliferation

6.1. ESTROGEN RECEPTOR
6.1.1. ER STRUCTURE AND FUNCTION
6.1.1.1. INTRODUCTION

Cell proliferation is driven by estradiol-estrogen receptor (E$_2$-ER)-mediated gene transcription and growth factor signaling.

The estrogen receptor (ER) is a member of the nuclear receptor superfamily for steroid and thyroid hormones, vitamin D, retinoids, and prostanoids (1, 2). The ESR1 gene at 6q25.1 extends over more than 140 kb, contains eight exons, and encodes the ERα protein, which consists of 595 amino acids with a predicted molecular weight of 66,182 daltons (Figure 6.1.1.). Comparison of the amino acid sequence of the human and chicken estrogen receptors indicated six regions of differing homology (A - F) of which only regions C and E are highly conserved (3). The N-terminal A/B region contains a ligand-independent transactivation function, AF-1, which is required for maximal ER function (4). Region C, which has a characteristic helix-loop-helix structure stabilized by two zinc atoms, contains the DNA-binding domain that determines the specificity of target gene activation (5). The hinge region D contains nuclear localization signals (6). The E and F domains are involved in ligand-binding and also contain a ligand-dependent transcription activation function, AF-2 (7-10).

The interaction of several domains is necessary for the estrogen-dependent activation of gene transcription which can be visualized as a series of steps (11). In the first step, estrogen binds to the ligand binding domain in region E and induces the dissociation of heat shock proteins and the formation of stable ER homodimers (12-14). In the second step, the hormone-activated ER dimer interacts via the DNA-binding domain with regulatory DNA enhancer sequences. Specific ER-binding DNA sequences have been identified in the 5' region of several estrogen-inducible target genes (15, 16). They are called estrogen response element (ERE) and are characterized as a 13-base pair palindrome with 5-bp stems separated by a 3-bp spacer and a consensus sequence of GGTCAnnnTGACC (17, 18). Additional proteins may facilitate the ER-ERE interaction (19-21). In the final step, the E$_2$-ER-ERE complex is postulated to promote the dissociation or inactivation of corepressor proteins and the recruitment of coactivator proteins, most of which interact with the AF-2 domain (22-24). This protein-DNA complex comprises the nucleus of a preinitiation complex to which RNA polymerase II and other basal transcription factors can bind efficiently to activate high levels of gene transcription (25, 26).

The transcriptional activity of the ER is regulated by several factors, such as the nature of the ligand, the phosphorylation state of the receptor, and interactions with coactivator proteins (27). The action of estrogens is antagonized by antiestrogens, which bind to ER in a manner competitive with E$_2$, but fail to effectively activate gene transcription.

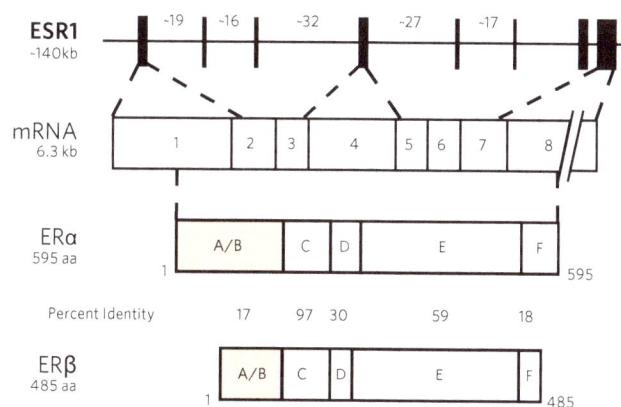

Figure 6.1.1. The ESR1 gene at 6q25.1 spans more than 140 kb and contains 8 exons.
The open reading frame begins in the middle of the first exon and is 1785 nucleotides in length. The ERα protein is composed of 595 amino acids and organized into functional domains A - F. The A/B region contains a ligand-independent transactivation function, AF-1, which is required for maximal ER function. Region C contains the DNA-binding domain that determines the specificity of target gene activation. Domain E contains the hormone-binding region as well as a ligand-dependent transcription activation function, AF-2. The ESR2 gene at 14q22-24 encodes the ERβ protein, which is organized into the same functional domains as ERα; the percent amino acid identity of each domain is indicated.

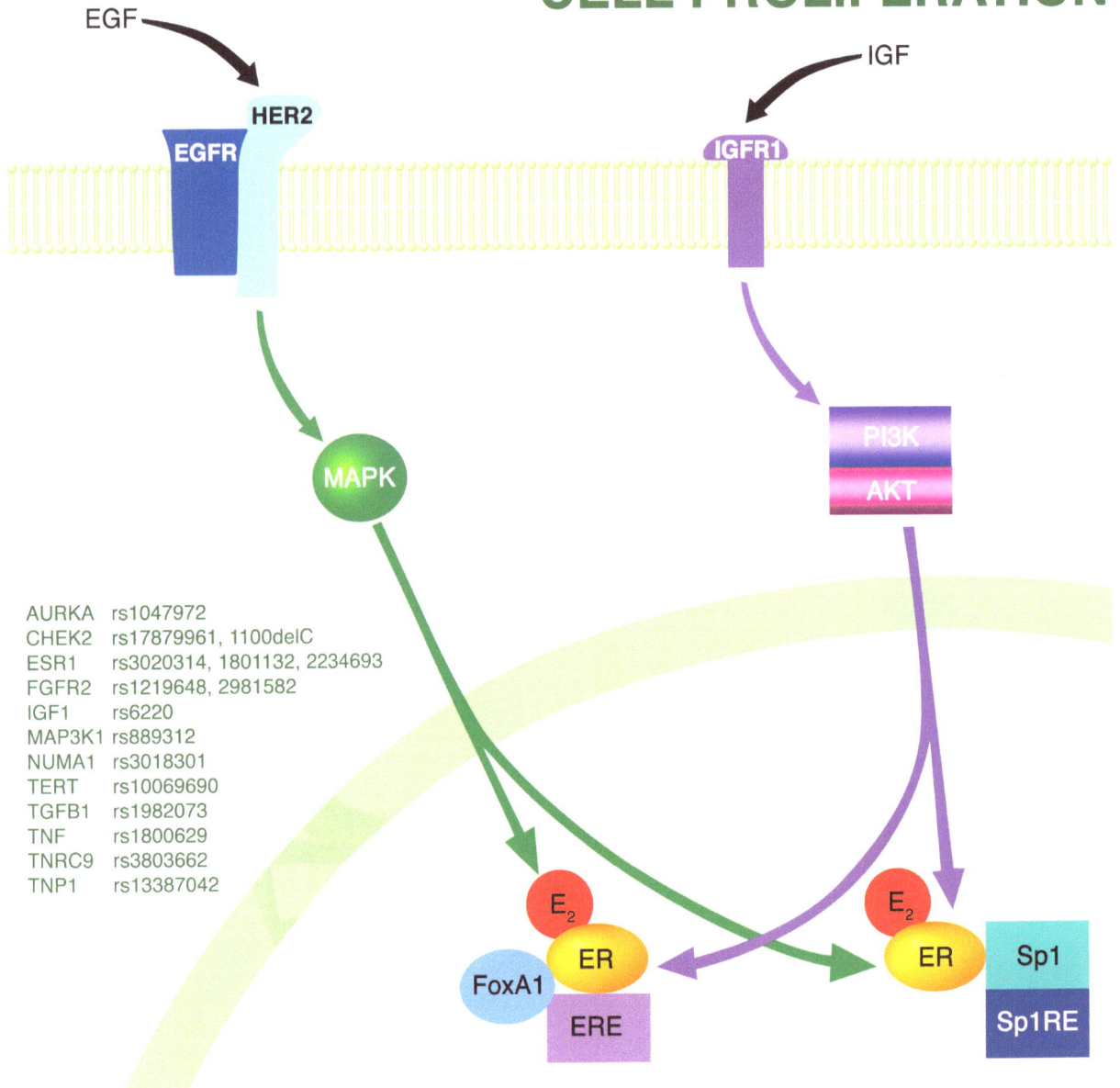

CELL PROLIFERATION

AURKA   rs1047972
CHEK2   rs17879961, 1100delC
ESR1   rs3020314, 1801132, 2234693
FGFR2   rs1219648, 2981582
IGF1   rs6220
MAP3K1   rs889312
NUMA1   rs3018301
TERT   rs10069690
TGFB1   rs1982073
TNF   rs1800629
TNRC9   rs3803662
TNP1   rs13387042

Figure 6.1.2. Overview of estradiol-estrogen receptor (E_2-ER)-mediated gene transcription and growth factor signaling in cell proliferation.

In the classical pathway of estrogen action, coactivators like FoxA1-facilitate binding of ER to estrogen response element (ERE), which triggers E_2-ER-induced gene transcription. While hormone and receptor are important, neither hormone binding nor receptor-DNA contact are essential. In an alternative pathway, ER interacts synergistically with another transcription factor, such as Sp1 that is bound to its respective response element (Sp1RE). This ERE-independent mechanism explains E_2-induced expression of certain genes without identifiable EREs. In a second alternative pathway, E_2-independent stimulation of ER-mediated gene transcription can be achieved by extracellular peptide growth factors (e.g., epidermal growth factor [EGF], insulin-like growth factor [IGF]) that interact with transmembrane growth factor receptors such as EGFR, HER2, and insulin-like growth factor receptor I (IGFR1) and other signaling molecules, including components of the mitogen-activated protein kinase (MAPK) and phosphatidylinositol 3-kinase (PI3K)/AKT pathways. These factors can activate nuclear ER through phosphorylation in the absence of hormone. Moreover, E_2 can act synergistically, promoting cell proliferation through 'cross-talk' between the signal transduction and steroid receptor pathways. The insert lists cell proliferation proteins with genetic variants implicated in breast cancer risk by genome-wide association studies. Details are discussed in subsequent sections.

Antiestrogens vary in their biological actions. Certain ones such as tamoxifen act as partial agonists/antagonists, while others such as ICI 164,384 act more as complete antagonists (28, 29). Altogether, these results can be explained by the tripartite model of steroid hormone receptor action (30). The model postulates that the biological response to a hormone signal is not just based on ligand binding and potency, but requires additional mechanistic events such as receptor conformational changes, dissociation of corepressors, and binding of coactivators in a promoter- and cell type-specific manner. Corepressors and coactivators are involved in chromatin remodeling by virtue of their enzymatic activities as histone deacetylases and acetyltransferases, histone demethylases and methyltransferases, respectively, which allows integration of nuclear receptors into the transcriptional machinery of gene expression (24, 31, 32). In this dynamic and in part cyclical process, hormone signaling may function by controlling the balance of competing enzymatic activities at a target promoter (33, 34). Figure 6.1.2. provides an overview of E_2-ER-mediated gene transcription in cell proliferation.

The ESR2 gene at 14q22-24 shows a similar functional organization as the ESR1 gene but is shorter (40 kb vs 140 kb) and encodes a shorter ERß protein (485 vs 595 amino acids) (35) (Figure 6.1.1.). The DNA-binding domain of ERß is highly homologous to ERα, differing by only two amino acids (97% identity). The high degree of conservation of the DNA-binding domain allows binding to ERE as ERß/ß homodimer or ERα/ß heterodimer (36, 37). The DNA-binding affinity of ERα/ß heterodimers was similar to that of ERα/α homodimers and greater than that of ERß/ß homodimers (38). The ligand-binding domain of ERß shows 59% amino acid identity to ERα. Not surprisingly, ERß displays similar ligand binding characteristics as ERα with high affinity for E_2 (K_d = 0.6 nM) and lower affinities for E_1 and E_3 (39). However, the shorter A/B domain, the hinge region D, and the F domain are not conserved. The dissimilarity in amino acid sequence of the A/B domain explains the difference observed in the transactivation function AF-1 of ERß and ERα.

References

1. Mangelsdorf DJ, Thummel C, Beato M, Herrlich P, Schutz G, Umesono K, et al. *The nuclear receptor superfamily: the second decade.* Cell. 1995;83:835-9.

2. Parl FF. *Estrogens, Estrogen Receptor and Breast Cancer.* Amsterdam: IOS Press; 2000.

3. Krust A, Green S, Argos P, Kumar V, Walter P, Bornert JM, et al. *The chicken oestrogen receptor sequence: homology with v-erbA and the human oestrogen and glucocorticoid receptors.* EMBO J. 1986;5:891-7.

4. Tora L, White J, Brou C, Tasset D, Webster N, Scheer E, et al. *The human estrogen receptor has two independent nonacidic transcriptional activation functions.* Cell. 1989;59:477-87.

5. Green S, Chambon P. *Oestradiol induction of a glucocorticoid-responsive gene by a chimaeric receptor.* Nature. 1987;325:75-8.

6. Ylikomi T, Bocquel MT, Berry M, Gronemeyer H, Chambon P. *Cooperation of proto-signals for nuclear accumulation of estrogen and progesterone receptors.* EMBO J. 1992;11:3681-94.

7. Danielian PS, White R, Lees JA, Parker MG. *Identification of a conserved region required for hormone dependent transcriptional activation by steroid hormone receptors.* EMBO J. 1992;11:1025-33.

8. Kumar V, Green S, Staub A, Chambon P. *Localisation of the oestradiol-binding and putative DNA-binding domains of the human oestrogen receptor.* EMBO J. 1986;5:2231-6.

9. Pierrat B, Heery DM, Chambon P, Losson R. *A highly conserved region in the hormone-binding domain of the human estrogen receptor functions as an efficient transactivation domain in yeast.* Gene. 1994;143:193-200.

10. Webster NJ, Green S, Jin JR, Chambon P. *The hormone-binding domains of the estrogen and glucocorticoid receptors contain an inducible transcription activation function.* Cell. 1988;54:199-207.

11. Truss M, Beato M. *Steroid hormone receptors: interaction with deoxyribonucleic acid and transcription factors.* Endocrine Rev. 1993;14:459-79.

12. Kumar V, Chambon P. *The estrogen receptor binds tightly to its responsive element as a ligand-induced homodimer.* Cell. 1988;55:145-56.

13. Linstedt AD, West NB, Brenner RM. *Analysis of monomeric-dimeric states of the estrogen receptor with monoclonal antiestrophilins.* J Steroid Biochem. 1986;24:677-86.

14. Smith D, Toft D. *Steroid receptors and their associated proteins.* Mol Endocrinol. 1993;7:4-11.

15. Martinez E, Wahli W. *Cooperative binding of estrogen receptor to imperfect estrogen-responsive DNA elements correlates with their synergistic hormone-dependent enhancer activity.* EMBO J. 1989;8:3781-91.

16. Seiler-Tuyns A, Walker P, Martinez E, Merillat A-M, Givel F, Wahli W. *Identification of estrogen-responsive DNA sequences by transient expression experiments in a human breast cancer cell line.* Nucleic Acids Res. 1986;14:8755-70.

17. Klein-Hitpass L, Schorpp M, Wagner U, Ryffel GU. *An estrogen-responsive element derived from the 5' flanking region of the Xenopus vitellogenin A2 gene functions in transfected human cells.* Cell. 1986;46:1053-61.

18. Klock G, Strahle U, Schutz G. *Oestrogen and glucocorticoid responsive elements are closely related but distinct.* Nature. 1987;329:734-6.

19. Landel CC, Kushner PJ, Greene GL. *The interaction of human estrogen receptor with DNA is modulated by receptor-associated proteins.* Mol Endocrinol. 1994;8:1407-19.

20. Romine LE, Wood JR, Lamia JA, Prendergast P, Edwards DP, Nardulli AM. *The high mobility group protein 1 enhances binding of the estrogen receptor DNA binding domain to the estrogen response element.* Mol Endocrinol. 1998;12:664-74.

21. Verrier CS, Roodi N, Yee CJ, Bailey R, Jensen RA, Bustin M, et al. *High-mobility group (HMG) protein HMG-1 and TATA-binding protein-associated factor TAFII30 affect estrogen receptor-mediated transcriptional activation.* Mol Endocrinol. 1997;11:1009-19.

22. Horwitz KB, Jackson TA, Bain DL, Richer JK, Takimoto GS, Tung L. *Nuclear receptor coactivators and corepressors.* Mol Endocrinol. 1996;10:1167-77.

23. Watson PJ, Fairall L, Schwabe JW. *Nuclear hormone receptor co-repressors: structure and function.* Mol Cell Endocrinol. 2012;348:440-9.

24. York B, O'Malley BW. *Steroid receptor coactivator (SRC) family: masters of systems biology.* J Biol Chem. 2010;285:38743-50.

25. Bagchi MK, Tsai M, O'Malley BW, Tsai SY. *Analysis of the mechanism of steroid hormone receptor-dependent gene activation in cell-free systems.* Endocrine Rev. 1992;13:525-35.

26. Ptashne M, Gann A. *Transcriptional activation by recruitment.* Nature. 1997;386:569-77.

27. Beato M, Herrlich P, Schutz G. *Steroid hormone receptors: many actors in search of a plot.* Cell. 1995;83:851-7.

28. Jordan VC, Murphy CS. *Endocrine pharmacology of antiestrogens as antitumor agents.* Endocrine Rev. 1990;11:578-610.

29. Montano MM, Ekena K, Krueger KD, Keller AL, Katzenellenbogen BS. *Human estrogen receptor ligand activity inversion mutants: receptors that interpret antiestrogens as estrogens and estrogens as antiestrogens and discriminate among different antiestrogens.* Mol Endocrinol. 1996;10:230-42.

30. Katzenellenbogen JA, O'Malley BW, Katzenellenbogen BS. *Tripartite steroid hormone receptor pharmacology: interaction with multiple effector sites as a basis for the cell-and promoter-specific action of these hormones.* Mol Endocrinol. 1996;10:119-31.

31. Chen JD, Li H. *Coactivation and corepression in transcriptional regulation by steroid/nuclear hormone receptors.* Crit Rev Eukaryotic Gene Expression. 1998;8:169-90.

32. Perlmann T, Evans RM. *Nuclear receptors in Sicily: all in the famiglia.* Cell. 1997;90:391-7.

33. Carlberg C, Seuter S. *Dynamics of nuclear receptor target gene regulation.* Chromosoma. 2010;119:479-84.

34. Metivier R, Penot G, Hubner MR, Reid G, Brand H, Kos M, et al. *Estrogen receptor-alpha directs ordered, cyclical, and combinatorial recruitment of cofactors on a natural target promoter.* Cell. 2003;115:751-63.

35. Enmark E, Pelto-Huikko M, Grandien K, Lagercrantz S, Langercrantz J, Fried G, et al. *Human estrogen receptor b-gene structure, chromosomal localization, and expression pattern.* J Clin Endocrinol Metabol. 1997;82:4258-65.

36. Pace P, Taylor J, Suntharalingam S, Coombes RC, Ali S. *Human estrogen receptor beta binds DNA in a manner similar to and dimerizes with estrogen receptor alpha.* J Biol Chem. 1997;272:25832-8.

37. Pettersson K, Grandien K, Kuiper GJM, Gustafsson J. *Mouse estrogen receptor b forms estrogen response element-binding heterodimers with estrogen receptor a.* Mol Endocrinol. 1997;11:1486-96.

38. Cowley SM, Hoare S, Mosselman S, Parker MG. *Estrogen receptors alpha and beta form heterodimers on DNA.* J Biol Chem. 1997;272:19858-62.

39. Kuiper GJM, Carlsson B, Grandien K, Enmark E, Haggblad J, Nilsson S, et al. *Comparison of the ligand binding specificity and transcript tissue distribution of estrogen receptors alpha and beta.* Endocrinology. 1997;138:863-70.

6.1.1.2. ER DOMAIN STRUCTURE AND FUNCTION

The ER binds E_2 with high affinity (K_d = 0.3 – 0.5 nM) at a single binding site in the **ligand binding domain (LBD)** of region E (Figure 6.1.1.). Binding occurs in a positive cooperative manner as a result of receptor dimerization (1). The cooperative binding mechanism involves site-site interactions between monomers of the dimeric ER in which E_2 binding by one monomer induces conformational change in the dimeric receptor that results in an increased affinity of the second monomer for E_2. The binding is stereospecific since 17α-estradiol is bound with two times lower affinity than E_2 (2). Most ER ligand binding studies done in the past actually involved mixtures of ERα and ERβ protein, such as those with rat uterus cytosol (3). Ligand binding analysis of *in vitro* synthesized ERα showed the following order of affinities for physiological estrogens: E_2 > E_1, 17α-estradiol > E_3 > catechol estrogens > E_1S. Stilbene estrogens, which consist of a composite diphenolic ring structure, bind with the following order: diethylstilbestrol > hexestrol > dienestrol > E_2. Triphenyl-ethylene antiestrogens (Figure 6.1.3.) bind with the following affinities: 4-hydroxytamoxifen >> nafoxidene > clomiphene > tamoxifen. Androgens have no significant affinity for ER except those with a hydroxyl group at C3 and C17 (5-androstenediol, 3β-androstanediol), consistent with the estrogen-like effect of the latter on uterus and mammary gland (4, 5). Progesterone and progestins derived from 19-nor-testosterone (norethynodrel, norethindrone) have negligible affinity for ER (3). The estrogenic potential of the latter on the induction of alkaline phosphatase activity in ER-positive Ishikawa endometrial cancer cells appears to result from *in vivo* conversion into more active metabolites by aromatization or hydroxylation at C3 (6).

It has been difficult to identify those amino acids that are critical for E_2 binding because region E contains overlapping domains not only for hormone binding but also for transcriptional activation (AF-2), receptor dimerization, and heat shock protein binding. In addition, the LBD is the largest of the ER domains, extending for about 250 amino acids from 302Lys to 553Thr in region E (Table 6.1.1.). The crystal structure of the ERα LBD has been determined in the presence of E_2, the agonist DES, and the antiestrogens 4-hydroxytamoxifen and raloxifene revealing a wedge-shaped molecule folded into a three-layered sandwich composed of 12α-helices (H1 - H12) and a single β-sheet (S1/S2) (Table 6.1.1.) (7-9). Two LBDs are arranged as homodimer with each monomer contributing about 15 percent of its solvent-accessible surface area to the dimer interface. Contacts between the monomers are made primarily through the H11 helices but also involve H8 from one monomer and H9 and H10 from the other monomer. The hormone binding cavity of 450 Å3 in the hydrophobic core of the LBD is completely partioned from the external environment. It is located at one end of the molecule and is formed by parts of H3 (342Met to 354Leu), H6 (383Trp to 394Arg), H8 and the preceding loop (418Val to 428Leu), H11 (517Met to 528Met), H12 (539Leu to 547His), and the S1/S2 hairpin (402Leu to 410Leu). Within the cavity the hormone is oriented by two types of contacts, i.e., hydrogen bonding at the two ends and hydrophobic van der Waals contacts along the body of the steroid molecule. The phenolic 3-hydroxyl group of the A-ring nestles between H3 and H6 and makes direct hydrogen bonds to the γ-carboxylate of 353Glu, the guanidinium group of 394Arg, and a structurally conserved water molecule that is part of a solvent channel extending from the A-ring to the exterior of the LBD. The 17β-hydroxyl of the D-ring makes a single hydrogen bond to the δ-nitrogen of 524His in H11. The hydrophobic contacts are concentrated over the A- and D-rings. The A-ring is sandwiched between the side chains of 350Ala and 387Leu on its β face and 404Phe on its α face. The D-ring makes nonpolar contacts with 424Ile, 521Gly, and 525Gly. The molecular volume of E_2 (245 Å3) is only about half the size of the hormone binding cavity. There are large, unoccupied regions opposite the α face of the B-ring and the β face of the C-ring that can accept a number of different hydrophopbic groups from a variety of nonsteroidal compounds such as DES, 4-hydroxytamoxifen, and raloxifene. For example, the ethyl groups of DES, which project perpendicular from the plane of the phenolic rings, fit snugly into these spaces. The resulting additional nonpolar contacts with the side chains of 350Ala, 384Leu, 404Phe, and 428Leu may account for the higher affinity of DES for the receptor (3).

Besides the dimerization interface and the hormone binding cavity, the LBD contains AF-2, the transcriptional activation function that interacts with a number of transcriptional coactivators in a ligand-dependent manner (10, 11). A subregion, AF-2a, in the N-terminal part of the LBD is formed by H3, H4, and the intervening loop which contains a 'signature sequence' with the highly conserved 362Lys (12, 13). The core of AF-2 is located in H12, which also forms the 'lid' of the hormone binding cavity in the E$_2$-LBD complex without actually making contact with E$_2$ (7). The inner hydrophobic surface of H12 projects toward the bound hormone while the charged surface with the highly conserved 542Glu is directed away from the body of the LBD on the side of the molecule lying perpendicular to the dimerization interface. This precise hormone-induced positioning of H12 seems to be a prerequisite for transcriptional activation since, by sealing the hormone binding cavity, it generates a competent AF-2 that is capable of interacting with coactivators. Both 362Lys and 542Glu are

highly conserved and mutation of either residue interfered with coactivator recognition (12, 14).

The LBD crystal structure changes drastically in the presence of antiestrogens such as 4-hydroxytamoxifen and raloxifene (Figure 6.1.3.) which are partially accommodated in the hormone binding cavity (7, 8). However, the long side chains of these antiestrogens cannot be accommodated; they protrude from the hormone binding cavity and thereby prevent H12 from covering the cavity as it does in the E$_2$-LBD complex. Instead, H12 is displaced and occupies the coactivator binding groove which, in turn, prevents binding of coactivators and transcriptional activation via AF-2. Thus, antiestrogen binding promotes a conformation of the LBD that is different from that stabilized by either E$_2$ or DES. Other structural features besides length of the side chain can determine antiestrogen activity. For example, the piperazine ring nitrogen of raloxifene must form a hydrogen bond with 351Asp in the LBD in order to achieve antagonistic action. The single substitution Asp351Tyr prevents hydrogen bonding and changes the pharmacology of raloxifene from antiestrogen to estrogen (15).

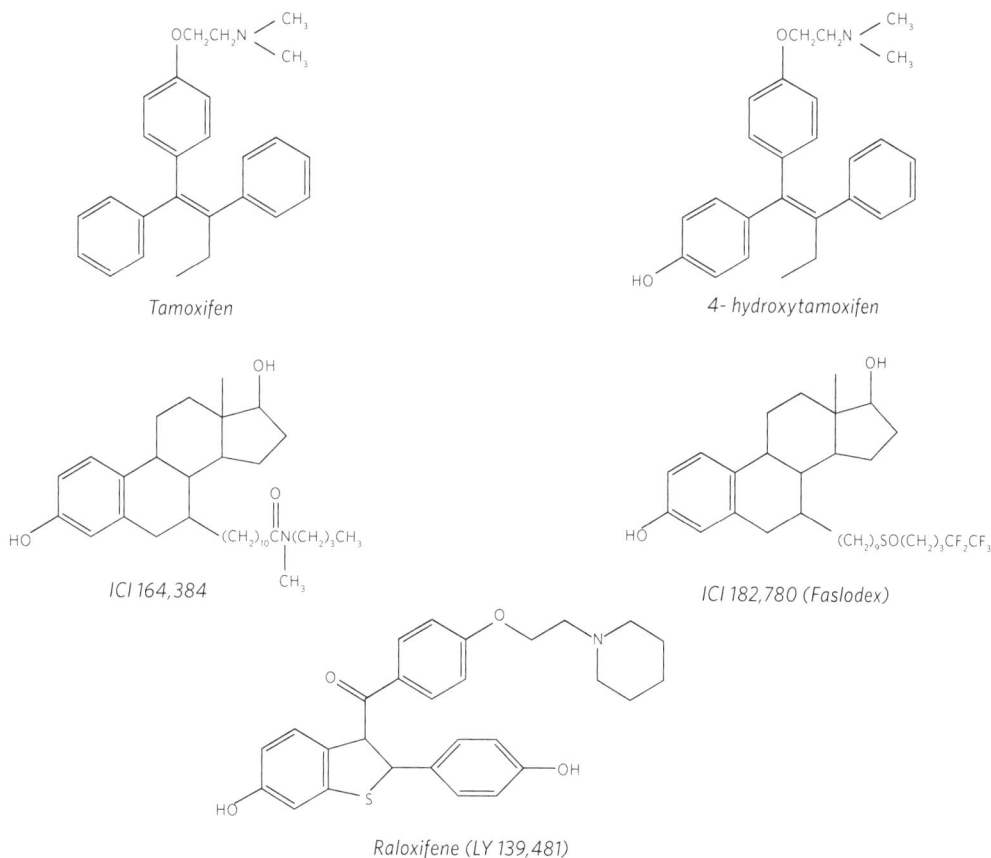

Figure 6.1.3. Chemical structures of selected antiestrogens

Table 6.1.1. Overview of Ligand Binding Domain

Crystal Structure	General Function	Codon	Specific Function
H3		350Ala	van der Waals contact to E_2 (β face of A-ring)
		351Asp	hydrogen bond to Raloxifene
		353Glu	hydrogen bond to phenolic 3-OH group of E_2
		358Ile	
H4		362Lys	
	AF-2a	364Val	coactivator binding
H5		376Val	
		379Leu	
		380Glu	coactivator binding
H6		387Leu	van der Waals contact to E_2 (β face of A-ring)
		394Arg	hydrogen bond to phenolic 3-OH group of E_2
		400Ala	Ala400Val decreases E_2 binding at room temperature
	S1/S2 hairpin	404Phe	van der Waals contact to E_2 (α face of A-ring)
H7			
H8	Dimerization	424Ile	van der Waals contact to E_2 (D-ring)
H9	interface	447Cys	palmitoylation for plasma membrane localization
H10			
H11		521Gly	van der Waals contact to E_2 (D-ring)
		524His	hydrogen bond to 17β-OH
		525Leu	van der Waals contact to E_2 (D-ring)
		530Cys	covalent binding site for tamoxifen aziridine
		537Tyr	phosphorylation site independent of E_2 binding
		538Asp	
H12	AF-2	542Glu	'lid' of hormone binding cavity and coactivator binding
		543Met	
		544Leu	

The binding of E_2 is regulated by phosphorylation of 537Tyr, the only phosphorylated tyrosine residue in the ER protein (16, 17). Although 537Tyr is located in the LBD, its phosphorylation is independent of E_2 binding, suggesting that the phosphorylation is regulated by cellular signaling pathways not involving its cognitive ligand, E_2 (18). Indeed, two src family tyrosine kinases, p60c-src and p56lck, were shown to phosphorylate 537Tyr in purified recombinant ER whereas two tyrosine phosphatases, protein tyrosine phosphatase-1B and src homology-2 protein tyrosine phosphatase-1, dephosphorylated phosphotyrosine 537 *in vitro*. Dephosphorylation of wild type ER with tyrosine phosphatase-1B severely reduced E_2 binding. Conversely, site-specific rephosphorylation of 537Tyr with p60c-src almost completely restored E_2 binding to wild type levels (19). It appears that 537Tyr phosphorylation enhances ER dimerization and site-site interaction resulting in cooperative E_2 binding (20). Replacement of 537Tyr by amino acids whose side chains have reduced size or decreased hydrophobicity (e.g., Ala, Ser, Asp) generated constitutively active ER, i.e., the same transcriptional activity in the absence or presence of E_2. These findings suggest that 537Tyr is required to maintain ER in a transcriptionally inactive state in the absence of hormone, possibly by stabilizing a conformation of the LBD in which the hormone binding cavity is collapsed

without E_2 (21, 22). Indeed, structural analysis suggests that 537Tyr participates in the shift of H12 towards the LBD core that creates a transcriptionally active receptor (13). In three-dimensional space, ER residues 537Tyr and 380Glu are within secondary structural elements that come together when ER is activated. The importance of 537Tyr and 380Glu was underlined by the finding that the mutation of either residue affected the transcriptional activity of ER (23). These studies emphasize the complexity of the LBD and illustrate the importance of individual amino acid residues located in this domain for functions other than hormone recognition.

The C-terminal F domain consists of approximately 40 amino acids (residues 554 - 595) that are not well conserved among ER proteins of different species. The F domain plays a role in maintaining the conformation of the AF-2 region and appears to have a modulatory function in regulating the transcriptional response to estrogens and antiestrogens which depend greatly on the cell examined (24, 25).

The **DNA binding domain (DBD**; amino acids 180 to 262, region C) is the most highly conserved domain among members of the nuclear receptor superfamily. The core of region C consists of two subregions CI (amino acids 185-215) and CII (amino acids 216-250), each containing

Table 6.1.2. Overview of DNA-Binding Domain

Codon	Domain	Function
185Cys		coordinates Zn in first zinc finger
188Cys		coordinates Zn in first zinc finger
202Cys		coordinates Zn in first zinc finger
203Glu		(1) ER-ERE binding: accepts hydrogen bond from C4
		(2) P-box ER vs GR half-site specificity
204Gly	CI	P-box ER vs GR half-site specificity
205Cys		coordinates Zn in first zinc finger
206Lys		ER-ERE binding: donates hydrogen bond to G-5
207Ala		P-box ER vs GR half-site specificity
210Lys		ER-ERE binding: forms hydrogen bonds with T-3 and G-4
211Arg		ER-ERE binding: donates one direct and one water-mediated hydrogen bond to G+2
221Cys		coordinates Zn in second zinc finger
222Pro		
223Ala		
224Thr		
225Asn		
226Gln	CII	D-box participates in dimerization interface
227Cys		coordinates Zn in second zinc finger
237Cys		coordinates Zn in second zinc finger
240Cys		coordinates Zn in second zinc finger
245Cys		ninth Cys in DNA-binding domain
256Arg		
260Arg		nuclear localization signal (NLS) 3
265Leu		
273Asp	D	NLS 2
299Lys		
303Lys		NLS 1

four cysteine residues which tetrahedrally coordinate zinc ion to form a 'zinc finger' (26) (Table 6.1.2.). The zinc finger motif is highly conserved between different nuclear receptors (27, 28). Removal of zinc by dialysis at low pH in the presence of chelating agents yields an apoprotein that is unable to bind DNA. Thus, the metal ion is an essential cofactor for DNA binding, apparently by maintaining the protein in its native, active form (29).

The three-dimensional structure of the DBD has been solved by nuclear magnetic resonance imaging and x-ray crystallography of the DBD-ERE complex (30, 31). While the two fingers share the tetrahedral zinc coordination structure, they differ in many other respects. Each finger is encoded by a separate exon, i.e. CI by ER exon 2 and CII by ER exon 3. While the first finger contains several hydrophobic residues, the second is more basic. The first finger contains two amino acids between each pair of four coordinating cysteines; the second has five residues between the first two coordinating cysteines. While the first finger has four cysteines, the second contains five absolutely conserved cysteines. Of the five cysteines in CII only the first four are part of the tetrahedral coordination complex. The fifth, 245Cys, is in the middle of an α-helix

critical for the proper folding of the DBD (31). All these differences suggest that the two zinc fingers represent more or less independent substructures within the DBD. Indeed, experiments with chimaeric ERs indicate that it is the first zinc finger which largely determines the specificity of ER binding to ERE (26). Mutagenesis experiments have identified three amino acids, 203Glu, 204Gly and 207Ala in the so-called P-box, at the C-terminal side of the first finger that play a key role in distinguishing between an ERE and GRE (32). A substitution of these three residues with the corresponding GR residues, 439Gly, 440Ser, 443Val, completely changes specificity so that this mutant transactivates strongly from a GRE-driven reporter and not at all from an ERE. Nuclear magnetic resonance and crystallographic studies of the ER and GR suggest that 203Glu, in particular, specifies ER-ERE recognition due to a direct side-chain contact to the amino group of cytosine (C+4), one of two bases within the ERE half-site (TGACC) that distinguishes it from a GRE (TGTTCT) (31, 33). Additional side-chain contacts are made by two residues in the α-helix of the first finger, 206Lys and 211Arg, both of which are conserved throughout the nuclear receptor family. In summary, amino acids at the C-terminal base of the first finger in ER, GR and probably all other nuclear receptors make direct contact with and discriminate between the bases of their respective response elements (34).

Experiments with ER and GR constructs containing only the DBD have shown that the domains, even at millimolar concentrations, are monomeric in solution (31, 35). Upon addition of the respective response element, the DBDs bind cooperatively as dimers, i.e., the interaction of one ER DBD monomer to one of two ERE half-sites facilitates the binding of the second monomer to the second half-site (36). The fact that the DBD alone cannot dimerize but requires association with a palindromic ERE suggests that the architecture of both the protein and DNA are important for dimerization. In fact, both the second zinc finger and the response element spacer contribute to the correct stereo-specificity and are essential for stabilizing the interactions required for the head-to-head symmetrical orientation of the two monomers on each half-site.

Mutagenesis experiments have identified five amino acids (termed the 'D box') at the N-terminal base of the second zinc finger, between coordinating 221Cys and 227Cys, as dimer interface (37). Crystallographic analysis of the GR-GRE complex demonstrates that amino acids in the D box of one monomer make direct contact, via salt bridges and hydrogen bonds, with residues in the D box of the opposing monomer. Thus, the dyads of protein and DNA are aligned by correct half-site spacing of three nucleotides, allowing specific contacts between first finger residues and bases and phosphates of the response element. If the half-sites are separated by two or four nucleotides, the monomers are not in complementary register with their binding sites. In summary, DBD and ERE form an intimately aligned unit with the first finger being the principal determinant of specificity and the second finger providing the dimer interface that ensures a twofold symmetrical orientation of ER monomers in register with properly spaced ERE half-sites.

The ER interacts not only with the nucleotide bases of the ERE but also with the phosphates of the DNA backbone. Phosphate ethylation interference analysis revealed a total of eight phosphate contact points for the perfect ERE (38, 39). These were symmetrically distributed about the palindromic sequence, producing a twofold axis of symmetry, consistent with the binding of an ER dimer. Furthermore, the footprint pattern displayed a 5' stagger, which is indicative of major groove binding. Accordingly, when the contact points are projected on a B-DNA helix model, the interfering phosphates are seen to straddle successive major grooves and all the contact points lie on one half of the cylindrical surface when viewed down the helical axis. This would suggest that the ER dimer is bound predominantly to one face of the DNA helix. Purine methylation interference experiments also indicate that the ERE contact sites are on one face of the DNA helix,

extending over one and one half successive turns of the major groove (40). When the ER dimer is bound to ERE, the DBD is inaccessible to site-specific antibodies (41).

In addition to the 66-amino acid zinc finger core (residues 185 – 240), amino acids C-terminal to the core in regions C and D play a role in DNA binding (42). Up to 30 adjacent C-terminal residues greatly stabilize DNA binding by the DBD to perfectly palindromic EREs and are absolutely required for formation of electrophoretic mobility shift complexes by the DBD on certain physiological imperfectly palindromic EREs. The three-dimensional topography of ER-ERE contacts is even more complex due to participation of the E region in ER dimerization and high affinity DNA binding (43). Analysis of the E region in mouse ER revealed a heptad repeat of hydrophobic residues that is conserved in all members of the nuclear receptor family. ER constructs that include this sequence have an affinity for DNA of about 10^{-9} M, more than 10-fold higher than the affinity of ER constructs containing the C domain alone. Thus, while the zinc finger region of the C domain is sufficient for specific recognition of ERE, the stability of the ER-ERE complex is increased by the heptad repeat sequence in region E. Specific point mutations introduced in the sequence at residues 507Arg, 511Leu, or 518Ile prevent ER dimerization and ERE binding suggesting that there is a direct correlation between the ability of mutants to dimerize and their ability to bind to DNA. Although the zinc finger region and the dimerization domain are separated by a distance of more than 250 amino acids, these data suggest that in the tertiary ER structure zinc fingers and dimerization domain are spatially juxtaposed to permit dimer formation and high affinity ERE binding.

Region C also contains one of three constitutive nuclear localization signals (NLSs), NLS 3, between residues 256 and 260 (Table 6.1.2.). The other two are located in the hinge region D, NLS 2 (residues 265 – 273) and NLS 1 (residues 299 – 303). A fourth NLS, which is estrogen-inducible, is present in the LBD (44). The three constitutive NLSs share lysine/arginine-rich clusters characteristic of the classical NLS motif Lys-Lys/Arg-X-Lys/Arg (45). All four NLSs cooperate in nuclear targeting, while none of them individually is sufficient for achieving wild-type nuclear ER localization (44). Unlike the highly conserved C and E regions, the D domain sequence varies, although its length is conserved across species (46). Maintaining the proper distance between DBD and LBD may be an important role of the hinge domain D in overall receptor function, necessary for distinguishing estrogen agonists from antagonists (47).

The ER stimulates transcription by means of two **transcriptional activation functions, AF-1 and AF-2** (Figure 6.1.1.). AF-1, in the N-terminal region A/B, is constitutively active (i.e., hormone independent), while the activity of AF-2 in region E requires binding of E$_2$ (36, 48-50). The activities of AF-1 and AF-2 vary depending upon the responsive promoter and cell type and, in some cases, both are required for full transcriptional activation (49, 51, 52).

The A/B region contains four serine residues (104Ser, 106Ser, 118Ser, and 167Ser) which become phosphorylated upon estrogen binding to the LBD in region E (53-56). The estrogen-inducible phosphorylation of these sites is associated with a conformational change in protein structure and important for the transcription activation function of AF-1. For example, a Ser118Ala mutation, which prevents phosphorylation caused a significant reduction in transcriptional activation by ER from reporter genes containing an ERE, but did not affect the DNA binding properties or nuclear localization of ER. Thus, phosphorylation of AF-1 may regulate the transcriptional activity of ER. The modulation of transcriptional activation by ligand-dependent phosphorylation appears to be residue- and cell specific. For example, 118Ser phosphorylation was observed in HeLa cells but not in chicken embryo fibroblasts (53). The five amino acids located around 118Ser (PQLSP, Pro-Gln-Leu-Ser-Pro) are conserved among vertebrate species (human, mouse, rat, chick, *Xenopus*, trout) and correspond to the consensus phosphorylation site [PXn(S or T)P] (where X is a neutral or basic amino acid and n = 1 or 2) for mitogen-activated protein kinase (MAPK). MAPK phosphorylates 118Ser *in vitro* and in cells treated with epidermal growth factor and insulin-like growth factor *in vivo* (57). Overexpression of MAPK kinase (MAPKK) or of the guanine nucleotide binding protein Ras, both of which activate MAPK, enhanced estrogen- and tamoxifen-induced transcriptional activity of wild-type ER, but not that of a mutant ER with Ser118Ala. Thus, the activity of AF-1 is modulated by phosphorylation of 118Ser through the membrane-associated receptor tyrosine kinase-Ras-Raf-MAPK cascade suggesting crosstalk between the E_2-nuclear ER signaling pathway and the growth factor signaling pathways (58, 59). Other kinases also phosphorylate the serine residues in the A/B region. Ku70/80, the catalytic subunit of the DNA-dependent protein kinase (DNA-PKcs), which plays a key role in the non-homologous end-joining (NHEJ) pathway of double-strand break repair, was shown to form a complex with ER and phosphorylate 118Ser (60). The phosphorylation stabilized the receptor, activated its transcriptional function and promoted E_2-induced proliferation in breast cancer cells. Casein kinase phosphorylated ER primarily at 167Ser (54). In summary, the site-specific phosphorylation of ERα on serine and tyrosine residues (104Ser, 106Ser, 118Ser, 167Ser, 537Tyr) influences receptor expression, stability, transcriptional activity, and protein-protein interaction (61).

The activity of AF-2 is hormone-dependent *in vivo* and may result from the cooperative function of two activating domains contained within the E region, one being located at the N-terminal (AF-2a) and the other at the C-terminal end (AF-2) of region E (10, 11, 62). The highly conserved AF-2a sequence interacts specifically with one of the TBP-associated factors, TAFII30, which is required for transcriptional activation by the ER (63). A unique ER mutant, Val364Glu, exhibits transcriptional superactivity (up to 250% of wild type ER activity), yet it is able to strongly inhibit wild type ER when both are coexpressed in cells (64). The fact that the mutant receptor can show transcriptional activity greater than that of wild type ER suggests that 364Val is in a region of the ER important in interactions with transcription factors and coactivators (Table 6.1.1.). Sixty amino acids at the far C-terminal end of the E-region contain part of the AF-2 activity and a 14-amino acid stretch within this region is conserved among steroid receptor family members (10, 48). The major feature of the sequence, which extends from amino acid 534 to 548 is an invariant 542Glu (Table 6.1.1.). Mutation of 542Glu, which is conserved throughout the steroid/thyroid receptor family, abolished transcription. 546Glu is flanked by two pairs of hydrophobic residues (539Leu, 540Leu, 543Met and 544Leu). Replacement of either pair of hydrophobic residues with alanines abolished transcriptional activation without significantly affecting steroid and DNA binding functions, indicating that these residues are specifically involved in transcriptional activation. AF-1 and AF-2 synergize intramolecularly (49). When AF-1 and AF-2 were expressed as separate polypeptides in mammalian cells, they exhibited functional interaction in response to E_2, evidence of the evolutionarily conserved modular structural and functional organization of the ER (65, 66). The interaction was transcriptionally productive only in response to E_2, and was eliminated by point or deletion mutations that destroyed AF-1 or AF-2 activity or E_2 binding. Thus, hormone binding has a definitive mechanistic role in the activity of ER, namely, to alter receptor conformation to promote an association of the N- and C-terminal regions, leading to transcriptional synergism between AF-1 and AF-2.

References

1. Notides AC, Lerner N, Hamilton DE. *Positive cooperativity of the estrogen receptor.* Proc Natl Acad Sci USA. 1981;78:4926-30.

2. Noteboom WD, Gorski J. *Stereospecific binding of estrogens in the rat uterus.* Arch Biochem Biophys. 1965;111:559-68.

3. Kuiper GJM, Carlsson B, Grandien K, Enmark E, Haggblad J, Nilsson S, et al. *Comparison of the ligand binding specificity and transcript tissue distribution of estrogen receptors alpha and beta.* Endocrinology. 1997;138:863-70.

4. Garcia M, Rochefort H. *Evidence and characterization of the binding of two 3H-labeled androgens to the estrogen receptor.* Endocrinology. 1979;104:1797-804.

5. van Doorn LG, Poortman J, Thijssen JH, Schwarz F. *Actions and interactions of delta 5-androstene-3 beta, 17 beta-diol and 17 beta-estradiol in the immature rat uterus.* Endocrinology. 1981;108:1587-93.

6. Botella J, Duranti E, Viader V, Duc I, Delansorne R, Paris J. *Lack of estrogenic potential of progesterone- or 19-nor-progesterone-derived progestins as opposed to testosterone or 19-nor-testosterone derivatives on endometrial Ishikawa cells.* J Steroid Biochem Mol Biol. 1995;55:77-84.

7. Brzozowski AM, Pike ACW, Dauter Z, Hubbard RE, Bonn T, Engstrom O, et al. *Molecular basis of agonism and antagonism in the oestrogen receptor.* Nature. 1997;389:753-8.

8. Shiau AK, Barstad D, Loria PM, Cheng L, Kushner PJ, Agard DA, et al. *The structural basis of estrogen receptor/coactivator recognition and the antagonism of this interaction by tamoxifen.* Cell. 1998;95:927-37.

9. Tanenbaum DM, Wang Y, Williams SP, Sigler PB. *Crystallographic comparison of the estrogen and progesterone receptor's ligand banding domains.* Proc Natl Acad Sci USA. 1998;95:5998-6003.

10. Danielian PS, White R, Lees JA, Parker MG. *Identification of a conserved region required for hormone dependent transcriptional activation by steroid hormone receptors.* EMBO J. 1992;11:1026-33.

11. Pierrat B, Heery DM, Chambon P, Losson R. *A highly conserved region in the hormone-binding domain of the human estrogen receptor functions as an efficient transactivation domain in yeast.* Gene. 1994;143:193-200.

12. Henttu PM, Kalkhoven E, Parker MG. *AF-2 activity and recruitment of steroid receptor coactivator 1 to the estrogen receptor depend on a lysine residue conserved in nuclear receptors.* Mol Cell Biol. 1997;17:1832-9.

13. Wurtz JM, Bourguet W, Renaud JP, Vivat V, Chambon P, Moras D, et al. *A canonical structure for the ligand-binding domain of nuclear receptors.* Nature Struct Biol. 1996;3:87-94.

14. Feng W, Riberio RCJ, Wagner RL, Nguyen H, Apriletti JW, Fletterick RJ, et al. *Hormone-dependent coactivator binding to a hydrophobic cleft on nuclear receptors.* Science. 1998;280:1747-9.

15. Levenson AS, Jordan VC. *The key to the antiestrogenic mechanism of raloxifene is amino acid 351 (asparate) in the estrogen receptor.* Cancer Res. 1998;58:1872-5.

16. Castoria G, Migliaccio A, Green S, Di Domenico M, Chambon P, Auricchio F. *Properties of a purified estradiol-dependent calf uterus tyrosine kinase.* Biochemistry. 1993;32:1740-50.

17. Migliaccio A, Di Domenico M, Green S, De Falco A, Kajtaniak EL, Blasi F, et al. *Phosphorylation on tyrosine of in vitro synthesized human estrogen receptor activates its hormone binding.* Mol Endocrinol. 1989;3:1061-9.

18. Arnold SF, Obourn JD, Jaffe H, Notides AC. *Phosphorylation of the human estrogen receptor on tyrosine 537 in vivo and by src family tyrosine kinases in vitro.* Mol Endocrinol. 1995;9,No.1:24-33.

19. Arnold SF, Melamed M,

Vorojeikina DP, Notides AC, Sasson S. *Estradiol-binding mechanism and binding capacity of the human estrogen receptor is regulated by tyrosine phosphorylation.* Mol Endocrinol. 1997;11:48-53.

20. Arnold SF, Vorojeikina DP, Notides AC. *Phosphorylation of tyrosine 537 on the human estrogen receptor is required for binding to an estrogen response element.* J Biol Chem. 1995;270:30205-12.

21. Carlson KE, Choi I, Gee A, Katzenellenbogen BS. *Altered ligand binging properties and enhanced stability of a constitutively active estrogen receptor: evidence that an open pocket conformation is required for ligand interaction.* Biochemistry. 1997;36:14897-905.

22. White R, Sjoberg M, Kalkhoven E, Parker MG. *Ligand-independent activation of the oestrogen receptor by mutation of a conserved tyrosine.* EMBO J. 1997;16:1427-35.

23. Weis KE, Ekena K, Thomas JA, Lazennec G, Katzenellenbogen BS. *Constitutively active human estrogen receptors containing amino acid substitutions for tyrosine 537 in the receptor protein.* Mol Endocrinol. 1996;10:1388-98.

24. Montano MM, Ekena K, Krueger KD, Keller AL, Katzenellenbogen BS. *Human estrogen receptor ligand activity inversion mutants: receptors that interpret antiestrogens as estrogens and estrogens as antiestrogens and discriminate among different antiestrogens.* Mol Endocrinol. 1996;10:230-42.

25. Montano MM, Muller V, Trobaugh A, Katzenellenbogen BS. *The carboxy-terminal F domain of the human estrogen receptor: Role in the transcriptional activity of the receptor and the effectiveness of antiestrogens as estrogen antagonists.* Mol Endocrinol. 1995;9:814-25.

26. Green S, Kumar V, Theulaz I, Wahli W, Chambon P. *The N-terminal DNA-binding 'zinc finger' of the oestrogen and glucocorticoid receptors determines target gene specificity.* EMBO J. 1988;7:3037-44.

27. Evans RM, Hollenberg SM. *Zinc fingers: gilt by association.* Cell. 1988;52:1-3.

28. Krust A, Green S, Argos P, Kumar V, Walter P, Bornert JM, et al. *The chicken oestrogen receptor sequence: homology with v-erbA and the human oestrogen and glucocorticoid receptors.* EMBO J. 1986;5:891-7.

29. Freedman LP, Luisi BF, Korszun ZR, Basavappa R, Sigler PB, Yamamoto KR. *The function and

structure of the metal coordination sites within the glucocorticoid receptor DNA binding domain.* Nature. 1988;334:543-6.

30. Schwabe JWR, Chapman L, Finch JT, Rhodes D. *The crystal structure of the estrogen receptor DNA-binding domain bound to DNA: how receptors discriminate between their response elements.* Cell. 1993;75:567-78.

31. Schwabe JWR, Neuhaus D, Rhodes D. *Solution structure of the DNA-binding domain of the oestrogen receptor.* Nature. 1990;348:458-61.

32. Mader S, Kumar V, de Verneuil H, Chambon P. *Three amino acids of the oestrogen receptor are essential to its ability to distinguish an oestrogen from a glucocorticoid-responsive element.* Nature. 1989;338:271-4.

33. Freedman LP. *Anatomy of the steroid receptor zinc finger region.* Endocrine Rev. 1992;13:129-45.

34. Berg JM. *DNA binding specificity of steroid receptors.* Cell. 1989;57:1065-8.

35. Hard R, Kellenback D, Boelens R, al. e. *Solution structure of the glucocorticoid receptor DNA-binding domain.* Science. 1990;249:157-60.

36. Kumar V, Green S, Stack G, Berry M, Jin J-R, Chambon P. *Functional domains of the human estrogen receptor.* Cell. 1987;51:941-51.

37. Umesono K, Evans RM. *Determinants of target gene specificity for steroid/thyroid hormone receptors.* Cell. 1989;57:1139-46.

38. Koszewski NJ, Notides AC. *Phosphate-sensitive binding of the estrogen receptor to its response elements.* Mol Endocrinol. 1991;5:1129-36.

39. Obourn JD, Koszewski NJ, Notides AC. *Hormone- and DNA-binding mechanisms of the recombinant human estrogen receptor.* Biochemistry. 1993;32:6229-36.

40. Klein-Hitpass L, Tsai SY, Greene GL, Clark JH, Tsai M, O'Malley BW. *Specific binding of estrogen receptor to the estrogen response element.* Mol Cell Biol. 1989;9:43-9.

41. Ikeda M, Ogata F, Curtis SW, Lubahn DB, French FS, Wilson EM, et al. *Characterization of the DNA-binding domain of the mouse uterine estrogen receptor using site-specific polyclonal antibodies.* J Biol Chem. 1993;268:10296-302.

42. Mader S, Chambon P, White JH. *Defining a minimal estrogen receptor DNA binding domain.* Nucleic Acids Res. 1993;21:1125-32.

43. Fawell SE, White R, Hoare S, Sydenham M, Page M, Parker MG. *Inhibition of estrogen receptor-DNA binding by the "pure" antiestrogen ICI 164,384 appears to be mediated by impairedreceptor dimerization.* Proc Natl Acad Sci USA. 1990;87:6883-7.

44. Ylikomi T, Bocquel MT, Berry M, Gronemeyer H, Chambon P. *Cooperation of proto-signals for nuclear accumulation of estrogen and progesterone receptors.* EMBO J. 1992;11:3681-94.

45. Picard D, Kumar V, Chambon P, Yamamoto KR. *Signal transduction by steroid hormones: nuclear localization is differentially regulated in estrogen and glucocorticoid receptors.* Cell Regulation. 1990;1:291-9.

46. Gronemeyer H, Laudet V. *Transcription factors 3: nuclear receptors.* Protein Profile. 1995;2:1173-308.

47. Nichols M, Rientjes JMJ, Logie C, Stewart AF. *FLP Recombinase/estrogen receptor fusion proteins require the receptor D Domain for responsivemess to antagonists, but not agonists.* Mol Endocrinol. 1997;11:950-61.

48. Lees JA, Fawell SE, Parker MG. *Identification of two transactivation domains in the mouse oestrogen receptor.* Nucleic Acids Res. 1989;17:5477-88.

49. Tora L, White J, Brou C, Tasset D, Webster N, Scheer E, et al. *The human estrogen receptor has two independent nonacidic transcriptional activation functions.* Cell. 1989;59:477-87.

50. Webster NJ, Green S, Tasset D, Ponglikitmongkol M, Chambon P. *The transcriptional activation function located in the hormone-binding domain of the human oestrogen receptor is not encoded in a single exon.* EMBO J. 1989;8:1441-6.

51. Metzger D, Losson R, Bornert JM, Lemoine Y, Chambon P. *Promoter specificity of the two transcriptional activation functions of the human oestrogen receptor in yeast.* Nucleic Acids Res. 1992;20:2813-7.

52. Tasset D, Tora L, Fromental C, Scheer E, Chambon P. *Distinct classes of transcriptional activating domains function by different mechanisms.* Cell. 1990;62:1177-87.

53. Ali S, Metzger D, Bornert J-M, Chambon P. *Modulation of transcriptional activation by ligand-dependent phosphorylation of the human oestrogen receptor A/B region.* EMBO J. 1993;12:1153-60.

54. Arnold SF, Obourn JD, Jaffe H, Notides AC. *Serine 167 is the major

estradiol-induced phosphorylation site on the human estrogen receptor. Mol Endocrinol. 1994;8:1208-14.

55. Joel PB, Traish AM, Lannigan DA. *Estradiol and phorbol ester cause phosphorylation of serine 118 in the human estrogen receptor.* Mol Endocrinol. 1995;9:1041-52.

56. LeGoff P, Montano MM, Schodin DJ, Katzenellenbogen BS. *Phosphorylation of the human estrogen receptor. Identification of hormone-regulated sites and examination of their influence on transcriptional activity.* J Biol Chem. 1994;269:4458-566.

57. Kato S, Endoh H, Masuhiro Y, Kitamoto T, Uchiyama S, Sasaki H, et al. *Activation of the estrogen receptor through phosphorylation by mitogen-activated protein kinase.* Science. 1995;270:1491-4.

58. Lannigan DA. *Estrogen receptor phosphorylation.* Steroids. 2003;68:1-9.

59. Smith CL. *Cross-talk between peptide growth factor and estrogen receptor signaling pathways.* Biol Reprod. 1998;58:627-32.

60. Medunjanin S, Weinert S, Schmeisser A, Mayer D, Braun-Dullaeus RC. *Interaction of the double-strand break repair kinase DNA-PK and estrogen receptor-alpha.* Mol Biol Cell. 2010;21:1620-8.

61. Weigel NL, Moore NL. *Steroid receptor phosphorylation: a key modulator of multiple receptor functions.* Mol Endocrinol. 2007;21:2311-9.

62. Norris JD, Fan D, Kerner SA, McDonnell DP. *Identification of a third autonomous activation domain within the human estrogen receptor.* Mol Endocrinol. 1997;11:747-54.

63. Jacq X, Brou C, Lutz Y, Davidson I, Chambon P, Tora L. *Human TAFII30 is present in a distinct TFIID complex and is required for transcriptional activation by the estrogen receptor.* Cell. 1994;79:107-17.

64. McInerney EM, Ince BA, Shapiro DJ, Katzenellenbogen BS. *A transcriptionally active estrogen receptor mutant is a novel type of dominant negative inhibitor of estrogen action.* Mol Endocrinol. 1996;10:1519-26.

65. Kraus WL, McInerney EM, Katzenellenbogen BS. *Ligand-dependent, transcriptionally productive association of the amino-and carboxyl-terminal regions of a steroid hormone nuclear receptor.* Proc Natl Acad Sci USA. 1995;92:12314-8.

66. Thornton JW, Need E, Crews D. *Resurrecting the ancestral steroid receptor: ancient origin of estrogen signaling.* Science. 2003;301:1714-7.

6.1.1.3. PROMOTER AND 3' UNTRANSLATED REGIONS

The promoter region of the ER gene exhibits a complex architecture that is characteristic of genes with heterogeneous transcription initiation sites. Sequence analysis reveals a panoply of *cis*-regulatory elements with the potential to bind several known transcription factors (Table 6.1.3.). The majority of promoter studies were performed with the ER-positive breast cancer cell line, MCF-7, with a more limited analysis in ER-negative MDA-MB-231 cells (1-4). The ER gene does not have a canonical TATA box, but a TATA-like element, TACTTAAA, located at nucleotide position -27. A consensus CAAT motif (CCAAT) is present at nucleotide -101, a typical position relative to the TATA box (5). Three consensus sequences of the cis-regulatory octamer motif, ATGCAAAT, are identified that bind the ubiquitous transcription factors Oct-1 and Oct-2. The 5' upstream region lacks a complete ERE response element but contains several ERE half-sites (Table 6.1.3.). The three ERE half-sites between nucleotides -420

and -892 have been implicated in the estrogen inducibility of ER expression (4). Baculovirus-expressed ER was able to bind each of the ERE half-sites but not their mutated counterparts.

The functional analysis of the promoter region has identified six ER mRNA isoforms (A - F), which are produced by usage of several promoters (1-4). All isoforms encode the same ER protein but differ in the length of their respective 5' untranslated region (5' UTR) as a consequence of alternative splicing of several upstream exons. The isoforms display a differential pattern of expression in tissues and cell types. Isoform A is the original 6.3 kb mRNA with its CAP site at +1 and a functional promoter with a noncanonical TATA element (TACTTAAA) at -27 and a consensus CAAT motif at -101 bp (5, 16). Exon 1 of isoform A ER mRNA contains the translation start site at +233 and a splice site at +163, which serves as a common acceptor splice site for five different upstream exons (1B - 1F) that extend over approximately 20 kb (1, 3). Since the other upstream exons are also spliced into the acceptor site at +163, all ER mRNA isoforms are identical in sequence downstream from the common splice site +163. The splice site at +163 is situated 70 nucleotides upstream of the translation start site at +233. Between nucleotides +163 and +233 is a stop codon in-frame with the translation initiation codon. Therefore, all ER mRNA isoforms encode the same ER protein.

RT-PCR and S1 nuclease analyses demonstrated ER mRNA isoforms A – F in MCF-7 breast cancer cells with the following relative concentrations: 48% (A), 18% (B), 8% (C), <1% (D), 2% (E), and 9% (F) (1). Both the absolute and relative amounts of these isoforms varied among breast cancer cell lines and benign and malignant breast tissues. However, A ER mRNA is the predominant isoform in all ER-positive breast tissues and breast cancer cell lines (1-3). The relative abundance of some of the other isoforms was higher in other tissues, e.g., isoform E was highest in liver (80%) (1). Computer modeling based on thermodynamic principles suggests that the 5' UTR sequences of the ER mRNA isoforms could be folded into more or less stable secondary structures, as indicated by free energy values ranging from ΔG -40 kcal/mol (isoform E) to -100 kcal/mol (isoform A) (1). Therefore, one possible function of the alternatively spliced 5' UTR exons could be the regulation of ER synthesis by controlling the turnover and/or translation efficiency of ER mRNA isoforms. In keeping with this hypothesis is the fact that the 5' UTR sequences of human ER mRNA isoforms A and C are conserved in 5' UTR sequences of ER genes of other species. The existence of several upstream exons and transcriptional start sites has been demonstrated in the 5' UTR of the chicken, monkey, mouse, rat, trout, and Xenopus ER genes (17-22). In summary, the ER gene is a complex genomic unit with multiple promoters, which produces several ER mRNA isoforms by alternative splicing. The isoforms are expressed

Table 6.1.3. Potential Regulatory Elements in ER Promoter Region

Element	Consensus	Nucleotide Position[1]	Reference
TATA	TATAAA	-1210 to -1205; -1375 to -1380	(6)
CAAT	CCAAT	-101 to -97; -2728 to -2724	(7)
Octamer	ATGCAAAT	-1441 to -1433; -1686 to -1686; -3067 to -3059	(8)
ERE	GGTCA half site	-424 to -420; -864 to -860; -2720 to -2716; -2780 to -2776;	(9)
	TGACC half site	-892 to -888; -3225 to -3219	
GRE	TGTTCT half site	-1757 to -1752	(10)
AP-1	GTGACTCA	-3773 to -3766	(11)
p53	PuPuPuC(T/A)(T/A)GPyPyPy	-1918 to -1885	(12)
GATA-1	TGATAG	-1263 to -1258	(13)
CTF-1	GCCAAT	-125 to -120	(14)
ERF-1/AP2g	CCCTGCGGGG	+189 to +198	(15)

[1]All nucleotide positions are in reference to the transcription start site of isoform 1 (5)

in a tissue-specific manner, which may allow differential ER expression in a wide range of physiological and pathological processes.

The majority of eukaryotic mRNAs are monocistronic; less than 10% contain more than one open reading frame. Upstream open reading frames (uORFs) carried by the 5' leader sequence of some mRNAs have been shown to impair the expression of the downstream protein coding sequence (23, 24). According to the scanning model for initiation of protein synthesis proposed by Kozak (25), the uORF would interfere with the initiation of the ribosome at the downstream AUG codon. A short open reading frame (ORF), which on translation could produce a small peptide of 18 amino acids in length, is encoded in the 5' UTR of the mouse ER mRNA (22). This same ORF is conserved in the human and rat ER genes, while the rainbow trout gene presents a similar element in the same position (26-28). The conservation in various species suggests that the uORFs may be functionally important in the regulation of ER expression. This is supported by experimental evidence, which indicates that the ER 5' UTR and especially the uORF within it may decrease the efficiency of translation (29, 30).

As is true for the 5' UTR, virtually everything known about the 3' untranslated region (3' UTR) of the ERα gene is based on sequence analysis of the ER-positive breast cancer cell line, MCF-7 (5, 31). The 3' UTR extends for 4304 bases from nucleotide 2018 to 6322. Such a long 3' UTR is unusual for eukaryotic genes, but is in common with other members of the steroid receptor gene family, e.g., the androgen and glucocorticoid receptors (32, 33). Moreover, the ER genes of other species, i.e., chicken, mouse, rat, and *Xenopus*, have equally long 3' UTRs (21, 22, 27, 34). This conservation among different receptors and species suggests a functional role for the 3' UTR, such as post-transcriptional regulation of ERα mRNA levels (35). Indeed, the expression of ERα is downregulated by base pairing of miRNAs to the 3' UTR of ERα mRNA. Several miRNAs have

been shown to repress ERα expression by targeting the ERα 3' UTR: miR-22, miR-206, miR-221, miR-222 (36-38). The levels of these miRNAs were higher in ER-negative than ER-positive breast cancer cell lines. The miR-17-92 cluster is a polycistronic gene at 13q31-q32 that encodes six miRNAs including miR-18a, miR-19b, and miR-20b, which downregulate ERα expression at the protein translational level (39). Interestingly, the miR-17-92 locus is upregulated by E_2-ERα via c-MYC, which interacts directly with the miR-17-92 promoter in an estrogen-dependent manner specific to breast cells. Thus, the E_2-ERα-c-MYC stimulatory mechanism sets up a negative feedback loop resulting in ERα suppression. Other mechanisms have been proposed to explain how E_2 downregulates ERα in an autoregulatory feedback loop. In pituitary lactotrope and MCF-7 cells, a non-genomic action of E_2 stimulated rapid degradation of nuclear ERα protein via a proteosome-mediated proteolytic mechanism (40). The half-life of ERα in MCF-7 cells decreased from greater than 24 hr to around 3 – 5 hr. The transcriptional activity of ERα and the proteasome-mediated degradation of ERα are linked in a continuous cyclic turnover of ERα on responsive promoters (41).

References

1. Flouriot G, Griffin C, Kenealy M, Sonntag-Buck V, Gannon F. *Differentially expressed messenger RNA isoforms of the human estrogen receptor-a gene are generated by alternative splicing and promoter usage.* Mol Endocrinol. 1998;12:1939-54.

2. Grandien K, Berkenstam A, Gustafsson JA. *The estrogen receptor gene: promoter organization and expression.* Int J Biochem Cell Biol. 1997;29:1343-69.

3. Thompson DA, McPherson LA, Carmeci C, deConinck EC, Weigel RJ. *Identification of two estrogen receptor transcripts with novel 5' exons isolated from a MCF7 CDNA library.* J Steroid Biochem Mol Biol. 1997;62:143-53.

4. Treilleux I, Peloux N, Brown M, Sergeant A. *Human estrogen receptor (ER) gene promoter-P1: estradiol-independent activity and estradiol inducibility in ER+ and ER- cells.* Mol Endocrinol. 1997;11:1319-31.

5. Green S, Walter P, Kumar V, Krust A, Bornert J-M, Argos P, et al. *Human oestrogen receptor cDNA: sequence, expression and homology to v-erb-A.* Nature. 1986;320:134-9.

6. Breathnach R, Chambon P. *Organization and expression of eucaryotic split genes coding for proteins.* Ann Rev Biochem. 1981;50:349-83.

| *The Etiology of Breast Cancer*

7. McKnight SL. *Discovery of a transcription factor that catalyzes terminal cell differentiation.* Harvey Lectures. 1991-92;87:57-68.

8. Aurora R, Herr W. *Segments of the POU domain influence one another's DNA-binding specificity.* Mol Cell Biol. 1992;12:455-67.

9. Klein-Hitpass L, Schorpp M, Wagner U, Ryffel GU. *An estrogen-responsive element derived from the 5' flanking region of the Xenopus vitellogenin A2 gene functions in transfected human cells.* Cell. 1986;46:1053-61.

10. Klock G, Strahle U, Schutz G. *Oestrogen and glucocorticoid responsive elements are closely related but distinct.* Nature. 1987;329:734-6.

11. Bohmann D, Bos TJ, Admon A, Nishimura T, Vogt PK, Tjian R. *Human proto-oncogene c-jun encodes a DNA binding protein with structural and functional properties of transcription factor AP-1.* Science. 1987;238:1386-92.

12. El-Deiry WS, Kern SE, Pietenpol JA, Kinzler KW, Vogelstein B. *Definition of a consensus binding site for p53.* Nature. 1992;1:45-9.

13. Orkin SH. *GATA-binding transcription factors in hematopoietic cells.* Blood. 1992;80:575-81.

14. Mitchell PJ, Tjian R. *Transcriptional regulation in mammalian cells by sequence-specific DNA binding proteins.* Science. 1989;245:371-8.

15. McPherson LA, Baichwal VR, Weigel RJ. *Identification of ERF-1 as a member of the AP2 transcription factor family.* Proc Natl Acad Sci USA. 1997;94:4342-7.

16. Greene GL, Gilna P, Waterfield M, Baker A, Hort Y, Shine J. *Sequence and expression of human estrogen receptor complementary DNA.* Science. 1986;231:1150-4.

17. Freyschuss B, Grandien K. *The 5' flank of the rat estrogen receptor gene: structural characterization and evidence for tissue-and species-specific promoter utilization.* J Mol Endocrinol. 1996;17:197-206.

18. Griffin C, Flouriot G, Sonntag-Buck V, Gannon F. *Two functionally different protein isoforms are produced from the chicken estrogen receptor-alpha gene.* Mol Endocrinol. 1999;13:1571-87.

19. Hirata S, Koh T, Yamada-Mouri N, Hoshi K, Kato J. *The untranslated first exon 'exon OS' of the rat estrogen receptor (ER) gene.* FEBS Lett. 1996;394:371-3.

20. Pakdel F, Le Guellec C, Vaillant C, Le Roux MG, Valotaire Y. *Identification and estrogen induction of two estrogen receptors (ER) messenger ribonucleic acids in the rainbow trout liver: sequence homology with other ERs.* Mol Endocrinol. 1989;3:44-51.

21. Weiler IJ, Lew D, Shapiro DJ. *The Xenopus laevis estrogen receptor: sequence homology with human and avian receptors and identification of multiple estrogen receptor messenger ribonucleic acids.* Mol Endocrinol. 1987;1:355-62.

22. White R, Lees JA, Needham M, Ham J, Parker M. *Structural organization and expression of the mouse estrogen receptor.* Mol Endocrinol. 1987;1:735-44.

23. Kozak M. *Selection of initiation sites by eucaryotic ribosomes: effect of inserting AUG triplets upstream from the coding sequence for preproinsulin.* Nucleic Acids Res. 1984;12:3873-93.

24. Mueller PP, Hinnebusch AG. *Multiple upstream AUG codons mediate translational control of GCN4.* Cell. 1986;45:201-7.

25. Kozak M. *Evaluation of the "scanning model" for initiation of protein synthesis in eucaryotes.* Cell. 1980;22:7-8.

26. Keaveney M, Klug J, Dawson MT, Nestor PV, Neilan JG, Forde RC, et al. *Evidence for a previously unidentified upstream exon in the human oestrogen receptor gene.* J Mol Endocr. 1991;6:111-5.

27. Koike S, Sakai M, Muramatsu M. *Molecular cloning and characterization of rat estrogen receptor cDNA.* Nucleic Acids Res. 1987;15:2499-513.

28. Le Roux MG, Theze N, Wolff J, Le Pennec JP. *Organization of a rainbow trout estrogen receptor gene.* Biochim Biophys Acta. 1993;1172:226-30.

29. Claret F, Chapel S, Garces J, Tsai-Pflugfelder M, Bertholet C, Shapiro DJ, et al. *Two functional forms of the Xenopus laevis estrogen receptor translated from a single mRNA species.* J Biol Chem. 1994;269:14047-55.

30. Wang Y, Miksicek RJ. *Characterization of estrogen receptor cDNAs from human uterus: identification of a novel PvuII polymorphism.* Mol Cell Endocrinol. 1994;101:101-10.

31. Keaveney M, Parker MG, Gannon F. *Identification of a functional role for the 3' region of the human oestrogen receptor gene.* J Mol Endocrinol. 1993;10:

32. Charest NJ, Zhou ZX, Lubahn DB, Olsen KL, Wilson EM, French FS. *A frameshift mutation destabilizes androgen receptor messenger RNA in the Tfm mouse.* Mol Endocrinol. 1991;5:573-81.

33. Hollenberg SM, Weinberger C, Ong ES, et al. *Primary structure and expression of a functional human glucocorticoid receptor cDNA.* Nature. 1985;318:635-41.

34. Krust A, Green S, Argos P, Kumar V, Walter P, Bornert JM, et al. *The chicken oestrogen receptor sequence: homology with v-erbA and the human oestrogen and glucocorticoid receptors.* EMBO J. 1986;5:891-7.

35. Kenealy MR, Flouriot G, Pope C, Gannon F. *The 3'untranslated region of the human estrogen receptor gene post-transcriptionally reduces mRNA levels.* Biochem Soc Transact. 1996;24:107S.

36. Adams BD, Furneaux H, White BA. *The micro-ribonucleic acid (miRNA) miR-206 targets the human estrogen receptor-alpha (ERalpha) and represses ERalpha messenger RNA and protein expression in breast cancer cell lines.* Mol Endocrinol. 2007;21:1132-47.

37. Pandey DP, Picard D. *miR-22 inhibits estrogen signaling by directly targeting the estrogen receptor alpha mRNA.* Mol Cell Biol. 2009;29:3783-90.

38. Zhao JJ, Lin J, Yang H, Kong W, He L, Ma X, et al. *MicroRNA-221/222 negatively regulates estrogen receptor alpha and is associated with tamoxifen resistance in breast cancer.* J Biol Chem. 2008;283:31079-86.

39. Castellano L, Giamas G, Jacob J, Coombes RC, Lucchesi W, Thiruchelvam P, et al. *The estrogen receptor-alpha-induced microRNA signature regulates itself and its transcriptional response.* Proc Natl Acad Sci USA. 2009;106:15732-7.

40. Alarid ET, Bakopoulos N, Solodin N. *Proteasome-mediated proteolysis of estrogen receptor: a novel component in autologous down-regulation.* Mol Endocrinol. 1999;13:1522-34.

41. Reid G, Hubner MR, Metivier R, Brand H, Denger S, Manu D, et al. *Cyclic, proteasome-mediated turnover of unliganded and liganded ERalpha on responsive promoters is an integral feature of estrogen signaling.* Mol Cell. 2003;11:695-707.

6.1.1.4. ER-ERE INTERACTION AND ER-MEDIATED GENE TRANSCRIPTION

The minimal functional ERE is a double-stranded 13-base pair palindrome with 5-base pair half-sites and a 3-base pair spacer (1, 2). The consensus sequence for the two half-sites is 5GGTCA1 and 1TGACC5, respectively, while the sequence of the spacer can accommodate any nucleotide, n, provided its length of three base pairs is invariant. The ERE differs in several respects from the response elements for the PR and GR which are identical, PRE/GRE. The ERE half-sequence has five base pairs, 1TGACC5, while the PRE/GRE half-sequence has six, 1TGTYCY6 (Y = T or C). Whereas positions 1,2, and 5 are identical, positions 3 and 4 differ and form the basis for the discrimination between ERE and GRE/PRE (3). The half-site sequence 6AGGTCA1 is recognized not only by the ER but also by the thyroid hormone receptor (TR), RAR, RXR, and vitamin D3 receptor (VDR). However, the ER binds to this sequence only in the form of a palindrome or inverted repeat with a 3-base pair spacer. In contrast, the TR and RAR recognize it either as an inverted repeat without spacer or, together with the other receptors, as a direct repeat (4). To change the binding from an inverted to a direct repeat would require a 180° rotational inversion of the DNA binding domain of the receptor dimer.

Thus, while the sequence of the consensus half-site is identical, its orientation and spacing assures discrimination between ER and other receptors. Like the other receptors, the ER can accommodate a great deal of asymmetry between the half-sites. Provided one half of the palindrome corresponds to the conserved element, the other half can vary considerably and tolerate bases in this position that are incompatible with binding to symmetric response elements (5-8). Thus, imperfect EREs containing changes in some of the half-site residues may still confer estrogen respon-siveness, particularly when two or more of these EREs are found in proximity to each other. An example is the PR gene, whose expression is induced by estrogen in several target tissues as well as in breast cancer and breast cancer cell lines (9-11). Estrogen responsiveness was shown to be dependent on PR gene sequences extending from about -2000 to +750, containing five widely spaced, imperfect EREs within this sequence including one intragenic ERE located between nucleotides +698 and +723 that overlaps the translation initiation site (12, 13).

In vitro saturation analysis of the ER-ERE interaction showed that one dimeric ER binds one ERE with a K_d of 0.24 nM (14). Steric constraints do not inhibit binding of ER to multiple tandem EREs as reflected by a similar K_d of 0.23 nM for ER binding to tandem double EREs. When the test plasmid contained three or more tandem copies of ERE, ER bound in a cooperative manner as indicated by convex Scatchard plots and Hill coefficients greater than 1.5. Adjacent EREs separated by 4 or 5 integral helical turns stimulated ER-mediated transcription about three-fold more strongly than EREs spaced 3.5 or 4.5 turns apart (7). Thus, synergism between two ER molecules bound to adjacent EREs requires proper stereoalignment of receptors and template. The number of nucleotides separating the EREs also proved critical for high-affinity ER binding. When the ERE centers of symmetry were spaced 1.5, 2.0, 3.0, 6.4, and 6.7 helical turns apart, the affinity of ER binding decreased as the distance between adjacent EREs increased, suggesting that ER binding to closely spaced EREs is more stable (K_d = 0.38, 0.58, 0.83, 1.23, and 0.96 nM, respectively, for the above spacings). In fact, two overlapping EREs, spaced less than one helical turn apart (5 base pairs center-to-center) were four- to six-fold more active in transfection experiments due to cooperative ER binding and synergistic induction of transcription (15). The sequence flanking the ERE may also affect ER binding. In particular, the presence of AT-rich sequences flanking the ERE increased both binding capacity and affinity of ER, the latter reflected in slower dissociation rates of ER from ERE (16). The induction capacity of EREs located immediately upstream of the promoter decreases slowly with distance from the promoter (17). For distant EREs, presumed to interact by a looping mechanism at the promoter, the length of DNA between the EREs and the promoter is not critical. These results suggest that ERE sequence, number, spacing,

flanking sequence, and distance to promoter determine the amount of ER binding and the induction capacity of EREs in estrogen-responsive genes. Furthermore, the nature of the ligand appears important for the ER-ERE interaction. For example, E_3 not only binds with lower affinity to ER than E_2, but the association constant of the resulting E_3-ER complex with ERE is only half that of the E_2-ER complex (18).

The interaction of ER and ERE leads to conformational changes of both protein and DNA. Circular permutation and DNA phasing analyses demonstrated that ER binding induces a distortion of ERE-containing fragments and directs bending of the DNA helix toward the major groove (19, 20). Since the receptor undergoes major changes in its conformation upon hormone binding it is not surprising that ER induced a different DNA bend in the absence of E_2 (15.6°) than in the presence of E_2 (7.3° to 8.3°) (21, 22). On the other hand, different EREs do not simply bind ER but induce allosteric alterations in receptor conformation, which in turn influence recruitment of specific coactivator proteins and lead to differential ligand-dependent expression of genes containing divergent ERE sequences (23, 24).

Genome-wide searches for ER binding sites have been performed by a combination of methods (chromatin immunoprecipitation, genomic expression microarrays, DNA sequencing) (25-30). When stringent selection criteria were applied, a total of 1,234 ER binding sites were identified in the whole genome of which 71% harbored full EREs, 25% bore ERE half sites and 4% had no recognizable ERE sequence (28). Only 5% of the ER binding sites are located within 5 kb upstream of the transcriptional start sites of adjacent genes typically targeted by transcription factors such as TFIID (26). In contrast to TFIID binding to RNA polymerase II sites in proximal promoters, the majority of ER binding mapped to intronic or distal locations (>5 kb from 5' and 3' ends of adjacent transcripts) suggesting transcriptional regulatory mechanisms over significant physical distances. For example, the ESR1 gene contains an ERE within the 1-kb proximal promoter region and three additional binding sites 150 to 192 kb upstream of the gene, which are important for autoregulation of ERα expression (26, 31). E_2 also regulates gene expression with distinct time-course patterns, i.e., some genes are induced early (3 h) and others late (12 h). Moreover, E_2 down-regulates the expression of many genes (26, 27). Thus, estrogen-regulated genes can be classified by (i) ERE sequence, (ii) time course, (iii) up- or down-regulation. The genome-wide search showed a high degree of overlap between ERα- and ERβ-binding sites but also selective binding regions for either receptor, which might explain differences in gene expression mediated by the respective receptors (32). Detailed sequence analysis revealed an overrepresentation of TA-rich motifs in ERα- and a predominance of GC-rich motifs in ERβ-binding sites. In general, the estrogen-regulated genes encode receptors, ligands, transcription

Table 6.1.4. Genome-wide Distribution of Selected E_2-Induced Genes

Locus	Gene	Protein	Function	Induction Mechanism	Reference
1q21	MUC1	mucin 1			(33)
2p21	COX7RP	COX7RP	oxidative metabolism	perfect ERE palindrome	(29)
2p13.3	TGFA	TGFα		GGTCAgctGTGCC at -215 to -203	(34)
				GGTGAcggTAGCC at -249 to -237	
3p21.31	LTF	lactotransferrin		GGTCAaggCGATC	(35)
4q22	ABCG2	breast cancer resistance protein	transporter protein	CGGCAgggTGACC	(36)
5p15.33	TERT	TERT	cell cycle	GGTCA plus Sp1 and AP-1 sites	(37)
6q25	ESR1	ERα	transcription		(38)
7p12	EGFR	HER1			(39)
8q23	EBAG6 (RCAS1)			perfect ERE palindrome	(40)
10q26.3	myc			GGGCA plus Sp1 site	(41)
11q22-23	PR	progesterone receptor		GGTCGacaTGACT, intragenic	(13)
12q13	CCND1	cyclin D1		ERE plus CRE and Sp1	(42)
13q34	CUL4A	cullin 4A	cell cycle	no ERE	(43)
14q24.3	FOS	fos	oncogene	CGGCAgcgTGACC	(44)
17q11-q21	HER2				
17q23.1	EFP	EFP	ring finger protein	perfect ERE palindrome	(29)
18q21.3	BCL2	Bcl-2			
19p13.3	C3	complement C3		GGTGGcccTGACC	(45)
21q22.3	pS2	trefoil factor 1		GGTCAcggTGGCC	(5)
22q12.1	XBP1		transcription		(46)

factors, enzymes, structural and other proteins involved in all aspects of cell physiology. As expected, the expression profile of ERα-positive breast cancers was enriched with genes with ERα-binding sites and significantly different from ERα-negative tumors (26, 28). Table 6.1.4. lists selected E_2-induced genes identified in genome-wide searches and characterized in experimental studies.

Research on estrogen-regulated transcription has generally focused on upregulated genes. However, downregulated genes constitute a significant fraction of all estrogen-dependent expression changes in cell lines and tumor samples (27, 47). An expression array analysis of MCF-7 cells revealed that 51.2% of early (3 h) gene changes after E_2 treatment were downregulated events (26). For example, E_2 inhibits the expression of the interleukin 6 and glycoprotein hormone α-subunit genes. Examination of the 5' flanking sequences of these genes failed to show ERE motifs or ER binding despite the responsiveness of the respective promoter regions to E_2-induced transcriptional repression (48, 49). In the case of interleukin 6, the repression appears to be due to antagonism of ER to the p65 subunit of NF-kB heterodimer, preventing binding of the latter to the NF-kB site of the interleukin 6 promoter. Conversely, NF-kB p65 can repress ER activity, much like it represses GR activity (50). The ER also participates in a protein complex that binds to the electrophile/antioxidant response element (EpRE/ARE) in the promoter of the quinone reductase gene. Paradoxically, antiestrogens (4-hydroxytamoxifen,

ICI 182,780) stimulate and E_2 represses expression of quinone reductase (51).

The hepatocyte nuclear factor (HNF) superfamily of transcriptional regulators represents a unique subclass of DNA binding proteins that regulates organ development and differentiation across multiple species. The DNA binding occurs via the conserved forkhead box domain of about 100 amino acids found across species from Drosophila to mammals. Crystal structure analysis of the domain revealed a central helix-turn-helix (HTH) motif with two flanking loops reminiscent of wings (52). The HTH makes direct contact with the major groove in a base-specific manner, crucial for target site recognition: 5'-[AC]A[AT]T[AG]TT[GT][AG][CT]T[CT]-3 (53). In recognition of the forkhead box, the vertebrate HNF family was renamed 'FOX', with a letter representing the subclass, e.g., FoxA, which, in turn, consists of FoxA1, FoxA2, and FoxA3. FoxA1 is involved in the global regulation of gene expression by acting as a 'pioneer' factor that opens the compacted chromatin for other transcription factors through interactions with nucleosomal core histones (54). Specifically, FoxA1 binding at hypomethylated DNA regions induces histone H3 lysine 4 methylation (H3K4me2) (55, 56). This epigenetic change facilitates the recruitment of other transcription factors to enhancer regions in a cell type-specific manner, including nuclear hormone receptors such as ERα in mammary and AR in prostate epithelial cells. In the developing mammary gland, FoxA1 and ERα are co-expressed within luminal epithelial cells, with strong expression in the terminal end buds (57). FoxA1 deletion resulted in a mammary phenotype that mirrors that of the ERα null phenotype suggesting that the influence of FoxA1 deletion on mammary ductal development is due to the loss of ERα expression in the luminal progenitor cells that give rise to the ductal lineage. Hence, Fox A1 appears to be necessary for acquisition of ERα expression in the mammary ducts. FoxA1 is a key determinant of ERα function as reflected by the enrichment of FoxA1 motifs near 56% of ERα binding sites in a genome-wide analysis (46). A limited analysis of the apoptosis pathway in MCF-7 cells revealed the frequent proximity of FoxA1 and ERα binding sites in apoptosis-related genes (58). Knockdown of FoxA1 in MCF-7 cells blocked the association of ERα with chromatin and E_2-induced gene expression. The activity of the antagonist tamoxifen was similarly dependent on FoxA1 (59). Thus, FoxA1 translates epigenetic signatures into cell-specific enhancer-driven transcription and is a major determinant of E_2-ERα activity and endocrine response in normal and malignant mammary epithelial cells. The binding of FoxA1 to chromatin is inhibited by an upstream negative regulator, CCCTC-binding factor (CTCF), a conserved zinc finger transcription factor involved in both intra- and inter-chromosomal looping (59, 60).

Clinical studies of FOXA1 expression in over 3500 primary invasive breast cancers demonstrated positive immunohistochemical staining in 86% of tumors (61). The expression of FoxA1 correlated positively with ERα and PR and was seen especially in luminal subtype A breast cancers (62). As tumors develop resistance to hormonal therapies and spread to distant sites, FoxA1 appears to induce ERα recruitment to different genomic sites (63). The FoxA1-mediated reprogramming of ERα binding resulted in a parallel redistribution of both proteins, e.g., ERα and FoxA1 were co-expressed in metastatic samples. The differential ERα binding was associated with clinical outcome.

The classical pathway of estrogen-induced gene transcription involves FoxA1-facilitated association of ER with ERE and E_2 binding to ER (Figure 6.1.2.). While hormone and receptor are important, experiments revealed that neither hormone binding nor receptor-DNA contact are essential and that alternate pathways exist for ER-mediated gene transcription (Figure 6.1.2.). First, ER can interact synergistically with another transcription factor, such as Sp1 or AP-1 that is bound to its respective response element (26, 29, 64, 65). E_2 stimulates gene transcription indirectly through protein-protein contact of ER with the adjacent transcription factor. This mechanism may explain E_2-induced expression of certain genes without identifiable EREs. Second, E_2-independent stimulation of ER-mediated gene transcription can be achieved by signals from tyrosine kinase-linked cell surface receptors initiated by peptide

growth factors and protein kinase activators (66-68). These factors can activate nuclear ER through phosphorylation in the absence of hormone. Addition of E_2 acts synergistically, providing an example of 'cross-talk' between the signal transduction and steroid receptor pathways (69). In addition to nuclear-initiated (genomic) responses, E_2 can elicit non-genomic (non-transcriptional) action in seconds or minutes by binding to extranuclear ER, which is embedded in the plasma membrane and interacts in a cooperative fashion with transmembrane growth factor receptor signaling pathways (70-73). Thus, nuclear and extranuclear ERs act in concert with growth factor receptor pathways to promote downstream signaling for cell proliferation and survival (68).

The synergistic effect of nuclear receptors with other transcription factors can be explained by two mechanisms: (1) cooperative binding of both proteins to DNA and (2) intermediary factors that interact with receptor and transcription factor and transfer the activation signal to the transcription initiation complex (74). An example of the first mechanism is the interaction of ER with the ubiquitous transcription factor Sp1 (29). Several estrogen-inducible genes contain ERE half-sites adjacent to Sp1 sites in their promoters, e.g., cathepsin D, c-myc, heat shock protein 27, and RARα (41, 75-77). Formation of Sp1/ER complexes on oligonucleotides with adjacent Sp1 and ERE half-sites was demonstrated by electrophoretic mobility shift assays and transient transfection studies. These findings suggest that an Sp1/ER complex may be involved in E_2-induced expression of genes containing only one ERE half site. In fact, it appears that ER can enhance Sp1 binding to its response element even in the absence of an ERE site, due to direct Sp1/ER interaction (78-80). The interaction of Sp1 and ER proteins and the enhancement of Sp1-DNA binding by ER were observed in the presence or absence of E_2, whereas transactivation was hormone-dependent (Figure 6.1.2.).

ER also interacts with activator protein-1 (AP-1), a ubiquitous transcription factor composed of homodimers of Jun or heterodimers of Jun and Fos family members. One example is the chicken ovalbumin gene which is expressed by administration of E_2. Analysis of the proximal promoter revealed two ERE half-sites, one of which (GGTCA, located between positions -47 and -43) is preceded by a TG dinucleotide, resulting in the sequence TGGGTCA, which differs only in one nucleotide from a canonical AP-1 recognition site, known as 12-O-tetradecanoylphorbol-13-acetate (TPA) response element (81). Transcription experiments showed synergistic activation of the ovalbumin promoter by c-Jun, c-Fos, and ER (82). Thus, ER can bind to DNA directly via ERE or indirectly by interacting with AP-1, which binds AP-1 recognition sites to induce estrogen-mediated gene transcription, e.g., insulin-like growth factor I and E2F transcription factor 1 (83, 84). Interestingly, ERα and ERβ send opposite signals through the AP-1 pathway (85, 86).

E_2 activated ERα- but inhibited ERβ-mediated transcription. The pharmacological response to ligands was also reversed. The antiestrogens tamoxifen, raloxifene, and ICI 164,384 were potent transcriptional activators with ERβ contacting Jun/Fos at the AP-1 site while they acted as inhibitors in the presence of ERα. The opposite ligand activation of ERα and ERβ at AP-1 sites is due to differential binding of coactivator and corepressor proteins to AP-1 and both ERα and ERβ thereby leading to up- or down-regulation of gene expression (26, 87). The importance of the interplay between ERα and AP-1 in regulating gene expression was demonstrated by Fos knockdown experiments, which reduced the expression of 37% of all estrogen-regulated genes in MCF-7 cells (83). The mechanism described for Sp1 and AP-1 also applies to other DNA-binding transcription factors, e.g., C/EBP, Oct, and Forkhead with their respective recognition motifs, i.e., they can be tethered to the ER and allow ER-mediated ERE-independent gene transcription. A genome-wide analysis has shown that response elements for AP-1, C/EBP, Oct, and Forkhead frequently cluster within the same DNA region as the ERE allowing interaction of these cooperating transcription factors (26). These data indicate that the context in which an ERE is placed has a profound effect on its effectiveness and that a receptor can cooperate functionally both with itself and other transcription factors bound to their respective response elements.

Another group of proteins that interacts with ER, the estrogen-related receptors (ERR) –α, –β, -γ, are so-called orphan nuclear receptors because they have no known ligand (88). The DBDs of ERα and ERRα have 68% amino acid identity and the latter can bind to EREs as well as to the recognition sequence TCAAGGTCA (89). Binding of both ER and ERRα to adjacent binding sites was shown to be required for maximal transactivation of the lactoferrin gene promoter (90). Thus, ERRs may play a role in the response of certain estrogen target genes by heterodimerizing with ER. For example, ERRα and ERRγ stimulate the expression of CYP19 (aromatase) and thereby may promote the growth of breast cancers by increasing local estrogen synthesis (91, 92). Interestingly, E_2 induces ERRα expression in both ER-positive and –negative breast cancer cells (93).

In addition to the traditional estrogen-dependent ER activation there is an estrogen-independent pathway of ER activation that involves (i) extracellular peptide growth factors acting through tyrosine kinase-linked cell surface receptors such as epidermal growth factor (EGF), transforming growth factor alpha (TGF-α), and insulin-like growth factor I (IGF-I) and (ii) protein kinase A and C activators such as cyclic AMP (cAMP) and phorbol ester (12-O-tetradecanoylphorbol-13-acetate, TPA), respectively (Figure 6.1.3.). Equally effective were agents which increased intracellular cAMP levels, such as cholera toxin plus 3-isobutyl-1-methyl-xanthine (CT + IBMX) and

8-bromo-cAMP (8-Br-cAMP). The peptide growth factors reproduce many of the effects of estrogen on human breast cancer cell lines and the murine female reproductive tract (94). For example, EGF administered to adult ovariectomized mice mimicked the stimulatory effects of E_2 on uterine DNA synthesis, lactoferrin gene expression, and phosphatidyl lipid metabolism (95, 96). In ER-positive cell lines, EGF and TGF-α stimulated transcription from a reporter plasmid containing an ERE and a TATA box linked to a chloramphenicol acetyltransferase (CAT) gene in a dose-dependent manner. Combinations of E_2 with EGF, TGF-α or IGF-I induced synergistic activation of transcription, whereas an additive response was observed with combinations of EGF and TGF-α or IGF-I (97). A synergistic effect on ER-mediated transcription was also achieved by combining E_2 and protein kinase activators (98). E_2 alone evoked a 25-fold increase in transcriptional activation of an ERE-CAT reporter plasmid transfected into ER-positive MCF-7 cells. Treatment with CT + IBMX or TPA elicited a weak (<2-fold) rise in CAT activity while cotreatment with E_2 increased CAT activity 60-fold. The steroid-independent stimulation of ER-mediated transcription by peptide growth factors and protein kinase activators is dependent upon the presence of functional ER. Cotransfection of ERE-reporter genes with truncated forms of ER revealed that peptide growth factors and protein kinase activators require different regions of the receptor to produce their effects on transcription. EGF acts primarily via AF-1, whereas cAMP acts via AF-2 (99). The synergistic stimulation of ER-mediated transcription by E_2 and protein kinase activators or peptide growth factors does not appear to result from changes in cellular ER content, but rather from changes in phosphorylation of ER since the latter increased 3- to 5-fold upon exposure to E_2, CT + IBMX, 8-Br-cAMP, or IGF-I (100). ER activation by EGF was shown to involve the MAPK pathway and phosphorylation of 118Ser in AF-1 (101). On the other hand, ER activation by Src/JNK kinase did not require 118Ser but enhanced AF-1 activity by modification of ER AF-1 associated proteins (102). These findings suggest that the synergistic effect of estrogen and tyrosine kinase activators results from phosphorylation of ER and/or other proteins involved in the ER-mediated transcriptional response. In summary, these data support the importance of 'cross-talk' between the steroid receptor, tyrosine kinase receptor, and protein kinase signal transduction pathways and suggest that the ER provides a site for the integration of signals activating these pathways, e.g., E_2, IGF-I, and cAMP (103, 104).

References

1. Driscoll MD, Sathya G, Saidi LF, DeMott MS, Hilf R, Bambara RA. *An explanation for observed estrogen receptor binding to single-stranded estrogen-responsive element DNA.* Mol Endocrinol. 1999;13:958-68.

2. Klein-Hitpass L, Ryffel GU, Heitlinger E, Cato ACB. *A 13 bp palindrome is a functional estrogen responsive element and interacts specifically with estrogen receptor.* Nucleic Acids Res. 1988;16:647-63.

3. Truss M, Chalepakis G, Slater EP, Mader S, Beato M. *Functional*

interaction of hybrid response elements with wild-type and mutant steroid hormone receptors. Mol Endocrinol. 1991;11:3247-58.

4. Umesono K, Murakami KK, Thompson CC, Evans RM. *Direct repeats as selective response elements for the thyroid hormone, retinoic acid, and vitamin D3 receptors.* Cell. 1991;65:1255-66.

5. Berry M, Nunez A, Chambon P. *Estrogen-responsive element of the human pS2 gene is an imperfectly palindromic sequence.* Proc Natl Acad Sci USA. 1989;86:1218-22.

6. Martinez E, Wahli W. *Cooperative binding of estrogen receptor to imperfect estrogen-responsive DNA elements correlates with their synergistic hormone-dependent enhancer activity.* EMBO J. 1989;8:3781-91.

7. Ponglikitmongkol M, White JH, Chambon P. *Synergistic activation of transcription by the human estrogen receptor bound to tandem responsive elements.* EMBO J. 1990;9:2221-31.

8. Slater EP, Redeuilh G, Theis K, Suske G, Beato M. *The uteroglobin promoter contains a noncanonical estrogen responsive element.* Mol Endocrinol. 1990;4:604-10.

9. Horwitz KB, McGuire WL. *Estrogen control of progesterone receptor in human breast cancer. Correlation with nuclear processing of estrogen receptor.* J Biol Chem. 1978;253:2223-8.

10. Nardulli AM, Greene GL, O'Malley BW, Katzenellenbogen BS. *Regulation of progesterone receptor messenger ribonucleic acid and protein levels in MCF-7 cells by estradiol: analysis of estrogen's effect on progesterone receptor synthesis and degradation.* Endocrinology. 1988;122:935-44.

11. Scott REM, Wu-peng XS, Yen PM, Chin WW, Pfaff DW. *Interactions of estrogen- and thyroid hormone receptors on a progesterone receptor estrogen response element (ERE) sequence: a comparison with the vitellogenin A2 consensus ERE.* Mol Endocrinol. 1997;11:1581-92.

12. Kraus WL, Montano MM, Katzenellenbogen BS. *Identification of multiple, widely spaced estrogen-responsive regions in the rat progesterone receptor gene.* Mol Endocrinol. 1994;8:952-69.

13. Savouret JF, Bailly A, Misrahi M, Rauch C, Redeuilh G, Chauchereau A, et al. *Characterization of the hormone responsive element involved in the regulation of the progesterone receptor gene.* EMBO J. 1991;10:1875-83.

14. Klinge CM, Peale FVJ, Hilf R, Bambara RA, Zain S. *Cooperative estrogen receptor interaction with consensus or variant estrogen responsive elements in vitro.* Cancer Res. 1992;52:1073-81.

15. Massaad C, Coumoul X, Sabbah M, Garlatti M, Redeuilh G, Barouki R. *Properties of overlapping EREs: Synergistic activation of transcription and cooperative binding of ER.* Biochemistry. 1998;37:6023-32.

16. Anolik JH, Klinge CM, Bambara RA, Hilf R. *Differential impact of flanking sequences on estradiol-vs-4-hydroxytamoxifen-liganded estrogen receptor binding to estrogen responsive element DNA.* J Steroid Biochem. 1993;46:713-30.

17. Sathya G, Li W, Klinge CM, Anolik JH, Hilf R, Bambara RA. *Effects of multiple estrogen responsive elements, their spacing, and location on estrogen response of reporter genes.* Mol Endocrinol. 1997;11:1994-2003.

18. Melamed M, Castano E, Notides AC, Sasson S. *Molecular and kinetic basis for the mixed agonist/antagonist activity of estriol.* Mol Endocrinol. 1997;11:1868-78.

19. Nardulli AM, Grobner C, Cotter D. *Estrogen receptor-induced DNA bending: Orientation of the bend and replacement of an estrogen response element with an intrinsic DNA bending sequence.* Mol Endocrinol. 1995;9:1064-76.

20. Nardulli AM, Shapiro DJ. *Binding of the estrogen receptor DNA-binding domain to the estrogen response element induces DNA bending.* Mol Cell Biol. 1992;12:2037-42.

21. Lazennec G, Ediger TR, Petz LN, Nardulli AM, Katzenellenbogen BS. *Mechanistic aspects of estrogen receptor activation probed with constitutively active estrogen receptors: correlations with DNA and coregulator interactions and receptor conformational changes.* Mol Endocrinol. 1997;11:1375-86.

22. Potthoff SJ, Romine LE, Nardulli AM. *Effects of wild type and mutant estrogen receptors on DNA flexibility, DNA bending, and transcription activation.* Mol Endocrinol. 1996;10:1095-106.

23. Hall JM, McDonnell DP, Korach KS. *Allosteric regulation of estrogen receptor structure, function, and coactivator recruitment by different estrogen response elements.* Mol Endocrinol. 2002;16:469-86.

24. Wood JR, Likhite VS, Loven MA, Nardulli AM. *Allosteric modulation of estrogen receptor conformation by different estrogen response elements.*

Mol Endocrinol. 2001;15:1114-26.

25. Bourdeau V, Deschenes J, Metivier R, Nagai Y, Nguyen D, Bretschneider N, et al. *Genome-wide identification of high-affinity estrogen response elements in human and mouse.* Mol Endocrinol. 2004;18:1411-27.

26. Carroll JS, Meyer CA, Song J, Li W, Geistlinger TR, Eeckhoute J, et al. *Genome-wide analysis of estrogen receptor binding sites.* Nature Genet. 2006;38:1289-97.

27. Frasor J, Danes JM, Komm B, Chang KC, Lyttle CR, Katzenellenbogen BS. *Profiling of estrogen up- and down-regulated gene expression in human breast cancer cells: insights into gene networks and pathways underlying estrogenic control of proliferation and cell phenotype.* Endocrinology. 2003;144:4562-74.

28. Lin CY, Vega VB, Thomsen JS, Zhang T, Kong SL, Xie M, et al. *Whole-genome cartography of estrogen receptor alpha binding sites.* PLoS genetics. 2007;3:e87.

29. O'Lone R, Frith MC, Karlsson EK, Hansen U. *Genomic targets of nuclear estrogen receptors.* Mol Endocrinol. 2004;18:1859-75.

30. Tang Q, Chen Y, Meyer C, Geistlinger T, Lupien M, Wang Q, et al. *A comprehensive view of nuclear receptor cancer cistromes.* Cancer Res. 2011;71:6940-7.

31. Le Drean Y, Lazennec G, Kern L, Saligaut D, Pakdel F, Valotaire Y. *Characterization of an estrogen-responsive element implicated in regulation of the rainbow trout estrogen receptor gene.* J Mol Endocrinol. 1995;15:37-47.

32. Liu Y, Gao H, Marstrand TT, Strom A, Valen E, Sandelin A, et al. *The genome landscape of ERalpha- and ERbeta-binding DNA regions.* Proc Natl Acad Sci USA. 2008;105:2604-9.

33. Zaretsky JZ, Barnea I, Aylon Y, Gorivodsky M, Wreschner DH, Keydar I. *MUC1 gene overexpressed in breast cancer: structure and transcriptional activity of the MUC1 promoter and role of estrogen receptor alpha (ERalpha) in regulation of the MUC1 gene expression.* Mol Cancer. 2006;5:57.

34. Saeki T, Cristiano A, Lynch MJ, Brattain M, Kim N, Normanno N, et al. *Regulation by estrogen through the 5'-flanking region of the transforming growth factor alpha gene.* Mol Endocrinol. 1991;5:1955-63.

35. Teng CT, Liu Y, Yang N, Walmer D, Panella T. *Differential molecular mechanism of the estrogen action*

that regulates lactoferrin gene in human and mouse. Mol Endocrinol. 1992;6:1969-81.

36. Ee PL, Kamalakaran S, Tonetti D, He X, Ross DD, Beck WT. *Identification of a novel estrogen response element in the breast cancer resistance protein (ABCG2) gene.* Cancer Res. 2004;64:1247-51.

37. Misiti S, Nanni S, Fontemaggi G, Cong YS, Wen J, Hirte HW, et al. *Induction of hTERT expression and telomerase activity by estrogens in human ovary epithelium cells.* Mol Cell Biol. 2000;20:3764-71.

38. Holst F, Stahl PR, Ruiz C, Hellwinkel O, Jehan Z, Wendland M, et al. Estrogen receptor alpha (ESR1) gene amplification is frequent in breast cancer. Nature Genet. 2007;39:655-60.

39. Zaczek A, Welnicka-Jaskiewicz M, Bielawski KP, Jaskiewicz J, Badzio A, Olszewski W, et al. *Gene copy numbers of HER family in breast cancer.* J Cancer Res Clin Oncol. 2008;134:271-9.

40. Tsuneizumi M, Emi M, Nagai H, Harada H, Sakamoto G, Kasumi F, et al. *Overrepresentation of the EBAG9 gene at 8q23 associated with early-stage breast cancers.* Clin Cancer Res. 2001;7:3526-32.

41. Dubik D, Shiu RP. *Mechanism of estrogen activation of c-myc onogene expression.* Onogene. 1992;7:1587-94.

42. Castro-Rivera E, Samudio I, Safe S. *Estrogen regulation of cyclin D1 gene expression in ZR-75 breast cancer cells involves multiple enhancer elements.* J Biol Chem. 2001;276:30853-61.

43. Chen LC, Manjeshwar S, Lu Y, Moore D, Ljung BM, Kuo WL, et al. *The human homologue for the Caenorhabditis elegans cul-4 gene is amplified and overexpressed in primary breast cancers.* Cancer Res. 1998;58:3677-83.

44. Weisz A, Rosales R. *Identification of an estrogen response element upstream of the human c-fos gene that binds the estrogen receptor and the AP-1 transcription factor.* Nucleic Acids Res. 1990;18:5097-6106.

45. Norris JD, Fan D, Wagner BL, McDonnell DP. *Identification of the sequences within the human complement 3 promoter required for estrogen responsiveness provides insight into the mechanism of tamoxifen mixed agonist activity.* Mol Endocrinol. 1996;10:1605-16.

46. Carroll JS, Liu XS, Brodsky AS, Li W, Meyer CA, Szary AJ, et al. *Chromosome-wide mapping of estrogen receptor binding reveals* long-range regulation requiring the forkhead protein FoxA1. Cell. 2005;122:33-43.

47. Nishidate T, Katagiri T, Lin ML, Mano Y, Miki Y, Kasumi F, et al. *Genome-wide gene-expression profiles of breast-cancer cells purified with laser microbeam microdissection: identification of genes associated with progression and metastasis.* Int J Oncol. 2004;25:797-819.

48. Keri RA, Andersen B, Kennedy GC, Hamernik DL, Clay CM, Brace AD, et al. *Estradiol inhibits transcription of the human glycoprotein hormone a- subunit gene despite the absence of a high affinity binding site for estrogen receptor.* Mol Endocrinol. 1991;5:7250733.

49. Ray P, Ghosh SK, Zhang DH, Ray A. *Repression of interleukin-6 gene expression by 17 beta-estradiol: inhibition of the DNA-binding activity of the transcription factors NF-IL6 and NF-kappa B by the estrogen receptor.* FEBS Lett 1997;409:79-85.

50. McKay LI, Cidlowski JA. *Cross-talk between nuclear factor-kB and the steroid hormone receptors: mechanisms of mutal antagonism.* Mol Endocrinol. 1998;12:45-56.

51. Montano MM, Jaiswal AK, Katzenellenbogen BS. *Transcriptional regulation of the human quinone reductase gene by antiestrogen-liganded estrogen receptor-a and estrogen receptor-b.* J Biol Chem. 1998;273:25443-9.

52. Clark KL, Halay ED, Lai E, Burley SK. *Co-crystal structure of the HNF-3/fork head DNA-recognition motif resembles histone H5.* Nature. 1993;364:412-20.

53. Cirillo LA, Zaret KS. *Specific interactions of the wing domains of FOXA1 transcription factor with DNA.* J Mol Biol. 2007;366:720-4.

54. Augello MA, Hickey TE, Knudsen KE. *FOXA1: master of steroid receptor function in cancer.* EMBO J. 2011;30:3885-94.

55. Lupien M, Eeckhoute J, Meyer CA, Wang Q, Zhang Y, Li W, et al. *FoxA1 translates epigenetic signatures into enhancer-driven lineage-specific transcription.* Cell. 2008;132:958-70.

56. Serandour AA, Avner G, Percevault F, Demay F, Bizot M, Lucchetti-Miganeh C, et al. *Epigenetic switch involved in activation of pioneer factor FOXA1-dependent enhancers.* Genome Res. 2011;21:555-65.

57. Bernardo GM, Lozada KL, Miedler JD, Harburg G, Hewitt SC, Mosley JD, et al. *FOXA1 is an essential determinant of ERalpha expression and mammary ductal morphogenesis.* Development. 2010;137:2045-54.

58. Liu Z, Chen S. *ER regulates an evolutionarily conserved apoptosis pathway.* Biochem Biophys Res Commun. 2010;400:34-8.

59. Hurtado A, Holmes KA, Ross-Innes CS, Schmidt D, Carroll JS. *FOXA1 is a key determinant of estrogen receptor function and endocrine response.* Nature Genet. 2011;43:27-33.

60. Ross-Innes CS, Brown GD, Carroll JS. *A co-ordinated interaction between CTCF and ER in breast cancer cells.* BMC Genomics. 2011;12:593.

61. Mehta RJ, Jain RK, Leung S, Choo J, Nielsen T, Huntsman D, et al. *FOXA1 is an independent prognostic marker for ER-positive breast cancer.* Breast Cancer Res Treat. 2012;131:881-90.

62. Badve S, Turbin D, Thorat MA, Morimiya A, Nielsen TO, Perou CM, et al. *FOXA1 expression in breast cancer--correlation with luminal subtype A and survival.* Clin Cancer Res. 2007;13:4415-21.

63. Ross-Innes CS, Stark R, Teschendorff AE, Holmes KA, Ali HR, Dunning MJ, et al. *Differential oestrogen receptor binding is associated with clinical outcome in breast cancer.* Nature. 2012;481:389-93.

64. Klinge CM. *Estrogen receptor interaction with co-activators and co-repressors.* Steroids. 2000;65:227-51.

65. Sabbah M, Courilleau D, Mester J, Redeuilh G. *Estrogen induction of the cyclin D1 promoter: involvement of a cAMP response-like element.* Proc Natl Acad Sci USA. 1999;96:11217-22.

66. Levin ER. *Bidirectional signaling between the estrogen receptor and the epidermal growth factor receptor.* Mol Endocrinol. 2003;17:309-17.

67. Stoica A, Saceda M, Fakhro A, Solomon HB, Fenster BD, Martin MB. *The role of transforming growtrh factor-beta in the regulation of estrogen receptor expression in the MCF-7 breast cancer cell line.* Endocrinology. 1997;138:1498-505.

68. Stoica GE, Franke TF, Moroni M, Mueller S, Morgan E, Iann MC, et al. *Effect of estradiol on estrogen receptor-alpha gene expression and activity can be modulated by the ErbB2/PI 3-K/Akt pathway.* Oncogene. 2003;22:7998-8011.

69. Lange CA. *Making sense of cross-talk between steroid hormone receptors and intracellular signaling pathways: who will have the last word?* Mol Endocrinol. 2004;18:269-78.

70. Moriarty K, Kim KH, Bender JR. *Minireview: estrogen receptor-mediated rapid signaling.* Endocrinology. 2006;147:5557-63.

71. Murphy E. *Estrogen signaling and cardiovascular disease.* Circulation Res. 2011;109:687-96.

72. Pedram A, Razandi M, Levin ER. *Nature of functional estrogen receptors at the plasma membrane.* Mol Endocrinol. 2006;20:1996-2009.

73. Pietras RJ, Marquez-Garban DC. *Membrane-associated estrogen receptor signaling pathways in human cancers.* Clin Cancer Res. 2007;13:4672-6.

74. Bastian LS, Nordeen SK. *Concerted stimulation of transcription by glucocorticoid receptors and basal transcription factors: limited transcriptional synergism suggests mediation by coactivators/adaptors.* Mol Endocrinol. 1991;5:619-27.

75. Krishnan V, Wang X, Safe S. *Estrogen receptor-Spl complexes mediate estrogen-induced cathepsin D gene expression in MCF-7 human breast cancer cells.* J Biol Chem. 1994;269:15912-7.

76. Porter W, Wang F, Wang W, Duan R, Safe S. *Role of estrogen receptor/Sp1 complexes in estrogen-induced heat shock protein 27 gene expression.* Mol Endocrinol. 1996;10:1371-8.

77. Rishi AK, Shao ZM, Baumann RG, Li XS, Sheikh S, Kimura S, et al. *Estradiol regulation of the human retinoic acid receptor a gene in human breast carcinoma cells is mediated via an imperfect half-palindromic estrogen response element and Sp1 motifs.* Cancer Res. 1995;55:4999-5006.

78. Porter W, Saville B, Hoivik D, Safe S. *Functional synergy between the transcription factor Sp1 and the estrogen receptor.* Mol Endocrinol. 1997;11:1569-80.

79. Sun G, Porter W, Safe S. *Estrogen-induced retinoic acid receptor a1 gene expression: role of estrogen receptor-Sp1 complex.* Mol Endocrinol. 1998;12:882-90.

80. Wang W, Dong L, Saville B, Safe S. *Transcriptional activation of E2F1 gene expression by 17beta-estradiol in MCF-7 cells is regulated by NF-Y-Sp1/ estrogen receptor interactions.* Mol Endocrinol. 1999;13:1373-87.

81. Lee W, Mitchell P, Tjian R. *Purified transcription factor AP-1 interacts with TPA-inducible enhancer elements.* Cell. 1987;49:741-52.

81. Lee W, Mitchell P, Tjian R. *Purified transcription factor AP-1 interacts with TPA-inducible enhancer elements.* Cell. 1987;49:741-52.

82. Gaub MP, Bellard M, Scheuer I, ChambonP, Sassone-Corsi P. *Activation of the ovalbumin gene by the estrogen receptor involves the fos-jun complex.* Cell. 1990;63:1267-76.

83. Dahlman-Wright K, Qiao Y, Jonsson P, Gustafsson JA, Williams C, Zhao C. *Interplay between AP-1 and estrogen receptor alpha in regulating gene expression and proliferation networks in breast cancer cells.* Carcinogenesis. 2012;33:1684-91.

84. Umayahara Y, Kawamori R, Watada H, Imano E, Iwama N, Morishima T, et al. *Estrogen-regulation of the insulin-like growth factor I gene transcription involves an AP-1 enhancer.* J Biol Chem. 1994;269:16433-42.

85. Kushner PJ, Agard DA, Greene GL, Scanlan TS, Shiau AK, Uht RM, et al. *Estrogen receptor pathways to AP-1.* J Steroid Biochem Mol Biol. 2000;74:311-7.

86. Webb P, Lopez GN, Uht RM, Kushner PJ. *Tamoxifen activation of the estrogen receptor/AP-1 pathway: potential origin for the cell-specific estrogen-like effects of antiestrogens.* Mol Endocrinol. 1995;9:443-56.

87. Paech K, Webb P, Kuiper GJM, Nilsson S, Gustafsson J, Kushner PJ, et al. *Differential ligand activation of estrogen receptors ERa and ERb at AP1 sites.* Science. 1997;277:1508-10.

88. O'Malley BW, Conneely OM. *Orphan receptors: In search of a unifying hypothesis for activation.* Mol Endocrinol. 1992;6:1359-61.

89. Johnston SD, Liu X, Zuo F, Eisenbraun TL, Wiley SR, Kraus RJ, et al. *Estrogen-related receptor a1 functionally binds as a monomer to extended half-site sequences including ones contained within estrogen-response elements.* Mol Endocrinol. 1997;11:342-52.

90. Yang N, Shigeta H, Shi H, Teng CT. *Estrogen-related receptor, hERR1, modulates estrogen receptor-mediated response of human lactoferrin gene promoter.* J Biol Chem. 1996;271:5795-804.

91. Kumar P, Mendelson CR. *Estrogen-related receptor gamma (ERRgamma) mediates oxygen-dependent induction of aromatase (CYP19) gene expression during human trophoblast differentiation.* Mol Endocrinol. 2011;25:1513-26.

92. Yang C, Zhou D, Chen S. *Modulation of aromatase expression in the breast tissue by ERRα-1 orphan receptor.* Cancer Res. 1998;58:5695-700.

93. Hu P, Kinyamu HK, Wang L, Martin J, Archer TK, Teng C. *Estrogen induces estrogen-related receptor alpha gene expression and chromatin structural changes in estrogen receptor (ER)-positive and ER-negative breast cancer cells.* J Biol Chem. 2008;283:6752-63.

94. Dickson RB, Lippman ME. *Growth factors in breast cancer.* Endocr Rev. 1995;16:559-89.

95. Ignar-Trowbridge DM, Nelson KG, Bidwell MC, Curtis SW, Washburn TF, McLachlan JA, et al. *Coupling of dual signaling pathways: Epidermal growth factor action involves the estrogen receptor.* Proc Natl Acad Sci USA. 1992;89:4658-62.

96. Nelson KG, Takahashi T, Bossert NL, Walmer DK, McLachlan JA. *Epidermal growth factor replaces estrogen in the stimulation of female genital-tract growth and differentiation.* Proc Natl Acad Sci USA. 1991;88:21-5.

97. Ignar-Trowbridge DM, Pimentel M, Parker MG, McLachlan JA, Korach KS. *Peptide growth factor cross-talk with the estrogen receptor requires the A/B domain and occurs independently of protein kinase C or estradiol.* Endocrinology. 1996;137:1735-44.

98. Cho H, Katzenellenbogen BS. *Synergistic activation of estrogen receptor-mediated transcription by estradiol and protein kinase activators.* Mol Endocrinol. 1993;7:441-52.

99. El-Tanani MKK, Green CD. *Two separate mechanisms for ligand-independent activation of the estrogen receptor.* Mol Endocrinol. 1997;11:928-37.

100. Aronica SM, Katzenellenbogen BS. *Stimulation of estrogen receptor-mediated transcription and alteration in the phosphorylation state of the rat uterine estrogen receptor by estrogen, cyclic adenosine monophosphate, and insulin-like growth factor-1.* Mol Endocrinol. 1993;7:743-52.

101. Bunone G, Briand PA, Miksicek RJ, Picard D. *Activation of the unliganded estrogen receptor by EGF involves the MAP kinase pathway and direct phosphorylation.* EMBO J. 1996;15:2174-83.

102. Feng W, Webb P, Nguyen P, Liu X, Li J, Karin M, et al. *Potentiation of estrogen receptor activation function 1 (AF-1) by Src/JNK through a serine 118-independent pathway.* Mol Endocrinol. 2001;15:32-45.

103. Katzenellenbogen BS. *Estrogen receptors: bioactivities and interactions with cell signaling pathways.* Biol Reprod. 1996;54:287-93.

104. Migliaccio A, Di Domenico M, Castoria G, de Falco A, Bontempo P, Nola E, et al. *Tyrosine kinase/p21ras/ MAP-kinase pathway activation by estradiol-receptor complex in MCF-7 cells.* EMBO J. 1996;15:1292-300.

6.1.1.5. EXTRANUCLEAR ER

In addition to nuclear ER, there is a subset of extranuclear ERs that associate either with the plasma membrane or mitochondria (1). ERα associates with caveolae or lipid raft domains in the plasma membrane and interacts with transmembrane growth factor receptors, such as EGFR, HER2, and insulin-like growth factor receptor receptor I (2). The plasma membrane-associated ERα was detected by controlled homogenization procedures with quantitative subcellular fractionation to limit extraction artifacts and by antibodies directed to different domains of nuclear ERα (2, 3). Transfection of ERα cDNA into Chinese hamster ovary cells resulted in a single transcript by Northern blot, specific binding of labeled E_2, and expression of ER in both nuclear and membrane cell fractions (4). Membrane-bound ER made up less than 3% of the protein product from the nuclear ER transcript. Mass spectrometry analysis of peptides isolated from plasma membranes confirmed that plasma membrane and nuclear ER are identical proteins (5). Thus, the plasma membrane-associated ER derives from the same gene as nuclear ERα but may undergo alternative splicing in endothelial and other cells resulting in an N-terminal truncated 46-kD variant (6). The receptor localization to the membrane involves post-translational palmitoylation of 447Cys in the LBD (Table 6.1.2.), which interacts with caveolin-1 (6-8). In cancer cells, E_2 reduced both ERα palmitoylation and its interaction with caveolin-1 in a time- and dose-dependent manner. The membrane-associated ERα also interacts with PELP1 (proline-, glutamic acid-, leucine-rich protein 1; also known as modulator of the non-genomic actions of the ER, MNAR), coupling the receptor to signaling cascades such as the MAPK and PI3K/ Akt pathways, Shc, Src kinases, and nitric oxide synthase (9-11). In summary, the plasma membrane-associated ERα mediates rapid non-genomic responses of the cell to E_2, including activation of signaling pathways. However, the non-genomic E_2 action can be linked to its genomic action, e.g., PI3K/Akt may interact with nuclear ERα, altering its expression and genomic activity (12). Other membrane proteins, such as G-protein-coupled receptor GPR30, are capable of binding estrogen with high affinity (K_d 2.7 nM) and may contribute to promiscuous estrogen signaling in cells with or without ER expression (13, 14). The location of GPR30 in the cell membrane allows the interaction with other membrane-associated proteins, such as the epidermal growth factor receptor (15).

Finally, ERα and ERβ have been localized in mitochondria of MCF-7 breast cancer cells and various other cells and tissues (16). ER is transcriptionally active in mitochondria and E_2 treatment enhanced mitochondrial DNA (mtDNA)-encoded gene transcript levels (17). In addition, E_2 stimulated ERα-mediated transcription of nuclear respiratory factor-1 (NRF-1) via an imperfect ERE in the promoter (18). In turn, NRF-1 increased transcription of the mitochondrial transcription factor Tfam, which then increased the transcription of mtDNA-encoded genes. Thus, E_2 has a direct and indirect effect on mitochondrial gene expression, e.g., cytochrome c oxidase, via mitochondrial and nuclear ERα (19). Besides the effect on ER-mediated gene transcription, E_2 has non-transcriptional effects on the production of reactive oxygen species by directly interacting with the mitochondrial respiratory complexes (20).

References

1. Yager JD, Davidson NE. *Estrogen carcinogenesis in breast cancer.* N Engl J Med. 2006;354:270-82.

2. Pietras RJ, Marquez-Garban DC. *Membrane-associated estrogen receptor signaling pathways in human cancers.* Clin Cancer Res. 2007;13:4672-6.

3. Pappas TC, Gametchu B, Watson CS. *Membrane estrogen receptors identified by multiple antibody labeling and impeded-ligand binding.* FASEB J. 1995;9:404-10.

4. Razandi M, Pedram A, Greene GL, Levin ER. *Cell membrane and nuclear estrogen receptors (ERs) originate from a single transcript: studies of ERalpha and ERbeta expressed in Chinese* *hamster ovary cells.* Mol Endocrinol. 1999;13:307-19.

5. Pedram A, Razandi M, Levin ER. *Nature of functional estrogen receptors at the plasma membrane.* Mol Endocrinol. 2006;20:1996-2009.

6. Li L, Haynes MP, Bender JR. *Plasma membrane localization and function of the estrogen receptor alpha variant (ER46) in human endothelial cells.* Proc Natl Acad Sci U S A. 2003;100:4807-12.

7. Acconcia F, Ascenzi P, Bocedi A, Spisni E, Tomasi V, Trentalance A, et al. *Palmitoylation-dependent estrogen receptor alpha membrane localization: regulation by 17beta-estradiol.* Mol Biol Cell. 2005;16:231-7.

8. Razandi M, Oh P, Pedram A, Schnitzer J, Levin ER. *ERs associate with and regulate the production of caveolin: implications for signaling and cellular actions.* Mol Endocrinol. 2002;16:100-15.

9. Barletta F, Wong CW, McNally C, Komm BS, Katzenellenbogen B, Cheskis BJ. *Characterization of the interactions of estrogen receptor and MNAR in the activation of cSrc.* Mol Endocrinol. 2004;18:1096-108.

10. Murphy E. *Estrogen signaling and cardiovascular disease.* Circ Res. 2011;109:687-96.

11. Vadlamudi RK, Manavathi B, Balasenthil S, Nair SS, Yang Z, Sahin AA, et al. *Functional implications of altered subcellular localization of PELP1 in breast cancer cells.* Cancer Res. 2005;65:7724-32.

12. Stoica GE, Franke TF, Moroni M, Mueller S, Morgan E, Iann MC, et al. *Effect of estradiol on estrogen receptor-alpha gene expression and activity can be modulated by the ErbB2/PI 3-K/Akt pathway.* Oncogene. 2003;22:7998-8011.

13. Prossnitz ER, Arterburn JB, Sklar LA. *GPR30: A G protein-coupled receptor for estrogen.* Mol Cell Endocrinol. 2007;265-266:138-42.

14. Thomas P, Pang Y, Filardo EJ, Dong J. *Identity of an estrogen membrane receptor coupled to a G protein in human breast cancer cells.* Endocrinology. 2005;146:624-32.

15. Filardo EJ, Quinn JA, Sabo E. *Association of the membrane estrogen receptor, GPR30, with breast tumor metastasis and transactivation of the epidermal growth factor receptor.* Steroids. 2008;73:870-3.

16. Chen JQ, Eshete M, Alworth WL, Yager JD. *Binding of MCF-7 cell mitochondrial proteins and recombinant human estrogen receptors alpha and beta to human mitochondrial DNA estrogen response elements.* J Cell Biochem. 2004;93:358-73.

17. Chen JQ, Yager JD. *Estrogen's effects on mitochondrial gene expression: mechanisms and potential contributions to estrogen carcinogenesis.* Ann N Y Acad Sci. 2004;1028:258-72.

18. Mattingly KA, Ivanova MM, Riggs KA, Wickramasinghe NS, Barch MJ, Klinge CM. *Estradiol stimulates transcription of nuclear respiratory factor-1 and increases mitochondrial biogenesis.* Mol Endocrinol. 2008;22:609-22.

19. Klinge CM. *Estrogenic control of mitochondrial function and biogenesis.* J Cell Biochem. 2008;105:1342-51.

20. Felty Q, Xiong WC, Sun D, Sarkar S, Singh KP, Parkash J, et al. *Estrogen-induced mitochondrial reactive oxygen species as signal-transducing messengers.* Biochemistry. 2005;44:6900-9.

box-binding protein (TBP) and TFIIB. The interaction of ER with TBP results in enhanced E_2-mediated transcription (5). While ER alone activated a simple promoter consisting of ERE and TATA box, TBP overexpression potentiated transcription two-fold as measured by CAT activity. TBP potentiated only estrogen-induced but not basal transcription and was independent of spacing between response element and TATA box. TBP overexpression also reduced autoinhibition by overexpressed ER, suggesting that one target of the autoinhibition may be TBP itself. The ER was also shown to interact with TFIIB (6). A truncated ER construct containing only the N-terminal AF-1 did not associate with TFIIB, suggesting that the domain responsible for TFIIB-ER interaction can be narrowed to the AF-2 region. TFIID is a large complex consisting of at least 11 protein subunits, TBP and ten TBP-associated factors or TAFs ranging in size from 15 to 250 kD. In addition to the common TAFs there is a distinct subunit, TAFII30, which appears to be necessary for transcriptional activation by ER (7). Neither E_2 nor estrogen antagonists influenced the association of TBP, TFIIB, or TAFII30 with ER. *In vitro*, TAFII30 stimulated E_2-dependent transcription initiation 20-fold in the presence of the ubiquitous high-mobility group protein 1, which promotes ER-ERE binding (28).

The mechanism by which the hormone-dependent AF-2 stimulates transcription was examined with an AF2-gluta-thione-S-transferase chimeric fusion protein, GST-AF2 (29). Two different assays were utilized to analyze AF2-protein interaction *in vitro*, i.e., far-Western blotting and GST pull-down assay. By far-Western blotting, three receptor-interacting proteins (RIPs) of 160, 140, and 80 kD were detected in extracts of different mammalian cells (ZR-75, HeLa, COS-1). The GST pull-down assay identified 160, 100, and 50 kD RIPs. RIPs interacted with GST-AF2 only in the presence of estrogen; the interaction was abolished by antiestrogens (4-hydroxytamoxifen, ICI 164,384, ICI 182,780). Thus, ligands specify the binding affinity of RIPs to ER (30). The interaction was also abolished by a GST-AF2 mutant containing two altered residues, Met547Ala and Leu548Ala that had little effect on E_2 binding but interfered with RIP binding, possibly by virtue of their location in a putative amphipathic helix flanked by prolines.

6.1.1.6. ER, COACTIVATORS, AND CHROMATIN

To accomplish estrogen-dependent transcription, the ER has to interact not only with DNA but also with a network of proteins that help to integrate the hormone response into diverse regulatory pathways (1). Among the interacting proteins are basal transcription factors, transcriptional coactivators, and epigenetic regulators (Table 6.1.5.). The C-terminal AF-2 domain of the ER has been established as main contact site for interacting proteins, which may also include RNA-binding proteins (2-4). Examples of basal transcription factors interacting with ER are the TATA

'Coactivators' are a diverse group of non-DNA binding proteins that function as bridging proteins between nuclear receptors and other transcription factors. The coactivators induce structural changes in 'activator proteins' (i.e., nuclear receptors or other transcription factors) with a DNA-binding domain resulting in enhanced transcription. Examples of coactivators include receptor interacting proteins (RIPs), receptor associated proteins (RAPs) and steroid receptor coactivator (SRC) family. The 140 kD receptor-interacting protein (RIP140) interacts with the ERα domain AF-2 in the presence of E_2; no signal was observed in the absence of ligand or in the presence of antiestrogens (8). The

Table 6.1.5. ER Interaction with Network of Transcription Factors, Coactivators, and Epigenetic Regulators

Protein	Function	ER Contact	E_2 Effect	References
TBP	basal transcription factor	AF-2	-	(5)
TFIIB	basal transcription factor	AF-2	-	(6)
TAFII30	basal transcription factor	AF-2a	-	(7)
RIP 140	coactivator	AF-2	+	(8)
ERAP140	coactivator	AF2	+	(9)
SRC family				(10,11)
SRC-1, ERAP 160	coactivator, HAT*	AF-2	+	(10,12)
SRC-2, TIF2, GRIP1	coactivator, HAT	AF-2	+	(13,14)
SRC-3, also known as	coactivator, HAT	AF-2	+	(15-18)
AIB1,ACTR,RAC3,p/CIP				
CBP/p300	coactivator, HAT		+	(16,19)
PCAF	coactivator, HAT			
Cyclin D1	coactivator	AF2	+	(20,21)
MNAR (PELP1)	coactivator	AF2	+	(22)
TIF1 family				
TIF1a(TRIM24)	epigenetic regulator	AF-2	+	(23)
TIF1b		AF-2	+	(24)
SWI/SNF complex	chromatin remodeling			(25-27)
BRG1 (SMARCA4)		AF-2	+	
BRM (SMARCA2)		AF-2	+	

* HAT = histone acetyltransferase

dependence of RIP140 interactions on estrogen binding is consistent with the fact that the hormone induces a conformational change in the LBD. However, the observation that AF-2 is not sufficient by itself to stimulate transcription when fused to a heterologous DBD suggests that other regions are necessary to bind RIP140.

In vivo, RIP140 is recruited by E_2-bound ER to the promoter region of ER target genes and enhances their transcription. RIP140 also regulates the E2F pathway and represses E2F1 transactivation, thereby inhibiting the expression of E2F1 target genes, such as cyclin E and cyclin B2 (31). In breast cancer cell lines, increasing RIP140 levels resulted in a reduction in the proportion of cells in S phase. In breast cancer, low RIP140 mRNA expression was associated with high E2F1 expression and basal-like tumors. The unrelated 140 kD nuclear protein ERAP140 interacts with ER like RIP140 (9). ERAP is primarily expressed in the brain but also detected in ER-positive (MCF-7) and ER-negative (MDA-231) breast cancer cells.

The coactivator paradigm is the **steroid receptor coactivator (SRC) family,** which consists of three members (SRC-1, SRC-2, and SRC-3; Table 6.1.5.) that have been called masters of systems biology because of their extensive regulatory effects on multiple transcription factors (e.g., nuclear receptors, AP-1, NFκB, HIF1α, STATs, and E2F1) in normal tissues and cancer (11, 32). The SRCs function

in transcription initiation, elongation, mRNA splicing and translation as well as receptor and coactivator turnover. They share both functional and structural similarities. For example, they bind *in vitro* to the LBD of nuclear receptors in a ligand- and AF-2 integrity-dependent manner and they enhance in vivo ligand-dependent transcriptional activity. The structural similarity extends to overall amino acid identity and conserved sequence motifs. The N-terminal bHLH-PAS (basic helix-loop-helix-PER/ARNT/Sim) domain is the most highly conserved and is necessary for several protein-protein interactions with other co-regulators. The midregion contains LXXLL (where L is leucine and X is any amino acid), which is repeated several times in each member of the SRC family, e.g., four times in SRC-1. Functional analysis demonstrated that the ability of these coactivators to bind nuclear receptors and enhance their transcriptional activity is dependent on the integrity of the LXXLL motifs. Mutations in the second and third LXXLL motifs in SRC-2 (TIF2, GRIP1) eliminated functional and binding interactions with ER, AR, and GR (33, 34). Interactions with ER were more strongly affected by mutations in the second motif, whereas interactions with AR and GR were more strongly affected by mutations in the third motif. The non-leucine residues within the LXXLL motifs and adjacent residues may also determine which nuclear receptors assemble with which coactivators following hormonal induction (35, 36). Structural analysis predicts that the LXXLL motifs are α-helical with the conserved leucines forming a hydrophobic face interacting with key hydrophobic residues in helix 12 of the ER LBD that are essential for ligand-dependent AF-2 (18, 37). Weaker contacts of the C-terminal end of SRC-1 may also exist with AF-1, thereby explaining the synergistic interaction of AF-1 and AF-2 observed in some cells (38, 39). Binding of estrogens induces a conformation in ER that stabilizes SRC-1 binding and, in a reciprocal fashion, SRC-1 binding stabilizes the estrogen-ER complex (40). Increasing the concentration of SRC-1 in Chinese hamster ovary cells by an expression plasmid encoding SRC-1 caused a 17-fold increase in the potency of E_2 in an estrogen-responsive reporter gene transcription assay.

SRC-1 and SRC-3 are expressed in normal and malignant breast tissue and cell lines, such as MCF-7 and MDA-MB-231. SRC-1 expression is low in normal mammary epithelial cells but increased in breast cancer, especially HER2-positive tumors (41-43). SRC-1 overexpression in breast cancer cell lines enhanced the stimulatory effect of E_2 and the agonistic effect of 4-hydroxytamoxifen in a dose-dependent manner (44). SRC-1 null mice exhibited decreased growth and development of mammary glands in response to estrogens (45).

Nuclear receptors can compete for coactivators and thereby interfere in the activity of another receptor. For example, the ERα-E_2 complex interferes with the action of the PR bound to PRE even though the ERα does not bind to the PRE (46). This interference was attributed to the titration by ERα-E_2 of a common and limiting target factor that is required by PRE-bound PR. Thus, the cross-interference or squelching is defined as the sequestering of limiting factors, e.g., coactivators, from the transcription complex by an overabundance of another coactivator-binding protein, i.e., nuclear receptor. ER squelching was found to involve AF-2 and be cell specific, occurring in human breast and CV1 cells, but not HepG2 liver cells (47-49).

In breast cancer cells SRC-1 also enhanced the ligand-independent activation of ER by agents that raise the intracellular concentration of cyclic AMP, e.g., forskolin and IBMX (44). Cyclic AMP is capable of eliciting a transcriptional response through the activation of protein kinase A and the subsequent phosphorylation of the cAMP response element binding protein (CREB). This phosphorylation has been shown to result in the interaction between CREB and the CREB-binding protein (CBP). In addition to CREB, CBP serves as coactivator for several, diverse transcription factors, e.g., c-Fos, c-Jun, c-Myb, and MyoD (50, 51). CBP and the closely related p300 protein also interact with components of the basal transcription apparatus including RNA polymerase II, TBP, and TFIIB (52). Finally, CBP/p300 interact with several nuclear hormone receptors, including ER (19, 53, 54). Hormone binding greatly stimulates this interaction which involves the respective LBD and the

N-terminal 101 amino acids of CBP/p300. The N-terminal region of CBP/p300 also interacts with other coactivators, such as SRC-1 and p/CIP (18, 54, 55). Thus, nuclear hormone receptors can bind simultaneously to more than one coactivator, e.g., CBP/p300 and SRC-1, in principle forming a ternary complex bound to DNA. Individually, CBP and SRC-1 stimulated ER to an equivalent extent (6- to 7-fold). However, when they were cotransfected into the same cell, the increase in ER transcriptional activity was synergistic (about 38-fold) (44). This suggests that multiple coactivators may jointly modulate and, in some instances, synergistically stimulate transactivation mediated by a single hormone receptor (50). Since CBP/p300 coordinate the transcriptional effects of simultaneous signals emanating from cell surface and nuclear receptors, they serve as 'cointegrators' of extracellular and intracellular signaling pathways (53).

Cyclin D1 and MNAR (modulator of non-genomic activity of estogen receptor; also known as proline-, glutamic acid-, and leucine-rich protein 1, PELP1) are examples of coactivators that connect ER with specific regulatory pathways, such as cell cycle progression. Cyclin D1, a key regulator of the G1/S cell cycle transition, has been shown to activate ER-mediated transcription independent of complex formation with cyclin-dependent kinases (20, 21). The activation, which does not occur with other cyclins, is by direct binding of cyclin D1 to the LBD of the receptor rather than by enhanced ER phosphorylation. The cyclin D1-ER interaction results in increased ER-ERE binding in the absence of E_2. Addition of E_2 further enhances ER-ERE binding and upregulates ER-mediated transcription. The dual activity of cyclin D1 as activator of ER and mediator of cell cycle progression defines this particular cyclin as 'cointegrator' for estrogen-mediated proliferation of hormone-responsive cells, such as breast epithelium. MNAR is another coactivator that interacts with ERα and enhances the ERα-mediated action on G1 progression (22). MNAR also interacts with the retinoblastoma tumor suppressor protein (pRb), which, in turn, forms a complex with Cdk4/6-cyclin D and the transcription factors E2F. MNAR is phosphorylated by Cdk4 and enhances the expression of pRb-E2F target genes, such as cyclin D1 and Cdc25C (56). Thus, MNAR connects estrogen signaling to the activation of E2F target genes, which may explain the oncogenic potential of overexpressed MNAR (57).

A central question in understanding nuclear hormone receptor-mediated transcription is how each receptor gains access to DNA tightly packed in chromatin. Nuclear receptors have been shown to induce chromatin remodelling at hormone response elements to gain access to their respective DNA binding site (58, 59). The effect of ER on chromatin structure has been examined in yeast containing chromosomally integrated ERE-reporter genes. ER deletion mutants lacking either AF-1 or AF-2 generated a smaller area of disrupted chromatin than wild type ER, suggesting that regions of the receptor containing trans-activation functions also play a role in mediating local chromatin structure disruption (60). The receptor-induced alterations in chromatin structure were dependent on the presence of E_2, indicating that ER requires ligand to bind to ERE in vivo and that DNA binding per se, i.e., independent of transcription, disrupts chromatin structure (61). Tran-scriptional activation by ER or GR expressed in yeast was dependent on the SWI1, SWI2/SNF2 and SWI3 proteins of the SWI/SNF (SWItch/Sucrose Non-Fermentable) complex (62). SWI/SNF is a multi-subunit complex uses the energy of ATP hydrolysis to modify chromatin structure in order to regulate gene expression (63, 64). The complex is evolu-tionarily conserved and is comprised of a catalytic subunit with helicase/ATPase activity that may be either BRG1 (Brahma-related gene 1, homologous to both yeast SWI2/SNF2 and Drosophila brahma; also known as SMARCA4) or BRM (SMARCA2) and several associated proteins, known as BAF (BRG1- or BRM-associated factors) that modulate its activity and the recruitment of the complex to specific promoters (65). Although the SWI/SNF complex has no intrinsic ability to bind DNA it controls transcriptional initiation at the chromatin level by mediating nucleosomal reorganization and by facilitating binding of transcription factors to DNA targets. ER interacts with both BRG1 and BRM and transcriptional activation by ER was enhanced by coexpression of either BRG1 (hSNF2ß; 9.4-fold) or BRM (hSNF2α; 3-fold) cDNA (26, 66). Mutagenesis experiments showed that the N-terminal regions of BRG1 and BRM interact with the AF-2, specifically amino acids 543Met and 544Leu in helix 12 of the LBD (27). E_2 increased the interaction whereas 4-hydroxytamoxifen had no effect.

BRG1 is mutated in cancer cell lines of several types, including breast cancer cell lines ALAB and Hs578T (65, 67). Knockdown of BRG1 by RNA interference in MDA-MB-231 and BT-549 breast cancer cell lines markedly inhibited cell proliferation due to G1 phase arrest (68). High expression of BRG1 in 437 breast cancers was inversely correlated with overall and disease-specific survival (68).

Another nucleosome assembly factor, the yeast protein SPT6, also stimulated ER transcriptional activity by interacting in vivo with the AF-2 (25). The binding of the SWI/SNF complex and ySPT6 to AF-2 suggests that at least part of the transcriptional activation brought about by ER AF-2 involves E_2-dependent remodelling of the chromatin template of estrogen target genes.

The fundamental building block of chromatin is the nucleosome, which is composed of an octamer of histone proteins (two molecules each of histones H2A, H2B, H3, and H4) and 146 bp of DNA in 1.75 turns tightly wrapped around the octamer. Histones possess a globular core domain, required for histone-histone interactions central to

nucleosome formation, and highly charged, unstructured N-terminal tail domains that protrude from the octamer. These tails are important both for histone-DNA interactions and for interactions with other non-histone proteins. Consecutive nucleosomes are separated by unwrapped linker DNA, typically between 20 and 50 bp in length. Since wrapped nucleosomal DNA is inherently less accessible than linker DNA, the genomic positioning and compaction of nucleosomes strongly influences the ability of nonhistone proteins to bind target sequences within DNA. While the SWI/SNF complex influences global chromatin architecture, epigenetic regulators influence local chromatin structure by covalent modifications of either DNA or histones (69-71). **Epigenetic regulators** include proteins capable of adding (epigenetic 'writers') or removing chemical modifications (epigenetic 'erasers') as well as epigenetic or chromatin 'readers' that recognize different covalent modifications of the nucleosome and assemble functional complexes onto specific loci to facilitate DNA-templated processes, e.g., transcription, repair, and replication (70, 72-74). At least four different DNA modifications and 16 distinct classes of histone modification have been described (75-77). DNA methyltransferases (DNMT1, DNMT3a, DNMT3b) catalyze the addition of a methyl group to the 5' position of cytosine (5mC), predominantly within the CpG dinucleotides of CpG islands in the promoter region of genes. The histone tails of all four core histones are subject to a variety of post-translational modifications including methylation, acetylation, phosphorylation, ubiquitylation, sumoylation, crotonylation, and proline isomerization, all of which occur at the site of a specific amino acid, e.g., lysine 4 or 9 on the histone 3 tail (H3K4, H3K9). Neutralization of positive charges associated with these highly conserved lysine residues within the N-terminal histone tails by acetylation provides a reversible mechanism for regulating interactions with DNA and non-histone proteins. Acetylation causes an allosteric change in nucleosome conformation that renders nucleosomal DNA more accessible to transcription factors and thereby facilitates gene expression. Conversely, histone deacetylation stabilizes histone-DNA interactions that lead to transcriptional repression. Several of the transcrip-tional coactivators function as histone acetyltransferases whereas several transcriptional corepressors act as histone deacetylases (Table 6.1.5.). Histone methyltransferases and demethylases also introduce reversible post-translational modifications of histone proteins. Thus, histones and chromatin structure are fully integrated components of the transcriptional machinery of gene expression (71).

Several coactivator proteins are histone acetyltransferases (HATs) including SRC-1, SRC-2, SRC-3, CBP/p300, the p300/CBP-associated protein (PCAF), and the TBP-associ-ated factor TAFII250 (16, 17, 78-81) (Table 6.1.5.). Members of the SRC family acetylate primarily histones H3 and H4, the HAT activity is located in their C-terminal region, they recruit other HATs such as CBP/p300, and they interact with many members of the nuclear receptor superfamily (16, 81). With regard to ERα, SRC-1, SRC-3, CBP/p300, and PCAF were shown to synergistically activate transcription driven by ERα (16, 44). They are targeted to the promoter regions of estrogen-responsive genes by interaction with ERα bound to specific EREs in response to E_2 binding. Thus, the E_2-induced recruitment of ER-associated coactivators with intrinsic HAT activity may bias the existing equilibrium between histone acetylation and histone deacetylation toward increasing transcription of estrogen-dependent genes from transcriptionally repressed chromatin (16, 81). The expression of SRC-3 is downregulated by miRNAs miR-17-5p, miR-20a, miR-20b, and miR-106b, the first two originating from the miR-17-92 cluster that also inhibits protein translation of ERα (82).

Chromatin readers that simultaneously recognize histones with multiple marks allow transduction of complex modification patterns into the expression of specific target genes. An example of a chromatin reading regulator is the tripartitite motif-containing 24 (TRIM24; also known as transcriptional intermediary factor 1α, TIF1α) protein, which contains a tandem plant homeodomain (PHD) and bromodomain region that recognizes unmodified H3K4 (histone H3 unmodified at lysine 4, H3K4me0) and acetylated H3K23 (histone H3 acetylated at lysine 23, H3K23ac) within the same histone tail (83). TRIM24 also interacts with the AF2 domain of ERα and other nuclear receptors (84). Genome-wide analysis of chromatin interactions revealed E_2-dependent binding of TRIM 24 and ERα at sites that exhibited E_2-activated loss of H3K4me2 and gain of histone acetylation. These findings stand in contrast to a model of chromatin accessibility at ERα binding sites, facilitated by FOXA1 and H3K4me2 enrichment in response to E_2 treatment (85), but are in agreement with findings that H3K4me3 is not present at a majority of distal ERE regions (86). The number of target sites shared by TRIM24 and ERα (1,677 sites) was similar to ERα and FOXA1, with little overlap among all three (263 sites) (83, 85). Thus, ERα-regulated genes may be divided into multiple classes, defined by specific coactivators and their dependence on H3K4 methylation. Aberrant overexpression of TRIM24 occurred frequently in breast cancers and correlated with poor survival of patients (83).

While coactivators are recruited to ligand-bound nuclear receptors to enhance gene expression, **corepressors** fulfill the opposite role, i.e., they bind to un-liganded receptors and repress transcription (87). The best studied corepressors are two homologous ~275 kD proteins, silencing mediator of retinoic acid and thyroid hormone receptors (SMRT) and nuclear corepressor (N-CoR) (88, 89). SMRT and N-CoR have an overall sequence identity of 40% and conserved regions such as the deacetylase activation domain (DAD) and the histone interaction domain (HID). The DAD both recruits and activates histone

deacetylase enzymes (e.g., HDAC1, HDAC3) whereas the HID interacts directly with histone tails, i.e., the enzymatic substrate of HDACs. In addition, SMRT and N-CoR contain two conserved amphipathic co-repressor motifs, φxxφφ (where φ is a hydrophobic amino acid and x any amino acid), that interact with the LBD in the AF2 domain of the nuclear receptor. In some cases the corepressor may also interact with the DBD of the nuclear receptor (90). ER was shown to bind to SMRT and N-CoR regardless of the presence of estrogen or antiestrogen (44, 91). Contrary to its assumed role as an obligatory corepressor, SMRT stimulated E_2-induced ER activity in MCF-7 but not HepG2 cells and was required for full expression of the target gene cyclin D1 but not pS2 (92). Whereas N-CoR acted as a corepressor, SMRT behaved as a coactivator in a gene- and cell-specific manner. Yet another member of the repressor complex is the transcriptional regulatory protein Sin3 (93, 94). Thus, a quaternary complex (HDAC, Sin3, SMRT/N-Co-R, nuclear receptor) would tether the histone deacetylase activity of HDAC to a specific promoter determined by a nuclear receptor, e.g., TR, RXR, RAR, ER. Ligand binding causes a conformational change in the LBD, resulting in dissociation of SMRT/N-Co-R from the receptor and thereby release of Sin3 and HDAC as well (95). At the same time, the displacement of the corepressor is associated with the formation of a suitable coactivator binding surface. In summary, one can envision a competition/chromatin structure model, in which multiprotein deacetylase repressor complexes compete with acetyltransferase transcriptional activator complexes to assemble an active or inactive chromatin structure. Taking the thyroid hormone receptor TR as an example, transcriptional regulation by nuclear receptors would occur in three steps: (i) precise rotational positioning of the DNA sequence containing the TRE on the surface of the histone octamer allows the TR/RXR dimer access to chromatin; (ii) once in place, the unliganded receptor recruits a deacetylase complex (HDAC, Sin3, N-CoR) to augment repression; (iii) subsequent addition of thyroid hormone leads to dissociation of the deacetylase complex and recruitment of acetyltransferases (CBP/p300, PCAF, TAFII250) that can disrupt these repressive histone-DNA interactions and permit transcription of thyroid hormone-dependent genes (96). Time-resolved monitoring of the interaction of ER with the pS2 (trefoil factor 1) gene revealed that ERα directed an ordered, cyclical, and combinatorial recruitment and release of coactivators, corepressors, histone acetyltransferases and deacetylases as well as histone methyltransferases and demethylases at the chromatin region of the pS2 promoter (97). Thus, nuclear receptor-mediated regulation of target gene transcription involves the dynamic and in part cyclical interaction with multiple proteins in regulatory chromatin regions (98).

The regulation of transcription requires complex interactions between proteins bound to DNA sequences that are often separated by hundreds of base pairs. Three models have been proposed to explain the mechanism by which transacting factors act at a distance: the scanning model, the structural transmission model, and the **DNA looping model** (99). Experimental evidence supports the looping model in which long-range interactions are accomplished by looping out of the intervening DNA sequence between DNA-bound transacting factors, allowing the factors to come into close proximity with one another (100). Pertinent to E_2-mediated gene transcription, the looping model is supported by data obtained with the *Xenopus* vitellogenin B1, rat prolactin, and human pS2 and BRCA1 genes. The formation of loops during E_2-induced transcription not only brings together far upstream or downstream regions but also establishes contacts between the 5' and 3' ends of genes permitting interaction of 3' end-processing factors with components of the transcriptional machinery (101). The E_2-dependent transcription of the *Xenopus* vitellogenin B1 gene is enhanced through positioning of a nucleosome between -300 and -140 relative to the start site of transcription. The DNA wraps around the histone core of the nucleosome and through this static loop brings the distal ER binding site close to proximal promoter elements. This, in turn, appears to facilitate interactions between transacting factors (e.g., ER, NF1) and/or RNA polymerase II (102). The transcription of the rat prolactin gene is controlled by ER and other transacting factors that interact with two DNA elements, the distal enhancer and the proximal promoter, which are separated by nearly 1500 base pairs of DNA (103, 104). E_2 stimulates transcription through interaction of ER with an ERE located in the distal enhancer between -1550 and -1578 base pairs from the transcription start site. Mutation of the ERE abolished the response of the prolactin gene to estrogen (105). E_2 treatment of rat pituitary GH3 cells stimulated formation of a chromatin loop between the distal enhancer and proximal promoter of the prolactin gene, juxtaposing these two transcriptional control regions (106, 107). These data suggest that the activation of prolactin gene transcription by E_2 involves the stabilization of a chromatin loop that facilitates protein-protein interactions between transcription factors that are associated with the distal enhancer and the proximal promoter. It appears that some of the factors modifying the chromatin structure are specific for the pituitary (108). Transcription of the pS2 promoter is activated by ER, which binds to an imperfect ERE located at nucleotide positions -405 to -393, approximately 375 base pairs from the TATA box at -30 to -24 (109). *In vivo* digestion by nucleases followed by ligation-mediated PCR and footprinting analysis revealed two nucleosomes within this promoter region, one containing the ERE at its 5' edge and the other encompassing the TATA box at its 3' edge (110). Rotational phasing of the two nucleosomes would juxtapose ERE and TATA box within 8 to 14 nm of each other and thereby dramatically reduce the distance between these binding sites relative to their locations on naked DNA.

Binding of ER could facilitate subsequent binding of TBP by coordinating an opening of the chromatin structure around the TATA box, previously inaccessible within the chromatin fiber. Gene conformation analysis in mouse mammary tissue and MCF-7 breast cancer cells demonstrated the existence of BRCA1 gene loop structures between the promoter and sequences including the introns and the termination region (111). E_2 stimulation resulted in a loss of association between the 5' and 3' ends of the BRCA1 gene. Although the 'gene-looping mediator' factor is unknown, the data suggest that the 5' and 3' ends of BRCA1 are juxtaposed when gene expression is repressed and released upon transcriptional induction by E_2. Thus, ER regulates the transcription of many genes by dynamic long-range chromatin interactions (101). The interactions are intrachromosomal as well as interchromosomal associating with interchromatin granules that harbor key factors for transcriptional elongation and splicing (112). A genome-wide analysis using paired-end tag sequencing (ChIA-PET) identified 689 ERα-bound chromatin interaction regions, in which distal ERα-binding sites interact with proximal sites, forming chromatin loops (86). ERα dimers are recruited to multiple and primarily distal EREs, which interact with one another and possibly other factors such as FoxA1 and RNA polymerase II to form chromatin looping structures around target genes. This three-dimensional architecture may partition individual genes into subcompartments of nuclear space such as interaction-anchor-associated genes and interaction-loop-associated genes for differential transcriptional activation or repression.

How many genes are regulated by ERα? The answer depends largely on the technique of investigation. Large scale chromatin immunoprecipitation (ChIP) profiling revealed 5000 to 10000 ERα binding sites whereas gene expression analysis by microarrays identified 100 to 1500 E_2-responsive genes (58, 113-119). The apparent discrepancy between binding sites and regulated genes is in part due to the fact that not all ERα-binding sites are active under all conditions while mRNA levels do not necessarily reflect gene activity because they are subject to degradation and regulation. Moreover, ERα-target genes respond differently to E_2, tamoxifen, and faslodex (fulvestrant, ICI 182,780) (119).

References

1. Robyr D, Wolffe AP, Wahli W. *Nuclear hormone receptor coregulators in action: diversity for shared tasks.* Mol Endocrinol. 2000;14:329-47.

2. Norris JD, Fan D, Sherk A, McDonnell DP. *A negative coregulator for the human ER.* Mol Endocrinol. 2002;16:459-68.

3. Powers CA, Mathur M, Raaka BM, Ron D, Samuels HH. *TLS (translocated-in-liposarcoma) is a high-affinity interactor for*

steroid, thyroid hormone, and retinoid receptors. Mol Endocrinol. 1998;12:4-18.

4. Watanabe M, Yanagisawa J, Kitagawa H, Takeyama K, Ogawa S, Arao Y, et al. *A subfamily of RNA-binding DEAD-box proteins acts as an estrogen receptor alpha coactivator through the N-terminal activation domain (AF-1) with an RNA coactivator, SRA.* EMBO J. 2001;20:1341-52.

5. Sadovsky Y, Webb P, Lopez G,

Baxter JD, Fitzpatrick PM, Gizang-Ginsberg E, et al. *Transcriptional activators differ in their responses to overexpression of TATA-box binding protein.* Mol Cell Biol. 1995;15:1554-63.

6. Ing NH, Beekman MM, Tsai SY, Tsai MJ, O'Malley BW. *Members of the steroid hormone receptor superfamily interact with TFIIB (S300-II).* J Biol Chem. 1992;267:17617-23.

7. Jacq X, Brou C, Lutz Y, Davidson I, Chambon P, Tora L. *Human TAFII30 is present in a distinct TFIID complex and is required for transcriptional activation by the estrogen receptor.* Cell. 1994;79:107-17.

8. Cavailles V, Dauvois S, L'Horset F, Lopez G, Hoare S, Kushner PJ, et al. *Nuclear factor RIP140 modulates transcriptional activation by the estrogen receptor.* EMBO J. 1995;14:3741-51.

9. Shao W, Halachmi S, Brown M. *ERAP140, a conserved tissue-specific nuclear receptor coactivator.* Mol Cell Biol. 2002;22:3358-72.

10. Onate SA, Tsai SY, Tsai M, O'Malley BW. *Sequence and characterization of a coactivator for the steroid hormone receptor superfamily.* Science. 1995;270:1354-7.

11. York B, O'Malley BW. Steroid receptor coactivator (SRC) family: masters of systems biology. J Biol Chem. 2010;285:38743-50.

12. Halachmi S, Marden E, Martin G, MacKay H, Abbondanza C, Brown M. *Estrogen receptor-associated proteins: possible mediators of hormone-induced transcription.* Science. 1994;264:1455-8.

13. Hong H, Kohli K, Trivedi A, Johnson DL, Stallcup MR. *GRIP1, a novel mouse protein that serves as a transcriptional coactivator in yeast for the hormone binding domains of steroid receptors.* Proc Natl Acad Sci USA. 1996;93:4948-52.

14. Voegel JJ, Heine MJ, Zechel C, Chambon P, Gronemeyer H. *TIF2, a 160 kDa transcriptional mediator for the ligand-dependent activation function AF-2 of nuclear receptors.* EMBO J. 1996;15:3667-75.

15. Anzick SL, Kononen J, Walker RL, Azorsa DO, Tanner MM, Guan XY, et al. *AIB1, a steroid receptor coactivator amplified in breast and ovarian cancer.* Science. 1997;277:965-8.

16. Chen H, Lin RJ, Schiltz RL, Chakravarti D, Nash A, Nagy L, et al. *Nuclear receptor coactivator ACTR is a novel histone acetyltransferase and forms a multimeric activation complex with P/CAF and CBP/p300.* Cell. 1997;90:569-80.

17. Li H, Gomes PJ, Chen JD. *RAC3, a steroid/nuclear receptor-associated coactivator that is related to SRC-1 and TIF2.* Proc Natl Acad Sci USA. 1997;94:8479-84.

18. Torchia J, Rose DW, Inostroza J, Kamei Y, Weston S, Glass CK, et al. *The transcriptional co-activator p/CIP binds CBP and mediates nuclear-receptor function.* Nature. 1997;387:677-84.

19. Hanstein B, Eckner R, DiRenzo J, Halachm S, Liu H, Searcy B, et al. *p300 is a component of an estrogen receptor coactivator complex.* Proc Natl Acad Sci USA. 1996;93:11540-5.

20. Neuman E, Ladha MH, Lin N, Upton TM, Miller SJ, DiRenzo J, et al. *Cyclin D1 stimulation of estrogen receptor transcriptional activity dependent of cdk4.* Mol Cell Biol. 1997;17:5338-47.

21. Zwijsen RML, Wientjens E, Klompmaker R, van der Sman J, Bernards R, Michalides RJA. *CDK-independent activation of estrogen receptor by cyclin D1.* Cell. 1997;88:405-15.

22. Balasenthil S, Vadlamudi RK. *Functional interactions between the estrogen receptor coactivator PELP1/MNAR and retinoblastoma protein.* J Biol Chem. 2003;278:22119-27.

23. Le Douarin BL, Zechel C, Garnier J, Lutz Y, Tora L, Pierrat B, et al. *The N-terminal part of TIF1, a putative mediator of the ligand-dependent activation function (AF-2) of nuclear receptors, is fused to B-raf in the oncogenic protein T18.* EMBO J. 1995;14:2020-33.

24. Le Douarin B, Nielsen AL, Garnier JM, Ichinose H, Jeanmougin F, Losson R, et al. *A possible involvement of TIF1 alpha and TIF1 beta in the epigenetic control of transcription by nuclear receptors.* EMBO J. 1996;15:6701-15.

25. Baniahmad C, Nawaz Z, Baniahmad A, Gleeson MAG, Tsai M, O'Malley BW. *Enhancement of human estrogen receptor activity by SPT6: a potential coactivator.* Mol Endocrinol. 1995;9:34-42.

26. Chiba H, Muramatsu M, Nomoto A, Kato H. *Two human homologues of Saccharomyces cerevisiae SW12/SNF2 and drosophila brahma are transcriptional coactivators cooperating with the estrogen receptor and the retinoic acid receptor.* Nucleic Acids Res. 1994;22:1815-20.

27. Ichinose H, Garnier J, Chambon P, Losson R. *Ligand-dependent interaction between the estrogen receptor and the human homologues of SWI2/SNF2.* Gene. 1997;188:95-100.

28. Verrier CS, Roodi N, Yee CJ, Bailey R, Jensen RA, Bustin M, et al. *High-mobility group (HMG) protein HMG-1 and TATA-binding protein-associated factor TAFII30 affect estrogen receptor-mediated transcriptional activation.* Mol Endocrinol. 1997;11:1009-19.

29. Cavailles V, Dauvois S, Danielian PS, Parker MG. *Interaction of proteins with transcriptionally active estrogen receptors.* Proc Natl Acad Sci USA. 1994;91:10009-13.

30. Bramlett KS, Wu Y, Burris TP. *Ligands specify coactivator nuclear receptor (NR) box affinity for estrogen receptor subtypes.* Mol Endocrinol. 2001;15:909-22.

31. Docquier A, Harmand PO, Fritsch S, Chanrion M, Darbon JM, Cavailles V. *The transcriptional coregulator RIP140 represses E2F1 activity and discriminates breast cancer subtypes.* Clin Cancer Res. 2010;16:2959-70.

32. Walsh CA, Qin L, Tien JC, Young LS, Xu J. *The function of steroid receptor coactivator-1 in normal tissues and cancer.* Int J Biol Sci. 2012;8:470-85.

33. Ding XF, Anderson CM, Ma H, Hong H, Uht RM, Kushner PJ, et al. *Nuclear receptor-binding sites of coactivators glucocorticoid receptor interacting protein 1 (GRIP1) and steroid receptor coactivator 1 (SRC-1): Multiple motifs with different binding specificities.* Mol Endocrinol. 1998;12:302-13.

34. Voegel JJ, Heine MJS, Tini M, Vivat V, Chambon P, Gronemeyer H. *The coactivator TIF2 contains three nuclear receptor-binding motifs and mediates transactivaton through CPB binding-dependent and -independent pathways.* EMBO J. 1998;17:507-19.

35. Ko L, Cardona GR, Iwasaki T, Bramlett KS, Burris TP, Chin WW. *Ser-884 adjacent to the LXXLL motif of coactivator TRBP defines selectivity for ERs and TRs.* Mol Endocrinol. 2002;16:128-40.

36. Montminy M. *Something new to hang your HAT on.* Nature. 1997;387:654-5.

37. Heery DM, Kalkhoven E, Hoare S, Parker MG. *A signature motif in transcriptional co-activators mediates binding to nuclear receptors.* Nature. 1997;387:733-6.

38. Metivier R, Penot G, Flouriot G, Pakdel F. *Synergism between ERalpha transactivation function 1 (AF-1) and AF-2 mediated by steroid receptor coactivator protein-1: requirement for the AF-1 alpha-helical core and for a direct interaction between the N- and C-terminal domains.* Mol Endocrinol. 2001;15:1953-70.

39. Webb P, Nguyen P, Shinsako J, Anderson C, Feng W, Nguyen MP, et al. *Estrogen receptor activation function 1 works by binding p160 coactivator proteins.* Mol Endocrinol. 1998;12:1605-18.

40. Gee AC, Carlson KE, Martini PG, Katzenellenbogen BS, Katzenellenbogen JA. *Coactivator peptides have a differential stabilizing effect on the binding of estrogens and antiestrogens with the estrogen receptor.* Mol Endocrinol. 1999;13:1912-23.

41. Fleming FJ, Hill AD, McDermott EW, O'Higgins NJ, Young LS. *Differential recruitment of coregulator proteins steroid receptor coactivator-1 and silencing mediator for retinoid and thyroid receptors to the estrogen receptor-estrogen responseeelement by beta-estradiol and 4-hydroxytamoxifen in human breast cancer.* J Clin Endocrinol Metab. 2004;89:375-83.

42. Hudelist G, Czerwenka K, Kubista E, Marton E, Pischinger K, Singer CF. *Expression of sex steroid receptors and their co-factors in normal and malignant breast tissue: AIB1 is a carcinoma-specific co-activator.* Breast Cancer Res Treat. 2003;78:193-204

43. Myers E, Hill AD, Kelly G, McDermott EW, O'Higgins NJ, Buggy Y, et al. *Associations and interactions between Ets-1 and Ets-2 and coregulatory proteins, SRC-1, AIB1, and NCoR in breast cancer.* Clin Cancer Res. 2005;11:2111-22.

44. Smith CL, Nawaz Z, O'Malley BW. *Coactivator and corepressor regulation of the agonist/antagonist activity of the mixed antiestrogen, 4-hydroxytamoxifen.* Mol Endocrinol. 1997;11:657-66.

45. Xu J, Qiu Y, DeMayo FJ, Tsai SY, Tsai MJ, O'Malley BW. *Partial hormone resistance in mice with disruption of the steroid receptor coactivator-1 (SRC-1) gene.* Science. 1998;279:1922-5.

46. Meyer ME, Gronemeyer H, Turcotte B, Bocquel MT, Tasset D, Chambon P. *Steroid hormone receptors compete for factors that mediate their enhancer function.* Cell. 1989;57:433-42.

47. Lopez GN, Webb P, Shinsako JH, Baxter JD, Greene GL, Kushner PJ. *Titration by estrogen receptor activation function-2 of targets that* are downstream from coactivators. Mol Endocrinol. 1999;13:897-909.

48. Pfitzner E, Sak A, Ulber V, Ryffel GU, Klein-Hitpass L. *Recombinant activation domains of virion protein 16 and human estrogen receptor generate transcriptional interference in vitro by distinct mechanisms.* Mol Endocrinol. 1993;7:1061-71.

49. Tasset D, Tora L, Fromental C, Scheer E, Chambon P. *Distinct classes of transcriptional activating domains function by different mechanisms.* Cell. 1990;62:1177-87.

50. Janknecht R, Hunter T. *A growing coactivator network.* Nature. 1996;383:22-3.

51. Lundblad JR, Kwok RPS, Laurance ME, Harter ML, Goodman RH. *Adenoviral E1A-associated protein p300 as a functional homologue of the transcriptional co-activator CBP.* Nature. 1995;374:85-8.

52. Nakajima T, Uchida C, Anderson SF, Parvin JD, Montminy M. *Analysis of a cAMP-responsive activator reveals a two-component mechanism for transcriptional induction via signal-dependent factors.* Genes Dev. 1997;11:738-47.

53. Chakravarti D, LaMorte VJ, Nelson MC, Nakajima T, Schulman IG, Juguilon H, et al. *Role of CBP/ P300 in nuclear receptor signalling.* Nature. 1996;383:99-103.

54. Kamei Y, Xu L, Heinzel T, Torchia J, Kurokawa R, Gloss B, et al. *A CBP integrator complex mediates transcriptional activation and AP-1 inhibition by nuclear receptors.* Cell. 1996;85:403-14.

55. Yao TP, Ku G, Zhou N, Scully R, Livingston DM. *The nuclear hormone receptor coactivator SRC-1 is a specific target of p300.* Proc Natl Acad Sci USA. 1996;93:10626-31.

56. Nair BC, Nair SS, Chakravarty D, Challa R, Manavathi B, Yew PR, et al. *Cyclin-dependent kinase-mediated phosphorylation plays a critical role in the oncogenic functions of PELP1.* Cancer Res. 2010;70:7166-75.

57. Rajhans R, Nair S, Holden AH, Kumar R, Tekmal RR, Vadlamudi RK. *Oncogenic potential of the nuclear receptor coregulator proline-, glutamic acid-, leucine-rich protein 1/ modulator of the nongenomic actions of the estrogen receptor.* Cancer Res. 2007;67:5505-12.

58. Biddie SC, John S, Hager GL. *Genome-wide mechanisms of nuclear receptor action.* Trends Endocrinol Metab: TEM. 2010;21:3-9.

59. Kadonaga JT. *Eukaryotic*

transcription: an interlaced network of transcription factors and chromatin-modifying machines. Cell. 1998;92:307-13.

60. Pham TA, Hwung Y, Santiso-Mere D, McDonnell DP, O'Malley BW. *Ligand-dependent and -independent function of the transacti-vation regions of the human estrogen receptor in yeast.* Mol Endocrinol. 1992;6:1043-50.

61. Gilbert DM, Losson R, Chambon P. *Ligand dependence of estrogen receptor induced changes in chromatin structure.* Nucleic Acids Res. 1992;20,No.17:4525-31.

62. Yoshinaga SK, Peterson CL, Herskowitz I, Yamamoto KR. *Roles of SWI1, SWI2, and SWI3 proteins for transcriptional enhancement by steroid receptors.* Science. 1992;258:1598-604.

63. Peterson CL, Tamkun JW. *The SWI-SNF complex: a chromatin remodeling machine.* Trends Biochem Sci. 1995;20:143-6.

64. Schnitzler G, Sif S, Kingston RE. *Human SWI/SNF interconverts a nucleosome between its base state and a stable remodeled state.* Cell. 1998;94:17-27.

65. Medina PP, Sanchez-Cespedes M. *Involvement of the chromatin-remodeling factor BRG1/SMARCA4 in human cancer.* Epigenetics. 2008;3:64-8.

66. DiRenzo J, Shang Y, Phelan M, Sif S, Myers M, Kingston R, et al. *BRG-1 is recruited to estrogen-respon-sive promoters and cooperates with factors involved in histone acetylation.* Mol Cell Biol. 2000;20:7541-9.

67. Wong AK, Shanahan F, Chen Y, Lian L, Ha P, Hendricks K, et al. *BRG1, a component of the SWI-SNF complex, is mutated in multiple human tumor cell lines.* Cancer Res. 2000;60:6171-7.

68. Bai J, Mei P, Zhang C, Chen F, Li C, Pan Z, et al. *BRG1 is a prognostic marker and potential therapeutic target in human breast cancer.* PloS one. 2013;8:e59772.

69. Dawson MA, Kouzarides T, Huntly BJ. *Targeting epigenetic readers in cancer.* N Engl J Med. 2012;367:647-57.

70. Strahl BD, Allis CD. *The language of covalent histone modifications.* Nature. 2000;403:41-5.

71. Umov FD, Wolffe AP. *Above and within the genome: Epigenetics past and present.* J Mammary Gland Biol Neoplasia. 2001;6:153-67.

72. Berger SL. *The complex language of chromatin regulation during transcription.* Nature. 2007;447:407-12.

73. Jenuwein T, Allis CD. *Translating the histone code.* Science. 2001;293:1074-80.

74. Ruthenburg AJ, Li H, Patel DJ, Allis CD. *Multivalent engagement of chromatin modifications by linked binding modules.* Nature Rev Mol Cell Biol. 2007;8:983-94.

75. Baylin SB, Jones PA. *A decade of exploring the cancer epigenome - biological and translational implications.* Nature Rev Cancer. 2011;11:726-34.

76. Tan M, Luo H, Lee S, Jin F, Yang JS, Montellier E, et al. *Identification of 67 histone marks and histone lysine crotonylation as a new type of histone modification.* Cell. 2011;146:1016-28.

77. Wu H, Zhang Y. *Mechanisms and functions of Tet protein-mediated 5-methylcytosine oxidation.* Genes Dev. 2011;25:2436-52.

78. Bannister AJ, Kouzarides T. *The CBP co-activator is a histone acetyl-transferase.* Nature. 1996;384:641-3.

79. Mizzen CA, Yang X, Kokubo T, Brownell JE, Bannister AJ, Owen-Hughes T, et al. *The TAFII250 subunit of TFIID has histone acetyltransferase activity.* Cell. 1996;87:1261-70.

80. Ogryzko VV, Schiltz RL, Russanova V, Howard BH, Nakatani Y. *The transcriptional coactivators p300 and CBP are histone acetyltrans-ferases.* Cell. 1996;87:953-9.

81. Spencer TE, Jenster G, Burcin MM, Allis CD, Zhou J, Mizzen CA, et al. *Steroid receptor coactivator-1 is a histone acetyltransferase.* Nature. 1997;389:194-8.

82. Castellano L, Giamas G, Jacob J, Coombes RC, Lucchesi W, Thiruchelvam P, et al. *The estrogen receptor-alpha-induced microRNA signature regulates itself and its tran-scriptional response.* Proc Natl Acad Sci USA. 2009;106:15732-7.

83. Tsai WW, Wang Z, Yiu TT, Akdemir KC, Xia W, Winter S, et al. *TRIM24 links a non-canonical histone signature to breast cancer.* Nature. 2010;468:927-32.

84. Thenot S, Bonnet S, Boulahtouf A, Margeat E, Royer CA, Borgna JL, et al. *Effect of ligand and DNA binding on the interaction between human transcription intermediary factor 1alpha and estrogen receptors.* Mol Endocrinol. 1999;13:2137-50.

85. Lupien M, Eeckhoute J, Meyer CA, Wang Q, Zhang Y, Li W, et al. *FoxA1 translates epigenetic signatures into enhancer-driven lineage-specific transcription.* Cell. 2008;132:958-70.

86. Fullwood MJ, Liu MH, Pan YF, Liu J, Xu H, Mohamed YB, et al. *An oestrogen-receptor-alpha-bound human chromatin interactome.* Nature. 2009;462:58-64.

87. Watson PJ, Fairall L, Schwabe JW. *Nuclear hormone receptor co-repressors: structure and function.* Mol Cell Endocrinol. 2012;348:440-9.

88. Chen JD, Evans RM. *A transcrip-tional co-repressor that interacts with nuclear hormone receptors.* Nature. 1995;377:454-7.

89. Horlein AJ, Naar AM, Heinzel T, Torchia J, Gloss B, Kurokawa R, et al. *Ligand-independent repression by the thyroid hormone receptor mediated by a nuclear receptor co-repressor.* Nature. 1995;377:397-403.

90. Varlakhanova N, Snyder C, Jose S, Hahm JB, Privalsky ML. *Estrogen receptors recruit SMRT and N-CoR corepressors through newly recognized contacts between the corepressor N terminus and the receptor DNA binding domain.* Mol Cell Biol. 2010;30:1434-45.

91. Zhang X, Jeyakumar M, Petukhov S, Bagchi MK. *A nuclear receptor corepressor modulates transcriptional activity of antagonist-occupied steroid hormone receptor.* Mol Endocrinol. 1998;12:513-24.

92. Peterson TJ, Karmakar S, Pace MC, Gao T, Smith CL. *The silencing mediator of retinoic acid and thyroid hormone receptor (SMRT) corepressor is required for full estrogen receptor alpha transcriptional activity.* Mol Cell Biol. 2007;27:5933-48.

93. Alland L, Muhle R, Hou HJ, Potes J, Chin L, Schreiber-Agus N, et al. *Role for N-CoR and histone deacetylase in Sin3-mediated transcriptional repression.* Nature. 1997;387:49-55.

94. Heinzel T, Lavinsky RM, Mullen T, Soderstrom M, Laherty CD, Torchia J, et al. *A complex containing N-CoR, mSin3 and histone deacetylase mediates transcriptional repression.* Nature. 1997;387:43-8.

95. Nagy L, Kao H, Chakravarti D, Lin RJ, Hassig CA, Ayer DE, et al. *Nuclear receptor repression mediated by a complex containing SMRT, mSin3A, and histone deacetylase.* Cell. 1997;89:373-80.

96. Wolffe AP. *Sinful repression.* Nature. 1997;387:16-7.

97. Metivier R, Penot G, Hubner MR, Reid G, Brand H, Kos M, et al. *Estrogen receptor-alpha directs ordered, cyclical, and combinatorial recruitment of cofactors on a natural target promoter.* Cell. 2003;115:751-63.

98. Carlberg C, Seuter S. *Dynamics of nuclear receptor target gene regulation.* Chromosoma. 2010;119:479-84.

99. Ptashne M. *Gene regulation by proteins acting nearby and at a distance.* Nature. 1986;322:697-701.

100. Bulger M, Groudine M. *Looping versus linking: toward a model for long-distance gene activation.* Genes Dev. 1999;13:2465-77.

101. Abbondanza C, De Rosa C, Ombra MN, Aceto F, Medici N, Altucci L, et al. *Highlighting chromosome loops in DNA-picked chromatin (DPC).* Epigenetics. 2011;6:979-86.

102. Schild C, Claret F-X, Wahli W, Wolffe AP. *A nucleosome-dependent static loop potentiates estrogen-regulated transcription from the Xenopus vitellogenin B1 promoter in vitro.* EMBO J. 1993;12,No.2:423-33.

103. Day RN, Maurer RA. *The distal enhancer region of the rat prolactin gene contains elements conferring response to multiple hormones.* Mol Endocrinol. 1989;3:3-9.

104. Lufkin T, Jackson AE, Pan WT, Bancroft C. *Proximal rat prolactin promoter sequences direct optimal, pituitary cell-specific transcription.* Mol Endocrinol. 1989;3:559-66.

105. Day RN, Maurer RA. *Thyroid hormone-responsive elements of the prolactin gene: evidence for both positive and negative regulation.* Mol Endocrinol. 1989;3:931-8.

106. Cullen KE, Kladde MP, Seyfred MA. *Interaction between transcription regulatory regions of prolactin chromatin.* Science. 1993;261:203-6.

107. Gothard LQ, Hibbard JC, Seyfred MA. *Estrogen-mediated induction of rat prolactin gene transcription requires the formation of a chromatin loop between the distal enhancer and proximal promoter regions.* Mol Endocrinol. 1996;10:185-95.

108. Willis SD, Seyfred MA. *Pituitary-specific chromatin structure of the rat prolactin distal enhancer element.* Nucleic Acids Res. 1996;24:1065-72.

109. Berry M, Nunez A, Chambon P. *Estrogen-responsive element of the human pS2 gene is an imperfectly palindromic sequence.* Proc Natl Acad Sci USA. 1989;86:1218-22.

110. Sewack GF, Hansen U. *Nucleosome positioning and transcription-associated chromatin alterations on the human estrogen-responsive pS2 promoter.* J Biol Chem. 1997;272:31118-29.

111. Tan-Wong SM, French JD, Proudfoot NJ, Brown MA. *Dynamic interactions between the promoter and terminator regions of the mammalian BRCA1 gene.* Proc Natl Acad Sci USA. 2008;105:5160-5.

112. Hu Q, Kwon YS, Nunez E, Cardamone MD, Hutt KR, Ohgi KA, et al. *Enhancing nuclear receptor-induced transcription requires nuclear motor and LSD1-dependent gene networking in interchromatin granules.* Proc Natl Acad Sci USA. 2008;105:19199-204.

113. Bourdeau V, Deschenes J, Metivier R, Nagai Y, Nguyen D, Bretschneider N, et al. *Genome-wide identification of high-affinity estrogen response elements in human and mouse.* Mol Endocrinol. 2004;18:1411-27.

114. Carroll JS, Meyer CA, Song J, Li W, Geistlinger TR, Eeckhoute J, et al. *Genome-wide analysis of estrogen receptor binding sites.* Nature Genet. 2006;38:1289-97.

115. Frasor J, Danes JM, Komm B, Chang KC, Lyttle CR, Katzenellenbogen BS. *Profiling of estrogen up- and down-regulated gene expression in human breast cancer cells: insights into gene networks and pathways underlying estrogenic control of proliferation and cell phenotype.* Endocrinology. 2003;144:4562-74.

116. Lin CY, Vega VB, Thomsen JS, Zhang T, Kong SL, Xie M, et al. *Whole-genome cartography of estrogen receptor alpha binding sites.* PLoS Genetics. 2007;3:e87.

117. Liu Z, Chen S. *ER regulates an evolutionarily conserved apoptosis pathway.* Biochem Biophys Res Commun. 2010;400:34-8.

118. O'Lone R, Frith MC, Karlsson EK, Hansen U. *Genomic targets of nuclear estrogen receptors.* Mol Endocrinol. 2004;18:1859-75.

119. Welboren WJ, van Driel MA, Janssen-Megens EM, van Heeringen SJ, Sweep FC, Span PN, et al. *ChIP-Seq of ERalpha and RNA polymerase II defines genes differentially responding to ligands.* EMBO J. 2009;28:1418-28.

6.1.2. ER IN BREAST CANCER
6.1.2.1. INTRODUCTION

Unlike most types of cancer, breast cancer is known to be under hormonal control. Estrogens stimulate the growth of breast cancer cells expressing the ER by affecting the cell cycle machinery and by inducing specific growth factors and their receptors (1). The ER-mediated effects of estrogens on tumor promotion have been examined in numerous studies and summarized in many reviews including my earlier monograph (2). Since the breast is a target organ for estrogens and the physiologic response to estrogens is dependent on the presence of ER, one would expect normal mammary epithelial cells to express ER. However, immunohistochemical examination revealed a striking variation in intensity and distribution of stainable ER in nuclei of benign epithelial cells. The absence of ER in a sizable percentage of benign cells challenged the view that ER is expressed in all normal mammary epithelial cells. In fact, several immunohistochemical studies revealed that only 4 to 17% of normal epithelial cells are ER-positive (3-6). The ER-positive cells are distributed as scattered single cells, with the highest frequency and intensity of staining in lobules as compared to interlobular ducts. Clearly, the epithelial cell component itself is complex, derived from mammary stem cells that can both self-renew and propagate the full spectrum of cell types to generate functional lobulo-alveolar units during pregnancy (7, 8). Thus, mammary epithelial cell development is thought to progress in a hierarchical process from undifferentiated stem cells into at least two differentiated cell types, basal/myoepithelial and luminal cells, which can be distinguished by cytokeratin markers. In turn, the basal/myoepithelial and luminal cells are progenitor cells, which give rise to at least four distinct breast cancer subtypes: basal-like cells expressing human epidermal growth factor receptor 1 (EGFR; HER1), human epidermal growth factor receptor 2 (HER2)-enriched cells, and luminal A and luminal B cells typically expressing ERα and PR (Table 6.1.6.) (9, 10). Thus, ER-positive breast cancers are divided into two broad categories of luminal types A and B, largely depending on whether the tumor has low or high proliferation as measured by the expression of Ki67, a nuclear marker of cell proliferation (11-13). Utilization of the immunohistochemical Ki67 index allowed distinction of the luminal A subgroup with a good prognosis from the luminal B subgroup with a poor prognosis.

Interestingly, both mouse and human mammary stem cells exhibit a triple-negative phenotype for ERα, PR, and HER2, which is found in 15 – 20% of primary invasive breast cancers (15, 16). Another subtype of ER-negative breast cancer is called apocrine carcinoma because of the enrichment of apocrine histological features. The molecular signature of apocrine cancer is characterized by lack of

Table 6.1.6. Genetic Abnormalities and Expression Patterns of Breast Cancer Subtypes (9, 14)

Genes	Luminal A	Luminal B	Basal-like	HER2-enriched
ER	ER +	ER +	ER -	ER -
PR	PR +	PR +	PR -	PR -
HER2	HER2 -	HER2 +	HER2 -	HER2 +++
HER1		HER1 +	HER1 +	
Ki67 index	low	high		
P53 mutations	12%	29%	72%	80%
BRCA1	wild type	wild type		mutated BRCA1
Cytokeratin-5, -6 (basal)			CK5/6 ++	
Cytokeratin-8, -18 (luminal)	CK8/18 ++	CK8/18 ++		

ER expression but high levels of AR and FoxA1 expression (17-19). The frequency of apocrine breast cancer varies depending on the method of assessment. When histopathological criteria are used, apocrine cancer is rare, accounting for 1 – 4% of breast cancers (20). In contrast, when molecular signatures are used, 8 – 12% of breast cancers are considered apocrine subtype (13, 19). HER2 overexpression was observed in 50% of apocrine tumors (19, 20).

The ESR2 gene encodes ERβ, which is expressed in some breast cancers and breast cancer cell lines, either alone or together with ERα (21-23). For example, T47D cells are ERα- and ERβ-positive, whereas MDA-MB-231 cells are ERα-negative but ERβ-positive (24). While the concentration of ERα mRNA is usually higher in breast cancer than in adjacent normal tissue, ERβ mRNA levels did not differ or were actually lower in the malignant compartment (25, 26). The level of expression of ERβ in breast tissues and cell lines varies widely, but is considerably lower than that of ERα. Thus, ERβ expression may be detectable as ERβ mRNA by reverse transcriptase-PCR but not as ERβ protein. Quantitation showed that ERβ mRNA levels were 3- to 30-fold lower than those of ERα mRNA in breast cancers that were ER-positive by ligand binding assay (25). Moreover, the ER status of 40 breast cancers analyzed by ligand binding assay failed to show a significant correlation with ERβ mRNA levels (27). Clearly, ERα is the predominant receptor isoform in breast tissue, which accounts for most of the hormone binding activity in ER-positive tumors. The ESR2 gene contains sequence variants including the synonymous Leu392Leu (C33390G; rs1256054), which is located in a splicing enhancer motif in exon 7 (28). In the Shanghai Breast Cancer Study the variant was associated with high levels of circulating hormones (testosterone, estradiol, estrone sulfate, DHEA sulfate) and an increased risk of breast cancer in postmenopausal women. The effect was synergistic resulting in a 3-4-fold elevated risk among women with a CG or GG genotype.

References

1. Yue W, Yager JD, Wang JP, Jupe ER, Santen RJ. *Estrogen receptor-dependent and independent mechanisms of breast cancer carcinogenesis.* Steroids. 2013;78:161-70.

2. Parl FF. *Estrogens, Estrogen Receptor and Breast Cancer.* Amsterdam: IOS Press; 2000.

3. Clarke RB, Howell A, Potten CS, Anderson E. *Dissociation between steroid receptor expression and cell proliferation in the human breast.* Cancer Res. 1997;57:4987-91.

4. Jacquemier JD, Hassoun J, Torrente M, Martin PM. *Distribution of estrogen and progesterone receptors in healthy tissue adjacent to breast lesions at various stages - immunohistochemical study of 107 cases.* Breast Cancer Res Treat. 1990;15:109-17.

5. Petersen OW, Hoyer PE, Van Deurs B. *Frequency and distribution of estrogen receptor-positive cells in normal, nonlactating human breast tissue.* Cancer Res. 1987;47:5748-51.

6. Williams G, Anderson E, Howell A, Watson R, Coyne J, Roberts SA, et al. *Oral contraceptive (OCP) use increases proliferation and decreases oestrogen receptor content of epithelial cells in the normal human breast.* Int J Cancer. 1991;48:206-10.

7. Stingl J, Eirew P, Ricketson I, Shackleton M, Vaillant F, Choi D, et al. *Purification and unique properties of mammary epithelial stem cells.* Nature. 2006;439:993-7.

8. Shackleton M, Vaillant F, Simpson KJ, Stingl J, Smyth GK, Asselin-Labat ML, et al. *Generation of a functional mammary gland from a single stem cell.* Nature. 2006;439:84-8.

9. Nielsen TO, Hsu FD, Jensen K, Cheang M, Karaca G, Hu Z, et al. *Immunohistochemical and clinical characterization of the basal-like subtype of invasive breast carcinoma.* Clin Cancer Res. 2004;10:5367-74.

10. Prat A, Perou CM. *Mammary development meets cancer genomics.* Nat Med. 2009;15:842-4.

11. Cheang MC, Chia SK, Voduc D, Gao D, Leung S, Snider J, et al. *Ki67 index, HER2 status, and prognosis of patients with luminal B breast cancer.* J Natl Cancer Inst. 2009;101:736-50.

12. Loi S, Haibe-Kains B, Desmedt C, Lallemand F, Tutt AM, Gillet C, et al. *Definition of clinically distinct molecular subtypes in estrogen receptor-positive breast carcinomas through genomic grade.* J Clin Oncol. 2007;25:1239-46.

13. Perou CM, Sorlie T, Eisen MB, van de Rijn M, Jeffrey SS, Rees CA, et al. *Molecular portraits of human breast tumours.* Nature. 2000;406:747-52.

14. Cancer Genome Atlas N. *Comprehensive molecular portraits of human breast tumours.* Nature. 2012;490:61-70.

15. Asselin-Labat ML, Shackleton M, Stingl J, Vaillant F, Forrest NC, Eaves CJ, et al. *Steroid hormone receptor status of mouse mammary stem cells.* J Natl Cancer Inst. 2006;98:1011-4.

16. Lim E, Vaillant F, Wu D, Forrest NC, Pal B, Hart AH, et al. *Aberrant luminal progenitors as the candidate target population for basal tumor development in BRCA1 mutation carriers.* Nat Med. 2009;15:907-13.

17. Augello MA, Hickey TE, Knudsen KE. *FOXA1: master of steroid receptor function in cancer.* EMBO J. 2011;30:3885-94.

18. Doane AS, Danso M, Lal P, Donaton M, Zhang L, Hudis C, et al. *An estrogen receptor-negative* breast cancer subset characterized by a hormonally regulated transcriptional program and response to androgen. Oncogene. 2006;25:3994-4008.

19. Farmer P, Bonnefoi H, Becette V, Tubiana-Hulin M, Fumoleau P, Larsimont D, et al. *Identification of molecular apocrine breast tumours by microarray analysis.* Oncogene. 2005;24:4660-71.

20. O'Malley FP, Bane A. *An update on apocrine lesions of the breast.* Histopathology. 2008;52:3-10.

21. Enmark E, Pelto-Huikko M, Grandien K, Lagercrantz S, Langercrantz J, Fried G, et al. *Human estrogen receptor b-gene structure, chromosomal localization, and expression pattern.* J Clin Endocrinol Metabol. 1997;82:4258-65.

22. Hu YF, Lau KM, Ho SM, Russo J. *Increased expression of estrogen receptor beta in chemically transformed human breast epithelial cells.* Int J Oncol. 1998;12:1225-8.

23. Kuiper GJM, Carlsson B, Grandien K, Enmark E, Haggblad J, Nilsson S, et al. *Comparison of the ligand binding specificity and transcript tissue distribution of estrogen receptors alpha and beta.* Endocrinology. 1997;138:863-70.

24. Dotzlaw H, Leygue E, Watson PH, Murphy LC. *Expression of estrogen receptor-beta in human breast tumors.* J Clin Endocrinol Metab. 1996;82:2371-4.

25. Leygue E, Dotzlaw H, Watson PH, Murphy LC. *Altered estrogen receptor alpha and beta messenger RNA expression during human breast tumorigenesis.* Cancer Res. 1998;58:3197-201.

26. Speirs V, Parkes AT, Kerin MJ, Walton DS, Carleton PJ, Fox JN, et al. *Coexpression of estrogen receptor alpha and beta: poor prognostic factors in human breast cancer.* Cancer Res. 1999;59:525-8.

27. Dotzlaw H, Leygue E, Watson PH, Murphy LC. *Estrogen receptor-beta messenger RNA expression in human breast tumor biopsies: relationship to steroid receptor status and regulation by progestins.* Cancer Res. 1999;59:529-32.

28. Zheng SL, Zheng W, Chang BL, Shu XO, Cai Q, Yu H, et al. *Joint effect of estrogen receptor beta sequence variants and endogenous estrogen exposure on breast cancer risk in Chinese women.* Cancer Res. 2003;63:7624-9.

6.1.2.2. ER GENE, mRNA AND PROTEIN ABNORMALITIES

Twenty to 30% of breast cancers are ER-negative at the time of diagnosis. The remaining 70 to 80% are ER-positive but only two third respond to endocrine therapy. Despite the sustained presence of the receptor, over time an increasing number of ER-positive cancers becomes therapy-resistant, although few convert to ER-negativity. Many investigators have attempted to identify the molecular basis for the ER-negative status and the resistance of ER-positive tumors to endocrine therapy. In principle, these studies have investigated the ER at several levels of cellular expression: (i) at the genomic level, (ii) as mRNA transcriptional product, (iii) as protein and (iv) at the functional level via induction of estrogen-dependent genes.

ERα is encoded by the ESR1 gene located at 6q25.1 (1). Since chromosome 6 is frequently abnormal in breast cancer (2, 3), the question arises whether structural alterations at 6q could affect ER expression. A study of 95 cytogenetically characterized breast cancers showed a significant association between loss of 6q and lower ER concentrations, suggesting a gene-dosage effect (4). However, another study of aneusomy of chromosome 6,

measured by both nondisomy and chromosomal gain in 55 tumors, showed no correlation with ER concentration (5). Deletion of one allele (loss of heterozygosity) of the ESR1 gene could also diminish production of functional ER protein. Genomic DNA from 67 primary breast cancers was examined for loss of heterozygosity using a dinucleotide repeat $[TA]_n$ marker positioned 1 kb of the ESR1 gene (6). Only nine cases (19.1%) tested positive for loss of heterozygosity; three cases showed total loss and six showed marked reduction in the signal intensity from one allele. There was no significant correlation between loss of heterozygosity and ER status. Thus, both gross disruption and loss of heterozygosity of the ESR1 gene appear to be uncommon in breast cancer and do not explain the lack of ER expression in ER-negative tumors. Southern blot analysis also failed to show major rearrangements or deletions of the ESR1 gene in ER-negative breast cancers (7, 8). A more detailed analysis of the coding region of the ER gene by PCR amplification and restriction endonuclease digestion revealed a complete set of eight exons of normal sizes in ER-negative tumors, which were indistinguishable from those seen in ER-positive breast cancers (9). We performed a detailed search for point **mutations** of the ESR1 gene in 118 ER-positive and 70 ER-negative primary breast cancers and found only two missense mutations in the same ER-negative tumor (10). One mutation, Asn69Lys, was in exon 1 in the N-terminal transactivation domain AF1 of the receptor. The second mutation, Met396Val, was in exon 5 in a region of the LBD that is highly conserved among members of the steroid receptor family. Both human and chicken ER contain methionine at codon 396, whereas the corresponding amino acid in *Xenopus* and rainbow trout ER is valine and isoleucine, respectively (11, 12). Since wild type *Xenopus* ER contains valine in this position, it is unlikely that a mutant 396Val would alter the receptor function. Analysis of normal peripheral lymphocyte DNA from the patient showed wild type ER sequence indicating that the tumor contained two somatic mutations. The patient had node-negative, stage II breast cancer and was alive and well ten years post-mastectomy. The fact that we observed only two mutations in 188 tumors (1%) indicates that missense mutations in the ESR1 gene are rare in primary breast cancer. The figure of 1% is far lower than that observed for p53 gene mutations (22%) with the same screening technique in the same group of tumors (13). Since size and complexity of the ESR1 and p53 genes are similar, the question arises as how to explain the difference in mutational frequency between the two genes. It has been shown that genes of similar size and complexity do not necessarily exhibit similar frequencies of mutational events (14). This apparent non-randomness in the spectrum of genetic alterations indicates that the frequency of mutational lesions may be influenced by the local DNA sequence environment. Stated in more general terms, the non-randomness of mutations suggests the existence of endogenous mechanisms of mutagenesis as distinct from the better characterized exogenous causes, such as radiation or chemical mutagens (15). Alternatively, the rate of repair of damaged DNA at individual nucleotides is highly variable and sequence-dependent, suggesting that DNA repair efficiency may also contribute to the difference in mutational frequency between the ESR1 and p53 genes (16). Another study identified a missense mutation, Leu296Pro, in two breast cancers and Lys303Arg in another tumor (17). A missense mutation, Ala86Val, was subsequently reported to represent a sequencing error (18-20). Examination of subgroups of breast cancers did not detect mutations in the DBD of ER-positive, PR-negative tumors or in the LBD of ER-negative, PR-positive tumors (21, 22). The Cancer Genome Atlas Research Network did not identify any ESR1 mutations in sequencing 390 primary ER-positive breast cancers (23). In summary, ESR1 mutations are infrequent in primary breast cancers.

In contrast, ESR1 mutations are relatively frequent in metastatic lesions arising from advanced ER-positive hormone-resistant breast cancer (24). An analysis of the ESR1 gene in 20 tamoxifen-sensitive and 20 tamoxifen-resistant breast cancers revealed both synonymous and non-synonymous mutations (25). A missense mutation, Glu352Val, was detected in a tamoxifen-sensitive tumor. Since the tumor responded to the antiestrogen, the amino acid change did not appear to affect the receptor function. Two mutations affecting ER structure were detected in the group of tamoxifen-resistant tumors. The first was a single base pair deletion in codon 432 [TCA/CA], which was found in a tamoxifen-resistant metastatic tumor but not in the primary tumor from the same patient. The second mutation involved a substitution of 47 nucleotides (1503 - 1550) of exon 6 by 42 nucleotides (1380 - 1422) of exon 5 in a metastatic tumor. If translated, both mutations would generate truncated receptors with an intact DBD and a defective LBD that could constitutively activate transcription of restrogen-responsive genes. Analysis of two tamoxifen-stimulated breast cancer cell lines (MCF-7/TAM, MCF-7/MT2) and two tamoxifen-stimulated endometrial cancer cell lines revealed a single nonsynonymous mutation, Asp351Tyr, in MCF-7/MT2 cells, resulting in increased estrogenicity of 4-hydroxytamoxifen (26-28). Examination of the entire ESR1 coding region in 30 breast cancer metastases identified three non-synonymous mutations, Ser47Thr, Lys531Glu, and Tyr537Asn (29). Functional analysis of the first two mutants showed transcriptional activity similar to wild type ER. The 537Asn mutant, which was found in an ER-negative bone metastasis, possessed strong E_2-independent transcriptional activity, probably due to abolition of the 537Tyr phosphorylation site that has been implicated in ligand-dependent transcriptional activation of AF2 (30). A substitution of the neighboring amino acid, Asp538Gly, was identified in liver metastases from 5 of 13 patients (38%) resistant to endocrine treatment (31). Importantly, the mutation was not detected in the

Table 6.1.7. Somatic Mutations of ESR1 Coding Region in Primary and Metastatic Breast Cancer

Exon	Condon	Tumor	Comment	References
1	Ser47Thr	metastasis	wild type transcriptional activity	(29)
1	Asn69Lys	primary		(10)
4	Leu296Pro	primary		(17)
4	Lys303Arg	primary		(17)
4	Glu352Val	primary	tamoxifen-sensitive	(25)
5	Glu380Gln	primary		(33)
5	Val392Ile	primary		(33)
5	Met396Val	primary		(10)
6	432: frame shift	metastasis	tamoxifen-sensitive	(25)
6	truncation at codon 455	metastasis	tamoxifen-sensitive	(25)
6	Ser463Pro	metastasis	constitutive transcriptional activity	(33)
8	Lys531Glu	metastasis	wild type transcriptional activity	(29)
8	Val534Glu	metastasis		(33)
8	Pro535His	metastasis		(33)
8	Leu536Arg	primary, metastasis		(33)
8	Leu536Gln	metastasis	constitutive transcriptional activity	(32)
8	Tyr537Asn	primary, metastasis	constitutive transcriptional activity	(32, 33)
8	Tyr537Cys	primary, metastasis	constitutive transcriptional activity	(32, 33)
8	Tyr537Ser	metastasis	constitutive transcriptional activity	(32, 33)
8	Asp538Gly	primary, metastasis	constitutive transcriptional activity	(31-33)

primary tumor of these patients, obtained at diagnosis prior to commencing endocrine treatment. Another study identified ESR1 mutations in six of 11 patients (55%) with ER-positive hormone-resistant metastatic breast cancer (32). The non-synonymous mutations were localized in the LBD (Leu536Gln, Tyr537Ser, Tyr537Cys, Tyr537Asn, Asp538Gly) and shown experimentally to result in constitutive activity and continued responsiveness to antiestrogen therapies *in vitro*. The clustering of somatic mutations in the LBD with a hot spot at amino acids 534 – 538 was also observed in a two-part study involving nine of 36 (25%) ER-positive metastatic lessions and five of 44 (11%) ER-positive metastatic breast cancers, which had progressed during treatment with aromatase inhibitors (33).

One group has examined the 5′ upstream regulatory region of the ESR1 gene to determine whether sequence alterations in this region could account for the ER-negative phenotype (34). Sequence comparison of ER-negative MDA-MB-231 with ER-positive MCF-7 breast cancer cells revealed eight alterations between nucleotide positions -2497 and -226, seven of which were G or C insertions/deletions. None of the alterations were located within putative cis-acting regulatory elements. Functional analysis showed that two alterations (G insertion at -1976 and C/T transition at -1752) in combination resulted in a 50% decrease in CAT activity. Either alteration alone did not change CAT activity.

In summary, ESR1 mutations are rare in newly diagnosed, untreated breast cancers. Thus, they do not account for the ER phenotype of the majority of ER-negative ER-positive breast cancers. Whether or not ESR1 gene alterations are causally related to tumor proliferation or dissemination is uncertain. The incidence of ESR1 mutations is also low in endometrial cancer, ranging from 2 to 8% (35, 36). In contrast to the rare occurrence in primary breast cancers, somatic ESR1 mutations occur in about 25% of metastases, apparently acquired during progression to hormone resistance, especially in the context of estrogen deprivation therapy (24). Table 6.1.7. summarizes the somatic mutations identified in primary and metastatic breast cancers.

Germline mutations of the ER gene could conceivably affect ER expression and function. Several **polymorphisms** of the ER gene have been identified (Figure 6.1.4.). A *PvuII* polymorphism was initially implicated in lack of ER expression in breast cancer, but a larger study failed to show a convincing correlation with the ER-negative phenotype (9, 37). The polymorphic site was localized to the first intron, 0.4 kb upstream of exon 2 and about 100 kb apart from the hormone-binding domain of ER. The location in the intron and the distance made it unlikely that hormone binding was influenced by the polymorphism. A second polymorphism in

intron 1, identified as *XbaI* site, was shown to be separated from the *PvuII* site by less than 50 bp (38). The ER coding region contains several neutral polymorphisms in codon 10 [TCT/TCC (Ser)], codon 87 [GCG/GCC (Ala)], codon 240 [TGC/TGT (Cys)], codon 243 [CGC/CGT (Arg)], codon 325 [CCC/CCG (Pro)], codon 377 [CAC/CAT (His)], codon 545 [GAC/GAT (Asp)], and codon 594 [ACA/ACG (Thr)] (10, 18, 29, 39). There was no correlation of any of the polymorphic alleles with the ER phenotype. Finally, a polymorphic dinucleotide repeat, $[TA]_n$, in the 5′ upstream region of the ER gene also did not show any correlation with ER expression (40).

Since the ER protein mediates the effect of estrogen on growth and differentiation of normal mammary tissue, germline mutations of the ER gene could conceivably play a role in the development of breast cancer. Indeed, several polymorphisms have been associated with breast cancer risk. For example, the *XbaI* site in intron 1 has been linked to increased breast cancer susceptibility and inheritance (42, 43). The Shanghai Breast Cancer Study (1069 cases, 1166 controls) observed an increased risk associated with the *PvuII* polymorphism (44). The odds ratios for genotypes *Pp* and *pp* were 1.3 (95% CI1.0 – 1.7) and 1.4 (95% CI 1.1. – 1.8), respectively, compared to genotype *PP*. There was no risk association of the *XbaI* polymorphism nor a synergistic effect of the two polymorphisms. In the coding region, the frequency of C alleles in codon 87 [GCG/GCC; (Ala)] was reported to be higher in breast cancer patients than in female controls above the age of 50 (45). Exon 2 has a polymorphic site in codon 160 [GGT/TGT; (Gly160Cys)], which was initially described in a cohort of 75 Norwegian patients with familial clustering of breast and/or ovarian cancer (39). This substitution was identified in two breast cancer patients and one ovarian cancer patient. However, the T allele was also present in eight of 729 healthy controls indicating that the germline G/T transversion was likely to represent a polymorphism. The allele frequency did not differ significantly between patients with familial breast cancer and controls although one female control carrying the T allele had a sister with breast cancer, leaving open the possibility of an association of the codon 160 variant with breast cancer susceptibility. The T allele was not observed in American, British, or Swedish cohorts with familial breast cancer (46). Finally, exon 4 contains a polymorphic site in codon 325 which is a silent third base alteration [CCC/CCG; (Pro)]. In a clinic-based study of 188 patients, the codon 325 polymorphism was associated with familial breast cancer (10). However, a larger, population-based case-control study (388 case and 294 control subjects) found no difference in the frequency of the polymorphism between cases and controls or between patients with and without a family history of breast cancer (47).

Several studies of primary breast cancers used Northern or dot blot hybridization to relate ER mRNA and protein levels. In general, these studies showed an excellent correlation between the presence and absence of ER mRNA and the ER-positive and -negative phenotype as determined by ligand binding assay (48, 49). However, the correlation between ER transcript and protein levels deteriorated when more sensitive techniques of mRNA detection were used. For example, an RNase protection assay detected ER mRNA in eight of 12 ER-negative breast cancers and a sensitive cRNA probe detected ER mRNA in 15 of 21 ER-negative tumors (50, 51). To determine whether the ER-negative phenotype is due to a deficiency of transcriptional or post-transcriptional regulation of ER expression, we examined the presence and concentration of ER mRNA and protein in five ER-negative (BT-20, HBL-100, MDA-MB-157, MDA-MB-231, MDA-MB-468) and five ER-positive human breast cancer cell lines (MCF-7, T47D, ZR-75-1, MDA-MB-134, MDA-MB-361) (52). We determined the steady-state levels of ER mRNA by Northern blot using a full-length cRNA probe. High levels of ER mRNA were seen in MCF-7 and MDA-MB-134 cells. Lower amounts were observed in BT-20, ZR-75-1, T47D, and MDA-MB-361 cells. The remaining cell lines had no detectable ER mRNA. The level of ER protein was determined by Western blot analysis using anti-ER monoclonal antibodies H222 and D547 and by hormone binding assay. High levels of ER protein were seen in MCF-7, T47D and MDA-MB-134 cells. Low levels of ER protein were seen in ZR-75-1 and MDA-MB-361 cells. ER protein was not detectable in the remaining cell lines. The hormone binding assays yielded high ER levels for MCF-7, T47D and MDA-MB-134 cells, lower levels for ZR-75-1 and MDA-MB-361 cells and values below 10 fmol/mg for the remaining cell lines. There was not a good quantitative correlation between ER transcript and protein levels although the general relation between the presence or absence of ER mRNA and protein was maintained in all cell lines, except BT-20. BT-20 cells contained clearly detectable ER mRNA, but no ER protein by Western blot or ligand binding assay. This finding suggests a post-transcriptional defect in ER expression in BT-20 cells. ER mRNA was detectable in ER-negative cell lines by reverse transcription-PCR amplification but only segments of ER cDNA could be amplified and not the complete full-length cDNA as in ER-positive cell lines. This result reflects the difference in ER mRNA template concentration between ER-negative and -positive cell lines and confirms the data obtained by Northern blot analysis. In summary, ER-negative breast cancers and cell lines do not completely lack the ability to synthesize ER transcripts although the level of

Site	Nucleotide	Codon	Frequency of Variant Allele (%)
5' UTR	[TA]n		
Exon 1	278T→C	10Ser	45
Exon 1	493G→C	87Ala	6
Intron 1	C→T (PvuII)		50
Intron 1	G→A (XbaI)		
Exon 3	952C→T	240Cys	
Exon 3	961C→T	243Arg	2
Exon 4	1270C→G	325Pro	14
Exon 5	1363C→T	377His	
Exon 8	1867C→T	545Asp	
Exon 8	2014A→G	594Thr	81

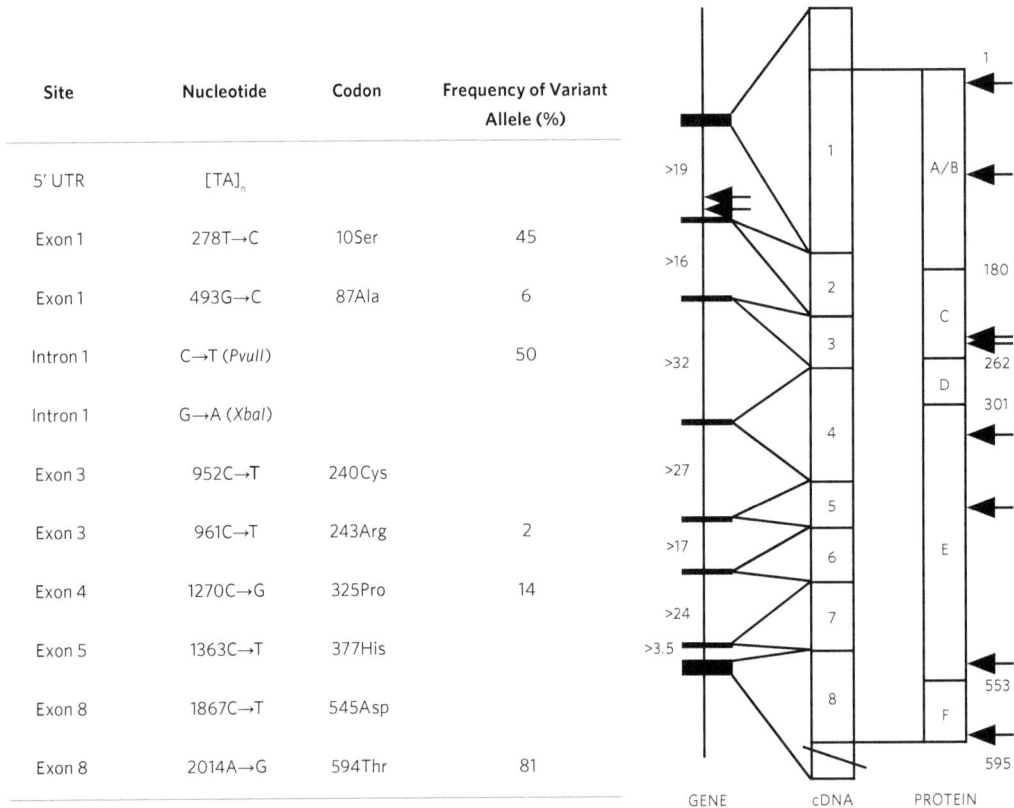

Figure 6.1.4. Diagram of ERα gene, cDNA, and protein with polymorphic sites indicated by arrows. *Nucleotides are numbered from transcription start site (41).*

Table 6.1.8. ER mRNA Splice Variants

Variant	Missing ER Domain (partial or complete)	Structural Effect (Frameshift/Truncation)	Functional Effect of Putative Variant ER Protein
$\sum 2$	A/B, C	yes	none
$\sum 3$	C	no	dominant negative effect; transcriptional repressor
$\sum 4$	D,E	no	none
$\sum 5$	E	yes	dominant positive effect; constitutively active, ligand-independent transcriptional activator
$\sum 6$	E	yes	?
$\sum 7$	E	yes	dominant negative effect

$\sum$ = devoid of exon as a result of alternative splicing, in contrast to genomic deletions which are generally designated by the symbol 'Δ' (62)

transcription is severely reduced resulting in the lack of ER protein and the ER-negative phenotype.

The interpretation of ER transcript studies based on sensitive RNase protection and reverse transcription PCR assays is complicated by the presence of **ER mRNA variant species**, which have been observed in breast tumors and breast cancer cell lines (48, 49, 53, 54). Most of the variants were shown to result from alternative splicing and precise deletion of individual exons 2 - 7 (Table 6.1.8.). For example, ER-positive T47D cells contained three ER cDNA variants with deletions of exons 2, 3, or 7, respectively (55). With the exception of the skipped exons, the remainder of the sequence of each of the three ER cDNA variants was identical to wild-type ER mRNA. An ER mRNA variant with a precise deletion of exon 5 was identified in ER-negative BT-20 cells (56, 57). However, the same variant species was also observed in ER-positive MCF-7 cells. Other investigators identified ER mRNA variants lacking exon 4 in MCF-7 and ZR-75 cells and exon 6 in T47D cells (58-60). A systematic analysis of six ER-positive cell lines (MCF-7, T47D, ZR-75, LCC1, LCC2 and LCC3) showed different patterns of variant transcripts for each cell line, including deletion of one, two, three or even four exons (61). All cell lines contained normal, full-length message in addition to the variants.

Functional studies of recombinant proteins corresponding to these mRNA variants confirmed the altered functional activity predicted from the domain structure of wild type ER (17, 60, 63) (Table 6.1.8.). The $\sum 2$ ER mRNA variant would encode a truncated protein due to introduction of an out-of-frame stop codon. Since the putative variant protein would lack both DBD and LBD, it plays no significant functional role. Deletion of exon 3 from wild type ER mRNA is in-frame and generates a protein of approximately 61 kDa that lacks the second zinc finger of the DBD. Recombinant $\sum 3$ ER variant was unable to form an ER-ERE complex but inhibited the ability of wild type ER to bind to ERE. When transfected into HeLa cells, $\sum 3$ ER cDNA exhibited a dominant negative effect, inhibiting wild type ER transcriptional activity without any intrinsic activity of its own (55,

64). The $\sum 4$ ER mRNA contains an in-frame deletion and is predicted to encode a protein of approximately 54 kDa which would be missing portions of the D and E domains of wild type ER. Functional studies with expression vectors showed the variant protein to be nonfunctional, i.e., it did not bind E_2 or an ERE and had no transcriptional activity of its own nor any dominant negative activity against wild type ER (58). Although the $\sum 5$ ER mRNA lacks only exon 5, the predicted protein would miss the C-terminal third of wild type ER due to introduction of an out-of-frame stop codon. The truncated protein of approximately 40 kDa would be identical to wild type ER up to amino acid 366 (exons 1 - 4), followed by five novel amino acids unique to the variant after which the stop codon terminates translation. A yeast expression system containing the $\sum 5$ ER cDNA construct yielded a variant ER protein with ligand-independent transcriptional activity (54). The $\sum 6$ ER mRNA results in a truncated protein whose function has not yet been reported. Finally, the predicted protein encoded by the $\sum 7$ ER mRNA is identical to wild type ER up to amino acid 456 followed by 10 novel amino acids and truncation resulting from an out-of-frame stop codon. In a yeast expression vector system $\sum 7$ ER cDNA had a dominant negative effect, inhibiting wild type ER transcriptional activity (63). Since the variant ER mRNA species were never found in the absence of normal, full-length ER transcript, attempts to explain the ER-negative phenotype have to take into account the relative proportion of wild type and variant ER mRNA. Two reasons make it unlikely that a dominant-negative mechanism will readily explain the ER-negative phenotype. First, the concentration of the variant is usually less than that of the wild type transcript. Second, the $\sum 3$ and $\sum 7$ ER mRNA variants which encode putative proteins with dominant negative effects are also detected in ER-positive tumors and cell lines.

$\sum 5$ ER mRNA has received attention because of its constitutive dominant positive effect. The $\sum 5$ ER mRNA was identified in 19 of 27 breast cancers, always coexpressed with wild type ER mRNA (65). When MCF-7 cells were transfected with $\sum 5$ ER cDNA to express equal levels of variant and wild type receptor, the cells became resistant to the growth-inhibitory effects of tamoxifen (66). The amount of $\sum5$ ER mRNA relative to wild type ER mRNA was measured in 70 tamoxifen-resistant and 50 primary breast carcinomas to determine whether tumors with constitutively active ER would be unresponsive to endocrine therapy (67). Both wild type and $\sum5$ ER mRNA were detected in the majority of tumors, although $\sum5$ ER mRNA was detected only in the presence of wild type ER mRNA. Overall, no significant difference was seen in the $\sum5$/ wild type ER mRNA ratio between tamoxifen-resistant and primary control tumors. However, tumors in both control and resistant groups, which expressed PR and pS2 in the absence of measurable ER protein (ER-negative, PR-positive, pS2-positive), had significantly higher $\sum5$ ER mRNA levels compared with other phenotypes (p <0.002). These findings indicate that $\sum5$ ER mRNA is unlikely to be responsible for tamoxifen resistance in breast cancer, but that elevated $\sum5$ ER mRNA levels may be important in tumors which continue to express high levels of PR and pS2. However, stable and transient transfection of the exon 5 variant in ER-positive MCF-7 cells had no effect on expression of the estrogen target genes pS2 and PR (68). The stimulatory effects of estrogen and growth-inhibitory effects of tamoxifen and ICI 182,780 were unchanged. Thus, expression of the variant alone was not sufficient to give rise to hormone independence or tamoxifen resistance. Transfection experiments with the ER-negative human breast epithelial cell line HMT-3522S1 showed ten-fold higher expression of $\sum5$ ER mRNA compared to that of wild type ER mRNA indicating increased stability of the former (69). However, $\sum5$ ER protein was significantly less stable than wild type ER protein.

To complicate matters further, another ER mRNA variant lacking the entire exon 5 and part of exons 4 and 6 was identified by reverse transcriptase PCR in 96 % of 102 breast cancers, of which 62 were tamoxifen-resistant tumors (70). The variant/wild type ER mRNA ratio was significantly higher in ER-positive, PR-negative, pS2-negative tumors. Examination of ten tamoxifen-resistant, ER-positive MCF-7 sublines revealed the presence of $\sum2$, $\sum3$, $\sum4$, $\sum5$, $\sum7$, and $\sum4,7$ ER mRNA in addition to wild type ER mRNA in all sublines and in parental MCF-7 cells (71). Quantitation by RNase protection assay showed no significant difference between the tamoxifen-resistant sublines and sensitive parental MCF-7 cells, indicating that ER splice variants are not involved in antiestrogen resistance, at least in this model system. A study of 100 breast cancers revealed deletions of two ($\sum3,4$, $\sum4,7$) and three exons ($\sum2$-4, $\sum2,3,7$) in many of the ER-positive, but only in three of the 30 ER-negative tumors (72). The DNA sequence surrounding the splice junctions in these variants conformed to wild type ER sequence. A study of 126 primary breast cancers failed to show a significant association between the presence of ER mRNA variants and prognosis (73).

Finally, mRNA splice variants of ERβ lacking exons 5 or 6 or both were detected in normal breast tissues, breast cancers, and ERα-positive and -negative cell lines (MCF-7, T47D, MDA-MB-231) (74, 75). Wild type ERβ mRNA was present in greater amounts than the ERβ splice variants. Two other types of ERβ mRNA variants have been identified. The first type is characterized by deletions in exon 8, while exons 1 - 7 remain intact (76, 77). The truncated receptor failed to exhibit ligand binding and ligand-dependent transactivation. The second type is characterized by an in-frame insertion of 54 nucleotides that results in the predicted insertion of 18 amino acids in the LBD (78). The recombinant variant ERβ protein bound E_2 with eight-fold lower affinity than wild type ERβ and required 100- to 1000-fold higher concentrations of E_2 to stimulate transcription to the same extent as wild type ERβ.

Although many investigators have identified ER mRNA splice variants in breast cancer cells, expression of the corresponding variant proteins has only been demonstrated by two groups (56, 79). The majority of studies did not detect variant ER proteins by Western blotting with anti-ER antibodies directed against epitopes in the N-terminal A/B domain which should recognize all putative variant ER proteins (71). A monoclonal anti-ER antibody raised against the C-terminal amino acids of $\sum5$ ER recognized the variant, but not wild type ER protein in breast cancer (79). The authors observed a positive correlation between the presence of $\sum5$ ER and disease-free survival (p = 0.05), making it unlikely that the variant is responsible for endocrine resistance in breast cancer.

After the discovery of ER mRNA splice variants in breast cancer, several groups decided to search for their presence in normal breast tissue. Multiple variant ER mRNA species lacking one or more exons were detected identical to those in breast cancer and breast cancer cell lines (62, 72, 80). Given the expression of the variants in normal tissue, and given the expression of potentially dominant positive variants in conjunction with potentially dominant negative ones, it appears unlikely that the ER mRNA splicing variants are involved in the neoplastic process or account for hormone antagonist resistance. No splicing variants of GR or RAR could be detected in MCF-7 cells, which normally express these members of the nuclear receptor superfamily. Thus, the ER mRNA variants do not appear to be the result of a leaky splicing machinery but may have a physiological role in mammary epithelial cells (81).

In summary, ER mRNA variants with precise deletions of individual exons [2 - 7] occur in breast tumors and breast cancer cell lines (82, 83). The level of ER mRNA splice variants is generally higher in breast cancer than in normal breast tissue supporting the notion that variants represent abnormal ER mRNA species resulting from a derangement of physiologically regulated splicing pathways (21, 84). The significance of these abnormal ER transcripts is uncertain for two reasons (i) they are found simultaneously with normal, full-length ER mRNA, and (ii) they occur in both ER-negative and -positive tumors and cell lines. The fact that ER-positive cancers and cell lines also harbor ER mRNA variants makes it less likely that the ER-negative status results from ER splice variants. Moreover, ER mRNA variants have also been found in normal endometrium, ovary, liver, and brain, indicating that their occurrence is not restricted to breast tissue or limited to malignant cells (54, 85-89).

Another type of variant ER mRNA is created by transsplicing, a process where ER mRNA is truncated and linked to unrelated sequences during the splicing process (90). An examination of transsplicing clone 4 variant ER and wild type ER by RNase protection assay in 106 breast cancers revealed wild type and clone 4 variant ER mRNAs in ER-positive, PR-positive and ER-positive, PR-negative tumors but not in ER-negative, PR-negative cancers (91). Significantly higher levels of clone 4 variant relative to wild type ER mRNA were found in tumors with markers of poor prognosis (node positivity, tumor size >2 cm, high percentage S-phase fraction) compared to those with markers of good prognosis (p = 0.0004).

The ER mRNA variants described so far are shorter than wild type ER mRNA due to truncation or exon skipping. ER transcripts that are longer than wild type ER mRNA have also been identified in breast cancer. The increase in length is due to duplication of exons or insertion of intronic sequence. For example, a 69-nucleotide sequence, which is normally present in intron 5 of the ER gene, was precisely inserted between exon 5 and 6 sequences (92). Cloning and sequencing of the corresponding region in genomic DNA isolated from the breast tumor expressing ER transcript with the 69-nucleotide insert revealed an A/G point mutation immediately 3' to the 69 nucleotide sequence. This point mutation resulted in the generation of a new consensus splice donor site. A consensus splice acceptor site sequence is normally present immediately 5' to the 69-nucleotide sequence. These data are consistent with the 69-nucleotide sequence being recognized as an exon by the splicing machinery, resulting in processing of a mature ER mRNA containing the 69-nucleotide insert as a new exon. An 80 kDa ER was detected by Western blotting in an estrogen-independent subclone of the MCF-7 breast cancer cell line, MCF-7A:2A (93). Structural analysis of ER mRNA revealed a lengthened ER transcript that was not derived from the transsplicing of two ER mRNAs but resulted from a genomic rearrangement in which exons 6 and 7 were duplicated in an in-frame fashion (94). The MCF-7:2A cells also expressed wild type ER transcript and 66 kDa protein. Analysis of abnormally large ER cDNA products obtained by reverse transcriptase PCR from breast cancer RNA has also shown complete duplication of exons 3, 4, or 6, the latter in 7.5%

of 212 ER-positive and -negative tumors (94). It is unknown if these ER mRNA variants are stably translated *in vivo*. Any resulting protein would be structurally different from wild type ER and therefore likely to differ in function.

Abnormal ER cDNAs were cloned from T47D$_{CO}$, a subline of T47D breast cancer cells (95). One transcript contained an insertion of approximately 130 nucleotides in exon 5. Sequence analysis displayed similarity of the insert to the human *Alu* family of repetitive sequences. Another transcript contained two T residues inserted in exon 3, resulting in a frame shift. The predicted protein would be truncated just beyond the last cysteine of the second zinc finger and completely lack the LBD. MCF-7 cells were also shown to contain ER cDNA variants with inserts, either in form of a 6 bp insertion between exons 2 and 3 or a 39 bp insertion between exons 3 and 4 (60).

Finally, several authors detected abnormal species of ER protein in breast cancer cytosol. For example, a 47 kDa species was identified by sodium dodecyl sulfate-polyacrylamide gel electrophoresis in over 50% of 41 primary breast cancers (96). A truncated 50 kDa receptor form was detected by Western analysis in tumor cytosol treated with protease inhibitors to minimize proteolysis (97). Although a posttranslational modification has been proposed to account for the truncated ER variant, the molecular mechanism has not been determined and the effect on ER function is uncertain.

After extensive molecular ER studies, how can we explain the ER-negative phenotype of one third of primary breast cancers? The critical analysis of data from this and other laboratories allows the following conclusions: (i) The overall structure of the ER gene is intact, i.e., in the majority of tumors there is no evidence for rearrangements or deletions/insertions. All eight exons are present and of normal length. (ii) Point mutations are rare; they occur in only 1% of primary tumors. Most are missense mutations that do not necessarily explain the ER-negative phenotype. (iii) The ER gene contains several polymorphic sites; none of the alleles correlates with the ER-negative phenotype. (iv) Most ER-negative breast cancers lack ER mRNA by Northern blot analysis. More sensitive techniques, such as RNase protection assays and reverse transcriptase PCR, may reveal ER mRNA, albeit at extremely low levels. (v) Variant species of ER mRNA can be detected by reverse transcriptase PCR that result from alternative splicing. The variant species are never found in the absence of normal, full-length ER transcript. The concentration of the latter generally exceeds that of the variant species. Variant ER mRNA is found in ER-positive and -negative breast cancers as well as normal breast. Most investigators fail to detect the corresponding variant ER proteins. (vi) Routine analysis for ER mutants and variants is not justified in primary breast cancers at the time of diagnosis because of lack of evidence that they predict resistance to endocrine therapy (98, 99). (vii) In contrast, sequence analysis may be indicated in metastases of ER-positive tumors, which harbor somatic ESR1 mutations in about 25% of lesions, apparently acquired during progression to hormone resistance (24). (viii) The ER-negative phenotype is characterized by the absence of ER protein as determined by hormone binding assay, Western blot, or immunohistochemical analysis.

References

1. Menasce LP, White GR, Harrison CJ, Boyle JM. *Localization of the estrogen receptor locus (ESR) to chromosome 6q25.1 by FISH and a simple post-FISH banding technique.* Genomics. 1993;17:263-5.

2. Devilee P, van Vliet M, van Sloun P, Kuipers Dijkshoorn N, Hermans J, Pearson PL, et al. *Allelotype of human breast carcinoma: a second major site for loss of heterozygosity is on chromosome 6q.* Oncogene. 1991;6:1705-11.

3. Thompson F, Emerson J, Dalton W, Yang JM, McGee D, Villar H, et al. *Clonal chromosome abnormalities in human breast carcinomas. I. Twenty-eight cases with primary disease.* Genes Chromosom Cancer. 1993;7:185-93.

4. Magdelenat H, Gerbault-Seureau M, Dutrillaux B. *Relationship between loss of estrogen and progesterone receptor expression and of 6q and 11q chromosome arms in breast cancer.* Int J Cancer. 1994;57:63-6.

5. Persons DL, Robinson RA, Hsu PH, Seelig SA, Borell TJ, Hartmann LC, et al. *Chromosome-specific aneusomy in carcinoma of the breast.* Clin Cancer Res. 1996;2:883-8.

6. Iwase H, Greenman JM, Barnes DM, Bobrow L, Hodgson S, Mathew CG. *Loss of heterozygosity of the oestrogen receptor gene in breast cancer.* Br J Cancer. 1995;71:448-50.

7. Koh EH, Ro J, Wildrick DM, Hortobagyi GN, Blick M. *Analysis of the estrogen receptor gene structure in human breast cancer.* Anticancer Res. 1989;9:1841-5.

8. Parl FF, Cavener DR, Dupont WD. *Genomic DNA analysis of the estrogen receptor gene in breast cancer.* Breast Cancer Res Treat. 1989;14:57-64.

9. Yaich L, Dupont WD, Cavener DR, Parl FF. *Analysis of the PvuII restriction fragment-length polymorphism and exon structure of the estrogen receptor gene in breast cancer and peripheral blood.* Cancer Res. 1992;52:77-83.

10. Roodi N, Bailey LR, Kao W-Y, Verrier CS, Yee CJ, Dupont WD, et al. *Estrogen receptor gene analysis in estrogen receptor-positive and receptor-negative primary breast cancer.* J Natl Cancer Inst. 1995;87:446-51.

11. Krust A, Green S, Argos P, Kumar V, Walter P, Bornert JM, et al. *The chicken oestrogen receptor sequence: homology with v-erbA and the human oestrogen and glucocorticoid receptors.* EMBO J. 1986;5:891-7.

12. Pakdel F, Le Guellec C, Vaillant C, Le Roux MG, Valotaire Y. *Identification and estrogen induction of two estrogen receptors (ER) messenger ribonucleic acids in the rainbow trout liver: sequence homology with other ERs.* Mol Endocrinol. 1989;3:44-51.

13. Caleffi M, Teague MW, Jensen RA, Vnencak-Jones CL, Dupont WD, Parl FF. *p53 gene mutations and steroid receptor status in breast cancer.* Cancer. 1994;73:2147-56.

14. Cooper DN, Krawczak M. *The mutational spectrum of single base-pair substitutions causing human genetic disease: patterns and predictions.* Hum Genet. 1990;85:55-74.

15. Krawczak M, Cooper DN. *Gene deletions causing human genetic disease: mechanisms of mutagenesis and the role of the local DNA sequence environment.* Hum Genet. 1991;86:425-41.

16. Tornaletti S, Pfeifer GP. *Slow repair of pyrimidine dimers at p53 mutation hotspots in skin cancer.* Science. 1994;263:1436-8.

17. McGuire WL, Chamness GC, Fuqua SAW. *Estrogen receptor variants in clinical breast cancer.* Mol Endocrinol. 1991;5:1571-7.

18. Garcia T, Sanchez M, Cox JL, Shaw PA, Ross JBA, Lehrer S, et al. *Identification of a variant form of the human estrogen receptor with an amino acid replacement.* Nucleic Acids Res. 1989;17:8364.

19. Macri P, Khoriaty G, Lehrer S, Karurunaratne A, Milne C,

Schachter BS. *Sequence of a human estrogen receptor variant allele.* Nucleic Acids Res. 1992;20:2008.

20. Taylor JA, Li Y, You M, Wilcox AJ, Liu E. *B region variant of the estrogen receptor gene.* Nucleic Acids Res. 1992;20:2895.

21. Fuqua SAW, Chamness GC, McGuire WL. *Estrogen receptor mutations in breast cancer.* J Cell Biochem. 1993;51:135-9.

22. Iwase H, Greenman JM, Barnes DM, Hodgson S, Bobrow L, Mathew CG. *Sequence variants of the estrogen receptor (ER) gene found in breast cancer patients with ER negative and progesterone receptor positive tumors.* Cancer Lett. 1996;108:179-84.

23. Cancer Genome Atlas N. *Comprehensive molecular portraits of human breast tumours.* Nature. 2012;490:61-70.

24. Oesterreich S, Davidson NE. *The search for ESR1 mutations in breast cancer.* Nature Genet. 2013;45:1415-6.

25. Karnik PS, Kulkarni S, Liu XP, Budd GT, Bukowski RM. *Estrogen receptor mutations in tamoxifen-resistant breast cancer.* Cancer Res. 1994;54:349-53.

26. Bilimoria MM, Assikis VJ, Muenzner HD, Wolf DM, Satyaswaroop PG, Jordan VC. *An analysis of tamoxifen-stimulated human carcinomas for mutations in the AF-2 region of the estrogen receptor.* J Steroid Biochem Mol Biol. 1996;58:479-88.

27. Catherino WH, Wolf DM, Jordan VC. *A naturally occurring estrogen receptor mutation results in increased estrogenicity of a tamoxifen analog.* Mol Endocrinol. 1995;9:1053-63.

28. Wolf DM, Jordan VC. *The estrogen receptor from a tamoxifen stimulated MCF-7 tumor variant contains a point mutation in the ligand binding domain.* Breast Cancer Res Treat. 1994;31:129-38.

29. Zhang Q, Borg A, Wolf DM, Oesterreich S, Fuqua SAW. *An estrogen receptor mutant with strong hormone-independent activity from a metastatic breast cancer.* Cancer Res. 1997;57:1244-9.

30. Arnold SF, Melamed M, Vorojeikina DP, Notides AC, Sasson S. *Estradiol-binding mechanism and binding capacity of the human estrogen receptor is regulated by tyrosine phosphorylation.* Mol Endocrinol. 1997;11:48-53.

31. Merenbakh-Lamin K, Ben-Baruch N, Yeheskel A, Dvir A, Soussan-Gutman L, Jeselsohn R,

et al. *D538G Mutation in Estrogen Receptor-alpha: A Novel Mechanism for Acquired Endocrine Resistance in Breast Cancer.* Cancer Res. 2013.

32. Robinson DR, Wu YM, Vats P, Su F, Lonigro RJ, Cao X, et al. *Activating ESR1 mutations in hormone-resistant metastatic breast cancer.* Nature Genet. 2013;45:1446-51.

33. Toy W, Shen Y, Won H, Green B, Sakr RA, Will M, et al. *ESR1 ligand-binding domain mutations in hormone-resistant breast cancer.* Nature Genet. 2013;45:1439-45.

34. Sullivan JA, Cohn CS, Hill SM. *Identification of sequence alterations in the upstream regulatory region of the estrogen receptor gene in an ER-negative breast cancer cell line.* Cancer Lett. 1997;113:131-9.

35. Assikis V, Bilimoria MM, Muenzner HD, Lurain JR, Jordan VC. *Mutations of the estrogen receptor in endometrial carcinoma: Evidence of an association with high tumor grade.* Gynec Oncol. 1996;63:192-9.

36. Kohler MF, Berkholz A, Risinger JI, Elbendary A, Boyd J, Berchuck A. *Mutational analysis of the estrogen-receptor gene in endometrial carcinoma.* Obstet Gynecol. 1995;86:33-7.

37. Hill SM, Fuqua SAW, Chamness GC, Greene GL, McGuire WL. *Estrogen receptor expression in human breast cancer associated with an estrogen receptor gene restriction fragment length polymorphism.* Cancer Res. 1989;49:145-8.

38. Kobayashi S, Inoue S, Hosoi T, Ouchi Y, Shiraki M, Orimo H. *Association of bone mineral density with polymorphism of the estrogen receptor gene.* J Bone Miner Res. 1996;11.

39. Anderson T, Wooster R, Laake K, Collins N, Warren W, Skrede M, et al. *Screening for ESR mutations in breast and ovarian cancer patients.* Hum Mutat. 1997;9:531-6.

40. Piva R, Bianchi N, Aguiari GL, Gambari R, Del Senno L. *Sequencing of an RNA transcript of the human estrogen receptor gene: evidence for a new transcriptional event.* J Steroid Biochem Molec Biol. 1993;46:531-8.

41. Green S, Walter P, Kumar V, Krust A, Bornert J-M, Argos P, et al. *Human oestrogen receptor cDNA: sequence, expression and homology to v-erb-A.* Nature. 1986;320:134-9.

42. Andersen TI, Heimdal KR, Skrede M, Tveit K, Berg K, Borresen A. *Oestrogen receptor (ESR) polymorphisms and breast* *cancer susceptibility.* Hum Genet. 1994;94:665-70.

43. Zuppan P, Hall JM, Lee MK, Ponglikitmongkol M, King MC. *Possible linkage of the estrogen receptor gene to breast cancer in a family with late-onset disease.* Am J Hum Genet. 1991;48:1065-8.

44. Cai Q, Shu XO, Jin F, Dai Q, Wen W, Cheng JR, et al. *Genetic polymorphisms in the estrogen receptor alpha gene and risk of breast cancer: results from the Shanghai Breast Cancer Study.* Cancer Epidemiol Biomarkers Prev 2003;12:853-9

45. Schmutzler RK, Sanchez M, Lehrer S, Chaparro CA, Phillips C, Rabin J, et al. *Incidence of an estrogen receptor polymorphism in breast cancer patients.* Breast Cancer Res Treat. 1991;19:111-7.

46. Zelada-Hedman M, Borresen-Dale A, Lindblom A. *Screening of 229 family cancer patients for a germline estrogen receptor gene (ESR) base mutation.* Hum Mutat. 1997;9:289.

47. Southey MC, Batten LE, McCredie MRE, Giles GG, Dite G, Hopper JL, et al. *Estrogen receptor polymorphism at codon 325 and risk of breast cancer in women before age forty.* J Natl Cancer Inst. 1998;90:532-6.

48. Barrett-Lee PJ, Travers MT, McClelland RA, Luqmani Y, Coombes RC. *Characterization of estrogen receptor messenger RNA in human breast cancer.* Cancer Res. 1987;47:6653-9.

49. Murphy LC, Dotzlaw H. *Variant estrogen receptor mRNA species detected in human breast cancer biopsy samples.* Mol Endocrinol. 1989;3:687-93.

50. Garcia T, Lehrer S, Bloomer WD, Schachter B. *A variant estrogen receptor messenger ribonucleic acid is associated with reduced levels of estrogen binding in human mammary tumors.* Mol Endocrinol. 1988;2:785-91.

51. Henry JA, Nicholson S, Farndon JR, Westley BR, May FEB. *Measurement of oestrogen receptor mRNA levels in human breast tumours.* Br J Cancer. 1988;58:600-5.

52. Yaich LE, Roodi N, Bailey LR, Verrier CS, Yee CJ, Cavener DR, et al. *Analysis of the estrogen receptor (ER) gene, transcript and protein in ER-positive and -negative breast cancer cell lines.* Endocrine-Related Cancer. 1995;2:293-309.

53. Fasco MJ. *Quantitation of estrogen receptor nRNA and its*

alternatively spliced mRNA's in breast tumor cells and tissues. Anal Biochem. 1997;245:167-78.

54. Fuqua SAW, Fitzgerald SD, Chamness GC, Tandon AK, McDonnell DP, Nawaz Z, et al. *Variant human breast tumor estrogen receptor with constitutive transcriptional activity.* Cancer Res. 1991;51:105-9.

55. Wang Y, Miksicek RJ. *Identification of a dominant negative form of the human estrogen receptor.* Mol Endocrinol. 1991;5:1707-15.

56. Castles CG, Fuqua ASW, Klotz DM, Hill SM. Expression of a constitutively active estrogen receptor variant in the estrogen receptor-negative BT-20 human breast cancer cell line. Cancer Res. 1993;53:5934-9.

57. Daffada AA, Johnston SR, Nicholls J, Dowsett M. *Detection of wild type and exon 5-deleted splice variant oestrogen receptor (ER) mRNA in ER-positive and-negative breast cancer cell lines by reverse transcription/polymerase chain reaction.* J Mol Endocrinol. 1994;13:265-73.

58. Koehorst G, Cox JJ, Donker GH, Lopes da Silva S, Burbach JP, Thijssen JH, et al. *Functional analysis of an alternatively spliced estrogen receptor lacking exon 4 isolated from MCF-7 breast cancer cells and meningioma tissue.* Mol Cell Endocrinol. 1994;101:237-45.

59. Pfeffer U, Fecarotta E, Castagnetta L, Vidali G. *Estrogen receptor variant messenger RNA lacking exon 4 in estrogen-responsive human breast cancer cell lines.* Cancer Res. 1993;53:741-3.

60. van Dijk MAJ, Floore AN, Kloppenborg KIM, van't Veer LJ. *A functional assay in yeast for the human estrogen receptor displays wild-type and variant estrogen receptor messenger RNAs present in breast carcinoma.* Cancer Res. 1997;57:3478-85.

61. Poola I, Koduri S, Chatra S, Clarke R. *Identification of twenty alternatively spliced estrogen receptor alpha mRNAs in breast cancer cell lines and tumors using splice targeted primer approach.* J Steroid Biochem Mol Biol. 2000;72:249-58.

62. Pfeffer U, Fecarotta E, Vidali G. *Coexpression of multiple estrogen receptor variant messenger RNAs in normal and neoplastic breast tissues and in MCF-7 cells.* Cancer Res. 1995;55:2158-65.

63. Fuqua SAW, Fitzgerald SD, Allred DC, Elledge RM, Nawaz Z, McDonnell DP, et al. *Inhibition of*

estrogen receptor action by a naturally occurring variant in human breast tumors. Cancer Res.1992;52:483-6.

64. Miksicek RJ, Lei Y, Wang Y. *Exon skipping gives rise to alternatively spliced forms of the estrogen receptor in breast tumor cells.* Breast Cancer Res Treat. 1993;26:163-74.

65. Zhang Q, Borg A, Fuqua SAW. *An exon 5 deletion variant of the estrogen receptor frequently coexpressed with wild-type estrogen receptor in human breast cancer.* Cancer Res. 1993;53:5882-4.

66. Fuqua SAW. *Estrogen receptor mutagenesis and hormone resistance.* Cancer. 1994;74:1026-9.

67. Daffada AAI, Johnston SRD, Smith IE, Detre S, King N, Dowsett M. *Exon 5 deletion variant estrogen receptor messenger RNA expression in relation to tamoxifen resistance and progesterone receptor/pS2 status in human breast cancer.* Cancer Res. 1995;55:288-93.

68. Rea D, Parker MG. *Effects of an exon 5 variant of the estrogen receptor in MCF-7 breast cancer cells.* Cancer Res.1996;56:1556-63.

69. Ohlsson H, Lykkesfeldt AE, Madsen MW, Briand P. *The estrogen receptor variant lacking exon 5 has dominant negative activity in the human breast epithelial cell line HMT-3522S1.* Cancer Res. 1998;58:4264-8.

70. Chan CM, Dowsett M. *A novel estrogen receptor variant mRNA lacking exons 4 to 6 in breast carcinoma.* J Steroid Biochem Mol Biol. 1997;62:419-30.

71. Madsen MW, Reiter BE, Larsen SS, Briand P, Lykkesfeldt AE. *Estrogen receptor messenger RNA splice variants are not involved in antiestrogen resistance in sublines of MCF-7 human breast cancer cells.* Cancer Res.1997;57:585-9.

72. Leygue E, Huang A, Murphy LC, Watson PH. *Prevalence of estrogen receptor variant messenger RNAs in human breast cancer.* Cancer Res.1996;56:4324-7.

73. Rennie PS, Mawji NR, Coldman AJ, Godolphin W, Jones EC, Vielking JR, et al. *Relationship between variant forms of estrogen reeptor RNA and an apoptosis-related RNA, TRPM-2, with survival in patients with breast cancer.* Cancer. 1993;72:3648-54.

74. Lu B, Leygue E, Dotzlaw H, Murphy LJ, Murphy LC, Watson PH. *Estrogen receptor-beta mRNA variants in human and murine tissues.* Mol Cell Endocrinol. 1998;138:199-203.

75. Vladusic EA, Hornby AE, Guerra-Vladusic FK, Lupu R. *Expression of estrogen receptor b messenger RNA variant in breast cancer.* Cancer Res. 1998;58:210-4.

76. Leygue E, Dotzlaw H, Watson PH, Murphy LC. *Expression of estrogen receptor beta1, beta2, and beta5 messenger RNAs in human breast tissues.* Cancer Res.1999;59:1175-9.

77. Ogawa S, Inoue S, Watanabe T, Orimo A, Hosoi T, Ouchi Y, et al. *Molecular cloning and characterization of human estrogen receptor betacx: a potential inhibitor of estrogen action in human.* Nucleic Acids Res. 1998;26:3505-12.

78. Hanstein B, Liu H, Yancisin MC, Brown M. *Functional analysis of a novel estrogen receptor-b isoform.* Mol Endocrinol. 1999;13:129-37.

79. Desai AJ, Luqmani YA, Walters JE, Coope RC, Dagg B, Gomm JJ, et al. *Presence of exon 5-deleted oestrogen receptor in human breast cancer: functional analysis and clinical significance.* Br J Cancer. 1997;75:1173-84.

80. Gotteland M, Desauty G, Delarue JC, Liu L, May E. *Human estrogen receptor messenger RNA variants in both normal and tumor breast tissues.* Mol Cell Endocrinol. 1995;112:1-13.

81. Pfeffer U, Fecarotta E, Arena G, Forlani A, Vidali G. *Alternative splicing of the estrogen receptor primary transcript normally occurs in estrogen receptor positive tissues and cell lines.* J Steroid Biochem Mol Biol. 1996;56:99-105.

82. Murphy LC, Leygue E, Dotzlaw H, Douglas D, Coutts A, Watson PH. *Oestrogen receptor variants and mutations in human breast cancer.* Ann Med. 1997;29:221-34.

83. Sluyser M. *Mutations in the estrogen receptor gene.* Hum Mutat. 1995;6:97-103.

84. Leygue ER, Watson PH, Murphy LC. *Estrogen receptor variants in normal human mammary tissue.* J Natl Cancer Inst. 1996;88:284-90.

85. Daffada AA, Dowsett M. *Tissue-dependent expression of a novel splice variant of the human oestrogen receptor.* J Steroid Biochem Mol Biol. 1995;55:413-21.

86. Friend KE, Ang LW, Shupnik MA. *Estrogen regulates the expression of several different estrogen receptor mRNA isoforms in rat pituitary.* Proc NatlAcad Sci USA. 1995;92:4367-71.

87. Park W, Choi J, Hwang E, Lee J. *Identification of a variant estrogen receptor lacking exon 4 and its coexpression with wild-type estrogen receptor in ovarian carcinomas.* Clin Cancer Res. 1996;2:2029-35.

88. Skipper JK, Young LJ, Bergeron JM, Tetzlaff MT, Osborn CT, Crews D. *Identification of an isoform of the estrogen receptor messenger RNA lacking exon four and present in the brain.* Proc Natl Acad Sci USA. 1993;90:7172-5.

89. Villa E, Camellini L, Dugani A, Zucchi F, Grottola A, Merighi A, et al. *Variant estrogen receptor messenger RNA species detected in human primary hepatocellular carcinoma.* Cancer Res. 1995;55:498-500.

90. Dotzlaw H, Alkhalaf M, Murphy LC. *Characterization of estrogen receptor variant mRNAs from human breast cancers.* Mol Endocrinol. 1992;6:773-85.

91. Murphy LC, Hilsenbeck SG, Dotzlaw H, Fuqua SAW. *Relationship of clone 4 estrogen receptor variant messenger RNA expression to some known prognostic variables in human breast cancer.* Clin Cancer Res. 1995;1:155-9.

92. Wang M, Dotzlaw H, Fuqua SAW, Murphy LC. *A point mutation in the human estrogen receptor gene is associated with the expression of an abnormal estrogen receptor mRNA containing a 69 novel nucleotide insertion.* Breast Cancer Res Treat. 1997;44:145-51.

93. Pink JJ, Jiang SY, Fritsch M, Jordan VC. *An estrogen-independent MCF-7 breast cancer cell line which contains a novel 80-kilodalton estrogen receptor-related protein.* Cancer Res. 1995;55:2583-90.

94. Murphy LC, Wang M, Coutt A, Dotzlaw H. *Novel mutations in the estrogen receptor messenger RNA in human breast cancers.* J Clin Endocrinol Metab. 1996;81:1420-7.

95. Graham ML, Krett NL, Miller LA, Leslie KK, Gordon DF, Wood WM, et al. *T47Dco cell, genetically unstable and containing estrogen receptor mutations, are a model for the progression of breast cancers to hormone resistance.* Cancer Res.1990;50:6208-17.

96. Jozan S, Julia A-M, Carretie A, Eche N, Maisongrosse V, Fouet B, et al. *65 and 47 kDa forms of estrogen receptor in human breast cancer: relation with estrogen responsiveness.* Breast Cancer Res Treat. 1991;19:103-9.

97. Scott GK, Kishner P, Vigne JL, Benz CC. *Truncated forms of DNA-binding estrogen receptors in human breast cancer.* J Clin Invest. 1991;88:700-6.

98. Dowsett M, Daffada A, Chan CMW, Johnston SRD. *Oestrogen receptor mutants and variants in breast cancer.* Eur J Cancer. 1997;33:1177-83.

99. Tonetti DA, Jordan VC. *The role of estrogen receptor mutations in tamoxifen-stimulated breast cancer.* J Steroid Biochem Mol Biol. 1997;62:119-28.

6.1.2.3. MECHANISMS AFFECTING ER EXPRESSION

ESR1 (ERα) and ESR2 (ERβ) are epigenetically regulated genes. Stable changes in gene expression result not only from genetic but also epigenetic mechanisms, especially the addition of a methyl group to the 5' position of cytosine (5mC) catalyzed by DNA methyltransferases (DNMT1, DNMT3a, DNMT3b), predominantly within the CpG dinucleotides of CpG islands in the 5' regulatory region of some genes (1, 2). CpG island hypermethylation may inhibit transcription by interfering with the recruitment of basal transcription factors or transcriptional coactivators. Active gene promoters, particularly those that are CpG-rich and that normally lack DNA methylation, have nucleosome-depleted regions just upstream of their transcription start sites. The nucleosomes that flank these sites are marked by modified histone H3 trimethylated on lysine 4 (H3K4me3), extensive lysine acetylation and the histone variant H2A.Z, which may destabilize nucleosomes to facilitate transcriptional initiation (3-5). By contrast, DNA methylation stabilizes epigenetic gene silencing in promoters that lack H2A.Z, that have nucleosomes positioned over the transcriptional start site and that harbour repressive histone modifications, such as H2K9me2 or H2K9me3 marks (6). Besides posttranslational histone modifications and CpG methylation a third mechanism of gene silencing involves recruitment of proteins with a methyl-CpG-binding domain (MBD) that recognize methylated CpGs but have negligible affinity for unmethylated DNA (7, 8). At least five MBD proteins have been identified, including MeCP2 and MBD1 – 4 that interact with histone deacetylase (HDAC)-containing repressor complexes. Since cancer cells often display anomalous patterns of DNA methylation (9-11), methylation of CpG islands within the 5' region of the ER gene might impair ER gene transcription and result in the ER-negative phenotype.

The initial studies of ER gene CpG island methylation used Southern analysis of genomic DNA with methylation-sensitive restriction enzymes, which is limited to the examination of CpG dinucleotides within restriction sites, such as *HpaII, MspI,* and *NotI* (12-18). In contrast, a methylation-sensitive PCR assay permits the assessment of the methylation state of all 29 CpGs across the entire ER gene CpG island, extending from nucleotides 44 to 529, relative to the transcription start site (19). The island was completely unmethylated in normal breast tissue and ER-positive breast cancer cell lines, but extensively methylated in all ER-negative cell lines and 13 of 13 ER-negative, PR-negative breast cancers. In addition, ER-negative tumors that were unmethylated by Southern analysis of a single restriction site, were positive at other sites by the methylation-specific PCR assay, demonstrating the greater sensitivity of the latter technique. Based on these results, methylation of the ER gene CpG island may account for loss of ER transcription in ER-negative breast cancer cell lines and tumors. Interestingly, analysis of 11 ER-positive, PR-negative and 11 ER-positive, PR-positive breast cancers showed evidence of ER gene methylation in 8 (70%) and 4 (35%) of cases, respectively, possibly reflecting tumor cell heterogeneity. ER-negative cells had higher levels of DNA methyltransferase than ER-positive cells suggesting an increased capacity to methylate DNA, which might account for the low level of ER mRNA in the former cells. Demethylation of the CpG island by the DNMT inhibitor 5-azacytidine or 5-aza-2'-deoxycytidine reactivated ER gene expression with increased ER mRNA levels in ER-negative MDA-MB-231 cells (20). Treatment with trichostatin A, an inhibitor of histone deacetylase also reactivated ER gene transcription as indicated by a dose- and time-dependent re-expression of ER mRNA without alteration in CpG island methylation (21). The combination of DNMT and HDAC inhibitors had a synergistic effect on ER re-expression in MDA-MB-231 cells (22). The pharmacological intervention induced a complex chromatin remodeling process that included the release of a repressor complex containing MBD proteins (MeCP2, MBD1, MBD2), DNMTs (DNMT1, DNMT3b), and HDAC1 from the ER promoter and concomitant enrichment of acetyl-H4, acetyl-H3, K4-dimethylated H3 and diminished methyltaton at K4-H3 (23). Thus, the reactivated ER promoter in MDA-MB-231 cells treated with both drugs acquired a profile similar to that of the innately active promoter in MCF-7 cells providing proof that DNMTs, HDACs, and MBD proteins participate in the transcriptional control of ER.

ER expression in breast cancer reflects not only the basic biology of the tumor but also the influence of extratumoral factors. For example, breast cancers in the postmenopausal age group are more commonly ER-positive, and at higher concentrations, than those in premenopausal patients (24-26). This increase in mean ER tumor concentration with age is generally attributed to the decrease in circulating estrogen levels in post-menopausal patients. During normal development, ER is expressed at low levels in breast tissue of infants, prepubertal children, and perimenarchal girls (27). In the premenopausal age group, mammary ER expression appears to vary during the menstrual cycle in normal breast but not in breast cancer. In a fine needle aspiration study of 68 premenopausal women with normal breasts, 21 specimens (31%) were ER-positive (28). All 21 were obtained from 35 women who were in the first half of their menstrual cycles. None of the 33 samples obtained during the second half of the cycle expressed ER. In contrast, 51 of 83 (61%) carcinomas were ER-positive, 24 in the first and 27 in the second half of the cycle. Thus, ER production is suppressed at the time of ovulation in the normal breast of premenopausal women (28, 29). In contrast, cancer cells in ER-positive tumors synthesize ER continuously throughout the menstrual cycle or even display increasing ER expression with progression of the cycle (30).

Several studies investigated the relation of ER content to either plasma or tumor E_2 levels. There is no consistent relationship between plasma E_2 and tumor ER concentrations (31-33). The reason for this is the highly variable blood-tissue gradient of E_2 between different individuals and ages, with tissue concentrations exceeding plasma levels in the postmenopausal age group. In other words, plasma E_2 levels generally do not reflect tissue E_2 levels (34, 35). A closer correlation exists between tissue E_2 concentration and tumor ER content. Several authors observed higher E_2 concentrations in cytosol of ER-positive tumors than in cytosol of ER-negative tumors (35-38). This finding suggests that ER-positive breast cancers retain greater amounts of E_2 than ER-negative cancers.

At the molecular level, the effects of estrogen on ER gene expression are not clear, since both positive and negative regulation have been described (39, 40). ER is downregulated by E_2 in some estrogen-responsive tissues as well as in cell lines such as MCF-7 and GH3 (41, 42). However, this effect is not true under all culture conditions of MCF-7 cells or for all breast cancer cell lines (43). In T47D cells, estrogen treatment has been shown to either have no effect on ER (44) or to result in a 2.5-fold increase in ER mRNA (41). The model devised to explain these findings invokes an interaction of ER with its own gene, similar to the mechanism by which this same transcription factor regulates any other target gene. Accordingly, the activated ER would bind selected regions in the promoter of its own gene. There are two aspects of this model that require explanation. First, the promoter of the ER gene lacks a canonical or imperfect ERE suggesting that a novel mechanism of ER-DNA interaction may be involved. Second, estrogen causes autologous down- as well as up-regulation of ER expression, suggesting that ER can interact with DNA in different ways, at different binding sites, and possibly with accessory proteins (45).

Only one *trans*-acting factor, ER factor 1 (ERF-1), has been identified that may play a role in the differential expression of ER in ER-positive and -negative breast cancer (46). ERF-1 is identical to AP-2γ, a member of the AP-2 family of developmentally regulated transcription factors that includes AP-2α, AP-2β, and AP-2γ (47). The three related AP-2 proteins bind as homo- or heterodimers to a GC-rich response element with the palindromic sequence GCCNNNGGC. Two imperfect versions of this palindrome were identified in the 5' untranslated region of the ER gene between nucleotides +135 and +210. Binding of AP-2γ (ERF-1) to this 75-base pair sequence augmented *in vivo* expression from the ER promoter (46). AP-2γ is abundantly expressed in ER-positive breast (MCF-7, T47D, BT20) and endometrial (ECC-1, RL95-2) cancer cell lines. ER-negative cell lines, such as MDA-MB-231, HBL-100, HEC1A, and HEC1B, express low levels or no AP-2 γ (ERF-1). An immunohistochemical comparison of AP-2α and AP-2γ expression in breast tissue revealed significantly higher levels of AP-2γ in malignant than in benign breast epithelial cells, whereas AP-2α showed no difference (48). These findings suggest up-regulation of AP-2γ expression in breast cancer. However, the presence of AP-2γ protein did not correlate with ER-positivity, whereas AP-2α and ER expression showed a significant positive correlation (p = 0.018). Several ER-positive tumors did not express either AP-2α or AP-2γ, indicating that the presence of AP-2 proteins was not essential for ER gene transcription. A 35-base pair enhancer element located between nucleotides -3778 and -3744 was shown to be critical for the high level of ER expression in ER-positive breast cancer cell lines (49). This enhancer element, termed ER-EHO, consists of an AP-1 site, which binds c-fos and c-jun, and adjacent sequences which bind another unknown factor(s). Mutation of ER-EHO reduced activity 60% in ER-positive cells but had little effect in ER-negative cells, while mutation of the downstream ERF-1 binding element had only a minimal effect in any of the cell lines. This result suggests that ER-EH0 is the dominant *cis*-acting element in differential ER expression. Both c-fos and c-jun are present in breast cancer cell lines. In fact, the proto-oncogene c-fos is rapidly and transiently induced by E_2 in MCF-7 cells and an ERE has been detected in the promoter region of the gene (50, 51). However, E_2 does not induce c-jun in MCF-7 cells (52, 53). ER-positive MCF-7 cells contain a cell-specific protein that acted in cooperation with ER to stimulate transcriptional activation from an ERE-containing 200-base pair enhancer region of the rainbow trout ER gene (54). This protein was not found in ER-negative MDA-MB-231 breast cancer cells. Finally, the coding region of ER cDNA possesses a sequence(s) necessary for ER downregulation of both protein and mRNA (55). E_2 did not appear to influence the stability of the ER transcript, which implies that the autologous downregulation is occurring at the transcriptional level. The authors speculated that negative regulation by E_2 may involve ER interactions with repressor proteins at composite regulatory elements.

Interestingly, ER-mediated transcriptional activity can be inhibited by PR in the presence of progestin, an example of inhibitory crosstalk between steroid hormone receptors (56, 57). In fact, the synthetic progestin, ORG 2058, was shown to decrease ER mRNA 35% 6 hours after treatment of T47D cells (58). Examination of the effects of retinoids on ER expression yielded contradictory results. One group reported that all-*trans* retinoic acid increased ER mRNA and protein two-fold in MCF-7 cells (59, 60). Another group reported that both all-*trans* and 9-*cis* retinoic acid induced downregulation of ER mRNA and protein in MCF-7 cells (61). The ER activity was also affected by phorbol esters such as 12-O-tetradecanoylphorbol-13-acetate (TPA). In short term treatment, TPA synergistically activated ER in the presence of E_2 (62). In contrast to the short term effect, long term treatment with TPA downregulated ER expression by facilitating rapid degradation of ER mRNA (63-65). A half-life study in MCF-7 cells demonstrated that TPA decreased ER mRNA half-life from 4 hours in control cells to 40 minutes in TPA-treated cells (66). ER protein declined by about 80% in 24 hours from 236 fmol/mg to 50 fmol/mg. Following removal of TPA from the culture medium, the levels of ER protein and mRNA returned to control values. However, the receptor failed to bind E_2, suggesting that TPA induces a factor which interacts with ER and blocks E_2 binding. The protein kinase C inhibitors, H-7 and bryostatin, prevented the effects of TPA on ER expression and binding, indicating that activation of the protein kinase C signal transduction pathway inhibits ER function (67).

Peptide hormones, growth factors, and cytokines can also affect ER expression, although the underlying mechanisms are in most cases unclear. For example, insulin and prolactin were shown to change ER levels in MCF-7 cells in opposite directions. High prolactin concentrations caused a 50% increase in ER levels (68). In contrast, ER levels decreased from over 100 fmol/mg in the presence of low insulin to less than 30 fmol/mg in the presence of high insulin concentrations (69, 70). Similarly, insulin-like growth factor 1 decreased ER levels in MCF-7 cells (71). TGFβ caused a 75% reduction in steady-state levels of ER mRNA at 6 hours (72). A putative ligand of the erbB-2 receptor, gp30, also reduced ER mRNA levels in MCF-7 and BT474 cells (73). Similarly, heregulin-ß2 (HRG), ligand for the HER4 (erbB-4) receptor, is involved in the acquisition of the hormone-independent phenotype of breast cancer cells (74). Stably HRG-transfected MCF-7 cells lost ER expression due to decreased ER gene transcription. The mechanism for the loss of ER expression in these MCF-7/T6 cells is unknown. The cytokine interleukin-1a downregulated ER protein in MCF-7 cells by about 40% within 3 hours (75). This occurred without altering the steady-state concentration of ER mRNA, suggesting that the site of action is at the posttranscriptional level. The antiviral agent distamycin also inhibited ER gene transcription in MCF-7 cells (76).

Oral contraceptives and antiestrogens can also affect the level of ER expression. Use of oral contraceptives significantly reduced the percentage of ER-positive cells in histologically normal breast tissue removed during operation for fibroadenoma or reduction mammoplasty (29, 77). Studies of ER expression in sequential breast cancer biopsies from patients on tamoxifen have reported inconsistent results. Early studies reported a decrease in ER expression based on ligand-binding assay results (78, 79). Subsequent studies determined that tamoxifen competes with labeled E_2 in the ligand-binding assay, giving a false ER-negative result (80). Application of immunological ER assays showed either no change or an increase in ER expression during tamoxifen therapy (81-83). The up-regulation of ER expression was attributed to the estrogen-agonist effect of tamoxifen. In contrast, steroidal antiestrogens such as ICI 164,384, ICI 182,780, and RU 58,668, which are free of estrogenic properties, consistently caused a decrease in ER expression (82, 84, 85). In the case of ICI 164,384, the decrease in ER concentration appeared to be due to an increase in its turnover with a reduction in half-life of the receptor from five hours in the presence of E_2 to less than one hour by ICI 164,384 (86).

Tamoxifen is the most common antiestrogen in clinical use and the majority of breast cancer patients receive the drug at some stage. It is estimated that 30 to 40% of ER-positive breast cancers fail to respond to tamoxifen therapy (87, 88). Almost all of those who initially respond eventually develop resistance to tamoxifen. Thus, resistance may be due to primary insensitivity or acquired after initial sensitivity of the tumor. Potential mechanisms for acquired tamoxifen resistance can be divided into three broad categories. (i) The pharmaco-dynamics of tamoxifen may be altered, resulting in lower drug levels within the tumor or in an altered balance of agonistic and antagonistic metabolites (89-91). (ii) Changes in ER expression may be responsible for the development of tamoxifen resistance (87, 92, 93). The change in expression could be due to marked downregu-lation of ER in previously ER-positive cells or a shift in the balance of ER-positive and -negative cells, for example, by apoptosis of the former (94). (iii) The effector pathway of cell proliferation downstream of the ER may be altered, resulting in independence from estrogenic control. As described in subsequent sections of Chapter 6, there is substantial evidence for the involvement of growth factor pathways in the mediation of the proliferative signal from estrogen.

With regard to changes in ER expression being responsible for the development of tamoxifen resistance in breast cancer, this notion is contradicted by sequential studies of ER expression in tamoxifen-resistant breast cancer (95). The initial clinical response to tamoxifen was strongly correlated with ER status; 89% of responders compared to only 15% of nonresponders were ER-positive at presentation. Following relapse or progression on tamoxifen, 61% of tumors remained ER-positive despite the acquired tamoxifen-resistance. Thus, a basic biological difference exists between tumors with acquired tamoxifen resistance and those with intrinsic resistance to the drug. The former appear to remain ER-positive while the latter are ER-negative to begin with. About 60% of ER-positive breast cancers were deficient in DNA binding as determined by ER-ERE complex formation in the electrophoretic mobility shift assay (96, 97). To assess whether ER DNA-binding function is altered during the development of antiestrogen resistance, investigators compared *in vitro* ER-ERE formation in ER-positive breast cancers that were either untreated primary tumors or treated tumors with acquired tamoxifen resistance (98). The resistant tumors continued to express ER of normal size and DNA-binding ability suggesting that the failure of antiestrogens to arrest tumor growth during emergence of clinical resistance results from altered gene-regulatory mechanisms other than ER-ERE complex formation.

The progression from hormone dependence to independence, leading to resistance to hormonal therapy, can be recapitulated experimentally to some extent in breast cancer cell lines (99). Long-term growth of estrogen-responsive human breast cancer cell lines in estrogen-free media inevitably leads to the development of estrogen-

independent growth. After an initial period of two to three months of slowed growth, rapid, steroid-independent growth rates were consistently obtained in MCF-7 cells (100, 101). These cells contained three-fold higher levels of ER, which was functional as determined by induction of PR and transactivation of a transiently transfected estrogen-responsive gene construct. Antiestrogen still effectively suppressed cell proliferation, although estrogens only minimally increased the proliferation rate. Breast cancer cells can adapt to low levels of estrogens by enhancing their sensitivity to E_2 (102). After depriving MCF-7 cells of estrogens in tissue culture medium for periods of one to six months, cell proliferation could be stimulated maximally by 10^{-14} to 10^{-15} M E_2. In contrast, wild type MCF-7 cells not exposed to estrogen deprivation required 10^{-10} M E_2 to grow at the same rate.

MCF-7 are the prototype of ER-positive cells, which display estrogen-dependent growth *in vitro* and estrogen-dependent tumorigenicity *in vivo*. Several MCF-7 sublines have been isolated, which are capable of *in vitro* growth in the absence of estrogen, e.g., MCF-7 clone 5C, MCF-7:2A, MCF7/MIII, and MCF7/LCC1 (103-106). Additional phenotypic changes such as acquired resistance to sytemic therapy were elicited in response to appropriate selective pressure, i.e., growth in the presence of anti-estrogen. These observations are consistent with the concept of clonal selection and expansion in the process of malignant progression. A complete lineage of MCF-7 cell sublines was selected by *in vivo* passage in ovariectomized, athymic, nude mice. The first of these sublines, MCF7/MIII, was isolated from an MCF-7 tumor grown in an ovariectomized, athymic, nude mouse, i.e., without estrogen stimulation (104). MCF7/MIII cells no longer required hormonal supplementation to form proliferating tumors and were more invasive and metastatic *in vivo* than the parental MCF-7 cells (107). Further selection of MCF7/MIII cells in ovariectomized, athymic, nude mice produced the MCF7/LCC1 cell line with an even more malignant phenotype as judged by the shortened lag time to the appearance of tumors in the animals (103). The MCF7/MIII and MCF7/LCC1 cells proliferate both *in vivo* and *in vitro* without estrogen but retain sensitivity to nonsteroidal and steroidal antiestrogens such as tamoxifen and ICI 182,780, respectively. To determine the consequences of acquired resistance to antiestrogens, the MCF7/LCC1 cells were further selected against either nonsteroidal or steroidal antiestrogens. *In vitro* selection of MCF/LCC1 cells against increasing concentrations of 4-hydroxytamoxifen produced the MCF7/LCC2 cells, which retain estrogen-independent growth and sensitivity to ICI 182,780, but have acquired resistance to tamoxifen *in vitro* and *in vivo* (108). A similar selection process was used against increasing concentrations of ICI 182,780 to produce the MCF7/LCC9 cell line (109). MCF7/LCC9 cells are not only resistant to ICI 182,780, but also exhibit full cross-resistance to tamoxifen, despite never having been exposed to this drug. The mechanism that confers tamoxifen resistance to MCF7/LCC2 and tamoxifen/ICI 182,780 cross-resistance to MCF7/LCC9 cells is unclear. Despite their antiestrogen resistance, MCF7/LCC2 and MCF7/LCC9 cells retain a level of ER expression comparable to that of their parental MCF7/LCC1 cells which, in turn, is similar to that of the original MCF-7 cell line. Functional studies showed that the dissociation constant for ER ligand binding was unchanged and structural analysis of ER mRNA revealed wild-type message without evidence of ER mRNA variants. All sublines displayed PR expression but differed either in the level of expression or the degree of E_2 regulation. Specifically, the E_2 induction of PR expression in MCF-7 cells was 5.8-fold compared to 2.5- and 2.6-fold in MCF7/LCC1 and MCF7/LCC2 cells, respectively. In MCF7/LCC9 cells, PR was not inducible by E_2, but overexpressed to levels higher than those in E_2-treated MCF-7 cells. The constitutive up-regulation of PR in MCF7/LCC9 cells implicates events downstream of the ER in acquired estrogen resistance (109). In summary, the collective evidence of clinical and experimental data supports the notion that ER expression is a stable phenotype in breast cancer (110).

To determine whether the estrogen independence of ER-negative breast cancer cells can be reversed by introduction of functional ER, MDA-MB-231 cells have been stably transfected with ER cDNA. As expected, wild type ER cDNA transfection resulted in the expression of functional ER as determined by estrogen inducibility of ERE reporter plasmids. Mutant ER cDNA (Asp351Tyr, Gly400Val) resulted in enhanced estrogenic activity of the antiestrogens keoxifene and 4-hydroxytamoxifen, respectively (111, 112). However, contrary to the E_2-stimulated proliferation of ER-positive MCF-7 cells, treatment of the transfected MDA-MB-231 cells with E_2 inhibited DNA replication and decreased the ability to invade *in vitro* and to metastasize in athymic nude mice (113, 114). These results indicate that factors other than the ER are involved in the progression toward hormone independence. Post-receptor alterations have been postulated in steroid receptor coactivator or transcription factor interactions that could affect the transcriptional activity of ER (115). An example is the coactivator RIP140, which enhances ER-mediated transcription (116). RIP140 also regulates the E2F pathway and represses E2F1 transactivation, thereby inhibiting the expression of E2F1 target genes, such as cyclin E and cyclin B2 (117). Increasing RIP140 levels in breast cancer cell lines resulted in a reduction in the proportion of cells in S phase.

Estrogen independence of ER-positive breast cancer cells may also arise from changes in growth factor production. This was demonstrated experimentally in MCF-7 cells, which possess a single copy of the HER2 growth factor receptor gene. Introduction of two to five copies of HER2 cDNA resulted in HER2 overexpression and the development of estrogen-independent *in vitro* and *in vivo*

growth with insensitivity to both estrogen and tamoxifen (118). Overexpression of HER2 receptor also resulted in E_2-independent down-regulation of ER from 275 to 105 fmol/mg. Addition of the HER receptor ligand heregulin further reduced ER binding capacity to 35 fmol/mg with no change in affinity of hormone binding. Overexpression and activation of the HER2 receptor also elicited downregulation of ER mRNA. Based on these results, the authors propose that long-term hormone-independent downregulation of ER by a HER2-mediated pathway may lead to estrogen resistance, providing experimental evidence for the clinical observation that overexpression of the HER2 gene in breast cancer is associated with an ER-negative phenotype.

Finally, the expression of ERα is influenced by microRNAs some of which can downregulate ERα via negative feedback loops while others upregulate ERα via positive feedback (119). For example, the miR-17-92 cluster at 13q31-q32 encodes six miRNAs including miR-18a, miR-19b, and miR-20b, which downregulate ERα expression at the protein translational level (120). Pre-miR-18a derived from DROSHA-pre-miR-17-92 cleavage was significantly more expressed in ERα-positive than ERα-negative breast cancers. Several other miRNAs have been shown to repress ERα expression by targeting the ERα 3' UTR: miR-22, miR-206, miR-221, miR-222 (121-123). The levels of these miRNAs were higher in ER-negative than ER-positive breast cancer cell lines. In contrast, miR375 upregulated ERα through the repression of Ras dexamethasone-induced 1 (RASD1), an anti-proliferative factor in MCF-7 cells (119, 124). This is consistent with the finding that a number of miRNAs show differential expression between ERα-positive and -negative tumors (125, 126).

References

1. Baylin SB, Jones PA. *A decade of exploring the cancer epigenome - biological and translational implications.* Nat Rev Cancer. 2011;11:726-34.

2. Bird AP. *CpG-rich islands and the function of DNA methylation.* Nature. 1986;321:209-13.

3. Berger SL. *The complex language of chromatin regulation during transcription.* Nature. 2007;447:407-12.

4. Dawson MA, Kouzarides T, Huntly BJ. *Targeting epigenetic readers in cancer.* N Engl J Med. 2012;367:647-57.

5. Kelly TK, Miranda TB, Liang G, Berman BP, Lin JC, Tanay A, et al. *H2A.Z maintenance during mitosis reveals nucleosome shifting on mitotically silenced genes.* Mol Cell. 2010;39:901-11.

6. Lin JC, Jeong S, Liang G, Takai D, Fatemi M, Tsai YC, et al. *Role of nucleosomal occupancy in the epigenetic silencing of the MLH1 CpG island.* Cancer Cell. 2007;12:432-44.

7. Roloff TC, Ropers HH, Nuber UA. *Comparative study of methyl-CpG-binding domain proteins.* BMC Genomics. 2003;4:1.

8. Wade PA. *Methyl CpG binding proteins: coupling chromatin architecture to gene regulation.* Oncogene. 2001;20:3166-73.

9. Baylin SB, Herman JG. *DNA hypermethylation in tumorigenesis. Epigenetics joins genetics.* Trends Genet. 2000;16:168-74.

10. Esteller M. *Epigenetics in cancer.* N Engl J Med. 2008;358:1148-59.

11. Laird PW, Jaenisch R. *The role of DNA methylation in cancer genetics* *and epigenetics.* Ann Rev Genet. 1996;30:441-64.

12. Chen Z, Ko A, Yang J, Jordan VC. *Methylation of CpG island is not a ubiquitous mechanism for the loss of oestrogen receptor in breast cancer cells.* Br J Cancer. 1998;77:181-5.

13. Falette NS, Fuqua AW, Chamness GC, Cheah MS, Greene GL, McGuire WL. *Estrogen receptor gene methylation in human breast tumors.* Cancer Res. 1990;50:3974-8.

14. Lapidus RG, Ferguson AT, Ottaviano YL, Parl FF, Smith HS, Weitzman SA, et al. *Methylation of estrogen and progesterone receptor gene 5' CpG islands correlates with lack of estrogen and progesterone receptor gene expression in breast tumors.* Clin Cancer Res. 1996;2:805-10.

15. Ottaviano YL, Issa J-P, Parl FF, Smith HS, Baylin SB, Davidson NE. *Methylation of the estrogen receptor gene CpG island marks loss of estrogen receptor expression in human breast cancer cells.* Cancer Res. 1994;54:2552-5.

16. Piva R, Rimondi AP, Hanau S, et al. *Different methylation of oestrogen receptor DNA in human breast carcinomas with and without oestrogen receptor.* Br J Cancer. 1990;61:270-5.

17. van Agthoven T, van Agthoven TLA, Dekker A, Foekens JA, Dorssers LCJ. *Induction of estrogen independence of ZR-75-1 human breast cancer cells by epigenetic alterations.* Mol Endocrinol. 1994;8:1474-83.

18. Watts CKW, Handel ML, King RJB, Sutherland RL. *Oestrogen receptor gene structure and function in breast cancer.* J Steroid Biochem Mol Biol. 1992;41:529-36.

19. Lapidus RG, Nass SJ, Butash KA, Parl FF, Wietzman SA, Graff JG, et al. *Mapping of ER gene CpG island methylation by methylation-specific polymerase chain reaction.* Cancer Res. 1998;58:2515-9.

20. Ferguson AT, Lapidus RG, Baylin SB, Davidson NE. *Demethylation of the estrogen receptor gene in estrogen receptor-negative breast cancer cells can reactivate estrogen receptor gene expression.* Cancer Res. 1995;55:2279-83.

21. Yang X, Ferguson AT, Nass SJ, Phillips DL, Butash KA, Wang SM, et al. *Transcriptional activation of estrogen receptor alpha in human breast cancer cells by histone deacetylase inhibition.* Cancer Res. 2000;60:6890-4.

22. Yang X, Phillips DL, Ferguson AT, Nelson WG, Herman JG, Davidson NE. *Synergistic activation of functional estrogen receptor (ER)-alpha by DNA methyltransferase and histone deacetylase inhibition in human ER-alpha-negative breast cancer cells.* Cancer Res. 2001;61:7025-9.

23. Sharma D, Blum J, Yang X, Beaulieu N, Macleod AR, Davidson NE. *Release of methyl CpG binding proteins and histone deacetylase 1 from the Estrogen receptor alpha (ER) promoter upon reactivation in ER-negative human breast cancer cells.* Mol Endocrinol. 2005;19:1740-51.

24. Fisher ER, Redmond CK, Liu H, Rockette H, Fisher B. *Correlation of estrogen receptor and pathologic characteristics of invasive breast cancer.* Cancer. 1980;45:349-53.

25. Johansson R, Vanharanta R, Soderholm J. *Estrogen receptors in mammary cancer: correlation with age, menopausal status, and response to therapy.* J Cancer Res Clin Oncol. 1984;107:221-4.

26. Wilking N, Rutqvist LE, Nordenskjold B, Skoog L. *Steroid receptor levels in breast cancer. Relationships with age and menopausal status.* Acta Oncol. 1989;28:807-10.

27. Boyd MT, Hildebrandt RH, Bartow SA. *Expression of the estrogen receptor gene in developing and adult human breast.* Breast Cancer Res Treat. 1996;37:243-51.

28. Markopoulos C, Berger U, Wilson P, Gazet JC, Coombes RC. *Oestrogen receptor content of normal breast cells and breast carcinomas throughout the menstrual cycle.* Br Med J. 1988;296:1349-51.

29. Williams G, Anderson E, Howell A, Watson R, Coyne J, Roberts SA, et al. *Oral contraceptive (OCP) use increases proliferation and decreases oestrogen receptor content of epithelial cells in the normal human breast.* Int J Cancer. 1991;48:206-10.

30. Khan SA, Rogers MAM, Khurana KK, Meguid MM, Numann PJ. *Estrogen receptor expression in benign breast epithelium and breast cancer risk.* J Natl Cancer Inst. 1998;90:37-42.

31. Maass H, Engel B, Hohmeister H, Lehmann F, Trams G. *Estrogen receptors in human breast cancer tissue.* Am J Obstet Gynecol. 1972;113:377--82.

32. Mason RC, Miller WR, Hawkins RA. *Plasma oestrogens and oestrogen receptors in breast cancer patients.* Br J Cancer. 1985;52:793-6.

33. Theve NO, Carlstrom K, Gustafsson JA, Gustafsson S, Nordenskjold B, Skoldefors H, et al. *Oestrogen receptors and peripheral serum levels of oestradiol -17b in patients with mammary carcinoma.* Eur J Cancer. 1978;14:1337-40.

34. Nagai R, Kataoka M, Kobayashi S, Ishihara K, Tobioka N, Nakashima J, et al. *Estrogen and progesterone receptors in human breast cancer with concomitant assay of plasma 17b-estradiol, progesterone, and prolactin levels.* Cancer Res. 1979;39:1835-40.

35. van Landeghem AA, Poortman J, Nabuurs M, Thijssen JH. *Endogenous concentration and subcellular distribution of estrogens in normal and malignant human breast tissue.* Cancer Res. 1985;45:2900-6.

36. Edery M, Goussard J, Dehennin L, Scholler R, Reiffsteck J, Drosdowsky MA. *Endogenous oestradiol-17beta concentration in breast tumours determined by mass fragmentography and by radioimmunoassay: relationship to receptor content.* Eur J Cancer. 1981;17:115-20.

37. Fishman J, Nisselbaum JS, Menendez-Botet CJ, Schwartz MK. *Estrone and estradiol content in human breast tumors: relationship to estradiol receptors.* J Steroid Biochem. 1977;8:893-6.

38. Vermeulen A, Deslypere JP, Paridaens R, Leclercq G, Roy F, Heuson JC. *Aromatase, 17 beta-hydroxysteroid dehydrogenase and intratissular sex hormone concentrations in cancerous and normal glandular breast tissue in postmenopausal women.* Eur J Cancer Clin Oncol. 1986;22:515-25.

39. Cicatiello L, Cobellis G, Addeo R, Papa M, Altucci L, Sica V, et al. *In vivo functional analysis of the mouse estrogen receptor gene promoter: a transgenic mouse model to study tissue-specific and developmental regulation of estrogen receptor gene transcription.* Mol Endocrinol. 1995;9:1077-90.

40. Pink JJ, Jordan VC. *Models of estrogen receptor regulation by estrogens and antiestrogens in breast cancer cell lines.* Cancer Res. 1996;56:2321-30.

41. Read LD, Greene GL, Katzenellenbogen BS. *Regulation of estrogen receptor messenger ribonucleic acid and protein levels in human breast cancer cell lines by sex steroid hormones, their antagonists, and growth factors.* Mol Endocrinol. 1989;3:295-304.

42. Saceda M, Lippman ME, Chambon P, Lindsey RL, Ponglikitmongkol M, Puente M, et al. *Regulation of the estrogen receptor in MCF-7 cells by estradiol.* Mol Endocrinol. 1988;2:1157-62.

43. Martin MB, Saceda M, Garcia-Morales P, Gottardis MM. *Regulation of estrogen receptor expression.* Breast Cancer Res Treat. 1994;31:183-9.

44. Berkenstam A, Glaumann H, Martin M, Gustafsson A, Norstedt G. *Hormonal regulation of estrogen receptor messenger ribonucleic acid in T47DCO and MCF-7 breast cancer cells.* Mol Endocrinol. 1989;3:22-8.

45. Santagati S, Gianazza E, P. A, Vegeto E, Patrone C, Pollio G, et al. *Oligonucleotide squelching reveals the mechanism of estrogen receptor autologous down-regulation.* Mol Endocrinol. 1997;11:938-49.

46. DeConinck EC, McPherson LA, Weigel RJ. *Transcriptional regulation of estrogen receptor in breast carcinomas.* Mol Cell Biol. 1995;15:2191-6.

47. McPherson LA, Baichwal VR, Weigel RJ. *Identification of ERF-1 as a member of the AP2 transcription factor family.* Proc Natl Acad Sci USA. 1997;94:4342-7.

48. Turner BC, Zhang J, Gumbs AA, Maher MG, Kaplan L, Carter D, et al. *Expression of AP-2 transcription factors in human breast cancer correlates with the regulation of multiple growth factor signalling pathways.* Cancer Res. 1998;58:5466-72.

49. Tang Z, Treilleux I, Brown M. *A transcriptional enhancer required for the differential expression of the human estrogen receptor in breast cancers.* Mol Cell Biol. 1997;17:1274-80.

50. Weisz A, Bresciani F. *Estrogen induces expression of c-fos and c-myc protooncogenes in rat uterus.* Mol Endocrinol. 1988;2:816-24.

51. Weisz A, Rosales R. *Identification of an estrogen response element upstream of the human c-fos gene that binds the estrogen receptor and the AP-1 transcription factor.* Nucleic Acids Res. 1990;18:5097-6106.

52. Chalbos D, Philips A, Rochefort H. *Genomic cross-talk between the estrogen receptor and growth factor regulatory pathways in estrogen target tissues.* Semin Cancer Biol. 1994;5:361-8.

53. Davidson NE, Prestigiacomo LJ, Hahm HA. *Induction of jun gene family members by transforming growth factor a but not 17b-estradiol in human breast cancer cells.* Cancer Res. 1993;53:291-7.

54. Lazennec G, Kern L, Salbert G, Saligaut D, Valotaire Y. *Cooperation between the human estrogen receptor (ER) and MCF-7 cell-specific transcription factors elicits high activity of an estrogen-inducible enhancer from the trout ER gene promoter.* Mol Endocrinol. 1996;10:1116-26.

55. Kaneko KJ, Furlow JD, Gorski J. *Involvement of the coding sequence for the estrogen receptor gene in autologous ligand-dependent down-regulation.* Mol Endocrinol. 1993;7:879-88.

56. Kraus WL, Weis KE, Katzenellenbogen BS. *Inhibitory cross-talk between steroid hormone receptors: differential targeting of estrogen receptor in the repression of its transcriptional activity by agonist-and antagonist-occupied progestin receptors.* Mol Cell Biol. 1995;15:1847-57.

57. McDonnell DP, Goldman ME. *RU486 exerts antiestrogenic activities through a novel progesterone receptor A form-mediated mechanism.* J Biol Chem. 1994;269:11945-9.

58. Alexander IE, Shine J, Sutherland RL. *Progestin regulation of estrogen receptor messenger RNA in human breast cancer cells.* Mol Endocrinol. 1990;4:821-8.

59. Butler WB, Fontana JA. *Responses to retinoic acid of tamoxifen-sensitive and -resistant sublines of human breast cancer cell line MCF-7.* Cancer Res. 1992;52:6164-7.

60. Fontana JA, Nervi C, Shao Z, Jetten AM. *Retinoid antagonism of estrogen-responsive transforming growth factor a and pS2 gene expression in breast carcinoma cells.* Cancer Res. 1992;52:3938-45.

61. Rubin M, Fenig E, Rosenauer A, Menendez-Botet C, Achkar C, Bentel JM, et al. *9-Cis retinoic acid inhibits growth of breast cancer cells and down-regulates estrogen receptor RNA and protein.* Cancer Res. 1994;54:6549-56.

62. Cho H, Ng PA, Katzenellenbogen BS. *Differential regulation of gene expression by estrogen in estrogen growth-independent and -dependent MCF-7 human breast camcer cell sublines.* Mol Endocrinol. 1991;5:1323-30.

63. Gierthy JF, Spink BC, Figge HL, Pentecost BT, Spink DC. *Effects of 2,3,7,8-tetrachlorodibenzo-p-dioxin, 12-O-tetradecanoylphorbol-13 acetate and 17 beta-estradiol on estrogen receptor regulation in MCF-7 human breast cancer cells.* J Cell Biochem. 1996;60:173-84.

64. Ree AH, Knutsen HK, Landmark BF, Eskild W, Hansson V. *Down-regulation of messenger ribonucleic acid (mRNA) for the estrogen receptor (ER) by phorbol ester requires ongoing RNA synthesis but not protein synthesis. Is hormonal control of ER mRNA degradation mediated by an RNA molecule?* Endocrinology. 1992;131:1810-14.

65. Tzukerman M, Zhang XK, Pfahl M. *Inhibition of estrogen receptor activity by the tumor promoter 12-O-tetradeconylphorbol-13 acetate: a molecular analysis.* Mol Endocrinol. 1991;5:1983-92.

66. Saceda M, Knabbe C, Dickson RB, Lippman ME, Bronzert D, Lindsey RK, et al. *Post-transcriptional destabilization of estrogen receptor mRNA in MCF-7 cells by 12-O-tetra-decanoylphorbol-13-acetate.* J Biol Chem. 1991;266:17809-14.

67. Martin MB, Garcia-Morales P, Stoica A, Solomon HB, Pierce MB, Katz D, et al. *Effects of 12-O-tetradec-anoylphorbol-13-acetate on estrogen receptor activity in MCF-7 cells.* J Biol Chem. 1995;270:25244-51.

68. Shafie S, Brooks SC. *Effect of prolactin on growth and the estrogen receptor level of human breast cancer cells (MCF-7).* Cancer Res. 1977;37:792-9.

69. Butler WB, Kelsey WH, Goran N. *Effects of serum and insulin on the sensitivity of the human breast cancer cell line MCF-7 to estrogen and antiestrogens.* Cancer Res. 1981;41:82-8.

70. Moore MR. *An insulin effect on cytoplasmic estrogen receptor in the human breast cancer cell line MCF-7.* J Biol Chem. 1981;256:3637-40.

71. Clayton SJ, May FE, Westley BR. *Insulin-like growth factors control the regulation of oestrogen and progesterone receptor expression by oestrogens.* Mol Cell Endocrinol. 1997;128:57-68.

72. Stoica A, Saceda M, Fakhro A, Solomon HB, Fenster BD, Martin MB. *The role of transforming growth factor-beta in the regulation of estrogen receptor expression in the MCF-7 breast cancer cell line.* Endocrinology. 1997;138:1498-505.

73. Saceda M, Grunt TW, Colomer R, Lippman ME, Lupu R, Martin MB. *Regulation of estrogen receptor concentration and activity by an erbB/HER ligand in breast carcinoma cell lines.* Endocrinology. 1996;137:4322-30.

74. Tang CK, Perez C, Grunt T, Waibel C, Cho C, Lupu R. *Involvement of heregulin-b2 in the acquisition of the hormone-independent phenotype of breast cancer cells.* Cancer Res. 1996;56:3350-8.

75. Danforth DNJ, M.K. S. *Interleulin 1a blocks estradiol-stimulated growth and down-regulates the estrogen receptor in MCDF-7 breast cancer cells in vitro.* Cancer Res. 1991;51:1488-93.

76. Bianchi N, Passadore M, Feriotto G, Mischiati C, Gambari R, Piva R. *Alteration of the expression of human estrogen receptor gene by distamycin.* J Steroid Biochem Mol Biol. 1995;54:211-5.

77. Battersby S, Robertson BJ, Anderson TJ, King RJB, McPherson K. *Influence of menstrual cycle, parity and oral contraceptive use on steroid hormone receptors in normal breast.* Br J Cancer. 1992;65:601-7.

78. Nomura Y, Tashiro H, Shinozuka K. *Changes of steroid hormone receptor content by chemotherapy and/or endocrine therapy in advanced breast cancer.* Cancer. 1985;55:546-51.

79. Taylor RE, Powles TJ, Humphreys J, Bettelheim R, Dowsett M, Casey AJ, et al. *Effects of endocrine therapy on steroid-receptor content of breast cancer.* Br J Cancer. 1982;45:80-5.

80. Encarnacion CA, Ciocca DR, McGuire WL, Clark GM, Fuqua SAW, Osborne CK. *Measurement of steroid hormone receptors in breast cancer patients on tamoxifen.* Breast Cancer Res Treat. 1993;26:237-46.

81. Clarke RB, Laidlaw IJ, Jones LJ, Howell A, Anderson E. *Effect of tamoxifen on Ki67 labelling index in human breast tumours and its relationship to estrogen and progesterone receptor status.* Br J Cancer. 1993;67:606-11.

82. McClelland RA, Manning DL, Gee JM, Anderson E, Clarke R, Howell A, et al. *Effects of short-term antiestrogen treatment of primary breast cancer on estrogen receptor mRNA and protein expression and on estrogen-regulated genes.* Breast Cancer Res Treat. 1996;41:31-41.

83. Noguchi S, Motomura K, Inaji H, Imaoka S, Koyama H. *Up-regulation of estrogen receptor by tamoxifen in human breast cancer.* Cancer. 1993;71:1266-72.

84. Gibson MK, Nemmers LA, Beckman WCJ, Davis VL, Curtis SW, Korach KS. *The mechanism of ICI 164,384 antiestrogenicity involves rapid loss of estrogen receptor in uterine tissue.* Endocrinology. 1991;129:2000-10.

85. Muller V, Jensen EV, Knabbe C. *Partial antagonism between steroidal and nonsteroidal antiestrogens in human breast cancer cell lines.* Cancer Res. 1998;58:263-7.

86. Dauvois S, Danielian PS, White R, Parker MG. *Antiestrogen ICI 164,384 reduces cellular estrogen receptor content by increasing its turnover.* Proc Natl Acad Sci USA. 1992;89:4037-41.

87. Maass H, Jonat W, Stolzenbach G, Trams G. *The problem of nonresponding estrogen receptor-positive patients with advanced breast cancer.* Cancer. 1980;46:2835-7.

88. McGuire WL. *Steroid hormone receptor in breast cancer treatment strategy.* Recent Prog Horm Res. 1980;36:135-56.

89. Johnston SR, Haynes BP, Smith IE, Jarman M, Sacks NP, Ebbs SR, et al. *Acquired tamoxifen resistance in human breast cancer and reduced intra-tumoral tumoral concentration.* Lancet. 1993;342:1521-2.

90. Morrow J, Jordan VC. *Molecular mechanisms of resistance to tamoxifen therapy in breast cancer.* Arch Surg. 1993;128:1187-91.

91. Osborne CK, Wiebe VJ, McGuire WL, Ciocca DR, DeGregorio MW. *Tamoxifen and the isomers of 4-hydroxytamoxifen in tamoxifen-resistant tumors from breast cancer patients.* J Clin Oncol. 1992;10:304-10.

92. Kuukasjarvi T, Kononen J, Helin H, Holli K, Isola J. *Loss of estrogen receptor in recurrent breast cancer is associated with poor response to endocrine therapy.* J Clin Oncol. 1996;14:2584-9.

93. van Netten JP, Algard FT, Coy P, Carlyle SJ, Brigden ML, Thornton KR, et al. *Heterogeneous estrogen receptor levels detected via multiple microsamples from individual breast cancers.* Cancer. 1985;56:2019-24.

94. Ellis PA, Saccani-Jotti G, Clarke R, Johnston SR, Anderson E, Howell A, et al. *Induction of apoptosis by tamoxifen and ICI 182780 in primary breast cancer.* Int J Cancer. 1997;72:608-13.

95. Johnston SR, Saccani-Jotti G, Smith IE, Salter J, Newby J, Coppen M, et al. *Changes in estrogen receptor, progesterone receptor, and pS2 expression in tamoxifen-resistant human breast cancer.* Cancer Res. 1995;55:3331-8.

96. Foster BD, Cavener DR, Parl FF. *Binding analysis of the estrogen receptor to its specific DNA target site in human breast cancer.* Cancer Res. 1991;51:3405-10.

97. Montgomery PA, Scott GK, Luce MC, Kaufmann M, Benz CC. *Human* breast tumors containing non-DNA-bindig immunoreactive (67 kDa) estrogen receptor. Breast Cancer Res Treat. 1993;26:181-9.

98. Johnston SRD, Lu B, Dowsett M, Liang X, Kaufmann M, Scott GK, et al. *Comparison of estrogen receptor DNA binding in untreated and acquired antiestrogen-resistant human breast tumors.* Cancer Res. 1997;57:3723-7.

99. Darbre PD, King RJB. *Progression to steroid insensitivity can occur irrespective of the presence of functional steroid receptors.* Cell. 1987;51:521-8.

100. Herman ME, Katzenellenbogen BS. *Alterations in transforming growth factor-a and -b production and cell responsiveness during the progression of MCF-7 human breast cancer cells to estrogen-autonomous growth.* Cancer Res. 1994;54:5867-74.

101. Katzenellenbogen BS, Kendra KL, NormanMJ, Berthois Y. *Proliferation, hormonal responsiveness, and estrogen receptor content of MCF-7 human breast cancer cells grown in the short-term and long-term absence of estrogens.* Cancer Res. 1987;47:4355-60.

102. Masamura S, Santner SJ, Heijan DF, Santen RJ. *Estrogen deprivation causes estradiol hypersensitivity in human breast cancer cells.* J Clin Endocrinol Metab. 1995;80:2918-25.

103. Brunner N, Boulay V, Fojo A, Freter CE, Lippmn ME, Clarke R. *Acquistion of hormone-independent growth in MCF-7 cells is accompanied by increased expression of estrogen-regulated genes but without detectable DNA amplifications.* Cancer Res. 1993;53:283-90.

104. Clarke R, Brunner N, Katzenellenbogen BS, Thompson EW, Norman MJ, Koppi C, et al. *Progression of human breast cancer cells from hormone-dependent to hormone-independent growth both in vitro and in vivo.* Proc Natl Acad Sci USA. 1989;86:3649-53.

105. Jiang SY, Wolf DM, Yingling JM, Chang C, Jordan VC. *An estrogen receptor positive MCF-7 clone that is resistant to antiestrogens and estradiol.* Mol Cell Endocrinol. 1992;90:77-86.

106. Pink JJ, Jiang SY, Fritsch M, Jordan VC. *An estrogen-independent MCF-7 breast cancer cell line which contains a novel 80-kilodalton estrogen receptor-related protein.* Cancer Res. 1995;55:2583-90.

107. Thompson EW, Brunner N, Torri J, Johnson MD, Boulay V, Wright A, et al. *The invasive and metastatic properties of hormone-independent but hormone-responsive variants of MCF-7 human breast cancer cells.* Clin Exper Metastasis. 1993;11:15-26.

108. Brunner N, Frandsen TL, Holst-Hansen C, Bei M, Thompson EW, Wakeling AE, et al. *MCF7/LCC2: a 4-hydroxytamoxifen resistant human breast cancer variant that retains sensitivity to the steroidal antiestrogen ICI 182, 780.* Cancer Res. 1993;53:3229-32.

109. Brunner N, Boysen B, Jirus S, Skaar TC, Holst-Hansen C, Lippman J, et al. *MCF7/LCC9: An antiestrogen-resistant MCF-7 variant in which acquired resistance to the steroidal antiestrogen ICI 182,780 confers an early cross-resistance to the nonsteroidal antiestrogen tamoxifen.* Cancer Res. 1997;57:3486-93.

110. Robertson JFR. *Oestrogen receptor: a stable phenotype in breast cancer.* Br J Cancer. 1996;73:5-12.

111. Jiang SY, Langan-Fahey SM, Stella AL, McCague R, Jordan VC. *Point mutation of estrogen receptor (ER) in the ligand-binding domain changes the pharmacology of antiestrogens in ER-negative breast cancer cells stably expressing complementary DNAs for ER.* Mol Endocrinol. 1992;6:2167-74.

112. Levenson AS, Catherino WH, Jordan VC. *Estrogenic activity is increased for an antiestrogen by a natural mutation of the estrogen receptor.* J Steroid Biochem Mol Biol. 1997;60:261-8.

113. Garcia M, Derocq D, Freiss G, Rochefort H. *Activation of estrogen receptor transfected into a receptor-negative breast cancer cell line decreases the metastatic and invasive potential of the cells.* Proc Natl Acad Sci USA. 1992;89:11538-42.

114. Jiang SY, Jordan VC. *Growth regulation of estrogen receptor-negative breast cancer cells transfected with complementary DNAs for estrogen receptor.* J Natl Cancer Inst. 1992;84:580-91.

115. Katzenellenbogen BS, Montano MM, Ekna K, Herman ME, McInerney EM. *Antiestrogens: Mechanisms of action and resistance in breast cancer.* Breast Cancer Res Treat. 1997;44:23-38.

116. Cavailles V, Dauvois S, L'Horset F, Lopez G, Hoare S, Kushner PJ, et al. *Nuclear factor RIP140 modulates transcriptional activation by the estrogen receptor.* EMBO J. 1995;14:3741-51.

117. Docquier A, Harmand PO, Fritsch S, Chanrion M, Darbon JM, Cavailles V. *The transcriptional coregulator RIP140 represses E2F1 activity and discriminates breast cancer subtypes.* Clin Cancer Res. 2010;16:2959-70.

118. Pietras RJ, Arboleda J, Reese DM, Wongvipat N, Pegram MD, Ramos L, et al. *HER-2 tyrosine kinase pathway targets estrogen receptor and promotes hormone-independent growth in human breast cancer cells.* Oncogene. 1995;10:2435-46.

119. de Souza Rocha Simonini P, Breiling A, Gupta N, Malekpour M, Youns M, Omranipour R, et al. *Epigenetically deregulated microRNA-375 is involved in a positive feedback loop with estrogen receptor alpha in breast cancer cells.* Cancer Res. 2010;70:9175-84.

120. Castellano L, Giamas G, Jacob J, Coombes RC, Lucchesi W, Thiruchelvam P, et al. *The estrogen receptor-alpha-induced microRNA signature regulates itself and its transcriptional response.* Proc Natl Acad Sci U S A. 2009;106:15732-7.

121. Adams BD, Furneaux H, White BA. *The micro-ribonucleic acid (miRNA) miR-206 targets the human estrogen receptor-alpha (ERalpha) and represses ERalpha messenger RNA and protein expression in breast cancer cell lines.* Mol Endocrinol. 2007;21:1132-47.

122. Pandey DP, Picard D. *miR-22 inhibits estrogen signaling by directly targeting the estrogen receptor alpha mRNA.* Mol Cell Biol. 2009;29:3783-90.

123. Zhao JJ, Lin J, Yang H, Kong W, He L, Ma X, et al. *MicroRNA-221/222 negatively regulates estrogen receptor alpha and is associated with tamoxifen resistance in breast cancer.* J Biol Chem. 2008;283:31079-86.

124. Vaidyanathan G, Cismowski MJ, Wang G, Vincent TS, Brown KD, Lanier SM. *The Ras-related protein AGS1/RASD1 suppresses cell growth.* Oncogene. 2004;23:5858-63.

125. Iorio MV, Ferracin M, Liu CG, Veronese A, Spizzo R, Sabbioni S, et al. *MicroRNA gene expression deregulation in human breast cancer.* Cancer Res. 2005;65:7065-70.

126. Mattie MD, Benz CC, Bowers J, Sensinger K, Wong L, Scott GK, et al. *Optimized high-throughput microRNA expression profiling provides novel biomarker assessment of clinical prostate and breast cancer biopsies.* Mol Cancer. 2006;5:24.

6.1.2.4. ER-POSITIVE AND –NEGATIVE PHENOTYPES

ER expression is positively correlated with breast cancer differentiation, i.e., highly differentiated tumors are more frequently ER-positive than anaplastic ones, regardless of examined differentiation parameter (Table 6.1.9.). Many studies have documented a positive correlation between histopathological tumor grade and ER status (1-3). Certain histologic subtypes are usually ER-positive (lobular, tubular, mucinous) or ER-negative (medullary, apocrine) (4-8). In carcinoma *in situ* (CIS) of the breast well-differentiated subtypes such as lobular and non-comedo ductal CIS are usually ER-positive, whereas poorly differentiated comedo ductal CIS are ER-negative in about 80% (9-12). Neither tumor stage nor the degree of nodal involvement appear to affect ER status except that very large cancers are more likely to be ER-negative (13, 14). ER expression is also significantly correlated with the proliferative rate of tumors, assessed by [^{3}H]thymidine labeling index, flow cytometric S phase fraction (SPF), mitotic index calculated from direct microscopic counts, or immunohistochemical staining of the proliferation-associated nuclear antigen Ki67 (Mib1). Early studies used thymidine labeling to determine the proliferative rate of breast cancers and observed an inverse correlation between ER-positivity and thymidine labeling index (15, 16). DNA flow cytometry studies confirmed and extended these findings (14, 17-19). One study of more than 100,000 breast cancers demonstrated significantly lower SPFs in tumors containing both ER and PR (median SPFs for diploid and aneuploid tumors were 3.1 and 8.5, respectively) as compared to tumors lacking both receptors (median SPFs 5.1 and 15.3, respectively) (20). An inverse relation was also established between positivity of ER, mitotic index, and Ki67 protein which is expressed only in cycling cells, i.e., it is present in late G1, S, G2, and M phases of the cell cycle (17, 21, 22). ER is also inversely related to another proliferation marker, mitosin, a 350 kD nuclear phosphoprotein that associates with the mitotic apparatus during M phase (23). Finally, ER expression has been related to ploidy (DNA content) of cancer cells as yet another marker of tumor aggressiveness. Several studies reported that ER-negative breast cancers are more frequently aneuploid than ER-positive tumors, while the converse applies to diploidy (18, 24-27). Other investigators did not find such a consistent relationship (17, 28). Combination of ER status with DNA flow cytometric results in a multivariate analysis offered greater statistical power to predict the clinical behavior of primary breast cancers than either factor alone (29).

Khan and associates (30) performed an immunohistochemical examination of ER expression in benign breast epithelium in relation to breast cancer risk. Semiquantitative analysis of ER expression in normal breast epithelium from 376 women undergoing diagnostic or therapeutic breast surgery showed that ER levels were higher in patients with breast cancer than in control patients.

Moreover, the odds of a woman with ER-positive normal epithelium having breast cancer were significantly elevated with an adjusted odds ratio of 2.63 (95%CI 1.47 – 4.90) controlled for known breast cancer risk factors. The effect of ER positivity was stronger for postmenopausal than premenopausal women. However, the use of hormone replacement therapy resulted in a drastic increase in the proportion of ER-positive control subjects, so that there was no difference between ER-positivity rates between case and control subjects in this subset. Overall, these findings suggest that overexpression of ER in normal breast epithelium may augment estrogen sensitivity and hence the risk of breast cancer.

Breast cancers in African American women are more likely to occur at a younger age, to be more poorly differentiated, and to exhibit high grade nuclear atypia and higher SPF (31). Associated with this more aggressive tumor phenotype is a higher percentage of ER-negativity in African American than Caucasian patients (32, 33). A large study (4885 Caucasian, 1016 African American, 777 Hispanic) found that tumors of Hispanic women were intermediate between tumors of the two other groups in terms of ER and PR positivity (32). A small study of 135 American women (48 Caucasian, 44 African American, 43 Asian) detected no significant differences in ER breast cancer expression between the three groups (34). Several investigators examined the ER phenotype of tumors in relation to family history of breast cancer and found no significant association (35, 36). The incidence of ER expression in breast cancer from pregnant women was not significantly different from that of tumors from nonpregnant age-matched patients (37). This finding is

Table 6.1.9. Correlation of ER Phenotype with Tumor Differentiation, Biomarker Expression, and Molecular Subtype

	ER-positive	ER-negative
Histology	well differentiated	poorly differentiated
	low nuclear grade	high nuclear grade
	low S-phase fraction	high S-phase fraction
	low proliferative index	high proliferative index
Biomarkers	PR	Ki67 (Mib1)
	pS2	vimentin
	hsp 27	
	tissue-type plasminogen activator	urokinase-type plasminogen activator
		Loss of heterozygosity at 17q21
		HER2 amplification
Molecular Subtype		p53 mutation
	Luminal A	Basal-like
	Luminal B	HER2-enriched

consistent with data indicating similar prognosis, stage for stage, for pregnant and nonpregnant patients with breast cancer (38). ER is also expressed in male breast cancer with a higher incidence of ER-positivity (ranging from 73 to 87%) than in female tumors (39-41).

Multiple **chromosomal anomalies** are present in breast cancer cells, both in terms of qualitative and quantitative aberrations (42), which are discussed in detail in Chapter 6.5. Both types of aberrations are more frequent in ER-negative than -positive tumors (43, 44). A case in point is chromosome 17, which exhibits more genetic abnormalities than other chromosomes in breast cancer. A study by the European Breast Cancer Linkage Consortium of 42 different loci on chromosome 17 in 1280 breast cancers showed significantly more loss of heterozygosity (LOH) in ER-negative than -positive tumors at several loci along the chromosome, especially on the q arm (45). A Japanese study of 616 primary breast cancers revealed that LOH at 17q21 was significantly correlated with absence of ER ($p < 0.0003$) or PR ($p < 0.0001$), and with the absence of both ($p < 0.0001$) (46). The HER2 gene at 17q12 (see Chapter 6.2.2.) is amplified in about 30% of breast cancers and consistently associated with the ER-negative phenotype, leading to the distinction of the HER-2 enriched subgroup of tumors (Table 6.1.6.) (43, 47-52). The p53 gene at 17p13.1 (see Chapter 8.4) is mutated in approximately 25% of breast cancers, mostly in form of missense mutations, which lead to inactivation of the p53 protein (53). A meta-analysis showed a significant correlation between p53 and ER status, i.e., wild type p53 was associated with ER-positivity, while mutant p53 was associated with the ER-negative phenotype ($p < 0.001$) (54). These data suggest a relationship between specific genetic changes on chromosome 17 and lack of ER expression.

As described in the preceding Section 6.1.1., estrogens induce the transcription of a number of genes. Several of these genes have been examined in breast cancer, e.g., PR, pS2, cathepsin D, c-myc, and hsp 27. Although these genes are estrogen-induced via the ER, their expression in breast cancer does not necessarily parallel the ER status. For example, cathepsin D and c-myc show inconsistent or no correlation with ER expression. The most likely explanation for this lack of coexpression, despite estrogen inducibility, is the influence of other factors, e.g., growth factors, on the expression of these genes. The expression of other genes, e.g., BRCA1, may parallel the expression of ER. However, instead of being directly estrogen-responsive, the coexpression simply reflects coordinate regulation with the effect of estrogen on the cell cycle (55). A third group of genes, e.g., Ki67, uPA, and vimentin, consistently shows an inverse correlation with ER expression. The mechanism of the transcriptional repression is presently unknown but does

not appear to directly involve E_2. In summary, ER-positive and -negative breast cancers display distinct phenotypes of gene expression and chromosomal anomalies summarized in Table 6.1.9. These studies have allowed to further differentiate ER-positive and -negative tumors resulting in four molecular subgroups of breast cancer with prognostic and therapeutic implications: luminal A and luminal B subtypes express ERα and PR whereas basal-like and HER2-enriched subtypes lack ERα and PR expression (49, 50).

The **pS2 gene** is the prototype of an estrogen-inducible gene, whose expression is regulated at the transcriptional level via binding of ER to an ERE in the pS2 promoter (56, 57). However, other factors such as protein kinase A and C activators also stimulate pS2 expression and may prevail, as shown in an estrogen-independent subline of MCF-7 cells (58, 59). pS2 is also known as trefoil factor 1 (TFF1), belonging to the family of trefoil factors that includes spasmolytic polypeptide (SP, TFF2) and intestinal trefoil factor (ITF, TFF3). The structurally and evolutionarily related TFFs are normally expressed in the gastrointestinal tract (60, 61). Experimentally, they were shown to affect *in vitro* cell motility and spreading. pS2 and ITF are also present in normal and neoplastic breast epithelium as well as ER-positive breast cancer cell lines (62-65). In breast cancers, pS2 expression has been assessed by immunohistochemical, Northern, and reverse transcriptase-PCR analysis (66, 67). Premenopausal but not postmenopausal breast cancers show a highly significant association between pS2 expression and ER positivity (68). Overall, 96% of pS2-positive tumors were ER-positive and 76% of ER-positive tumors were pS2-positive (69, 70).

The expression of **vimentin** intermediate filaments shows a significant correlation with ER-negativity. ER-positive breast cancers are generally vimentin-negative whereas about half of ER-negative tumors express vimentin, especially those that are poorly differentiated (21, 71). Nearly 80% of lobular breast cancer are ER-positive but only 5% express vimentin, in contrast to ductal cancer of which about 60% are ER-positive and close to 20% vimentin-positive (72). Vimentin expression is also correlated with poor prognosis in breast cancer patients and increased invasiveness of breast cancer cell lines (73-75).

The **heat shock protein 27 (hsp 27)** is estrogen-inducible at the transcriptional level via ER/Sp1 complexes binding to its promoter (76, 77). Hsp 27 belongs to the family of heat shock proteins and is thought to be involved in cell differentiation, drug resistance, and chaperone processes (78, 79). As expected from the estrogen-inducibility, there is a significant positive correlation between hsp 27 and ER expression in breast cancer (80-82). In spite of the association with ER status, the majority of studies involving a combined total of over 1500 breast cancers did not find any correlation between hsp 27 expression and prognosis in terms of disease-free interval, overall survival, or response to tamoxifen therapy (83-85).

Breast cancer invasion is accomplished by the concerted action of four different classes of tumor-associated proteases: (i) serine proteases including **tissue-type plasminogen activator (tPA)** and **urokinase-type plasminogen activator (uPA)**, (ii) aspartyl protease cathepsin D, (iii) cysteine proteases including cathepsins B and L, and (iv) matrix metalloproteases (MMP) including collagenases, gelatinases, and stromelysins (86). Both types of PA have been identified within normal breast and breast cancer and higher uPA activity was found in malignant compared to normal breast tissue (87, 88). The two types of PA differ in other respects. tPA is induced by E_2 in ER-positive breast cancer cells and several studies involving over 500 breast cancers showed significantly higher tPA activity in ER-positive than -negative tumors (89-92). In contrast, determination of uPA activity in over 1000 breast cancers yielded significantly lower values in ER-positive than -negative tumors (93-95). Furthermore, high tPA levels were associated with good prognosis (91, 96, 97), while high uPA levels were associated with poor prognosis and poor response to tamoxifen therapy (95, 97, 98). uPA can form complexes with endogenous molecules such as its two inhibitors, PAI-1 and PAI-2, and its receptor, uPAR. Several studies have shown that patients with high tumor levels of PAI-1 and uPAR and low levels of PAI-2 have a worse prognosis than patients with low PAI-1 and uPAR and high PAI-2 levels (99-102). uPAR levels showed a weak or no correlation with ER content (101, 103). Two groups showed an inverse correlation between PAI-1 and ER levels (104, 105). **Cathepsin D** is a lysosomal aspartyl protease that is expressed at low levels in all cells, but overexpressed in breast cancer cells (106, 107). Cathepsin D transcription is stimulated by estrogen in ER-positive breast cancer cells via EREs in the cathepsin D promoter and in ER-negative cells by an unknown mechanism (108-110). Although cathepsin D expression is induced by estrogen, there is no correlation between ER and cathepsin D levels in breast cancer, probably because the latter is also regulated by growth factors (58, 111-113). Like cathepsin D, the cysteine proteases cathepsin B and L are lysosomal enzymes expressed at higher levels in breast cancer than in normal breast tissue (114-116). Cathepsin B activity showed no correlation with ER status whereas cathepsin L showed an inverse correlation (90, 117). The expression of several MMPs has also been examined in breast cancer. Levels of MMP-8 and MMP-9 in breast cancer were shown to be inversely related to ER levels (118). The MMP activity is regulated by at least three different tissue inhibitors of metalloproteases (TIMP). The expression of TIMP-1 and TIMP-2 showed no correlation with ER status (119).

The expression of the **c-myc** oncogene is enhanced by E_2 in cultured normal mammary epithelium and breast cancer cell lines (120-123). E_2 treatment of MCF-7 cells increased transcription of c-myc, as measured by accumulation of mRNA within 30 minutes and peak protein levels within 1 – 3 hours, placing c-myc induction among the earliest transcriptional responses to E_2. The c-myc gene is amplified in about 15% of breast cancers, especially ER-negative tumors of the basal-like subtype (43, 48, 124-126).

The **c-myb** oncogene encodes a nuclear phosphoprotein that plays a role in proliferation and differentiation of hematopoietic precursor cells (127). c-myb is expressed in hematopoietic malignancies as well as a wide variety of solid tumors including breast cancer. Addition of E_2 to the culture medium of MCF-7 cells induced a 20-fold increase in c-myb mRNA and protein (128). Estrogen withdrawal downregulated c-myb mRNA and protein. c-myb mRNA was also expressed and modulated by estrogen in ER-negative MDA-MB-231 cells stably transfected with ER cDNA. Two promoters for the c-myb gene have been identified and sequenced. Because there are no ERE consensus sequences in either promoter, estrogen may not affect c-myb mRNA accumulation at the transcriptional level but rather at a posttranscriptional level (128). A study of 169 primary breast cancers showed a positive correlation between ER positivity and c-myb mRNA expression (129). The c-myb expression also correlated with PR and pS2 positivity, but showed an inverse association with HER2 overexpression.

Estrogens have been shown to influence angiogenesis (130, 131). Immunohistochemical analysis identified ERα-positive stromal cells in the microenvironment of breast cancer cells in both ER-positive and –negative tumors (132). In animal models of ERα-negative cancer cells grafted subcutaneously into syngeneic ovariectomized immunocompetent mice, E_2 enhanced tumor growth and increased intratumoral blood vessel density. Thus, E_2 may promote the growth of ER-negative tumors through the activation of stromal ERα-positive cells, which may adapt tumor angiogenesis to prevent hypoxia and necrosis. E_2 rapidly up-regulates production of vascular endothelial growth factor (VEGF) mRNA. The estrogen effect is blocked by the antiestrogen ICI 182,780 in a dose-dependent manner (133). Nevertheless, the estrogen-mediated induction of the VEGF gene is likely to occur indirectly as the promoter lacks an apparent ERE (134). Three studies involving over 300 breast cancers failed to show a significant correlation between ER status and VEGF expression or angiogenesis assessed by quantification of intratumoral microvessel density (135-137).

References

1. Maynard PV, Davies CJ, Blamey RW, Elston CW, Johnson J, Griffiths K. *Relationship between oestrogen-receptor content and histological grade in human primary breast tumours.* Br J Cancer. 1978;38:745-8.

2. McCarty KS, Barton TK, Fetter BF, Woodard BH, Mossler JA, Reeves W, et al. *Correlation of estrogen and progesterone receptors with histologic differentiation in mammary carcinoma.* Cancer. 1980;46:2851-8.

3. Parl FF, Wagner RK. *The histo-pathological evaluation of human breast cancers in correlation with estrogen receptor values.* Cancer. 1980;46:362-7.

4. Helin HJ, Helle MJ, Kallioniemi OP, Isola JJ. *Immunohistochemi-cal determination of estrogen and progesterone receptors in human breast carcinoma.* Cancer. 1989;63:1761-7.

5. O'Malley FP, Bane A. *An update on apocrine lesions of the breast.* Histopathology. 2008;52:3-10.

6. Reiner A, Reiner G, Spona J, Schemper M, Holzner JH. *Histopathologic characterization of human breast cancer in correlation with estrogen receptor status. A comparison of immunocytochemical and biochemical analysis.* Cancer. 1988;61:1149-54.

7. Rosen PP, Menendez-Botet CJ, Nisselbaum JS, Urban JA, Mike V, Fracchia A, et al. *Pathological review of breast lesions analyzed for estrogen receptor protein.* Cancer Res. 1975;35:3187-94.

8. Shousha S, Coady AT, Stamp T, James KR, Alaghband-Zadeh J. *Oestrogen receptors in mucinous carcinoma of the breast: an immuno-histological study using paraffin wax sections.* J Clin Pathol. 1989;42:902-5.

9. Barnes R, Masood S. *Potential value of hormone receptor assay in carcinoma in situ of breast.* Am J Clin Pathol. 1990;94:533-7.

10. Bur ME, Zimarowski MJ, Schnitt SJ, Baker S, Lew R. *Estogen receptor immunohistochemistry in carcinoma in situ of the breast.* Cancer. 1992;69:1176-81.

11. Malafa M, Chaudhuri T, Thomford NR, Chaudhuri PK. *Estrogen receptors in ductal carcinoma in situ of breast.* Am Surg. 1990;56:436-9.

12. Pallis L, Wilking N, Cedermark B, Rutqvist LE, Skoog L. *Receptors for estrogen and progesterone in breast carcinoma in situ.* Anticancer Res. 1992;12:2113-5.

13. Gordon NH. *Association of education and income with estrogen receptor status in primary breast cancer.* Am J Epidemiol. 1995;142:796-803.

14. Pegoraro RJ, Kaman V, Nirmul D, Joubert SM. *Estrogen and progesterone receptors in breast cancer among women of different racial groups.* Cancer Res. 1986;46:2117-20.

15. Meyer JS, Friedman E, McCrate MM, Bauer WC. *Prediction of early course of breast carcinoma by thymidine labeling.* Cancer. 1983;51:1879-86.

16. Silvestrini R, Daidone MG, Luisi A, Mastore M, Leutner M, Salvadori B. *Cell proliferation in 3,800 node-negative breast cancers: consistency over time of biological and clinical information provided by 3H-thymidine labelling index.* Int J Cancer. 1997;74:122-7.

17. Helin M, Helle M, Helin H, Isola J. *Proliferative activity and steroid receptors determined by immunohisto-chemistry in adjacent frozen sections of 102 breast carcinomas.* Arch Pathol Lab Med. 1989;113:854-7.

18. Kallioniemi OP, Blanco G, Alavaikko M, Hietanen T, Mattila J, Lauslahti K, et al. *Improving the prognostic value of DNA flow cytometry in breast cancer by combining DNA index and S-phase fraction. A proposed classification of DNA histograms in breast cancer.* Cancer. 1988;62:2183-90.

19. Lykkesfeldt AE, Balslev I, Christensen IJ, Larsen JK, Molgaard H, Rasmussen BB, et al. *DNA ploidy and S-phase fraction in primary breast carcinomas in relation to prognostic factors and survival for premenopausal patients at high risk for recurrent disease.* Acta Oncol. 1988;27:749-56.

20. Wenger CR, Beardslee S, Owens MA, Pounds G, Oldaker T, Vendely P, et al. *DNA ploidy, S-phase, and steroid receptors in more than 127,000 breast cancer patients.* Breast Cancer Res Treat. 1993;28:9-20.

21. Domagala W, Lasota J, Bartkowiak J, Weber K, Osborn M. *Vimentin is preferentially expressed in human breast carcinomas with low estrogen receptor and high Ki-67 growth fraction.* Am J Path. 1990;136:219-27.

22. Gaffney EV, Venz-Williamson TL, Hutchinson G, Biggs PJ, Nelson KM. *Relationship of standardized mitotic indices to other prognostic factors in breast cancer.* Arch Pathol Lab Med. 1996;120:473-7.

23. Clark GM, Allred DC, Hilsenbeck SG, Chamness GC, Osborne CK, Jones D, et al. *Mitosin (a new proliferation marker) correlates with clinical outcome in node-negative breast cancer.* Cancer Res. 1997;57:5505-8.

24. Dressler LG, Seamer LC, Owens MA, Clark GM, McGuire WL. *DNA flow cytometry and prognostic factors in 1331 frozen breast cancer specimens.* Cancer. 1988;61:420-7.

25. Horsfall DJ, Tilley WD, Orell SR, Marshall VR, Cant EL. *Relationship between ploidy and steroid hormone receptors in primary invasive breast cancer.* Br J Cancer. 1986;53:23-8.

26. Olszewski W, Darzynkiewicz Z, Rosen PP, Schwartz MK, Melamed MR. *Flow cytometry of breast carcinoma: I. Relation of DNA ploidy level to histology and estrogen receptor.* Cancer. 1981;48:980-4.

27. Visscher DW, Zarbo RJ, Jacobsen G, Kambouris A, Talpos G, Sakr W, et al. *Multiparamet-ric deoxyribonucleic acid and cell cycle analysis of breast carcinomas by flow cytometry. Clinicopatho-logic correlations.* Lab Invest. 1990;62:370-8.

28. Muss HB, Kute TE, Case LD, Smith LR, Booher C, Long R, et al. *The relation of flow cytometry to clinical and biologic characteris-tics in women with node negative primary breast cancer.* Cancer. 1989;64:1894-900.

29. Clark GM, Wenger CR, Beardslee S, Owens MA, Pounds G, Oldaker T, et al. *How to integrate steroid hormone receptor, flow cytometric, and other prognostic information in regard to primary breast cancer.* Cancer. 1993;71:2157-62.

30. Khan SA, Rogers MAM, Khurana KK, Meguid MM, Numann PJ. *Estrogen receptor expression in benign breast epithelium and breast cancer risk.* J Natl Cancer Inst. 1998;90:37-42.

31. Trock BJ. *Breast cancer in African American women: Epidemiology and tumor biology.* Breast Cancer Res Treat. 1996;40:11-24.

32. Elledge RM, Clark GM, Chamness GC, Osborne CK. *Tumor biologic factors and breast cancer prognosis among white, Hispanic, and black women in the United States.* J Natl Cancer Inst. 1994;86:705-12.

33. Mohla S, Sampson CC, Khan T, Enterline JP, Leffall LJ, White JE. *Estrogen and progesterone receptors in breast cancer in Black Americans: Correlation of receptor data with tumor differentiation.* Cancer. 1982;50:552-9.

34. Krieger N, van den Eden SK, Zava D, Okamoto A. *Race/ethnicity, social class, and prevalence of breast cancer prognostic biomarkers: a study of white, black, and Asian women in the San Francisco bay area.* Ethn Dis. 1997;7:137-49.

35. Israeli D, Tartter PI, Brower ST, Mizrachy B, Bratton J. *The significance of family history for patients with carcinoma of the breast.* J Am Coll Surg. 1994;179:29-32.

36. Tutera AM, Sellers TA, Potter JD, Drinkard CR, Wiesner GL, Folsom AR. *Association between family history of cancer and breast cancer defined by estrogen and progesterone receptor status.* Genet Epidemiol. 1996;13:207-21.

37. Elledge RM, Ciocca DR, Langone G, McGuire WL. *Estrogen receptor, progesterone receptor, and HER-2/neu protein in breast cancers from pregnant patients.* Cancer. 1993;71:2499-506.

38. Petrek JA, Dukoff R, Rogatko A. *Prognosis of pregnancy-associated breast cancer.* Cancer. 1991;67:869-72.

39. Borgen PI, Senie RT, McKinnon WM, Rosen PP. *Carcinoma of the male breast: analysis of prognosis compared with matched female patients.* Ann Surg Oncol. 1997;4:385-8.

40. Dawson PJ, Paine TM, Wolman SR. *Immunocytochemical character-ization of male breast cancer.* Modern Pathol. 1992;5:651-25.

41. Joshi MG, Lee AK, Loda M, Damus MG, Pedersen C, Heatley GJ, et al. *Male breast carcinoma: an evaluation of prognostic factors contributing to a poorer outcome.* Cancer. 1996;77:490-8.

42. Devilee P, Cornelisse CJ. *Somatic geneticchanges in human breast cancer.* Biochim Biophys Acta. 1994;1198:113-30.

43. Berns EMJJ, Klijn JGM, van Staveren IL, Portengen H, Noordegraaf E, Foekens JA. *Prevalence of amplification of the oncogenes c-myc, HER2/neu, and int-2 in one thousand human breast tumours: Correlation with steroid receptors.* Eur J Cancer. 1992;28:697-700.

44. Magdelenat H, Gerbault-Seureau M, Laine-Bidron C, Prieur M, Dutrillaux B. *Genetic evolution of breast cancer: II. Relationship with estrogen and progesterone receptor expression.* Breast Cancer Res Treat. 1992;22:119-27.

45. Phelan CM, Borg A, Cuny M, Crichton DN, Baldersson T, Andersen TI, et al. *Consortium study on 1280 breast carcinomas: allelic loss on chromosome 17 targets subregions associated with family history and clinical parameters.* Cancer Res. 1998;58:1004-12.

46. Ito I, Yoshimoto M, Iwase T, Watanabe S, Katagiri T, Harada Y, et al. *Association of genetic alterations on chromosome 17 and loss of hormone receptors in breast cancer.* Br J Cancer. 1995;71:438-41.

47. Adnane J, Gaudray P, Simon MP, Simony-Lafontaine J, Jeanteur P, Theillet C. *Proto-oncogene amplification and human breast tumor phenotype.* Oncogene. 1989;4:1389-95.

48. Courjal F, Cuny M, Simony-Lafontaine J, Louason G, Speiser P, Zeillinger R, et al. Mapping of DNA amplifictions at 15 chromosomal localizations in 1875 breast tumors: *Definition of phenotypic groups.* Cancer Res. 1997;57:4360-7.

49. Nielsen TO, Hsu FD, Jensen K, Cheang M, Karaca G, Hu Z, et al. *Immunohistochemical and clinical characterization of the basal-like subtype of invasive breast carcinoma.* Clin Cancer Res. 2004;10:5367-74.

50. Prat A, Perou CM. *Mammary development meets cancer genomics.* Nat Med. 2009;15:842-4.

51. Slamon DJ, Clark GM, Wong SG, Levin WJ, Ullrich A, McGuire WL. *Human breast cancer: correlation of relapse and survival with amplification of the HER-2/neu oncogene.* Science. 1987;235:177-82.

52. Valeron PF, Chirino R, Vega V, Falcon O, Rivero JF, Torres S, et al. *Quantitative analysis of p185 (HER-2/neu) protein in breast cancer and its association with other prognostic factors.* Int J Cancer. 1997;74:175-9.

53. Greenblatt MS, Bennett WP, Hollstein M, Harris CC. *Mutations in the p53 tumor suppressor gene: Clues to cancer etiology and molecular pathogenesis.* Cancer Res. 1994;54:4855-78.

54. Caleffi M, Teague MW, Jensen RA, Vnencak-Jones CL, Dupont WD, Parl FF. *p53 gene mutations and steroid receptor status in breast cancer.* Cancer. 1994;73:2147-56.

55. Kuang WW, Thompson DA, Hoch RV, Weigel RJ. *Differential screening and suppression subtractive hybridization identified genes differentially expressed in an estrogen receptor-positive breast carcinoma cell line.* Nucleic Acids Res. 1998;26:1116-23.

56. Berry M, Nunez A, Chambon P. *Estrogen-responsive element of the human pS2 gene is an imperfectly palindromic sequence.* Proc Natl Acad Sci USA. 1989;86:1218-22.

57. Sewack GF, Hansen U. *Nucleosome positioning and transcription-associated chromatin alterations on the human estrogen-responsive pS2 promoter.* J Biol Chem. 1997;272:31118-29.

58. Cavailles V, Garcia M, Rochefort H. *Regulation of cathepsin-D and pS2 gene expression by growth factors in MCF7 human breast cancer cells.* Mol Endocrinol. 1989;3:552-8.

59. Cho H, Ng PA, Katzenellenbogen BS. *Differential regulation of gene expression by estrogen in estrogen growth-independent and -dependent MCF-7 human breast camcer cell sublines.* Mol Endocrinol. 1991;5:1323-30.

60. Gott P, Beck S, Machado JC, Carneiro F, Schmitt H, Blin N. *Human trefoil peptides: genomic structure in 21q22.3 and coordinated expression.* Eur J Hum Genet. 1996;4:308-15.

61. Plaut AG. *Trefoil peptides in the defense of the gastrointestinal tract.* N Engl J Med. 1997;336:506-7.

62. Brown AM, Jeltsch JM, Roberts M, Chambon P. *Activation of pS2 gene transcription is a primary response to estrogen in the human breast cancer cell line MCF-7.* Proc Natl Acad Sci USA. 1984;81:6344-8.

63. May FE, Westley BR. *Expression of human intestinal trefoil factor in malignant cells and its regulation by oestrogen in breast cancer cells.* J Pathol. 1997;182:404-13.

64. Poulsom R, Hanby AM, Lalani EN, Hauser F, Hoffmann W, Stamp GW. *Intestinal trefoil factor (TFF3) and pS2 (TFF1), but not spasmolytic polypeptide (TFF2) and mRNAs are co-expressed in normal, hyperplastic, and neoplastic human breast epithelium.* J Pathol. 1997;183:30-8.

65. Willard ST, Faught WJ, Frawley LS. *Real-time monitoring of estrogen-regulated gene expression in single, living breast cancer cells: A new paradigm for the study of molecular dynamics.* Cancer Res. 1997;57:4447-50.

66. Knowlden JM, Gee JMW, Bryant S, McClelland RA, D.L. M, Mansel R, et al. *Use of reverse transcription-polymerase chain reaction methodology to detect estrogen-regulated gene expression in small breast cancer specimens.* Clin Cancer Res. 1997;3:2165-72.

67. Piggott NH, Henry JA, May FE, Westley BR. *Antipeptide antibodies against the pNR-2 oestrogen-regulated protein of human breast cancer cells and detection of pNR-2 expression in normal tissues by immunohistochemistry.* J Pathol. 1991;163:95-104.

68. Detre S, King N, Salter J, MacLennan K, McKinna JA, Dowsett M. *Immunohistochemical and biochemical analysis of the oestrogen regulated protein pS2, and its relation with oestrogen receptor and progesterone receptor in breast cancer.* J Clin Pathol. 1994;47:240-4.

69. Gion M, Mione R, Pappagallo GL, Gatti C, Nascimben O, Bari M, et al. *PS2 in breast cancer-alternative or complementary tool to steroid receptor status? Evaluation of 446 cases.* Br J Cancer. 1993;68:374-9.

70. Rio MC, Bellocq JP, Gairard B, Rasmussen UB, Krust A, Koehl C, et al. *Specific expression of the pS2 gene in subclasses of breast cancers in comparison with expression of the estrogen and progesterone receptors and the oncogene ERBB2.* Proc Natl Acad Sci U S A. 1987;84:9243-7.

71. Cattoretti G, Andreola S, Clemente C, D'Amato L, Rilke F. *Vimentin and p53 expression on epidermal growth factor receptor-positive, oestrogen receptor-negative breast carcinomas.* Br J Cancer. 1988;57:353-7.

72. Domagala W, Markiewski M, Kubiak R, Bartkowiak J, Osborn M. *Immunohistochemical profile of invasive lobular carcinoma of the breast: predominantly vimentin and p53 protein negative, cathepsin D and oestrogen receptor positive.* Virchows Archiv A Pathol Anat. 1993;423:497-502.

73. Domagala W, Markiewski M, Harezga B, Dukowicz A, Osborn M. *Prognostic significance of tumor cell proliferation rate as determined by the MIB-1 antibody in breast carcinoma: Its relationship with vimentin and p53 protein.* Clin Cancer Res. 1996;2:147-54.

74. Russo A, Bazan V, Morello V, Tralongo V, Nagar C, Nuara R, et al. *Vimentin expression, proliferating cell nuclear antigen and flow cytometric factors. Prognostic role in breast cancer.* Anal Quant Cytol Histol. 1994;16:365 - 74.

75. Thompson EW, Paik S, Brunner N, Sommers CL, Zugmaier G, Clarke R, et al. *Association of increased basement membrane invasiveness with absence of estrogen receptor and expression of vimentin in human breast cancer cell lines.* J Cell Physiol. 1992;150:534-44.

76. Fuqua SA, Blum-Salingaros M, McGuire WL. *Induction of the estrogen-regulated "24K" protein by heat shock.* Cancer Res. 1989;49:4126-9.

77. Porter W, Wang F, Wang W, Duan R, Safe S. *Role of estrogen receptor/Sp1 complexes in estrogen-induced heat shock protein 27 gene expression.* Mol Endocrinol. 1996;10:1371-8.

78. Ciocca DR, Oesterreich S, Chamness GC, McGuire WL, Fuqua SA. *Biological and clinical implications of heat shock protein 27,000 (Hsp27): a review.* J Natl Cancer Inst. 1993;85:1558-70.

79. Vargas-Roig LM, Fanelli MA, Lopez LA, Gago FE, Tello O, Aznar JC, et al. *Heat shock-proteins and cell proliferation in human breast cancer biopsy samples.* Cancer Detect Prev. 1997;21:441-51.

80. Adams DJ, McGuire WL. *Quantitative enzyme-linked immunosorbent assay for the estrogen-regulated Mr 24,000 protein in human breast tumors: correlation with estrogen and progesterone receptors.* Cancer Res. 1985;45:2445-9.

81. Love S, King RJ. *A 27 kDa heat shock protein that has anomalous prognostic powers in early and advanced breast cancer.* Br J Cancer. 1994;69:743-8.

82. Thor A, Benz C, Moore D, Goldman E, Edgerton S, Landry J, et al. *Stress response protein(srp-27) determination in primary human breast carcinomas: clinical, histologic, and prognostic correlations.* J Natl Cancer Inst. 1991;83:170-8.

83. Ciocca DR, Green S, Elledge RM, Clark GM, Pugh R, Ravdin P, et al. *Heat shock proteins hsp27 and hsp70: Lack of correlation with response to tamoxifen and clinical course of disease in estrogen receptor-postive metastatic breast cancer (A Southwest Oncology Group Study).* Clin Cancer Res. 1998;5:1263-6.

84. Damstrup L, Andersen J, Kufe DW, Hayes DF, Poulsen HS. *Immunocytochemical determination of the estrogen-regulated proteins Mr 24,000, Mr 52,000 and DF3 breast cancer associated antigen: clinical value in advanced breast cancer and correlation with estrogen receptor.* Ann Oncol. 1992;3:71-7.

85. Oesterreich S, Hilsenbeck SG, Ciocca DR, Allred DC, Clark GM, Chamness GC, et al. *The small heat shock protein HSP27 is not an independent prognostic marker in axillary lymph node-negative breast cancer patients.* Clin Cancer Res. 1996;2:1199-206.

86. Duffy MJ. *Proteases as prognostic markers in cancer.* Clin Cancer Res. 1996;2:613-8.

87. O'Grady P, Lijnen HR, Duffy MJ. *Multiple forms of plasminogen activator in human breast tumors.* Cancer Res. 1985;45:6216-8.

88. Tissot JD, Hauert J, Bachmann F. *Characterization of plasminogen activators from normal human breast and colon and from breast and colon carcinomas.* Int J Cancer. 1984;34:295-302.

89. Dickerman HW, Martinez HL, Seeger JI, Kumar SA. *Estrogen regulation of human breast cancer cell line MCF-7 tissue plasminogen activator.* Endocrinology. 1989;125:492-500.

90. Duffy MJ, O'Grady P, Simon J, Rose M, Lijnen HR. *Tissue-type plasminogen activator in breast cancer: relationship with estradiol and progesterone receptors.* J Natl Cancer Inst. 1986;77:621-3.

91. Needham GK, Nicholson S, Angus B, Farndon JR, Harris AL. *Relationship of membrane-bound tissue type and urokinase type plasminogen activators in human breast cancers to estrogen and epidermal growth factor receptors.* Cancer Res. 1988;48:6603-7.

92. Rella C, Coviello M, Quaranta M, Paradiso A. *Tissue-type plasminogen activator as marker of functional steroid receptors in human breast cancer.* Thrombosis Res. 1993;69:209-20.

93. Duffy MJ, Reilly D, O'Sullivan C, O'Higgins N, Fennelly JJ, Andreasen P. *Urokinase-plasminogen activator, a new and independent prognostic marker in breast cancer.* Cancer Res. 1990;50:6827-9.

94. Ferno M, Bendahl PO, Borg A, Brundell J, Hirschberg L, Olsson H, et al. *Urokinase plasminogen activator, a strong independent prognostic factor in breast cancer, analysed in steroid receptor cytosols with a luminometric immunoassay.* Eur J Cancer. 1996;32A:793-801.

95. Foekens JA, Look MP, Peters HA, van Putten WL, Portengen H, Klijn JG. *Urokinase-type plasminogen activator and its inhibitor PAI-1: predictors of poor response to tamoxifen therapy in recurrent

breast cancer.* J Natl Cancer Inst. 1995;87:751-6.

96. Duffy MJ, O'Grady P, Devaney D, O'Siorain L, Fennelly JJ, Lijnen HR. *Tissue-type plasminogen activator, a new prognostic marker in breast cancer.* Cancer Res. 1988;48:1348-9.

97. Kim SJ, Shiba E, Kobayashi T, Yayoi E, Furukawa JF, Takatsuka Y, et al. *Prognostic impact of urokinase-type plasminogen activator (PA), PA inhibitor type-1, and tissue-type PA antigen levels in node-negative breast cancer: A prospective study on multicenter basis.* Clin Cancer Res. 1998;4:177-82.

98. Grondahl-Hansen J, Christensen IJ, Rosenquist C, Brunner N, Mouridsen HT, Dano K, et al. *High levels of urokinase-type plasminogen activator and its inhibitor PAI-1 in cytosolic extracts of breast carcinomas are associated with poor prognosis.* Cancer Res. 1993;53:2513-21.

99. Bouchet C, Spyratos F, Martin PM, Hacene K, Gentile A, Oglobine J. *Prognostic value of urokinase-type plasminogen activator (uPA) and plasminogen activator inhibitors PAI-1 and PAI-2 in breast carcinomas.* Br J Cancer. 1994;69:398-405.

100. Duggan C, Kennedy S, Kramer MD, Barnes C, Elvin P, McDermott E, et al. *Plasminogen activator inhibitor type 2 in breast cancer.* Br J Cancer. 1997;76:622-7.

101. Duggan C, Maguire T, McDermott E, O'Higgins N, Fennelly JJ, Duffy MJ. *Urokinase plasminogen activator and urokinase plasminogen activator receptor in breast cancer.* Int J Cancer. 1995;61:597-600.

102. Schmitt M, Thomssen C, Ulm K, Seiderer A, Harbeck N, Hofler H, et al. *Time-varying prognostic impact of tumour biological factors urokinase (uPA), PAI-1 and steroid hormone receptor status in primary breast cancer.* Br J Cancer. 1997;76:306-11.

103. Grondahl-Hansen J, Peters HA, van Putten WLJ, M.P. L, Pappot H, Ronne E, et al. *Prognostic significance of the receptor for urokinase plasminogen activator in breast cancer.* Clin Cancer Res. 1995;1:1079-87.

104. Foekens JA, Schmitt M, van Putten WL, Peters HA, Kramer MD, Janicke F, et al. *Plasminogen activator inhibitor-1 and prognosis in primary breast cancer.* J Clin Oncol. 1994;12:1648-58.

105. Janicke F, Schmitt M, Pache L, Ulm K, Harbeck N, Hofler H, et al.

Urokinase (uPA) and its inhibitor PAI-1 are strong independent prognostic factors in node-negative breast cancer. Breast Cancer Res Treat. 1993;24:195-208.

106. Rochefort H. *Oestrogens, proteases and breast cancer. From cell lines to clinical applications.* Eur J Cancer. 1994;30A:1583-6.

107. Schultz DC, Bazel S, Wright LM, Tucker S, Lange MK, Tachovsky T, et al. Western blotting and enzymatic activity analysis of cathepsin D in breast tissue and sera of patients with breast cncer and benign breast disease and of normal controls. Cancer Res. 1994;54:48-54.

108. Augereau P, Miralles F, Cavailles V, Gaudelet C, Parker M, Rochefort H. *Characterization of the proximal estrogen-responsive element of human cathepsin D gene.* Mol Endocrinol. 1994;8:693-703.

109. Garcia M, Platet N, Liaudet E, V. L, Derocq D, Brouillet JP, et al. *Biological and clinical significance of cathepsin D in breast cancer metastasis.* Stem Cells. 1996;14:642-50.

110. Krishnan V, Wang X, Safe S. *Estrogen receptor-Spl complexes mediate estrogen-induced cathepsin D gene expression in MCF-7 human breast cancer cells.* J Biol Chem. 1994;269:15912-7.

111. Cavailles V, Garcia M, Salazar G, Domergue J, Simony J, Pujol H, et al. *Immunodetection of estrogen receptor and 52,000-dalton protein in fine needle aspirates of breast cancer tumors.* J Natl Cancer Inst. 1987;79:245-52.

112. Marsigliante S, Biscozzo L, Correale M, Paradiso A, Leo G, Abbate I, et al. *Immunoradiometric detection of pS2 and total cathepsin D in primary breast cancer biopsies: their correlation with steroid receptors.* Br J Cancer. 1994;69:550-4.

113. Tandon AK, Clark GM, Chamness GC, Chirgwin JM, McGuire WL. *Cathepsin D and prognosis in breast cancer.* N Engl J Med. 1990;322:297-302.

114. Castiglioni T, Merino MJ, Elsner B, Lah TT, Sloane BF, Emmert-Buck MR. *Immunohistochemical analysis of cathepsins D, B, and L in human breast cancer.* Hum Pathol. 1994;25:857-62.

115. Lah TT, Kokalj-Kunovar M, Strukelj B, Pungercar J, Barlic-Maganja D, M. D-K, et al. *Stefins and lysosomal cathepsins, B.L and D in human breast carcinoma.* Int J Cancer. 1992;50:36-44.

116. Santamaria I, Velasco G,

Cazorla M, Fueyo A, Campo E, Lopez-Otin C. *Cathepsin L2, a novel human cysteine proteinase produced by breast and colorectal carcinomas.* Cancer Res. 1998;58:1624-30.

117. Thomssen C, Schmitt M, Goretzki L, Oppelt P, Pache L, Dettmar P, et al. *Prognostic value of the cysteine proteases cathepsin B and cathepsin L in human breast cancer.* Clin Cancer Res. 1995;1:741-6.

118. Duffy MJ, Blaser J, Duggan C, McDermott E, O'Higgins N, Fennelly JJ, et al. *Assay of matrix metalloproteases type 8 and 9 by ELISA in human breast cancer.* Br J Cancer. 1995;7:1025-8.

119. Ree AH, Florenes VA, Berg JP, Maelandsmo GM, Nesland JM, Fodstad O. *High levels of messenger RNAs for tissue inhibitors of metal-loproteinases (TIMP-1 and TIMP-2) in primary breast carcinomas are associated with development of distant metastases.* Clin Cancer Res. 1997;3:1623-8.

120. Dubik D, Dembinski TC, Shiu RPC. *Stimulation of c-myc oncogene expression associated with estrogen-induced proliferation of human breast cancer cells.* Cancer Res. 1987;47:6517-21.

121. Dubik D, Shiu RP. *Mechanism of estrogen activation of c-myc onogene expression.* Onogene. 1992;7:1587-94.

122. Leygue E, Gol-Winkler R, Gompel A, Louis-Sylvestre C, Soquet L, Staub S, et al. *Estradiol stimulates c-myc proto-oncogene expression in normal human breast epithelial cells in culture.* J Steroid Biochem Mol Biol. 1995;52:299-305.

123. Nass SJ, Dickson RB. *Defining a role for c-myc in breast tumorigenesis.* Breast Cancer Res Treat. 1997;44:1-22.

124. Chandriani S, Frengen E, Cowling VH, Pendergrass SA, Perou CM, Whitfield ML, et al. *A core MYC gene expression signature is prominent in basal-like breast cancer but only partially overlaps the core serum response.* PLoS One. 2009;4:e6693.

125. Chin K, DeVries S, Fridlyand J, Spellman PT, Roydasgupta R, Kuo WL, et al. *Genomic and transcri-tional aberrations linked to breast cancer pathophysiologies.* Cancer Cell. 2006;10:529-41.

126. Persons DL, Borelli KA, Hsu PH. *Quantitation of HER-2/ neu and c-myc gene amplification in breast carcinoma using fluorescence in situ hybridization.* Mod Pathol. 1997;10:720-7.

127. Wolff L. *Myb-induced transformation.* Crit Rev Oncogenesis. 1996;7:245-60.

128. Gudas JM, Klein RC, Oka M, Cowan KH. *Posttranscriptional regulation of the c-myb proto-oncogene in estrogen receptor-positive breast cancer cells.* Clin Cancer Res. 1995;1:235-43.

129. Guerin M, Sheng Z, Andrieu N, Riou G. *Strong association between c-myb and oestrogen-receptor expression in human breast cancer.* Oncogene. 1990;5:131-5.

130. Gupta PB, Proia D, Cingoz O, Weremowicz J, Naber SP, Weinberg RA, et al. *Systemic stromal effects of estrogen promote the growth of estrogen receptor-negative cancers.* Cancer Res. 2007;67:2062-71.

131. Losordo DW, Isner JM. *Estrogen and angiogenesis: A review.* Arterioscler Thromb Vasc Biol. 2001;21:6-12.

132. Pequeux C, Raymond-Letron I, Blacher S, Boudou F, Adlanmerini M, Fouque MJ, et al. *Stromal estrogen receptor-alpha promotes tumor growth by normalizing an increased angiogenesis.* Cancer Res. 2012;72:3010-9.

133. Hyder SM, Chiappetta C, Murthy L, Stancel GM. *Selective inhibition of estrogen-regulated gene expression in vivo by the pure antiestrogen ICI 182,780.* Cancer Res. 1997;57:2547-9.

134. Kolch W, Martiny-Baron G, Kieser A, Marme D. Regulation of the expression of the VEGF/VPS and its receptors: role in tumor angiogenesis. Breast Cancer Res Treat. 1995;36:139-55.

135. Gasparini G, Fox SB, Verderio P, Bonoldi E, Bevilacqua P, Boracchi P, et al. *Determination of angiogenesis adds information to estrogen receptor status in predicting the efficacy of adjuvant tamoxifen in node-positive breast cancer patients.* Clin Cancer Res. 1996;2:1191-8.

136. Relf M, LeJeune S, Scott PA, Fox S, Smith K, Leek R, et al. *Expression of the angiogenic factors vascular endothelial cell growth factor, acidic and basic fibroblast growth factor, tumor growth factor beta-1, platelet-derived endothelial cell growth factor, placenta growth factor, and pleiotrophin in human primary breast cancer and its relation to angiogenesis.* Cancer Res. 1997;57:963-9.

137. Toi M, Inada K, Hoshina S, Suzuki H, Kondo S, Tominaga T. *Vascular endothelial growth factor and platelet-derived endothelial cell growth factor are frequently coexpressed in highly vascularized human breast cancer.* Clin Cancer Res. 1995;1:961-4.

6.2. GROWTH FACTORS AND GROWTH FACTOR RECEPTORS

6.2.1. INTRODUCTION

Numerous growth factors regulate cell growth, survival, adhesion, migration, and differentiation functions that are amplified or weakened in cancer cells (1-5). Most growth factors activate cell proliferation and differentiation through binding their attendant receptors (Table 6.2.1.), which are tyrosine kinase enzymes embedded in the cell membrane that communicate the growth factor signals from outside the cell to pathways inside the cell, turning genes on and off in the process (Figure 6.1.2.) (6-8). As a result of the growth factor binding, the receptor tyrosine kinases (RTKs) undergo dimerization and conformational changes that result in transphosphorylation of discrete tyrosine residues. The predominant substrate is the receptor itself and this autophosphorylation creates the binding sites for proteins containing src homology 2 (SH2) domains. These SH2 domains recognize and bind to phosphotyrosine residues in a sequence-specific context (9). SH2 domains are found in proteins that either have enzymatic activity, such as

PLC-γ, or serve as adapters that, in turn, link to proteins with enzymatic activity, e.g., the regulatory subunit of PI3-kinase. Thus, the phosphorylation provides binding sites for signaling or linker/adapter molecules and the recruitment of additional signal molecules. Signal cascades are triggered, dependent upon the translocation, membrane association, and activation of tyrosine, serine/threonine, and lipid kinases, including mitogen-activated protein kinases (MAPKs), ras, protein kinase C, and PI3-kinase (Figure 6.1.2.). In turn, these kinases phosphorylate substrate proteins in the cytoplasm, altering target protein function. For example, ligand binding to RTKs leads to the activation of ras, which allows the recruitment of raf to the membrane and mediates the sequential phosphorylation and activation of raf, MEK1/2 and ERK1/2 (10). RTKs also translocate to the nucleus, where they phosphorylate and activate transcription factors that induce a variety of genes (11).

Different types of RTK alterations, including abnormal expression levels, gene amplifications, mutations, and SNPs have been described in cancer (5). Since RTKs activate multiple pathways, each of these alterations can cause deregulated RTK signaling with constitutive activation of several downstream pathways and aberrant cell proliferation and increased cell survival.
The development of antibodies directed against overexpressed or amplified RTKs and inhibitors targeted to proteins encoded by mutated RTK genes permits optimized therapy for individual tumors (3, 12-14). The ability of one RTK pathway to compensate for another to maintain tumor cell viability is emerging as a common resistance mechanism to antitumor agents targeting individual RTKs (15, 16). In addition, numerous feedback systems regulate these pathways and inhibition of one pathway may lead to activation of a parallel pathway (17).

Cell proliferation and differentiation are modulated by both estrogens and growth factors (6, 18). Growth factor responses can be modulated by ER ligands and, conversely, estrogen responses can be modulated by growth factors and RTKs. For example, TGFα is regulated by estradiol via an estrogen response element at the 5' end of the TGFα gene (19). PI3K/AKT signaling is upregulated via an interaction between ER and PI3K (20). Although estrogens and growth factors interact with receptors localized in different compartments of the cell, their nuclear end points are often on the same gene.

The majority of growth factors bind to tyrosine-kinase linked cell surface receptors. However, there are also non-receptor tyrosine kinases such as Src, which can activate multiple signal transduction cascades, e.g., Raf/MEK/ERK and MEKK/JNKK/JNK (21). The non-receptor Src tyrosine kinase is relevant for breast cancer because over 80% of tumors show increased Src activity compared with normal breast tissue (22).

Table 6.2.1. Growth Factors and Growth Factor Receptors

Gene	Abbreviation	Locus	Genetic Alteration in Breast Cancer
Epidermal growth factor receptor family			
Epidermal growth factor receptor	EGFR (HER1, ErbB1)	7p12	amplification/deletion
Human epidermal growth factor receptor 2	HER2/neu (ErbB2)	17q12	amplification
Human epidermal growth factor receptor 3	HER3 (ErbB3)	12q13	amplification
Human epidermal growth factor receptor 4	HER4 (ErbB4)	2q33.3-q34	amplification
Fibroblast growth factor receptor family			
Fibroblast growth factor receptor 1	FGFR1	8p11.2	amplification
Fibroblast growth factor receptor 2	FGFR2	10q26.3	amplification, SNPs
Fibroblast growth factor receptor 3	FGFR3	11q13	
Fibroblast growth factor receptor 4	FGFR4	5q35.2	
Transforming growth factor			
Transforming growth factor beta 1	TGFβ1	19q13.1	
Transforming growth factor beta 2	TGFβ2	1q41	
Transforming growth factor beta 3	TGFβ3	14q24	
Transforming growth factor receptor			
Transforming growth factor beta receptor 1	TGFBR1	9q22	
Transforming growth factor beta receptor 2	TGFBR2	3p22	
Transforming growth factor beta receptor 3	TGFBR3	1p33-p32	

References

1. Benson JR. *Role of transforming growth factor beta in breast carcinogenesis.* Lancet Oncol. 2004;5:229-39.

2. Dickson RB, Lippman ME. *Growth factors in breast cancer.* Endocr Rev. 1995;16:559-89.

3. Klein S, Levitzki A. *Targeting the EGFR and the PKB pathway in cancer.* Curr Opinion Cell Biol. 2009;21:185-93.

4. Stern DF. *ERBB3/HER3 and ERBB2/HER2 duet in mammary development and breast cancer.* J Mammary Gland Biol Neoplasia. 2008;13:215-23.

5. Turner N, Grose R. *Fibroblast growth factor signalling: from development to cancer.* Nature Rev Cancer. 2010;10:116-29.

6. Levin ER. *Bidirectional signaling between the estrogen receptor and the epidermal growth factor receptor.* Mol Endocrinol. 2003;17:309-17.

7. Massague J. *Receptors for the TGF-b family.* Cell. 1992;69:1067-70.

8. Siegel PM, Massague J. *Cytostatic and apoptotic actions of TGF-beta in homeostasis and cancer.* Nature Rev Cancer. 2003;3:807-21.

9. Pawson T, Gish GD. *SH2 and SH3 domains: From structure to function.* Cell. 1992;71:359-62.

10. Raman M, Chen W, Cobb MH. *Differential regulation and properties of MAPKs.* Oncogene. 2007;26:3100-12.

11. Wang SC, Lien HC, Xia W, Chen IF, Lo HW, Wang Z, et al. *Binding at and transactivation of the COX-2 promoter by nuclear tyrosine kinase receptor ErbB-2.* Cancer Cell. 2004;6:251-61.

12. Gradishar WJ. *HER2 therapy--an abundance of riches.* N Engl J Med. 2012;366:176-8.

13. Hynes NE, Dey JH. *Potential for targeting the fibroblast growth factor receptors in breast cancer.* Cancer Res. 2010;70:5199-202.

14. McDermott U, Downing JR, Stratton MR. *Genomics and the continuum of cancer care.* N Engl J Med. 2011;364:340-50.

15. Abramson V, Arteaga CL. *New strategies in HER2-overexpressing breast cancer: many combinations of targeted drugs available.* Clin Cancer Res. 2011;17:952-8.

16. Buck E, Gokhale PC, Koujak S, Brown E, Eyzaguirre A, Tao N, et al. *Compensatory insulin receptor (IR) activation on inhibition of insulin-like growth factor-1 receptor (IGF-1R): rationale for cotargeting IGF-1R and IR in cancer.* Mol Cancer Therapeutics. 2010;9:2652-64.

17. Turke AB, Song Y, Costa C, Cook R, Arteaga CL, Asara JM, et al. *MEK Inhibition Leads to PI3K/AKT Activation by Relieving a Negative Feedback on ERBB Receptors.* Cancer Res. 2012;72:3228-37.

18. Chalbos D, Philips A, Rochefort H. *Genomic cross-talk between the estrogen receptor and growth factor regulatory pathways in estrogen target tissues.* Semin Cancer Biol. 1994;5:361-8.

19. Saeki T, Cristiano A, Lynch MJ, Brattain M, Kim N, Normanno N, et al. *Regulation by estrogen through the 5'-flanking region of the transforming growth factor alpha gene.* Mol Endocrinol. 1991;5:1955-63.

20. Sun M, Paciga JE, Feldman RI, Yuan Z, Coppola D, Lu YY, et al. *Phosphatidylinositol-3-OH Kinase (PI3K)/AKT2, activated in breast cancer, regulates and is induced by estrogen receptor alpha (ERalpha) via interaction between ERalpha and PI3K.* Cancer Res. 2001;61:5985-91.

21. Parsons JT, Parsons SJ. *Src family protein tyrosine kinases: cooperating with growth factor and adhesion signaling pathways.* Curr Opinion Cell Biol. 1997;9:187-92.

22. Muthuswamy SK, Muller WJ. *Activation of the Src family of tyrosine kinases in mammary tumorigenesis.* Adv Cancer Res. 1994;64:111-23.

6.2.2. EPIDERMAL GROWTH FACTOR RECEPTORS
6.2.2.1. INTRODUCTION

The epidermal growth factor receptor (EGFR) family consists of four members, which contain a conserved epidermal growth factor (EGF) domain: (i) epidermal growth factor receptor (EGFR; also known as human epidermal growth factor receptor 1, HER1, or ErbB1; the gene symbol, ErbB, is derived from the name of a viral oncogene to which the receptor is homologous: Erythro-blastic Leukemia Viral Oncogene), (ii) HER2 (ErbB2; also named neu in rodents), (iii) HER3 (ErbB3), and (iv) HER4 (ErbB4) (Table 6.2.1.). The EGFRs consist of an extracellular region or ectodomain, a single transmembrane-spanning region, and a cytoplasmic tyrosine kinase domain. Each ectodomain is composed of four subdomains containing leucine- and cysteine-rich regions. Upon ligand binding, the ectodomain adopts the so-called open conformation, which triggers EGFRs dimerization (1). The formation of homo- or heterodimers results in further conformational changes and transphosphorylation of discrete tyrosine residues that, in turn, trigger various intracellular signal cascades (2, 3). Ligands include EGF, TGFα, amphiregulin and neuregulins.

EGFR (HER1). The EGFR gene at 7p12 encodes a 134 kDa protein, which is embedded in the cell membrane and activated by binding of its specific ligands, including EGF and TGFα. Upon activation by its growth factor ligands, EGFR undergoes a transition from an inactive monomeric form to an active homodimer. In addition to forming homodimers after ligand binding, EGFR may pair with another member of the ErbB receptor family, such as HER2, to create an activated heterodimer. The consequence of the EGFR ectodomain dimerization is the positioning of the two cytoplasmic domains in such a way that transphosphorylation of specific tyrosine, serine, and threonine residues can occur within the C-terminal tyrosine kinase portion of the receptor. The autophosphorylation includes several tyrosine residues including 992Tyr, 1045Tyr, 1068Tyr, and 1173Tyr. In turn, this autophosphorylation elicits downstream activation and signaling by several other proteins that associate with the phosphorylated tyrosines through their own phosphotyrosine-binding SH2 domains (4). These downstream signaling proteins initiate several signal transduction cascades, principally the MAPK, PI3K-AKT, PLC-γ-PKC and JNK pathways, leading to DNA synthesis and cell proliferation. For example, EGFR phosphorylates PCNA, an essential component for DNA synthesis, at 211Tyr, thereby protecting PCNA from polyubiquitination at 164Lys by cullin 4a E3 ligase and ubiquitin-mediated proteolysis (5). Blocking the phosphorylation induced ubiquitin-mediated degradation of the chromatin-bound PCNA results in suppression of cell proliferation.

HER2. The HER2 gene at 17q12 encodes a 185 kDa protein with extensive sequence homology to other members of the EGFR family. Despite not binding any direct ligand, HER2 is readily activated through heterodimerization with other EGFR family members because its ectodomain is constitutively in an open conformation (6). Thus, HER2 serves as a co-receptor and facilitates signal transduction as part of a heterodimer complex, e.g., HER2/3 that forms after ligand binding to HER3. The intracellular domain of HER2 has tyrosine kinase activity that regulates growth and differentiation of cells via activation of multiple signal transduction pathways including MAPK, PI3K, and phospholipase C-γ (3, 7). In turn, these pathways stimulate other transcription factors, e.g., PI3K/AKT induces NF-κB, which activates the Arf/p53 pathway via Dmp1, thereby linking HER2 and p53 signaling (8-10). HER2 gene transcription is under the control of at least two promoters (11). AP-2 and ETS factor family members are required for maximal promoter activity and associated with HER2 overexpression in breast cancer. However, other transactivators, such as estrogen-related receptor α (ERRα) and its coregulator peroxisome proliferator-activated receptor γ coactivator 1β (PGC-1β) have been shown to enhance HER2 expression (12). In contrast, HER2 expression is inhibited by several transcription factors including GATA, PAX2, and ERα, which can repress HER2 expression in cooperation with PAX2 through binding to a *cis*-regulatory element in the presence of either 17β-estradiol or 4-hydroxytamoxifen (13, 14). HER2 expression can also be regulated by nutrient availability, such that reduced glucose concentrations resulted in diminished Neu2 protein levels (15). PGC-1α was shown to promote the growth of HER2-induced mammary tumors by regulating the glucose supply. Treatment of HER2-positive tumor cells with HER2-specific antagonistic antibodies or kinase inhibitors blocks these cells in the G1 phase of the cell cycle (16).

HER3. The HER3 gene at 12q13 encodes a 148-kDa protein, which lacks intrinsic tyrosine kinase activity, requiring heterodimerization with other EGFR family members to activate and transphosphorylate HER3 (7). Thus, HER2 and HER3 homodimers are inactive, while HER2/HER3 heterodimers are fully functional. Although they cooperate in heterodimerization and transphosphorylation, each receptor activates distinct signaling pathways. HER2 recruits adapter proteins that primarily function through the RAS/MAPK pathway, whereas HER3 recruits the PI3K pathway by direct interaction with the p85 regulatory subunit of PI3K (16-18). HER3 heterodimerization with overexpressed HER2 can also result in up-regulation of PI3K/AKT signaling (19).

HER4. The HER4 gene at 2q33.3-q34 encodes a 146-kDa protein but variants can result from alternative splicing (20). In contrast to other HER family members, HER4 functions as a proapoptotic protein, suppressing the growth of malignant cells (21). Ligand activation and subsequent proteolytic processing of HER4 results in mitochondrial accumulation of the HER4 intracellular domain and cytochrome c efflux, the essential and committed step of mitochondrial-regulated apoptosis (22).

Mammary epithelial cells co-express multiple EGF ligands (e.g., EGF, TGFα, amphiregulin, neuregulins) and receptors (18, 23-25). Studies in rodent models indicate that each of the EGFRs has unique functions in mammary development ranging from early patterning to terminal differentiation during lactation (18, 26). For example, HER4 functions as a nuclear chaperone for signal transducer and activator of transcription (STAT) 5A, thereby stimulating β-casein gene expression during the initiation of lactation at parturition (21). The receptors also cooperate by forming heterodimers. For example, the formation of HER2/HER3 heterodimers plays a critical role in HER2-mediated signaling in both normal mammary development and tumorigenesis (3, 27, 28). HER3 ablation in a mouse mammary model decreased HER2-induced tumor formation from 93% to 7% (29). Because of the functional interdependence among HER receptors it is not always possible to clearly define the extent to which different signaling pathways interact in breast carcinogenesis. Finally, single-cell induction of HER2 overexpression in a three-dimensional acinar culture of non-transformed human mammary MCF10A cells resulted in luminal outgrowth characteristic of early-stage breast cancer (30). In contrast, overexpression of Myc, Akt, or cyclin D1 was not sufficient to drive clonal outgrowth.

References

1. Burgess AW, Cho HS, Eigenbrot C, Ferguson KM, Garrett TP, Leahy DJ, et al. *An open-and-shut case? Recent insights into the activation of EGF/ErbB receptors.* Mol Cell. 2003;12:541-52.

2. Chang L, Karin M. *Mammalian MAP kinase signalling cascades.* Nature. 2001;410:37-40.

3. Reese DM, Slamon DJ. *HER-2/neu signal transduction in human breast and ovarian cancer.* Stem Cells. 1997;15:1-8.

4. Pawson T, Gish GD. *SH2 and SH3 domains: From structure to function.* Cell. 1992;71:359-62.

5. Lo YH, Ho PC, Wang SC. *Epidermal Growth Factor Receptor Protects Proliferating Cell Nuclear Antigen from Cullin 4A Protein-mediated Proteolysis.* J Biol Chem. 2012;287:27148-57.

6. Garrett TP, McKern NM, Lou M, Elleman TC, Adams TE, Lovrecz GO, et al. *The crystal structure of a truncated ErbB2 ectodomain reveals an active conformation, poised to interact with other ErbB receptors.* Mol Cell. 2003;11:495-505.

7. Citri A, Skaria KB, Yarden Y. *The deaf and the dumb: the biology of ErbB-2 and ErbB-3.* Exp Cell Res. 2003;284:54-65.

8. Liu M, Ju X, Willmarth NE, Casimiro MC, Ojeifo J, Sakamaki T, et al. *Nuclear factor-kappaB enhances ErbB2-induced mammary tumorigenesis and neoangiogenesis in vivo.* Am J Pathol. 2009;174:1910-20.

9. Pratt MA, Tibbo E, Robertson SJ, Jansson D, Hurst K, Perez-Iratxeta C, et al. *The canonical NF-kappaB pathway is required for formation of luminal mammary neoplasias and is activated in the mammary progenitor population.* Oncogene. 2009;28:2710-22.

10. Taneja P, Maglic D, Kai F, Sugiyama T, Kendig RD, Frazier DP, et al. *Critical roles of DMP1 in human epidermal growth factor receptor 2/neu-Arf-p53 signaling and breast cancer development.* Cancer Res. 2010;70:9084-94.

11. Hurst HC. *Update on HER-2 as a target for cancer therapy: the ERBB2 promoter and its exploitation for cancer treatment.* Breast Cancer Res. 2001;3:395-8.

12. Deblois G, Chahrour G, Perry MC, Sylvain-Drolet G, Muller WJ, Giguere V. *Transcriptional control of the ERBB2 amplicon by ERRalpha and PGC-1beta promotes mammary gland tumorigenesis.* Cancer Res. 2010;70:10277-87.

13. Hua G, Zhu B, Rosa F, Deblon N, Adelaide J, Kahn-Perles B, et al. *A negative feedback regulatory loop associates the tyrosine kinase receptor ERBB2 and the transcription factor GATA4 in breast cancer cells.* Mol Cancer Res. 2009;7:402-14.

14. Hurtado A, Holmes KA, Geistlinger TR, Hutcheson IR, Nicholson RI, Brown M, et al. *Regulation of ERBB2 by oestrogen receptor-PAX2 determines response to tamoxifen.* Nature. 2008;456:663-6.

15. Klimcakova E, Chenard V, McGuirk S, Germain D, Avizonis D, Muller WJ, et al. *PGC-1alpha promotes the growth of ErbB2/Neu-induced mammary tumors by regulating nutrient supply.* Cancer Res. 2012;72:1538-46.

16. Holbro T, Beerli RR, Maurer F, Koziczak M, Barbas CF, 3rd, Hynes NE. *The ErbB2/ErbB3 heterodimer functions as an oncogenic unit: ErbB2 requires ErbB3 to drive breast tumor cell proliferation.* Proc Natl Acad Sci U S A. 2003;100:8933-8.

17. Hellyer NJ, Cheng K, Koland JG. *ErbB3 (HER3) interaction with the p85 regulatory subunit of phosphoinositide 3-kinase.* Biochem J. 1998;333:757-63.

18. Stern DF. *ERBB3/HER3 and ERBB2/HER2 duet in mammary development and breast cancer.* J Mammary Gland Biol Neoplasia. 2008;13:215-23.

19. Bacus SS, Altomare DA, Lyass L, Chin DM, Farrell MP, Gurova K, et al. *AKT2 is frequently upregulated in HER-2/neu-positive breast cancers and may contribute to tumor aggressiveness by enhancing cell survival.* Oncogene. 2002;21:3532-40.

20. Sawyer C, Hiles I, Page M, Crompton M, Dean C. *Two erbB-4 transcripts are expressed in normal breast and in most breast cancers.* Oncogene. 1998;17:919-24.

21. Vidal GA, Clark DE, Marrero L, Jones FE. A constitutively active ERBB4/HER4 allele with *enhanced transcriptional coactivation and cell-killing activities.* Oncogene. 2007;26:462-6.

22. Naresh A, Long W, Vidal GA, Wimley WC, Marrero L, Sartor CI, et al. *The ERBB4/HER4 intracellular domain 4ICD is a BH3-only protein promoting apoptosis of breast cancer cells.* Cancer Res. 2006;66:6412-20.

23. Dunn M, Sinha P, Campbell R, Blackburn E, Levinson N, Rampaul R, et al. *Co-expression of neuregulins 1, 2, 3 and 4 in human breast cancer.* J Pathol. 2004;203:672-80.

24. Panico L, D'Antonio A, Salvatore G, Mezza E, Tortora G, De Laurentiis M, et al. *Differential immunohistochemical detection of transforming growth factor alpha, amphiregulin and Cripto in human normal and malignant breast tissues.* Int J Cancer. 1996;65:51-6.

25. Sheffield LG. *Epidermal growth factor as an autocrine modulator of stress response in mammary epithelial cells.* J Endocrinol. 1998;159:111-6.

26. Muthuswamy SK, Li D, Lelievre SA, Bissell MJ, Brugge JS. *Erbb2, but not ErbB1, reinitiates proliferation and induces luminal repopulation in epithelial acini.* Nature Cell Biol. 2001;3:785-92.

27. Lahlou H, Muller T, Sanguin-Gendreau V, Birchmeier C, Muller WJ. *Uncoupling of PI3K from ErbB3 Impairs Mammary Gland Development but Does Not Impact on ErbB2-Induced Mammary Tumorigenesis.* Cancer Res. 2012;72:3080-90.

28. Siegel PM, Ryan ED, Cardiff RD, Muller WJ. *Elevated expression of activated forms of Neu/ErbB-2 and ErbB-3 are involved in the induction of mammary tumors in transgenic mice: implications for human breast cancer.* EMBO J. 1999;18:2149-64.

29. Vaught DB, Stanford JC, Young C, Hicks DJ, Wheeler F, Rinehart C, et al. *HER3 is required for HER2-induced preneoplastic changes to the breast epithelium and tumor formation.* Cancer Res. 2012;72:2672-82.

30. Leung CT, Brugge JS. *Outgrowth of single oncogene-expressing cells from suppressive epithelial environments.* Nature. 2012;482:410-3.

6.2.2.2. CLINICAL STUDIES

Clinical evidence implicates the involvement of HER1 – 4 in breast cancer development. Abnormal copy numbers (i.e., amplifications, deletions) of all HER genes are frequently present in breast cancers. One study detected amplifications of HER1, HER2, HER3, and HER4 in 15, 26, 10, and 15% tumors, respectively, in a consecutive series of 225 primary breast cancers (1). Deletions occurred in 31, 2, 2, and 7%, respectively. Taken together, 65% of tumors had at least one HER gene with abnormal copy number. The majority of single gene alterations included HER2 amplifications and EGFR amplifications/deletions, whereas HER3 and HER4 abnormalities typically occurred in combination with other HER anomalies. HER abnormalities including two or more genes were found in 31% cases, most frequently co-amplifications of HER2 with other members of the HER family.

EGFR (HER1). Amplification of the EGFR (HER1) gene at 7p12 is found in 5 to 15% of breast cancers except in the basal-like subtype, which shows overexpression in 50% of tumors (1-3). The EGFR gene has a polymorphic dinucleotide (CA) repeat in intron 1 whose length ranges from 8 to 23. The basal transcription activity of the gene was inversely related to the number of repeats (4). There was no association of the allelic length with breast cancer risk except in women younger than age 50 with a first-degree family history of breast cancer (5). In these women, the presence of two long alleles (≥19 CA) was associated with a significantly increased odds ratio of 10.4 (95% CI 1.85 – 58.7). A risk increase associated with high red meat consumption (OR 10.68; 95% CI 1.57 – 72.6) and a protective effect of high vegetable intake (OR 0.07; 95% CI 0.004 – 1.07) was also most pronounced among carriers of two long alleles (≥19 CA). Thus, the length of the EGFR repeat may increase the risk for familial breast cancer and its effect could be modulated by dietary factors. A variant of the EGFR gene, named EGFRvIII, is characterized by an in-frame 801 base pair deletion (exons 2 through 7), which gives rise to a truncated receptor that lacks a portion of the extracellular ligand-binding domain and is ligand-independent and constitutively active (6). The EGFR variant occurs in the absence of EGFR amplification and is thought to arise by genomic rearrangement or alternative splicing (7). The variant is frequently present in glioblastoma multiforme brain tumors and also found in breast cancers with percentages ranging from 5 to 80% depending on the method of detection. Functional studies revealed that the EGFR variant mediated its effects through the Wnt/β-catenin pathway, leading to increased β-catenin expression and a stem cell phenotype. A proteomic mass spectrometry analysis of >500 plasma proteins combined with ELISA assays identified elevated EGFR levels in the circulation as a predictor of breast cancer risk in current users of estrogen plus progestin menopausal hormone therapy (8).

HER2. Gene expression profiling has distinguished HER2-positive tumors as one of the main molecular classes of breast cancer (9, 10). The HER2 gene at 17q12 is amplified from 2- to greater than 20-fold in about 30% of breast cancers (11). Generally, HER2 amplification determined by fluorescence *in situ* hybridization (FISH) correlates with HER2 overexpression determined by immunohistochemistry (12, 13). However, in about 10% of breast cancers, there is clear overexpression at the mRNA and protein levels despite a single HER2 gene copy by DNA analysis (14). In particular, lobular breast cancers exhibit HER2 overexpression without amplification (15). Thus, similar to other genes, HER2 overexpression may result from mechanisms other than gene amplification (16). One such mechanism may involve the estrogen-related receptor α (ERRα) and its coregulator PGC-1β, which enhance not only HER2 expression but act as global transcriptional regulator for several genes in the HER2 amplicon at 17q12 (17). ERα has the opposite effect, inhibiting HER2 expression in cooperation with PAX2 through binding to a cis-regulatory element in the presence of either 17β-estradiol or 4-hydroxytamoxifen (18). Based on the ER-mediated transcriptional repression of HER2, lack of ER expression may be a cause of HER2 overexpression, even in the absence of HER2 amplification. Indeed, amplification and/or overexpression of HER2 have been associated with the ER-negative phenotype in breast cancer (11, 19, 20). In two large series of 1052 and 1875 breast cancers, HER2 amplification was observed in 18.7% and 15.9% of tumors, respectively, with amplification levels ranging from 3 to 43 gene copies (21, 22). The correlation between HER2 amplification and ER-negativity was highly significant (p < 0.0001) in both studies. HER2 may be co-amplified with other members of the HER family, enabling heterodimerization, e.g., HER2/3 (1). HER2 may also be co-amplified with other genes in the 17q12 amplicon (e.g., HSD17B1) (23). Finally, intron 23 of the HER2 gene contains a microRNA (miR-4728) of unknown significance, which is co-amplified with HER2 amplification (24).

HER2 overexpression elicits multiple molecular changes by activating cytoplasmic signaling cascades and nuclear targets. For example, HER2 overexpression averts TGFβ-mediated tumor suppression by switching expression of two functionally distinct isoforms of the transcription factor C/EBPβ, i.e., LAP and LIP (25). Specifically, HER2 signaling activates the translational regulatory factor CUGP1, which favors the transcriptionally inhibitory isoform LIP over that of the active isoform LAP. In turn, LIP overexpression prevents the assembly of LAP/SMAD transcriptional repressor complexes on the

MYC promoter in response to TGFβ silencing its anti-prolif-erative effect. HER2 also induces the expression of COX-2 and prostaglandin E$_2$ prostaglandin (PGE2) production (26). In turn, PGE2 stimulates the aromatase CYP19 gene in mammary tissue (27). Moreover, HER2 overexpression elicits a proinflammatory autocrine loop of IL-6/Stat3 expression (28). HER2 overexpression was accompanied by the presence of nuclear HER2, which enhanced rRNA synthesis, protein translation, and cell growth (29).

Nineteen of 45 (42%) DCIS had HER2 overexpression as indicated by cell membrane staining (30). These 19 were all of the large-cell, comedo growth type. None of 16 DCIS of small-cell, papillary, or cribriform growth type exhibited HER2 overexpression. These findings suggest that HER2 overexpression may be an early step in the development of a distinct histologic type of carcinoma of the breast, i.e., comedo DCIS. On the other hand, there was no evidence of HER2 amplification/overexpression in atypical hyperplasia, hyperplasia, or benign breast disease (31, 32).

HER2-positive breast cancers express not only the full-length 185 kDa protein composed of 1,255 amino acids but also a series of C-terminal fragments (CTFs), also known as p95HER2, which arise through at least two different mechanisms (33). Proteolytic shedding of the ectodomain is carried out by the metalloprotease ADAM10 at codon 648, generating a 95 to 100 kDa membrane-anchored fragment, also known as 648-CTF (34, 35). Alternative translation from two internal initiation codons at positions 611 and 678, located upstream and downstream of the transmembrane domain, generates fragments of 100 to 115 kDa and 90 to 95 kDa, also known as 611-CTF and 678-CTF, respectively. The soluble 678-CTF was inactive, whereas the activity of the membrane-bound fragments was comparable (648-CTF) or even exceeded (611-CTF) that of full-length HER2 (36). Tumors expressing p95HER2 appear to be intrinsically resistant to treatment with trastuzumab, which is directed against the extracellular domain of HER2 (33).

HER3. The formation of HER2/HER3 heterodimers plays a critical role in both normal mammary development and tumorigenesis (37-39). Like the HER2 gene, the HER3 gene at 12q13 is amplified and over-expressed in 15 to 30% of breast cancers (40). Thus, HER2 and HER3 are frequently co-overexpressed and cause breast cells to reproduce uncontrollably.

HER4 is expressed in normal breast and in most breast cancers (41, 42). Amplification and overexpression of the HER4 gene were observed in about 15% of tumors (43, 44). Down-regulation of HER4 in MCF-7 and T47D breast cancer cell lines inhibited tumor formation in athymic nude mice.

References

1. Zaczek A, Welnicka-Jaskiewicz M, Bielawski KP, Jaskiewicz J, Badzio A, Olszewski W, et al. *Gene copy numbers of HER family in breast cancer.* J Cancer Res Clin Oncol. 2008;134:271-9.

2. Kersting C, Tidow N, Schmidt H, Liedtke C, Neumann J, Boecker W, et al. *Gene dosage PCR and fluorescence in situ hybridization reveal low frequency of egfr amplications despite protein overexpression in invasive breast carcinoma.* Lab Invest. 2004;84:582-7.

3. Nielsen TO, Hsu FD, Jensen K, Cheang M, Karaca G, Hu Z, et al. *Immunohistochemical and clinical characterization of the basal-like subtype of invasive breast carcinoma.* Clin Cancer Res. 2004;10:5367-74.

4. Gebhardt F, Zanker KS, Brandt B. *Modulation of epidermal growth factor receptor gene transcription by a polymorphic dinucleotide repeat in intron 1.* J Biol Chem. 1999;274:13176-80.

5. Brandt B, Hermann S, Straif K, Tidow N, Buerger H, Chang-Claude J. *Modification of breast cancer risk in young women by a polymorphic sequence in the egfr gene.* Cancer Res. 2004;64:7-12.

6. Moscatello DK, Montgomery RB, Sundaresh an P, McDanel H, Wong MY, Wong AJ. *Transformational and altered signal transduction by a naturally occurring mutant EGF receptor.* Oncogene. 1996;13:85-96.

7. Del Vecchio CA, Jensen KC, Nitta RT, Shain AH, Giacomini CP, Wong AJ. *Epidermal growth factor receptor variant III contributes to cancer stem cell phenotypes in invasive breast carcinoma.* Cancer Res. 2012;72:2657-71.

8. Pitteri SJ, Amon LM, Busald Buson T, Zhang Y, Johnson MM, Chin A, et al. *Detection of elevated plasma levels of epidermal growth factor receptor before breast cancer diagnosis among hormone therapy users.* Cancer Res. 2010;70:8598-606.

9. Perou CM, Sorlie T, Eisen MB, van de Rijn M, Jeffrey SS, Rees CA, et al. *Molecular portraits of human breast tumours.* Nature. 2000;406:747-52.

10. Sorlie T, Perou CM, Tibshirani R, Aas T, Geisler S, Johnsen H, et al. *Gene expression patterns of breast carcinomas distinguish tumor subclasses with clinical implications.* Proc Natl Acad Sci U S A. 2001;98:10869-74.

11. Slamon DJ, Clark GM, Wong SG, Levin WJ, Ullrich A, McGuire WL. *Human breast cancer: correlation of relapse and survival with amplification of the HER-2/neu oncogene.* Science. 1987;235:177-82.

12. Bloom KJ, Cote RJ. *Counterpoint: Both immunohistochemistry and fluorescence in situ hybridization play important roles for HER2 evaluation.* Clin Chem. 2011;57:983-5.

13. Ross JS. Point: *Fluorescence in situ hybridization is the preferred approach over immunohistochemistry for determining HER2 status.* Clin Chem. 2011;57:980-2.

14. Slamon DJ, Godolphin W, Jones LA, Holt JA, Wong SG, Keith DE, et al. *Studies of HER-2neu proto-oncogene in human breast and ovarian cancer.* Science. 1989;244:707-12.

15. Smith CA, Police AA, Gu LP, Brown KA, Singh SG, Janocko LE, et al. *Correlations among p53, Her-2/neu, and ras overexpression and aneuploidy by multiparameter flow cytometry in human breast cancer: evidence for a common phenotypic evolutionary pattern in infiltrating ductal carcinomas.* Clin Cancer Res. 2000;6:112-26.

16. Hurst HC. *Update on HER-2 as a target for cancer therapy: the ERBB2 promoter and its exploitation for cancer treatment.* Breast Cancer Res. 2001;3:395-8.

17. Deblois G, Chahrour G, Perry MC, Sylvain-Drolet G, Muller WJ, Giguere V. *Transcriptional control of the ERBB2 amplicon by ERRalpha and PGC-1beta promotes mammary gland tumorigenesis.* Cancer Res. 2010;70:10277-87.

18. Hurtado A, Holmes KA, Geistlinger TR, Hutcheson IR, Nicholson RI, Brown M, et al. *Regulation of ERBB2 by oestrogen receptor-PAX2 determines response to tamoxifen.* Nature. 2008;456:663-6.

19. Adnane J, Gaudray P, Simon MP, Simony-Lafontaine J, Jeanteur P, Theillet C. *Proto-oncogene amplification and human breast tumor phenotype.* Oncogene. 1989;4:1389-95.

20. Valeron PF, Chirino R, Vega V, Falcon O, Rivero JF, Torres S, et al. *Quantitative analysis of p185 (HER-2/neu) protein in breast cancer and its association with other prognsotic factors.* Int J Cancer. 1997;74:175-9.

21. Berns EMJJ, Klijn JGM, van Staveren IL, Portengen H, Noordegraaf E, Foekens JA. *Prevalence of amplification of the oncogenes c-myc, HER2/neu, and int-2 in one thousand human*

breast tumours: Correlation with steroid receptors. Eur J Cancer. 1992;28:697-700.

22. Courjal F, Cuny M, Simony-Lafontaine J, Louason G, Speiser P, Zeillinger R, et al. *Mapping of DNA amplifictions at 15 chromosomal localizations in 1875 breast tumors: Definition of phenotypic groups.* Cancer Res. 1997;57:4360-7.

23. Gunnarsson C, Ahnstrom M, Kirschner K, Olsson B, Nordenskjold B, Rutqvist LE, et al. *Amplification of hsd17b1 and erbb2 in primary breast cancer.* Oncogene. 2003;22:34-40.

24. Persson H, Kvist A, Rego N, Staaf J, Vallon-Christersson J, Luts L, et al. *Identification of new microRNAs in paired normal and tumor breast tissue suggests a dual role for the ERBB2/Her2 gene.* Cancer Res. 2011;71:78-86.

25. Arnal-Estape A, Tarragona M, Morales M, Guiu M, Nadal C, Massague J, et al. *HER2 silences tumor suppression in breast cancer cells by switching expression of C/EBPss isoforms.* Cancer Res. 2010;70:9927-36.

26. Wang SC, Lien HC, Xia W, Chen IF, Lo HW, Wang Z, et al. *Binding at and transactivation of the COX-2 promoter by nuclear tyrosine kinase receptor ErbB-2.* Cancer Cell. 2004;6:251-61.

27. Subbaramaiah K, Howe LR, Port ER, Brogi E, Fishman J, Liu CH, et al. *HER-2/neu status is a determinant of mammary aromatase activity in vivo: evidence for a cyclooxygenase-2-dependent mechanism.* Cancer Res. 2006;66:5504-11.

28. Hartman ZC, Yang XY, Glass O, Lei G, Osada T, Dave SS, et al. *HER2 overexpression elicits a proinflammatory IL-6 autocrine signaling loop that is critical for tumorigenesis.* Cancer Res. 2011;71:4380-91.

29. Li LY, Chen H, Hsieh YH, Wang YN, Chu HJ, Chen YH, et al. *Nuclear ErbB2 enhances translation and cell growth by activating transcription of ribosomal RNA genes.* Cancer Res. 2011;71:4269-79.

30. Van de Vijver M, Van de Bersselaar R, Devilee P, Cornelisse C, Peterse J, Nusse R. *Amplification of the neu (c-erbB-2) oncogene in human mammary tumors is relatively frequent and is often accompanied by amplification of the linked c-erbA oncogene.* Mol Cell Biol. 1987;7:2019-23.

31. Aubele M, Werner M, Hofler H. *Genetic alterations in presumptive precursor lesions of breast carcinomas.* Anal Cell Pathol. 2002;24:69-76.

32. Lizard-Nacol S, Lidereau R, Collin F, Arnal M, Hahnel L, Roignot P, et al. *Benign breast disease: Absence of genetic alterations of several loci implicated in breast cancer malignancy.* Cancer Res. 1995;55:4416-9.

33. Arribas J, Baselga J, Pedersen K, Parra-Palau JL. *p95HER2 and breast cancer.* Cancer Res. 2011;71:1515-9.

34. Christianson TA, Doherty JK, Lin YJ, Ramsey EE, Holmes R, Keenan EJ, et al. *NH2-terminally truncated HER-2/neu protein: relationship with shedding of the extracellular domain and with prognostic factors in breast cancer.* Cancer Res. 1998;58:5123-9.

35. Liu PC, Liu X, Li Y, Covington M, Wynn R, Huber R, et al. *Identification of ADAM10 as a major source of HER2 ectodomain sheddase activity in HER2 overexpressing breast cancer cells.* Cancer Biol Ther. 2006;5:657-64.

36. Pedersen K, Angelini PD, Laos S, Bach-Faig A, Cunningham MP, Ferrer-Ramon C, et al. *A naturally occurring HER2 carboxy-terminal fragment promotes mammary tumor growth and metastasis.* Mol Cell Biol. 2009;29:3319-31.

37. Lahlou H, Muller T, Sanguin-Gendreau V, Birchmeier C, Muller WJ. *Uncoupling of PI3K from ErbB3 Impairs Mammary Gland Development but Does Not Impact on ErbB2-Induced Mammary Tumorigenesis.* Cancer Res. 2012;72:3080-90.

38. Siegel PM, Ryan ED, Cardiff RD, Muller WJ. *Elevated expression of activated forms of Neu/ErbB-2 and ErbB-3 are involved in the induction of mammary tumors in transgenic mice: implications for human breast cancer.* EMBO J. 1999;18:2149-64.

39. Vaught DB, Stanford JC, Young C, Hicks DJ, Wheeler F, Rinehart C, et al. *HER3 is required for HER2-induced preneoplastic changes to the breast epithelium and tumor formation.* Cancer Res. 2012;72:2672-82.

40. Naidu R, Yadav M, Nair S, Kutty MK. *Expression of c-erbB3 protein in primary breast carcinomas.* Br J Cancer. 1998;78:1385-90.

41. Koutras AK, Fountzilas G, Kalogeras KT, Starakis I, Iconomou G, Kalofonos HP. *The upgraded role of HER3 and HER4 receptors in breast cancer.* Crit Rev Oncol Hematol. 2010;74:73-8.

42. Sawyer C, Hiles I, Page M, Crompton M, Dean C. *Two erbB-4 transcripts are expressed in normal breast and in most breast cancers.* Oncogene. 1998;17:919-24.

43. Tang CK, Concepcion XZ, Milan M, Gong X, Montgomery E, Lippman ME. *Ribozyme-mediated down-regulation of ErbB-4 in estrogen receptor-positive breast cancer cells inhibits proliferation both in vitro and in vivo.* Cancer Res. 1999;59:5315-22.

44. Vogt U, Bielawski K, Schlotter CM, Bosse U, Falkiewicz B, Podhajska AJ. *Amplification of erbB-4 oncogene occurs less frequently than that of erbB-2 in primary human breast cancer.* Gene. 1998;223:375-80.

6.2.2.3. TRIPLE-NEGATIVE BREAST CANCER

Breast cancers lacking expression of HER2 and hormone receptors (ERα-negative, PR-negative), so-called triple-negative breast cancers (TNBCs), account for approximately 15% of primary breast cancers (1, 2). Due to the lack of ERα and HER2 expression, TNBCs cannot be effectively treated with endocrine therapy or trastuzumab (Herceptin®), a humanized monoclonal anti-HER2 antibody. Thus, TNBCs have a relatively poor prognosis and contribute to a disproportionately high percentage of breast cancer deaths (3-6). TNBCs occur more frequently in young African-American and Hispanic women than in young women of other racial or ethnic group (1). Other risk associations include early menarche and high body mass index. Use of oral contraceptives was associated with a 2.5-fold increased risk for TNBC and no significantly increased risk for non-triple negative breast cancer (7).

Gene expression studies using DNA microarrays have identified subtypes of breast cancer that were not apparent using traditional histopathological methods (8, 9). These studies suggest that the basal/myoepithelial and luminal cells are progenitor cells, which give rise to at least four distinct tumor subtypes of which two are ER-positive (luminal A and luminal B cells) and two ER-negative (basal-like cells and HER2-positive) (10, 11). The majority of basal-like cancers do not express ERα, PR, or HER2, but generally do express EGFR (HER1) and the basal cytokeratins-5 and -17, markers usually found in normal basal/myoepithelial breast cells (8-10). Thus, it has been claimed that the triple-negative and basal-like phenotypes are effectively synonymous and the two designations have been used interchangeably (12). However, both basal-like and TNBCs are heterogeneous. Histologically, the majority of TNBCs are grade 3 invasive ductal carcinomas of no special type although most medullary, metaplastic, and adenoid cystic carcinomas also display a triple-negative phenotype (13). Over 75% of breast cancers carrying a BRCA1 mutation have a triple-negative phenotype. While TNBCs are ER- and HER2-negative by definition, up to 20% of basal-like tumors express ER or overexpress HER2 (13, 14). TNBCs include so-called claudin-low tumors enriched with breast cancer stem cells (1). Lactoferrin, a secretory protein expressed in mammary epithelial cells and present in the circulation was shown to down-regulate ERα, PR, and HER2 in a proteasome-dependent manner in breast cancer cells (15). Lactoferrin directly stimulated the transcription of endothelin-1 (ET-1), a secreted proinvasive polypeptide that acts through a specific receptor, ET(A)R, leading to secretion of the bioactive ET-1 peptide. Patients with TNBCs had elevated plasma and tissue lactoferrin and ET-1 levels compared with patients with ER-positive tumors.

A genome-wide association study of the Breast Cancer Association Consortium (BCAC; 48,869 cases and 49,787 controls) showed a significant association of TNBC with a SNP (rs8170) at 19p13.1: OR 1.22; 95% CI 1.13 – 1.31; p = 2.22 x 10^{-7} (16). This association was confirmed by the Triple Negative Breast Cancer Consortium (2,980 cases and 4,978 controls), which examined 22 common SNPs including rs8170 (17). However, while rs8170 was associated with increased risk of TNBC, another SNP (rs8100241) at 19p13.11 was associated with decreased risk: OR 0.84; 95% CI 0.78 – 0.90. Both SNPs had a multiplicative effect on TNBC risk. Two SNPs in the ERα gene at 6q25.1 (ESR1 rs2046210 and 12662670) also showed an increased association with TNBC (17).

A whole-genome analysis of 104 TNBCs observed a wide and continuous spectrum of somatic mutations and copy number alterations indicating that TNBCs are heterogeneous from the outset (18). Some tumors contained few mutations and a small number of implicated pathways, whereas other tumors exhibited extensive mutation burdens and involvement of multiple pathways. The most frequently mutated gene was p53 (62% of basal TNBC, 43% of non-basal TNBC). Other frequently mutated genes included PIK3CA (10.2%; see Chapter 6.2.5.), PTEN (7.7%; see Chapter 8.6.5.) and RB1 (7.7%). RNA sequencing revealed that only 36% of somatic single nucleotide mutations were in the transcriptome sequence, i.e., expressed in proteins. Other mutations were identified in splice junctions with evidence for an impact on splicing patterns and in non-coding sequences of regulatory regions with known effects on protein expression. Gene copy number alterations were mostly amplifications of tumor suppressor and oncogenes: PARK2 (6%), EGFR (5%), RB1 (5%), and PTEN (3%). Overall, the highest frequencies of genetic alterations involved the p53 and PIK3CA pathways, consistent with their roles in early tumorigenesis, whereas lower frequencies were observed in pathways with cytoskeletal genes, such as myosins, laminins, collagens, and integrins, suggesting that somatic mutations in these genes are acquired much later. Notably, the median clonal frequency was 73% for 'p53 pathway feedback loops' (including 46 mutations in ATM, ATR, NRAS, PIK3CA, PTEN, SIAH1 and p53) compared to 42% for 'integrin cell surface interactions' (including 23 mutations in integrin, laminin, and collagen genes; Wilcoxon, q = 0.0007 versus q = 0.9569). For example, immunohistochemical studies revealed a reduction of ATM protein expression in TNBCs (19).

References

1. Foulkes WD, Smith IE, Reis-Filho JS. *Triple-negative breast cancer.* N Engl J Med. 2010;363:1938-48.

2. Schneider BP, Winer EP, Foulkes WD, Garber J, Perou CM, Richardson A, et al. *Triple-negative breast cancer: risk factors to potential targets.* Clin Cancer Res. 2008;14:8010-8.

3. Carey LA, Dees EC, Sawyer L, Gatti L, Moore DT, Collichio F, et al. *The triple negative paradox: primary tumor chemosensitivity of breast cancer subtypes.* Clin Cancer Res. 2007;13:2329-34.

4. Dent R, Trudeau M, Pritchard KI, Hanna WM, Kahn HK, Sawka CA, et al. *Triple-negative breast cancer: clinical features and patterns of recurrence.* Clin Cancer Res. 2007;13:4429-34.

5. Montanger M, Enzo E, Forcato M, Zanconato F, Parenti A, Rampazzo E, et al. *SHARP1 suppresses breast cancer metastasis by promoting degradation of hypoxia-inducible factors.* Nature. 2012;487:380-4.

6. Turner N, Lambros MB, Horlings HM, Pearson A, Sharpe R, Natrajan R, et al. *Integrative molecular profiling of triple negative breast cancers identifies amplicon drivers and potential therapeutic targets.* Oncogene. 2010;29:2013-23.

7. Dolle J, Daling J, White E, Brinton L, Doody D, Porter P, et al. *Risk factors for triple-negative breast cancer in women under the age of 45 years.* Cancer Epidemiol Biomarkers Prev. 2009;18:1157-66.

8. Perou CM, Sorlie T, Eisen MB, van de Rijn M, Jeffrey SS, Rees CA, et al. *Molecular portraits of human breast tumours.* Nature. 2000;406:747-52.

9. Sorlie T, Perou CM, Tibshirani R, Aas T, Geisler S, Johnsen H, et al. *Gene expression patterns of breast carcinomas distinguish tumor subclasses with clinical implications.* Proc Natl Acad Sci U S A. 2001;98:10869-74.

10. Nielsen TO, Hsu FD, Jensen K, Cheang M, Karaca G, Hu Z, et al. *Immunohistochemical and clinical characterization of the basal-like subtype of invasive breast carcinoma.* Clin Cancer Res. 2004;10:5367-74.

11. Prat A, Perou CM. *Mammary development meets cancer genomics.* Nat Med. 2009;15:842-4.

12. Sotiriou C, Pusztai L. *Gene-expression signatures in breast cancer.* N Engl J Med. 2009;360:790-800.

13. Reis-Filho JS, Tutt AN. *Triple negative tumours: a critical review.* Histopathology. 2008;52:108-18.

14. Bertucci F, Finetti P, Cervera N, Esterni B, Hermitte F, Viens P, et al. *How basal are triple-negative breast cancers?* Int J Cancer. 2008;123:236-40.

15. Ha NH, Nair VS, Reddy DN, Mudvari P, Ohshiro K, Ghanta KS, et al. *Lactoferrin-endothelin-1 axis contributes to the development and invasiveness of triple-negative breast cancer phenotypes.* Cancer Res. 2011;71:7259-69.

16. Stevens KN, Fredericksen Z, Vachon CM, Wang X, Margolin S, Lindblom A, et al. *19p13.1 is a triple-negative-specific breast cancer susceptibic locus.* Cancer Res. 2012;72:1795-803.

17. Stevens KN, Vachon CM, Lee AM, Slager S, Lesnick T, Olswold C, et al. *Common breast cancer susceptibility loci are associated with triple-negative breast cancer.* Cancer Res. 2011;71:6240-9.

18. Shah SP, Roth A, Goya R, Oloumi A, Ha G, Zhao Y, et al. *The clonal and mutational evolution spectrum of primary triple-negative breast cancers.* Nature. 2012;486:395-9.

19. Tommiska J, Bartkova J, Heinonen M, Hautala L, Kilpivaara O, Eerola H, et al. *The DNA damage signalling kinase ATM is aberrantly reduced or lost in BRCA1/BRCA2-deficient and ER/PR/ERBB2-triple-negative breast cancer.* Oncogene. 2008;27:2501-6.

6.2.3. FIBROBLAST GROWTH FACTOR RECEPTORS

Four genes encode fibroblast growth factor receptors (FGFRs), which have an extracellular ligand-binding domain, a transmembranous domain, and an intracellular tyrosine kinase domain (1). There are 18 FGF ligands, which bind specific FGFR isoforms and induce receptor dimerization, leading to kinase activation and autophosphorylation of Tyr residues in the intracellular kinase domain. The major FGFR effector, the adaptor protein FRS2, is constitutively associated with the receptor and following activation becomes phosphorylated on Tyr and Ser residues. In turn, phosphorylated FRS2 acts as a docking site for Grb2, which interacts with SOS and Gab1 to activate the ERK and AKT pathways for cell proliferation and survival. Different types of FGF/FGFR alterations, including abnormal expression levels, SNPs, mutations, and amplifications have been described in cancer (2). Each of these alterations can cause deregulated FGFR signaling with constitutive activation of downstream pathways and aberrant cell proliferation and increased cell survival. In breast cancer, FGFR gene amplifications are more common than activating mutations. Overall, the data provide evidence for distinct roles of specific FGF/FGFR family members in the malignant process.

FGFR1: Germline mutations of FGFR1 at 8p11.2 cause Kallmann syndrome, which leads to hypogonadotrophic hypogonadism whereas somatic gain-of-function mutations have been identified in glioblastoma brain tumors (1).
The FGFR1 gene is preferentially activated by FGF1 and FGF2 and expressed as cell membrane-bound protein in epithelial cells of normal breast tissue, fibroadenomas, and invasive breast carcinomas. About 10% of breast cancers exhibited overexpression of FGFR1 protein due to amplification of the FGFR1 gene (3-5). The tumor phenotype was characterized as ER-positive, PR-negative, HER2-negative, highly proliferative, luminal B-type with tamoxifen resistance and poor outcome (6, 7).

FGFR2: Germline mutations in the kinase domain of FGFR2 at 10q26.3 cause craniosynostosis with premature closure of cranial sutures before the completion of brain growth (1).
FGFR2 amplification was identified in a small subgroup (6 of 165 = 4%) of triple-negative breast cancers (8). Interestingly, no case of FGFR2 amplification was found in >200 tumors that were not triple-negative.

Genome-wide screens in European women have identified several SNPs in intron 2 of FGFR2 to be significantly associated with breast cancer risk (9, 10). The SNP rs2981582 is prevalent with 40% of the population carrying at least one copy, but the increase in risk was relatively small with a relative risk of 1.26 (95% CI 1.23 – 1.30) in a heterozygote and 1.63-fold in a homozygote. The association with the FGFR2 gene was corroborated in African-American, Chinese, and Jewish women although the specific locus within intron 2 remained in question (11-14). Fine-scale mapping of intron 2 identified another SNP, rs2981578, to be associated with breast cancer risk (15). Rs2981578 mapped to highly accessible

chromatin by DNase I hypersensitive site analysis and additional experiments identified rs2981578 and another cis-regulatory SNP (rs7895676) to alter binding affinity for transcription factors Oct-1/Runx2 and C/EBPβ (15, 16). In transient transfection experiments, the two SNPs synergized to lead to increased FGFR2 expression. The FGFR2 rs2981582 was associated with increased breast cancer risk in BRCA2 mutation carriers suggesting that the FGFR2 SNP contributes to the genetic variability seen in familial BRCA2 mutation-carrying breast cancers (17). Interestingly, amplification of FGF3 (also known as int-2, at 11q13), a FGFR2 ligand, has been observed in 15% of breast cancers and shown to correlate with increased aggressiveness in node-negative tumors (18-20). Clinical correlation studies yielded inconsistent evidence of an interaction of FGFR2 variants with hormone replacement therapy or ER-positive tumors (14, 21). Eight SNPs in intron 2 of the FGFR2 gene were examined in the Women's Health Initiative Dietary Modification Trial, which included 48,835 postmenopausal women, aged 50 to 79 years (22). The women were randomly assigned to a dietary modification group or a comparison group. After an average of 8.1 years of intervention and follow-up, the incidence of invasive breast cancer was 9% lower in the intervention than the comparison group (95% CI, 0.83 – 1.01; p = 0.09). Case-only analyses showed that odds ratios for the dietary intervention did not vary significantly with the genotype for any of the eight polymorphisms. However, among women whose baseline percent of energy from fat was in the upper quartile there was a significant risk association of the intervention with polymorphism rs3570817 (T/C), ranging from OR 1.06 to 0.53 and 0.62 at 0, 1, and 2 minor alleles (p = 0.03). Thus, women having one or two C alleles of rs3570817 may benefit from reduction from a high-fat to a lower fat dietary pattern.

FGFR3: Germline mutations of FGFR3 at 11q13 cause achondroplasia, a common inherited form of dwarfism. Somatic FGFR3 mutations have been described in multiple myeloma and bladder cancer but not in breast tumors (1).

FGFR4: The breast cancer cell line MDA-MB-453 contains an activating mutation (Tyr367Cys) in the FGFR4 gene at 5q35.2 that causes constitutive receptor dimerization (23). A non-synonymous polymorphism (rs351855; minor allele frequency ~50%) encodes Gly388Arg, which is located in the transmembrane region of the receptor (24). The 388Arg variant exhibited markedly decreased degradation and increased phosphorylation after ligand binding compared to the 388Gly variant, resulting in enhanced stability of activated FGFR4 (25). Gly388Arg acts as an activity switch of a membrane type 1 matrix metalloproteinase (MT1-MMP) – FGFR4 complex (26). The 388Arg variant was shown to induce MT1-MMP and collagen invasion whereas the alternative

388Gly variant downregulated MT1-MMP. In a transgenic animal model, the 388Arg allele promoted more rapid mammary tumor development than the 388Gly allele (27). In humans, Gly388Arg does not seem to influence the incidence of breast cancer but the 388Arg allele was associated with accelerated disease progression (24). Overexpression of FGFR4 was also correlated with poor response to endocrine therapy (28).

References

1. Beenken A, Mohammadi M. *The FGF family: biology, pathophysiology and therapy.* Nature Rev Drug Discovery. 2009;8:235-53.

2. Turner N, Grose R. *Fibroblast growth factor signalling: from development to cancer.* Nature Rev Cancer. 2010;10:116-29.

3. Blanckaert VD, Hebbar M, Louchez MM, Vilain MO, Schelling ME, Peyrat JP. *Basic fibroblast growth factor receptors and their prognostic value in human breast cancer.* Clin Cancer Res. 1998;4:2939-47.

4. Cuny M, Kramar A, Courjal F, Johannsdottir V, Iacopetta B, Fontaine H, et al. *Relating genotype and phenotype in breast cancer: An analysis of prognostic significance of amplification at eight different genes or loci and of p53 mutations.* Cancer Res. 2000;60:1077-83.

5. Ugolini F, Adelaide J, Charafe-Jauffret E, Nguyen C, Jacquemier J, Jordan B, et al. *Differential expression assay of chromosome arm 8p genes indentifies frizzled-related (FRP1/FRZB) and fibroblast growth factor receptor 1 (FGFR1) as candidate breast cancer genes.* Oncogene. 1999;18:1903-10.

6. Elbauomy Elsheikh S, Green AR, Lambros MB, Turner NC, Grainge MJ, Powe D, et al. *FGFR1 amplification in breast carcinomas: a chromogenic in situ hybridisation analysis.* Breast Cancer Res: BCR. 2007;9:R23.

7. Turner N, Pearson A, Sharpe R, Lambros M, Geyer F, Lopez-Garcia MA, et al. *FGFR1 amplification drives endocrine therapy resistance and is a therapeutic target in breast cancer.* Cancer Res. 2010;70:2085-94.

8. Turner N, Lambros MB, Horlings HM, Pearson A, Sharpe R, Natrajan R, et al. *Integrative molecular profiling of triple negative breast cancers identifies amplicon drivers and potential therapeutic targets.* Oncogene. 2010;29:2013-23.

9. Easton DF, Pooley KA, Dunning AM, Pharoah PDP, Thompson D, Ballinger DG, et al. *Genome-wide association study indentifies novel breast cancer susceptibility loci.* Nature. 2007;447:1087-95.

10. Hunter DJ, Kraft P, Jacobs KB, Cox DG, Yeager M, Hankinson SE, et al. *A genome-wide association study identifies alleles in FGFR2 associated with risk of sporadic postmenopausal breast cancer.* Nature Genet. 2007;39:870-4.

11. Gold B, Kirchhoff T, Stefanov S, Lautenberger J, Viale A, Garber J, et al. *Genome-wide association study provides evidence for a breast cancer risk locus at 6q22.33.* Proc Natl Acad Sci USA. 2008;105:4340-5.

12. Liang J, Chen P, Hu Z, Zhou X, Chen L, Li M, et al. *Genetic variants in fibroblast growth factor receptor 2 (FGFR2) contribute to susceptibility of breast cancer in Chinese women.* Carcinogenesis. 2008;29:2341-6.

13. Raskin L, Pinchev M, Arad C, Lejbkowicz F, Tamir A, Rennert HS, et al. *FGFR2 is a breast cancer susceptibility gene in Jewish and Arab Israeli populations.* Cancer Epidemiol Biomarkers Prev. 2008;17:1060-5.

14. Rebbeck TR, DeMichele A, Tran TV, Panossian S, Bunin GR, Troxel AB, et al. *Hormone-dependent effects of FGFR2 and MAP3K1 in breast cancer susceptibility in a population-based sample of postmenopausal African-American and European-American women.* Carcinogenesis. 2009;30:269-74.

15. Udler MS, Meyer KB, Pooley KA, Karlins E, Struewing JP, Zhang J, et al. *FGFR2 variants and breast cancer risk: fine-scale mapping using African American studies and analysis of chromatin conformation.* Hum Mol Genetics. 2009;18:1692-703.

16. Meyer KB, Maia AT, O'Reilly M, Teschendorff AE, Chin SF, Caldas C, et al. *Allele-specific up-regulation of FGFR2 increases susceptibility to breast cancer.* PLoS Biology. 2008;6:e108.

17. Antoniou AC, Beesley J, McGuffog L, Sinilnikova OM, Healey S, Neuhausen SL, et al. *Common breast cancer susceptibility alleles and the risk of breast cancer for BRCA1 and BRCA2 mutation carriers: implications for risk prediction.* Cancer Res. 2010;70:9742-54.

18. Berns EMJJ, Klijn JGM, van Staveren IL, Portengen H, Noordegraaf E, Foekens JA. *Prevalence of amplification of the oncogenes c-myc, HER2/neu, and int-2 in one thousand human breast tumours: Correlation with steroid receptors.* Eur J Cancer. 1992;28:697-700.

19. Fioravanti L, Cappelletti V, Coradini D, Miodini P, Borsani G, Daidone MG, et al. *int-2 oncogene amplification and prognosis in node-negative breast carcinoma.* Int J Cancer. 1997;74:620-4.

20. Hui R, Campbell DH, Lee CS, McCaul K, Horsfall DJ, Musgrove EA, et al. *EMS1 amplification can occur independently of CCND1 and INT-2 amplification at 11q13 and may identify different phenotypes in primary breast cancer.* Oncogene. 1997;15:1617-23.

21. Travis RC, Reeves GK, Green J, Bull D, Tipper SJ, Baker K, et al. *Gene-environment interactions in 7610 women with breast cancer: prospective evidence from the Million Women Study.* Lancet. 2010;375:2143-51.

22. Prentice RL, Huang Y, Hinds D, Peters U, Cox DR, Beilharz E, et al. *Variation in the FGFR2 gene and the effect of a low-fat dietary pattern on invasive breast cancer.* Cancer Epidemiol Biomarkers Prev. 2010;19:74-9.

23. Roidl A, Foo P, Wong W, Mann C, Bechtold S, Berger HJ, et al. *The FGFR4 Y367C mutant is a dominant oncogene in MDA-MB453 breast cancer cells.* Oncogene. 2010;29:1543-52.

24. Bange J, Prechtl D, Cheburkin Y, Specht K, Harbeck N, Schmitt M, et al. *Cancer progression and tumor cell motility are associated with the FGFR4 Arg(388) allele.* Cancer Res. 2002;62:840-7.

25. Wang J, Yu W, Cai Y, Ren C, Ittmann MM. *Altered fibroblast growth factor receptor 4 stability promotes prostate cancer progression.* Neoplasia. 2008;10:847-56.

26. Sugiyama N, Varjosalo M, Meller P, Lohi J, Chan KM, Zhou Z, et al. *FGF receptor-4 (FGFR4) polymorphism acts as an activity switch of a membrane type 1 matrix metalloproteinase-FGFR4 complex.* Proc Natl Acad Sci USA. 2010;107:15786-91.

27. Seitzer N, Mayr T, Streit S, Ullrich A. *A single nucleotide change in the mouse genome accelerates breast cancer progression.* Cancer Res. 2010;70:802-12.

28. Meijer D, Sieuwerts AM, Look MP, van Agthoven T, Foekens JA, Dorssers LC. *Fibroblast growth factor receptor 4 predicts failure on tamoxifen therapy in patients with recurrent breast cancer.* Endocrine-Related Cancer. 2008;15:101-11.

6.2.4. TRANSFORMING GROWTH FACTOR BETA

The transforming growth factor β (TGF-β) superfamily has >40 members including three TGF-β isoforms (TGF-β1, TGF-β2, TGF-β3) activins, anti-Müllerian hormone, bone morphogenetic proteins (BMPs), growth differentiation factors (GDFs), inhibins, and nodals, which play pivotal roles in the regulation of cell proliferation, cell differentiation, and apoptosis (1-4). The three TGF-β genes are located on different chromosomes (TGF-β1 at 19q13.1, TGF-β2 at 1q41, TGF-β3 at 14q24), but encode isoforms with similar structural features, i.e., a large protein precursor consisting of an N-terminal signal peptide of 20 – 30 amino acids required for secretion from a cell, a pro-domain region (called latency-associated peptide, LAP) required for proper folding and dimerization of the 112 - 114 amino acid C-terminal region that becomes the mature TGF-β. TGF-β is secreted by many cell types in an inactive form in which it is complexed with latent TGF-β binding protein (LTBP; four isoforms LTBP1 – 4). The latent TGF-β becomes activated by binding of integrin and cleavage from LTBP by various proteinases (5). The mature TGF-β dimerizes to form a 25 kDa active molecule that binds to a TGF-β receptor type II (TGFBR2) dimer, which recruits and phosphorylates a TGF-β receptor type I (TGFBR1) dimer. Both receptors have a cysteine-rich extracellular domain, a transmembrane domain, and a cytoplasmic serine/threonine-rich kinase domain. The binding of TGF-β ligand to the hetero-tetra-meric TGFBR complex causes the rotation of the receptors so that their cytoplasmic kinase domains are arranged in a catalytically favorable orientation to initiate downstream signal transduction by phosphorylation of transcription factors, especially members of the SMAD family (SMAD1 – 5). The SMAD complex migrates to the nucleus and acts as a transcription factor to regulate cell-specific patterns of gene expression. SMAD6 and SMAD7 block phosphoryla-tion of SMAD2 or SMAD3 and act as inhibitors of the TGF-β signaling pathway (3). In contrast, Smad anchor for receptor activation (SARA) stabilizes the SMAD-TGFBR interaction and thereby facilitates TGF-β signaling. SMAD-independent TGF-β signaling can occur via the activation of Ras, phos-phoinositide 3-kinase (PI3K), mitogen-activated protein kinase (MAPK), and other pathways.

TGF-β is a pre-eminent negative growth factor that exerts strong inhibitory effects on the growth of normal, premalignant, and early transformed mammary epithelial cells (6). In normal cells, TGF-β stops the cell cycle at G1 to inhibit proliferation, induce differentiation, and promote apoptosis (1, 4, 7). Experimentally, TGF-β inhibits the development of the normal mouse mammary gland and the *in vitro* growth of normal human mammary epithelial cells (8, 9). Treatment of a mouse mammary carcinoma cell line with TGF-β and TGF-α generated breast cancer stem cells by epithelial-mesenchymal transition with a claudin-low phenotype (10). In carcinogenesis, the TGF-β signaling pathway plays a dual role. In early stages of carcinogen-esis, TGF-β acts as antiproliferative factor and as tumor suppressor. However, once a primary tumor has been established and the tumor suppressor activity is overridden by oncogenic mutations in other pathways, TGF-β may act to enhance tumor progression (2, 11, 12).

Mutations of the TGFBR1 and TGFBR2 genes in breast cancer are rare in primary tumors but more common in recurrent or metastatic lesions (13, 14). Similarly, loss of expression in components of the TGF-β signaling pathway (e.g., SMAD2, SMAD3, SMAD4) appears to be more common in advanced cancer (15, 16). High levels of TGF-β1 in breast cancer tissue and circulating plasma were associated with shorter survival (17, 18).

Several SNPs in genes of the TGF-β signaling pathway have been associated with increased breast cancer

risk (19). The TGF-β1 gene contains several common polymorphisms: G-800A (minor allele frequency ~15%), C-509T (MAF ~40%), and a non-synony-mous polymorphism in the signal peptide, Leu10Pro (rs1982073, merged into rs1800471; MAF ~40%) that is in strong linkage disequilibrium with C-509T. The 10Pro variant was associated with a 2.8-fold increase in TGF-β1 secretion compared with 10Leu and an increased risk of breast cancer relative to the 10Leu allele (OR 1.08; 95% 1.04 – 1.11) (20, 21). However, other studies observed no or even a reduced association with breast cancer risk (19).

The TGFBR1 gene at 9q22 contains a repeat length polymorphism of either 9 or 6 GCG triplets in exon 1 encoding either 9 or 6 Ala residues (22). The 6 Ala variant was shown experimentally to decrease TGF-β signaling activity. A meta-analysis of 15 breast cancer studies (10,826 cases, 12,964 controls) showed a significant association for allelic effect (6Ala versus 9Ala) with increased breast cancer risk: OR 1.16; 95% CI 1.01 – 1.34 (23). The TGFBR2 gene at 3p22 contains several SNPs in noncoding regions including rs4522809 (A/G) in intron 2 and rs1078985 (A/G) in intron 3 (24, 25). The latter is part of a putative binding site for transcription factors such as NFκB.

Two comprehensive analyses of SNPs in TGF-β signaling pathway genes have been performed. The first analysis of 354 SNPs tagging 17 genes in the pathway including TGF-β1 and TGFBR1 was performed on three large breast cancer study populations: Studies of Epidemiology and Risk Factors in Cancer Heredity (SEARCH; 6,703 cases and 6,840 controls), NCI Polish Breast Cancer Study (PBCS; 1,966 cases and 2,347 controls), and Breast Cancer Association Consortium (BCAC) (25). The TGF-β1 gene Leu10Pro (rs1982073, merged into rs1800471) was associated with increased breast cancer risk: OR 1.05 (95% CI 1.02 – 1.09; p = 0.002). The TGFBR1 gene contains a SNP in intron 1 (A/G; rs10512263) in a region conserved between species but of unknown functional significance. The G allele (MAF ~7%) showed a protective effect: OR 0.87 (95% CI 0.81 – 0.95; p = 0.001). The 9/6 Ala repeat length polymorphism in exon 1 was not examined and its linkage with rs10512263 in intron 1 has not been investigated. A SNP (A/G; rs4522809) in intron 2 of the TGFBR2 gene showed a protective effect of the minor G allele (MAF ~35%): OR 0.95; 95% CI 0.91 – 0.99; p = 0.022). SNPs in the LTBP, SMAD, and TGFBR3 genes showed no association with breast cancer risk. The second comprehensive analysis evaluated 341 SNPs tagging 11 pathway genes in the Shanghai Breast Cancer Genetics Study (7,291 cases; 6,723 controls) and Asian Breast Cancer Consortium (5,077 cases; 5,384 controls) (24). Pooled analysis of all data indicated that minor allele homozygotes (GG) of TGFBR2 rs1078985 had a 24% reduced risk of breast cancer risk compared with major allele carriers (AA or AG; OR 0.76; 95% CI 0.65 – 0.89; p = 8.42 x 10^{-4}). SNPs in other genes encoding ligands (TGFβ1 – 3), receptors (TGFBR1, TGFBR3), and cofactors (SMAD, SARA) showed no association with breast cancer risk. It is of interest that both analyses found a decrease in breast cancer risk association for SNPs in the TGFBR2 gene: rs4522809 in intron 2 (25) and rs1078985 in intron 3 (24); the linkage of these alleles has not been examined.

References

1. Alexandrow MG, Moses HL. *Transforming growth factor b and cell cycle regulation.* Cancer Res. 1995;55:1452-7.

2. Bachman KE, Park BH. *Duel nature of TGF-beta signaling: tumor suppressor vs. tumor promoter.* Curr Opinion Oncol. 2005;17:49-54.

3. Schmierer B, Hill CS. *TGFbeta-SMAD signal transduction: molecular specificity and functional flexibility.* Nat Rev Mol Cell Biol. 2007;8:970-82.

4. Siegel PM, Massague J. *Cytostatic and apoptotic actions of TGF-beta in homeostasis and cancer.* Nat Rev Cancer. 2003;3:807-21.

5. Shi M, Zhu J, Wang R, Chen X, Mi L, Walz T, et al. *Latent TGF-beta structure and activation.* Nature. 2011;474:343-9.

6. Benson JR. *Role of transforming growth factor beta in breast carcinogenesis.* Lancet Oncol. 2004;5:229-39.

7. Dickson RB, Lippman ME. *Growth factors in breast cancer.* Endocr Rev. 1995;16:559-89.

8. Bronzert DA, Bates SE, Sheridan JP, Lindsey R, Valverius EM, Stampfer MR, et al. *Transforming growth factor-beta induces platelet-derived growth factor (PDGF) messenger RNA and PDGF secretion while inhibiting growth in normal human mammary epithelial cells.* Mol Endocrinol. 1990;4:981-9.

9. Silberstein GB, Daniel CW. *Reversible inhibition of mammary gland growth by transforming growth factor-beta.* Science. 1987;237:291-3.

10. Asiedu MK, Ingle JN, Behrens MD, Radisky DC, Knutson KL. *TGFbeta/TNF(alpha)-mediated epithelial-mesenchymal transition generates breast cancer stem cells with a claudin-low phenotype.* Cancer Res. 2011;71:4707-19.

11. Akhurst RJ, Balmain A. *Genetic events and the role of TGF beta in epithelial tumour progression.* J Pathol. 1999;187:82-90.

12. Xu Y, Pasche B. *TGF-beta signaling alterations and susceptibility to colorectal cancer.* Hum Mol Genet. 2007;16 Spec No 1:R14-20.

13. Chen T, Carter D, Garrigue-Antar L, Reiss M. *Transforming growth factor beta type I receptor kinase mutant associated with metastatic breast cancer.* Cancer Res. 1998;58:4805-10.

14. Lucke CD, Philpott A, Metcalfe JC, Thompson AM, Hughes-Davies L, Kemp PR, et al. *Inhibiting mutations in the transforming growth factor beta type 2 receptor in recurrent human breast cancer.* Cancer Res. 2001;61:482-5.

15. Jeruss JS, Sturgis CD, Rademaker AW, Woodruff TK. *Down-regulation of activin, activin receptors, and Smads in high-grade breast cancer.* Cancer Res. 2003;63:3783-90.

16. Xie W, Mertens JC, Reiss DJ, Rimm DL, Camp RL, Haffty BG, et al. *Alterations of Smad signaling in human breast carcinoma are associated with poor outcome: a tissue microarray study.* Cancer Res. 2002;62:497-505.

17. Desruisseau S, Palmari J, Giusti C, Romain S, Martin PM, Berthois Y. *Determination of TGFbeta1 protein level in human primary breast cancers and its relationship with survival.* Br J Cancer. 2006;94:239-46.

18. Grau AM, Wen W, Ramroopsingh DS, Gao YT, Zi J, Cai Q, et al. *Circulating transforming growth factor-beta-1 and breast cancer prognosis: results from the Shanghai Breast Cancer Study.* Breast Cancer Res Treat. 2008;112:335-41.

19. Zheng W. *Genetic polymorphisms in the transforming growth factor-beta signaling pathways and breast cancer risk and survival.* Methods Mol Biol. 2009;472:265-77.

20. Cox A, Dunning AM, Garcia-Closas M, Balasubramanian S, Reed MW, Pooley KA, et al. *A common coding variant in CASP8 is associated with breast cancer risk.* Nat Genet. 2007;39:352-8.

21. Dunning AM, Ellis PD, McBride S, Kirschenlohr HL, Healey CS, Kemp PR, et al. *A transforming growth factorbeta1 signal peptide variant increases secretion in vitro and is associated with increased incidence of invasive breast cancer.* Cancer Res. 2003;63:2610-5.

22. Pasche B, Kolachana P, Nafa K, Satagopan J, Chen YG, Lo RS, et al. *TbetaR-I(6A) is a candidate tumor susceptibility allele.* Cancer Res. 1999;59:5678-82.

23. Liao RY, Mao C, Qiu LX, Ding H, Chen Q, Pan HF. *TGFBR1*6A/9A polymorphism and cancer risk: a meta-analysis of 13,662 cases and 14,147 controls.* Mol Biol Reports. 2010;37:3227-32.

24. Ma X, Beeghly-Fadiel A, Lu W, Shi J, Xiang YB, Cai Q, et al. *Pathway Analyses Identify TGFBR2 as Potential Breast Cancer Susceptibility Gene: Results from a Consortium Study among Asians.* Cancer Epidemiol Biomarkers Prev. 2012;21:1176-84.

25. Scollen S, Luccarini C, Baynes C, Driver K, Humphreys MK, Garcia-Closas M, et al. *TGF-beta signaling pathway and breast cancer susceptibility.* Cancer Epidemiol Biomarkers Prev. 2011;20:1112-9.

6.2.5. PHOSPHATIDYLINOSITOL-3 KINASES (PI3Ks)

Mammalian phosphatidylinositol-3 kinases (PI3Ks) are divided into three classes (I – III) that differ in structure, substrate preference, and mechanism of activation (1, 2). Class I PI3Ks are heterodimers composed of a regulatory subunit (p85α, p55α, p50α, p85β, p55γ) and a catalytic subunit (p110α, β, δ) (3). PI3K phosphorylates phosphatidylinositol 4,5-biphosphate (PIP2) at the 3 position of the inositol ring. The product, phosphatidylinositol 3,4,5-triphosphate (PIP3), recruits the serine-threonine kinases AKT (cellular homolog of murine thymoma virus Akt8 oncoprotein) and phosphatidylinositol-dependent kinase 1 (PDK1) (2). In turn, PDK1 phosphorylates and thereby activates AKT, initiating the PI3K/AKT signaling pathway, which plays a key role in cell cycle progression and survival.

The importance of the PI3K/AKT pathway is highlighted in cancer where several mechanisms can cause inappropriate activation including (i) somatic mutations of the PIK3CA gene at 3q26 that encodes the α-catalytic subunit of the kinase (p110 α), (ii) loss of expression of the PTEN phosphatase that reverses PI3K action (see Section 8.6.5. Cowden Syndrome), (iii) activation of upstream oncogenic receptor tyrosine kinases, such as HER3 (ErbB3), which directly interacts with the p85 regulatory subunit and (iv) mutation or amplification of AKT (1, 4-7). Somatic mutations of PIK3CA are found in various types of cancer including 8 – 40% of breast cancers (7, 8). PIK3CA mutations were reported to be more frequent in ER-positive and HER2-negative tumors but other studies found no association with these variables (1, 9). An inverse association of PIK3CA mutations and loss of PTEN expression was observed by some authors but not others (9, 10). PIK3CA mutations have been detected in breast ductal carcinoma *in situ* (11). The mutation frequencies in pure DCIS, DCIS adjacent to invasive ductal carcinoma, and invasive ductal carcinoma were similar, suggesting that PIK3CA mutations occur early in breast tumorigenesis. About 80% of PIK3CA mutations occur in three hotspots in the helical (exon 9) or kinase (exon 20) domains, resulting in substitutions of three amino acids (Glu542Lys, Glu545Lys, and His1047Arg), which induce a gain of PI3KCA function with constitutive signaling activity. The mutant PIK3CA isoforms are oncogenic in cell culture and animal model systems (12-14). Amplification of the PI3KCB gene, which encodes p110 β has also been observed in breast cancer (15).

References

1. Carvalho S, Schmitt F. *Potential role of PI3K inhibitors in the treatment of breast cancer.* Future Oncol. 2010;6:1251-63.

2. Engelman JA, Luo J, Cantley LC. *The evolution of phosphatidylinositol 3-kinases as regulators of growth and metabolism.* Nature Rev Genetics. 2006;7:606-19.

3. Lempiainen H, Halazonetis TD. *Emerging common themes in regulation of PIKKs and PI3Ks.* EMBO J. 2009;28:3067-73.

4. Carpten JD, Faber AL, Horn C, Donoho GP, Briggs SL, Robbins CM, et al. *A transforming mutation in the pleckstrin homology domain of AKT1 in cancer.* Nature. 2007;448:439-44.

5. Hellyer NJ, Cheng K, Koland JG. *ErbB3 (HER3) interaction with the p85 regulatory subunit of phosphoinositide 3-kinase.* Biochem J. 1998;333 (Pt 3):757-63.

6. Sansal I, Sellers WR. *The biology and clinical relevance of the PTEN tumor suppressor pathway.* J Clin Oncol 2004;22:2954-63.

7. Vogt PK, Kang S, Elsliger MA, Gymnopoulos M. *Cancer-specific mutations in phosphatidylinositol 3-kinase.* Trends Biochem Sci. 2007;32:342-9.

8. Paradiso A, Mangia A, Azzariti A, Tommasi S. *Phosphatidylinositol 3-kinase in breast cancer: where from here?* Clin Cancer Res. 2007;13:5988-90.

9. Saal LH, Holm K, Maurer M, Memeo L, Su T, Wang X, et al. *PIK3CA mutations correlate with hormone receptors, node metastasis, and ERBB2, and are mutually exclusive with PTEN loss in human breast carcinoma.* Cancer Res. 2005;65:2554-9.

10. Yuan TL, Cantley LC. *PI3K pathway alterations in cancer: variations on a theme.* Oncogene. 2008;27:5497-510.

11. Miron A, Varadi M, Carrasco D, Li H, Luongo L, Kim HJ, et al. *PIK3CA mutations in in situ and invasive breast carcinomas.* Cancer Res. 2010;70:5674-8.

12. Isakoff SJ, Engelman JA, Irie HY, Luo J, Brachmann SM, Pearline RV, et al. *Breast cancer-associated PIK3CA mutations are oncogenic in mammary epithelial cells.* Cancer Res. 2005;65:10992-1000.

13. Meyer DS, Brinkhaus H, Muller U, Muller M, Cardiff RD, Bentires-Alj M. *Luminal expression of PIK3CA mutant H1047R in the mammary gland induces heterogeneous tumors.* Cancer Res. 2011;71:4344-51.

14. Zhao L, Vogt PK. *Class I PI3K in oncogenic cellular transformation.* Oncogene. 2008;27:5486-96.

15. Crowder RJ, Phommaly C, Tao Y, Hoog J, Luo J, Perou CM, et al. *PIK3CA and PIK3CB inhibition produce synthetic lethality when combined with estrogen deprivation in estrogen receptor-positive breast cancer.* Cancer Res. 2009;69:3955-62.

6.2.6. MITOGEN-ACTIVATED PROTEIN KINASE (MAPK) PATHWAY

MAPKs are important signal transducing enzymes that connect cell-surface RTKs to critical regulatory targets within cells (1-4). MAPK activity is regulated through three-tiered cascades composed of (i) a MAPK, (ii) a MAPK kinase (also known as MAPKK, MKK, or MEK), and (iii) a MAPKK or MEK kinase (MAPKKK or MEKK). The cascades transmit and amplify extracellular signals into cell proliferation. Mammals express at least four distinctly regulated groups of MAPKs: extracellular regulated kinase (ERK), Jun NH2-terminal kinase (JNK), p38 protein, and ERK5. Several of these kinases have isoforms, e.g., ERK exists as ERK1 and ERK2 isoforms, which are 85% homologous with molecular weights of 42-44 kDa. Similarly, JNK exists as three isoforms, JNK1/2/3, and p38 as four isoforms, p38α/β/γ/δ. Each MAPK is activated by a specific MAPKK: MEK1/1 for ERK1/2, MKK4/7 (JNKK1/2) for the JNKs, MKK3/6 for the p38, and MEK5 for ERK5. Each MAPKK, however, can be activated by more than one MAPKKK, increasing the complexity and diversity of MAPK signaling (1). For example, MAP3K1 (MEKK1) encodes the MAP kinase kinase that phosphorylates and activates the MAP kinase (MAPK2) that, in turn, phosphorylates MAPK/ERK to produce downstream signaling effects on cMyc, cJun, cFos and other cancer genes involved in cell proliferation (5). Alternatively, Raf-1 can act as the initial MAPKKK in the same pathway (2).

MAPKs play a role in mammary gland development and carcinogenesis. Normal breast development requires the coordinated growth and movement of individual cells and cell sheets. Ductal epithelial cells asymmetrically organize into an apical pole oriented toward the lumen and a basal pole that interfaces with the stroma and vasculature (6). The adaptor protein AMOT (for angiomotin) regulates cellular asymmetry and promotes cell proliferation via activation of ERK1/2 (7). JNK signaling is required for normal branching of mammary ducts (8).

In ER-positive breast cancers, MAPK pathways can exert "cross talk" effects at the level of ER-induced transcription while estradiol can activate MAPK either through rapid, non-transcription effects or by increasing growth factor production and consequently MAPK (4, 9). Specifically, MAPK (ERK) can directly catalyze the phosphorylation of 118Ser of ERα and increase its transcriptional activity (10). Estradiol can utilize cell membrane-associated ER to activate MAPK through non-genomic effects (11, 12). The rapid effect of E_2 was mediated via interaction of ERα with the adapter protein Shc, which interacts with the MAPK pathway (13). E_2 also stimulates the expression of several growth factors such as TGFα and IGF-1, which leads to up-regulation of MAPK (4, 14, 15). Approximately 50% of breast cancers express more activated MAPK than the surrounding benign tissue (16, 17). The up-regulation of MAPK activity was not due to Ras mutations but appeared to result from expression of EGFR and HER2 (18). Interestingly, hyperactivation of MAPK in EGFR- or HER2-overexpressing breast cancer cells resulted in down-regulation of ERα (19). ERα expression was restored by MAPK inhibition. Estrogen has also been shown to activate MAPK in ER-negative SKBR3 breast cancer cells by an alternative mechanism involving the G-protein-coupled receptor homolog, GPR30 (20, 21).

A genome-wide screen of European women identified a SNP in MAP3K1 to be significantly associated with breast cancer risk (22). The high-risk allele of SNP rs2981582 carried a relative risk of 1.13 (95% CI 1.09 – 1.18) per allele as compared with the low-risk allele. This was corroborated in African American but not European American women (5). However, other genome-wide screens found no significant associations with MAP3K1 (23, 24). Clinical correlation studies yielded inconsistent evidence of an interaction of MAP3K1 variants with hormone replacement therapy or ER-positive tumors (5, 25). In African American, but not European American women, a significant risk interaction was observed between the MAP3K1 rs889312 and FGFR2 rs2981582 variants (5).

References

1. Chang L, Karin M. *Mammalian MAP kinase signalling cascades.* Nature. 2001;410:37-40.

2. Pearson G, Robinson F, Beers Gibson T, Xu BE, Karandikar M, Berman K, et al. *Mitogen-activated protein (MAP) kinase pathways: regulation and physiological functions.* Endocr Rev. 2001;22:153-83.

3. Raman M, Chen W, Cobb MH. *Differential regulation and properties of MAPKs.* Oncogene. 2007;26:3100-12.

4. Santen RJ, Song RX, McPherson R, Kumar R, Adam L, Jeng MH, et al. *The role of mitogen-activated protein (MAP) kinase in breast cancer.* J Steroid Biochem Mol Biol. 2002;80:239-56.

5. Rebbeck TR, DeMichele A, Tran TV, Panossian S, Bunin GR, Troxel AB, et al. *Hormone-dependent effects of FGFR2 and MAP3K1 in breast cancer susceptibility in a population-based sample of post-menopausal African-American and European-American women.* Carcinogenesis. 2009;30:269-74.

6. Itoh M, Bissell MJ. *The organization of tight junctions in epithelia: implications for mammary gland biology and breast tumorigenesis.* J Mammary Gland Biol Neoplasia. 2003;8:449-62.

7. Ranahan WP, Han Z, Smith-Kinnaman W, Nabinger SC, Heller B, Herbert BS, et al. *The adaptor protein AMOT promotes the proliferation of mammary epithelial cells via the prolonged activation of the extracellular signal-regulated kinases.* Cancer Res. 2011;71:2203-11.

8. Cellurale C, Girnius N, Jiang F, Cavanagh-Kyros J, Lu S, Garlick DS, et al. *Role of JNK in mammary gland development and breast cancer.* Cancer Res. 2012;72:472-81.

9. Migliaccio A, Di Domenico M, Castoria G, de Falco A, Bontempo P, Nola E, et al. *Tyrosine kinase/p21ras/MAP-kinase pathway activation by estradiol-receptor complex in MCF-7 cells.* EMBO J. 1996;15:1292-300.

10. Kato S, Endoh H, Masuhiro Y, Kitamoto T, Uchiyama S, Sasaki H, et al. *Activation of the estrogen receptor through phosphorylation by mitogen-activated protein kinase.* Science. 1995;270:1491-4.

11. Pietras RJ, Marquez-Garban DC. *Membrane-associated estrogen receptor signaling pathways in human cancers.* Clin Cancer Res. 2007;13:4672-6.

12. Watson CS, Gametchu B, Morfleet AM, Campbell CH, THomas ML. *Rapid, nongenomic actions of estrogens.* Women and Cancer. 1998;1:21-8.

13. Song RX, McPherson RA, Adam L, Bao Y, Shupnik M, Kumar R, et al. *Linkage of rapid estrogen action to MAPK activation by ERalpha-Shc association and Shc pathway activation.* Mol Endocrinol. 2002;16:116-27.

14. Chalbos D, Philips A, Rochefort H. *Genomic cross-talk between the estrogen receptor and growth factor regulatory pathways in estrogen target tissues.* Semin Cancer Biol. 1994;5:361-8.

15. Levin ER. *Bidirectional signaling between the estrogen receptor and the epidermal growth factor receptor.* Mol Endocrinol. 2003;17:309-17.

16. Mueller H, Flury N, Eppenberger-Castori S, Kueng W, David F, Eppenberger U. *Potential prognostic value of mitogen-activated protein kinase activity for disease-free survival of primary breast cancer patients.* Int J Cancer. 2000;89:384-8.

17. Salh B, Marotta A, Matthewson C, Ahluwalia M, Flint J, Owen D, et al. *Investigation of the Mek-MAP kinase-Rsk pathway in human breast cancer.* Anticancer Res. 1999;19:731-40.

18. von Lintig FC, Dreilinger AD, Varki NM, Wallace AM, Casteel DE, Boss GR. *Ras activation in human breast cancer.* Breast Cancer Res Treat. 2000;62:51-62.

19. Oh AS, Lorant LA, Holloway JN, Miller DL, Kern FG, El-Ashry D. *Hyperactivation of MAPK induces loss of ERalpha expression in breast cancer cells.* Mol Endocrinol. 2001;15:1344-59.

20. Filardo EJ, Quinn JA, Bland KI, Frackelton AR, Jr. *Estrogen-induced activation of Erk-1 and Erk-2 requires the G protein-coupled receptor homolog, GPR30, and occurs via trans-activation of the epidermal growth factor receptor through release of HB-EGF.* Mol Endocrinol. 2000;14:1649-60.

21. Filardo EJ, Quinn JA, Frackelton AR, Jr., Bland KI. *Estrogen action via the G protein-coupled receptor, GPR30: stimulation of adenylyl cyclase and cAMP-mediated attenuation of the epidermal growth factor receptor-to-MAPK signaling axis.* Mol Endocrinol. 2002;16:70-84.

22. Easton DF, Pooley KA, Dunning AM, Pharoah PDP, Thompson D, Ballinger DG, et al. *Genome-wide association study identifies novel breast cancer susceptibility loci.* Nature. 2007;447:1087-95.

23. Gold B, Kirchhoff T, Stefanov S, Lautenberger J, Viale A, Garber J, et al. *Genome-wide association study provides evidence for a breast cancer risk locus at 6q22.33.* Proc Natl Acad Sci USA. 2008;105:4340-5.

24. Hunter DJ, Kraft P, Jacobs KB, Cox DG, Yeager M, Hankinson SE, et al. *A genome-wide association study identifies alleles in FGFR2 associated with risk of sporadic post-menopausal breast cancer.* Nature Genet. 2007;39:870-4.

25. Travis RC, Reeves GK, Green J, Bull D, Tipper SJ, Baker K, et al. *Gene-environment interactions in 7610 women with breast cancer: prospective evidence from the Million Women Study.* Lancet. 2010;375:2143-51.

6.2.7. MYC

The myc oncogene family consists of three members, c-myc at 8q24, L-myc at 1p32, and N-myc at 2p24. Deregulated expression of each myc gene is associated with specific malignancies, e.g., N-myc amplification is associated with neuroblastoma whereas elevated expression of c-myc is associated with breast and other cancers (1). C-myc is a basic helix-loop-helix (bHLH) transcription factor that forms a heterodimer with MAX and binds to E-box sequences near the core promoter elements of actively transcribed genes (2). In normal cells, c-myc links growth factor stimulation and cellular proliferation (3). Mitogenic growth factor signaling induces c-myc expression, which then enhances the transcription of proliferation-associated genes. In tumor cells, c-myc expression is elevated through multiple mechanisms, including gene amplification, chromosomal translocation, single nucleotide polymorphism in regulatory regions, mutation of upstream signaling pathways, and mutations that enhance the stability of the protein (3). In tumor cells that express high levels of c-myc, cellular proliferation is no longer dependent on growth factor stimulation and this uncoupling leads to the uncontrolled proliferation characteristic of cancer cells. Elevated expression of c-myc also causes changes in chromatin structure, metabolic pathways, cell adhesion, apoptosis, and angiogenesis (4). How elevated levels of c-myc cause such a broad spectrum of cellular effects was not understood since c-myc target signatures showed little overlap to ascribe c-myc's oncogenic properties to any one set of target genes (5). Rather than acting as an on-off specifier of particular transcriptional programs, c-myc acts universally at active genes, except for immediate early genes that are strongly induced before c-myc (4, 6). In tumor cells with elevated levels of c-myc, the transcription factor occupies both the core promoters and enhancers of all active genes. The increase in c-myc occupancy leads to increased transcription elongation by RNA polymerase II and increased levels of transcripts per cell. Thus, c-myc causes universal transcriptional amplification, producing elevated levels of transcripts from the existing gene expression program of tumor cells.

The c-myc gene is involved in the stimulation of cell proliferation and its expression is enhanced by estradiol in cultured normal mammary epithelium and breast cancer cell lines (7-10). Estrogen-induced renal cancer in the Syrian hamster exhibited c-myc amplification and overexpression (11). Amplification of the c-myc gene may promote cell replication and thereby accelerate breast cancer. In general, clinical studies found a significant association between c-myc amplification and the ER-negative phenotype (12-14). An immunohistochemical analysis of 206 breast cancers detected a significant correlation between ER-negative phenotype and nuclear, but not cytoplasmic c-myc staining (15). Gene expression analysis revealed a higher average c-myc expression in basal-like breast cancers that frequently show 8q24 amplification (5, 16). In contrast, a lower average expression was observed in the luminal A subtype.

References

1. Nesbit CE, Tersak JM, Prochownik EV. *MYC oncogenes and human neoplastic disease.* Oncogene. 1999;18:3004-16.

2. Blackwood EM, Eisenman RN. *Max: a helix-loop-helix zipper protein that forms a sequence-specific DNA-binding complex with Myc.* Science. 1991;251:1211-7.

3. Dang CV. *MYC on the path to cancer.* Cell. 2012;149:22-35.

4. Lin CY, Loven J, Rahl PB, Paranal RM, Burge CB, Bradner JE, et al. *TTranscriptional amplification in tumor cells with elevated c-myc.* Cell. 2012;151:56-67.

5. Chandriani S, Frengen E, Cowling VH, Pendergrass SA, Perou CM, Whitfield ML, et al. *A core MYC gene expression signature is prominent in basal-like breast cancer but only partially overlaps the core serum response.* PLoS One. 2009;4:e6693.

6. Nie Z, Hu G, Wei G, Cui K, Yamane A, Resch W, et al. *c-myc is a universal amplifier of expressed genes in lymphocytes and embryonic stem cells.* Cell. 2012;151:68-79.

7. Dubik D, Dembinski TC, Shiu RPC. *Stimulation of c-myc oncogene expression associated with estrogen-induced proliferation of human breast cancer cells.* Cancer Res. 1987;47:6517-21.

8. Dubik D, Shiu RP. *Mechanism of estrogen activation of c-myc onogene expression.* Onogene. 1992;7:1587-94.

9. Leygue E, Gol-Winkler R, Gompel A, Louis-Sylvestre C, Soquet L, Staub S, et al. *Estradiol stimulates c-myc proto-oncogene expression in normal human breast epithelial cells in culture.* J Steroid Biochem Mol Biol. 1995;52:299-305.

10. Nass SJ, Dickson RB. *Defining a role for c-myc in breast tumorigenesis.* Breast Cancer Res Treat. 1997;44:1-22.

11. Li JJ, Hou X, Banerjee SK, Liao DJ, Maggouta F, Norris JS, et al. *Overexpression and amplification of c-myc in the syrian hamster kidney during estrogen carcinogenesis: A probable critical role in neoplastic transformation.* Cancer Res. 1999;59:2340-6.

12. Berns EMJJ, Klijn JGM, van Putten WLJ, van Staveren IL, Portengen H, Foekens JA. *c-myc amplification is a better prognostic factor than HER2/neu amplification in primary breast cancer.* Cancer Res. 1992;52:1107-13.

13. Courjal F, Cuny M, Simony-Lafontaine J, Louason G, Speiser P, Zeillinger R, et al. *Mapping of DNA amplifications at 15 chromosomal localizations in 1875 breast tumors: Definition of phenotypic groups.* Cancer Res. 1997;57:4360-7.

14. Persons DL, Borelli KA, Hsu PH. *Quantitation of HER-2/ neu and c-myc gene amplification in breast carcinoma using fluorescence in situ hybridization.* Mod Pathol. 1997;10:720-7.

15. Pietilainen T, Lipponen. P, Aaltomaa S, Eskelinen M, Kosma VM, Syrjanen K. *Expression of c-myc proteins in breast cancer as related to established prognostic factors and survival.* Anticancer Res. 1995;15:959-64.

16. Chin K, DeVries S, Fridlyand J, Spellman PT, Roydasgupta R, Kuo WL, et al. *Genomic and transcriptional aberrations linked to breast cancer pathophysiologies.* Cancer Cell. 2006;10:529-41.

6.3. INSULIN-LIKE GROWTH FACTOR FAMILY
6.3.1. INTRODUCTION

Three hormones or growth factors constitute the insulin-like growth factor (IGF) family – insulin, IGF1 and IGF2 (Table 6.3.1.). Insulin is synthesized in the beta cells of the pancreas as proinsulin, which is cleaved to form insulin and C peptide. The IGFs, which are synthesized primarily by the liver, retain the C peptide and have an extended carboxy terminus. Insulin circulates at picomolar concentrations and has a half-life of minutes. The IGFs, on the other hand, circulate at much higher (nanomolar) concentrations and are largely (99%) bound to one of six IGF-binding proteins (IGFBP1 – 6) that modulate IGF activity (1). Like the IGFs, the IGFBPs are synthesized primarily in the liver but also produced locally by most tissues, including the breast, where they act in an autocrine or paracrine manner (2-4). A key role in the IGF network is played by pituitary growth hormone (GH), which is under the regulation of the hypothalamic hormones somatostatin (SST, an inhibitor) and GH-releasing hormone (GHRH, a stimulator). GH receptor, GHRH receptor, and five somatostatin receptors (SSTR1 – 5) bind their respective ligands and regulate their function (1). Thus, GH interacts with its hepatic receptor to stimulate the expression of the IGF1 and IGFBP3 genes and secretion of the mature peptides. In turn, IGF1 acts on the pituitary to inhibit the secretion of GH (2).

The IGF family has been shown to play a role in normal mammary gland development as well as mammary carcinogenesis (4-6). Insulin acts primarily on the liver, muscle, and adipose tissue to regulate carbohydrate metabolism but also influences cell growth, e.g., it exerts a mitogenic effect in breast cancer cells (7, 8). The mitogenic action of insulin is mediated by the membrane-associated insulin receptor (IR; Table 6.3.1.), a tyrosine kinase (9). Activated insulin receptor phosphorylates insulin receptor substrates (IRS1 – 4), which bind the p85 subunit of phosphoinositide 3-kinase (PI3K). In turn, PI3K activates downstream effectors including AKT-TOR. Both IGFs are essential for embryonic development but after birth IGF1 appears to have the predominant role in regulating growth by stimulating mitosis and inhibiting apoptosis (2, 3). Both IGFs bind specifically to the IR and the highly homologous IGF1 receptor (IGF1R; Table 6.3.1.) as well as IR-IGF1R heterodimers. Following ligand binding to IGF1R, its tyrosine kinase activity is activated and this stimulates signaling through downstream intracellular pathways, including the PI3K-AKT-TOR and RAF-MAPK systems that regulate cell proliferation and survival (1, 9). A third receptor, the IGF2 receptor (IGF2R, also known as the mannose-6-phosphate receptor) binds IGF2 but has no tyrosine kinase domain and therefore no signaling activity (3, 10). Instead, IGF2R internalizes IGF2, which is transported to lysosomes for degradation. By removing IGF2 from the extracellular environment and precluding its activation of IGF1R, IGF2R is believed to reduce the mitogenic effects of IGF2. Transgenic mice that overexpress IGF1 and IGF2 specifically in the mammary gland have an increased incidence of breast adenocarcinoma (11). Similarly, overexpression of either IR or IGF1R is tumorigenic in mouse tumor models (12). In the circulation, the IGFBPs limit the access of the IGFs to tissues and to the insulin and IGF1 receptors. Of

the six binding proteins, IGFBP3 is the most abundant and accounts for 80% of bound IGFs in serum. The IGF-IGFBP3 dimer forms a complex with another hepatic protein, the acid-labile subunit (IGFALS), and in this ternary complex the IGFs have a serum half-life of 12 hours (2). There is considerable biological variation of IGF1 levels in the circulation with 3 – 36% within-individual biological imprecision of IGF1 assays (13). Genetic variation, multiple physiological variables and clinical conditions can affect the concentrations of IGF1. Two common SNPs in IGF1 (rs1520220, rs10735380) and four in SSTR5 were significantly associated with circulating IGF1 levels (14). Puberty, pregnancy, and extremes of body mass index influence circulating levels but there is little variation during the menstrual cycle (13, 15, 16). IGF1 is inversely associated with age, with no additional decline in concentration after age 50, suggesting that menopause itself does not have a marked effect on IGF1. IGF1 was highest in women with a BMI of 25 to 27.4 kg/m^2 than in thinner or more overweight women. In adults, sex and ethnicity are minor contributors of variation in IGF1 concentrations. Circadian and meal-related changes do not appear to affect IGF1 measurements. Chronic hyperinsulinemia decreases concentrations of IGFBPs, which leads to an increase in free IGF1 (17). Some medications, such as oral estrogen, can also affect IGF1 levels. Considerable analytical differences exist between currently available IGF1 assays (13). An individual IGF1 sample measured by different assays can yield very different results because assays differ with respect to the epitope specificity of the antibodies used and because the different types of antibodies vary in their ability to bind to the different molecular forms of IGF1. Moreover, IGFBPs may interfere with IGF1 assays producing falsely low values. SNPs in the IGFBP3 and IGFALS genes showed a significant association with circulating IGFBP3 concentrations (14, 18). In particular the positive association of SNPs Gly32Ala in exon 1 (rs2854746) and C-202A (rs2854744) in the promoter region of the IGFBP3 gene with circulating IGFBP3 levels has been noted in several studies, the latter resulting from higher promoter activity of the A allele.

Table 6.3.1. The Insulin-Like Growth Factor Family

Protein	Chromosome Locus	Size (kDa, Amino Acids)
Proinsulin/Insulin	11p15.5	82/51 AA
IGF1	12q23.2	70 AA
IGF2	11p15.5	180 AA
Insulin Receptor	19p13.2	320 kDa
IGF1 Receptor	19p13.2	300 kDa
IGF2 Receptor	6q26-27	300 kDa
IGFBP3	7p12.3	32 kDa
IGF-ALS	16p13.3	85 kDa
Somatostatin	3q28	14 AA
SSTR5	16p13.3	39 kDa

References

1. Pollak M. *Insulin and insulin-like growth factor signalling in neoplasia.* Nature Rev Cancer. 2008;8:915-28.

2. Le Roith D. Seminars in medicine of the Beth Israel Deaconess Medical Center. *Insulin-like growth factors.* N Engl J Med. 1997;336:633-40.

3. Pollak MN, Schernhammer ES, Hankinson SE. *Insulin-like growth factors and neoplasia.* Nature Rev Cancer. 2004;4:505-18.

4. Sachdev D, Yee D. *The IGF system and breast cancer.* Endocrine-Related Cancer. 2001;8:197-209.

5. Clayton PE, Banerjee I, Murray PG, Renehan AG. *Growth hormone, the insulin-like growth factor axis, insulin and cancer risk.* Nature Rev Endocrinol. 2011;7:11-24.

6. Hadsell DL, Bonnette SG. *IGF and insulin action in the mammary gland: lessons from transgenic and knockout models.* J Mammary Gland Biol Neoplasia. 2000;5:19-30.

7. Chappell J, Leitner JW, Solomon S, Golovchenko I, Goalstone ML, Draznin B. *Effect of insulin on cell cycle progression in MCF-7 breast cancer cells. Direct and potentiating influence.* J Biol Chem.

2001;276:38023-8.

8. Osborne CK, Bolan G, Monaco ME, Lippman ME. *Hormone responsive human breast cancer in long-term tissue culture: effect of insulin.* Proc Natl Acad Sci USA. 1976;73:4536-40.

9. Belfiore A, Frasca F. *IGF and insulin receptor signaling in breast cancer.* J Mammary Gland Biol Neoplasia. 2008;13:381-406.

10. Oates AJ, Schumaker LM, Jenkins SB, Pearce AA, DaCosta SA, Arun B, et al. *The mannose 6-phosphate/insulin-like growth factor 2 receptor (M6P/IGF2R), a putative breast tumor suppressor gene.* Breast Cancer Res Treat. 1998;47:269-81.

11. Hadsell DL, Murphy KL, Bonnette SG, Reece N, Laucirica R, Rosen JM. *Cooperative interaction between mutant p53 and des(1-3) IGF-I accelerates mammary tumorigenesis.* Oncogene. 2000;19:889-98.

12. Buck E, Gokhale PC, Koujak S, Brown E, Eyzaguirre A, Tao N, et al. *Compensatory insulin receptor (IR) activation on inhibition of insulin-like growth factor-1 receptor (IGF-1R): rationale for cotargeting IGF-1R and IR in cancer.* Mol Cancer Therapeutics. 2010;9:2652-64.

13. Clemmons DR. *Consensus statement on the standardization and evaluation of growth hormone and insulin-like growth factor assays.* Clin Chem. 2011;57:555-9.

14. Gu F, Schumacher FR, Canzian F, Allen NE, Albanes D, Berg CD, et al. *Eighteen insulin-like growth factor pathway genes, circulating levels of IGF-I and its binding protein, and risk of prostate and breast cancer.* Cancer Epidemiol Biomarkers Prev. 2010;19:2877-87.

15. Key TJ, Appleby PN, Reeves GK, Roddam AW. *Insulin-like growth factor 1 (IGF1), IGF binding protein 3 (IGFBP3), and breast cancer risk: pooled individual data analysis of 17 prospective studies.* Lancet Oncol. 2010;11:530-42.

16. Missmer SA, Spiegelman D, Bertone-Johnson ER, Barbieri RL, Pollak MN, Hankinson SE. *Reproducibility of plasma steroid hormones, prolactin, and insulin-like growth factor levels among premenopausal women over a 2- to 3-year period.* Cancer Epidemiol Biomarkers Prev. 2006;15:972-8.

17. Renehan AG, Frystyk J, Flyvbjerg A. *Obesity and cancer risk: the role of the insulin-IGF axis.* Trends Endocrinol Metab. 2006;17:328-36.

18. Fletcher O, Gibson L, Johnson N, Altmann DR, Holly JM, Ashworth A, et al. *Polymorphisms and circulating levels in the insulin-like growth factor system and risk of breast cancer: a systematic review.* Cancer Epidemiol Biomarkers Prev. 2005;14:2-19.

6.3.2. EXPERIMENTAL AND CLINICAL STUDIES

Crosstalk between the IGF and estrogen-mediated signaling pathways results in synergistic effects on cell growth (1, 2). The IGF family interacts with estrogens at multiple levels. Hyperinsulinemia is associated with decreased plasma SHBG, thereby increasing E_2 bioavailability (3). Conversely, a reduction in insulin levels increased SHBG, with a decrease in E_2 bioavailability. Stimulation of IGF1R increases the phosphorylation and activity of ERα (4). Reciprocally, estrogen treatment of breast cancer cells alters the expression of several IGF family members including IGF1, IGF2, IGFBPs, IGF1R, IRS1, and IRS2 (2, 5). For example, E_2 treatment increased IGF1R levels in ER-positive breast cancer cells but IGF1R levels in ER-negative cells remained decreased and IGF1 was non-mitogenic (4, 5). Part of the estrogen effect on IGF1R expression is mediated through activation of the Sp1 transcription factor (6). An analysis of plasma levels of IGFs and sex steroid hormones in the Shanghai Breast Cancer study showed evidence of synergy in association with breast cancer risk (7). Pre- and postmenopausal women with high circulating levels of IGF1 and estrone were at much higher risk for breast cancer compared with high levels of only one hormone. Similar patterns of association were also seen for IGF1 with testosterone as well as for IGFBP3 with estrone or testosterone. No joint effects were found for IGF1 or IGFBP3 with estradiol or estrone sulfate. The insulin and IGF1 receptors were shown to be essential for the growth of ER-positive MCF-7 cells that became resistant to long-term estrogen deprivation (8).

Several epidemiologic studies suggest a positive association between levels of circulating IGF1 and risk of breast cancer. A systematic review in 2004 of 21 studies observed an increased risk of premenopausal breast cancer with high concentrations of IGF1 and IGFBP3 (9). A 2005 review found in 5 of 8 studies that premenopausal women in the highest quartile of circulating IGF1 levels had more than twice the risk of developing breast cancer than those in the lowest quartile (10). Similarly, 4 of 6 studies showed an increased cancer risk of premenopausal women with higher IGFBP3 levels. In postmenopausal women, however, there was no consistent association of either IGF1 or IGFBP3 with breast cancer risk (9, 10). A 2010 analysis of the Breast and Prostate Cancer Cohort Consortium observed significantly higher circulating IGF1 and IGFBP3 levels in men with prostate cancer than in controls but found no significant differences in either IGF1 or IGFBP3 levels between women with breast cancer and controls (11). In contrast, a 2010 pooled analysis by the Endogenous Hormones and Breast Cancer Collaborative Group of 17 prospective studies from 12 countries obtained an odds ratio for breast cancer of 1.28 (95% CI 1.14 – 1.44; $p < 0.0001$) for women in the highest versus the lowest fifth of IGF1 concentrations (12). This association was not altered by adjusting for IGFBP3 and did not vary significantly by menopausal status. However, the odds ratios were 1.38 (95% CI 1.14 – 1.68) for ER-positive tumors and 0.80 (95% 0.57 – 1.13) for ER-negative tumors, indicating that the association between circulating IGF1 levels and breast cancer risk is independent of menopausal status but confined to ER-positive cancers.

Insulin and IGF1 receptors as well as IR-IGF1R heterodimers are overexpressed in breast cancers (13). Phosphorylated receptors were detected in different breast cancer subtypes (ER-positive luminal, HER2-positive, and ER/PR/HER2-triple negative) and high levels correlated with poor survival (14). Loss of heterozygosity at the IGF2R locus has been identified in breast cancers and somatic mutations of the remaining allele resulted in altered ligand binding (15, 16). The IGF2R has been proposed to be a tumor suppressor gene given its antagonist role on cellular growth and evidence of loss of heterozygosity and loss of function mutations (17). An extensive investigation of the IGF2R gene in the Multiethnic Cohort study found no evidence for an association of 12 nonsynonymous polymorphisms with breast cancer risk (18).

What is the biological basis for the observed association between higher levels of IGF1 and breast cancer risk? A model has been presented that suggests subtle influences of higher IGF1 levels on renewal dynamics of epithelial cell populations (19). Mammary epithelial cells of women with higher levels of IGF1 might show slightly higher proliferation rates and have a slightly increased chance of survival in the presence of genetic damage because of the anti-apoptotic effects of IGF1. This would facilitate stepwise accumulation of genetic damage leading to carcinogenesis. Higher IGF1 levels might also reduce the time interval between emergence of a transformed clone of cells and clinically significant cancer. Are there genetic markers to allow the identification of women with increased IGF1 levels at high risk to develop breast cancer? A systematic analysis of 302 SNPs in 18 genes in the IGF signaling pathway (including all those listed in Table 6.3.1.) was performed in 5,500 Caucasian women and >5,500 Caucasian men from the Breast and Prostate Cancer Cohort Consortium (11). SNPs in the IGF1 and SSTR5 genes were significantly associated with circulating IGF1 levels. Similarly, SNPs in the IGFBP3 and IGFALS genes showed a significant association with circulating IGFBP3 concentrations. However, these variants explain only a small percentage of the variation in IGF1 (<0.7%) or IGFBP3 (<4%) levels. This may explain the finding that there was no significant association of these SNPs with risk of breast or prostate cancer (10, 11). An even larger analysis of the Breast and Prostate Cancer Cohort Consortium (1416 SNPs in 24 genes of the IGF1 pathway; 6292 postmenopausal breast cancer cases, 8135 controls) reached the same conclusion that none of the SNPs was significantly associated with breast cancer risk after correction for multiple comparisons (20). Instead of a conventional p-value threshold (e.g., p < 0.05, p < 0.01) for single candidate gene studies, the authors chose a threshold of p < 0.00005 to minimize false positives among the large number of SNPs tested. Among ER-negative cases, a SNP (rs719756) located 3' of the growth hormone receptor gene was marginally associated with increased risk (p = 1.5 x 10^{-4}) (20). Results from the Breast Cancer Association Consortium also yielded null results with the exception of a borderline statistical significance (p = 0.06) of SNP C-202A (rs2854744) in the promoter region of the IGFBP3 gene with breast cancer risk (21). One study observed a significant association of linkage disequilibrium blocks in the IGF2 gene with breast cancer risk in BRCA1 and BRCA2 mutation carriers (22).

Obesity is an established risk factor for postmenopausal breast cancer as well as type 2 diabetes. Some epidemiological studies have reported diabetics to have a greater risk of breast cancer than nondiabetics, independent of body weight (23, 24). However, two large analyses of 31 and 40 common genetic markers for type 2 diabetes and obesity showed no association with breast cancer (25, 26).

References

1. Dupont J, Le Roith D. *Insulin-like growth factor 1 and oestradiol promote cell proliferation of MCF-7 breast cancer cells: new insights into their synergistic effects.* Mol Pathol: MP. 2001;54:149-54.

2. Hamelers IH, Steenbergh PH. *Interactions between estrogen and insulin-like growth factor signaling pathways in human breast tumor cells.* Endocrine-Related Cancer. 2003;10:331-45.

3. Kaaks R. *Nutrition, hormones, and breast cancer: is insulin the missing link?* Cancer Causes Control. 1996;1996:605-25.

4. Surmacz E, Bartucci M. *Role of estrogen receptor alpha in modulating IGF-I receptor signaling and function in breast cancer.* J Exper Clin Cancer Res: CR. 2004;23:385-94.

5. Lee AV, Jackson JG, Gooch JL, Hilsenbeck SG, Coronado-Heinsohn E, Osborne CK, et al. *Enhancement of insulin-like growth factor signaling in human breast cancer: estrogen regulation of insulin receptor substrate-1 expression in vitro and in vivo.* Mol Endocrinol. 1999;13:787-96.

6. Maor S, Mayer D, Yarden RI, Lee AV, Sarfstein R, Werner H, et al. *Estrogen receptor regulates insulin-like growth factor-I receptor gene expression in breast tumor cells: involvement of transcription factor Sp1.* J Endocrinol. 2006;191:605-12.

7. Yu H, Shu XO, Li BD, Dai Q, Gao YT, Jin F, et al. *Joint effect of insulin-like growth factors and sex steroids on breast cancer risk.* Cancer Epidemiol Biomarkers Prev. 2003;12:1067-73.

8. Fox EM, Miller TW, Balko JM, Kuba MG, Sanchez V, Smith RA, et al. *A kinome-wide screen identifies the insulin/IGF-I receptor pathway as a mechanism of escape from hormone dependence in breast cancer.* Cancer Res. 2011;71:6773-84.

9. Renehan AG, Zwahlen M, Minder C, O'Dwyer ST, Shalet SM, Egger M. *Insulin-like growth factor (IGF)-I, IGF binding protein-3, and cancer risk: systematic review and meta-regression analysis.* Lancet. 2004;363:1346-53.

10. Fletcher O, Gibson L, Johnson N, Altmann DR, Holly JM, Ashworth A, et al. *Polymorphisms and circulating levels in the insulin-like growth factor system and risk of breast cancer: a systematic review.* Cancer Epidemiol Biomarkers Prev. 2005;14:2-19.

11. Gu F, Schumacher FR, Canzian F, Allen NE, Albanes D, Berg CD, et al. *Eighteen insulin-like growth factor pathway genes, circulating levels of IGF-I and its binding protein, and risk of prostate and breast cancer.* Cancer Epidemiol Biomarkers Prev. 2010;19:2877-87.

12. Key TJ, Appleby PN, Reeves GK, Roddam AW. *Insulin-like growth factor 1 (IGF1), IGF binding protein 3 (IGFBP3), and breast cancer risk: pooled individual data analysis of 17 prospective studies.* Lancet Oncol. 2010;11:530-42.

13. Pandini G, Vigneri R, Costantino A, Frasca F, Ippolito A, Fujita-Yamaguchi Y, et al. *Insulin and insulin-like growth factor-I (IGF-I) receptor overexpression in breast cancers leads to insulin/IGF-I hybrid receptor overexpression: evidence for a second mechanism of IGF-I signaling.* Clin Cancer Res. 1999;5:1935-44.

14. Law JH, Habibi G, Hu K, Masoudi H, Wang MY, Stratford AL, et al. *Phosphorylated insulin-like growth factor-i/insulin receptor is present in all breast cancer subtypes and is related to poor survival.* Cancer Res. 2008;68:10238-46.

15. Byrd JC, Devi GR, de Souza AT, Jirtle RL, MacDonald RG. *Disruption of ligand binding to the insulin-like growth factor II/mannose 6-phosphate receptor by cancer-associated missense mutations.* J Biol Chem. 1999;274:24408-16.

16. Chappell SA, Walsh T, Walker RA, Shaw JA. *Loss of heterozygosity at the mannose 6-phosphate insulin-like growth factor 2 receptor gene correlates with poor differentiation in early breast carcinomas.* Br J Cancer. 1997;76:1558-61.

17. Oates AJ, Schumaker LM, Jenkins SB, Pearce AA, DaCosta SA, Arun B, et al. *The mannose 6-phosphate/insulin-like growth factor 2 receptor (M6P/IGF2R), a* putative breast tumor suppressor gene. Breast Cancer Res Treat. 1998;47:269-81.

18. Cheng I, Stram DO, Burtt NP, Gianniny L, Garcia RR, Pooler L, et al. *IGF2R missense single-nucleotide polymorphisms and breast cancer risk: the multiethnic cohort study.* Cancer Epidemiol Biomarkers Prev. 2009;18:1922-4.

19. Pollak MN, Schernhammer ES, Hankinson SE. *Insulin-like growth factors and neoplasia.* Nature Rev Cancer. 2004;4:505-18.

20. Canzian F, Cox DG, Setiawan VW, Stram DO, Ziegler RG, Dossus L, et al. *Comprehensive analysis of common genetic variation in 61 genes related to steroid hormone and insulin-like growth factor-I metabolism and breast cancer risk in the NCI breast and prostate cancer cohort consortium.* Hum Mol Genet. 2010;19:3873-84.

21. *Commonly studied single-nucleotide polymorphisms and breast cancer: results from the Breast Cancer Association Consortium.* J Natl Cancer Inst. 2006;98:1382-96.

22. Neuhausen SL, Brummel S, Ding YC, Steele L, Nathanson

KL, Domchek S, et al. *Genetic variation in IGF2 and HTRA1 and breast cancer risk among BRCA1 and BRCA2 carriers.* Cancer Epidemiol Biomarkers Prev. 2011;20:1690-702.

23. Clayton PE, Banerjee I, Murray PG, Renehan AG. *Growth hormone, the insulin-like growth factor axis, insulin and cancer risk.* Nature Rev Endocrinol. 2011;7:11-24.

24. Novosyadlyy R, Lann DE, Vijayakumar A, Rowzee A, Lazzarino DA, Fierz Y, et al. *Insulin-mediated acceleration of breast cancer development and progression in a nonobese model of type 2 diabetes.* Cancer Res. 2010;70:741-51.

25. Chen F, Wilkens LR, Monroe KR, Stram DO, Kolonel LN, Henderson BE, et al. *No association of risk variants for diabetes and obesity with breast cancer: the Multiethnic Cohort and PAGE studies.* Cancer Epidemiol Biomarkers Prev. 2011;20:1039-42.

26. Hou N, Zheng Y, Gamazon ER, Ogundiran TO, Adebamowo C, Nathanson KL, et al. *Genetic susceptibility to type 2 diabetes and breast cancer risk in women of European and African ancestry.* Cancer Epidemiol Biomarkers Prev. 2012;21:552-6.

6.4. CELL CYCLE

6.4.1. INTRODUCTION

Each of the 100 trillion (10^{14}) cells in the human body contains a complete set of 46 chromosomes that is required for normal viability and development. Each cell contains the complete genome, which is 3.3 billion (10^9) bases long. Thus, there are 46 x 10^{14} chromosomes and 3.3 x 10^{23} bases in the human body. Within each cell, more than 1 m of stretched DNA fiber is packaged into a 10 μm nucleus. The correct distribution of the full complement of chromosomes and bases to each cell occurs during the cell cycle, which depends on faithful replication of DNA during S phase and equal segregation of sister chromatids into two daughter cells during mitosis (M phase). Interspersed are two gap phases, G1 and G2. The G1 phase allows responses to extracellular cues that induce either commitment to another round of cell division or withdrawal from the cell cycle (G0 resting phase or quiescence) to embark on a cell-specific differentiation pathway (1). The G2 phase is interspersed between the S and M phases to ensure the completion of DNA replication and genomic integrity before starting mitosis. Progression through the cell cycle is driven by

cyclin-dependent kinases (Cdks), whose catalytic activity and substrate specificity depend on their association with regulatory cyclins (Table 6.4.1.). Cdks and cyclins interact in a tightly controlled sequence of protein phosphorylation and degradation reactions (2-5). The proteolytic degradation reactions are mediated by the ubiquitin ligase activity of the anaphase-promoting complex, also called cyclosome (APC/C), which contains cell division cycle (Cdc) proteins Cdc16, 23, and 27 as well as a cullin and RING subunit (6, 7). Other Cdc proteins have phosphatase activity, i.e., Cdc14 and Cdc25 activate target Cdks by dephosphorylation, thereby establishing a network of kinases and phosphatases. The intrinsic cell cycle machinery is controlled by external signals such as hormones, growth factors, and anti-mitogens that integrate cell division with environmental and developmental stimuli.

Cdk-Cyclins. While yeast employ a single Cdk, higher eukaryotes use distinct Cdks to coordinate cell cycle progression through the G1/S and G2/M transitions (5, 8). Cdk2 interacts with cyclin E at the beginning of S phase to induce the initiation of DNA synthesis, and then binds cyclin A for the progression through S phase. Cdk1 interacts with cyclin B to initiate mitosis. Cdk4 and Cdk6 interact with members of the cyclin D family (D1 - D3) for progression through G1. The D-type cyclins and cyclin E are collectively referred to as "G1 cyclins" because both are expressed during the G1 phase of the cell cycle, with cyclin E levels peaking in late G1 (9, 10). The mitogen-dependent cyclin D-Cdk4/6 complexes are responsible for cell cycle progression through G1 while the mitogen-independent cyclin E and its partner Cdk2 are responsible for entry into S phase (11). Independent of its Cdk-associated cell cycle activity, cyclin D1 participates in DNA repair by directly binding RAD51, a recombinase that cooperates with BRCA2 on repairing double-stranded DNA breaks by homologous recombination (12).

Cdc25 Phosphatases regulate key transitions between cell cycle phases during normal cell division and, in the event of DNA damage, they are key targets of the checkpoint machinery that ensures genetic stability (13). The Cdc25 family consists of three genes, Cdc25A at 3p21, Cdc25B at 20p13, and Cdc25C at 5q31 (14, 15). The three Cdc25 homologs share structural and functional features including a conserved C-terminal catalytic domain (16). In spite of the similarities, they differ in their physiological substrates. Cdc25A is expressed during the G1/S transition and controls entry into S phase by activating Cdk2-cyclin E and Cdk2-cyclin A (17, 18). Cdc25B and CdcC are primarily required for entry into mitosis (19, 20). During G2/M transition, Cdc25B initiates the activation of Cdk1-cyclin B complexes at the centrosome, followed by complete activation of the complexes in the nucleus by Cdc25C at the onset of mitosis (13). In order to ensure the timely progression through G1/S and G2/M, Cdc25 phosphatases are themselves regulated by multiple mechanisms,

Table 6.4.1. Major Cell Cycle and Checkpoint Control Proteins

Proteins	Complex	Function
Cdk1	cyclin B	M phase
Cdk2	cyclin E	G1/S
	cyclin A	S and G2 phases
Cdk4	cyclin D	G1 phase
Cdk6	cyclin D	G1 phase
Cdc14 phosphatase		exit from mitosis
APC/C	Cdc16, Cdc 23, Cdc27, cullin, RING	anaphase initiation, exit from mitosis
Cdc25A - C phosphatases		entry into S phase, G2/M transition
p16^{ink4a}, p15^{ink4b}, p18^{ink4c}, p19^{ink4d}		inhibit Cdk4 and Cdk6
p21$^{CIP1/WAF1}$, p27^{KIP1}, p57^{KIP2}		universal Cdk inhibitor
ATM	Chk2	G1/S checkpoint control
ATR	Chk1	S and G2/M checkpoint control
Polo-like & Aurora kinases		spindle assembly checkpoint

including inhibitory and activating phosphorylations. For example, Cdc25B is phosphorylated and activated by Aurora A at the centrosome and Cdc25C by polo-like kinase 1 for nuclear translocation (21, 22). Cdc25 overexpression, particularly Cdc25A and Cdc25B, has been reported in various cancers, including breast cancer.

Cdk inhibitors. While cyclins and Cdks serve as positive regulators, there are two families of Cdk inhibitors that act as negative cell cycle regulators by binding to and inactivating Cdk-cyclin complexes (23-25). The Ink4 (Inhibitors of Cdk4) family includes the p16^{INK4A}, p15^{INK4B}, p18^{INK4C}, and p19^{INK4D} proteins, which contain a related series of ankyrin repeats and specifically bind Cdk4 and Cdk6, thereby blocking the assembly of catalytically active Cdk4/6-cyclin D complexes (26, 27). The second family includes p21$^{CIP1/WAF1}$, p27^{KIP1}, and p57^{KIP2}, which share a common Cdk-binding domain (28). In contrast to the Ink4 family, the Cip/Kip family members are promiscuous and interact with all Cdks. The CDKN1A gene at 6p21.2 encodes p21$^{CIP/WAF1}$, which has a Cdk-binding domain in the mid-molecule region and two cyclin-binding domains, one near the N-terminus and the other in the C-terminal region. The latter overlaps with yet another binding domain for PCNA (28-32). These binding domains accommodate interactions with cyclins, Cdks, and PCNA. Moreover, p21 can directly bind and inhibit the transcription factors E2F1, STAT3, and Myc. Multiple transcription factors, ubiquitin ligases, and protein kinases regulate the transcription, stability, and cellular localization of p21 (31, 33). In particular, the tumor suppressor protein p53 controls the expression of p21 at the transcriptional and post-transcriptional levels (34, 35).

Cell Cycle Phases and Checkpoint Controls. The onset of each phase of the cell cycle is dependent on successful completion of previous cell cycle events and

these dependencies are maintained by checkpoints, whose activation leads to cell cycle arrest (36). Several checkpoints have been identified that monitor completion of DNA replication, repair of damaged DNA, correct spindle assembly, and attainment of adequate cell size. Damage to DNA can delay the start of a new cell cycle (G1/S checkpoint), DNA synthesis (S phase progression checkpoint), or mitosis (G2/M checkpoint), while the presence of misaligned chromosomes or spindle damage can delay anaphase (spindle assembly or mitotic checkpoint) (37-39). There may be additional checkpoints, e.g., during S phase to prevent conflicts between DNA replication and transcription, especially rapidly induced transcription in response to extracellular stimuli (40). Activation of any of these checkpoint pathways arrests the cell cycle to provide time to repair and induce the transcription of genes that facilitate repair.

The cell cycle checkpoints are inactive in normal tissue but they become activated in precancerous lesions, leading to cell cycle blockade or apoptosis and thereby containing tumor progression (41, 42). Failure to activate checkpoint pathways can result in the propagation of mutations or damaged chromosomes and therefore may contribute to genetic instability and cancer (36, 43-47).

The ataxia-teleangiectasia mutated (ATM) and ATM and Rad3-related (ATR) protein kinases act as master regulators of the DNA-damage response by signalling to control cell cycle transitions and DNA repair (48). ATM and ATR share structural and functional similarities but differ in important aspects. ATM responds to double-stranded DNA breaks by recruiting the trimeric MRN (MRE11, RAD50, NBS1) protein complex leading to the phosphorylation of checkpoint kinase 2 (Chk2) and other targets, such as the variant histone H2AX, yielding γH2AX, which interacts with mediator of DNA damage checkpoint 1 (MDC1) (49,

50). In contrast to ATM, ATR responds to single-stranded DNA breaks and therefore becomes activated during every S phase to regulate the firing of replication origins from single-stranded DNA. Replication protein A (RPA)-ssDNA recruit ATR, ATR-interacting protein (ATRIP) and the trimeric complex RAD9-RAD1-HUS1 (also known as 9-1-1) leading to the phosphorylation of checkpoint kinase 1 (Chk1) and other downstream substrates (49). ATR is involved not only in regular DNA replication but also reponds to stalled replication forks and damaged DNA (48). Thus, the checkpoint kinases Chk2 and Chk1 are downstream effectors of ATM and ATR, respectively (51-53). A third checkpoint effector is the mitogen activated protein kinase p38MAPK/MK2 (46, 52).

Germ line mutations in the ATM gene result in defects in DNA repair, loss of cell cycle checkpoints, increased sensitivity to ionizing radiation and predisposition to malignancy, including breast cancer (see Chapter 8.5. Ataxia teleangiectasia). Germ line mutations in the ATR gene are rare and not associated with cancer.

G1 phase and G1/S checkpoint. The activity of Cdk4/6-cyclin D during G1 is regulated by forming a ternary complex with the retinoblastoma tumor suppressor protein (pRb), which, in turn, binds the transcription factor E2F, resulting in repression of E2F-mediated gene transcription. The Rb gene product is a nuclear phosphoprotein that undergoes cell cycle-dependent changes in its phosphorylation status (54). In G0 and early G1, pRb is unphosphorylated, allowing it to bind to cyclin D via the sequence Leu-X-Cys-X-Glu, which is present near the N-terminus of each of the D-type cyclins (55, 56). pRb is not only bound to cyclin D-Cdk4/6, but also becomes the target of phosphorylation by Cdk4/6 (57, 58). The phosphorylation of pRb destabilizes the entire pRb-cyclin D-Cdk4/6 complex and leads to the release of pRb during middle and late G1 (59). In turn, the release of pRb removes the repression from E2F and allows the expression of genes necessary for entry into S phase. E2F is a heterodimer composed of a subunit encoded by the E2F gene family and a subunit encoded by the DP family of genes. Six E2F genes (E2F1 - E2F6) and two DP genes (DP1, DP2) have been identified in mammalian cells and all E2F and DP proteins contain highly conserved dimerization- and DNA-binding domains (60). In addition, E2F1 – 5, but not E2F6, contain a binding site for pRb family proteins (61-63). In its hypophosphorylated form, pRb binds and inhibits E2F transcription factors (64, 65). The E2F group of transcription factors are all targeted to variants of the consensus nucleotide sequence TTTCGCGC, which is present in the promoter regions of several genes involved in cell cycle regulation, nucleotide synthesis, and DNA replication, e.g., cyclin E, cyclin A, Cdk1, Cdc25A, dihydrofolate reductase, thymidine kinase, and DNA polymerase α (62, 66, 67). By using already bound E2Fs as docking sites, pRb can repress E2F-mediated transcription

and silence specific genes that are active in the S phase of the cell cycle. E2F is released from the pRb-E2F complex as a result of pRb phosphorylation. Free E2F then becomes an active transcription factor that promotes the transcription of genes required for DNA replication and drives cells from G1 into S phase. Thus, pRb main role is as a signal transducer during G1, connecting the cell cycle clock with the transcriptional machinery, whereas E2F may be considered the ultimate transcription factor activated by the cyclin D-Cdk4/6-pRb pathway (60, 68).

pRb is phosphorylated first by cyclin D-cdk4/6 and then by cyclin E-cdk2 (69). The sequential phosphorylation of pRb, in turn, affects its interaction with other proteins, such as the histone deacetylase HDAC and BRG1, a component of the SWI/SNF nucleosome remodeling complex. HDAC and BRG1 can bind to different sites on pRb, resulting in formation of a ternary complex, HDAC-pRb-SWI/SNF, which inhibits transcription of cyclin E and arrests cells in G1 (70). Phosphorylation of pRb by cyclin D-cdk4/6 disrupts the HDAC-pRb interaction (71). Removal of HDAC from pRb allows expression of cyclin E and progression into S phase. However, the pRb-SWI/SNF complex persists and is sufficient to maintain repression of the cyclin A and cdc2 genes, thereby inhibiting exit from S phase. Phosphorylation of pRb by cyclin E-cdk2 finally disrupts the pRb-SWI/SNF complex, allowing cyclin A and cdc2 expression and exit from S phase. Thus, the sequential phosphorylation by cyclin D-cdk4/6 and cyclin E-cdk2 of HDAC-pRb-SWI/SNF and pRb-SWI/SNF, respectively, maintains the order of cyclin E and A expression during the cell cycle, which in turn regulates exit from G1 and S phase, respectively (70).

Diverse signals (e.g., serum starvation, TGFβ treatment, bovine papillomavirus) can cause G1/S arrest via activation of the ATM-Chk2 pathway. Activated Chk2 phosphorylates Cdc25A, which triggers ubiquitylation and proteasome-dependent degradation of Cdc25A in the cytoplasm away from Cdk2-cyclin E (67, 72-74). Another downstream target of Chk2 is p53, which induces CDKN1A gene transcription, resulting in increased expression of p21[CIP1/WAF1]. p21 inhibits the activity of G1 phase Cdks and thereby mediates G1/S growth arrest (75). p21 may also control DNA replication during S phase by blocking the ability of PCNA to activate DNA polymerase δ, the principal replicative DNA polymerase (76). Thus, p21 can inhibit cell proliferation by two separate mechanisms involving either its Cdk- or PCNA-binding domains. By virtue of its capacity to inhibit cell cycle progression at major transitions points, p21 becomes, in principle, a universal inhibitor of cell proliferation. However, the effect of p21 on cell cycle regulation is complex because p21 can activate rather than inhibit Cdk4/6-cyclin D, thereby promoting progression through G1 and cell proliferation (33). p53 controls the expression of p21 at the transcriptional and post-transcriptional levels (34, 35). Through increased p53-induced CDKN1A gene transcription, p21 becomes the key effector of p53-dependent G1/S arrest in response to a variety of growth inhibitory stimuli, including DNA damage (75, 77-80). Alternatively, p53 can down-regulate the expression of p21 by inducing a specific microRNA gene, miR-22, which causes increased degradation of p21 mRNA and reduction of p21 protein levels. As a consequence, p21 will not be available for G1/S cell cycle arrest and cells can accumulate more DNA damage resulting in the activation of p53-mediated apoptosis. However, p21 is also stimulated by p53-independent pathways allowing the protein to participate not only in growth arrest but also cell senescence and differentiation (33). To add to the complexity, p21 can have oncogenic properties and is required for HER2-mediated proliferation of breast cancer cells and mammary carcinogenesis (81-84). Finally, p53-deficient tumor cells do not block G1/S but instead rely on Chk1 to arrest cell cycle progression in the S and G2 phases (85).

S phase and intra-S phase progression checkpoint. The chromosomal DNA is replicated from 10^3 to 10^5 replication origins, which are typically spaced 30 – 100 kb apart (86). Factors other than DNA sequence, such as chromatin structure, play a role in determining the position of the replication origins and the replication timing (87). It is important that origins are replicated once and only once in each cell cycle to ensure genomic stability. The origins become 'licensed' for one round of replication during S phase by the assembly onto chromatin of prereplica-tive complexes containing the <u>o</u>rigin <u>r</u>ecognition <u>c</u>omplex (ORC), Cdc6, Cdt1, and <u>mini</u>chromosome <u>m</u>aintenance (Mcm) 2-7 helicases (86). The binding of Mcm2-7 helicases to replication origins is essential for the origins to fire, i.e., initiate a pair of replication forks for DNA synthesis. The licensing is withdrawn by geminin, which acts by competitively binding to Ctd1, thereby blocking Mcm2-7 re-loading onto chromatin.

Even in normal cells, replication forks encounter DNA lesions that must be repaired prior to replication, leading to fork stalling (88). When replication forks encounter DNA damage, the Mcm2-7 helicases continue to unwind the DNA template, resulting in exposure of streches of single-stranded DNA, which are rapidly coated by ssDNA-binding replication protein A (RPA). RPA-ssDNA recruit ATR, ATR-interacting protein (ATRIP) and the trimeric complex RAD9-RAD1-HUS1 (also known as 9-1-1) leading to the phos-phorylation of Chk1 and Mcm2, which creates a docking site for polo-like kinase 1 (Plk1) (48, 89). The ATR-Mcm2-Plk1 pathway acts locally to control replication at the stalled fork, whereas the ATR-Chk1 pathway operates throughout the nucleus to slow overall rates of DNA synthesis.

G2 phase and G2/M checkpoint. DNA damaging agents (e.g., UV light, gamma radiation, alkylating agents) can cause G2/M arrest by setting in motion the sequential interaction of several proteins: ATR, Chk1 kinase, Cdc25C, 14-3-3 (13, 90-95). The DNA damage is detected by ATR, which phosphorylates Chk1. In turn, Chk1 phosphorylates Cdc25C on 216Ser, which potentiates 14-3-3 binding to Cdc25C leading to its sequestration in the cytoplasm away from Cdk1-cyclin B. Thus, 14-3-3 binding prevents Cdc25C from dephosphorylating Cdk1 thereby blocking activation of the Cdk1-cyclin B complex and mitotic entry.

Mitosis and spindle assembly or mitotic checkpoint. The main purpose of mitosis is to segregate sister chromatids into two nascent cells, such that each daughter cell inherits one complete set of chromosomes (3). This requires the temporal and spatial coordination of chromosome segregation with mitotic spindle disassembly and cytokinesis. In yeast, the Cdc14 phosphatase (there are two human homologs, Cdc14A and Cdc14B) is a key regulator of the late mitotic events (96). Cdc14 is sequestered in an inactive state in the nucleolus during all phases of the cell cycle except anaphase (97, 98). At the metaphase-to-anaphase transition, Cdc20 associates with APC/C to induce APC/C-mediated proteolysis of the anaphase inhibitor Pds1, the chromatid cohesion protein Scc1, and the Cfi1 protein, which anchors Cdc14 in the nucleolus. The liberated Cdc14 dephosphorylates Sic1, Swi5, and Cdh1, which, in turn, allows these proteins to inactivate Cdk1 and degrade the spindle components Ase1 and polo-like kinase 1, resulting in exit from mitosis and reentry in G1 phase (96, 99). The APC/C also degrades M cyclins and securin. When securin undergoes ubiquination by APC/C, it releases a protease, separase, which in turn triggers the release of cohesin, the protein complex that binds sister chromatids together during metaphase, allowing them to move to opposite poles for anaphase. The APC/C also targets the mitotic cyclins A and B for degradation, resulting in the inactivation of Cdk1-M cyclin complexes, which promotes exit from mitosis and cytokinesis. Thus, proteolysis mediated by APC/C triggers the transition from metaphase to anaphase with separation of sister chromatids, cytokinesis, and re-duplication of chromosomes in the subsequent cell cycle (100). Together, APC/C and Cdc14 may be considered the ultimate effectors of the mitotic exit pathway. Cells that lack Cdc14 are unable to exit from mitosis with impaired movement of chromosomes to the spindle poles. In human cells, Cdc14B is also involved in centrosome duplication (101).

The spindle assembly checkpoint arrests the cell cycle in mitosis when even a single chromosome is not properly attached to the mitotic spindle (38). The signal generated by an unattached kinetochore recruits a group of proteins, including Mad1, Mad2, Bub1, Bub3, BubR1/Bub1B, and Mps1, which inhibit the Cdc20-dependent recognition of cyclin B and securin by APC/C, thereby preventing advance to anaphase. Two families of mitotic kinases, namely the polo-like kinases (Plks) and Aurora kinases, regulate the formation of a bipolar spindle, accurate segregation of chromosomes, and the completion of cytokinesis (102). Abnormal expression of these mitotic checkpoint genes leads to chromosomal instability and aneuploidy (44). For details see Chapter 6.5.2.

References

1. Caldon CE, Sutherland RL, Musgrove E. *Cell cycle proteins in epithelial cell differentiation: implications for breast cancer.* Cell Cycle. 2010;9:1918-28.

2. King RW, Jackson PK, Kirschner MW. *Mitosis in transition.* Cell. 1994;79:563-71.

3 Nigg EA. *Mitotic kinases as regulators of cell division and its checkpoints.* Nat Rev Mol Cell Biol. 2001;2:21-32.

4. Sherr CJ. *Mammalian G1 Cyclins.* Cell. 1993;73:1059-65.

5. Sherr CJ. *G1 phase progression: cycling on cue.* Cell. 1994;79:551-5.

6. Barford D. *Structural insights into anaphase-promoting complex function and mechanism.* Philos Trans R Soc Lond B Biol Sci. 2011;366:3605-24.

7. Koepp DM, Harper JW, Elledge SJ. *How the cyclin became a cyclin: regulated proteolysis in the cell cycle.* Cell. 1999;97:431-4.

8. Morgan DO. *Cyclin-dependent kinases: engines, clocks, and micro-processors.* Annu Rev Cell Dev Biol. 1997;13:261-91.

9. Dulic V, Lees E, Reed SI. *Association of human cyclin E with a periodic G1-S phase protein kinase.* Science. 1992;257:1958-61.

10. Koff A, Giordano A, D. D, Yamashita K, Harper JW, Elledge S, et al. *Formation and activation of cyclin-E-cdk2 complex during the G1 phase of the human cell cycle.* Science. 1992;257:1689-94.**11.** Resnitzky D, Reed SI. *Different roles for cyclins D1 and E in regulation of the G1-to-S transition.* Mol Cell Biol. 1995;15:3463-9.

12. Jirawatnotai S, Hu Y, Michowski W, Elias JE, Becks L, Bienvenu F, et al. *A function for cyclin D1 in DNA repair uncovered by protein interactome analyses in human cancers.* Nature. 2011;474:230-4.

13. Boutros R, Lobjois V, Ducommun B. *CDC25 phosphatases in cancer cells: key players? Good targets? Nat Rev Cancer. 2007;7:495-507.

14. Demetrick DJ, Beach DH. *Chromosome mapping of human CDC25A and CDC25B phosphatases.* Genomics. 1993;18:144-7.

15. Draetta G, Eckstein J. *Cdc25 protein phosphatases in cell proliferation.* Biochim Biophys Acta. 1997;1332:M53-M63.

16. Fauman EB, Cogswell JP, Lovejoy B, Rocque WJ, Holmes W, Montana VG, et al. *Crystal structure of the catalytic domain of the human cell cycle control phosphatase, Cdc25A.* Cell. 1998;93:617-25.

17. Hoffmann I, Draetta G, Karsenti E. *Activation of the phosphatase activity of human cdc25A by a cdk2-cyclin E dependent phosphorylation at the G1/S transition.* EMBO J. 1994;13:4302-10.

18. Jinno S, Suto K, Nagata A, Igarashi M, Kanaoka Y, Nojima H, et al. *Cdc25A is a novel phosphatase functioning early in the cell cycle.* EMBO J. 1994;13:1549-56.

19. Lammer C, Wagerer S, Saffrich R, Mertens D, Ansorge W, Hoffmann I. *The cdc25B phosphatase is essential for the G2/M phase transition in human cells.* J Cell Sci. 1998;111:2445-53.

20. Millar JB, Blevit J, Gerace L, Sadhu K, Featherstone C, Russell P. *p55CDC25 is a nuclear protein required for the initiation of mitosis in human cells.* Proc Natl Acad Sci USA. 1991;88:10500-4.

21. Barr FA, Silljé HH, Nigg EA. *Polo-like kinases and the orchestration of cell division.* Nat Rev Mol Cell Biol. 2004;5:429-40.

22. Dutertre S, Cazales M, Quaranta M, Froment C, Trabut V, Dozier C, et al. *Phosphoryla-tion of CDC25B by Aurora-A at the centrosome contributes to the G2-M transition.* J Cell Sci. 2004;117:2523-31.

23. Hunter T, Pines J. *Cyclins and cancer II: cyclin D and CDK inhibitors come of age.* Cell. 1994;79:573-82.

24. Peter M, Herskowitz I. *Joining the complex: cyclin-dependent kinase inhibitory proteins and the cell cycle.* Cell. 1994;79:181-4.

25. Sherr CJ, Roberts JM. *Inhibitors of mammalian G1 cyclin-dependent kinases.* Genes Dev. 1995;9:1149-63.

26. Roussel MF. *The INK4 family of cell cycle inhibitors in cancer.* Oncogene. 1999;18:5311-7.

27. Serrano M, Hannon GJ, Beach D. *A new regulatory motif in cell-cycle control causing specific inhibition of cyclin D/CDK4.* Nature. 1993;366:704-7.

28. Luo Y, Hurwitz J, Massague J. *Cell-cycle inhibition by independent CDK and PCNA binding domains in p21Cip1.* Nature. 1995;375:159-61.

29. Cai K, Dynlacht BD. *Activity and nature of p21WAF1 complexes during the cell cycle.* Proc Natl Acad Sci USA. 1998;95:12254-9.

30. Chen J, Jackson PK, Kirschner MW, Dutta A. *Separate domains of p21 involved in the inhibition of cdk kinase and PCNA.* Nature. 1995;374:386-8.

31. Child ES, Mann DJ. *The intricacies of p21 phosphorylation: protein/protein interactions, subcellular localization and stability.* Cell Cycle. 2006;5:1313-9.

32. Warbrick E, D.P. L, Glover DM, Cox LS. *A small peptide inhibitor of DNA replication defines the site of interaction between the cyclin-depen-dent kinase inhibitor p21WAF1 and proliferating cell nuclear antigen.* Curr Biol. 1995;5:275-82.

33. Abbas T, Dutta A. *p21 in cancer: intricate networks and multiple activities.* Nat Rev Cancer. 2009;9:400-14.

34. El-Deiry WS, Tokino T, Velculescu VE, Levy DB, Parsons R, Trent JM, et al. *WAF1, a potential mediator of p53 tumor suppression.* Cell. 1993;75:817-25.

35. Tsuchiya N, Izumiya M, Ogata-Kawata H, Okamoto K, Fujiwara Y, Nakai M, et al. *Tumor suppressor miR-22 determines p53-dependent cellular fate through post-transcriptional regulation of p21.* Cancer Res. 2011;71:4628-39.

36. Hartwell LH, Kastan MB. *Cell Cycle Control and Cancer.* Science. 1994;266:1821-8.

37. Elledge SJ. *Cell cycle checkpoints:preventing an identity crisis.* Science. 1996;274:1664-72.

38. Musacchio A, Salmon ED. *The spindle-assembly checkpoint in space and time.* Nat Rev Mol Cell Biol. 2007;8:379-93.

39. Stillman B. *Cell cycle control of DNA replication.* Science. 1996;274:1659-64.

40. Duch A, Felipe-Abrio I, Barroso S, Yaakov G, Garcia-Rubio M, Aguilera A, et al. *Coordinated control of replication and transcription by a SAPK protects genomic integrity.* Nature. 2013;493:116-9.

41. Bartkova J, Horejsi Z, Koed K, Kramer A, Tort F, Zieger K, et al. *DNA damage response as a candidate anti-cancer barrier in early human tumorigenesis.* Nature. 2005;434:864-70.

42. Gorgoulis VG, Vassiliou LV, Karakaidos P, Zacharatos P, Kotsinas A, Liloglou T, et al. *Activation of the DNA damage checkpoint and genomic instability in human precancerous lesions.* Nature. 2005;434:907-13.

43. Kastan MB, Bartek J. *Cell-cycle checkpoints and cancer.* Nature. 2004;432:316-23.

44. Kops GJ, Weaver BA, Cleveland DW. *On the road to cancer: aneuploidy and the mitotic checkpoint.* Nat Rev Cancer. 2005;5:773-85.

45. Paulovich AG, Toczyski DP, Hartwell LH. *When checkpoints fail.* Cell. 1997;88:315-21.

46. Poehlmann A, Roessner A. *Importance of DNA damage checkpoints in the pathogenesis of human cancers.* Pathol Res Pract. 2010;206:591-601.

47. Sherr CJ. *Cancer cell cycles.* Science. 1996;274:1672-7.

48. Cimprich KA, Cortez D. *ATR: an essential regulator of genome integrity.* Nat Rev Mol Cell Biol. 2008;9:616-27.

49. Matsuoka S, Ballif BA, Smogorzewska A, McDonald ER, 3rd, Hurov KE, Luo J, et al. *ATM and ATR substrate analysis reveals extensive protein networks responsive to DNA damage.* Science. 2007;316:1160-6.

50. Yu B, Dalton WB, Yang VW. *CDK1 regulates mediator of DNA damage checkpoint 1 during mitotic DNA damage.* Cancer Res. 2012;72:5448-53.

51. Bartek J, Falck J, Lukas J. *CHK2 kinase--a busy messenger.* Nat Rev Mol Cell Biol. 2001;2:877-86.

52. Reinhardt HC, Yaffe MB. *Kinases that control the cell cycle in response to DNA damage: Chk1, Chk2, and MK2.* Curr Opin Cell Biol. 2009;21:245-55.

53. Stracker TH, Usui T, Petrini JH. *Taking the time to make important decisions: the checkpoint effector kinases Chk1 and Chk2 and the DNA damage response.* DNA Repair (Amst). 2009;8:1047-54.

54. Goodrich DW, Wang NP,

Qian YW, Lee EHP, Lee WH. *The retinoblastoma gene product regulates progression through the G1 phase of the cell cycle.* Cell. 1991;67:293-302.

55. Dowdy SF, Hinds PW, Louie K, Reed SI, Arnold A, Weinberg RA. *Physical interaction of the retinoblastoma protein with human D cyclins.* Cell. 1993;73:499-511.

56. Ewen ME, Sluss HK, Sherr CJ, Matsushime H, Kato J, Livingston DM, et al. *Functional interactions of the retinoblastoma protein with mammalian D-type cyclins.* Cell. 1993;73:487-97.

57. Lees JA, Buchkovich K, Marshak DR, Anderson CW, Harlow E. *The retinoblastoma protein is phosphorylated on multiple sites by human cdc2.* EMBO J. 1991;10:4279-90.

58. Lin BT, Gruenwald S, Morla AO, Lee WH, Wang JY. *Retinoblastoma cancer suppressor gene product is a substrate of the cell cycle regulator cdc2 kinase.* EMBO J. 1991;10:857-64.

59. Kato J, Matsushime H, Hiebert SW, Ewen ME, Sherr CJ. *Direct binding of cyclin D to the retinoblastoma gene product (pRb) and pRb phosphorylation by the cyclin D-dependent kinase CDK4.* Genes Dev. 1993;7:331-42.

60. Dyson N. T*he regulation of E2F by pRB-family proteins.* Genes Dev. 1998;12:2245-62.

61. Cartwright P, Muller H, Wagener C, Holm K, Helin K. *E2F-6: a novel member of the E2F family is an inhibitor of E2F-dependent transcription.* Oncogene. 1998;17:611-23.

62. Helin K. *Regulation of cell proliferation by the E2F transcription factors.* Curr Opin Genet Dev. 1998;8:28-35.

63. Trimarchi JM, Fairchild B, Verona R, Moberg K, Andon N, Lees JA. *E2F-6, a member of the E2F family that can behave as a transcriptional repressor.* Proc Natl Acad Sci USA. 1998;95:2850-5.

64. Chellappan SP, Hiebert SP, Mudryj M, Horowitz JM, Nevins JR. *The E2F transcritpion factor is a cellular target for the RB protein.* Cell. 1991;65:1053-61.

65. Chittenden T, Livingston D, Kaelin W. *The T/E1A-binding domain of the retinoblastoma product can interact selectively with a sequence-specific DNA-binding protein.* Cell. 1991;65:1073-82.

66. Nevins JR. *E2F: a link betwen the Rb tumor suppressor protein and viral oncoproteins.* Science. 1992;258:424-9.

67. Vigo E, Muller H, Prosperini E, Hateboer G, Cartwright P, Moroni MC, et al. *CDC25A phosphatase is a target of E2F and is required for efficient E2F-induced S phase.* Mol Cell Biol. 1999;19:6379-95.

68. Nevins JR. *Toward an understanding of the functional complexity of the E2F and retinoblastoma families.* Cell Growth Diff. 1998;9:585-93.

69. Lundberg AS, Weinberg RA. *Functional inactivation of the retinoblastoma protein requires sequential modification of at least two distinct cyclin-cdk complexes.* Mol Cell Biol. 1998;18:753-61.

70. Zhang HS, Gavin M, Dahiya A, Postigo AA, Ma D, Luo RX, et al. *Exit from G1 and S phase of the cell cycle is regulated by repressor complexes containing HDAC-Rb-hSWI/SNF and Rb-hSWI/SNF.* Cell. 2000;101:79-89.

71. Harbour JW, Luo RX, Dei Santi A, Postigo AA, Dean DC. *Cdk phosphorylation triggers sequential intramolecular interactions that progressiviely block Rb functions as cells move through G1.* Cell. 1999;98:859-69.

72. Chen X, Prywes R. *Serum-induced expression of the cdc25A gene by relief of E2F-mediated repression.* Mol Cell Biol. 1999;19:4695-702.

73. Iavarone A, Massague J. *E2F and histone decetylase mediate transforming growth factor beta repression of cdc25A during keratinocyte cell cycle arrest.* Mol Cell Biol. 1999;19:916-22.

74. Wu L, Goodwin EC, Naeger LK, Vigo E, Galaktionov K, Helin K, et al. *E2F-Rb complexes assemble and Inhibit cdc25A transcription in cervical carcinoma cells following repression of human papillomavirus oncogene expression.* Mol Cell Biol. 2000;20:7059-67.

75. Dulic V, Kaufmann WK, Wilson SJ, Tisty TD, Lees E, Harper JW, et al. *p53-dependent inhibition of cyclin-dependent kinase activities in human fibroblasts during radiation-induced G1 arrest.* Cell. 1994;76:1013-23.

76. Waga S, Hannon GJ, Beach D, Stillman B. *The p21 inhibitor of cyclin-dependent kinases controls DNA replication by interaction with PCNA.* Nature. 1994;369:574-8.

77. Brugarolas J, Chandrasekaran C, Gordon JI, Beach D, T. J, Hannon GJ. *Radiation-induced cell cycle arrest compromised by p21 deficiency.* Nature. 1995;377:552-7.

78. Deng C, Zhang P, Harper JW, Elledge SJ, Leder P. *Mice lacking p21CIP1/WAF1 undergo normal development, but are defective in G1 checkpoint control.* Cell. 1995;82:675-84.

79. Terada Y, Tatsuka M, Jinno S, Okayama H. *Requirement for tyrosine phosphorylation of Cdk4 in G1 arrest induced by ultraviolet irradiation.* Nature. 1995;376:358-62.

80. Xiong Y, Zhang H, Beach D. *D-type cyclins associate with multiple protein kinases and the DNA replication and repair factor PCNA.* Cell. 1992;71:505-14.

81. Roninson IB. *Oncogenic functions of tumour suppressor p21(Waf1/Cip1/Sdi1): association with cell senescence and tumour-promoting activities of stromal fibroblasts.* Cancer Lett. 2002;179:1-14.

82. Winters ZE, Leek RD, Bradburn MJ, Norbury CJ, Harris AL. *Cytoplasmic p21WAF1/CIP1 expression is correlated with HER-2/neu in breast cancer and is an independent predictor of prognosis.* Breast Cancer Res. 2003;5:R242-9.

83. Xia W, Chen JS, Zhou X, Sun PR, Lee DF, Liao Y, et al. *Phosphorylation/cytoplasmic localization of p21Cip1/WAF1 is associated with HER2/neu overexpression and provides a novel combination predictor for poor prognosis in breast cancer patients.* Clin Cancer Res. 2004;10:3815-24.

84. Zhou BP, Liao Y, Xia W, Spohn B, Lee MH, Hung MC. *Cytoplasmic localization of p21Cip1/WAF1 by Akt-induced phosphorylation in HER-2/neu-overexpressing cells.* Nat Cell Biol. 2001;3:245-52.

85. Ma CX, Cai S, Li S, Ryan CE, Guo Z, Schaiff WT, et al. *Targeting Chk1 in p53-deficient triple-negative breast cancer is therapeutically beneficial in human-in-mouse tumor models.* J Clin Invest. 2012;122:1541-52.

86. Blow JJ, Dutta A. *Preventing re-replication of chromosomal DNA.* Nat Rev Mol Cell Biol. 2005;6:476-86.

87. Hansen RS, Thomas S, Sandstrom R, Canfield TK, Thurman RE, Weaver M, et al. *Sequencing newly replicated DNA reveals widespread plasticity in human replication timing.* Proc Natl Acad Sci U S A. 2010;107:139-44.

88. Cox MM, Goodman MF, Kreuzer KN, Sherratt DJ, Sandler SJ, Marians KJ. *The importance of repairing stalled replication forks.* Nature. 2000;404:37-41.

89. Wang J, Han X, Zhang Y. *Auto-regulatory mechanisms of phosphorylation of checkpoint kinase 1.* Cancer Res. 2012;72:3786-94.

90. Furnari B, Rhind N, Russell P. *Cdc25 mitotic inducer targeted by Chk1 DNA damage checkpoint kinase.* Science. 1997;277:1495-7.

91. Hermeking H, Benzinger A. *14-3-3 proteins in cell cycle regulation.* Semin Cancer Biol. 2006;16:183-92.

92. Nurse P. *Checkpoint pathways come of age.* Cell. 1997;91:865-7.

93. Peng C, Graves PR, Thoma RS, Wu Z, Shaw AS, Piwnica-Worms H. *Mitotic and G2 checkpoint control: regulation of 14-3-3 protein binding by phosphorylation of Cdc25C on serine-216.* Science. 1997;277:1501-5.

94. Sanchez Y, Wong C, Thoma RS, Richman R, Wu Z, Piwnica-Worms H, et al. *Conservation of the Chk1 checkpoint pathway in mammals: linkage of DNA damage to Cdk regulation through Cdc25.* Science. 1997;277:1497-501.

95. Walworth NC, Bernards R. *Rad-dependent response of the chk1-encoded protein kinase at the DNA damage checkpoint.* Science. 1996;271:353-6.

96. Stegmeier F, Amon A. *Closing mitosis: the functions of the Cdc14 phosphatase and its regulation.* Annu Rev Genet. 2004;38:203-32.

97. Shou W, Seol JH, Shevchenko A, Baskerville C, Moazed D, Chen ZWS, et al. *Exit from mitosis is treggered by Tem1-dependent release of the protein phosphatase Cdc14 from nucleolar RENT complex.* Cell. 1999;97:233-44.

98. Visintin R, Hwang ES, Amon A. *Cfi1 prevents premajure exit from mitosis by anchoring Cdc14 phosphatase in the nucleolus.* Nature. 1999;398:818-23.

99. Shirayama M, Toth A, Galova M, Nasmyth K. *APC cdc20 promotes exit from mitosis by destroying the anaphase inhibitor Pds1 and cyclin Clb5.* Nature. 1999;402:203-7.

100. Yu H. *Regulation of APC-Cdc20 by the spindle checkpoint.* Curr Opin Cell Biol. 2002;14:706-14.

101. Wu J, Cho HP, Rhee DB, Johnson DK, Dunlap J, Liu Y, et al. *Cdc14B depletion leads to centriole amplification, and its overexpression prevents unscheduled centriole duplication.* J Cell Biol. 2008;181:475-83.

102. Lens SM, Voest EE, Medema RH. *Shared and separate functions of polo-like kinases and aurora kinases in cancer.* Nat Rev Cancer. 2010;10:825-41.

6.4.2. CELL CYCLE IN BREAST CANCER

Germ line defects in three key cell-cycle arrest genes are associated with familial breast cancer, namely ATM (see Chapter 8.5. Ataxia teleangiectasia), Chk2 (also known as CHEK2; Chapter 8.6.3.), and p53 (Chapter 8.4. Li-Fraumeni syndrome).

If the cell cycle has a beginning, it starts in G1 with the recruitment of noncycling cells from a quiescent G0 state in response to mitogenic stimuli. One such stimulus is provided by estrogens, which are mitogenic for several cell types within the major estrogen target tissues, particularly epithelial cells of the mammary gland, uterus, and vagina. Studies of ER-positive breast cancer cell lines demonstrated that 17β-estradiol (E_2) increased the rate of cell proliferation by two mechanisms, i.e., by recruiting noncycling cells from G0 into the cell cycle and by shortening the overall cell cycle time due predominantly to a reduction in length of the G1 phase (1, 2). Studies with antiestrogens support this conclusion. Antiestrogen treatment of MCF-7 cells leads to growth arrest, with a reduction in the proportion of cells in S phase and a concomitant increase in the percentage of cells in G1 phase (3-6). This decrease in cell proliferation rate can be negated by simultaneous or subsequent addition of E_2, providing support for the concept that antiestrogens inhibit cell proliferation by competitive inhibition of estrogen-induced mitogenic signaling. Cells are sensitive to growth inhibition by antiestrogens in a limited segment of the G1 phase, extending from soon after mitosis until mid-G1 (7). Cells elsewhere in the cell cycle are essentially insensitive, and proceed through S phase and mitosis at the same rate as untreated cells. In summary, estrogens and antiestrogens regulate cell proliferation by their actions on a cell cycle control point in early to mid-G1 phase (1).

D-type cyclins. Estrogens independently regulate the expression of c-myc and cyclin D and the induction of either c-myc or cyclin D is sufficient to recapitulate the effects of estrogen on cell cycle progression (8, 9). C-myc links growth factor stimulation and cell proliferation (10) (see Section 6.2.7.). The molecular mechanism by which estrogens control cell cycle progression is linked to the induction of D-type cyclins (11). The D-type cyclins assemble with their catalytic partners, Cdk4 and Cdk6, as cells progress through G1 phase (12). There are three types of D cyclin (D1, D2, and D3), each encoded by a separate gene. Thus, cyclin D1 is encoded by the CCND1 gene at 11q13, cyclin D2 by the CCND2 gene at 12p13, and cyclin D3 by the CCND3 gene at 6p21 (13, 14). The D-type cyclins are induced in a cell lineage-specific manner (15, 16). For example, E_2 treatment of growth-arrested MCF-7 cells caused an increase in cyclin D1 mRNA and protein, the latter rising fourfold within six hours (17, 18). E_2 also caused a fivefold increase in the activity of cyclin D1-Cdk4 as a result of the enhanced cyclin D1 expression (19). Antiestrogens have the opposite effects, i.e., ICI 182,780 treatment of MCF-7 cells caused a decrease in cyclin D1 concentration and Cdk4 activities (20, 21). The proliferative effect of estrogens is mediated by ERα, whereas ERβ inhibits proliferation by repressing cyclin D, cyclin A, and c-myc gene transcription and increasing the expression of p21[CIP1/WAF1] and p27[Kip1], which leads to G2 arrest (22). The cyclin D1 promoter does not contain a classical estrogen response element, ERE. Instead, induction by ERα has been mapped to a <u>c</u>AMP <u>r</u>esponse <u>e</u>lement (CRE) close to the transcription start site, suggesting that E_2 activates the cyclin D1 gene via c-Jun-ATF2 binding at the CRE (23, 24). In addition, a more distal Sp1 site bound both ERα and Sp1, indicating that estrogen regulation of cyclin D1 expression involves multiple enhancer elements (25). Cyclin D1 is also induced by the Ras-mitogen-activated protein kinase (MAPK) signaling pathway (26-29). The Ras signaling pathway not only acts transcriptionally to induce the cyclin D1 gene but also functions posttranslationally to regulate cyclin D1 assembly with Cdk4 (30). Inactivation of Ras in cycling cells caused a decline in cyclin D1 protein levels and G1 arrest (31). In summary, the D-type cyclins provide the link between extracellular mitogenic stimuli, such as E_2 and Ras, and the cell cycle machinery. E_2 and Ras stimulate cyclin D expression and increase cyclin D-cdk4/6 activity, resulting in recruitment of noncycling G0 cells and progression through G1.

The D-type cyclins exhibit half-lives of less than 30 min, whether or not bound to Cdk4 or Cdk6. Thus, D-type cyclins coprecipitating in complexes with Cdk4 are degraded almost as rapidly as the free regulatory subunits (12). Since Cdk4 has a half-life of about 4 hr, D-type cyclins associate only transiently with the catalytic subunit, allowing for cyclin exchange and formation of multiple cyclin D/Cdk heterodimers in response to different extracellular stimuli. The D-type cyclins are synthesized as long as mitogen stimulation persists and exhibit only moderate oscillations during the cell cycle with peak levels near G1-S. However, the cyclin D levels rapidly decrease when mitogens are withdrawn, regardless of the position of the cell in the cycle. Their decline in response to mitogen deprivation during G1 results in the failure of cells to enter S phase. Thus, cells can withdraw from the cell cycle with an unduplicated DNA content and persist in the quiescent G0 state in which macromolecular synthesis is reduced. This mechanism provides an explanation for the involution of estrogen-dependent tissues in menopause.

The CCND1 gene at 11q13 is amplified and overexpressed in approximately 15% of primary breast cancers (32, 33). CCND1 amplification does not invariably lead to cyclin D1 overexpression and, conversely, cyclin D1 overexpression can occur with and without CCND1 amplification (34). A study of 94 biopsies used *in situ* hybridization to compare cyclin D mRNA levels in benign lesions (18%), premalignant atypical ductal hyperplasias (18%), ductal carcinoma *in situ* (76% low grade, 87% high grade) and invasive ductal carcinomas (83%) (35). However, an immunohistochemical analysis of 471 breast tissue samples showed a gradual progression of cyclin D1 protein overexpression from normal breast epithelium (11.7%), to proliferative disease without atypia (25.0%), atypical ductal hyperplasia (39.4%), low-grade DCIS (43.6%), high-grade DCIS (47.9%), and infiltrating carcinoma (48.3%) (36). Thus, the conclusion of the first study that overexpression of cyclin D mRNA distinguishes invasive and *in situ* breast cancers from non-malignant lesions was not borne out by the second study, which showed that cyclin D1 protein overexpression is important in the earliest stages of breast oncogenesis and increases significantly with the progression to invasive cancer. Clinical and experimental evidence indicates a direct link between cyclin D1 and estrogen action in breast cancer. Several clinical studies involving a combined total of over 1500 breast cancers reported a significant positive correlation between cyclin D1 amplification/overexpression and ER positivity (34, 37-41). Experimentally, cyclin D1 was shown to activate transcription of ER-regulated genes independent of complex formation to a Cdk partner (42, 43). The activation involved direct binding of cyclin D1 to the ligand-binding domain of the receptor, resulting in increased ER-ERE interaction. Activation occurred in the absence of estrogen, but was enhanced in the presence of E_2 and regulated in a protein kinase A-dependent manner (44). The synergism between E_2 and cyclin D1 in ER activation results from the cooperative recruitment of steroid receptor coactivators such as SRC-1 to the cyclin D1-ER complex (45). Thus, cyclin D1 contributes to mammary epithelial cell proliferation in a dual fashion, i.e., as activator of ER-mediated transcription and as mediator of cell cycle progression via Cdk kinase activity. In contrast to cyclin D1, cyclin D2 transcript and protein levels are decreased in approximately 50% of primary breast cancers. The loss of expression results from promoter hypermethylation of the CCND2 gene (46, 47). In comparison to cyclin D1, cyclin D2 and D3 only minimally stimulated ER-mediated gene transcription. Inactivation of cyclin D1 by injected antibodies or antisense oligonucleotides severely delayed entry of MCF-7 cells into S phase (48, 49). On the other hand, overexpression of cyclin D1 in growth-arrested breast cancer cells was sufficient to allow cells to reenter and progress through the cell cycle in the absence of estrogens, overriding even the growth-inhibitory effects of antiestrogens (1, 50, 51). These findings suggest that deregulated expression of cyclin D1 enables cells to bypass the requirement for estrogens and thereby provides a mechanism for estrogen-independent growth of cyclin D1-overexpressing breast cancer cells. Constitutive activation of the D cyclin pathway can also bypass the requirement for Ras signaling in cell proliferation and thereby contribute

to oncogenic transformation (52-55). For example, targeted expression of cyclin D1 in mammary epithelial cells of transgenic mice resulted in ductal hyperproliferation and eventual tumor formation (56). On the other hand, mice nullizygous for cyclin D1 presented with defects in mammary lobuloalveolar development during pregnancy, indicating that cyclin D1 plays a critical role in normal gestational maturation of breast tissue (57, 58). Cyclin D1 regulates mammary epithelial differentiation via C/EBPβ and PPARγ as both a regulator and target of these transcription factors (59-61). Mouse strains engineered to allow ablation of individual cyclins in a living animal showed that ubiquitous shutdown of cyclin D1 or inhibition of cyclin D-associated kinase activity triggered tumor cell senescence in HER2-driven mammary carcinomas (62). The loss of cyclin D1 had no apparent effect on healthy adult mice.

PELP1 (proline-, glutamic acid-, and leucine-rich protein 1) is a nuclear receptor coregulator that interacts with ERα and enhances the ERα-mediated action on G1 progression (63). PELP1 also interacts with the retinoblastoma tumor suppressor protein (pRb), which, in turn, forms a complex with Cdk4/6-cyclin D and the transcription factors E2F. PELP1 is phosphorylated by Cdk4 and enhances the expression of pRb-E2F target genes, such as cyclin D1 and Cdc25C (64). Thus, PELP1 connects estrogen signaling to the activation of E2F target genes, which may explain the oncogenic potential of overexpressed PELP1 (65). In breast cancers, PELP1 overexpression was associated with poor prognosis (66). The E2F transcription factors are key regulators of cell proliferation as demonstrated by the observation that overexpression of E2F1 can induce quiescent cells to enter S phase, whereas deficiency of E2F3 can decrease the rate of proliferation (67-69). Interestingly, overexpression of E2F1 not only promotes cell proliferation but may also lead to apoptosis via induction of the tumor suppressor p14[ARF] (70). Immunohistochemical analysis of E2F1 expression in breast tissue revealed an increase in the percentage of epithelial cells with E2F1-positive nuclear staining from 1.9% in normal breast to 6.3% in ductal carcinoma *in situ* and to 15.3% in invasive breast carcinomas (71). The breast carcinoma cell lines T47D and MCF-7 displayed higher levels of E2F1 expression than the non-tumorigenic cell line MCF-10F by immunohistochemistry and Western blot analysis. E2F1 expression correlated well with the expression of Mib-1 (Ki67), a proliferation marker that is present in all phases of the cell cycle except G0 and early G1.

Cyclin E. Functional knockout of cyclin E by injection of anti-cyclin E antibodies into fibroblasts caused cell arrest in the G1 phase (72). Conversely, overexpression of cyclin E protein caused acceleration through G1 along with a decrease in cell size, similar to the effect of overexpressed cyclin D (73, 74). Expression of the human cyclin E gene in transgenic mice under the control of the ovine β-lactoglobulin promoter induced mammary gland hyperplasia and carcinoma (75). Not surprisingly, abnormal regulation of cyclin E expression has been associated with a number of human malignancies, including breast cancer (76). Immunohistochemical examination revealed lack of cyclin E in normal mammary tissue and benign breast lesions. In contrast, positive nuclear cyclin E immunoreactivity was observed in over 20% of invasive breast cancers and the cyclin E gene was amplified eight-fold in MDA-MB-157 breast cancer cells (77-80). However, cyclin E could be detected by Western immunoblot in homogenate of normal breast tissue suggesting that the low level of cyclin E expression in individual normal nuclei is the reason for the lack of immunohistochemical staining (79). By contrast, the positive immunohistochemical staining in malignant nuclei reflects the overexpression of cyclin E and its abnormal persistence throughout the cell cycle. The cyclin E overexpression was associated with increased cyclin E-Cdk2 activity (81, 82). Biochemical analysis of the overexpressed cyclin E revealed the presence of the 51-kD wild type protein, several variants arising by alternative splicing, and several low molecular isoforms (34 - 49 kD) (72, 81). The splice variants are found in both normal and malignant tissues, in contrast to the low molecular isoforms, which are formed by posttrans-lational N-terminal proteolytic processing by a protease (elastase) that is more active in tumor cells than in normal cells (83-85). Since the N-terminally truncated isoforms retain the ability to bind Cdk2, cyclin E-Cdk2 kinase activity was ten times higher in tumor cells than in normal cells. The oncogenic potential of low-molecular-weight cyclin E and the requirement for Cdk2 were demonstrated in transgenic mice, which exhibited an increased incidence of mammary tumors and distant-metastasis when compared with mice with full-length cyclin E (86, 87). Expression of low-molecular-weight cyclin E resulted in genomic instability due to premature inactivation of Cdc25C and polo-like kinase 1, leading to shortened mitosis and centrosome amplification (88-90). Thus, overexpression of cyclin E, especially of the low-molecular-weight isoform may lead to uncoordinated upregulation of many genes involved in cell cycle control, resulting in uncoordinated entry into S phase with aberrant activation of nucleotide biosynthesis pathways (91). The uncoordinated activation of factors regulating cell proliferation can lead to nucleotide deficiency, which promotes genomic instability in early stages of cancer development. In summary, deregulated cyclin E induces chromosomal instability in breast epithelial cells whereas analogous expression of cyclin D or A does not increase the frequency of chromosomal instability (92).

Cdc25 phosphatase overexpression, particularly Cdc25A and Cdc25B, has been reported in various malignancies,

including breast cancer. One study did not detect any Cdc25B mRNA in normal breast but observed transcript expression in 32% of 124 node-negative invasive breast cancers, most of which exhibited poor histologic differ-entiation (93). Although it is clear that both Cdc25A and Cdc25B are commonly overexpressed in breast carcinomas, the correlation between mRNA and protein levels and the association with clinicopathological features have been inconsistent (94-96).The Cdc25 overexpression in cancer occurs at the transcriptional and post-translational levels. Both Cdc25A and Cdc25B are transcriptional targets of the c-myc oncogene (97). A significant correlation between increased Cdc25A and myc expression has been observed in breast cancer (98). The half-life of Cdc25A protein was prolonged in breast cancer cell lines that overexpress Cdc25A compared to cells that express lower levels suggesting that the increase in protein stability contributes to the elevation in phosphatase activity (99). Constitutive overexpres-sion of Cdc25A in hTERT-immortalized primary human mammary epithelial cells resulted in defective DNA damage response (100). Cdc25A and Cdc25B have oncogenic properties but are insufficient to cause cancer suggesting that increased Cdc25 phosphatase activity alone does not override checkpoint controls and promote inappropriate progression through the cell cycle (101). The targeting of Cdc25B overexpression to the mammary glands of mice resulted in hyperplasia but required additional challenge with the carcinogen 9,10-dimethyl-1,2,-benzanthracene to induce tumor growth (102, 103). Similarly, transgenic expression of Cdc25A alone resulted in alveolar hyperplasia in mammary tissue of mice but only the cooperation with H-ras or neu/HER2 induced tumor growth and chromosomal instability (104). Thus, Cdc25A may be a determining factor in tumorigenesis induced by neu/Her2 or Ras (105).

Cell cycle checkpoints are inactive in normal, nonpro-liferating breast tissue but they are activated in ductal carcinoma *in situ* and invasive breast cancer (106). The **Cdk inhibitor p16**[Ink4a] is a key regulator of cell cycle progression and senescence (107, 108). Elevated expression of p16[Ink4a] causes a G1/S arrest while inactivation of p16[Ink4a] allows cell cycle progression (109-112). Both overexpression and inactivation occur in breast cancer. The p16[Ink4a] inactivation is due to heterozygous or homozygous deletion of the Ink4 gene at 9p21 and/or silencing of the gene by promoter hypermethylation (113-115). In a study of 100 primary breast carcinomas, promoter hypermeth-ylation was the major mechanism of decreased p16[INK4A] expression, followed by hemizygous deletion of the Ink4 gene. Homozygous deletion was rare and mutation was absent (116). In most cases, p16[Ink4a] was silenced and overexpressed concomitantly with p14[ARF], not surprising since both proteins are encoded by the same Ink4 (CDNK2A) gene. Interestingly, Ink4 promoter hypermethylation was observed throughout all early stages of intraepithelial mammary neoplasia and even in morphologically normal appearing epithelial cells (111, 117). Similarly, p16^{Ink4a} overexpression was observed in atypical hyperplasia but not associated with increased risk of invasive breast cancer (118).

The **Cdk inhibitor p21**$^{CIP1/WAF1}$ acts as a tumor suppressor by promoting cell cycle arrest in response to many stimuli but also has oncogenic properties promoting tumor progression (119, 120). p21 null mice exhibit no increased tumor incidence, but p21 deficiency had opposite effects on mammary tumor incidence and age of onset in MMTV-ras and MMTV-myc transgenic mice, resulting in increased and decreased cell proliferation, respectively (121-123). In normal human breast, p21 was localized immunohistochemically in the nuclei of rare luminal cells and in occasional myoepithelial cells (124). Well-differentiated ductal carcinoma *in situ* lesions showed few p21-reactive cells with many more positive cells in poorly differentiated DCIS. Invasive breast carcinomas express p21 in over 50% of tumors but with varying degrees of intensity (124-128). The p21 expression was inversely related to p53 expression in some studies but not at all in others. Similarly, there was no consistent association of p21 expression with clinicopathological data, including estrogen receptor status and relapse-free survival. Several single nucleotide polymorphisms have been identified in the CDKN1A gene, which encodes p21 including Ser31Arg (rs1801270) but none were associated with breast cancer (129-132). In contrast to p53, p21 mutations are rare in breast cancer (131, 133, 134). The growth inhibiting activity of p21 is strongly correlated with its nuclear localization, whereas the oncogenic activity is associated with cytoplasmic accumulation, which may be induced by other oncogenes. For example, HER2 overexpression activates the phosphatidyl-3 kinase (PI3K)/Akt1 pathway leading to the phosphorylation of p21 residue 145 Thr in the PCNA binding site, which disrupts binding with PCNA and induces its cytoplasmic localization (135). HER2-positive breast cancers showed proportionately higher cytoplasmic p21 staining compared with HER2-negative tumors and a worse prognosis (136-138). Similarly, overexpression of the IKKβ (inhibitor of nuclear factor κB kinase β), which is seen in some breast cancers, is associated with Akt1 phosphorylation and the cytoplasmic accumulation of p21 (139).

p27^{KIP1} exhibits sequence identity in its N-terminus to p21$^{CIP1/WAF1}$, which explains their common ability to inhibit Cdk activity (140). However, p27 and p21 respond to different signals, and their levels change reciprocally as cells progress through G1, with those of p27 being elevated in quiescent cells and declining during the G0 to S phase interval (141). Thus, p27 is expressed at high levels in normal mammary epithelial cells in contrast to

low p21 levels (142). Conversely, breast cancers frequently show loss of p27 expression through accelerated proteolysis while p21 expression is increased (143). Most studies showed an inverse correlation between p27 expression and degree of malignancy and poor prognosis (142-145). p27 mutations are infrequent in tumors (142-144).

The **minichromosome maintenance (Mcm) 2-7 helicases** are key components of the S phase progression checkpoint (146). They are part of the licensing machinery that regulates initiation of DNA replication. On the other hand, geminin is a licensing repressor that prevents replication of DNA replication during the S, G2, and M phases by blocking the reloading of Mcm2-7 at replication origins. The importance of the helicases was illustrated by a Mcm4 mutant that caused chromosomal instability and mammary adenocarcinoma in over 80% of female mice (147).

Immunohistochemical analysis of human breast tissue showed that a large proportion of epithelial cells of the terminal duct lobular unit reside in a primed 'replication licensed' but not proliferating state, which was characterized by Mcm2 expression and absence of geminin and Ki67 (148). Mcm2 expression increased progressively from atypical lesions to high-grade invasive breast cancer (149). Increasing tumor grade was associated with increased Mcm2, geminin and Ki67 (148). Another protein required for the S phase progression checkpoint is BRIT1 (BRCT-repeat inhibitor of hTERT, also known as microcephalin, MCPH1), which acts in the ATM/ATR DNA damage response pathways (150-152). Silencing of BRIT1 expression in normal human mammary epithelial cells induced chromosomal instability in the form of chromosomal breaks and polyploidy (152). Breast cancer cell lines showed decreased BRIT1 mRNA and protein levels compared to non-transformed breast epithelial cells.

References

1. Musgrove EA, Sutherland RL. *Cell cycle control by steroid hormones.* Sem Cancer Biol. 1994;5:381-9.

2. Sutherland RL, Watts CKW, Musgrove EA. *Cell cycle control by steroid hormones in breast cancer: implications for endocrine resistance.* Endocr Relat Cancer. 1995;2:87-96.

3. Sutherland RL, Green MD, Hall RE, Reddel RR, Taylor IW. *Tamoxifen induces accumulation of MCF 7 human mammary carcinoma cells in the G0/G1 phase of the cell cycle.* Eur J Cancer Clin Oncol. 1983;19:615-21.

4. Sutherland RL, Hall RE, Taylor IW. *Cell proliferation kinetics of MCF-7 human mammary carcinoma cells in culture and effects of tamoxifen on exponentially growing and plateau-phase cells.* Cancer Res. 1983;43:3998-4006.

5. Taylor IW, Hodson PJ, Green MD, Sutherland RL. *Effects of tamoxifen on cell cycle progression of synchronous MCF-7 human mammary carcinoma cells.* Cancer Res. 1983;43:4007-10.

6. Wakeling AE, Newboult E, Peters SW. *Effects of antioestrogens on the proliferation of MCF-7 human breast cancer cells.* J Mol Endocrinol. 1989;2:225-34.

7. Musgrove EW, Wakeling AE, Sutherland RL. *Points of action of estrogen antagonist and a calmodulin antagonist within the MCF-7 human breast cancer cell cycle.* Cancer Res. 1989;49:2398-404.

8. Doisneau-Sixou SF, Sergio CM, Carroll JS, Hui R, Musgrove EA, Sutherland RL. *Estrogen and antiestrogen regulation of cell cycle* progression in breast cancer cells. Endocr Relat Cancer. 2003;10:179-86.

9. Foster JS, Henley DC, Ahamed S, Wimalasena J. Estrogens and cell-cycle regulation in breast cancer. Trends Endocrinol Metab. 2001;12:320-7.

10. Dang CV. MYC on the path to cancer. Cell. 2012;149:22-35.

11. Musgrove EA, Hamilton JA, Lee CS, Sweeney KJ, Watts CK, Sutherland RL. Growth factor, steroid, and steroid antagonist regulation of cyclin gene expression associated with changes in T-47D human breast cancer cell cycle progression. Mol Cell Biol. 1993;13:3577-87.

12. Sherr CJ. Mammalian G1 Cyclins. Cell. 1993;73:1059-65.

13. Inaba T, Matsushine H, Valentine M, Roussel MF, Sherr CJ, Look AT. Genomic organization, chromosomal localization, and independent expression of human cyclin D genes. Genomics. 1992;13:565-74.

14. Xiong Y, Menninger J, Beach D, Ward D. Molecular cloning and chromosomal mapping of CCND tgenes encoding human D-type cyclins. Genomics. 1992;13:575-84.

15. Ajchenbaum F, Ando K, DeCaprio JA, Griffin JD. Independent regulation of human D-type cyclin gene expression during G1 phase in primary human T lymphocytes. J Biol Chem. 1993;268:4113-9.

16. Bartkova J, Lukas J, Strauss M, Bartek J. Cyclin D3: requirement for G1/S transtion and high abundance in quiescent tissues suggest a dual role in proliferation and differentiation. Oncogene. 1998;17:1027-37.

17. Altucci L, Addeo R, Cicatiello L, Dauvois S, Parker MG, Truss M, et al. 17b - Estradiol induces cyclin D1 gene transcription, p36D1 -p34 cdk4 complex activation and p105 Rb phosphorylatiuon during mitogenic stimulation of G1-arrested human breast cancer cells. Oncogene. 1996;12:2315-24.

18. Foster JS, Wimalasena J. Estrogen regulates activity of cyclin-dependent kinases and retinoblastoma protein phosphorylation in breast cancer cells. Mol Endocrinol. 1996;10:488-98.

19. Prall OWJ, Sarcevic B, Musgrove EA, Watts CKW, Sutherland RL. Estrogen-induced activation of Cdk4 and Cdk2 during G1-S phase progression is accompanied by increased cyclin D1 expression and decreased cyclin-dependent kinase inhibitor association with

cyclin E-Cdk2. J Biol Chem. 1997;272:10882-94.

20. Watts CKW, Brady A, Sarcevic B, deFazio A, Musgrove EA, Sutherland RL. Antiestrogen inhibition of cell cycle progression in breast cancer cells is associated with inhibition of cyclin-dependent kinase activity and decreased retinoblastoma protein phosphorylation. Mol Endocrinol. 1995;9:1804-13.

21. Watts CKW, Sweeney KJE, Warlters A, Musgrove EA, Sutherland RL. Antiestrogen regulation of cell cycle progression and cyclin D1 gene expression in MCF-7 human breast cancer cells. Breast Cancer Res Treat. 1994;31:95-105.

22. Paruthiyil S, Parmar H, Kerekatte V, Cunha GR, Firestone GL, Leitman DC. Estrogen receptor beta inhibits human breast cancer cell proliferation and tumor formation by causing a G2 cell cycle arrest. Cancer Res. 2004;64:423-8.

23. Planas-Silva MD, Donaher JL, Weinberg RA. Functional activity of ectopically expressed estrogen receptor is not sufficient for estrogen-mediated cyclin D1 expression. Cancer Res. 1999;59:4788-92.

24. Sabbah M, Courilleau D, Mester J, Redeuilh G. Estrogen induction of the cyclin D1 promoter: involvement of a cAMP response-like element. Proc Natl Acad Sci U S A. 1999;96:11217-22.

25. Castro-Rivera E, Samudio I, Safe S. Estrogen regulation of cyclin D1 gene expression in ZR-75 breast cancer cells involves multiple enhancer elements. J Biol Chem. 2001;276:30853-61.

26. Filmus J, Robles AI, Shi W, Wong MJ, Colombo LL, Conti CJ. Induction of cyclin D1 overexpression by activated ras. Oncogene. 1994;9:3627-33.

27. Lavoie JN, Rivard N, L'Allemain G, Pouyssegur J. A temporal and biochemical link between growth factor-activated MAP kinases, cyclin D1 induction and cell cycle entry. Prog Cell Cycle Res. 1996;2:49-58.

28. Weber JD, Raben DM, Phillips PJ, Baldassare JJ. Sustained activation of extracellular-signal-regulated kinase 1 (ERK1) is required for the continued expression of cyclin D1 in G1 phase. Biochem J. 1997;326:61-8.

29. Winston JT, Coats SR, Wang YZ, Pledger WJ. Regulation of the ell cyle machinery by oncogenic ras. Oncogene. 1996;12:127-34.

30. Cheng M, Sexl V, Sherr CJ, Roussel MF. Assembly of cyclin D-dependent kinase and titration of

p27Kip1 regulated by mitogen-activated protein kinase kinase (MEK1). Proc Natl Acad Sci USA. 1998;95:1091-6.

31. Peeper DS, Upton TM, Ladha MH, Neuman E, Zalvide J, Bernards R, et al. Ras signalling linked to the cell-cycle machinery by the retinoblastoma protein. Nature. 1997;386:177-81.

32. Gillett C, Fantl V, Smith R, Fisher C, Bartek J, Dickson C, et al. Amplification and overexpression of cyclin D1 in breast cancer detected by immunohistochemical staining. Cancer Res. 1994;54:1812-7.

33. Peters G, Fantl V, Smith R, Brookes S, Dickson C. Chromosome 11q13 markers and D-type cyclins in breast cancer. Breast Cancer Res Treat. 1995;33:125-35.

34. Buckley MF, Sweeney KJ, Hamilton JA, Sini RL, D.L. M, Nicholson RI, et al. Expression and amplification of cyclin genes in human breast cancer. Oncogene. 1993;8:2127-33.

35. Weinstat-Saslow D, Merino MJ, Manrow RE, Lawrence JA, Bluth RF, Wittenbel KD, et al. Overexpression of cyclin D mRNA distinguishes invasive and in situ breast carcinomas from non-malignant lesions. Nat Med. 1995;1:1257-60.

36. Alle KM, Henshall SM, Field AS, Sutherland RL. Cyclin D1 protein is overexpressed in hyperplasia and intraductal carcinoma of the breast. Clin Cancer Res. 1998;4:847-54.

37. Adnane J, Gaudray P, Simon MP, Simony-Lafontaine J, Jeanteur P, Theillet C. Proto-oncogene amplification and human breast tumor phenotype. Oncogene. 1989;4:1389-95.

38. Hui R, Cornish AL, McClelland RA, Robertson JFR, Blamey RW, Musgrove EA, et al. Cyclin D1 and estrogen receptor messenger RNA levels are positively correlated in primary breast cancer. Clin Cancer Res. 1996;2:923-8.

39. Michalides R, Hagemen P, van Tinteren H, Houben L, Wientjens E, Klompmaker R, et al. A clinicopathological study on overexpression of cyclin D1 and of p53 in a series of 248 patients with operable breast cancer. Br J Cancer. 1996;73:728-34.

40. Oyama T, Kashiwabara K, Yoshimoto K, Arnold A, Koerner F. Frequent overexpression of the cyclin D1 oncogene in invasive lobular carcinoma of the breast. Cancer Res. 1998;58:2876-80.

41. Seshadri R, Lee CSL, Hui R, McCaul K, Horsfall DJ, Sutherland RL. Cyclin D1 amplification is not

associated with reduced overall survival in primary breast cancer but may predict early relapse in patients with features of good prognosis. Clinical Cancer Res. 1996;2:1177-84.

42. Neuman E, Ladha MH, Lin N, Upton TM, Miller SJ, DiRenzo J, et al. Cyclin D1 stimulation of estrogen receptor transcriptional activity dependent of cdk4. Mol Cell Biol. 1997;17:5338-47.

43. Zwijsen RML, Wientjens E, Klompmaker R, van der Sman J, Bernards R, Michalides RJA. CDK-independent activation of estrogen receptor by cyclin D1. Cell. 1997;88:405-15.

44. Lamb J, Ladha MH, McMahon C, Sutherland RL, Ewen ME. Regulation of the functional interaction between cyclin D1 and the estrogen receptor. Mol Cell Biol. 2000;20:8667-75.

45. Zwijsen RML, Buckle RS, Hijmans EM, Loomans CJM, Bernards R. Ligand-independent recruitment of steroid receptor coactivators to estrogen receptor by cyclin D1. Genes Dev. 1998;12:3488-98.

46. Evron E, Umbricht CB, Korz D, Raman V, Loeb DM, Niranjan B, et al. Loss of cyclin D2 expression in the majority of breast cancers is associated with promoter hypermethylation. Cancer Res. 2001;61:2782-7.

47. Lewis CM, Cler LR, Bu DW, Zochbauer-Muller S, Milchgrub S, Naftalis EZ, et al. Promoter hypermethylation in relation to predicted breast cancer risk. Clin Cancer Res. 2005;11:166-72.

48. Bartkova J, Lukas J, Muller H, Lutzhoft D, Strauss M, Bartek J. Cyclin D1 protein expression and function in human breast cancer. Int J Cancer. 1994;57:353-61.

49. Lukas J, Pagano M, Staskova Z, Draetta G, Bartek J. Cyclin D1 protein oscillates and is essential for cell cycle progression in human tumour cell lines. Oncogene. 1994;9:707-18.

50. Wilcken NRC, Prall OWJ, Musgrove EA, Sutherland RL. Inducible overexpression of cyclin D1 in breast cancer cells reverses the growth-inhibitory effects of antiestrogens. Clin Cancer Res. 1997;3:849-54.

51. Zwijsen RM, Klompmaker R, Wientjens EB, Kristel PM, van der Burg B, Michalides RJ. Cyclin D1 triggers autonomous growth of breast cancer cells by governing cell cycle exit. Mol Cell Biol. 1996;16:2554-60.

52. Aktas H, Cai H, Cooper GM. *Ras links growth factor signaling to the cell cycle machinery via regulation of cyclin D1 and the Cdk inhibitor p27KIP1.* Mol Cell Biol. 1997;17:3850-7.

53. Bartkova J, Lukas J, Bartek J. *Aberrations of the G1- and G1/S-regulating genes in human cancer.* Prog Cell Cycle Res. 1997;3:211-20.

54. Sherr CJ. *Cancer cell cycles.* Science. 1996;274:1672-7.

55. Weinberg RA. *The retinoblastoma protein and cell cycle control.* Cell. 1995;81:323-30.

56. Wang TC, Cardiff RD, Zukerberg L, Lees E, Arnold A, Schmidt EV. *Mammary hyperplasia and carcinoma in MMTV-cyclin D1 transgenic mice.* Nature. 1994;369:669-71.

57. Fantl V, Stamp G, Andrews A, Rosewell I, Dickson C. *Mice lacking cyclin D1 are small and show defects in eye and mammary gland development.* Genes Dev. 1995;9:2364-72.

58. Sicinski P, Donaher J, Parker SB, Li T, Fazeli A, Gardner H, et al. *Cyclin D1 provides a link between development and oncogenesis in the retina and breast.* Cell. 1995;82:621-30.

59. Caldon CE, Sutherland RL, Musgrove E. *Cell cycle proteins in epithelial cell differentiation: implications for breast cancer.* Cell Cycle. 2010;9:1918-28.

60. Grimm SL, Rosen JM. *The role of C/EBPbeta in mammary gland development and breast cancer.* J Mammary Gland Biol Neoplasia. 2003;8:191-204.

61. Mueller E, Sarraf P, Tontonoz P, Evans RM, Martin KJ, Zhang M, et al. *Terminal differentiation of human breast cancer through PPAR gamma.* Mol Cell. 1998;1:465-70.

62. Choi YJ, Li X, Hydbring P, Sanda T, Stefano J, Christie AL, et al. *The requirement for cyclin d function in tumor maintenance.* Cancer Cell. 2012;22:438-51.

63. Balasenthil S, Vadlamudi RK. *Functional interactions between the estrogen receptor coactivator PELP1/MNAR and retinoblastoma protein.* J Biol Chem. 2003;278:22119-27.

64. Nair BC, Nair SS, Chakravarty D, Challa R, Manavathi B, Yew PR, et al. *Cyclin-dependent kinase-mediated phosphorylation plays a critical role in the oncogenic functions of PELP1.* Cancer Res. 2010;70:7166-75.

65. Rajhans R, Nair S, Holden AH, Kumar R, Tekmal RR, Vadlamudi RK. *Oncogenic potential of the nuclear receptor coregulator proline-, glutamic acid-, leucine-rich protein 1/modulator of the nongenomic actions of the estrogen receptor.* Cancer Res. 2007;67:5505-12.

66. Habashy HO, Powe DG, Rakha EA, Ball G, Macmillan RD, Green AR, et al. *The prognostic significance of PELP1 expression in invasive breast cancer with emphasis on the ER-positive luminal-like subtype.* Breast Cancer Res Treat. 2010;120:603-12.

67. Humbert PO, Verona R, Trimarchi JM, Rogers C, Dandapani S, Lees JA. *E2F3 is critical for normal cellular proliferation.* Genes Dev. 2000;14:690-703.

68. Johnson DG, Schwarz KJ, Cress WD, Nevins JR. *Expression of transcription factor E2F1 induces quiescent cells to enter S phase.* Nature. 1993;365:349-52.

69. Muller H, Helin K. *The E2F transcription factors: key regulators of cell proliferation.* Biochim Biophys Acta. 2000;1470:M1-12.

70. DeGregori J, Leone G, Miron A, Jakoi L, Nevins JR. *Distinct roles for E2F proteins in cell growth control and apoptosis.* Proc Natl Acad Sci USA. 1997;94:7245-50.

71. Zhang SY, Liu SC, Al-Saleem LF, Holloran D, Babb J, Guo X, et al. *E2F-1: A proliferative marker of breast neoplasia.* Cancer Epidemiol Biomarkers Prev. 2000;9:395 - 401.

72. Ohtsubo M, Theodoras AM, Schumacher J, Roberts JM, Pagano M. *Human cyclin E, a nuclear protein essential for the G1-to-S phase transition.* Mol Cell Biol. 1995;15:2612-24.

73. Quelle DE, Ashmun RA, Shurtleff SA, Kato JY, Bar-Sagi D, Roussel MF, et al. *Overexpression of mouse D-type cyclins acccelerates G1 phase in rodent fibroblasts.* Genes Dev. 1993;7:1559-71.

74. Resnitzky D, Gossen M, Bujard H, Reed S. *Acceleration of the G1/S phase transition by expression of cyclins D1 and E with an inducible system.* Mol Cell Biol. 1994;14:1669-79.

75. Bortner DM, Rosenberg MP. *Induction of mammary gland hyperplasia and carcinomas in transgenic mice expressing human cyclin E.* Mol Cell Biol. 1997;17:453-9.

76. Sherr CJ, Roberts JM. *Living with or without cyclins and cyclin-dependent kinases.* Genes Dev. 2004;18:2699-711.

77. Dutta A, Chandra R, Leiter LM, Lester S. *Cyclins as markers of tumor proliferation: immunocytochemical studies in breast cancer.* Proc Natl Acad Sci USA. 1995;92:5386-90.

78. Keyomarsi K, O'Leary N, Molnar G, Lees E, Fingert HJ, Pardee AB. *Cyclin E, a potential prognostic marker for breast cancer.* Cancer Res. 1994;54:380-5.

79. Keyomarsi K, Pardee AB. *Redundant cyclin overexpression and gene amplification in breast cancer cells.* Proc Natl Acad Sci USA. 1993;90:1112-6.

80. Scott KA, Walker RA. *Lack of cyclin E immunoreactivity in non-malignant breast and association with proliferation in breast cancer.* Br J Cancer. 1997;76:1288-92.

81. Keyomarsi K, Conte DJ, Toyofuku W, Fox MP. *Deregulation of cyclin E in breast cancer.* Oncogene. 1995;11:941-50.

82. Loden M, Nielsen NH, Roos G, Emdin SO, Landberg G. *Cyclin E dependent kinase activity in human breast cancer in relation to cyclin E, p27 and p21 expression and retinoblastoma protein phosphorylation.* Oncogene. 1999;18:2557-66.

83. Akli S, Keyomarsi K. *Cyclin E and its low molecular weight forms in human cancer and as targets for cancer therapy.* Cancer Biol Ther. 2003;2:S38-47.

84. Harwell RM, Porter DC, Danes C, Keyomarsi K. *Processing of cyclin E differs between normal and tumor breast cells.* Cancer Res. 2000;60:481-9.

85. Porter DC, Zhang N, Danes C, McGahren MJ, Harwell RM, Faruki S, et al. *Tumor-specific proteolytic processing of cyclin E generates hyperactive lower-molecular-weight forms.* Mol Cell Biol. 2001;21:6254-69.

86. Akli S, Van Pelt CS, Bui T, Meijer L, Keyomarsi K. *Cdk2 is required for breast cancer mediated by the low-molecular-weight isoform of cyclin E.* Cancer Res. 2011;71:3377-86.

87. Akli S, Van Pelt CS, Bui T, Multani AS, Chang S, Johnson D, et al. *Overexpression of the low molecular weight cyclin E in transgenic mice induces metastatic mammary carcinomas through the disruption of the ARF-p53 pathway.* Cancer Res. 2007;67:7212-22.

88. Akli S, Zheng PJ, Multani AS, Wingate HF, Pathak S, Zhang N, et al. *Tumor-specific low molecular weight forms of cyclin E induce genomic instability and resistance to p21, p27, and antiestrogens in breast cancer.* Cancer Res. 2004;64:3198-208.

89. Bagheri-Yarmand R, Biernacka A, Hunt KK, Keyomarsi K. *Low molecular weight cyclin E overexpression shortens mitosis, leading to chromosome missegregation and centrosome amplification.* Cancer Res. 2010;70:5074-84.

90. Bagheri-Yarmand R, Nanos-Webb A, Biernacka A, Bui T, Keyomarsi K. *Cyclin E deregulation impairs mitotic progression through premature activation of Cdc25C.* Cancer Res. 2010;70:5085-95.

91. Bester AC, Roniger M, Oren YS, Im MM, Sarni D, Chaoat M, et al. *Nucleotide deficiency promotes genomic instability in early stages of cancer development.* Cell. 2011;145:435-46.

92. Spruck CH, Won KA, Reed SI. *Deregulated cyclin E induces chromosome instability.* Nature. 1999;401:297-300.

93. Galaktionov K, Lee AK, Eckstein J, Draetta G, Meckler J, Loda M, et al. *CDC25 phosphatases as potential human oncogenes.* Science. 1995;269:1575-7.

94. Bonin S, Brunetti D, Benedetti E, Gorji N, Stanta G. *Expression of cyclin-dependent kinases and CDC25a phosphatase is related with recurrences and survival in women with peri- and post-menopausal breast cancer.* Virchows Arch. 2006;448:539-44.

95. Cangi MG, Cukor B, Soung P, Signoretti S, Moreira G, Jr., Ranashinge M, et al. *Role of the Cdc25A phosphatase in human breast cancer.* J Clin Invest. 2000;106:753-61.

96. Ito Y, Yoshida H, Uruno T, Takamura Y, Miya A, Kuma K, et al. *Expression of cdc25A and cdc25B phosphatase in breast carcinoma.* Breast Cancer. 2004;11:295-300.

97. Galaktionov K, Chen X, Beach D. *Cdc25 cell-cycle phosphatase as a target of c-myc.* Nature. 1996;382:511-7.

98. Ben-Yosef T, Yanuka O, Halle D, Benvenisty N. *Involvement of Myc targets in c-myc and N-myc induced human tumors.* Oncogene. 1998;17:165-71.

99. Loffler H, Syljuasen RG, Bartkova J, Worm J, Lukas J, Bartek J. *Distinct modes of deregulation of the proto-oncogenic Cdc25A phosphatase in human breast cancer cell lines.* Oncogene. 2003;22:8063-71.

100. Cangi MG, Piccinin S, Pecciarini L, Talarico A, Dal Cin E, Grassi S, et al. *Constitutive overexpression of CDC25A in primary human*

mammary epithelial cells results in both defective DNA damage response and chromosomal breaks at fragile sites. Int J Cancer. 2008;123:1466-71.

101. Boutros R, Lobjois V, Ducommun B. CDC25 phosphatases in cancer cells: key players? Good targets? Nat Rev Cancer. 2007;7:495-507.

102. Ma ZQ, Chua SS, DeMayo FJ, Tsai SY. Induction of mammary gland hyperplasia in transgenic mice over-expressing human Cdc25B. Oncogene. 1999;18:4564-76.

103. Yao Y, Slosberg ED, Wang L, Hibshoosh H, Zhang YJ, Xing WQ, et al. Increased susceptibility to carcinogen-induced mammary tumors in MMTV-Cdc25B transgenic mice. Oncogene. 1999;18:5159-66.

104. Ray D, Terao Y, Fuhrken PG, Ma ZQ, DeMayo FJ, Christov K, et al. Deregulated CDC25A expression promotes mammary tumorigenesis with genomic instability. Cancer Res. 2007;67:984-91.

105. Ray D, Kiyokawa H. CDC25A phosphatase: a rate-limiting oncogene that determines genomic stability. Cancer Res. 2008;68:1251-3.

106. Bartkova J, Horejsi Z, Koed K, Kramer A, Tort F, Zieger K, et al. DNA damage response as a candidate anti-cancer barrier in early human tumorigenesis. Nature. 2005;434:864-70.

107. Garbe JC, Holst CR, Bassett E, Tlsty T, Stampfer MR. Inactivation of p53 function in cultured human mammary epithelial cells turns the telomere-length dependent senescence barrier from agonescence into crisis. Cell Cycle. 2007;6:1927-36.

108. Roussel MF. The INK4 family of cell cycle inhibitors in cancer. Oncogene. 1999;18:5311-7.

109. Braig M, Schmitt CA. Onco-gene-induced senescence: Putting the brakes on tumor development. Cancer Res. 2006;66:2881-4.

110. Foster SA, Wong DJ, Barrett MT, Galloway DA. Inactivation of p16 in human mammary epithelial cells by CpG island methylation. Mol Cell Biol. 1998;18:1793-801.

111. Holst CR, Nuovo GJ, Esteller M, Chew K, Baylin SB, Herman JG, et al. Methylation of p16(INK4a) promoters occurs in vivo in histologically normal human mammary epithelia. Cancer Res. 2003;63:1596-601.

112. Stampfer MR, Yaswen P. Human epithelial cell immortalization as a step in carcinogenesis. Cancer Lett. 2003;194:199-208.

113. Cairns P, Polascik TJ, Eby Y, Tokino K, Califano J, Merlo A, et al. Frequency of homozygous deletion at p16/CDKN2 in primary human tumours. Nat Genet. 1995;11:210-2.

114. Esteller M, Fraga MF, Guo M, Garcia-Foncillas J, Hedenfalk I, Godwin AK, et al. DNA methylation patterns in hereditary human cancers mimic sporadic tumorigenesis. Hum Mol Genet. 2001;10:3001-7.

115. Witcher M, Emerson BM. Epigenetic silencing of the p16(INK4a) tumor suppressor is associated with loss of CTCF binding and a chromatin boundary. Mol Cell. 2009;34:271-84.

116. Silva J, Silva JM, Dominguez G, Garcia JM, Cantos B, Rodriguez R, et al. Concomitant expression of p16INK4a and p14ARF in primary breast cancer and analysis of inactivation mechanisms. J Pathol. 2003;199:289-97.

117. Bean GR, Bryson AD, Pilie PG, Goldenberg V, Baker JC, Jr., Ibarra C, et al. Morphologically normal-appearing mammary epithelial cells obtained from high-risk women exhibit methylation silencing of INK4a/ARF. Clin Cancer Res. 2007;13:6834-41.

118. Radisky DC, Santisteban M, Berman HK, Gauthier ML, Frost MH, Reynolds CA, et al. p16(INK4a) expression and breast cancer risk in women with atypical hyperplasia. Cancer Prev Res (Phila). 2011;4:1953-60.

119. Abbas T, Dutta A. p21 in cancer: intricate networks and multiple activities. Nat Rev Cancer. 2009;9:400-14.

120. Roninson IB. Oncogenic functions of tumour suppressor p21(Waf1/Cip1/Sdi1): association with cell senescence and tumour-promoting activities of stromal fibroblasts. Cancer Letters. 2002;179:1-14.

121. Adnane J, Jackson RJ, Nicosia SV, Cantor AB, Pledger WJ, Sebti SM. Loss of p21WAF1/CIP1 accelerates Ras oncogenesis in a transgenic/knockout mammary cancer model. Oncogene. 2000;19:5338-47.

122. Bearss DJ, Lee RJ, Troyer DA, Pestell RG, Windle JJ. Differential effects of p21(WAF1/CIP1) deficiency on MMTV-ras and MMTV-myc mammary tumor properties. Cancer Res. 2002;62:2077-84.

123. Jones JM, Cui XS, Medina D, Donehower LA. Heterozygosity of p21WAF1/CIP1 enhances tumor cell proliferation and cyclin D1-associated kinase activity in a murine mammary cancer model. Cell Growth Differ. 1999;10:213-22.

124. Barbareschi M, Caffo O, Doglioni C, Fina P, Marchetti A, Buttitta F, et al. p21WAF1 immunohistochemical expression in breast carcinoma: correlations with clinicopathological data, oestrogen receptor status, MIB1 expression, p53 gene and protein alterations and relapse-free survival. Br J Cancer. 1996;74:208-15.

125. Caffo O, Doglioni C, Veronese S, Bonzanini M, Marchetti A, Buttitta F, et al. Prognostic value of p21(WAF1) and p53 expression in breast carcinoma: an immunohisto-chemical study in 261 patients with long-term follow-up. Clin Cancer Res. 1996;2:1591-9.

126. Ellis PA, Lonning PE, Borresen-Dale A, Aas T, Geisler S, Akslen LA, et al. Absence of p21 expression is associated with abnormal p53 in human breast carcinomas. Br J Cancer. 1997;76:480-5.

127. Pellikainen MJ, Pekola TT, Ropponen KM, Kataja VV, Kellokoski JK, Eskelinen MJ, et al. p21WAF1 expression in invasive breast cancer and its association with p53, AP-2, cell proliferation, and prognosis. J Clin Pathol. 2003;56:214-20.

128. Wakasugi E, Kobayashi T, Tamaki Y, Ito Y, Miyashiro I, Komoike Y, et al. p21(Waf1/Cip1) and p53 protein expression in breast cancer. Am J Clin Pathol. 1997;107:684-91.

129. Chedid M, Michieli P, Lengel C, Huppi K, Givol D. A single nucleotide substitution at codon 31 (Ser/Arg) defines a polymorphism in a highly conserved region of the p53-inducible gene WAF1/CIP1. Oncogene. 1994;9:3021-4.

130. Keshava C, Frye BL, Wolff MS, McCanlies EC, Weston A. Waf-1 (p21) and p53 polymorphisms in breast cancer. Cancer Epidemiol Biomarkers Prev. 2002;11:127-30.

131. Lukas J, Groshen S, Saffari B, Niu N, Reles A, Wen WH, et al. WAF1/Cip1 gene polymorphism and expression in carcinomas of the breast, ovary, and endometrium. Am J Pathol. 1997;150:167-75.

132. Wang N, Wang S, Zhang Q, Lu Y, Wei H, Li W, et al. Association of p21 SNPs and risk of cervical cancer among Chinese women. BMC Cancer. 2012;12:589.

133. Balbin M, Hannon GJ, Pendas AM, Ferrando AA, Vizoso F, Fueyo A, et al. Functional analysis of a p21WAF1,CIP1,SDI1 mutant (Arg94 --> Trp) identified in a human breast carcinoma. Evidence that the mutation impairs the ability of p21 to inhibit cyclin-dependent kinases. J Biol Chem. 1996;271:15782-6.

134. McKenzie KE, Siva A, Maier S, Runnebaum IB, Seshadri R, Sukumar S. Altered WAF1 genes do not play a role in abnormal cell cycle regulation in breast cancers lacking p53 mutations. Clin Cancer Res. 1997;3:1669-73.

135. Zhou BP, Liao Y, Xia W, Spohn B, Lee MH, Hung MC. Cytoplasmic localization of p21Cip1/WAF1 by Akt-induced phosphorylation in HER-2/neu-overexpressing cells. Nat Cell Biol. 2001;3:245-52.

136. Winters ZE, Hunt NC, Bradburn MJ, Royds JA, Turley H, Harris AL, et al. Subcellular localisation of cyclin B, Cdc2 and p21(WAF1/CIP1) in breast cancer. association with prognosis. Eur J Cancer. 2001;37:2405-12.

137. Winters ZE, Leek RD, Bradburn MJ, Norbury CJ, Harris AL. Cytoplasmic p21WAF1/CIP1 expression is correlated with HER-2/neu in breast cancer and is an independent predictor of prognosis. Breast Cancer Res. 2003;5:R242-9.

138. Xia W, Chen JS, Zhou X, Sun PR, Lee DF, Liao Y, et al. Phosphorylation/cytoplasmic localization of p21Cip1/WAF1 is associated with HER2/neu overexpression and provides a novel combination predictor for poor prognosis in breast cancer patients. Clin Cancer Res. 2004;10:3815-24.

139. Ping B, He X, Xia W, Lee DF, Wei Y, Yu D, et al. Cytoplasmic expression of p21CIP1/WAF1 is correlated with IKKbeta overexpression in human breast cancers. Int J Oncol. 2006;29:1103-10.

140. Polyak K, Lee MH, Erdjument-Bromage H, Koff A, Roberts JM, Tempst P, et al. Cloning of p27Kip1, a cyclin-dependent kinase inhibitor and a potential mediator of extracellular antimitogenic signals. Cell. 1994;78:59-66.

141. Slingerland JM, Hengst L, Pan CH, Alexander D, Stampfer MR, Reed SI. A novel inhibitor of cyclin-Cdk activity detected in transforming growth factor beta-arrested epithelial cells. Mol Cell Biol. 1994;14:3683-94.

142. Porter PL, Malone KE, Heagerty PJ, Alexander GM, Gatti LA, Firpo EJ, et al. Expression of cell-cycle regulators p27Kip1 and cyclin E, alone and in combination, correlate with survival in young breast cancer patients. Nat Med. 1997;3:222-5.

143. Alkarain A, Slingerland J.

Deregulation of p27 by oncogenic signaling and its prognostic significance in breast cancer. Breast Cancer Res. 2004;6:13-21.

144. Catzavelos C, Bhattacharya N, Ung YC, Wilson JA, Roncari L, Sandhu C, et al. *Decreased levels of the cell-cycle inhibitor p27Kip1 protein: prognostic implications in primary breast cancer.* Nat Med. 1997;3:227-30.

145. Fredersdorf S, Burns J, Milne AM, Packham G, Fallis L, Gillett CE, et al. *High level expression of p27(kip1) and cyclin D1 in some human breast cancer cells: inverse correlation between the expression of p27(kip1) and degree of malignancy in human breast and colorectal cancers.* Proc Natl Acad Sci U S A. 1997;94:6380-5.

146. Blow JJ, Dutta A. *Preventing re-replication of chromosomal DNA.* Nat Rev Mol Cell Biol. 2005;6:476-86.

147. Shima N, Alcaraz A, Liachko I, Buske TR, Andrews CA, Munroe RJ, et al. *A viable allele of Mcm4 causes chromosome instability and mammary adenocarcinomas in mice.* Nat Genet. 2007;39:93-8.

148. Shetty A, Loddo M, Fanshawe T, Prevost AT, Sainsbury R, Williams GH, et al. *DNA replication licensing and cell cycle kinetics of normal and neoplastic breast.* Br J Cancer. 2005;93:1295-300.

149. Nasir A, Chen DT, Gruidl M, Henderson-Jackson EB, Venkataramu C, McCarthy SM, et al. *Novel molecular markers of malignancy in histologically normal and benign breast.* Patholog Res Int. 2011;2011:489064.

150. Liang Y, Gao H, Lin SY, Peng G, Huang X, Zhang P, et al. *BRIT1/MCPH1 is essential for mitotic and meiotic recombination DNA repair and maintaining genomic stability in mice.* PLoS Genet. 2010;6:e1000826.

151. Lin SY, Rai R, Li K, Xu ZX, Elledge SJ. *BRIT1/MCPH1 is a DNA damage responsive protein that regulates the Brca1-Chk1 pathway, implicating checkpoint dysfunction in microcephaly.* Proc Natl Acad Sci U S A. 2005;102:15105-9.

152. Rai R, Dai H, Multani AS, Li K, Chin K, Gray J, et al. *BRIT1 regulates early DNA damage response, chromosomal integrity, and cancer.* Cancer Cell. 2006;10:145-57.

6.5. GENOMIC ABNORMALITIES
6.5.1. INTRODUCTION

Like other malignancies, breast cancer presents with a bewildering array of genetic abnormalities that have been documented in numerous studies by a variety of techniques ranging from traditional karyotype analysis to interphase fluorescence *in situ* hybridization (FISH), comparative genomic hybridization (CGH), DNA sequence analysis of individual genes, and whole genome sequencing. The results of these studies can be summarized in the following general comments:

• All invasive breast cancers contain genetic abnormalities

• Invasive breast cancers display intratumoral heterogeneity of genetic changes

• Genetic abnormalities are already present in DCIS and continue to be acquired as tumors progress from primary invasive to metastatic breast cancers

• Distinct patterns of genomic abnormalities are associated with different subtypes of breast cancer defined by expression profiling.

These findings allow several conclusions. About 80% of primary breast cancers contain an abnormal number of chromosomes referred to as aneuploidy (Section 6.5.2.). The 20% tumors that do not display aneuploidy harbor less conspicuous genetic abnormalities in form of gains or losses of chromosome parts. The acquisition of additional copies of a chromosome segment is referred to as gene amplification, which can result in the presence of over 100 copies of a particular gene (Section 6.5.3.). Aneuploidy is associated with basal-like breast cancers whereas amplification is more frequent in luminal B tumors (1-3). The loss of genetic material can take the form of loss of heterozygosity (LOH; Section 6.5.2.). Genetic instability also involves short DNA sequence repeats known as microsatellites (Section 6.5.2.). Finally, malignant tumors exhibit chromosomal rearrangements in form of reciprocal translocations, inversions and insertions. Microsatellite instability and chromosomal rearrangements are not common in breast cancer but frequently found in colorectal and hematological malignancies, respectively (4-6).

Breast cancers are among the most heterogeneous neoplasms (7). Cytogenetic, FISH, and LOH studies have demonstrated a striking intratumoral heterogeneity of genetic abnormalities, which suggests that an unknown number of the genetic abnormalities are not important for tumor development but merely result from genetic instability (8-11). One study examined the clonal relationship of macroscopically distinct, ipsilateral breast carcinomas by cytogenetic analysis of 26 malignant lesions from 12 patients (12). Seven patients had two ipsilateral lesions with clonal chromosome abnormalities. Four of these cases had cytogenetically related clonal abnormalities in the two ipsilateral lesions, whereas the remaining three cases had completely different clonal aberrations in the two ipsilateral foci. The authors noted an association between the proximity of the foci and the likelihood of them being clonally related. These findings indicate that multiple synchronous breast tumors may arise by two mechanisms: (i) through intramammary spreading of a single primary carcinoma or (ii) as the result of simultaneous emergence of pathogenetically independent carcinomas within the same breast (Figure 6.5.1.). Cytogenetic analysis of bilateral breast cancers revealed a similar pattern. A study of 16 bilateral tumors showed the same clonal abnormalities in samples from both breasts of two patients indicating that the bilaterality had arisen through a metastatic process (13). However, the absence of similarities between the two sides in the remaining 14 cases indicated an independent origin of the two carcinomas. The majority of bilateral cancers occur asynchronously, but even when they present synchronously they are likely to have a multicentric origin as shown by their different LOH patterns (14). A cytogenetic analysis of 29 primary breast

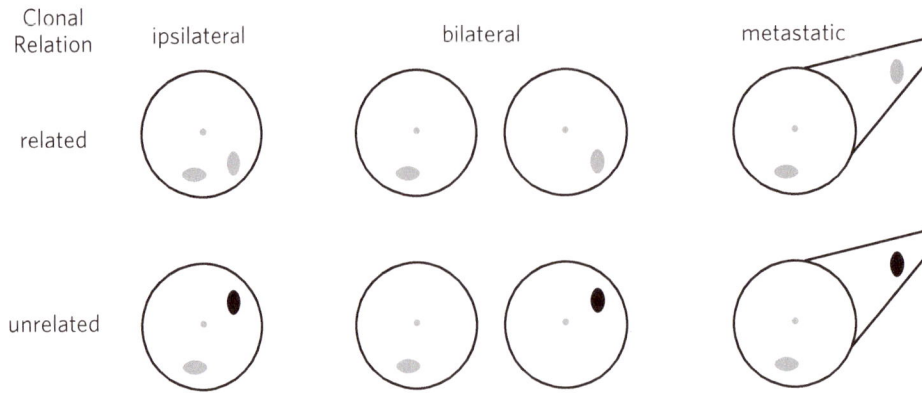

Figure 6.5.1. Clonal relationship of ipsilateral, bilateral, and metastatic breast cancer

cancers and their paired asynchronous metastases also revealed a high degree of clonal difference in 9 (31%) metastases (15). At times a lymph node metastasis may be genetically less heterogeneous than the corresponding primary tumor, apparently because only a few of the multiple clones in the primary tumor metastasize (16).

References

1. Bergamaschi A, Kim YH, Wang P, Sorlie T, Hernandez-Boussard T, Lonning PE, et al. *Distinct patterns of DNA copy number alteration are associated with different clinicopathological features and gene-expression subtypes of breast cancer.* Genes Chromosomes Cancer. 2006;45:1033-40.

2. Chin K, DeVries S, Fridlyand J, Spellman PT, Roydasgupta R, Kuo WL, et al. *Genomic and transcriptional aberrations linked to breast cancer pathophysiologies.* Cancer Cell. 2006;10:529-41.

3. Natrajan R, Weigelt B, Mackay A, Geyer FC, Grigoriadis A, Tan DS, et al. *An integrative genomic and transcriptomic analysis reveals molecular pathways and networks regulated by copy number aberrations in basal-like, HER2 and luminal cancers.* Breast Cancer Res Treat. 2010;121:575-89.

4. Frohling S, Dohner H. *Chromosomal abnormalities in cancer.* N Engl J Med. 2008;359:722-34.

5. Peltomaki P, Aaltonen LA, P. S, Pylkkanen L, Medklin J, Jarvinen H, et al. *Genetic mapping of a locus predisposing to human colorectal cancer.* Science. 1993;260:810-2.

6. Thibodeau SN, Gren G, Schaid D. *Microsatellite instability in cancer of the proximal colon.* Science. 1993;260:816-9.

7. Wolman SR, Heppner GH. *Genetic heterogeneity in breast cancer.* J Natl Cancer Inst. 1992;84:469-70.

8. Balazs M, Matsumura K, Moore D, Pinkel D, Gray JW, Waldman FM. *Karyotypic heterogeneity and its relation to labeling index in interphase beast tumor cells.* Cytometry. 1995;20:62-73.

9. Devilee P, Cornelisse CJ. *Somatic genetic changes in human breast cancer.* Biochim Biophys Acta. 1994;1198:113-30.

10. Fiegl M, Tueni C, Schenk T, Jakesz R, Gnant M, Reiner A, et al. *Interphase cytogenetics reveals a high incidence of aneuploidy and intra-tumour heterogeneity in breast cancer.* Br J Cancer. 1995;72:51-5.

11. Teixeira MR, Pandis N, Bardi G, Andersen JA, Mandahl N, Mitelman F, et al. *Cytogenetic analysis of multifocal breast carcinomas: detection of karyotypically unrelated clones as well as clonal similarities between tumour foci.* Br J Cancer. 1994;70:922-7.

12. Teixeira MR, Pandis N, Bardi G, Andersen JA, Bohler PJ, Qvist H, et al. *Discrimination between multicentric and multifocal breast carcinoma by cytogenetic investigation of macroscopically distinct ipsilateral lesions.* Genes Chromosomes Cancer. 1997;18:170-4.

13. Pandis N, Teixeira MR, Gerdes AM, Limon J, Bardi G, Andersen JA, et al. *Chromosome abnormalities in bilateral breast carcinomas. Cytogenetic evaluation of the clonal origin of multiple primary tumors.* Cancer. 1995;76:250-8.

14. Tsuda H, Hirohashi S. *Identification of multiple breast cancers of multicentric origin by histological observations and distribution of allele loss on chromosome 16q.* Cancer Res. 1995;55:3395-8.

15. Kuukasjarvi T, Karhu R, Tanner M, Kahkonen M, Schaffer A, Nupponen N, et al. *Genetic heterogeneity and clonal evolution underlying development of asynchronous metastasis in human breast cancer.* Cancer Res. 1997;57:1597-604.

16. Teixeira MR, Pandis N, Bardi G, Andersen JA, Heim S. *Karyotypic comparisons of multiple tumorous and macroscopically normal surrounding tissue samples from patients with breast cancer.* Cancer Res. 1996;56:855-9.

6.5.2. CHROMOSOME INSTABILITY
6.5.2.1. INTRODUCTION

Chromosomal Instability (CIN) is a characteristic of malignancy and cancers often display abnormal mitotic figures and karyotypes as a hallmark of malignancy. Defects in chromosome segregation, cell cycle checkpoints, telomere function and DNA damage response can cause CIN (1-3). CIN can contribute to tumor progression by the following two mechanisms: (i) Loss of chromosomes that harbor genes encoding negative regulators of cell cycle progression or proteins involved in apoptosis or senescence, and (ii) gain of chromosomes that harbor genes encoding positive regulators of cell cycle progression or anti-apoptosis or anti-senescence proteins. Central to the equal distribution of chromosomes to daughter cells is the mitotic apparatus, which is composed of mitotic spindle, centrosome, and centromere. The **mitotic spindle** is a dynamic bipolar array of microtubules that forms during mitosis to move the duplicated chromosomes apart. The **centrosome** in animal cells and the equivalent spindle pole body (SPB) in yeast cells are also called microtubule organizing centers (MTOCs) because they direct the polymerization of microtubules from free γ-tubulin subunits (a process called 'microtubule nucleation') and organize those microtubules into arrays that form the mitotic spindle (4). Each centrosome contains a pair of orthogonally oriented centrioles surrounded by pericentriolar material. Each centriole is a short cylinder of nine triplet microtubules. Centrins are a family of calcium-binding proteins (CETN1, CETN2, CETN3) found in centrosomes that play an essential role in centriole duplication (4-6). Centrosomes are often found in the center of the cell in the vicinity of, but physically separate from, the nucleus. For many cells, including epithelial cells, the position of the centrosome relative to the nucleus defines the structural 'cell axis' that indicates the overall functional polarity of the cell (7). The **centromere** is a unique region on each chromosome where the kinetochore assembles for the attachment of spindle microtubules. The **kinetochore** contains multiple centromere proteins (CENPs) that are organized into a trilaminar structure to mediate the spindle-centromere association (8, 9). Human centromeric DNA contains repeated 171 bp sequences (called alpha satellite DNA) typically arranged in tandem arrays of 2 to 4 Mb (8, 10).

To keep the number of chromosomes (ploidy) in a cell constant, chromosomes and centrosomes are duplicated once in every cell cycle (11). Because the centrosome plays an important role in the maintenance of cellular polarity and chromosome segregation during mitosis, the loss of cell polarity and abnormal chromosome number (polyploidy, aneuploidy) could result from defects in centrosome function. Indeed, carcinoma cells frequently show 'centrosome amplification', characterized by increased size and number of centrosomes and >4 centrioles in abnormal orientation, which drive CIN and aneuploidy (12). The centrosomal abnormalities may affect both cell polarity and genomic stability. First, the cytoplasmic architecture and directional vesicular trafficking become disorganized in a cell with multiple MTOCs. In the extreme case, the polarity of a cell may become completely inverted, with the apical membrane domain facing the stroma rather than the lumen. Such a complete loss of cellular organization, particularly with regard to orientation of adjacent cells relative to one another, is evident in anaplastic carcinomas. Second, the centrosomal abnormalities may increase the incidence of multipolar mitoses, leading to chromosomal mis-segregation and aneuploidy.

An abnormal chromosome number or **aneuploidy** results from the gain or loss of an entire chromosome. Cancer cells become aneuploid as a result of aberrant mitotic divisions (13). There are numerous proteins that play

structural and functional roles in maintaining the normal karyotype (14, 15). For example, the kinetochore is a multiprotein complex composed of centromere protein (CENP)-A through CENP-T plus a variety of other proteins (14). Any unattached kinetochore recruits a group of mitotic checkpoint proteins, including Mad1, Mad2, Bub1, Bub3, Mad3 (BubR1/Bub1B in humans), and Mps1 (TTK), which inhibit the Cdc20-dependent recognition of cyclin B and securin (PTTG1) by APC/C, thereby preventing advance to anaphase (14). When securin undergoes ubiquination by APC/C, it releases a protease, separase, which in turn triggers the release of cohesin, the protein complex that binds sister chromatids together during metaphase, allowing them to move to opposite poles for anaphase. Thus, securin is required for efficient chromosome segregation in anaphase and, at the same time, securing is needed to prevent anaphase in response to spindle damage (16). Abnormal expression of these mitotic checkpoint genes leads to CIN and aneuploidy (13). However, extensive analysis has revealed that these genes are only rarely mutated in cancer (17-20).

Several mitotic kinases regulate M phase progression by protein phosphorylation (15). The most prominent is the cyclin-dependent kinase Cdk1. Others include members of the Polo and Aurora families (21). The family of **Polo-like kinases (Plks)** comprises four isoenzymes, Plk1 to Plk4, which function in spindle duplication and entry into mitosis (22-24). Immunofluorescent confocal laser scanning microscopy revealed that Plk1 binds to components of the spindle at all stages of mitosis but is not associated with chromosomes at any stage (22). In case of DNA damage Plk1 is inactivated and depleted as part of the DNA damage-induced G2/M checkpoint. The Plk1 down-regulation was shown to be mediated via p53-dependent transcriptional repression, which involves direct interaction of p53 with the Plk1 promoter (25). p53-null cells were unable to down-regulate Plk1 levels in response to genotoxic drugs. Various tumor types, including breast cancer, express high levels of Plk1 mRNA (26). Mutations in polo, the homologous *Drosophila* gene, caused abnormal mitotic and meiotic divisions due to abnormal spindle formation. The spindle abnormalities included monopolar spindles, highly branched bipolar spindles, and overcondensed chromosomes (27). Plk mutations have not been described in human tumors.

The **Aurora family** of centrosome-associated serine/threonine kinases is composed of three members, Aurora A, B, and C, which share similar N-terminal regulatory domains and highly related C-terminal kinase domains (15, 28-33). The Aurora kinases are pivotal to the successful execution of cell division (34). Together they ensure the formation of a bipolar spindle, accurate segregation of chromosomes and the completion of cytokinesis. Aurora A was shown to phosphorylate centrin and localize with centrin at the centrosome (35). Loss or dysfunction of the *Drosophila* gene aurora (homolog of Aurora A) caused failure of the centrosomes to separate and form a bipolar spindle (36). On the other hand, overexpression of Aurora A induced centrosome amplification and aneuploidy (37). Disruption of Aurora B caused chromosome misalignment and cytokinesis failure in both *Drosophila* and HeLa Cells (29). Aurora B and Aurora C are part of the chromosomal passenger complex, which also comprises survivin, borealin, and the inner centromere protein, INCENP (29, 38).

Experiments in yeast have revealed many potential causes of aneuploidy reflecting the participation of multiple genes in chromosome condensation, sister-chromatid cohesion, kinetochore structure and function and centrosome/ microtubule formation and dynamics as well as cell cycle checkpoints (3, 14, 15). However, as pointed out by Duesberg (39), mutations in any of these genes are found in only a minority of cancers. Instead, he proposes that aneuploidy is the cause rather than a consequence of cancer (40). Because aneuploidy unbalances numerous genes, it may explain, by abnormal dosages of normal genes and thus independent of gene mutation, not only abnormal centrosomes but also the many other dominant, cancer-specific phenotypes (41-43). Duesberg argues further that aneuploidy can change cellular phenotypes much better than gene mutations, which are typically recessive. However, it remains unclear whether aneuploidy causes cancer (20). Do cancer cells divide uncontrollably because they are aneuploid or because they have accumulated mutations that allow them to tolerate aneuploidy (44, 45)?

References

1. Albertson DG, Collins C, McCormick F, Gray JW. *Chromosome aberrations in solid tumors.* Nat Genet. 2003;34:369-76.

2. Feldser DM, Hackett JA, Greider CW. *Telomere dysfunction and the initiation of genome instability.* Nat Rev Cancer. 2003;3:623-7.

3. Lengauer C, Kinzler KW, Vogelstein B. *Genetic instabilities in human cancers.* Nature. 1998;396:643-9.

4. Stearns T, Winey M. *The cell center at 100.* Cell. 1997;91:303-9.

5. Salisbury JL. *Centrin, centrosomes, and mitotic spindle poles.* Curr Opin Cell Biol. 1995;7:39-45.

6. Salisbury JL, Suino KM, Busby R, Springett M. *Centrin-2 is required for centriole duplication in mammalian cells.* Curr Biol. 2002;12:1287-92.

7. Kellogg DR, M. M, Alberts BM. *The centrosome and cellular organization.* Ann Rev Biochem. 1994;63:639-74.

8. Murphy TD, Karpen GH. *Centromeres take flight: alpha satellite and the quest for the human centromere.* Cell. 1998;93:317-20.

9. Pluta AF, Mackay AM, Ainsztein AM, Goldberg IG, Earnshaw WC. *The centromere: hub of chromosomal activities.* Science. 1995;270:1591-4.

10. Wiens GR, Sorger PK. *Centromeric chromatin and epienetic effects in kinetochore assembly.* Cell. 1998;93:313+6.

11. Blow JJ, Dutta A. *Preventing re-replication of chromosomal DNA.* Nat Rev Mol Cell Biol. 2005;6:476-86.

12. Lingle WL, Lutz WH, Ingle JN, Maihle NJ, Salisbury JL. *Centrosome hypertrophy in human breast tumors: implications for genomic stability and cell polarity.* Proc Natl Acad Sci USA. 1998;95:2950-5.

13. Kops GJ, Weaver BA, Cleveland DW. *On the road to cancer: aneuploidy and the mitotic checkpoint.* Nat Rev Cancer. 2005;5:773-85.

14. Musacchio A, Salmon ED. *The spindle-assembly checkpoint in space and time.* Nat Rev Mol Cell Biol. 2007;8:379-93.

15. Nigg EA. *Mitotic kinases as regulators of cell division and its checkpoints.* Nat Rev Mol Cell Biol. 2001;2:21-32.

16. Hornig NC, Knowles PP, McDonald NQ, Uhlmann F. *The dual mechanism of separase regulation by securin.* Curr Biol. 2002;12:973-82.

17. Langerod A, Stromberg M, Chin K, Kristensen VN, Borresen-Dale AL. *BUB1 infrequently mutated in human breast carcinomas.* Hum Mutat. 2003;22:420.

18. Myrie KA, Percy MJ, Azim JN, Neeley CK, Petty EM. *Mutation and expression analysis of human BUB1 and BUB1B in aneuploid breast cancer cell lines.* Cancer Lett. 2000;152:193-9.

19. Percy MJ, Myrie KA, Neeley CK, Azim JN, Ethier SP, Petty EM. *Expression and mutational analyses of the human MAD2L1 gene in breast cancer cells.* Genes Chromosomes Cancer. 2000;29:356-62.

20. Weaver BA, Cleveland DW. *Does aneuploidy cause cancer?* Curr Opin Cell Biol. 2006;18:658-67.

21. Lens SM, Voest EE, Medema RH. *Shared and separate functions of polo-like kinases and aurora kinases in cancer.* Nat Rev Cancer. 2010;10:825-41.

22. Golsteyn RM, Mundt KE, Fry AM, Nigg EA. *Cell cycle regulation of the activity and subcellular localization of PLK1, a human protein kinase implicated in mitotic spindle fuction.* J Cell Biol. 1995;129:1617-28.

23. Lee KS, Yuan YO, Kuriyama R, Erikson RL. *Plk is an M-phase-specific protein kinase and interacts with a kinesin-like protein, CHO1/MKLP-1.* Mol Cell Biol. 1995;15:7143-51.

24. Nigg EA. *Polo-like kinases: positive regulators of cell division from start to finish.* Current Opin Cell Biol. 1998;10:776-83.

25. McKenzie L, King S, Marcar L, Nicol S, Dias SS, Schumm K, et al. *p53-dependent repression of polo-like kinase-1 (PLK1).* Cell Cycle. 2010;9:4200-12.

26. Holtrich U, Wolf G, Brauninger A, Karn T, Bohme B, Rubsamen-Waigmann H, et al. *Induction and down-regulation of PLK, a human serine/threonine kinase expressed in proliferating cells and tumors.* Proc Natl Acad Sci USA. 1994;91:1736-40.

27. Llamazares S, Moreira A, Tavares A, Girdham C, Spruce BA, Gonzalez C, et al. *Polo encodes a protein kinase homolog required for mitosis in Drosophila.* Genes Dev. 1991;5:2153-65.

28. Bernard M, Sanseau P, Henry C, Couturier A, Prigent C. *Cloning of STK13, a third human protein kinase related to Drosophila aurora and budding yeast Ipl1 that maps on chromosome 19q13.3-ter.* Genomics. 1998;53:406-9.

29. Fu J, Bian M, Jiang Q, Zhang C. *Roles of Aurora kinases in mitosis and tumorigenesis.* Mol Cancer Res. 2007;5:1-10.

30. Kimura M, Matsuda Y, Eki T, Yoshioska T, Okumera K, Okano Y. *Assignment of STK6 to human chromosome 20q13.2-q13.3 and a pseudogene STK6 to 1q41--q42.* Cytogenet Cell Genet. 1997;79:201-3.

31. Kimura M, Matsuda Y, Yoshioka T, Okano Y. *Cell cycle-dependent expession and centrosome localization of a third human aurora /Ipl1-related protein kinase, AIK3.* J Biol Chem. 1999;274:7334-40.

32. Kimura M, Matsuda Y, Yoshioka T, Sumi N, Okano Y. *Identification and characterization of STK12/Aik2: a human gene related to aurora of Drosophila and yeast IPL1.* Cytogenet Cell Genet. 1998;82:147-52.

33. Shindo M, Nakano H, Kuroyanagi H, Shirasawa T, Mihara M, Gilbert DJ, et al. *cDNA cloning, expression, subcellular localization, and chromosomal assignment of mammalian aurora homologues, aurora-related kinase (ARK) 1 and 2.* Biochem Biophys Res Commun. 1998;244:285-92.

34. Crane R, Gadea B, Littlepage L, Wu H, Ruderman JV. *Aurora A, meiosis and mitosis.* Biol Cell. 2004;96:215-29.

35. Lukasiewicz KB, Greenwood TM, Negron VC, Bruzek AK, Salisbury JL, Lingle WL. *Control of centrin stability by Aurora A.* PLoS One. 2011;6:e21291.

36. Glover DM, Leibowitz MH, McLean DA, Parry H. *Mutations in aurora prevent centrosome separation leading to the formation of monopolar spindles.* Cell. 1995;81:95-105.

37. Zhou H, Kuang J, Zhong L, Kuo W, Gray JW, Sahin A, et al. *Tumour amplified kinase STK15/BTAK induces centrosome amplification, aneuploidy* and transformation. Nature Genet. 1998;20:189-93.

38. Ruchaud S, Carmena M, Earnshaw WC. *Chromosomal passengers: conducting cell division.* Nat Rev Mol Cell Biol. 2007;8:798-812.

39. Duesberg P. *Are centrosomes or aneuploidy the key to cancer?* Science. 1999;284:2091-2.

40. Duesberg P, Li R, Fabarius A, Hehlmann R. *The chromosomal basis of cancer.* Cell Oncol. 2005;27:293-318.

41. Duesberg P, Rausch C, Rasnick D, Hehlmann R. *Genetic instability of cancer cells is proportional to their degree of aneuploidy.* Proc Natl Acad Sci USA. 1998;95:13692-7.

42. Li R, Yerganian G, P. D, Kramer A, Willer A, Hehlmann R. *Aneuploidy correlated 100% with chemical transformation of Chinese hamster cells.* Proc Natl Acad Sci USA. 1997;94:14506-11.

43. Rasnick D, Duesberg PH. *How aneuploidy affects metabolic control and causes cancer.* Biochem J. 1999;340:621-30.

44. Berman J. *Evolutionary genomics: When abnormality is beneficial.* Nature. 2010;468:183-4.

45. Williams BR, Amon A. *Aneuploidy: cancer's fatal flaw?* Cancer Res. 2009;69:5289-91.

6.5.2.2. CHROMOSOME INSTABILITY IN BREAST CANCER

Mitotic figures are rarely found in normal breast epithelium although a population of proliferating cells is maintained at all times. The greatest likelihood of finding mitoses is in breast tissue from premenopausal women in the luteal phase of the menstrual cycle (1-3). Several techniques have been employed to assess the extent of proliferation, e.g., *in vitro* labeling with tritiated thymidine, *in vivo* labeling with 5'-bromodeoxyuridine, and immunohistochemical staining with anti-Ki67 or anti-PCNA antibodies (3-6). The proliferative index, calculated as the percentage of labeled or stained cells out of the total number of epithelial cells, was technique-dependent, ranging from 0.2 to 12%. The fraction of proliferating epithelial cells declined while the turnover time lengthened with increasing age of the woman (2-4). Based on a postulated duration of DNA synthesis of 12 hours, the mean turnover of the cells of normal terminal ducts was calculated to be approximately 22 days for 20-year old women with regular menstrual cycles and increase to 147 days at age 40.

An ultrastructural study of over 100 epithelial cells undergoing mitosis in normal breast showed that the mitotic cells were luminally positioned polarized cells with two specific orientations of the mitotic spindle (7). In one orientation, the spindle formed parallel to the lumen and division resulted in two luminally positioned daughter cells. This orientation would increase the number of luminal epithelial cells by one while maintaining the normal architecture. In the majority of mitotic cells, the spindle was oriented at right angles to the lumen, which resulted in a luminally and a basally positioned daughter cell. Although the fate of the basal daughter cells cannot be determined with certainty, the data suggest that the basal daughter cell

could develop into a myoepithelial cell or undergo apoptosis. Thus, the two orientations of mitosis would allow a single luminally positioned proliferative cell to give rise to both epithelial and myoepithelial cells without disrupting the integrity of the glandular architecture. The deletion of basal daughter cells by apoptosis could represent an adaptation to prevent cellular proliferation in the 'resting' breast, which shows little evidence of growth of the lobular units despite repeated proliferative stimulation during the menstrual cycle. Another ultrastructural study showed that normal mammary epithelium contains two apically positioned centrioles with sparse electron-dense pericentriolar material. In contrast, breast cancers showed supernumerary centrioles, excess pericentriolar material, disrupted centriole barrel structure, unincorporated microtubule complexes, centrioles of unusual length, centrioles functioning as ciliary basal bodies, and mispositioned centrosomes (5). Centrins are required for the duplication of centrioles (8). In normal mammary epithelial cells, centrins are phosphorylated only during mitosis but not during interphase. In contrast, in malignant epithelial cells, centrins were phosphorylated at inappropriate times during the cell cycle and phosphocentrin accumulated in the unusually large tumor centrosomes. Immunohistochemical staining with anti-centrin antibody showed that normal mammary epithelial cells contain distinct pairs (1.5 ± 0.3) of spots corresponding to centrioles. In contrast, malignant mammary epithelial cells contain two-to-four times (4.3 ± 1.2) this number (5, 9). One study found a positive correlation between centrosome defects and aneuploidy with copy numbers of chromosomes 1 and 8 ranging from 1 to 22 (10).

About 80% of primary breast cancers are aneuploid. Analysis of these aneuploid tumors by flow cytometry reveals that the majority has an excess number of chromosomes referred to as polyploidy (Table 6.5.1.). Near-tetraploid karyotypes result from mis-segregation of single chromosomes before or after doubling of the genome, usually from failure of cytokinesis, the process of cytoplasmic division. Less than 2% of tumors are hypodiploid.

The spindle assembly or mitotic checkpoint combines the action of multiple proteins, including Mad1, Mad2, Bub1, Bub3, Mad3 (BubR1/Bub1B in humans), and Mps1 (TTK), which inhibit the Cdc20-dependent recognition of cyclin B and securin (PTTG1) by APC/C, thereby preventing advance to anaphase (13). Abnormal expression of these mitotic checkpoint genes leads to chromosomal instability and aneuploidy (14). A comprehensive mRNA expression analysis of 76 mitotic spindle checkpoint genes in 9 normal breast tissues, 14 benign breast tumors, 14 DCIS, 11 invasive grade I and 12 invasive grade III breast cancers revealed abnormal expression of 49 genes (64.5%) (15). Compared to normal breast tissues, the cancers showed upregulation of 40 genes and down-

Table 6.5.1. Aneuploidy in Breast Cancer (11, 12)

Stemline*	DNA Index	Frequency (%)
Hypo-diploid	<0.95	1 – 2
Near-diploid	1.1 – 1.4	5 – 15
Triploid	1.4 – 1.6	5 – 15
Hypotetraploid	1.7 – 1.8	10 – 25
Tetraploid	1.9 – 2.1	5 – 15
Hypertetraploid	>2.1	5 – 15

* 5 to 10% of breast cancers contain more than one aneuploid stemline

regulation of 9 genes. Most of these expression changes were observed in epithelial cells and several already detected in benign tumors (e.g., Bub1, BubR1/Bub1B, NDC80, cyclin B1, cyclin A2, Cdk1, Cdc20), suggesting that they may be involved in the malignant transformation. A study of invasive breast cancers showed that 15 of 25 tumors expressed Mad1 at fivefold greater levels than normal breast samples. To determine the relevance of Mad1 overexpression a microarray analysis of 242 breast cancers was performed, which showed an association of Mad1 overexpression with poor prognosis and p53 mutation but not estrogen receptor status (16). A study of 19 breast cancers showed an increase in Mad2 and BubR1/Bub1B transcripts and a positive association of BubR1/Bub1B expression with CIN (17). Moreover, BubR1/Bub1B expression increased progressively from atypical lesions to high-grade invasive breast cancer (18, 19). Synuclein gamma (also known as breast cancer-specific gene 1, BCSG1), which is not expressed in normal breast but abundantly expressed in poorly differentiated breast cancer, was shown to bind and degrade BubR1/Bub1B (20, 21). Increased Mps1/TTK expression was observed in poorly differentiated breast cancer suggesting that high Mps1 levels permit cancer cells to tolerate aneuploidy (22).

The mitotic checkpoint genes contain polymorphisms which have been examined for possible association with breast cancer, yielding inconsistent results. Two nonsynonymous polymorphisms in Mad1 (Arg558His; rs1801368) and Bub1B (Gln348Arg; rs1801376) genes were identified in twelve breast cancer cell lines but not associated with apparent functional protein changes (19). An analysis of nine SNPs in six spindle checkpoint genes (Mad2, Mad2L2, Bub3, Bub1B, CENPE) or any of the haplotypes showed no association with breast cancer risk in a study of 441 patients with familial breast cancer and 552 controls (23). An analysis of 13 SNPs in five genes (Mad2, Bub1, Bub1B, Mps1/TTK, securin/PTTG1) in a Taiwanese case-control study (698 cancers, 1492 controls) showed an increased association with breast cancer risk for two SNPs in Mps1/TTK (intron; rs151658; OR 1.40; 95%CI 1.03 – 1.89) and securin/PTTG1 (intron; rs2910203; OR 1.33; 95% CI 1.00 – 1.78) (24). The

combination of the two SNPs was associated with a greater risk which was further increased by inclusion of reproductive risk factors. Only one of 22 breast cancer cell lines contained a mutation in the Mad2 gene causing a frameshift that creates a truncated Mad2 protein product (25). No Bub1 mutations were detected in 19 aneuploid breast cancer cell lines or in 20 breast cancers with CIN (25, 26).

The families of polo-like kinases (Plks) and Aurora kinases regulate the formation of a bipolar spindle, accurate segregation of chromosomes, and the completion of cytokinesis (27-30). Various tumor types, including breast cancer, express high levels of Plk1 mRNA (31). Since Plk1 mRNA was undetectable in surrounding tissue, its expression appears to be associated with cell proliferation. Indeed, experimental Plk1 depletion/inhibition induced an increase in G2/M arrest and apoptosis as well as reduced viability of breast cancer cell lines (32). An immunohistochemical analysis of 135 breast carcinomas revealed strong Plk isoform overexpression in 42.2% (Plk1) and 47.7% (Plk3) of tumors when compared to normal breast tissue (33). Another study of 215 breast cancers revealed immunohistochemical staining of Plk1 in 23 (11%) tumors and showed a significant association (p = 0.0063) with the presence of mutated p53 (34). Triple-negative breast cancers had the highest levels of Plk1 expression compared with other breast cancer subtypes (32, 34).

Aurora A activity can be controlled at several levels, including phosphorylation, ubiquitin-dependent proteolysis and interaction with both positive regulators, such as TPX2, and negative ones, like p53 (35). After four months of estradiol treatment, levels of Aurora A and centrosomal proteins, γ-tubulin and centrin, rose significantly in female ACI rat mammary glands and remained elevated in mammary tumors at five or six months of estrogen treatment (35). The estrogen-induced Aurora A overexpression was associated with centrosome amplification and aneuploidy. Transient transfection of the near diploid human breast epithelial cell line MCF10A with a STK15 (Aurora A) expression vector induced aneuploidy as shown by an increase in centromere number for chromosomes 13, 21, and X, which are diploid in MCF10A. Approximately 12% STK15 transfectants revealed more than two centrosomes, compared with less than 3% of the vector-transfected control cells (36). These results suggest that specific gene abnormalities can directly influence chromosome ploidy in tumor cells. Loss of p53 function due to mutational inactivation, deletion, or Mdm2 overexpression may also cause centrosome abnormalities (37, 38). However, analysis of breast cancer cell lines and tumors failed to show a strict correlation between p53 immunohistochemistry labeling index and CIN or centrosome alterations (9, 39).

Table 6.5.2. Mitotic checkpoint proteins and kinases in breast cancer

Protein	Function	Alteration in Breast Cancer	Reference
Centrin	Duplication of centrioles	Aberrant centrin phosphorylation	(9)
Mad1	Mad1 recruits Mad2 to kinetochore, catalyzes formation of Mad2-Cdc20 complexes	Mad1 overexpression	(16)
Mad2	Mad2 binds to Cdc20 and inhibits APC/C	Mad2 overexpression	(17, 25)
BubR1/Bub1B	Inhibition of APC/C	BubR1/Bub1B expression progressively increasing from atypical lesions to high grade breast cancer	(17-19)
Mps1/TTK	Accurate segregation of chromosomes	Mps1/TTK overexpression	(22)
Plk1	Formation of bipolar spindle	Plk1 overexpression associated with mutant p53 and triple-negative subtype	(32-34)
Aurora A	Formation of bipolar spindle, accurate segregation of chromosomes	Aurora A amplification and overexpression	(36, 40)

Immunohistochemical analysis revealed that Aurora A was overexpressed in 94% (31 of 33) breast cancers irrespective of histopathological type, whereas the protein was not detected in normal ductal and lobular cells (40). The staining in the tumor cells was predominantly cytoplasmic. Benign breast lesions including fibrocystic disease and fibroadenoma (epithelial component) displayed weakly detectable Aurora A expression in parts of the lesions. The molecular mechanisms by which Aurora A protein is overexpressed in breast cancer cells have not been defined. The Aurora A gene at 20q13.2-13.3 is frequently amplified in breast cancers and breast cancer cell lines such as BT474 and MDA-MB-231 (36, 41-44). However, compared to the 94% of breast cancers with Aurora A protein overexpression, the proportion of cases with Aurora A gene amplification was low, i.e., between 12 and 25% (36, 40). The Aurora A gene at 20q13.2-13.3 contains a nonsynonymous single nucleotide polymorphism, Phe31Ile (1712T>A), which showed no association with familial breast cancer risk (45). The Aurora B gene at 17p13.1 contains a nonsynonymous polymorphism, Thr298Met (893G>A), which showed no association whereas the synonymous Ser295Ser (885A>G) polymorphism resulted in an increased risk for carriers of the homozygous 885G genotype (OR = 1.45; 95% CI 1.05 – 2.0; p = 0.02).

In summary, centrosome abnormalities are a common feature of breast cancers and these defects contribute causally to CIN of mammary tumors. Abnormal expression of mitotic checkpoint genes is frequently detected in breast cancers (Table 6.5.2.). However, mutational inactivation of these genes occurs rarely in malignant tumors including breast cancer because complete inhibition of mitotic checkpoint genes is deleterious for normal cells as well as cancer cells (13, 14, 46).

References

1. Ferguson DJ, Anderson TJ. *Morphological evaluation of cell turnover in relation to the menstrual cycle in the "resting" human breast.* Br J Cancer. 1981;44:177-81.

2. Going JJ, Anderson TJ, Battersby S, MacIntyre CC. *Proliferative and secretory activity in human breast during natural and artificial menstrual cycles.* Am J Pathol. 1988;130:193-204.

3. Meyer JS. *Cell proliferation in normal human breast ducts, fibroadenomas and other ductal hyperplasias measured by nuclear laeling with tritiated thymidine. Effects of menstrual phase, age, and oral contraceptive hormones.* Hum Pathol. 1977;8:67-81.

4. Christov K, Chew KL, Ljung BM, Waldman FM, Duarte LA, Goodson WHD. *Proliferation of normal breast epithelial cellls as shown by in vivo labeling with bromodeoxyuridine.* Am J Pathol. 1991;138:1371-7.

5. Lingle WL, Salisbury JL. *Altered centrosome structure is associated with abnormal mitoses in human breast tumors.* Am J Pathol. 1999;155:1941-51.

6. Olsson H, Jernstrom H, Alm P, Kreipe H, Ingvar C, Jonnson P-E, et al. *Proliferation of the breast epithelium in relation to menstrual cycle phase, hormonal use, and reproductive factors.* Breast Cancer Res Treat. 1996;40:187-96.

7. Ferguson DJP. *An ultrastructural study of mitosis and cytokinesis in normal resting human breast.* Cell Tissue Res. 1988;252:581-7.

8. Salisbury JL, Suino KM, Busby R, Springett M. *Centrin-2 is required for centriole duplication in mammalian cells.* Curr Biol. 2002;12:1287-92.

9. Lingle WL, Lutz WH, Ingle JN, Maihle NJ, Salisbury JL. *Centrosome hypertrophy in human breast tumors: implications for genomic stability and cell polarity.* Proc Natl Acad Sci U S A. 1998;95:2950-5.

10. Pihan GA, Purohit A, Wallace J, Knecht H, Woda B, Quesenberry P, et al. *Centrosome defects and genetic instability in malignant tumors.* Cancer Res. 1998;58:3974-93985.

11. Carcangiu ML, Casalini P, Menard S. *Breast tumors: an Overview.* Atlas Genet Cytogenet Oncol Haematol; 2005.

12. Mitelman F, Johansson B, Mertens F. *Mitelman Database of Chromosome Aberrations and Gene Fusions in Cancer.* 2012.

13. Musacchio A, Salmon ED. *The spindle-assembly checkpoint in space and time.* Nat Rev Mol Cell Biol. 2007;8:379-93.

14. Kops GJ, Weaver BA, Cleveland DW. *On the road to cancer: aneuploidy and the mitotic checkpoint.* Nat Rev Cancer. 2005;5:773-85.

15. Bieche I, Vacher S, Lallemand F, Tozlu-Kara S, Bennani H, Beuzelin M, et al. *Expression analysis of mitotic spindle checkpoint genes in breast carcinoma: role of NDC80/HEC1 in early breast tumorigenicity, and a two-gene signature for aneuploidy.* Mol Cancer. 2011;10:23.

16. Ryan SD, Britigan EM, Zasadil LM, Witte K, Audhya A, Roopra A, et al. *Up-regulation of the mitotic checkpoint component Mad1 causes chromosomal instability*

and resistance to microtubule poisons. Proc Natl Acad Sci U S A. 2012;109:E2205-14.

17. Scintu M, Vitale R, Prencipe M, Gallo AP, Bonghi L, Valori VM, et al. *Genomic instability and increased expression of BUB1B and MAD2L1 genes in ductal breast carcinoma.* Cancer Lett. 2007;254:298-307.

18. Nasir A, Chen DT, Gruidl M, Henderson-Jackson EB, Venkataramu C, McCarthy SM, et al. *Novel molecular markers of malignancy in histologically normal and benign breast.* Pathol Res Int. 2011;2011:489064.

19. Yuan B, Xu Y, Woo JH, Wang Y, Bae YK, Yoon DS, et al. *Increased expression of mitotic checkpoint genes in breast cancer cells with chromosomal instability.* Clin Cancer Res. 2006;12:405-10.

20. Gupta A, Inaba S, Wong OK, Fang G, Liu J. *Breast cancer-specific gene 1 interacts with the mitotic checkpoint kinase BubR1.* Oncogene. 2003;22:7593-9.

21. Inaba S, Li C, Shi YE, Song DQ, Jiang JD, Liu J. *Synuclein gamma inhibits the mitotic checkpoint function and promotes chromosomal instability of breast cancer cells.* Breast Cancer Res Treat. 2005;94:25-35.

22. Daniel J, Coulter J, Woo JH, Wilsbach K, Gabrielson E. *High levels of the Mps1 checkpoint protein are protective of aneuploidy in breast cancer cells.* Proc Natl Acad Sci U S A. 2011;108:5384-9.

23. Vaclavicek A, Bermejo JL, Wappenschmidt B, Meindl A, Sutter C, Schmutzler RK, et al. *Genetic variation in the major mitotic checkpoint genes does not affect familial breast cancer risk.* Breast Cancer Res Treat.. 2007;106:205-13.

24. Lo YL, Yu JC, Chen ST, Hsu GC, Mau YC, Yang SL, et al. *Breast cancer risk associated with genotypic polymorphism of the mitotic checkpoint genes: a multigenic study on cancer susceptibility.* Carcinogenesis. 2007;28:1079-86.

25. Myrie KA, Percy MJ, Azim JN, Neeley CK, Petty EM. *Mutation and expression analysis of human BUB1 and BUB1B in aneuploid breast cancer cell lines.* Cancer Lett. 2000;152:193-9.

26. Langerod A, Stromberg M, Chin K, Kristensen VN, Borresen-Dale AL. *BUB1 infrequently mutated in human breast carcinomas.* Hum Mutat. 2003;22:420.

27. Golsteyn RM, Mundt KE, Fry AM, Nigg EA. *Cell cycle regulation of the activity and subcellular localization of PLK1, a human protein kinase implicated in mitotic spindle function.* J Cell Biol. 1995;129:1617-28.

28. Lee KS, Yuan YO, Kuriyama R, Erikson RL. *Plk is an M-phase-specific protein kinase and interacts with a kinesin-like protein, CHO1/MKLP-1.* Mol Cell Biol. 1995;15:7143-51.

29. Lens SM, Voest EE, Medema RH. *Shared and separate functions of polo-like kinases and aurora kinases in cancer.* Nat Rev Cancer. 2010;10:825-41.

30. Nigg EA. *Polo-like kinases: positive regulators of cell division from start to finish.* Curr Opin Cell Biol. 1998;10:776-83.

31. Holtrich U, Wolf G, Brauninger A, Karn T, Bohme B, Rubsamen-Waigmann H, et al. *Induction and down-regulation of PLK, a human serine/threonine kinase expressed in proliferating cells and tumors.* Proc Natl Acad Sci USA. 1994;91:1736-40.

32. Maire V, Nemati F, Richardson M, Vincent-Salomon A, Tesson B, Rigaill G, et al. *Polo-like Kinase 1: A Potential Therapeutic Option in Combination with Conventional Chemotherapy for the Management of Patients with Triple-Negative Breast Cancer.* Cancer Res. 2013;73:813-23.

33. Weichert W, Kristiansen G, Winzer KJ, Schmidt M, Gekeler V, Noske A, et al. *Polo-like kinase isoforms in breast cancer: expression patterns and prognostic implications.* Virchows Arch. 2005;446:442-50.

34. King SI, Purdie CA, Bray SE, Quinlan PR, Jordan LB, Thompson AM, et al. *Immunohistochemical detection of Polo-like kinase-1 (PLK1) in primary breast cancer is associated with TP53 mutation and poor clinical outcome.* Breast Cancer Res. 2012;14:R40.

35. Crane R, Gadea B, Littlepage L, Wu H, Ruderman JV. *Aurora A, meiosis and mitosis.* Biol Cell. 2004;96:215-29.

36. Zhou H, Kuang J, Zhong L, Kuo W, Gray JW, Sahin A, et al. *Tumour amplified kinase STK15/BTAK induces centrosome amplification, aneuploidy and transformation.* Nature Genet. 1998;20:189-93.

37. Carroll PE, Okuda M, Horn HR, Biddinger P, P.J. S, Gleich LL, et al. *Centrosome hyperamplification in human cancer: chromosome instability induced by p53 mutation and/or Mdm2 overexpression.* Oncogene. 1999;18:1935-44.

38. Fukasawa K, Choi T, Kuriyama R, Rulong S, Woude GFV. *Abnormal centrosome amplification in the absence of p53.* Science. 1996;271:1744-7.

39. Yoon DS, Wersto RP, Zhou W, Chrest FJ, Garrett ES, Kwon TK, et al. *Variable levels of chromosomal instability and mitotic spindle checkpoint defects in breast cancer.* Am J Pathol. 2002;161:391-7.

40. Tanaka T, Kimura M, Matsunaga KR, Fukada D, Mori H, Okano Y. *Centrosomal kinase AIK1 is overexpressed in invasive ductal carcinoma of the breast.* Cancer Res. 1999;59:2041-4.

41. Isola JJ, Kallioniemi OP, Chu LW, Fuqua SA, Hilsenbeck SG, Osborne CK, et al. *Genetic aberrations detected by comparative genomic hybridization predict outcome in node-negative breast cancer.* Am J Pathol. 1995;147:905-11.

42. Kallioniemi A, Kallioniemi OP, Piper J, Tanner M, Stokke T, Chen L, et al. *Detection and mapping of amplified DNA sequences in breast cancer by comparative genomic hybridization.* Proc Natl Acad Sci USA. 1994;91:2156-60.

43. Kimura M, Matsuda Y, Eki T, YoshioskaT, Okumera K, Okano Y. *Assignment of STK6 to human chromosome 20q13.2-q13.3 and a pseudogene STK6 to 1q41--q42.* Cytogenet Cell Genet. 1997;79:201-3.

44. Sen S, Zhou H, White RA. *A putative serine/threoine kinase encoding gene BTAK on chromosome 20q13 is amplified and overexpressed in human breast cancer cell lines.* Oncogene. 1997;14:2195-200.

45. Tchatchou S, Wirtenberger M, Hemminki K, Sutter C, Meindl A, Wappenschmidt B, et al. *Aurora kinases A and B and familial breast cancer risk.* Cancer Lett. 2007;247:266-72.

46. Weaver BA, Cleveland DW. *Does aneuploidy cause cancer?* Curr Opinion Cell Biol. 2006;18:658-67.

6.5.2.3. LOSS OF HETEROZYGOSITY AND MICRO-SATELLITE INSTABILITY

Loss of heterozygosity (LOH) is defined as the loss of normal function of one allele of a gene in which the other allele was already inactivated. In malignancies, LOH occurs when the remaining functional allele in a tumor cell becomes inactivated by mutation. This could cause a normal tumor suppressor to no longer be produced, which could result in tumorigenesis. Thus, LOH has been classically viewed as indirect evidence for the possible existence of a tumor suppressor gene within a region affected by the allele loss. For example, the genes BRCA1 and BRCA2 show LOH in breast cancers from patients who have germline mutations. LOH can be identified in cancers by noting the presence of heterozygosity at a genetic locus in germline DNA and the absence of heterozygosity at that locus in the cancer cells. This can be accomplished by using polymorphic markers, such as microsatellites and SNPs for which the two parents contributed different alleles. LOH is the most frequent genetic alteration found in solid tumors and breast cancer is no exception to this observation (1). However, the majority of LOH loci are outside known genes limiting the development of clinical applications for LOH.

The number of chromosomal arms that show LOH in DCIS is smaller than in invasive breast cancer, i.e. the median FAL [fractional allelic loss = number of arms showing LOH divided by total number of informative arms/tumor] is 0.037 for DCIS and 0.05 for invasive cancer (2). The comedo subtype and high nuclear grade DCIS are associated with higher FAL in agreement with the more aggressive clinical behavior of these phenotypes (2, 3). This would be expected if one considers DCIS to be a preinvasive landmark on one of the pathways to invasion. The accumulation of additional mutations and deletions on other chromosomal regions may then result in the invasive phenotype. In DCIS, the highest percentage of LOH was shown for loci on 8p (18.7%), 13q (18%), 16q (28.6%), 17p (37.5%), 17q (15.9%), accompanied by amplification of HER2 and INT2 genes as well as mutation of the p53 gene (3, 4).

The genomes of all eukaryotes contain microsatellites and minisatellites. **Microsatellites** (also known as short tandem repeats or simple sequence repeats) are sequences of DNA comprising multiple copies of a repeat unit of 1 – 6 base pairs. A particularly common microsatellite consists of multiple copies of the dinucleotide GT, known as poly(GT). The human genome has about 100,000 poly(GT) microsatellites, each 20 or more base pairs in length. Microsatellites are particularly prone to frameshift mutation by insertion-deletion loop formation during replication resulting from DNA polymerase slippage (see Chapter 5.4. DNA Mismatch Repair, MMR). For example, poly(GT) tracts alter at rates about 10^{-4} or 10^{-5} per division in mammalian cells (5). Mutation rates favor insertions, longer repeats, and heterozygous sites (6, 7). In summary, microsatellites are common, polymorphic, and distributed widely throughout the genome.

The insertion or deletion of microsatellite repeats is called **microsatellite instability (MSI)** (8). Cells deficient in MMR are prone to mismatch and recombination errors resulting in a replication error-positive [RER+] or mutator phenotype and promiscuity in inter- and intragenomic recombination, the recombinator phenotype (9-14). These phenotypes are characterized by up to 1000-fold increased rates of spontaneous mutation including MSI, which is observed in nucleotide repeats of non-coding as well as coding regions of genes such as transforming growth factor β1 receptor II (TGFβ1RII), insulin-like growth factor II receptor (IGFIIR), BCL-2-associated X (BAX), retinoblastoma-protein-interacting zinc finger (RIZ), and transcription factor 4 (TCF 4) (15-19). The genes with coding microsatellites accumulate frame-shift mutations and thereby lose function. Deficiency of MMR proteins plays a causative role in the hereditary non-polyposis colorectal cancer (HNPCC) syndrome, which includes colorectal, endometrial, and ovarian cancers. MSI also occurs in sporadic colorectal cancers (about 15%) and has been detected in a variety of other cancer types (8, 20, 21). Investigations of MSI in breast cancer have yielded inconsistent results, which may in part be due to the choice of microsatellite markers, i.e., the number of loci and the type of sequence repeats

assayed that can influence the MSI detection rate (15, 22-26). Overall, MSI is present in less than 5% of primary breast cancers. There is no consistent association of MMR defects with breast cancer risk.

Minisatellites are tandem arrays of a locus-specific consensus sequence that varies between 14 and 100 bp in length. Minisatellites are often polymorphic in the number of tandem repeats of the consensus, hence the alternative designations, **variable number of tandem repeats (VNTRs)** or variable tandem repetitions (VTRs). Dispersed throughout the genome, minisatellites are often situated just upstream or downstream of genes; many occur within introns. For example, the H-ras VNTR at 11p15.5 is 1000 bp downstream from the polyadenylation signal. The VNTR is composed of 30 to 100 units of a 28-bp consensus sequence (27). Detailed analysis revealed about 130 alleles differing in length from 1000 to 3000 bp (28). The four most common alleles (a1, a2, a3, a4) represent approximately 90% of all alleles in Caucasians. Allele 3.5 is rare in Caucasians but common in African Americans (29). The rare alleles (individual frequencies < 0.5%) are thought to derive from germline mutations of the nearest common alleles (30, 31). Molecular epidemiological investigations have shown that rare H-ras VNTR alleles occur more commonly in cancer patients. A meta-analysis of 23 studies found that aggregation of rare alleles appeared twice as frequently in the genomes of cancer patients as in normal controls (32). The association between mutant alleles and risk of cancer varied with the type of cancer. A highly significant association was observed for sporadic breast cancer (29, 33). One of 11 patients with sporadic breast cancer has rare H-ras VNTR alleles (OR 2.29; 95% CI 1.18 – 4.46) (32). There was no association between rare H-ras VNTR alleles and familial breast cancer (34, 35).

Both micro- and minisatellites are extraordinarily hyperallelic. Many loci display dozens of alleles. As a consequence, the heterozygosity rate (het rate), or fraction of individuals in the population with two different alleles, can approach 100%. For example, the insulin gene at 11p15.5 contains a VNTR upstream of the transcriptional start site that has a het rate in excess of 90%. The mucin 1 (MUC1) gene at 1q21 contains both a VNTR region within exon 2 and a microsatellite (CA repeat) sequence within intron 6. A study of the MUC1 gene in 118 women with breast cancer found no association between CA repeat micro- and VNTR minisatellite suggesting independent mechanisms for these changes (36). The CA repeat was altered at a higher frequency than two other microsatellites at 1q21.

References

1. Devilee P, Schuuring E, van de Vijver MJ, Cornelisse CJ. *Recent developments in the molecular genetic understanding of breast cancer.* Crit Rev Oncog. 1994;5:247-70.

2. Radford DM, Phillips NJ, Fair KL, Ritter JH, Holt M, Donis-Keller H. *Allelic loss and the progression of breast cancer.* Cancer Res. 1995;55:5180-3.

3. Tsuda H, Fukutomi T, Hirohashi S. *Pattern of gene alterations in intraductal breast neoplasms associated with histological type and grade.* Clin Cancer Res. 1995;1:261-7.

4. Radford DM, Fair KL, Phillips NJ, Ritter JH, Steinbrueck T, Holt MS, et al. *Allelotyping of ductal carcinoma in situ of the breast: deletion of loci on 8p, 13q, 17p and 17q.* Cancer Res. 1995;55:3399-405.

5. Farber RA, Petes TD, Dominska M, Hudgens SS, Liskay RM. *Instability of simple sequence repeats in a mammalian cell line.* Hum Mol Genet. 1994;3:253-6.

6. Amos W, Flint J, Xu X. *Heterozygosity increases micro-satellite mutation rate, linking it to demographic history.* BMC Genet. 2008;9:72.

7. Yamada NA, Smith GA, Castro A, Roques CN, Boyer JC, Farber RA. *Relative rates of insertion and deletion mutations in dinucleotide repeats of various lengths in mismatch repair proficient mouse and mismatch repair deficient human cells.* Mutat Res. 2002;499:213-25.

8. Ionov Y, Peinado MA, Malkhosyan S, Shibata D, Perucho M. *Ubiquitous somatic mutations in simple repeated sequences reveal a new mechanism for colonic carcinogenesis.* Nature. 1993;363:558-61.

9. Aaltonen LA, Peltomaki P, Leach FS, Sistonen P, Pylkkanen L, Mecklin J, et al. *Clues to the pathogenesis of familial colorectal cancer.* Science. 1993;260:812-6.

10. Fishel R, Ewel A, Lescoe MK. *Purified human MSH2 protein binds to DNA containing mismatched nucleotides.* Cancer Res. 1994;54:5539-42.

11. Loeb LA. *Microsatellite instability: Marker of a mutator phenotype in cancer.* Cancer Res. 1994;54:5059-63.

12. Loeb LA. *A mutator phenotype in cancer.* Cancer Res. 2001;61:3230-9.

13. MacPhee DG. *Mismatch repair, somatic mutations and the origins of cancer.* Cancer Res. 1995;55:5489-92.

14. Strand M, Prolla TA, Liskay RM, Petes TD. *Destabilization of tracts of simple repetitive DNA in yeast by mutations affecting DNA mismatch repair.* Nature. 1993;365:274-6.

15. Anbazhagan R, Fujii H, Gabrielson E. *Microsatellite instability is uncommon in breast cancer.* Clin Cancer Res. 1999;5:839-44.

16. Lynch HT, de la Chapelle A. *Hereditary colorectal cancer.* N Engl J Med. 2003;348:919-32.

17. Markowitz S, Wang J, Myeroff L, Parsons R, Sun L, Lutterbaugh J, et al. *Inactivation of the type II TGF-b receptor in colon cancer cells with micrsatellite instability.* Science. 1995;268:1336-8.

18. Rampino N, Yamamoto H, Ionov Y, Li Y, Sawai H, Reed JC, et al. *Somatic frameshift mutations in the BAX gene in colon cancers of the microsatellite mutator phenotype.* Science. 1997;275:967-9.

19. Seitz S, Waßmuth P, Plaschke J, Schackert HK, Karsten U, Santibanez-Koref MF, et al. *Identification of microsatellite instability and mismatch repair gene mutations in breast cancer cell lines.* Genes Chromosomes Cancer. 2003;37:29-35.

20. Peltomaki P, Aaltonen LA, P. S, Pylkkanen L, Medklin J, Jarvinen H, et al. *Genetic mapping of a locus predisposing to human colorectal cancer.* Science. 1993;260:810-2.

21. Thibodeau SN, Gren G, Schaid D. *Microsatellite instability in cancer of the proximal colon.* Science. 1993;260:816-9.

22. Aldaz CM, Chen T, Sahin A, Cunningham J, Bondy M. *Comparative allelotype of in situ and invasive human breast cancer: high frequency of microsatellite instability in lobular breast carcinomas.* Cancer Res. 1995;55:3976-81.

23. Dillon EK, de Boer WB, Papadimitriou JM, Turbett GR. Br J Cancer. 1997;76:156-62.

24. Siah SP, Quinn DM, Bennett GD, Casey G, Flower RL, Suthers G, et al. *Microsatellite instability markers in breast cancer: a review and study showing MSI was not detected at 'BAT 25' and 'BAT 26' microsatellite markers in early-onset breast cancer.* Breast Cancer Res Treat. 2000;60:135-42.

25. Wooster R, Neuhausen SL, Mangion J, Quirk Y, Ford D, Collins N, et al. *Localization of a breast cancer susceptibility gene, BRCA2, to chromosome 13q12-13.* Science. 1994;265:2088-90.

26. Yee CJ, Roodi N, Verrier CS, Parl FF. *Microsatellite instability and loss of heterozygosity in breast cancer.* Cancer Res. 1994;54:1641-4.

27. Capon DJ, Chen EY, Levinson AD, Seeburg PH, Goeddel DV. *Complete nucleotide sequences of the T24 human bladder carcinoma oncogene and its normal homologue.* Nature. 1983;302:33-7.

28. Ding S, Larson GP, Foldenauer K, Zhang G, Krontiris TG. *Distinct mutation patterns of breast cancer-associated alleles of the HRAS1 minisatellite locus.* Hum Mol Genet. 1999;8:515-21.

29. Weston A, Godbold JH. *Polymorphisms of H-ras-1 and p53 in breast cancer and lung cancer: a meta-analysis.* Environ Health Perspect. 1997;105 Suppl:919-26.

30. Kasperczyk A, DiMartino NA, Krontiris TG. *Minisatellite allele diversification: the origin of rare alleles at the HRAS1 locus.* Am J Hum Genet. 1990;47:854-9.

31. Krontiris TG. *Minisatellites and human disease.* Science. 1995;269:1682-3.

32. Krontiris TG, Devlin B, Karp Dd, Robert NJ, Risch N. *An association between the risk of cancer and mutations if the HRAS1 minisatellite locus.* New Engl J Med. 1993;329:517-23.

33. Lidereau R, Escot C, Theillet C, Champeme MH, Brunet M, Gest J. *High frequenc y of rare alleles of the human c-Ha-ras-1 proto-oncogene in breast cancer patients.* J Natl Cancer Inst. 1986;77:697-701.

34. Barkardottir RB, Johannsson OT, Arason A, Gudnason V, Egilsson V. *Polymorphism of the c-Ha-ras-1 proto-oncogene in sporadic and familial breast cancer.* Int J Cancer. 1989;44:251-5.

35. Hall JM, Huey B, Morrow J, Newman B, Lee M, Jones E, et al. *Rare HRAS alleles and susceptibility to human breast cancer.* Genomics. 1990;6:188-91.

36. Waltz MR, Pandelidis SM, Pratt W, Barnes D, Swallow DM, Gendler SJ, et al. *A microsatellite within the MUC1 locus at 1q21 is altered in the neoplastic cells of breast cancer patients.* Cancer Genet Cytogenet. 1998;100:63-7.

6.5.2.4. TELOMERE DYSFUNCTION

Telomeres are specialized structures at the end of chromosomes that are made up of tandem 5'-TTAGGG repeats with a single-stranded G-rich 3' overhang of about 50 – 210 bases and several telomere-associated proteins (1). The single-stranded G-rich 3' overhang invades the duplex repeats to form a large duplex telomere loop (T-loop) and a smaller single-stranded displacement loop (D-loop). This configuration, together with a complex of six telomere-associated proteins called shelterin, creates a telomere cap that protects the chromosome end and distinguishes it from a double-strand break (2). The shelterin complex is composed of telomeric repeat binding factors 1 and 2 (TRF1, TRF2), protection of telomeres 1 (POT1), RAP1, TIN2, and TPP1. TRF1, TRF2, and POT1 directly recognize TTAGGG repeats, i.e., TRF1 and TRF2 form homodimers binding to the double-stranded region, whereas POT1 binds to the single-stranded 3' overhang and D-loop (3). TIN2 and TPP1 bridge the TRF1 and TRF2 subcomplexes (4). Shelterin hides the chromosome ends from nonhomologous end-joining DNA repair and represses the ATM and ATR kinase signaling pathways (5). Thus, telomeres are structures that prevent the ends of a linear chromosome to be mistaken for a double strand break. A special enzyme called telomerase adds telomeres to existing telomeres at the end of chromosomes and maintains the 3' overhang (1, 6). Telomerase is not the usual protein enzyme but is instead a ribonucleoprotein complex composed of telomerase reverse transcriptase (hTERT; 130 kD), an RNA component (hTERC; 153 kD), and the associated proteins dyskerin (57 kD), NOP10 (10 kD), NHP2 (22 kD), and GAR1 (25 kD) (7-9). The hTERT gene at 5p15.33 includes 16 exons, which encode the catalytic subunit of telomerase that uses as its own template the hTERC component encompassing 11 nucleotides (5'-CUAACCCUAAC), complementary to the telomere sequence TTAGGG (10, 11).

Human telomeres are 10 – 15 kb in germ cells and substantially shorter in normal somatic cells. Telomere lengths differ between individual chromosomes and even between chromosome arms of the same chromosome (12, 13). During DNA replication, both DNA strands serve as templates for the synthesis of two complementary strands of DNA. However, DNA polymerase is not capable to replicate linear DNA to its very end. As a result, telomeres shorten progressively with cell divisions. Although telomerase can add telomeres onto the end of chromosomes, telomerase activity is weak or undetectable in normal somatic cells. Therefore, the intrinsic end replication problem cannot be solved and telomere shortening ensues. *In vivo* studies indicate that the average telomere lengths in normal somatic cells shorten at an estimated rate of 15 – 40 bp per year and that telomere erosion declines with age (13-16). It has also been demonstrated that oxidative stress contributes to telomere shortening by inducing DNA damage, specifically by causing cleavage at the central guanine of 5'-GGG in the TTAGGG sequence (17-19). Conversely, decreasing the level of reactive oxygen species in the cell slowed the rate of telomere shortening (20).

Telomere shortening leads to structural telomere changes and triggers DNA damage signals, which lead to cell cycle arrest and so-called replicative senescence that prevents normal somatic cells from replicating indefinitely (1, 3, 21). Specifically, the double-stranded TTAGGG repeats may become too short to bind enough telomere binding proteins for T-loop formation and the single-stranded 3' overhang too short to form a D-loop for the appropriate sealing of the overhang. Several DNA double-strand break repair proteins including ATM, γ-H2AX, 53BP1, MDC1, and NBS1 and checkpoint factors, such as phospho-CHK1 and CHK2, assemble at exceptionally short telomeres in senescent cells. The resulting DNA damage response activates primarily the p53 but also p16[INK4A]/Rb pathways, which can induce both G1 and G2 phase arrest or apoptosis (21-23). In cells with defective p53 or p16[INK4A]/Rb pathways or defective cell cycle checkpoints, telomere dysfunction resulting from excessive telomere attrition or disruption of telomere structure may lead to broken chromosome ends, which tend to fuse with their sister chromatids or other broken chromosomes (1, 24, 25). The propagation of breakage-fusion-bridge cycles generates chromosomal instability in form of aneuploidy, loss of heterozygosity, and gene amplification. In summary, the cell responds

to dysfunctional telomeres by undergoing senescence, apoptosis, or chromosomal instability (26, 27).

Since breast cancers arise from mammary epithelial cells, normal human mammary epithelial cells (HMEC) have been studied extensively for barriers to indefinite growth (28-30). A first senescence barrier, stasis, is stress-associated and mediated by the retinoblastoma (Rb) protein. In cultured HMEC, this barrier is enforced by increasing expression of p16[INK4A]. Stasis can be overcome by inactivation of the p16[INK4A]/Rb pathways, e.g., through loss of p16[INK4A] expression. A second senescence barrier is due to telomere dysfunction resulting from ongoing telomere attrition. This barrier can be overcome by exposure to pathologically relevant agents, such as chemical carcinogens, oncogenes associated with breast cancer, and/or p53 inactivation, to generate immortally transformed HMEC from normal cells. Unlike rodent cells, spontaneous transformation to immortality of human cells cultured from normal tissues is virtually nonexistent. The non-malignant immortally transformed HMEC lines display several properties that resemble DCIS *in vivo*, such as short telomeres, genomic instability, and specific gene expression patterns (28, 29, 31-37). Thus, telomere shortening occurs early during breast carcinogenesis. Direct telomere FISH analysis revealed telomere shortening in 22 of 29 (78%) ductal carcinoma *in situ* lesions and 80 of 114 (70%) invasive cancers (38). Normal telomere lengths were observed in 21% of invasive carcinomas, while only 5% contained elongated telomeres. Surprisingly, moderate telomere shortening was observed in benign secretory cells in approximately 50% of histologically normal lobular units, while such shortening was not seen in myoepithelial cells or epithelial cells lining large ducts.

Since extensive chromosomal instability is detrimental to cell survival, the telomere length must be stabilized to alleviate the instability to a level permitting tumor progression. This is accomplished by telomerase activation, which is required to maintain telomere length for long-term proliferation of cancer cells. Telomerase activity is weak or undetectable in normal somatic cells but strong in the majority of cancers and immortalized cell lines (39). The generation of transgenic mice that express mTERT at high levels in a variety of tissues was associated with the spontaneous development of *in situ* and invasive mammary carcinomas in a significant proportion of aged females (40). Forced expression of hTERT contributed to the malignant transformation of normal human epithelial cells *in vitro* in the presence of cooperating oncogenes (41). A number of transcription factors, tumor suppressors, cell cycle inhibitors, and hormone receptors have been implicated in the control of hTERT expression (42). In ER-positive MCF-7 breast cancer cells, estradiol treatment up-regulated telomerase expression via an estrogen response element in the hTERT promoter (43). Genomic footprinting indicated that one ERE, 950 bp upstream of the translation start site, was occupied in ERα-positive but not ERα-negative cells (44). Estradiol also activates c-Myc expression, which has its own binding site on the hTERT promoter, resulting in a direct and indirect positive effect of estrogen on telomerase activation. Retrovirally introduced c-Myc cDNA resulted in immortalization of normal human mammary epithelial cells in which p16^{INK4A} was inactivated (45). However, while c-Myc introduction immediately resulted in increased activity of transiently transfected hTERT promoter reporter constructs, endogenous hTERT mRNA levels did not change until about 60 population doublings after c-Myc transfection. These results suggest that telomerase activation by c-Myc in nHMEC requires additional genomic changes, such as the removal of repressive chromatin structures around the hTERT locus. It has been proposed that cancer cells can be killed by inducing critical shortening of telomeres with subsequent senescence or apoptosis through inhibition of telomerase. Support for this hypothesis was provided by transfecting HeLa cells with antisense hTERC (10). The cells lost telomeric DNA and began to die after 23 to 26 doublings. Thus, telomerase is a critical enzyme for the long-term growth of tumor cells. In general, telomerase activation correlates with oncogenic transformation. However, some tumors (sarcomas, astrocytomas) can use an alternative telomere lengthening (ALT) mechanism to stabilize the existing telomeres and alleviate chromosomal instability (46).

The crucial role of telomere attrition in cell turnover is highlighted by patients with germ line mutations in telomere biology genes. These patients have very short telomeres and develop dyskeratosis congenita, an inherited bone marrow failure and cancer predisposition syndrome with a high incidence of head and neck squamous cell carcinomas (7, 47-50). Since short telomeres can result in chromosomal instability, a number of epidemiological studies have examined the association between telomere length in surrogate tissues (e.g., white blood cells) and cancer. The strongest evidence for such an association was observed in esophageal, gastric, urinary bladder, and renal cancers (51). The majority were case-control studies, which obtained DNA from the cases after the diagnosis. This could result in reverse causation bias, where changes in surrogate tissue telomere length could be a consequence of the presence of malignant disease rather than an etiologic marker. There were few studies in which samples were collected months or years prior to cancer diagnosis. Studies of breast cancer risk and telomere length in white blood cells have been inconsistent. Three prospective studies found no association between breast cancer risk and telomere length (52-54) and 6 retrospective studies were inconsistent, i.e., shorter telomeres were associated with increased, decreased, or no risk (54-59). A study of 103 breast cancers reported significantly shorter telomere length in more aggressive subtypes, such as luminal B

(n = 28; p = 0.002), HER2-positive (n = 20; p = 0.011), and triple-negative (n = 37; p = 0.0003) tumors (60). Overall, analyses of telomere length in cases and controls by menopausal status were inconsistent. Oxidative biomarkers such as urinary 15-F(2)-isoprostanes or 8-oxo-7,8-dihydroxyguanosine did not modify the risk association (56). The prospective Nurses' Health Study found no association of telomere length with postmenopausal breast cancer risk but observed that plasma estrone and estradiol levels were inversely associated (p = 0.02) with telomere length in white blood cells (52). Another study observed that exogenous hormonal exposure was inversely correlated with telomere length, i.e., increased duration of hormone replacement therapy was associated with reduced telomere length (61).

Over 500 SNPs have been identified in the hTERT gene at 5p15.33 but only two were reported with potential functional effects. A T/C polymorphism (rs2735940; TERT-08) at -1381 in the promoter region was associated with lower telomerase activity and shorter telomere length (62). A population-based breast cancer study (1,995 cases, 2,296 controls) of this SNP found no association with breast cancer risk overall but a reduced risk in women with a family history of breast cancer (63). However, the Long Island Breast Cancer Study Project (1,067 cases, 1,110 controls) found no significant genotype-breast cancer association or genotype-family history interaction for the TERT-08 SNP (64). Another T/C polymorphism (rs2853669; TERT-67) at -244 was associated with lower telomerase activity and marginally longer telomere length for the CC compared with the TT genotype (65). The association of TERT-67 with familial and sporadic breast cancer risk was inconsistent (63, 64, 66). No functional data have been reported for an A/G polymorphism (rs2736109; TERT-07), which is in strong linkage disequilibrium to TERT-67 in the same haplotype block. The Long Island Breast Cancer Study Project observed an association of the G-allele with increased risk in Caucasian women (OR 1.60; 95% CI 1.24 – 2.05; p = 0.0002) (64). However, the same investigators were unable to replicate these results in a sister study (333 cases, 409 unaffected sisters) from the New York site of the Breast Cancer Family Registry; i.e., no significant association was observed between TERT-07 and breast cancer risk (61). A minisatellite tandem repeat (MNS16A) in the putative promoter region of the antisense RNA transcript has been reported to affect telomerase activity (67). A study of Chinese women (1,029 cases, 1,107 controls) observed an increased breast cancer risk for carriers of the shorter MNS16A allele (68). However, no significant correlation was found in the Long Island Breast Cancer Study Project (64). Polymorphisms in other telomere-related genes have been reported but associations with breast cancer risk were inconsistent (61, 63, 64, 69). Somatic mutations of the hTERT gene have been identified in hematological malignancies but not in solid tumors including breast cancer (7). In contrast, hTERT amplification has been observed in 5 of 19 (26%) breast carcinomas (70).

References

1. Cheung AL, Deng W. *Telomere dysfunction, genome instability and cancer.* Front Biosci. 2008;13:2075-90.

2. de Lange T. *Shelterin: the protein complex that shapes and safeguards human telomeres.* Genes Dev. 2005;19:2100-10.

3. Crabbe L, Karlseder J. *In the end, it's all structure.* Curr Mol Med. 2005;5:135-43.

4. O'Connor MS, Safari A, Xin H, Liu D, Songyang Z. *A critical role for TPP1 and TIN2 interaction in high-order telomeric complex assembly.* Proc Natl Acad Sci U S A. 2006;103:11874-9.

5. Palm W, de Lange T. *How shelterin protects mammalian telomeres.* Annu Rev Genet. 2008;42:301-34.

6. Stewart SA, Ben-Porath I, Carey VJ, O'Connor BF, Hahn WC, Weinberg RA. *Erosion of the telomeric single-strand overhang at replicative senescence.* Nat Genet. 2003;33:492-6.

7. Carroll KA, Ly H. *Telomere dysfunction in human diseases: the long and short of it!* Int J Clin Exp Pathol. 2009;2:528-43.

8. Cohen SB, Graham ME, Lovrecz GO, Bache N, Robinson PJ, Reddel RR. *Protein composition of catalytically active human telomerase from immortal cells.* Science. 2007;315:1850-3.

9. Greider CW, Blackburn EH. *A telomeric sequence in the RNA of Tetrahymena telomerase required for telomere repeat synthesis.* Nature. 1989;337:331-7.

10. Feng J, Funk WD, Wang S, Weinrich SL, Avilion AA, Chiu E, et al. *The RNA component of human telomerase.* Science. 1995;269:1236-41.

11. Kilian A, Bowtell DD, Abud HE, Hime GR, Venter DJ, Keese PK, et al. *Isolation of a candidate human telomerase catalytic subunit gene, which reveals complex splicing patterns in different cell types.* Hum Mol Genet. 1997;6:2011-9.

12. Lansdorp PM, Verwoerd NP, van de Rijke FM, Dragowska V, Little MT, Dirks RW, et al. *Heterogeneity in telomere length of human chromosomes.* Hum Mol Genet. 1996;5:685-91.

13. Londono-Vallejo JA, DerSarkissian H, Cazes L, Thomas G. *Differences in telomere length between homologous chromosomes in humans.* Nucleic Acids Res. 2001;29:3164-71.

14. de Lange T, Shiue L, Myers RM, Cox DR, Naylor SL, Killery AM, et al. *Structure and variability of human chromosome ends.* Mol Cell Biol. 1990;10:518-27.

15. Rhyu MS. *Telomeres, telomerase and immortality.* J Natl Cancer Inst. 1995;87:884-94.

16. Unryn BM, Cook LS, Riabowol KT. *Paternal age is positively linked to telomere length of children.* Aging Cell. 2005;4:97-101.

17. Kawanishi S, Oikawa S. *Mechanism of telomere shortening by oxidative stress.* Ann N Y Acad Sci. 2004;1019:278-84.

18. Oikawa S. *Sequence-specific DNA damage by reactive oxygen species: Implications for carcinogenesis and aging.* Environ Health Prev Med. 2005;10:65-71.

19. von Zglinicki T. *Role of oxidative stress in telomere length regulation and replicative senescence.* Ann NY Acad Sci. 2000;908:99-110.

20. Serra V, von Zglinicki T, Lorenz M, Saretzki G. *Extracellular superoxide dismutase is a major antioxidant in human fibroblasts and slows telomere shortening.* J Biol Chem. 2003;278:6824-30.

21. d'Adda di Fagagna F, Reaper PM, Clay-Farrace L, Fiegler H, Carr P, Von Zglinicki T, et al. *A DNA damage checkpoint response in telomere-initiated senescence.* Nature. 2003;426:194-8.

22. Beausejour CM, Krtolica A, Galimi F, Narita M, Lowe SW, Yaswen P, et al. *Reversal of human cellular senescence: roles of the p53 and p16 pathways.* EMBO J. 2003;22:4212-22.

23. Herbig U, Jobling WA, Chen BP, Chen DJ, Sedivy JM. *Telomere shortening triggers senescence of human cells through a pathway involving ATM, p53, and p21(CIP1), but not p16(INK4a).* Mol Cell. 2004;14:501-13.

24. Baird DM. *Mechanisms of telomeric instability.* Cytogenet Genome Res. 2008;122:308-14.

25. McClintock B. *The fusion of broken ends of chromosones following nuclear fusion.* Proc Natl Acad Sci USA. 1942;28:458-63.

26. Murnane JP. *Telomere dysfunction and chromosome instability.* Mutat Res. 2012;730:28-36.

27. Titen SW, Golic KG. *Telomere loss provokes multiple pathways to apoptosis and produces genomic instability in Drosophila melanogaster.* Genetics. 2008;180:1821-32.

28. Garbe JC, Holst CR, Bassett E, Tlsty T, Stampfer MR. *Inactivation of p53 function in cultured human mammary epithelial cells turns the telomere-length dependent senescence barrier from agonescence into crisis.* Cell Cycle. 2007;6:1927-36.

29. Romanov SR, Kozakiewicz BK, Holst CR, Stampfer MR, Haupt LM, Tisty TD. *Normal human mammary epithelial cells spontaneously escape senescence and acquire genomic changes.* Nature. 2001;409:633-7.

30. Stampfer MR, Yaswen P. *Human epithelial cell immortalization as a step in carcinogenesis.* Cancer Lett. 2003;194:199-208.

31. Braig M, Schmitt CA. *Oncogene-induced senescence: Putting the brakes on tumor development.* Cancer Res. 2006;66:2881-4.

32. Chin K, de Solorzano CO, Knowles D, Jones A, Chou W, Rodriguez EG, et al. *In situ analyses of genome instability in breast cancer.* Nat Genet. 2004;36:984-8.

33. Holst CR, Nuovo GJ, Esteller M, Chew K, Baylin SB, Herman JG, et al. *Methylation of p16INK4a promoters occurs in vivo in histologically normal human mammary epithelia.* Cancer Res. 2003;63:1596-601.

34. Kim H, Farris J, Christman SA, Kong BW, Foster LK, O'Grady SM, et al. *Events in the immortalizing process of primary human mammary epithelial cells by the catalytic subunit of human telomerase.* Biochem J. 2002;365:765-72.

35. Li Y, Pan J, Li JL, Lee JH, Tunkey C, Saraf K, et al. *Transcriptional changes associated with breast cancer occur as normal human mammary epithelial cells overcome senescence barriers and become immortalized.* Mol Cancer. 2007;6:7.

36. Stampfer MR, Bodnar A, Garbe J, Wong M, Pan A, Villeponteau B, et al. *Gradual phenotypic conversion associated with immortalization of cultured human mammary epithelial cells.* Mol Biol Cell. 1997;8:2391-405.

37. Stampfer MR, Garbe J, Nijjar T, Wigington D, Swisshelm K, Yaswen P. *Loss of p53 function accelerates acquisition of telomerase activity in indefinite lifespan human mammary epithelial cell lines.* Oncogene. 2003;22:5238-51.

38. Meeker AK, Hicks JL, Gabrielson E, Strauss WM, De Marzo AM, Argani P. *Telomere shortening occurs in subsets of normal breast epithelium as well as in situ and invasive carcinoma.* Am J Pathol. 2004;164:925-35.

39. Kim NW, Piatyszek MA, Prowse KR, Harley CB, West MD, Ho PL, et al. *Specific association of human telomerase activity with immortal cells and cancer.* Science. 1994;266:2011-5.

40. Artandi SE, Alson S, Tietze MK, Sharpless NE, Ye S, Greenberg RA, et al. *Constitutive telomerase expression promotes mammary carcinomas in aging mice.* Proc Natl Acad Sci U S A. 2002;99:8191-6.

41. Hahn WC, Counter CM, Lundberg AS, Beijersbergen RL, Brooks MW, Weinberg RA. *Creation of human tumour cells with defined genetic elements.* Nature. 1999;400:464-8.

42. Ducrest AL, Szutorisz H, Lingner J, Nabholz M. *Regulation of the human telomerase reverse transcriptase gene.* Oncogene. 2002;21:541-52.

43. Kyo S, Takakura M, Kanaya T, Zhuo W, Fujimoto K, Nishio Y, et al. *Estrogen activates telomerase.* Cancer Res. 1999;59:5917-21.

44. Misiti S, Nanni S, Fontemaggi G, Cong YS, Wen J, Hirte HW, et al. *Induction of hTERT expression and telomerase activity by estrogens in human ovary epithelium cells.* Mol Cell Biol. 2000;20:3764-71.

45. Bazarov AV, Hines WC, Mukhopadhyay R, Beliveau A, Melodyev S, Zaslavsky Y, et al. *Telomerase activation by c-Myc in human mammary epithelial cells requires additional genomic changes.* Cell Cycle. 2009;8:3373-8.

46. Muntoni A, Reddel RR. *The first molecular details of ALT in human tumor cells.* Hum Mol Genet. 2005;14 Spec No. 2:R191-6.

47. Alter BP, Giri N, Savage SA, Rosenberg PS. *Cancer in dyskeratosis congenita.* Blood. 2009;113:6549-57.

48. Aubert G, Lansdorp PM. *Telomeres and aging.* Physiol Rev. 2008;88:557-79.

49. Campisi J, Kim SH, Lim CS, Rubio M. *Cellular senescence, cancer and aging: the telomere connection.* Exp Gerontol. 2001;36:1619-37.

50. Gilley D, Tanaka H, Herbert BS. *Telomere dysfunction in aging and cancer.* Int J Biochem Cell Biol. 2005;37:1000-13.

51. Wentzensen IM, Mirabello L, Pfeiffer RM, Savage SA. *The association of telomere length and cancer: a meta-analysis.* Cancer Epidemiol Biomarkers Prev. 2011;20:1238-50.

52. De Vivo I, Prescott J, Wong JY, Kraft P, Hankinson SE, Hunter DJ. *A prospective study of relative telomere length and postmenopausal breast cancer risk.* Cancer Epidemiol Biomarkers Prev. 2009;18:1152-6.

53. Kim S, Sandler DP, Carswell G, De Roo LA, Parks CG, Cawthon R, et al. *Telomere length in peripheral blood and breast cancer risk in a prospective case-cohort analysis: results from the Sister Study.* Cancer Causes Control. 2011;22:1061-6.

54. Pooley KA, Sandhu MS, Tyrer J, Shah M, Driver KE, Luben RN, et al. *Telomere length in prospective and retrospective cancer case-control studies.* Cancer Res. 2010;70:3170-6.

55. Gramatges MM, Telli ML, Balise R, Ford JM. *Longer relative telomere length in blood from women with sporadic and familial breast cancer compared with healthy controls.* Cancer Epidemiol Biomarkers Prev. 2010;19:605-13.

56. Shen J, Gammon MD, Terry MB, Wang Q, Bradshaw P, Teitelbaum SL, et al. *Telomere length, oxidative damage, antioxidants and breast cancer risk.* Int J Cancer. 2009;124:1637-43.

57. Shen J, Terry MB, Gurvich I, Liao Y, Senie RT, Santella RM. *Short telomere length and breast cancer risk: a study in sister sets.* Cancer Res. 2007;67:5538-44.

58. Svenson U, Nordfjall K, Stegmayr B, Manjer J, Nilsson P, Tavelin B, et al. *Breast cancer survival is associated with telomere length in peripheral blood cells.* Cancer Res. 2008;68:3618-23.

59. Zheng YL, Ambrosone C, Byrne C, Davis W, Nesline M, McCann SE. *Telomere length in blood cells and breast cancer risk: investigations in two case-control studies.* Breast Cancer Res Treat. 2010;120:769-75.

60. Heaphy CM, Subhawong AP, Gross AL, Konishi Y, Kouprina N, Argani P, et al. *Shorter telomeres in luminal B, HER-2 and triple-negative breast cancer subtypes.* Mod Pathol. 2011;24:194-200.

61. Shen J, Terry MB, Liao Y, Gurvich I, Wang Q, Senie RT, et al. *Genetic variation in telomere maintenance genes, telomere length and breast cancer risk.* PLoS One. 2012;7:e44308.

62. Matsubara Y, Murata M, Yoshida T, Watanabe K, Saito I, Miyaki K, et al. *Telomere length of normal leukocytes is affected by a functional polymorphism of hTERT.* Biochem Biophys Res Commun. 2006;341:128-31.

63. Savage SA, Chanock SJ, Lissowska J, Brinton LA, Richesson D, Peplonska B, et al. *Genetic variation in five genes important in telomere biology and risk for breast cancer.* Br J Cancer. 2007;97:832-6.

64. Shen J, Gammon MD, Wu HC, Terry MB, Wang Q, Bradshaw PT, et al. *Multiple genetic variants in telomere pathway genes and breast cancer risk.* Cancer Epidemiol Biomarkers Prev. 2010;19:219-28.

65. Hsu CP, Hsu NY, Lee LW, Ko JL. *Ets2 binding site single nucleotide polymorphism at the hTERT gene promoter--effect on telomerase expression and telomere length maintenance in non-small cell lung cancer.* Eur J Cancer. 2006;42:1466-74.

66. Varadi V, Brendle A, Grzybowska E, Johansson R, Enquist K, Butkiewicz D, et al. *A functional promoter polymorphism in the TERT gene does not affect inherited susceptibility to breast cancer.* Cancer Genet Cytogenet. 2009;190:71-4.

67. Wang L, Soria JC, Chang YS, Lee HY, Wei Q, Mao L. *Association of a functional tandem repeats in the downstream of human telomerase gene and lung cancer.* Oncogene. 2003;22:7123-9.

68. Wang Y, Hu Z, Liang J, Wang Z, Tang J, Wang S, et al. *A tandem repeat of human telomerase reverse transcriptase (hTERT) and risk of breast cancer development and metastasis in Chinese women.* Carcinogenesis. 2008;29:1197-201.

69. Varadi V, Brendle A, Brandt A, Johansson R, Enquist K, Henriksson R, et al. *Polymorphisms in telomere-associated genes, breast cancer susceptibility and prognosis.* Eur J Cancer. 2009;45:3008-16.

70. Zhang A, Zheng C, Lindvall C, Hou M, Ekedahl J, Lewensohn R, et al. *Frequent amplification of the telomerase reverse transcriptase gene in human tumors.* Cancer Res. 2000;60:6230-5.

6.5.3. AMPLIFICATION
6.5.3.1. INTRODUCTION

Genetic alterations in human cancer include DNA sequence alterations (mutations) together with copy number gains and losses. The term gene amplification or better DNA amplification refers to the genetic alteration through which a cell gains additional copies of a small part of its genome (1). The size of the amplified unit (amplicon) ranges from ~50 kb to >10 Mb (2-4). The amplified sequences may appear as two abnormal structures that can be visualized by cytogenetic analysis. The first type consists of expanded chromosomal regions that fail to exhibit trypsin-Giemsa bands, and hence were termed homogeneously staining regions (HSRs) (5). The second type consists of extra-chromosomal, self-replicating structures referred to as double minute chromosomes (DMs), which are composed of paired chromatin bodies that lack centromeric regions. Numerous experimental studies of drug-resistance genes (e.g., dihydrofolate reductase [DHFR] and carbamyl-P-synthetase aspartate transcarbamylase dihydroorotase [CAD] in methotrexate [MTX] and N-phosphonacetyl-L-aspartate [PALA]-treated cells, respectively) confirmed that HSRs and DMs were indeed the repositories of amplified genes (6-8). These studies also shed light on molecular events leading to gene amplification.

Stepwise Selection. MTX resistance in cultured cells is acquired by overproduction of DHFR, which results from a dose-related amplification of the DHFR gene. Cells with amplified DHFR genes are generally selected by stepwise selection and cell variants can be obtained eventually with as many as 100 to 1000 DHFR genes. Thus, it appears that gene amplification occurs incrementally in response to multiple, small-step rather than single, large-step selection protocols. However, the trigger for the initial amplification remains unknown. In the 1980s, investigators suggested that overexpression of multidrug resistance genes can precede gene amplification in the development of multidrug resistance (9). Similarly, overexpression of HER2 would confer an initial selective growth advantage to a tumor cell and subsequent gene amplification cause an additional step where the selective growth advantage is enhanced and stabilized. However, experimental data do not provide convincing proof that overexpression precedes amplification (10). On the other hand, there is clear evidence that gene copy number abnormalities are correlated with gene expression levels (11).

A stepwise selection process may lead to amplification of the androgen receptor (AR) gene at Xq11-12, which has been implicated in the resistance of prostate cancer to endocrine therapy (12). Patients with advanced prostate cancer are treated either by classical androgen deprivation (orchiectomy or luteinizing hormone-releasing hormone agonists) or by maximal androgen blockade (castration combined with anti-androgens). About 75% of patients respond favorably and disease palliation is achieved for several months or years. Eventually, however, the disease progresses despite the therapy (13). One study used FISH to examine AR amplification in tumor specimens collected from 54 prostate cancer patients at the time of local recurrence following therapy failure (14). Fifteen (28%) of the recurrent therapy-resistant tumors, but none of the untreated primary tumors showed AR amplification. The mean AR copy number ranged from 2.7 to 28 per cell, with individual cells containing much higher copy numbers (>60 copies/cell). AR amplification was most likely to occur in recurrent tumors from patients who had initially responded well to androgen deprivation and whose endocrine therapy response had lasted for more than 12 months. AR gene amplification detected in tumors during androgen deprivation monotherapy was associated with favorable treatment response to second line combined androgen blockade (15). These are typical features of androgen-sensitive tumors whose proliferation is critically dependent on androgens. Any increase of AR copy number is likely to provide a survival or proliferative advantage to the cells by allowing them to resume androgen-dependent growth. In fact, none of the recurrent tumors from patients who showed no initial response to androgen deprivation or whose disease recurred less than 12 months after therapy contained AR amplification. These results suggest that failure of conventional androgen deprivation therapy in prostate cancer may be caused by clonal expansion of tumor cells that are able to continue androgen-dependent growth in the presence of low, residual levels of circulating androgens originating from the adrenals. Thus, AR amplification is not involved in the genesis of prostate cancer but the increased expression of a wild-type AR gene may play a key role in the progression of prostate cancer to endocrine therapy resistance.

Initial Amplicons are Genetically Unstable. Cytogenetic and molecular studies of drug-selected gene amplification within 20 cell doublings of the initial treatment revealed substantial heterogeneity in number, size, and location of amplicons (16). For example, individual clones contained DHFR amplicons ranging in size from <1 Mb to >30 Mb. Descendants of the same parental cell could contain amplified DHFR genes either in DMs or HSRs (4). Many examples of gene amplification point to a unidirectional progression from populations dominated by cells with extrachromosomal elements at an early time to those dominated by cells with chromosomally amplified regions at a later time (17). In general, amplicons residing in DMs appear less stable than those present in HSRs or, stated in another way, when the amplified genes are stable, they are present on chromosomes.

A cytogenetic analysis determined the percentage of primary tumors with amplified genes on extra- or

intrachromosomal DNA (18). Of 200 primary tumors, 91% contained only DMs, 6.5% contained only HSRs, and 2.5% contained both. In a parallel review of 109 cell lines with cytogenetic or molecular evidence of gene amplification, 60.6% contained DMs, 20.6% contained HSRs, and 12.8% contained both. These data indicate that DMs are the predominant cytogenetic marker for gene amplification in primary tumors, but occur less frequently in established cell lines.

Amplification Models. In current models of DNA amplification, a double-strand break (DSB) is regarded as the principal initiator but other models have been described to account for the variety in amplicon configurations (17, 19, 20). One mechanism that does not require DSBs involves replication fork stalling and template switching while another mechanism is driven by small DNA fragments (21-23). The prevailing model, known as the **breakage-fusion-bridge (BFB)** cycle, is initiated by a double strand break telomeric to the amplifiable gene (24-27). After replication, the two broken sister chromatids fuse. During mitosis, the centromeres of the dicentric chromatid move to opposite poles of the spindle, creating a bridge at anaphase. If this bridge is broken asymmetrically, one daughter cell will have a terminal deletion while the other has an inverted duplication. Each broken chromosome can fuse again after replication, perpetuating the BFB cycle until the broken end is healed. Cells with the terminal deletion are likely to perish. In contrast, cells with the inverted duplication may accumulate higher gene copy numbers with each additional BFB cycle, leading to intrachromosomal amplification. The BFB model invokes the presence of two different sites of chromosome breakage relative to the selected gene: a telomeric site which is involved in initiation and a centromeric site which defines the size and organization of the amplicon. Some of these sites may correspond to 'fragile sites' characterized by a high frequency of gaps and breaks in metaphase chromosomes (19, 24, 28-33).

The BFB model leads to amplicons with an inverted orientation but does not explain tandem repeats. Other models allow different arrangements of amplicons. For example, amplification could start if DNA undergoes unscheduled replication during cell division or if part of the DNA is excised following loop formation (20). Yet another possibility involves unequal segregation at mitosis of extrachromosomal elements, which are also initiated by chromosome breaks (34). If the extrachromosomal DNA contains a replication origin, multiple copies can be generated to form DMs, which eventually reintegrate into chromosomes either at the native site or at a new site. A comprehensive study of human cancer amplicons at sequence-level resolution analyzed 133 different genomic rearrangements of c-myc, N-myc, and HER2 (35). The observed architectures of rearrangement were diverse and highly distinctive, with evidence for sister chromatid

BFB cycles, formation and reinsertion of double minutes, and the presence of bizarre clusters of small genomic fragments. There were characteristic features of sequences at the breakage-fusion junctions, indicating roles for non-homologous end joining and homologous recombination-mediated repair mechanisms together with nontemplated DNA synthesis. The large number of breakpoints sequenced in this study also provided answers to the question whether there are structural features of the genome that are particularly prone to breakage or repair. Sequences surrounding breakpoints were GC-rich (p < 0.01) with a slight excess of polypurine and polypyrimidine runs (p < 0.01) providing evidence for sequence-dependent variation in susceptibility of the genome to somatic rearrangement.

Abnormal cell cycle progression may enhance gene amplification. Specifically, loss of G1-S checkpoint control by either p53 loss or cyclin D1 overexpression enhanced PALA-induced gene amplification (36-38). Transfection of wild type p53 restored G1 control and reduced the frequency of amplification to undetectable levels (39). The source of the amplified DNA can in principle be (i) DNA donated by a sister chromatid (unequal sister chromatid exchange) or (ii) additional DNA from "overreplication" or "rereplication" within a cell cycle. While most studies have focused on the first process, there is evidence that dissociation of the normal progression from S through M phase may result in multiple initiations of DNA replication, generating cells with additional DNA prior to mitosis (7). For example, transient inhibition of DNA synthesis may cause replication of some chromosomal DNA segments more than once in a single S phase (40, 41). Theoretically, this would create a complex structure consisting of replication bubbles within bubbles, also referred to as "onionskin" bubble with free DNA strands of rereplicated DNA (6). Although experimental evidence supports the occurrence of overreplication, it is unlikely to be the source of amplicons. The main argument against amplification by overreplication is the excessive time required for producing amplicons. It would require over 40 hours of replication to generate a 10 Mb amplicon by multiple initiations from a single origin in a single S phase.

The process of gene amplification involves recombination-rearrangement of DNA sequences. Since the recombination events are predominantly nonhomologous, the number of potential recombination "joints" in DNA appears unlimited and the structure of amplified DNA sequences should be highly variable. Indeed, amplicons differ from one cell type to another in terms of lengths of amplified sequences and variety of chromosomal arrangements. They are configured as either direct tandem repeats (head-to-tail orientation) or, more commonly, as a mixture of tandem and inverted repeats (head-to-head and tail-to-tail orientation) (35, 42-45). These latter findings indicate that the structure of amplicons can be generated by a variety of mechanisms. In some instances, the amplicon appears at or near the

site of the normal, non-amplified gene. However, in the majority of cases, amplicons occur on a number of different chromosomes, presumably resulting from secondary translocation rearrangements when cells are subjected to progressive selections. To account for the diverse cytogenetic consequences of gene amplification, it has been proposed that chromosome breakage plays a central role in the amplification process by (i) generating intermediates that are initially acentric and lead to copy number increase primarily by unequal segregation, (ii) creating atelomeric ends that are either incompletely replicated or resected by exonucleases to generate deletions, and (iii) producing recombinogenic ends that provide preferred sites for amplicon relocalization (4). The head-to-tail orientation of amplified genes has counterparts in multigene families such as the drug metabolizing cytochrome P450 (CYP) family. For example, the CYP2D6 gene can be duplicated, multiduplicated or amplified, resulting in ultra-rapid metabolism of CYP2D6 substrates (46). Alleles with two, three, four, five and 13 gene copies in tandem have been reported (47). Those with two-to-five gene copies in tandem have likely been formed by unequal recombination between two homologous but non-allelic sequences flanking the gene. The outcome is one locus that contains two genes in tandem and a deleted gene on the other locus. By contrast, the allele containing 13 copies of the gene has probably occurred by unequal segregation and extrachromosomal replication of acentric DNA (48).

While amplification is a preferred mechanism for cultured cell lines to acquire drug resistance, substantial amplification of drug resistance-related genes appears to occur infrequently in primary or even metastatic tumors. For example, the multiple drug resistance (MDR1) gene at 1q26, which encodes a P-glycoprotein that is responsible for acquired multidrug resistance is more frequently overexpressed as a result of (post)transcriptional events than amplification (10, 49).

DNA amplification is rare in normal cells. For example, exposure of normal mammary epithelial cells, foreskin keratinocytes or diploid fibroblasts to MTX or hydroxyurea failed to generate a single amplification of the DHFR and ribonucleotide reductase genes in a total of 5×10^8 cells (50). Similarly, the frequency of CAD and DHFR gene amplification in seven diploid normal cell populations was below the limit of detection at 10^{-8}, whereas the same genes were amplified 10^{-3} to 10^{-7} in 22 transformed cell lines (51). Why does amplification only occur in tumor cells? If amplification is initiated by a chromosome break, it stands to reason that one of the early, rate-limiting steps in tumor development is a mutation that affects the ability to repair single- or double-strand breaks. Breaks might occur naturally as a consequence of occasional failures in the rejoining reaction carried out by topoisomerases I or II during replication and/or transcription. A normal cell will repair these breaks whereas a transformed cell that is not capable of repair may initiate DNA sequence amplification.

It has been shown that increased levels of normal cellular oncogene products are capable of transforming cells (52, 53). For example, the normal RAS genes can each be turned into oncogenes by two mechanisms: a 5- to 50-fold amplification of the wild type gene or point mutations in codons 12, 13, or 61 (54-57). Each of these missense mutations lock p21ras in its active, GTP-bound state by altering the conformation of p21ras in such a way that the p21ras-GDP/p21ras -GTP equilibrium is shifted toward the latter. Amplification of the wild type RAS gene or mutations at codons 12, 13, or 61 stimulate cell proliferation, induce cell transformation, and are oncogenic, although the mechanism of transformation by amplification of the normal gene is poorly understood. Although transgenic mouse models have illustrated the role of oncogenes in the induction of tumors, oncogene expression in these tumors is driven by strong viral promoters of questionable relevance to human malignancies. Expression of activated HER2/neu under the control of the intact endogenous *neu* promoter in a mouse model of mammary tumorigenesis resulted in accelerated lobulo-alveolar development and formation of focal mammary tumors after a long latency period (52). All tumors carried amplified copies (2 – 22 copies) of the activated *neu* allele and expressed highly elevated levels of *neu* transcript and protein. However, the *neu* amplification was not sufficient for the initiation of mammary carcinogenesis.

Summary. Based on experimental studies of drug-resistance genes, amplification is considered to be the consequence of selective pressure. Presumably, amplification is exploited by cancer cells to increase copy number and hence expression of dominantly acting cancer genes (35). Although such adaptive response of cells to environmental stress may be at work in primary cancers as well, it is much more difficult to clearly identify the associated growth advantage (58). This means that in most malignancies we are faced with a list of amplified genes without a clear indication of the underlying reason. Breast cancer is no exception to this situation as described in Section 6.5.3.2. At the same time, it is clear that distinctive gene amplification and expression patterns provide sufficient proliferative advantage of specific cell types to achieve immortality and clonal dominance resulting in defined breast cancer subtypes; e.g., HER2-enriched tumors (11, 59, 60).

References

1. Schwab M. *Oncogene amplification in solid tumors.* Semin Cancer Biol. 1999;9:319-25.

2. Brodeur GM, Seeger RC. *Gene amplification in human neuroblastomas: Basic mechanisms and clinical implications.* Cancer Genet Cytogenet. 1986;19:101-11.

3. Kinzler KW, Zehnbauer BA, Brodeur GM, Seeger RC, Trent JM, Meltzer PS, et al. *Amplification units containing human n-myc and c-myc genes.* Proc Natl Acad Sci USA. 1986;83:1031-5.

4. Windle B, Draper BW, Yin Y, O'Gorman S, Wahl GM. *A central role for chromosome breakage in gene amplification, deletion formation, and amplicon integration.* Genes Dev. 1991;5:160-74.

5. Biedler JL. *Drug resistance: genotype versus phenotype - Thirty-second G.H.A. Clowes Memorial Award Lecture.* Cancer Res. 1994;54:666-78.

6. Schimke RT. *Gene amplification in cultured animal cells.* Cell. 1984;37:705-13.

7. Schimke RT. *Gene amplification in cultured cells.* J Biol Chem. 1988;263:5989-92.

8. Stark GR, Wahl GM. *Gene amplification.* Ann Rev Biochem. 1984;53:447-91.

9. Kraus MH, Popescu NC, Amsbaugh SC, King CR. *Overexpression of the EGF receptor-related proto-oncogene erbB-2 in human mammary tumor cell lines by different molecular mechanisms.* EMBO J. 1987;6:605-10.

10. Shen DW, Fojo A, Chin JE, Roninson IB, Richert N, Pastan I, et al. *Human multidrug-resistant cell lines: increased mdr1 expression can precede gene amplification.* Science. 1986;232:643-5.

11. Chin K, DeVries S, Fridlyand J, Spellman PT, Roydasgupta R, Kuo WL, et al. *Genomic and transcriptional aberrations linked to breast cancer pathophysiologies.* Cancer Cell. 2006;10:529-41.

12. Koivisto P, Kononen J, Palmberg C, Tammela T, Hyytinen E, Isola J, et al. *Androgen receptor gene amplification: a possible molecular mechanism for androgen deprivation therapy failure in prostate cancer.* Cancer Res. 1997;57:314-9.

13. Gittes RF. *Carcinoma of the prostate.* N Engl J Med. 1991;324:236-45.

14. Koivisto P, Kolmer M, Visakorpi T, Kallioniemi OP. *Androgen receptor gene and hormonal therapy failure of prostate cancer.* Am J Pathol. 1998;152:1-9.

15. Palmberg C, Koivisto P, Kakkola L, Tammela TL, Kallioniemi OP, Visakorpi T. *Androgen receptor gene amplification at primary progression predicts response to combined androgen blockade as second line therapy for advanced prostate cancer.* J Urol. 2000;164:1992-5.

16. Smith KA, Gorman PA, Stark MB, Groves RP, Stark GR. *Distinctive chromosomal structures are formed very early in the amplification of CAD genes in Syrian hamster cells.* Cell. 1990;63:1219-27.

17. Windle BE, Wahl GM. *Molecular dissection of mammalian gene amplification: New mechanistic insights revealed by analyses of very early events.* Mutation Res. 1992;276:199-224.

18. Benner SE, Wahl GM, Von Hoff DD. *Double minute chromosomes and homogeneously staining regions in tumors taken directly from patients versus in human tumor cell lines.* Anti-Cancer Drugs. 1991;2:11-25.

19. Myllykangas S, Knuutila S. *Manifestation, mechanisms and mysteries of gene amplifications.* Cancer Lett. 2006;232:79-89.

20. Savelyeva L, Schwab M. *Amplification of oncogenes revisited: from expression profiling to clinical application.* Cancer Lett. 2001;167:115-23.

21. Hastings PJ, Ira G, Lupski JR. *A microhomology-mediated break-induced replication model for the origin of human copy number variation.* PLoS Genet. 2009;5:e1000327.

22. Mizuno K, Miyabe I, Schalbetter SA, Carr AM, Murray JM. *Recombination-restarted replication makes inverted chromosome fusions at inverted repeats.* Nature. 2013;493:246-9.

23. Mukherjee K, Storici F. *A mechanism of gene amplification driven by small DNA fragments.* PLoS Genet. 2012;8:e1003119.

24. Debatisse M, Coquelle A, Toledo F, Buttin G. *Gene amplification mechanisms: The role of fragile sites.* Rec Results Cancer Res. 1998;154:216-26.

25. Ma C, Martin S, Trask B, Hamlin JL. *Sister chromatid fusion initiates amplification of the dihydrofolate reductase gene in Chinese hamster cells.* Genes Dev. 1993;7:605-20.

26. McClintock B. *The fusion of broken ends of chromosones following nuclear fusion.* Proc Natl Acad Sci USA. 1942;28:458-63.

27. Poupon MF, Smith KA, Chernova OB, Gilbert C, Stark GR. *Inefficient growth arrest in response to dNTP starvation stimulates gene amplification through bridge-breakage-fusion cycles.* Mol Biol Cell. 1996;7:345-54.

28. Calin GA, Sevignani C, Dumitru CD, Hyslop T, Noch E, Yendamuri S, et al. *Human microRNA genes* are frequently located at fragile sites and genomic regions involved in cancers. Proc Natl Acad Sci USA.2004;101:2999-3004.

29. Coquelle A, Pipiras E, Toledo F, Buttin G, Debatisse M. *Expression of fragile sites triggers intrachromosomal mammalian gene amplification and sets boundaries to early amplicons.* Cell. 1997;89:215-25.

30. Sutherland GR. *Heritable fragile sites on human chromosomes I. Factors affecting expression in lymphocyte culture.* Am J Hum Genet. 1979;31:125-35.

31. Sutherland GR, Richards RI. *The molecular basis of fragile sites in human chromosomes.* Curr Opin Genet Dev. 1995;5:323-7.

32. Yunis JJ, Soreng AL. *Constitutive fragile sites and cancer. Science.* 1984;226:1199-204.

33. Yunis JJ, Soreng AL, Bowe AE. *Fragile sites are targets of diverse mutagens and carcinogens.* Oncogene. 1987;1:59-69.

34. Singer MJ, Mesner LD, Friedman CL, Trask BJ, Hamlin JL. *Amplification of the human dihydrofolate reductase gene via double minutes is initiated by chromosome breaks.* Proc Natl Acad Sci USA. 2000;97:7921-6.

35. Bignell GR, Santarius T, Pole JC, Butler AP, Perry J, Pleasance E, et al. *Architectures of somatic genomic rearrangement in human cancer amplicons at sequence-level resolution.* Genome Res. 2007;17:1296-303.

36. Ishizaka Y, Chernov MV, Burns CM, Stark GR. *p53-dependent growth arrest of REF52 cells containing newly amplified DNA.* Proc Natl Acad Sci USA. 1995;92:3224-8.

37. Livingstone LR, White A, Sprouse J, Livanos E, Jacks T, Tisty TD. *Altered cell cycle arrest and gene amplification potential accompany loss of wild-type p53.* Cell. 1992;70:923-35.

38. Zhou P, Jiang W, Weghorst CM, Weinstein IB. *Overexpression of Cyclin D1 enhances gene amplification.* Cancer Res. 1996;56:36-9.

39. Yin Y, Tainsky MA, Bischoff FZ, Strong LC, Wahl GM. *Wild-type p53 restores cell cycle control and inhibits gene amplification in cells with mutant p53 alleles.* Cell. 1992;70:937-48.

40. Woodcock DM, Cooper IA. *Evidence for double replication of chromosomal DNA segments as a general consequence of DNA replication inhibition.* Cancer Res. 1981;41:2483-90.

41. Zannis-Hadjopoulos M, Persico M, Martin RG. *The remarkable instability of replication loops provides a general method for the isolation of origins of DNA replication.* Cell. 1981;27:155-63.

42. Ford M, Fried M. *Large inverted duplications associated with gene amplification.* Cell. 1986;45:425-30.

43. Ma C, Leu TH, Hamlin JL. *Multiple origins of replication in the dihydrofolate reductase amplicons of a methotrexate-resistant chinese hamster cell line.* Mol Cell Biol. 1990;10:1338-46.

44. Ma C, Looney JE, Leu TH, Hamlin JL. *Organization and genesis of dihydrofolate reductase amplicons in the genome of a methotrexate-resistant Chinese hamster ovary cell line.* Mol Cell Biol. 1988;8:2316-27.

45. Schneider SS, Hiemstra JL, Zehnbauer BA, Taillon-Miller P, LePaslier DL, Vogelstein B, et al. *Isolation and structural analysis of a 1.2 Megabase n-myc amplicon from a human neuroblastoma.* Mol Cell Biol. 1992;12:5563-70.

46. Johansson I, Lundqvist E, Bertilsson L, Dahl ML, Sjoqvist F, Ingelman-Sundberg M. *Inherited amplification of an active gene in the cytochrome P450 CYP2D locus as a cause of an ultrarapid metabolism of debrisoquin.* Proc Natl Acad Sci USA. 1993;90:11825-9.

47. Ingelman-Sundberg M, Oscarson M, McLellan RA. *Polymorphic human cytochrome P450 enzymes: an opportunity for individualized drug treatment.* Trends Pharmacol Sci. 1999;20:342-9.

48. Lundqvist E, Johansson I, Ingelman-Sundberg M. *Genetic mechanisms for duplication and multi-duplication of the human CYP2D6 gene and methods for detection of duplicated CYP2D6 genes.* Gene. 1999;226:327-38.

49. Riordan JR, Deuchars K, Kartner N, Alon N, Trent J, Ling V. *Amplification of P-glycoprotein genes in multidrug-resistant mammalian cell lines.* Nature. 1985;316:817-9.

50. Wright JA, Smith HS, Watt FM, Hancock MC, Hudson DL, Stark GR. *DNA amplification is rare in normal human cells.* Proc Natl Acad Sci USA. 1990;87:1791-5.

51. Tlsty TD. *Normal diploid human and rodent cells lack of detectable frequency of gene amplification.* Proc Natl Acad Sci USA. 1990;87:3132-6.

52. Andrechek ER, Hardy WR, Siegel PM, Rudnicki MA, Cardiff RD, Muller WJ. *Amplification of the neu/erbB-2 oncogene in a mouse model of mammary tumorigenesis.* Proc Natl Acad Sci U S A. 2000;97:3444-9.

53. DeFeo D, Gonda MA, Young HA, Chang EH, Lowy DR, Scolnick EM, et al. *Analysis of two divergent rat genomic clones homologous to the transforming gene of Harvey murine sarcoma virus.* Proc Natl Acad Sci USA. 1981;78:3328-32.

54. Bos JL. *The ras gene family and human carcinogenesis.* Mutation Res. 1988;195:255-71.

55. Pulciani S, Santos E, Long LK, Sorrentino V, Barbacid M. *Ras gene amplification and malignant transformation.* Mol Cell Biol. 1985;5:2836-41.

56. Sekiya T, Fushimi M, Hori H, Hirohashi S, Nishimura S, Sugimura T. *Molecular cloning and the total nucleotide sequence of the human c-Ha-ras-1 gene activated in a melanoma from a Japanese patient.* Proc Natl Acad Sci USA. 1984;81:4771-5.

57. Spandidos DA, Wilkie NM. *Malignant transformation of early passage rodent cells by a single mutated human oncogene.* Nature. 1984;310:469-75.

58. Santarius T, Shipley J, Brewer D, Stratton MR, Cooper CS. *A census of amplified and overexpressed human cancer genes.* Nature Rev Cancer. 2010;10:59-64.

59. Perou CM, Sorlie T, Eisen MB, van de Rijn M, Jeffrey SS, Rees CA, et al. *Molecular portraits of human breast tumours.* Nature. 2000;406:747-52.

60. Sorlie T, Perou CM, Tibshirani R, Aas T, Geisler S, Johnsen H, et al. *Gene expression patterns of breast carcinomas distinguish tumor subclasses with clinical implications.* Proc Natl Acad Sci U S A. 2001;98:10869-74.

6.5.3.2. AMPLIFICATION IN BREAST CANCER

A comprehensive analysis of genetic alterations in breast cancers identified a median of 86 point mutations and a median of 24 genes altered by major copy changes, i.e., deletion of all copies or amplification to at least 12 copies per cell (1). Amplification can be detected by different techniques. Until the mid-1990s, the Southern blot technique was the principal tool to determine amplification. For example, amplification of the HER-2 oncogene was detected by Southern blotting with the probe HER-2 and quantitated by comparing the signal strength between tumor sample and a normal control tissue (2). A control probe for a different marker was used to compare the amount of DNA loaded on the gel. Density ratios of the tumor probe and the reference probe were determined in each lane. Ratios were normalized by calculating a mean value of several normal DNAs. Ratios exceeding two were considered amplified. The Southern blot technique has been replaced by FISH of interphase chromosomes, which allows detection of amplification in single cells (3). The sensitivity and resolution of this method allow the identification of single-copy loci with a resolution of approximately 100 kb in interphase chromosomes and 1 Mb in metaphase chromosomes (4-6). The lower resolution in the latter is due to compacted DNA in the condensed sister chromatids. FISH also permits screening of gene amplification in multiple tumors arranged in a tissue microarray (7). In general, amplification values below 5 should be interpreted with caution because they could be the result of polyploidy (8). The combination of gene locus-specific and centromeric enumeration probes allows correlation of amplification with polysomy. For example, combination of the HER-2 probe with a centromeric probe specific for chromosome 17 revealed that the increase in HER2 gene copy number may in some instances be due to chromosome 17 polysomy (9). Quantitative measurement of DNA copy number across amplified chromosome regions by array comparative genomic hybridization (CGH) may allow precise localization of amplicon boundaries and amplification maxima (10). This technique was applied to resolve two regions of amplification within an approximately 2-Mb region of recurrent aberration at 20q13.2 in breast cancer (11). The putative oncogene ZNF217 mapped to one peak and CYP24 (encoding vitamin D 24-hydroxylase) to the other. Quantification of gene amplification in tumor DNA can also be accomplished by real-time PCR (12). Precise studies of amplicon architecture can be performed by genome sequencing (13). Amplification in small and precancerous lesions containing as few as 50 cells can be detected by combining laser-assisted microdissection of archival paraffin-embedded tissue with real-time PCR (14).

FISH and CGH studies of breast cancers revealed amplification at >40 different regions of the genome (15-19). The critical genes giving tumor cells a growth advantage are still unknown at many amplification sites. Typically, a candidate gene is the focus of interest. However, given the large size of the average amplicon, it is likely that it contains not only the candidate gene but also adjacent genes which are coamplified. In breast cancer, coamplifications frequently involve regions of chromosomes 8, 11, 12, 17, and 20, i.e., 8q24, 11q13-14, 12q13-14, 17q11-12, 17q21-24, and 20q13 (Table 6.5.3.). An example is the coamplification of the HER2 gene with the c-erbA, topoisomerase IIα, and MLN 64 genes in chromosomal segment 17q11–q21. Coamplification of the c-erbA gene was observed in 6 of 10 breast cancers with HER2 amplification (20). The topoisomerase IIα gene was amplified in 5 – 15% of breast cancers, usually together with HER2 (21). The MLN 64 gene was coamplified with HER2 in 22 of 25 breast cancers (22). Both genes were always amplified to a similar level. The long arm of chromosome 20 contains three distinct nonsyntenic regions between 20q11 and 20q13.2 that are frequently amplified in breast cancer either independently or together (23). The involvement of three distinct loci in the coamplifications as well as their independent amplification suggests extensive chromosomal rearrangements and multiple breaks at 20q.

Densitometric quantification of Southern blots revealed a 3- to 7-fold increase in copy number of the **mucin 1 (MUC1) gene at 1q21** in breast cancer DNA (24). MUC1 mRNA levels were higher than those in normal breast tissue in 20 of the 32 (62%) cancers. Densitometric quantification of the mRNA levels showed that the MUC1 expression was increased 2- to 5-fold in nine tumors and 6- to more than 11-fold in eleven tumors. However, 12 cancers had similar or lower MUC1 mRNA levels relative to those in normal breast tissue. Nevertheless, there was a significant correlation (p <0.0001) between acquisition of additional copies of the MUC1 gene and high MUC1 mRNA levels. The acquisition of additional copies of an arm or a whole chromosome (trisomy, tetrasomy, or polysomy) may be due to mitotic nondisjunction. This type of genetic alteration, which affects hundreds of genes, is usually associated with a 2- to 5-fold overexpression. Different genetic mechanisms are involved in 5- to 100-fold amplification of a small number of genes. Clearly, acquisition of one or two extra copies of 1q is not the only mechanism responsible for the overexpression of MUC1 in breast cancer. Indeed, MUC1 mRNA was also increased in seven of 14 tumors with normal MUC1 gene copy number (53). Overexpression of the MUC1 gene in these cancers must therefore be due to mechanisms other than a gene dosage effect, e.g., mutations in regulatory sequences or altered patterns of methylation.

Amplification of the **telomerase reverse transcriptase (TERT) gene at 5p15.33** was observed in 5 of 19 (26%) breast carcinomas (25). The **MYB oncogene at 6q22-q24** plays a role in the tumorigenesis of ERα-positive/HER2-positive cancers (54). MYB was amplified in 5 of 17 (29%) BRCA1 breast cancers, 2 of 100 sporadic breast cancers, and none of 8 BRCA2 tumors and 13 breast cancer cell lines (26). The MYB amplification was accompanied by mRNA and protein overexpression. The **ESR1 gene at 6q25** encodes the ERα. A tissue microarray analysis showed ESR1 amplification in 358 of 1,739 (20.6%) breast cancers (27). Ninety-nine percent of tumors with ESR1 amplification showed ERα protein overexpression, compared with 66.6% cancers without ESR1 amplification (P <0.0001). ESR1 amplification was observed in all histological subtypes of cancer as well as in premalignant lesions (hyperplasia, papilloma, ductal and lobular carcinoma *in situ*). ESR1 amplification was not found in normal breast tissue and fibrocystic disease. Amplification of the **EGFR (HER1) gene at 7p12** was found in 5 to 15% of breast cancers except in the basal-like subtype, which showed overexpression in 50% of tumors (28, 29, 55).

The **8p11.2–p12 amplicon** is complex, comprising at least four cores, which can be amplified independently (30, 31). The second core encompasses the **fibroblast growth factor receptor 1 (FGFR1) gene at 8p11.2**, which is amplified in about 10% of primary breast cancers (56-59). The telomeric core contains the **ZNF703 oncogene** which regulates the proliferation of luminal epithelium and is predominantly amplified in luminal

B breast cancers (32). Other genes in the amplicon influence receptor trafficking (RAB11FIP1) and regulate the histone code (WHSC1L1), RNA metabolism (LSM1), or endoplasmic reticulum stress pathway (ERLIN2) (60). These genes may cooperate in mediating neoplastic transformation. Amplification of the **FGFR2 gene at 10q26.3** was identified in 4% (6 of 165) of triple-negative breast cancers (36).

The myc oncogene family consists of three members, **c-myc at 8q24**, L-myc at 1p32, and N-myc at 2p24. Deregulated expression of each myc gene is associated with specific malignancies, e.g., N-myc amplification is associated with neuroblastoma whereas c-myc amplification is associated with breast and other cancers (61). In two large series of 1052 and 1875 breast cancers, c-myc amplification was observed in 17.1% and 12.3% of tumors, respectively, with amplification levels ranging from 2 to 18 gene copies (30, 35). A smaller series of 121 breast cancers showed amplification in 38 (32%) and rearrangement of the c-myc locus in 5 (4%) tumors (62). With three exceptions, all tumors containing altered c-myc were invasive ductal carcinomas. Amplification levels in these studies were determined by Southern

Table 6.5.3. DNA Amplification in Breast Cancer

Locus	Genes	Frequency	References
1q21	MUC1	38 – 62%	(24)
5p15.33	hTERT	5/19 (26%)	(25)
6q22-q24	c-myb	5/17 (29%) BRCA1-pos.	(26)
6q25	ESR1 (ERα)	358/1,739 (20.6%)	(27)
7p12	EGFR (HER1)	5 – 10%	(28, 29)
8p11.2-p12	FGFR1	5 – 12%	(30, 31)
	ZNF703	3 – 13%	(33)
8q21.1-21.3	E2F5	19/442 (4.2%)	(33)
8q23	EIF3S3	20%	(34)
8q24	c-myc	15%	(30, 35)
10q26.3	FGFR2	6/165 (4% of triple neg.)	(36)
11q13	CCND1	15%	(37)
12q13	MDM 2	5.8%	(30)
13q34	CUL4A	4.5 – 20%	(38, 39)
15q25-qter	IGF-1-R	19/975 (2%)	(40)
17q12	HER2	15 – 30%	(41, 42)
17q11-q21	c-erbA	10/95 (10%)	(20)
	Topoisomerase IIα	25/97 (26%)	(21)
	MLN 64	22/98 (22%)	(22)
17q21.3	CRD-BP	14/40 (35%)	(43)
17q22-23	RAD51C	3 – 8%	(44, 45)
	PS6K	8%	(44, 45)
	SIGMA1B	11/92 (11.9%)	(45)
	PAT1	9 – 19%	(45, 45)
	TBX2	5 – 9%	(44, 47)
20q11	AIB3	29/335 (8.7%)	(47)
20q12	AIB1	56/1157 (4.8%)	(48)
20q13.1	MYBL2	25/372 (6.7%)	(49)
20q13.2	ZNF217, NABC1	20%	(50)
20q13.2-13.3	Aurora A kinase	12 – 25%	(51, 52)

blotting. However, a real-time PCR method detected extra copies of c-myc in only 10% of 108 breast cancers (12). Another study showed that c-myc gene overrepresentation in breast carcinoma is mainly due to polysomy of chromosome 8 and/or genomic endoreduplication rather than specific c-myc gene amplification (63). Furthermore, quantitation of c-myc mRNA by real-time PCR showed c-myc overexpression in 29 of 134 (22%) breast cancers, ranging from 3.2 to 19 times the level in normal breast tissues (64). Interestingly, the c-myc overexpression in these tumors was rarely due to an increase in c-myc gene copy number. These findings indicate a dissociation of c-myc amplification and overexpression in breast cancer. There was no evidence of c-myc amplification in benign breast disease (37 fibroadenomas, 8 cystosacoma phyllodes, 5 fibrocystic disease lesions) (65). The **c-myc mRNA coding region determinant (CRD-BP) gene at 17q21.3** was amplified in 14 of 40 primary breast cancers (43). CRD-BP binds *in vitro* to c-myc mRNA and is thought to stabilize the mRNA and increase c-myc protein abundance. Therefore, amplification of the CRD-BP gene may upregulate c-myc abundance.

Located centromerically of the c-myc gene is the **EIF3S3 (eIF3-p40) gene at 8q23**, which is amplified and overexpressed in about 20% of primary breast cancers (34). The EIF3S3 protein is a subunit of translation factor 3, which is required for initiation of protein synthesis in eukaryotic cells (66). The c-myc and EIF3S3 genes are frequently coamplified, indicating a large size of the amplicon. However, several breast cancers exhibited higher copy numbers of EIF3S3 than c-myc. Detailed mapping revealed that the EIF3S3 amplicon in these cases did not include c-myc (34). The breast cancer cell line SK-BR-3 showed a high copy number for both EIF3S3 and c-myc, but a low copy number for the intermediate marker D8S1484. This discontinuity indicates a complex amplification pattern at 8q23-24. Located further centromerically of c-myc at **8q21.1-21.3** is the **E2F5 gene**, which was amplified in 19 of 442 (4.2%) primary breast cancers (33). The c-myc gene was coamplified in 6 of the 19 cases. There was no evidence for amplification of other members of the E2F/DP family of transcription factors. Approximately 15% of primary breast cancers show amplification of genomic DNA centered on chromosome band **11q13** (67). The 11q13 locus encompasses several driver genes that could play a role in mammary carcinogenesis. For example, the GALN gene encodes the preprogalanin protein, which is cleaved to liberate galanin, a neuropeptide involved in pituitary secretion of prolactin and a tumor cell mitogen. However, the increase in GALN copy number was not matched by a proportional increase in preprogalanin mRNA levels, making GALN an unlikely candidate oncogene for mammary carcinogenesis (68). Other candidate genes include **fibroblast growth factors 3 (FGF3 or INT2)** and 4 (FGF4 or HST1), two established

oncogenes (69). However, RNase protection assays demonstrated that FGF3 and FGF4 are generally not expressed in human breast cancers irrespective of gene copy number (70). About 1.5 Mb telomeric of FGF3 is the **EMS1 gene**, which encodes an actin binding protein. The EMS1 gene is either coamplified with the FGF3 gene or amplified alone (71-73). The strongest candidate for a potential role in mammary tumorigenesis is the **CCND1 gene**, which encodes the cell cycle regulator cyclin D1. The CCND1 gene is amplified and overexpressed in approximately 15% of primary breast cancers (37, 74). Further studies showed that overexpression of cyclin D1 can occur in both presence and absence of CCND1 gene amplification. Conversely, amplification does not invariably lead to overexpression (75). Thus, overexpression may result from mechanisms other than gene amplification. Cyclin D1 amplification occurs three to four times more often in hypodiploid than in unselected breast cancers (76). FISH revealed cyclin D1 amplification in tumor cells with either one or two copies of chromosome 11, suggesting that amplification in these hypodiploid tumors occurred before losing one copy of chromosome 11. The prognostic significance of cyclin D1 amplification/overexpression in breast cancer is uncertain as long-term follow-up studies showed reduced survival, no effect, or improved survival (57, 77, 78).

Amplification of the **MDM 2 gene at 12q13** appears to be more common in tumors of non-epithelial origin. However, MDM 2 amplification was observed in 5 – 10% of primary breast cancers (30, 57, 79). Amplification of the **cullin 4A (CUL4A) gene at 13q34** was observed in 4.5% of breast cancers, but the frequency increased to 8.1% in BRCA1-associated tumors and to 20% in basal like tumors (38, 39). Cullins are a family of structural proteins that provide a scaffold for ubiquitin ligases involved in many protein degradation reactions. Amplification of the epidermal growth factor receptor **HER2 gene at 17q12** is discussed in Section 6.2.2.

About 20% of primary breast cancers exhibit amplification of the **17q22-23** region and an even higher percentage (about 40%) is observed in breast cancer metastases (18, 80). Several investigators have performed a systematic analysis of this region and identified several target genes in breast cancer cell lines and tumors. One study employed FISH and tissue microarray to examine amplification and overexpression of **RAD51C, PS6K, PAT1, and TBX2** in 372 breast cancers and found amplification of at least one of the genes in 10.5% of tumors (44). Another study used Southern blotting to identify amplification of RAD51C, PS6K, PAT1, and **SIGMA1B** in 26 of 94 (28%) breast cancers (45). Of the 26 tumors with amplification, only 2 (2%) were amplified for all of the four genes, whereas 14 (54%) were amplified for only one of the genes. Thus, the 17q23 amplicons appear to be heterogeneous in composition despite the short 14-cR interval (corresponding to about 3.5 Mb) harboring these genes (81). Based on biological function each of the five amplified genes has the potential to play a role in breast cancer development or progression. For example, the ribosomal protein PS6K is a serine-threonine kinase involved in cell cycle progression from G1 to S phase, which may also play a role in the control of cell size (82, 83). Amplification of the PS6K gene has been observed in both breast cancer tissues and cell lines (84, 85). Amplification/overexpression of PS6K at 17q23 was associated with poor prognosis independent of HER-2 amplification at 17q12. TBX2 is a member of the T-box family of transcription factors, which downregulates Cdkn2a (p19^{ARF}) and prevents induction of Cdkn2a by oncogenic signals such as myc or H-ras. Amplification and overexpression of TBX2 enhanced cell proliferation by inhibiting senescence (46).

Detailed FISH analysis revealed the independent amplification and frequent coamplification of three distinct nonsyntenic regions on 20q (23, 86). The **AIB1 gene at 20q12** was amplified 2- to 8-fold in 4.8% (56/1157) of primary breast cancers (48). The related genes **AIB3 and AIB4 at 20q11** were amplified with similar frequencies and levels (47). The AIB1 protein is a member of the SRC-1 family of nuclear receptor coactivators, which was shown to interact with the estrogen receptor in a ligand-dependent manner (87). AIB1 overexpression in transient transfection assay resulted in increased levels of estrogen-dependent transcription. In a series of 1157 breast cancers, AIB1 amplification showed a positive correlation with both ER positivity and tumor size (p = 0.02), suggesting a role in the growth of hormone-dependent tumors (48). The oncogene **MYBL2 at 20q13.1** was amplified in 6.7% of 372 primary breast cancers (49). The **NABC1 and ZNF217 genes at 20q13.2** are frequently coamplified and overexpressed in breast cancer (50). The NABC1 gene encodes a 585-amino acid protein of unknown function. The ZNF217 gene encodes an alternatively spliced transcription factor of 1,062 and 1,108 amino acids, each having a DNA binding domain with eight C2H2 Kruppel-like motifs and a proline-rich transcription activation domain. When finite life span mammary epithelial cells from reduction mammoplasty tissue were transfected with ZNF217, the cultures gave rise to immortalized cells (88). The cells that overcame senescence exhibited continued telomere erosion, followed by increasing telomerase activity, stabilization of telomere length, and resistance to TGFβ growth inhibition. These experiments suggest that aberrant expression of ZNF217 may be selected for during tumorigenesis because it allows mammary epithelial cells to overcome senescence and attain immortality. Also located at 20q13.2, but approximately 2 Mb apart from the ZNF217 gene is the **CYP24 gene**, which encodes

vitamin D24 hydroxylase. The amplification and overexpression of CYP24 is suspected to lead to abrogation of growth control mediated by vitamin D (11). The **Aurora A kinase gene at 20q13.2-13.3** encodes a centrosome-associated serine-threonine kinase that is essential for successful execution of cell division. Aurora A kinase is amplified in 12 to 25% of breast cancers (51, 52, 89). Compared to the amplification percentage, the proportion of tumors with Aurora A kinase protein overexpression is much higher (94%), indicating the existence of other overexpression pathways. Ectopic expression of the mouse homolog STK15 in mouse NIH 3T3 cells led to the appearance of multiple centrosomes and *in vitro* transformation. Transient transfection of STK15 into the near diploid human breast epithelial cell line MCF-10A revealed an increase in centrosome number as well as induction of aneuploidy (52). These results suggest that specific gene abnormalities can directly influence chromosome ploidy in tumor cells. Gene amplification already occurs in the early preinvasive stages of breast cancer. For example, amplification of the CCND1 gene was observed in 6 to 18 % of DCIS using Southern blot, FISH, and microdissection of paraffin-embedded tissue combined with real-time PCR (90, 91). Amplifications increase in frequency and level with DCIS grade (92-94). In a study of 83 DCIS classified according to the Van Nuys grading system (95), HER-2 amplification was detected in 0 of 6 low grade, 1 of 4 (25%) intermediate grade, and 11 of 27 (40%) high-grade lesions (90). These findings suggest that HER2 amplification may be an early step in the development of DCIS and that the frequency and level of HER-2 amplification in high-grade DCIS is indistinguishable from invasive tumors, which is consistent with the progression of *in situ* to invasive cancer. There was no evidence of HER2 amplification/ overexpression in atypical hyperplasia, hyperplasia or benign breast disease (65, 96). On the other hand, there is evidence that the progression from primary breast cancer to metastasis is associated with the acquisition of further genetic changes including amplification. For example, comparative genomic hybridization analysis of 16 pairs of primary cancer and metastasis/recurrence revealed amplification in four metastases (two lymph node and two distant metastases) but not in the corresponding primary tumors (97).

Investigators have examined the amplification of multiple genes in breast cancer and defined complex amplification patterns for subgroups of tumors (30). One study examined the prognostic significance of gene coamplification at eight different loci in 640 breast cancer patients (57). Both disease-free interval and overall survival were significantly shorter in patients whose tumors contained two or more amplified genes compared with patients showing only one or none at all. The prognostic effect was observed with coamplification of CCND1 and FGFR1 as well as HER-2 and c-myc while other combinations showed no additional effect compared to single gene amplification. This cooperative effect suggests the existence of a selective advantage associated with the coamplification, which may lead to the formation of hybrid CCND1/FGFR1 and HER-2/c-myc amplification units (57, 98). It has been shown that coamplified oncogenes from different regions of the genome may localize to the same HSRs (99). For example, the breast cancer cell line MDA-MB-134 carries a single hybrid HSR in which amplified sequences originating from chromosomes 8 (FGFR1 at 8p12) and 11 (CCND1 at 11q13) were arranged as alternating repeats (100). In general, the recurrent genomic and transcriptional characteristics of 51 breast cancer cell lines mirror those of 145 primary breast tumors (101). FISH analysis of primary breast cancers showed the formation of a hybrid amplified structure of 8p12 and 11q13 sequences in 3 of 125 tumors (98). Analysis of centromeres indicated that the 11q13 sequence became rearranged onto chromosome 8 during the amplification process. A detailed investigation of the amplicon organization was performed by pretreatment of nuclei with a mild alkaline solution resulted in decondensation of chromatin. Analysis of the elongated chromatin revealed that the amplicon in one tumor consisted of alternating portions of 8p12 and 11q13 sequences while another tumor contained alternating clusters of five to six 8p12 and five to six 11q13 sequences. In neuroblastoma, N-myc (2p24) and MDM 2 (12q13) were coamplified first in extrachromosomal DMs and then integrated into chromosomes to develop to HSRs (102). This observation points to molecular mechanisms that predispose to coamplification of multiple important genes in tumor cells and to localization of such initially nonsyntenic sequences into the same chromosomal structure. A study combining gene copy number and gene expression analyses revealed 66 genes, including many of those listed in Table 6.5.3 ., whose expression levels were deregulated by high-level amplification (15). Interestingly, there were also frequent low-level copy-number changes in genes involved in cellular metabolism. Presumably, these low-level copy-number alterations are selected during early cancer formation because they increase basal metabolism, thereby providing a net survival/proliferative advantage to the cells that carry them. Whole-genome sequencing of 2000 breast cancers with integrative cluster analysis of gene copy number variation and gene expression data also revealed amplification of many of the genes listed in Table 6.5.3. (103). The cluster analysis identified 10 subgroups with distinct copy number profiles.

References

1. Leary RJ, Lin JC, Cummins J, Boca S, Wood LD, Parsons DW, et al. *Integrated analysis of homozygous deletions, focal amplifications, and sequence alterations in breast and colorectal cancers.* Proc Natl Acad Sci U S A. 2008;105:16224-9.

2. Eyfjord JE, Thorlacius S, Steinarsdottir M, Valgardsdottir R, Ogmundsdottir HM, Anamthawat-Jonsson K. *p53 Abnormalities and genomic instability in primary human breast carcinomas.* Cancer Res. 1995;55:646-51.

3. Pinkel D, Straume T, Gray JW. *Cytogenetic analysis using quantitative, high-sensitivity, fluorescence hybridization.* Proc Natl Acad Sci USA. 1986;83:2934-838.

4. Trask BJ. *Fluorescence in situ hybridization: applications in cytogenetics and gene mapping.* Trends Genet. 1991;7:149-54.

5. Trask BJ, Allen S, Massa H, Fertitta A, Sachs R, van den Engh G, et al. *Studies of metaphase and interphase chromosomes using fluorescence in situ hybridization.* Cold Spring Harbor Symp Quant Biol. 1993;58:767-75.

6. Yokota H, van den Engh G, Mostert M, Trask BJ. *Treatment of cells with alkaline borate buffer extends the capability of interphase FISH mapping.* Genomics. 1995;25:485-91.

7. Schraml P, Kononen J, Bubendorf L, Moch H, Bissig H, Nocito A, et al. *Tissue microarrays for gene amplification surveys in many different tumor types.* Clin Cancer Res. 1999;5:1966-75.

8. Schwab M. *Oncogene amplification in solid tumors.* Semin Cancer Biol. 1999;9:319-25.

9. Szollosi J, Balazs M, Feuerstein BG, Benz CC, Waldman FM. *ERBB-2 (HER2/neu) gene copy number, p185Her-2 overexpression, and intratumor heterogeneity in human breast cancer.* Cancer Res. 1995;55:5400-7.

10. Pinkel D, Segraves R, Sudar D, Clark S, Poole J, Kowbel D, et al. *High resolution analysis of DNA copy number variation using comparative genomic hybridization to microarrays.* Nature Genet. 1998;20:207-11.

11. Albertson DG, Ylstra B, Segraves R, Collins C, Dairkee SH, Kowbel D, et al. *Quantitative mapping of amplicon structure by array CGH identifies CYP24 as a candidate oncogene.* Nature Genet. 2000;25(2):144-6.

12. Bieche I, Olivi M, Champeme MH, Vidaud D, Lidereau R, Vidaud M. *Novel approach to quantitative polymerase chain reaction using real-time detection: application to the detection of gene amplification in breast cancer.* Int J Cancer. 1998;78:661-6.

13. Bignell GR, Santarius T, Pole JC, Butler AP, Perry J, Pleasance E, et al. *Architectures of somatic genomic rearrangement in human cancer amplicons at sequence-level resolution.* Genome Res. 2007;17:1296-303.

14. Lehmann U, Glockner S, Kleeberger W, von Wasielewski HF, Kreipe H. *Detection of gene amplification in archival breast cancer specimens by laser-assisted microdissection and quantitative real-time polymerase chain reaction.* Am J Pathology. 2000;156:1855-64.

15. Chin K, DeVries S, Fridlyand J, Spellman PT, Roydasgupta R, Kuo WL, et al. *Genomic and transcriptional aberrations linked to breast cancer pathophysiologies.* Cancer Cell. 2006;10:529-41.

16. Forozan F, Mahlamaki EH, Monni O, Chen Y, Veldman R, Jiang Y, et al. *Comparative genomic hybridization analysis of 38 breast cancer cell lines: A basis for interpreting complementary DNA microarray data.* Cancer Res. 2000;60:4519-25.

17. Kallioniemi A, Kallioniemi O-P, Sudar D, Rutovitz D, Gray JW, Waldman F, et al. *Comparative genomic hybridization for molecular cytogenetic analysis of solid tumors.* Science. 1992;258:818-21.

18. Kallioniemi A, Kallioniemi OP, Piper J, Tanner M, Stokke T, Chen L, et al. *Detection and mapping of amplified DNA sequences in breast cancer by comparative genomic hybridization.* Proc Natl Acad Sci USA. 1994;91:2156-60.

19. Knuutila S, Bjorkqvist AM, Autio K, Tarkkanen M, Wolf M, Monni O, et al. *DNA copy number amplifications in human neoplasms.* Am J Pathol. 1998;152:1107-23.

20. Van de Vijver M, Van de Bersselaar R, Devilee P, Cornelisse C, Peterse J, Nusse R. *Amplification of the neu (c-erbB-2) oncogene in human mammary tumors is relatively frequent and is often accompanied by amplification of the linked c-erbA oncogene.* Mol Cell Biol. 1987;7:2019-23.

21. Jarvinen T, Tanner M, Rantanen V, Barlund M, Borg A, Grenman S, et al. *Amplification and deletion of topoisomerase IIa associate with erbB-2 amplification and affect sensitivity to topoisomerase 11 inhibitor doxorubicin in breast cancer.* Am J Pathol. 2000;156:839-47.

22. Bieche I, Tomasetto C, Regnier CH, Moog-Lutz C, Rio MC, Lidereau R. *Two distinct amplified regions at 17q11-q21 involved in human primary breast cancer.* Cancer Res. 1996;56:3886-90.

23. Tanner MM, Tirkkonen M, Kallioniemi A, Isola J, Kuukasjarvi T, Collins C, et al. Independent amplification and frequent *co-amplification of three neosyntenic regions on the long arm of chromosome 20 in human breast cancer.* Cancer Res. 1996;56:3441-5.

24. Bieche I, Lidereau R. *A gene dosage effect is responsible for high overexpression of the MUC1 gene observed in human breast tumors.* Cancer Genet Cytogenet. 1997;98:75-80.

25. Zhang A, Zheng C, Lindvall C, Hou M, Ekedahl J, Lewensohn R, et al. *Frequent amplification of the telomerase reverse transcriptase gene in human tumors.* Cancer Res. 2000;60:6230-5.

26. Kauraniemi P, Hedenfalk I, Persson K, Duggan DJ, Tanner M, Johannsson O, et al. *MYB oncogene amplification in hereditary BRCA1 breast cancer.* Cancer Res. 2000;60:5323-8.

27. Holst F, Stahl PR, Ruiz C, Hellwinkel O, Jehan Z, Wendland M, et al. *Estrogen receptor alpha (ESR1) gene amplification is frequent in breast cancer.* Nature Genet. 2007;39:655-60.

28. Nielsen TO, Hsu FD, Jensen K, Cheang M, Karaca G, Hu Z, et al. *Immunohistochemical and clinical characterization of the basal-like subtype of invasive breast carcinoma.* Clin Cancer Res. 2004;10:5367-74.

29. Zaczek A, Welnicka-Jaskiewicz M, Bielawski KP, Jaskiewicz J, Badzio A, Olszewski W, et al. *Gene copy numbers of HER family in breast cancer.* J Cancer Res Clin Oncol. 2008;134:271-9.

30. Courjal F, Cuny M, Simony-Lafontaine J, Louason G, Speiser P, Zeillinger R, et al. *Mapping of DNA amplifications at 15 chromosomal localizations in 1875 breast tumors: Definition of phenotypic groups.* Cancer Res. 1997;57:4360-7.

31. Gelsi-Boyer V, Orsetti B, Cervera N, Finetti P, Sircoulomb F, Rouge C, et al. *Comprehensive profiling of 8p11-12 amplification in breast cancer.* Mol Cancer Res. 2005;3:655-67.

32. Holland DG, Burleigh A, Git A, Goldgraben MA, Perez-Mancera PA, Chin SF, et al. *ZNF703 is a common Luminal B breast cancer oncogene that differentially regulates luminal and basal progenitors in human mammary epithelium.* EMBO Mol Med. 2011;3:167-80.

33. Polanowska J, Le Cam L, Orsetti B, Valles H, Fabbrizio E, Fajas L, et al. *Human E2F5 gene is oncogenic in primary rodent cells and is amplified in*

human breast tumors. Genes Chrom Cancer. 2000;28:126-30.

34. Nupponen NN, Isola J, Visakorpi T. *Mapping the amplification of EIF3S3 in breast and prostate cancer.* Genes. 2000;28:203-10.

35. Berns EMJJ, Klijn JGM, van Putten WLJ, van Staveren IL, Portengen H, Foekens JA. *c-myc amplification is a better prognostic factor than HER2/neu amplification in primary breast cancer.* Cancer Res. 1992;52:1107-13.

36. Turner N, Lambros MB, Horlings HM, Pearson A, Sharpe R, Natrajan R, et al. *Integrative molecular profiling of triple negative breast cancers identifies amplicon drivers and potential therapeutic targets.* Oncogene. 2010;29:2013-23.

37. Gillett C, Fantl V, Smith R, Fisher C, Bartek J, Dickson C, et al. *Amplification and overexpression of cyclin D1 in breast cancer detected by immunohistochemical staining.* Cancer Res. 1994;54:1812-7.

38. Chen LC, Manjeshwar S, Lu Y, Moore D, Ljung BM, Kuo WL, et al. *The human homologue for the Caenorhabditis elegans cul-4 gene is amplified and overexpressed in primary breast cancers.* Cancer Res. 1998;58:3677-83.

39. Melchor L, Saucedo-Cuevas LP, Munoz-Repeto I, Rodriguez-Pinilla SM, Honrado E, Campoverde A, et al. *Comprehensive characterization of the DNA amplification at 13q34 in human breast cancer reveals TFDP1 and CUL4A as likely candidate target genes.* Breast Cancer Res. 2009;11:R86.

40. Berns EMJJ, Klijn JGM, van Staveren IL, Portengen H, Foekens JA. *Sporadic amplification of the insulin-like growth factor 1 receptor gene in human breast tumors.* Cancer Res. 1992;52:1036-9.

41. Persons DL, Borelli KA, Hsu PH. *Quantitation of HER-2/ neu and c-myc gene amplification in breast carcinoma using fluorescence in situ hybridization.* Mod Pathol. 1997;10:720-7.

42. Slamon DJ, Clark GM, Wong SG, Levin WJ, Ullrich A, McGuire WL. *Human breast cancer: correlation of relapse and survival with amplification of the HER-2/neu oncogene.* Science. 1987;235:177-82.

43. Doyle GA, Bourdeau-Heller JM, Coulthard S, Meisner LF, Ross J. *Amplification in human breast cancer of a gene encoding a c-myc mRNA-binding protein.* Cancer Res. 2000;60:2756-9.

44. Barlund M, Monni O, Kononen

J, Cornelison R, Torhorst J, Sauter G, et al. *Multiple Genes at 17q23 undergo amplification and overexpression in beast cancer.* Cancer Res. 2000;60:5340-4.

45. Wu GJ, Sinclair CS, Paape J, Ingle JN, Roche PC, James CD, et al. *17q23 amplifications in breast cancer involve the PAT1, RAD51C, PS6K and SIGMA1B genes.* Cancer Res. 2000;60:5371-5.

46. Jacobs JJL, Keblusek P, Robanus-Maandag E, Kristel P, Lingbeek M, Nederlof PM, et al. *Senescence bypass screen identified TBX2, which represses Cdkn2a (p19ARF) and is amplified in a subset of human breast cancers.* Nature Genet. 2000;26:291-9.

47. Lee SK, Anzick SL, Choi JE, Bubendorf L, Guan XY, Jung YK, et al. *A nuclear factor, ASC-2 as a cancer-amlified transcriptional coactivator essential for ligand-dependent transactivation by nuclear receptors in vivo.* J Biol Chem. 1999;274:34283-93.

48. Bautista S, Valles H, Walker RL, Anzick S, Zeillinger R, Meltzer P, et al. *In breast cancer, amplification of the steroid receptor coactivator gene AIB1 is correlated with estrogen and progesterone receptor positivity.* Clin Cancer Res. 1998;4:2925-9.

49. Kononen J, Bubendorf L, Kallioniemi A, Barlund M, Schraml P, Leighton S, et al. *Tissue microarrays for high-throughput molecular profiling of tumor specimens.* Nature Med. 1998;4:844-7.

50. Collins C, Rommens JM, Kowbel D, Godfrey T, Tanner M, Hwang S, et al. *Positional cloning of ZNF217 and NABC1: Genes amplified at 20q13.2 and overexpressed in breast carcinoma.* Proc Natl Acad Sci USA. 1998;95:8703-8.

51. Tanaka T, Kimura M, Matsunaga KR, Fukada D, Mori H, Okano Y. *Centrosomal kinase AIK1 is overexpresssed in invasive ductal carcinoma of the breast.* Cancer Res. 1999;59:2041-4.

52. Zhou H, Kuang J, Zhong L, Kuo W, Gray JW, Sahin A, et al. *Tumour amplified kinase STK15/BTAK induces centrosome amplification, aneuploidy and transformation.* Nature Genet. 1998;20:189-93.

53. Zrihan-Licht S, Weiss M, Keydar I, Wreschner DH. *DNA methylation status of the MUC1 gene coding for a breast-cancer-associated protein.* Int J Cancer. 1995;62:245-51.

54. Miao RY, Drabsch Y, Cross RS, Cheasley D, Carpinteri S, Pereira L, et al. *MYB is essential for mammary*

tumorigenesis. Cancer Res. 2011;71:7029-37.

55. Kersting C, Tidow N, Schmidt H, Liedtke C, Neumann J, Boecker W, et al. *Gene dosage PCR and fluorescence in situ hybridization reveal low frequency of egfr amplifications despite protein overexpression in invasive breast carcinoma.* Lab Invest. 2004;84:582-7.

56. Blanckaert VD, Hebbar M, Louchez MM, Vilain MO, Schelling ME, Peyrat JP. *Basic fibroblast growth factor receptors and their prognostic value in human breast cancer.* Clin Cancer Res. 1998;4:2939-47.

57. Cuny M, Kramar A, Courjal F, Johannsdottir V, Iacopetta B, Fontaine H, et al. *Relating genotype and phenotype in breast cancer: An analysis of prognostic significance of amplification at eight different genes or loci and of p53 mutations.* Cancer Res. 2000;60:1077-83.

58. Reis-Filho JS, Tutt AN. *Triple negative tumours: a critical review.* Histopathology. 2008;52:108-18.

59. Ugolini F, Adelaide J, Charafe-Jauffret E, Nguyen C, Jacquemier J, Jordan B, et al. *Differential expression assay of chromosome arm 8p genes indentifies frizzled-related (FRP1/FRZB) and fibroblast growth factor receptor 1 (FGFR1) as candidate breast cancer genes.* Oncogene. 1999;18:1903-10.

60. Yang ZQ, Liu G, Bollig-Fischer A, Giroux CN, Ethier SP. *Transforming properties of 8p11-12 amplified genes in human breast cancer.* Cancer Res. 2010;70:8487-97.

61. Nesbit CE, Tersak JM, Prochownik EV. *MYC oncogenes and human neoplastic disease.* Oncogene. 1999;18:3004-16.

62. Escot C, Theillet C, Lidereau R, Spyratos F, Champeme MH, Gest J, et al. *Genetic alteration of the c-myc protooncogene (MYC) in human primary breast carcinomas.* Proc Natl Acad Sci USA. 1986;83:4834-8.

63. Visscher DW, Wallis T, Awussah S, Mohamed A, Crissman JD. *Evaluation of MYC and chromosome 8 copy number in breast carcinoma by interphase cytogenetics.* Genes Chrom Cancer. 1997;18:1-7.

64. Bieche I, Laurendeau I, Tozlu S, Olivi M, Vidaud D, Lidereau R, et al. *Quantitation of MYC gene expression in sporadic breast tumors with a real-time reverse transcription-PCR assay.* Cancer Res. 1999;59:2759-65.

65. Lizard-Nacol S, Lidereau R, Collin F, Arnal M, Hahnel L, Roignot P, et al. *Benign breast disease: Absence of genetic alterations of* several loci implicated in breast cancer malignancy. Cancer Res. 1995;55:4416-9.

66. Asano K, Goss Kinzy T, Merrick WC, Hershey JWB. *Conservation and diversity of eukaryotic translation initiation factor eIF3.* J Biol Chem. 1997;272:1101-9.

67. Fantl V, Richards MA, Smith R, Lammie GA, Johnstone G, Allen D, et al. *Gene amplification on chromosome band 11q13 and estrogen receptor status in breast cancer.* Eur J Cancer. 1990;26:423-9.

68. Ormandy CJ, Lee CSL, Ormandy HF, V. F, Shine J, Peters G, et al. *Amplification, expression, and steroid regulation of the preprogalanin gene in human breast cancer.* Cancer Res. 1998;58:1353-7.

69. Lammie GA, Peters G. *Chromosome 11q13 abnormalities in human cancer.* Cancer Cells. 1991;3:413-20.

70. Fantl V, Smith R, Brookes S, Dickson C, Peters G. *Chromosome 11q13 abnormalities in human breast cancer.* Cancer Surv. 1993;18:77-94.

71. Brookes S, Lammie GA, Schuuring E, de Boer C, Michalides R, Dickson C, et al. *Amplified region of chromosome band 11q13 in breast and squamous cell carcinomas encompasses three CpG islands telomeric of FGF3, including the expressed gene EMS1.* Genes Chrom Cancer. 1993;6:222-31.

72. Hui R, Campbell DH, Lee CS, McCaul K, Horsfall DJ, Musgrove EA, et al. *EMS1 amplification can occur independently of CCND1 or INT-2 amplification at 11q13 and may identify different phenotypes in primary breast cancer.* Oncogene. 1997;15:1617-23.

73. Karlseder J, Zeillinger R, Schneeberger C, Czerwenka K, Speiser P, Kubista E, et al. *Patterns of DNA amplification at band q13 of chromosome 11 in human breast cancer.* Genes Chrom Cancer. 1994;9:42-8.

74. Peters G, Fantl V, Smith R, Brookes S, Dickson C. *Chromosome 11q13 markers and D-type cyclins in breast cancer.* Breast Cancer Res Treat. 1995;33:125-35.

75. Buckley MF, Sweeney KJ, Hamilton JA, Sini RL, D.L. M, Nicholson RI, et al. *Expression and amplification of cyclin genes in human breast cancer.* Oncogene. 1993;8:2127-33.

76. Tanner MM, Karhu RA, Nupponen NN, Borg A, Baldetorp B, Pejovic T, et al. *Genetic abberations in hypodiploid breast cancer*

frequent loss of chromosome 4 and amplification of cyclin D1 oncogene. Am J Pathol. 1998;153:191-9.

77. Gillett C, Smith P, Gregory W, Richards M, Millis R, Peters G, et al. *Cyclin D1 and prognosis in human breast cancer.* Int J Cancer. 1996;69:92-9.

78. Seshadri R, Lee CSL, Hui R, McCaul K, Horsfall DJ, Sutherland RL. *Cyclin D1 amplification is not associated with reduced overall survival in primary breast cancer but may predict early relapse in patients with features of good prognosis.* Clinical Cancer Res. 1996;2:1177-84.

79. Momand J, Zambetti GP. *Mdm-2: "big brother" of p53.* J Cell Biochem. 1997;64:343-52.

80. Kuukasjarvi T, Karhu R, Tanner M, Kahkonen M, Schaffer A, Nupponen N, et al. *Genetic heterogeneity and clonal evolution underlying development of asynchronous metastasis in human breast cancer.* Cancer Res. 1997;57:1597-604.

81. Wu GJ, Sinclair C, Hinson S, Ingle JN, Roche PC, Couch FJ. *Structural analysis of the 17q22-23 amplicon identifies several independent targets of amplification in breast cancer cell lines and tumors.* Cancer Res. 2001;61:4951-5.

82. Chou MM, Blenis J. *The 70 kDa S6 kinase: regulation of a kinase with multiple roles in mitogenic signaling.* Curr Opinion Cell Biol. 1995;7:806-14.

83. Montagne J, Stewart MJ, Stocker H, Hafen E, Kozma SC, Thomas G. *Drosophila S6 kinase: a regulator of cell size.* Science. 1999;285:2126-9.

84. Barlund M, Forozan F, Kononen J, Bubendorf L, Chen Y, Bittner ML, et al. *Detecting activation of ribosomal protein S6 kinase by complementary DNA and tissue microarray analysis.* J Natl Cancer Inst. 2000;92:1252-9.

85. Couch FJ, Wang XY, Wu GJ, Qian J, Jenkins RB, James CD. *Localization of PS6K to chromosomal region 17q23 and determination of its amplification in breast cancer.* Cancer Res. 1999;59:1408-11.

86. Tanner MM, Tirkkonen M, Kallioniemi A, Collins C, Stokke T, Karhu R, et al. *Increased copy number at 20q13 in breast cancer: defining the critical region and exclusion of candidate genes.* Cancer Res. 1994;54:4257-60.

87. Anzick SL, Kononen J, Walker RL, Azorsa DO, Tanner MM, Guan X, et al. *AIB1, a steroid receptor coactivator amplified in*

breast and ovarian cancer. Science. 1997;277:965-7.

88. Nonet GH, Stampfer MR, Chin K, Gray JW, Collins CC, Yaswen P. *The ZNF217 gene amplified in breast cancers promotes immortalization of human mammary epithelial cells.* Cancer Res. 2001;61:1250-4.

89. Sen S, Zhou H, White RA. *A putative serine/threoine kinase encoding gene BTAK on chromosome 20q13 is amplified and overexpressed in human breast cancer cell lines.* Oncogene. 1997;14:2195-200.

90. Glockner S, Lehmann U, Wilke N, Kleeberger W, Langer F, Kreipe H. *Amplification of growth regulatory genes in intraductal breast cancer is associated with higher nuclear grade but not with the progression to invasiveness.* Lab Invest. 2001;81:565-71.

91. Simpson JF, D.E. Q, O'Malley F, Odom-Maryon T, Clarke PE. *Amplification of CCND1 and expession of its protein product, cyclin D1, in ductal carcinoma in situ of the breast.* Am J Pathol. 1997;151:161-8.

92. Buerger H, Otterbach F, Simon R, Poremba C, Diallo R, Decker T, et al. *Comparative genomic hybridization of ductal carcinoma in situ of the breast-evidence of multiple genetic pathways.* J Pathol. 1999;187:396-402.

93. Kuukasjarvi T, Tanner M, Pennanen S, Karhu R, Kallioniemi OP, Isola J. *Genetic changes in intraductal breast cancer detected by comparative genomic hybridization.* Am J Pathol. 1997;150:1465-71.

94. Moore E, Hagee H, Coyne J, Gorey T, Dervan PA. *Widespread chromosomal abnormalities in high-grade ductal carcinoma in situ of the breast. Comparative genomic hybridization study of pure high-grade DCIS.* J Pathol. 1999;187:403-9.

95. Silverstein MJ, Poller DN, Waisman JR, Colburn WJ, Barth A, Gierson ED, et al. *Prognostic classification of breast ductal carcinoma-in-situ.* Lancet. 1995;345:1154-7.

96. Aubele M, Werner M, Hofler H. *Genetic alterations in presumptive precursor lesions of breast carcinomas.* Anal Cell Pathol. 2002;24:69-76.

97. Nishizaki T, DeVries S, Chew K, Goodson WH, Ljung B, Thor A, et al. *Genetic alterations in primary breast cancers and their metastases: Direct comparison using modified comparative genomic hybridization.* Genes Chrom Cancer. 1997;19:267-72.

98. Bautista S, Theillet C. *CCNDI and FGFRI coamplification results in the colocalization of 11q13 and 8p12 sequences in breast tumor nuclei.* Genes

Chrom Cancer. 1998;22:268-77.

99. Guan XY, Meltzer PS, Dalton WS, Trent JM. *Identification of cryptic sites of DNA sequence amplification in human breast cancer by chromosome microdissection.* Nature Genet. 1994;8:155-61.

100. Lafage M, Pedeutour F, Marchetto S, Simonetti J, Prosperi MT, Gaudray P, et al. *Fusion and amplification of two originally non-syntenic chromosomal regions in a mammary carcinoma cell line.* Genes Chrom Cancer. 1992;5:40-9.

101. Neve RM, Chin K, Fridlyand J, Yeh J, Baehner FL, Fevr T, et al. *A collection of breast cancer cell lines for the study of functionally distinct cancer subtypes.* Cancer Cell. 2006;10:515-27.

102. Corvi R, Savelyeva L, Schwab M. *Duplication of N-MYC at its resident site 2p24 may be a mechanism of activation alternative to amplification in human neuroblastoma cells.* Cancer Res. 1995;55:3471-4.

103. Curtis C, Shah SP, Chin SF, Turashvili G, Rueda OM, Dunning MJ, et al. *The genomic and transcriptomic architecture of 2,000 breast tumours reveals novel subgroups.* Nature. 2012;486:346-52.

CHAPTER 7

Apoptosis

7.1. INTRODUCTION

Apoptosis is a multi-step cell-death program that is inherent in every cell of the body. The average adult human body contains approximately 10^{13} cells of which an estimated 50 billion (10^9) cells die each day due to apoptosis.

Of course, the level of apoptosis varies between tissues. In the female breast, mammary epithelial cells undergo multiplication (mitosis) and death (apoptosis) during the menstrual cycle. The cell turnover is cyclical with the mitotic and apoptotic peaks occurring at days 25 and 28, respectively, of a 28-day menstrual cycle (1). Thus, normal tissue homeostasis is maintained through a balance of cell proliferation and apoptosis, which represents a physiological mechanism that eliminates damaged cells and thereby controls cell numbers and tissue size. A balance of proliferation and apoptosis is also maintained through all stages of mammary gland development (2). Apoptosis of mammary epithelial and other cells is characterized by specific morphological and biochemical changes, such as overall shrinkage, chromatin condensation, nuclear fragmentation, formation of apoptotic bodies, and DNA fragmentation at inter-nucleosomal sites.

Apoptosis can be initiated by a wide variety of intra- and extracellular factors, such as reactive oxygen species, c-jun-NH2-kinases (JNK), p53, adiponectin, cytotoxic chemicals, UV light and ionizing radiation (3-7). However, the execution of apoptosis follows evolutionarily conserved pathways involving a cascade of protease reactions carried out by caspases and regulated by the Bcl-2 (B-cell CLL/lymphoma 2) family of proteins (8-10). All caspases possess an active-site cysteine that is part of a conserved pentapeptide motif, QACXG (where X is R, Q or G), and cleaves substrates at Asp-Xxx bonds, i.e., after aspartic acid residues. Thus, caspases are aspartate-specific cysteine proteases. Two main pathways have been identified that activate caspases. The extrinsic pathway is activated by ligand-bound death receptors in the cell membrane, while the intrinsic pathway is initiated by release of mitochondrial proteins (Figure 7.1.).

The extrinsic pathway is initiated via binding of members of the tumor necrosis factor (TNF) family to their plasma membrane-bound receptors. Thus, the cytokine TNF itself binds to TNF receptors 1 and 2, also known as death receptors 4 and 5 (DR4 and DR5). Alternatively, Fas ligand (FasL or CD95L) binds to the Fas receptor, also known as CD95. The DR4 and DR5 proteins are composed of two extracellular cysteine-rich, ligand-binding pseudorepeats, a single transmembrane helix, and an intracellular death domain (11, 12). The interaction between ligands and

receptors leads to binding of the adaptor protein, Fas-associated death domain (FADD), which in turn recruits and activates the death-inducing signaling complex (DISC) composed of FADD, caspase-8 and caspase-10, and a regulator of caspase-8 activity, FLIP.

The intrinsic pathway is used extensively in response to DNA damage. It begins with mitochondrial permeabilization, release of cytochrome c from the intermembrane compartment into the cytosol, and activation of procaspase-9 (9). The cytochrome c release is regulated by members of the Bcl-2 family, which consists of over a dozen proteins classified into three functional groups (13, 14). Members of the first group, such as Bcl-2 and Bcl-xL, are characterized by four short, conserved Bcl-2 homology domains, BH1 – BH4. They also possess a C-terminal hydrophobic tail, which localizes the proteins to the outer surface of mitochondria. The key feature of group I members is that they possess anti-apoptotic activity and protect cells from death. In contrast, group II members, such as Bax and Bak, have pro-apoptotic activity. Group II proteins contain the same structural elements as group I proteins except the BH4 domain at the N-terminal. Group III members, such as Bid, contain the BH3 domain, but otherwise share little sequence similarity. Pro- and anti-apoptotic Bcl-2 members meet at the surface of mitochondria, where they compete to regulate the exit of cytochrome c. Cytochrome c associates with another protein cofactor, Apaf-1, and ATP, which then bind to pro-caspase-9 to form a complex called the apoptosome. This complex dimerizes and activates caspase-9, which in turn activates the effector caspase-3.

Both pathways converge and lead to the activation of caspase 3, which initiates controlled degradation of the cell through proteolytic cleavage of the cytoskeleton and nuclear envelope, resulting in cell shrinkage and chromatin condensation or pyknosis. Under most conditions the two pathways operate independently of each other. Crosstalk is provided by Bid, a member of Bcl-2 group III. Caspase-8-mediated cleavage of Bid greatly increases its pro-apoptotic activity and results in its translocation to mitochondria, where it promotes cytochrome c exit (15). The apoptotic cell death can be distinguished from necrotic cell death, which is characterized morphologically by rapid cell swelling and lysis and involves some proteins instrumental in apoptosis, such as caspase-8 (16). Apoptosis can also be blocked by a wide variety of factors, including estrogens. In addition to their well-known proliferative effect on mammary epithelial cells, estrogens have been shown to prevent the induction of apoptosis by altering the expression of the Bcl-2 family of proteins. Breast epithelium undergoes cyclic apoptosis and fluctuation in Bcl-2 protein levels during the menstrual cycle with maximal expression at midcycle and a sharp decrease at the end of the cycle (17). Treatment of MCF-7 breast cancer cells with E2 resulted in increased

APOPTOSIS

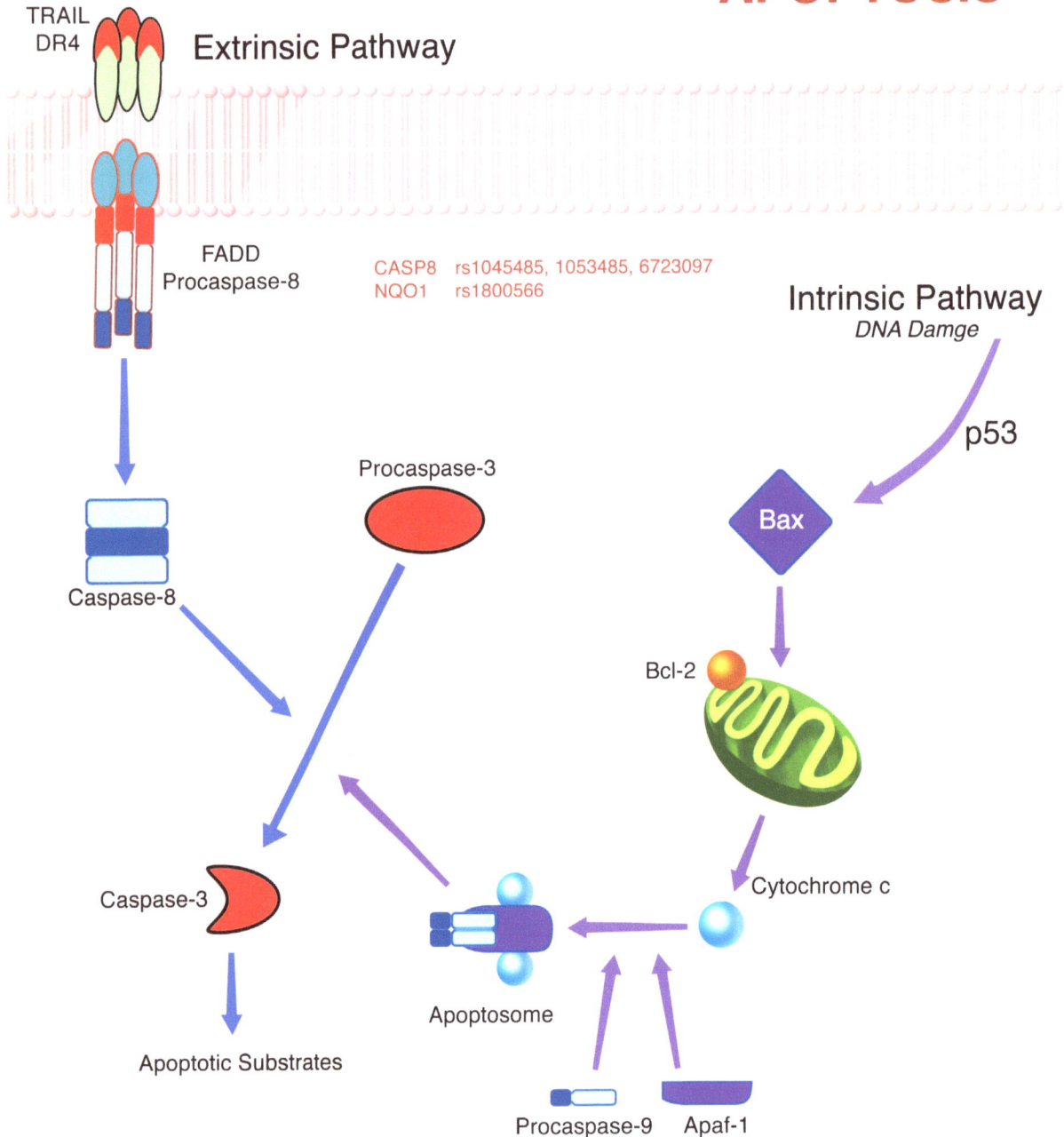

TRAIL
DR4

Extrinsic Pathway

FADD
Procaspase-8

CASP8 rs1045485, 1053485, 6723097
NQO1 rs1800566

Intrinsic Pathway
DNA Damge

p53

Bax

Procaspase-3

Bcl-2

Caspase-8

Caspase-3

Cytochrome c

Apoptosome

Apoptotic Substrates

Procaspase-9   Apaf-1

Figure 7.1. Schematic overview of apoptosis shows extrinsic and intrinsic pathways.

The extrinsic pathway is initiated via binding of ligands to membrane-bound death receptors, such as TNF-related apoptosis-inducing ligand (TRAIL) to the extracellular domain of death receptor 4 (DR4). Structural analysis revealed three receptors and three ligands assembled as a hexameric complex. The intracellular death domain of DR4 interacts with an adaptor protein, Fas-associated death domain (FADD). In turn, FADD recruits and activates procaspase-8 through self-cleavage. Caspase-8 then activates downstream effector caspase-3, committing the cell to apoptosis. The intrinsic pathway is regulated by members of the Bcl-2 family, such as proapoptotic Bax and anti-apoptotic Bcl-2. Bax and Bcl-2 meet at the surface of mitochondria, where they compete to regulate the exit of cytochrome c. Cytochrome c associates with another protein cofactor, Apaf-1, and ATP, which then bind to pro-caspase-9 to form a complex called the apoptosome. This complex dimerizes and activates caspase-9, which in turn activates the effector caspase-3. Genome-wide association studies have implicated genetic variants of CASP8 and NQO1 in breast cancer risk. Details are discussed in Chapters 7.2. and 4.5.

The Etiology of Breast Cancer

anti-apoptotic Bcl-2 mRNA and protein levels and increased Bcl-2/Bax protein ratio (18). Conversely, aromatase inhibitors and antiestrogens stimulated apoptosis by down-regulation of Bcl-2, up-regulation of Bax and activation of caspase 9, caspase 6, and caspase 7 (19). 2-methoxyestradiol, an estrogen metabolite, was also shown to induce apoptosis of MCF-7 cells through several mechanisms, including increased formation of ROS, up-regulation of the intrinsic pathway via inactivation of anti-apoptotic Bcl-2, and up-regulation of the extrinsic pathway via increased expression of death receptor 5 leading to sequential activation of caspase-8, caspase-9, and caspase-3 (20-22).

Apoptosis contributes to the pathogenesis of a number of diseases, including cancer (23, 24). In general, tumors result from cellular transformation leading to an inappropriate increase in cell number. However, the failure of dividing cells to initiate apoptosis after sustaining severe DNA damage also contributes to cancer. In principle, a cell with DNA damage has two fates; one is apoptosis, and the other is mutation leading to carcinogenesis. The cells that incur strong DNA damage and undergo apoptosis are no longer candidates for producing cancer cells. When weak DNA damage is induced, the cellular response allows repair of the damage. However, if the damage fails to be repaired, mutagenic lesions could be propagated and might lead to cancer. Thus, the fate of cells may be dependent on the intensity of the DNA damage and the ability to repair DNA (25, 26). The balance between repair and apoptosis appears to be influenced by the phosphorylation of histone H2AX (27). In this context, the suppression of apoptosis can lead to deregulated cell proliferation and tumor formation.

References

1. Ferguson DJ, Anderson TJ. *Morphological evaluation of cell turnover in relation to the menstrual cycle in the "resting" human breast.* Br J Cancer. 1981;44:177-81.

2. Humphreys RC, Krajewska M, Krnacik S, Jaeger R, Weiher H, Krajewski S, et al. *Apoptosis in the terminal endbud of the murine mammary gland: a mechanism of ductal morphogenesis.* Development. 1996;122:4013-22.

3. Chen F. *JNK-Induced Apoptosis, Compensatory Growth, and Cancer Stem Cells.* Cancer Research. 2012;72:379-86.

4. Hiraku Y, Kawanishi S. *Oxidative DNA damage and apoptosis induced by benzene metabolites.* Cancer Research. 1996;56:5172-8.

5. Kang JH, Lee YY, Yu BY, Yang BS, Cho KH, Yoon DK, et al. *Adiponectin induces growth arrest and apoptosis of MDA-MB-231 breast cancer cell.* Arch Pharm Res. 2005;28:1263-9.

6. Shimamura A, Fisher DE. *p53 in life and death.* Clin Cancer Res. 1996;2:435-40.

7. Ueda S, Masutani H, Nakamura H, Tanaka T, Ueno M, Yodoi J. *Redox control of cell death.* Antiox Redox Signal. 2002;4:405-14.

8. Cohen GM. *Caspases: the executioners of apoptosis.* Biochem J. 1997;326 (Pt 1):1-16.

9. Hengartner MO. *The biochemistry of apoptosis.* Nature. 2000;407:770-6.

10. Thorburn A. *Death receptor-induced cell killing.* Cell Signal. 2004;16:139-44.

11. Hymowitz SG, Christinger HW, Fuh G, Ultsch M, O'Connell M, Kelley RF, et al. *Triggering cell death: the crystal structure of Apo2L/TRAIL in a complex with death receptor 5.* Mol Cell. 1999;4:563-71.

12. Pan G, O'Rourke K, Chinnaiyan AM, Gentz R, Ebner R, Ni J, et al. *The receptor for the cytotoxic ligand TRAIL.* Science. 1997;276:111-3.

13. Adams JM, Cory S. *The Bcl-2 protein family: arbiters of cell survival.* Science. 1998;281:1322-6.

14. Antonsson B, Martinou JC. *The Bcl-2 protein family.* Exp Cell Res. 2000;256:50-7.

15. Gross A, Yin XM, Wang K, Wei MC, Jockel J, Milliman C, et al. *Caspase cleaved BID targets mitochondria and is required for cytochrome c release, while BCL-XL prevents this release but not tumor necrosis factor-R1/Fas death.* J Biol Chem. 1999;274:1156-63.

16. Peter ME. *Programmed cell death: Apoptosis meets necrosis.* Nature. 2011;471:310-2.

17. Sabourin JC, Martin A, Baruch J, Truc JB, Gompel A, Poitout P. *bcl-2 expression in normal breast tissue during the menstrual cycle.* Int J Cancer. 1994;59:1-6.

18. Wang TT, Phang JM. *Effects of estrogen on apoptotic pathways in human breast cancer cell line MCF-7.* Cancer Research. 1995;55:2487-9.
19. Thiantanawat A, Long BJ, Brodie AM. *Signaling pathways of apoptosis activated by aromatase inhibitors and antiestrogens.* Cancer Research. 2003;63:8037-50.

20. Fukui M, Zhu BT. *Mechanism of 2-methoxyestradiol-induced apoptosis and growth arrest in human breast cancer cells.* Mol Carcinog. 2009;48:66-78.

21. LaVallee TM, Zhan XH, Johnson MS, Herbstritt CJ, Swartz G, Williams MS, et al. *2-methoxyestradiol up-regulates death receptor 5 and induces apoptosis through activation of the extrinsic pathway.* Cancer Research. 2003;63:468-75.

22. Stander BA, Marais S, Vorster CJ, Joubert AM. *In vitro effects of 2-methoxyestradiol on morphology, cell cycle progression, cell death and gene expression changes in the tumorigenic MCF-7 breast epithelial cell line.* J Steroid Biochem Mol Biol. 2010;119:149-60.

23. Evan GI, Vousden KH. *Proliferation, cell cycle and apoptosis in cancer.* Nature. 2001; 411:342-8.

24. Thompson CB. *Apoptosis in the pathogenesis and treatment of disease.* Science. 1995; 267:1456-62.

25. Jackson SP, Bartek J. *The DNA-damage response in human biology and disease.* Nature. 2009;461:1071-8.

26. Oberle C, Blattner C. *Regulation of the DNA Damage Response to DSBs by Post-Translational Modifications.* Curr Genomics. 2010;11:184-98.

27. Cook PJ, Ju BG, Telese F, Wang X, Glass CK, Rosenfeld MG. *Tyrosine dephosphorylation of H2AX modulates apoptosis and survival decisions.* Nature. 2009;458:591-6.

7.2. EXPERIMENTAL AND CLINICAL STUDIES

It is generally accepted that mammary gland involution after the end of lactation is the result of an increase in apoptosis. When compared to lactating tissue, Bcl-2 protein levels decrease during mammary gland involution whereas Bax levels remain high (1). These findings suggest that Bcl-2 might serve as an intracellular mediator of signals that influence the apoptotic activity of mammary epithelial cells and mammary gland remodeling. Reduced Bcl-2 expression was also observed in 16% of breast biopsies with ductal hyperplasia (2). There was no correlation with the degree of proliferation. An immunohistochemical analysis of breast cancer specimens showed that loss of Bcl-2 expression was associated with a high rate of apoptosis whereas Bax expression showed no correlation (3).

Experimental studies suggest that apoptosis plays a role in breast cancer development. Apoptosis repressor with caspase recruitment domain (ARC) is an endogenous inhibitor of both the intrinsic and extrinsic pathways (4). The effects of ARC on breast tumorigenesis were examined in the polyoma middle T-antigen (PyMT) transgenic mouse model of breast cancer, in which endogenous ARC is strongly upregulated and in a xenograft model using MDA-MB-231-derived LM2 metastatic breast cancer cells. Deletion of the ARC-encoding gene nol3 decreased primary tumor burden in the PyMT model and RNAi-mediated knockdown of ARC levels reduced tumor volume in the LM2 xenograft model. These results indicate that ARC promotes breast carcinogenesis by driving primary tumor growth (4). Other endogenous compounds, such as conjugated linoleic acid, have the opposite effect and induced apoptosis in a rat mammary epithelial cell line and inhibited the formation of premalignant lesions by reducing the expression of Bcl-2 (5). Natural products, such as the alkaloid piperlongumine, have similar effects and induce apoptosis selectively in cancer cells resulting in growth inhibition of murine breast tumors (6).

Estrogens inhibit apoptosis in many cells while stimulating apoptosis in other cells (7). The inhibitory effect involves the interaction of E_2 with its membrane receptor, ERm, activating the MAP kinase ERK and inhibiting JNK, which leads to the induction of bcl-2 and inhibition of apoptosis in MCF-7 breast cancer cells (8). In addition, E_2 inhibits the cleavage of procaspase-9 to caspase-9 (9). The stimulating effect of estrogen was shown experimentally in MCF-7 cells that were adapted to grow in an estrogen-free environment for prolonged periods, i.e., 6 – 24 months. E_2 treatment of these cells resulted in a 7-fold increase in apoptosis accompanied by a marked increase of Fas expression compared with the parental MCF-7 cells (10). E_2 treatment of long-term estrogen-deprived cells also increased the expression of several pro-apoptotic proteins in the intrinsic pathway, including Bax and Bak (11). The expression of NAD(P)H:quinone oxidoreductase 1 (NQO1), which prevents redox cycling and oxidative DNA damage is suppressed by E_2 and up-regulated by antiestrogens such as tamoxifen (12).

The estrogen-mediated activation of both extrinsic and intrinsic apoptotic pathways may explain the finding that high doses of estrogen can promote tumor regression in postmenopausal women with hormone-dependent breast cancer (13).

The DR4 gene at 8p21-22 contains two common non-synonymous polymorphisms, Arg209Thr (626G/C; rs4871857; minor allele frequency over 40%) and Glu228Ala (683A/C; rs17088993; MAF 18%). A case-control study (999 cases, 996 controls) showed no association of Arg209Thr with breast cancer risk (14).

Another case-control study (521 cases, 1100 controls) also found no association of Arg209Thr or Glu228Ala with breast cancer risk (15). However, haplotype analysis revealed a 3.5-fold risk for carriers of the rare 626C-683C haplotype (OR 3.52; 95% CI 1.45 – 8.52; p = 0.003). DR4 gene mutations have been identified in breast cancer cell lines and metastatic breast cancers (16, 17).

The CASP8 gene at 2q33 contains a polymorphism (rs1045485) Asp302His in exon 12, with a minor allele frequency of 4 – 16%. Two studies observed an association of the variant allele with reduced breast cancer risk (14, 18). The larger of these studies (16,423 cases, 17,109 controls) determined odds ratios of 0.89 (95% CI 0.85 – 0.94) and 0.74 (0.62 – 0,87) for heterozygotes and rare homozygotes, respectively, compared with wild-type homozygotes (18). There was no association between the variant and age of onset, grade, stage, or hormone receptor status. An investigation of over 7,000 BRCA1 and BRCA2 mutation-carriers also showed an association of the variant 302His allele with reduced risk of breast cancer (per allele hazard ratio 0.85; 95% CI 0.76 – 0.97; p_{trend} 0.011) and ovarian cancer for BRCA1 but not BRCA2 carriers (19). Another variant, a 6-bp deletion in the CASP8 promoter (-652 6N del; rs3834129) was shown to abolish an Sp1 transcription factor binding site resulting in decreased mRNA expression in lymphocytes and lower caspase-8 and T lymphocytes apoptotic activity (20). The 6N del was associated with reduced risk of breast cancer in a Chinese population but not in two other studies consisting of a European and an American multiethnic cohort (20-22). Two independent data sets from the UK and US, including 3,888 breast cancer cases and controls, were genotyped for 45 tagging single nucleotide polymorphisms in the expanded CASP8 region (23, 24). A three-SNP haplotype across rs3834129, rs6723097, and rs3817578 was significantly associated with increased breast cancer risk (risk ratio 1.28; 95% CI 1.21 – 1.35; p <5 x 10^{-6}) and frequency of 0.29 in controls. This haplotype was substantially more significant than any individual SNP and consistent with previous results, i.e., the increased risk associated with the common 302Asp allele at rs1045485 and the ins allele of rs3834129.

The CASP10 gene is adjacent to the CASP8 gene and caspase-10 cooperates with caspase-8 in the transduction of death receptor-mediated apoptotic signals (25-27). In analogy with caspase 8, caspase 10 has two DEDs (death effector domains) that interact with the DED of FADD. The CASP10 gene contains two non-synonymous polymorphisms. Val410Ile (rs13010627; minor allele frequency 5 – 9%) is located seven amino acids downstream of the active site (28). Ile522Leu (rs13006529; minor allele frequency over 40%) is the last amino acid of the protein in exon 10. There is no association of Ile522Leu with breast cancer risk (14). A large study (over 30,000 cases and 30,000 controls) also found no evidence of an association of Val410Ile with breast cancer (29). However, in a case-control study (511 cases, 547 controls) of familial breast cancer, which did not carry BRCA1 or BRCA2 mutations, carriers of the variant 410Ile allele had a reduced risk (30). There was a mutual effect of CASP 10 Val410Ile with CASP 8 Asp302His resulting in a highly decreased familial breast cancer risk (OR 0.35; p_{trend} 0.007), pointing to the interaction between the CASP 10 and CASP 8 polymorphisms in breast carcinogenesis. A separate investigation of over 7,000 BRCA1 and BRCA2 mutation carriers showed no association of the variant 410Ile allele with breast or ovarian risk in either BRCA1 or BRCA1 carriers (19). There were also no significant interactions between CASP 8 and CASP 10.

In summary, apoptosis plays in role in breast cancer development and molecular epidemiological studies implicate primarily the extrinsic pathway.

References

1. Merto GR, Cella N, Hynes NE. *Apoptosis is accompanied by changes in Bcl-2 and Bax expression, induced by loss of attachment, and inhibited by specific extracellular matrix proteins in mammary epithelial cells.* Cell Growth Differ. 1997;8:251-60.

2. Mommers EC, van Diest PJ, Leonhart AM, Meijer CJ, Baak JP. *Expression of proliferation and apoptosis-related proteins in usual ductal hyperplasia of the breast.* Hum Pathol. 1998;29:1539-45.

3. van Slooten HJ, van de Vijver MJ, van de Velde CJ, van Dierendonck JH. *Loss of Bcl-2 in invasive breast cancer is associated with high rates of cell death, but also with increased proliferative activity.* Br J Cancer. 1998;77:789-96.

4. Medina-Ramirez CM, Goswami S, Smirnova T, Bamira D, Benson B, Ferrick N, et al. *Apoptosis inhibitor ARC promotes breast tumorigenesis, metastasis, and chemoresistance.* Cancer Res. 2011;71:7705-15.

5. Ip C, Ip MM, Loftus T, Shoemaker S, Shea-Eaton W. *Induction of apoptosis by conjugated linoleic acid in cultured mammary tumor cells and premalignant lesions of the rat mammary gland.* Cancer Epidemiol Biomarkers Prev. 2000;9:689-96.

6. Raj L, Ide T, Gurkar AU, Foley M, Schenone M, Li X, et al. *Selective killing of cancer cells by a small molecule targeting the stress response to ROS.* Nature. 2011;475:231-4.

7. Lewis-Wambi JS, Jordan VC. *Estrogen regulation of apoptosis: how can one hormone stimulate and inhibit?* Breast Cancer Res. 2009;11:206.

8. Razandi M, Pedram A, Greene GL, Levin ER. *Cell membrane and nuclear estrogen receptors (ERs) originate from a single transcript: studies of ERα and ERb expressed in chinese hamster ovary cells.* Mol Endocrinol. 1999;13:307-19.

9. Razandi M, Pedram A, Levin ER. *Plasma membrane estrogen receptors signal to antiapoptosis in breast cancer.* Mol Endocrinol. 2000; 14:1434-47.

10. Song RX, Mor G, Naftolin F, McPherson RA, Song J, Zhang Z, et al. *Effect of long-term estrogen deprivation on apoptotic responses of breast cancer cells to 17beta-estradiol.* J Natl Cancer Inst. 2001;93:1714-23.

11. Lewis JS, Meeke K, Osipo C, Ross EA, Kidawi N, Li T, et al. *Intrinsic mechanism of estradiol-induced apoptosis in breast cancer cells resistant to estrogen deprivation.* J Natl Cancer Inst. 2005;97:1746-59.

12. Bianco NR, Perry G, Smith MA, Templeton DJ, Montano MM. *Functional implications of antiestrogen induction of quinone reductase: inhibition of estrogen-induced deoxyribonucleic acid damage.* Mol Endocrinol. 2003;17:1344-55.

13. Jordan VC, Ford LG. *Paradoxical clinical effect of estrogen on breast cancer risk: a "new" biology of estrogen-induced apoptosis.* Cancer Prev Res (Phila). 2011;4:633-7.

14. MacPherson G, Healey CS, Teare MD, Balasubramanian SP, Reed MW, Pharoah PD, et al. *Association of a common variant of the CASP8 gene with reduced risk of breast cancer.* J Natl Cancer Inst. 2004;96:1866-9.

15. Frank B, Hemminki K, Shanmugam KS, Meindl A, Klaes R, Schmutzler RK, et al. *Association of death receptor 4 haplotype 626C-683C with an increased breast cancer risk.* Carcinogenesis. 2005;26:1975-7.

16. Seitz S, Wassmuth P, Fischer J, Nothnagel A, Jandrig B, Schlag PM, et al. *Mutation analysis and mRNA expression of trail-receptors in human breast cancer.* Int J Cancer. 2002;102:117-28.

17. Shin MS, Kim HS, Lee SH, Park WS, Kim SY, Park JY, et al. *Mutations of tumor necrosis factor-related apoptosis-inducing ligand receptor 1 (TRAIL-R1) and receptor 2 (TRAIL-R2) genes in metastatic breast cancers.* Cancer Res. 2001;61:4942-6.

18. Cox A, Dunning AM, Garcia-Closas M, Balasubramanian S, Reed MW, Pooley KA, et al. *A common coding variant in CASP8 is associated with breast cancer risk.* Nat Genet. 2007;39:352-8.

19. Engel C, Versmold B, Wappenschmidt B, Simard J, Easton DF, Peock S, et al. *Association of the variants CASP8 D302H and CASP10 V410I with breast and ovarian cancer risk in BRCA1 and BRCA2 mutation carriers.* Cancer Epidemiol Biomarkers Prev. Ann Rev Public Health. 2010;19:2859-68.

20. Sun T, Gao Y, Tan W, Ma S, Shi Y, Yao J, et al. *A six-nucleotide insertion-deletion polymorphism in the CASP8 promoter is associated with susceptibility to multiple cancers.* Nat Genet. 2007;39:605-13.

21. Frank B, Rigas SH, Bermejo JL, Wiestler M, Wagner K, Hemminki K, et al. *The CASP8 -652 6N del promoter polymorphism and breast cancer risk: a multicenter study.* Breast Cancer Res Treat. 2008;111:139-44.

22. Haiman CA, Garcia RR, Kolonel LN, Henderson BE, Wu AH, Le Marchand L. *A promoter polymorphism in the CASP8 gene is not associated with cancer risk.* Nat Genet. 2008;40:259-60; author reply 60-1.

23. Camp NJ, Parry M, Knight S, Abo R, Elliott G, Rigas SH, et al. *Fine-Mapping CASP8 Risk Variants in Breast Cancer.* Cancer Epidemiol Biomarkers Prev. 2012;21:176-81.

24. Shephard ND, Abo R, Rigas SH, Frank B, Lin WY, Brock IW, et al. *A breast cancer risk haplotype in the caspase-8 gene.* Cancer Res. 2009; 69:2724-8.

25. Cohen GM. *Caspases: the executioners of apoptosis.* Biochem J. 1997;326 (Pt 1):1-16.

26. Milhas D, Cuvillier O, Therville N, Clave P, Thomsen M, Levade T, et al. *Caspase-10 triggers Bid cleavage and caspase cascade activation in FasL-induced apoptosis.* J Biol Chem. 2005;280:19836-42.

27. Wang J, Chun HJ, Wong W, Spencer DM, Lenardo MJ. *Caspase-10 is an initiator caspase in death receptor signaling.* Proc Natl Acad Sci U S A. 2001;98:13884-8.

28. Wang J, Zheng L, Lobito A, Chan FK, Dale J, Sneller M, et al. *Inherited human Caspase 10 mutations underlie defective lymphocyte and dendritic cell apoptosis in autoimmune lympho-proliferative syndrome type II.* Cell. 1999;98:47-58.

29. Gaudet MM, Milne RL, Cox A, Camp NJ, Goode EL, Humphreys MK, et al. *Five polymorphisms and breast cancer risk: results from the Breast Cancer Association Consortium.* Cancer Epidemiol Biomarkers Prev. 2009;18:1610-6.

30. Frank B, Hemminki K, Wappenschmidt B, Meindl A, Klaes R, Schmutzler RK, et al. *Association of the CASP10 V410I variant with reduced familial breast cancer risk and interaction with the CASP8 D302H variant.* Carcinogenesis. 2006;27:606-9.

CHAPTER 8

Familial Breast Cancer

8.1. INTRODUCTION

Breast cancers can be divided into sporadic and familial tumors. A family history of breast cancer is associated with an increased risk of breast cancer.

The risk was quantified by The Collaborative Group on Hormonal Factors in Breast Cancer, which combined and reanalyzed data from 52 studies including 58,209 women with breast cancer and 101,986 women without the disease (1). Altogether 7,496 (12.9%) women with breast cancer and 7,438 (7.3%) controls reported that one or more first-degree relative (mother, sister, or daughter) had a history of breast cancer. Among cases, 12% had one affected relative and 1% had two or more. Risk ratios for breast cancer increased with increasing numbers of affected first-degree relatives. Compared with women who had no affected relative, the ratios were 1.80 (99% CI 1.69 – 1.91), 2.93 (2.36 – 3.64), and 3.90 (2.03 – 7.49), respectively, for one, two, and three or more affected first-degree relatives (p < 0.0001 each). The risk ratios were greatest at young ages and, for women of a given age, were greater the younger the relative was when diagnosed. The results did not differ substantially between women reporting an affected mother or sister. Other factors, such as childbearing history, did not significantly alter the risk ratios associated with a positive family history. The risk estimates translate into the following incidence percentages. The probability of developing breast cancer between ages 20 and 80 is 7.8% (1 in 13 women) with no affected first-degree relative, 13.3% (1 in 8) with one affected relative, and 21.1% (1 in 5) with two affected relatives. In other words, the large majority of women who develop breast cancer have no mother, sister, or daughter with a history of the disease. In more-developed countries, 8 out of 9 breast cancers occur in women with no history of breast cancer in a first-degree relative (1).

The contribution of family history to breast cancer risk has been assessed by analyzing large cohorts of twins in Scandinavia (44,788 pairs) and the nationwide Swedish family-cancer database, which encompasses about 10 million individuals (2, 3). This analysis indicates that genetic factors account for about 25% of the risk for breast cancer. Familial aggregation may be due to either genetic or environmental factors shared within families. A sub-analysis of the Swedish database addressed familial effects (genetic + shared environmental + childhood shared environmental effects) and found that shared environment and shared environment during childhood contributed about 9% and 6%, respectively, of the overall risk (2). Thus, the main contribution to risk was assigned to non-shared environmental effects, accounting for

about 60%, which means that the environment plays the principal causative role in breast cancer.

Several empirical models have been developed to estimate the breast cancer risk of patients with a positive family history. The main models by Gail, Claus, and Tyrer-Cuzick differ in the risk factors included. The Gail model includes age, age at menarche, age at first live birth, number of previous breast biopsies, atypical hyperplasia, and number of first-degree relatives with breast cancer ((4); www. cancer.gov/bcrisktool; www4.utsouthwestern.edu/ breasthealth/cagene). The Claus model includes age and family history, i.e., number and type (mother, sister, or aunt) of first- and second-degree relatives with breast cancer and their age at onset ((5); www4.utsouthwestern.edu/ breasthealth/cagene). The Tyrer-Cuzick model includes age at menarche, parity, age at first child birth, age at menopause, atypical hyperplasia, lobular carcinoma *in situ*, height, body mass index, presence of BRCA1/2 mutation and family history, i.e., breast/ovarian cancer in first/ second/third-degree relatives and their age at onset ((6); www.ems-trials.org/riskevaluator/). The strengths and limitations of these models have been discussed extensively (7-13). The predictive accuracy of a model to discriminate between women who will develop breast cancer and those who will not is generally expressed as the area under the receiver operating characteristics curve (AUC). The AUCs of the breast cancer risk models mentioned range from 0.558 to 0.762 (14).

Population-based estimates of the prevalence of family history among U.S. women indicate that ~6% of women have a first-degree relative with breast cancer, whereas 13% have one or more second-degree relatives (including maternal or paternal aunts, uncles, nieces, nephews, grandparents, and grandchildren) with breast cancer (15). Several studies have found that cancer in second-degree relatives is often underestimated (15). Failure to extend information of family history to male relatives may also lead to underestimation because a family history of prostate cancer may increase the risk of breast cancer (16, 17). Early age at diagnosis of breast cancer is a useful marker of genetic susceptibility to breast cancer (18). Among women diagnosed with breast cancer under age 40, 50% had family members with breast and/or ovarian cancer (19).

A positive family history, often used as a proxy for an unmeasured, common genetic background within a family, is one of the most important risk factors in predicting breast cancer risk (15). It has been estimated that there is a 40-fold range of genetic risk for breast cancer in the general population with half of all tumors occurring in the 12% of women at the highest genetic risk (20-22). By taking risk level into account as well as prevalence in the population, three reasonably

Table 8.1. Genes Associated with a Hereditary Predisposition to Breast Cancer (11, 21, 23, 26-28)

Gene	Syndrome	Risk Penetrance	Contribution to Familial Breast Cancer (%)
BRCA1		high	~8
BRCA2		high	~8
p53	Li-Fraumeni	high	<1
PTEN	Cowden	high	<1
ATM	Ataxia teleangiectasia	moderate	2
BACH1 (BRIP1)	Fanconi's Anemia	moderate	<1
BARD1		moderate	<1
CHK2	Li-Fraumeni variant	moderate	5
NBS1	Nijmegen Breakage Syndrome	moderate	<1
PALB2	Fanconi's Anemia	moderate	<1
STK11	Peutz-Jeghers	moderate	<1

well-defined classes of breast cancer susceptibility alleles have been defined according to their penetrance, which is the proportion of genetically similar individuals that show any phenotypical manifestation (i.e., breast cancer) of a mutation that they have in common: rare high-penetrance alleles, rare moderate-penetrance alleles and common low-penetrance alleles (23). The two most important breast cancer susceptibility genes, BRCA1 and BRCA2, were identified by linkage analysis and positional cloning in the 1990s (24, 25). In addition to BRCA1 and BRCA2, there are nine other genes, which can be considered well established breast cancer susceptibility genes (11, 21, 26-28). Table 8.1. provides an overview of the genes associated with inherited breast cancer, including BRCA1, BRCA2, p53, and ATM, which are discussed in detail in the following sections. All of these genes play crucial roles in DNA repair, cell cycle checkpoints, and genetic stability but they differ in disease frequency and risk.

BRCA1 and BRCA2 are high-penetrance susceptibility genes. A combined analysis of 22 studies showed that BRCA1 and BRCA2 germ-line mutations confer a cumulative lifetime risk of invasive breast cancer of ~65% in BRCA1 mutation carriers and 45% in BRCA2 carriers up to the age of 70 (29). However, the proportion of the familial risk of breast cancer that is accounted for by BRCA1 and BRCA2 mutations together is no more than 17% (30-33). Even in a referral clinic specializing in screening women from high-risk families, the majority of tests for BRCA1 mutations will be negative and therefore uninformative (34). Mutations in p53 and PTEN also confer high risks of breast cancer but are far less common than those in BRCA1 or BRCA2 (26). ATM, BACH1/BRIP1, CHEK2, and PALB2 encode proteins that interact with BRCA1, BRCA2, and/or p53. However, compared with mutations in BRCA1, BRCA2, and p53, the relative risks associated with mutations in ATM, BACH1/BRIP1, CHEK2, and PALB2 are much lower, 2.0 – 2.5 (35, 36).

Moreover, mutations in these moderate-level risk genes are rare, with less than 1% of the population being heterozygous. Mutations in NBS1 and STK11 are also associated with lower risks of developing breast cancer (27). Collectively, mutations in ATM, BACH1/BRIP1, CHEK2, NBS1, PALB2, and STK11 are estimated to account for no more than approximately 5% of the familial clustering of breast cancer, indicating that the majority of the familial clustering is still unexplained (26, 27, 37).

A third component of breast cancer susceptibility is comprised of common low-penetrance alleles, which have been discovered through genome-wide association studies (GWAS) (23). GWAS based on stringent levels of statistical significance revealed seven genes encoding known or unidentified proteins (37). The increased risk of breast cancer associated with carrying a single copy of each risk allele is small, ranging from 1.07 to 1.26. However, the population prevalence of each risk allele is high, ranging from 28 to 87%. Because the predisposing alleles are common, despite the low risks they confer, their contribution to the familial risk of breast cancer is not insignificant, estimated to account for 3.9% of the familial risk of breast cancer in European populations (23). In summary, no more than 5% of the genetic attributable risk of breast cancer can be explained by these low-level susceptibility alleles. Even allowing for the effect of these alleles, it appears that the known genes account for no more than 25% of the familial aggregation of breast cancer (26, 27). Thus, the underlying cause of roughly three-quarters of familial breast cancer cases is still unclear. Inherited breast cancer is highly genetically heterogeneous with respect to both loci and alleles involved. All evidence to date is that the model that best reflects this heterogeneity is not a "common disease-common allele" model but instead a "common disease-multiple rare alleles" model (38).

In current practice, family history is used to separate hereditary from sporadic breast cancer cases. However, family history is an unsatisfactory criterion, even when based on first-degree relatives. Many genetically determined cases give a false negative family history whereas false positive histories occur by chance, especially in large families (39, 40). Thus, at the present time neither the true proportion of breast cancers that are hereditary nor the proportionate role of specific genetic determinants is precisely known. Twin studies have been used to determine the contribution of inheritance. Monozygotic twins are genetically identical and develop in a single placenta, whereas dizygotic twins share only half their genetic material and develop in two placentas. Disease in monozygotic twins who are both affected is considered largely to represent hereditary breast cancer, whereas disease in only one twin of a monozygotic pair is believed to represent sporadic, or less heritable, disease (41). A prospective study showed a constant and much higher age-specific incidence of breast cancer throughout adulthood in the identical twins of women with cancer than in similar women in the general population (42). A comparison of mono- with dizygotic twins showed that 20% of monozygotic twins were disease-concordant (i.e., both twins developed breast cancer) compared to 12% in dizygotic twins (40). Such a high incidence in monozygotic twins appears incompatible with a Mendelian model of inheritance with only two suscepti- bility categories. The results are more consistent with the hypothesis that genetic susceptibility to breast cancer is due to multiple low-penetrance alleles coexisting in high- penetrance combinations, i.e., a type of polygenic model (26, 40, 42). In this polygenic model for breast cancer susceptibility, a large number of genes that confer low risk individually would act in combination, resulting in a wide spectrum of risk in the population.

References

1. Beral V, Bull D, Doll R, Peto R, Reeves G. *Familial breast cancer: collaborative reanalysis of individual data from 52 epidemiological studies including 58,209 women with breast cancer and 101,986 women without the disease.* Lancet. 2001;358:1389-99.

2. Czene K, Lichtenstein P, Hemminki K. *Environmental and heritable causes of cancer among 9.6 million individuals in the Swedish Family-Cancer Database.* Int J Cancer. 2002;99:260-6.

3. Lichtenstein P, Holm NV, Verkasalo PK, Iliadou A, Kaprio J, Koskenvuo M, et al. *Environmental and heritable factors in the causation of cancer--analyses of cohorts of twins from Sweden, Denmark, and Finland.* N Engl J Med. 2000;343:78-85.

4. Gail MH, Brinton LA, Byar DP, Corle DK, Green SB, Schairer C, et al. *Projecting individualized probabilities of developing breast cancer for white females who are being examined annually.* J Natl Cancer Inst. 1989;81:1879-86.

5. Claus EB, Risch N, Thompson WD. *Autosomal dominant inheritance of early-onset breast cancer. Implications for risk prediction.* Cancer. 1994;73:643-51.

6. Tyrer J, Duffy SW, Cuzick J. *A breast cancer prediction model incorporating familial and personal risk factors.* Stat Med. 2004;23:1111-30.

7. Armstrong K, Eisen A, Weber B. *Assessing the risk of breast cancer.* N Engl J Med. 2000;342:564-71.

8. Bellcross C. *Approaches to applying breast cancer risk prediction models in clinical practice.* Commun Oncol. 2009;6:373-9.

9. Costantino JP, Gail MH, Pee D, Anderson S, Redmond CK, Benichou J, et al. *Validation studies for models projecting the risk of invasive and total breast cancer incidence.* J Natl Cancer Inst. 1999;91:1541-8.

10. Pankratz VS, Hartmann LC, Degnim AC, Vierkant RA, Ghosh K, Vachon CM, et al. *Assessment of the accuracy of the Gail model in women with atypical hyperplasia.* J Clin Oncol. 2008;26:5374-9.

11. Robson M, Offit K. *Clinical practice. Management of an inherited predisposition to breast cancer.* N Engl J Med. 2007;357:154-62.

12. Spiegelman D, Colditz GA, Hunter D, Hertzmark E. *Validation of the Gail et al. model for predicting individual breast cancer risk.* J Natl Cancer Inst. 1994;86:600-7.

13. van Asperen CJ, Jonker MA, Jacobi CE, van Diemen-Homan JE, Bakker E, Breuning MH, et al. *Risk estimation for healthy women from breast cancer families: new insights and new strategies.* Cancer Epidemiol Biomarkers Prev. 2004;13:87-93.

14. van Zitteren M, van der Net JB, Kundu S, Freedman AN, van Duijn CM, Janssens AC. *Genome-based prediction of breast cancer risk in the general population: a modeling study based on meta-analyses of genetic associations.* Cancer Epidemiol Biomarkers Prev. 2011;20:9-22.

15. Hall IJ, Burke W, Coughlin S, Lee NC. *Population-based estimates of the prevalence of family history of cancer among women.* Community Genet. 2001;4:134-42.

16. McCahy PJ, Harris CA, Neal DE. *Breast and prostate cancer in the relatives of men with prostate cancer.* Br J Urol. 1996;78:552-6.

17. Sellers TA, Potter JD, Rich SS, Drinkard CR, Bostick RM, Kushi LH, et al. *Familial clustering of breast and prostate cancers and risk of postmeno- pausal breast cancer.* J Natl Cancer Inst. 1994;86:1860-5.

18. FitzGerald MG, MacDonald DJ, Krainer M, Hoover I, O'Neil E, Unsal H, et al. *Germ-line BRCA1 mutations in Jewish and non-Jewish women with early-onset breast cancer.* N Engl J Med. 1996;334:143-9.

19. Samphao S, Wheeler AJ, Rafferty E, Michaelson JS, Specht MC, Gadd MA, et al. *Diagnosis of breast cancer in women age 40 and younger: delays in diagnosis result from underuse of genetic testing and breast imaging.* Am J Surg. 2009;198:538-43.

20. Bellcross CA, Lemke AA, Pape LS, Tess AL, Meisner LT. *Evaluation of a breast/ovarian cancer genetics referral screening tool in a mammography population.* Genet Med. 2009;11:783-9.

21. Easton DF. *How many more breast cancer predisposition genes are there?* Breast Cancer Res. 1999;1:14-7.

22. Pharoah PD, Antoniou A, Bobrow M, Zimmern RL, Easton DF, Ponder BA. *Polygenic susceptibility to breast cancer and implications for prevention.* Nature Genet. 2002;31:33-6.

23. Stratton MR, Rahman N. *The emerging landscape of breast cancer susceptibility.* Nature Genet. 2008;40:17-22.

24. Miki Y, Swensen J, Shattuck- Eidens D, Futreal PA, Harshman K, Tavtigian S, et al. *A strong candidate for the breast and ovarian cancer susceptibility gene BRCA1.* Science. 1994;266:66-71.

25. Wooster R, Bignell G, Lancaster J, Swift S, Seal S, Mangion J, et al. *Identification of the breast cancer susceptibility gene BRCA2.* Nature. 1995;378:789-92.

26. Antoniou AC, Easton DF. *Models of genetic susceptibil- ity to breast cancer.* Oncogene. 2006;25:5898-905.

27. Hollestelle A, Wasielewski M, Martens JW, Schutte M. *Discovering moderate-risk breast cancer suscep- tibility genes.* Curr Opin Genet Dev. 2010;20:268-76.

28. Wooster R, Weber BL. *Breast and ovarian cancer.* N Engl J Med. 2003;348:2339-47.

29. Antoniou A, Pharoah PD, Narod S, Risch HA, Eyfjord JE, Hopper JL, et al. *Average risks of breast and ovarian cancer associated with BRCA1 or BRCA2 mutations detected in case Series unselected for family history: a combined analysis of 22 studies.* Am J Hum Genet. 2003;72:1117-30.

30. Anglian Breast Cancer Study Group. *Prevalence and penetrance of BRCA1 and BRCA2 mutations in a population-based series of breast cancer cases. Anglian Breast Cancer Study Group.* Br J Cancer. 2000;83:1301-8.

31. Martin AM, Weber BL. *Genetic and hormonal risk factors in breast cancer.* J Natl Cancer Inst. 2000;92:1126-35.

32. Struewing JP, Hartge P, Wacholder S, Baker SM, Berlin M, McAdams M, et al. *The risk of cancer associated with specific mutations of BRCA1 and BRCA2 among Ashkenazi*

Jews. N Engl J Med. 1997;336:1401-8.

33. Szabo CI, King MC. *Population genetics of BRCA1 and BRCA2.* Am J Hum Genet. 1997;60:1013-20.

34. Couch FJ, DeShano ML, Blackwood MA, Calzone K, Stopfer J, Campeau L, et al. *BRCA1 mutations in women attending clinics that evaluate the risk of breast cancer.* N Engl J Med. 1997;336:1409-15.

35. Renwick A, Thompson D, Seal S, Kelly P, Chagtai T, Ahmed M, et al. *ATM mutations that cause ataxia-telangiectasia are breast cancer susceptibility alleles.* Nature Genet. 2006;38:873-5.

36. Seal S, Thompson D, Renwick A, Elliott A, Kelly P, Barfoot R, et al. *Truncating mutations in the Fanconi anemia J gene BRIP1 are low-penetrance breast cancer susceptibility alleles.* Nature Genet. 2006;38:1239-41.

37. Pharoah PD, Antoniou AC, Easton DF, Ponder BA. *Polygenes,*

risk prediction, and targeted prevention of breast cancer. N Engl J Med. 2008;358:2796-803.

38. Walsh T, King MC. *Ten genes for inherited breast cancer.* Cancer Cell. 2007;11:103-5.

39. Cui J, Hopper JL. *Why are the majority of hereditary cases of early-onset breast cancer sporadic? A simulation study.* Cancer Epidemiol Biomarkers Prev. 2000;9:805-12.

40. Mack TM, Hamilton AS, Press MF, Diep A, Rappaport EB. *Heritable breast cancer in twins.* Br J Cancer. 2002;87:294-300.

41. Hamilton AS, Mack TM. *Puberty and genetic susceptibility to breast cancer in a case-control study in twins.* N Engl J Med. 2003;348:2313-22.

42. Peto J, Mack TM. *High constant incidence in twins and other relatives of women with breast cancer.* Nature Genet. 2000;26:411-4.

8.2. BRCA1
8.2.1. INTRODUCTION

Breast cancer susceptibility protein 1 (BRCA1) plays multiple roles in the cell, primarily in the nucleus where it is involved in DNA repair, control of cell cycle checkpoints, and transcription regulation (1-6). Extranuclear roles of BRCA1 include the regulation of centrosome replication and cytokinesis. Germ-line mutations in the BRCA1 gene are associated with a highly penetrant, autosomal dominant predisposition to early-onset breast and/or ovarian cancer (7, 8). Mutations in BRCA1 most commonly induce frameshifts that cause protein truncation and loss of function (9, 10; http://research.nhgri.nih.gov/bic/). Given the multitude of cellular roles played by BRCA1, it is difficult to determine which ones are compromised by its absence to the point of causing malignant growth, and even more difficult to explain the tissue predilection leading to breast and/or ovarian cancer. Tumors of afflicted individuals show loss of heterozygosity for the wild-type BRCA1 allele with retention of the inherited mutant allele indicating that BRCA1 is a tumor suppressor gene.

The 100-kb BRCA1 gene at 17q21 is composed of 24 exons (there is no exon 4) of which exon 11 is longer than all others combined. Translation starts in exon 2 resulting in a protein of 1,863 amino acids (8) (Figure 8.1.). The N-terminal 109 residues contain the so-called RING finger domain, a conserved motif composed of 40 – 60 amino acids including eight cysteine residues that coordinate two ions of zinc.

Nuclear localization sequences have been identified at residues 503 – 508 and 606 – 615 (2). The C-terminal end of BRCA1 contains two tandem repeats of ~100 amino acids named BRCT (for BRCA1 C-terminus). The BRCT repeats are conserved domains that interact with several phosphorylated proteins involved in DNA repair and transcription regulation by specifically targeting the motif pSer-x-x-Phe (11-14). Apart from the conserved RING domain and BRCT repeats, the central region of BRCA1 encompassing ~1,500 residues does not have any identifiable domains but is known to interact with numerous proteins involved in DNA repair, control of cell cycle checkpoints, and transcription regulation (Figure 8.1.). The central region of BRCA1 (amino acids 452 – 1,079) is also capable of binding DNA directly with preference for branched structures but without sequence specificity (15). The main binding partner of BRCA1 is BARD1 (for BRCA1-associated RING domain), a protein of 777 amino acids (1). BARD1 contains a RING domain and BRCT repeats and thereby resembles BRCA1 more than BRCA2, which is structurally unrelated to BRCA1. BRCA1 and BARD1 form a stable heterodimeric complex connected via their individual RING fingers. Structural analysis revealed that the zinc-binding elements of each RING motif are flanked by long α-helices that combine to form a four-helix bundle and stabilize the heterodimer (16). RING-domain proteins can self-assemble into large complexes and the BRCA1-BARD1 RING domains have been shown to form 30 nm ring-shaped superstructures visible by electron microscopy (17). After DNA damage, BARD1 colocalizes with BRCA1 and the homologous DNA repair protein RAD51 to participate as a complex in damage repair (18, 19). The BRCA1-BARD1 heterodimer also functions as an E2-dependent E3 ubiquitin ligase (20, 21). The cellular consequences of ubiquination are diverse, including protein degradation, transcription regulation, cell cycle control, and DNA repair activation (1, 5). In addition, the presence of BARD1 stimulates the binding of BRCA1 to DNA (22). Interestingly, the BARD1 gene is also targeted by germ-line mutations in a subset of breast and ovarian cancers and BARD1 mutations have been identified in rare hereditary breast cancers negative for BRCA1/BRCA2 mutations (23, 24).

In addition to BARD1, BRCA1 interacts with numerous proteins in response to various cellular challenges (2). The long list of binding partners includes Abraxas (forming the BRCA1 A complex), BACH1 (for BRCA1-associated C-terminal helicase; forming the BRCA1 B complex), CtIP (for CtBP-interacting protein; forming the BRCA1 C complex), as well as histone H2AX, MDC1 (for mediator of DNA damage checkpoint protein 1), and 53BP1 (for p53-binding protein) (3, 25, 26). Each of the binding proteins interacts via the pSer-x-x-Phe motif of the BRCT domain of BRCA1 in a mutually exclusive manner. In case of DNA damage, BRCA1 is involved in more than one repair pathway, i.e., homologous recombination, non-homologous

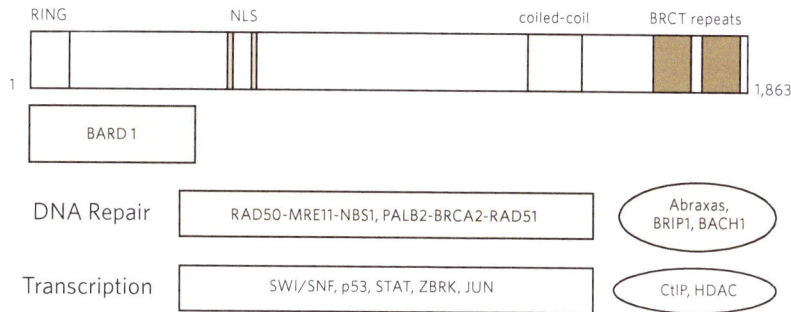

Figure 8.1. Diagram of BRCA1 protein.
The ~220-kDa BRCA1 protein contains conserved RING and BRCT repeat domains at the N- and C-termini, respectively. The RING domain is endowed with E3 ubiquitin ligase activity, while the BRCT repeats interact with a variety of phosphorylated proteins including Abraxas, BACH1/ BRIP1, and CtIP. Nuclear localization signals (NLS) are consistent with the presence of BRCA1 in the nucleus and its roles in DNA damage repair, especially homologous recombination and non-homologous end-joining repair, which are accomplished by interaction with RAD50-MRE-11-NBS1 and PALB2-BRCA2-RAD51 complexes. BRCA1 is also involved in transcription regulation by interacting with multiple proteins e.g., SWI/SNF, p53, STAT, ZBRK, JUN. Germ-line mutations associated with early-onset breast cancer are mostly missense or frame-shift mutations, e.g., a deletion of adenine and guanine in codon 185 (185delAG).

end-joining, and mismatch repair. In the repair process, BRCA1 is not the first protein to arrive at the damage site but rather acts downstream to recruit other repair factors, resulting in the assembly of so-called DNA damage-induced repair foci at DSBs (11, 12, 14, 27, 28). Thus, BRCA1 has been identified as part of a larger protein complex called BRCA1-associated genome surveillance complex (BASC), which contains ATM and several proteins playing structural and functional roles in different DNA repair pathways, i.e., the RAD50-MRE11-NBS1 complex, the MSH2-MSH6 heterodimer, and replication factor C (29). Moreover, BRCA1 interacts indirectly with BRCA2 via PALB2 (for partner and localizer of BRCA2). The interaction is established by binding of a coiled-coil region upstream of the BRCT repeats in BRCA1 to a coiled-coil region in the N-terminus of PALB2, which then recruits BRCA2 and RAD51 (30-32). Abolishing the interaction of PALB2 with either BRCA2 or BRCA1 has been shown to impair DSB repair by homologous recombination (30, 32, 33). These results highlight the functional linkage of all three proteins in maintaining genomic stability and explain how familial breast cancer can result from germ-line mutations in each of the three genes (3, 4, 34).

At ionizing radiation-induced DSBs, the histone H2AX is the first protein to arrive at the damage site, where it is phosphorylated by ATM. The phosphorylated form of H2AX, termed γH2AX, sets in motion the recruitment of several proteins starting with MDC1 and followed by RNF8, an E3 ubiquitin ligase, which interacts with the E2 ubiquitin-conjugating enzyme UBC13 to ubiquinate γH2AX (27, 28, 35, 36). In turn, the ubiquitylation of γH2AX allows the attachment of ubiquitylated receptor-associated protein 80 (RAP80), which recruits the phosphoprotein Abraxas

that finally binds BRCA1 via its BRCT repeats (26, 37). In the process, ATM also phosphorylates two BRCA1 residues, 1423Ser and 1524Ser, as well as the BARD1 residue 714Thr, which are important for the DNA damage function of the BRCA1-BARD1 complex (38, 39). To complicate matters, BRCA1 is endowed with E3 ubiquitin ligase activity in its RING domain, which is even more pronounced in the BRCA1-BARD1 complex, allowing ubiquitylation of other proteins as well as possible auto-ubiquitylation of BRCA1-BARD1 heterodimers (22). Finally, BRCA1 is covalently modified by attachment of small ubiquitin-like modifier (SUMO) proteins to lysine residues such as 119Lys (40). Thus, BRCA1 plays a key role in the DNA damage response pathways by interacting with multiple proteins with modification reactions of phosphorylation, ubiquitylation, sumoylation, and acetylation (3, 4, 40-44). Moreover, BRCA1 appears to influence the choice of repair pathway by promoting error-free HR over error-prone NHEJ (45-47). This occurs through at least two mechanisms involving 53BP1 and CtIP. In BRCA1-deficient cells, 53BP1 inhibits HR by blocking resection of DNA breaks leading to error-prone repair by NHEJ (46, 47). In wild-type cells, BRCA1 overcomes the inhibitory effect of 53BP1 by presently unknown mechanisms and thereby promotes active DSB resection and accurate repair by HR. The other process involves another binding partner of BRCA1, CtIP (yeast homolog Sae2), which undergoes phosphorylation on residue 327Ser and recruits BRCA1 to shift the balance of DSB repair from NHEJ to HR (48). CtIP is also acetylated by histone acetyltransferases resulting in its degradation by autophagy (49, 50). Instead of Abraxas or CtIP, BRCA1 can interact with BACH1 (also known as BRIP1 or FancJ) and is then joined by RAD51 in HR or by RAD50-MRE11-NBS1 in NHEJ repair (18, 51-53). Mutations in BACH1 cause Fanconi anemia and early-onset breast cancer (54).

Besides its role in DNA repair, BRCA1 is important in transcription and cell cycle regulation (2, 5). For example, the BRCA1-BARD1 complex functions as an adaptor for p53, enabling the latter to be targeted for phosphorylation by ATM at 15Ser (55). In turn, this phosphorylation event is essential for enhancing p53 activity as a transcription factor to induce the CDK2-cyclin E inhibitor, p21, which causes G1/S cell cycle arrest. Thus, cellular depletion of BRCA1 not only disrupted p53 phosphorylation but also compromised p21 induction and G1/S checkpoint arrest following ionizing radiation (55). BRCA1 participates in transcription regulation at several levels by interacting with sequence-specific transcription factors, RNA polymerase II, and enzymes involved in chromatin remodeling (56, 57). BRCA1 interacts with two members of the p53 family of transcription factors, namely p53 and p63, upregulating the latter by recruiting AP-2γ and ΔNp63γ to an intronic enhancer region within the p63 gene (58, 59).

BRCA1 is a serine phosphoprotein localized predominantly in the nucleus where its expression is tied to the cell cycle, phosphorylation, and ubiquination (37, 60-62). Steady-state levels are low in resting G0 and cycling G1 cells but expression and phosphorylation increase as the cell enters S-phase. Levels of BRCA1 remain elevated throughout mitosis and then rapidly decrease at the beginning of the G1 phase as result of ubiquination and proteasome-dependent degradation (61). Thus, BRCA1 is subject to ubiquitin-mediated degradation while simultaneously promoting the ubiquination of other proteins via the E3 ligase function of the BRCA1-BARD1 heterodimer. One of the proteins ubiquinated by BRCA1-BARD1 is γ-tubulin in centrosomes (63). On the other hand, two separate enzymes, the E2 ubiquitin-conjugating UBE2T and the E3 ligase HERC2, have been shown to ubiquinate and downregulate BRCA1 (64, 65). Subcellular localization studies have shown that BRCA1 protein is also present in mitochondria and in the cytoplasm segregating with centrosomes (2, 63, 66).

The tissue-specific association of BRCA1 germ-line mutations with breast and/or ovarian cancer suggests a connection with steroid hormone-responsive pathways (67). One study reported that estradiol induced the transcription of BRCA1 via recruitment of an estrogen receptor (ERα)-p300 complex to a region of the proximal BRCA1 promoter containing an AP-1 response element (68). At the same time, activation of BRCA1 transcription by E_2 required occupancy of the BRCA1 promoter by the unliganded aromatic hydrocarbon receptor (AhR) (69). To complicate matters, BRCA1 was shown to inhibit ERα signaling by direct interaction with the transcriptional activation domain (AF-2) of the receptor resulting in down-regulation of estrogen-responsive genes (70). BRCA1 also interacted with the progesterone receptor

(PR-A, PR-B) resulting in reduced receptor activity (71). Most BRCA1-mutant breast cancers are ER- and PR-negative leaving open the question how to explain the tissue predilection of BRCA1 germ-line mutations for mammary cancer (72, 73).

References

1. Baer R, Ludwig T. *The BRCA1/BARD1 heterodimer, a tumor suppressor complex with ubiquitin E3 ligase activity.* Curr Opin Genet Dev. 2002;12:86-91.

2. Henderson BR. *Regulation of BRCA1, BRCA2 and BARD1 intracellular trafficking.* BioEssays. 2005;27:884-93.

3. Huen MS, Sy SM, Chen J. *BRCA1 and its toolbox for the maintenance of genome integrity.* Nature Rev Mol Cell Biol. 2010;11:138-48.

4. Moynahan ME, Jasin M. *Mitotic homologous recombination maintains genomic stability and suppresses tumorigenesis.* Nature Rev Mol Cell Biol. 2010;11:196-207.

5. Starita LM, Parvin JD. *The multiple nuclear functions of BRCA1: transcription, ubiquitination and DNA repair.* Curr Opin Cell Biol. 2003;15:345-50.

6. Venkitaraman AR. *Cancer susceptibility and the functions of BRCA1 and BRCA2.* Cell. 2002;108:171-82.

7. Futreal PA, Liu Q, Shattuck-Eidens D, Cochran C, Harshman K, Tavtigian S, et al. *BRCA1 mutations in primary breast and ovarian carcinomas.* Science. 1994;266:120-2.

8. Miki Y, Swensen J, Shattuck-Eidens D, Futreal PA, Harshman K, Tavtigian S, et al. *A strong candidate for the breast and ovarian cancer susceptibility gene BRCA1.* Science. 1994;266:66-71.

9. Couch FJ, DeShano ML, Blackwood MA, Calzone K, Stopfer J, Campeau L, et al. *BRCA1 mutations in women attending clinics that evaluate the risk of breast cancer.* N Engl J Med. 1997;336:1409-15.

10. Langston AA, Malone KE, Thompson JD, Daling JR, Ostrander EA. *BRCA1 mutations in a population-based sample of young women with breast cancer.* N Engl J Med. 1996;334:137-42.

11. Callebaut I, Mornon JP. *From BRCA1 to RAP1: a widespread BRCT module closely associated with DNA repair.* FEBS Lett. 1997;400:25-30.

12. Manke IA, Lowery DM, Nguyen A, Yaffe MB. BRCT repeats as phosphopeptide-binding modules

involved in protein targeting. Science. 2003;302:636-9.

13. Yu X, Chen J. *DNA damage-induced cell cycle checkpoint control requires CtIP, a phosphorylation-dependent binding partner of BRCA1 C-terminal domains.* Mol Cell Biol. 2004;24:9478-86.

14. Yu X, Chini CC, He M, Mer G, Chen J. The BRCT domain is a phospho-protein binding domain. Science. 2003;302:639-42.

15. Paull TT, Cortez D, Bowers B, Elledge SJ, Gellert M. *Direct DNA binding by Brca1.* Proc Natl Acad Sci U S A. 2001;98:6086-91.

16. Brzovic PS, Rajagopal P, Hoyt DW, King MC, Klevit RE. Structure of a BRCA1-BARD1 heterodimeric RING-RING complex. Nature Struct Biol. 2001;8:833-7.

17. Kentsis A, Gordon RE, Borden KL. Self-assembly properties of a model RING domain. Proc Natl Acad Sci U S A. 2002;99:667-72.

18. Scully R, Chen J, Ochs RL, Keegan K, Hoekstra M, Feunteun J, et al. *Dynamic changes of BRCA1 subnuclear location and phosphorylation state are initiated by DNA damage.* Cell. 1997;90:425-35.

19. Westermark UK, Reyngold M, Olshen AB, Baer R, Jasin M, Moynahan ME. BARD1 participates with BRCA1 in homology-directed repair of chromosome breaks. Mol Cell Biol. 2003;23:7926-36.

20. Hashizume R, Fukuda M, Maeda I, Nishikawa H, Oyake D, Yabuki Y, et al. *The RING heterodimer BRCA1-BARD1 is a ubiquitin ligase inactivated by a breast cancer-derived mutation.* J Biol Chem. 2001;276:14537-40.

21. Lorick KL, Jensen JP, Fang S, Ong AM, Hatakeyama S, Weissman AM. *RING fingers mediate ubiquitin-conjugating enzyme (E2)-dependent ubiquitination.* Proc Natl Acad Sci U S A. 1999;96:11364-9.

22. Simons AM, Horwitz AA, Starita LM, Griffin K, Williams RS, Glover JN, et al. *BRCA1 DNA-binding activity is stimulated by BARD1.* Cancer Res. 2006;66:2012-8.

23. Ghimenti C, Sensi E, Presciuttini

S, Brunetti IM, Conte P, Bevilacqua G, et al. *Germline mutations of the BRCA1-associated ring domain (BARD1) gene in breast and breast/ovarian families negative for BRCA1 and BRCA2 alterations.* Genes Chromosomes Cancer. 2002;33:235-42.

24. Karppinen SM, Heikkinen K, Rapakko K, Winqvist R. *Mutation screening of the BARD1 gene: evidence for involvement of the Cys557Ser allele in hereditary susceptibility to breast cancer.* J Med Genet. 2004;41:e114.

25. Sobhian B, Shao G, Lilli DR, Culhane AC, Moreau LA, Xia B, et al. *RAP80 targets BRCA1 to specific ubiquitin structures at DNA damage sites.* Science. 2007;316:1198-202.

26. Wang B, Matsuoka S, Ballif BA, Zhang D, Smogorzewska A, Gygi SP, et al. *Abraxas and RAP80 form a BRCA1 protein complex required for the DNA damage response.* Science. 2007;316:1194-8.

27. Goldberg M, Stucki M, Falck J, D'Amours D, Rahman D, Pappin D, et al. *MDC1 is required for the intra-S-phase DNA damage checkpoint.* Nature. 2003;421:952-6.

28. Stewart GS, Wang B, Bignell CR, Taylor AM, Elledge SJ. *MDC1 is a mediator of the mammalian DNA damage checkpoint.* Nature. 2003;421:961-6.

29. Wang Y, Cortez D, Yazdi P, Neff N, Elledge SJ, Qin J. *BASC, a super complex of BRCA1-associated proteins involved in the recognition and repair of aberrant DNA structures.* Genes Dev. 2000;14:927-39.

30. Sy SM, Huen MS, Chen J. *PALB2 is an integral component of the BRCA complex required for homologous recombination repair.* Proc Natl Acad Sci U S A. 2009;106:7155-60.

31. Zhang F, Fan Q, Ren K, Andreassen PR. *PALB2 functionally connects the breast cancer susceptibility proteins BRCA1 and BRCA2.* Mol Cancer Res. 2009;7:1110-8.

32. Zhang F, Ma J, Wu J, Ye L, Cai H, Xia B, et al. *PALB2 links BRCA1 and BRCA2 in the DNA-damage response.* Curr Biol. 2009;19:524-9.

33. Xia B, Sheng Q, Nakanishi K, Ohashi A, Wu J, Christ N, et al. *Control of BRCA2 cellular and clinical functions by a nuclear partner, PALB2.* Mol Cell. 2006;22:719-29.

34. Rahman N, Seal S, Thompson D, Kelly P, Renwick A, Elliott A, et al. *PALB2, which encodes a BRCA2-interacting protein, is a breast cancer susceptibility gene.* Nature Genet. 2007;39:165-7.

35. Paull TT, Rogakou EP, Yamazaki V, Kirchgessner CU, Gellert M, Bonner WM. *A critical role for histone H2AX in recruitment of repair factors to nuclear foci after DNA damage.* Curr Biol. 2000;10:886-95.

36. Wang B, Elledge SJ. *Ubc13/Rnf8 ubiquitin ligases control foci formation of the Rap80/Abraxas/Brca1/Brcc36 complex in response to DNA damage.* Proc Natl Acad Sci U S A. 2007;104:20759-63.

37. Kim H, Chen J, Yu X. *Ubiquitin-binding protein RAP80 mediates BRCA1-dependent DNA damage response.* Science. 2007;316:1202-5.

38. Cortez D, Wang Y, Qin J, Elledge SJ. *Requirement of ATM-dependent phosphorylation of brca1 in the DNA damage response to double-strand breaks.* Science. 1999;286:1162-6.

39. Kim HS, Li H, Cevher M, Parmelee A, Fonseca D, Kleiman FE, et al. *DNA damage-induced BARD1 phosphorylation is critical for the inhibition of messenger RNA processing by BRCA1/BARD1 complex.* Cancer Res. 2006;66:4561-5.

40. Morris JR, Boutell C, Keppler M, Densham R, Weekes D, Alamshah A, et al. *The SUMO modification pathway is involved in the BRCA1 response to genotoxic stress.* Nature. 2009;462:886-90.

41. Bergink S, Jentsch S. *Principles of ubiquitin and SUMO modifications in DNA repair.* Nature. 2009;458:461-7.

42. Foulkes WD. *Traffic control for BRCA1.* N Engl J Med. 2010;362:755-6.

43. Galanty Y, Belotserkovskaya R, Coates J, Polo S, Miller KM, Jackson SP. *Mammalian SUMO E3-ligases PIAS1 and PIAS4 promote responses to DNA double-strand breaks.* Nature. 2009;462:935-9.

44. Potenski CJ, Klein HL. *Molecular biology: The expanding arena of DNA repair.* Nature. 2011;471:48-9.

45. Boulton SJ. *DNA repair: Decision at the break point.* Nature. 2010;465:301-2.

46. Bouwman P, Aly A, Escandell JM, Pieterse M, Bartkova J, van der Gulden H, et al. *53BP1 loss rescues BRCA1 deficiency and is associated with triple-negative and BRCA-mutated breast cancers.* Nature Struct Mol Biol. 2010;17:688-95.

47. Bunting SF, Callen E, Wong N, Chen HT, Polato F, Gunn A, et al. *53BP1 inhibits homologous recombination in Brca1-deficient cells by blocking resection of DNA breaks.* Cell. 2010;141:243-54.

48. Yun MH, Hiom K. *CtIP-BRCA1 modulates the choice of DNA double-strand-break repair pathway throughout the cell cycle.* Nature. 2009;459:460-3.

49. Kaidi A, Weinert BT, Choudhary C, Jackson SP. *Human SIRT6 promotes DNA end resection through CtIP deacetylation.* Science. 2010;329:1348-53.

50. Robert T, Vanoli F, Chiolo I, Shubassi G, Bernstein KA, Rothstein R, et al. *HDACs link the DNA damage response, processing of double-strand breaks and autophagy.* Nature. 2011;471:74-9.

51. Cantor S, Drapkin R, Zhang F, Lin Y, Han J, Pamidi S, et al. *The BRCA1-associated protein BACH1 is a DNA helicase targeted by clinically relevant inactivating mutations.* Proc Natl Acad Sci U S A. 2004;101:2357-62.

52. Cousineau I, Abaji C, Belmaaza A. *BRCA1 regulates RAD51 function in response to DNA damage and suppresses spontaneous sister chromatid replication slippage: implications for sister chromatid cohesion, genome stability, and carcinogenesis.* Cancer Res. 2005;65:11384-91.

53. Zhong Q, Chen CF, Li S, Chen Y, Wang CC, Xiao J, et al. *Association of BRCA1 with the hRad50-hMre11-p95 complex and the DNA damage response.* Science. 1999;285:747-50.

54. D'Andrea AD. *Susceptibility pathways in Fanconi's anemia and breast cancer.* N Engl J Med. 2010;362:1909-19.

55. Fabbro M, Savage K, Hobson K, Deans AJ, Powell SN, McArthur GA, et al. *BRCA1-BARD1 complexes are required for p53Ser-15 phosphorylation and a G1/S arrest following ionizing radiation-induced DNA damage.* J Biol Chem. 2004;279:31251-8.

56. Anderson SF, Schlegel BP, Nakajima T, Wolpin ES, Parvin JD. *BRCA1 protein is linked to the RNA polymerase II holoenzyme complex via RNA helicase A.* Nature Genet. 1998;19:254-6.

57. Bochar DA, Wang L, Beniya H, Kinev A, Xue Y, Lane WS, et al. *BRCA1 is associated with a human SWI/SNF-related complex: linking chromatin remodeling to breast cancer.* Cell. 2000;102:257-65.

58. Buckley NE, Conlon SJ, Jirstrom K, Kay EW, Crawford NT, O'Grady A, et al. *The {Delta}Np63 Proteins Are Key Allies of BRCA1 in the Prevention of Basal-Like Breast Cancer.* Cancer Res. 2011;71:1933-44.

Cell. 2010;141:243-54.

59. Zhang H, Somasundaram K, Peng Y, Tian H, Bi D, Weber BL, et al. *BRCA1 physically associates with p53 and stimulates its transcriptional activity.* Oncogene. 1998;16:1713-21.

60. Chen CF, Chen PL, Zhong Q, Sharp ZD, Lee WH. *Expression of BRC repeats in breast cancer cells disrupts the BRCA2-Rad51 complex and leads to radiation hypersensitivity and loss of G(2)/M checkpoint control.* J Biol Chem. 1999;274:32931-5.

61. Choudhury AD, Xu H, Baer R. *Ubiquitination and proteasomal degradation of the BRCA1 tumor suppressor is regulated during cell cycle progression.* J Biol Chem. 2004;279:33909-18.

62. Ruffner H, Verma IM. *BRCA1 is a cell cycle-regulated nuclear phospho-protein.* Proc Natl Acad Sci U S A. 1997;94:7138-43.

63. Sankaran S, Starita LM, Simons AM, Parvin JD. *Identification of domains of BRCA1 critical for the ubiquitin-dependent inhibition of centrosome function.* Cancer Res. 2006;66:4100-7.

64. Ueki T, Park JH, Nishidate T, Kijima K, Hirata K, Nakamura Y, et al. *Ubiquitination and downregulation of BRCA1 by ubiquitin-conjugating enzyme E2T overexpression in human breast cancer cells.* Cancer Res. 2009;69:8752-60.

65. Wu W, Sato K, Koike A, Nishikawa H, Koizumi H, Venkitaraman AR, et al. *HERC2 is an E3 ligase that targets BRCA1 for degradation.* Cancer Res. 2010;70:6384-92.

66. Coene ED, Hollinshead MS, Waeytens AA, Schelfhout VR, Eechaute WP, Shaw MK, et al. *Phosphorylated BRCA1 is predominantly located in the nucleus and mitochondria.* Mol Biol Cell. 2005;16:997-1010.

67. Hilakivi-Clarke L. *Estrogens, BRCA1, and breast cancer.* Cancer Res. 2000;60:4993-5001.

68. Jeffy BD, Hockings JK, Kemp MQ, Morgan SS, Hager JA, Beliakoff J, et al. *An estrogen receptor-alpha/p300 complex activates the BRCA-1 promoter at an AP-1 site that binds Jun/Fos transcription factors: repressive effects of p53 on BRCA-1 transcription.* Neoplasia. 2005;7:873-82.

69. Hockings JK, Thorne PA, Kemp MQ, Morgan SS, Selmin O, Romagnolo DF. *The ligand status of the aromatic hydrocarbon receptor modulates transcriptional activation of BRCA-1 promoter by estrogen.* Cancer Res. 2006;66:2224-32.

70. Fan S, Wang J, Yuan R, Ma Y, Meng Q, Erdos MR, et al. *BRCA1 inhibition of estrogen receptor signaling in transfected cells.* Science. 1999;284:1354-6.

71. Ma Y, Katiyar P, Jones LP, Fan S, Zhang Y, Furth PA, et al. *The breast cancer susceptibility gene BRCA1 regulates progesterone receptor signaling in mammary epithelial cells.* Mol Endocrinol. 2006;20:14-34.

72. Karp SE, Tonin PN, Begin LR, Martinez JJ, Zhang JC, Pollak MN,

et al. *Influence of BRCA1 mutations on nuclear grade and estrogen receptor status of breast carcinoma in Ashkenazi Jewish women.* Cancer. 1997;80:435-41.

73. Lakhani SR, Van De Vijver MJ, Jacquemier J, Anderson TJ, Osin PP, McGuffog L, et al. *The pathology of familial breast cancer: predictive value of immunohistochemical markers estrogen receptor, progesterone receptor, HER-2, and p53 in patients with mutations in BRCA1 and BRCA2.* J Clin Oncol. 2002;20:2310-8.

8.2.2. EXPERIMENTAL AND CLINICAL STUDIES

Animal experiments have shown that disruption of the BRCA1 gene in mice led to embryonic lethality whereas breast-specific inactivation of the gene resulted in mammary tumor formation after long latency (1, 2). Gene expression profiling of these murine carcinoma models revealed a basal-like pattern characteristic of human tumors with BRCA1 germ-line mutations (3). Interestingly, the basal-like murine tumors originated from luminal epithelial progenitor rather than basal stem cells (4). Analysis of the BRCA1 coding sequence in 41 human breast cancer cell lines revealed mutations in HCC1937, MDA-MB-436, SUM149PT, and SUM1315MO2 cells (5). In each of the four cell lines, the BRCA1 mutation was accompanied by loss of the other BRCA1 allele and absence of nuclear BRCA1 protein expression. Similarly, immunohistochemical studies of breast cancers have shown a loss of nuclear BRCA1 staining in tumors with germ-line BRCA1 mutations but also in high-grade, non-inherited tumors (6). Isoforms of BRCA1 have also been identified, such as an exon-11 splice variant and a truncated form, BRCA1-IRIS (7-9).

The majority of BRCA1 germ-line mutations associated with early-onset breast cancer are frame-shift or missense mutations (10-12; http://research.nhgri.nih.gov/bic). The frame-shift mutations result in protein truncation with loss of the C-terminal BRCT repeats. The missense mutations frequently occur in evolutionarily conserved amino acids in either the RING domain or BRCT repeats, indicating the importance of these two domains in the tumor suppressor role of BRCA1 (13). Other types of mutations, e.g., genomic rearrangements (exon deletions or duplications) have also been identified and account for almost 30% of BRCA1 mutations in non-Ashkenazi Jewish women (14-16). Germ-line mutations in BRCA1 are rare in the general population with an estimated prevalence of 0.1 – 0.2%. However, a few mutations occur more frequently in selected populations where they can be traced to founder

mutations. For example, BRCA1 185delAG and 5382insC occur in about 2% of the Ashkenazi Jewish population (17, 18). The frequency of BRCA1 mutations in women with breast cancer varies accordingly. In most populations, BRCA1 and BRCA2 mutations together account for 5 – 10% of breast and ovarian cancer unselected for family history compared to 15% in Israel (17-19). The frequency of BRCA1 mutations is 1.5- to 2.0-fold higher than the frequency of BRCA2 mutations (19-21). A multicenter, population-based, case-control study of African-American and Caucasian women observed that BRCA1 mutations were significantly more common in White than Black cases (2.9 versus 1.4%) (22).

Several models have been developed to predict the likelihood that an individual will carry a BRCA1/2 mutation. Online programs, such as BRCAPRO, BOADICEA and the Myriad II prevalence tables (23-25) require computer entry of a full three-generation pedigree and are most useful in a referral clinic specializing in screening women from high-risk families. Even in a referral clinic, the majority of tests for BRCA1 mutations will be negative and therefore uninformative (11).

A combined analysis of 22 studies showed that BRCA1 germ-line mutations confer a cumulative lifetime risk of invasive breast cancer of ~65% in mutation carriers up to the age of 70 (26). The risk begins to increase around the age of 25 years. Among first-degree relatives of carriers, risk was significantly associated with younger age at diagnosis in the cancer patient (20). A large series of 1,187 BRCA1 carriers from the International BRCA1/2 Carrier Cohort Study (IBCCS) found no evidence of an association between breast cancer risk and ages at menarche or menopause (27). This may reflect the different natural history of the disease in BRCA1 carriers compared to sporadic breast cancer.

Several non-synonymous polymorphisms have also been identified in the BRCA1 gene with minor allele frequencies (MAFs) in the general population ranging from 1.2 to 34.3%: Gln356Arg, Pro871Leu, Glu1038Gly, Ser1613Gly, Met1652Ile (28). Although these non-conservative amino acid substitutions located in known interaction domains of the BRCA1 protein are potentially biologically relevant polymorphisms, they were not associated with increased risk of breast and/or ovarian cancer. Variants of unknown pathogenic potential include rare missense mutations, mutations in regulatory regions, and synonymous mutations in splicing enhancer regions (29). Since these BRCA1 variants are of uncertain clinical significance, *in vitro* assays have been designed to assess their effect on BRCA1 protein function. One assay analyzed the effect of BRCA1 point mutations on homologous recombination DNA repair and identified about 300 residues at both termini of the BRCA1 protein as being essential for

HRR (30). Another study employed biochemical and cell-based transcription assays to identify 50 pathogenic variants in the BRCT repeats (31). Multifactorial *in silico* likelihood models have been developed to classify these BRCA1 variants as neutral or pathogenic (32, 33). BRCA1 interacts with multiple other proteins, including ATM, p53, and BRCA2 (34). The interaction was examined in a study of 1,037 non-synonymous polymorphisms in various candidate cancer genes in 473 women with two breast cancers and 2463 controls (35). There was a significant trend in risk with increasing numbers of variant alleles for 25 SNPs in BRCA1, BRCA2, ATM, p53 and CHEK2 (p_{trend} = 0.005). The trend was due almost entirely to SNPs with MAFs <10%, including two SNPs in BRCA1: Gln356Arg (MAF 5.4%) and Asp693Asn (MAF 7.4%). A case-control study (469 cases, 740 controls) examined the risk association of the Glu1038Gly polymorphism with SNPs in five proteins involved in the non-homologous end-joining DNA repair pathway (Ku70, Ku80, DNA-PKcs, XRCC4, ligase IV). Women with at least one variant BRCA1 allele and high-risk genotypes of the NHEJ genes had a significantly increased risk (36). An investigation of over 4,844 BRCA1 mutation carriers observed an association of the CASP8 polymorphism (rs1045485) Asp302His with reduced risk of breast cancer (per allele hazard ratio 0.85; 95% CI 0.76 – 0.97; p_{trend} 0.011) (37).

Breast cancers with BRCA1 germ-line mutations have a distinctive morphologic and molecular phenotype. The morphological appearance is characterized by a high histopathological grade (defined by number of mitoses, degree of cellular pleomorphism, and extent of tubule formation) (38, 39). Immunohistochemical analysis revealed that the majority of tumors lack ERα, PR, and HER2 expression, so-called "triple-negative" cancers, which are associated with a poor prognosis (40, 41). The majority of BRCA1-mutated breast cancers also contained acquired p53 mutations and showed reduced expression of 53BP1 (42, 43). Chromosomal analysis identified a high prevalence of 4p, 4q, and 5q deletions (44, 45). Gene expression studies suggest that breast cancers with BRCA1 germ-line mutations arise from basal stem cells based on their "basal-like" histopathological appearance with expression of cytokeratins-5 and -6 (CK-5, CK-6) found in normal basal mammary cells (46, 47). However, other data indicate that BRCA1 mutant tumors arise from aberrant luminal progenitor cells (4, 48). The basal-like morphology is also seen in a subgroup of sporadic breast cancers with high histopathological grade, which contain lower expression levels of BRCA1 in the absence of BRCA1 mutations (6, 41). These findings suggest that a dysfunction of the BRCA1 pathway plays a role in the development of all basal-like cancers, inherited as well as sporadic.

References

1. Gowen LC, Johnson BL, Latour AM, Sulik KK, Koller BH. *Brca1 deficiency results in early embryonic lethality characterized by neuroepithelial abnormalities.* Nature Genet. 1996;12:191-4.

2. Xu X, Wagner KU, Larson D, Weaver Z, Li C, Ried T, et al. *Conditional mutation of Brca1 in mammary epithelial cells results in blunted ductal morphogenesis and tumour formation.* Nature Genet. 1999;22:37-43.

3. Herschkowitz JI, Simin K, Weigman VJ, Mikaelian I, Usary J, Hu Z, et al. *Identification of conserved gene expression features between murine mammary carcinoma models and human breast tumors.* Genome Biol. 2007;8:R76.

4. Molyneux G, Geyer FC, Magnay FA, McCarthy A, Kendrick H, Natrajan R, et al. *BRCA1 basal-like breast cancers originate from luminal epithelial progenitors and not from basal stem cells.* Cell Stem Cell. 2010;7:403-17.

5. Elstrodt F, Hollestelle A, Nagel JH, Gorin M, Wasielewski M, van den Ouweland A, et al. *BRCA1 mutation analysis of 41 human breast cancer cell lines reveals three new deleterious mutants.* Cancer Res. 2006;66:41-5.

6. Taylor J, Lymboura M, Pace PE, A'Hern R P, Desai AJ, Shousha S, et al. *An important role for BRCA1 in breast cancer progression is indicated by its loss in a large proportion of non-familial breast cancers.* Int J Cancer. 1998;79:334-42.

7. ElShamy WM, Livingston DM. *Identification of BRCA1-IRIS, a BRCA1 locus product.* Nature Cell Biol. 2004;6:954-67.

8. Lu M, Conzen SD, Cole CN, Arrick BA. *Characterization of functional messenger RNA splice variants of BRCA1 expressed in nonmalignant and tumor-derived breast cells.* Cancer Res. 1996;56:4578-81.

9. Wilson CA, Ramos L, Villasenor MR, Anders KH, Press MF, Clarke K, et al. *Localization of human BRCA1 and its loss in high-grade, non-inherited breast carcinomas.* Nature Genet. 1999;21:236-40.

10. Collins FS. *BRCA1--lots of mutations, lots of dilemmas.* N Engl J Med. 1996;334:186-8.

11. Couch FJ, DeShano ML, Blackwood MA, Calzone K, Stopfer J, Campeau L, et al. *BRCA1mutations in women attending clinics that evaluate the risk of breast cancer.* N Engl J Med. 1997;336:1409-15.

12. Langston AA, Malone KE, Thompson JD, Daling JR, Ostrander EA. *BRCA1 mutations in a population-based sample of young women with breast cancer.* N Engl J Med. 1996;334:137-42.

13. Figge MA, Blankenship L. *Missense mutations in the BRCT domain of BRCA-1 from high-risk women frequently perturb strongly hydrophobic amino acids conserved among mammals.* Cancer Epidemiol Biomarkers Prev. 2004;13:1037-41.

14. Hogervorst FB, Nederlof PM, Gille JJ, McElgunn CJ, Grippeling M, Pruntel R, et al. *Large genomic deletions and duplications in the BRCA1 gene identified by a novel quantitative method.* Cancer Res. 2003;63:1449-53.

15. Palma MD, Domchek SM, Stopfer J, Erlichman J, Siegfried JD, Tigges-Cardwell J, et al. *The relative contribution of point mutations and genomic rearrangements in BRCA1 and BRCA2 in high-risk breast cancer families.* Cancer Res. 2008;68:7006-14.

16. Walsh T, Casadei S, Coats KH, Swisher E, Stray SM, Higgins J, et al. *Spectrum of mutations in BRCA1, BRCA2, CHEK2, and TPS3 in families at high risk of breast cancer.* JAMA. 2006;295:1379-88.

17. Martin AM, Weber BL. *Genetic and hormonal risk factors in breast cancer.* J Natl Cancer Inst. 2000;92:1126-35.

18. Struewing JP, Hartge P, Wacholder S, Baker SM, Berlin M, McAdams M, et al. *The risk of cancer associated with specific mutations of BRCA1 and BRCA2 among Ashkenazi Jews.* N Engl J Med. 1997;336:1401-8.

19. Szabo CI, King MC. *Population genetics of BRCA1 and BRCA2.* Am J Hum Genet. 1997;60:1013-20.

20. Begg CB, Haile RW, Borg A, Malone KE, Concannon P, Thomas DC, et al. *Variation of breast cancer risk among BRCA1/2 carriers.* JAMA. 2008;299:194-201.

21. Krainer M, Silva-Arrieta S, FitzGerald MG, Shimada A, Ishioka C, Kanamaru R, et al. *Differential contributions of BRCA1 and BRCA2 to early-onset breast cancer.* N Engl J Med. 1997;336:1416-21.

22. Malone KE, Daling JR, Doody DR, Hsu L, Bernstein L, Coates RJ, et al. *Prevalence and predictors of BRCA1 and BRCA2 mutations in a population-based study of breast cancer in white and black American* *women ages 35 to 64 years.* Cancer Res. 2006;66:8297-308.

23. Antoniou AC, Pharoah PP, Smith P, Easton DF. *The BOADICEA model of genetic susceptibility to breast and ovarian cancer.* Br J Cancer. 2004;91:1580-90.

24. Berry DA, Iversen ES, Jr., Gudbjartsson DF, Hiller EH, Garber JE, Peshkin BN, et al. *BRCAPRO validation, sensitivity of genetic testing of BRCA1/BRCA2, and prevalence of other breast cancer susceptibility genes.* J Clin Oncol. 2002;20:2701-12.

25. Frank B, Hemminki K, Wappenschmidt B, Meindl A, Klaes R, Schmutzler RK, et al. *Association of the CASP10 V410I variant with reduced familial breast cancer risk and interaction with the CASP8 D302H variant.* Carcinogenesis. 2006;27:606-9.

26. Antoniou A, Pharoah PD, Narod S, Risch HA, Eyfjord JE, Hopper JL, et al. *Average risks of breast and ovarian cancer associated with BRCA1 or BRCA2 mutations detected in case Series unselected for family history: a combined analysis of 22 studies.* Am J Hum Genet. 2003;72:1117-30.

27. Chang-Claude J, Andrieu N, Rookus M, Brohet R, Antoniou AC, Peock S, et al. *Age at menarche and menopause and breast cancer risk in the International BRCA1/2 Carrier Cohort Study.* Cancer Epidemiol Biomarkers Prev. 2007;16:740-6.

28. Dombernowsky SL, Weischer M, Freiberg JJ, Bojesen SE, Tybjaerg-Hansen A, Nordestgaard BG. *Missense polymorphisms in BRCA1 and BRCA2 and risk of breast and ovarian cancer.* Cancer Epidemiol Biomarkers Prev. 2009;18:2339-42.

29. Monteiro AN, Couch FJ. *Cancer risk assessment at the atomic level.* Cancer Res. 2006;66:1897-9.

30. Ransburgh DJ, Chiba N, Ishioka C, Toland AE, Parvin JD. *Identification of breast tumor mutations in BRCA1 that abolish its function in homologous DNA recombination.* Cancer Res. 2010;70:988-95.

31. Lee MS, Green R, Marsillac SM, Coquelle N, Williams RS, Yeung T, et al. *Comprehensive analysis of missense variations in the BRCT domain of BRCA1 by structural and functional assays.* Cancer Res. 2010;70:4880-90.

32. Chenevix-Trench G, Healey S, Lakhani S, Waring P, Cummings M, Brinkworth R, et al. *Genetic and histopathologic evaluation of BRCA1 and BRCA2 DNA sequence variants of unknown clinical significance.* Cancer Res. 2006;66:2019-27.

33. Easton DF, Deffenbaugh AM, Pruss D, Frye C, Wenstrup RJ, Allen-Brady K, et al. *A systematic genetic assessment of 1,433 sequence variants of unknown clinical significance in the BRCA1 and BRCA2 breast cancer-predisposition genes.* Am J Hum Genet. 2007;81:873-83.

34. Venkitaraman AR. *Cancer susceptibility and the functions of BRCA1 and BRCA2.* Cell. 2002;108:171-82.

35. Johnson N, Fletcher O, Palles C, Rudd M, Webb E, Sellick G, et al. *Counting potentially functional variants in BRCA1, BRCA2 and ATM predicts breast cancer susceptibility.* Hum Mol Genet. 2007;16:1051-7.

36. Bau DT, Fu YP, Chen ST, Cheng TC, Yu JC, Wu PE, et al. *Breast cancer risk and the DNA double-strand break end-joining capacity of nonhomologous end-joining genes are affected by BRCA1.* Cancer Res. 2004;64:5013-9.

37. Engel C, Versmold B, Wappenschmidt B, Simard J, Easton DF, Peock S, et al. *Association of the variants CASP8 D302H and CASP10 V410I with breast and ovarian cancer risk in BRCA1 and BRCA2 mutation carriers.* Cancer Epidemiol Biomarkers Prev. 2010;19:2859-68.

38. Eisinger F, Stoppa-Lyonnet D, Longy M, Kerangueven F, Noguchi T, Bailly C, et al. *Germ line mutation at BRCA1 affects the histoprognostic grade in hereditary breast cancer.* Cancer Res. 1996;56:471-4.

39. Lakhani SR, Gusterson BA, Jacquemier J, Sloane JP, Anderson TJ, van de Vijver MJ, et al. *The pathology of familial breast cancer: histological features of cancers in families not attributable to mutations in BRCA1 or BRCA2.* Clin Cancer Res. 2000;6:782-9.

40. Lakhani SR, Van De Vijver MJ, Jacquemier J, Anderson TJ, Osin PP, McGuffog L, et al. *The pathology of familial breast cancer: predictive value of immunohistochemical markers estrogen receptor, progesterone receptor, HER-2, and p53 in patients with mutations in BRCA1 and BRCA2.* J Clin Oncol. 2002;20:2310-8.

41. Turner NC, Reis-Filho JS. *Basal-like breast cancer and the BRCA1 phenotype.* Oncogene. 2006;25:5846-53.

42. Bouwman P, Aly A, Escandell JM, Pieterse M, Bartkova J, van der Gulden H, et al. *53BP1 loss rescues BRCA1 deficiency and is associated with triple-negative and BRCA-mutated breast cancers.* Nature Struct Mol Biol. 2010;17:688-95.

43. Phillips KA, Nichol K, Ozcelik H, Knight J, Done SJ, Goodwin PJ, et al. *Frequency of p53 mutations in breast carcinomas from Ashkenazi Jewish carriers of BRCA1 mutations.* J Natl Cancer Inst. 1999;91:469-73.

44. Jonsson G, Naylor TL, Vallon-Christersson J, Staaf J, Huang J, Ward MR, et al. *Distinct genomic profiles in hereditary breast tumors identified by array-based comparative genomic hybridization.* Cancer Res. 2005;65:7612-21.

45. Tirkkonen M, Johannsson O, Agnarsson BA, Olsson H, Ingvarsson S, Karhu R, et al. *Distinct somatic genetic changes associated with tumor progression in carriers of BRCA1 and BRCA2 germ-line mutations.* Cancer Res. 1997;57:1222-7.

46. Foulkes WD. *BRCA1 functions as a breast stem cell regulator.* J Med Genet. 2004;41:1-5.

47. Perou CM, Sorlie T, Eisen MB, van de Rijn M, Jeffrey SS, Rees CA, et al. *Molecular portraits of human breast tumours.* Nature. 2000;406:747-52.

48. Lim E, Vaillant F, Wu D, Forrest NC, Pal B, Hart AH, et al. *Aberrant luminal progenitors as the candidate target population for basal tumor development in BRCA1 mutation carriers.* Nature Med. 2009;15:907-13.

8.3. BRCA2

8.3.1. INTRODUCTION

Breast cancer susceptibility protein 2 (BRCA2) is a key mediator in repairing chromosomal breaks by homologous recombination, which is essential for maintaining genomic integrity (1). In addition, BRCA2 plays a pivotal role in centrosome duplication and cell division (2). Germline mutations in the BRCA2 gene are associated with a highly penetrant, autosomal dominant predisposition to early-onset breast and/or ovarian cancer (3-5). Tumors of afflicted individuals show loss of heterozygosity for the wild-type BRCA2 allele with retention of the inherited mutant allele indicating that BRCA2 is a tumor suppressor gene. Like BRCA1, mutations in BRCA2 most commonly induce frameshifts that cause protein truncation and loss of function (3, 4). Loss of BRCA2 results in increased sensitivity to cross-linking agents, compromised repair of DNA double-strand breaks (DSBs) by homologous recombination, and defects in replication and cell cycle checkpoint controls (6-8). These deficiencies provide a molecular basis for the severe chromosomal instability observed in cells with BRCA2 mutations, which exhibit an accumulation of chromosomal breaks, translocations, and other abnormal structures (9).

The 70-kb BRCA2 gene at 13q12.3 contains 27 exons (exon 1 is untranslated), which encode a ~410 kDa protein composed of 3,418 amino acids that interacts with DNA and several other proteins (Figure 8.2.). The C-terminal half of BRCA2 contains a single-stranded DNA (ssDNA)-binding region that consists of four globular domains arranged in a linear manner and a fifth domain with a coiled-coil structure extending like a tower. Three globular domains are oligonucleotide-oligosaccharide binding folds found in ssDNA-binding proteins such as replication protein A (RPA), whereas the tower domain has a three-helix bundle at the apex that is similar in structure to some dsDNA-binding domains (1, 10). In addition, BRCA2 interacts with RAD51 recombinase, the central protein of homologous recombination. BRCA2 contains eight conserved BRC repeats and a C-terminal region that bind up to six RAD51 molecules with varying affinity in the presence of DSS1, a 70-amino acid polypeptide (11-13). The DNA- and RAD51 binding domains interact to deliver RAD51 to ss-DNA-dsDNA junctions that are formed by 5′ to 3′ end resection of DSBs. Although BRCA2 can bind to both ssDNA and RAD51, a ternary complex of BRCA2-RAD51-ssDNA was not detected, suggesting that BRCA2 delivers RAD51 onto ssDNA, but does not become a stable part of the RAD51-ssDNA filament (14, 15). Thus, BRCA2 stimulates the binding of RAD51 to ssDNA over double-stranded DNA, enabling RAD51 to displace RPA from ssDNA. Additionally, BRCA2 stabilizes RAD51-ssDNA filaments by blocking ATP hydrolysis (13, 16). BRCA2 also interacts with microcephalin (MCPH1), a BRCT-domain containing protein that enhances binding of BRCA2 and RAD1 to DSBs and another recombinase, the meiosis-specific DMC1, in a 26 amino acid sequence (2386 – 2411) distinct from the BRC repeat region (17, 18). An important partner in localizing BRCA2 and RAD51 at the DNA damage site is PALB2 (for partner and localizer of BRCA2), which contains a C-terminal β-propeller domain that binds to the N-terminal third of BRCA2 (19). At its N-terminus PALB2 also contains a coiled-coil region, which interacts with a coiled-coil region upstream of the BRCA1 repeats in BRCA1, thereby serving as an important link between BRCA2 and BRCA1 (20-22). Abolishing the interaction of PALB2 with either BRCA2 or BRCA1 has been shown to impair DSB repair by homologous recombination (19, 20, 22). These results highlight the functional linkage of all three proteins in maintaining genomic stability

and provide a molecular basis for understanding how germ-line mutations in BRCA1, BRCA2, or PALB2 result in early-onset breast and ovarian cancer (1, 23, 24).

The N-terminal third of BRCA2 protein contains binding sites for nucleophosmin (NPM), a nucleolar phosphoprotein that regulates centrosome duplication and accurate cell division (2, 19). The interaction with additional proteins (e.g., EMSY) has been described but is of unknown significance (25, 26). The expression of BRCA2 fluctuates during the cell cycle, reaching maximum levels in late G1 and S phase, similar to BRCA1 (27, 28). This expression profile is consistent with the role of BRCA2 in replication and repair of DNA during S phase (29). Animal experiments have shown that disruption of the BRCA2 gene in mice led to embryonic lethality whereas breast-specific inactivation of the gene resulted in mammary adenocarcinomas after long latency (30, 31).

References

1. Moynahan ME, Jasin M. *Mitotic homologous recombination maintains genomic stability and suppresses tumorigenesis.* Nat Rev Mol Cell Biol. 2010;11:196-207.

2. Wang HF, Takenaka K, Nakanishi A, Miki Y. *BRCA2 and nucleophosmin coregulate centrosome amplification and form a complex with the Rho effector kinase ROCK2.* Cancer Res. 2011;71:68-77.

3. Phelan CM, Lancaster JM, Tonin P, Gumbs C, Cochran C, Carter R, et al. *Mutation analysis of the BRCA2 gene in 49 site-specific breast cancer families.* Nature Genet. 1996;13:120-2.

4. Tavtigian SV, Simard J, Rommens J, Couch F, Shattuck-Eidens D, Neuhausen S, et al. *The complete BRCA2 gene and mutations in chromosome 13q-linked kindreds.* Nature Genet. 1996;12:333-7.

5. Wooster R, Bignell G, Lancaster J, Swift S, Seal S, Mangion J, et al. *Identification of the breast cancer susceptibility gene BRCA2.* Nature. 1995;378:789-92.

6. Chen CF, Chen PL, Zhong Q, Sharp ZD, Lee WH. *Expression of BRC repeats in breast cancer cells disrupts the BRCA2-Rad51 complex and leads to radiation hypersensitivity and loss of G(2)/M checkpoint control.* J Biol Chem. 1999;274:32931-5.

7. Lomonosov M, Anand S, Sangrithi M, Davies R, Venkitaraman AR. *Stabilization of stalled DNA replication forks by the BRCA2 breast cancer*

BRCA2 Protein

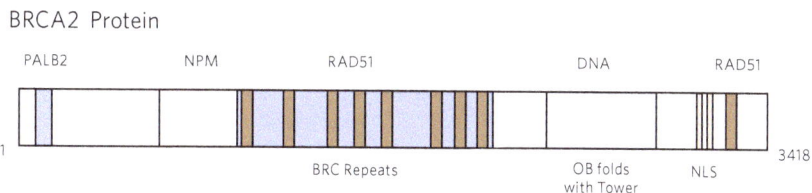

Figure 8.2. Diagram of BRCA2 protein.
The central region of the protein (corresponding to exon 11, amino acids 1009 – 2083) contains a series of eight short degenerate motifs called the BRC repeats, which bind RAD51 recombinase. An unrelated RAD51 interaction domain in exon 27 at the C terminus is required for RAD51 localization to sites of DNA damage. In addition, BRCA2 contains a DNA binding region, which consists of four globular domains arranged in a linear manner and a fifth domain with a coiled-coil structure extending like a tower. Three of the globular domains are oligonucleotide-oligosaccharide–binding (OB) folds found in ssDNA-binding proteins. The N-terminal end of BRCA2 interacts with PALB2, which also binds BRCA1 and serves as a link between the two BRCAs. A nuclear localization sequence is present between the OB folds and the C-terminal RAD51 binding domain. Binding of the nucleolar phosphoprotein NPM involves amino acids 639 – 1000. Most BRCA2 mutations induce frameshifts that cause protein truncation and loss of function. For example, the common mutation 6174delT encodes a truncated protein of 2002 amino acids, including 1981 BRCA2 amino acids and 21 amino acids that are generated by the frameshift. The truncation of BRCA2 sequences at amino acid 1981 occurs within BRC repeat 7.

susceptibility protein. Genes Dev. 2003;17:3017-22.

8. Moynahan ME, Pierce AJ, Jasin M. *BRCA2 is required for homology-directed repair of chromosomal breaks.* Mol Cell. 2001;7:263-72.

9. Yu VP, Koehler M, Steinlein C, Schmid M, Hanakahi LA, van Gool AJ, et al. *Gross chromosomal rearrangements and genetic exchange between nonhomologous chromosomes following BRCA2 inactivation.* Genes Dev. 2000;14:1400-6.

10. Yang H, Jeffrey PD, Miller J, Kinnucan E, Sun Y, Thoma NH, et al. *BRCA2 function in DNA binding and recombination from a BRCA2-DSS1-ssDNA structure.* Science. 2002;297:1837-48.

11. Esashi F, Galkin VE, Yu X, Egelman EH, West SC. *Stabilization of RAD51 nucleoprotein filaments by the C-terminal region of BRCA2.* Nature Struct Mol Biol. 2007;14:468-74.

12. Galkin VE, Esashi F, Yu X, Yang S, West SC, Egelman EH. *BRCA2 BRC motifs bind RAD51-DNA filaments.* Proc Natl Acad Sci U S A. 2005;102:8537-42.

13. Liu J, Doty T, Gibson B, Heyer WD. *Human BRCA2 protein promotes RAD51 filament formation on RPA-covered single-stranded DNA.* Nature Struct Mol Biol. 2010;17:1260-2.

14. Thorslund T, McIlwraith MJ, Compton SA, Lekomtsev S, Petronczki M, Griffith JD, et al. *The breast cancer tumor suppressor BRCA2 promotes the specific targeting of RAD51 to single-stranded DNA.* Nature Struct Mol Biol. 2010;17:1263-5.

15. Zou L. DNA repair: *A protein giant in its entirety.* Nature. 2010;467:667-8.

16. Jensen RB, Carreira A, Kowalczykowski SC. *Purified human BRCA2 stimulates RAD51-mediated recombination.* Nature. 2010;467:678-83.

17. Thorslund T, Esashi F, West SC. *Interactions between human BRCA2 protein and the meiosis-specific recombinase DMC1.* EMBO J. 2007;26:2915-22.

18. Wu X, Mondal G, Wang X, Wu J, Yang L, Pankratz VS, et al. *Microcephalin regulates BRCA2 and Rad51-associated DNA double-strand break repair.* Cancer Res. 2009;69:5531-6.

19. Xia B, Sheng Q, Nakanishi K, Ohashi A, Wu J, Christ N, et al.

Control of BRCA2 cellular and clinical functions by a nuclear partner, PALB2. Mol Cell. 2006;22:719-29.

20. Sy SM, Huen MS, Chen J. *PALB2 is an integral component of the BRCA complex required for homologous recombination repair.* Proc Natl Acad Sci U S A. 2009;106:7155-60.

21. Zhang F, Fan Q, Ren K, Andreassen PR. *PALB2 functionally connects the breast cancer susceptibility proteins BRCA1 and BRCA2.* Mol Cancer Res. 2009;7:1110-8.

22. Zhang F, Ma J, Wu J, Ye L, Cai H, Xia B, et al. *PALB2 links BRCA1 and BRCA2 in the DNA-damage response.* Curr Biol. 2009;19:524-9.

23. Huen MS, Sy SM, Chen J. *BRCA1 and its toolbox for the maintenance of genome integrity.* Nature Rev Mol Cell Biol. 2010;11:138-48.

24. Rahman N, Seal S, Thompson D, Kelly P, Renwick A, Elliott A, et al. *PALB2, which encodes a BRCA2-interacting protein, is a breast cancer susceptibility gene.* Nature Genet. 2007;39:165-7.

25. Hughes-Davies L, Huntsman D, Ruas M, Fuks F, Bye J, Chin SF, et al. *EMSY links the BRCA2 pathway to sporadic breast and ovarian cancer.* Cell. 2003;115:523-35.

26. King MC. *A novel BRCA2-binding protein and breast and ovarian tumorigenesis.* N Engl J Med. 2004;350:1252-3.

27. Rajan JV, Wang M, Marquis ST, Chodosh LA. *Brca2 is coordinately regulated with Brca1 during proliferation and differentiation in mammary epithelial cells.* Proc Natl Acad Sci U S A. 1996;93:13078-83.

28. Vaughn JP, Cirisano FD, Huper G, Berchuck A, Futreal PA, Marks JR, et al. *Cell cycle control of BRCA2.* Cancer Res. 1996;56:4590-4.

29. Esashi F, Christ N, Gannon J, Liu Y, Hunt T, Jasin M, et al. *CDK-dependent phosphorylation of BRCA2 as a regulatory mechanism for recombinational repair.* Nature. 2005;434:598-604.

30. Ludwig T, Fisher P, Murty V, Efstratiadis A. *Development of mammary adenocarcinomas by tissue-specific knockout of Brca2 in mice.* Oncogene. 2001;20:3937-48.

31. Sharan SK, Morimatsu M, Albrecht U, Lim DS, Regel E, Dinh C, et al. *Embryonic lethality and radiation hypersensitivity mediated by Rad51 in mice lacking Brca2.* Nature. 1997;386:804-10.

8.3.2. CLINICAL STUDIES

Germline mutations in BRCA2 most commonly induce frameshifts that cause protein truncation and loss of function (1, 2; http://research.nhgri.nih.gov/bic/). For example, a 1-bp thymidine deletion at nucleotide 6174 in codon 1982 (6174delT) introduces a stop codon at 2003 resulting in premature termination of translation (3). In addition, tumors from individuals with BRCA2 mutations may display loss of the wild-type allele (loss of heterozygosity), suggesting that the absence of functional BRCA2 protein causes early onset breast cancer (4). While monoallelic inheritance of BRCA2 mutations leads to breast cancer in adults, biallelic inheritance results in Fanconi anemia of the D1 subtype in children; for this reason, BRCA2 is also referred to as FANCD1 gene (5). The majority of BRCA2 germline mutations are point mutations (http://research.nhgri.nih.gov/bic). They may be traced as founder mutations, e.g., BRCA2 6174delT occurs in 1.2% of the Ashkenazi Jewish population and 999del5 is carried by 0.6% of the Icelandic population (6). In non-Ashkenazi Jewish women, 6% of BRCA2 mutations are genomic rearrangements (exon duplications or deletions) (7). In BRCA2 probands, partial (exons 12 and 13) and complete gene deletions have been identified (8). Several missense polymorphisms have also been identified in the BRCA2 gene with minor allele frequencies in the general population ranging from 1.3 to 28.2%: Asn289His, Asn372His, Asp1420Tyr, Tyr1915Met (9). Although these non-conservative amino acid substitutions located in known interaction domains of the BRCA2 protein are potentially biologically relevant polymorphisms, they were not associated with increased risk of breast and/or ovarian cancer. Rare missense variants of unknown pathogenic potential have been termed "unclassified variants". A multifactorial likelihood model has been developed to classify these BRCA2 variants as neutral or pathogenic (10).

Germ-line mutations in BRCA1 or BRCA2 are rare in the general population with an estimated prevalence of 0.1 – 0.2%. In most populations identification of BRCA1 and BRCA2 mutations together account for 5 – 10% of breast and ovarian cancer unselected for family history (6). In Israel, the attributable fraction is higher (15%) due to the presence of BRCA1 and BRCA2 mutations in over 2% of Ashkenazi Jews (11, 12). In most populations, the frequency of BRCA1 mutations is 1.5- to 2.0-fold higher than the frequency of BRCA2 mutations (12-14).

The BRCA2 mutation frequency is slightly but non-significantly greater (2.6 versus 2.1%) in African American than Caucasian women with breast cancer (15). Breast cancer is rare in men but when it occurs in families with multiple breast cancer cases, there is a high likelihood that the tumors are due to BRCA2 mutations (8, 16-18).

A combined analysis of 22 studies showed that BRCA1 and BRCA2 germline mutations confer a cumulative lifetime risk of invasive breast cancer of ~65% in BRCA1 mutation carriers and 45% in BRCA2 carriers up to the age of 70 (19). The risk begins to increase around the age of 25 years. Among first-degree relatives of carriers, risk was significantly associated with younger age at diagnosis in the cancer patient (13). A large series of 1,187 BRCA1 and 414 BRCA2 carriers from the International BRCA1/2 Carrier Cohort Study (IBCCS) found no evidence of an association between breast cancer risk and ages at menarche or menopause (20). This may reflect the different natural history of the disease in BRCA1 and BRCA2 carriers compared to sporadic breast cancer. The Consortium of Investigators of Modifiers of BRCA1/2 (CIMBA) analyzed the interaction of BRCA2 mutations with SNPs identified in breast cancer susceptibility genome-wide association studies (GWAS). In a study population of 7,409 BRCA2 carriers, seven of nine GWAS SNPs were associated with breast cancer risk: FGFR2 rs2981582, TOX3 rs3803662, MAP3K1 rs889312, LSP1 rs3817198, 2q35 rs13387042, SLC4A7 rs4973768, 5p12 rs10941679 (21). The seven SNPs appear to interact multiplicatively so that the 5% of BRCA2 mutation carriers at highest risk will have a lifetime risk of developing breast cancer of ≥80% whereas the 5% at lowest risk will have a lifetime risk of ≤50%. Overall, the seven SNPs were estimated to account for approximately 4% of the genetic variability of breast cancer in BRCA2.

The BRCA2 gene contains several non-synonymous polymorphisms: Asn289His (rs766173; minor allele frequency, MAF 3.1%), Asn372His (rs144848; MAF 27.9%), Thr1915Met (rs4987117; MAF 3.5%), Arg2034Cys (rs1799954; MAF 0.8%), and Glu2856Ala (rs11571747; MAF 0.3%). A 2006 meta-analysis of 10 studies (13,314 cases, 13,032 controls) observed a weak association of the homozygous variant 372His/His genotype with increased breast cancer risk (odds ratio 1.13; 95% CI 1.01 – 1.28; p = 0.04) (22). However, the Breast Cancer AssociationConsortium found no risk association of the Asn372His polymorphism in an analysis of 15,627 cases and 15,968 controls (23). Another study examined the interaction of non-synonymous SNPs in BRCA2 and other candidate genes in 473 women with two breast cancers and 2463 controls (24). There was a significant trend in risk with increasing numbers of variant alleles for 25 SNPs in BRCA1, BRCA2, ATM, p53 and CHEK2 (p_{trend} = 0.005). The trend was due almost entirely to SNPs with MAFs <10%, including four SNPs in BRCA2: Asn289His, Thr1915Met, Arg2034Cys, and Glu2856Ala.

Breast cancers with BRCA1 mutations differ from those with BRCA2 mutations in several ways. The histopathological grade (composed of a number of mitoses, degree of cellular pleomorphism, and extent of tubule formation) is high in both types of tumors but for different reasons;

tumors with BRCA1 mutations have a high mitotic index, whereas those with BRCA2 mutations exhibit less tubule formation (25). The former are generally ER- and PR-negative while the latter express both receptors (26). These differences imply that the mutant BRCA1 and BRCA2 genes induce the development of breast cancer through separate pathways. Indeed, gene expression studies have shown that significantly different groups of genes are expressed by breast cancers with BRCA1 and BRCA2 mutations (27, 28). On the other hand, there are genes like ATM, whose protein expression was reduced in both BRCA1- and BRCA2-deficient tumors (29).

Prophylactic bilateral total mastectomy reduced the incidence of breast cancer by at least 90% in women with a BRCA1 or BRCA2 mutation (30, 31). In contrast, prophylactic oophorectomy reduced the risk of breast cancer only by 25% in premenopausal carriers (32). Breast and ovarian cancers with BRCA2 mutations are also treated with drugs that exploit the defect in homologous recombination repair, such as cisplatin and poly(ADP-ribose) polymerase (PARP) inhibitors (33, 34). The tumors respond initially but ultimately develop resistance to both drugs. The mechanism underlying the resistance appears to be the emergence of new BRCA2 isoforms that arise from secondary intragenic mutations of BRCA2, which restore the wild-type BRCA2 reading frame (35, 36). Thus, in a subset of tumors, the cancer cells survive when exposed to cisplatin or PARP inhibitors and the tumors recur.

References

1. Phelan CM, Lancaster JM, Tonin P, Gumbs C, Cochran C, Carter R, et al. *Mutation analysis of the BRCA2 gene in 49 site-specific breast cancer families.* Nature Genet. 1996;13:120-2.

2. Tavtigian SV, Simard J, Rommens J, Couch F, Shattuck-Eidens D, Neuhausen S, et al. *The complete BRCA2 gene and mutations in chromosome 13q-linked kindreds.* Nature Genet. 1996;12:333-7.

3. Berman DB, Costalas J, Schultz DC, Grana G, Daly M, Godwin AK. *A common mutation in BRCA2 that predisposes to a variety of cancers is found in both Jewish Ashkenazi and non-Jewish individuals.* Cancer Res. 1996;56:3409-14.

4. Cleton-Jansen AM, Collins N, Lakhani SR, Weissenbach J, Devilee P, Cornelisse CJ, et al. *Loss of heterozygosity in sporadic breast tumours at the BRCA2 locus on chromosome 13q12-q13.* Br J Cancer. 1995;72:1241-4.

5. D'Andrea AD. *Susceptibility pathways in Fanconi's anemia and breast cancer.* N Engl J Med. 2010;362:1909-19.

6. Martin AM, Weber BL. *Genetic and hormonal risk factors in breast cancer.* J Natl Cancer Inst. 2000;92:1126-35.

7. Palma MD, Domchek SM, Stopfer J, Erlichman J, Siegfried JD, Tigges-Cardwell J, et al. The relative contribution of point mutations and genomic rearrangements in BRCA1 and BRCA2 in high-risk breast cancer families. Cancer Res. 2008;68:7006-14.

8. Tournier I, Paillerets BB, Sobol H, Stoppa-Lyonnet D, Lidereau R, Barrois M, et al. *Significant contribution of germline BRCA2 rearrangements in male breast cancer families.* Cancer Res. 2004;64:8143-7.

9. Dombernowsky SL, Weischer M, Freiberg JJ, Bojesen SE, Tybjaerg-Hansen A, Nordestgaard BG. *Missense polymorphisms in BRCA1 and BRCA2 and risk of breast and ovarian cancer.* Cancer Epidemiol Biomarkers Prev. 2009;18:2339-42.

10. Chenevix-Trench G, Healey S, Lakhani S, Waring P, Cummings M, Brinkworth R, et al. *Genetic and histopathologic evaluation of BRCA1* and BRCA2 DNA sequence variants of unknown clinical significance. Cancer Res. 2006;66:2019-27.

11. Struewing JP, Hartge P, Wacholder S, Baker SM, Berlin M, McAdams M, et al. The risk of cancer associated with specific mutations of BRCA1 and BRCA2 among Ashkenazi Jews. N Engl J Med. 1997;336:1401-8.

12. Szabo CI, King MC. Population genetics of BRCA1 and BRCA2. Am J Hum Genet. 1997;60:1013-20.

13. Begg CB, Haile RW, Borg A, Malone KE, Concannon P, Thomas DC, et al. Variation of breast cancer risk among BRCA1/2 carriers. JAMA. 2008;299:194-201.

14. Krainer M, Silva-Arrieta S, FitzGerald MG, Shimada A, Ishioka C, Kanamaru R, et al. Differential contributions of BRCA1 and BRCA2 to early-onset breast cancer. N Engl J Med. 1997;336:1416-21.

15. Malone KE, Daling JR, Doody DR, Hsu L, Bernstein L, Coates RJ, et al. Prevalence and predictors of BRCA1 and BRCA2 mutations in a population-based study of breast cancer in white and black American women ages 35 to 64 years. Cancer Res. 2006;66:8297-308.

16. Ford D, Easton DF, Stratton M, Narod S, Goldgar D, Devilee P, et al. Genetic heterogeneity and penetrance analysis of the BRCA1 and BRCA2 genes in breast cancer families. The Breast Cancer Linkage Consortium. Am J Hum Genet. 1998;62:676-89.

17. Hakansson S, Johannsson O, Johansson U, Sellberg G, Loman N, Gerdes AM, et al. Moderate frequency of BRCA1 and BRCA2 germ-line mutations in Scandinavian familial breast cancer. Am J Hum Genet. 1997;60:1068-78.

18. Haraldsson K, Loman N, Zhang QX, Johannsson O, Olsson H, Borg A. BRCA2 germ-line mutations are frequent in male breast cancer patients without a family history of the disease. Cancer Res. 1998;58:1367-71.

19. Antoniou A, Pharoah PD, Narod S, Risch HA, Eyfjord JE, Hopper JL, et al. Average risks of breast and ovarian cancer associated with BRCA1 or BRCA2 mutations detected in case Series unselected for family history: a combined analysis of 22 studies. Am J Hum Genet. 2003;72:1117-30.

20. Chang-Claude J, Andrieu N, Rookus M, Brohet R, Antoniou AC, Peock S, et al. Age at menarche and menopause and breast cancer risk in the International BRCA1/2 Carrier Cohort Study. Cancer Epidemiol Biomarkers Prev. 2007;16:740-6.

21. Antoniou AC, Beesley J, McGuffog L, Sinilnikova OM, Healey S, Neuhausen SL, et al. Common breast cancer susceptibility alleles and the risk of breast cancer for BRCA1 and BRCA2 mutation carriers: implications for risk prediction. Cancer Res. 2010;70:9742-54.

22. Garcia-Closas M, Egan KM, Newcomb PA, Brinton LA, Titus-Ernstoff L, Chanock S, et al. Polymorphisms in DNA double-strand break repair genes and risk of breast cancer: two population-based studies in USA and Poland, and meta-analyses. Hum Genet. 2006;119:376-88.

23. Commonly studied single-nucleotide polymorphisms and breast cancer: results from the Breast Cancer Association Consortium. J Natl Cancer Inst. 2006;98:1382-96.

24. Johnson N, Fletcher O, Palles C, Rudd M, Webb E, Sellick G, et al. Counting potentially functional variants in BRCA1, BRCA2 and ATM predicts breast cancer susceptibility. Hum Mol Genet. 2007;16:1051-7.

25. Lakhani SR, Gusterson BA, Jacquemier J, Sloane JP, Anderson TJ, van de Vijver MJ, et al. The pathology of familial breast cancer: histological features of cancers in families not attributable to mutations in BRCA1 or BRCA2. Clin Cancer Res. 2000;6:782-9.

26. Karp SE, Tonin PN, Begin LR, Martinez JJ, Zhang JC, Pollak MN, et al. Influence of BRCA1 mutations on nuclear grade and estrogen receptor status of breast carcinoma in Ashkenazi Jewish women. Cancer. 1997;80:435-41.

27. Armes JE, Trute L, White D, Southey MC, Hammet F, Tesoriero A, et al. Distinct molecular pathogeneses of early-onset breast cancers in BRCA1 and BRCA2 mutation carriers: a population-based study. Cancer Res. 1999;59:2011-7.

28. Hedenfalk I, Duggan D, Chen Y, Radmacher M, Bittner M, Simon R, et al. Gene-expression profiles in hereditary breast cancer. N Engl J Med. 2001;344:539-48.

29. Tommiska J, Bartkova J, Heinonen M, Hautala L, Kilpivaara O, Eerola H, et al. The DNA damage signalling kinase ATM is aberrantly reduced or lost in BRCA1/BRCA2-deficient and ER/PR/ERBB2-triple-negative breast cancer. Oncogene. 2008;27:2501-6.

30. Hartmann LC, Schaid DJ, Woods JE, Crotty TP, Myers JL, Arnold PG, et al. Efficacy of bilateral prophylactic mastectomy in women with a family history of breast cancer.

N Engl J Med. 1999;340:77-84.

31. Meijers-Heijboer H, van Geel B, van Putten WL, Henzen-Logmans SC, Seynaeve C, Menke-Pluymers MB, et al. Breast cancer after prophylactic bilateral mastectomy in women with a BRCA1 or BRCA2 mutation. N Engl J Med. 2001;345:159-64.

32. Rebbeck TR, Lynch HT, Neuhausen SL, Narod SA, Van't Veer L, Garber JE, et al. Prophylactic oophorectomy in carriers of BRCA1 or BRCA2 mutations. N Engl J Med. 2002;346:1616-22.

33. Fong PC, Boss DS, Yap TA, Tutt A, Wu P, Mergui-Roelvink M, et al. Inhibition of poly(ADP-ribose) polymerase in tumors from BRCA mutation carriers. N Engl J Med. 2009;361:123-34.

34. Iglehart JD, Silver DP. Synthetic lethality--a new direction in cancer-drug development. N Engl J Med. 2009;361:189-91.

35. Edwards SL, Brough R, Lord CJ, Natrajan R, Vatcheva R, Levine DA, et al. Resistance to therapy caused by intragenic deletion in BRCA2. Nature. 2008;451:1111-5.

36. Sakai W, Swisher EM, Karlan BY, Agarwal MK, Higgins J, Friedman C, et al. Secondary mutations as a mechanism of cisplatin resistance in BRCA2-mutated cancers. Nature. 2008;451:1116-20.

8.4. P53 AND LI-FRAUMENI SYNDROME

8.4.1. INTRODUCTION

The p53 gene is the most commonly mutated gene identified in human cancers with somatic mutations ranging in frequency from 5 to 70% depending on cancer type and stage (1, 2; www.iarc.fr/P53/). The high rate of somatic mutations suggests that p53 plays a major role in tumor suppression. Indeed, p53 plays a crucial role in the prevention of malignant transformation by maintaining the integrity of the genome threatened by DNA damage, activation of oncogenes, and hypoxic conditions (3). The protective response of p53 depends on the degree of DNA damage and ranges from the induction of cell cycle arrest at the G1 checkpoint to the initiation of apoptosis (4-6). The G1 arrest prevents replication of a potentially damaged genome during S phase and enables DNA repair and cell survival whereas apoptosis completely eliminates the damaged cell if the DNA damage proves to be irreparable. Thus, DNA damage is the principal trigger that elicits a p53 response, which in turn arrests or kills damaged cells, thereby suppressing the emergence of mutant cells that have oncogenic potential. An alternative model suggests that the p53 response is not to DNA damage but to cellular oncogenic stress (cell-cycle disruption) mediated by p14[ARF], a protein that facilitates p53 stabilization (7-9).

The p53 gene at 17p13.1 consists of 11 exons, which encode a nuclear protein with an apparent molecular weight of 53 kDa. The p53 protein is composed of 393 amino acids organized into distinct functional domains (Figure 8.3.) (6, 10). Exon 1 is non-coding. Exons 2 – 4 encode the N-terminal domain of p53 that functions as a transcriptional activator of a range of genes in response to cellular stress (3, 11). Specific amino acids (e.g., 22Leu, 23Trp) interact with the TATA-associated factors TAFII70 and TAFII31, subunits of the TFIID complex (12). Thus, p53 uses a hydrophobic interface in its N-terminal domain to interact with the transcriptional machinery of the cell. P53 can also recruit the histone acetyltransferases (HATs) CBP, p300, and PCAF to the promoter-enhancer region of genes through high-affinity protein-protein binding (13, 14). These HATs acetylate lysine residues of histones in chromatin, increasing transcriptional activity (15). Exons 5 – 9 encode the sequence-specific DNA-binding domain in the central portion of the molecule (residues 102 – 292) (16). Consistent with its tetrameric oligomerization state, the p53 protein binds DNA sites that contain the so-called p53 response element, 5'-Pu-Pu-Pu-C-(A/T)(T/A)-G-Py-Py-Py-3', typically in the context of two such decameric motifs separated by 0 - 20 bp (15, 17, 18). This degenerate sequence is present in a large number of genes (e.g., p21, MDM2, GADD45, Bax, IGF-BP3) all of which are transcriptionally activated by p53 (3, 4). More than 90% of the non-synonymous mutations in p53 occur in the DNA-binding domain involving residues that contact DNA directly or play a role in the structural integrity of the domain (2, 19). For example, the two most frequently altered residues in the protein, 273Arg and 248Arg, make contact with the phosphate backbone in the major groove and via hydrogen bonds with the minor groove of the DNA helix, respectively (16). Predictably, mutations in these residues result in defective DNA binding and loss of the ability of p53 to act as a transcription factor. Finally, exons 10 and 11 encode the oligomerization domain (residues 325 – 356) and the lysine-rich basic C-terminal domain (residues 363 – 393) (20, 21). The native p53 forms homotetrameric complexes in solution. The crystal structure of the oligomerization domain in a monomer consists of a ß-strand and α-helix, which associate with a second monomer across an antiparallel ß sheet and an antiparallel helix-helix interface to form a dimer

(20). Two of these dimers associate across a second and distinct parallel helix-helix interface to form the tetramer as a four-helix bundle. In summary, each cognate half-site in DNA binds two molecules of p53 and two such dimers assemble into tetramers (22). The tetramerization domain is linked to the DNA-binding domain by a flexible linker of 37 amino acids (residues 287 – 323) (20). Lastly, the C-terminal domain contains nine basic amino acid residues that can interact with DNA to regulate the ability of p53 to bind to its cognate recognition sequence via the core DNA-binding domain (23).

The transcriptional network of p53-responsive genes produces proteins that interact with a large number of other signal transduction pathways in the cell and a number of positive and negative autoregulatory feedback loops act upon the p53 response (4). In addition, the p53 protein lies at the hub of a complex signaling network interacting with multiple upstream and downstream proteins. The explanation for its promiscuity lies in the versatile structure of p53, which features every possible conformation from order to disorder (24). The core domain is globular and binds to diverse DNA targets containing two decameric p53 response elements. Its two flanking wings are mostly disordered and can bind to hundreds of signaling protein partners. Indeed, a segment within one wing can flip between four different ordered states, depending on the binding partner. Thus, the p53 protein uses different disordered regions to bind to different partners but also has several individual disordered regions that each bind to multiple partners. In addition, post-translational modification of p53 by phosphorylation, acetylation, or ubiquitination results in a specific cellular outcome in response to p53 activation (e.g., cell cycle arrest versus apoptosis) that depends on preferential activation of target genes (6, 25, 26).

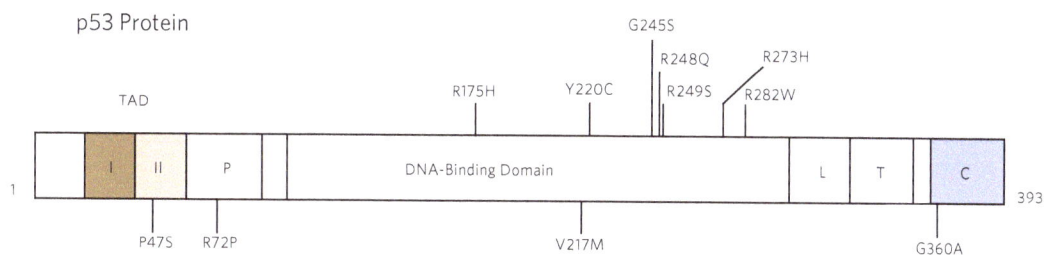

Figure 8.3. Diagram of p53 protein and its three major functional domains.
The N-terminus contains a transcriptional activation domain, the core domain contains a sequence-specific DNA-binding domain, and the C terminus incorporates oligomerization and regulatory domains. The transcriptional activation domain can be subdivided into TAD I (residues 20 – 40) and TAD II (residues 40 – 60), followed by the proline domain (P; residues 60 – 90). The DNA-binding domain (residues 100 – 300) is linked to the tetramerization domain (T; residues 325 - 356) by a flexible linker (L; residues 301 – 324). The C-terminal domain (C; residues 363 - 393) can interact with DNA to regulate the ability of p53 to bind to its cognate recognition sequence via the core DNA-binding domain. The p53 gene is the most commonly mutated gene identified in human cancers. Over 95% of the mutations are clustered in the core DNA-binding domain and >75% are non-synonymous mutations. The most common missense mutations are listed on top of the diagram: Arg175His, Tyr220Cys, Gly245Ser, Arg248Gln, Arg249Ser, Arg273His, Arg282Trp. Alternate mutations lead to Arg248Trp or Arg273Cys. Non-synonymous SNPs are listed below the diagram: Pro47Ser, Arg72Pro, Val217Met, and Gly360Ala.

To give one example of protein-protein interaction, p53 interacts with BRCA1 which interacts with the p53-binding protein 1 (53BP1). P53 mediates the export of BRCA1 from the nucleus to the cytoplasm by binding to the breast cancer carboxy-terminal (BRCT) region of BRCA1, displacing its usual binding partner, BARD1 (27). Since nuclear BRCA1 plays a key role in DNA repair, p53-dependent BRCA1 nuclear export controls cellular susceptibility to DNA damage, which is increased by augmentation of cytosolic BRCA1. In sporadic breast cancers, mutant p53 strongly correlated with nuclear retention of BRCA1 as a mechanism to increase cellular resistance to DNA damage. 53BP1 plays a key role in the activation of cell cycle checkpoints and DNA repair. It co-localizes at DSBs with the MRN complex, γH2AX, MDC1, and BRCA1. The interaction with the MRN complex occurs via the tandem breast cancer carboxy-terminal (BRCT) repeats in 53BP1 and is accompanied by ATM-mediated phosphorylation of both 53BP1 and NBS1 (28, 29). Immunohistochemical analysis revealed that the majority of sporadic breast cancers that lack ERα, PR, and HER2 expression (so-called "triple-negative" tumors) and BRCA1/2-mutated inherited cancers also showed reduced expression of 53BP1 (30).

In normal, unstressed cells, p53 levels are kept low through continuous proteolytic degradation. This is accomplished through the action of a ubiquitin ligase, Mdm2, which binds to the transactivation domain of p53 and ubiquitinates the protein targeting it for degradation by the proteasome (31-34). Mdm2 itself is induced by p53 and its binding to p53 establishes a negative feedback loop. Cellular stress leads to the activation of various protein kinases (e.g., MAPK family, ATM, CHEK2), which phosphorylate p53 in its N-terminal domain, thereby disrupting Mdm2-binding and stopping p53 degradation. The half-life of p53 increases drastically followed by rapid accumulation of p53. The Mdm2 gene is frequently amplified in sarcomas and overexpressed in a p53-independent manner in ERα-positive breast cancer cells MCF-7 and T47D (35).

P53 protein variants exist that differ from full-length p53 in their N- or C-terminal region, but generally conserve the central DNA-binding domain (36). These shorter isoforms are formed through alternative splicing, internal initiation of translation, or proteolytic cleavage. The variants are expressed at low levels, restricted to specific cell types and physiological conditions suggesting a limited role in cell function. In addition, there are two other members of the p53 family, p63 and p73, which show homology with p53, not so much in terms of primary sequence but in overall domain structure and conformation (36, 37). Despite the structural similarities with p53, the role of these proteins in tumorigenesis is controversial (38). Unlike the p53 gene, mutations in p63 and p73 are rare. Germline mutations in p63 are linked to developmental abnormalities rather than tumor formation.

Cell Cycle. Cell experiments with ionizing radiation have shown that p53 regulates cell cycle progression. Cells containing wild-type p53 undergo G1 arrest following radiation-induced DNA damage, whereas cells lacking functional p53 fail to arrest in G1 (39, 40). By holding the cell cycle at the G1/S checkpoint, p53 permits DNA repair to take place and thereby prevents replication of a potentially damaged genome during S phase. DNA damage induced by ionizing radiation leads to rapid increases in p53 protein by posttranslational changes in protein stability. The posttranslational modification of p53 is mediated by ATM, which binds p53 directly and is responsible for its phosphorylation at multiple residues including 15Ser, 20Ser, and 46Ser (41-43). The tight control exerted by ATM on p53 is enhanced by phosphorylating two additional proteins, namely CHEK2 kinase, which in turn phosphorylates p53 on a different site, and Mdm2, which controls p53 degradation. The abundant p53 protein binds to the DNA of several target genes, including p21$^{CIP1/WAF1}$, which encodes p21, a cyclin-dependent kinase inhibitor (44). p21 binds to CDK2 and inhibits the kinase activity necessary for transition into S phase. Thus, the p53-induced G1/S cell cycle arrest is mediated by p21. Alternatively, p53 can also down-regulate the expression of p21 by inducing a specific microRNA gene, miR-22 at 17p13.3 (45). The target of miR-22 is p21 mRNA and miR-22 overexpression leads to increased degradation of p21 mRNA and reduction of p21 protein levels. As a consequence, p21 will not be available for G1/S cell cycle arrest and cells can accumulate more DNA damage resulting in the activation of p53-mediated apoptosis. These results suggest that miR-22 is a molecular switch or sensor for the determination of the p53-dependent cellular fate in response to distinct stresses. Low cellular stress activates the p53-p21 axis to induce cell cycle arrest and high extensive stress activating the p53-miR-22-p21 axis to initiate apoptosis. Almost all of the mutant p53 proteins associated with malignancies lose the ability to regulate transcription and experiments have shown an excellent correlation between transcriptional activation and growth suppression in p53-null human cells (46, 47). Thus, a mutant p53 will result in decreased tran-scriptional expression of p21. As a consequence, p21 will not be available for G1/S cell cycle arrest and cells can divide uncontrollably resulting in tumor formation.

Apart from regulating the cell cycle at G1 in response to DNA damage, p53 regulates an apparent spindle checkpoint, which ensures the maintenance of chromosomal diploidy. Fibroblasts from p53$^{-/-}$ mouse embryos exposed to spindle inhibitors (nocodazole, colcemid) underwent multiple rounds of DNA synthesis without completing chromosome segregation, thus forming tetraploid and octaploid cells (48). Human fibroblasts lacking p53 activity arrested with a 4n DNA content upon exposure to colcemid (49). p53 also associates with the centrosome during interphase, but not during mitosis (50, 51). In contrast to p53$^{+/+}$ mouse embryo fibroblasts, >30%

of the p53$^{-/-}$ fibroblasts at interphase contained more than two centrosomes (3 to 10 per cell), and at mitosis >50% of the cells contained spindles organized by multiple spindle poles (52). The supernumerary centrosomes retained microtubule-nucleating activity and localized to the bipolar axis in >90% of the p53$^{-/-}$ cells at metaphase, giving rise to balanced chromosome segregation and the generation of viable polyploid daughter cells. However, in a few cells, there was an unequal distribution of chromosomes to daughter cells, perhaps caused by unequal numbers of centrosomes and resultant differences in the mitotic force exerted at each spindle pole. These observations implicate p53 in the regulation of centrosome duplication and identify yet one more mechanism by which the loss of p53 function may cause chromosome instability. p53 also plays a role in preserving the integrity of telomeres at chromosome ends. Telomere shortening results in p53 activation, which, in turn, suppresses telomerase activity (53, 54).

Apoptosis. In addition to its nuclear function as a transcription factor, p53 possesses biological activities that are cytoplasmic and transcription-independent (55). One example is apoptosis, which is regulated by Bcl-2 and Bax, homologous proteins that have opposing effects on cell life and death, with Bcl-2 serving to prolong cell survival and Bax acting as an accelerator of apoptosis. p53 is a direct transcriptional activator of the *bax* and *PUMA* (p53-upregulated modulator of apoptosis) genes (56, 57). Under a variety of cell death-inducing conditions, cytosolic p53 rapidly moves to the mitochondria and induces mitochondrial outer membrane permeabilization (MOMP), thereby triggering the release of pro-apoptotic factors from the mitochondrial intermembrane space, such as cytochrome c, leading to the downstream activation of caspases and apoptosis. MOMP is usually inhibited by anti-apoptotic proteins of the Bcl-2 family (e.g., Bcl-X$_L$). p53 is sequestered by Bcl-X$_L$ until PUMA disrupts the Bcl-X$_L$–p53 interaction, releasing p53 to trigger MOMP and apoptosis by activating Bax (58). Thus, there is interplay of nuclear and cytoplasmic functions of p53 in apoptosis. In addition to bax and PUMA, p53 induces the activation of other pro-apoptotic genes, such as NOXA to initiate the intrinsic apoptotic pathway (59). Another cytoplasmic effect of p53 is the inhibition of autophagy, a programmed cell-survival strategy that allows cells to sustain themselves by digesting parts of the cytoplasm and remove damaged, potentially harmful, cytoplasmic organelles (55).

References

1. Hollstein M, Sidransky D, Vogelstein B, Harris CC. *p53 mutations in human cancers.* Science. 1991;253:49-53.

2. Olivier M, Eeles R, Hollstein M, Khan MA, Harris CC, Hainaut P. *The IARC TP53 database: new online mutation analysis and recommendations to users.* Hum Mutat. 2002;19:607-14.

3. Levine AJ. *p53, the cellular gatekeeper for growth and division.* Cell. 1997;88:323-31.

4. Harris SL, Levine AJ. *The p53 pathway: positive and negative feedback loops.* Oncogene. 2005;24:2899-908.

5. Sablina AA, Budanov AV, Ilyinskaya GV, Agapova LS,

Kravchenko JE, Chumakov PM. *The antioxidant function of the p53 tumor suppressor.* Nature Med. 2005;11:1306-13.

6. Vousden KH, Prives C. *Blinded by the Light: The Growing Complexity of p53.* Cell. 2009;137:413-31.

7. Christophorou MA, Ringshausen I, Finch AJ, Swigart LB, Evan GI. *The pathological response to DNA damage does not contribute to p53-mediated tumour suppression.* Nature. 2006;443:214-7.

8. Efeyan A, Garcia-Cao I, Herranz D, Velasco-Miguel S, Serrano M. *Tumour biology: Policing of oncogene activity by p53.* Nature. 2006;443:159.

9. Van Dyke T. *p53 and tumor suppression.* N Engl J Med. 2007;356:79-81.

10. Soussi T, de Fromentel CC, May P. *Structural aspects of the p53 protein in relation to gene evolution.* Oncogene. 1990;5:945-52.

11. Lin J, Chen J, Elenbaas B, Levine AJ. *Several hydrophobic amino acids in the p53 amino-terminal domain are required for transcriptional activation, binding to mdm-2 and the adenovirus 5 E1B 55-kD protein.* Genes Dev. 1994;8:1235-46.

12. Thut CJ, Chen J, Klemm R, Tjian R. *p53 Transcriptional activation mediated by coactivators TAF 40 and TAF 60.* Science. 1995;267:100-4.

13. Gu W, Roeder RG. *Activation of p53 sequence-specific DNA binding by acetylation of the p53 c-terminal domain.* Cell. 1997;90:595-606.

14. Gu W, Shi X, Roeder RG. *Synergistic activation of transcription by CBP and p53.* Nature. 1997;387:819-22.

15. Riley T, Sontag E, Chen P, Levine A. *Transcriptional control of human p53-regulated genes.* Nature Rev Mol Cell Biol. 2008;9:402-12.

16. Cho Y, Gorina S, Jeffrey PD, Pavletich NP. *Crystal structure of a p53 tumor suppressor-DNA complex: understanding tumorigenic mutations.* Science. 1994;265:346-55.

17. El-Deiry WS, Kern SE, Pietenpol JA, Kinzler KW, Vogelstein B. *Definition of a consensus bindingsite for p53.* Nature. 1992;1:45-9.

18. Kern SE, Pietenpol JA, Thiagalingam S, Seymour A, Kinzler KW, Vogelstein B. *Oncogenic forms of p53 inhibit p53-regulated gene expression.* Science. 1992;256:827-30.

19. Hollstein M, Rice K, Greenblatt MS, Soussi T, Fuchs R, Sorlie T, et

al. *Database of p53 gene somatic mutations in human tumors and cell lines.* Nucleic Acids Res. 1994;22:3551-5.

20. Jeffrey PD, Gorina S, Pavletich NP. *Crystal structure of the tetramerization domain of the p53 tumor suppressor at 1.7 Angstroms.* Science. 1995;267:1498-502.

21. Sturzbecher H-W, Brain R, Addison C, Rudge K, Remm M, Grimaldi M, et al. *A c-terminal alpha-helix plus basic region motif is the major structural determinant of p53 tetramerization.* Oncogene. 1992;7:1513-23.

22. Kitayner M, Rozenberg H, Kessler N, Rabinovich D, Shaulov L, Haran TE, et al. *Structural basis of DNA recognition by p53 tetramers.* Molecular Cell. 2006;22:741-53.

23. Lee S, Elenbaas B, Levine A, Griffith J. *p53 and its 14 kDa C-terminal domain recognize primary DNA damage in the form of insertion/deletion mismatches.* Cell. 1995;81:1013-20.

24. Oldfield CJ, Meng J, Yang JY, Yang MQ, Uversky VN, Dunker AK. *Flexible nets: disorder and induced fit in the associations of p53 and 14-3-3 with their partners.* BMC Genomics. 2008;9 Suppl 1:S1.

25. Jayaraman L, Prives C. *Covalent and noncovalent modifiers of the p53 protein.* Cellular and Molecular Life Sciences : CMLS. 1999;55:76-87.

26. Kruse JP, Gu W. SnapShot: *p53 posttranslational modifications.* Cell. 2008;133:930-30 e1.

27. Jiang J, Yang ES, Jiang G, Nowsheen S, Wang H, Wang T, et al. *p53-dependent BRCA1 nuclear export controls cellular susceptibility to DNA damage.* Cancer Res. 2011;71:5546-57.

28. Lee JH, Goodarzi AA, Jeggo PA, Paull TT. *53BP1 promotes ATM activity through direct interactions with the MRN complex.* EMBO J. 2010;29:574-85.

29. Matsuoka S, Ballif BA, Smogorzewska A, McDonald ER, 3rd, Hurov KE, Luo J, et al. *ATM and ATR substrate analysis reveals extensive protein networks responsive to DNA damage.* Science. 2007;316:1160-6.

30. Bouwman P, Aly A, Escandell JM, Pieterse M, Bartkova J, van der Gulden H, et al. *53BP1 loss rescues BRCA1 deficiency and is associated with triple-negative and BRCA-mutated breast cancers.* Nature Struct Mol Biol. 2010;17:688-95.

31. Bond GL, Hu W, Levine AJ. *MDM2 is a central node in the p53 pathway: 12 years and counting.* Curr Cancer Drug Targets. 2005;5:3-8.

32. Haupt Y, Maya R, Kazaz A, Oren M. *Mdm2 promotes the rapid degradation of p53.* Nature. 1997;387:296-9.

33. Kubbutat MHG, Jones SN, Vousden KH. *Regulation of p53 stability by Mdm2.* Nature. 1997;387:299-303.

34. Momand J, Zambetti GP, Olson DC, George D, Levine AJ. *The mdm-2 oncogene product forms a complex with the p53 protein and inhibits p53-mediated transactivation.* Cell. 1992;69:1237-45.

35. Phelps M, Darley M, Primrose JN, Blaydes JP. *p53-independent activation of the hdm2-P2 promoter through multiple transcription factor response elements results in elevated hdm2 expression in estrogen receptor alpha-positive breast cancer cells.* Cancer Res. 2003;63:2616-23.

36. Courtois S, Caron de Fromentel C, Hainaut P. *p53 protein variants: structural and functional similarities with p63 and p73 isoforms.* Oncogene. 2004;23:631-8.

37. Kaghad M, Bonnet H, Yang A, L. C, Biscan J, Valent A, et al. *Monoallelically expressed gene related to p53 at 1p36, a region frequently deleted in neuroblastoma and other human cancers.* Cell. 1997;90:809-19.

38. Tomkova K, Tomka M, Zajac V. *Contribution of p53, p63, and p73 to the developmental diseases and cancer.* Neoplasma. 2008;55:177-81.

39. Kastan MB, Zhan Q, El-Deiry WS, Carrier F, Jacks T, Walsh WV, et al. *A mammalian cell cycle checkpoint pathway utilizing p53 and GADD45 is defective in ataxia-telangiectasia.* Cell. 1992;71:587-97.

40. Kuerbitz SJ, Plunkett BS, Walsh WV, Kastan MB. *Wild-type p53 is a cell cycle checkpoint determinant following irradiation.* Proc Natl Acad, Sci USA. 1992;89:7491-5.

41. Banin S, Moyal L, Shieh S, Taya Y, Anderson CW, Chessa L, et al. *Enhanced phosphorylation of p53 by ATM in response to DNA damage.* Science. 1998;281:1674-7.

42. Khanna KK, Keating KE, Kozlov S, Scott S, Gatei M, Hobson K, et al. *ATM associates with and phosphorylates p53: mapping the region of interaction.* Nature Genet. 1998;20:398-400.

43. Saito S, Goodarzi AA, Higashimoto Y, Noda Y, Lees-Miller SP, Appella E, et al. *ATM mediates phosphorylation at multiple p53 sites, including Ser(46), in response to ionizing radiation.* J Biol Chem. 2002;277:12491-4.

44. El-Deiry WS, Tokino T, Velculescu VE, Levy DB, Parsons R, Trent JM, et al. *WAF1, a potential mediator of p53 tumor suppression.* Cell. 1993;75:817-25.

45. Tsuchiya N, Izumiya M, Ogata-Kawata H, Okamoto K, Fujiwara Y, Nakai M, et al. *Tumor suppressor miR-22 determines p53-dependent cellular fate through post-transcriptional regulation of p21.* Cancer Res. 2011;71:4628-39.

46. Crook T, Marston NJ, Sara EA, Vousden KH. *Transcriptional activation by p53 correlates with suppression of growth but not transformation.* Cell. 1994;79:818-27.

47. Harris CC, Hollstein M. *Clinical Implications of the p53 tumor-suppressor gene.* N Engl J Med. 1993;329:1318-127.

48. Cross SM, Sanchez CA, Morgan CA, Schimke MK, Ramel S, Idzerda RL, et al. *A p53-dependent mouse spindle checkpoint.* Science. 1995;267:1353-6.

49. Gualberto A, Aldape K, Kozakiewicz K, Tlsty TD. *An oncogenic form of p53 confers a dominant, gain-of-function phenotype that disrupts spindle checkpoint control.* Proc Natl Acad Sci USA. 1998;95:5166-71.

50. Brown CR, Doxsey SJ, White E, Welch WJ. *Both viral (adenovirus E1B) and cellular (hsp 70, p53) components interact with centrosomes.* J Cell Physiol. 1999;160:4760.

51. Zaidel B, Blair GE. *The intracellular distribution of the transformation-associated protein p53 in adenovirus-transformed rodent cells.* Oncogene. 1988;2:579-84.

52. Fukasawa K, Choi T, Kuriyama R, Rulong S, Woude GFV. *Abnormal centrosome amplification in the absence of p53.* Science. 1996;271:1744-7.

53. Sahin E, Colla S, Liesa M, Moslehi J, Muller FL, Guo M, et al. *Telomere dysfunction induces metabolic and mitochondrial compromise.* Nature. 2011;470:359-65.

54. Stampfer MR, Garbe J, Nijjar T, Wigington D, Swisshelm K, Yaswen P. *Loss of p53 function accelerates acquisition of telomerase activity in indefinite lifespan human mammary epithelial cell lines.* Oncogene. 2003;22:5238-51.

55. Green DR, Kroemer G. *Cytoplasmic functions of the tumour suppressor p53.* Nature. 2009;458:1127-30.

56. Miyashita T, Reed JC. *Tumor suppressor p53 is a direct transcriptional activator of the human bax gene.* Cell. 1995;80:293-9.

57. Yu J, Zhang L, Hwang PM, Kinzler KW, Vogelstein B. *PUMA induces the rapid apoptosis of colorectal cancer cells.* Mol Cell. 2001;7:673-82.

58. Chipuk JE, Bouchier-Hayes L, Kuwana T, Newmeyer DD, Green DR. *PUMA couples the nuclear and cytoplasmic proapoptotic function of p53.* Science. 2005;309:1732-5.

59. Oda E, Ohki R, Murasawa H, Nemoto J, Shibue T, Yamashita T, et al. *Noxa, a BH3-only member of the Bcl-2 family and candidate mediator of p53-induced apoptosis.* Science. 2000;288:1053-8.

8.4.2. CLINICAL STUDIES

Three groups of sequence alterations can affect the p53 gene at 17p13.1 (i) germline mutations cause the Li-Fraumeni syndrome, (ii) somatic mutations occur in 5 to 70% of human cancers including about 25% of breast cancers, and (iii) germline polymorphisms of which more than 90% occur in non-coding sequences (1-3).

The Li-Fraumeni syndrome is a rare multicancer familial syndrome, which includes early-onset breast cancer, sarcoma, leukemia, brain tumors, and adrenocortical carcinoma. The syndrome was linked to the p53 gene by the presence of germline mutations in approximately 50% of pedigrees (4, 5). Although breast cancer is a common component of the Li-Fraumeni syndrome, germline p53 mutations are rare in breast cancer-only pedigrees because the syndrome is usually associated with other tumors (6). The contribution of p53 germline mutations to breast cancer predisposition overall is thought to be small (much less than 1%) and certainly less than that of BRCA1 or BRCA2. However, recent studies indicate a larger role than previously thought in very-early onset breast cancer, i.e., women diagnosed before age 30 years (7).

Somatic p53 mutations are present in approximately 25% of breast cancers (www.iarc.fr/P53/). A higher prevalence of p53 mutations was found in breast tumors of current smokers (36.5%; p = 0.02) compared with never smokers (23.6%) (8). A higher frequency of somatic p53 mutations was also found in breast cancers from carriers of BRCA1 mutations than from non-carriers suggesting that loss of p53 function is an important step in the pathogenesis of BRCA1 mutation-associated breast tumors (9). The International Agency for Research on Cancer (IARC) p53 mutation database contains >17,000 somatic mutations and 225 germline mutations (www.iarc.fr/p53/). Of all p53 mutations, >95% are clustered in the core DNA-binding domain. Of the mutations in this domain, >75% are non-synonymous mutations and about 30% fall within seven 'hotspot' residues: Arg175His,

Tyr220Cys, Gly245Ser, Arg248Gln, Arg249Ser, Arg273His, and Arg282Trp (Figure 8.3.). Alternate mutations lead to Arg248Trp and Arg273Cys. The existence of these hotspot residues could be explained by the susceptibility of these codons to carcinogen-induced alterations and by positive selection of mutations that provide the cell with growth and survival advantages (10). Deletions and chain-termination mutations occur less frequently in the p53 gene (11, 12), which is opposite to BRCA1 and BRCA2 gene alterations that are usually deletions and truncating mutations and rarely missense mutations. In total, over 1,100 distinct amino acid substitutions caused by missense mutations have been documented in the p53 gene (13). The majority of germline and somatic mutations occurring in tumors are pathogenic missense mutations, which encode defective proteins that compromise normal activities of p53.

The high frequency of somatic mutations has allowed studies of mutational spectra at the p53 gene locus in different tumor types. These studies revealed characteristic fingerprints of DNA alterations reflecting the involvement of specific environmental agents in inducing these mutations; e.g., skin carcinoma and UV light, lung carcinoma and cigarette smoking, and hepatocellular carcinoma and aflatoxin (8, 14). However, no specific exogenous carcinogens could be conclusively linked with the development of breast cancer. The analysis of American and European studies revealed somatic p53 mutations in 25% of breast cancers; 15% of these were G:C/T:A transversions, 16% G:C/A:T transitions at non-CpG sites, and 11% A:T/G:C transitions, with nontranscribed strand bias of 89, 63, and 75%, respectively (1, 14, 15). In contrast, G:C/A:T transitions constituted the majority (over 50%) of mutations in breast cancers from Japanese women (16, 17). An analysis of the IARC p53 mutation database (1392 mutations) confirmed an excess of transversions on G bases in tumors from Western (USA and Europe) as compared to Eastern (Japan) countries (18). Moreover, the patterns of inherited p53 mutations associated with breast cancer differed from those of somatic mutations. Thus, p53 mutational spectra in breast cancer appear to differ between populations suggesting that different carcinogens or environmental exposures may play a role. Alternatively, there may be population differences in carcinogen metabolism or DNA repair.

Over 200 SNPs have been identified in the p53 gene with over 90% in non-coding sequences (3). The best characterized is a 16-bp insertion in intron 3, which has been associated with increased breast cancer risk in one study (19, 20). Nineteen SNPs have been reported in exons of which 8 are synonymous. Although these silent SNPs do not change the amino acid sequence or structure of the protein, the changes in base sequence

and codon usage could modify protein expression and microRNA miR-122 has been shown to target p53 mRNA and regulate p53 translation (21). In contrast, most of the p53 SNPs are intronic with no or at best subtle biological effects. Of the 11 reported non-synonymous SNPs, four have been validated by multiple submissions to the IARC p53 database and not observed as somatic mutations in tumors: Pro47Ser, Arg72Pro, Val217Met, and Gly360Ala (Figure 8.3.). Pro47Ser is only present in Africans (minor allele frequency about 5%) and associated with inconsistent phenotypic consequences (3). Arg72Pro in exon 4 is common worldwide with minor allele frequencies ranging from 25% in Caucasians to 60% in Asians. Functional studies have shown that the 72Arg allele is more effective at inducing apoptosis and protecting cells from neoplastic development than 72Pro (22). However, associations with cancer risk have been inconsistent and do not support a role for this SNP in developing breast cancer (3, 19, 23). Val217Met and Gly360Ala are rare and of uncertain phenotypic consequences. Finally, none of the susceptibility loci for breast cancer identified in genome-wide association studies have been close to the p53 gene (24, 25).

The p53 status of mammary cancers has been assessed by two techniques, at the protein level by immunohistochemistry and at the gene level by DNA sequencing. The percentage of tumors reported to have abnormal p53 immunostaining has been consistently higher than the rate of tumors with mutations. There are several reasons for the discordance between the two techniques, which may be as high as 38% (26). The immunohistochemical studies are based on the assumption that visible nuclear immunostaining reflects accumulation of mutant p53 protein, which is then quantitated by examining the intensity of staining and the percentage of stained cells. The immunohistochemical technique has two major limitations. It probes at the protein rather than at the DNA level and therefore cannot verify the presence or define the site or nature of a mutation. This can only be accomplished by DNA sequencing. The second weakness of the immunohistochemical technique lies in the arbitrary decision level of calling a tumor p53-positive which may vary from staining of any cell (>0%) to >20% or even >75% of cells (27, 28). Thus, in some studies tumors may be classified as p53-positive that contain only solitary cell clusters with positive nuclear staining. In contrast, the parallel genetic test on such heterogeneous tumors may lack the sensitivity to detect a mutation that is present in only a minor fraction of tumor cells. In short, the immunohistochemical assay may overdiagnose and result in false positives while the genetic tests may underdiagnose and result in false negatives. Consequently, the frequency of p53-positive cases is clearly dependent on the methodology.

It is still an open question whether p53 mutations are involved in the initiation of malignant transformation or only at a more advanced stage of cancer, leading to additional growth and aggressiveness (10). p53 overexpression has been detected immunohistochemically in ductal hyperplasia of the breast (29). p53 mutations have been identified in ductal carcinoma *in situ* (30). A study of 1,794 invasive breast cancers revealed that p53 mutations were more frequent in medullary and ductal tumors with an aggressive phenotype (high grade, large size, lymph node positive cases, and low hormone receptor content) (31). The p53 mutation status, tumor size, and lymph node status were the strongest predictors of breast cancer survival (32). The accumulation of p53 mutations leads to general genomic instability and additional genetic aberrations. One study examined genetic instability in 183 primary human breast cancers with and without p53 abnormalities and observed a significant association between p53 mutations (17.6% of tumors) and genetic instability detected by gene amplification, allele loss, karyotype analysis, and fluorescent *in situ* hybridization (33).

References

1. Hollstein M, Sidransky D, Vogelstein B, Harris CC. *p53 mutations in human cancers*. Science. 1991;253:49-53.

2. Olivier M, Eeles R, Hollstein M, Khan MA, Harris CC, Hainaut P. *The IARC TP53 database: new online mutation analysis and recommendations to users*. Hum Mutat. 2002;19:607-14.

3. Whibley C, Pharoah PD, Hollstein M. *p53 polymorphisms: cancer implications*. Nature Rev Cancer. 2009;9:95-107.

4. Malkin D. *p53 and the Li-Fraumeni syndrome*. Cancer Genet Cytogenet. 1993;66:83-92.

5. Malkin D, Li FP, Strong LC, Fraumeni JFJ, Nelson CE, Kim DH, et al. *Germ line p53 mutations in a familial syndrome of breast cancer, sarcomas, and other neoplasms*. Science. 1990;250:1233-8.

6. Sidransky D, Tokino T, Helzlsouer K, Zehnbauer B, Rausch G, Shelton B, et al. *Inherited p53 gene mutations in breast cancer*. Cancer Res. 1992;52:2984-6.

7. Mouchawar J, Korch C, Byers T, Pitts TM, Li E, McCredie MR, et al. *Population-based estimate of the contribution of TP53 mutations to subgroups of early-onset breast cancer: Australian Breast Cancer Family Study*. Cancer Res. 2010;70:4795-800.

8. Conway K, Edmiston SN, Cui L, Drouin SS, Pang J, He M, et al. *Prevalence and spectrum of p53 mutations associated with smoking in breast cancer*. Cancer Res. 2002;62:1987-95.

9. Phillips KA, Nichol K, Ozcelik H, Knight J, Done SJ, Goodwin PJ, et al. *Frequency of p53 mutations in breast carcinomas from Ashkenazi Jewish carriers of BRCA1 mutations*. J Natl Cancer Inst. 1999;91:469-73.

10. Rivlin N, Brosh R, Oren M, Rotter V. Mutations in the p53 Tumor Suppressor Gene: *Important Milestones at the Various Steps of Tumorigenesis*. Genes Cancer. 2011;2:466-74.

11. Harris CC, Hollstein M. *Clinical Implications of the p53 tumor-suppressor gene*. N Engl J Med. 1993;329:1318-127.

12. Hollstein M, Rice K, Greenblatt MS, Soussi T, Fuchs R, Sorlie T, et al. *Database of p53 gene somatic mutations in human tumors and cell lines*. Nucleic Acids Res. 1994;22:3551-5.

13. Kato S, Han SY, Liu W, Otsuka K, Shibata H, Kanamaru R, et al. *Understanding the function-structure and function-mutation relationships of p53 tumor suppressor protein by high-resolution missense mutation analysis*. Proc Natl Acad Sci U S A. 2003;100:8424-9.

14. Greenblatt MS, Bennett WP, Hollstein M, Harris CC. *Mutations in the p53 tumor suppressor gene: Clues to cancer etiology and molecular pathogenesis*. Cancer Res. 1994;54:4855-78.

15. Coles C, Condie A, Chetty U, Steel CM, Evans HJ, Prosser J. *p53 mutations in breast cancer*. Cancer Res. 1992:5291-8.

16. Sasa M, Kondo K, Komaki K, Uyama T, Morimoto T, Monden Y. *Frequency of spontaneous p53 mutations (CpG site) in breast cancer in Japan*. Breast Cancer Res Treat. 1993;27,No.3:247-52.

17. Tsuda H, Iwaya K, Fukutomi T, Hirohashi S. *p53 mutations and c-erbB-2 amplification in intraductal and invasive breast carcinomas of high histologic grade*. Jpn J Cancer Res. 1993;84:394-401.

18. Olivier M, Hainaut P. TP53 mutation patterns in breast cancers: *searching for clues of environmental carcinogenesis*. Semin Cancer Biol. 2001;11:353-60.

19. Costa S, Pinto D, Pereira D, Rodrigues H, Cameselle-Teijeiro J, Medeiros R, et al. *Importance of TP53 codon 72 and intron 3 duplication 16bp polymorphisms in prediction of susceptibility on breast cancer*. BMC Cancer. 2008;8:32.

20. Lazar V, Hazard F, Bertin F, Janin N, Bellet D, Bressac B. *Simple sequence repeat polymorphism within the p53 gene*. Oncogene. 1993;8:1703-5.

21. Burns DM, D'Ambrogio A, Nottrott S, Richter JD. *CPEB and two poly(A) polymerases control miR-122 stability and p53 mRNA translation*. Nature. 2011;473:105-8.

22. Dumont P, Leu JI, Della Pietra AC, 3rd, George DL, Murphy M. *The codon 72 polymorphic variants of p53 have markedly different apoptotic potential*. Nature Genet. 2003;33:357-65.

23. Commonly studied single-nucleotide polymorphisms and breast cancer: results from the Breast Cancer Association Consortium. J Natl Cancer Inst. 2006;98:1382-96.

24. Easton DF, Pooley KA, Dunning AM, Pharoah PDP, Thompson D, Ballinger DG, et al. *Genome-wide association study indentifies novel breast cancer susceptibility loci*. Nature. 2007;447:1087-95.

25. Hunter DJ, Kraft P, Jacobs KB, Cox DG, Yeager M, Hankinson SE, et al. *A genome-wide association study identifies alleles in FGFR2 associated with risk of sporadic post-menopausal breast cancer*. Nature Genet. 2007;39:870-4.

26. Allred DC, Clark GM, Elledge R, Fuqua SAW, Brown RW, Chamness GC, et al. *Association of p53 protein expression with tumor cell proliferation rate and clinical outcome in node-negative breast cancer*. Natl Cancer Inst. 1993;85:200-6.

27. Barnes DM, Dublin EA, Fisher CJ, Levison DA, Millis RR. *Immunohistochemical detection of p53 protein in mammary carcinoma: An important new independent indicator of prognosis?* Hum Pathol. 1993;24:469-76.

28. Thor AD, Moore II DH, Edgerton SM, Kawasaki ES, Reihsaus E, Lynch HT, et al. *Accumulation of p53 tumor suppressor gene protein: An independent marker of prognosis in breast cancers*. J Natl Cancer Inst. 1992;84:845-55.

29. Mommers EC, van Diest PJ, Leonhart AM, Meijer CJ, Baak JP. *Expression of proliferation and apoptosis-related proteins in usual ductal hyperplasia of the breast*. Human Pathol. 1998;29:1539-45.

30. Done SJ, Eskandarian S, Bull S, Redston M, Andrulis IL. *p53 missense mutations in microdissected high-grade ductal carcinoma in situ of the breast*. J Natl Cancer Inst. 2001;93:700-4.

31. Olivier M, Langerod A, Carrieri P, Bergh J, Klaar S, Eyfjord J, et al. *The clinical value of somatic TP53 gene mutations in 1,794 patients with breast cancer*. Clin Cancer Res. 2006;12:1157-67.

32. Langerod A, Zhao H, Borgan O, Nesland JM, Bukholm IR, Ikdahl T, et al. *TP53 mutation status and gene expression profiles are powerful prognostic markers of breast cancer*. Breast Cancer Res : BCR. 2007;9:R30.

33. Eyfjord JE, Thorlacius S, Steinarsdottir M, Valgardsdottir R, Ogmundsdottir HM, Anamthawat-Jonsson K. *p53 Abnormalities and genomic instability in primary human breast carcinomas*. Cancer Res. 1995;55:646-51.

8.5. ATAXIA-TELEANGIECTASIA MUTATED (ATM)

8.5.1. INTRODUCTION

Ataxia-teleangiectasia (AT) is an autosomal recessive syndrome characterized by progressive cerebellar ataxia, oculocutaneous teleangiectases, immunodeficiency, radiation sensitivity, chromosomal instability, and a predisposition to malignancy (1). Lymphoid cancers predominate in childhood whereas epithelial cancers, including breast cancer, are seen in adults. The syndrome is caused by mutations in the ataxia-teleangiectasia mutated (ATM) gene at 11q22.3, which is composed of 66 exons, 62 of which encode a 350-kDa protein composed of 3056 amino acids (Figure 8.4.) (2, 3). ATM is a nuclear protein kinase that plays a key role in the cellular response to DNA double-strand breaks (DSBs) by phosphorylating multiple substrates, including BRCA1, Chk2, H2AX, Mdm2, NBS1, p53, 53BP1, SMC1, SSB1, resulting in the activation of DSB repair, cell cycle checkpoints or apoptosis (4-9). Consequently, cells lacking functional ATM protein exhibit defects in DNA repair and loss of cell cycle checkpoints, resulting in increased sensitivity to ionizing radiation and predisposition to malignancy.

ATM belongs to the family of phosphatidylinositol-3 kinase-related kinases (PIKKs) with sequence similarity to phosphatidylinositol-3 kinases (PI3Ks; Figure 8.4.) (10-12). Other members of the PIKK family are ATR (ATM- and RAD3-related), DNA-PKcs (DNA-dependent protein kinase catalytic subunit) and mTOR (mammalian target of rapamycin). ATR and DNA-PK also participate in DNA damage response pathways and, like ATM, phosphorylate multiple downstream substrates (8, 13-15). The phosphorylation targets for all three kinases are serines or threonines followed by glutamine, i.e., SQ or TQ motifs (5).

ATM can be activated by several mechanisms involving the interaction with different proteins. One mechanism involves the autophosphorylation of ATM on multiple sites including 1981Ser, which leads to the dissociation of inactive ATM dimers into active monomers with accessibility of substrates to the kinase domain (16). The autophosphorylation appears to be triggered by the interaction of ATM with NBS1, a component of the trimeric MRN (MRE11, RAD50, NBS1) protein complex that functions as primary sensor of DSBs (17, 18). The nuclease MRE11 processes DSBs producing overhanging DSB ends required for efficient homologous recombination and generates short single-strand oligonucleotides that stimulate ATM activity (19). NBS1 and MRE11 gene mutations cause the Nijmegen breakage syndrome (NBS) and ataxia teleangiectasia-like disorder (ATLD), respectively, which are associated with multisystem defects, including immunodeficiency and increased lymphoid malignancies (20, 21). NBS cells are hypersensitive to ionizing radiation, defective in intra-S and G2/M checkpoints, and deficient in ATM substrate phosphorylation despite the presence of wild-type ATM. A second mechanism involves the acetylation of ATM on 3016 Lys by Tip60 histone acetyltransferase whose activity is increased by DNA damage (22). Finally, ATM can be activated in the absence of DNA damage by agents that relax chromatin, such as chloroquine (16). The isoflavenoid genistein that is abundant in soybeans has also been shown to phosphorylate ATM (23). In summary, once ATM is activated, a positive feedback loop is initiated resulting in the phosphorylation of multiple ATM targets of which several, in turn, phosphorylate ATM, thereby amplifying ATM-dependent signaling.

ATM interacts directly with two tumor suppressor proteins, BRCA1 and p53. ATM forms a complex with

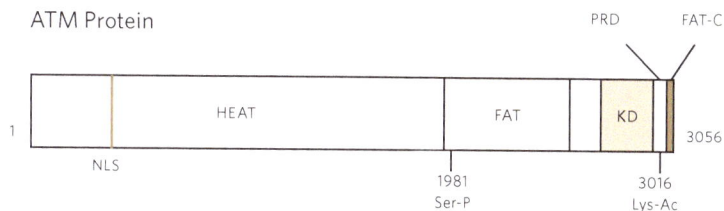

Figure 8.4. Diagram of ATM protein domains.
The 350-kDa ATM is a Mn^{2+}-dependent protein kinase composed of 3056 amino acids. The N-terminal two-thirds contain a nuclear localization signal (NLS, amino acids 385 – 388) and multiple HEAT sequence repeats (so called because they are present in huntingtin, elongation factor 3, subunit A of protein phosphatase 2A and TOR1), which are predicted to form a massive α-helical scaffold. The C-terminal third contains four domains that share features with other members of the PIKK family, including a FAT domain (acknowledging a conserved amino acid sequence in FRAP, ATR, and TRRAP), a PIKK-regulatory domain (PRD), and the FAT-C terminal (FATC) domain. All three domains regulate the activity of the kinase domain (KD). Over 300 distinct mutations have been reported in AT patients of which >80% are base substitutions, insertions, or deletions that generate premature termination codons or splicing abnormalities. In addition, there are several non-synonymous polymorphisms: Ser49Cys (rs1800054), Ser707Pro (rs4986761; MAF 1.3%), Phe858Leu (rs1800056; MAF 1.1%), Pro1054Arg (rs1800057; MAF 2.4%), Leu1420Phe (rs1800058; MAF 1.8%). Activation of ATM involves auto-phosphorylation of 1981Ser and acetylation of 3016Lys.

BRCA1 and phosphorylates two residues, 1423Ser and 1524Ser, important for BRCA1 function (24). The biochemical link between ATM and BRCA1 may partially explain why heterozygous carriers of a dysfunctional ATM gene are at increased risk of breast cancer. Activated ATM also binds p53 directly and is responsible for its 15Ser phosphorylation, thereby contributing to the activation and stabilization of p53, chief regulator of the G1/S checkpoint (25-27). The tight control exerted by ATM on p53 is enhanced by phosphorylating two additional proteins, namely checkpoint kinase 2 (Chk2), which in turn phosphorylates p53 on a different residue (20Ser), and Mdm2, which controls p53 degradation. Thus, activated ATM mediates a two-step response to DSBs. In the rapid response, ATM phosphorylates Chk2, which phosphorylates CDC25A, targeting it for ubiquitination and degradation. Therefore phosphorylated CDK2-Cyclin accumulates and progression through the cell cycle is blocked. In the delayed response, ATM phosphorylates the inhibitor of p53, Mdm2, and p53 itself, which is also phosphorylated by Chk2. The resulting activation and stabilization of p53 leads to an increased expression of Cdk inhibitor p21, which further helps to keep Cdk activity low and to maintain long-term cell cycle arrest.

References

1. Swift M, Reitnauer PJ, Morrell D, Chase CL. *Breast and other cancers in families with ataxia-telangiectasia.* N Engl J Med. 1987;316:1289-94.

2. Lee JH, Paull TT. *Activation and regulation of ATM kinase activity in response to DNA double-strand breaks.* Oncogene. 2007;26:7741-8.

3. Savitsky K, Bar-Shira A, Gilad S, Rotman G, Ziv Y, Vanagaite L, et al. *A single ataxia telangiectasia gene with a product similar to PI-3 kinase.* Science. 1995;268:1749-53.

4. Barzilai A, Rotman G, Shiloh Y. *ATM deficiency and oxidative stress: a new dimension of defective response to DNA damage.* DNA Repair. 2002;1:3-25.

5. Kim ST, Lim DS, Canman CE, Kastan MB. *Substrate specificities and identification of putative substrates of ATM kinase family members.* J Biol Chem. 1999;274:37538-43.

6. Kitagawa R, Bakkenist CJ, McKinnon PJ, Kastan MB. *Phosphorylation of SMC1 is a critical downstream event in the ATM-NBS1-BRCA1 pathway.* Genes Dev. 2004;18:1423-38.

7. Kurz EU, Lees-Miller SP. *DNA damage-induced activation of ATM and ATM-dependent signaling pathways.* DNA Repair (Amst). 2004;3:889-900.

8. Matsuoka S, Ballif BA, Smogorzewska A, McDonald ER, 3rd, Hurov KE, Luo J, et al. *ATM and ATR substrate analysis reveals extensive protein networks responsive to DNA damage.* Science. 2007;316:1160-6.

9. Richard DJ, Bolderson E, Cubeddu L, Wadsworth RI, Savage K, Sharma GG, et al. *Single-stranded DNA-binding protein hSSB1 is critical for genomic stability.* Nature. 2008;453:677-81.

10. Keith CT, Schreiber SL. *PIK-related kinases: DNA repair, recombination, and cell cycle checkpoints.* Science. 1995;270:50-1.

11. Lempiainen H, Halazonetis TD. *Emerging common themes in regulation of PIKKs and PI3Ks.* EMBO J. 2009;28:3067-73.

12. Perry J, Kleckner N. *The ATRs, ATMs, and TORs are giant HEAT repeat proteins.* Cell. 2003;112:151-5.

13. Bao S, Tibbetts RS, Brumbaugh KM, Fang Y, Richardson DA, Ali A, et al. *ATR/ATM-mediated phosphorylation of human Rad17 is required for genotoxic stress responses.* Nature. 2001;411:969-74.

14. Chen J. *Ataxia telangiectasia-related protein is involved in the phosphorylation of BRCA1 following deoxyribonucleic acid damage.* Cancer Res. 2000;60:5037-9.

15. Gatei M, Zhou BB, Hobson K, Scott S, Young D, Khanna KK. *Ataxia telangiectasia mutated (ATM) kinase and ATM and Rad3 related kinase mediate phosphorylation of Brca1 at distinct and overlapping sites. In vivo assessment using phospho-specific antibodies.* J Biol Chem. 2001;276:17276-80.

16. Bakkenist CJ, Kastan MB. *DNA damage activates ATM through intermolecular autophosphorylation and dimer dissociation.* Nature. 2003;421:499-506.

17. Kang J, Ferguson D, Song H, Bassing C, Eckersdorff M, Alt FW, et al. *Functional interaction of H2AX, NBS1, and p53 in ATM-dependent DNA damage responses and tumor suppression.* Mol Cell Biol. 2005;25:661-70.

18. You Z, Chahwan C, Bailis J, Hunter T, Russell P. *ATM activation and its recruitment to damaged DNA require binding to the C terminus of Nbs1.* Mol Cell Biol. 2005;25:5363-79.

19. Jazayeri A, Balestrini A, Garner E, Haber JE, Costanzo V. *Mre11-Rad50-Nbs1-dependent processing of DNA breaks generates oligonucle-otides that stimulate ATM activity.* EMBO J. 2008;27:1953-62.

20. Lee JH, Xu B, Lee CH, Ahn JY, Song MS, Lee H, et al. *Distinct functions of Nijmegen breakage syndrome in ataxia telangiectasia mutated-dependent responses to DNA damage.* Mol Cancer Res. 2003;1:674-81.

21. Varon R, Vissinga C, Platzer M, Cerosaletti KM, Chrzanowska KH, Saar K, et al. *Nibrin, a novel DNA double-strand break repair protein, is mutated in Nijmegen breakage syndrome.* Cell. 1998;93:467-76.

22. Sun Y, Xu Y, Roy K, Price BD. *DNA damage-induced acetylation of lysine 3016 of ATM activates ATM kinase activity.* Mol Cell Biol. 2007;27:8502-9.

23. Ye R, Goodarzi AA, Kurz EU, Saito S, Higashimoto Y, Lavin MF, et al. *The isoflavonoids genistein and quercetin activate different stress signaling pathways as shown by analysis of site-specific phosphorylation of ATM, p53 and histone H2AX.* DNA Repair (Amst). 2004;3:235-44.

24. Cortez D, Wang Y, Qin J, Elledge SJ. *Requirement of ATM-dependent phosphorylation of brca1 in the DNA damage response to double-strand breaks.* Science. 1999;286:1162-6.

25. Banin S, Moyal L, Shieh S, Taya Y, Anderson CW, Chessa L, et al. *Enhanced phosphorylation of p53 by ATM in response to DNA damage.* Science. 1998;281:1674-7.

26. Canman CE, Lim DS, Cimprich KA, Taya Y, Tamai K, Sakaguchi K, et al. *Activation of the ATM kinase by ionizing radiation and phosphorylation of p53.* Science. 1998;281:1677-9.

27. Khanna KK, Keating KE, Kozlov S, Scott S, Gatei M, Hobson K, et al. *ATM associates with and phosphorylates p53: mapping the region of interaction.* Nat Genet. 1998;20:398-400.

8.5.2. CLINICAL STUDIES

Over 300 distinct ATM mutations have been reported of which >80% are base substitutions, insertions, or deletions that generate premature termination codons or splicing abnormalities. The truncated species is usually unstable and results in absent or severely reduced ATM protein expression (1, 2). Most patients with AT have compound heterozygous rather than homozygous ATM mutations, with the latter present in consanguineous families or in the case of a few population-specific founders. Biallelic (compound heterozygous or homozygous) individuals develop cancers at a rate approximately 100 times higher than unaffected age-matched controls. Women heterozygous for truncating mutations in ATM are healthy, but have a 2-fold higher risk of developing breast cancer, which increases to about 5-fold in those younger than age 50 years (3, 4). Thus, ATM mutations that cause

AT in biallelic individuals are breast cancer susceptibility alleles in monoallelic carriers. Certain missense mutations can also cause AT but only account for approximately 10% of ATM mutations identified in AT patients (1). Overall, it has been estimated that 0.5 – 1% of women in Western populations are heterozygous carriers of ATM mutations (2).

Non-synonymous polymorphisms in ATM are common and it can therefore be difficult to deduce whether a specific base substitution is causally associated with disease, particularly if the ATM protein expression is not abnormal. Thus, the association of missense ATM variants with breast cancer has been contentious (5-9). The uncertainty was resolved by a comprehensive analysis of 26,101 cases and 29,842 controls from 23 studies in the Breast Cancer Association Consortium (10). Five non-synonymous polymorphisms (minor allele frequency, 0.9 – 2.6%) were genotyped: Ser49Cys (rs1800054), Ser707Pro (rs4986761), Phe858Leu (rs1800056), Pro1054Arg (rs1800057), Leu1420Phe (rs1800058). No single missense variant was significantly associated with breast cancer risk but there was a significant trend in risk with increasing numbers of variant ATM polymorphisms. The odds ratio was 1.05 for being heterozygous for any of the SNPs and 1.51 for being a rare homozygote for any of the SNPs with an overall OR of 1.06 (p_{trend} = 0.04). Among bilateral and familial cases the OR was 1.12 (95% CI 1.02 – 1.23; p_{trend} = 0.02). Overall, the five rare missense SNPs explain an estimated 0.03% of excess familial risk of breast cancer. Four of the five ATM SNPs together with SNPs in BRCA1 and BRCA2 contributed to the increased risk observed in a study of 473 women with two breast cancers and 2463 controls (11). In addition to the 5 SNPs, there are 78 other rare missense SNPs in ATM listed in dbSNP (10). The distribution of ATM missense mutations and polymorphisms varies widely across ethnic groups (6). Interestingly, one study found that carriers of common ATM variants, such as Pro1054Arg (rs1800057), had a reduced risk of a second primary cancer in the contralateral breast (12). The authors speculate that certain ATM alleles may exert an antineoplastic effect by altering the activity of ATM as initiator of DNA damage response or regulator of p53. Immunohistochemical studies have shown that ATM protein expression is reduced in ER/PR/ERBB2-triple-negative breast cancers as well as in BRCA1- and BRCA2-deficient tumors (13).

References

1. Ahmed M, Rahman N. *ATM and breast cancer susceptibility.* Oncogene. 2006;25:5906-11.

2. Renwick A, Thompson D, Seal S, Kelly P, Chagtai T, Ahmed M, et al. *ATM mutations that cause ataxia-telangiectasia are breast cancer susceptibility alleles.* Nature Genet. 2006;38:873-5.

3. Swift M, Reitnauer PJ, Morrell D, Chase CL. *Breast and other cancers in families with ataxia-telangiectasia.* N Engl J Med. 1987;316:1289-94.

4. Thompson D, Duedal S, Kirner J, McGuffog L, Last J, Reiman A, et al. *Cancer risks and mortality in heterozygous ATM mutation carriers.* J Natl Cancer Inst. 2005;97:813-22.

5. Angele S, Romestaing P, Moullan N, Vuillaume M, Chapot B, Friesen M, et al. *ATM haplotypes and cellular response to DNA damage: association with breast cancer risk and clinical radiosensitivity.* Cancer Res. 2003;63:8717-25.

6. Bretsky P, Haiman CA, Gilad S, Yahalom J, Grossman A, Paglin S, et al. *The relationship between twenty missense ATM variants and breast cancer risk: the Multiethnic Cohort.* Cancer Epidemiol Biomarkers Prev. 2003;12:733-8.

7. Chen J, Birkholtz GG, Lindblom P, Rubio C, Lindblom A. *The role of ataxia-telangiectasia heterozygotes in familial breast cancer.* Cancer Res. 1998;58:1376-9.

8. Cox A, Dunning AM, Garcia-Closas M, Balasubramanian S, Reed MW, Pooley KA, et al. *A common coding variant in CASP8 is associated with breast cancer risk.* Nature Genet. 2007;39:352-8.

9. Lee KM, Choi JY, Park SK, Chung HW, Ahn B, Yoo KY, et al. *Genetic polymorphisms of ataxia telangiectasia mutated and breast cancer risk.* Cancer Epidemiol Biomarkers Prev. 2005;14:821-5.

10. Fletcher O, Johnson N, dos Santos Silva I, Orr N, Ashworth A, Nevanlinna H, et al. *Missense variants in ATM in 26,101 breast cancer cases and 29,842 controls.* Cancer Epidemiol Biomarkers Prev. 2010;19:2143-51.

11. Johnson N, Fletcher O, Palles C, Rudd M, Webb E, Sellick G, et al. *Counting potentially functional variants in BRCA1, BRCA2 and ATM predicts breast cancer susceptibility.* Hum Mol Genet. 2007;16:1051-7.

12. Concannon P, Haile RW, Borresen-Dale AL, Rosenstein BS, Gatti RA, Teraoka SN, et al. *Variants in the ATM gene associated with a reduced risk of contralateral breast cancer.* Cancer Res. 2008;68:6486-91.

13. Tommiska J, Bartkova J, Heinonen M, Hautala L, Kilpivaara O, Eerola H, et al. *The DNA damage signalling kinase ATM is aberrantly reduced or lost in BRCA1/BRCA2-deficient and ER/PR/ERBB2-triple-negative breast cancer.* Oncogene. 2008;27:2501-6.

8.6. OTHER GENES
8.6.1. BACH1 (BRIP1, FANCJ) AND PALB2 (FANCN)

Fanconi's anemia (FA) is a rare disorder of chromosomal instability associated with aplastic anemia in childhood, multiple congenital abnormalities, susceptibility to leukemia and other cancers (1). Cultured cells from FA patients treated with interstrand DNA cross-linking agents (e.g., cisplatin) become more susceptible to chromosome breakage, a response that forms the basis of the definitive diagnostic test for FA. There are 13 genes involved in FA (1, 2). The encoded proteins fall into several classes of enzymes and structural proteins that cooperate in the recognition and repair of damaged DNA, including a ubiquitin ligase, monoubiquinated proteins, a helicase, and one protein with both helicase and nuclease motifs. The genes are inactivated not only in FA but also in a variety of malignancies including breast cancer. Mutations in three FA genes have been associated with familial breast cancer: FANCD1, FANCJ, and FANCN. Surprisingly, FANCD1 is identical to BRCA2 (3). Biallelic germline mutations in BRCA2/FANCD1 cause FA whereas heterozygote carriers of a BRCA2/FANCD1 mutation have a high risk of familial breast or ovarian cancer. The tumors in heterozygotes result from loss of the second (wild-type) allele, leading to biallelic extinction of BRCA2/FANCD1. Paradoxically, breast or ovarian cancer rarely, if ever, develops in FA patients, most likely due to hypogonadism with decreased estrogen levels in women with FA (2). In families that carry a BRCA2/FANCD1 mutation, the peak age at the onset of breast or ovarian cancer among heterozygote carriers is in the fourth decade, whereas homozygotes (BRCA2/FANCD1-/-) often die from complications of aplastic anemia well before this age. Thus, timing of the bialleleic loss of BRCA2/FANCD1 determines the specific cancer spectrum with germline loss of both alleles leading to FA and secondary loss of the second allele later in life leading to breast/ovarian cancer (4).

BRCA1 recruits another BRCT-containing protein termed **BACH1 (for BRCA1-associated C-terminal helicase; also known as BRIP1 or FANCJ)** and is joined by RAD51 in homologous recombination repair or by RAD50-MRE11-NBS1 in non-homologous end-joining repair (5-7). Inactivating truncating mutations of BACH1/BRIP1/FANCJ have been identified in British women with breast cancer from BRCA1/BRCA2 mutation-negative families (8, 9). Similar to those in BRCA2, biallelic germline mutations of BACH1/BRIP1/FANCJ cause FA and monoallelic carriers are susceptible to early-onset breast cancer with a relative risk of 2.0 (95% CI 1.2 – 3.2; p = 0.012) (9). Mutational analysis of the BACH1/BRIP1/FANCJ gene in Chinese and French-Canadian women did not find truncating mutations but instead observed variants in the promoter and a non-synonymous mutation, Gln944Glu, with uncertain association to breast cancer risk (10, 11).

FANCN is identical with **PALB2 (partner and localizer of BRCA2)**, which independently interacts with BRCA1 and BRCA2 through its N- and C-termini, respectively. The N-terminus of PALB2 contains a coiled-coil region, which interacts with a coiled-coil region upstream of the BRCA1 repeats in BRCA1 (12-14). The C-terminus of PALB2 contains a β-propeller domain that binds to the N-terminal third of BRCA2 (15). Thus, BRCA1 can recruit PALB2, which in turn organizes BRCA2 and RAD51 (12-14). Abolishing the interaction of PALB2 with either BRCA2 or BRCA1 has been shown to impair double-stranded break repair by homologous recombination (12, 14, 15). These results highlight the physical and functional linkage of all three proteins in maintaining genomic stability and explain how familial breast cancer can result from germ-line mutations in each of the three genes (16-18). Inactivating truncating mutations of PALB2 have been identified in women with breast cancer from BRCA1/BRCA2 mutation-negative families (18-22). For example, a mutation in exon 4, 1592delT, results in a frame-shift at 531Leu, with the new reading frame progressing for 28 codons before termination. The truncated PALB2 protein had a markedly decreased BRCA2-binding affinity without affecting endogenous BRCA2 abundance upon transient

overexpression (21). Heterozygosity for loss-of-function PALB2 mutations increases risk of developing breast cancer 2- to 6-fold and also contributes to familial prostate cancer (20, 21, 23, 24). Tumors carrying the 1592delT mutation exhibited an aggressive phenotype with higher grade and more frequent triple negative (ER-, PR-, HER2-) status than those of other familial or sporadic breast cancer patients (25).

References

1. D'Andrea AD. *Susceptibility pathways in Fanconi's anemia and breast cancer.* N Engl J Med. 2010;362:1909-19.

2. Kennedy RD, D'Andrea AD. *The Fanconi Anemia/BRCA pathway: new faces in the crowd.* Genes Dev. 2005;19:2925-40.

3. Howlett NG, Taniguchi T, Olson S, Cox B, Waisfisz Q, De Die-Smulders C, et al. *Biallelic inactivation of BRCA2 in Fanconi anemia.* Science. 2002;297:606-9.

4. Berwick M, Satagopan JM, Ben-Porat L, Carlson A, Mah K, Henry R, et al. *Genetic heterogeneity among Fanconi anemia heterozygotes and risk of cancer.* Cancer Res. 2007;67:9591-6.

5. Cantor S, Drapkin R, Zhang F, Lin Y, Han J, Pamidi S, et al. *The BRCA1-associated protein BACH1 is a DNA helicase targeted by clinically relevant inactivating mutations.* Proc Natl Acad Sci U S A. 2004;101:2357-62.

6. Scully R, Chen J, Ochs RL, Keegan K, Hoekstra M, Feunteun J, et al. *Dynamic changes of BRCA1 subnuclear location and phosphorylation state are initiated by DNA damage.* Cell. 1997;90:425-35.

7. Zhong Q, Chen CF, Li S, Chen Y, Wang CC, Xiao J, et al. *Association of BRCA1 with the hRad50-hMre11-p95 complex and the DNA damage response.* Science. 1999;285:747-50.

8. De Nicolo A, Tancredi M, Lombardi G, Flemma CC, Barbuti S, Di Cristofano C, et al. *A novel breast cancer-associated BRIP1 (FANCJ/BACH1) germ-line mutation impairs protein stability and function.* Clin Cancer Res. 2008;14:4672-80.

9. Seal S, Thompson D, Renwick A, Elliott A, Kelly P, Barfoot R, et al. *Truncating mutations in the Fanconi anemia J gene BRIP1 are low-penetrance breast cancer susceptibility alleles.* Nature Genet. 2006;38:1239-41.

10. Cao AY, Huang J, Hu Z, Li WF, Ma ZL, Tang LL, et al. *Mutation analysis of BRIP1/BACH1 in BRCA1/*

BRCA2 negative Chinese women with early onset breast cancer or affected relatives. Breast Cancer Res Treat. 2009;115:51-5.

11. Guenard F, Labrie Y, Ouellette G, Joly Beauparlant C, Simard J, Durocher F. *Mutational analysis of the breast cancer susceptibility gene BRIP1/BACH1/FANCJ in high-risk non-BRCA1/BRCA2 breast cancer families.* J Hum Genet. 2008;53:579-91.

12. Sy SM, Huen MS, Chen J. *PALB2 is an integral component of the BRCA complex required for homologous recombination repair.* Proc Natl Acad Sci U S A. 2009;106:7155-60.

13. Zhang F, Fan Q, Ren K, Andreassen PR. *PALB2 functionally connects the breast cancer susceptibility proteins BRCA1 and BRCA2.* Mol Cancer Res. 2009;7:1110-8.

14. Zhang F, Ma J, Wu J, Ye L, Cai H, Xia B, et al. *PALB2 links BRCA1 and BRCA2 in the DNA-damage response.* Curr Biol. 2009;19:524-9.

15. Xia B, Sheng Q, Nakanishi K, Ohashi A, Wu J, Christ N, et al. *Control of BRCA2 cellular and clinical functions by a nuclear partner, PALB2.* Mol Cell. 2006;22:719-29.

16. Huen MS, Sy SM, Chen J. *BRCA1 and its toolbox for the maintenance of genome integrity.* Nature Rev Mol Cell Biol. 2010;11:138-48.

17. Moynahan ME, Jasin M. *Mitotic homologous recombination maintains genomic stability and suppresses tumorigenesis.* Nature Rev Mol Cell Biol. 2010;11:196-207.

18. Rahman N, Seal S, Thompson D, Kelly P, Renwick A, Elliott A, et al. *PALB2, which encodes a BRCA2-interacting protein, is a breast cancer susceptibility gene.* Nature Genet. 2007;39:165-7.

19. Cao AY, Huang J, Hu Z, Li WF, Ma ZL, Tang LL, et al. *The prevalence of PALB2 germline mutations in BRCA1/BRCA2 negative Chinese women with early onset breast cancer or affected relatives.* Breast Cancer Res Treat. 2009;114:457-62.

20. Casadei S, Norquist BM, Walsh T, Stray S, Mandell JB, Lee MK, et al. *Contribution of inherited mutations in the BRCA2-interacting protein PALB2 to familial breast cancer.* Cancer Res. 2011;71:2222-9.

21. Erkko H, Xia B, Nikkila J, Schleutker J, Syrjakoski K, Mannermaa A, et al. *A recurrent mutation in PALB2 in Finnish cancer families.* Nature. 2007;446:316-9.

22. Garcia MJ, Fernandez V, Osorio A, Barroso A, Llort G, Lazaro C, et al. *Analysis of FANCB and FANCN/ PALB2 fanconi anemia genes in BRCA1/2-negative Spanish breast cancer families.* Breast Cancer Res Treat. 2009;113:545-51.

23. Erkko H, Dowty JG, Nikkila J, Syrjakoski K, Mannermaa A, Pylkas K, et al. *Penetrance analysis of the PALB2 c.1592delT founder mutation.* Clin Cancer Res. 2008;14:4667-71.

24. Hollestelle A, Wasielewski M, Martens JW, Schutte M. *Discovering moderate-risk breast cancer susceptibility genes.* Curr Opin Genet Dev. 2010;20:268-76.

25. Heikkinen T, Karkkainen H, Aaltonen K, Milne RL, Heikkila P, Aittomaki K, et al. *The breast cancer susceptibility mutation PALB2 1592delT is associated with an aggressive tumor phenotype.* Clin Cancer Res. 2009;15:3214-22.

8.6.2. BARD1

The main binding partner of BRCA1 is BARD1 (for BRCA1-associated RING domain) (1). BRCA1 and BARD1 form a stable heterodimeric complex connected via their individual RING fingers. After DNA damage, BARD1 colocalizes with BRCA1 and the homologous DNA repair protein RAD51 to participate as a complex in damage repair (2, 3). Interestingly, germ-line BARD1 mutations (e.g., Cys557Ser) have been identified in rare hereditary breast cancers negative for BRCA1/BRCA2 mutations (4, 5). In patients positive for BRCA1/BRCA2 mutations, neither the Cys557Ser SNP nor additional BARD1 haplotypic variation not captured by Cys557Ser modified the BRCA1/BRCA2-associated cancer risk (6). Since BARD1 and BRCA1 are bound together and act in the same pathway, an inactivating mutation in BARD1 will not confer an additional breast cancer risk.

References

1. Baer R, Ludwig T. *The BRCA1/ BARD1 heterodimer, a tumor suppressor complex with ubiquitin E3 ligase activity.* Curr Opin Genet Dev. 2002;12:86-91.

2. Scully R, Chen J, Ochs RL, Keegan K, Hoekstra M, Feunteun J, et al. *Dynamic changes of BRCA1 subnuclear location and phosphorylation state are initiated by DNA damage.* Cell. 1997;90:425-35.

3. Westermark UK, Reyngold M, Olshen AB, Baer R, Jasin M, Moynahan ME. *BARD1 participates with BRCA1 in homology-directed repair of chromosome breaks.* Mol Cell Biol. 2003;23:7926-36.

4. Ghimenti C, Sensi E, Presciuttini S, Brunetti IM, Conte P, Bevilacqua G, et al. *Germline mutations of the BRCA1-associated ring domain (BARD1) gene in breast and breast/ovarian families negative for BRCA1 and BRCA2 alterations.* Genes Chromosomes Cancer. 2002;33:235-42.

5. Karppinen SM, Heikkinen K, Rapakko K, Winqvist R. *Mutation screening of the BARD1 gene: evidence for involvement of the Cys557Ser allele in hereditary susceptibility to breast cancer.* J Med Genet. 2004;41:e114.

6. Spurdle AB, Marquart L, McGuffog L, Healey S, Sinilnikova O, Wan F, et al. *Common genetic variation at BARD1 is not associated with breast cancer risk in BRCA1 or BRCA2 mutation carriers.* Cancer Epidemiol Biomarkers Prev. 2011;20:1032-8.

8.6.3. CHEK2

CHEK2 (also known as CHK2) is a cell cycle G2 checkpoint kinase involved in cell cycle control and DNA repair through its ability to phosphorylate p53, Cdc25c, and BRCA1 (1-3). A mutation, 1100delC, leads to a truncated CHEK2 that abrogates the kinase activity of the protein. CHEK2 1100delC was identified as a founder mutation in northern European countries and associated with a twofold increase in breast cancer risk (4-6). The frequency of the 1100delC allele varies widely, being highest in northern Europeans, rare in Mediterranean countries, and nonexistent in Asians (7). In northern Europeans, the variant allele was observed in 1.1% of healthy individuals and in 5.1% of breast cancer cases with a positive family history in which BRCA1 and BRCA2 mutations had been excluded (4). Other deleterious mutations including Ser428Phe have been identified in other populations (2).

References

1. Bartek J, Falck J, Lukas J. *CHK2 kinase--a busy messenger.* Nature Rev Mol Cell Biol. 2001;2:877-86.

2. Nevanlinna H, Bartek J. *The CHEK2 gene and inherited breast cancer susceptibility.* Oncogene. 2006;25:5912-9.

3. Wang HC, Chou WC, Shieh SY, Shen CY. *Ataxia telangiectasia mutated and checkpoint kinase 2 regulate BRCA1 to promote the fidelity of DNA end-joining.* Cancer Res. 2006;66:1391-400.

4. Meijers-Heijboer H, van den Ouweland A, Klijn J, Wasielewski M, de Snoo A, Oldenburg R, et al. *Low-penetrance susceptibility to breast cancer due to CHEK2(*)1100delC in noncarriers of BRCA1 or BRCA2 mutations.* Nature Genet. 2002;31:55-9.

5. Oldenburg RA, Kroeze-Jansema K, Kraan J, Morreau H, Klijn JG, Hoogerbrugge N, et al. *The CHEK2*1100delC variant acts as a breast cancer risk modifier in non-BRCA1/BRCA2 multiple-case families.* Cancer Res. 2003;63:8153-7.

6. *The CHEK2 Breast Cancer Case-Control Consortium. CHEK2*1100delC and susceptibility to breast cancer: a collaborative analysis involving 10,860 breast cancer cases and 9,065 controls from 10 studies.* Am J Hum Genet. 2004;74:1175-82.

7. Zhang S, Phelan CM, Zhang P, Rousseau F, Ghadirian P, Robidoux A, et al. *Frequency of the CHEK2 1100delC mutation among women with breast cancer: an international study.* Cancer Res. 2008;68:2154-7.

8.6.4. NBS1 AND NIJMEGEN BREAKAGE SYNDROME

The MRE11-RAD50-NBS1 (MRN) protein complex plays a central role in the cellular response to DNA damage by serving as primary sensor of double-stranded DNA breaks (1, 2). Homozygous or compound heterozygous germline mutations in the Nijmegen Breakage Syndrome (NBS1) gene predispose to the chromosomal instability Nijmegen breakage syndrome, which is associated with lymphoid malignancies and various types of cancer including breast cancer (3, 4). The NBS1 657del5 mutation is the most prevalent pathogenic NBS1 variant, implicated in 90% of NBS patients and 50% of NBS1 heterozygotes that develop cancer (4, 5). Heterozygous carriers of the 657del5 allele have an estimated threefold increased breast cancer risk, with higher risks for women younger than 50 years (6). Mutations have also been identified in the MRE11 (Arg633Stop) and RAD50 (687delT) genes in breast cancer cases from non-BRCA1/BRCA2 families (7-9). However, conclusive confirmation of mutations in these genes in other studies has not been reported (10).

References

1. Kang J, Ferguson D, Song H, Bassing C, Eckersdorff M, Alt FW, et al. *Functional interaction of H2AX, NBS1, and p53 in ATM-dependent DNA damage responses and tumor suppression.* Mol Cell Biol. 2005;25:661-70.

2. You Z, Chahwan C, Bailis J, Hunter T, Russell P. *ATM activation and its recruitment to damaged DNA require binding to the C terminus of Nbs1.* Mol Cell Biol. 2005;25:5363-79.

3. Lee JH, Xu B, Lee CH, Ahn JY, Song MS, Lee H, et al. *Distinct functions of Nijmegen breakage syndrome in ataxia telangiectasia mutated-dependent responses to DNA damage.* Mol Cancer Res. 2003;1:674-81.

4. Seemanova E, Jarolim P, Seeman P, Varon R, Digweed M, Swift M, et al. *Cancer risk of heterozygotes with the NBN founder mutation.* J Natl Cancer Inst. 2007;99:1875-80.

5. di Masi A, Antoccia A. *NBS1 Heterozygosity and Cancer Risk.* Curr Genomics. 2008;9:275-81.

6. Steffen J, Nowakowska D, Niwinska A, Czapczak D, Kluska A, Piatkowska M, et al. *Germline mutations 657del5 of the NBS1 gene contribute significantly to the incidence of breast cancer in Central Poland.* Int J Cancer. 2006;119:472-5.

7. Bartkova J, Tommiska J, Oplustilova L, Aaltonen K, Tamminen A, Heikkinen T, et al. *Aberrations of the MRE11-RAD50-NBS1 DNA damage sensor complex in human breast cancer: MRE11 as a candidate familial cancer-predisposing gene.* Mol Oncol. 2008;2:296-316.

8. Heikkinen K, Rapakko K, Karppinen SM, Erkko H, Knuutila S, Lundan T, et al. *RAD50 and NBS1 are breast cancer susceptibility genes associated with genomic instability.* Carcinogenesis. 2006;27:1593-9.

9. Tommiska J, Seal S, Renwick A, Barfoot R, Baskcomb L, Jayatilake H, et al. *Evaluation of RAD50 in familial breast cancer predisposition.* Int J Cancer. 2006;118:2911-6.

10. Hollestelle A, Wasielewski M, Martens JW, Schutte M. *Discovering moderate-risk breast cancer susceptibility genes.* Curr Opin Genet Dev. 2010;20:268-76.

8.6.5. PTEN AND COWDEN SYNDROME

Phosphatidylinositol-3 kinases (PI3Ks) phosphorylate phosphatidylinositol 4,5-biphosphate (PIP$_2$) at the 3 position of the inositol ring (1). The product, phosphatidylinositol 3,4,5-triphosphate (PIP$_3$), recruits the serine-threonine kinases AKT (cellular homolog of murine thymoma virus Akt8 oncoprotein) and phosphatidylinositol-dependent kinase 1 (PDK1) (2). In turn, PDK1 phosphorylates and thereby activates AKT, initiating the PI3K/AKT signaling pathway, which plays a key role in cell cycle progression and survival. The tumor suppressor gene PTEN (for Phosphatase and Tensin homologue deleted on chromosome Ten) at 10q23 encodes a phosphatase that hydrolyzes PIP$_3$, thus acting as catalytic antagonist of PI3K and opposing activation of the PI3K/AKT signaling pathway (3-5).

Both somatic and germline mutations have been identified in the PTEN gene. Somatic PTEN mutations occur in multiple tumors, including breast cancer and glioblastomas. Heterozygous germline mutations in PTEN are responsible for most cases of Cowden syndrome, a rare familial trait characterized by benign hamartomas (skin, breast, thyroid, oral mucosa, intestinal epithelium) and predisposition to breast and thyroid cancers (6, 7). Mutational analysis identified truncating or non-synonymous mutations in one allele, which were accompanied by loss of the wild-type allele. Single nucleotide polymorphisms and common haplotypes of PTEN were not associated with breast cancer risk (8). However, sporadic breast cancers frequently have somatic mutations or loss of at least one copy of the PTEN gene, which results in the activation of the proto-oncogenic phosphatidylinositol 3-kinase/AKT signaling pathway, supporting the concept of 'obligate haploinsufficiency' in which partial loss of a tumor suppressor gene is more tumorigenic than complete loss (9). The gene dose and level of PTEN expression also affects tumor progression and therapeutic response to trastuzumab (Herceptin), a humanized monoclonal antibody against HER2/neu, a membrane-receptor kinase that activates the AKT pathway (10).

References

1. Lempiainen H, Halazonetis TD. *Emerging common themes in regulation of PIKKs and PI3Ks.* EMBO J. 2009;28:3067-73.

2. Engelman JA, Luo J, Cantley LC. *The evolution of phosphatidylinositol 3-kinases as regulators of growth and metabolism.* Nature Rev Genet. 2006;7:606-19.

3. Di Cristofano A, Pandolfi PP. *The multiple roles of PTEN in tumor suppression.* Cell. 2000;100:387-90.

4. Li J, Yen C, Liaw D, Podsypanina K, Bose S, Wang SI, et al. *PTEN, a putative protein tyrosine phosphatase gene mutated in human brain, breast, and prostate cancer.* Science. 1997;275:1943-7.

5. Sansal I, Sellers WR. *The biology and clinical relevance of the PTEN tumor suppressor pathway.* J Clin Oncol. 2004;22:2954-63.

6. FitzGerald MG, Marsh DJ, Wahrer D, Bell D, Caron S, Shannon KE, et al. *Germline mutations in PTEN are an infrequent cause of genetic predisposition to breast cancer.* Oncogene. 1998;17:727-31.

7. Lynch ED, Ostermeyer EA, Lee MK, Arena JF, Ji H, Dann J, et al. *Inherited mutations in PTEN that are associated with breast cancer, Cowden disease, and juvenile polyposis.* Am J Hum Genet. 1997;61:1254-60.

8. Haiman CA, Stram DO, Cheng I, Giorgi EE, Pooler L, Penney K, et al. *Common genetic variation at PTEN and risk of sporadic breast and prostate cancer.* Cancer Epidemiol Biomarkers Prev. 2006;15:1021-5.

9. Berger AH, Knudson AG, Pandolfi PP. *A continuum model for tumour suppression.* Nature. 2011;476:163-9.

10. Pandolfi PP. *Breast cancer--loss of PTEN predicts resistance to treatment.* N Engl J Med. 2004;351:2337-8.

8.6.6. STK11 AND PEUTZ-JEGHERS SYNDROME

Germline mutations in the STK11 gene cause the Peutz-Jeghers syndrome, an autosomal dominant disease characterized by the development of characteristic polyps throughout the gastrointestinal tract, mucocutaneous pigmentation, and increased risk of malignancy, particularly breast and gastrointestinal cancer (1, 2).

References

1. Beggs AD, Latchford AR, Vasen HF, Moslein G, Alonso A, Aretz S, et al. *Peutz-Jeghers syndrome: a systematic review and recommendations for management.* Gut. 2010;59:975-86.

2. Boardman LA, Thibodeau SN, Schaid DJ, Lindor NM, McDonnell SK, Burgart LJ, et al. *Increased risk for cancer in patients with the Peutz-Jeghers syndrome.* Ann Int Med. 1998;128:896-9.

AK, et al. *Familial cancer associated with a polymorphism in ARLTS1.* N Engl J Med. 2005;352:1667-76.

3. Frank B, Hemminki K, Meindl A, Wappenschmidt B, Klaes R, Schmutzler RK, et al. *Association of the ARLTS1 Cys148Arg variant with familial breast cancer risk.* Int J Cancer. 2006;118:2505-8.

4. Egan KM, Newcomb PA, Ambrosone CB, Trentham-Dietz A, Titus-Ernstoff L, Hampton JM, et al. *STK15 polymorphism and breast cancer risk in a population-based study.* Carcinogenesis. 2004;25:2149-53.

5. Fletcher O, Johnson N, Palles C, dos Santos Silva I, McCormack

V, Whittaker J, et al. *Inconsistent association between the STK15 F31I genetic polymorphism and breast cancer risk.* J Natl Cancer Inst. 2006;98:1014-8.

6. Ding SL, Yu JC, Chen ST, Hsu GC, Shen CY. *Genetic variation in the premature aging gene WRN: a case-control study on breast cancer susceptibility.* Cancer Epidemiol Biomarkers Prev. 2007;16:263-9.

7. Wirtenberger M, Frank B, Hemminki K, Klaes R, Schmutzler RK, Wappenschmidt B, et al. *Interaction of Werner and Bloom syndrome genes with p53 in familial breast cancer.* Carcinogenesis. 2006;27:1655-60.

8.6.7. OTHER LOW-PENETRANCE BREAST CANCER SUSCEPTIBILITY GENES

There are several other putative breast cancer susceptibility genes, which have been implicated by some studies but not confirmed by others. These genes have in common variant alleles that are associated with moderate- or low- penetrance of breast cancer risk.   To prove the risk association conclusively would require mutational screening of large numbers of breast cancer cases and controls and require p-values of 0.01 or better (1). The following are examples of candidate breast cancer susceptibility genes. **ARLTS1** (ADP-ribosylation factor-like tumor suppressor gene 1) is a member of the ADP-ribosylation factor (ARF) family. A polymorphism, Trp149Stop, results in a truncated ARLTS1 protein, which has been shown to induce lower levels of apoptosis and predispose to familial cancer including familial breast cancer (2, 3). The **STK15** gene encodes a serine/threonine kinase that acts as a key regulator of mitotic chromosome segregation. There is an inconsistent association of a common genotype, combining two non-synonymous polymorphisms, Phe31Ile (rs2273535) and Val57Ile (rs1047972), with increased risk of breast cancer (4, 5). Germline mutations in the **WRN** gene (encoding RECQL2, a DNA helicase), cause Werner syndrome, a premature aging disease associated with genomic instability and increased cancer incidence. There is an inconsistent association of single nucleotide polymorphisms (rs1346044, Cys1367Arg; rs9649886, 46729A/C) with increased risk of breast cancer (6, 7).

References

1. Hollestelle A, Wasielewski M, Martens JW, Schutte M. *Discovering moderate-risk breast cancer susceptibility genes.* Curr Opinion Genet

Develop. 2010;20:268-76.

2. Calin GA, Trapasso F, Shimizu M, Dumitru CD, Yendamuri S, Godwin

8.7. GENOME-WIDE ASSOCIATION STUDIES

In contrast to studies of known candidate genes, the primary goal of genome-wide association studies (GWAS) is the discovery of novel genetic associations, possibly in genes with unknown functions. The identification of DNA sequence variants in GWAS typically involves single nucleotide polymorphisms (SNPs). Usually >3000 SNPs spread across the genome fairly evenly are tested without an existing hypothesis about a particular gene or locus (1). The selected SNPs are relatively common with minor allele frequencies above 5% and show low-penetrance, which is reflected in relative risks of 1.05 – 1.3 (2). SNPs showing an association with disease are scrutinized in an independent sample set and those select SNPs showing an unequivocal association in replication experiments are subjected to fine mapping of the locus to determine the pathologically relevant gene. Only 3% of disease-associated polymorphisms in GWAS are non-synonymous SNPs, which can be subjected to functional follow-up studies to establish disease causality (3). Synonymous SNPs are commonly ignored because these silent polymorphisms do not change the amino acid sequence or structure of a protein. However, changes in base sequence and codon usage may affect mRNA splicing and folding or modify protein expression whereas microRNA has been shown to target mRNA and regulate its translation (4, 5). The great majority of SNPs identified in GWAS are not in the coding region of a gene. Approximately 40% of disease-associated SNPs fall in intergenic regions and another 40% are located in non-coding introns (2).   It is difficult to prove conclusively that a SNP located in an intron has a deleterious effect and is responsible for a disease. An example is a SNP (rs2981582) identified by GWAS in intron 2 of the fibroblast growth factor receptor 2 (FGFR2) gene to be associated with breast cancer risk

(6-8). The high-risk allele of SNP rs2981582 carried a relative risk of 1.26 per allele as compared with the low-risk allele. However, another study found no association of rs2981582 with breast cancer risk, except in subgroups of ER+, PR+, HER- tumors (9). Fine-scale mapping of intron 2 identified another SNP, rs2981578, to be associated with breast cancer risk (10). Rs2981578 has been implicated in up-regulating FGFR2 expression and mapped to highly accessible chromatin by DNase I hypersensitive site analysis (8, 10). Genome-wide association studies have also been performed to identify DNA methylation changes in genomic CpG islands to discover associations between epigenetic DNA changes and breast cancer characteristics (11).

Several GWAS have been performed to discover new genes relevant to breast cancer (6, 7, 12-20). Over 40 SNPs have been identified (21, 22). Three GWAS based on stringent levels of statistical significance have revealed SNPs in six genes encoding four known and two unidentified proteins (6, 7, 17, 23). The four known genes were FGFR2, TNRC9, MAP3K1, and LSP1. Overall, the increased risk of breast cancer associated with carrying a single copy of any of the risk alleles was small, ranging from 1.07 to 1.26. However, the population prevalence of each risk allele is high, ranging from 28 to 87%. Because the predisposing alleles are common, despite the low risks they confer, their contribution to the familial risk of breast cancer is not insignificant, estimated to account for 3.9% of the familial risk of breast cancer in European populations (24).

Overall, the GWAS results provide support for the existence of polygenic influences on breast cancer risk, in which the inherited susceptibility is due to a large number of cumulative weak effects contributed by many variants. At the same time, the genetic contribution to breast cancer risk estimated in a GWAS from statistically significant SNPs reveals only a small fraction of the heritability, not because heritability is missing but because most causal variants are not appropriately tagged by measured SNPs and hidden in the random noise contributed by the numerous SNPs that are irrelevant for breast cancer (25). A consequence of this finding is that larger studies will always identify new associations, while the contribution of newly identified loci decreases as study size increases (26). Additional reasons for the relatively small contribution of individual SNPs to heritability could be gene-environment interactions or more complex pathways involving multiple genes and exposures (27). A pathway-based analysis of SNPs identified in GWAS revealed three pathways that were highly enriched with association signals: syndecan-1-, hepatocyte growth factor receptor-, and growth hormone-mediated signaling (28). Clustering analysis revealed that pathways containing key components of the RAS/RAF/mitogen-activated protein kinase canonical signaling

cascade were significantly more likely to have an excess of association signals than expected by chance.

Breast cancer risk models based solely on genetic information obtained from GWAS yielded receiver operator characteristics (ROC) AUCs of 0.55 for a combination of seven polymorphisms associated with breast cancer risk (23, 29). A study comparing the Gail model alone and in combination with seven SNPs yielded AUCs of 0.557 and 0.594, respectively, showing that the combined model was more discriminating (30). A similar comparison study with ten SNPs obtained AUCs of 0.58 and 0.618, respectively (19). Yet another study assessed multiple SNPs associated with breast cancer risk and obtained AUCs of 0.67 for 41 polymorphisms and 0.68 for 96 polymorphisms (31). Thus, newly discovered genetic factors alone or included in existing risk models modestly improved the performance of the latter.

References

1. Hardy J, Singleton A. *Genomewide association studies and human disease.* N Engl J Med. 2009;360:1759-1768.

2. Manolio TA. *Genomewide association studies and assessment of the risk of disease.* N Engl J Med. 2010;363:166-176.

3. Freedman ML, Monteiro AN, Gayther SA, Coetzee GA, Risch A, Plass C, et al. *Principles for the post-GWAS functional characterization of cancer risk loci.* Nature Genet. 2011;43:513-518.

4. Burns DM, D'Ambrogio A, Nottrott S, Richter JD. *CPEB and two poly(A) polymerases control miR-122 stability and p53 mRNA translation.* Nature. 2011;473:105-108.

5. Hurst LD. *Molecular genetics: The sound of silence.* Nature. 2011;471:582-583.

6. Easton DF, Pooley KA, Dunning AM, Pharoah PDP, Thompson D, Ballinger DG, et al. *Genome-wide association study indentifies novel breast cancer susceptibility loci.* Nature. 2007;447:1087-1095.

7. Hunter DJ, Kraft P, Jacobs KB, Cox DG, Yeager M, Hankinson SE, et al. *A genome-wide association study identifies alleles in FGFR2 associated with risk of sporadic postmenopausal breast cancer.* Nature Genet. 2007;39:870-874.

8. Meyer KB, Maia AT, O'Reilly M, Teschendorff AE, Chin SF, Caldas C, et al. *Allele-specific up-regulation of FGFR2 increases susceptibility to breast cancer.* PLoS Biol. 2008;6:e108.

9. Rebbeck TR, DeMichele A, Tran TV, Panossian S, Bunin GR, Troxel AB, et al. *Hormone-dependent effects of FGFR2 and MAP3K1 in breast cancer susceptibility in a population-based sample of post-menopausal African-American and European-American women.* Carcinogenesis. 2009;30:269-274.

10. Udler MS, Meyer KB, Pooley KA, Karlins E, Struewing JP, Zhang J, et al. *FGFR2 variants and breast cancer risk: fine-scale mapping using African American studies and analysis of chromatin conformation.* Hum Mol Genet. 2009;18:1692-1703.

11. Hill VK, Ricketts C, Bieche I, Vacher S, Gentle D, Lewis C, et al. *Genome-wide DNA methylation profiling of CpG islands in breast cancer identifies novel genes associated with tumorigenicity.* Cancer Res. 2011;71:2988-2999.

12. Cai Q, Wen W, Qu S, Li G, Egan KM, Chen K, et al. *Replication and functional genomic analyses of the breast cancer susceptibility locus at 6q25.1 generalize its importance in women of Chinese, Japanese, and European ancestry.* Cancer Res. 2011;71:1344-1355.

13. Higginbotham KS, Breyer JP, Bradley KM, Schuyler PA, Plummer WD, Freudenthal ME, et al. *A multistage association study identifies a breast cancer genetic locus at NCOA7.* Cancer Res. 2011;71:3881-3888.

14. Hutter CM, Young AM, Ochs-Balcom HM, Carty CL, Wang T, Chen CT, et al. *Replication of breast cancer GWAS susceptibility loci in the Women's Health Initiative*

African American SHARe Study. Cancer Epidemiol Biomarkers Prev. 2011;20:1950-1959.

15. Long J, Shu XO, Cai Q, Gao YT, Zheng Y, Li G, et al. *Evaluation of breast cancer susceptibility loci in Chinese women.* Cancer Epidemiol Biomarkers Prev. 2010;19:2357-2365.

16. Song H, Koessler T, Ahmed S, Ramus SJ, Kjaer SK, Dicioccio RA, et al. *Association study of prostate cancer susceptibility variants with risks of invasive ovarian, breast, and colorectal cancer.* Cancer Res. 2008;68:8837-8842.

17. Stacey SN, Manolescu A, Sulem P, Rafnar T, Gudmundsson J, Gudjonsson SA, et al. *Common variants on chromosomes 2q35 and 16q12 confer susceptibility to estrogen receptor-positive breast cancer.* Nature Genet. 2007;39:865-869.

18. Turnbull C, Ahmed S, Morrison J, Pernet D, Renwick A, Maranian M, Seal S, et al. *Genome-wide association study identifies five new breast cancer susceptibility loci.* Nature Genet. 2010;42:504-507.

19. Wacholder S, Hartge P, Prentice R, Garcia-Closas M, Feigelson HS, Diver WR, et al. *Performance of common genetic variants in breast-cancer risk models.* N Engl J Med. 2010;362:986-993.

20. Zheng W, Wen W, Gao YT, Shyr Y, Zheng Y, Long J, et al. *Genetic and clinical predictors for breast cancer risk assessment and stratification among Chinese women.* J Natl Cancer Inst. 2010;102:972-981.

21. Peng S, Lu B, Ruan W, Zhu Y, Sheng H, Lai M. *Genetic polymorphisms and breast cancer risk: evidence from meta-analyses, pooled analyses, and genome-wide association studies.* Breast Cancer Res Treat. 2011;127:309-324.

22. Zhang B, Beeghly-Fadiel A, Long J, Zheng W. *Genetic variants associated with breast-cancer risk: comprehensive research synopsis, meta-analysis, and epidemiological evidence.* Lancet Oncol. 2011;12:477-488.

23. Pharoah PD, Antoniou AC, Easton DF, Ponder BA. *Polygenes, risk prediction, and targeted prevention of breast cancer.* N Engl J Med. 2008;358:2796-2803.

24. Stratton MR, Rahman N. *The emerging landscape of breast cancer susceptibility.* Nature Genet. 2008;40:17-22.

25. Frazer KA, Murray SS, Schork NJ, Topol EJ. *Human genetic variation and its contribution to complex traits.* Nature Rev Genet. 2009;10:241-251.

26. Cambien F. *Heritability, weak effects, and rare variants in genomewide association studies.* Clin Chem. 2011;57:1263-1266.

27. Thomas D. *Methods for investigating gene-environment interactions in candidate pathway and genome-wide association studies.* Annu Rev Public Health. 2010;31:21-36.

28. Menashe I, Maeder D, Garcia-Closas M, Figueroa JD, Bhattacharjee S, Rotunno M, et al. *Pathway analysis of breast cancer genome-wide association study highlights three pathways and one canonical signaling cascade.* Cancer Res. 2010;70: 4453-4459

29. Gail MH. *Discriminatory accuracy from single-nucleotide polymorphisms in models to predict breast cancer risk.* J Natl Cancer Inst. 2008;100:1037-1041.

30. Mealiffe ME, Stokowski RP, Rhees BK, Prentice RL, Pettinger M, Hinds DA. *Assessment of clinical validity of a breast cancer risk model combining genetic and clinical information.* J Natl Cancer Inst. 2010;102:1618-1627.

31. van Zitteren M, van der Net JB, Kundu S, Freedman AN, van Duijn CM, Janssens AC. *Genome-based prediction of breast cancer risk in the general population: a modeling study based on meta-analyses of genetic associations.* Cancer Epidemiol Biomarkers Prev. 2011;20:9-22.

Epilog

EPILOG

Breast cancer has more than one etiology and is caused by endogenous and exogenous, phenotypic and genotypic factors. It is not a single river but a stream fed by many tributaries and hence difficult to control once it reaches the surface. In the 1960s, estrogens were identified as risk factors for the development of breast cancer (1). While the identification of estrogens and estrogen-related reproductive factors (e.g., early menarche, late menopause, hormone replacement therapy) helped in defining groups of women at risk, little progress was made in narrowing the risk prediction for individual women. Little progress was also made in the definition of estrogen-unrelated risk factors until the 1990s when BRCA1 and 2 were discovered (2, 3). However, even the discovery of additional genes (e.g., p53, ATM, CHK2, NBS1, PALB2, PTEN) did not fulfill the expectation to at least explain the etiology of familial breast cancer. It is now widely accepted that breast cancer suscep-tibility is partly due to low-penetrance genetic variants and gene-environment interactions. However, an estimated 70% of inherited tumors have no known cause and an even higher percentage of sporadic breast cancers lack clearly defined risk factors. In other words, the majority of breast cancers occur unexpectedly, contributing to the frightful diagnosis of breast cancer.

To remedy this situation, risk assessment has to cast a wide net and consider multiple etiologic factors at the same time to provide a useful risk prediction for the individual woman. This presents the challenge of combining etiologic factors, which we have attempted in a genotypic-phenotypic model of breast cancer risk determination (4). To cast the net even wider, I propose to employ a systems biology approach based on the assumption that breast cancer is caused by an imbalance of DNA damage, DNA repair, cell proliferation, and apoptosis summarized in the following simplistic 'equation' of its etiology:

estrogens, and stimulating cell proliferation and gene expression via the estrogen receptor (5-7). Thus, estrogens and their oxidative metabolites are unique carcinogens that affect both tumor initiation and promotion. As illustrated in Figure 1, genotypic variants of key enzymes involved in the generation and detoxification of $4\text{-}OHE_2$ are incorporated into a traditional cumulative estrogen exposure model (phenotypic factors) to predict breast cancer risk.

The current model (4) used experimentally determined rate constants to simulate kinetic conversion of E_2 to $4\text{-}OHE_2$. Using rate constants of genetic variants of CYP1A1, CYP1B1, and COMT, the model further allowed examination of the kinetic impact of enzyme polymorphisms on the entire estrogen metabolic pathway. The levels of $4\text{-}OHE_2$, represented by calculated areas under the curve (AUC), serve as an indicator of exposure to the oxidative metabolites of E_2. This model was applied to two popula-tion-based breast cancer case-control groups with the goal of determining whether the calculated level of $4\text{-}OHE_2$ was associated with relative risk. The two population studies were the German GENICA study (967 cases, 971 controls) and the Nashville Breast Cohort (NBC; 465 cases, 885 controls). In the GENICA study, premenopausal women at the 90th percentile of $4\text{-}OHE_2$-AUC among control subjects had a risk of breast cancer that was 2.30 times that of women at the 10th control $4\text{-}OHE_2$-AUC percentile (95% CI: 1.7–3.2, p = 2.9×10^{-7}). This relative risk was 1.89 (95% CI: 1.5–2.4, p = 2.2×10^{-8}) in postmenopausal women. In the NBC, this relative risk in postmenopausal women was 1.81 (95% CI: 1.3–2.6, p = 7.6×10^{-4}), which increased to 1.83 (95% CI: 1.4–2.3, p = 9.5×10^{-7}) when a history of proliferative breast disease was included in the model. These findings demonstrate the potential of this genotype-phenotype model for predicting breast cancer risk and provide additional evidence for a role of the

$$(DNA\ damage - DNA\ repair) + (cell\ proliferation - apoptosis) = breast\ cancer$$

The goal will be to meld the genotypic-phenotypic model with the systems biological approach (DNA damage, DNA repair, cell proliferation, apoptosis) for a truly comprehensive risk assessment. In the following, I will sketch out the components of this new over-riding model starting with a brief description of the phenotypic-genotypic component centered on estrogens. Estrogens are the most important and best understood risk factor warranting an 'estrocentric' approach to risk modeling. What explains the predominant role of estrogens in mammary carcino-genesis? They play a dual role as enzyme substrate and receptor ligand, simultaneously causing DNA damage via their oxidation products, the 2-OH and 4-OH catechol

oxidative metabolism of E_2/E_1 via the 4-OH-catechol/4-quinone pathway in breast cancer. Other enzymes and their polymorphisms can readily be incorporated into the model. An example is CYP19A1 (aromatase), which catalyzes the formation of E_2 from testosterone and E_1 from androstene-dione (Figure 1). Additional support for the carcinogenic role of the 4-OH-catechol/4-quinone pathway is provided by a prospective case-control study nested within the Prostate, Lung, Colorectal, and Ovarian Cancer Screening Trial (PLCO; 277 breast cancer cases, 423 controls) (8). An increased ratio of the 4-hydroxylation pathway catechols to their inactive methylated derivatives was associated with an increased risk of breast cancer (1.34, 95% CI 1.04-1.72).

Reproductive and Lifestyle Factors

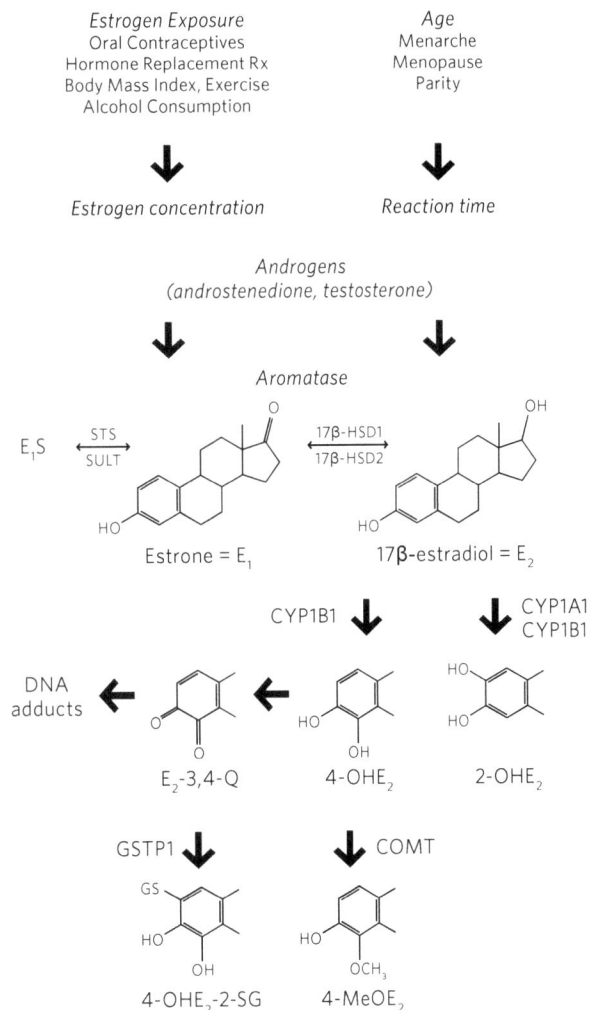

Estrogen Exposure
Oral Contraceptives
Hormone Replacement Rx
Body Mass Index, Exercise
Alcohol Consumption

Age
Menarche
Menopause
Parity

Estrogen concentration

Reaction time

Androgens
(androstenedione, testosterone)

Aromatase

E_1S ⇄ STS / SULT

Estrone = E_1

17β-HSD1
17β-HSD2

17β-estradiol = E_2

CYP1B1

CYP1A1
CYP1B1

DNA adducts

E_2-3,4-Q

4-OHE$_2$

2-OHE$_2$

GSTP1

COMT

4-OHE$_2$-2-SG

4-MeOE$_2$

Haplotype Configuration	in silico Model	
Aromatase $h_{1,1}/h_{1,2}$	Aromatase variable expression	
17β-HSD1 $h_{2,1}/h_{2,2}$	17β-HSD1 kat_1, Km_1 kat_2, Km_2	
SULT1E1 $h_{3,1}/h_{3,2}$	SULT1E1 variable expression	
CYP1A1 $h_{4,1}/h_{4,2}$	CYP1A1 kat_1, Km_1 kat_2, Km_2	4-OHE$_2$ production
CYP1B1 $h_{5,1}/h_{5,2}$	CYP1B1 kat_1, Km_1 kat_2, Km_2	
COMT $h_{6,1}/h_{6,2}$	COMT kat_1, Km_1 kat_2, Km_2	
GSTP1 $h_{7,1}/h_{7,2}$	GSTP1: kat_1, Km_1 kat_2, Km_2	

On the other hand, the ratio of the 2-hydroxylation pathway to parent estrogens (E_1/E_2) was associated with a decreased risk of breast cancer (0.66, 95% CI 0.51–0.87).

Although these data support the carcinogenic role of estrogen metabolites, the development of breast cancer is more complex than the 4-hydroxylation pathway portrays. Tumor initiation depends not only on DNA damage but also on DNA repair, two concurrent processes that are linked rather than separate as illustrated in Figure 2.

The principal enzyme controlling estrogen metabolism in breast tissue is CYP1B1, which sequentially oxidizes E_2 to 4-OHE$_2$ and the quinone E_2-3,4-Q. The highly reactive quinone nonenzymatically attacks DNA and forms covalent adducts with bases, such as 4-OHE$_2$-N7-Gua (Gua$_{adduct}$), which undergoes depurination by hydrolysis of the N-glycosylic bond between base and sugar, leaving an apurinic (AP) site in the double-stranded DNA. AP sites are repaired by the base excision repair (BER) pathway in a sequence of enzymatic reactions. AP endonuclease APE1 cleaves 5' to the AP site to generate a nick with 3'-OH and

Figure 1. Overview of breast cancer risk prediction model based on genotypic-phenotypic assessment of mammary estrogen metabolism and exposure to carcinogenic estrogen metabolites.
In the center is the estrogen metabolism pathway used to generate the proteomic-genomic component of the model. Androgens are converted by aromatase (CYP19A1) to the parent estrogens, E_1 and E_2. E_1 is also produced from estrone sulfate (E_1S) by STS and further converted to E_2 by 17β-HSD1. The reverse pathway is catalyzed by 17β-HSD2 and SULT1E1. In the main pathway, E_1 and E_2 are oxidized by CYP1A1 and CYP1B1 to catechol estrogens (e.g., 2-OHE$_2$ and 4-OHE$_2$). The catechol estrogens are either methylated by COMT to methoxyestrogens (e.g., 4-MeOE$_2$) or further oxidized by the CYPs to quinones, e.g., E_2-3,4-Q, the main quinone that forms depurinating estrogen DNA adducts. Alternatively, the quinones are conjugated by GSTs, primarily GSTP1, thereby preventing adduct formation. Each of the genes encoding variant enzymes is genotyped for all subjects and the SNP genotype data used to derive the haplotype configuration for each subject. The model then uses the kinetic constants in a system of nonlinear differential equations to calculate the production of the main carcinogenic estrogen metabolite, 4-OHE$_2$, for each haplotype configuration as well as the weighted average of all 4-OH-E$_2$ production values; the haplotype configuration probabilities are used as weights in these calculations. The genotype model is influenced by traditional risk factors, which are thought to affect hormone concentration (OC, HRT, BMI, exercise, alcohol consumption) and exposure time (ages at menarche and menopause, parity). The weight of each phenotypic factor on hormone concentration is determined by maximum likelihood estimation. Thus, the combined genotypic-phenotypic model allows the calculation of 4-OHE$_2$ produced by each woman, expressed as 4-OHE$_2$ area under the curve metric (4-OHE$_2$-AUC), for a personalized risk estimate of developing breast cancer. See reference (4) for details.

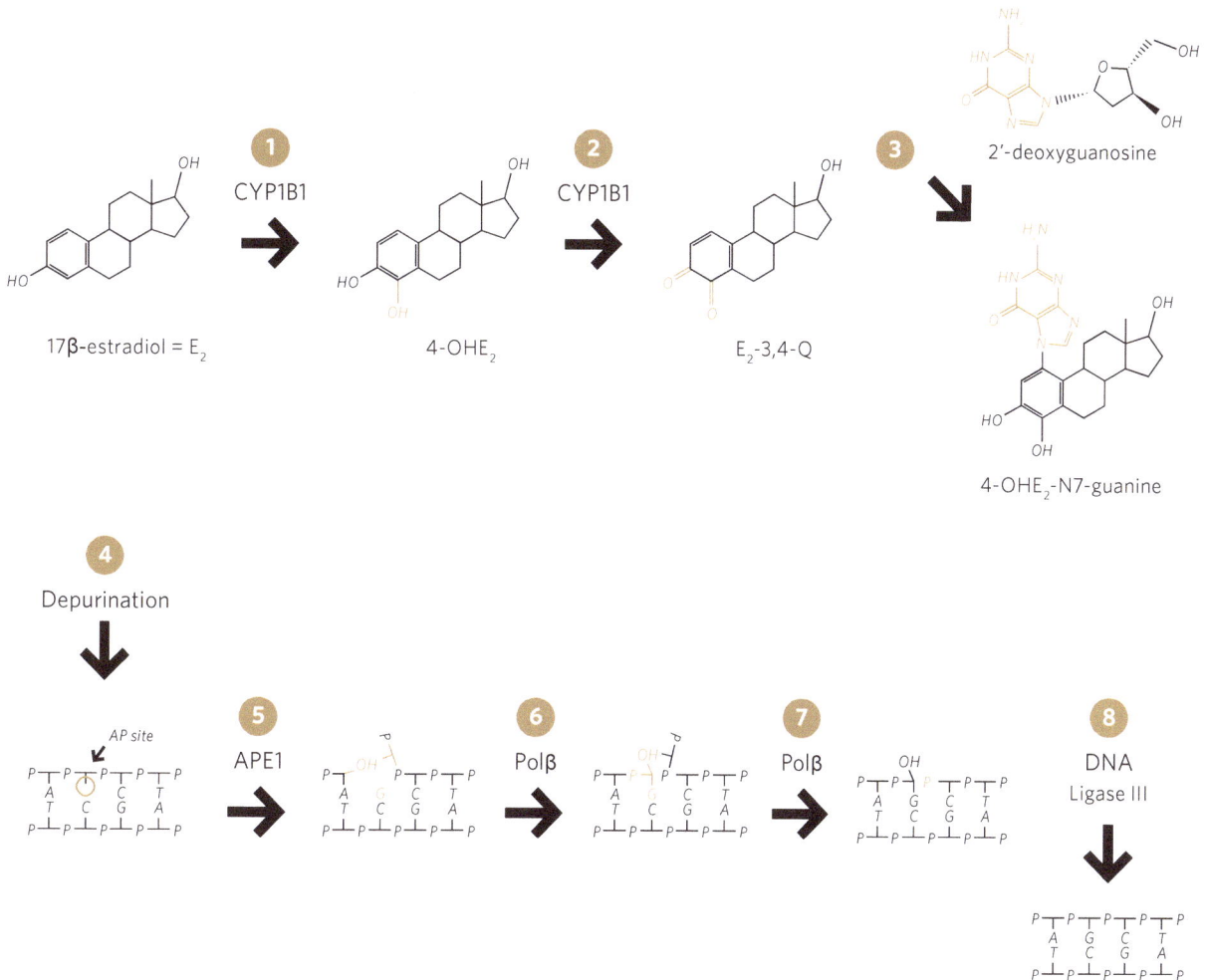

Figure 2. Overview of estrogen metabolism-induced DNA adduct formation and subsequent DNA base excision repair (BER). *The top sequence of reactions shows the CYP1B1-mediated oxidative estrogen metabolism resulting in estrogen-deoxyribonucleoside adduct formation. (1) CYP1B1 catalyzes the oxidation of E_2 to the catechol estrogen 4-OHE_2 and (2) further to the estrogen quinone, E_2-3,4-Q. (3) The highly reactive quinone nonenzymatically attacks DNA and forms covalent adducts with bases such as 4-OHE_2-N7-guanine, which (4) undergoes depurination by hydrolysis of the N-glycosylic bond between base and sugar, leaving an apurinic (AP) site in the double-stranded DNA. (5) AP endonuclease Ape1 cleaves 5' to the AP site, creating a single-strand break. The following sequence of reactions shows the BER with its predominant short-patch pathway (6) DNA polymerase β (Polβ) attaches dGTP to the newly generated 3'-OH and thereby displaces the baseless sugar-phosphate, which (7) Polβ subsequently removes by deploying its inherent AP-lyase activity. (8) DNA ligase III seals the remaining nick to restore the original DNA sequence. Not shown is the formation of the catechol estrogen 2-OHE_2, which yields stable adducts, and the redox cycling of catechol estrogens and estrogen quinones, which generates reactive oxygen species capable of producing oxidative DNA adducts. Also not shown are the alternate long-patch BER pathway and a variant of the short-patch pathway involving a bifunctional DNA glycosylase/AP-lyase.*

5'-deoxyribose phosphate termini. DNA polymerase (Pol) catalyzes the release of the deoxyribose phosphate residue and DNA synthesis to fill the gap, which is then sealed by DNA ligase.

We have developed a model for the formation of DNA adducts from oxidative estrogen metabolism followed by BER of these adducts (9). The model represents the sequence of enzymatic reactions in both damage and repair pathways. By combining both pathways, we could simulate the overall process by starting from a given time-dependent concentration of E_2 and 2'-deoxyguanosine, determine the extent of adduct formation and the correction by BER required to preserve the integrity of DNA. The model allowed us to examine the effect of phenotypic and genotypic factors such as different concentrations of estrogen and variant enzyme haplotypes on the formation and repair of DNA adducts (Figure 3).

Of course, the development of breast cancer is more complex than envisioned in the two-stage process of initiation and promotion. As initiation is a composite of DNA damage and concurrent DNA repair, promotion is the sum of cell proliferation and cell death or apoptosis. Moreover, all four processes are linked rather than separate. The maintenance of genomic integrity following DNA damage depends on the coordination of DNA repair with cell cycle checkpoint controls (10, 11). The integrity of the DNA damage response pathway is crucial for the prevention of neoplastic proliferation, as suggested by the fact that many proteins involved in these pathways are tumor suppressors such as BRCA1, BRCA2, ATM, and p53 and play roles in both DNA repair and apoptosis.

Genome-wide association studies (GWAS) have identified over 40 SNPs related to breast cancer risk (12-15). I have categorized the affected genes in Table 1 according to the role of the encoded proteins in each of the four processes, i.e., DNA damage, DNA repair, cell proliferation, and apoptosis with the understanding that a protein (e.g., ATM) may be involved in more than one pathway. Some SNPs are categorized as 'other genes' although they could be assigned to DNA repair (e.g., MTR, MTHFR, and TYMS play key roles in folate metabolism and DNA synthesis). Overall, the increased risk of breast cancer associated with carrying a single copy of any of the risk alleles was small, ranging from 1.07 to 1.26. However, the population prevalence of each risk allele is high, ranging from 28 to 87%. Because the predisposing alleles are common, despite the low risks they confer, their contribution to breast cancer is not insignificant (16). Moreover, the GWAS results provide support for the existence of polygenic influences on breast cancer risk, in which the inherited susceptibility is due to a large number of cumulative weak effects contributed by many variants (17-19). This is illustrated in Figure 3, Panel C for the augmented effect of combined CYP1B1 and APE1 variants on the formation and repair of DNA adducts.

Contemporary ideas of carcinogenesis envisage a series of stochastic genetic changes that confer a selective growth advantage over healthy cells. The concurrent changes brought about by multiple SNPs in proteins involved in DNA damage, DNA repair, cell proliferation, and apoptosis are expected to lead to fundamental change in cellular behavior with progressive dysregulation and acquisition of a malignant phenotype (20). We propose a comprehensive model that combines these genetic characteristics based on systems biology with phenotypic parameters to predict breast cancer risk for the individual woman.

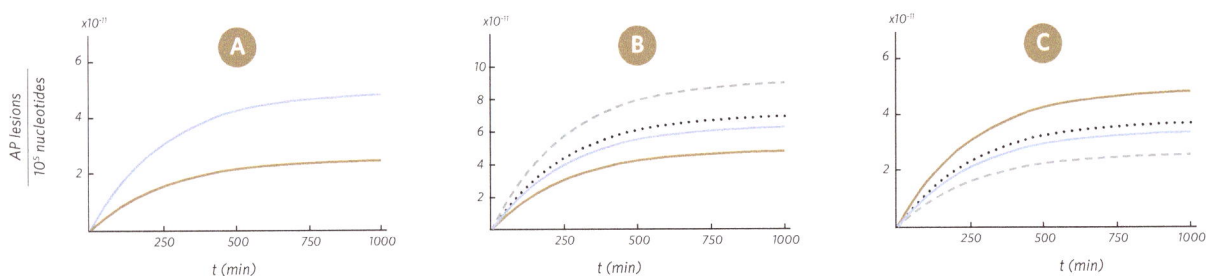

Figure 3. Effects of different concentrations of estrogen and variant haplotypes of CYP1B1 and APE1 on the formation and repair of DNA adducts.
Panel A shows the effect of doubling the concentration of E_2 on the number of AP lesions per 10^5 nucleotides over time using $E_2 = 1$ (gold curve) and $E_2 = 2$ (blue curve). Panel B shows the effect of variant haplotypes of CYP1B1 and APE1 leading to an increase in the number of AP lesions per 10^5 nucleotides relative to wild-type enzymes. Each curve uses different assumptions about the kinetic constant, k_{cat}, for the CYP1B1 and APE1 enzymes. The gold curve depicts the wild-types with $k_{cat1} = 1.6$ (CYP1B1) and $k_{cat5} = 192.168$ (APE1). The blue curve uses a 30% increase in k_{cat1}, keeping $k_{cat5} = 192.168$. The dotted curve employs a 30% reduction in k_{cat5}, keeping $k_{cat5} = 192.168$. Finally, the dashed curve uses both a 30% increase in k_{cat1} and a 30% reduction in k_{cat5}. Panel C shows the effect of different variant haplotypes of CYP1B1 and APE1 that cause a decrease in the number of AP lesions per 10^5 nucleotides relative to wild-type enzymes. The wild-type is shown in gold, a 30% decrease of CYP1B1 activity with APE1 activity held at its wild-type level in blue, 30% increase of APE1 activity with CYP1B1 activity held at its wild-type level (dotted), and a 30% decrease in CYP1B1 activity with a 30% increase of APE1 activity (dashed). See reference (9) for details.

Table 1. Genetic Variants with a Significant Association with Breast Cancer Risk (12-15)

DNA Damage	DNA Repair	Cell Proliferation	Apoptosis	Other
CAT 511895*	ATM 1800057	AURKA 1047972	CASP8 1045485 6723097	MTR 1805087
COMT 4680	BABAM1 8170	CHEK2 17879961 1100delC	NQO1 1800566	MTHFR 1801133
CYP17A1 743572	ERCC2 13181	ESR1 3020314 1801132 2234693		TYMS 28 bp repeat
CYP19A1 700519, $(TTTA)_{10}$	NBN 657del5	FGFR2 1219648		DYL2 13329835
CYP1A1 1048943	P53 12947788 12951053 17878362	IGF1 6220		CTLA4 231775
CYB1B1 1056836, 1800440	RAD51L1 999737	MAP3K1 889312		FTO 17817449
GSTP1 1695	WRN 1346044	NUMA1 3018301		IFNG 2430561
GSTM1 deletion	XRCC3 861539	TERT 10069690		INBB 4849887
GSTT1 deletion		TNF 1800629		LRTOMT 673478
HSD17B1 606059, 676387		TNRC9 3803662		LSP1 3817198
TXN1 4135179		TNP1 13387042		VDR 731236
TXNRD2 5748469 756661				

* *The numbers are reference SNP IDs, rs#*

Every year, scientists publish thousands of papers on breast cancer. Many are worthwhile, high quality studies, but they all have a limited focus, e.g. one gene, one protein, this pathway, that response, or that outcome. Although studies of pathways are quite sophisticated, as a rule they do not include processes leading to DNA damage, i.e., they omit cancer initiation. My concern is that we will not understand breast cancer unless we develop an integrated view of the disease. It is my hope that this book will contribute to a better understanding of breast cancer and its etiology allowing better risk prediction.

References

1. MacMahon B, Cole P, Brown J. *Etiology of human breast cancer: a review.* J Natl Cancer Inst. 1973;50:21-42.

2. Futreal PA, Liu Q, Shattuck-Eidens D, Cochran C, Harshman K, Tavtigian S, et al. *BRCA1 mutations in primary breast and ovarian carcinomas.* Science. 1994;266:120-2.

3. Miki Y, Swensen J, Shattuck-Eidens D, Futreal PA, Harshman K, Tavtigian S, et al. *A strong candidate for the breast and ovarian cancer susceptibility gene BRCA1.* Science. 1994;266:66-71.

4. Crooke PS, Justenhoven C, Brauch H, Dawling S, Roodi N, Higginbotham KS, et al. *Estrogen metabolism and exposure in a genotypic-phenotypic model for breast cancer risk prediction.* Cancer Epidemiol Biomarkers Prev. 2011;20:1502-15.

5. Parl FF, Dawling S, Roodi N, Crooke PS. *Estrogen metabolism and breast cancer: a risk model.* Ann N Y Acad Sci. 2009;1155:68-75.

6. Yager JD, Davidson NE. *Estrogen carcinogenesis in breast cancer.* New Engl J Med. 2006;354:270-82.

7. Yue W, Yager JD, Wang JP, Jupe ER, Santen RJ. *Estrogen receptor-dependent and independent mechanisms of breast cancer carcinogenesis.* Steroids. 2013;78:161-70.

8. Fuhrman BJ, Schairer C, Gail MH, Boyd-Morin J, Xu X, Sue LY, et al. *Estrogen metabolism and risk of breast cancer in postmenopausal women.* J Natl Cancer Inst. 2012;104:326-39.

9. Crooke PS, Parl FF. *A mathematical model for DNA damage and repair.* J Nucl Acids. 2010.doi: 10.4061/2010/352603

10. Kastan MB, Bartek J. *Cell-cycle checkpoints and cancer.* Nature. 2004;432:316-23.

11. Sherr CJ. *Cancer cell cycles.* Science. 1996;274:1672-7.

12. Cebrian A, Pharoah PD, Ahmed S, Smith PL, Luccarini C, Luben R, et al. *Tagging single-nucleotide polymorphisms in antioxidant defense enzymes and susceptibility to breast cancer.* Cancer Res. 2006;66:1225-33.

13. Long J, Shu XO, Cai Q, Gao YT, Zheng Y, Li G, et al. *Evaluation of breast cancer susceptibility loci in Chinese women.* Cancer Epidemiol Biomarkers Prev. 2010;19:2357-65.

14. Peng S, Lu B, Ruan W, Zhu Y, Sheng H, Lai M. *Genetic polymorphisms and breast cancer risk: evidence from meta-analyses, pooled analyses, and genome-wide association studies.* Breast Cancer Res Treat. 2011;127:309-24.

15. Zhang B, Beeghly-Fadiel A, Long J, Zheng W. *Genetic variants associated with breast-cancer risk: comprehensive research synopsis, meta-analysis, and epidemiological evidence.* Lancet Oncol. 2011;12:477-88.

16. Stratton MR, Rahman N. *The emerging landscape of breast cancer susceptibility.* Nat Genet. 2008;40:17-22.

17. Cambien F. *Heritability, weak effects, and rare variants in genomewide association studies.* Clin Chemistry. 2011;57:1263-6.

18. Frazer KA, Murray SS, Schork NJ, Topol EJ. *Human genetic variation and its contribution to complex traits.* Nat Rev Genet. 2009;10:241-51.

19. Thomas D. *Methods for investigating gene-environment interactions in candidate pathway and genome-wide association studies.* Annl Rev Public Health. 2010;31:21-36.

20. Benson JR. *Role of transforming growth factor beta in breast carcinogenesis.* Lancet Oncol. 2004;5:229-39.

www.ingramcontent.com/pod-product-compliance
Lightning Source LLC
Chambersburg PA
CBHW080018240326
41598CB00075B/62

* 9 7 8 0 6 1 5 9 9 3 7 3 7 *